CRC Series in Modern Nutrition Science

Phytopharmaceuticals in Cancer Chemoprevention

CRC Series in Modern Nutrition Science

Series Editor
Stacey J. Bell
Ideasphere, Inc.
Grand Rapids, Michigan

Phytopharmaceuticals in Cancer Chemoprevention
Edited by Debasis Bagchi and Harry G. Preuss

Handbook of Minerals as Nutritional Supplements
Robert A. DiSilvestro

Intestinal Failure and Rehabilitation: A Clinical Guide
Edited by Laura E. Matarese, Ezra Steiger,
and Douglas L. Seidner

CRC Series in Modern Nutrition Science

Phytopharmaceuticals in Cancer Chemoprevention

EDITED BY

Debasis Bagchi
Harry G. Preuss

CRC Press
Taylor & Francis Group
Boca Raton   London   New York

CRC Press is an imprint of the
Taylor & Francis Group, an **informa** business

CRC Press
Taylor & Francis Group
6000 Broken Sound Parkway NW, Suite 300
Boca Raton, FL 33487-2742

First issued in paperback 2019

ISBN-13: 978-0-8493-1560-2 (hbk)
ISBN-13: 978-0-367-39368-7 (pbk)

Library of Congress Cataloging-in-Publication Data

Phytopharmaceuticals in cancer chemoprevention / edited by Debasis Bagchi, Harry G. Preuss.
      p. cm. (Modern nutrition science)
   ISBN 0-8493-1560-3 (alk. paper)
    1. Cancer--Chemoprevention. 2. Carcinogenesis. 3. Cancer--Pathophysiology. 4. Materia medica, Vegetable. 5. Dietary supplements. I. Bagchi, Debasis. II. Preuss, Harry G. III. Series.

RC268.15.P48 2004
616.99′4061—dc22   2004051918

Library of Congress Card Number 2004051918

Visit the Taylor & Francis Web site at
http://www.taylorandfrancis.com

and the CRC Press Web site at
http://www.crcpress.com

Dedication

Dedicated to My Beloved Parents, the late Tarak Chandra Bagchi and Pratima Bagchi, My Beloved Wife, Manashi, and My Sweet Daughter, Deepanjali

— Debasis Bagchi

Dedicated to My Beloved Parents, the late Harry G. Preuss Jr. and Mary K. Preuss, My Beloved Wife, Bonnie, and My Children, Mary Beth, Jeffrey, Christopher and Michael

— Harry G. Preuss

Series Preface

When I was asked to serve as the series editor of *Modern Nutrition Science*, I jumped at the opportunity. I have worked in the field of nutrition for 30 years both as a researcher and as a dietitian. This role gave me the opportunity to select expert colleagues, many of whom are key thought leaders in the field, to be volume editors for the individual books in the series. I am happy to report that in each case, my first choice for book editor accepted the job.

The books in this series are geared for two audiences: nutritional scientists and health care professionals, such as dietitians, physicians, nurses, and pharmacists. Each book contains chapters that are scholarly, extensive, scientific reviews covering their topics in a comprehensive way. My hope is that the volumes in this series will stimulate new topics of research for nutritional scientists. I also hope that busy health care practitioners will use these books as reliable resources when providing clinical care.

Many patients today are using alternative therapies to manage their diseases, and it is difficult for health care practitioners to keep up with this evolving field. Moreover, especially in the area of dietary supplements, the science behind the perceived benefits is lacking. This book is presented as an effort to provide a scientific perspective on the use of phytopharmaceuticals in battling cancer.

The books in this series are written with a definitive structure to the chapters so that specific information is easily found, without having to read the entire chapter. For example, if a patient with breast cancer asks about the use of selenium, one can easily find cancer and efficacious doses without reading the entire chapter.

I am pleased that Drs. Bagchi and Preuss agreed to edit this book, *Phytopharmaceuticals in Cancer Chemoprevention*. I have known them both for many years and have been impressed by their excellent research in this field. They have assembled a team of well-known and well-published researchers and clinicians in the field of cancer and nutrition to create a unique book. I hope that researchers who read this book will discover a myriad of topics to be studied. For the clinicians, I am sure that the book can be used as a guide to answering questions posed by anxious patients.

This book and others in the series offer a fresh and scholarly approach to many of the cutting-edge topics in the field of nutritional science. I am grateful to CRC Press for its support in publishing this series.

Stacey J. Bell, Ph.D.
Series Editor

Preface

In Western medicine, professionals dealing with cancer rely heavily on aggressive treatment through the use of pharmaceutical intervention. The use of plant-based phytopharmaceuticals for prevention, or as adjuncts to conventional therapy, is largely ignored. Considering the billions of dollars spent annually on research and development of pharmaceutical cancer drugs, progress of effective new treatments has been marginal, while serious side effects remain an ongoing problem. Consequently, safe and effective alternatives, primarily from natural sources, have been sought. Even though a number of natural products may lessen and/or ameliorate some cancers without causing any significant adverse effects, more often than not these are ignored, or at least not understood, by practicing physicians. This is unfortunate because a significant amount of research on phytopharmaceuticals has validated their efficacy and usefulness in the prevention, as well as treatment, of cancer. As a result, many patients are deprived of these life improving, if not life saving, compounds.

This book was developed to expand the knowledge of practicing medical professionals, clinical nutritionists and the public in general on the usefulness of phytopharmaceuticals in the prevention and treatment of cancer. We are deeply grateful to the highly renowned scientists and cancer biologists who have contributed their knowledge and understanding of the subject matter contained in the chapters that follow. Our contributors begin with a discussion of the epidemiology of cancer, molecular mechanisms of carcinogenesis, pathophysiology and mechanistic roles of metals, metalloids, tobacco, asbestos, ozone and oxidative stress, radiation therapy, xenobiotics, *Helicobacter pylori*, obesity, estrogen, and non-steroidal anti-inflammatory drugs in the induction of cancer. Important sections were added to discuss DNA methylation, mechanisms behind tumor formation, and the etiology of selected cancers.

Following this setting, our experts describe the broad range of phytopharmaceuticals available to practicing healthcare professionals, which hold the greatest promise in the prevention and treatment of cancer. Three chapters are dedicated to describing the potential mechanistic roles of natural products in cancer prevention and therapy, while individual chapters summarize information on natural, cancer-fighting alternatives, including vitamins C and E and carotenoids; polyphenolic and flavonoid antioxidants; grape seed proanthocyanidins; curcumin; tea; phytoestrogens; resveratrol; berry anthocyanins and fruits; tocotrienols; maitake mushroom; taxol; lycopene; NADH; astaxanthin; selected spice ingredients; coenzyme Q_{10}; and probiotics.

We extend our sincere thanks to our eminent contributors for their invaluable contributions.

Debasis Bagchi and Harry G. Preuss

Editors

Debasis Bagchi, Ph.D., FACN, CNS, MAIChE, received his Ph.D. in medicinal chemistry in 1982. He is a professor of toxicology in the Department of Pharmacy Sciences at Creighton University Medical Center, Omaha, NE. Dr. Bagchi is also the vice president of research and development at InterHealth Nutraceuticals, Inc., in Benicia, CA. His research interests include free radicals, human diseases, carcinogenesis pathophysiology, mechanistic aspects of cytoprotection by antioxidants, regulatory pathways in obesity, and gene expression.

Dr. Bagchi has written 186 papers that have been published in peer-reviewed journals. He has delivered invited lectures and served as session chairperson in various national and international scientific conferences, organized workshops, and group discussion sessions. Dr. Bagchi is a fellow of the American College of Nutrition, member of the Society of Toxicology, member of the New York Academy of Sciences, fellow of the Nutrition Research Academy, and member of the TCE Stakeholder Committee of the Wright Patterson Air Force Base, OH. Dr. Bagchi is a member of the Study Section and Peer Review Committee of the National Institutes of Health, Bethesda, MD. Dr. Bagchi also serves as a reviewer of U.S. Army Research Grants on Gulf War Illness. Dr. Bagchi is also serving as editorial board member of numerous peer-reviewed journals, including *Antioxidants and Redox Signaling, Toxicology Letters, The Original Internist, Research Communications in Pharmacology and Toxicology*, and the *International Journal of Geriatric Urology and Nephrology*.

Dr. Bagchi has received research funding from various institutions and agencies, including the U.S. Air Force Office of Scientific Research; National Institutes of Health (NIH); National Cancer Institute (NCI); National Heart, Lung, and Blood Institute (NHLBI); American Heart Association; Nebraska State Department of Health, Biomedical Research Support Grant; Health Future Foundation; The Procter & Gamble Company; and Abbott Laboratories.

Harry G. Preuss, MD, MACN, CNS, received his BA and MD from Cornell University, Ithaca, NY, and New York City, NY, trained for 3 years in internal medicine at Vanderbilt University Medical Center under Dr. David E. Rogers, studied for 2 years as a fellow in renal physiology at Cornell University Medical Center under Dr. Robert F. Pitts, and spent 2 years in clinical and research training in nephrology at Georgetown University Medical Center under Dr. George E. Schreiner. During his training years, he was a special research fellow of the National Institutes of Health (NIH) and an established investigator of the American Heart Association. Following 5 years as an assistant and associate (tenured) professor of medicine at the University of Pittsburgh Medical Center, he returned to Georgetown Medical Center and is now a professor of physiology, medicine, and pathology (tenured). He subsequently performed a 6-month sabbatical in molecular biology at the NIH in the laboratories of Dr. Maurice Burg.

His bibliography includes over 300 medical papers and more than 200 abstracts. Dr. Preuss has edited or co-edited six books and three symposia published in well-established journals. He is the co-author of two books written for the lay public, *The Prostate Cure* and *Maitake Magic*. In 1976, Dr. Preuss was elected to membership in the American Society of Clinical Investigations. He is currently an advisory editor for six journals. His previous government appointments included 4 years on the Advisory Council for the National Institute on Aging, 2 years on the Advisory Council of the director of the NIH, and 2 years on the Advisory Council for the Office of Alternative Medicine of the NIH. He has been a member of many other peer research review committees for the NIH and American Heart Association and is now a member of the National Cholesterol Education Program of the NHLBI.

Dr. Preuss was elected the ninth Master of the American College of Nutrition (ACN). He is a former chairman of two ACN councils — the Cardiovascular and Aging Council and the Council on Dietary Supplements, Nutraceuticals, and Functional Foods. After a brief stint on the Board of Directors of the ACN, Dr. Preuss spent 3 years as secretary-treasurer and 3 consecutive years as vice president, president-elect, and president. He is currently chairman of the Nutritional Policy Institute of the ACN. Dr. Preuss was a member of the board of directors for the American Association for Health Freedom (AAHF) and their treasurer. Dr. Preuss wrote the nutrition section for the *Encyclopedia Americana* and recently was elected president of the Certification Board for Nutrition Specialists (CBNS) that gives the CNS certification. He is at present co-chairman of the Institutional Review Board (IRB) at Georgetown University, which reviews all clinical protocols at Georgetown University Medical Center. His research centers on the use of dietary supplements and nutraceuticals to favorably influence or even prevent a variety of medical perturbations, especially those related to obesity, insulin resistance, and cardiovascular perturbations. Lately, he has also researched the ability of many oils and fats to overcome various infections, including those resistant to antibiotics.

Contributors

Vaqar Mustafa Adhami
University of Wisconsin
Madison, Wisconsin

Bharat B. Aggarwal
University of Texas M.D. Anderson Cancer
  Center
Houston, Texas

Manoj S. Aggarwal
University of Texas M.D. Anderson Cancer
  Center
Houston, Texas

Nihal Ahmad
University of Wisconsin
Madison, Wisconsin

R. Ambra
National Institute for Food and Nutrition
  Research (INRAN)
Rome, Italy

Moammir Hasan Aziz
University of Wisconsin
Madison, Wisconsin

Angelo Azzi
University of Bern
Bern, Switzerland

Manashi Bagchi
InterHealth Research Center
Benicia, California

Bhaskar Banerjee
Washington University School of Medicine
St. Louis, Missouri

Snigdha Banerjee
VA Medical Center
Kansas City, Missouri
and
University of Kansas Medical Center
Kansas City, Kansas

Sushanta K. Banerjee
VA Medical Center
Kansas City, Missouri
and
University of Kansas Medical Center
Kansas City, Kansas

W. Vanden Berghe
Ghent University
Ghent, Belgium

Hemmi N. Bhagavan (Deceased)
Biomedical Research Consultants

George D. Birkmayer
University of Graz
and
Birkmayer Laboratories
Vienna, Austria

Emile G. Bliznakov
Biomedical Research Consultant
Pompano Beach, Florida

Jeffrey Blumberg
Tufts University
Boston, Massachusetts

R. Canali
National Institute for Food and Nutrition
  Research (INRAN)
Rome, Italy

Jean Carper
Key West, Florida

Doug Case
Wake Forest University Baptist Medical Center
Winston-Salem, North Carolina

Shampa Chatterjee
University of Pennsylvania Medical Center
Philadelphia, Pennsylvania

Fei Chen
National Institute for Occupational Safety and
  Health
Morgantown, West Virginia

Raj K. Chopra
Tishcon Corp.
Westbury, New York

Marco d'Ischia
University of Naples "FedericaII"
Naples, Italy

A. De Naeyer
Ghent University
Ghent, Belgium

John E. Dore
Cyanotech Corp.
Kailua-Kona, Hawaii

Rama S. Dwivedi
Northwestern University
Chicago, Illinois

Aron B. Fisher
University of Pennsylvania Medical Center
Philadelphia, Pennsylvania

Balz Frei
Oregon State University
Corvallis, Oregon

Gayle M. Gordillo
The Ohio State University
Columbus, Ohio

Lone Gothard
Institute of Cancer Research and Roy Marsden
  Hospital
Sutton, United Kingdom

O. Gulati
Horphag Research Ltd.
Geneva, Switzerland

René, Gysin
University of Bern
Bern, Switzerland

G. Haegeman
Ghent University
Ghent, Belgium

Gabriel Keith Harris
National Institute for Occupational Safety and
  Health
Morgantown, West Virginia

Jane Higdon
Oregon State University
Corvallis, Oregon

Aminul Islam
VA Medical Center
Kansas City, Missouri
and
University of Kansas Medical Center
Kansas City, Kansas

Jim Joseph
Tufts University
Boston, Massachusetts

G. R. Kaats
Health and Medical Research Foundation
San Antonio, Texas

D. De Keukeleire
Ghent University
Ghent, Belgium

Sensuke Konno
New York Medical College
New York, New York

Anushree Kumar
University of Texas M.D. Anderson Cancer
  Center
Houston, Texas

Hyong Joo Lee
Seoul National University
Seoul, Korea

Henry T. Lynch
Creighton University School of Medicine
Omaha, Nebraska

Jane F. Lynch
Creighton University School of Medicine
Omaha, Nebraska

Karen Madsen
University of Alberta
Edmonton, Alberta, Canada

Gautam Maulik
Harvard Medical School
Boston, Massachusetts

Nilanjana Maulik
University of Connecticut
Farmington, Connecticut

Rajendra G. Mehta
University of Illinois
Chicago, Illinois

Bernard L. Mirkin
Northwestern University
Chicago, Illinois

Hasan Mukhtar
University of Wisconsin
Madison, Wisconsin

Hye-Kyung Na
Seoul National University
Seoul, Korea

Kalanithi Nesaretnam
Malaysian Palm Oil Board
Kuala Lumur, Malaysia

Giovanni Pagano
Italian National Cancer Institute
Mercogliano, Italy

John M. Pezzuto
Purdue University
West Lafayette, Indiana

Ronald L. Prior
USDA Arkansas Children's Nutrition Center
Little Rock, Arkansas

Y-Yong Qiu
Northwestern University
Chicago, Illinois

Sidhartha D. Ray
Long Island University
Brooklyn, New York

Henry Rodriguez
National Institute of Standards and Technology
Gaithersburg, Maryland

Hemant K. Roy
Evanston-Northwestern Healthcare
Evanston, Illinois

Chandan K. Sen
The Ohio State University
Columbus, Ohio

Ginette Serrero
A & G Pharmaceutical Inc.
Columbus, Maryland
and
University of Maryland
Baltimore, Maryland

Trudy G. Shaw
Creighton University School of Medicine
Omaha, Nebraska

Xianglin Shi
National Institute for Occupational Safety and
  Health
Morgantown, West Virginia

Shishir Shishodia
University of Texas M.D. Anderson Cancer
  Center
Houston, Texas

Clinton Snedegar
Washington University School of Medicine
St. Louis, Missouri

Vernon E. Steele
National Cancer Institute
Bethesda, Maryland

Sidney J. Stohs
Creighton University Medical School
Omaha, Nebraska

Gottumukkala V. Subbaraju
Laila Impex R&D Center
Jawahar Autonagar
Vijayawada, India

Young-Joon Surh
Seoul National University
Seoul, Korea

Wisit Tangkeangsirisin
A & G Pharmaceutical Inc.
Columbus, Maryland
and
University of Maryland
Baltimore, Maryland

Golakoti Trimurtulu
Laila Impex R&D Center
Jawahar Autonagar
Vijayawada, India

Sandeep K. Tripathy
Washington University School of Medicine
St. Louis, Missouri

Shinya Toyokuni
Kyoto University
Kyoto, Japan

Val Vallyathan
National Institute for Occupational Safety and
  Health
Morgantown, West Virginia

Fabio Virgili
National Institute for Food and Nutrition
  Research (INRAN)
Rome, Italy

Theresa Visarius
University of Bern
Bern, Switzerland

John Walker
University of Alberta
Edmonton, Alberta, Canada

John Yarnold
Institute of Cancer Research and Roy Marsden
  Hospital
Sutton, United Kingdom

Jiren Zhang
Zhujiang Hospital
Guangzhou, China

Table of Contents

Section I
Introduction 1

Chapter 1 Epidemiology of Cancer: an Overview 3
Henry Rodriguez and Doug Case

Section II
Pathophysiology of Cancer...15

Chapter 2 Role of DNA Methylation in Cancer and Chemotherapy 17
Y-Yong Qiu, Bernard L. Mirkin, and Rama S. Dwivedi

Chapter 3 Metals, Metalloids, and Cancer 29
Gabriel Keith Harris and Xianglin Shi

Chapter 4 Molecular Mechanisms of Asbestos- and
              Silica-Induced Lung Cancer..41
Fei Chen and Val Vallyathan

Chapter 5 Involvement of Receptor Tyrosine Kinases in Lung
              Cancer: Clinical Importance 63
Nilanjana Maulik and Gautam Maulik

Chapter 6 Free Radicals, Environmental Radiation, and Cancer................................79
Okezie I. Aruoma

Chapter 7 Smokeless Tobacco, Oxidative Stress, and Oral Cancer 89
Manashi Bagchi, Sidney J. Stohs, and Debasis Bagchi

Chapter 8 The Regulatory Roles of Estrogen in Carcinogenesis:
              an Overview...105
Sushanta K. Banerjee, Aminul Islam, and Snigdha Banerjee

Chapter 9 Nonsteroidal Anti-Inflammatory Drugs (NSAID)
              and Colorectal Cancer 123
Henry T. Lynch, Trudy G. Shaw, Jane F. Lynch,
and Hemant K. Roy

Chapter 10 Current Perspectives in Gastric Adenocarcinoma 143

Clinton Snedegar and Bhaskar Banerjee

Chapter 11 Adenocarcinoma of the Esophagus..151

Sandeep K. Tripathy and Bhaskar Banerjee

Chapter 12 *Helicobacter pylori* and Gastric Cancer 163

Shinya Toyokuni

Chapter 13 Free Radicals, Oxidative Stress, and Cancer..171

Shampa Chatterjee and Aron B. Fisher

Chapter 14 Biotransformation and Mechanism of Action
of Xenobiotics: What Lessons from the Past 40 Years? 187

Giovanni Pagano and Marco d'Ischia

Chapter 15 Obesity and Cancer 197

*Harry G. Preuss, Manashi Bagchi, Debasis Bagchi,
and G.R. Kaats*

Chapter 16 Hemangioendothelioma as a Model to Study
the Antiangiogenic Effects of Dietary Chemopreventive
Agents *In Vivo* 205

Gayle M. Gordillo and Chandan K. Sen

Chapter 17 Modulation of Late Adverse Effects of Curative Radiation
Therapy for Cancer..215

John Yarnold and Lone Gothard

**Section III
Phytopharmaceuticals and Chemoprevention**..**227**

Chapter 18 Development of Selected Phytochemicals for
Cancer Chemoprevention 229

Vernon E. Steele

Chapter 19 Phytochemicals as Potential Cancer
Chemopreventive Agents..237

Rajendra G. Mehta and John M. Pezzuto

Chapter 20 History of Natural Supplements in Cancer
Therapy and Prevention..247

Gottumukkala V. Subbaraju and Golakoti Trimurtulu

Chapter 21 Vitamin C, Vitamin E, and β-Carotene in
Cancer Chemoprevention 271

Jane Higdon and Balz Frei

Chapter 22 Roles of Polyphenols, Flavonoids, and Oligomeric
Proanthocyanidins in Cancer Chemoprevention 311

Sidhartha D. Ray and Debasis Bagchi

Chapter 23 Curcumin Derived from Turmeric (*Curcuma longa*):
a Spice for All Seasons 349

*Bharat B. Aggarwal, Anushree Kumar, Manoj S. Aggarwal,
and Shishir Shishodia*

Chapter 24 Tea in Chemoprevention of Cancer 389

*Vaqar Mustafa Adhami, Moammir Hasan Aziz, Nihal Ahmad,
and Hasan Mukhtar*

Chapter 25 Phytoestrogens in Cancer Prevention: Characterization
and Beneficial Effects of Kurarinone, a New Flavanone
and a Major Phytoestrogen Constituent
of *Sophora flavescens* Ait..427

*A. De Naeyer, W. Vanden Berghe, D. De Keukeleire,
and G. Haegeman*

Chapter 26 Resveratrol in the Chemoprevention and Chemotherapy
of Breast Cancer 449

Wisit Tangkeangsirisin and Ginette Serrero

Chapter 27 Berries and Fruits in Cancer Chemoprevention 465

Ronald L. Prior and Jim Joseph

Chapter 28 Palm Tocotrienols and Cancer 481

Kalanithi Nesaretnam

Chapter 29 Pycnogenol in Cancer Chemoprevention.......................491

Fabio Virgili, R. Ambra, R. Canali, and O. Gulati

Chapter 30 Overview of the Use of Maitake Mushroom
and Fraction D in Cancer...509

Harry Preuss, Sensuke Konno, and Debasis Bagchi

Chapter 31 Taxol in Cancer Treatment and Chemoprevention 519

Sidney J. Stohs

Chapter 32 Lycopene and Cancer 525

Theresa Visarius, René Gysin, and Angelo Azzi

Chapter 33 NADH in Cancer Prevention and Therapy 541

George D. Birkmayer and Jiren Zhang

Chapter 34 Astaxanthin and Cancer Chemoprevention.............................555

John E. Dore

Chapter 35 Chemopreventive Effects of Selected Spice Ingredients 575

Young-Joon Surh, Hye-Kyung Na, and Hyong Joo Lee

Chapter 36 Coenzyme Q_{10} and Neoplasia: Overview of Experimental
and Clinical Evidence 599

Emile G. Bliznakov, Raj K. Chopra, and Hemmi N. Bhagavan

Chapter 37 Probiotics in the Prevention of Cancer 623

Karen Madsen and John Walker

Section IV
Concluding Remarks...637

Commentary 1 Do Dietary Antioxidants Really Help
Prevent or Treat Cancer?...639

Jeffrey Blumberg

Commentary 2 Who Gets Cancer? Do Healthy Foods, Healthy Living,
and Natural Antioxidants Really Help?.................................643

Jean Carper

Concluding Remarks...645

Index 647

Section I

Introduction

1 Epidemiology of Cancer: an Overview

Henry Rodriguez and Doug Case

CONTENTS

1.1 Epidemiology 3
1.2 Cancer: Basic Facts 4
    1.2.1 What is Cancer? 4
    1.2.2 Who Gets Cancer? 4
1.3 Cancer Terms 4
    1.3.1 Benign and Malignant 4
    1.3.2 Types of Cancer 5
    1.3.3 Primary and Secondary 5
    1.3.4 Grading and Staging of Cancer 5
1.4 Cancer: Statistics 6
    1.4.1 Incidence and Mortality 6
    1.4.2 Trends over Time 7
    1.4.3 Age at Diagnosis 9
    1.4.4 Cancer Prevalence 10
    1.4.5 Cancer Worldwide 12
1.5 Conclusion 12
References 14

1.1 EPIDEMIOLOGY

Epidemiology is the study of how often diseases occur in different groups of people and why. Epidemiological information is used to plan and evaluate strategies to prevent illness and as a guide to the management of patients in whom disease has already developed. Like clinical findings and pathology, the epidemiology of a disease is an integral part of its basic description.

A key feature of epidemiology is the measurement of disease outcomes in relation to a population at risk. The population at risk is the group of people, healthy or sick, who would be counted as cases if they developed the disease being studied. Implicit in any epidemiological investigation is the notion of a target population about which conclusions are to be drawn. Occasionally, measurements can be made on the full target population. More often, observations can only be made on a study sample, which is selected in some way from the target population. How well the sample observations describe the target population depends on how representative the sample is of the target population.

Another task of epidemiology is monitoring or surveillance of time trends to show which diseases are increasing or decreasing in incidence and which are changing in their distribution. This information is needed to identify emerging problems and also to assess the effectiveness of measures to control old problems.

0-8493-1560-3/05/$0.00+$1.50
© 2005 by CRC Press LLC

1.2 CANCER: BASIC FACTS

This section deals with an overview of the biology of cancer and the terminology associated with this field of science/medicine.

1.2.1 WHAT IS CANCER?

Cancer is derived from the Latin word for crab. Like a crab, cancer cells attach themselves to tissues and organs and hang on. The study of cancer is called "oncology," named after *oncos*, which means "tumor" in Greek. Although cancer is often referred to as a single condition, it actually consists of more than 100 different diseases, all characterized by the uncontrolled growth and spread of abnormal cells causing masses of tissues called tumors.[1]

Cancer can arise in many sites and behaves differently depending on its organ of origin. For example, breast cancer has different characteristics than lung cancer, and breast cancer that spreads to the lungs should not be confused with lung cancer. Even in the lungs, breast cancer continues to behave like breast cancer and, under the microscope, continues to look like breast cancer.

While there are many different types of cells with as many different functions, all cells have some basic similarities. We now know that cells in a tumor descend from just one initial cell that at some point (usually decades before the tumor is detected) starts a program of unsuitable growth.[1] This is because some of the information carried in the cell's genes has been altered or damaged. The transformation into a cancer cell comes about through the step-by-step accumulation of a series of genetic alterations or mutations in those particular genes that replicate this growth advantage. Hundreds of genes have a role to play in this process in the normal cell, and thus hundreds of genes have the potential to be damaged in the cancer cell.[1]

Cancers spread by invading the surrounding tissue until they reach a blood or lymph vessel. Small groups of cells may then break off the original tumor and be carried by the blood or lymph to another part of the body. A cancer that has spread is said to have metastasized.

1.2.2 WHO GETS CANCER?

Cancer is a very common disease. Based on incidence data from the National Cancer Institute's Surveillance, Epidemiology, and End Results (SEER) program,[2] two in five persons will be diagnosed with cancer some time in their life, and it will eventually affect three out of every four families. One in five deaths in the U.S. is due to cancer. Cancer develops in people of all ages, but most often in the middle-aged and the elderly. The number of cancer cases has risen dramatically over the past 40 years, but much of this increase is a reflection of the increase in population, especially in the older age groups. Cancers of the prostate, lung, and colon are the most common types diagnosed among adult males. Breast, lung, and colon cancers are the most common types among females. Common sites of metastatic cancer (cancer that has spread from the primary site) include the brain and liver.

Researchers do not fully understand why some people develop cancer while others do not. Susceptibility to cancer-causing agents probably varies among individuals due to genetic factors. Other factors, as yet unknown, may also play a part in causing this disease.

1.3 CANCER TERMS

This section deals with the terminology used by doctors and medical staff in the diagnosis and treatment of cancer. The following provides some information on commonly used terms and their meanings.

1.3.1 BENIGN AND MALIGNANT

The mass of abnormal cells formed when cells start dividing uncontrollably is called a tumor. There are two types of tumors: benign and malignant.

Benign tumors contain cells that are not able to spread to a different site in the body. This means they are not cancerous. They are generally contained within a covering of normal cells. They usually grow slowly and are often harmless, not requiring any treatment. However, if they grow large and begin to encroach on the space of surrounding organs, they can cause problems. In this circumstance they do require treatment.

Malignant tumors contain cells that are capable of spreading beyond the original tumor to another part of the body. This can be dangerous because as the cells invade surrounding tissues, they can damage them and stop them from working properly. This type of tumor is cancerous. Doctors can tell whether a tumor is benign or malignant by examining a small sample of cells from the tumor under the microscope. The removal of tissues or cells from the body for examination is called a biopsy.

1.3.2 TYPES OF CANCER

All of the various cell types of the body can grow into a cancer. In general, cancers fall into four major groups, classified according to the body tissues in which they arise. All types can spread to other types of tissues of the body while retaining their original cellular characteristics. Cancers that derive from different cell types have distinct names to denote the type of cell involved:

Carcinoma: a cancer that has developed from epithelial cells. These cells line the digestive tract and make up organs such as the liver, kidney, and pancreas. These are the most common type of cancers.
Sarcoma: a cancer that has developed from cells making up muscle, nerves, or blood vessels.
Lymphoma: a cancer that has developed from cells in a lymph gland.
Leukemias: cancers of the blood-forming tissue.

Cancers from different cell types can behave very differently. They may:

- Grow faster or slower
- Produce different symptoms
- Respond differently to the same treatment
- Be more or less likely to spread to a specific part of the body

All of these factors are important in the diagnosis and treatment of each type of cancer.

1.3.3 PRIMARY AND SECONDARY

The original site of a tumor is called the primary cancer. It refers to the place at which the cancer originated. When cells from malignant tumors break away and travel to other parts of the body, they form new groups of cells that are called secondary tumors. This indicates that these cells have traveled from a different site and did not originate at the site where the secondary tumor is located. A secondary tumor can also be referred to as a metastasis.

1.3.4 GRADING AND STAGING OF CANCER

Physicians use a system called "staging" to describe the size of a tumor and whether it has spread beyond the area in which it started. Each type of cancer has a different system of staging that helps to describe the progress of that cancer. In general, there are a number of stages that begin with a small, localized cancer, right through to one that has spread extensively into surrounding areas of the body, a so-called advanced cancer. Knowing the stage is important in deciding on the most appropriate treatment.

After the determination is made as to the type of cancer, the cancer is graded — a measurement of the aggressiveness of the tumor. Most cancer cells are graded by how much they look like normal cells. Grading is done in the lab using cancerous cells taken during biopsy. There are many different types of grading systems used by doctors. These vary depending on the cancer. In general, however, lower grades mean a less aggressive behavior, and higher grades predict for a relatively more aggressive cancer. The most commonly used grading system is called the Gleason System, which is based on a number range from 2 to 10. The lower the number, the lower the grade. Grades under 4 mean that the cancer cells look similar to your normal cells and the cancer is likely to be less aggressive. Grades in the 5 to 7 range are intermediate, which means that the cancer cells do not look like normal cells and are more likely to be aggressive and grow faster. Grades of 8 to 10 indicate that the cancer cells are more likely to be very aggressive in growth.

Once cancer is diagnosed, more tests will be done to find out if the cancer cells have spread to other parts of the body. This testing is called staging. To plan treatment, a physician needs to know the stage of the disease. Stage refers to the extent, or the size, of the cancer. Each cancer, by organ, has its own staging system.

>*Stage 0* (carcinoma *in situ*): Carcinoma *in situ* is very early cancer. The abnormal cells are found only in the first layer of cells of the primary site and do not invade the deeper tissues.
>*Stage I:* Cancer involves the primary site, but has not spread to nearby tissues.
>>*Stage IA:* A very small amount of cancer that is visible under a microscope and is found deeper in the tissues.
>>*Stage IB:* A larger amount of cancer is found in the tissues.
>*Stage II:* Cancer has spread to nearby areas but is still inside the primary site.
>>*Stage IIA:* Cancer has spread beyond the primary site.
>>*Stage IIB:* Cancer has spread to other tissue around the primary site.
>*Stage III:* Cancer has spread throughout the nearby area.
>*Stage IV:* Cancer has spread to other parts of the body.
>>*Stage IVA:* Cancer has spread to organs close to the pelvic area.
>>*Stage IVB:* Cancer has spread to distant organs, such as the lungs.
>*Recurrent:* Cancer has come back (recurred) after it has been treated and supposedly eliminated. The recurrence may be in the same location as the original tumor or in a different location.

Once a stage is assigned and treatment given, the stage is never changed. The important thing about staging is that it determines the appropriate treatment, provides a prognosis, and allows for comparison of treatment results between different treatments.

1.4 CANCER: STATISTICS

The data in this section are based on information from the SEER Cancer Statistics Review[2] and the American Cancer Society.[3] Note that the values presented in this chapter are estimates and are offered as a guide and should be interpreted with caution.

1.4.1 INCIDENCE AND MORTALITY

Cancer incidence refers to the number of new cases of cancer diagnosed in a given year, and the incidence rate is usually expressed as the number of new cancers diagnosed per year per 100,000 persons at risk. The cancer mortality rate is the number of deaths per year per 100,000 persons at risk. In the U.S., it is expected that over 1.3 million new cases of cancer will be diagnosed in 2003. In that same year, more than 550,000 deaths will occur due to cancer. The lifetime risk of developing cancer is approximately 44% for men and 39% for women, and 24% of men and 20% of women will die from cancer. Cancer is the second leading cause of death in the U.S., next to heart disease. Figure 1.1 shows the 10 leading causes of death and the percent of total deaths attributable to each cause in the year 2000. One sees that cancer accounts for approximately 23% of the total number of deaths and accounts for more deaths than the next five leading causes combined. In the quarter century between 1973 and 1999, the percentage of deaths due to heart disease decreased from 38.4% to 29.6%, while

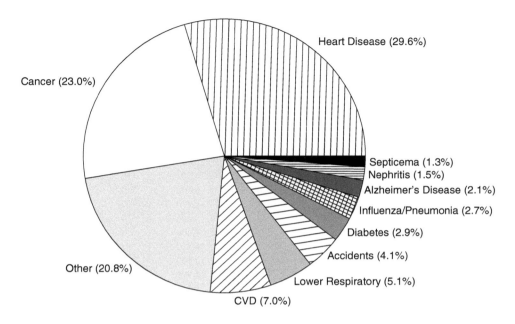

FIGURE 1.1 Ten leading causes of death in the U.S., 2000.
Source: Anderson, R.N. Deaths: leading causes for 2000, National Vital Statistics Report, 50, 1–88, 2002.

the percentage of deaths due to cancer increased from 17.7% to 23%. More than 8 million person-years of life were lost in 1999 due to cancer deaths, making it the leading cause of lost years of life.

Table 1.1 and Table 1.2 show the number of new cancer cases and the number of deaths expected in 2003 for the most common types of cancer for men and women, respectively. Approximately 16,500 more cases of cancer are expected to be diagnosed for males compared with females, and males are expected to suffer about 15,000 more deaths. Prostate cancer is the most common type of cancer in men, accounting for 33% of the new diagnoses. Lung and colorectal cancer account for 14 and 11% of the new diagnoses, respectively, while the other cancers each account for less than 10%. Although prostate cancer accounts for 33% of the new diagnoses, it is responsible for only 10% of the deaths. The 5-year relative survival for patients newly diagnosed with prostate cancer is 97%. Lung cancer, while accounting for 14% of the cancers, is responsible for almost one-third of the deaths. The 5-year relative survival for men newly diagnosed with lung cancer is only 13%, making it one of the most lethal cancers. Colorectal cancer is responsible for 10% of the cancer deaths, while the other cancers are each responsible for less than 6% of the deaths in men.

For women, breast cancer accounts for almost one-third of the new diagnoses. Lung cancer and colorectal cancer account for 12 and 11% of the new diagnoses, respectively, and the other cancers account for 6% or less of the new diagnoses. Breast cancer, the most common cancer in women, has a 5-year relative survival of 86% and is responsible for 15% of the cancer deaths. As with men, lung cancer is responsible for the most deaths, accounting for 25% of the total number of deaths. The 5-year relative survival for women diagnosed with lung cancer is 17%, just slightly better than that for men. Colorectal cancer accounts for 11% of the cancer deaths, and the other cancers are each responsible for less than 10% of the deaths. Pancreatic cancer is the most lethal of the major cancers for both men and women, with a 5-year relative survival of only 4% for both genders.

1.4.2 Trends over Time

Age-adjusted incidence rates are shown by gender in Figure 1.2 and Figure 1.3 for the years between 1973 and 1999 for some of the most common cancers.[2] For men, one notes that there has been a

TABLE 1.1

Expected Incidence and Mortality for the Most Common Types of Cancer for Men of All Races, 2003

Cancer	Number of New Cases	Cancers (%)	Deaths	Deaths (%)	Relative 5-Year Survival (%) [a]
Total	675,300	100	285,900	100	62
Prostate	220,900	33	28,900	10	97
Lung/bronchus	91,800	14	88,400	31	13
Colon/rectum	72,800	11	28,300	10	62
Urinary/bladder	42,200	6	8,600	3	84
Melanoma of skin	29,900	4	4,700	2	87
Non-Hodgkin's lymphoma	28,300	4	12,200	4	51
Kidney/renal pelvis	19,500	3	7,400	3	62
Oral cavity	18,200	3	4,800	2	55
Leukemia	17,900	3	12,100	4	47
Pancreas	14,900	2	14,700	5	4
Stomach	13,400	2	7,000	2	21
Liver/bile duct	11,700	2	9,200	3	6
Esophagus	10,600	2	9,900	3	13
Brain/CNS	10,200	2	7,300	3	33
Multiple myeloma	7,800	1	5,400	2	32
Testis	7,600	1	400	<1	95

[a] All stages.

Source: Ries, L. et al., SEER Cancer Statistics Review, 1973–1999, National Cancer Institute, Bethesda, MD, 2002.

steady increase in the incidence of prostate cancer, with a sharp spike from the mid-1980s to the mid-1990s, when prostate-specific antigen (PSA) testing was becoming more commonplace. Lung and colorectal cancer have been gradually declining during the last decade. The incidence of urinary/bladder cancer has been fairly steady over the last quarter century, while the incidence of melanoma of the skin has approximately tripled over that time period. For women, the incidence of breast cancer has increased approximately 30 to 40% since the 1970s. During that same time, the incidence of lung cancer has doubled. The incidences of colorectal cancer and that of uterine corpus cancer have declined slightly over the last two decades, while the incidence of ovarian cancer has remained relatively stable.

Age-adjusted mortality rates are shown by gender in Figure 1.4 and Figure 1.5 for the years between 1973 and 1999 for some of the cancers that currently result in the most deaths.[2] For men, the overall death rate declined by approximately 1.5% per year between 1992 and 1999. One notes that the mortality rate for lung cancer increased during the 1970s and 1980s, a trend begun early in the century, and finally began to decrease during the 1990s. The mortality rate for prostate cancer increased in the late 1980s and early 1990s and decreased subsequently. The death rate for colorectal cancer has been decreasing slightly over the last two decades. The death rate for pancreas cancer has also decreased slightly, while that of non-Hodgkin's lymphoma has increased, and that of leukemia has remained relatively stable.

For women, the overall death rate decreased by 0.6% per year between 1992 and 1999, this despite the fact that the death rate for lung cancer continued to rise. The death rate for lung cancer has risen sharply in women and only just recently appears possibly to be leveling off. The death rate for breast cancer declined by approximately 6% during the 1990s, from 33.2 per 100,000 in 1989 to 27.0 per 100,000 in 1999. The death rate for colorectal cancer has decreased steadily since the early 1970s. The death rates for pancreas cancer and non-Hodgkin's lymphoma have increased slightly during the last two decades, while that for ovarian cancer has remained relatively stable.

TABLE 1.2
Expected Incidence and Mortality for the Most Common Types of Cancer for Women of All Races, 2003

Cancer	Number of New Cases	Cancers (%)	Deaths	Deaths (%)	Relative 5-Year Survival (%) [a]
Total	658,800	100	270,600	100	63
Breast	211,300	32	39,800	15	86
Lung/bronchus	80,100	12	68,800	25	17
Colon/rectum	74,700	11	28,800	11	62
Uterine corpus	40,100	6	6,800	3	85
Ovary	25,400	4	14,300	5	53
Non-Hodgkin's lymphoma	25,100	4	11,200	4	60
Melanoma of skin	24,300	4	2,900	1	92
Thyroid	16,300	2	800	<1	97
Pancreas	15,800	2	15,300	6	4
Urinary/Bladder	15,200	2	3,900	1	75
Leukemia	12,700	2	9,800	4	44
Kidney/renal pelvis	12,400	2	4,500	2	62
Cervix	12,200	2	4,100	2	71
Oral cavity	9,500	1	2,400	1	60
Stomach	9,000	1	5,100	2	24
Brain/CNS	8,100	1	5,800	2	32

[a] All stages.

Source: Ries, L. et al., SEER Cancer Statistics Review, 1973–1999, National Cancer Institute, Bethesda, MD, 2002.

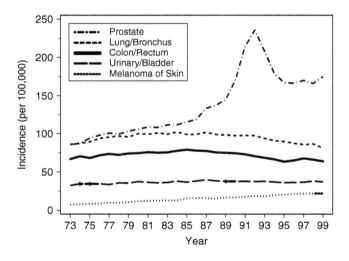

FIGURE 1.2 Age-adjusted incidence over time of the most common cancers in men.

1.4.3 AGE AT DIAGNOSIS

In general, cancer is a disease of the elderly, with a median age at diagnosis across all cancer sites of 67 years and a median age at death of 72 years. However, cancers can be diagnosed at any age, and some cancers are more common among younger adults or children. Indeed, cancer is the second

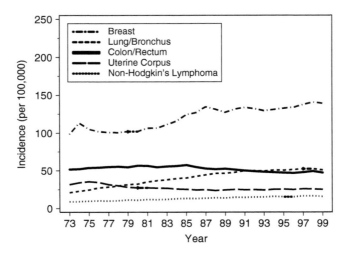

FIGURE 1.3 Age-adjusted incidence over time of the most common cancers in women.

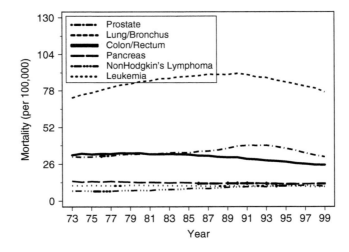

FIGURE 1.4 Age-adjusted mortality rates over time of the leading causes of cancer deaths in men.

leading cause of death among children between 1 and 15 years of age. Table 1.3 shows the median age at diagnosis for some of the more common cancers and some cancers that are more common among younger individuals. In general, one sees that the median age is 60 or above for the majority of the cancers. However, for Hodgkin's disease and cancers of the testis and bones and joints, the median age at diagnosis is in the mid-30s. For acute lymphocytic leukemia, the median age at diagnosis is only 10 years old. Fortunately, the 5-year relative survival for this latter disease has increased from 53% in the mid-1970s to 85% in the late 1990s.

1.4.4 CANCER PREVALENCE

Prevalence is the number of people currently alive who have been diagnosed with the disease (cancer survivors). It includes new and old cases of the disease and is a function of incidence and survival. As of January 1, 1999, there were approximately 8.9 million cancer survivors in the U.S., 5 million of whom were women and 3.9 million of whom were men.[7] Over half the survivors (59%)

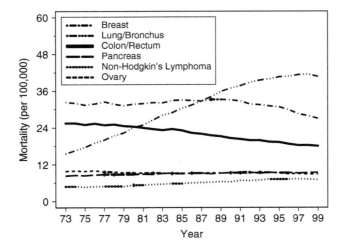

FIGURE 1.5 Age-adjusted mortality rates over time of the leading causes of cancer deaths in women.

TABLE 1.3
Age at Diagnosis for Some Common Cancers and Some Cancers That Are More Common among Younger Individuals[a]

Cancer	Males	Females
Total	68	67
Urinary/bladder	71	73
Pancreas	70	74
Colon/rectum	70	74
Lung/bronchus	70	70
Prostate	69	—
Breast	68	62
Non-Hodgkin's lymphoma	63	69
Uterine corpus	—	65
Melanoma of skin	60	52
Brain/CNS	54	57
Hodgkin's lymphoma	37	34
Bones and joints	34	42
Testis	34	—
Acute lymphocytic leukemia	10	10

[a] All stages

Source: Ries, L. et al., SEER Cancer Statistics Review, 1973–1999, National Cancer Institute, Bethesda, MD, 2002.

were 65 years of age or older. The most common prevalent cancer sites are the breast, prostate, and colon/rectum. Figure 1.6 shows the distribution of cancer diagnoses for cancer survivors in 1999. The three common sites mentioned above account for slightly more than 50% of the cancers among survivors. One notes that although lung cancer is the second most common type of cancer among men and women, only 4% of the current survivors have this type of cancer.

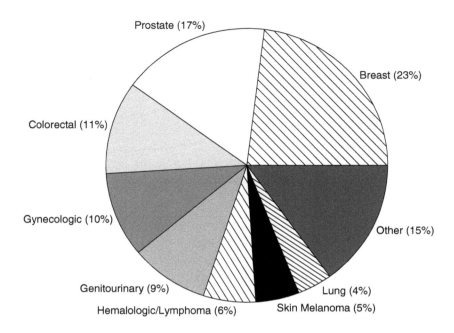

FIGURE 1.6 Most common types of prevalent cancer in the U.S.

1.4.5 CANCER WORLDWIDE

All the data presented above pertain to the U.S., but cancer is certainly not only a problem in our country. Indeed, cancer is a global problem, and while some cancers pose a greater risk in our country, other cancers pose a much greater risk in other countries. Table 1.4 shows the incidence and mortality of the four most common types of cancer worldwide[4] for several countries around the world. Worldwide, lung cancer is the most commonly diagnosed and the most common cause of cancer death; over 1.2 million people were diagnosed with lung cancer in 2000, and over 900,000 died from the disease. Women in the U.S. lead the world in lung cancer incidence and mortality. The incidence and mortality rates for men in the U.S. are almost twice that of women. However, men in many other countries have much greater incidence and mortality rates. Hungary has the highest rates, and these are approximately 60% greater than those in the U.S. Breast cancer is the leading type of cancer among women worldwide and the second most common type of cancer overall, with over 1 million new cases diagnosed in 2000. The Netherlands and the U.S. have the greatest incidence of breast cancer, while Denmark and the Netherlands have the greatest mortality. Approximately 945,000 cases of colorectal cancer were diagnosed in 2000, making it the third most common type of cancer. For men, the Czech Republic has the greatest colorectal cancer incidence and mortality, while for women, New Zealand has the greatest incidence, and Hungary has the greatest mortality. Stomach cancer is the fourth most common cancer type, with approximately 875,000 new cases diagnosed in 2000. Stomach cancer incidence and mortality in the U.S. have decreased dramatically since the early part of the 20th century, and now the U.S. rates are among the lowest in the world. The incidence of stomach cancer for men is greatest in Korea and Japan, with rates that are about nine times higher than those in the U.S. For women, the incidence is greatest in Japan, with a rate that is about eight times higher than that in the U.S.

1.5 CONCLUSION

Cancer is a major public health problem. The outlook for Americans with cancer has improved steadily since the beginning of the 20th century, when few cancer victims survived for very long.[1,5]

TABLE 1.4
Worldwide Incidence and Mortality for the Most Common Types of Cancers [a]

Country	Lung Male	Lung Female	Breast Female	Colon/Rectum Male	Colon/Rectum Female	Stomach Male	Stomach Female
All							
New cases	34.9	11.0	35.7	19.1	14.4	21.5	10.4
Deaths	31.4	9.5	12.5	9.8	7.6	15.6	7.8
United States							
New cases	58.6	34.0	91.4	40.6	30.7	7.6	3.6
Deaths	53.2	27.2	21.2	15.9	12.0	4.5	2.3
Canada							
New cases	55.1	30.2	81.8	40.8	29.8	9.1	4.2
Deaths	50.4	25.0	22.8	16.4	11.6	6.4	3.2
Mexico							
New cases	24.5	9.1	38.4	9.3	9.0	16.5	12.3
Deaths	22.1	8.2	12.2	4.7	4.6	13.2	9.8
South America							
New cases	25.3	8.3	45.1	15.6	14.2	23.1	11.7
Deaths	22.6	7.4	14.8	8.0	7.3	18.3	9.3
United Kingdom							
New cases	47.6	21.8	74.9	35.4	25.3	12.4	5.5
Deaths	48.6	21.1	26.8	18.7	13.8	10.1	4.8
Germany							
New cases	50.2	11.4	73.6	45.0	32.0	16.2	9.2
Deaths	46.2	9.6	23.7	21.7	17.0	12.9	7.8
Hungary							
New cases	95.4	22.6	67.2	59.8	34.6	25.0	12.0
Deaths	86.2	20.0	25.3	33.5	20.9	21.0	10.0
Italy							
New cases	59.4	9.0	64.9	35.3	24.0	19.9	10.3
Deaths	52.6	8.2	20.7	16.4	11.3	14.6	7.6
Turkey							
New cases	40.1	4.0	20.7	9.1	5.3	10.5	5.6
Deaths	37.0	3.7	9.2	5.9	3.4	9.0	4.8
Russian Federation							
New cases	74.9	7.6	48.8	31.8	22.1	42.9	18.0
Deaths	68.2	6.8	16.7	17.5	12.7	35.6	15.2
India							
New cases	9.0	2.0	19.1	4.7	3.2	5.7	2.8
Deaths	8.4	1.8	9.9	3.1	2.1	5.3	2.6
China							
New cases	38.5	15.7	16.4	13.0	9.8	36.1	17.5
Deaths	33.2	13.4	4.5	7.2	5.3	27.0	13.0

[a] Adjusted to the WHO (World Health Organization) world standard population.

Source: Ferlay, J. et al., in *IARC Cancer Base*, IARC Press, Lyon, France, 2001, No. 5.

By the 1930s, only one out of five cancer patients survived 5 or more years after treatment and were considered "cured." Since then, the cure rate has climbed in almost every decade.[3,6] During the 1940s, the 5-year survival improved to one out of four; in the 1960s, it was one out of three; and in the 1970s, 38% of cancer patients were cured. Today, 51% of cancer patients survive for 5 years or more, and the American Cancer Society estimates that an additional 25 to 30% of cancer deaths could be prevented with earlier diagnosis and treatment.[7]

Yet, despite gains in treating many cancers and improved survival rates, cancer deaths continue to mount.[7,8] While 143 people per 100,000 died of cancer in 1930, today the figure is nearly 180 per 100,000. Much of this increase is due to increased life expectancy (cancer incidence increases with age) and to marked increases in lung cancer. If current trends continue, cancer is expected to be the leading cause of death in the United States by the year 2010.[7]

Many cancer myths persist, exaggerating its worst aspects. Although cancer is reputed to be a hopeless condition, nearly half of all cancer patients can expect to be alive and free of any sign of the disease in 5 years, a much better outlook than that facing most heart attack patients.

Intense anxiety and dread cause some cancer patients to delay seeking a diagnosis until the disease reaches an advanced, less treatable stage. And while cancer treatment is feared for its supposed pain and risk of spreading the disease, recent advances in surgical, radiologic, and chemotherapeutic therapies, and concurrent improvements in medicines to prevent nausea and pain (supportive care), have greatly lessened the uncomfortable side effects of cancer treatment.

Basic and applied research into the causes and cures for cancer continues, including investigations designed to dramatically change screening, diagnosis, and treatment. Two of the most promising areas include the study of cancer on the molecular level and the role of genetics in the development of cancer.

Although cancer therapy is often rigorous, debilitating, and uncomfortable, many people successfully undergo treatment with a minimal disruption of their normal lives. While others are quite ill during the intensive treatment stage, as is true for many diseases, they then recover to resume their careers and other pursuits. And while most people assume that cancer produces intense discomfort, many cancers are associated with little or no pain, or with pain that can often be controlled or minimized. Indeed, less dreaded diseases such as arthritis and certain neurological disorders cause more pain than most forms of cancer.

Overcoming the common cancer fears and misconceptions is important both for individual cancer patients and society as a whole. Cancer must be detected early to obtain the highest probability of cure, but even persistent cancer is often a chronic disease that, although unfortunate, can be effectively treated for many years. Increased information about cancer allows cancer patients and their families to make better decisions about where to go for treatment, how to pick the most appropriate treatments, and what to expect at each step in diagnosis and treatment.

REFERENCES

1. American Cancer Society, *SEER Cancer Statistics Review,* 1973–1999, Atlanta, 2003.
2. Ries, L. et al., SEER Cancer Statistics Review, 1973–1999, National Cancer Institute, Bethesda, MD, 2001.
3. American Cancer Society, Cancer Facts and Figures 2003, American Cancer Society, Atlanta, 2003.
4. Ferlay, J. et al., Globocan 2000: cancer incidence, mortality and prevalence worldwide, in *IARC Cancer Base*, IARC Press, Lyon, France, 2001, No. 5.
5. National Center for Health Statistics, *Health, United States 2001 with Urban and Rural Health Chartbook*, National Center for Health Statistics, Hyattsville, MD, 2001.
6. Reis, L. et al., Annual report to the nation on the status of cancer, 1973–1997, with a special section on colorectal dancer, *Cancer*, 88, 2398–2424, 2000.
7. National Cancer Institute, *Trends in Screening for Colorectal Cancer—United States, 1997 and 1999*, JAMA 285, 1570–1571, 2001. Bethesda, MD, 2003.

Section II

Pathophysiology of Cancer

2 Role of DNA Methylation in Cancer and Chemotherapy

Y-Yong Qiu, Bernard L. Mirkin, and Rama S. Dwivedi

CONTENTS

2.1 Introduction..17
2.2 Maintenance of Methylation Patterns in Genomic DNA 18
2.3 Role of DNA Hypermethylation in Tumorigenesis 18
2.4 DNA Hypomethylation in Tumor Development..21
2.5 DNA Methylation and Chemotherapeutic Response..22
2.6 Conclusions...23
Acknowledgment..24
References 24

2.1 INTRODUCTION

Alterations in the DNA methylation patterns of CpG (cytosine phosphoguanosine) dinucleotides, which are clustered in CpG islands, are very common in neoplastic cells.[1,2] These islands, approximately 29,000 in the human genome,[3,4] are very small stretches of DNA between 0.2 to 2 kb that are mostly found near the transcription start site of almost 60% of the promoters in human genes. They generally exist in an unmethylated state except for those genes located in inactive X-chromosomes of the female and/or silenced alleles of imprinted genes. The unmethylated status of CpG islands is very important in maintaining the transcriptional activity of genes that are expressed in tissue-specific manner.[5] A change in the methylation pattern of CpG islands alters the expression and transcriptional activity of critical genes involved in carcinogenesis and chemotherapy, leading to abnormal phenotypes by epigenetic mechanisms.

Epigenetic processes affect gene expression by evoking changes in DNA methylation and histone modification (acetylation), but they do not alter nucleotide sequences in genomic DNA. The enzymes responsible for maintaining the methylation patterns of CpG islands throughout the genome are DNA methyltransferases (DNMT), which transfer methyl groups from S-adenosylmethionine to cytosine bases. DNMT expression has been shown to be required for c-fos transformation *in vitro*, and mice carrying disrupted DNMT1 alleles have been reported to develop fewer intestinal polyps *in vivo*.[6] DNA hypermethylation leading to transcriptional silencing and/or functional inactivation of tumor suppressor genes has been postulated to be one of the key causative events in development of primary leukemia as well as prostate, lung, head and neck, breast, and colon tumors.[7,8]

A recent report has shown that DNA hypomethylation may contribute to the induction of tumors in mice via an epigenetic mechanism causing chromosomal instability.[9] In addition to facilitating neoplastic transformation, alterations in DNA methylation patterns have been shown to exert effects on embryonic development[10] and acquired drug resistance to cytotoxic agents.[11–13] Recent data from our laboratory have demonstrated the reversal of cisplatin-induced drug resistance in human neuroblastoma cell lines following an increase in DNA methylation.[12] This chapter discusses our current understanding of DNA methylation and its potential role in tumor initiation and cellular response to chemotherapy.

2.2 MAINTENANCE OF METHYLATION PATTERNS IN GENOMIC DNA

Normal embryonic development is regulated by the methylation pattern of genomic DNA. Loss of methylation leads to apoptosis in *Xenopus* embryos[10] and to abnormalities in embryonic stem and human cancer cells.[14] The methylation patterns of genomic DNA are primarily maintained by DNA methyltransferase (DNMT1), known as the maintenance methyltransferase, as it converts hemimethylated DNA to a fully methylated state.[15] This enzyme acts to maintain a stable methylation pattern of DNA following replication in differentiated cell types. The DNMT1 catalyzes the transfer of a methyl group to the cytosine base in CpG dinucleotides. DNMT3a and DNMT3b,[16] which belong to the methyltransferase family, are significantly expressed during embryogenesis and sustain global *de novo* methylation after implantation.[17]

Recent data have provided evidence that in addition to DNMT1, DNMT3a and 3b are required to establish and maintain the methylation patterns of DNA in tumor cells.[18] Another member of the methyltransferase family, DNMT3l does not have DNMT activity by itself. However, in coexpression with DNMT3a and 3b, its function is to establish the methylation pattern in female germ cells.[19] Similarly, DNMT2, another member of the DNMT family, does not possess any DNMT activity or specific phenotype in mice.[20] DNMT2 has been found to be highly conserved in *Drosophila* and expressed during embryogenesis,[21] where it may be responsible for some non-CpG methylation events.

The molecular mechanism(s) by which DNA methylation regulates gene expression and their transcriptional activity is not yet fully understood. Methyl cytosine binding proteins such as MeCP1, MeCP2,[22] MBD1, MBD2, MBD3, and MBD4 have been characterized as transcriptional repressors and implicated in methylation-dependent transcriptional regulation of gene expression.[23] These methyl cytosine-binding proteins have been shown to have a preferential binding to methylated promoters and to repress the transcription process.[24] Studies using Mbd$^{-/-}$ knock-out mice demonstrated the expression of the IL4 gene encoding interleukin-4 in a subset of T-helper cells in which it is normally repressed.[25] IL4 is generally expressed at higher levels in differentiating type 2 T-helper cells, suggesting its role in methylation-mediated repression processes.[26] Interference with the binding of transcriptional factors in the major groove of the DNA helix due to the presence of methylated cytosine could be another possible mechanism of DNA methylation repression and loss of gene expression.[27]

2.3 ROLE OF DNA HYPERMETHYLATION IN TUMORIGENESIS

The CpG islands of a number of genes that are usually nonmethylated in normal cell types undergo methylation in a variety of human tumors.[28] Methylation of CpG islands leads to inactivation or complete loss of expression and silencing of tumor suppressor genes involved in the determination of tumor cell phenotype. Transcriptional silencing and functional inactivation of tumor suppressor genes due to DNA hypermethylation provides an advantage of selective growth to neoplastic cells over normal cell types. DNA hypermethylation resulting in loss of transcriptional activity and function of a number of tumor-associated genes has been strongly implicated in malignant transformation (Table 2.1).

The promoter region of the p15^{INK4B} tumor suppressor gene is commonly hypermethylated in gliomas and to some extent in leukemia, leading to the loss of its transcriptional activity.[7]

TABLE 2.1
Silencing of Cancer-Related Critical Genes due to DNA Hypermethylation

Gene	Tumor Type	Reference
p16^{INK4A}	melanoma	[81]
p16^{INK4A} and p14^{ARF}	pancreatic carcinoma	[82]
p15^{INK4A}	AML	[46]
E-cadherin	diffused gastric carcinoma	[52]
VHL	clear cell renal carcinoma	[83]
hMLH1	colon cancer	[84]
RASSF1A	lung, breast, kidney, and neuroblastoma	[85]
RAR-β	breast cancer	[40]
GSTP1	prostate cancer	[8, 86]
O^6-MGMT	colon cancer	[87]
ER	breast cancer	[88]
BRCA1	breast and ovarian cancer	[35]
HIC1	brain tumor	[89]
PAX6	colon cancer	[90]

Transcriptional loss of the p15^{INK4B} gene has also been observed in conjunction with the p16^{INK4B} gene in lung, head and neck, breast, prostate, and colon cancers. Functional inactivation of the p16^{INK4A} gene due to DNA hypermethylation has been demonstrated in colon cancer.[29] The p16^{INK4A} gene expression is extremely critical in maintaining the retinoblastoma protein (Rb) in its active form. In esophageal adenocarcinoma, functional inactivation of p16^{INK4A} gene due to methylation is a common finding, leading to the loss of p16^{INK4A} gene expression.[30] Hypermethylation of p15^{INK4b} (4/10) and p16^{INK4a} gene (2/10) has been implicated in the transformation of AMM (agiogenic myloid metaplasia, also known as myelofibrosis) into leukemia.[31]

The HIC-1 (hypermethylated in cancer) gene is expressed in wild-type mammary cancer cells (MCF7), while its expression is lost in mutant p53 cells (MDA-MB-231). Treatment of p53 mutant cells with 5-aza-2′-deoxycytidine, a DNA methylation inhibitor, completely restores the loss of HIC-1 gene expression, suggesting an active involvement of DNA methylation.[32] Loss of the Cx 26 (connexin 26) gene due to methylation of its promoter region has also been reported in MDA-MDA-MB-453 breast cancer cell line.[33] Yuan et al.[34] have recently reported the loss of another tumor suppressor gene, the SYK, in breast cancer patients (12/37) and breast cancer cell lines (6/20) due to DNA hypermethylation.

The BRCA (the breast cancer predisposition gene 1) located on chromosome 17q21 was uniformly present in normal breast and ovarian tissues but was significantly reduced or undetectable in high-grade carcinoma. Increased methylation of the BRCA1 gene promoter was found to be a major cause for the decreased expression of this gene in many breast carcinomas (11/84). Loss-of-heterozygosity studies indicated that 20% of tumors had BRCA1 hypermethylation in ovarian cancer and subsequent silencing and functional inactivation of the BRCA1 gene.[35] Loss of E-cadherin gene expression by promoter hypermethylation has been reported in human epithelial cancers,[36] gastric cancer,[37] and in some breast carcinomas.[38] Germline mutations in the E-cadherin gene has been described in families with an inherited predisposition to gastric carcinoma.[38]

Inactivation of the helicase-like transcription factor (HLTF) gene, which encodes a SW1/SNF family protein, due to hypermethylation of its promoter region, was found in 43% of colon cancer tissues in contrast to breast and lung cancer, where this gene was not methylated.[39] The methylation and silencing of retinoic acid receptor-β (RAR-β), a tumor suppressor gene, has been shown to be responsible for its decreased expression in several malignant breast

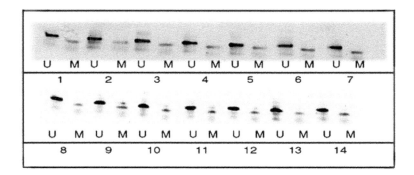

FIGURE 2.1 Methylation-specific PCR (MSP)[7] analysis of the MDR-1 gene promoter in neuroblastoma specimens procured from Children's Oncology Group (COG), Philadelphia, PA. This study shows that the MDR-1 gene promoter is hypomethylated (U) in neuroblastoma specimens with a variable degree of methylation.

cancer specimens, indicating its role in tumor progression and the pathogenesis of breast cancer.[40] Aberrant methylation of RAR-β, RASSF1A (ras association domain family 1A), DAP (death-associated protein)-kinase, p16, p15, p14, MGMT, and GSTp1 genes were present in primary nasopharyngeal carcinomas when compared with normal nasopharyngeal specimens. Silencing and transcriptional inactivation of the RASSF1A gene due to promoter hypermethylation was frequently noticed in 37 of 67 neuroblastoma biopsies.[41] Recent studies from our laboratory using the methylation-specific polymerase chain reaction (MSP)[7] to determine the methylation status of the multidrug resistance (MDR-1) gene promoter in neuroblastoma specimens suggest that variable degrees of methylation of the MDR-1 gene promoter are present in neuroblastoma specimens obtained from the Children's Oncology Group (COG, Philadelphia, PA) (Figure 2.1).

The p57KIP2 promoter, a cycline-dependent kinase inhibitor, has been reported to be highly methylated in gastric (5/35), hepatocellular (6/20), and pancreatic cancers (2/18) as well as AML (acute myeloid leukemia, 7/25).[42] The p15 and p16 gene promoters were frequently found to be methylated in AML. However, no correlation was established between the degree of promoter methylation and increased methyltransferase levels in some patients with AML.[43] Herman et al.[44] have proposed the characterization of AML by determining the methylation profile of p15 only, not of p16. Restriction landmark genomic scanning analysis showed a wide variation in the amount of aberrant methylation among different AML patients; however, a frequent hypermethylated phenotype has been confirmed in AML.[45] Methylation analyses of patients with myelodysplastic syndrome, which has a tendency to progress to AML, have shown the presence of p15 methylation in 50% (16/32) of the patients examined, without any sign of the p16 methylation.[46]

Increased methylation of the calcitonin,[47] HIC1,[48] ER,[49] and ABL1[50] genes with decreased gene expression has been reported in chronic myeloid leukemia patients. A variable level of methylation was noticed in p15,[51] ER,[49] HICI,[48] and E-cadherin[52] genes in B and T lineage samples obtained from acute lymphocytic leukemia patients. Promoter methylation and down-regulation of p21, a cycline-dependent kinase inhibitor and candidate tumor suppressor gene, have recently been reported in acute lymphocytic leukemia.[53] Decreased levels of the hTERT (the catalytic unit of telomerase) gene expression due to methylation of the hTERT promoter region has been correlated with decreased levels of telomerase activity in chronic lymphocytic leukemia. Methylation of the p16 and p15 gene promoters,[54] along with the DAP kinase gene, may be responsible for the pathogenesis and progression of advanced disease in multiple myeloma patients.[55]

It has been observed that hypermethylation of the transcriptional regulatory region of GSTP1, a gene encoding the Π-class glutathione-s-transferase (GST P-Π) protein, leads to silencing of GSTP1 expression in the most common forms of human prostate cancer.[8] Silencing of GSTP1 expression was reversed in LNCaP human prostate cancer cells following treatment with inhibitors of methyltransferase, such as 5-aza-cytidine and procainamide. The results of this study suggest the participation of DNA methylation in silencing GSTP1 expression and facilitating the progression of human prostate tumors.

2.4 DNA HYPOMETHYLATION IN TUMOR DEVELOPMENT

DNA hypomethylation and its association with human cancer was first described by Feinberg and Vogelstein in 1983[56] in animal models of carcinogenesis; hyper- and hypo-methylation of DNA were both reported.[57] Similar to DNA hypermethylation, alterations in DNA hypomethylation play a significant role in the progression and pathogenesis of tumors.[58,59] Cell lines with low metastatic tendency were converted to high metastatic cell lines following treatment with 5-azadeoxycytidine, an inhibitor of DNA methyltransferase (DNMT1), causing genome-wide DNA hypomethylation. Rats and mice treated with methyl-deficient diets showed an increased degree of genomic demethylation and activation of proto-oncogenes, leading to hepatocarcinogenesis.[9]

DNA hypomethylation was reported in DNMT1+/− heterozygous knock-out mice, and treatment of these mice with low doses of 5-azadeoxycytidine developed thymic lymphoma, while normal mice were not affected.[60] To further elucidate the role of DNA hypomethylation in the pathogenesis of tumor development, Jaenisch's laboratory has studied genomic hypomethylation in DNMT1 hypomorphic mice (DNMT1/chip/c), a heterozygous mouse model with significantly lower levels of DNA methyltransferase activity.[9] The DNMT1 hypomorphic mice developed aggressive T-cell lymphomas at 4 to 8 months of age with a high frequency of chromosome 15 trisomy. These results support the involvement of DNA hypomethylation in mouse tumorigenesis by producing chromosomal instability and increased loss of heterozygosity in regions harboring tumor suppressor genes. Genomic instability as a result of DNA hypomethylation has been reported in some colorectal tumor cell lines treated with 5-azacytidine and 5-azadeoxycytidine.[61]

Aberrant gene expression due to DNA hypomethylation has also been observed in some breast cancers,[62,63] in contrast to other findings that did not reveal any significant changes in DNA methylation patterns.[64] In another study, global DNA methylation analyses were performed in breast lesions and respective adjacent parenchyma; DNA hypomethylation was significantly increased in breast carcinomas and proved to be a causative factor in tumor development.[65]

A potential role for DNA hypomethylation as the cause of chromosomal abnormalities in patients with ICF (immunodeficiency, centromeric instability, and facial abnormality) syndrome has also been proposed. ICF syndrome is a rare recessive disorder caused by diverse mutations in the human DNMT3b DNA methyltransferase gene.[66] These mutations include two missense substitutions and a 3-aa insertion resulting from creation of a unique 3′ splice acceptor in the human DNMT3b gene due to DNA hypomethylation. Variable immunodeficiency, mild facial abnormalities, and chromosomal abnormalities due to hypomethylation of pericentromeric satellite 2 and satellite 3 DNA regions of chromosomes 1, 9, and 16 have been reported in patients affected with ICF syndrome.[60,66] Recent data from our laboratory demonstrate a variable degree of methylation of the MDR-1 gene promoter in neuroblastoma specimens and human neuroblastoma cell lines (Figure 2.1 and Figure 2.2) obtained from COG (Children's Oncology Group, Philadelphia, PA). Hypomethylation of the MDR1 gene promoter and its association with acute myeloid leukemia has been previously reported.[67]

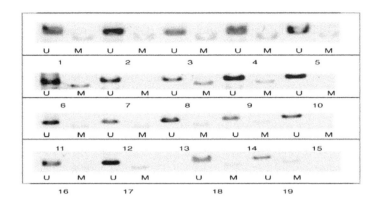

FIGURE 2.2 Methylation-specific PCR (MSP) analysis of the MDR-1 gene promoter methylation in human neuroblastoma (NB) cell lines (procured from Children's Oncology Group [COG] Philadelphia, PA) established from patients after chemotherapy.[80] Methylation status of MDR-1 gene is represented as unmethylated (U) and methylated (M) in each cell type. A total of 19 neuroblastoma cell lines[80] were examined by MSP and data indicate hypomethylation of MDR-1 gene in all cell lines tested. Hypomethylation (U) of MDR-1 gene indicates its increased expression and development of resistant phenotype in neuroblastoma cells after intensive chemotherapy.

2.5   DNA METHYLATION AND CHEMOTHERAPEUTIC RESPONSE

Alteration in DNA methylation patterns due to drug-mediated response may be capable of creating drug-resistant phenotypes by inactivating genes that generate products required for the pharmacologic actions of cytotoxic drugs. Change in DNA methylation patterns can also produce drug-resistant tumor cells, characterized by overexpression of specific gene products, which potentiate gene amplification.[68] In Chinese hamster V-79 cells, the DNA synthesis inhibitor 3′-azido-3′-deoxythymidine (zidovudine, AZT) was shown to induce resistance to AZT due to genome-wide DNA hypermethylation and low-frequency silencing of thymidine kinase (TK) gene expression.[69]

AZT has also been reported to produce drug resistance in Jurkat-T cells due to hypermethylation of the 5′ end of the TK gene and its decreased expression.[70] DNA methylation of the CpG-rich 5′ region of the deoxycytidine kinase (dCK) gene is potentially involved in its suppression of the gene and the resistance of tumor cells to arabinosylcytosine (ara-C), 2-chlorodeoxyadenosine (cladribine, CdA) and 2-chloro-2′-arabino-fluoro-2′-deoxyadenosine (CAFdA), which are purine nucleoside analogues that undergo phosphorylation.[71] Despite growing interest in the methylation-mediated silencing/inactivation of tumor-related genes, the signaling mechanisms for silencing remains largely unknown in oncogenesis.[72,73] Experimental evidence from several studies has identified a link between DNA mismatch repair (MMR) deficiency and cytotoxic drug resistance. A decreased expression of the MGMT gene, which repairs the DNA damage caused by alkylating agents such as cyclophosphamide, has been correlated with the hypermethylation of MGMT gene promoter and the survival of lymphoma patients.[74]

Drug resistance in the human ovarian cancer cell line, A2780, results in loss of expression of the mismatch repair (MMR) protein hMLH1.[75] Expression of hMLH1 gene is not detectable in ovarian, gastric, and endometrial tumors that are hypermethylated, in contrast to all unmethylated tumors that still express the hMLH1 protein. DNA analysis of the hMLH1 gene suggests that hypermethylation of the hMLH1 promoter is a common mechanism for loss of hMLH1 expression and cisplatin resistance in these cancers. Treatment of tumor-bearing mice with the demethylating agent 2′-deoxy-5-azacytidine (DAC) at a nontoxic dose induces hMLH1 expression and sensitizes the xenografts to cisplatin, carboplatin, temozolomide, and epirubicin treatment. DAC treatment does not sensitize the xenografts that lack MMR because of hMLH1 mutation.

These observations further suggest that another potential mechanism of drug resistance could be drug-mediated alterations in DNA methylation levels, causing transcriptional inactivation of cellular genes (switching on/off) whose products are required for drug reactivity. MMR proteins provide a link between DNA damage and apoptotic response effectors such as p[53] and p[57]. Abnormal DNA methylation of the ER (estrogen receptor) gene has been implicated in loss of estrogen receptor expression and subsequent hormone resistance in some human breast cancers.

Absence of a NotI-EcoRV fragment in the ALX3 gene, the human ortholog of the hamster homeobox gene Alx3 and its suppression in neuroblastoma tumors due to DNA hypermethylation has been recently observed by using the restriction landmark genome scanning (RLGS) and virtual genome scan (VGS) analyses, which very precisely identify amplification, deletion, or methylation changes in CpG islands in genomic DNA. Hypermethylation-associated silencing and inactivation of the caspase-8 gene has been reported in tissue biopsies obtained from neuroblastoma tumors.[76] Disruption of apoptotic pathways, caspase-8 inactivation, and resistance to TNF-related apoptosis-inducing ligand (TRAIL) in neuroblastoma cells suggest the involvement of gene hypermethylation in these events.[77] Loss of heterozygosity (LOH) on chromosomes 2q, 9p, and 18q due to hypermethylation has frequently been observed in neuroblastoma.[78] Recent data from our laboratory have demonstrated a significant increase in DNMT1 activity, global DNA methylation, and expression of DNMT1 and DNMT3b genes in drug-resistant human neuroblastoma (SK-N-SH) cell lines.[12,13]

2.6 CONCLUSIONS

Aberrant methylation patterns in genomic DNA may be one of the causative factors responsible for the progression and pathogenesis of a variety of human cancers (Figure 2.3). Alterations in methylation status due to DNA hypomethylation may produce chromosomal instability, rearrangement, and mutations in critical genes, whereas DNA hypermethylation may lead to silencing of

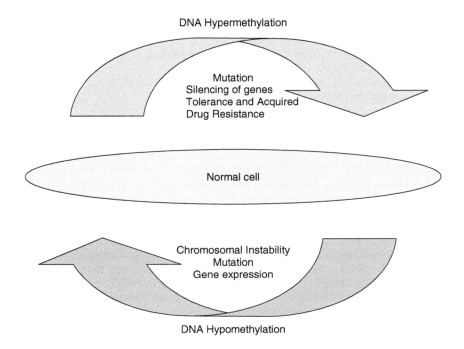

FIGURE 2.3 Effects of genome-wide DNA hypermethylation and hypomethylation on chromosomal instability, mutation, and gene expression.

tumor suppressor genes responsible for checking the uncontrolled growth of cancer cells and acquired drug resistance against cytotoxic agents. Due to its reversible nature, DNA methylation phenomena could be exploited as a "turn off and turn on" switch for gene expression in cancer pathogenesis and the reversal of drug resistance.[13] Alterations in genome-wide methylation may be prevented by providing balanced dietary supplements such as folate, which is an important carbon donor for methylation reaction. Folate deficiency leading to abnormal methylenetetrahydrofolate reductase gene (1p36) expression and altered DNA methylation status has been correlated with an increased risk of cancers.[79]

ACKNOWLEDGMENT

We thank the Children's Oncology Group (COG) Philadelphia, PA, for providing the neuroblastoma specimens and C. Patrick Reynolds, M.D., Ph.D. (Children's Hospital Los Angeles, CA), for DNA samples from neuroblastoma cell lines established from neuroblastoma patients. Assistance from Dr. James Herman, M.D. (Johns Hopkins University, Baltimore, MD), in providing the MDR-1 gene primers for MSP analysis is gratefully acknowledged. This investigation was partially supported by grants from the Anderson Foundation, North Suburban Medical Research Junior Board, Medical Research Institute Council, Children's Memorial Institution for Education and Research Program in Cancer Biology and Therapeutics, and the R. Wile Neuroblastoma and Chemotherapeutic Fund. The editorial assistance of Roberta Gerard and A. Woodworth and the technical support of Sandra Clark are gratefully appreciated.

REFERENCES

1. Jones, P.A. and Laird, P.W., Cancer epigenetics comes of age, *Nature Genetics*, 21, 163–167, 1999.
2. Leegwater, P.A., De Abreu, R.A., and Albertioni, F., Analysis of DNA methylation of the 5' region of the deoxycytidine kinase gene in CCRF-CEM-sensitive and cladribine (CdA)- and 2-chloro-2'-arabino-fluoro-2'-deoxyadenosine (CAFdA)-resistant cell, *Cancer Lett.*, 130 (1–2), 169–173, 1998.
3. Venter, J.C., Adams, M.D., Myers, E.W., Li, P.W., Mural, R.J. et al., The sequence of the human genome, *Science*, 291 (5507), 1304–1351, 2001.
4. Lander, E.S., Linton, L.M., Birren, B., Nusbaum, C., Zody, M.C. et al., Initial sequencing and analysis of the human genome, *Nature*, 409 (6822), 860–921, 2001.
5. Bird, A., DNA methylation de novo, *Science*, 286 (5448), 2287–2288, 1999.
6. Laird, P.W., Jackson-Gusby, L., Fazeli, A., Dickinson, S.L., Jung, W.E., Li, E., Weinberg, R.A., and Jaenisch, R., Suppression of intestinal neoplasia by DNA hypomethylation, *Cell*, 81, 197–205, 1995.
7. Herman, J.G., Jen, J., Merlo, A., and Baylin, S.B., Hypermethylation associated inactivation indicates a tumor suppressor role for p15INK4B1, *Cancer Res.*, 56, 722–727, 1996.
8. Lin, X., Asgari, K., Putzi, M.J., Gage, W.L., Yu, X., Cornblatt, B.S., Kumar, A., Piantadosi, S., DeWeese, T.L., De Marzo, A.M., and Nelson, W.G., Reversal of GSTP1 CpG island hypermethylation and reactivation of p-class glutathione S-transferase (GSTP1) expression in human prostate cancer cells by treatment with procainamide, *Cancer Res.*, 61, 8611–8616, 2001.
9. Gaudet, F., Hodgson, G.A.E., Jackson-Grusby, L., Dausman, J., Gray, J.W., Leonhardt, H., and Jaenisch, R., Induction of tumors in mice by genomic hypomethylation, *Science*, 300, 489–492, 2003.
10. Stancheva, I., Hensey, C., and Meehan, R.R., Loss of the maintenance methyltransferase, xDnmt1, induces apoptosis in Xenopus embryos, *EMBO J.*, 20 (8), 1963–1973, 2001.
11. Nyce, J.W., Drug-induced DNA hypermethylation: a potential mediator of acquired drug resistance during cancer chemotherapy, *Mutat. Res.*, 386 (2), 153–161, 1997.
12. Qiu, Y.Y., Mirkin, B.L., and Dwivedi, R.S., Differential expression of DNA-methyltransferases in drug resistant murine neuroblastoma cells, *Cancer Detect. Prev.*, 26 (6), 444–453, 2002.
13. Dwivedi, R.S., Qiu, Y.Y., Devine, J., and Mirkin, B.L., Role of DNA methylation in acquired drug resistance in neuroblastoma tumors, *Proc. Indian Natl. Sci. Acad.*, B69, 111–120, 2003.

14. Lei, H., Oh, S.P., Okano, M., Juttermann, R., Goss, K.A., Jaenisch, R., and Li, E., De novo DNA cytosine methyltransferase activities in mouse embryonic stem cells, *Development*, 122, 3195–3205, 1996.

15. Rountree, M.R., Bachman, K.E., and Baylin, S.B., DNMT1 binds HDAC2 and a new co-repressor, DMAP1, to form a complex at replication foci, *Nat. Genet.*, 25 (3), 269–277, 2000.

16. Xie, S., Wang, Z., Okano, M., Nogami, M., Li, Y., He, W.W., Okumura, K., and Li, E., Cloning, expression and chromosome locations of the human DNMT3 gene family, *Gene*, 236 (1), 87–95, 1999.

17. Okano, M., Bell, D.W., Haber, D.A., and Li, E., DNA methyltransferases Dnmt3a and Dnmt3b are essential for de novo methylation and mammalian development, *Cell*, 99 (3), 247–257, 1999.

18. Rhee, I., Bachman, K.E., Park, B.H., Jair, K.W., Yen, R.W., Schuebel, K.E., Cui, H., Feinberg, A.P., Lengauer, C., Kinzler, K.W., Baylin, S.B., and Vogelstein, B., DNMT1 and DNMT3b cooperate to silence genes in human cancer cells, *Nature*, 416 (6880), 552–556, 2002.

19. Hata, K., Okano, M., Lei, H., and Li, E., Dnmt3L cooperates with the Dnmt3 family of de novo DNA methyltransferases to establish maternal imprints in mice, *Development*, 129 (8), 1983–1993, 2002.

20. Okano, M., Xie, S., and Li, E., Dnmt2 is not required for de novo and maintenance methylation of viral DNA in embryonic stem cells, *Nucleic Acids Res.*, 26 (11), 2536–2540, 1998.

21. Gowher, H., Leismann, O., and Jeltsch, A., DNA of Drosophila melanogaster contains 5-methylcytosine, *EMBO J.*, 19 (24), 6918–6923, 2000.

22. Lewis, J.D., Meehan, R.R., Henzel, W.J., Maurer-Fogy, I., Jeppesen, P., Klein, F., and Bird, A., Purification, sequence, and cellular localization of a novel chromosomal protein that binds to methylated DNA, *Cell*, 69 (6), 905–914, 1992.

23. Hendrich, B. and Bird, A., Identification and characterization of a family of mammalian methyl-CpG binding proteins, *Mol. Cell Biol.*, 18 (11), 6538–6547, 1998.

24. Boyes, J. and Bird, A., DNA methylation inhibits transcription indirectly via a methyl-CpG binding protein, *Cell*, 64 (6), 1123–1134, 1991.

25. Hendrich, B., Guy, J., Ramsahoye, B., Wilson, V.A., and Bird, A., Closely related proteins MBD2 and MBD3 play distinctive but interacting roles in mouse development, *Genes Dev.*, 15 (6), 710–723, 2001.

26. Hutchins, A.S., Mullen, A.C., Lee, H.W., Sykes, K.J., High, F.A., Hendrich, B.D., Bird, A.P., and Reiner, S.L., Gene silencing quantitatively controls the function of a developmental trans-activator, *Mol. Cell*, 10 (1), 81–91, 2002.

27. Kass, S.U., Pruss, D., and Wolffe, A.P., How does DNA methylation repress transcription? *Trends Genet.*, 13 (11), 444–449, 1997.

28. Jones, P.A. and Baylin, S.B., The fundamental role of epigenetic events in cancer, *Nat. Rev. Genet.*, 3 (6), 415–428, 2002.

29. Herman, J.G., Merlo, A., Mao, L., Lapidus, R.G., Issa, J.P., Davidson, N.E., Sidransky, D., and Baylin, S.B., Inactivation of the CDKN2/p16/MTS1 gene is frequently associated with aberrant DNA methylation in all common human cancers, *Cancer Res.*, 55 (20), 4525–4530, 1995.

30. Bian, Y.S., Osterheld, M.C., Fontolliet, C., Bosman, F.T., and Benhattar, J., p16 inactivation by methylation of the CDKN2A promoter occurs early during neoplastic progression in Barrett's esophagus, *Gastroenterology*, 122 (4), 1113–1121, 2002.

31. Wang, J.C., Chen, W., Nallusamy, S., Chen, C., and Novetsky, A.D., Hypermethylation of the P15[INK4b] and P16[INK4a] in agnogenic myeloid metaplasia (AMM) and AMM in leukaemic transformation, *Br. J. Haematol.*, 116 (3), 582–586, 2002.

32. Fujii, H., Biel, M.A., Zhou, W., Weitzman, S.A., Baylin, S.B., and Gabrielson, E., Methylation of the HIC-1 candidate tumor suppressor gene in human breast cancer, *Oncogene*, 16 (16), 2159–2164, 1998.

33. Tan, L.W., Bianco, T., and Dobrovic, A., Variable promoter region CpG island methylation of the putative tumor suppressor gene Connexin 26 in breast cancer, *Carcinogenesis*, 23 (2), 231–236, 2002.

34. Yuan, B.Z., Durkin, M.E., and Popescu, N.C., Promoter hypermethylation of DLC-1, a candidate tumor suppressor gene, in several common human cancers, *Cancer Genet. Cytogenet.*, 140 (2), 113–117, 2003.

35. Fearon, E.R., BRCA1 and E-cadherin promoter hypermethylation and gene inactivation in cancer-association or mechanism, *J. Natl. Cancer Inst.*, 92 (7), 515–517, 2000.

36. Hirohashi, S., Inactivation of the E-cadherin-mediated cell adhesion system in human cancers, *Am. J. Pathol.*, 153 (2), 333–339, 1998.

37. Tamura, G., Yin, J., Wang, S., Fleisher, A.S., Zou, T., Abraham, J.M., Kong, D., Smolinski, K.N., Wilson, K.T., James, S.P., Silverberg, S.G., Nishizuka, S., Terashima, M., Motoyama, T., and Meltzer, S.J., E-cadherin gene promoter hypermethylation in primary human gastric carcinomas, *J. Natl. Cancer Inst.*, 92 (7), 569–573, 2000.

38. Guilford, P.J., Hopkins, J.B., Grady, W.M., Markowitz, S.D., Willis, J., Lynch, H., Rajput, A., Wiesner, G.L., Lindor, N.M., Burgart, L.J., Toro, T.T., Lee, D., Limacher, J.M., Shaw, D.W., Findlay, M.P., and Reeve, A.E., E-cadherin germline mutations define an inherited cancer syndrome dominated by diffuse gastric cancer, *Hum. Mutat.*, 14 (3), 249–255, 1999.

39. Moinova, H.R., Chen, W.D., Shen, L., Smiraglia, D., Olechnowicz, J., Ravi, L., Kasturi, L., Myeroff, L., Plass, C., Parsons, R., Minna, J., Willson, J.K., Green, S.B., Issa, J.P., and Markowitz, S.D., HLTF gene silencing in human colon cancer, *Proc. Natl. Acad. Sci. USA*, 99 (7), 4562–4567, 2002.

40. Widschwendter, M., Berger, J., Hermann, M., Muller, H.M., Amberger, A., Zeschnigk, M., Widschwendter, A., Abendstein, B., Zeimet, A.G., Daxenbichler, G., and Marth, C., Methylation and silencing of the retinoic acid receptor-beta2 gene in breast cancer, *J. Natl. Cancer Inst.*, 92 (10), 826–832, 2000.

41. Astuti, D., Agathanggelou, A., Honorio, S., Dallol, A., Martinsson, T., Kogner, P., Cummins, C., Neumann, H.P., Voutilainen, R., Dahia, P., Eng, C., Maher, E.R., and Latif, F., RASSF1A promoter region CpG island hypermethylation in phaeochromocytomas and neuroblastoma tumours, *Oncogene*, 20 (51), 7573–7577, 2001.

42. Kikuchi, T., Toyota, M., Itoh, F., Suzuki, H., Obata, T., Yamamoto, H., Kakiuchi, H., Kusano, M., Issa, J.P., Tokino, T., and Imai, K., Inactivation of p57^{KIP2} by regional promoter hypermethylation and histone deacetylation in human tumors, *Oncogene*, 21 (17), 2741–2749, 2002.

43. Melki, J.R., Vincent, P.C., and Clark, S.J., Concurrent DNA hypermethylation of multiple genes in acute myeloid leukemia, *Cancer Res.*, 59 (15), 3730–3740, 1999.

44. Herman, J.G., Civin, C.I., Issa, J.P., Collector, M.I., Sharkis, S.J., and Baylin, S.B., Distinct patterns of inactivation of p15^{INK4B} and p16^{INK4A} characterize the major types of hematological malignancies, *Cancer Res.*, 57 (5), 837–841, 1997.

45. Rush, L.J. and Plass, C., Alterations of DNA methylation in hematologic malignancies, *Cancer Lett.*, 185 (1), 1–12, 2002.

46. Uchida, T., Kinoshita, T., Nagai, H., Nakahara, Y., Saito, H., Hotta, T., and Murate, T., Hypermethylation of the p15^{INK4B} gene in myelodysplastic syndromes, *Blood*, 90 (4), 1403–1409, 1997.

47. Nelkin, B.D., Przepiorka, D., Burke, P.J., Thomas, E.D., and Baylin, S.B., Abnormal methylation of the calcitonin gene marks progression of chronic myelogenous leukemia, *Blood*, 77 (11), 2431–2434, 1991.

48. Issa, J.P., Zehnbauer, B.A., Kaufmann, S.H., Biel, M.A., and Baylin, S.B., HIC1 hypermethylation is a late event in hematopoietic neoplasms, *Cancer Res.*, 57 (9), 1678–1681, 1997.

49. Issa, J.P., Zehnbauer, B.A., Civin, C.I., Collector, M.I., Sharkis, S.J., Davidson, N.E., Kaufmann, S.H., and Baylin, S.B., The estrogen receptor CpG island is methylated in most hematopoietic neoplasms, *Cancer Res.*, 56 (5), 973–977, 1996.

50. Asimakopoulos, F.A., Shteper, P.J., Krichevsky, S., Fibach, E., Polliack, A., Rachmilewitz, E., Ben-Neriah, Y., and Ben-Yehuda, D., ABL1 methylation is a distinct molecular event associated with clonal evolution of chronic myeloid leukemia, *Blood*, 94 (7), 2452–2460, 1999.

51. Sakashita, K., Koike, K., Kinoshita, T., Shiohara, M., Kamijo, T., Taniguchi, S., and Kubota, T., Dynamic DNA methylation change in the CpG island region of p15 during human myeloid development, *J. Clin. Invest.*, 108 (8), 1195–1204, 2001.

52. Graff, J.R., Greenberg, V.E., Herman, J.G., Westra, W.H., Boghaert, E.R., Ain, K.B., Saji, M., Zeiger, M.A., Zimmer, S.G., and Baylin, S.B., Distinct patterns of E-cadherin CpG island methylation in papillary, follicular, Hurthle's cell, and poorly differentiated human thyroid carcinoma, *Cancer Res.*, 58 (10), 2063–2066, 1998.

53. Roman-Gomez, J., Castillejo, J.A., Jimenez, A., Gonzalez, M.G., Moreno, F., Rodriguez Mdel, C., Barrios, M., Maldonado, J., and Torres, A., 5′ CpG island hypermethylation is associated with transcriptional silencing of the p21(CIP1/WAF1/SDI1) gene and confers poor prognosis in acute lymphoblastic leukemia, *Blood*, 99 (7), 2291–2296, 2002.

54. Mateos, M.V., Garcia-Sanz, R., Lopez-Perez, R., Balanzategui, A., Gonzalez, M.I., Fernandez-Calvo, J., Moro, M.J., Hernandez, J., Caballero, M.D., Gonzalez, M., and San Miguel, J.F., p16/INK4a gene inactivation by hypermethylation is associated with aggressive variants of monoclonal gammopathies, *Hematol. J.*, 2 (3), 146–149, 2001.

55. Ng, M.H., To, K.W., Lo, K.W., Chan, S., Tsang, K.S., Cheng, S.H., and Ng, H.K., Frequent death-associated protein kinase promoter hypermethylation in multiple myeloma, *Clin. Cancer Res.*, 7 (6), 1724–1729, 2001.

56. Feinberg, A.P. and Vogelstein, B., Hypomethylation distinguishes genes of some human cancers from their normal counterparts, *Nature*, 301, 89–92, 1983.

57. Lengauer, C., CANCER: An Unstable Liaison, *Science*, 300 (5618), 442–443, 2003.

58. Esteller, M., Avizienyte, E., Corn, P.G., Lothe, R.A., Baylin, S.B., Aaltonen, L.A., and Herman, J.G., Epigenetic inactivation of LKB1 in primary tumors associated with the Peutz-Jeghers syndrome, *Oncogene*, 19 (1), 164–168, 2000.

59. Chen, Z., Ko, A., Yang, J., and Jordan, V.C., Methylation of CpG island is not a ubiquitous mechanism for the loss of oestrogen receptor in breast cancer cells, *Br. J. Cancer*, 77 (2), 181–185, 1998.

60. Ehrlich, M., Jiang, G., Fiala, E., Dome, J.S., Yu, M.C., Long, T.I., Youn, B., Sohn, O.S., Widschwendter, M., Tomlinson, G.E., Chintagumpala, M., Champagne, M., Parham, D., Liang, G., Malik, K., and Laird, P.W., Hypomethylation and hypermethylation of DNA in Wilms tumors, *Oncogene*, 21 (43), 6694–6702, 2002.

61. Haff, T., The effects of 5-azacytidine and 5-azadeoxycytidine on chromosome structure and function: implication for methylation-associated cellular processes, *Pharmacol. Ther.*, 65, 19–46, 1995.

62. Ferguson, A.T., Lapidus, R.G., Baylin, S.B., and Davidson, N.E., Demethylation of the estrogen receptor gene in estrogen receptor-negative breast cancer cells can reactivate estrogen receptor gene expression, *Cancer Res.*, 55 (11), 2279–2283, 1995.

63. Petrangeli, E., Lubrano, C., Ravenna, L., Vacca, A., Cardillo, M.R., Salvatori, L., Sciarra, F., Frati, L., and Gulino, A., Gene methylation of oestrogen and epidermal growth factor receptors in neoplastic and perineoplastic breast tissues, *Br. J. Cancer*, 72 (4), 973–975, 1995.

64. Kass, D.H., Shen, M., Appel, N.B., Anderson, D.E., and Saunders, G.F., Examination of DNA methylation of chromosomal hot spots associated with breast cancer, *Anticancer Res.*, 13 (5A), 1245–1251, 1993.

65. Soares, J., Pinto, A.E., Cunha, C.V., Andre, S., Barao, I., Sousa, J.M., and Cravo, M., Global DNA hypomethylation in breast carcinoma: correlation with prognostic factors and tumor progression, *Cancer*, 85 (1), 112–118, 1999.

66. Hansen, L.H., Vester, B., and Douthwaite, S., Core sequence in the RNA motif recognized by the ErmE methyltransferase revealed by relaxing the fidelity of the enzyme for its target, *RNA*, 5 (1), 93–101, 1999.

67. Nakayama, M., Wada, M., Harada, T., Nagayama, J., Kusaba, H., Ohshima, K., Kozuru, M., Komatsu, H., Ueda, R., and Kuwano, M., Hypomethylation status of CpG sites at the promoter region and overexpression of the human MDR1 gene in acute myeloid leukemias, *Blood*, 92, 4296–4307, 1998.

68. Nyce, J.W., Drug-induced DNA hypermethylation and drug resistance in human tumors, *Cancer Res.*, 49, 5829–5836, 1989.

69. Nyce, J., Leonard, S., Canupp, D., Schulz, S., and Wong, S., Epigenetic mechanisms of drug resistance: drug-induced DNA hypermethylation and drug resistance, *Proc. Natl. Acad. Sci. USA*, 90 (7), 2960–2964, 1993.

70. Wu, S., Liu, X., Solorzano, M.M., Kwock, R., and Avramis, V.I., Development of zidovudine (AZT) resistance in Jurkat T cells is associated with decreased expression of the thymidine kinase (TK) gene and hypermethylation of the 5' end of human TK gene, *J. Acquir. Immune Defic. Syndr. Hum. Retrovirol.*, 8 (8), 1–9, 1995.

71. Leegwater, P.A., De Abreu, R.A., and Albertioni, F., Analysis of DNA methylation of the 5' region of the deoxycytidine kinase gene in CCRF-CEM-sensitive and cladribine (CdA)- and 2-chloro-2'-arabino-fluoro-2'-deoxyadenosine (CAFdA)-resistant cells, *Cancer Lett.*, 130 (1–2), 169–173, 1998.

72. Conway, K.E., McConnell, B.B., Bowring, C.E., Donald, C.D., Warren, S.T., and Vertino, P.M., TMS1, a novel proapoptotic caspase recruitment domain protein, is a target of methylation-induced gene silencing in human breast cancers, *Cancer Res.*, 60 (22), 6242, 2000.

73. Eads, C.A., Lord, R.V., Kurumboor, S.K., Wickramasinghe, K., Skinner, M.L., Long, T.I., Peters, J.H., DeMeester, T.R., Danenberg, K.D., Danenberg, P.V., Laird, P.W., and Skinner, K.A., Fields of aberrant CpG island hypermethylation in Barrett's esophagus and associated adenocarcinoma, *Cancer Res.*, 60 (18), 5021–5026, 2000.

74. Esteller, M., Guo, M., Moreno, V., Peinado, M.A., Capella, G., Galm, O., Baylin, S.B., and Herman, J.G., Hypermethylation-associated inactivation of the cellular retinol-binding-protein 1 gene in human cancer, *Cancer Res.*, 62 (20), 5902–5905, 2002.

75. Karran, P., Mechanisms of tolerance to DNA damaging therapeutic drugs, *Carcinogenesis*, 22 (12), 1931–1937, 2001.

76. Teitz, T., Wei, T., Valentine, M.B., Vanin, E.F., Grenet, J., Valentine, V.A., Behm, F.G., Look, A.T., Lahti, J.M., and Kidd, V.J., Caspase 8 is deleted or silenced preferentially in childhood neuroblastomas with amplification of MYCN [see comments], *Nature Med.*, 6 (5), 529–535, 2000.

77. Eggert, A., Grotzer, M.A., Zuzak, T.J., Wiewrodt, B.R., Ikegaki, N., and Brodeur, G., Resistance to TRAIL-induced apoptosis in neuroblastoma cells correlates with a loss of caspase-8 expression, *Medical Pediatric Oncol.*, 35 (6), 603–607, 2000.

78. Takita, J., Hayashi, Y., Kohno, T., Yamaguchi, N., Hanada, R., Yamamoto, K., and Yokota, J., Deletion map of chromosome 9 and p16 (CDKN2A) gene alterations in neuroblastoma, *Cancer Res.*, 57 (5), 907–912, 1997.

79. Heijmans, B.T., Boer, J.M., Suchiman, H.E., Cornelisse, C.J., Westendorp, R.G., Kromhout, D., Feskens, E.J., and Slagboom, P.E., A common variant of the methylenetetrahydrofolate reductase gene (1p36) is associated with an increased risk of cancer, *Cancer Res.*, 63 (6), 1249–1253, 2003.

80. Keshelava, N., Groshen, S., and Reynolds, C.P., Cross-resistance of topoisomerase I and II inhibitors in neuroblastoma cell lines, *Cancer Chemother. Pharmacol.*, 45 (1), 1–8, 2000.

81. Baylin, S.B., Herman, J.G., Graff, J.R., Vertino, P.M., and Issa, J.P., Alterations in DNA methylation: a fundamental aspect of neoplasia, *Adv. Cancer Res.*, 72, 141–196, 1998.

82. Klump, B., Hsieh, C.J., Nehls, O., Dette, S., Holzmann, K., Kiesslich, R., Jung, M., Sinn, U., Ortner, M., Porschen, R., and Gregor, M., Methylation status of p16INK4A and p14ARF as detected in pancreatic secretion, *Brit. J. Cancer*, 88, 217–222, 2003.

83. Herman, J.G., Latif, F., Weng, Y., Lerman, M.I., Zbar, B., Liu, S., Samid, D., Duan, D.S., Gnarra, J.R., Linehan, W.M. et al., Silencing of the VHL tumor-suppressor gene by DNA methylation in renal carcinoma, *Proc. Natl. Acad. Sci. USA*, 91 (21), 9700–9704, 1994.

84. Herman, J.G., Umar, A., Polyak, K., Graff, J.R., Ahuja, N., Issa, J.P., Markowitz, S., Willson, J.K., Hamilton, S.R., Kinzler, K.W., Kane, M.F., Kolodner, R.D., Vogelstein, B., Kunkel, T.A., and Baylin, S.B., Incidence and functional consequences of hMLH1 promoter hypermethylation in colorectal carcinoma, *Proc. Natl. Acad. Sci. USA*, 95 (12), 6870–6875, 1998.

85. Hesson, L., Dallol, A., Minna, J.D., Maher, E.R., and Latif, F., NORE1A, a homologue of RASSF1A tumor suppressor gene is inactivated in human cancers, *Oncogene*, 22, 947–954, 2003.

86. Harden, S.V., Guo, Z., Epstein, J.I., and Sidransky, D., Quantitative GSTP1 methylation clearly distinguishes benign prostatic tissue and limited prostate carcinoma, *J. Urol.*, 169, 1138–1142, 2003.

87. Watts, C.K., Handel, M.L., King, R.J., and Sutherland, R.L., Oestrogen receptor gene structure and function in breast cancer, *J. Steroid Biochem. Mol. Biol.*, 41 (3–8), 529–536, 1992.

88. Ottaviano, Y.L., Issa, J.P., Parl, F.F., Smith, H.S., Baylin, S.B., and Davidson, N.E., Methylation of the estrogen receptor gene CpG island marks loss of estrogen receptor expression in human breast cancer cells, *Cancer Res.*, 54 (10), 2552–2555, 1994.

89. Li, Q., Jedlicka, A., Ahuja, N., Gibbons, M.C., Baylin, S.B., Burger, P.C., and Issa, J.P., Concordant methylation of the ER and N33 genes in glioblastoma multiforme, *Oncogene*, 16 (24), 3197–3202, 1998.

90. Toyota, M., Ahuja, N., Ohe-Toyota, M., Herman, J.G., Baylin, S.B., and Issa, J.P., CpG island methylator phenotype in colorectal cancer, *Proc. Natl. Acad. Sci. USA*, 96 (15), 8681–8686, 1999.

3 Metals, Metalloids, and Cancer

Gabriel Keith Harris and Xianglin Shi

CONTENTS

3.1 Introduction: Metals and Metalloids 29
3.2 Epidemiology: the Case for Metals and Cancer 30
3.3 Carcinogenic Metals 31
    3.3.1 Arsenic 32
    3.3.2 Beryllium 32
    3.3.3 Cadmium 32
    3.3.4 Chromium 33
    3.3.5 Cobalt 33
    3.3.6 Lead 33
    3.3.7 Nickel 33
    3.3.8 Iron and Copper 34
3.4 Mechanisms of Cancer Causation by Metals 34
    3.4.1 DNA Damage 34
    3.4.2 Inhibition of DNA Repair 35
    3.4.3 Generation of Reactive Oxygen Species 35
    3.4.4 Effects on Signal Transduction and Apoptosis 36
3.5 Conclusions 37
References 38

3.1 INTRODUCTION: METALS AND METALLOIDS

Mankind has sought out and worked with metals for millennia. The earliest known metalwork dates to approximately 5000 B.C.[1] Metals are ubiquitous in modern society. They form the structures of our buildings and our automobiles, and they are vital components of our computers and appliances. Many chemical reactions, including those used to manufacture products ranging from foods to drugs, are catalyzed by metals. Metals, including iron, chromium, and cobalt, are essential components of our diet.[2] Not all exposures to metal are benign, however. For example, various forms of the metalloid arsenic have been used for centuries as a poison.[3] Exposure to specific forms of metal, especially in the workplace, where workers may be exposed to high concentrations for many hours per day, has been linked to cancers of the lung, skin, and other organs. Because metals do not biodegrade, the level of environmental metal exposure tends to increase over time in areas where they are utilized. Archaeological data indicate that humans today possess levels of skeletal lead and cadmium many times higher than those found in ancient humans.[4] The purpose of this chapter is to discuss metals and metalloids that have been identified as carcinogens in humans via epidemiological studies. Special emphasis will be given to the free-radical and molecular mechanisms by which metals may induce or promote cancer.

Metals are defined as elements that have an unoccupied space in their outer valence shell, possess a positive charge in solution, and are good conductors of electricity. Metalloids are defined less stringently than metals. They are semiconductor elements that may or may not display metal characteristics, depending on chemical conditions.[5] The metalloid of greatest interest to cancer research is arsenic. The majority of elements in the periodic chart (96 of 113) can be classified as either metals or metalloids. Of these, only seven have been positively associated with human cancer.

The complex chemistry of metals defies simple categorization with respect to carcinogenesis. Even metals recognized as carcinogens vary greatly with respect to carcinogenic potential, which is often a function of oxidation state. For example, exposure to chromium (VI) increases cancer risk, whereas chromium (III) exposure does not.[6] It is, therefore, more correct to refer to specific metal forms (metal salts versus pure metals, for example) or oxidation states than to categorize any single metal as carcinogenic.

3.2 EPIDEMIOLOGY: THE CASE FOR METALS AND CANCER

Numerous epidemiological studies have examined the effects of metals on human cancer. The first epidemiological studies linking metals and human cancer date from the early 1900s.[7] Based on available epidemiological data, a number of government agencies — including the U.S. Environmental Protection Agency, the Occupational Safety and Health Administration, the National Institute for Occupational Safety and Health, and the National Toxicology Program, as well as nongovernmental advisory organizations such as the American Conference of Governmental Industrial Hygienists and the International Agency for Research on Cancer — have classified metals with regard to their carcinogenicity in humans. Although there is some disagreement among these organizations as to the carcinogenic potential of specific metal compounds, there is general agreement among them that a given metal (chromium, for example) may occur in carcinogenic forms. Table 3.1 lists the metals and metalloids that have been positively identified as human carcinogens by one or more of these organizations, as well as the metallic form involved, the organ sites they affect, and the relative risk of lung cancer (the most common form of metal-induced cancer) associated with each metal/metalloid.

Relative risk is an epidemiological estimate of the fold increase in risk for a particular disease compared with the general population. Relative risk also serves as a means of comparison among factors associated with a particular disease. Based on current data, the relative carcinogenic potential of the seven metals implicated as human lung carcinogens is as follows: arsenic > chromium > nickel > beryllium and cadmium > lead > cobalt.

One of the greatest difficulties for the epidemiologist studying metal-related cancer is the determination of the risk associated with a single metal or metal compound. Multiple types of metals, or multiple forms of a single metal, are often present in an industrial work environment. For example, welders may be exposed to fumes containing mixtures of iron, chromium, nickel, and manganese, whereas workers in copper or lead smelting plants may also be exposed to significant quantities of arsenic. In addition, persons who work with metals may be exposed to other cancer-causing agents, such as asbestos and cigarette smoke.[13] These factors, especially cigarette smoking, may play an additive or synergistic role in cancer causation. It is interesting to note that cigarettes, the major cause of human lung cancer, contain a number of carcinogenic metals, including cadmium, arsenic, nickel, and chromium, and are associated with a five-fold increase in the risk of lung cancer, higher than any single metal.[15] The result of these multicarcinogen exposures is that it is generally possible to identify specific metals and metal compounds as human carcinogens, but more difficult to quantify the level of risk associated with a given metal species.

Animal studies have been used to determine the relative carcinogenic potential of various metal species. Because traditional *in vitro* methods of determining mutagenic or carcinogenic potential, such as Ames testing, do not consistently identify metals as mutagens, animal studies are especially important to determine the relative carcinogenic potentials of metals. An advantage of animal studies is that they allow for the use of single metals or mixtures of known composition and concentration. The disadvantage of animal studies of metal-induced cancer is that they do not always reflect what has been observed in human epidemiological studies. This is especially true for arsenic, which has been associated with human carcinogenesis for more than a century, but which, in animals, does not produce the same lesions

TABLE 3.1
Metal and Metalloid Forms Positively Identified as Human Carcinogens

Metals/Metalloids	Forms Considered Human Carcinogens	Target Organs	Relative Lung Cancer Risk [a]
Arsenic	elemental arsenic; inorganic arsenic compounds	lung, skin, kidney, bladder, liver, lymphoma, leukemia	3.69
Beryllium	pure metal; alloys	lung	1.49
Cadmium	dust and fumes; carbonate, chloride, fluoroborate, nitrate, oxide, sulfate, and sulfite forms	lung, prostate	1.49
Chromium	chromium metal; chromite (mineral form); chromic acid; hexavalent (chromate) compounds; carbamate, phosphate, and triacetate forms	lung, nasal passages	2.78
Cobalt	chromium/cobalt alloys	lung	1.20
Lead	chromate and phosphate compounds	respiratory tract, kidney	1.33
Nickel	nickel metal; inorganic nickel; carbonyl and sulfide forms; soluble and insoluble forms	lung, nasal passages	1.56

[a] Relative risk is an epidemiological estimate of the fold increase in risk for a particular disease compared with the general population.

Sources: EPA, Health Assessment Document for Cadmium, U.S. Environmental Protection Agency, Washington, DC, 1981; EPA, Evaluation of the Potential Carcinogenicity of Lead and Lead Compounds, U.S. Environmental Protection Agency, Washington, DC, 1989a; IARC, Monograph on the Evaluation of Carcinogenicity: an Update of IARC Monographs, Vol. 1–42, Suppl. 7, World Health Organization, International Agency for Research on Cancer, Lyon, France, 1987; NTP, National Toxicology Program Report on Carcinogens, Public Health Service, U.S. Department of Health and Human Services, Washington, DC, 2002; Fu, H. and Boffetta, P., *Occup. Environ. Med.*, 52, 73, 1995; Steenland, K. et al., *Am. J. Ind. Med.*, 29, 474, 1996; Tuchsen, F. et al., *Scand. J. Work Environ. Health*, 22, 444, 1996.

or multi-organ carcinogenesis observed in humans.[16] Despite their limitations, animal studies have proven invaluable for the study of the mechanisms of a number of metal-related cancers.

3.3 CARCINOGENIC METALS

As Table 3.1 indicates, at least seven metals have been identified as human carcinogens, primarily of the lung. This may lead to the conclusion that these metals are extremely dangerous, and that any contact may result in cancer. On the contrary, based on available epidemiological evidence, it appears that the majority of metal-associated cancers are the result of chronic overexposure over a period of years or decades. With the exception of arsenic and lead, metal-induced cancers are largely preventable through the use of proper environmental controls or respiratory protective equipment. Surprisingly, two of the seven metals listed in Table 3.1, chromium and cobalt, are required in minute quantities as essential nutrients. Three others (arsenic, cadmium, and lead) may also be essential dietary nutrients.[2] As Paracelsus indicated, it is not the substance that makes the poison, but the dose.

3.3.1 ARSENIC

Arsenic is a metalloid that occurs naturally in both organic and inorganic forms as either arsenic (III) or arsenic (V). Arsenic (III), or arsenite, is considered to be the most toxic and is also the form most closely associated with cancer, both as a carcinogen and as a chemotherapy. Arsenic exposure may be occupational or may result from drinking arsenate-contaminated water. Workers in pesticide manufacturing plants, smelting operations, and the art glass industry have the greatest potential for exposure. Persons living in countries including Bangladesh, India, Taiwan, Argentina, and Chile are at risk for overexposure due to high arsenic levels in ground water.[17,18] Shellfish consumption is a third source of arsenic exposure, although the primarily methylated forms contained in these foods are rapidly excreted in urine and do not appear to increase cancer risk.

Arsenic is the most potent of the metals, both as a carcinogen and as a chemotherapy. Chronic exposure to arsenic has been associated with cancers of the lung, skin, and other organs. Lung cancers are most common in arsenic smelter workers and are typically adenocarcinomas.[19] One of the unique characteristics of arsenic carcinogenesis relative to other carcinogenic metals is that either respiration or ingestion can result in lung cancer. Skin cancers resulting from arsenic exposure may be of either basal- or squamous-cell origin and are unrelated to sun exposure. Other cancers, including lymphoma, leukemia, kidney, bladder, and liver cancer, have also been linked to arsenic exposure. Prior to the advent of radiation and other forms of chemotherapy, arsenic was used as a cancer treatment. Studies in China during the 1970s renewed interest in the use of arsenic to treat specific forms of cancer. Today, arsenic trioxide (III) is considered one of the most effective treatments for promyelocytic leukemia.[20]

3.3.2 BERYLLIUM

Beryllium is a light, heat-resistant metal that is often alloyed with other metals to increase their strength. Beryllium is used in nuclear reactors, spacecraft, rocket fuel, and aircraft. There is some disagreement concerning the carcinogenicity of beryllium in humans. Groups including the National Institute for Occupational Safety and Health and the National Toxicology Program consider beryllium a confirmed human carcinogen. Other organizations, such as the U.S. Environmental Protection Agency, list it as a probable carcinogen, and the International Agency for Research on Cancer "reasonably anticipates" that beryllium is a carcinogen.[11,13,21,22] The greatest anthropogenic source of beryllium results from the burning of fossil fuels, particularly coal. Beryllium exposure can result from respiration of dust or ingestion in food or water. Occupational exposure has been reported to increase lung cancer risk.[23,24] Environmental exposure does not appear to significantly increase cancer risk, however.[25]

3.3.3 CADMIUM

Relative to other metals described in this chapter, cadmium has come into use only recently, during the second half of the 20th century. It is used in galvanizing applications (because of its resistance to corrosion) and forms the positive pole of nickel/cadmium and zinc/cadmium batteries. Cadmium (II) is produced as a by-product of zinc and lead refining. In areas where soil is contaminated with cadmium, plant foods such as rice are a major source of exposure. Divalent cadmium is also present in seafood and in cigarettes.[16] At normal exposure levels, cadmium is rendered harmless by binding to metallothionein, although it may be stored in the body for more than two decades. Excessive exposure to cadmium overwhelms the ability of metallothionein to bind cadmium, and organ damage occurs, mainly in the kidneys and liver.[26] Curiously, the organ sites most commonly associated with cadmium-induced cancer are the lung and prostate. Workers in nickel/cadmium battery factories were found to have an increased risk for both types of cancer.[27] Some authors have suggested that cadmium-associated cancers do not result from cadmium exposure alone but are, instead, related to coexposure to nickel or lead.[28] Unlike carcinogenic metal forms, such as chromium (IV) and cobalt (II), cadmium (II) does not directly participate in the formation of reactive oxygen species. Instead, it indirectly enhances free-radical formation through the deletion of glutathione and other cellular reducing agents.[29]

3.3.4 CHROMIUM

Although chromium can exist in at least four oxidation states, trivalent and hexavalent chromium are the forms of greatest interest to human health. Chromium compounds are used in steel production, in the construction of furnaces, in leather tanning, in pigments, and in welding. Chromium (III) is the form most commonly encountered in nature. Unlike chromium (VI), ingested or respired chromium (III) is neither toxic nor carcinogenic. Instead, chromium (III) is an essential dietary nutrient with a recommended daily intake of 50 to 200 µg for adults. A component of the "glucose tolerance factor," chromium is necessary for the proper function of insulin.[2]

Hexavalent forms of chromium are the most important for industrial applications. Chromium (VI) is associated with an increased risk of lung cancer. It is primarily utilized by the steel industry for the production of ferrochrome, which is, in turn, used for the production of stainless steel. Unlike trivalent chromium, hexavalent chromium easily crosses cellular and nuclear membranes. Once inside the cell, it is converted by glutathione reductase, ascorbic acid, or any of a number of other intracellular reducing agents to chromium (V), chromium (IV), or chromium (III). It is the formation of reactive oxygen species during reduction of chromium and the binding of chromium (III) and chromium (V) to DNA that are considered to be the major causes of chromium-induced carcinogenesis.[30] Hexavalent chromium differs from other metal carcinogens in that it is also a potent mutagen.

3.3.5 COBALT

Cobalt is a by-product of copper refining. Its industrial applications include uses in alloys, magnets, and pigments. Radioactive forms of cobalt, including cobalt-57 and cobalt-60, have been used for over 30 years as chemotherapy agents. Cobalt (II) is associated with human lung cancer. According to the most recent epidemiological data, it is the weakest of the metal carcinogens.[14] Like trivalent chromium, cobalt is an essential nutrient. Indeed, it forms the center of the vitamin B-12 molecule.[2]

3.3.6 LEAD

Lead is the most common of the carcinogenic metals in the environment. In the past, it has been used as an additive in gasoline and in paint. Industrial uses of lead today include lead/acid batteries, alloys, and the use of the pure metal to form nonreactive surfaces. At least two lead-containing compounds (lead phosphate and lead chromate) are associated with cancer.[31,32] Lead workers are at increased risk for cancers of the respiratory tract, the digestive tract, and to a lesser extent, the kidney. Recent data indicate that some of the cancers attributed to lead may have been the result of arsenic contamination.[33] Like cadmium, lead compounds do not participate in redox cycling, but they do indirectly increase the formation of free radicals via the depletion of glutathione and other cellular reducing agents.[34]

3.3.7 NICKEL

Nickel is used in stainless steel alloys, for metal plating, and in batteries. Sources of nickel exposure most closely associated with cancer include metal refining operations, oil fly ash, and cigarette smoke. Several forms of nickel, both soluble and insoluble, have been identified as human carcinogens. Nickel (II) chloride is a soluble form of nickel that has been identified as a weak lung carcinogen. Nickel subsulfide (Ni_3S_2) and nickel oxide (NiO) are sparingly soluble forms of nickel that are potent lung and nasal carcinogens. Nickel sulfides are more carcinogenic than other forms of nickel.[35] It has been theorized that this is due to the formation of radicals by both the metal and the sulfur components. Insoluble metallic nickel is also a lung and nasal carcinogen. One explanation for the greater carcinogenic potential of less soluble nickel compounds is that they are more persistent in the respiratory tract and are phagocytized. Inside phagocytes, in the low-pH environment of lysosomes, nickel species that are insoluble or sparingly soluble at neutral pH can be solubilized and released into the cytoplasm at high concentrations.[36]

3.3.8 IRON AND COPPER

Although not considered to be carcinogenic metals per se, iron and copper are considered by some authors to be the ultimate carcinogens in metal-induced cancer. The introduction of carcinogenic metals to the body and the resulting damage they induce may cause the release of iron and copper from the sequestrant proteins to which they are normally bound tightly. The resulting free-radical reactions and DNA damage may lead to an initiation of cancer.[37] In support of this hypothesis, experimental data indicate that iron-overload diseases increase the risk for liver cancer (where iron is concentrated) and other cancers as well.[38]

3.4 MECHANISMS OF CANCER CAUSATION BY METALS

In this section, we discuss the possible mechanisms by which metals may induce cancer. A review of the current literature reveals several common themes with regard to the effects of carcinogenic metals on the cellular level. These include DNA damage, the inhibition of DNA repair, generation of reactive oxygen species, and effects on apoptosis and signal transduction. These are not discrete phenomena but, rather, are interrelated in the pathology of metal-induced cancer.

3.4.1 DNA DAMAGE

DNA damage is considered to be the initiating step in cancer causation. If left unrepaired, this damage can result in cell death or a permanent mutation. Accumulated mutations in oncogenes or tumor suppressor genes may then result in cancer. Metals are capable of inducing a wide range of DNA damage, including strand breaks, depurinations, base modifications, and DNA cross-links. The effect of metals on DNA depends on the type of metal in question. Nickel (II) is capable of inducing DNA strand breaks at cytosine, thymine, and guanine residues, whereas chromium (VI) appears to target guanine selectively.[39,40] Strand breaks may also be caused by the attack of a DNA sugar by the hydroxyl radical. Chromium (VI) and nickel (II) can cause depurination at guanine or adenine residues. Metals are capable of inducing two base modifications in DNA that are associated with cancer, including 8-hydroxy-2′-deoxyguanosine and O-6-methylguanine. The formation of cancer-associated DNA-base alterations as a result of metal-induced free-radical reactions is, perhaps, the strongest link between metal-induced cancer and reactive oxygen species. Cross-links are the most common form of DNA damage induced by metals. Cross-links can result from noncovalent binding of DNA complexes or the formation of covalent bonds between strands.[37]

In addition to effects on DNA, metals can also affect protein expression, structure, and function. Damage to protein is generally not considered as serious as DNA mutations because protein is a "renewable resource." Protein alterations can have pronounced effects on essential cell functions, however, including the inhibition of DNA repair. Metals can oxidize sulfhydryl groups to disulfides, as well as bind to cysteine, histidine, arginine, lysine, or proline. Conformational and functional changes result from these interactions. Protein–protein and protein–DNA cross-links can also occur. In addition, binding of metals to protein may enhance their ability to form free radicals, thus causing further damage to the protein. Metals can also bind to the chromatin protein associated with DNA or to the DNA itself. Chromatin cross-links have been identified in workers exposed to nickel and chromium.[41]

3.4.2 INHIBITION OF DNA REPAIR

Arsenic, nickel, lead, cadmium, and cobalt, are weak mutagens. Since mutagenicity is related to carcinogenic potential, they might be expected to lack carcinogenic activity. In fact, they are classified as human carcinogens. This apparent paradox is partly explained by the fact that these metals also inhibit DNA repair. Metals can inhibit DNA repair in several ways: by interfering with mismatch repair, by base excision, or by nucleotide excision. Many DNA repair proteins contain a zinc-finger motif. DNA repair can be inhibited either through oxidative inactivation of these proteins or by the exchange of zinc for another metal.[42] Arsenic (III) inhibits DNA repair through oxidative mechanisms. Nickel (II) chloride inhibits the removal of oxidatively damaged DNA bases and inhibits the ligation of broken DNA strands. Both nickel (II) chloride and nickel oxide enhance the mutagenic activity of the carcinogen benzo(a)pyrene by inhibiting DNA repair. Other carcinogenic metals, including arsenic, cadmium, and cobalt, share this ability to enhance the mutagenic activity of directly acting (DNA-damaging) carcinogens.[38] In addition to the inhibition of DNA repair itself, metals including cadmium, nickel, and cobalt are capable of inactivating zinc-finger proteins that are involved in the detection of DNA repair. In this way, metals are capable of compromising not only DNA repair itself, but the ability to detect DNA damage as well. The mechanism by which these repair proteins are inactivated involves the replacement of metals essential to DNA-repair protein function (zinc and magnesium) with other metals. Cadmium (II), nickel (II) and cobalt (II) are all capable of this activity.[43]

3.4.3 GENERATION OF REACTIVE OXYGEN SPECIES

The formation of reactive oxygen species (ROS) is an inevitable consequence of oxidative metabolism. In the body, the levels of these potentially damaging chemicals are kept in check by a complex system of reducing agents and enzymes that serve to convert ROS into less harmful species. These systems include glutathione, antioxidant vitamins and their precursors (vitamins C and E, beta-carotene), and natural chelators and sequestrants, such as metallothionein. Not all ROS generation is undesirable, as some ROS serve to relay intracellular signals. Metals can upset this delicate balance between formation and disappearance of ROS, tipping the balance in favor of oxidation, a state known as oxidative stress.

Major forms of oxygen-centered ROS include hydrogen peroxide (H_2O_2), superoxide radical ($O_2^{\bullet-}$), and the hydroxyl radical ($^{\bullet}OH$). The hydroxyl radical is particularly toxic because it reacts with any biological molecule in its vicinity, including DNA, RNA, or protein.

$$M^{(n+)} + H_2O_2 \rightarrow M^{(n+1)} + {}^{-}OH + {}^{\bullet}OH \tag{3.1}$$

$$M^{(n+1)} + O_2^{\bullet-} \rightarrow M^{(n+)} + O_2 \tag{3.2}$$

$$H_2O_2 + O_2^{\bullet-} \rightarrow O_2 + {}^{-}OH + {}^{\bullet}OH \tag{3.3}$$

Equation 3.1 through Equation 3.3 demonstrate the effects of metals that possess multiple oxidation states (transition metals) on the formation of these radicals. In Fenton-type reactions (Equation 3.1), metal oxidation ($M^{(n+)} \rightarrow M^{(n+1)}$) is coupled to H_2O_2 disproportionation to form the hydroxyl ion and the hydroxyl radical. Equation 3.2 demonstrates the ability of superoxide to reduce an oxidized metal through the donation of a single electron. It is this reduction that can, in turn, catalyze Haber–Weiss type reactions (Equation 3.3), in which hydrogen peroxide reacts with the superoxide radical to form molecular oxygen and the hydroxyl radical. In fact, the reaction in Equation 3.3 will not proceed unless it is coupled to metal reduction.[44] The reactions described in Equation 3.1 through Equation 3.3 can function in a continuous cycle wherein metals are repeatedly oxidized and reduced (thus the term *redox cycling*). This oxidation–reduction cycle results in the formation of increasingly greater quantities of the highly toxic hydroxyl radical, as both the Fenton and Haber–Weiss reactions produce this radical. This continual supply of $^{\bullet}OH$ mimics the effects of one of the most potent known carcinogens, ionizing radiation.[45] Chromium (VI), nickel (II), and cobalt (II) are capable of producing hydroxyl radicals and inducing DNA damage via Fenton and Haber–Weiss reactions.

Metal chelation may enhance or inhibit the Fenton reaction, depending on the metal and the chelator in question. Chelation of iron (II) by EDTA enhances the formation of hydroxyl radical, while deferoxamine, another chelator, reduces its formation. This is significant because peptides or proteins can chelate metals in the body, thus influencing the resulting degree of damage. The formation of hydroxyl radicals by nickel (II) and cobalt (II) is enhanced by this type of chelation. In addition to the Fenton and Haber–Weiss reactions, metals can also catalyze the formation of the hydroxyl radical via reaction with hypochlorite (HOCl), which is produced by neutrophils.[37]

In addition to the hydroxyl radical, metals catalyze the formation of a wide variety of radical species through numerous types of reactions with lipids. These processes are collectively known as lipid peroxidation or autoxidation. Autoxidation occurs in three basic stages: initiation, propagation, and termination (Equation 3.4 through Equation 3.10). RH represents a lipid molecule; ROOH represents a lipid peroxide. Initiation reactions (Equation 3.4 and Equation 3.5) are typically slow, but metals can dramatically increase reaction rates by lowering the energies of activation and by the donation of electrons. In Equation 3.4, a lipid radical ($R^{\bullet}$) is formed. In Equation 3.5, the cleavage of two lipid molecules results in the formation of an alkoxy ($RO^{\bullet}$) and a peroxyl radical ($ROO^{\bullet}$). Equation 3.6 and Equation 3.7 illustrate propagation reactions. First, a lipid radical reacts with molecular oxygen to form a peroxyl radical. The peroxyl radical can then react with other lipids to re-form a lipid peroxide and to form a lipid radical (Equation 3.7), thus "propagating" free-radical reactions. In much the same way as the Fenton and Haber–Weiss reactions, the initiation and propagation reactions of autoxidation can form a cycle, resulting in the formation of ever greater quantities of lipid radicals. The final stage of autoxidation is termination (Equation 3.8 through Equation 3.10), in which lipid and peroxyl radicals react with one another to form nonradical compounds.[46]

$$RH \rightarrow R^{\bullet} + H^{\bullet} \qquad (3.4)$$

$$2ROOH \rightarrow RO^{\bullet} + ROO^{\bullet} + H_2O \qquad (3.5)$$

$$R^{\bullet} + O_2 \rightarrow ROO^{\bullet} \qquad (3.6)$$

$$ROO^{\bullet} + RH \rightarrow ROOH + R^{\bullet} \qquad (3.7)$$

$$R^{\bullet} + R^{\bullet} \rightarrow R\text{-}R \qquad (3.8)$$

$$R^{\bullet} + ROO^{\bullet} \rightarrow ROOR \qquad (3.9)$$

$$ROO^{\bullet} + ROO^{\bullet} \rightarrow ROOR + O_2 \qquad (3.10)$$

Autoxidation is theorized to be important to the pathology of cancer because of its ability to change the permeability of cell membranes and to form organic peroxides. Changes in membrane permeability affect sodium channels and proteins involved in signal transduction. Changes in mitochondrial membrane permeability can result in increased formation of ROS and/or apoptosis.[47] Organic peroxides formed during autoxidation reactions can form hydroxyl radicals via Fenton and Haber–Weiss types of reactions. Theories concerning the importance of autoxidation in cancer are supported by the increases in measures of oxidative stress and cell permeability in human cancer patients.[48]

3.4.4 Effects on Signal Transduction and Apoptosis

The presence of metals can affect signal transduction, especially for genes known to be responsive to the oxidative status of the cell, such as p53. This will, in turn, affect the actions of the cell controlled by these genes. One of the effects most commonly associated with cancer-causing metals is apoptosis. Metals can cause apoptosis via extrinsically (cell receptor) or intrinsically (mitochondrial or DNA damage) mediated mechanisms. Apoptosis can be induced through p53-dependent or p53-independent pathways.[49] Chromium (VI) can indirectly induce p53-mediated apoptosis in at least three ways: through direct DNA damage, through activation of mitogen-activated (MAP) kinases upstream of p53, and through the oxidative activation of p53 itself. Reduced forms of chromium or the reactive oxygen species produced by them both have the capacity to damage DNA. Damage, such as DNA strand breaks, activates upstream kinases, such as DNA protein kinase, *Ataxia telangiectasia* mutated protein kinase, and the *Ataxia telangiectasia* and rad3[+]-related protein kinase, all of which are capable of activating p53.[50] The binding of chromium (III) and chromium (V) to DNA can result in the activation of MAP kinases, such as c-Jun N-terminal kinase and p38, which, in turn, can activate p53 without DNA damage. Due to the presence of cysteine residues within the protein, p53 is sensitive to and can be activated by changes in oxidative conditions within the cell.[51]

Arsenic-induced apoptosis is independent of p53 and is mediated through direct mitochondrial damage and MAP kinase activation. Arsenic can cause M-phase cell-cycle arrest and apoptosis in p53-negative cells. Arsenic damages mitochondria via free-radical mechanisms. Activation of either p38 or c-Jun N-terminal kinase subsequently activates growth arrest and DNA damage (GADD) genes. Activation of GADD45 causes arsenic-induced cell-cycle arrest.[52]

Paradoxically, both apoptosis and the inhibition of apoptosis can be carcinogenic. Generalized apoptosis of a cell population may select for apoptosis-resistant cells. The inhibition of apoptosis may allow for the growth and replication of cells that contain substantial DNA damage, thus perpetuating potentially carcinogenic cell traits. Chromium (VI) appears to do both of these, as it initially activates and later inactivates p53.

Although many of the effects of metals appear to be mediated through free-radical production, this does not explain all of the effects of metals. For example, chromium (VI) is a stronger pro-oxidant and mutagen than arsenic (III); however, arsenic (III) is a more potent carcinogen. One explanation for this is the differing ways in which metals affect the signal transduction pathways of cancer-related genes. The nuclear transcription factor kappa-B (NFκB) is known to be up-regulated in many cancers. Arsenic (III) and Cr (VI), both human carcinogens, have been shown to both activate and inactivate NFκB.[53] The extent to which activation or inactivation occurs appears to be dependent both on metal concentration and cell type. At levels that are physiologically achievable *in vivo* (in the low micromolar range), arsenic (III) appears to induce NFκB.

A number of authors have proposed that this activation is due to the production of free radicals. However, three experimental findings indicate that the mechanism may also be independent of oxidation. First of all, the binding of the NFκB protein to DNA, which is necessary for its activity, is reduced when the NFκB protein itself is oxidized. In other words, the NFκB protein is more effective as a transcription factor when it has not been oxidized. Second, chromium (VI) generates more radicals than arsenic (III) but induces NFκB to a lesser extent. Finally, the signal transduction pathways known to control NFκB activation (LPS, IL-1, Toll, CD28) are not activated by oxidative stress. Together, these findings indicate that carcinogenic metals may have mechanisms that are independent of or in addition to free-radical formation, at least with regard to NFκB.[54]

3.5 CONCLUSIONS

Mankind has used metals for millennia, but the relation between metals and cancer has only been known for over a century. Only in the past three decades have the tools been available to analyze the molecular and cellular effects of metals on cancer, and only very recently has it been possible to examine the role of free radicals in normal and disease states. Because of their ability to produce free radicals, transition metals provide a unique means by which to study not only metal-related diseases, but the effects of free radicals on DNA damage, intracellular signaling, and cell-to-cell communication. With the completion of the human genome project and the use of novel technologies such as genomics and proteomics, it will soon be possible to examine the global effects of free radicals on genes and their expression.

REFERENCES

1. Biehl, P.F., The Archaeology of Europe, in *The International Encyclopedia of Social and Behavioral Sciences*, Smelsar, N.J. and Balts, P.B., Eds. Elsevier Science, Oxford, 2002, p. 4913.
2. Ziegler, E.E. and Filer, L.J., Eds., *Present Knowledge in Nutrition*, ILSI Press, Washington, DC, 1996.
3. Anonymous, Poisoning in history, *JAMA*, 287, 1500, 2002.
4. Gonzalez-Reimers, E. et al., Bone cadmium and lead in prehistoric inhabitants and domestic animals from Gran Canaria, *Sci. Total Environ.*, 301, 97, 2003.
5. Lewis, R.J., *Hawley's Condensed Chemical Dictionary*, John Wiley & Sons, New York, 1997.
6. Langard, S., One hundred years of chromium and cancer: a review of epidemiological evidence and selected case reports, *Am. J. Ind. Med.*, 17, 189, 1990.
7. Clemens, F. and Landolph, J.R., Genotoxicity of samples of nickel refinery dust, *Toxicol. Sci.*, 25, 25, 2003.
8. EPA, Health Assessment Document for Cadmium, U.S. Environmental Protection Agency, Washington, DC, 1981.
9. EPA, Evaluation of the Potential Carcinogenicity of Lead and Lead Compounds, U.S. Environmental Protection Agency, Washington, DC, 1989a.
10. IARC, Monograph on the Evaluation of Carcinogenicity: an Update of IARC Monographs, Vol. 1-42, Suppl. 7, World Health Organization, International Agency for Research on Cancer, Lyon, France, 1987.
11. NTP, National Toxicology Program Report on Carcinogens, Public Health Service, U.S. Department of Health and Human Services, Washington, DC, 2002.
12. Fu, H. and Boffetta, P., Cancer and occupational exposure to inorganic lead compounds: a meta-analysis of published data, *Occup. Environ. Med.*, 52, 73, 1995.
13. Steenland, K. et al., Review of occupational lung carcinogens, *Am. J. Ind. Med.*, 29, 474, 1996.
14. Tuchsen, F. et al., Incidence of lung cancer among cobalt-exposed women, *Scand. J. Work Environ. Health*, 22, 444, 1996.
15. Agudo, A. et al., Lung cancer and cigarette smoking in women: a multicenter case-control study in Europe, *Int. J. Cancer*, 88, 820, 2000.
16. Goyer, R.A. and Clarksen, T.W., Toxic effects of metals, in *Casarett and O'Doul's Toxicology: the Basic Science of Poisons*, 5th ed., Klaassen, C.D., Ed., McGraw-Hill, New York, 1996.
17. Mahata, J. et al., Chromosomal aberrations and sister chromatid exchanges in individuals exposed to arsenic through drinking water in West Bengal, India, *Mutat. Res.*, 534, 133, 2003.
18. Smith, A.H. et al., Marked increase in bladder and lung cancer mortality in a region of Northern Chile due to arsenic in drinking water, *Am. J. Epidemiol.*, 147, 660, 1998.
19. Wicks, M.J. et al., Arsenic exposure in a copper smelter as related to histological type of lung cancer, *Am. J. Ind. Med.*, 2, 25, 1981.
20. Waxman, S. and Anderson, K.C., History of the development of arsenic derivatives in cancer therapy, *Oncologist*, 6, 3, 2001.
21. EPA, Integrated Risk Information System on Beryllium, U.S. Environmental Protection Agency, Washington, DC, 1999.
22. IARC, IARC Monographs on the Evaluation of the Carcinogenic Risk to Humans: Beryllium, Cadmium, Mercury and Exposures in the Glass Manufacturing Industry, Vol. 58, World Health Organization, International Agency for Research on Cancer, Lyon, France, 1994.
23. Steenland, K. and Ward, E., Lung cancer incidence among patients with beryllium disease: a cohort mortality study, *J. Natl. Cancer Inst.*, 83, 1380, 1991.
24. Ward, E. et al., A mortality study of workers at seven beryllium processing plants, *Am. J. Ind. Med.*, 22, 885, 1992.
25. McGavran, P.D., Rood, A.S., and Till, J.E., Chronic beryllium disease and cancer risk estimates with uncertainty for beryllium released to the air from the Rocky Flats Plant, *Environ. Health Perspect.*, 107, 731, 1999.
26. Jin, T., Lu, J., and Nordberg, M., Toxicokinetics and biochemistry of cadmium with special emphasis on the role of metallothionein, *Neurotoxicology*, 19, 529, 1998.
27. Waalkes, M.P., Cadmium carcinogenesis in review, *J. Inorg. Biochem.*, 79, 241, 2000.
28. Ades, A.E. and Kazantzis, G., Lung cancer in a non-ferrous smelter: the role of cadmium, *Br. J. Ind. Med.*, 45, 435, 1988.

29. Stohs, S.J. et al., Oxidative mechanisms in the toxicity of chromium and cadmium ions, *J. Environ. Pathol. Toxicol. Oncol.*, 20, 77, 2001.

30. Shi, X. et al., Reduction of chromium (VI) and its relationship to carcinogenesis, *J. Toxicol. Environ. Health B Crit. Rev.*, 2, 87, 1999.

31. Neuberger, J.S. and Hollowell, J.G., Lung cancer excess in an abandoned lead-zinc mining and smelting area, *Sci. Total Environ.*, 25, 287, 1982.

32. Sheffet, A. et al., Cancer mortality in a pigment plant utilizing lead and zinc chromates, *Arch. Environ. Health*, 37, 44, 1982.

33. Steenland, K. and Boffetta, P., Lead and cancer in humans: where are we now? *Am. J. Ind. Med.*, 38, 295, 2000.

34. Tatrai, E. et al., Comparative *in vitro* toxicity of cadmium and lead on redox cycling in type II pneumocytes, *J. Appl. Toxicol.*, 21, 479, 2001.

35. Norseth, T., Cancer hazards caused by nickel and chromium exposure, *J. Toxicol. Environ. Health*, 6, 1219, 1980.

36. Costa, M. et al., Phagocytosis, cellular distribution, and carcinogenic activity of particulate nickel compounds in tissue culture, *Cancer Res.*, 41, 2868, 1981.

37. Kasprzak, K.S., Oxidative DNA and protein damage in metal-induced toxicity and carcinogenesis, *Free Radical Biol. Med.*, 32, 958, 2002.

38. Blanc, J.F., Bioulac-Sage, P., and Balabaud, C., Iron overload and cancer, *Bull. Acad. Natl. Med.*, 184, 355, 2000.

39. Kawanishi, S., Inoue, S., and Yamamoto, K., Site-specific DNA damage induced by nickel(II) ion in the presence of hydrogen peroxide, *Carcinogenesis*, 10, 2231, 1989.

40. Xu, J. et al., Chromium(VI) treatment of normal human lung cells results in guanine-specific DNA polymerase arrest, DNA-DNA cross-links and S-phase blockade of cell cycle, *Carcinogenesis*, 17, 1511, 1996.

41. Costa, M., Molecular targets of nickel and chromium in human and experimental systems, *Scand. J. Work Environ. Health*, 19, 71, 1993.

42. Hartwig, A. and Schwerdtle, T., Interactions by carcinogenic metal compounds with DNA repair processes: toxicological implications, *Toxicol. Lett.*, 127, 47, 2002.

43. Hartwig, A. et al., Interference by toxic metal ions with zinc-dependent proteins involved in maintaining genomic stability, *Food Chem. Toxicol.*, 40, 1179, 2002.

44. Haliwell, B. and Gutteridge, J.M.C., *Free Radicals in Biology and Medicine*, Oxford University Press, Oxford, 2000.

45. Halpern, H.J. et al., *In situ* detection, by spin trapping, of hydroxyl radical markers produced from ionizing radiation in the tumor of a living mouse, *Proc. Natl. Acad. Sci. USA*, 92, 796, 1995.

46. Jadhav, S.J., Nimbalkar, S.S., Kulkarni, A.D., and Madhavi, D.L., Lipid oxidation in biological and food systems, in *Food Antioxidants: Technological, Toxicological, and Health Perspectives*, Madhavi, D.L., D., S.S., and Salunkhe, D.K., Eds., Marcel Dekker, New York, 1996.

47. Dubin, M. and Stoppani, A.O., Programmed cell death and apoptosis: the role of mitochondria, *Medicina*, 60, 375, 2000.

48. Kolanjiappan, K., Manoharan, S., and Kayalvizhi, M., Measurement of erythrocyte lipids, lipid peroxidation, antioxidants and osmotic fragility in cervical cancer patients, *Clin. Chim. Acta*, 326, 143, 2002.

49. Chen, F. et al., Cell apoptosis induced by carcinogenic metals, *Mol. Cell. Biochem.*, 222, 183, 2001.

50. Pernin, D. et al., p53 Activation by PI-3K family kinases after DNA double-strand breaks, *Bull. Cancer*, 87, 635, 2000.

51. Meplan, C., Richard, M.J., and Hainaut, P., Redox signalling and transition metals in the control of the p53 pathway, *Biochem. Pharmacol.*, 59, 25, 2000.

52. Chen, F. et al., Opposite effect of NF-kappa B and c-Jun N-terminal kinase on p53- independent GADD45 induction by arsenite, *J. Biol. Chem.*, 276, 11414, 2001.

53. Chen, F. et al., Carcinogenic metals and NF-kappaB activation, *Mol. Cell. Biochem.*, 222, 159, 2001.

54. Chen, F. and Shi, X., Signaling from toxic metals to NF-kappaB and beyond: not just a matter of reactive oxygen species, *Environ. Health Perspect.*, 110 (Suppl. 5), 807, 2002.

4 Molecular Mechanisms of Asbestos- and Silica-Induced Lung Cancer

Fei Chen and Val Vallyathan

CONTENTS

4.1   Introduction..41
4.2   Basic Molecular Events in Carcinogenesis..42
    4.2.1   Genomic Damage, the Earliest Step in Carcinogenesis 43
        4.2.1.1   DNA Damage...43
        4.2.1.2   DNA Repair..43
        4.2.1.3   DNA Damage Response Signals 44
    4.2.2   Impaired Apoptotic Responses Speed Up Carcinogenesis.......................45
        4.2.2.1   E2F1 46
        4.2.2.2   AP-1 47
        4.2.2.3   NF-κB...47
4.3   Linkage between Silica and Lung Cancer 48
    4.3.1   Silica Induces DNA Strand Breaks...49
    4.3.2   Silica Activates NF-κB..49
    4.3.3   Silica and Cell Death 49
    4.3.4   Silica and ROS 50
4.4   Asbestos, a Potent and Established Carcinogen 51
    4.4.1   Asbestos Damages DNA 52
    4.4.2   Asbestos Activates NF-κB 52
    4.4.3   AP-1, a Molecular Switch for Asbestos-Induced Carcinogenesis.............53
4.5   Conclusions...54
References 54

4.1   INTRODUCTION

Lung cancer continues to be the leading cause of cancer death in the U.S. and on a worldwide basis. In the U.S., the annual deaths from lung cancer are estimated to be 161,900 in 2000.[1] It is widely recognized that human lung cancers result from accumulated genetic damage leading to the activation of oncogenes and the inactivation or loss of tumor suppressor genes. Hereditary predispositions, age, occupational/environmental influences, nutritional status, lifestyle, exposure to infectious agents, and chemicals are known to play a role in carcinogenesis.[2] Among these factors, the burden of occupationally related cancer is considered important, especially when it is known that the majority of blue-collar workers in high-risk occupations are smokers, and cigarette smoking has been shown to be the single most important factor known to cause lung cancer. Doll and Peto[3] estimated that at least 4% of the lung cancer deaths each year in the U.S. are related to occupational exposure. In this respect, both asbestos and silica are established occupational carcinogens that cause lung cancers in exposed populations.[4,5] In addition, both bronchogenic carcinomas and mesotheliomas of the pleura and peritoneum develop in workers exposed to asbestos, and the risk for carcinoma is increased tenfold with cigarette smoking. This increased incidence of lung cancer compared with cigarette smokers with no asbestos exposure suggests the potential of asbestos to cause an additive or synergistic interaction to induce lung cancer. In comparison with asbestos, the relationship between crystalline silica exposure and lung cancer is controversial. Results of epidemiologic studies in exposed workers are inconsistent and ambiguous. On the other hand, the carcinogenic potential of crystalline silica has been clearly demonstrated in experimental animal studies with rats of both sexes using a single intratracheal instillation.[6]

It is apparent from several studies reported in the literature that both asbestos and crystalline silica can induce lung cancer, probably through a multistage process, by certain common and/or divergent mechanisms. Both asbestos and crystalline silica can cause DNA damage, altered gene expression, and activation of signaling pathways important in cell proliferation and carcinogenesis. In this chapter, we will review and explore recent studies regarding the molecular mechanisms governing the carcinogenetic effect of asbestos and silica.

4.2 BASIC MOLECULAR EVENTS IN CARCINOGENESIS

It has been well recognized for several decades that carcinogenesis is initiated through multiple unique steps, including altered genetic changes, activation of intracellular signaling events, abnormal cell growth by evading apoptosis, and sustained angiogenesis[7] (Figure 4.1). Rapid advances in cancer research have revealed complex molecular events showing dynamic changes in the genome and protein functions. In addition, although all cancers share certain degrees of similarities in their cellular, biological, and biochemical characteristics, different types or subtypes of cancer usually exhibit distinct divergent intracellular signaling events that govern cell proliferation and homeostasis. Evidence to support this notion comes from the recent application of gene profiling studies of normal cell transformation into a neoplastic state.[8–11] A similar change in gene expression could be observed among several types of cancer, whereas considerable specific patterns occurred in a small set of cancer-cell types.

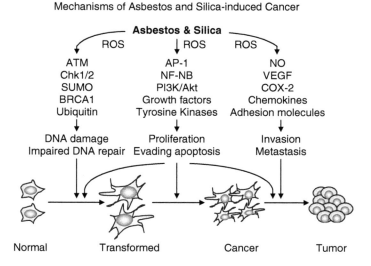

FIGURE 4.1 Mechanisms of asbestos- and silica-induced cancer.

4.2.1 Genomic Damage, the Earliest Step in Carcinogenesis

The hallmark of cancer is centered on mutations of genome that produce oncogenes with dominant gain in functions and inactivation of tumor suppressor genes. Earlier studies in mice and rodent cells suggested that a small number of mutations in the genome were sufficient to cause neoplastic transformation.[12,13] This conclusion, however, was not supported by the evidence, indicating that multiple mutations are necessary for the achievement of transformation.[14] Currently, it is generally accepted that a notable transformation is triggered by accumulation of a number of genomic mutations, some of which inactivate tumor suppressor genes and others activate oncogenes.

4.2.1.1 DNA Damage

It is evident from several studies that the genetic damage occurs in multiple steps that include small-scale insertions, deletions of DNA bases/segments, DNA base changes/modifications, microsatellite instability, chromosomal changes/loss, and translocation of segments.[15,16] A number of physical and chemical agents can trigger genomic damage by directly attacking DNA. Among these agents, reactive oxygen species (ROS) are the most documented. ROS-induced DNA damage is the most common mechanism for toxic occupational and environmental agents, such as asbestos, silica, ozone, and ultraviolet and ionizing radiations. During normal physiological or subtle pathophysiological processes, moderate overproduction of ROS is antagonized by the efficient endogenous antioxidant system. In contrast, enhanced ROS generation will overwhelm antioxidant defense, resulting in the interaction of ROS with a number of intracellular targets, especially DNA, lipids, and proteins.

Sustained, enhanced production of ROS can cause cumulative DNA damage, including base modifications, strand breaks, and large segment deletions. It can also convert cellular constituents into second-generation reactive intermediates capable of inducing further damage.[17] ROS damages DNA by attacking the nucleic acid bases, deoxyribose residues, or the phosphodiester backbone. The most important base attack of ROS is hydroxyl radical ($^{\bullet}$OH)-mediated purine oxidation.[18] $^{\bullet}$OH reacts with purines to form 8-oxo-7,8-dihydrodeoxyguanosine (8-oxo-dG) or 8-oxo-7,8 dihydrodeoxyadenosine (8-oxo-dA), which alters the base-pairing properties of guanines or adenine. $^{\bullet}$OH can also react with pyrimidines to generate pyramidine glycols by adding the 5,6 double bond of pyrimidines. In addition, by an indirect mechanism, $^{\bullet}$OH can induce DNA single-strand breaks through hydrogen abstraction from the 5-methyl group of thymine or the DNA sugar moieties.[19] The progressive accumulation of such oxidatively damaged DNA will obviously increase the incidence of cancer.

4.2.1.2 DNA Repair

The outcome of genomic damage is dependent on the type of DNA lesions and the ability of the cell to repair DNA efficiently. A limited extent of DNA damage usually triggers apoptosis to eliminate the damaged cells or arrests the cell cycle to have more time to repair the damaged DNA. If the damage of DNA surpasses the DNA-repair capacity of the cell, genomic mutations or instability will be inevitable.

Damage to the genome can block the movement of RNA polymerase II (RNAPII) on the transcribed strand of active genes and consequently inactivate the damaged genes.[20] Such damage includes ROS-induced 8-oxo-dG formation[21] and UV-induced cyclobutane pyrimidine dimers.[22] The stalled RNAPII on the damage site recruits Cockayne's Syndrome A (CSA) and CSB proteins[23] and initiates a DNA repair process, the so-called transcription-coupled repair (TCR) through the recruitment of other DNA repair factors, such as TFIIH, XPG, XPA, RPA, and ERCC1, to the damaged site. During the initiation of TCR, the RNAPII is phosphorylated and modified by polyubiquitin chains that target RNAPII into the proteasomal degradation pathway.[24] The transcription-coupled repair was considered a specific subpathway of the nucleotide excision repair (NER) pathway and base excision repair (BER) pathway.[25]

DNA damage may also interfere with the replication of DNA by blocking the activity of replicative DNA polymerase-δ/ε (polδ/ε).[26] The new types of DNA polymerase, polξ and polκ will act as alternative polymerases to replicate the DNA around the damaged sites.[27,28] Since these DNA polymerases have more flexible base-pairing properties, the replicated DNA contains a higher rate of error, which may be particularly relevant for the damage-induced point mutations seen in oncogenesis.[28,29]

Based on the types of DNA damage, eukaryotic cells developed a number of DNA repair systems that are conserved in prokaryotic cells.[30] In mammalian cells, at least three main types of DNA repair systems — excision repair, recombinational repair, and DNA post-replication repair (PRR)— function to repair damaged DNA. In the system of excision repair, there are three subtypes of repair mechanisms: nucleotide-excision repair (NER), base-excision repair (BER), and mismatch repair (MMR). The recombinational repair includes homologous recombination repair (HRR) and nonhomologous end joining (NHEJ). Although each type or subtype of DNA repair systems requires a unique subset of DNA repairing factors, a certain degree of overlapping in the damage repair pathways has been noted under many circumstances.

The NER repairs DNA by removing bulky DNA adducts that distort the DNA helix structure and interfere with the base pairing that affects transcription and normal replication. BER, on the other hand, predominantly deals with small chemical modifications or alterations of bases. Certain incorrect base pairs that do not follow the Watson-Crick pairing are corrected by MMR. When DNA double strand breaks or DNA interstrand cross-links occur, cells employ HRR or NHEJ to repair such damages. HRR repairs damage during DNA replication in S and early G2 phases of the cell cycle by using the second sequence of sister chromatid to align the DNA breaks. In contrast, NHEJ repairs damaged DNA by a less-accurate DNA repairing process in the G1 phase of the cell cycle where a second copy is not available.[20]

The DNA PRR has been studied mainly in prokaryotic cells or yeast. Little is known about the mechanisms of PRR in mammalian cells. PRR converts single-stranded gaps of DNA damage into large-molecular-weight DNA without eliminating the replication-blocking lesions. Two modes of PRR have been suggested. The error-free PRR repairs damage through the use of undamaged sister chromatid as template. The error-prone PRR, however, uses specific DNA polymerases that are able to read through sites of DNA damage. The initiation of PRR in yeast requires Rad18, a single stranded DNA binding protein, and Rad6/UbcH1, an ubiquitin-conjugating enzyme.[31] Human Rad6/UbcH1 has been demonstrated to be able to rescue yeast rad6 sensitivity to 8-methoxypsoralen plus UVA (32), indicating a possible evolutionary conservation of PRR within eukaryotes. The key regulatory mechanism of PRR is the modifications of PRR components by ubiquitin and SUMO. A recent study by Hoege et al.[33] indicated that Rad6/Rad18-dependent monoubiquitination of PCNA resulted in either the recruitment of specific DNA polymerases for error-prone PRR or polyubiquitination of PCNA by Rad5 and Mms2/Ubc13 complex to activate error-free PRR.

4.2.1.3 DNA Damage Response Signals

In general, different types of DNA damage activate unique signaling pathways. For instance, damage that triggers HRR usually activates ATM, BRCA1, and BRCA2, whereas damage that induces NHEJ predominantly activates DNA-PK.[34] Nevertheless, the various forms of DNA damage share common downstream signal transduction pathways, and cross-talk among different signaling pathways can occur even in the very early stages of DNA damage responses. One of the most important DNA damage responses is the activation of checkpoints that causes cell cycle arrest at specific phases and/or cell apoptosis. Extensive studies have been made in the last decade that revealed a number of checkpoint proteins that can sense various types of DNA lesions. These checkpoint proteins include ATM, ATR, RAD17, RAD9, BRCA1/2, checkpoint protein 1 (Chk1), and Chk2.

The most established signaling pathway of DNA damage is the cellular response to DNA double-strand breaks (DSB)[34] resulting from attacks by exogenous agents, such as ionizing radiation and ROS. In addition, DSB can also be derived from formation of improper DNA replication forks due to DNA single-strand breaks. DSB can be repaired by two major repair pathways: HRR and NHEJ systems. At the initiation of HRR, the DNA ends are recognized and bound by a complex containing RAD51, ATM, and BRCA1 or BRCA2 proteins.[35,36] Such recruitment of ATM to the DNA damage sites activates the kinase activity of ATM that consequently phosphorylates p53, Mdm2, Chk2, BRCA1, NBS1, and c-Abl. In NHEJ, the free DNA ends are bound with the Ku subunits of DNA-PK, another PIKK family member. DNA-PK consists of a heterodimeric DNA binding subunit (Ku70 and Ku80) and a catalytic serine/threonine kinase subunit whose activity is activated by the association with DNA. Activated DNA-PK is able to phosphorylate p53 or MDM2,[37,38] although this conclusion remains controversial.[39]

One of the most common outcomes of DNA damage response is cell cycle arrest through the activation of checkpoint pathways, which is considered a critical event for the cells to have more time to repair their damaged DNA. A number of checkpoints have been shown to be able to prevent the replication of damaged DNA or segregation of damaged chromosomes. The downstream targets of DNA damage-activated kinases, ATM and DNA-PK, are pivotal in executing the checkpoint signal transduction. Phosphorylation and activation of p53 by ATM or DNA-PK have been demonstrated as important processes of the G1 cell cycle checkpoint. The posttranslational modifications of p53 not only increase the stability of p53, but also enhance the transcriptional activity of p53 on its target genes, especially the G1 cyclin-dependent kinase inhibitor p21waf1.[40] The p21waf1 suppresses cyclin E- and cyclin A-associated Cdk2 kinase activity and, thereby, blocks the cell cycle transition from G1 to S phase. The G2/M phase cell cycle arrest may be mainly mediated by the activation of checkpoint protein kinases Chk1 and Chk2. Activated Chk1 or Chk2 phosphorylates a conserved site, Ser216, on the Cdc25C protein, a dual specific phosphatase that maintains the progress of the G2/M phase by dephosphorylating Tyr15 of cyclin-dependent kinase Cdc2. This phosphorylated Cdc25C is functionally inactive due to the binding of 14-3-3 protein that sequesters Cdc25C in cytoplasm and, therefore, prevents its phosphatase activity toward Tyr15 of Cdc2, an essential kinase for G2/M phase cell cycle transition.[41]

4.2.2 Impaired Apoptotic Responses Speed Up Carcinogenesis

Programmed cell death, or apoptosis, is one of the most extensively studied topics in the areas of basic biology and cancer. The apoptotic cell is characterized by membrane blebbing, cell shrinking, cytosolic and/or nuclear condensation, and breakdown of chromosomal DNA. Apoptosis is vital to life in multicellular organisms during embryonic development and to help the removal of damaged cells in an orderly way to shape the organs and maintain adequate immune and neural cell populations. Upon invasion of exogenous pathogens, apoptosis is an important process to eliminate pathogen-damaged cells. Therefore, the cellular apoptotic response has to be tightly regulated, as too little or too much apoptosis will lead to developmental deficiency, autoimmune response, or cancer.

Apoptosis differs from necrotic cell death not only morphologically but also biochemically. A key event in apoptosis is the cascade activation of cysteine aspartyl proteases, namely caspases, that implement cell apoptosis by cleaving substrates. More than one dozen caspases have been identified in mammalian cells. All caspases are synthesized as enzymatically inert zymogens under normal conditions. In the case of Fas ligand- or TNFα-induced "extrinsic" apoptosis, procaspase 8 is recruited to the intracellular domain of Fas or the TNFα receptor through death domain-containing proteins. Such recruitment activates caspase 8.[42] Many stimuli, especially ROS, induce "intrinsic" apoptosis by mitochondrial damage that releases cytochrome c and causes the subsequent activation of caspase 9 through the association of Apaf-1.[43,44] Both activated caspase 8 and caspase 9 are capable of activating caspase 3 by cleavage and removal of the N-terminal prodomain. The substrate specificity of distinct caspases is determined by the surrounding amino acids of the cleavage site after aspartate (Asp) residues. Although more than 100 cellular proteins have been identified as substrates for caspases, only a restricted set of proteins are cleaved by caspases at a specific stage of apoptosis.[45] The typical substrates of caspases include nuclear proteins, cytoskeleton proteins, protein kinases, proteases, cytokines, and several apoptotic regulatory proteins. The caspase cleavage usually results in loss of biological activity of target proteins. However, in many cases caspase cleavage will activate the function of proteins by removing a negative regulatory domain or inactivating a suppressive subunit.

Increase of survival signals and decrease of death signals are common strategies used by various types of tumors to escape apoptosis. Many tumor cells overexpress survival factors, such as insulin-like growth factor (IGF-I) and IGF-II, that promote cells for autonomous proliferation.[46] Similarly, a strong anti-apoptotic serine/threonine kinase, Akt, is overactivated in certain tumor cells due to the loss of the suppressor of Akt function, PTEN.[47] Furthermore, some anti-apoptotic proteins, such as Bcl-2 and Bcl-xl, which exhibit their roles through stabilization of mitochondria, are increased in several types of tumors. However, perhaps the most important factor for the avoidance of apoptosis of tumors is the improper function of transcription factors, including E2F1, AP-1, and NF-κB.

4.2.2.1 E2F1

E2F1, a key transcription factor originally discovered in the adenovirus early region 1A (E1A) transform cells, promotes proliferation by up-regulating a variety of genes, such as Cdc6, Cdc2, and cyclin A2. E2F1 is required for cell cycle transition from G1 to S phase. Enhanced E2F1 activity has been observed in a number of tumors as a result of impaired retinoblastoma tumor suppressor (pRb) function.[48] It has been demonstrated that pRb mutation or its absence occurs in at least one-third of all human tumors.[49] In normal cells, the binding of pRb to E2F1 is under the control of cell cycle-dependent phosphorylation. In G0 and early G1 phase, hypophosphorylated pRb binds to an 18-amino-acid motif within the transactivation domain of E2F1, resulting in the inhibition of transcriptional activity of E2F1.[50] In addition, a number of transcriptional repressors, such as histone deacetylase enzymes (HDACs),[51] human brahma related gene (BRG1),[52] and histone methyltransferase SUV39H1,[53] can associate with suppressed pRb-E2F1. Following the sequential activation of cell cycle-dependent kinase complexes cyclin D/Cdk4 and cyclin E/Cdk2 induced by mitogenic growth factors, pRb is hyperphosphorylated and dissociates from E2F1, leading to the activation of E2F1. Activated E2F1 can form a heterodimer with DP1 or DP2 and bind to various known E2F-responsive promoters of target genes.

The evidence indicating an anti-apoptotic effect of E2F1 was first provided by the studies of E2F1 overexpression.[54] Microinjection of the wild type-E2F1 cDNA, but not mutated E2F1 cDNA, into quiescent cells can activate DNA synthesis and induce S-phase entry.[54] Further studies indicated that overexpression of E2F1 can cause transformation of primary cells.[55] In contrast, mutation of E2F1 is sufficient to block cell proliferation completely.[48]

The target genes involved in cell cycle transition induced by E2F1 include DNA topoisomerase IIα, cyclines D and E, Cdc25A, replication protein A, replication factor C, DP1, and DNA polymerase α and δ.[56,57] In addition to the transcription of cell cycle regulatory genes, E2F1 can also stimulate the expression of genes involved in DNA repair and mitosis. These genes include MAP3 kinase 5, MSH2, MSH6, PCNA, BRCA1, BUB1b, AIM-1, KRP1, Cdc2, and serine threonine kinase SAK-α.[57,58]

Interestingly, E2F1 activation also increases the transcription of genes that facilitate cell apoptosis. These pro-apoptotic genes, regulated by E2F1, include caspase 3, AFAF-1, and the p53 family member p73.[57,59] Thus, although E2F1 appears to be an essential transcription factor required for cell cycle transition and proliferation, E2F1 is also a pro-apoptotic factor. Further support of this conclusion is the fact that E2F1 antagonizes the activity of NF-κB, a well-documented anti-apoptotic transcription factor, by direct association with p65, to abolish the NF-κB-dependent expression of SOD1 that is responsible for the elimination of ROS.[60] An alternative mechanism, that E2F1 negatively regulates NF-κB, is its inhibition on TRAF2, an activator of NF-κB upstream kinase, IKK.[61]

4.2.2.2 AP-1

AP-1 is one of the first gene-regulating transcription factors identified in mammalian cells. The key component of AP-1 complex is the Jun protein, which was originally identified as an oncoprotein encoded by a cellular insert in the genome of avian sarcoma virus 17.[62] Jun can form a homodimer or heterodimer with Fos or ATF members. The most common AP-1 dimer is the Jun/Fos complex that recognizes the palindromic DNA sequence TGAC/GTCA, the so-called TPA response element (TRE), in the promoter or intron region of a number of genes.[63] In addition, Jun can also form a complex with some activating transcription factor (ATF) members, such as ATF2, ATF3, and ATF4.[63] All Jun proteins from different species contain an N-terminal JNK docking domain (delta domain) adjacent to the JNK phosphorylating site Ser63/73. In the C-terminal, there is a basic domain for DNA binding, followed by a nuclear localization signal (NLS) and a leucine zipper motif for dimerization with partner proteins.

Emerging evidence indicates that AP-1 plays an important role in oncogenesis. A number of studies suggest that active Ras-induced oncogenic transformation requires constitutive activation of AP-1.[64,65] In immortalized rodent fibroblasts, experimental overexpression of c-jun is sufficient to induce oncogenic transformation in the absence of Ras. The transformational activity of AP-1 is mainly attributed to its anti-apoptotic effect, as evidenced by the fact that c-Jun protects cells against UV-induced cell death.[66,67] AP-1 appears to be able to suppress the expression of pro-apoptotic factor, Fas.[67] Gene knock-out and biochemical studies have suggested that the anti-apoptotic activity of an AP-1 component, especially the c-Jun subunit, is achieved by down-regulating the transcription of p53, a critical pro-apoptotic transcription factor.[68] Furthermore, ChIP experiments revealed a clear negative effect of c-Jun on association of p53 with the p21 promoter, resulting in a decreased expression of p21 and an increased G1 to S phase progression.[69]

It should be noted that a considerable number of reports indicated a pro-apoptotic effect of AP-1.[70] The pro- or anti-apoptosis of AP-1 may be highly dependent on tissue type, development stages, and extracellular stimuli that determine which simultaneous and asynchronous signaling pathways are activated. For example, in the cells where p53 is deficient, AP-1 may exert anti-apoptotic effect. In contrast, in the cells where p53 is functionally normal, activation of AP-1 may be pro-apoptotic. As a transcription factor, the activity of AP-1 is reflected by its transcriptional regulation of target genes. Several anti-apoptotic and transformational genes have been recently identified, such as Jun-activated gene in CEF (JAC)[71] and heparin-binding epidermal growth-factor-like growth factor (HB-EGF).[72]

4.2.2.3 NF-κB

NF-κB is an important transcription factor found in eukaryotic cells that governs the expression of genes involved in cell-to-cell communication, development, apoptosis, and carcinogenic transformation. NF-κB family members include proteins that are synthesized in mature form, such as p65 (RelA), c-Rel, and RelB, and the proteins that are synthesized as large precursors that undergo proteolytic processing, such as p50/p105 (NF-κB1) and p52/p100 (NF-κB2) that undergo proteolytic processing before being assembled as homodimers or heterodimers. Since p65, c-Rel, and RelB have an N-terminal transactivation domain, the NF-κB dimers containing these subunits exhibit great transcriptional activity. In contrast, the transcriptional activity of NF-κB homodimers p50/p50 and p52/p52 are less active. All members contain a conserved Rel homology domain (RHD) with a size about 300 amino acids. This domain is required for DNA binding, dimerization, nuclear translocation, and interaction with inhibitory proteins, IκB family members. All IκB family members, including IκBα, IκBβ, IκBγ, and IκBε, are characterized by six to seven ankyrin repeats responsible for binding to the RHD of Rel proteins. Upon stimulation of the cells, IκB is phosphorylated at the N-terminal serine residues by IκB kinase (IKK) complexes and subjected to ubiquitin-proteasome-mediated degradation. The degradation of IκB liberates NF-κB, resulting in its nuclear translocation and binding to the decameric κB site, "GGGRNNYYCC," found in the promoter, enhancer, or intron regions of target genes.

The signaling pathways that activate NF-κB have been extensively studied during the past several years. A major breakthrough is the identification of an IKK kinase complex containing two catalytic subunits, IKKα and IKKβ, and a structural component named NEMO/IKKγ/IKKAP. Both IKKα and IKKβ contain an amino-terminal kinase domain, a carboxyl-terminal region with a leucine zipper, and a helix-loop-helix domain. IKKβ appears to be more responsible for MEKK1- and pro-inflammatory signal-induced IκBα phosphorylation in the canonical pathway that uses NF-κB dimers composed of RelA, c-Rel, and/or p50.[73] In contrast, IKKα is preferentially involved in B-cell activating factor receptor (BAFF-R)-, NIK- and lymphotoxin β-induced IκBα phosphorylation, and p100 processing.[74–76]

The signaling pathway for the activation of IKK and NF-κB by TNFα, IL-1, LPS, and CpG DNA has been extensively studied and well documented. Yet little is known about the signaling events that trigger IKK and NF-κB activation following oxidative stress or intracellular stress, e.g., DNA damage. Oxidative stress is a common biological response in cells challenged with cytokines or bacterial and/or viral products. The evidence suggesting activation of IKK by ROS is from the studies using H_2O_2-treated human epithelial cells or mouse fibroblasts.[77, 78] However, the activation of IKK by ROS was challenged by observations indicating oxidative inactivation of IKK kinase by H_2O_2.[79] An alternative mechanism has been proposed recently regarding the oxidative stress-induced IKK and NF-κB activation.[80] In HeLa cells, H_2O_2 activates IKK through tyrosine kinases Src- and Abl-induced tyrosine 463 (Y463) phosphorylation of protein kinase D (PKD, also named as PKCμ), which phosphorylates serine 181 (S181) of the IKKβ subunit.[80]

The role of NF-κB in carcinogenesis and its enhancement in many types of tumors has been investigated in a number of studies during the past decade.[81] In transgenic mice expressing v-Rel oncogene, malignant T-cell lymphomas develop at a fast rate.[82] It is believed that the contribution of NF-κB in carcinogenesis is largely based on its transcriptional regulation on several anti-apoptotic genes encoding Bcl-x, A20, cIAP1, cIAP2, XIAP, and Blf-1. In addition, NF-κB is capable of antagonizing the pro-apoptotic signaling pathways elicited by p53 and c-Jun N-terminal kinase (JNK).[83–85]

4.3 LINKAGE BETWEEN SILICA AND LUNG CANCER

Crystalline silica (quartz) is a mineral found in many types of rocks, granite, mineral deposits, and sand. Silica is a well-recognized health hazard for workers engaged in mining, construction, sandblasting, farming, and other occupations. Environmental exposure to aerosolized fine silica is also a major public health issue. Although silica is now considered as a human carcinogen, the molecular mechanisms involved in silica-induced carcinogenesis remain to be answered, because evidence for carcinogenicity in humans and animals is still debated. Human epidemiologic studies have not conclusively documented evidence or a dose–response relationship regarding the role of silica as a potential carcinogen in many exposed population studies. Furthermore, animal exposure studies to silica are controversial, because silica shows species differences in carcinogenicity. In the rat, a single intratracheal instillation of silica induces lung tumors, i.e., mostly of adenocarcinomas associated with silicotic fibrosis. In mice and hamsters, on the other hand, similar exposure to silica does not produce characteristic fibrotic lesions or tumors.[86] However, silica can interact with human and animal pulmonary cells *in vitro* and *in vivo* to induce carcinogenesis through several mechanisms directly involving ROS generation, DNA damage, apoptosis, signal transduction, induction of oncogenes, and activation of transcription factors. Obviously, a sustained induction and activation of these silica-induced cellular responses may promote carcinogenesis.

4.3.1 Silica Induces DNA Strand Breaks

DNA strand break (DSB) is probably one of the earliest types of genetic damage that occurs in cells exposed to silica. Using the alkaline DNA unwinding and ethidium bromide fluorometric techniques, Vallyathan and colleagues showed that crystalline silica induced DSB of bovine DNA.[86] Silica-induced DNA damage in alveolar macrophages assayed using a single-cell gel electrophoresis Comet assay showed that N-acetylcysteine increased intracellular glutathione levels and protected against DNA strand breaks by inhibiting silica-induced ROS formation.[87] In addition, exposure of Rat (RLE) and human (A549) epithelial cells to DQ12 quartz causes •OH radical-dependent DNA lesions in cells.[88] These studies firmly document that silica causes DSBs, which are driven by •OH-radical-generating properties of the particles.

4.3.2 Silica Activates NF-κB

The first notion that silica causes NF-κB activation was identified in an experimental system using a mouse peritoneal macrophage cell line as a prototype model.[89] Since then, NF-κB activation by silica has been documented in a variety of cellular systems.[90–95] Chen et al.[96] postulated that silica activated NF-κB through proteolytic degradation of IκBα. However, degradation of IκBα in the cells treated with silica was not inhibited by a classic proteasome inhibitor, MG132. In contrast, an inhibitor for calpains attenuated the degradation of IκBα and NF-κB activation induced by silica. Recent studies by Kang et al.[90] reported that tyrosine phosphorylation of IκBα may contribute to silica-induced IκBα degradation.

The majority of studies on silica-induced NF-κB activation were conducted through *in vitro* silica exposure of selected cell types. Only a few studies have examined the effect of silica on NF-κB *in vivo*. Using bronchoalveolar lavage cells from rats exposed to silica, Porter et al.[97] demonstrated that NF-κB activation in the silica-exposed rats increased progressively throughout the 116-day silica inhalation period. The expression of two NF-κB-regulated cytokines, TNFα and IL-1, was correlated with the induction of NF-κB by silica. In transgenic mice carrying a NF-κB-dependent luciferase reporter gene, instillation of silica caused NF-κB activation in both alveolar macrophages and epithelial cells at 24 to 72 h postinstillation.[95]

The signal transduction pathways for NF-κB activation are well characterized for cellular responses to inflammatory cytokines, LPS, HTLV Tax protein, and CD28 signals.[98] It is unknown, however, whether silica activates NF-κB through similar signaling cascades. A number of reports indicated that silica was capable of inducing generation of ROS in the cells (see Section 4.3.4). Although it is still a debatable issue, ROS have long been assumed to be critical mediators for the NF-κB signaling.[99]

4.3.3 Silica and Cell Death

It has long been realized that exposure to silica causes alveolar macrophage and epithelial cell injury, leading to a necrotic or an apoptotic state.[100,101] Apoptosis is an integral part of homeostasis and development, which requires energy, *de novo* macromolecular synthesis, and the specific biochemical steps to initiate and execute apoptosis.[102] Under certain pathological circumstances, the cellular apoptotic program can be inappropriately implemented, leading to severe cellular destruction. Necrosis, usually occurring under high-exposure conditions, is dramatically different from apoptosis in many aspects.[103] Cells dying by necrosis are characterized by edema, loss of membrane integrity, and chaotic breakdown of cellular components, including protein and nuclear DNA. It has been generally believed that necrosis is a passive process, which does not require new protein synthesis and energy. Depending on the severity of pathological signals and the strength of the cellular protective system, cell death can be apoptotic, necrotic, or both. In fact, under many pathophysiological conditions, different forms of cell death can coexist in the same cell type and are switched on by specific stimuli. In addition, a number of reports revealed that shared pathways for apoptosis and necrosis coexist.[104]

The first notion that silica induces apoptosis was demonstrated in macrophages treated with silica, causing the processing of precursor IL-1β into mature IL-1β.[100,101] In human alveolar macrophages, silica-induced apoptosis was shown to be mediated through the scavenger receptor.[105] Silica-induced apoptosis in alveolar and granulomatous cells was present in rats exposed to silica after 10 to 56 days.[106] In an *in vivo* study using a mouse model, Srivastava et al.[107] demonstrated that apoptosis peaked between 1 and 6 weeks after silica exposure, i.e., before silicotic lesions are well organized. Recent experimental evidence suggests that apoptosis induced by silica may be through the Fas/Fas ligand system. First, an increased expression of soluble Fas was observed in peripheral blood mononuclear cells from silicosis patients.[108] Second, silica induced Fas ligand expression or apoptosis in lung macrophages *in vitro* and *in vivo*. Neither apoptosis nor silicosis could be induced by silica in Fas ligand-deficient mice.[109] Third, treatment of wild-type mice with an antibody against Fas ligand *in vivo* prevented silicosis.[109]

The notion of the silica-induced necrotic cell death was far earlier than that of the apoptotic induction by silica. In CFY rats, silica-induced necrosis occurred in a relatively earlier period of postinstillation, between 12 to 72 h.[101] In mice, necrotic epithelial cell death occurred within 24 h of silica instillation.[110] Driscoll et al.[111] reported that necrosis induced by silica was in a prolonged manner. The necrosis, as indicated by the presence of lactate dehydrogenase (LDH) in bronchoalveolar lavage fluid (BALF), could be detected from 7 to 28 days of intratracheal silica instillation.

It is believed that cell necrosis triggers inflammation but that apoptosis does not. Biochemical evidence suggests that necrotic cells release a number of intracellular lytic enzymes, mediators, cytokines, chemokines, and other proinflammatory factors that damage neighboring cells and elicit an inflammatory response. Recent studies by Scaffidi et al.[112] indicate that the release of chromatin protein, such as high mobility group 1 (HMGB1) from necrotic cells, could serve as a major diffusible signal for inflammatory responses. In cells without the *hmgb1* gene, the induction of inflammation is less efficient. In contrast, the apoptotic cells were unable to induce inflammation due to the tight association of HMGB1 with hypoacetylated intact or fragmented chromatin. However, several recent studies document that apoptotic cells could also be involved in the regulation of inflammation.[113] As discussed earlier, the apoptosis induced by silica might be through the Fas/FasL system. The possible involvement of Fas/FasL in inflammation has been investigated over the past two years. In hepatic cells, activation of Fas by Fas agonist causes induction of inflammatory chemokines.[114] In addition, phagocytes can produce inflammatory cytokines upon ingestion and clearance of apoptotic cells.[115] Inhibition of proinflammatory cytokines by apoptotic cells has also been presented.[116–118] It is obvious that the pro- or anti-inflammatory effect of apoptosis is likely dependent on the duration of apoptosis. Since inhaled silica is retained in the lung for a long time, the apoptotic response induced by silica may be prolonged. Therefore, chronic apoptosis in response to silica exposure may cause chronic inflammatory reaction that can lead to cancer.[119,120]

4.3.4 SILICA AND ROS

The importance of ROS in mediating silica-induced lung diseases has been well documented in recent years.[121,122] In aqueous media, silica has been shown to generate •OH radicals.[123] In addition, phagocytosis of silica activates cell membrane NADPH oxidase, inducing the respiratory burst and the generation of superoxide radicals •O_2^-, leading to lipoprotein oxidation.[124–126] In addition, a major source of •O_2^- in lung cells is also produced by the respiratory chain in mitochondria.

Nitric oxide (NO), another important free radical, and its metabolites have been implicated in the lung tissue damage associated with acute and chronic silica exposure. Van Dyke et al.[127] demonstrated that silica caused peroxynitrite-dependent increases in chemiluminescence (CL) of rat lung and bronchoalveolar lavage (BAL) cells. The CL reaction was markedly decreased by N-nitro-L-arginine methyl ester hydrochloride (L-NAME), indicating induction of NO by silica. Studies by Blackford et al.[128] suggested that silica was able to up-regulate inducible NO synthase (iNOS) gene expression in rat alveolar macrophages and neutrophils. Immunohistochemical double staining and *in situ* hybridization demonstrated silica-induced expression of iNOS not only in the alveolar macrophages and neutrophils, but also in the bronchiolar epithelial cells.[129] The importance of NO induction *in vivo* was demonstrated by Srivastava et al.[107] in a mouse silicosis model. Inhalation of silica resulted in iNOS gene expression in pulmonary cells and numerous silicotic lung lesions in wild-type mice. The silicotic lesions were significantly reduced in the iNOS gene knock-out mice exposed to silica.

Whereas the induction of NO *in vivo* by silica has been well documented,[130,131] the *in vitro* induction of NO by silica is controversial. Many studies failed to demonstrate that silica was able to induce NO generation *in vitro* using primary rat alveolar macrophages or macrophage cell lines derived from mice.[132,133] Using rat alveolar macrophages, Huffman et al.[134] suggested that the induction of NO *in vitro* by silica required certain cofactors derived from other cells, such as IL-1 and TNF-α.[107,135,136]

4.4 ASBESTOS, A POTENT AND ESTABLISHED CARCINOGEN

Animal and human epidemiologic studies indicate that inhalation of asbestos can cause serious respiratory diseases, including asbestosis, bronchogenic cancer, and mesothelioma.[137,138] Exposure to asbestos has also been linked to certain types of tumors that originate in the stomach, throat, and rectum.[139] People who are working with asbestos-containing products or who live near asbestos mining/manufacturing areas also have a high risk of contracting asbestos-induced lung diseases.[140]

Asbestos refers to a family of crystalline fibrous silicates that are relatively indestructible, heat resistant, and have a high commercial importance. Serpentines and amphiboles are the two distinct varieties of asbestos with succinct physical and chemical characteristics of commercial importance in different applications. The serpentine asbestos (chrysotile) is curly and flexible and accounts for most of the asbestos used commercially in North America. Crocidolite, amosite, tremolite, and anthophilite are the important commercial amphiboles, but these are used less frequently than chrysotile. Amphiboles are straight, rodlike fibers. The relative toxicity, pathogenicity, and carcinogenicity of these different types of asbestos have been extensively investigated both in animal and human studies. Epidemiologic data of human asbestos exposure suggest that amosite and crocidolite forms of asbestos are more pathogenic and carcinogenic than chrysotile, particularly in reference to induction of mesotheliomas.[141] Such potency may be largely due to their needlelike shape that hinders the clearance of these fibers from the lung. The chrysotile asbestos is considered less carcinogenic because of its curly shape, which is likely to cause impaction in the upper airways, which facilitates mucociliary clearance. Some studies of chrysotile workers have reported a few mesotheliomas. It has been suggested that these mesotheliomas are likely due to contamination of chrysotile with the amphibole, tremolite. However, animal studies indicate that all forms of asbestos are equally carcinogenic, and there is a clear dose–response relationship.[137]

The mechanism of asbestos-induced carcinogenesis is still a debatable issue. Inhaled asbestos fibers in alveoli can be phagocytized by alveolar macrophages.[4] For occupational exposures to long, thin fibers, several phagocytes are often found attempting to phagocytize a single long fiber. One plausible mechanism of asbestos-induced carcinogenesis is its ability of sustained induction of ROS that cause DNA strand breaks and deletion mutations. In addition, asbestos can also alter the normal intracellular signal transduction pathways by increased autophosphorylation of the EGF receptor, leading to activation of downstream kinases and transcription factors, such as AP-1 and NF-κB. Furthermore, asbestos is capable of depressing immune function by stimulating macrophages to produce immune suppressive lymphokines and other mediators.

4.4.1 ASBESTOS DAMAGES DNA

The first evidence indicating the DNA-damaging effect of asbestos was demonstrated in Chinese hamster and human cell lines exposed to asbestos.[142] Further studies indicated that asbestos can induce large deletions of DNA and consequent failure of chromosomal segregation and cell mitosis.[143,144] In workers exposed to asbestos, increased frequency of sister chromatid exchange and DNA double-strand breaks have been observed.[145,146] The DNA-damaging effect of asbestos was enhanced in the presence of H_2O_2.[147] The presence of surface-bound iron was considered a critical factor in the induction of •OH radicals and 8-OHdG DNA adduct formation.[148, 149]

Several studies suggested that cigarette smoking enhances the lifetime accumulation of asbestos in the lung and, consequently, increases the generation of ROS and the occurrence of lung cancer.[150] It was shown in *in vitro* studies that asbestos fibers absorb and retain benzo(a)pyrene, and thereby enhance direct delivery of this carcinogen to the nucleus of the cells.[151,152] This mechanism may be a reasonable explanation for the increased G to T transversions at codon 12 of the k-ras gene and G:C to T:A mutations of the p53 gene in lung cancer resulting from combined exposure to asbestos and cigarette smoke.[153,154]

Increased p53 expression or function has been frequently observed in the cellular response to DNA damage signals. If asbestos is capable of damaging DNA, one can speculate a change in p53 protein or its function following exposure to asbestos. Indeed, in an inhalation study of rats exposed to aerosolized asbestos, increases in the expression of p53 protein at the sites of asbestos fiber deposition were observed. The expression of p53 was detected by immunostaining after 24 h of exposure and peaked at 8 days.[155] A similar study reported on the accumulation of p53 in human lung cancer tissues, with a significant correlation with the asbestos content in the lung tissue.[156] In addition, an increased level of serum p53 protein was reported to be present in patients with asbestos-associated lung cancer.[157] The elevated level of p53 in asbestos-induced cancer is correlated with the enhanced downstream effects of p53, such as cell cycle arrest at G1 phase and expression of p53 target genes p21, gadd45, and gadd153.[158,159] It was proposed that the increased level of p53 protein may be largely because of the increased p53 stability resulting from mutation or post-transcriptional modification rather than the increase of p53 mRNA expression. A recent study by Matsuoka et al.[160] demonstrated that exposure of human pulmonary epithelial type II cells to asbestos increased the S15 phosphorylation of p53. This phosphorylation of p53 seems to be dependent on the activation PI3K family members, such as ATM, ATR, or DNA-PK, since the PI3K inhibitor wortmannin, but not MAPK inhibitors, blocked S15 phosphorylation induced by asbestos.[160]

4.4.2 ASBESTOS ACTIVATES NF-κB

The first evidence indicating NF-κB activation by asbestos was provided by Janssen et al.[161] in 1995. In this study, they reported that crocidolite asbestos caused a dose-dependent increase in the binding of p50 and p65 proteins to NF-κB-binding DNA elements in hamster tracheal epithelial (HTE) cells. Similarly, Simeonova and Luster[162] showed that asbestos-induced NF-κB activation is responsible for the increased expression of the inflammatory cytokine, IL-8, in human pulmonary epithelial cells. ROS scavengers as well as protein tyrosine kinase inhibitors partially decreased NF-κB activation and IL-8 induction by asbestos. In an asbestos inhalation model in the rat, increased NF-κB p65 immunofluorescence was observed in airway epithelial cells at 5 days after exposure, indicating the translocation of the p65 protein.[163]

Asbestos induces NF-κB and AP-1 transcription-factor activation, possibly through ROS-mediated lipid peroxidation and arachidonic acid metabolism.[164] In macrophages, asbestos caused increased release of arachidonic acid and the formation of prostaglandins E2 and F2α.[165] In alveolar macrophages, asbestos induced the releases of arachidonic acid metabolites as early as 1 h post-*in vitro* exposure. Persistent release of arachidonic acid metabolites at sites of asbestos deposition are implicated in the constitutive activation of NF-κB.[166,167]

Accumulating evidence suggests that NF-κB-dependent cytokine induction is one of the key events in asbestos-induced lung cancer and mesothelioma.[121] The cellular sources of asbestos-induced NF-κB-dependent cytokines include alveolar macrophages and type II epithelial cells. In asbestos-exposed individuals, alveolar macrophages produce increased levels of TNFα.[168] Biochemical studies indicate an involvement of iron-catalyzed ROS generation in asbestos-induced NF-κB activation and cytokine gene expression.[169] Both ROS scavengers and iron chelators blocked TNFα mRNA induction in rat alveolar macrophages stimulated with asbestos. TNFα itself is not only a cytokine whose expression is dependent on NF-κB, but also a potent NF-κB activator. Thus, activation of NF-κB by asbestos initiates a positive cytokine feedback loop, leading to the amplification of inflammation or carcinogenesis. Indeed, a report by Driscoll et al.[170] indicates that increased release of TNFα from macrophages was responsible for the subsequent induction of macrophage inflammatory protein-1α (MIP-1α), MIP-1, IL-1, IL-6, and IL-8 from either epithelial cells or macrophages. Interruption of this cytokine positive feedback loop, such as by TNFα receptor gene knock-out,[171] will be effective in blocking inflammation and cell proliferation.

4.4.3 AP-1, A Molecular Switch for Asbestos-Induced Carcinogenesis

Heintz et al.[172] first demonstrated that asbestos activates AP-1 transcription factor by inducing persistent expression of c-fos and c-jun genes in rat pleural mesothelial cells and hamster tracheal epithelial cells. In AP-1-luciferase reporter transgenic mice, asbestos activates AP-1 in both lung and bronchial tissues.[173] This activation of AP-1 by asbestos is persistent until day three after a single dose of crocidolite asbestos exposure. Further studies suggested that PKCα plays a critical role in asbestos-induced c-fos and c-jun expression, since a PKCα inhibitor substantially decreased c-fos and c-jun mRNA in asbestos-treated cells.[174] In addition, asbestos is also capable of inducing expression of fra-1, a gene encoding another AP-1 subunit, Fra, in rat mesothelial cells in a dose-dependent manner.[175] A recent study by Ramos-Nino et al.[176] demonstrated that the fra-1 induction by asbestos in mesothelial cell transformation is dependent on the activation of extracellular signal-regulated kinases (ERKs 1/2). Inhibition of ERK activation or fra expression reversed the transformed phenotype of mesothelioma cells, suggesting that fra-1 induction is the most important event in asbestos-induced carcinogenic transformation.

In addition to the induction of AP-1 subunit genes, asbestos activates AP-1 through the upstream kinases that are responsible for the phosphorylation of AP-1 subunits and enhancement of AP-1 transcriptional activity. Asbestos activates ERK1 and ERK2 through EGF-R.[177] Pretreatment of the cells with tyrphostin AG1478, a specific inhibitor of the tyrosine kinase activity of EGF-R, substantially decreased phosphorylation and activation of ERK1 and/or ERK2 induced by asbestos. The mechanism of how asbestos affects the function of EGF-R is elusive. Asbestos has been shown to be able to interrupt the interaction of EGF and its receptor, and at the same time to increase the protein expression of EGF-R.[178] The activation of ERK1 or ERK2 was generally in response to cell growth signals that induce cell proliferation. Intriguingly, asbestos is also capable of inducing cell apoptosis through the activation of ERK.[178] The ability of asbestos to cause kinase activation was largely attributed to its induction of ROS. However, a discrepancy was noted between asbestos and H_2O_2 in their stimulation on ERK and JNK, a stress and apoptotic MAP kinase.[179] In rat pleural mesothelial cells, H_2O_2 activates both ERK and JNK, whereas asbestos activates only ERK but not JNK.[179] The activation of p38 MAP kinase by asbestos appears to be responsible for asbestos-induced cytokine release from normal human alveolar macrophages.[180]

The DNA binding activity of AP-1 can be regulated by asbestos through the inducible nuclear translocation of redox factor-1 (Ref-1, APEX) in human alveolar macrophages or murine macrophage cells treated with low concentrations of asbestos.[181] In rat pleural mesothelial cells, asbestos induces increased levels of Ref-1 mRNA and protein in a dose-dependent manner.[182] Ref-1 is a nuclear protein possessing both DNA repair activity and redox regulatory activity, and the latter may be critical for the DNA binding activity of AP-1.[183] The DNA binding activity of AP-1 is partially dependent on the redox status of the conserved cysteine residues in the DNA binding domain of c-Jun and c-Fos proteins. Association of Ref-1 with AP-1 causes reduction of these cysteines and enhances the DNA binding activity of AP-1.[184]

4.5 CONCLUSIONS

In this chapter, we briefly reviewed some of the recent information regarding the molecular mechanisms underlying cancer biology of silica- and asbestos-induced carcinogenesis. Since there are rapid advances in these fields and new information is available on a daily basis using modern technology, it is difficult to summarize all aspects of related studies in the carcinogenesis of these two important minerals. We attempted to outline a few important molecular processes that are intensively investigated in our laboratories that are potentially critical and are associated with the carcinogenesis induced by silica and asbestos (Figure 4.1). Several decades have passed since the recognition of the health hazards of silica and asbestos exposure. Although silicosis and asbestosis are now mostly preventable by environmental dust control, these diseases in the workplace continue to occur on a worldwide basis. Because the latency period between exposure and development of silica- and asbestos-related cancers is long, there is an intense effort to identify markers of tumor or biomarker proteins as early indicators of the disease process. To achieve this goal, studies exploring the detailed molecular mechanisms of silica- and asbestos-induced cancer development in lung cells and animal models are our priorities.

REFERENCES

1. Greenlee, R.T., Murray, T., Bolden, S., and Wingo, P.A., Cancer statistics, 2000, *CA Cancer J. Clin.*, 50, 7–33, 2000.
2. Perera, F.P., Environment and cancer: who are susceptible? *Science*, 278, 1068–1073, 1997.
3. Doll, R. and Peto, R., The causes of cancer: quantitative estimates of avoidable risks of cancer in the United States today, *J. Natl. Cancer Inst.*, 66, 1191–1308, 1981.
4. Manning, C.B., Vallyathan, V., and Mossman, B.T., Diseases caused by asbestos: mechanisms of injury and disease development, *Int. Immunopharmacol.*, 2, 191–200, 2002.
5. Chen, F. and Shi, X., Signaling from toxic metals to NF-kappaB and beyond: not just a matter of reactive oxygen species, *Environ. Health Perspect.*, 110 (Suppl. 5), 807–811, 2002.
6. Saffiotti, U., Lung cancer induction by crystalline silica, *Prog. Clin. Biol. Res.*, 374, 51–69, 1992.
7. Hanahan, D. and Weinberg, R.A., The hallmarks of cancer, *Cell*, 100, 57–70, 2000.
8. Golub, T.R., Slonim, D.K., Tamayo, P., Huard, C., Gaasenbeek, M., Mesirov, J.P., Coller, H., Loh, M.L., Downing, J.R., Caligiuri, M.A., Bloomfield, C.D., and Lander, E.S., Molecular classification of cancer: class discovery and class prediction by gene expression monitoring, *Science*, 286, 531–537, 1999.
9. Bhattacharjee, A., Richards, W.G., Staunton, J., Li, C., Monti, S., Vasa, P., Ladd, C., Beheshti, J., Bueno, R., Gillette, M., Loda, M., Weber, G., Mark, E.J., Lander, E.S., Wong, W., Johnson, B.E., Golub, T.R., Sugarbaker, D.J., and Meyerson, M., Classification of human lung carcinomas by mRNA expression profiling reveals distinct adenocarcinoma subclasses, *Proc. Natl. Acad. Sci. USA*, 98, 13790–13795, 2001.
10. Garber, M.E., Troyanskaya, O.G., Schluens, K., Petersen, S., Thaesler, Z., Pacyna-Gengelbach, M., van de Rijn, M., Rosen, G.D., Perou, C.M., Whyte, R.I., Altman, R.B., Brown, P.O., Botstein, D., and Petersen, I., Diversity of gene expression in adenocarcinoma of the lung, *Proc. Natl. Acad. Sci. USA*, 98, 13784–13789, 2001.
11. Perou, C.M., Sorlie, T., Eisen, M.B., van de Rijn, M., Jeffrey, S.S., Rees, C.A., Pollack, J.R., Ross, D.T., Johnsen, H., Akslen, L.A., Fluge, O., Pergamenschikov, A., Williams, C., Zhu, S.X., Lonning, P.E., Borresen-Dale, A.L., Brown, P.O., and Botstein, D., Molecular portraits of human breast tumours, *Nature*, 406, 747–752, 2000.
12. Shih, C., Padhy, L.C., Murray, M., and Weinberg, R.A., Transforming genes of carcinomas and neuroblastomas introduced into mouse fibroblasts, *Nature*, 290, 261–264, 1981.

13. Balmain, A. and Pragnell, I.B., Mouse skin carcinomas induced *in vivo* by chemical carcinogens have a transforming Harvey-ras oncogene, *Nature*, 303, 72–74, 1983.
14. Renan, M.J., How many mutations are required for tumorigenesis? Implications from human cancer data, *Mol. Carcinog.*, 7, 139–146, 1993.
15. Tomlinson, I., Sasieni, P., and Bodmer, W., How many mutations in a cancer? *Am. J. Pathol.*, 160, 755–758, 2002.
16. Nyberg, K.A., Michelson, R.J., Putnam, C.W., and Weinert, T.A., Toward maintaining the genome: DNA damage and replication checkpoints, *Annu. Rev. Genet.*, 36, 617–656, 2002.
17. Marnett, L.J., Riggins, J.N., and West, J.D., Endogenous generation of reactive oxidants and electrophiles and their reactions with DNA and protein, *J. Clin. Invest.*, 111, 583–593, 2003.
18. Wei, H. and Frenkel, K., Relationship of oxidative events and DNA oxidation in SENCAR mice to *in vivo* promoting activity of phorbol ester-type tumor promoters, *Carcinogenesis*, 14, 1195–1201, 1993.
19. Casadevall, M., da Cruz Fresco, P., and Kortenkamp, A., Chromium(VI)-mediated DNA damage: oxidative pathways resulting in the formation of DNA breaks and abasic sites, *Chem. Biol. Interact.*, 123, 117–132, 1999.
20. Hoeijmakers, J.H., Genome maintenance mechanisms for preventing cancer, *Nature*, 411, 366–374, 2001.
21. Le Page, F., Kwoh, E.E., Avrutskaya, A., Gentil, A., Leadon, S.A., Sarasin, A., and Cooper, P.K., Transcription-coupled repair of 8-oxoguanine: requirement for XPG, TFIIH, and CSB and implications for Cockayne syndrome, *Cell*, 101, 159–171, 2000.
22. van Hoffen, A., Venema, J., Meschini, R., van Zeeland, A.A., and Mullenders, L.H., Transcription-coupled repair removes both cyclobutane pyrimidine dimers and 6-4 photoproducts with equal efficiency and in a sequential way from transcribed DNA in xeroderma pigmentosum group C fibroblasts, *EMBO J.*, 14, 360–367, 1995.
23. Henning, K.A., Li, L., Iyer, N., McDaniel, L.D., Reagan, M.S., Legerski, R., Schultz, R.A., Stefanini, M., Lehmann, A.R., Mayne, L.V. et al., The Cockayne syndrome group A gene encodes a WD repeat protein that interacts with CSB protein and a subunit of RNA polymerase II TFIIH, *Cell*, 82, 555–564, 1995.
24. Lee, K.B., Wang, D., Lippard, S.J., and Sharp, P.A., Transcription-coupled and DNA damage-dependent ubiquitination of RNA polymerase II *in vitro*, *Proc. Natl. Acad. Sci. USA*, 99, 4239–4244, 2002.
25. Svejstrup, J.Q., Mechanisms of transcription-coupled DNA repair, *Nat. Rev. Mol. Cell. Biol.*, 3, 21–29, 2002.
26. Goodman, M.F. and Tippin, B., Sloppier copier DNA polymerases involved in genome repair, *Curr. Opin. Genet. Dev.*, 10, 162–168, 2000.
27. Haracska, L., Prakash, L., and Prakash, S., Role of human DNA polymerase kappa as an extender in translesion synthesis, *Proc. Natl. Acad. Sci. USA*, 99, 16000–16005, 2002.
28. Lehmann, A.R., Replication of damaged DNA in mammalian cells: new solutions to an old problem, *Mutat. Res.*, 509, 23–34, 2002.
29. Kunkel, T.A. and Bebenek, K., DNA replication fidelity, *Annu. Rev. Biochem.*, 69, 497–529, 2000.
30. Bernstein, C., Bernstein, H., Payne, C.M., and Garewal, H., DNA repair/pro-apoptotic dual-role proteins in five major DNA repair pathways: fail-safe protection against carcinogenesis, *Mutat. Res.*, 511, 145–178, 2002.
31. Broomfield, S., Hryciw, T., and Xiao, W., DNA postreplication repair and mutagenesis in *Saccharomyces cerevisiae*, *Mutat. Res.*, 486, 167–184, 2001.
32. Kaiser, P., Mansour, H.A., Greeten, T., Auer, B., Schweiger, M., and Schneider, R., The human ubiquitin-conjugating enzyme UbcH1 is involved in the repair of UV-damaged, alkylated and cross-linked DNA, *FEBS Lett.*, 350, 1–4, 1994.
33. Hoege, C., Pfander, B., Moldovan, G.L., Pyrowolakis, G., and Jentsch, S., RAD6-dependent DNA repair is linked to modification of PCNA by ubiquitin and SUMO, *Nature*, 419, 135–141, 2002.
34. Khanna, K.K. and Jackson, S.P., DNA double-strand breaks: signaling, repair and the cancer connection, *Nat. Genet.*, 27, 247–254, 2001.
35. Abraham, R.T., Cell cycle checkpoint signaling through the ATM and ATR kinases, *Genes Dev.*, 15, 2177–2196, 2001.
36. Wang, Y., Cortez, D., Yazdi, P., Neff, N., Elledge, S.J., and Qin, J., BASC, a super complex of BRCA1-associated proteins involved in the recognition and repair of aberrant DNA structures, *Genes Dev.*, 14, 927–939, 2000.

37. Woo, R.A., McLure, K.G., Lees-Miller, S.P., Rancourt, D.E., and Lee, P.W., DNA-dependent protein kinase acts upstream of p53 in response to DNA damage, *Nature*, 394, 700–704, 1998.

38. Shieh, S.Y., Ikeda, M., Taya, Y., and Prives, C., DNA damage-induced phosphorylation of p53 alleviates inhibition by MDM2, *Cell*, 91, 325–334, 1997.

39. Jhappan, C., Yusufzai, T.M., Anderson, S., Anver, M.R., and Merlino, G., The p53 response to DNA damage *in vivo* is independent of DNA-dependent protein kinase, *Mol. Cell. Biol.*, 20, 4075–4083, 2000.

40. Jaiswal, A.S. and Narayan, S., SN2 DNA-alkylating agent-induced phosphorylation of p53 and activation of p21 gene expression, *Mutat. Res.*, 500, 17–30, 2002.

41. Piwnica-Worms, H., Cell cycle: fools rush in, *Nature*, 401, 535–537, 1999.

42. Krammer, P.H., CD95's deadly mission in the immune system, *Nature*, 407, 789–795, 2000.

43. Hengartner, M.O., The biochemistry of apoptosis, *Nature*, 407, 770–776, 2000.

44. Newmeyer, D.D. and Ferguson-Miller, S., Mitochondria: releasing power for life and unleashing the machineries of death, *Cell*, 112, 481–490, 2003.

45. Earnshaw, W.C., Martins, L.M., and Kaufmann, S.H., Mammalian caspases: structure, activation, substrates, and functions during apoptosis, *Annu. Rev. Biochem.*, 68, 383–424, 1999.

46. Yu, H. and Rohan, T., Role of the insulin-like growth factor family in cancer development and progression, *J. Natl. Cancer Inst.*, 92, 1472–1489, 2000.

47. Maehama, T. and Dixon, J.E., PTEN: a tumour suppressor that functions as a phospholipid phosphatase, *Trends Cell. Biol.*, 9, 125–128, 1999.

48. Wu, L., Timmers, C., Maiti, B., Saavedra, H.I., Sang, L., Chong, G.T., Nuckolls, F., Giangrande, P., Wright, F.A., Field, S.J., Greenberg, M.E., Orkin, S., Nevins, J.R., Robinson, M.L., and Leone, G., The E2F1-3 transcription factors are essential for cellular proliferation, *Nature*, 414, 457–462, 2001.

49. Weinberg, R.A., The retinoblastoma gene and gene product, *Cancer Surv.*, 12, 43–57, 1992.

50. Helin, K., Lees, J.A., Vidal, M., Dyson, N., Harlow, E., and Fattaey, A., A cDNA encoding a pRB-binding protein with properties of the transcription factor E2F, *Cell*, 70, 337–350, 1992.

51. Chen, T.T. and Wang, J.Y., Establishment of irreversible growth arrest in myogenic differentiation requires the RB LXCXE-binding function, *Mol. Cell. Biol.*, 20, 5571–5580, 2000.

52. Strobeck, M.W., Knudsen, K.E., Fribourg, A.F., DeCristofaro, M.F., Weissman, B.E., Imbalzano, A.N., and Knudsen, E.S., BRG-1 is required for RB-mediated cell cycle arrest, *Proc. Natl. Acad. Sci. USA*, 97, 7748–7753, 2000.

53. Nielsen, S.J., Schneider, R., Bauer, U.M., Bannister, A.J., Morrison, A., O'Carroll, D., Firestein, R., Cleary, M., Jenuwein, T., Herrera, R.E., and Kouzarides, T., Rb targets histone H3 methylation and HP1 to promoters, *Nature*, 412, 561–565, 2001.

54. Johnson, D.G., Schwarz, J.K., Cress, W.D., and Nevins, J.R., Expression of transcription factor E2F1 induces quiescent cells to enter S phase, *Nature*, 365, 349–352, 1993.

55. Johnson, D.G., Cress, W.D., Jakoi, L., and Nevins, J.R., Oncogenic capacity of the E2F1 gene, *Proc. Natl. Acad. Sci. USA*, 91, 12823–12827, 1994.

56. Kalma, Y., Marash, L., Lamed, Y., and Ginsberg, D., Expression analysis using DNA microarrays demonstrates that E2F-1 up-regulates expression of DNA replication genes including replication protein A2, *Oncogene*, 20, 1379–1387, 2001.

57. Muller, H., Bracken, A.P., Vernell, R., Moroni, M.C., Christians, F., Grassilli, E., Prosperini, E., Vigo, E., Oliner, J.D., and Helin, K., E2Fs regulate the expression of genes involved in differentiation, development, proliferation, and apoptosis, *Genes Dev.*, 15, 267–285, 2001.

58. Polager, S., Kalma, Y., Berkovich, E., and Ginsberg, D., E2Fs up-regulate expression of genes involved in DNA replication, DNA repair and mitosis, *Oncogene*, 21, 437–446, 2002.

59. Lissy, N.A., Davis, P.K., Irwin, M., Kaelin, W.G., and Dowdy, S.F., A common E2F-1 and p73 pathway mediates cell death induced by TCR activation, *Nature*, 407, 642–645, 2000.

60. Tanaka, H., Matsumura, I., Ezoe, S., Satoh, Y., Sakamaki, T., Albanese, C., Machii, T., Pestell, R.G., and Kanakura, Y., E2F1 and c-Myc potentiate apoptosis through inhibition of NF-kappaB activity that facilitates MnSOD-mediated ROS elimination, *Mol. Cell*, 9, 1017–1029, 2002.

61. Phillips, A.C., Ernst, M.K., Bates, S., Rice, N.R., and Vousden, K.H., E2F-1 potentiates cell death by blocking antiapoptotic signaling pathways, *Mol. Cell*, 4, 771–781, 1999.

62. Vogt, P.K., Fortuitous convergences: the beginnings of JUN, *Nat. Rev. Cancer*, 2, 465–469, 2002.

63. van Dam, H. and Castellazzi, M., Distinct roles of Jun:Fos and Jun:ATF dimers in oncogenesis, *Oncogene*, 20, 2453–2464, 2001.
64. Smeal, T., Binetruy, B., Mercola, D.A., Birrer, M., and Karin, M., Oncogenic and transcriptional cooperation with Ha-Ras requires phosphorylation of c-Jun on serines 63 and 73, *Nature*, 354, 494–496, 1991.
65. Suzuki, T., Murakami, M., Onai, N., Fukuda, E., Hashimoto, Y., Sonobe, M.H., Kameda, T., Ichinose, M., Miki, K., and Iba, H., Analysis of AP-1 function in cellular transformation pathways, *J. Virol.*, 68, 3527–3535, 1994.
66. Wisdom, R., Johnson, R.S., and Moore, C., c-Jun regulates cell cycle progression and apoptosis by distinct mechanisms, *EMBO J.*, 18, 188–197, 1999.
67. Ivanov, V.N., Bhoumik, A., Krasilnikov, M., Raz, R., Owen-Schaub, L.B., Levy, D., Horvath, C.M., and Ronai, Z., Cooperation between STAT3 and c-jun suppresses Fas transcription, *Mol. Cell*, 7, 517–528, 2001.
68. Schreiber, M., Kolbus, A., Piu, F., Szabowski, A., Mohle-Steinlein, U., Tian, J., Karin, M., Angel, P., and Wagner, E.F., Control of cell cycle progression by c-Jun is p53 dependent, *Genes Dev.*, 13, 607–619, 1999.
69. Shaulian, E., Schreiber, M., Piu, F., Beeche, M., Wagner, E.F., and Karin, M., The mammalian UV response: c-Jun induction is required for exit from p53-imposed growth arrest, *Cell*, 103, 897–907, 2000.
70. Jochum, W., Passegue, E., and Wagner, E.F., AP-1 in mouse development and tumorigenesis, *Oncogene*, 20, 2401–2412, 2001.
71. Bader, A.G., Hartl, M., and Bister, K., Conditional cell transformation by doxycycline-controlled expression of the ASV17 v-jun allele, *Virology*, 270, 98–110, 2000.
72. Fu, S., Bottoli, I., Goller, M., and Vogt, P.K., Heparin-binding epidermal growth factor-like growth factor, a v-Jun target gene, induces oncogenic transformation, *Proc. Natl. Acad. Sci. USA*, 96, 5716–5721, 1999.
73. Nakano, H., Shindo, M., Sakon, S., Nishinaka, S., Mihara, M., Yagita, H., and Okumura, K., Differential regulation of IkappaB kinase alpha and beta by two upstream kinases, NF-kappaB-inducing kinase and mitogen-activated protein kinase/ERK kinase kinase-1, *Proc. Natl. Acad. Sci. USA*, 95, 3537–3542, 1998.
74. Claudio, E., Brown, K., Park, S., Wang, H., and Siebenlist, U., BAFF-induced NEMO-independent processing of NF-kappa B2 in maturing B cells, *Nat. Immunol.*, 3, 958–965, 2002.
75. Yilmaz, Z.B., Weih, D.S., Sivakumar, V., and Weih, F., RelB is required for Peyer's patch development: differential regulation of p52-RelB by lymphotoxin and TNF, *EMBO J.*, 22, 121–130, 2003.
76. Senftleben, U., Cao, Y., Xiao, G., Greten, F.R., Krahn, G., Bonizzi, G., Chen, Y., Hu, Y., Fong, A., Sun, S.C., and Karin, M., Activation by IKKalpha of a second, evolutionary conserved, NF-kappa B signaling pathway, *Science*, 293, 1495–1499, 2001.
77. Jaspers, I., Zhang, W., Fraser, A., Samet, J.M., and Reed, W., Hydrogen peroxide has opposing effects on IKK activity and IkappaBalpha breakdown in airway epithelial cells, *Am. J. Respir. Cell. Mol. Biol.*, 24, 769–777, 2001.
78. Yin, Z., Ivanov, V.N., Habelhah, H., Tew, K., and Ronai, Z., Glutathione S-transferase p elicits protection against H2O2-induced cell death via coordinated regulation of stress kinases, *Cancer Res.*, 60, 4053–4057, 2000.
79. Korn, S.H., Wouters, E.F., Vos, N., and Janssen-Heininger, Y.M., Cytokine-induced activation of nuclear factor-kappa B is inhibited by hydrogen peroxide through oxidative inactivation of IkappaB kinase, *J. Biol. Chem.*, 276, 35693–35700, 2001.
80. Storz, P. and Toker, A., Protein kinase D mediates a stress-induced NF-kappaB activation and survival pathway, *EMBO J.*, 22, 109–120, 2003.
81. Chen, F., Castranova, V., Shi, X., and Demers, L.M., New insights into the role of nuclear factor-kappaB, a ubiquitous transcription factor in the initiation of diseases, *Clin. Chem.*, 45, 7–17, 1999.
82. Carrasco, D., Rizzo, C.A., Dorfman, K., and Bravo, R., The v-rel oncogene promotes malignant T-cell leukemia/lymphoma in transgenic mice, *EMBO J.*, 15, 3640–3650, 1996.
83. Chen, F., Castranova, V., and Shi, X., New insights into the role of nuclear factor-kappaB in cell growth regulation, *Am. J. Pathol.*, 159, 387–397, 2001.
84. Tang, G., Minemoto, Y., Dibling, B., Purcell, N.H., Li, Z., Karin, M., and Lin, A., Inhibition of JNK activation through NF-kappaB target genes, *Nature*, 414, 313–317, 2001.

85. De Smaele, E., Zazzeroni, F., Papa, S., Nguyen, D.U., Jin, R., Jones, J., Cong, R., and Franzoso, G., Induction of gadd45beta by NF-kappaB downregulates pro-apoptotic JNK signalling, *Nature*, 414, 308–313, 2001.

86. Vallyathan, V., Shi, X., and Castranova, V., Reactive oxygen species: their relation to pneumoconiosis and carcinogenesis, *Environ. Health Perspect.*, 106 (Suppl. 5), 1151–1155, 1998.

87. Zhang, Z., Shen, H.M., Zhang, Q.F., and Ong, C.N., Critical role of GSH in silica-induced oxidative stress, cytotoxicity, and genotoxicity in alveolar macrophages, *Am. J. Physiol.*, 277, L743–748, 1999.

88. Schins, R.P., Knaapen, A.M., Cakmak, G.D., Shi, T., Weishaupt, C., and Borm, P.J., Oxidant-induced DNA damage by quartz in alveolar epithelial cells, *Mutat. Res.*, 517, 77–86, 2002.

89. Chen, F., Sun, S.C., Kuh, D.C., Gaydos, L.J., and Demers, L.M., Essential role of NF-kappa B activation in silica-induced inflammatory mediator production in macrophages, *Biochem. Biophys. Res. Commun.*, 214, 985–992, 1995.

90. Kang, J.L., Pack, I.S., Hong, S.M., Lee, H.S., and Castranova, V., Silica induces nuclear factor-kappa B activation through tyrosine phosphorylation of I kappa B-alpha in RAW264.7 macrophages, *Toxicol. Appl. Pharmacol.*, 169, 59–65, 2000.

91. Porter, D.W., Barger, M., Robinson, V.A., Leonard, S.S., Landsittel, D., and Castranova, V., Comparison of low doses of aged and freshly fractured silica on pulmonary inflammation and damage in the rat, *Toxicology*, 175, 63–71, 2002.

92. Sacks, M., Gordon, J., Bylander, J., Porter, D., Shi, X.L., Castranova, V., Kaczmarczyk, W., Van Dyke, K., and Reasor, M.J., Silica-induced pulmonary inflammation in rats: activation of NF-kappa B and its suppression by dexamethasone, *Biochem. Biophys. Res. Commun.*, 253, 181–184, 1998.

93. Desaki, M., Takizawa, H., Kasama, T., Kobayashi, K., Morita, Y., and Yamamoto, K., Nuclear factor-kappa b activation in silica-induced interleukin 8 production by human bronchial epithelial cells, *Cytokine*, 12, 1257–1260, 2000.

94. Duffin, R., Gilmour, P.S., Schins, R.P., Clouter, A., Guy, K., Brown, D.M., MacNee, W., Borm, P.J., Donaldson, K., and Stone, V., Aluminium lactate treatment of DQ12 quartz inhibits its ability to cause inflammation, chemokine expression, and nuclear factor-kappaB activation, *Toxicol. Appl. Pharmacol.*, 176, 10–17, 2001.

95. Hubbard, A.K., Timblin, C.R., Shukla, A., Rincon, M., and Mossman, B.T., Activation of NF-kappaB-dependent gene expression by silica in lungs of luciferase reporter mice, *Am. J. Physiol. Lung Cell. Mol. Physiol.*, 282, L968–975, 2002.

96. Chen, F., Lu, Y., Kuhn, D.C., Maki, M., Shi, X., Sun, S.C., and Demers, L.M., Calpain contributes to silica-induced I kappa B-alpha degradation and nuclear factor-kappa B activation, *Arch. Biochem. Biophys.*, 342, 383–388, 1997.

97. Porter, D.W., Ye, J., Ma, J., Barger, M., Robinson, V.A., Ramsey, D., McLaurin, J., Khan, A., Landsittel, D., Teass, A., and Castranova, V., Time course of pulmonary response of rats to inhalation of crystalline silica: NF-kappa B activation, inflammation, cytokine production, and damage, *Inhal. Toxicol.*, 14, 349–367, 2002.

98. Ghosh, S. and Karin, M., Missing pieces in the NF-kappaB puzzle, *Cell*, 109 (Suppl.), S81–96, 2002.

99. Schreck, R., Rieber, P., and Baeuerle, P.A., Reactive oxygen intermediates as apparently widely used messengers in the activation of the NF-kappa B transcription factor and HIV-1, *EMBO J.*, 10, 2247–2258, 1991.

100. Sarih, M., Souvannavong, V., Brown, S.C., and Adam, A., Silica induces apoptosis in macrophages and the release of interleukin-1 alpha and interleukin-1 beta, *J. Leukoc. Biol.*, 54, 407–413, 1993.

101. Tatrai, E., Adamis, Z., Timar, M., and Ungvary, G., Comparative histopathological and biochemical analysis of early stages of exposure to non-silicogenic aluminium silicate- and strongly silicogenic quartz-dust in rats, *Exp. Pathol.*, 23, 163–171, 1983.

102. Reed, J.C., Apoptosis-regulating proteins as targets for drug discovery, *Trends Mol. Med.*, 7, 314–319, 2001.

103. Fiers, W., Beyaert, R., Declercq, W., and Vandenabeele, P., More than one way to die: apoptosis, necrosis and reactive oxygen damage, *Oncogene*, 18, 7719–7730, 1999.

104. Lemasters, J.J., V. Necrapoptosis and the mitochondrial permeability transition: shared pathways to necrosis and apoptosis, *Am. J. Physiol.*, 276, G1–6, 1999.

105. Iyer, R., Hamilton, R.F., Li, L., and Holian, A., Silica-induced apoptosis mediated via scavenger receptor in human alveolar macrophages, *Toxicol. Appl. Pharmacol.*, 141, 84–92, 1996.

106. Leigh, J., Wang, H., Bonin, A., Peters, M., and Ruan, X., Silica-induced apoptosis in alveolar and granulomatous cells *in vivo*, *Environ. Health Perspect.*, 105 (Suppl. 5), 1241–1245, 1997.
107. Srivastava, K.D., Rom, W.N., Jagirdar, J., Yie, T.A., Gordon, T., and Tchou-Wong, K.M., Crucial role of interleukin-1beta and nitric oxide synthase in silica-induced inflammation and apoptosis in mice, *Am. J. Respir. Crit. Care Med.*, 165, 527–533, 2002.
108. Otsuki, T., Sakaguchi, H., Tomokuni, A., Aikoh, T., Matsuki, T., Kawakami, Y., Kusaka, M., Ueki, H., Kita, S., and Ueki, A., Soluble Fas mRNA is dominantly expressed in cases with silicosis, *Immunology*, 94, 258–262, 1998.
109. Borges, V.M., Falcao, H., Leite-Junior, J.H., Alvim, L., Teixeira, G.P., Russo, M., Nobrega, A.F., Lopes, M.F., Rocco, P.M., Davidson, W.F., Linden, R., Yagita, H., Zin, W.A., and DosReis, G.A., Fas ligand triggers pulmonary silicosis, *J. Exp. Med.*, 194, 155–164, 2001.
110. Bowden, D.H. and Adamson, I.Y., The role of cell injury and the continuing inflammatory response in the generation of silicotic pulmonary fibrosis, *J. Pathol.*, 144, 149–161, 1984.
111. Driscoll, K.E., Maurer, J.K., Lindenschmidt, R.C., Romberger, D., Rennard, S.I., and Crosby, L., Respiratory tract responses to dust: relationships between dust burden, lung injury, alveolar macrophage fibronectin release, and the development of pulmonary fibrosis, *Toxicol. Appl. Pharmacol.*, 106, 88–101, 1990.
112. Scaffidi, P., Misteli, T., and Bianchi, M.E., Release of chromatin protein HMGB1 by necrotic cells triggers inflammation, *Nature*, 418, 191–195, 2002.
113. Chen, J.J., Sun, Y., and Nabel, G.J., Regulation of the proinflammatory effects of Fas ligand (CD95L), *Science*, 282, 1714–1717, 1998.
114. Faouzi, S., Burckhardt, B.E., Hanson, J.C., Campe, C.B., Schrum, L.W., Rippe, R.A., and Maher, J.J., Anti-Fas induces hepatic chemokines and promotes inflammation by an NF-kappa B-independent, caspase-3-dependent pathway, *J. Biol. Chem.*, 276, 49077–49082, 2001.
115. Uchimura, E., Kodaira, T., Kurosaka, K., Yang, D., Watanabe, N., and Kobayashi, Y., Interaction of phagocytes with apoptotic cells leads to production of pro-inflammatory cytokines, *Biochem. Biophys. Res. Commun.*, 239, 799–803, 1997.
116. Fadok, V.A., Bratton, D.L., Konowal, A., Freed, P.W., Westcott, J.Y., and Henson, P.M., Macrophages that have ingested apoptotic cells *in vitro* inhibit proinflammatory cytokine production through autocrine/paracrine mechanisms involving TGF-beta, PGE2, and PAF, *J. Clin. Invest.*, 101, 890–898, 1998.
117. McDonald, P.P., Fadok, V.A., Bratton, D., and Henson, P.M., Transcriptional and translational regulation of inflammatory mediator production by endogenous TGF-beta in macrophages that have ingested apoptotic cells, *J. Immunol.*, 163, 6164–6172, 1999.
118. Wahl, S.M., Orenstein, J.M., and Chen, W., TGF-beta influences the life and death decisions of T lymphocytes, *Cytokine Growth Factor Rev.*, 11, 71–79, 2000.
119. Kuempel, E.D., Tran, C.L., Bailer, A.J., Porter, D.W., Hubbs, A.F., and Castranova, V., Biological and statistical approaches to predicting human lung cancer risk from silica, *J. Environ. Pathol. Toxicol. Oncol.*, 20, 15–32, 2001.
120. Deshpande, A., Narayanan, P.K., and Lehnert, B.E., Silica-induced generation of extracellular factor(s) increases reactive oxygen species in human bronchial epithelial cells, *Toxicol Sci.*, 67, 275–283, 2002.
121. Mossman, B.T. and Churg, A., Mechanisms in the pathogenesis of asbestosis and silicosis, *Am. J. Respir. Crit. Care Med.*, 157, 1666–1680, 1998.
122. Shen, H.M., Zhang, Z., Zhang, Q.F., and Ong, C.N., Reactive oxygen species and caspase activation mediate silica-induced apoptosis in alveolar macrophages, *Am. J. Physiol. Lung Cell. Mol. Physiol.*, 280, L10–17, 2001.
123. Daniel, L.N., Mao, Y., and Saffiotti, U., Oxidative DNA damage by crystalline silica, *Free Radical Biol. Med.*, 14, 463–472, 1993.
124. Chvapil, M., Stankova, L., and Malshet, V., Lipid peroxidation as one of the mechanisms of silica fibrogenicity? I: study with erythrocytes, *Environ. Res.*, 11, 78–88, 1976.
125. Allison, A.C. and Ferluga, J., Cell membranes in cytotoxicity, *Adv. Exp. Med. Biol.*, 84, 231–246, 1977.
126. Shi, X.L., Dalal, N.S., Hu, X.N., and Vallyathan, V., The chemical properties of silica particle surface in relation to silica-cell interactions, *J. Toxicol. Environ. Health*, 27, 435–454, 1989.
127. Van Dyke, K., Antonini, J.M., Wu, L., Ye, Z., and Reasor, M.J., The inhibition of silica-induced lung inflammation by dexamethasone as measured by bronchoalveolar lavage fluid parameters and peroxynitrite-dependent chemiluminescence, *Agents Actions*, 41, 44–49, 1994.

128. Blackford, Jr., J.A., Antonini, J.M., Castranova, V., and Dey, R.D., Intratracheal instillation of silica up-regulates inducible nitric oxide synthase gene expression and increases nitric oxide production in alveolar macrophages and neutrophils, *Am. J. Respir. Cell. Mol. Biol.*, 11, 426–431, 1994.

129. Setoguchi, K., Takeya, M., Akaike, T., Suga, M., Hattori, R., Maeda, H., Ando, M., and Takahashi, K., Expression of inducible nitric oxide synthase and its involvement in pulmonary granulomatous inflammation in rats, *Am. J. Pathol.*, 149, 2005–2022, 1996.

130. Schapira, R.M., Wiessner, J.H., Morrisey, J.F., Almagro, U.A., and Nelin, L.D., L-arginine uptake and metabolism by lung macrophages and neutrophils following intratracheal instillation of silica *in vivo*, *Am. J. Respir. Cell. Mol. Biol.*, 19, 308–315, 1998.

131. Castranova, V., Huffman, L.J., Judy, D.J., Bylander, J.E., Lapp, L.N., Weber, S.L., Blackford, J.A., and Dey, R.D., Enhancement of nitric oxide production by pulmonary cells following silica exposure, *Environ. Health Perspect.*, 106 (Suppl. 5), 1165–1169, 1998.

132. Cunha, F.Q., Assreuy, J., Moncada, S., and Liew, F.Y., Phagocytosis and induction of nitric oxide synthase in murine macrophages, *Immunology*, 79, 408–411, 1993.

133. Claudio, E., Segade, F., Wrobel, K., Ramos, S., and Lazo, P.S., Activation of murine macrophages by silica particles *in vitro* is a process independent of silica-induced cell death, *Am. J. Respir. Cell. Mol. Biol.*, 13, 547–554, 1995.

134. Huffman, L.J., Judy, D.J., and Castranova, V., Regulation of nitric oxide production by rat alveolar macrophages in response to silica exposure, *J. Toxicol. Environ. Health A*, 53, 29–46, 1998.

135. Chen, F., Kuhn, D.C., Sun, S.C., Gaydos, L.J., and Demers, L.M., Dependence and reversal of nitric oxide production on NF-kappa B in silica and lipopolysaccharide-induced macrophages, *Biochem. Biophys. Res. Commun.*, 214, 839–846, 1995.

136. Kang, J.L., Lee, K., and Castranova, V., Nitric oxide up-regulates DNA-binding activity of nuclear factor-kappaB in macrophages stimulated with silica and inflammatory stimulants, *Mol. Cell. Biochem.*, 215, 1–9, 2000.

137. Tweedale, G., Asbestos and its lethal legacy, *Nat. Rev. Cancer*, 2, 311–315, 2002.

138. Mossman, B.T., Kamp, D.W., and Weitzman, S.A., Mechanisms of carcinogenesis and clinical features of asbestos-associated cancers, *Cancer Invest.*, 14, 466–480, 1996.

139. Ekstrom, A.M., Eriksson, M., Hansson, L.E., Lindgren, A., Signorello, L.B., Nyren, O., and Hardell, L., Occupational exposures and risk of gastric cancer in a population-based case-control study, *Cancer Res.*, 59, 5932–5937, 1999.

140. Rom, W.N., Hammar, S.P., Rusch, V., Dodson, R., and Hoffman, S., Malignant mesothelioma from neighborhood exposure to anthophyllite asbestos, *Am. J. Ind. Med.*, 40, 211–214, 2001.

141. Mossman, B.T., Bignon, J., Corn, M., Seaton, A., and Gee, J.B., Asbestos: scientific developments and implications for public policy, *Science*, 247, 294–301, 1990.

142. Sincock, A.M., Delhanty, J.D., and Casey, G., A comparison of the cytogenetic response to asbestos and glass fibre in Chinese hamster and human cell lines: demonstration of growth inhibition in primary human fibroblasts, *Mutat. Res.*, 101, 257–268, 1982.

143. Jaurand, M.C., Kheuang, L., Magne, L., and Bignon, J., Chromosomal changes induced by chrysotile fibres or benzo-3,4-pyrene in rat pleural mesothelial cells, *Mutat. Res.*, 169, 141–148, 1986.

144. Dopp, E., Saedler, J., Stopper, H., Weiss, D.G., and Schiffmann, D., Mitotic disturbances and micronucleus induction in Syrian hamster embryo fibroblast cells caused by asbestos fibers, *Environ. Health Perspect.*, 103, 268–271, 1995.

145. Rom, W.N., Livingston, G.K., Casey, K.R., Wood, S.D., Egger, M.J., Chiu, G.L., and Jerominski, L., Sister chromatid exchange frequency in asbestos workers, *J. Natl. Cancer Inst.*, 70, 45–48, 1983.

146. Marczynski, B., Czuppon, A.B., Marek, W., Reichel, G., and Baur, X., Increased incidence of DNA double-strand breaks and anti-ds DNA antibodies in blood of workers occupationally exposed to asbestos, *Hum. Exp. Toxicol.*, 13, 3–9, 1994.

147. Kasai, H. and Nishimura, S., DNA damage induced by asbestos in the presence of hydrogen peroxide, *Gann*, 75, 841–844, 1984.

148. Fubini, B. and Mollo, L., Role of iron in the reactivity of mineral fibers, *Toxicol. Lett.*, 82–83, 951–960, 1995.

149. Xu, A., Wu, L.J., Santella, R.M., and Hei, T.K., Role of oxyradicals in mutagenicity and DNA damage induced by crocidolite asbestos in mammalian cells, *Cancer Res.*, 59, 5922–5926, 1999.

150. Churg, A. and Stevens, B., Enhanced retention of asbestos fibers in the airways of human smokers, *Am. J. Respir. Crit. Care Med.*, 151, 1409–1413, 1995.

151. Brown, R.C., Fleming, G.T., and Knight, A.I., Asbestos affects the *in vitro* uptake and detoxification of aromatic compounds, *Environ. Health Perspect.*, 51, 315–318, 1983.

152. Gerde, P. and Scholander, P., Adsorption of benzo(a)pyrene on to asbestos and manmade mineral fibres in an aqueous solution and in a biological model solution, *Br. J. Ind. Med.*, 45, 682–688, 1988.

153. Husgafvel-Pursiainen, K., Hackman, P., Ridanpaa, M., Anttila, S., Karjalainen, A., Partanen, T., Taikina-Aho, O., Heikkila, L., and Vainio, H., K-ras mutations in human adenocarcinoma of the lung: association with smoking and occupational exposure to asbestos, *Int. J. Cancer*, 53, 250–256, 1993.

154. Wang, X., Christiani, D.C., Wiencke, J.K., Fischbein, M., Xu, X., Cheng, T.J., Mark, E., Wain, J.C., and Kelsey, K.T., Mutations in the p53 gene in lung cancer are associated with cigarette smoking and asbestos exposure, *Cancer Epidemiol. Biomarkers Prev.*, 4, 543–548, 1995.

155. Mishra, A., Liu, J.Y., Brody, A.R., and Morris, G.F., Inhaled asbestos fibers induce p53 expression in the rat lung, *Am. J. Respir. Cell. Mol. Biol.*, 16, 479–485, 1997.

156. Nuorva, K., Makitaro, R., Huhti, E., Kamel, D., Vahakangas, K., Bloigu, R., Soini, Y., and Paakko, P., p53 Protein accumulation in lung carcinomas of patients exposed to asbestos and tobacco smoke, *Am. J. Respir. Crit. Care Med.*, 150, 528–533, 1994.

157. Luo, J.C., Zehab, R., Anttila, S., Ridanpaa, M., Husgafvel-Pursiainen, K., Vainio, H., Carney, W., De Vivo, I., Milling, C., and Brandt-Rauf, P.W., Detection of serum p53 protein in lung cancer patients, *J. Occup. Med.*, 36, 155–160, 1994.

158. Johnson, N.F., Carpenter, T.R., Jaramillo, R.J., and Liberati, T.A., DNA damage-inducible genes as biomarkers for exposures to environmental agents, *Environ. Health Perspect.*, 105 (Suppl. 4), 913–918, 1997.

159. Isik, R., Metintas, M., Gibbs, A.R., Metintas, S., Jasani, B., Oner, U., Harmanci, E., Demircan, S., and Isiksoy, S., p53, p21 and metallothionein immunoreactivities in patients with malignant pleural mesothelioma: correlations with the epidemiological features and prognosis of mesotheliomas with environmental asbestos exposure, *Respir. Med.*, 95, 588–593, 2001.

160. Matsuoka, M., Igisu, H., and Morimoto, Y., Phosphorylation of p53 protein in A549 human pulmonary epithelial cells exposed to asbestos fibers, *Environ. Health Perspect.*, 111, 509–512, 2003.

161. Janssen, Y.M., Barchowsky, A., Treadwell, M., Driscoll, K.E., and Mossman, B.T., Asbestos induces nuclear factor kappa B (NF-kappa B) DNA-binding activity and NF-kappa B-dependent gene expression in tracheal epithelial cells, *Proc. Natl. Acad. Sci. USA*, 92, 8458–8462, 1995.

162. Simeonova, P.P. and Luster, M.I., Asbestos induction of nuclear transcription factors and interleukin 8 gene regulation, *Am. J. Respir. Cell. Mol. Biol.*, 15, 787–795, 1996.

163. Janssen, Y.M., Driscoll, K.E., Howard, B., Quinlan, T.R., Treadwell, M., Barchowsky, A., and Mossman, B.T., Asbestos causes translocation of p65 protein and increases NF-kappa B DNA binding activity in rat lung epithelial and pleural mesothelial cells, *Am. J. Pathol.*, 151, 389–401, 1997.

164. Faux, S.P. and Howden, P.J., Possible role of lipid peroxidation in the induction of NF, *Environ. Health Perspect.*, 105S, 1127–1130, 1997.

165. Brown, R.C. and Poole, A., Arachidonic acid release and prostaglandin synthesis in a macrophage-like cell line exposed to asbestos, *Agents Actions*, 15, 336–340, 1984.

166. Camandola, S., Leonarduzzi, G., Musso, T., Varesio, L., Carini, R., Scavazza, A., Chiarpotto, E., Baeuerle, P.A., and Poli, G., Nuclear factor kB is activated by arachidonic acid but not by eicosapentaenoic acid, *Biochem. Biophys. Res. Commun.*, 229, 643–647, 1996.

167. Bonizzi, G., Piette, J., Merville, M.P., and Bours, V., Distinct signal transduction pathways mediate nuclear factor-kappaB induction by IL-1beta in epithelial and lymphoid cells, *J. Immunol.*, 159, 5264–5272, 1997.

168. Zhang, Y., Lee, T.C., Guillemin, B., Yu, M.C., and Rom, W.N., Enhanced IL-1 beta and tumor necrosis factor-alpha release and messenger RNA expression in macrophages from idiopathic pulmonary fibrosis or after asbestos exposure, *J. Immunol.*, 150, 4188–4196, 1993.

169. Simeonova, P.P. and Luster, M.I., Iron and reactive oxygen species in the asbestos-induced tumor necrosis factor-alpha response from alveolar macrophages, *Am. J. Respir. Cell. Mol. Biol.*, 12, 676–683, 1995.

170. Driscoll, K.E., Carter, J.M., Hassenbein, D.G., and Howard, B., Cytokines and particle-induced inflammatory cell recruitment, *Environ. Health Perspect.*, 105 (Suppl. 5), 1159–1164, 1997.

171. Brody, A.R., Liu, J.Y., Brass, D., and Corti, M., Analyzing the genes and peptide growth factors expressed in lung cells *in vivo* consequent to asbestos exposure and *in vitro*, *Environ. Health Perspect.*, 105 (Suppl. 5), 1165–1171, 1997.

172. Heintz, N.H., Janssen, Y.M., and Mossman, B.T., Persistent induction of c-fos and c-jun expression by asbestos, *Proc. Natl. Acad. Sci. USA*, 90, 3299–3303, 1993.

173. Ding, M., Dong, Z., Chen, F., Pack, D., Ma, W.Y., Ye, J., Shi, X., Castranova, V., and Vallyathan, V., Asbestos induces activator protein-1 transactivation in transgenic mice, *Cancer Res.*, 59, 1884–1889, 1999.

174. Fung, H., Quinlan, T.R., Janssen, Y.M., Timblin, C.R., Marsh, J.P., Heintz, N.H., Taatjes, D.J., Vacek, P., Jaken, S., and Mossman, B.T., Inhibition of protein kinase C prevents asbestos-induced c-fos and c-jun proto-oncogene expression in mesothelial cells, *Cancer Res.*, 57, 3101–3105, 1997.

175. Sandhu, H., Dehnen, W., Roller, M., Abel, J., and Unfried, K., mRNA expression patterns in different stages of asbestos-induced carcinogenesis in rats, *Carcinogenesis*, 21, 1023–1029, 2000.

176. Ramos-Nino, M.E., Timblin, C.R., and Mossman, B.T., Mesothelial cell transformation requires increased AP-1 binding activity and ERK-dependent Fra-1 expression, *Cancer Res.*, 62, 6065–6069, 2002.

177. Zanella, C.L., Posada, J., Tritton, T.R., and Mossman, B.T., Asbestos causes stimulation of the extracellular signal-regulated kinase 1 mitogen-activated protein kinase cascade after phosphorylation of the epidermal growth factor receptor, *Cancer Res.*, 56, 5334–5338, 1996.

178. Zanella, C.L., Timblin, C.R., Cummins, A., Jung, M., Goldberg, J., Raabe, R., Tritton, T.R., and Mossman, B.T., Asbestos-induced phosphorylation of epidermal growth factor receptor is linked to c-fos and apoptosis, *Am. J. Physiol.*, 277, L684–693, 1999.

179. Jimenez, L.A., Zanella, C., Fung, H., Janssen, Y.M., Vacek, P., Charland, C., Goldberg, J., and Mossman, B.T., Role of extracellular signal-regulated protein kinases in apoptosis by asbestos and H2O2, *Am. J. Physiol.*, 273, L1029–1035, 1997.

180. Geist, L.J., Powers, L.S., Monick, M.M., and Hunninghake, G.W., Asbestos stimulation triggers differential cytokine release from human monocytes and alveolar macrophages, *Exp. Lung Res.*, 26, 41–56, 2000.

181. Flaherty, D.M., Monick, M.M., Carter, A.B., Peterson, M.W., and Hunninghake, G.W., Oxidant-mediated increases in redox factor-1 nuclear protein and activator protein-1 DNA binding in asbestos-treated macrophages, *J. Immunol.*, 168, 5675–5681, 2002.

182. Fung, H., Kow, Y.W., Van Houten, B., Taatjes, D.J., Hatahet, Z., Janssen, Y.M., Vacek, P., Faux, S.P., and Mossman, B.T., Asbestos increases mammalian AP-endonuclease gene expression, protein levels, and enzyme activity in mesothelial cells, *Cancer Res.*, 58, 189–194, 1998.

183. Xanthoudakis, S. and Curran, T., Identification and characterization of Ref-1, a nuclear protein that facilitates AP-1 DNA-binding activity, *EMBO J.*, 11, 653–665, 1992.

184. Abate, C., Patel, L., Rauscher, III, F.J., and Curran, T., Redox regulation of fos and jun DNA-binding activity *in vitro*, *Science*, 249, 1157–1161, 1990.

5 Involvement of Receptor Tyrosine Kinases in Lung Cancer: Clinical Importance

Nilanjana Maulik and Gautam Maulik

CONTENTS

5.1 Introduction...63
5.2 Receptor Tyrosine Kinases (RTKs)...64
    5.2.1 c-Met Receptor Tyrosine Kinase...64
    5.2.2 c-Kit Receptor Tyrosine Kinase...65
    5.2.3 EGFR Receptor Tyrosine Kinase...66
5.3 Role of c-Met, c-Kit, and EGFR 67
5.4 Role of Reactive Oxygen Species..73
5.5 Receptor Tyrosine Kinases as Target for Anti-Cancer Therapy 74
5.6 Conclusion 74
References 75

5.1 INTRODUCTION

Lung cancer is the most common cancer-related cause of death among men and women. The two main types of lung cancer are non-small-cell lung cancer (NSCLC) and small-cell lung cancer (SCLC). Each type affects different types of cells in the lung. NSCLC, generally known as the common type of lung cancer, can be divided further into three categories such as: squamous cell carcinoma, adenocarcinoma, and large-cell carcinoma. SCLC is highly associated with smoking and spreads very fast. Among all types of lung cancer, small-cell lung cancer is associated with the lowest rate of 5-year survival.

Lung cancer cells often express inappropriately high levels of receptor tyrosine kinases (RTKs) that accelerate cell proliferation upon binding of their ligands. Binding of ligands to the tyrosine kinase receptors (growth factors as a ligand) is one of the first steps in the internal signaling pathway that activates cell migration, proliferation, and growth.[1,2] Generally, growth factors and their receptors control cell growth in a rational way that initiates cell growth only when it is absolutely necessary. When a growth factor binds to its receptor, it activates the receptor.

We will restrict our interest only to three well-described RTKs in lung cancer, and they are: epidermal growth factor receptors (EGFR), c-Kit, and c-Met. The epidermal growth factor receptor, also known as the ERBB1 receptor, is expressed by approximately 30% of NSCLCs and has thus been the potential therapeutic target. At least 70% of small-cell lung cancer tumors coexpress the kit receptor tyrosine kinase and its ligand, stem cell factor (SCF). Overexpression of c-Met has also been documented in SCLC and NSCLC. Recently, it was demonstrated that c-Met is overexpressed and also activated in response to the binding of its ligand hepatocyte growth factor (HGF) in SCLC cells.[3] This RTK activation plays a crucial role in increasing cell proliferation and decreasing apoptosis. Thus, RTKs have been identified as important and very promising therapeutic targets in lung carcinogenesis.

5.2 RECEPTOR TYROSINE KINASES (RTKS)

5.2.1 c-Met Receptor Tyrosine Kinase

The proto-oncogene Met encodes the tyrosine kinase receptor for hepatocyte growth factor/scatter factor (HGF/SF).[4,5] Met was first detected as the product of a human oncogene, tpr-met, which represents the fusion of two distinct genetic loci: tpr, which contributes a protein-protein dimerization motif, and met, which contributes the intracellular portion of the met receptor containing the kinase domain. In addition to tpr-met, several other oncogenic forms of Met have been identified, and it is becoming increasingly apparent that understanding the cell biology of the HGF/SF-Met system is of great importance in elucidating the development and progression of human cancer.

Met receptor is initially synthesized as a partially glycosylated 170-kDa single-chain intracellular precursor. After additional glycosylation, it is cleaved, yielding a mature, cell surface-associated disulfide-linked heterodimer of 190 kDa. This dimer consists of an entirely extracellular α-chain (50 kDa) and a membrane-spanning β-chain (140 kDa) that possesses intrinsic, ligand-activated tyrosine kinase activity in its intracellular C-terminal domain. Receptor activation triggers a unique process of differentiation, which is known as "branching morphogenesis," that activates cell growth, inhibits apoptosis, controls cell dissociation, and also migrates into the extracellular matrix.[6,7] Met activation mediates invasive growth, which eventually leads to systemic metastases.[8–13] HGF binding to Met phosphorylates tyrosine kinase[5,14] at a unique docking site located in the Met carboxyl terminal tail, which contains the sequence $Y^{1349}VHVY^{1356}VNV$.[15] Thus the two phosphorylated tyrosines within this sequence couple the receptor and multiply intracellular effectors, among which are the Grb2/SOS complex, the p85 regulatory subunit of PI-3-kinase, Stat-3, Src, and the multiadaptor protein Gab1.[15–22] Besides Met there are four more prototypic members of a subfamily of heterodimeric receptor tyrosine kinases called Ron, Ryk, Sex, and Sea. However, only Met is being studied extensively. It is expressed in a variety of normal as well as malignant cells. Extensive expression of c-Met and activated mutation leads to carcinogenesis in multiple tumors. Overexpression of c-Met is already documented in small-cell lung cancer and non-small-cell lung cancer (Figure 5.1A and Figure 5.1B). When c-Met is activated by HGF/SF, it is phosphorylated along with a number of other proteins (Figure 5.1C).

Met also can be oncogenically activated via a number of specific point mutations. Missense mutations in Met were found to be associated with human papillary renal carcinoma,[23] and these mutations subsequently were shown to deregulate the enzymatic activity of this receptor. The missense mutations have been found to be located in the tyrosine kinase domain. Overexpression of c-Met tyrosine kinase has been reported in both SCLC and NSCLC; however, no mutations have been reported in either form of cancers[3,24,25] until now. Recently Ma et al.[26] demonstrated two different missense mutations in the juxtamembrane (JM) domain of c-Met, R988C and T1010I, which disrupt the wild-type c-Met RTK signaling pathway significantly. The JM domains of tyrosine kinase receptors are found to be responsible for the catalytic function of the kinases. According to this study, mutation of the c-Met JM domain (R988C or T1010I) was found to regulate cell morphology and tumorigenicity and cell adhesion of the H446 cells transfected with mutated c-Met receptor.

Overexpression of Met in some cancers can result in its own activation and does not depend on an autocrine mechanism.[27,28] Overexpression of HGF and Met has been observed in a large portion of human NSCLC,[29,30] and this has been clinically suggested to be a possible independent early predictor of poor survival of lung cancer patients.[29]

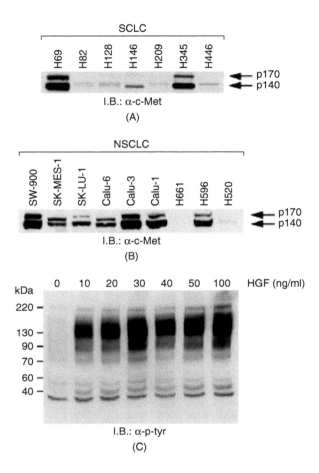

FIGURE 5.1 Expression of c-Met in lung cancer and tyrosine phosphorylation of several proteins in response to HGF/SF stimulation of small-cell lung cancer cells. (**A**) Immunoblot of c-Met in various small-cell lung cancer (SCLC) cell lines; (**B**) Immunoblot of c-Met in various non-small-cell lung cancer (NSCLC) cell lines. (**C**) Dose–response study of HGF/SF stimulation on tyrosine phosphorylation in H69 SCLC cell line. Tyrosine phosphorylated proteins were detected in cellular lysates by immunoblotting using an anti-phosphotyrosine antibody. (From Maulik, G. et al., *Cytokine Growth Factors Rev.*, 13, 41, 2002. With permission.)

5.2.2 c-Kit Receptor Tyrosine Kinase

At least 70% of small-cell lung cancers express the kit receptor tyrosine kinase and its ligand, stem cell factor (SCF). C-kit is a type III receptor tyrosine kinase. It is found that interactions between kit and its ligand, SCF, are important in the development and maintenance of hematopoietic cells, melanocytes, germ cells, and the interstitial cells. SCF acts either alone or in combination with other growth factors in promoting the survival and self-renewal of stem cells, and the proliferation, differentiation, and migration of various cell types. Numerous lines of evidence have demonstrated that this coexpression constitutes a functional autocrine loop, suggesting that inhibitors of kit tyrosine activity could have therapeutic efficacy in this disease. The kit receptor is a transmembrane protein of 145 kDa possessing an extracellular ligand-binding domain with five immunoglobulin-like motifs, a signal transmembrane region, and a cytoplasmic domain that exhibits protein tyrosine kinase activity.[31] SCF binding to the c-kit receptor results in autophosphorylation on tyrosines located in the COOH-terminal tail region, which then results in activation of its kinase activity.[32] When phosphorylated, these tyrosine residues become docking sites for several intracellular signaling molecules, leading to various cellular responses in different cell types.

Generally, kit is expressed in hematopoietic cells, melanocytes, other neural crest derivatives, and germ cells, as well as in a variety of solid tumors.[33] Abnormalities in melanogenesis, hematopoiesis, and gonadogenesis were observed in the mice during development with the c-kit or SCF mutations.[34] Until now, kit oncogenic mutations have been identified in exon 2 (myelofibrosis and chronic myelogenous leukemia), exon 11-juxtamembrane domain (mast cell tumors and GISTs), and also in the exon 17 phospho-transferase domain (mast cell tumors and germ cell tumors).[35]

A number of indolino compounds have shown inhibition of wild-type c-kit, but these were generally inefficient at inhibiting the kinase activity of c-kit with a mutation in the kinase domain (814 murine, 816 human, 817 rat); only SU6577 was effective at the high concentration of 40 μM.[36,37]

There are two major ways that c-kit mediates rapid cell growth in SCLC: (1) through the c-kit/SCF autocrine loop and (2) through the overexpression with increased sensitivity to endogenous SCF. The activation of RAS-MAPK cascade by c-kit in SCLC has been documented by Krystal et al.[38]

5.2.3 EGFR Receptor Tyrosine Kinase

Growth factor targets cell proliferation and differentiation, and it is now clear that tumor cells may overcome normal regulatory inhibition of proliferation by inappropriate activation of protein tyrosine kinases, such as the erbB receptor family. This particular family includes four distinct members: HER1 epidermal growth factor (EGF) receptor (c-erbB1), HER2 (c-erbB2), HER3 (c-erbB3), and HER4 (c-erbB4). These family members share an overall structure of two cysteine-rich regions in the extracellular domain and a cytoplasmic kinase pocket with a carboxy-terminal tail that is responsible for various downstream stimulation of signal transduction pathways. The overexpression of EGFR is found in many lung tumors.[39]

EGFR is expressed on the cell membrane of a variety of malignant cells.[40,41] Under various physiological conditions, EGF exerts its main action of growth stimulation and initiation of DNA by binding to the corresponding high-affinity cell surface receptor, EGFR, which leads to receptor tyrosine kinase activity, triggering a complex cascade of events that generally leads to cell proliferation. This process is enhanced by anti-apoptotic effects.[42,43] Both EGFR and c-erbB 2 are the members of the type I family of tyrosine kinase cell surface receptors, and activation of EGFR promotes cell proliferation or differentiation.[44] Upon binding with the ligand, the EGFR homodimerizes with the other member of the family.[45] Receptor dimerization leads to activation of the EGFR and, subsequently, to cross-phosphorylation of tyrosine residues in the cytoplasmic tail of the receptor. Phosphotyrosine residues in the carboxy-terminus of the receptor serve as high-affinity sites for proteins that, in turn, transmit the growth factor signal inside the cell.[46] Overexpression of the EGFR and also c-erbB2 has been found in various cancers such as bladder cancer,[47] colon carcinoma,[48] and lung cancer.[49]

In NSCLC the expression profile of EGFR was correlated with a very poor prognosis and also correlated with stage of disease. One study has documented median survival for 121 patients with squamous cell lung cancer, and patients with expression of EGFR had a shorter survival.[50]

Activation of EGFR was found to activate the phosphorylation of its C-terminus. Several SH2 domains (Src homology) containing proteins have been found to bind to the phosphotyrosine residues of the receptor. Several of these proteins activate different pathways that result in cell proliferation, differentiation, motility, and adhesion.[51]

EGFR expression was observed in >20% of cells in 11 of 16 (69%) epithelial malignant pleural mesothelioma (MPM) specimens by immunohistochemical analysis. MPM is a locally advancing disease with significant extension and growth into adjacent vital structures, such as esophagus, myocardium, and chest wall. MPM is found to be a very rare kind of aggressive malignancy and there is almost no therapeutic modality.

Squamous cell carcinomas overexpress EGFR very strongly (85%), whereas adenocarcinomas and large-cell carcinomas are positive in approximately 65% cases.[52] Some studies suggested that EGFR expression might be correlated with a decreased survival rate.[53,54] A few studies on lung cancer cell lines have demonstrated that, in NSCLC, a large portion of the examined cell lines expressed the EGF receptor, whereas SCLC frequently expressed the receptor.[55–57] The EGF receptor consists of an external ligand binding site and an internal tyrosine kinase complex.[58] The difference between SCLC and NSCLC could be accounted for if SCLC carried the truncated receptor. However, it was found that the monoclonal antibodies EGF-RF4 and EGF-RD10 recognize the internal domain of the EGF receptor.[59] The studies with these two antibodies have demonstrated that neither SCLC nor NSCLC carried the truncated receptor. Damstrup et al.[49] demonstrated that a relatively high proportion of SCLC cell lines carried high-affinity EGF receptors and expressed EGF receptor mRNA. This group also documented that in the SCLC cell lines, EGF receptors could not be expressed when the protein concentration was found to be lower than 75 µg/ml. It was also documented by this group that the binding capacity was stable in the range of 150 to 600 µg proteins.

5.3 ROLE OF c-MET, c-KIT, AND EGFR

c-Met signaling (Met-HGF/SF) has been well studied in various forms of human cancers. c-Met has been found to have a number of biological effects related to the cytoskeleton, for example scattering, cell motility, invasion, migration, and finally metastasis. HGF/SF-Met signaling plays an important role in lung cancer (solid tumor) and can invade and metastasize using this signaling pathway (Figure 5.2).[24] SCF/HGF-Met signaling significantly increases the motility of epithelial cells, which play a key role in physiology and various disease processes.[23]

Autophosphorylation-mediated activation of c-Met triggers multiple signal transducing intermediates such as Grb2, the p85 subunit of PI3-kinase, Stat-3, and Gab1.[60] The PI3-kinase pathway is assumed to control a number of cellular processes, including cytoskeletal organization, cell growth, and survival.[61–63] Significant c-Met phosphorylation was observed on several unique tyrosines in response to HGF in SCLC (Figure 5.3).[64]

Various phospho-specific c-Met antibodies (anti-phospho-tyrosine [pY]1003, pY1313, pY1349, pY1365, and pY1230/1234/1235-c-Met) were used to judge the functionality of the HGF/c-Met signaling pathway, as shown in Figure 5.4A.[64] H69 cell (lysates) exposed once to HGF (40 ng/ml, 7.5 min) immuno-precipitated with c-Met antibody. Membranes were immunoblotted with the particular phospho-specific c-Met and c-Met antibodies. Increased phosphorylation of tyrosine 1003 (c-Cb1 binding site), 1313 (PI3K binding site), 1230/1234/1235 (major ligand-induced autophosphorylation site), 1349 (SHC, Src, and Gab 1 binding site), and 1365 (site known to inhibit cell morphogenesis) were observed in response to HGF. The quantity of c-Met immunoprecipitate was found to be the same with or without HGF (Figure 5.4B).[64]

In the same study, examination of cell motility with time-lapse video microscopy documented that activation of c-Met leads to increased cell motility, as observed by the rate of formation of cellular clusters (Figure 5.5).[64] Cell motility is found to be regulated also by the focal adhesion molecules. It has been demonstrated that certain components of the focal adhesion, such as paxillin, p125FAK, and PYK2, are phosphorylated at specific sites due to HGF (Figure 5.6). Paxillin, a key focal adhesion cytoskeletal protein, was phosphorylated in R988C Met and T10101 Met, two different c-Met missense mutations in the juxtamembrane (JM) domain (R988C found in NCI-H69 and H 249 cell lines, and T10101 in SCLC tumor sample T31). This finding strongly supports a unique role of JM mutations in modulating SCLC cytoskeletal signaling and function.[26] JM domain alternative splicing is an important mechanism to modulate c-Met signaling, which includes at least three 8-kb c-Met mRNA variants generated by alternative splicing.[65]

HGF/SF-Met protects cells from DNA damage involving PI3K to c-Akt, resulting in enhanced DNA repair[66] and reduced apoptosis. In SCLC, PI3K is chronically activated because it was constitutively expressed. In MDA-MB-231 cancer cells, c-Met signaling has been found to increase

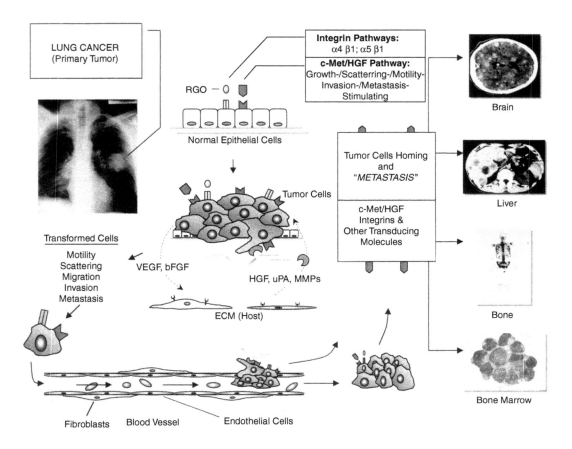

FIGURE 5.2 Mechanism of c-Met signaling and metastasis in a solid tumor. The role of c-Me/HGF and integrins signaling in the transformation and metastasis in solid-tumor lung cancer cells is shown here schematically. Increased motility, scattering, and migration allow tumor cells (such as lung cancer) to proceed eventually to invasion of the ECM (extracellular matrix). Metastasis of the circulating tumor cells to various distant organs (brain, liver, bone, and bone marrow shown here) is represented. Microenvironment of the tumor-host involves secretion of vascular endothelial growth factor (VEGF) and basic fibroblast growth factor (bFGF) by the tumor cells as well as secretion of HGF, uPA, and MMPs by the host stromal cells. (From Maulik, G. et al., *Cytokine Growth Factors Rev.*, 13, 41, 2002. With permission.)

cell adhesion on laminins 1 and 5 fibronectin and vitronectin through a PI3-kinase-dependent mechanism. Increased adhesion was found to increase invasiveness through reconstituted basement membranes. PI3K activation may also occur via integrin signaling. Integrin clustering is generally triggered by HGF-c-Met activation, at the actin-rich adhesive sites and lamellipodia, without interrupting the integrin expression level. Therefore, HGF/c-Met signaling promotes cell adhesion/invasion by triggering the integrin-signaling pathway.[67] Recently, it was demonstrated that PI3K is associated with SCF and SDF-1α, the natural ligands for c-kit and CXCR4, respectively.

Both c-Crk and CRKL members of the Crk family of adapter proteins are important candidates for HGF/c-Met signaling pathway. Several studies have implicated Gab1 (multisubstrate signaling adapter) as an essential mediator of HGF/SF-induced branching tubulogenesis of MDCK epithelial cells.[68,69]

The c-Kit that encodes the receptor for stem cell factor plays a significant role in signal transduction as well as in metastasis. More than 70% of SCLC cell lines and tumors coexpress the c-kit receptor and its ligand stem cell factor, compared with 40% in NSCLC. In SCLC, kit signaling promotes autocrine growth regulation and thus accelerates tumor proliferation. c-Kit is also found

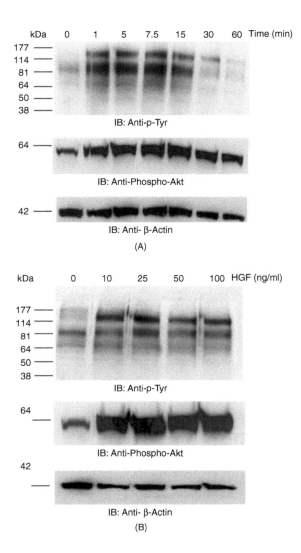

FIGURE 5.3. (**A**) Dose-response HGF stimulation on tyrosine phosphorylation in the H69 SCLC cell line. Cell lysates for the H69 cell line were applied on a 7.5% SDS-PAGE gel, transferred to an Immobilon-P membrane, and probed with the anti-phospho tyrosine, anti-phospho Akt, and β actin. Tyrosine phosphorylated proteins were detected in cellular lysates by immunoblotting using an anti-phospho tyrosine antibody (upper panel) and an anti-phospho-Akt antibody (middle panel). The β-actin control of the same membrane shows the similar protein loading (bottom panel); (**B**) Kinetics study of HGF stimulation on tyrosine phosphorylation in the H69 SCLC cell line. Tyrosine phosphorylated proteins were detected in cellular lysates by immunoblotting using an anti-phospho tyrosine antibody (upper panel) and an anti-phospho-Akt antibody (middle panel). The β-actin control of the same membrane shows the similar protein loading (bottom panel). (From Maulik, G. et al., *J. Cell. Mol. Med.*, 6, 539, 2002. With permission.)

to be responsible for the cellular cluster movement and the individual cell membrane ruffling/ blebbing documented in SCLC cell lines.[37,70,71] c-Kit demonstrates its downstream signaling via several intermediates, including that of cytoskeleton, in response to SCF binding. The binding leads to the phosphorylation of molecules involved in cell survival, such as PI3K, Akt, and S6K. The p85 PI3K subunit phosphorylation activates the p110 catalytic subunit of PI3K, which activates its downstream targets Akt, p70 ribosomal s6kinase. Akt has been found to play a significant role in

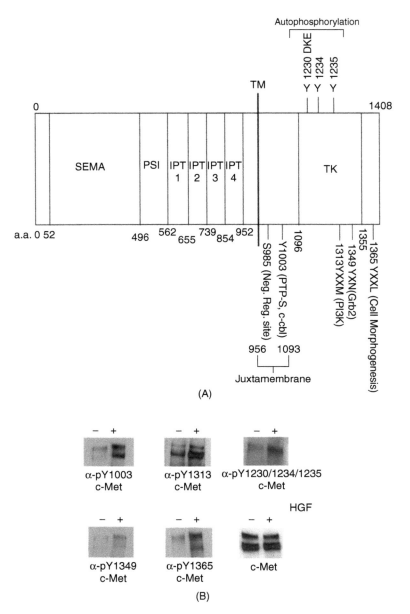

FIGURE 5.4 (A) Predicted functional domains of c-Met. Depicted here are the Sema domain (semaphorin-like), the PSI domain (found in plexins, semaphorins, and integrins), the IPT-repeat domains (found in Ig-like regions, plexins, and transcription factors), and the TK domain (tyrosine kinase located intracellularly). The various amino acid residues with regulatory functions for the c-Met/HGF pathway are illustrated here; **(B)** Phosphorylation of various tyrosine residues of c-Met in response to HGF in SCLC. Cell lysates for the H69 were treated with or without HGF (40 ng/ml; 7.5 min) and were immunoprecipitated with anti c-Met. The whole cell lysate and immunoprecipitations were immunoblotted with anti-phospho tyrosine pY1003, pY1313, 1230/1234/1235, pY1349, and pY1365 of c-Met and anti-c-Met antibody. (From Maulik, G. et al., *J. Cell. Mol. Med.*, 6, 539, 2002. With permission.)

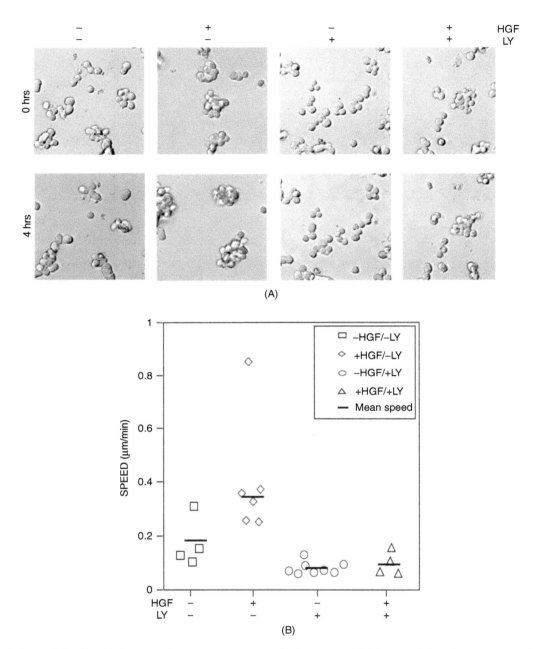

FIGURE 5.5 (**A**) SCLC morphology in response to HGF and PI3K inhibition. H69 cells were starved overnight and treated with or without HGF (100 ng/ml) and with or without LY294002 (50 μ*M*). Shown are representative pictures before and after 4 h with the conditions; (**B**) SCLC cellular speed in response to HGF and PI3K inhibition. Time-lapse video microscopy analysis in terms of speed of H69 cells in response to (−) or (+) HGF (100 ng/ml) and (−) or (+) LY294002 (50 μ*M*). Serum-starved H69 cells were visualized by time-lapse video microscopy and recorded for 4 h with or without HGF and with or without LY294002. These images were then analyzed by NIH image analysis, and each cell/cluster in every frame was traced every 5 min. The position of the cell centroid was measured, and the corresponding *X*- and *Y*-axis values were noted. The distance traversed by each cell/cluster was calculated using these values, and from this the speed was determined for each cell/cluster. These data are represented for each set of experiments, and the corresponding average speed has also been shown.

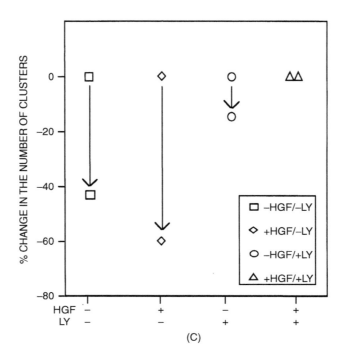

FIGURE 5.5 (CONTINUED) (C) SCLC cellular clusters in response to HGF and PI3K inhibition. Time-lapse video microscopy analysis in terms of the percentage change of total number of clusters of H69 cells in response to (–) or (+) HGF (100 ng/ml) and (–) or (+) LY294002 (50 μM) over a period of 4 h. Serum-starved H69 cells were visualized by time-lapse video microscopy and recorded for 4 h with or without HGF and with or without LY294002 (50 μM). These images were then analyzed by NIH image analysis, and each cell/cluster in every frame was traced every 5 min. The coming together of the smaller clusters to form larger ones was followed over a period of 4 h. These data have been plotted as the percent change in the total number of clusters formed for each of the four sets of experiments. (From Maulik, G. et al., *J. Cell. Mol. Med.*, 6, 539, 2002. With permission.)

mediating the balance between survival and apoptosis. Previous studies demonstrated that wild-type c-kit could promote multiple signal transduction pathways, including Src family members such as JAK/STAT pathway, MAPKinase cascade apart from PI3K.

c-Kit-mediated metastasis depends on the expression of chemokine CXCR4 receptor. CXCR4 is a seven-transmembrane G-protein coupled receptor and a co-receptor for HIV. CXCR4 and c-kit were found to induce morphological changes in the NCI-H69 SCLC cell line.[72] This study also demonstrated that SCF and SDF-1α are able to induce proliferation of the NCI-H69 SCLC line, and SDF-1α itself increased motility and adhesion.[72]

EGFR is expressed on the cell membrane of various malignant cells. Activation of EGFR leads to phosphorylation of c-terminus. Proteins in the SH-2 (Src homology) domain have been found to bind to the phosphotyrosine residues of the receptor. The binding of EGF or TGF-α to EGFR stimulates the formation of dimeric or oligomeric structures involving several receptors. The activated receptor initiates a series of signal-transduction events through the tyrosine phosphorylation of interacting proteins that belong to the SH-2 family. This results in a sequence of responses that are involved in the mitogenic signal-transduction pathways of cells, which leads to DNA replication and cell division.[73,74] Besides tyrosine phosphorylation, the EGF-activated receptors are phosphorylated *in vivo* on serine/threonine, a mechanism by which receptor functions are reduced. Protein kinase C (PKC), MAP kinase, p34 cdk2kinase, and casein kinase II are considered to be involved in this serine/threonine phosphorylation and receptor attenuation.[75] The activated EGF receptor/ligand complex is

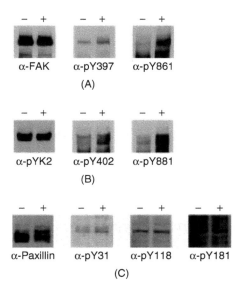

FIGURE 5.6 Phosphorylation of p125FAK, PYK2, and paxillin on specific amino acid residues on the stimulation of HGF in H69 cells. Shown are specific phosphorylations of p125FAK on tyrosines 397 and 861, PYK2 on tyrosines 402 and 881, and paxillin on tyrosine 31. Shown are WCLs of H69 cells with (+) or without (−) stimulation with HGF (7.5 min, 40 ng/ml) separated on SDS-PAGE gel, applied to membrane, and immunoblotted with various antibodies: (A) expression and specific phosphorylation of p125FAK, anti-p125FAK, and anti-phospho-FAK (tyrosine residues 397 or 861); (B) expression and specific phosphorylation of PYK2, anti-PYK2, and anti-phospho-PYK2 (tyrosine residues 402 or 881); (C) expression and specific phosphorylation of paxillin, antipaxillin clone 5H11, and anti-phospho-paxillin (tyrosine residues 31, 118, or 181). (From Maulik, G. et al., *Clin. Cancer Res.*, 8, 620, 2002. With permission.)

internalized and is eventually degraded in lysosomes or dissociated for EGF-receptor recycling. Several studies revealed that intrinsic tyrosine kinase activity and auto-phosphorylation sites of the EGFR are not required for all signaling pathways activated by EGF.

5.4 ROLE OF REACTIVE OXYGEN SPECIES

HGF was found to trigger reactive oxygen species (ROS) generation in SCLC. Cigarette smoke generates considerable toxic ROS, which leads to SCLC. ROS play a significant role as signaling molecules in the regulation of various cellular processes, including migration, adhesion, differentiation, and cell growth.[76] A recent study by Arakaki et al.[77] documented that N-acetylcysteine prevented HGF-suppressed growth of Sarcoma 180 and Meth A cells and HGF-induced apoptosis. In another study, Sattler et al.[76] demonstrated that ROS could induce cell cycle progression, increase cell migration, and inhibit protein tyrosine phosphatases that lead to increased phosphorylation of cellular proteins. These studies suggest a link between HGF/SCF-c-Met and ROS in SCLC. They also show that GM-CSF and other growth factors, including IL-3, SF, and TPO, are associated with increased levels of ROS in different hematopoietic cell lines compared with unstimulated cells. PDTC, a broad-range free-radical scavenger, was found to suppress ROS generation as well as GM-CSF-activated signal transduction, suggesting that ROS contribute to growth factor signal transduction.

Very recently, the signaling mechanism of the EGF receptor has been shown to involve generation of H_2O_2. Goldkorn et al.[78] reported H_2O_2-mediated stimulation of EGF-receptor tyrosine phosphorylation in lung epithelial cells. It was found that in all situations, phosphorylation was somewhat restricted to tyrosine residues. Serine and threonine residues of the EGF receptor were not phosphorylated, and the half-life of the receptor was prolonged after H_2O_2 exposure. It has been shown that MAP kinase down-regulates EGF-receptor tyrosine kinase activity via phosphorylation of the EGF-receptor Thr[669].[79]

5.5 RECEPTOR TYROSINE KINASES AS TARGET FOR ANTI-CANCER THERAPY

Overexpression and/or structural alteration of RTKs' family members are often associated with human cancers, and tumor cells are known to use RTK transduction pathways to achieve tumor growth, angiogenesis, and metastasis. Thus, RTKs represent a pivotal cancer therapy target. Several small molecules acting as RTK inhibitors have been synthesized by pharmaceutical companies and are under clinical trials. Some are being analyzed in animal models or already have been successfully marketed. Cancer-homing toxins are a group of man-made cytotoxic molecules targeting cancer cells. These molecules contain toxins, natural or usually derivatives, connected to a cancer-homing module, such as a monoclonal antibody or growth factor or their derivatives.

SUGEN, Inc., a biopharmaceutical company, focused on the discovery and development of small-molecule drugs that target specific cellular signal transduction pathways. Recent reports indicate that several small-molecule inhibitors are being used against the ATP-binding site of c-Met, which may be the new standard up-front treatment for patients with SCLC. Geldanamycin, of the anisamycin antibiotic family, has been found to inhibit c-Met in SCLC cells by down-regulating the expression of certain cell-signaling proteins.[24] This inhibition of c-Met RTK by Geldanamycin includes several tyrosine and serine/threonine kinases, as well as stress-related protein, HSP 90. Small peptides derived from the tail of the c-Met receptor also possess high affinity for the receptor, and after binding it they inhibit its activity.[24] Additionally, they were was found to block HGF-mediated invasive growth by 50% in the A549 cell line, as determined by invasion, cell migration, and branching morphogenesis. The Grb2-associated binder (Gab1) has recently been identified as a multisubstrate adapter protein of the insulin-responsive substrate 1 family that associates with the c-Met receptor and mediates epithelial morphogenesis. Gab1 is a potent inhibitor of HGF/SF-c-Met signaling pathways for cell survival and DNA repair downstream of PI3K.[66,80]

As mentioned earlier, 70% of SCLC express the kit receptor tyrosine kinase and its ligand, stem cell factor (SCF). Several studies determined that inhibitors of kit tyrosine kinase activity could have therapeutic efficacy in this disease. SCF-induced c-kit phosphorylation was found to be inhibited by PP2 (4-chlorophenyl-7-t-butyl-pyrazolo 3,4-d-pyrimidine) and STI571. When a small-cell lung cancer cell line (H526) was treated with STI571, SCF-mediated c-Kit activation was inhibited, as measured by decreased receptor tyrosine phosphorylation.[71] This drug can inhibit the kinase activity of c-kit, but it has no effect on the related receptor tyrosine kinases c-Fms and Flk2/Flt3. Based on these investigations, clinical trials are underway for the role of STI571 in SCLC.

Currently, there are many potent tyrosine kinase inhibitors against EGFR. It is an important potential therapeutic target, and the small-molecule, orally superactive EGFR-tyrosine kinase inhibitor (ZD1839) is currently under investigation as monotherapy in clinical trials in patients with a very advanced NSCLC. ZD1839 (up to 1000 mg/day) has been studied in four open-labeled, multicenter, phase I studies involving patients with a variety of solid malignant tumors, including NSCLC.[81–83] ZD1839 is already being used in clinics in combination with standard therapy for NSCLC, such as carboplatin and paclitaxel.

5.6 CONCLUSION

RTKs represent pivotal targets in approaches toward cancer therapy. Thus, c-Met, c-Kit, as well as EGFR are important receptor tyrosine kinases that have been implicated in the etiology of multiple tumor types, and they are important therapeutic targets.

REFERENCES

1. Hunter, T., Oncoprotein networks, *Cell*, 88, 333, 1997.
2. Marshall, C.J., Specificity of receptor tyrosine kinase signaling: transient versus sustained extracellular signal-regulated kinase activation, *Cell*, 80, 179, 1995.
3. Maulik, G. et al., Modulation of the c-Met/hepatocyte growth factor pathway in small cell lung cancer, *Clin. Cancer Res.*, 8, 620, 2002.
4. Bottaro, D.P. et al., Identification of the hepatocyte growth factor receptor as the c-met proto-oncogene product, *Science*, 251, 802, 1991.
5. Naldini, L. et al., Hepatocyte growth factor (HGF) stimulates the tyrosine kinase activity of the receptor encoded by the proto-oncogene c-MET, *Oncogene*, 6, 501, 1991.
6. Tamagnone, L. and Comoglio, P.M., Control of invasive growth by hepatocyte growth factor (HGF) and related scatter factors, *Cytokine Growth Factor Rev.*, 8, 129, 1997.
7. Birchmeier, C. and Gherardi, E., Developmental roles of HGF/SF and its receptor, the c-Met tyrosine kinase, *Trends Cell. Biol.*, 8, 404, 1998.
8. Bellusci, S. et al., Creation of an hepatocyte growth factor/scatter factor autocrine loop in carcinoma cells induces invasive properties associated with increased tumorigenicity, *Oncogene*, 9, 1091, 1994.
9. Giordano, S. et al., Transfer of motogenic and invasive response to scatter factor/hepatocyte growth factor by transfection of human MET protooncogene, *Proc. Natl. Acad. Sci. USA*, 90, 649, 1993.
10. Jeffers, M., Rong, S., and Vande Woude, G.F., Enhanced tumorigenicity and invasion-metastasis by hepatocyte growth factor/scatter factor-met signalling in human cells concomitant with induction of the urokinase proteolysis network, *Mol. Cell. Biol.*, 16, 1115, 1996.
11. Liang, T.J. et al., Transgenic expression of tpr-met oncogene leads to development of mammary hyperplasia and tumors, *J. Clin. Invest.*, 97, 2872, 1996.
12. Meiners, S. et al., Role of morphogenetic factors in metastasis of mammary carcinoma cells, *Oncogene*, 16, 9, 1998.
13. Rong, S. et al., Invasiveness and metastasis of NIH 3T3 cells induced by Met-hepatocyte growth factor/scatter factor autocrine stimulation, *Proc. Natl. Acad. Sci. USA*, 91, 4731, 1994.
14. Longati, P. et al., Tyrosines 1234-1235 are critical for activation of the tyrosine kinase encoded by the MET proto-oncogene (HGF receptor), *Oncogene*, 9, 49, 1994.
15. Ponzetto, C. et al., A multifunctional docking site mediates signaling and transformation by the hepatocyte growth factor/scatter factor receptor family, *Cell*, 77, 261, 1994.
16. Bardelli, A. et al., Gab1 coupling to the HGF/Met receptor multifunctional docking site requires binding of Grb2 and correlates with the transforming potential, *Oncogene*, 15, 3103, 1997.
17. Royal, I., Fournier, T.M., and Park, M., Differential requirement of Grb2 and PI3-kinase in HGF/SF-induced cell motility and tubulogenesis, *J. Cell. Physiol.*, 173, 196, 1997.
18. Nguyen, L. et al., Association of the multisubstrate docking protein Gab1 with the hepatocyte growth factor receptor requires a functional Grb2 binding site involving tyrosine 1356, *J. Biol. Chem.*, 272, 20811, 1997.
19. Fixman, E.D. et al., Pathways downstream of Shc and Grb2 are required for cell transformation by the tpr-Met oncoprotein, *J. Biol. Chem.*, 271, 13116, 1996.
20. Pelicci, G. et al., The motogenic and mitogenic responses to HGF are amplified by the Shc adaptor protein, *Oncogene*, 10, 1631, 1995.
21. Boccaccio, C. et al., Induction of epithelial tubules by growth factor HGF depends on the STAT pathway, *Nature*, 391, 285, 1998.
22. Weidner, K.M. et al., Interaction between Gab1 and the c-Met receptor tyrosine kinase is responsible for epithelial morphogenesis, *Nature*, 384, 173, 1996.
23. Jeffers, M. et al., The mutationally activated Met receptor mediates motility and metastasis, *Proc. Natl. Acad. Sci. USA*, 95, 14417, 1998.
24. Maulik, G. et al., Role of the hepatocyte growth factor receptor, c-Met, in oncogenesis and potential for therapeutic inhibition, *Cytokine Growth Factor Rev.*, 13, 41, 2002.
25. Ma, P.C. et al., c-Met: structure, functions and potential for therapeutic inhibition, *Cancer Metastasis Rev.*, 22, 309, 2003.

26. Ma, P.C. et al., c-Met mutational analysis in small cell lung cancer: novel juxtamembrane domain regulating cytoskeletal functions, *Cancer Res.*, 63, 2003 (in press).

27. Rusciano, D., Lorenzoni, P., and Burger, M.M., Regulation of c-met expression in B16 murine melanoma cells by melanocyte stimulating hormone, *J. Cell. Sci.*, 112 (Pt 5), 623, 1999.

28. Wang, R. et al., Activation of the Met receptor by cell attachment induces and sustains hepatocellular carcinomas in transgenic mice, *J. Cell. Biol.*, 153, 1023, 2001.

29. Olivero, M. et al., Overexpression and activation of hepatocyte growth factor/scatter factor in human non-small-cell lung carcinomas, *Br. J. Cancer*, 74, 1862, 1996.

30. Tsao, M.S. et al., Hepatocyte growth factor is predominantly expressed by the carcinoma cells in non-small-cell lung cancer, *Hum. Pathol.*, 32, 57, 2001.

31. Broudy, V.C., Stem cell factor and hematopoiesis, *Blood*, 90, 1345, 1997.

32. Lev, S. et al., Steel factor and c-kit protooncogene: genetic lessons in signal transduction. *Crit Rev Oncog*, 5, 141, 1994.

33. Nocka, K. et al., Expression of c-kit gene products in known cellular targets of W mutations in normal and W mutant mice: evidence for an impaired c-kit kinase in mutant mice, *Genes Dev.*, 3, 816, 1989.

34. Lux, M.L. et al., KIT extracellular and kinase domain mutations in gastrointestinal stromal tumors, *Am. J. Pathol.*, 156, 791, 2000.

35. Ashman, L.K., The biology of stem cell factor and its receptor C-kit, *Int. J. Biochem. Cell. Biol.*, 31, 1037, 1999.

36. Ma, Y. et al., Indolinone derivatives inhibit constitutively activated KIT mutants and kill neoplastic mast cells, *J. Invest. Dermatol.*, 114, 392, 2000.

37. Jafri, N.F. et al., *J. Environ. Pathol. Toxicol. and Oncol.*, 22, 147, 2003.

38. Krystal, G.W. et al., Lck associates with and is activated by Kit in a small cell lung cancer cell line: inhibition of SCF-mediated growth by the Src family kinase inhibitor PP1, *Cancer Res.*, 58, 4660, 1998.

39. Salomon, D.S. et al., Epidermal growth factor-related peptides and their receptors in human malignancies, *Crit. Rev. Oncol. Hematol.*, 19, 183, 1995.

40. Fontanini, G. et al., Evaluation of epidermal growth factor-related growth factors and receptors and of neoangiogenesis in completely resected stage I-IIIA non-small-cell lung cancer: amphiregulin and microvessel count are independent prognostic indicators of survival, *Clin. Cancer Res.*, 4, 241, 1998.

41. Hendler, F.J. and Ozanne, B.W., Human squamous cell lung cancers express increased epidermal growth factor receptors, *J. Clin. Invest.*, 74, 647, 1984.

42. Wells, A., EGF receptor, *Int. J. Biochem. Cell. Biol.*, 31, 637, 1999.

43. Grandis, J.R. et al., Requirement of Stat3 but not Stat1 activation for epidermal growth factor receptor-mediated cell growth *in vitro*, *J. Clin. Invest.*, 102, 1385, 1998.

44. Ullrich, A. and Schlessinger, J., Signal transduction by receptors with tyrosine kinase activity, *Cell*, 61, 203, 1990.

45. Yarden, Y. and Schlessinger, J., Self-phosphorylation of epidermal growth factor receptor: evidence for a model of intermolecular allosteric activation, *Biochemistry*, 26, 1434, 1987.

46. Downward, J., Parker, P., and Waterfield, M.D., Autophosphorylation sites on the epidermal growth factor receptor, *Nature*, 311, 483, 1984.

47. Neal, D.E. et al., Epidermal-growth-factor receptors in human bladder cancer: comparison of invasive and superficial tumours, *Lancet*, 1, 366, 1985.

48. Gross, M.E. et al., Cellular growth response to epidermal growth factor in colon carcinoma cells with an amplified epidermal growth factor receptor derived from a familial adenomatous polyposis patient, *Cancer Res.*, 51, 1452, 1991.

49. Damstrup, L. et al., Expression of the epidermal growth factor receptor in human small cell lung cancer cell lines, *Cancer Res.*, 52, 3089, 1992.

50. Volm, M., Rittgen, W., and Drings, P., Prognostic value of ERBB-1, VEGF, cyclin A, FOS, JUN and MYC in patients with squamous cell lung carcinomas, *Br. J. Cancer*, 77, 663, 1998.

51. Khazaie, K., Schirrmacher, V., and Lichtner, R.B., EGF receptor in neoplasia and metastasis, *Cancer Metastasis Rev.*, 12, 255, 1993.

52. Franklin, W.A. et al., Epidermal growth factor receptor family in lung cancer and premalignancy, *Semin. Oncol.*, 29, 3, 2002.

53. Cox, G., Jones, J.L., and O'Byrne, K.J., Matrix metalloproteinase 9 and the epidermal growth factor signal pathway in operable non-small cell lung cancer, *Clin. Cancer Res.*, 6, 2349, 2000.

54. Ohsaki, Y. et al., Epidermal growth factor receptor expression correlates with poor prognosis in non-small cell lung cancer patients with p53 overexpression, *Oncol. Rep.*, 7, 603, 2000.

55. Haeder, M. et al., Epidermal growth factor receptor expression in human lung cancer cell lines, *Cancer Res.*, 48, 1132, 1988.

56. Sherwin, S.A. et al., Expression of epidermal and nerve growth factor receptors and soft agar growth factor production by human lung cancer cells, *Cancer Res.*, 41, 3538, 1981.

57. Sakiyama, S., Nakamura, Y., and Yasuda, S., Expression of epidermal growth factor receptor gene in cultured human lung cancer cells, *Jpn. J. Cancer Res.*, 77, 965, 1986.

58. Cohen, S. et al., A native 170,000 epidermal growth factor receptor-kinase complex from shed plasma membrane vesicles, *J. Biol. Chem.*, 257, 1523, 1982.

59. Gullick, W.J. et al., Expression of epidermal growth factor receptors on human cervical, ovarian, and vulval carcinomas, *Cancer Res.*, 46, 285, 1986.

60. Stella, M.C. and Comoglio, P.M., HGF: a multifunctional growth factor controlling cell scattering, *Int. J. Biochem. Cell. Biol.*, 31, 1357, 1999.

61. Ueno, H. et al., The phosphatidylinositol 3′ kinase pathway is required for the survival signal of leukocyte tyrosine kinase, *Oncogene*, 14, 3067, 1997.

62. Kubota, Y. et al., Activation of phosphatidylinositol 3-kinase is necessary for differentiation of FDC-P1 cells following stimulation of type III receptor tyrosine kinases, *Cell Growth Differ.*, 9, 247, 1998.

63. Kennedy, S.G. et al., The PI 3-kinase/Akt signaling pathway delivers an anti-apoptotic signal, *Genes Dev.*, 11, 701, 1997.

64. Maulik, G. et al., Activated c-Met signals through PI3K with dramatic effects on cytoskeletal functions in small cell lung cancer, *J. Cell. Mol. Med.*, 6, 539, 2002.

65. Liu, Y., The human hepatocyte growth factor receptor gene: complete structural organization and promoter characterization, *Gene*, 215, 159, 1998.

66. Fan, S. et al., The multisubstrate adapter Gab1 regulates hepatocyte growth factor (scatter factor)-c-Met signaling for cell survival and DNA repair, *Mol. Cell. Biol.*, 21, 4968, 2001.

67. Trusolino, L. et al., Growth factor-dependent activation of alphavbeta3 integrin in normal epithelial cells: implications for tumor invasion, *J. Cell. Biol.*, 142, 1145, 1998.

68. Maroun, C.R. et al., The Gab1 PH domain is required for localization of Gab1 at sites of cell-cell contact and epithelial morphogenesis downstream from the met receptor tyrosine kinase, *Mol. Cell. Biol.*, 19, 1784, 1999.

69. Maroun, C.R. et al., A conserved inositol phospholipid binding site within the pleckstrin homology domain of the Gab1 docking protein is required for epithelial morphogenesis, *J. Biol. Chem.*, 274, 31719, 1999.

70. Wang, W.L. et al., Growth inhibition and modulation of kinase pathways of small cell lung cancer cell lines by the novel tyrosine kinase inhibitor STI 571, *Oncogene*, 19, 3521, 2000.

71. Krystal, G.W. et al., The selective tyrosine kinase inhibitor STI571 inhibits small cell lung cancer growth, *Clin. Cancer Res.*, 6, 3319, 2000.

72. Kijima, T. et al., Regulation of cellular proliferation, cytoskeletal function, and signal transduction through CXCR4 and c-Kit in small-cell lung cancer cells, *Cancer Res.*, 62, 6304, 2002.

73. Carpenter, G. and Cohen, S., Epidermal growth factor, *J. Biol. Chem.*, 265, 7709, 1990.

74. Carpenter, G., EGF: new tricks for an old growth factor, *Curr. Opin. Cell. Biol.*, 5, 261, 1993.

75. Heisermann, G.J. et al., Mutational removal of the Thr669 and Ser671 phosphorylation sites alters substrate specificity and ligand-induced internalization of the epidermal growth factor receptor, *J. Biol. Chem.*, 265, 12820, 1990.

76. Sattler, M. and Salgia, R., Role of the adapter protein CRKL in signal transduction of normal hematopoietic and BCR/ABL-transformed cells, *Leukemia*, 12, 637, 1998.

77. Arakaki, N. et al., Involvement of oxidative stress in tumor cytotoxic activity of hepatocyte growth factor/scatter factor, *J. Biol. Chem.*, 274, 13541, 1999.

78. Goldkorn, T. et al., EGF-receptor phosphorylation and signaling are targeted by H2O2 redox stress, *Am. J. Respir. Cell. Mol. Biol.*, 19, 786, 1998.

79. Countaway, J.L. et al., Multisite phosphorylation of the epidermal growth factor receptor: use of site-directed mutagenesis to examine the role of serine/threonine phosphorylation, *J. Biol. Chem.*, 265, 3407, 1990.

80. Atabey, N. et al., Potent blockade of hepatocyte growth factor-stimulated cell motility, matrix invasion and branching morphogenesis by antagonists of Grb2 Src homology 2 domain interactions, *J. Biol. Chem.*, 276, 14308, 2001.

81. Kris, M., Objective regressions in non-small cell lung cancer patients treated in phase I trials of oral ZD 1839('Iressa'), a selective tyrosine kinase inhibitor that blocks the epidermal growth factor receptor(EGFR), *Lung Cancer*, 29 (Suppl. 1), A233, 2000.

82. Negoro final results of a Phase 1 intermittent dose-escalation trial of ZD1839('Iressa') in Japanese patients with various solid tumors, *Proc. Am. Soc. Clin. Oncol.*, 20, 324, 2001.

83. Ranson, M. et al., ZD1839, a selective oral epidermal growth factor receptor-tyrosine kinase inhibitor, is well tolerated and active in patients with solid, malignant tumors: results of a phase I trial, *J. Clin. Oncol.*, 20, 2240, 2002.

6 Free Radicals, Environmental Radiation, and Cancer

Okezie I. Aruoma

CONTENTS

6.1 Free-Radical Chemistry 79
    6.1.1 Unimolecular Radical Reactions 80
        6.1.1.1 Radical–Molecule Interaction: Addition to Unsaturated Systems 80
        6.1.1.2 Radical–Molecule Interaction: S_H2 (Substitution
                 Homolytic Bimolecular) Reactions 80
        6.1.1.3 Radical–Molecule Interaction: Reaction with Oxidizing Agents 80
        6.1.1.4 Radical–Radical Interaction: Dimerization or Radical Coupling 81
    6.1.2 Radical Disproportionation 81
6.2 Radiation Biology and Cancer 81
6.3 Radiation and the Environment 83
References 86

6.1 FREE-RADICAL CHEMISTRY

Interest in free radicals began with the work of Gomberg (1900), who demonstrated the existence of the triphenylmethyl radical ($Ph_3C^\bullet$). The triphenylmethyl radical is stable enough to exist in solution at room temperature, often in equilibrium with a dimeric form. At room temperature, this concentration of $Ph_3C^\bullet$ in benzene is 2%. Although triphenylmethyl-type radicals are stabilized by resonance, it is steric hindrance to dimerization and not resonance that is the cause of their stability.

Free radicals are formed from molecules by breaking a bond such that each fragment keeps one electron (free radicals can also be formed by collision of the nonradical species by a reaction between a radical and a molecule, which must then result in a radical, since the total number of electron is odd), by cleavage of a radical to give another radical, and by oxidation or reduction reactions.

One technique widely used in free-radical research, electron spin resonance (ESR) or electron paramagnetic resonance, detects radicals by measuring the energy changes that occur when unpaired electrons change their direction of spin. Only free radicals can give an ESR spectrum, hence the technique is sensitive for detecting and determining the concentration of free radicals. This technique provides information concerning the electron distribution and hence the structure of the free radicals — the structure being deduced from the splitting pattern of the ESR spectrum.

Whereas ESR depends on reorientation of unpaired electrons in a magnetic field, NMR (nuclear magnetic resonance) depends on reorientation of magnetic nuclei in the presence of a magnetic field. In the mid 1960s, it was discovered that if an NMR spectrum is taken during the course of a reaction involving free-radical species, certain signals can be enhanced (positively or negatively) or reduced. The occurrence of this chemically induced dynamic nuclear polarization (CIDNP) is indicative that, in such reaction, a portion of the product that was formed was via a radical intermediate. However, this may not be an absolute proof, since the absence of CIDNP does not prove that a free-radical intermediate is necessarily absent. CIDNP results when protons in a reacting molecule become dynamically coupled to an unpaired electron while traversing the path from reactants to products.

So a free radical is any chemical species (capable of independent existence) possessing one or more unpaired electrons, an unpaired electron being one that is alone in an orbital. The simplest free radical is an atom of the element hydrogen. The hydrogen atom contains one proton and a single unpaired electron, which qualifies it as a free radical. The radical dot (which denotes the radical species) is inserted to indicate that one or more unpaired electrons is present. Electrons are more stable when paired together in orbitals: the two electrons in a pair have different directions of spin. Hence radicals are generally less stable than nonradicals, although their reactivity may vary. The reader is referred to the literature (Bensasson et al., 1993; Butler et al., 1984; Cadogan, 1993; Hay and Waters, 1937; and Moad and Solomon, 1995) for accounts of free-radical chemistry.

6.1.1 Unimolecular Radical Reactions

Unimolecular radical reactions result from the instability of the first formed radical. The radicals may completely decompose or they may rearrange before reaction with other molecules or radicals present. In decomposition reactions, the radical decomposes to give a stable molecule and a new radical. In rearrangement reactions, the breaking of an adjacent C–C bond in a cyclic system leads to the concomitant formation of a new bond, usually carbonyl, and a new isomeric radical. There could also be migration of an atom, via intramolecular abstraction by the radical center, thus creating a new, isomeric radical.

6.1.1.1 Radical–Molecule Interaction: Addition to Unsaturated Systems

This involves the addition of a radical to an olefinic double bond to give a new saturated radical. Typical reaction is the radical-induced polymerization of olefins. The intermediate step involving the addition of a radical to an aromatic double bond is widespread in free-radical chemistry, e.g., in the radical substitution of aromatic compounds (homolytic aromatic substitution). The net overall reaction is displacement of an aromatic substituent by a radical:

$$AR - X + Y^{\bullet} \rightarrow AR - Y + X^{\bullet}$$

6.1.1.2 Radical–Molecule Interaction: S_H2 (Substitution Homolytic Bimolecular) Reactions

The S_H2 reactions are bimolecular reactions involving homolytic attack of radicals on a molecule. The radical attacks a univalent atom, usually a terminal halogen or hydrogen in an abstraction reaction, to give rise to a new radical, e.g.:

$$Ph^{\bullet} + CBrCl_3 \rightarrow {}^{\bullet}CCl_3 + PhBr$$

Homolytic substitution at multivalent atoms also occurs, but neither normally occurs at saturated carbon centers. S_H2 reactions can be synchronous, occurring via a transition state, or stepwise via an intermediate that may exist for only a finite time.

6.1.1.3 Radical–Molecule Interaction: Reaction with Oxidizing Agents

Radicals readily undergo one-electron oxidations with oxidizing reagents of suitable redox potential to give positive ions. An example is the Meerwein reaction, which involves the oxidation of cynnamyl-derived radicals by cupric ions:

$$Ph^{\bullet}CHCHRCO_2Et + Cu^{2+} \rightarrow PhC^{+}HCHRCO_2Et + Cu^{+}$$

6.1.1.4 Radical–Radical Interaction: Dimerization or Radical Coupling

Here, localized radicals (methyl, phenyl radicals) react readily with little chance of dimerization. Only delocalized radicals have a high probability of dimerizing in solution. Thus:

$$R'^\bullet + R''^\bullet = R' - R''$$

When $R' = R''$, the reaction is dimerization; when $R' \neq R''$, the reaction is radical coupling or combination.

6.1.2 RADICAL DISPROPORTIONATION

This reaction involves collision of the radicals, resulting in the abstraction of an atom, usually hydrogen, by one radical from the other. This leads to the formation of two stable molecules, with the atom abstracted being β to the radical center, for example, the disproportionation of two phenylethyl radicals to give styrene and ethyl benzene. The disproportionation reaction derives its driving force from the formation of two new strong bonds and from the fact that the β–CH bonds in radicals are usually weak. The ratio of disproportionation to combination is dependent on the structural features of the radicals involved and may be affected, for example, by solvent, pressure, and temperature (reviewed in Gibian and Corley, 1993).

6.2 RADIATION BIOLOGY AND CANCER

The biological effect of radiation on living cells tends to vary and depends on dose. It is well known that radiation is known to cause cancer, but in clinical practice, radiotherapy is an accepted form of cancer treatment. Injured or damaged cells may self-repair, die, or be involved in a misrepair process (see discussion later). High-radiation doses tend to kill cells, while low doses tend to damage or alter the genetic code (DNA) of irradiated cells. High-radiation doses can kill so many cells that tissues and organs are damaged immediately and tend to initiate a rapid body response, often called "acute radiation syndrome." High-radiation dose manifests itself early. This was evident in many of the atomic bomb survivors in 1945 and in the emergency workers who responded to the 1986 Chernobyl nuclear power plant accident (Sali et al., 1996). Low doses of radiation over long periods may not cause immediate problems in body organs, but effects may occur at the cellular level. Genetic effects and the development of cancer are the primary health concerns attributed to radiation exposure (see below). The likelihood of cancer occurring after radiation exposure is about five times greater than a genetic effect (see review by Abrahamse, 2003).

Genetic effects are the result of a mutation produced in the reproductive cells of an exposed individual that are passed on to his or her offspring. Although these effects are widely suggested to appear in the exposed person's direct offspring or several generations later (depending on whether the altered genes are dominant or recessive), radiation-induced genetic effects are often observed in laboratory animals (given very high doses of radiation). However, according to the U.S. Nuclear Regulatory Commission (http://environment.about.com), no statistically significant increase in genetic effects has been observed among the children born to atomic bomb survivors from Hiroshima and Nagasaki.

The association between radiation exposure and the development of cancer is mostly based on populations exposed to relatively high levels of ionizing radiation (e.g., Japanese atomic bomb survivors and recipients of selected diagnostic or therapeutic medical procedures where radiation represents the major anticancer modality in terms of successful tumor care and patient survival) (Abrahamse, 2003). Those cancers that may develop as a result of radiation exposure are indistinguishable from those that occur naturally or as a result of exposure to other chemical carcinogens. Chemical and physical hazards and lifestyle factors (e.g., smoking, alcohol consumption, and diet) contribute to many of these same diseases. Cancers associated with high-dose exposure include leukemia, breast, bladder, colon, liver, lung, esophagus, ovarian, multiple myeloma, and stomach cancers. Radiation biology focuses on the understanding of the physical, biological, and chemical mechanisms of the interaction of radiation with living matter.

In living cells, there is a steady formation of DNA lesions. A substantial number of these lesions are formed by endogenous factors that damage DNA on a continuous basis. Free-radical attack upon DNA generates a series of modified purine and pyrimidine base products (Figure 6.1). One of the widely studied product is 8-oxo-7,8-dihydroguanine (8-oxoGua). To form 8-oxoGua, a hydroxyl radical first reacts with guanine to form a C8-OH adduct radical. The loss of an electron (e^-) and proton (H^+) generates 8-oxoG. The C8-OH adduct radical can also be reduced by uptake

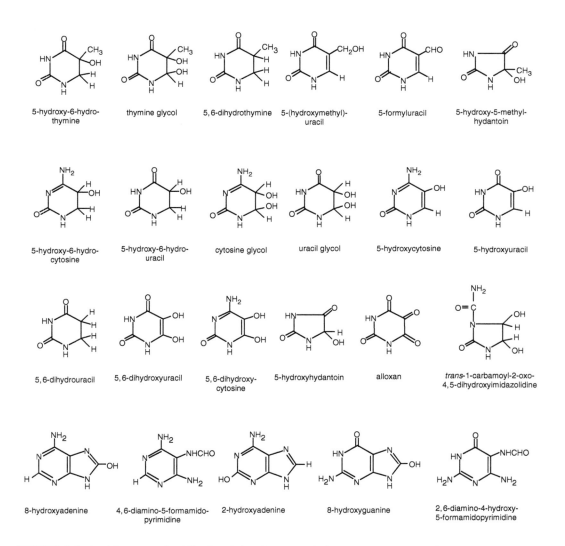

FIGURE 6.1 A wide range of oxidized and ring-fragmented nitrogen bases that are formed by endogenous reactive oxygen species or by ionizing or UV radiation. (Adapted from Dizdasoglu 2000 in *Free Radicals in Chemistry, Biology, and Medicine*. With permission from OICA International, Landon.)

of an electron and a proton to form 7-hydro-8-hydroxyguanine, which is subsequently converted to 2,6-diamino-4-hydroxy-5-formamidopyrimidine (FaPy). The formation of 8-oxoGua residues in DNA leads to GC → TA transversions unless repaired prior to DNA replication, and this may lead to point mutations (Aruoma and Halliwell 1998).

The most common damage to pyrimidines is the formation of thymine glycol (Tg), which results from attack by a hydroxyl radical to yield a 5-hydroxy-6-yl radical. In the absence of oxygen, loss of an electron followed by uptake of water and loss of a proton generates thymine glycol. In the presence of oxygen, uptake of oxygen at position 6 first yields a 5-hydroxy-6-peroxyl radical, which is converted to thymine glycol through loss of a proton and superoxide anion and reaction with water. Indeed 5-methylcytosine (5-meC) can also be converted to thymine glycol by ionizing radiation under aerobic conditions or H_2O_2, and thus may be an additional source of mutations in organisms that utilize 5-meC to regulate gene expression. The level of the modified bases *in vivo* can depend on oxidative DNA insult and can be reflective of an involvement of different repair mechanism(s). Increased levels of modified DNA bases may contribute to the genetic instability and metastatic potential of tumor cells in fully developed cancer. A direct correlation between 8-oxoGua formation and carcinogenesis *in vivo* as well as the induction of mutagenesis in hotspot codons of the human *p53* and *Ha-ras* genes are widely suggested.

The reader is referred to articles in volume 531 (Issue 1–2) of the journal *Mutation Research* for views on oxidative DNA damage and its repair, which is important for understanding the importance of oxidative stress in the development of cancer. These articles include some salient observations, including the fact that cancer patients show signs of extensive granulocyte activation with a release of ROS and show increases in the levels of 8-isoprostane, one of the biomarkers of oxidative stress. Tumors may stimulate the defense systems of the body so that they react against the tumor to produce cytokines (e.g., TNF, which is known to increase oxidative DNA damage of CD 34+ cells). It is also possible that a pro-oxidant environment is characteristic of advanced stages of cancer and that oxidative stress is a result of the disease development.

6.3 RADIATION AND THE ENVIRONMENT

Potential effects of UV-B radiation on the chemical environment of marine organisms is widely discussed. Quoting an abstract from Palenik et al. (Palenik et al., 1990):

> An increase in ultraviolet-B (UV-B) due to depletion of stratospheric ozone may affect growth of marine phytoplankton by altering the chemistry of their environment. Production of bioactive free radicals photodecomposition of organic matter, and availability of trace metals are likely to be altered by increased UV-B flux. Such changes to the chemical environment may be both deleterious and beneficial to marine phytoplankton. Extracellular free radicals such as OH, Br_2^-, and CO_3^- are predicted to have a negligible impact, but superoxide and its decomposition product hydrogen peroxide may react rapidly with cell surfaces and destroy membrane function and integrity. Increased UV-B will enhance the bioavailability of the redox active trace metals Fe and Cu. Thus, in the Fe-limited high latitude ocean, increased Fe availability may promote phytoplankton production, while in other parts of the ocean increased Cu availability may be toxic. Overall, the interdependent direct and indirect effects of UV-B on phytoplankton may compensate for each other and account for the ability of marine ecosystems to be subjected to widely variable UV-B flux without apparent damage.

Upon reviewing the biochemical aspects (antioxidants) for development of tolerance in plants growing at different low levels of ambient air pollutants, Rao and Dubey (Rao and Dubey, 1990) studied the role of antioxidants, which are thought to scavenge the oxygen free radicals in plants exposed to relatively low concentrations of ambient air pollutants for long durations, for a year. As a general response of all the plants in the polluted area, increases were observed in superoxide dismutase peroxidase activity, in sulfate, and in leaf area to dry weight ratio, and decreases were observed in stomatal conductance, ascorbic acid content, protein content, and total lipids.

This indicates that high peroxidase activity in the control plants and enhanced superoxide-dismutase activity in the polluted area might have enhanced the ability of *Cassia siamea* to tolerate stress better than *Dalbergia sissoo*. Similarly, enhanced activities in the polluted sites made *Calotropis procera* more tolerant of stress than *Ipomoea fistulosa*. Thus, it appears that monitoring of antioxidant activities offers a useful tool in understanding the mechanisms that make plants relatively tolerant in field conditions.

Indeed, plants also seem to have developed a strategy to preserve their cell integrity or to overcome the deleterious effects of ROS in situations of oxidative stress. They react by enhancing the expression of multiple, synergistic, antioxidant mechanisms, which include the synthesis of primary and secondary metabolites. Among these natural substances, polyphenolic compounds are a particularly attractive class commonly cited for their potent antioxidant activities. More than 4000 of these compounds, generally referred to as flavonoids, have been identified in both higher and lower plants. They can be subdivided into six classes: flavones, flavanones, isoflavones, flavonols, and anthocyanins. The basic ring skeletons of these classes are shown in Figure 6.2.

Classes	Structural Formula	Examples
Flavanones	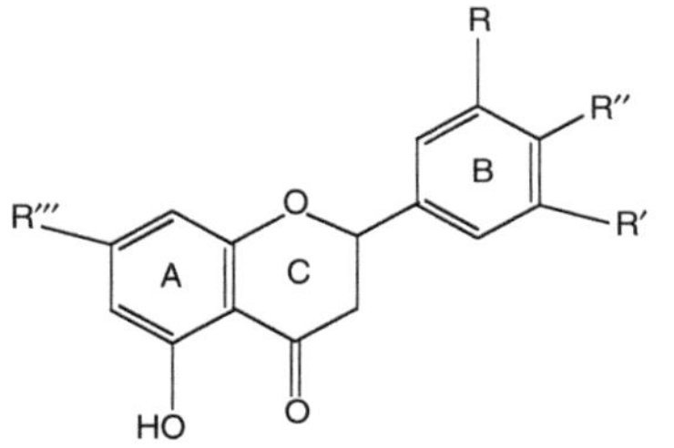	R = R′ = H, R″ = R‴ = OH; naringenin R = OH, R′ = H, R″ = R‴ = OH; eriodyctiol R = R′ = OH, R″ = R‴ = OH; 5′-OH-eriodyctiol.
Flavones	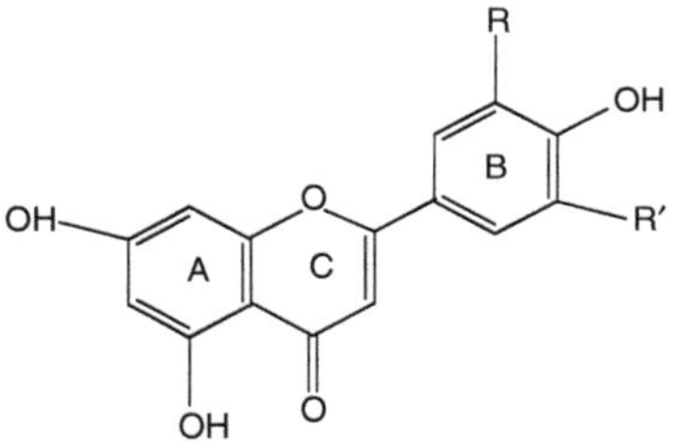	R = R′ = H; apigenin R = OH, R′ = H; luteolin R = R′ = OCH$_3$; tricetin
Isoflavones	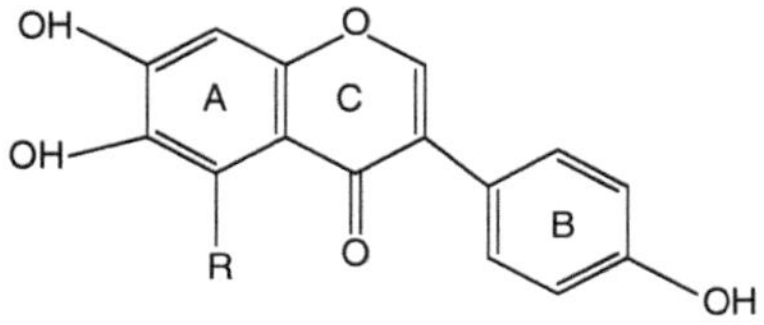	R = H; daidzein R = OH; genistein
Flavonols	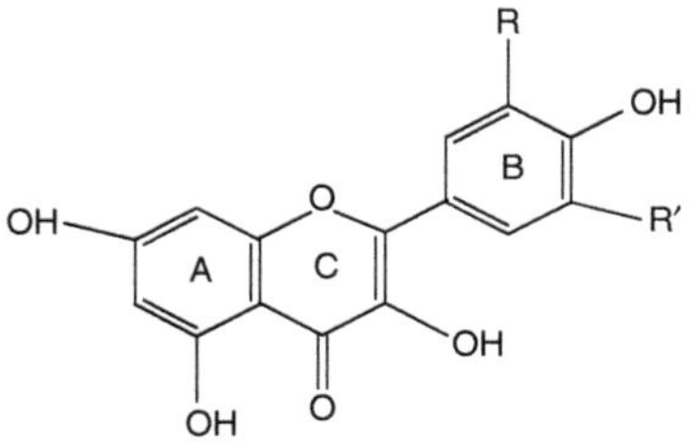	R = R′ = H; kaempferol R = OH, R′ = H; quercetin R = R′ = OH; myricetin

FIGURE 6.2. Chemical structures of flavonoids and some examples. (Courtesy of Dr. Theeshan Bahorun, University of Mauritius.)

Flavonols

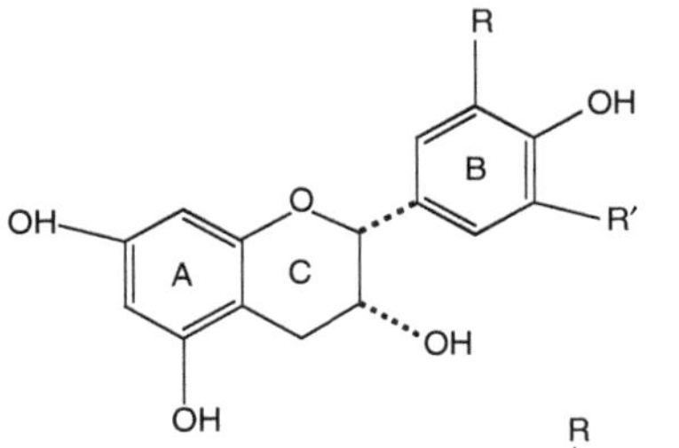

R = OH, R′ = H; (–) epicatechin
R = R′ = OH; (–) epigallocatechin
R = OH, R′ = H; (+) catechin
R = R′ = OH; (+) gallocatechin

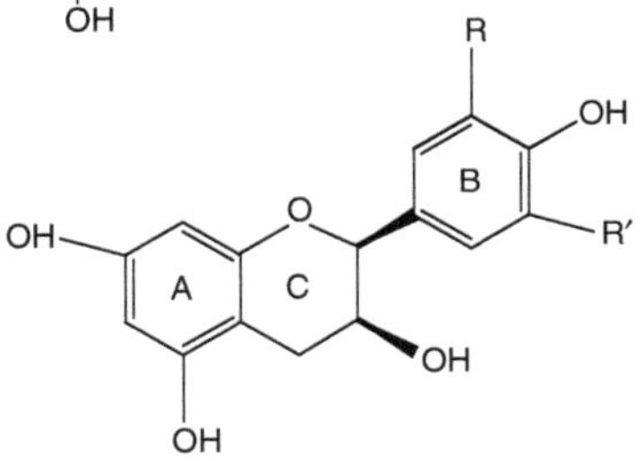

R = OH, R′ = H; (+) epicatechin
R = R′ = OH; (+) epigallocatechin
R = OH, R′ = H; (–) catechin
R = R′ = OH; (–) gallocatechin

Anthocyanidins

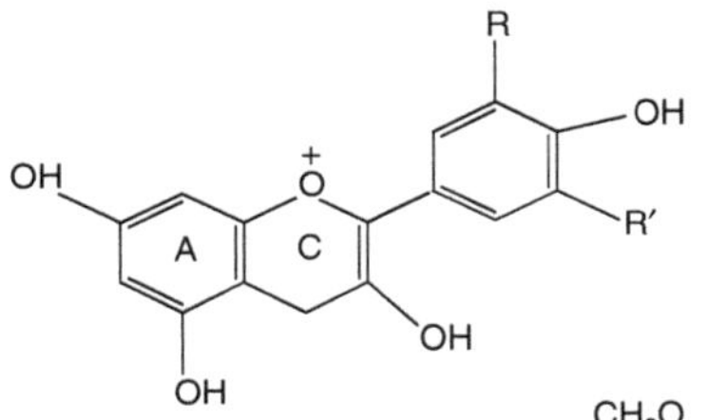

R = OH, R′ = H; cyanidin
R = R′ = OH; delphinidin
R = R′ = OCH₃; malvidin
R = R′ = H; pelargonidin

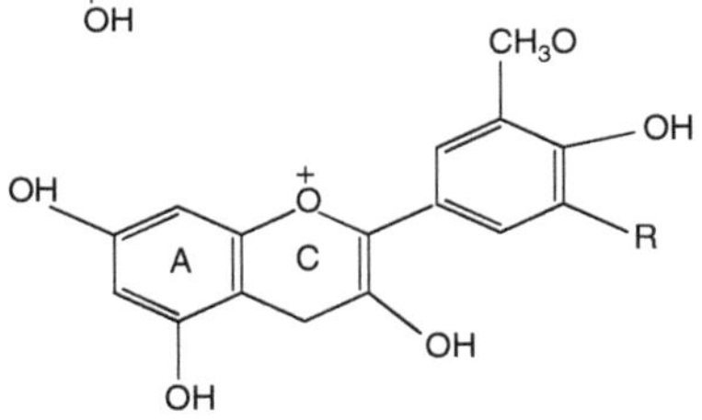

R = H; peonidin
R = OH; petunidin

FIGURE 6.2 (CONTINUED).

Many reports have highlighted the existing relationship between polyphenol contents of plant extracts and their antioxidant capacities. These phytochemicals, for example, including curcumin, epigallocatechin gallate, and resveratrol, have been shown to exert chemopreventive effects by blocking improper NF-κB activation, possibly through inhibiting the degradation of inhibitory unit IκBα. The antioxidant beverage EM-X inhibits H_2O_2- and TNF-mediated NF-κB activation and subsequent IL-8 release (Deiana et al., 2003).

Cancer prevention strategies are now being directed to understanding the potential for diet to modulate molecular signaling (Monson, 2003; Surh, 2003; Young et al., 2003). A comprehensive review in 1997 (World Cancer Research Fund/American Institute for Cancer Research, 1997) convincingly suggested that consumption of fruit and vegetables decreased the risk of cancers of the mouth, pharynx, esophagus, lung, stomach, colon, and rectum. This report also concluded that eating fruit and vegetables probably also reduces the risk of cancers of the larynx, pancreas, breast, and bladder. As a consequence, it was suggested that the adoption of "recommended diets together with maintenance of physical activity and appropriate body mass, could in time reduce cancer incidence by 30–40%." Hence, the dietary advice to individuals was to "choose predominantly plant-based diets rich in a variety of vegetables and fruit, pulses (legumes) and minimally processed starchy staple foods" (Monson, 2003).

The process by which a normal, healthy cell becomes a tumor involves multiple steps: initiation (single mutagenic event), promotion (chronic exposure to tumor promoters), formation of a benign tumor, and progression (benign tumor progression to carcinoma). In order to achieve full malignancy, cells must acquire certain transforming characteristics, including (1) self-sufficiency in growth signaling and limitless replicative potential, (2) becoming unresponsive to antiproliferative signals, (3) evading apoptosis, (4) inducing and sustaining angiogenesis, and (5) acquiring the ability to invade and metastasize. This sequence of events presents many opportunities for intervention, with the aim of preventing, slowing down, or reversing the transformation process. Thus chemoprevention (defined as use of agents to inhibit, reverse, or retard tumorigenesis) would be ideal if the strategy halts the carcinogenic process at an early stage, perhaps even preventing the formation of preneoplastic lesions.

Carcinogenesis requires several genetic modifications and, hence, blocking mechanisms could also be effective in slowing down the later stages of cancer progression and in preventing the development of second primary tumors following removal of the initial cancer. Indeed, tumors result from an imbalance between proliferative and apoptotic processes, so any mechanism that halts or slows down inappropriate cell division, or that induces damaged cells to undergo apoptosis, would represent a useful chemopreventive approach.

Extracellular tumor promoters activate signaling cascades from membrane receptors, resulting in altered gene transcription in the nucleus and altered translation in the cytoplasm. This leads to misregulation of protein expression (Young et al., 2003). The reader is referred to excellent reviews (Monson, 2003; Surh, 2003; and Young et al., 2003) for detailed accounts. Cancer involves chronic inflammatory responses. Thus the induction of inflammatory mediators can be regulated by the activation of redox-sensitive transcription factors such as activator protein-1 (AP-1), (c-Fos/c-Jun), and nuclear factor-kappa B (NF-κB) stimulated in response to oxidants and TNF. TNF increases AP-1 binding via the MAP kinase (stress-activated protein kinase/JNK) signaling pathway, activates NF-κB via the Iκ-B kinase pathway, and affects the local tissue oxidant/antioxidant balance. Transcription factors are redox-sensitive, but more research is needed to fully understand the molecular effects of antioxidants on the signal transduction pathways and how this could apply to modulation of cancer and diseases of chronic inflammation and the acute insult of stroke.

REFERENCES

Abrahamse, H., Radiation and cancer: a double edged sword, in *Molecular and Therapeutic Aspects of Redox Biochemistry*, Bahorun, T. and Gurib-Fakim, A., Eds., OICA International, London, 2003, pp. 103–116.

Aruoma, O.I. and Halliwell, B., *DNA and Free Radicals: Techniques, Mechanisms and Applications*, OICA International, London, 1998.

Bensasson, R.V., Land, E.J., and Truscott, T.G., *Excited States and Free Radicals in Biology and Medicine: Contribution from Flash Photolysis and Pulse Radiolysis*, Oxford University Press, Oxford, 1993.

Butler, J., Land, E.J., and Swallow, A.J., Mechanisms of the effects of high energy radiation on biological systems, *Radiat. Phys. Chem.*, 24 273–283, 1984.

Cadogan, J.I.G., *Principles of Free Radical Chemistry*, The Chemical Society, London, 1973.

Deiana M., Dessi, M.A., Ke, B., Liang, Y-F., Higa, T., Gilmour, P., Jen, L.S., Rahman, I., and Aruoma, O.I., The antioxidant cocktail, effective microorganism (EM-X) inhibits oxidant-induced interleukin-8 release and the peroxidation of phospholipids *in vitro*, *Biochemical Biophysical Res. Commn.*, 296, 1148–1151, 2002.

Gibian, M.J. and Corley, R.C., Organic radical-radical reactions: disproportionation versus combination, *Chem. Rev.*, 73, 441–464, 1973.

Goldstein, S. and Czapski, G., The reaction of NO• with O_2•$^-$ and H_2O_2: a pulse radiolysis study, *Free Radical Biol. Med.*, 19, 505–510, 1995.

Gomberg, M., An incidence of trivalent carbon trimethyl, *J. Am. Chem. Soc.*, 22, 757–771, 1900.

Hey, D.H. and Waters, W.A., Some organic reactions involving the occurrence of free radicals in solution, *Chem. Rev.*, 21, 169–208, 1937.

Moad, G. and Solomon, D.H., *The Chemistry of Free Radical Polymerization*, Pergamon Press, Oxford, 1995.

Monson, M.M., Cancer prevention: the potential for diet to modulate molecular signaling, *Trends Molecular Med.*, 9, 11–18, 2003.

Palenik, B., Price, N.M., and Morel, F.M.M., Potential effects of UV-B on the chemical environment of marine organisms: a review, *Environmental Pollut.*, 70, 117–130, 1990.

Rao, M.V. and Dubey, P.S., Biochemical aspects (antioxidants) for development of tolerance in plants growing at different low levels of ambient air pollutants, *Environmental Pollut.*, 64, 55–66, 1990.

Sali, D., Cardis, E., Sztanyik, L., Auvinen, A., Bairakova, A., Dontas, N., Grosche, B., Kerekes, A., Kusic, Z., Kusoglu, C., Lechpammer, S., Lyra, M., Michaelis, J., Petridou, E., and Szybinxki, Z., Cancer consequences of the Chernobyl accident in Europe outside the former USSR: a review, *Int. J. Cancer*, 67, 343–352, 1996.

Slater, T.F., Free radical mechanisms in tissue injury, *Biochemical J.*, 222, 1–15, 1984.

Surh, Y.-J., Chun, K.-S., Cha, H.-H., Han, S.-S., Keum, Y.-S., Park, K.-K., and Lee, S.-S., Molecular mechanisms underlying chemopreventive activities of anti-inflammatory phytochemicals: down-regulation of COX-2 and iNOS through suppression of NF-κB activation, *Mutation Res.*, 480–481, 243–268, 2001.

Surh, Y.-J., Cancer chemoprevention with dietary phytochemicals, *Nat. Rev. Cancer*, 3, 768–780, 2003.

United States Nuclear Regulatory Commission, fact sheet; available on-line at http://environment.about. comnewsissues/environment/library/weekly/blrwaste1/htm?iam.

World Cancer Research Fund/American Institute for Cancer Research, *Foad, Natrition, and the Prevention of Cancer: A Global Perspective,* American Institute for Cancer Research, Washington, DC, 1997.

Young, M.R., Yang, H.-S., and Colburn, N.H., Promising molecular targets for cancer prevention: AP-1, NF-κB and Pdcd4, *Trends Molecular Med.*, 9, 36–41, 2003.

7 Smokeless Tobacco, Oxidative Stress, and Oral Cancer

Manashi Bagchi, Sidney J. Stohs, and Debasis Bagchi

CONTENTS

7.1 Introduction..90
7.2 Preparation of Smokeless Tobacco Extract (STE)......................................91
7.3 Smokeless Tobacco Extract, Free-Radical Formation,
     and Lipid Peroxidation 91
7.4 Smokeless-Tobacco-Induced Excretion of Urinary Metabolites 92
7.5 Smokeless Tobacco Extract, Cytotoxicity, and LDH Leakage in J774A.1
     Macrophage and Peritoneal Macrophage Cells 92
7.6 Smokeless Tobacco Extract and Free-Radical Scavenging Effects
     of Common Antioxidants in Macrophage Cells 93
7.7 Subchronic Effects of Smokeless Tobacco Extract on Hepatic Lipid
     Peroxidation and the Increased Excretion of Urinary Metabolites 93
7.8 Chronic Effects of Smokeless Tobacco on the Histopathology
     of Rat Livers and Induction of HSP90 94
7.9 Cell Viability and Trypan Blue Exclusion Technique
     of Normal Human Oral Keratinocytes (NHOK) Treated with STE 94
7.10 Smokeless Tobacco and Superoxide Anion Production in
      Human Oral Keratinocytes..95
7.11 Smokeless Tobacco and Lipid Peroxidation of Human
      Oral Keratinocytes...95
7.12 Smokeless Tobacco and Change in Intracellular Oxidation
      States by Laser Scanning Confocal Microscopy 95
7.13 Smokeless Tobacco Induces Protein Kinase C (PKC) Activation
      in Human Oral Keratinocytes..96
7.14 Smokeless Tobacco and DNA Fragmentation in Human
      Oral Keratinocytes...97
7.15 Smokeless Tobacco and DNA Ladder Analysis in Human
      Oral Keratinocytes...97
7.16 Smokeless Tobacco and Apoptosis in Human Oral Keratinocytes
      by Flow Cytometry and DNA Cell Cycle Analysis 98
7.17 Effect of STE on Cell Viability as Determined by MTT,
      3-[4,5-Dimethylthiazol-2-yl]-2,5-Diphenyl Tetrazolium Bromide] Assay 98
7.18 STE-Induced Alterations in p53 and Bcl-2 Gene in NHOK Cells 99
7.19 Conclusion 101
Acknowledgment...101
References 102

7.1 INTRODUCTION

Approximately one-third of all cancers in the U.S. are believed to be caused by tobacco. The consumption of tobacco products is responsible for a significant portion of human cancers at various sites, in particular the lung, oral cavity, larynx, and esophagus.[1,2] Among the constituents of cigarette smoke and chewing tobacco, at least 40 carcinogens have been identified.[3] In 1989, the U.S. surgeon general released a list of carcinogens found in tobacco, which included polynuclear aromatic hydrocarbons, aromatic amines, aldehydes, alkenes, metals, and N-nitroso compounds. The most prevalent N-nitroso compounds in tobacco smoke and unburned tobacco products are known as tobacco-specific nitrosamines (TSNA), and these are responsible for the progression of cancer and leukoplakia (a white patch on the oral mucus membrane). Nitrosamines contain the organic functional group N–N=O, and are formed by the nitrosation (addition of an N=O group) of secondary and tertiary amines.[4] In tobacco, these amines are nicotine, nornicotine, and anabasine, which are formed from nicotine and other tobacco alkaloids during fermentation of tobacco. Three of the tobacco-specific nitrosamines[5] — 4-(methyl nitrosamino)-1-(3-pyridyl)-1-butanone (NNK) benzo[a]pyrene (BAP), and N-nitrosonornicotine (NNN) — are present in relatively high concentrations (1- to 100-ppm range).[4,6] Experimental data has indicated that NNK, a nicotine-derived lung carcinogen, is one of the most powerful carcinogens in smokeless tobacco and is known to cause oral and lung cancer. Researchers have also demonstrated that NNN is another nicotine-derived oral and respiratory carcinogen and that benzo[a]pyrene binds to nucleoproteins and is mutagenic as well as carcinogenic. Aryl hydrocarbon hydroxylase (AHH), an enzyme produced principally in human leukocytes in varying amounts, increases the carcinogenicity of benzo[a]pyrene. Nitrosamines and benzo(a)pyrene induce benign and malignant tumors in rats, mice, and hamsters and in the lungs, oral cavity, esophagus, pancreas, and livers of humans.[2,7,8]

The development of cancer can be divided into three stages. The first stage, initiation, is a persistent and heritable alteration of the cell. The second stage is promotion, which consists of the proliferation of the initiated cells. The third stage is progression, which is where the benign lesion becomes highly malignant and turns into rapidly growing neoplasm.[9,10] Free radicals produced by smokeless tobacco extract can act as mediators in the initiation and promotion stages of oral cancer. The oral use of moist smokeless tobacco products (snuff) is associated with cancers of the mouth, lip, nasal cavities, esophagus, and gut.[10] Winn demonstrated that chewing tobacco and snuff is carcinogenic and can induce tumors and lesions in the oral cavity of rats, mice, and hamsters and in the lungs, oral cavity, esophagus, pancreas, and livers of humans.[11]

The tumor suppressor p53 gene is a transcription factor regulating cell cycle progression, cell survival, and DNA repair in cells exposed to environmental stimuli. The p53 gene is also a component of the biochemical pathways central to human carcinogens, and p53 mutations provide a selective advantage for clonal expansion of preneoplastic and neoplastic cells. Loss or mutations in the p53 tumor suppressor gene are the most frequently observed genetic lesions in human cancer.[12,13] Some of the normal cellular functions of p53 can be modulated by interactions with noxious environmental stimuli. The normal physiological role of p53 appears to be in regulation of the cell cycle, while loss of this function may contribute to the emergence of the malignant phenotype. Furthermore, mutated forms of p53 may behave as oncogenes and immortalize primary cells, and they may cooperate with ras oncogenes to transform primary cells. In cells that lack p53, the loss of the ability to check cell cycle progression leads to increased rates of DNA damage, mutation, and translocations, which make these cells particularly dangerous. The p53 gene can also induce apoptosis as one of its modes of protecting the body against cells that are behaving in a discoordinated fashion or that have damaged DNA.[14,15]

Apoptosis (programmed cell death) is a selective process of physiological cell deletion and is characterized by nuclear chromatin condensation, cytoplasmic shrinking, dilated endoplasmic reticulum, membrane blebbing, fragmentation of nuclei, and extensive degradation of chromosomal DNA. It plays a major role in developmental biology and in the maintenance of homeostasis in vertebrates. Inducers of oxidative stress or stimulators of cellular oxidative metabolism are believed to induce apoptosis. Apoptotic cell death can be triggered by a wide variety of stimuli that cause oxidative stress. Among the most widely studied cell death stimuli is DNA damage, which in many cells leads to apoptotic death via a pathway dependent on p53 gene mutation.[15,16]

Another cellular regulatory gene, Bcl-2 protein, is a 25- to 26-kDa molecule that appears to be associated with a variety of membranes, notably the nuclear membrane and parts of the endoplasmic reticulum that has the ability to protect against and overcome the apoptosis mechanism.[17] Increased expression of Bcl-2 protein prevents the induction of apoptosis by a variety of oxidative stresses, including ionizing radiation, TNFα, heat shock, or inhibition of glutathione synthesis. It has been suggested that the Bcl-2 gene inhibits apoptosis by interacting with mitochondrial superoxide dismutase.[18,19]

Chronic snuff dipping has been associated with oral cancer in experimental animals and humans. Oral pharyngeal cancer accounts for 3% of these cancers and is the seventh most common form of cancer.[20–22] The oral use of moist smokeless tobacco products (chewing tobacco) has increased in recent years, resulting in leukoplakia, which is a clinical white patch or plaque on the oral mucus membrane that cannot be removed by scrapping. Most of these lesions can occur in all areas of the oral cavity.[23] Water-soluble smokeless tobacco extract has been shown to suppress lymphokine activated killer (LAK) cell activity by inhibiting DNA synthesis, and altered LAK function in the oral mucosa may promote the development of snuff-associated carcinogenesis.[23–25] The formation of free radicals, DNA damage, altered p53 gene mutation, increases in intracellular oxidized states, or alteration of Bcl-2 can occur with the use of smokeless tobacco, which may lead to apoptotic cell death and can play a major role in the development of oral cancer.[26–29]

7.2 PREPARATION OF SMOKELESS TOBACCO EXTRACT (STE)

Standardized smokeless tobacco (chewing tobacco) was purchased from the University of Kentucky, Tobacco and Health Research Institute (Lexington, KY).[29] Quantities of smokeless tobacco were mixed with 5 volumes (5 ml/g) of 0.10 M phosphate buffer, pH 7.0, and stirred at room temperature for 24 h. The pH of the extracts was readjusted as needed to pH 7.0 after 1 h of stirring to ensure a physiological pH, and the extracts were centrifuged at 40,000 g for 1 h. The extracts were reconstituted in phosphate buffer at a concentration of 1.0 mg freeze-dried material per milliliter. The STE was standardized from batch to batch by quantitating the nicotine content in a Perkin Elmer 200 gas chromatograph (Perkin Elmer Corp., Norwalk, CT) equipped with a hydrogen flame ionization detector and a 15- × 0.32-mm inside-diameter fused silica capillary column. The instrument was operated in a split mode, and the injector and detector temperatures were 225 and 227°C, respectively. Helium was used as the carrier gas at 25 psi.[29]

7.3 SMOKELESS TOBACCO EXTRACT, FREE-RADICAL FORMATION, AND LIPID PEROXIDATION

The mechanism by which smokeless tobacco constituents produce genetic damage and cause tissue damage is not clear. Lipid peroxidation is a common endpoint for assessing tissue damage due to the production of reactive oxygen species (ROS). The production of thiobarbituric acid reactive substances (TBARS) as an index of lipid peroxidation was determined according to Buege and Aust[30] and Bagchi and Stohs.[31] Malondialdehyde, a common lipid peroxidation product, was used as the standard, with a molar extinction coefficient of 1.56×10^5 $M^{-1}cm^{-1}$ at 535 nm. The effects of an aqueous smokeless tobacco extract at doses of 0, 125, 250, and 500 mg STE/kg in female Sprague–Dawley rats on the induction of hepatic mitochondrial and microsomal lipid peroxidation were examined at 24 h posttreatment.[29] Dose-dependent increases of 1.8-, 2.3-, and 4.4-fold in mitochondrial and 1.5-, 2.1-, and 3.6-fold in microsomal lipid peroxidation occurred at 125, 250, and 500 mg STE/kg, respectively, relative to control values. In order to determine whether STE could induce lipid peroxidation based on direct contact, hepatic mitochondria and microsomes from untreated female Sprague–Dawley rats were incubated at 37°C for 60 min in the presence of 0 to 500 µg STE/ml. Dose-dependent increases of 1.1- to 2.4-fold occurred in 35 mitochondria and microsomes.[32] These results strongly suggest that STE induces oxidative stress, which increases the production of ROS.

7.4 SMOKELESS-TOBACCO-INDUCED EXCRETION OF URINARY METABOLITES

Smokeless tobacco extract induces lipid peroxidation in the hepatic tissue of rats. When lipid peroxidation products accumulate, they leak from the tissue and may be excreted in the urine. The detection of lipid peroxidation products in the urine provides a noninvasive method of assessing lipid metabolism and oxidative stress.[29,33] High-pressure liquid chromatography–mass spectrometry (GC–MS) has provided conclusive identification and quantization of malondialdehyde (MDA), formaldehyde (FA), acetaldehyde (ACT), and acetone (ACON) excretion in the urine of rats.[29,33] Female Sprague–Dawley rats were placed in metabolism cages for urine collection. The animals were allowed free access to tap water but received no food during urine collection to avoid possible contamination of the urine with food particles. The urine-collecting vessels were positioned over Styrofoam containers filled with dry ice in order to collect the urine in the frozen state over a 4.5-h period, with the midpoints of collection times occurring at 0, 12, 24, 48, and 72 h after STE administration. The 2,4-dinitrophenylhydrazine derivatives of the four metabolic products of urine were quantitated by high-pressure liquid chromatography (HPLC) on a Waters µ-Bondapak C_{18} column, eluting with acetonitrile–water (51:49 v/v) mobile phase and using a UV absorbance detector at 330 nm, as previously described.[29,33] Urinary excretion of the four lipid metabolites malondialdehyde (MDA), formaldehyde (FA), acetaldehyde (ACT), and acetone (ACON) was monitored by HPLC for 72 h after treatment of rats with 125 and 250 mg STE/kg. Increases occurred in the excretion of the four lipid metabolites at every dose and time point, with maximum increases in the excretion of all lipid metabolites being observed between 12 and 24 h posttreatment. The results suggest the involvement of oxidative stress in rats in the toxicity of STE. The increased excretion of MDA, FA, ACT, and ACON after STE treatment correlates well with the increase in hepatic lipid peroxidation contributing to the cytotoxicity of STE.[29]

7.5 SMOKELESS TOBACCO EXTRACT, CYTOTOXICITY, AND LDH LEAKAGE IN J774A.1 MACROPHAGE AND PERITONEAL MACROPHAGE CELLS

The *in vitro* incubation of cultured J774A.1 macrophage cells with STE on the release of the enzyme lactate dehydrogenase (LDH) into the medium was assessed. LDH is an indicator of cellular membrane damage and cytotoxicity.[32,34] The LDH assay was performed according to the method of Moss et al.[35] The changes in absorbance of reaction media were measured for 3 min at 340 nm for LDH in a Perkin Elmer Lambda 6 spectrophotometer. The amount of LDH released by smokeless tobacco extract was both concentration- and time-dependent. The cytotoxicity of STE to macrophage J774A.1 cells in culture was further determined from the percent of viability after various periods of incubation. The addition of 250 µg smokeless tobacco/ml to the cultured J774A.1 cells resulted in a 2.9-fold increase in the release of LDH. The increased production of oxygen free radicals with STE may be responsible for tissue-damaging effects, including membrane damage.[32,34]

7.6 SMOKELESS TOBACCO EXTRACT AND FREE-RADICAL SCAVENGING EFFECTS OF COMMON ANTIOXIDANTS IN MACROPHAGE CELLS

Antioxidants are potent scavengers of free radicals, serve as inhibitors of neoplastic processes and cancer at both initiation and promotion/progression transformation stages of carcinogenesis, and provide protection from oxidative damage. A large number of synthetic and natural antioxidants have been demonstrated to induce beneficial effects on human health and disease prevention. In order to determine the possible sources of reactive oxygen species in response to smokeless tobacco, rat peritoneal macrophages (3×10^6/ml) and hepatic mitochondria and microsomes (1 mg protein/ml) from untreated female Sprague–Dawley rats were incubated with STE (200 µg/ml). STE resulted in rapid increases in chemiluminescence, with maximum increases occurring at approximately 6 min for the macrophages and 8 min for mitochondria and microsomes. Maximum increases in chemiluminescence of 1.4-, 3.2-, and 3.1-fold relative to control values occurred for macrophages, mitochondria, and microsomes, respectively. Hepatic mitochondria and microsomes (1 mg protein/ml) from female Sprague–Dawley rats were incubated at 37°C for 60 min in the presence of 0 to 500 µg/ml STE.[32] Potential tissue damage was measured as lipid peroxidation, and dose-dependent increases of 1.1- to 2.4-fold occurred in mitochondria and microsomes. Preincubation with various oxygen free-radical scavengers, including superoxide dismutase (SOD) (100 µg/ml), catalase (100 µg/ml), SOD + catalase (100 µg/ml each), mannitol (1.25 mmol/ml), and allopurinol (100 µg/ml), inhibited STE (200 µg/ml)-induced lipid peroxidation by 15 to 70%.[32] Vitamin E succinate also protected against STE-induced oxidative damage in rat peritoneal macrophages and cultured J774A.1 macrophage cells.[36] The results demonstrate that (1) oral cells, peritoneal macrophages, and hepatic mitochondria and microsomes produce reactive oxygen species following *in vitro* incubation with an aqueous extract of smokeless tobacco, and (2) antioxidants can attenuate these tissue-damaging effects.

7.7 SUBCHRONIC EFFECTS OF SMOKELESS TOBACCO EXTRACT ON HEPATIC LIPID PEROXIDATION AND THE INCREASED EXCRETION OF URINARY METABOLITES

Since the use of tobacco is a chronic process, the effects of an aqueous extract of STE in rats following low-dose exposure were examined. Female Sprague–Dawley rats were treated orally with 25 mg STE/kg every other day for 105 days. The effects of subchronic treatment of STE on hepatic microsomal and mitochondrial lipid peroxidation were assessed.[37] Urinary excretion of the four lipid metabolites malondialdehyde, formaldehyde, acetaldehyde, and acetone was monitored by HPLC, with maximum increases being observed between 60 and 75 days of treatment. The assessment of lipid peroxidation was based on the formation of thiobarbituric acid-reactive substances (TBARS) over the 105 days of the study.

An age-dependent increase in lipid peroxidation was observed in control animals. Over the 105 days of the study, lipid peroxidation increased by 1.9-fold in hepatic mitochondria and 1.6-fold in hepatic microsomes of control animals. In rats treated subchronically with STE, time-dependent increases in lipid peroxidation associated with hepatic mitochondria and microsomes were observed relative to the control animals. In hepatic mitochondria, a 2.7-fold increase in lipid peroxidation was observed after 45 days of treatment and remained constant thereafter, relative to the control animals. In hepatic microsomes, a time-dependent increase in lipid peroxidation was observed, with the greatest increase (3.3-fold) occurring after 90 days of STE treatment.[37]

The concurrent detection of lipid metabolites by HPLC provides a rapid, convenient, noninvasive method for assessing oxidative tissue damage and cytotoxicity by foreign chemicals.[29,37] Urine samples were collected after STE treatment over dry ice for 6.0-h time intervals on the indicated days, and the results are expressed as nmol/kg body weight/6.0 h. In control animals, time-dependent increases occurred in the urinary excretion of the four lipid metabolites over the 105 days of the study. Increases of 1.9-, 1.5-, 1.6-, and 1.5-fold occurred in the urinary excretion of MDA, FA, ACT, and ACON, respectively, between days 0 and 105 by control rats. MDA excretion in STE-treated rats was significantly different from control animals by the 10th day of treatment, with a 1.9-fold increase occurring by day 30, and a maximum increase of 2.3-fold occurring by day 45. No further increases in MDA urinary excretion occurred thereafter during the remainder of the 105-day study. The urinary excretion of FA in STE-treated rats relative to control animals significantly increased by day 20, with a 1.3-fold increase on day 30 and a maximum increase of 2.0-fold on day 60.

The urinary excretion of FA remained relatively constant thereafter in the STE-treated animals throughout the remainder of the 105-day study as compared with the control rats. The urinary excretion of ACT in STE-treated rats was significantly greater than in control animals by the 10th day of the study, with a maximum increase in ACT of 1.9-fold occurring relative to control animals on day 50 of the study. The excretion of ACT by STE-treated rats remained relatively constant thereafter as compared with the control animals. The urinary excretion of ACON was significantly elevated in the urine of STE-treated animals by the 10th day of the study, relative to the control animals. A maximum increase in the urinary excretion of ACON in STE-treated rats of 1.6-fold occurred by day 50 of the study and remained relatively constant thereafter as compared with the control animals during the remainder of the study. The results clearly demonstrate that a chronic low dose of STE treatment induces oxidative stress, resulting in tissue-damaging effects, which contributes to the toxicity and carcinogenicity of STE.[37]

7.8 CHRONIC EFFECTS OF SMOKELESS TOBACCO ON THE HISTOPATHOLOGY OF RAT LIVERS AND INDUCTION OF HSP90

Female Sprague–Dawley rats were treated orally with 25 mg STE/kg every other day for 90 days.[38] In order to obtain information regarding the cytotoxicity of STE, the ultrastructural changes occurring in livers of rats following administration of STE were examined under light and electron microscopy. Groups of rats were killed on days 30, 45, 60, and 90 by decapitation, and livers were fixed for light and electron microscopy.[38] Samples of liver from each animal were fixed in paraformaldehyde and glutaraldehyde, pH 7.4, for electron microscopy. A tissue buffer ratio of at least 1:10 was used in all cases. Sections approximately 5 μm thick were cut. Samples of livers from each control and experimental animal were fixed in 10% buffered formalin for light microscopy. The samples were dehydrated by standard procedures and embedded in paraffin. Sections approximately 5 μm thick were cut, stained with hematoxylin and eosin (H & E), and examined under a light microscope. Electron microscopy revealed that in the perisinusoidal spaces an accumulation of indistinct filamentous material occurred following 60 days of treatment, occupying most of the sinusoids. Moreover, the lipids were in a state of disintegration. Significant increases in 90-kDa protein expression were also observed due to chronic treatment with STE. Western blot analysis using a polyclonal mouse antibody against heat shock/stress protein 90 (HSP90) confirmed that the overexpressed proteins were heat shock/stress proteins (HSPs).[38] The HSPs are believed to serve as adaptive or survival functions involving a rapid but transient reprogramming of cellular metabolic activity to protect cells from oxidative damage.[39]

7.9 CELL VIABILITY AND TRYPAN BLUE EXCLUSION TECHNIQUE OF NORMAL HUMAN ORAL KERATINOCYTES (NHOK) TREATED WITH STE

Both concentration- and time-dependent effects of STE were assessed on cellular viability in a primary culture of human oral keratinocytes (NHOK), derived from oral cavities of normal human subjects, by the Trypan blue exclusion technique.[40] The control and STE-treated NHOK cells were trypsinized, centrifuged, and resuspended in culture medium. Each cell suspension (0.10 ml) was mixed with 0.1 ml Trypan blue solution (0.2% PBS). Live and dead cells were counted using a hemocytometer, and blue-stained cells were counted as nonviable. The percentage viability was calculated based on the percentage of unstained cells. A concentration-dependent effect (0 to 400 µg/ml) on cell death was observed with treatment of STE, and most cells lost their adhesiveness within 24 h when incubated with high concentrations of STE (>250 µg/ml). However, 55 to 70% of the cells were still viable.[28,40]

7.10 SMOKELESS TOBACCO AND SUPEROXIDE ANION PRODUCTION IN HUMAN ORAL KERATINOCYTES

Superoxide anion production by NHOK after treatment with various concentrations of STE and/or antioxidants was measured according to our previously published procedure.[40] Human oral keratinocytes were treated with 0 to 300 µg/ml of STE for 24 h. The cells (1×10^6 cells/ml) were incubated at 37°C for 15 min in a reaction mixture containing 0.05 mM cytochrome c (Type III, Sigma). The reactions were stopped by placing the reaction tubes in an ice bath. The mixtures were centrifuged for 10 min at 1500 g at 4°C, and the absorbance values of the supernatant fractions were measured at 550 nm.[40] To verify that the reactive oxygen species were produced in response to STE, cytochrome c reduction was measured. This assay is a specific test for the production of superoxide anion. Following treatment of the NHOK with 100, 200, and 300 µg/ml of STE, 1.6-, 2.0-, and 2.7-fold increases in superoxide anion production were observed. Preincubation of the cells with vitamin C (75 µM), vitamin E (75 µM), a combination of vitamins C plus E (75 µM each), and grape seed proanthocyanidin extract (GSPE) (100 µg/ml) for 4 h decreased STE (300 µg/ml)-induced cytochrome c reduction by 37, 39, 44, and 48%, respectively, as compared with the STE-treated cells.[28]

7.11 SMOKELESS TOBACCO AND LIPID PEROXIDATION OF HUMAN ORAL KERATINOCYTES

Normal human oral keratinocytes (NHOK) were incubated with STE (0 to 300 µg/ml) for 24 h at 37°C. Keratinocytes were also preincubated with vitamin C (75 µM), vitamin E (75 µM), a combination of vitamins C and E (75 µM each), and grape seed extract (GSPE) (100 µg/ml) for 4 h at 37°C. These optimal concentrations for vitamins C and E were selected based on our earlier experiments. Production of thiobarbituric acid reactive substances (TBARS), a commonly used technique that has been related to lipid peroxidation and tissue damage, was determined according to the method of Buege and Aust[30] and as described previously.[40] The production of TBARS was determined with NOHK cells following STE incubation. Concentration-dependent increases of 3.5-, 5.7-, and 7.6-fold in lipid peroxidation occurred at 100, 200, and 300 µg/ml of STE, respectively, for 24-h posttreatment relative to control values. Preincubation of the cells with vitamin C, vitamin E, a combination of vitamins C plus E, and GSPE for 4 h decreased STE (300 µg/ml)-induced lipid peroxidation by 22, 34, 43, and 54%, respectively, as compared with the corresponding STE-treated cells.[28,40]

7.12 SMOKELESS TOBACCO AND CHANGE IN INTRACELLULAR OXIDATION STATES BY LASER SCANNING CONFOCAL MICROSCOPY

The concentration-dependent effects of STE on the intracellular oxidized states of human keratinocytes were determined using confocal microscopy.[40] The intensity of fluorescence reflects the intracellular oxidized state. When cells were treated with concentrations of STE at 250 and 300 µg/ml, the color and fluorescence intensity of nuclei changed, possibly because of a change

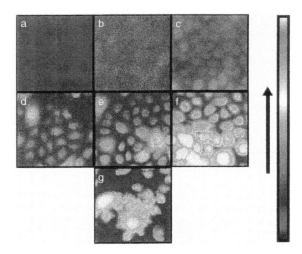

FIGURE 7.1 Changes in intracellular redox states of oral human keratinocytes after treatment with smokeless tobacco extract (STE). Cells (70% confluent) were treated with STE for 24 h. After 24 h, medium was replaced with fresh medium containing 5 μm 2,7-DCF, and fluorescence intensity was measured 5 min later at 513 nm with a confocal laser scanning microscope. A: control; B: 100 μg/ml; C: 150 μg/ml; D: 200 μg/ml; E: 250 μg/ml; F: 300 μg/ml; G: 3 mM H_2O_2 (positive control).

in the oxidized states of the cells and the accumulation of the dye in the nuclear fractions. A concentration-dependent effect was observed. Approximately 70% confluent primary NHOK grown in six-well plates were treated with different concentrations (0 to 300 μg/mg) of STE and incubated at 37°C in an atmosphere of 95% air and 5% carbon dioxide for 24 h. After 24 h of incubation with STE, the overall intracellular oxidized states of cells were measured by laser scanning confocal microscopy using the dye 2,7-dichlorofluorescin diacetate (2,7-DCF) as the fluorescent probe. The cells were incubated with STE for 24 h prior to exposure to 2,7-DCF. Cells were also treated with hydrogen peroxide as a positive control. The dye is incorporated into the cells and is converted to a fluorescent metabolite by oxidation. The fluorescence intensity for each point was measured by confocal laser scanning microscopy according to the method of Ohba et al.[41] An excitation wavelength of 513 nm was used. Relative fluorescence intensity was calculated using untreated control cells as the standard.[40]

The intensity of fluorescence increased with increasing concentrations of STE. Intensities were approximately 3.0-, 3.2-, 3.9-, and 4.6-fold higher than the control cells when the cells were treated with 150, 200, 250, and 300 μg/ml of STE, respectively (Figure 7.1). Intracellular oxidized state increased with increasing concentration of STE treatment. The highest level of intracellular fluorescence produced by STE was comparable with that which was observed in cells treated with 3 mM H_2O_2. The dye 2,7-DCF can be oxidized by any peroxidase and hyperperoxide (including H_2O_2), which indicates production of highly reactive forms of oxygen, which in turn leads to DNA damage, cell death, and may lead to cancers of the oral cavity.[40,41]

7.13 SMOKELESS TOBACCO INDUCES PROTEIN KINASE C (PKC) ACTIVATION IN HUMAN ORAL KERATINOCYTES

Changes in the redox state of cells are thought to induce modifications of cellular signaling molecules, including protein kinase.[40] PKC is involved in signaling pathways that mediate the regulation of many cell processes, including cell differentiation, cell survival, cytoskeletal function, gene expression, secretion, and cell–cell interactions. Studies by Fox et al.[42] have demonstrated that STE causes time- and concentration-dependent loss of cellular adhesiveness of HT 1080 cells to a variety of matrices. Cells that lost their adhesiveness could regain that function by removal of the STE. Our studies have shown that incubation of keratinocytes in culture with STE results in the increased activity of PKC.[40] Concentrations in protein kinase C activity in cytosol were measured after keratinocytes were exposed to STE, and time-dependent variations were observed. The cells were treated with 0 to 250 μg/ml of STE for 1, 3, and 24 h. After exposure of keratinocytes to STE for 1 h, increases of 113, 121, and 146% occurred in cytosolic PKC activity with respect to control values for STE treatments of 100, 150, and 250 μg/ml, respectively. Similarly, after exposure to STE for 3 h, increases of 113, 122, and 170% occurred in cytosolic PKC activity with respect to control values at these three same concentrations, respectively. Keratinocytes exposed to STE for 24 h exhibited no significant changes in cytosolic PKC activity with respect to controls. PKC activity was also measured in solubilized particulate fractions for the three time points, but no significant differences were observed with respect to the control values. A growing body of evidence indicates that free radicals and ROS may be involved in mediating signal transduction through interaction with PKC. Thus, the current results support a role for both ROS and PKC with respect to the tissue-damaging effects of smokeless tobacco.[40–43]

7.14 SMOKELESS TOBACCO AND DNA FRAGMENTATION IN HUMAN ORAL KERATINOCYTES

DNA fragmentation is a biological hallmark of oxidative DNA and cellular injury.[44] The cytotoxicity of STE was examined by determining DNA fragmentation (DNA damage) 24 h after treatment of normal human oral keratinocytes (NHOK) in culture with 100, 250, and 350 μg/ml of STE. STE induced significant DNA damage, as measured by DNA fragmentation.[40] Effects of STE treatment on keratinocytes were reported as percent of fragmentation relative to untreated cells. DNA fragmentation in samples is expressed as percent of total DNA appearing in the supernatant fractions.[40,44] The absorbencies were measured spectrometrically at 600 nm. A concentration-dependent increase in DNA fragmentation was observed with increases of 1.5-, 1.6-, and 2.9-fold following exposure of the cells to the three concentrations of STE, respectively. Pretreatment of the NHOK with vitamin C (75 μM), vitamin E (75 μM), a combination of vitamins C plus E (75 μM each), and GSPE (100 μg/ml) for 4 h decreased STE (300 μg/ml)-induced DNA fragmentation by 9, 12, 20, and 36%, respectively, as compared with the STE-treated cells. The results demonstrate that increased production of ROS with STE treatment causes oxidative DNA damage and cell death.

7.15 SMOKELESS TOBACCO AND DNA LADDER ANALYSIS IN HUMAN ORAL KERATINOCYTES

The status of genomic DNA (qualitative and quantitative) during STE treatment of keratinocytes was assessed. Agarose gel electrophoresis was used to provide evidence that NHOK treated with increasing concentrations of STE (0 to 300 μg/ml) for 24 h exhibited DNA damage. STE induced a ladderlike pattern of endonucleolytic DNA degradation at all concentrations. The characteristic ladder pattern of internucleosomal DNA cleavage that was observed is consistent with activation of an endogenous endonuclease and is readily evident in STE-treated cells.[28,45]

The fragmentation pattern for DNA from cells treated with STE indicated significant DNA damage caused by STE, while DNA from the control cells remained totally unfragmented. At a concentration of 100 μg STE/ml, the ladder formation was not very prominent.[28] However, at higher concentrations, the loss of higher-molecular-weight genomic DNA was observed, and a concentration-dependent response occurred. Partial reversal of STE-induced oligonucleosome length DNA-laddering was observed in cells treated with the various antioxidants. Greatest protection against STE-induced DNA fragmentation was provided by grape seed proanthocyanidin extract (GSPE).[28]

7.16 SMOKELESS TOBACCO AND APOPTOSIS IN HUMAN ORAL KERATINOCYTES BY FLOW CYTOMETRY AND DNA CELL CYCLE ANALYSIS

Programmed cell death (apoptosis) is a selective process of physiological cell deletion.[15] During apoptosis, cells shrink, rapidly display an altered plasma membrane, and suffer cytoplasmic damage. Apoptosis has been linked to several chronic conditions, including cancer. Induction of apoptosis has been demonstrated in a variety of cell types following exposure to low doses of hydrogen peroxide. Some agents that induce apoptosis are not free radicals themselves, but may elicit free-radical formation.[15,16,28]

Concentration-dependent effects of STE on the death of human keratinocytes were determined using flow cytometry.[28] The cells were incubated with STE for 24 h prior to labeling with propidium iodide. The percentages of the cells exhibiting apoptotic cell death increased with increasing concentration of STE (Figure 7.2). The correlation between the flow cytometric data measuring the Ao peak and DNA laddering has been described earlier by Telford et al.[46] and Darzynkiewicz.[47] A distinct subpopulation of cells is formed below the Go/G region during a poptotic cell death, which is denoted as the Ao region. All experiments were performed in triplicate, and concentration-dependent effects of STE on the death of human keratinocytes were determined using flow cytometry.[28] The cells were incubated with STE for 24 h prior to labeling with propidium iodide.

The percentages of the apoptotic peak (cell death) increased with increasing concentration of STE. The percentages of the cells exhibiting apoptotic cell death as analyzed by ModFit cell cycle analysis software were approximately 0, 9, 29, and 35% when keratinocytes were treated with 0, 100, 200, and 300 mg/ml of STE, respectively (Figure 7.2). A distinct loss of cells in the G2-M phase coincides with the appearance of the apoptotic cell population. Significant protection was observed against STE-induced cell death with the antioxidants used in this study. Following pretreatment of the NHOK with a combination of vitamins C plus E or GSPE decreased STE (300 μg/ml)-induced apoptotic cell death by approximately 46 and 85%, respectively, as compared with the STE-treated cells that received no antioxidants (Figure 7.3). The results indicate that STE treatment of oral keratinocytes enhanced the production of free radicals, which may trigger apoptosis as a result of oxidative damage and may initiate the pathways of oral cancer.[28]

7.17 EFFECT OF STE ON CELL VIABILITY AS DETERMINED BY MTT, 3-[4,5-DIMETHYLTHIAZOL-2-YL]-2,5-DIPHENYL TETRAZOLIUM BROMIDE, ASSAY

The cell viability measured by MTT assay is a very sensitive colorimetric assay, and only viable cells with intact mitochondria can reduce the dye tetrazolium (MTT). There is a direct proportionality between the formazan produced (expressed as Optical Density) and the number of viable cells.[48]

The cleavage of the tetrazolium salt MTT into a blue-colored formazan by the mitochondrial enzyme succinate dehydrogenase was used for assaying cell survival and proliferation, an assay technique that is used extensively.[49] Respiratory effects of selected antioxidants on STE (0 to 300 μg/ml)-exposed NHOK cells (24 h) were assessed by determining the ability of the cells to reduce tetrazolium dye MTT based on the activation of the enzyme succinate dehydrogenase, as described by Mosmann[48] and by Bagchi et al.[50] Cells were plated at a density of 1.0×10^6 cells. After treatment, 100 μl of dye (4 mg/ml in Dulbecco's phosphate buffered saline) was added to each well and incubated for 2 h at 37°C. The optical density of each well was read at 570 nm. Concentration-dependent effects of STE were studied on the reduction of the dye MTT in NHOK cells. Incubation of the NHOK cells with 100, 200, and 300 μg/ml of STE for 24 h at 37°C resulted in 15, 21, and 46% decreases in succinate dehydrogenase activities, respectively, as compared with untreated cells.[51]

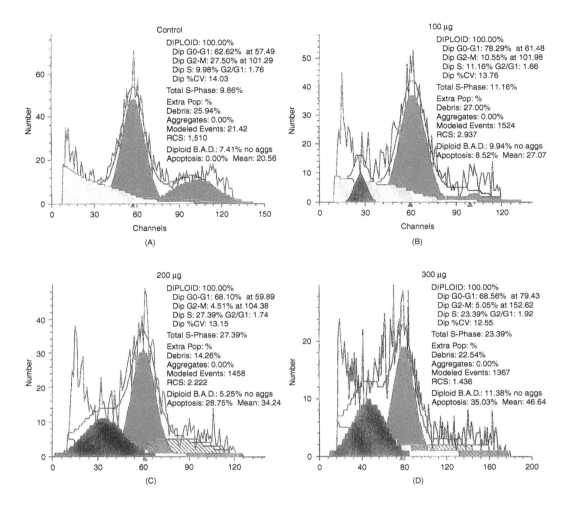

FIGURE 7.2 Flow cytometric analysis of cell cycle distribution and apoptosis in human oral keratinocytes in response to increasing concentrations of smokeless tobacco extract. Primary keratinocytes were grown to approximately 70% confluence and then treated with various concentrations of STE for 25 h. Following incubation, the cells were removed from the culture surfaces by trypsinization, and DNA content and apoptosis were determined using the method of Telford et al.[46] Raw data were modeled using ModFit cell cycle analysis software (Verity Software House, Topsham, ME). The profiles are representative histograms of triplicate assays. A: control; B: STE (100 μg/ml); C: STE (200 μg/ml); and D: STE (300 μg/ml).

7.18 STE-INDUCED ALTERATIONS IN p53 AND Bcl-2 GENE IN NHOK CELLS

To determine the effect of STE on Bcl-2 and p53 genes in NHOK cells, NHOK cells were treated with 0 to 200 μg/ml of STE.[51] Keratinocytes were also preincubated with GSPE (50 μg/ml and 100 μg/ml) before treatment with STE. A reverse-transcription polymerase chain reaction (RT-PCR) technique was used to determine the expression of p53 and Bcl-2 genes in NHOK treated with STE and GSPE.[51] The differences in the levels of expression of p53 and Bcl-2 in STE treated with and without GSPE was determined using densitometric analysis of the PCR product bands. The data were digitized using BioRad's Gel Analysis System. The density of each band was then determined using Image Pro Plus (Media Cybernetics, Inc., Silver Spring, MD) software and compared with the control values.[51]

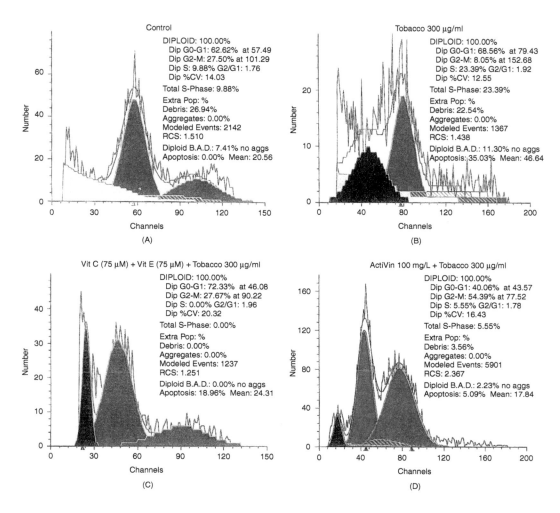

FIGURE 7.3 Flow cytometric analysis of cell cycle distribution and apoptosis in human oral keratinocytes in response to STE (300 µg/ml) and selected antioxidants. Primary keratinocytes were grown to approximately 70% confluence and then treated with STE alone or with the antioxidants for 24 h. Following incubation, the cells were removed and treated and described as stated in Figure 7.2. The profiles are representative histograms of triplicate assays. A: control; B: STE (300 µg/ml); C: STE (300 µg/ml) plus vitamins C plus E (75 µ*M* each); and D: STE (300 µg/ml) plus GSPE (100 µg/ml).

The RT-PCR analysis of STE-treated NHOK cells showed a remarkable decrease in Bcl-2 antiapoptotic gene expression following treatment with STE (200 µg/ml) for 24 h (Figure 7.4). NHOK cells preincubated with GSPE (50 µg/ml and 100 µg/ml) for 4 h prior to STE treatment showed an increase in Bcl-2 mRNA expression, indicating that GSPE ameliorates the toxic effects of STE.[51] A concentration-dependent increase in Bcl-2 mRNA expression was observed following preincubation with GSPE. At 100 µg/ml concentration of GSPE, preincubation completely reversed the cytotoxic effect of STE, as demonstrated by near-complete restoration of Bcl-2 mRNA expression in these cells. Thus, NHOK cells preincubated with GSPE for 4 h followed by STE treatment showed an increase in Bcl-2 mRNA expression,[51] which corresponds to our previous apoptosis study.[28]

The RT-PCR analysis of the STE-treated NHOK cells for the expression of proapoptotic gene p53 revealed that there was an increase/alteration of two-fold in p53 mRNA expression following

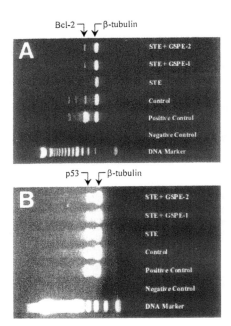

FIGURE 7.4 Modulation of Bcl-2, p53 mRNA expression in normal human oral keratinocytes (NHOK) following treatment with STE and/or GSPE (GSPE-1: 50 μg/ml; GSPE-2: 100 μg/ml). A: Bcl-2 mRNA expression; B: p53 mRNA expression.

STE (200 μg/ml) treatment. The two-fold increase of the band was quantified using Image Pro Plus software, as described previously. The expression of p53 reduces dramatically beyond 200 μg/ml STE, confirming increased apoptotic cell death with higher concentration of STE. Thus, expression of p53 and Bcl-2 genes in NHOK cells is affected following incubation with STE[28,51,52] (Figure 7.4).

7.19 CONCLUSION

In summary, the results of this foregoing study strongly suggest that oxidative stress and reactive oxygen species play a significant role in the cytotoxic effects of smokeless tobacco on hepatic mitochondria, microsomes, peritoneal macrophages, and cultured normal human oral keratinocyte cells.[28,29,32,34,36–38,40,45,51,53] Modulation of Bcl-2 and p53 genes in human oral keratinocyte cells by STE plays a significant role in oxidative stress, leading to cell death and oral cancer.[51] Furthermore, these results clearly demonstrate production of ROS oxidative DNA damage and enhanced intracellular oxidized states in NHOK cells in response to STE treatment, as demonstrated by cytochrome c reduction, increases in DNA damage with ladderlike fragmentation of genomic DNA, and increases in lipid peroxidation, ultimately leading to apoptotic cell death.[28,29,40,53]

Taken together, all of these cytotoxic effects leading to DNA damage and alterations in the expression of p53 and Bcl-2 in the presence of STE may constitute the defining stages of oral carcinogenesis and metastases.[54–58]

ACKNOWLEDGMENT

The authors thank Ms. Kristine Strong and Ms. Jessica De Leon for technical assistance.

REFERENCES

1. Wynder, E.L. and Hoffmann, D., *Tobacco and Tobacco Smoke: Studies in Experimental Carcinogenesis*, Academic Press, New York, 1967.
2. Brinton, L.A. et al., A case control study of cancers of the nasal cavity and paranoid sinuses, *Am. J. Epidemiol.*, 119, 896, 1984.
3. Chamberlain, W.J., Schlotzhauer, W.S., and Chortyk, O.T., Chemical composition of commercial tobacco products, *J. Agric. Food Chem.*, 36, 48, 1998.
4. Spielgelhalder, B. and Fischer, S., Formation of tobacco-specific nitrosamines, *Crit. Rev. Toxicol.*, 21, 241, 1991.
5. Andersen, R.A. et al., Effect of storage conditions on nitrosated, acetylated and oxidized pyridine alkaloid derivatives in smokeless tobacco products, *Cancer Res.*, 49, 5895, 1989.
6. Malkinson, A.M. et al., Inhibition of 4-(methylnitrosamino)-1-(3-pyridyl)-1-butanone-induced mouse lung tumor formation by FGN-1 (sulindac sulfone), *Carcinogenesis*, 19, 1353, 1998.
7. Hoffmann, D., Rivenson, A., and Hecht, S.S., The biological significance of tobacco-specific N-nitrosamines: smoking and adenocarcinoma of the lung, *Crit. Rev. Toxicol.*, 26, 199, 1996.
8. Johanssen, S.L. et al., Snuff induced carcinogenesis: effect of snuff in rats initiated with 4-nitroquinoline-N-oxide, *Cancer Res.*, 49, 303, 1989.
9. Bagchi, D., Cellular protection with proanthocyanidins derived from grape seeds, *Ann. N.Y. Acad. Sci.*, 957, 260, 2002.
10. Thun, M.J., Henley, S.J., and Calle, E.E., Tobacco use and cancer: an epidemiologic perspective for geneticists, *Oncogene*, 21, 7307, 2002.
11. Winn, D.M., Surveillance of knowledge about cancer associated with smokeless tobacco use, in *Smokeless Tobacco on Health: an International Perspective*, U.S. Department of Health and Human Services, National Institutes of Health, Bethesda, MD, 1992, pp. 11–17.
12. Kerdpon, D., Sriplung, H., and Kietthubthew, S., Expression of p53 in oral squamous cell carcinoma and its association with risk habits in southern Thailand, *Oral Oncol.*, 37, 553, 2001.
13. Kasten, M.M. and Giordano, A., pRb and the cdks in apoptosis and the cell cycle, *Cell Death Differ.*, 5, 132, 1998.
14. Hollstein, M. et al., p53 Mutations in human cancers, *Science*, 253, 49, 1991.
15. Choisy-Rossi, C. and Yonish-Rouach, E., Apoptosis and the cell cycle, the p53 connection, *Cell Death Differ.*, 5, 129, 1998.
16. Choisy-Rossi, C., Reisdorf, P., and Yonish-Rouach, E., Mechanisms of p53-induced apoptosis: in search of genes which are regulated during p53-mediated cell death, *Toxicol. Lett.*, 28 (102), 491, 1998.
17. McDonnell, T.J. et al., Importance of the Bcl-2 family in cell death regulation, *Experntia*, 52, 1008, 1996.
18. Beham, A. et al., Bcl-2 inhibits p53 nuclear import following DNA damage, *Oncogene*, 15, 2767, 1997.
19. Teni, T., Expression of Bcl-2 and bax in chewing tobacco-induced oral cancers and oral lesions from India, *Pathol. Oncol. Res.*, 8, 109, 2002.
20. Department of Health and Human Services, Health Benefits of Smoking Cessation: a Report of the Surgeon General, Centers for Disease Control, Center for Chronic Disease Prevention and Health Promotion, Office on Smoking and Health, Rockville, MD, 1992.
21. Hoffmann, D. et al., Carcinogenic agents in snuff, *J. Natl. Cancer Inst.*, 76, 435, 1986.
22. Farber, E., Cancer development and its natural history: a cancer prevention perspective, *Cancer*, 62, 1676, 1988.
23. Einhorn J. and Wersall, J., Incidence of oral carcinoma in patients with leukoplakia of the oral mucosa, *Cancer*, 20, 2189, 1967.
24. Pindborg, J.J., Studies in oral leukoplakia: a preliminary report on the period prevalence of malignant transformation in leukoplakia based on a follow-up study of 248 patients, *J. Am. Dent. Assoc.*, 76, 767, 1968.
25. Silverman, Jr., S. and Rozen, R.D., Observations on the clinical characteristics and natural history of oral leukoplakia, *J. Am. Dent. Assoc.*, 76, 772, 1968.
26. Schildt, E.B., Expression of p53, PCNA, Ki-67 and bcl-2 in relation to risk factors in oral cancer: a molecular epidemiological study, *Int. J. Oncol.*, 22, 861, 2003.

27. Li, P.F., Dietz, R., and von Harsdorf, R., p53 Regulates mitochondrial membrane potential through reactive oxygen species and includes cytochrome c independent apoptosis blocked by Bcl-2, *EMBO J.*, 18, 6027, 1999.

28. Bagchi, M. et al., Smokeless tobacco, oxidative stress, apoptosis, and antioxidants in human oral keratinocytes, *Free Radical Biol. Med.*, 26, 992, 1999.

29. Bagchi, M. et al., Smokeless tobacco induced increases in hepatic lipid peroxidation, DNA damage and excretion of urinary lipid metabolites, *Int. J. Exp. Path.*, 75, 197, 1994.

30. Buege, J.A. and Aust, S.A., Microsomal lipid peroxidation, in *Methods of Enzymology*, Fleisher, S. and Packer, L., Eds., Academic Press, New York, 1978, p. 302.

31. Bagchi, M. and Stohs, S.J., *In vitro* induction of reactive oxygen species by 2,3,7,8-tetrachlorodibenzo-p-dioxin, endrin and lindane in rat peritoneal macrophages, and hepatic mitochondria and microsomes, *Free Radical Biol. Med.*, 14, 11, 1993.

32. Bagchi, M., Bagchi, D., and Stohs, S.J., *In vitro* effects of a smokeless tobacco extract on the production of reactive oxygen species by human oral epidermal cells and rat hepatic mitochondria and microsomes, and peritoneal macrophages, *Arch. Environ. Contam. Toxicol.*, 30, 418, 1996.

33. Bagchi, D. et al., Time-dependent effects of 2,3,7,8-tetrachlorodibenzo-p-dioxin on serum and urine levels of malondialdehyde, formaldehyde, acetaldehyde and acetone in rats, *Toxicol. Appl. Pharmacol.*, 123, 83, 1993.

34. Bagchi, D. et al., Protective effects of free radical scavengers and antioxidants against smokeless tobacco extract (STE)-induced oxidative stress in macrophage J774A.1 cell cultures, *Arch. Environ. Contam. Toxicol.*, 29, 424, 1995.

35. Moss, D.W., Henderson, A.R., and Kachmar, J.R., Enzymes, in *Tietz N.W. Textbook of Clinical Chemistry*, Tietz, N.W. Ed., Saunders Co., Philadelphia, 1986, p. 619.

36. Hassoun, E.A. et al., Effect of vitamin E succinate on smokeless tobacco-induced production of nitric oxide by rat peritoneal macrophages and J774A.1 macrophage cells in culture, *Free Radical Biol. Med.*, 18, 577, 1995.

37. Bagchi, M. et al., Subchronic effects of smokeless tobacco extract (STE) on hepatic lipid peroxidation: DNA damage and excretion of urinary metabolites in rats, *Toxicology*, 127, 29, 1998.

38. Bagchi, M. et al., Chronic effects of smokeless tobacco extract on rat liver histopathology and production of HSP-90, *J. Environ. Pathol. Toxicol. Oncol.*, 14, 61, 1995.

39. Donati, Y.R., Slosman, D.O., and Polla, B.S., Oxidative injury and the heat shock response, *Biochem. Pharmacol.*, 15, 2571, 1990.

40. Bagchi, M., *In vitro* free radical production in human oral keratinocytes induced by smokeless tobacco extract, *In Vitro Toxicol.*, 10, 263, 1997.

41. Ohba, M., Production of hydrogen peroxide by transforming growth factor-beta 1 and its involvement in induction of egr-1 in mouse osteoblastic cells, *J. Cell Biol.*, 126, 1079, 1994.

42. Fox, J.W., Gyorfi, T., and Dragulev, B., Effect of smokeless tobacco extract on HT 1080 cell adhesion, pp 125[FAK] phosphorylation and apoptosis, *In Vitro Toxicol.*, 8, 339, 1995.

43. Brawn, M.K., Chiou, W.J., and Leach, K.L., Oxidant-induced activation of protein kinase C in UC11MG cells, *Free Radical Res.*, 22, 23, 1995.

44. Bagchi, D., *Helicobacter pylori*-induced oxidative stress and DNA damage in a primary culture of human gastric mucosal cells, *Dig. Dis. Sci.*, 47, 1405, 2002.

45. Bagchi, M. et al., Role of reactive oxygen species in the development of cytotoxicity with various forms of chewing tobacco and pan masala, *Toxicology*, 179, 247, 2002.

46. Telford, W.G., King, L.E., and Fraker, P.J., Evaluation of glucocorticoid-induced DNA fragmentation in mouse thymocytes by flow cytometry, *Cell Prolif.*, 24, 447, 1991.

47. Darzynkiewicz, Z., Features of apoptotic cells measured by flow cytometry, *Cytometry*, 13, 795, 1992.

48. Mosmann, T., Rapid colorimetric assay for cellular growth and survival: application to proliferation and cytotoxicity assays, *J. Immunol. Methods*, 55, 1983.

49. Gerlier, D. and Thomasset, N., Use of MTT colorimetric assay to measure cell activation, *J. Immunol. Methods*, 94, 57, 1986.

50. Bagchi, D. et al., Chromium- and cadmium-induced oxidative stress and apoptosis in cultured J774A.1 macrophage cell, *In Vitro Mol. Toxicol.*, 11, 171, 1998.

51. Bagchi, M. et al., Protective effects of antioxidants against smokeless tobacco-induced oxidative stress and modulation of Bcl-2 and p53 genes in human oral keratinocytes, *Free Radical Res.*, 35, 181, 2001.

52. Li, S.L. et al., Low p53 level in immortal non-tumorigenic oral keratinocytes harboring HPV-16 DNA, *Eur. J. Cancer Biol. Oral Oncol.*, 28B, 129, 1992.

53. Stohs, S.J., Bagchi, D., and Bagchi, M., Toxicity of trace elements in tobacco smoke, *Inhalation Toxicol.*, 9, 867, 1997.

54. Girja, K.P. et al., Biochemical changes of saliva in tobacco chewers, tobacco smokers, alcohol consumers, leukoplakia and oral cancer patients, *Indian J. Dent. Res.*, 13, 102, 2002.

55. Ernster, V.L. et al., Smokeless tobacco use and health effects among baseball players, *JAMA*, 264, 218, 1990.

56. Banoczy, J., Gintner, Z., and Dombi, C., Tobacco use and oral leukoplakia, *J. Dent. Educ.*, 65, 322, 2001.

57. Hockenbery, Z.N. et al., Bcl-2 functions in an antioxidant pathway to prevent apoptosis, *Cell*, 75, 241, 1993.

58. Lindemann, R.A. and Park, N.H., Inhibition of human lymphokine-activated killer activity by smokeless tobacco snuff extract, *Arch. Oral Biol.*, 33, 317, 1988.

8 The Regulatory Roles of Estrogen in Carcinogenesis: an Overview

Sushanta K. Banerjee, Aminul Islam, and Snigdha Banerjee

CONTENTS

8.1 Introduction..105
8.2 Roles of Estrogen in Tumorigenesis: from Tissue Culture to Humans 107
8.3 Roles of Estrogen in Angiogenesis..108
8.4 Roles of Estrogen in Controlling the Positive and Negative
    Regulators of Angiogenesis...109
8.5 Mechanism of Actions of Estrogen in Carcinogenesis:
    Genomic and Nongenomic Actions 110
8.6 Role of Coactivators and Cosuppressors in the Mediation of Estrogen Action 112
8.7 Phytoestrogens and Carcinogenesis 112
8.8 Conclusions and Future Directions...113
Acknowledgment..114
References 114

8.1 INTRODUCTION

Estrogens are the most important sex steroidal hormones in the female because, along with the other hormones, they play significant roles in the development of the female sex organs and secondary sex characters and in regulation of the menstrual cycle and pregnancy.[1,2] They are derived by the structural alteration of cholesterol to produce typically an 18-carbon steroid that has an unsaturated aromatic A ring with a phenolic hydroxyl group at C-3 and a methyl group at the C-13 position.[1] In the circulation, there are three distinct types of estrogens, termed E2 for estradiol-17β, E1 for estrone, and E3 for estriol. E2 and E1 are mainly secreted by follicular cells of mature Graafian follicles and are the most potent estrogens. E3 is present in large quantities during pregnancy and is mainly a metabolite of E2 with reduced activity.[1,3]

Estrogen not only plays a crucial role in the differentiation in fetal life[4] and maintaining the normal development and reproductive physiology, but also sustains the physiological balance in the maintenance of skeletal, cardiovascular, and central nervous systems throughout reproductive life.[5] Thus this reproductive hormone has some major nonreproductive sites of action. In fact, the physiological role of estrogen is particularly evident when Graafian follicles fail to secrete the precious hormone after menopause.

Normally, the biological actions of estrogen are regulated by its concentration in the serum and in the target tissue. In premenopausal women the source of estrogen is mainly the ovary, but in postmenopausal women it is mainly secreted by peripheral tissues, essentially adipose tissues. In the breast, the plasma/tissue ratio of E2 is 1:1 in premenopausal women, increasing to 1:10–50 in postmenopausal women.[3] Thus the risks of different cancers in postmenopausal women mainly owe to the tissue-specific production of estrogen in the absence of an ovarian source.[2] Many of the postmenopausal women become prone to osteoporosis, coronary artery diseases, cardiac stroke, and perhaps neurodegenerative diseases like Alzheimer's.[5,6] Despite the controversy as to the relationship of these diseases to lack of estrogen, many women continue to use estrogen therapy to maintain vigor and reduce these risk factors.[5,7–10] Moreover, the use of estrogen is not only restricted to diseases; natural or synthetic estrogens are the main component of the oral contraceptives that have been in use by large numbers of women for decades. The potential tragedy lies in the fact that this hormone, which is so important for normal reproductive life, has proved to be a major risk factor for the development of cancer in different estrogen-responsive organs, including breast, uterine endometrium, ovary, liver, lungs, and pituitary.[11,12]

Literature on this aspect reveals that estrogen is directly associated with the development of neoplastic processes in a variety of animal models, including rat pituitaries, hamster kidneys, and mammary tumors in virus-c-erbB-2 transgenic mice.[13–21] Therefore, estrogens are now referred to as "carcinogenic hormones." This unique property has drawn attention of the present-day research workers, not only because of the basic molecular biological mechanisms involved in these actions, but also because of immense public interest for its direct impact on women's health. The impetus for the growing interest mainly owes to the use of estrogens that ultimately devastate the balance of natural levels of estrogen in the body, which contributes significantly to the etiology of these endocrine-related cancers.

Estrogen exerts its effect through a multistep process involving initiation or promotion followed by progression to a series of steps in the carcinogenic processes.[2,16,22,23] Estrogens are regarded as promoters or initiators of the carcinogenic processes because of their well-recognized growth-promoting effects on target tissues by accelerating the rate of cell proliferation.[24] Stimulation of cell proliferation is mainly mediated through binding with estrogen receptors. However, there is evidence that metabolites, such as catechol estrogens, which are less estrogenic than E2, can also induce carcinogenesis by directly or indirectly damaging proteins, lipids, and DNA.[15,25]

Tumorigenesis is intimately associated with angiogenesis, a biological and pathobiological process of sprouting and configuring new blood vessels from preexisting blood vessels.[26] Normally, estrogen induces angiogenesis in the uterus during the menstrual cycle, which suggests its potentiality to induce neovascularization.[27–29] This is further evidenced by its critical role in disruption of uterine angiogenesis in estrogen-receptor knock-out mice.[30] The positive correlation between estrogen-receptor expression, angiogenic activity, and breast tumor invasiveness also supports the angiogenic role of estrogen.[31,32] Several additional observations suggest that estrogen, and other sex steroids, play positive roles in physiological and pathological angiogenesis.[13,29,33] Moreover, the inhibition of angiogenesis by antiestrogens provides further corroboration for its angiogenic roles.[34,35]

Humans are constantly exposed to an array of environmental chemicals that can disrupt the hormonal balance in the body. These chemicals are referred to as "endocrine disrupters" and include many pesticides such as DDT, lindane, and vinclozolin as well as industrial chemicals such as phthalates, bisphenol A, and alkylphenols. Pesticide residues can be found in foods; phthalates are in many PVC pipes; and bisphenol A is present in the linings of many food containers.[36–45] This relatively new area of investigation has drawn major attention to environmental endocrine science and has focused upon those chemicals that mimic the actions of estrogen. Many constituents of plants are known to potentiate or inhibit estrogenic effects.[45–53] The most commonly studied are the phytoestrogens, especially flavonoids, including isoflavones and flavones represented by genistein and luteolin, respectively. Leguminous plants like soybean and alfalfa contain flavonoids that are known to bind with estrogen receptor α (ERα) or ERβ and alter the transcription of estrogen-responsive genes.[46,54–57] These phytoestrogens are also reported to induce carcinogenesis in hamsters and mice.[20]

With a view to the multiphasic roles of estrogen in inducing carcinogenesis, this review attempts to focus on the molecular actions of estrogens for stimulation of multiple signaling cascades for carcinogenesis in estrogen-responsive cells.

8.2 ROLES OF ESTROGEN IN TUMORIGENESIS: FROM TISSUE CULTURE TO HUMANS

The role of 17β estradiol (E2) in the induction of tumorigenesis was first reported in the early 1940s by Lipschutz and Vargus for the guinea pig and by Gardner for the mouse.[58] Subsequently, many reports became available testifying to the carcinogenic role of estrogens *in vitro* and *in vivo*. Tsutsui and Barrett[59] have shown that acute exposure to estrogen is able to directly transform Syrian hamster embryo (SHE) fibroblasts. Successive studies from these authors have shown that catechol estrogens of E2 and estrone are also competent to transform SHE cells by forming DNA adducts,[60–62] which is also the parameter for determining genotoxic lesions and carcinogenic potentials of a chemical.[63,64]

Independent *in vivo* studies manifested that exogenous administration of estrogens caused mammary, pituitary, uterine, cervical, vaginal, testicular, lymphoid, and bone tumors in mice;[65–67] mammary or pituitary tumors in rats[13,68,69] and guinea pigs;[66] and renal tumors in hamsters.[18,25,70,71] Knock-out mice, which have their estrogen receptors genetically deleted, have been developed to aid in understanding the role of estrogen in regulation of tumorigenesis. Although defined mechanisms underlying estrogen carcinogenesis in various animal models have not yet been fully elucidated, decades of study have provided evidence that they involve interplay and cross talk between genotoxic and epigenetic alterations (Figure 8.1) potentiated by chronic exposure to estrogens.[21,25,72,73] For example, estrogens are able to form DNA adducts and chromosomal aberrations, including chromosome breaks and aneuploidy,[15–17,21,25,72,74] as well as to modulate DNA methylation during the early stages of estrogen-induced carcinogenesis in various target organs.[73,75–80] Although, there is no direct evidence indicating intimate association between these genetic errors and estrogen carcinogenesis, the development of malignant phenotypes is dependent upon somatic mutations and unscheduled cellular proliferation,[81] and both of these can be enhanced by genotoxic and epigenetic errors associated with estrogen exposure.

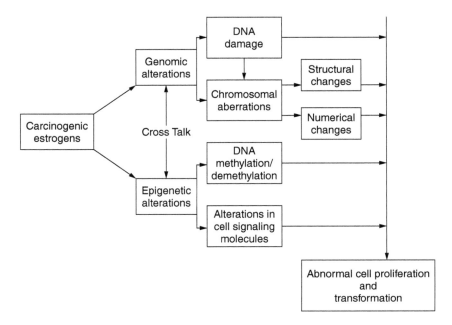

FIGURE 8.1 Multiple actions of estrogens on target tissues. Chronic exposure to carcinogenic estrogens enhances genotoxic and epigenetic alterations in target tissues and cells. It is anticipated that these two events are interdependent and cross-talk with each other to enhance the abnormal cell proliferation and carcinogenic process. This commentary discusses our current understanding of these interactions, along with some of the answered questions.

Because antiestrogens are able to block the estrogen-induced carcinogenesis, it is reasonable to believe that these genotoxic and epigenetic events may be mediated through estrogen receptors. Further studies are warranted to reconcile this issue.

Apart from these animal models, estrogen administration is considered to be a risk factor for carcinoma in different organs in humans, including uterine endometrium,[81] breast,[2,81] lungs,[82] and prostate.[83] Among all of these carcinomas, a role for endogenous estrogen as a risk factor for breast cancer has been among the most widely studied.[2] Increased concentrations of endogenous estrogens are strongly associated with an increased risk of breast cancer in postmenopausal women,[2,84] and clinical trials have indicated that anti-estrogen therapy can reduce this risk.[85] Exogenous estrogen alone or in combination with progestin can elevate the risk of breast cancer.[2] Constant use of oral contraceptives adds to the risk of uterine and breast cancer among women under the age of 45 years. Regular use of oral contraceptives for 10 years increases the risk of breast cancer by 1.36% per year in comparison with nonusers.[24] Such a relationship between estradiol and breast cancer has posed extreme problems in hormone-replacement therapy in postmenopausal women.

Breast cancer risk in the presence of endogenous estrogen is determined by several factors, such as early menarche, late menopause, and obesity in postmenopausal women.[2] Epidemiological studies have shown that despite a weak relationship, relative risk is about 1.2 for women in whom menarche occurred before the age of 12 when compared with menarche at age 14. This is probably because early menarche exposes the breast epithelium to more endogenous estrogen from regular ovulatory cycles,[11,12,81] and this increases the rate of proliferation of these cells. Late menopause could pose a greater risk of breast cancer because these women would have a greater number of ovulatory cycles and accumulate a greater exposure to estrogen during the whole reproductive period.[11] Obese postmenopausal women also show a higher risk to breast cancer because of higher levels of circulating estrogens, due to conversion of androgen to estrogen in adipose tissue.[86]

Nulliparity and late age of first child delivery both increase the risk of breast cancer. The risk is about 50% less in women having first delivery before the age of 20 compared with those with first delivery at age 30. Subsequent births, especially at an early age, further reduce the risk.[87,88] Lactation has been shown to be inversely related to breast cancer risk, presumably because it delays the reestablishment of the ovulatory cycles.[89] Women who occasionally used the synthetic estrogen diethylstilbestrol during pregnancy to reduce the risk of miscarriage have been reported to have an increased risk of breast cancer.[90–93] Those mothers who have a dizygotic twin pregnancy have been shown to have a higher risk of breast cancer when compared with those with a normal single-baby pregnancy, presumably because the former have higher levels of circulating estrogen.

Taken together, the data clearly indicate an association between estrogen exposure and carcinogenesis in target organs. Therefore, estrogen is now considered to be a potent carcinogen in animals as well as in humans. However, defining the mechanisms of carcinogenesis in human target tissues has proved elusive, and it is not yet known whether similar events are required to develop cancer in human as those found in animal models. The *in vitro* studies from various laboratories, including ours, have shown that estrogen modulates several oncogenes/tumor suppressor genes, and signaling pathways in human tumor cell lines.[94–99] These modulations are likely to be associated with cell proliferation and an increased opportunity for genetic errors and transformation.

8.3 ROLES OF ESTROGEN IN ANGIOGENESIS

Angiogenesis is a multistep process by which new blood vessels are formed from existing blood vessels.[26–100] It is a highly regulated process and essential for reproduction, morphogenesis, development, and wound repair.[101,102] Angiogenesis is also critical for patients with coronary artery diseases in order to maintain blood flow to bypass the blocked coronary artery and protect from heart ischemia and attacks. Uncontrolled and persistent changes in the angiogenic process "switch" can occur at different stages of tumor progression, depending on tumor type and the surrounding environment.[103] Physiological and pathophysiological angiogenesis can be regulated by various positive and negative regulators, including vascular endothelial growth factor (VEGF), fibroblast growth factors (FGFs), platelet-derived growth factors (PDGFs), epidermal growth factors (EGFs), thrombospondin-1 (TSP-1) and various types of statins, and inhibitors of HMG-Co and A-reductase.[103] Although, the molecular mechanisms associated with physiological and pathophysiological angiogenesis have not been fully elucidated, it is assumed that angiogenesis is the result of imbalanced expression and function of positive and negative regulators.[103–105] The angiogenic switch induced by imbalanced functions of positive and negative regulators can be modulated by various internal and external stimuli.[106]

In healthy adults, angiogenesis is a quiescent process, with the exception that in the female reproductive tract, this event is modulated in each menstrual cycle in response to the estrogen level in the blood. In premenopausal women, estrogen promotes angiogenesis in association with some diseases like Takayasu's arteritis and lupus erythematous.[27] Together, the studies indicate that angiogenesis can be regulated by estrogen. A role for estrogen in angiogenesis is further supported by our studies on the rat pituitary that indicated estrogen as the sole etiological agent for the development of tumor angiogenesis during the development of estrogen-induced tumorigenesis.[13] Some additional observations, such as inhibition of angiogenesis by antiestrogens, also support estrogen as a positive regulator of angiogenesis.[107–110]

8.4 ROLES OF ESTROGEN IN CONTROLLING THE POSITIVE AND NEGATIVE REGULATORS OF ANGIOGENESIS

Changes in the relative balance of positive and negative regulators are well-accepted manifestations of the angiogenic "switch," as discussed in the previous section. As normal cells proceed toward malignancy, they become increasingly angiogenic by either decreasing the secretion of negative regulators or by increasing the production of one or more positive regulators.[103–105] Our studies, as well as those of others, have (1) shown that VEGF-A, a positive regulator of angiogenesis, can be regulated by the natural estrogen 17β-estradiol in rat pituitaries, GH3 rat pituitary tumor cells, human breast tumor cells, and in human endocrine-related tumors via estrogen-receptor-mediated pathways,[13,111–117] and (2) suggested that estrogen augments angiogenesis in target tissues by enhancing expression and activities of VEGF-A. Moreover, this concept was further supported by our *in vivo* and *in vitro* studies, which have shown that an antitumor and antiangiogenic agent, 2-methoxyestradiol, is able to suppress tumor growth and tumor angiogenesis in rat pituitaries induced by chronic exposure of estrogen through the inhibition of VEGF-A.[118,119] In addition to its effect on VEGF, estradiol has been shown to up-regulate a variety of molecules associated with the positive control of switching of tumor angiogenesis during the development of tumorigenesis.[106,120–124] Endothelial cells have been shown to express adhesion molecules, including E-selectin. Despite some controversy, it is assumed that the changes in expression patterns of adhesion molecules and matrix proteins have a strong correlation with the level of estrogen that ultimately influences the angiogenic switch.[125–128]

Very little is known about a role for estrogen in controlling negative regulators, or angioinhibitory factors, in tumorigenesis. Recent studies from our laboratory have shown that the expression of thrombospondin-1 (TSP-1), a negative regulator of angiogenesis, is down-regulated by estrogen in endothelial or tumor cells. Reduced expression of TSP-1 by estrogen is associated with estrogen-induced endothelial cell proliferation.[129] The proliferative effect of estrogen can also be partially diminished by the addition of recombinant TSP-1 protein. A similar effect was found during the development of estrogen-induced pituitary tumors in Fisher 344 rats.[129] Taken together, these studies

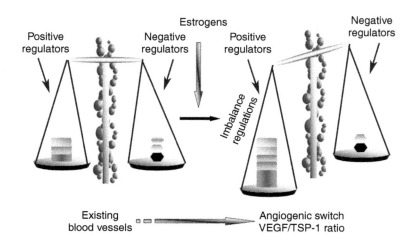

FIGURE 8.2 Estrogen-dependent activation of angiogenic switch. Imbalance of positive and negative regulators of angiogenesis is the crucial event in induction of physiological and pathophysiological angiogenesis. Estrogen has been considered as an angiogenesis inducer because it activates the angiogenic switch by enhancing the imbalance regulation of positive and negative regulators in both tumor cells and endothelial cells.

indicate that estrogens have the ability to perturb the angiogenic balance by activating both positive regulators, such as VEGF-A, as well as by inhibiting a negative regulator like TSP-1 through complex mechanism(s) (Figure 8.2). This perturbation eventually mediates the angiogenic "switch."

8.5 MECHANISM OF ACTIONS OF ESTROGEN IN CARCINOGENESIS: GENOMIC AND NONGENOMIC ACTIONS

How are estrogens involved in carcinogenesis? Despite several decades of investigation, an absolute answer for this question is unavailable. From various studies, it is obvious that "estrogen carcinogenesis" is a complex phenomenon involving multiple pathways, and these have proved to be very elusive. Petrangeli et al.[130] proposed that the imbalance between proliferative and differentiative estrogenic actions, which eventually causes an alteration in estrogen receptor expression and activities, may play a determinant role in mammary carcinogenesis. Subsequent studies came to similar conclusions and proposed that the estrogen receptor may be involved in this multifactorial and multi-event process.[131] The biological effects of 17β-estradiol (E2) are normally mediated by two distinct intracellular receptors, estrogen receptor α and β (ERα and ERβ), each encoded by separate genes, though they have some common domains.[132] Consequently, they exhibit similar binding-affinity profiles for natural and synthetic estrogens. The two receptors may play redundant roles in estrogen signaling, though tissue-localization studies have revealed distinct expression patterns. Both ERα and ERβ show close functional homology in the DNA binding domain (ligand/AF-2), but they exhibit less homology in the ligand-independent activation domain (AF-1). Both receptors are necessary to mediate unique biological actions in mammalian cell culture and show direct convergence in their pathways, though ERβ interacts in a ligand-independent manner and can attenuate the ligand-activated transcriptional activity of ERα.[133] Thus in cells where both receptors are expressed, the overall estrogen responsiveness is determined by the ERα:ERβ ratio. It is known that the two isoforms of ERβ, ERβcx and ERβ2, heterodimerize and inhibit the activity of ERα, which suggests a modulatory role for ERβ.[132]

Estrogens activate the target organs by two alternative pathways.[133,134] Like other steroid hormones, E2 usually binds to ERα or ERβ and makes an active E+ER complex in the nucleus. This active complex eventually binds DNA directly at classical inverse palindromic or nonclassical

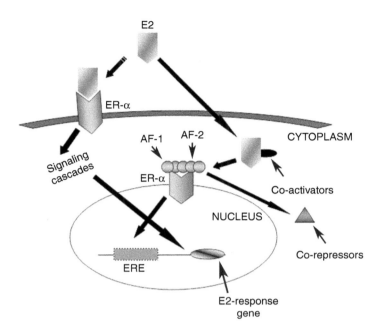

FIGURE 8.3 Mechanism of genomic and nongenomic estrogen-receptor-dependent action of estrogen in target cells. Estrogens interact with either membrane receptor or nuclear receptor and subsequently activate estrogen-response genes. Classically, estrogen, in association with co-activator, binds with the AF-2 ligand binding domain of nuclear receptor and forms an estrogen-receptor complex, which eventually interacts with estrogen response element (ERE) in the target genes and makes those genes transcriptionally active. Alternatively, estrogen interacts with membrane receptors and forms estrogen-receptor complexes, which sequentially activate signaling molecules in the cytoplasm, resulting in transcriptional activation of estrogen-response genes.

estrogen-response elements (ERE) within the promoter of the target genes, or it endorses transcription factor binding to DNA to make the genes transcriptionally active or inactive for specific functions (see Figure 8.3).[135–136] Alternatively, E2–ER complexes transactivate several genes by protein–protein interactions, and this transactivation is differently regulated by two different receptors.[133] This classical cell-signaling mechanism has been considered its "genomic action," and it can take hours or more to modulate specific genes. Estrogens may also bind with a small population of receptors on the plasma membrane of the target cells, from which signaling spreads through the cytoplasm, where it leads to the posttranslational modification of important structural and functional proteins in the cell. This is known as its "nongenomic action" (see the Figure 8.3).[135,137,138] This cell-signaling mechanism is rapid and takes minutes to achieve its final downstream effects.[139] Although two alternative pathways of estrogen action through its receptors have received much attention in endocrine research, it is still debatable whether the carcinogenic actions of estrogen that include cell proliferation and genomic and nongenomic alterations are explainable by these pathways.

Some biological actions of E2 in bone, breast, nervous system, and other tissues are reported to be mediated by cell surface receptors that are linked with intracellular signal-transduction proteins.[140–142] In a variety of cell types, E2 has been shown to activate mitogen-activated protein kinase (MAPK) signaling pathways, which point out a close linkage of ER with membrane-coupled tyrosine kinase pathways.[133,135] The membrane originating multiple signaling pathways is rapidly stimulated and becomes linked with the ERα and ERβ transcription factors, depending upon the cell type.[133] In breast tumor, a specific protein, MTA-1, is reported to keep the ER away from the nucleus to (1) strongly reduce ER-activated transcription and (2) promote increased MAPK (ERK) signaling, which is assumed to be responsible for the aggressive behavior of the tumor. Thus, by the nature of membrane-bound receptors and their signaling capacity, they actually regulate the physiological processes of the tumor.[143] Here E2 acts as a cell survival factor by preventing chemotherapy-induced activation of the tyrosine kinase JNK. JNK activates phosphorylation of the tumor suppressors Bcl-2/Bcl-XL, and as a result apoptosis or cell death fails to follow. In this way, E2 allows survival of the breast cancer cells and enhances malignancy.[142] In endothelial cells (EC), E2 activates the p38 MAPKAP-2 kinase pathway. MAPKAP-2 phosphorylates and modifies the functions of heat-shock protein-27, which leads to the morphological preservation and survival of EC and stimulates primitive capillary tube formation.

8.6 ROLE OF COACTIVATORS AND COSUPPRESSORS IN THE MEDIATION OF ESTROGEN ACTION

As described in an earlier section, estrogen induces its physiological action by binding with specific receptors in the plasma membrane and eventually triggering multiple gene activations to carry out cell proliferation and differentiation. It is now established that estrogen receptors themselves cannot carry out the transcription alone, but need an interaction with a complex of coregulatory proteins called coactivators or corepressors that act as signaling intermediates between ER and gene transcription machinery.[133] The replacement of cosuppressor and simultaneous binding of coactivator and ligand to the hormone-dependent activation domain (AF-2) of ER is essential in this transcription machinery.

On the basis of their ability to interact with ER, several coactivators are known. Of these proteins, the most notable are ERAP160, RIP140, and SRC1. RIP 140 interacts with the ligand binding domain (LBD) of ERα. SRC1 (also called p160/NcoA-1/ERAP-160), SRC2 (TIF2/GRIP-1), and SRC-3 (AIBI/RAC-3/TRAM-1) are not only transcription mediators for ligand-dependent AF2 of ER, but they are also involved in ligand-independent interaction with activating factor-1 (AF-1) of ERα and with ERβ through phosphorylation of its AF-1 via MAPK. The SRC-1 complex also contains SRA, which mediates transactivation via AF-1. ERα/ERβ can also activate gene transcription via SRC 1 (128). Another protein remaining in combination with SRC1 is p300/CBP. SRC1 and p300/CBP contain intrinsic acetyl transferase activity and can interact with other histone acetyl tranferases. Acetylation of histone brings about the activation of the transcription of specific genes by AF-1 and AF-2 of ER. The specific peptides of SRC1 are known to interact with ERα and ERβ.[144] Another coactivator is the estrogen-receptor-related receptors (ERRs). These are of three types, α, β, and γ. They identify with both ERs and exhibit similar, but distinct, biochemical and transcriptional activities as ERα and ERβ. However, ERRs do not bind natural estrogen but, rather, interact with transcriptional coactivators inside the cell in the absence of ligands.[145]

8.7 PHYTOESTROGENS AND CARCINOGENESIS

Almost all animals are exposed to an array of environmental chemicals, some of which can mimic the activities of endogenous hormones. Different plant products and synthetic compounds are known to have similar chemical compositions and act as estrogens.[146] These estrogenlike compounds (so-called phytoestrogens), under laboratory conditions or in the natural environment, might be capable of causing a spectrum of adverse effects on the endocrine system that eventually cause several disorders, including tumorigenesis.[147–150] Many in vitro studies using breast cancer cells indicate that phytoestrogens alone imitate the estrogenic effect and induce estrogen-dependent protein (pS2) and cell proliferation in ER-positive MCF-7 cells.[146] However, in the presence of estrogen, the effect of phytoestrogens was reversed and the effect was not ER dependent.[146] This issue remains a source of debate.[151]

Despite the adverse effects of naturally occurring phytoestrogens, these compounds also exhibit beneficial effects.[146] Both epidemiological and experimental studies have shown that the consumption of phytoestrogen-rich diets may reduced the risk of cardiovascular diseases, meno-pausal disorder, osteoporosis, as well as breast and prostate cancers,[146,152–15] suggesting that their use may be worthwhile.[155] There are two principal categories of phytoestrogens, isoflavones and lignanes. Isoflavones and flavones, represented by genistein, luteolin, and daidzein, are produced by leguminous plants (soybean, alfalfa), while lignanes are found in high concentration in linseed.[146] The commonly available lignanes are enterolactone and enterodiol. These phytoestro-gens have been reported to be weak estrogen-receptor agonists in the uterus and vagina of female rats.[155] However, more genistein is bound to ERβ than to ERα, so it could have an antiestrogen effect in systems with higher levels of ERβ than ERα. It shows an inhibitory effect on estrogen-mediated endometrial carcinogenesis in mice, possibly by suppressing the estrogen-responsive genes, such as *c-fos* and *c-jun* and the internal cytokines IL-1a and TNFα.[156]

From the results of several independent studies, flavonoids could be considered valuable as anticancer therapeutic agents. With the use of soy phytoestrogens and the avoidance of hormonal replacement therapy (HRT), encouraging results have been obtained with breast cancer survivors. Furthermore, administration of phytoestrogens to postmenopausal women, along with the anti-estrogen tamoxifen, has provided encouraging results. Dietary phytophenols are now also con-sidered to be anticarcinogenic. Tea, isoflavones, catechins, quercetin, resveratrol, and lignan are known to inhibit carcinogenesis by affecting estrogen-related activities.[157–161] Consumption of soy milk substantially lowered serum estrogen levels, which was assumed to reduce breast cancer risk.[146]

Together, these studies indicate that the phytoestrogens exhibit a complex biphasic effect on the development of carcinogenesis. These compounds have both stimulatory and chemopreventive actions on mammary tumor cells. Although, the pathways of their actions are not clear, some researchers predict that timing of exposure will prove to play a critical role in determining the positive or negative actions of these compounds. The protective effect of genistein on chemically (DMBA) induced breast tumorigenesis is more pronounced when rats are exposed to it during the neonatal or prepubertal period compared with exposure during a later period of life, suggesting an important developmental effect.[154,162]

8.8 CONCLUSIONS AND FUTURE DIRECTIONS

Multiple lines of evidence indicate that estrogens can be potent carcinogens in animals as well as in humans, and estrogens are directly associated with tumorigenesis and tumor angiogenesis in various models. The basis for this appears to be their ability to induce many genetic and epigenetic alterations, many of which could provide normal cells with advantages in terms of increased rates of proliferation, which ultimately facilitate transformations. Moreover, estrogens are capable of inducing the angiogenesis needed for tumor survival by modulating both positive and negative regulator of angiogenesis. In our opinion, therefore, estrogens not only promote tumorigenesis by helping the tumor cells proliferate, but also by nourishing the tumor cells through formation of new blood vessels. Many questions remain to be answered with regard to the mechanisms of estrogen involvement in carcinogenesis, and particularly the role of phytoestrogens in these pro-cesses. The fundamental question remains as to how estrogen actually promotes tumorigenesis and tumor angiogenesis *in vivo*. Which estrogen receptors play essential roles in potentiating estrogen-induced tumorigenesis and tumor angiogenesis? Are supplementary pathways also involved? What molecular pathways are recruited or compromised by carcinogenic estrogens for these events to occur?

Like natural and synthetic estrogens, the phytoestrogens that are abundantly found in nature are also associated with tumor growth. Ironically, these phytoestrogens can also exhibit enormous preventive roles in tumor growth and proliferation. However, the modes of biphasic actions of these compounds remain elusive, raising several questions regarding the benefit of these compounds.

Translation of this basic information into a nutritional preventive protocol for the preclusion of certain cancers requires further intensive studies and an intensive multidisciplinary approach.

ACKNOWLEDGMENT

This work is supported by NIH/NCI CA87680 (SKB); a V.A. Merit Review grant, the Midwest Biomedical Research Foundation grant, NIH COBRE award P20 RR15563 (SB), and a University of Kansas Medical Center Departmental Research Award grant. We are extremely grateful to members of the Cancer Research Unit and Professor Donald Johnson, Ph.D., University of Kansas Medical Center, for their helpful discussions and comments on the manuscript.

REFERENCES

1. Norman, A.W. and Litwack, G., *Estrogens and Progestins, Hormones*, Academic Press, New York, 1997, pp. 361–386.
2. Travis, R.C. and Key, T.J., Oestrogen exposure and breast cancer risk, *Breast Cancer Res.*, 5, 239–247, 2003.
3. Parl, F.F., *Estrogen Synthesis and Metabolism: Estrogens, Estrogen Receptor and Breast Cancer*, IOS Press, Washington, DC, 2002, pp. 21–56.
4. Beato, M., Gene regulation by steroid hormones, *Cell*, 56, 335–344, 1989.
5. Manolagas, S.C. and Kousteni, S., Perspective: nonreproductive sites of action of reproductive hormones, *Endocrinology*, 142, 2200–2204, 2001.
6. Kousteni, S., Bellido, T., Plotkin, L.I., O'Brien, C.A., Bodenner, D.L., Han. L., Han, K., DiGregorio, G.B., Katzenellenbogen, J.A., Katzenellenbogen, B.S., Roberson, P.K., Weinstein, R.S., Jilka, R.L., and Manolagas, S.C., Nongenotropic, sex-nonspecific signaling through the estrogen or androgen receptors: dissociation from transcriptional activity, *Cell*, 104, 719–730, 2001.
7. Norbury, R., Cutter, W.J., Compton, J., Robertson, D.M., Craig, M., Whitehead, M., and Murphy, D.G., The neuroprotective effects of estrogen on the aging brain, *Exp. Gerontol.*, 38, 109–117, 2003.
8. Hodis, H.N., Mack, W.J., Azen, S.P., Lobo, R.A., Shoupe, D., Mahrer, P.R., Faxon, D.P., Cashin-Hemphill, L., Sanmarco, M.E., French, W.J., Shook, T.L., Gaarder, T.D., Mehra, A.O., Rabbani, R., Sevanian, A., Shil, A.B., Torres, M., Vogelbach, K.H., and Selzer, R.H., Hormone therapy and the progression of coronary-artery atherosclerosis in postmenopausal women, *New England J. Med.*, 349, 535–545, 2003.
9. Hodis, H.N., Mack, W.J., and Lobo, R., What is the cardioprotective role of hormone replacement therapy? *Curr. Atheroscler. Rep.*, 5, 56–66, 2003.
10. Mack, W.J., Hameed, A.B., Xiang, M., Roy, S., Slater, C.C., Stanczyk, F.Z., Lobo, R.A., Liu, C., Liu, C., and Hodis, H.N., Does elevated body mass modify the influence of postmenopausal estrogen replacement on atherosclerosis progression: results from the estrogen in the prevention of atherosclerosis trial, *Atherosclerosis*, 168, 91–98, 2003.
11. Henderson, B.E., Ross, R.K., Lobo, R.A., Pike, M.C., and Mack, T.M., Re-evaluating the role of progestogen therapy after the menopause, *Fertil. Steril.*, 49, 9S–15S, 1988.
12. Henderson, B.E., Ross, R., and Bernstein, L., Estrogens as a cause of human cancer: the Richard and Hinda Rosenthal Foundation award lecture, *Cancer Res.*, 48, 246–253, 1988.
13. Banerjee, S.K., Sarkar, D.K., Weston, A.P., De, A., and Campbell, D.R., Over expression of vascular endothelial growth factor and its receptor during the development of estrogen-induced rat pituitary tumors may mediate estrogen-initiated tumor angiogenesis, *Carcinogenesis*, 18, 1155–1161, 1997.
14. Li, J.J., Gonzalez, A., Banerjee, S., Banerjee, S.K., and Li, S.A., Estrogen carcinogenesis in the hamster kidney: role of cytotoxicity and cell proliferation, *Environ. Health Perspect.*, 101 (Suppl. 5), 259–264, 1993.
15. Liehr, J.G., Is estradiol a genotoxic mutagenic carcinogen? *Endocr. Rev.*, 21, 40–54, 2000.
16. Yager, J.D. and Liehr, J.G., Molecular mechanisms of estrogen carcinogenesis, *Annu. Rev. Pharmacol. Toxicol.*, 36, 203–232, 1996.

17. Liehr, J.G., Genotoxic effects of estrogens, *Mutat. Res.*, 238, 269–276, 1990.
18. Roy, D. and Liehr, J.G., Inhibition of estrogen-induced kidney carcinogenesis in Syrian hamsters by modulators of estrogen metabolism, *Carcinogenesis*, 11, 567–570, 1990.
19. Li, J.J. and Li, S.A., Estrogen carcinogenesis in the hamster kidney: a hormone-driven multistep process, *Prog. Clin. Biol. Res.*, 394, 255–267, 1996.
20. Li, J.J. and Li, S.A., Estrogen carcinogenesis in hamster tissues: a critical review, *Endocr. Rev.*, 11, 524–531, 1990.
21. Cavalieri, E., Frenkel, K., Liehr, J.G., Rogan, E., and Roy, D., Estrogens as endogenous genotoxic agents: DNA adducts and mutations, *J. Natl. Cancer Inst. Monogr.*, 75–94, 2000.
22. Moolgavkar, S.H., Carcinogenesis modeling: from molecular biology to epidemiology, *Annu. Rev. Public Health*, 7, 151–169, 1986.
23. Moolgavkar, S.H., Hormones and multistage carcinogenesis, *Cancer Surv.*, 5, 635–648, 1986.
24. Key, T.J. and Pike, M.C., The role of oestrogens and progestagens in the epidemiology and prevention of breast cancer, *Eur. J. Cancer Clin. Oncol.*, 24, 29–43, 1988.
25. Banerjee, S.K., Banerjee, S., Li, S.A., and Li, J.J., Induction of chromosome aberrations in Syrian hamster renal cortical cells by various estrogens, *Mutat. Res.*, 311, 191–197, 1994.
26. Browder, T., Folkman, J., and Pirie-Shepherd, S., The hemostatic system as a regulator of angiogenesis, *J. Biol. Chem.*, 275, 1521–1524, 2000.
27. Shelhamer, J.H., Volkman, D.J., Parrillo, J.E., Lawley, T.J., Johnston, M.R., and Fauci, A.S., Takayasu's arteritis and its therapy, *Ann. Intern. Med.*, 103, 121–126, 1985.
28. Ansar, A.S., Penhale, W.J., and Talal, N., Sex hormones, immune responses, and autoimmune diseases: mechanisms of sex hormone action, *Am. J. Pathol.*, 121, 531–551, 1985.
29. Cullinan-Bove, K. and Koos, R.D., Vascular endothelial growth factor/vascular permeability factor expression in the rat uterus: rapid stimulation by estrogen correlates with estrogen-induced increases in uterine capillary permeability and growth, *Endocrinology*, 133, 829–837, 1993.
30. Johns, A., Freay, A.D., Fraser, W., Korach, K.S., and Rubanyi, G.M., Disruption of estrogen receptor gene prevents 17 beta estradiol-induced angiogenesis in transgenic mice, *Endocrinology*, 137, 4511–4513, 1996.
31. Toi, M., Kashitani, J., and Tominaga, T., Tumor angiogenesis is an independent prognostic indicator in primary breast carcinoma, *Int. J. Cancer*, 55, 371–374, 1993.
32. Vartanian, R.K. and Weidner, N., Correlation of intratumoral endothelial cell proliferation with microvessel density (tumor angiogenesis) and tumor cell proliferation in breast carcinoma, *Am. J. Pathol.*, 144, 1188–1194, 1994.
33. Banerjee, S.K., Campbell, D.R., Weston, A.P., and Banerjee, D.K., Biphasic estrogen response on bovine adrenal medulla capillary endothelial cell adhesion, proliferation and tube formation, *Mol. Cell. Biochem.*, 177, 97–105, 1997.
34. Iruela-Arispe, M.L., Porter, P., Bornstein, P., and Sage, E.H., Thrombospondin-1, an inhibitor of angiogenesis, is regulated by progesterone in the human endometrium, *J. Clin. Invest.*, 97, 403–412, 1996.
35. Gagliardi, A.R., Hennig, B., and Collins, D.C., Antiestrogens inhibit endothelial cell growth stimulated by angiogenic growth factors, *Anticancer Res.*, 16, 1101–1106, 1996.
36. Fox, J.E., Starcevic, M., Kow, K.Y., Burow, M.E., and McLachlan, J.A., Nitrogen fixation: endocrine disrupters and flavonoid signalling, *Nature*, 413, 128–129, 2001.
37. Welshons, W.V., Thayer, K.A., Judy, B.M., Taylor, J.A., Curran, E.M., and Vom Saal, F.S., Large effects from small exposures, I: mechanisms for endocrine-disrupting chemicals with estrogenic activity, *Environ. Health Perspect.*, 111, 994–1006, 2003.
38. Daston, G.P., Cook, J.C., and Kavlock, R.J., Uncertainties for endocrine disrupters: our view on progress, *Toxicol. Sci.*, 74, 245–252, 2003.
39. Buchner, V., Environmental endocrine disrupting chemicals, *Rev. Environ. Health*, 17, 249–252, 2002.
40. Singleton, D.W. and Khan, S.A., Xenoestrogen exposure and mechanisms of endocrine disruption, *Front Biosci.*, 8, s110–s118, 2003.
41. Quesada, I., Fuentes, E., Viso-Leon, M.C., Soria, B., Ripoll, C., and Nadal, A., Low doses of the endocrine disruptor bisphenol-A and the native hormone 17beta-estradiol rapidly activate transcription factor CREB, *FASEB J.*, 16, 1671–1673, 2002.
42. Takeshita, A., Koibuchi, N., Oka, J., Taguchi, M., Shishiba, Y., and Ozawa, Y., Bisphenol A, an environmental estrogen, activates the human orphan nuclear receptor, steroid and xenobiotic receptor-mediated transcription, *Eur. J. Endocrinol.*, 145, 513–517, 2001.

43. Gore, A.C., Environmental toxicant effects on neuroendocrine function, *Endocrine*, 14, 235–246, 2001.
44. Janssens, J.P., Van Hecke, E., Geys, H., Bruckers, L., Renard, D., and Molenberghs, G., Pesticides and mortality from hormone-dependent cancers, *Eur. J. Cancer Prev.*, 10, 459–467, 2001.
45. Cooper, R.L. and Kavlock, R.J., Endocrine disruptors and reproductive development: a weight-of-evidence overview, *J. Endocrinol.*, 152, 159–166, 1997.
46. Safe, S.H., Pallaroni, L., Yoon, K., Gaido, K., Ross, S., and McDonnell, D., Problems for risk assessment of endocrine-active estrogenic compounds, *Environ. Health Perspect.*, 110 (Suppl. 6), 925–929, 2002.
47. Horn-Ross, P.L., John, E.M., Canchola, A.J., Stewart, S.L., and Lee, M.M., Phytoestrogen intake and endometrial cancer risk, *J. Natl. Cancer Inst.*, 95, 1158–1164, 2003.
48. Watanabe, S., Uesugi, S., Zhuo, X., and Kimira, M., Phytoestrogen and cancer prevention, *Gan To Kagaku Ryoho*, 30, 902–908, 2003.
49. Tsutsui, T., Tamura, Y., Yagi, E., Someya, H., Hori, I., Metzler, M., and Barrett, J.C., Cell-transforming activity and mutagenicity of 5 phytoestrogens in cultured mammalian cells, *Int. J. Cancer*, 105, 312–320, 2003.
50. Adlercreutz, H., Phytoestrogens and breast cancer, *J. Steroid Biochem. Mol. Biol.*, 83, 113–118, 2002.
51. Mishra, S.I., Dickerson, V., and Najm, W., Phytoestrogens and breast cancer prevention: what is the evidence? *Am. J. Obstet. Gynecol.*, 188, S66–S70, 2003.
52. Albertazzi, P. and Purdie, D., The nature and utility of the phytoestrogens: a review of the evidence, *Maturitas*, 42, 173–185, 2002.
53. Chen, W.F., Huang, M.H., Tzang, C.H., Yang, M., and Wong, M.S., Inhibitory actions of genistein in human breast cancer (MCF-7) cells, *Biochim. Biophys. Acta*, 1638, 187–196, 2003.
54. Fox, J.E., Burow, M.E., and McLachlan, J.A., Symbiotic gene activation is interrupted by endocrine disrupting chemicals, *Scientific World J.*, 1, 653–655, 2001.
55. Levenson, A.S., Gehm, B.D., Pearce, S.T., Horiguchi, J., Simons, L.A., Ward, III, J.E., Jameson, J.L., and Jordan, V.C., Resveratrol acts as an estrogen receptor (ER) agonist in breast cancer cells stably transfected with ER alpha, *Int. J. Cancer*, 104, 587–596, 2003.
56. Mueller, S.O., Kling, M., Arifin, F.P., Mecky, A., Duranti, E., Shields-Botella, J., Delansorne, R., Broschard, T., and Kramer, P.J., Activation of estrogen receptor alpha and ERbeta by 4-methylbenzylidene-camphor in human and rat cells: comparison with phyto- and xenoestrogens, *Toxicol. Lett.*, 142, 89–101, 2003.
57. Rickard, D.J., Monroe, D.G., Ruesink, T.J., Khosla, S., Riggs, B.L., and Spelsberg, T.C., Phytoestrogen genistein acts as an estrogen agonist on human osteoblastic cells through estrogen receptors alpha and beta, *J. Cell. Biochem.*, 89, 633–646, 2003.
58. IARC, Monographs on the Evaluation of Carcinogenic Risks to Human: Hormonal Contraception and Post Menopausal Hormone Therapy, International Agency for Research on Cancer, Lyons, France, 1999.
59. Tsutsui, T. and Barrett, J.C., Neoplastic transformation of cultured mammalian cells by estrogens and estrogenlike chemicals, *Environ. Health Perspect.*, 105 (Suppl. 3), 619–624, 1997.
60. Yagi, E., Barrett, J.C., and Tsutsui, T., The ability of four catechol estrogens of 17beta-estradiol and estrone to induce DNA adducts in Syrian hamster embryo fibroblasts, *Carcinogenesis*, 22, 1505–1510, 2001.
61. Tsutsui, T., Tamura, Y., Yagi, E., and Barrett, J.C., Involvement of genotoxic effects in the initiation of estrogen-induced cellular transformation: studies using Syrian hamster embryo cells treated with 17beta-estradiol and eight of its metabolites, *Int. J. Cancer*, 86, 8–14, 2000.
62. Tsutsui, T., Tamura, Y., Hagiwara, M., Miyachi, T., Hikiba, H., Kubo, C., and Barrett, J.C., Induction of mammalian cell transformation and genotoxicity by 2-methoxyestradiol, an endogenous metabolite of estrogen, *Carcinogenesis*, 21, 735–740, 2000.
63. Kyrtopoulos, S.A., Anderson, L.M., Chhabra, S.K., Souliotis, V.L., Pletsa, V., Valavanis, C., and Georgiadis, P., DNA adducts and the mechanism of carcinogenesis and cytotoxicity of methylating agents of environmental and clinical significance, *Cancer Detection Prev.*, 21, 391–405, 1997.
64. Kyrtopoulos, S.A., DNA adducts in humans after exposure to methylating agents, *Mutat. Res.*, 405, 135–143, 1998.
65. Huseby, R.A., Demonstration of a direct carcinogenic effect of estradiol on Leydig cells of the mouse, *Cancer Res.*, 40, 1006–1013, 1980.

66. Highman, B., Roth, S.I., and Greenman, D.L., Osseous changes and osteosarcomas in mice continuously fed diets containing diethylstilbestrol or 17beta-estradiol, *J. Natl. Cancer Inst.*, 67, 653–662, 1981.

67. Son, D.S., Roby, K.F., Rozman, K.K., and Terranova, P.F., Estradiol enhances and estriol inhibits the expression of CYP1A1 induced by 2,3,7,8-tetrachlorodibenzo-p-dioxin in a mouse ovarian cancer cell line, *Toxicology*, 176, 229–243, 2002.

68. Han, B.S., Fukamachi, K., Takasuka, N., Ohnishi, T., Maeda, M., Yamasaki, T., and Tsuda, H., Inhibitory effects of 17beta-estradiol and 4-n-octylphenol on 7,12-dimethylbenz[a]anthracene-induced mammary tumor development in human c-Ha-ras proto-oncogene transgenic rats, *Carcinogenesis*, 23, 1209–1215, 2002.

69. Shull, J.D., Spady, T.J., Snyder, M.C., Johansson, S.L., and Pennington, K.L., Ovary-intact, but not ovariectomized female ACI rats treated with 17beta-estradiol rapidly develop mammary carcinoma, *Carcinogenesis*, 18, 1595–1601, 1997.

70. Liehr, J.G., Stancel, G.M., Chorich, L.P., Bousfield, G.R., and Ulubelen, A.A., Hormonal carcinogenesis: separation of estrogenicity from carcinogenicity, *Chem. Biol. Interact.*, 59, 173–184, 1986.

71. Bhat, H.K., Calaf, G., Hei, T.K., Loya, T., and Vadgama, J.V., Critical role of oxidative stress in estrogen-induced carcinogenesis, *Proc. Natl. Acad. Sci. USA*, 100, 3913–3918, 2003.

72. Cavalieri, E.L., Stack, D.E., Devanesan, P.D., Todorovic, R., Dwivedy, I., Higginbotham, S., Johansson, S.L., Patil, K.D., Gross, M.L., Gooden, J.K., Ramanathan, R., Cerny, R.L., and Rogan, E.G., Molecular origin of cancer: catechol estrogen-3,4-quinones as endogenous tumor initiators, *Proc. Natl. Acad. Sci. USA*, 94, 10937–10942, 1997.

73. Li, S., Washburn, K.A., Moore, R., Uno, T., Teng, C., Newbold, R.R., McLachlan, J.A., and Negishi, M., Developmental exposure to diethylstilbestrol elicits demethylation of estrogen-responsive lactoferrin gene in mouse uterus, *Cancer Res.*, 57, 4356–4359, 1997.

74. Liehr, J.G., Role of DNA adducts in hormonal carcinogenesis, *Regul. Toxicol. Pharmacol.*, 32, 276–282, 2000.

75. Berger, J. and Daxenbichler, G., DNA methylation of nuclear receptor genes: possible role in malignancy, *J. Steroid Biochem. Mol. Biol.*, 80, 1–11, 2002.

76. Li, S., Ma, L., Chiang, T., Burow, M., Newbold, R.R., Negishi, M., Barrett, J.C., and McLachlan, J.A., Promoter CpG methylation of Hox-a10 and Hox-a11 in mouse uterus not altered upon neonatal diethylstilbestrol exposure, *Mol. Carcinog.*, 32, 213–219, 2001.

77. Li, S., Hursting, S.D., Davis, B.J., McLachlan, J.A., and Barrett, J.C., Environmental exposure, DNA methylation, and gene regulation: lessons from diethylstilbestrol-induced cancers, *Ann. N.Y. Acad. Sci.*, 983, 161–169, 2003.

78. Jost, J.P. and Saluz, H.P., Steroid hormone dependent changes in DNA methylation and its significance for the activation or silencing of specific genes, *EXS*, 64, 425–451, 1993.

79. Ginger, M.R., Gonzalez-Rimbau, M.F., Gay, J.P., and Rosen, J.M., Persistent changes in gene expression induced by estrogen and progesterone in the rat mammary gland, *Mol. Endocrinol.*, 15, 1993–2009, 2001.

80. Castro-Rivera, E., Samudio, I., and Safe, S., Estrogen regulation of cyclin D1 gene expression in ZR-75 breast cancer cells involves multiple enhancer elements, *J. Biol. Chem.*, 276, 30853–30861, 2001.

81. Henderson, B.E. and Feigelson, H.S., Hormonal carcinogenesis, *Carcinogenesis*, 21, 427–433, 2000.

82. Mollerup, S., Jorgensen, K., Berge, G., and Haugen, A., Expression of estrogen receptors alpha and beta in human lung tissue and cell lines, *Lung Cancer*, 37, 153–159, 2002.

83. Steiner, M.S. and Raghow, S., Antiestrogens and selective estrogen receptor modulators reduce prostate cancer risk, *World J. Urol.*, 21, 31–36, 2003.

84. Endogenous Hormones and Breast Cancer Collaborative Group, Endogenous sex hormones and breast cancer in post-menopausal women: reanalysis on nine prospective studies, *J. Natl. Cancer Inst.*, 94, 606–616, 2002.

85. Cuzick, J., Powles, T., Veronesi, U., Forbes, J., Edwards, R., Ashley, S., and Boyle, P., Overview of the main outcomes in breast-cancer prevention trials, *Lancet*, 361, 296–300, 2003.

86. MacDonald, P.C., Edman, C.D., Hemsell, D.L., Porter, J.C., Siiteri, and P.K., Effect of obesity on conversion of plasma androstenedione to estrone in postmenopausal women with and without endometrial cancer, *Am. J. Obstet. Gynecol.*, 130, 448–455, 1978.

87. Kelsey, J.L. and Horn-Ross, P.L., Breast cancer: magnitude of the problem and descriptive epidemiology, *Epidemiol. Rev.*, 15, 7–16, 1993.
88. Kelsey, J.L., Breast cancer epidemiology: summary and future directions, *Epidemiol. Rev.*, 15, 256–263, 1993.
89. Hoffman, S., Grisso, J.A., Kelsey, J.L., Gammon, M.D., and O'Brien, L.A., Parity, lactation and hip fracture, *Osteoporos. Int.*, 3, 171–176, 1993.
90. Lipsett, M.B., Estrogen use and cancer risk, *J. Am. Med. Assoc.*, 237, 1112–1115, 1977.
91. Russo, J., Hu, Y.F., Tahin, Q., Mihaila, D., Slater, C., Lareef, M.H., and Russo, I.H., Carcinogenicity of estrogens in human breast epithelial cells, *APMIS*, 109, 39–52, 2001.
92. Laitman, C.J., DES exposure and the aging woman: mothers and daughters, *Curr. Womens Health Rep.*, 2, 390–393, 2002.
93. Titus-Ernstoff, L., Hatch, E.E., Hoover, R.N., Palmer, J., Greenberg, E.R., Ricker, W., Kaufman, R., Noller, K., Herbst, A.L., Colton, T., and Hartge, P., Long-term cancer risk in women given diethylstilbestrol (DES) during pregnancy, *Br. J. Cancer*, 84, 126–133, 2001.
94. Duan, R., Xie, W., Burghardt, R.C., and Safe, S., Estrogen receptor-mediated activation of the serum response element in MCF-7 cells through MAPK-dependent phosphorylation of Elk-1, *J. Biol. Chem.*, 276, 11590–11598, 2001.
95. Perillo, B., Sasso, A., Abbondanza, C., and Palumbo, G., 17Beta-estradiol inhibits apoptosis in MCF-7 cells, inducing bcl-2 expression via two estrogen-responsive elements present in the coding sequence, *Mol. Cell. Biol.*, 20, 2890–2901, 2000.
96. Urban, G., Golden, T., Aragon, I.V., Scammell, J.G., Dean, N.M., Honkanen, R.E., Identification of an estrogen-inducible phosphatase (PP5) that converts MCF-7 human breast carcinoma cells into an estrogen-independent phenotype when expressed constitutively, *J. Biol. Chem.*, 276, 27638–27646, 2001.
97. Burow, M.E., Weldon, C.B., Tang, Y., McLachlan, J.A., and Beckman, B.S., Oestrogen-mediated suppression of tumour necrosis factor alpha-induced apoptosis in MCF-7 cells: subversion of Bcl-2 by anti-oestrogens, *J. Steroid Biochem. Mol. Biol.*, 78, 409–418, 2001.
98. Krepinsky, J., Ingram, A.J., James, L., Ly, H., Thai, K., Cattran, D.C., Miller, J.A., and Scholey, J.W., 17Beta-estradiol modulates mechanical strain-induced MAPK activation in mesangial cells, *J. Biol. Chem.*, 277, 9387–9394, 2002.
99. Banerjee, S., Saxena, N., Sengupta, K., Tawfik, O., Mayo, M.S., and Banerjee, S.K., WISP-2 gene in human breast cancer: estrogen and progesterone inducible expression and regulation of tumor cell proliferation, *Neoplasia*, 5, 63–73, 2003.
100. Folkman, J., Angiogenesis and angiogenesis inhibition: an overview, *EXS*, 79, 1–8, 1997.
101. Folkman, J., Angiogenesis: initiation and modulation, *Symp. Fundam. Cancer Res.*, 36, 201–208, 1983.
102. Risau, W., Angiogenic growth factors, *Prog. Growth Factor Res.*, 2, 71–79, 1990.
103. Bergers, G. and Benjamin, L.E., Angiogenesis: tumorigenesis and the angiogenic switch, *Nat. Rev. Cancer*, 3, 401–410, 2003.
104. Liotta, L.A. and Stetler-Stevenson, W.G., Tumor invasion and metastasis: an imbalance of positive and negative regulation, *Cancer Res.*, 51, 5054s–5059s, 1991.
105. Hanahan, D. and Folkman, J., Patterns and emerging mechanisms of the angiogenic switch during tumorigenesis, *Cell*, 86, 353–364, 1996.
106. Cockerill, G.W., Gamble, J.R., and Vadas, M.A., Angiogenesis: models and modulators, *Int. Rev. Cytol.*, 159, 113–160, 1995.
107. Soares, R., Guo, S., Russo, J., and Schmitt, F., Role of the estrogen antagonist ICI 182,780 in vessel assembly and apoptosis of endothelial cells, *Ultrastruct. Pathol.*, 27, 33–39, 2003.
108. McNamara, D.A., Harmey, J., Wang, J.H., Kay, E., Walsh, T.N., and Bouchier-Hayes, D.J., Tamoxifen inhibits endothelial cell proliferation and attenuates VEGF-mediated angiogenesis and migration *in vivo*, *Eur. J. Surg. Oncol.*, 27, 714–718, 2001.
109. Calabrese, E.J., Estrogen and related compounds: biphasic dose responses, *Crit. Rev. Toxicol.*, 31, 503–515, 2001.
110. Marson, L.P., Kurian, K.M., Miller, W.R., and Dixon, J.M., The effect of tamoxifen on breast tumour vascularity, *Breast Cancer Res. Treat.*, 66, 9–15, 2001.
111. Banerjee, S., Saxena, N., Sengupta, K., and Banerjee, S.K., 17Alpha-estradiol-induced VEGF-A expression in rat pituitary tumor cells is mediated through ER independent but PI3K-Akt dependent signaling pathway, *Biochem. Biophys. Res. Commun.*, 300, 209–215, 2003.

112. Banerjee, S.K., Zoubine, M.N., Tran, T.M., Weston, A.P., and Campbell, D.R., Overexpression of vascular endothelial growth factor164 and its co-receptor neuropilin-1 in estrogen-induced rat pituitary tumors and GH3 rat pituitary tumor cells, *Int. J. Oncol.*, 16, 253–260, 2000.

113. Sengupta, K., Banerjee, S., Saxena, N., and Banerjee, S.K., Estradiol-induced vascular endothelial growth factor-A expression in breast tumor cells is biphasic and regulated by estrogen receptor-alpha dependent pathway, *Int. J. Oncol.*, 22, 609–614, 2003.

114. Maity, A., Sall, W., Koch, C.J., Oprysko, P.R., and Evans, S.M., Low pO2 and beta-estradiol induce VEGF in MCF-7 and MCF-7-5C cells: relationship to *in vivo* hypoxia, *Breast Cancer Res. Treat.*, 67, 51–60, 2001.

115. Ruohola, J.K., Valve, E.M., Karkkainen, M.J., Joukov, V., Alitalo, K., and Harkonen, P.L., Vascular endothelial growth factors are differentially regulated by steroid hormones and antiestrogens in breast cancer cells, *Mol. Cell. Endocrinol.*, 149, 29–40, 1999.

116. Mueller, M.D., Vigne, J.L., Minchenko, A., Lebovic, D.I., Leitman, D.C., and Taylor, R.N., Regulation of vascular endothelial growth factor (VEGF) gene transcription by estrogen receptors alpha and beta, *Proc. Natl. Acad. Sci. USA*, 97, 10972–10977, 2000.

117. Ochoa, A.L., Mitchner, N.A., Paynter, C.D., Morris, R.E., and Ben-Jonathan, N., Vascular endothelial growth factor in the rat pituitary: differential distribution and regulation by estrogen, *J. Endocrinol.*, 165, 483–492, 2000.

118. Banerjeei, S.K., Zoubine, M.N., Sarkar, D.K., Weston, A.P., Shah, J.H., and Campbell, D.R., 2-Methoxyestradiol blocks estrogen-induced rat pituitary tumor growth and tumor angiogenesis: possible role of vascular endothelial growth factor, *Anticancer Res.*, 20, 2641–2645, 2000.

119. Banerjee, S.N., Sengupta, K., Banerjee, S., Saxena, N., and Banerjee, S.K., 2-Methoxyestradiol exhibits a biphasic effect on VEGF-A in tumor cells and upregulation is mediated through ER-α: a possible signaling pathway associated with the impact of 2-ME$_2$ on proliferative cell$_s$, *Neoplasia*, 5, 417–426, 2003.

120. Menendez, J.A., Mehmi, I., Griggs, D.W., and Lupu, R., The angiogenic factor CYR61 in breast cancer: molecular pathology and therapeutic perspectives, *Endocr. Relat. Cancer*, 10, 141–152, 2003.

121. Lincoln, D.W., Phillips, P.G., and Bove, K., Estrogen-induced Ets-1 promotes capillary formation in an *in vitro* tumor angiogenesis model, *Breast Cancer Res. Treat.*, 78, 167–178, 2003.

122. Albrecht, E.D. and Pepe, G.J., Steroid hormone regulation of angiogenesis in the primate endometrium, *Front Biosci.*, 8, d416–d429, 2003.

123. Dabrosin, C., Palmer, K., Muller, W.J., and Gauldie, J., Estradiol promotes growth and angiogenesis in polyoma middle T transgenic mouse mammary tumor explants, *Breast Cancer Res. Treat.*, 78, 1–6, 2003.

124. Fukuro, H., Mogi, C., Yokoyama, K., and Inoue, K., Change in expression of basic fibroblast growth factor mRNA in a pituitary tumor clonal cell line, *Endocr. Pathol.*, 14, 145–149, 2003.

125. Millauer, B., Shawver, L.K., Plate, K.H., Risau, W., and Ullrich, A., Glioblastoma growth inhibited *in vivo* by a dominant-negative Flk-1 mutant, *Nature*, 367, 576–579, 1994.

126. Jilma, B., Eichler, H.G., Breiteneder, H., Wolzt, M., Aringer, M., Graninger, W., Rohrer, C., Veitl, M., and Wagner, O.F., Effects of 17 beta-estradiol on circulating adhesion molecules, *J. Clin. Endocrinol. Metab.*, 79, 1619–1624, 1994.

127. Aziz, K.E. and Wakefield, D., Modulation of endothelial cell expression of ICAM-1, E-selectin, and VCAM-1 by beta-estradiol, progesterone, and dexamethasone, *Cell. Immunol.*, 167, 79–85, 1996.

128. Brooks, P.C., Cell adhesion molecules in angiogenesis, *Cancer Metastasis Rev.*, 15, 187–194, 1996.

129. Sengupta, K., Banerjee, S., Saxena, N., and Banerjee, S.K., Estrogen-induced down regulation of thrombospondin-1 in rat pituitary tumor cells and in rat pituitaries may be associated with estrogen-induced tumor cell growth and angiogenesis, *Proc. Am. Assoc. Cancer Res.*, 44, 103, 2003.

130. Petrangeli, E., Lubrano, C., Ortolani, F., Ravenna, L., Vacca, A., Sciacchitano, S., Frati, L., and Gulino, A., Estrogen receptors: new perspectives in breast cancer management, *J. Steroid Biochem. Mol. Biol.*, 49, 327–331, 1994.

131. Parl, F.F., *Estrogen Receptor Expression in Breast Cancer: Estrogens, Estrogen Receptor and Breast Cancer*, IOS Press, Washington, DC, 2002, pp. 135–204.

132. Parl, F.F., *Estrogen Receptor: Estrogens, Estrogen Receptor and Breast Cancer*, IOS Press, Ohmsha, Washington, DC, 2002, pp. 57–110.

133. Levin, E.R., Cell localization, physiology, and nongenomic actions of estrogen receptors, *J. Appl. Physiol.*, 91, 1860–1867, 2001.

134. Collins, P. and Webb, C., Estrogen hits the surface, *Nat. Med.*, 5, 1130–1131, 1999.
135. Levin, E.R., Bidirectional signaling between the estrogen receptor and the epidermal growth factor receptor, *Mol. Endocrinol.*, 17, 309–317, 2003.
136. Valentine, J.E., Kalkhoven, E., White, R., Hoare, S., and Parker, M.G., Mutations in the estrogen receptor ligand binding domain discriminate between hormone-dependent transactivation and transrepression, *J. Biol. Chem.*, 275, 25322–25329, 2000.
137. Pedram, A., Razandi, M., Aitkenhead, M., Hughes, C.C., and Levin, E.R., Integration of the nongenomic and genomic actions of estrogen: membrane-initiated signaling by steroid to transcription and cell biology, *J. Biol. Chem.*, 277, 50768–50775, 2002.
138. Kelly, M.J. and Levin, E.R., Rapid actions of plasma membrane estrogen receptors, *Trends Endocrinol. Metab.*, 12, 152–156, 2001.
139. Goetz, R.M., Thatte, H.S., Prabhakar, P., Cho, M.R., Michel, T., and Golan, D.E., Estradiol induces the calcium-dependent translocation of endothelial nitric oxide synthase, *Proc. Natl. Acad. Sci. USA*, 96, 2788–2793, 1999.
140. Le, M.V., Grosse, B., and Lieberherr, M., Phospholipase C beta and membrane action of calcitriol and estradiol, *J. Biol. Chem.*, 272, 11902–11907, 1997.
141. Razandi, M., Alton, G., Pedram, A., Ghonshani, S., Webb, P., and Levin, E.R., Identification of a structural determinant necessary for the localization and function of estrogen receptor alpha at the plasma membrane, *Mol. Cell. Biol.*, 23, 1633–1646, 2003.
142. Razandi, M., Pedram, A., Greene, G.L., and Levin, E.R., Cell membrane and nuclear estrogen receptors (ERs) originate from a single transcript: studies of ERalpha and ERbeta expressed in Chinese hamster ovary cells, *Mol. Endocrinol.*, 13, 307–319, 1999.
143. Kumar, R., Wang, R.A., Mazumdar, A., Talukder, A.H., Mandal, M., Yang, Z., Bagheri-Yarmand, R., Sahin, A., Hortobagyi, G., Adam, L., Barnes, C.J., and Vadlamudi, R.K., A naturally occurring MTA1 variant sequesters oestrogen receptor-alpha in the cytoplasm, *Nature*, 418, 654–657, 2002.
144. Gee, A.C., Carlson, K.E., Martini, P.G., Katzenellenbogen, B.S., and Katzenellenbogen, J.A., Coactivator peptides have a differential stabilizing effect on the binding of estrogens and antiestrogens with the estrogen receptor, *Mol. Endocrinol.*, 13, 1912–1923, 1999.
145. Greschik, H., Wurtz, J.M., Sanglier, S., Bourguet, W., van Dorsselaer, A., Moras, D., and Renaud, J.P., Structural and functional evidence for ligand-independent transcriptional activation by the estrogen-related receptor 3, *Mol. Cell*, 9, 303–313, 2002.
146. This, P., De La, R.A., Clough, K., Fourquet, A., and Magdelenat, H., Phytoestrogens after breast cancer, *Endocr. Relat. Cancer*, 8, 129–134, 2001.
147. McLachlan, J.A. and Newbold, R.R., Estrogens and development, *Environ. Health Perspect.*, 75, 25–27, 1987.
148. Jones, L.A. and Hajek, R.A., Effects of estrogenic chemicals on development, *Environ. Health Perspect.*, 103 (Suppl. 7), 63–67, 1995.
149. Kohlmeier, L., Simonsen, N., and Mottus, K., Dietary modifiers of carcinogenesis, *Environ. Health Perspect.*, 103 (Suppl. 8), 177–184, 1995.
150. Coffey, D.S., Similarities of prostate and breast cancer: evolution, diet, and estrogens, *Urology*, 57, 31–38, 2001.
151. Golden, R.J., Noller, K.L., Titus-Ernstoff, L., Kaufman, R.H., Mittendorf, R., Stillman, R., and Reese, E.A., Environmental endocrine modulators and human health: an assessment of the biological evidence, *Crit. Rev. Toxicol.*, 28, 109–227, 1998.
152. Anderson, J.W., Dietary fibre, complex carbohydrate and coronary artery disease, *Can. J. Cardiol.*, 11 (Suppl. G), 55G–62G, 1995.
153. Potter, J.D. and Steinmetz, K., Vegetables, fruit and phytoestrogens as preventive agents, *IARC Sci Publ.*, 139, 61–90, 1996.
154. Barnes, S., Phytoestrogens and breast cancer, *Baillieres Clin. Endocrinol. Metab.*, 12, 559–579, 1998.
155. Diel, P., Smolnikar, K., Schulz, T., Laudenbach-Leschowski, U., Michna, H., and Vollmer, G., Phytoestrogens and carcinogenesis-differential effects of genistein in experimental models of normal and malignant rat endometrium, *Hum. Reprod.*, 16, 997–1006, 2001.
156. Lian, Z., Niwa, K., Tagami, K., Hashimoto, M., Gao, J., Yokoyama, Y., Mori, H., and Tamaya, T., Preventive effects of isoflavones, genistein and daidzein, on estradiol-17beta-related endometrial carcinogenesis in mice, *Jpn. J. Cancer Res.*, 92, 726–734, 2001.

157. Chhabra, S.K. and Yang, C.S., Tea and prostate cancer, *Epidemiol. Rev.*, 23, 106–109, 2001.
158. Hong, J., Smith, T.J., Ho, C.T., August, D.A., and Yang, C.S., Effects of purified green and black tea polyphenols on cyclooxygenase- and lipoxygenase-dependent metabolism of arachidonic acid in human colon mucosa and colon tumor tissues, *Biochem. Pharmacol.*, 62, 1175–1183, 2001.
159. Yang, C.S., Prabhu, S., and Landau, J., Prevention of carcinogenesis by tea polyphenols, *Drug Metab. Rev.*, 33, 237–253, 2001.
160. Yang, C.S., Inhibition of carcinogenesis and toxicity by dietary constituents, *Adv. Exp. Med. Biol.*, 500, 541–550, 2001.
161. Yang, C.S., Yang, G.Y., Chung, J.Y., Lee, M.J., and Li, C., Tea and tea polyphenols in cancer prevention, *Adv. Exp. Med. Biol.*, 492, 39–53, 2001.
162. Lamartiniere, C.A., Moore, J.B., Brown, N.M., Thompson, R., Hardin, M.J., and Barnes, S., Genistein suppresses mammary cancer in rats, *Carcinogenesis*, 16, 2833–2840, 1995.

9 Nonsteroidal Anti-Inflammatory Drugs (NSAID) and Colorectal Cancer

Henry T. Lynch, Trudy G. Shaw, Jane F. Lynch, and Hemant K. Roy

CONTENTS

9.1   Introduction..123
9.2   Definition of Chemoprevention..124
9.3   Epidemiology...124
    9.3.1   Animal Studies 125
    9.3.2   Human NSAID Intervention Studies 125
9.4   Aspirin 125
    9.4.1   Possible Mechanisms of CRC Chemoprevention by Aspirin....................127
9.5   COX Inhibitors 129
    9.5.1   Chemistry and COX Activity 129
9.6   Sulindac 131
    9.6.1   CRC Occurrence while Undergoing Sulindac Treatment 131
    9.6.2   Periampullary Polyps...132
    9.6.3   Exisulind (Sulindac Sulfone, FGN-1) in FAP 132
9.7   Chemoprevention and CRC: Genetic Models 132
    9.7.1   Adenoma Regression in FAP 133
    9.7.2   Angiogenesis and COX-2...133
9.8   Curcumin 133
9.9   Discussion..133
    9.8.1   Will NSAIDs Replace Screening and Prophylactic Colectomy in FAP? 134
9.10 Conclusion 135
Acknowledgment..135
References 135

9.1   INTRODUCTION

Chemoprevention has emerged as an exceedingly important prevention objective for many forms of cancer. An increasing variety of chemicals and hormones have received major attention as potential chemopreventive agents during the past several decades. Noteworthy among these is tamoxifen in the prevention of breast cancer.[1-4] Usage of oral contraceptives (OC) may also be protective against ovarian cancer.

With respect to breast cancer susceptibility settings, OC use is controversial in that it may be influenced significantly by host factors, including possible differences in the effect of tumor suppressor genes, such as *BRCA1* and *BRCA2*. For example, in a matched case-control study on 1311 pairs of women with known deleterious *BRCA1* or *BRCA2* mutations who were recruited from 52 centers in 11 countries, Narod et al.[5] found that among *BRCA2* mutation carriers, the use of OC was not associated with an increased risk of breast cancer (OR = 0.94, 95% CI = 0.72 to 1.24). However, among *BRCA1* mutation carriers, the use of OC was associated with a modestly increased risk of breast cancer (OR = 1.20, 95% CI = 1.02 to 1.40).

Why are we discussing hormonal factors, genetics, and *BRCA1* and *BRCA2* mutations for hereditary breast cancer in the introduction to a chapter on chemoprevention in colorectal cancer (CRC)? The reason is simply to stress the importance of understanding genotypic and phenotypic heterogeneity in the differential response to the testing of any drug, including chemopreventive agents, given the possibility of their pharmacogenetic effects. These pharmacogenetic modulations may conceivably occur in hereditary cancer-prone models for chemoprevention, such as familial adenomatous polyposis (FAP) and hereditary nonpolyposis colorectal cancer (HNPCC), as well as in sporadic forms of CRC.

9.2 DEFINITION OF CHEMOPREVENTION

Chemoprevention is defined as the use of natural or synthetic agents for the reversal of carcinogenesis. In their review of the subject, Gwyn and Sinicrope[6] discuss epidemiologic studies that have consistently shown that the chronic intake of a nonsteroidal anti-inflammatory drug (NSAID), principally aspirin, can successfully reduce the incidence of colorectal adenomas and possibly CRCs. They note that NSAIDs such as cyclooxygenase-2 inhibitors celecoxib and sulindac, in addition to other varieties of chemopreventive agents, are being studied in patients at increased genetic risk of CRC, as well as those with sporadic colonic adenomas.

Prevention of the initiation of colonic polyps and/or CRC can be considered to fall under the rubric of chemoprevention. Hong and Sporn's[7] statement that "the use of pharmacological or natural agents that inhibit development of invasive cancer either by blocking the DNA damage that initiates carcinogenesis or by arresting or reversing the progression of premalignant cells in which such damage has already occurred" is consistent with the definition of chemoprevention. While the outcomes of some chemopreventive studies will likely impact patient management practices, such chemopreventive agents cannot be recommended at the present time for average-risk individuals or for those with sporadic colorectal neoplasia. However, hereditary CRC-prone patients, such as those with FAP and HNPCC, are ideal candidates for studying the efficacy of chemopreventive agents.[8]

Hawk et al.[9] note that the goals of chemoprevention are to reduce cancer risk and improve quality of life. Studies of hereditary cohorts may hold promise for the chemoprevention of CRC in the sporadic setting as well. King et al.[10] stress the importance of chemoprevention as part of treatment strategies still in development for FAP patients, in order to improve the management of extracolonic neoplasms and desmoid tumors.

Chemopreventive agents must be safe and well tolerated for chronic usage, and ideally, they should be relatively cost effective. While the field of chemoprevention is still in its infancy, it has become a rapidly advancing area for clinical and basic science investigation, where it holds promise for producing a paradigm shift in the management of CRC.[6]

9.3 EPIDEMIOLOGY

It is estimated that approximately 70% of cancer may be due to environmental factors, i.e., occupational, dietary, smoking, alcohol consumption and other lifestyle factors.[11] Obesity appears to be a significant risk factor for CRC when controlling for the effects of age, race, education, smoking, exercise, alcohol, parental history of CRC, fat intake, vegetable and grain intake, as well as aspirin use. In women, estrogen-replacement therapy may add to these risk factors. When considering these risk factors in a prospective study, Murphy et al.[12] found that the hypothesis that obesity increases the risk of CRC death appeared to be stronger and more linear in men than in women. It logically follows that appropriate modification of these exogenous cancer-causing factors may be useful in avoiding cancer.

In the case of CRC, NSAIDs have become the focus of research investigating their potential for prevention. What is the motivation for conducting such research? In humans, there is strong epidemiologic evidence suggesting that NSAIDs reduce the risk of CRC and CRC death,[13–28] summarized by Thun.[29] Case-control studies have almost uniformly found an inverse association between the use of aspirin and incident colorectal adenomas or cancers.[13–19] Two prospective studies of persons with rheumatoid arthritis receiving therapy revealed a reduced risk for CRC and stomach cancer associated with their therapies (presumably involving the use of NSAIDs).[22,23] Six prospective studies have found a lower risk of CRC and lower mortality from the disease among regular aspirin users.[21,24–27,30] Only one observational study, conducted in an elderly retirement community, showed an elevated CRC risk with daily aspirin use.[31] Two additional studies showed little or no effect. In one, aspirin taken at a dose of 325 mg every other day produced no evidence of reduced risk in a relatively low-risk population.[28,32] In another study, the apparent benefit did not reach significance until self-medication with aspirin had been employed for over 20 years.[26] The results of these studies indicated the need for a formal, randomized, controlled trial of sufficient power to elicit a significant result, but this would be a large undertaking. The only known primary prevention study of which we are aware is the Women's Health Initiative. This study involved 16,608 patients who were followed over an average of 5.2 years. The result showed a 37% risk reduction of CRC, comparable to that anticipated with NSAIDs. The increased cardiovascular risk, however, makes this option nonviable.[33]

9.3.1 ANIMAL STUDIES

The chemopreventive efficacy of NSAIDs against colon cancer has been supported by a large range of animal studies (reviewed by Thun[29]). The NSAIDs aspirin, piroxicam, sulindac, naproxen, ketoprofen, indomethacin, and ibuprofen reduced both the number and size of carcinogen-induced colon tumors in rodent models, even in some cases when administered months after exposure to carcinogen and when microscopic tumors were already present. The NSAIDs piroxicam and sulindac have also been shown to inhibit colon carcinogenesis in the Min mouse. Davis and Patterson[34] reported that aspirin reduces the incidence of colonic carcinoma in the dimethylhydrazine rat model. Oshima et al.[35] discuss the redistribution of beta-catenin with rofecoxib. COX-2 (cyclooxygenase-2) inhibition may not, however, be the only mode of action of NSAIDs.[36] For example, sulindac sulfone is a derivative of sulindac that is effective in polyp suppression but does not inhibit COX-2.[37]

9.3.2 HUMAN NSAID INTERVENTION STUDIES

Human intervention studies of FAP, and of sporadic adenoma development, have shown efficacy associated with NSAID administration, manifested primarily by regression of existing adenomas.[38] One report suggests that NSAIDs may be ineffective for primary polyp prevention in FAP patients,[39] although this appears discordant with the vast majority of studies that evaluated patients with established polyps. Efficacy appears to exist within all regions of the colorectum, and possibly in other sites as well (esophagus, stomach, prostate, head and neck, pancreas, and breast).

9.4 ASPIRIN

Aspirin and other NSAIDs are all effective to some extent, though analgesics without antiprostaglandin activities, such as acetaminophen, are ineffective. As might be anticipated, activity of NSAIDs seems to be related to the dose, frequency, and duration of administration. However, there have been suggestions that aspirin, particularly in high doses, may incite rectal bleeding earlier due to its antiplatelet effects. However, Ahnen found no evidence for this in his review of the literature in 1998,[36] and he concluded that there were no confounding variables such as positive fecal occult blood tests (FOBs) or healthier lifestyles in the aspirin studies.

Sjödahl[40] notes that the relative risk of CRC with aspirin use is about 0.6 in large cohort studies (40% risk reduction). Attention was called to the fact that tumors in both human and experimental animal models contain increased amounts of prostaglandin E_2 which is believed to have a role in the accelerated proliferation that occurs in tumor tissue and that may be attributable to the activation of COX-2

> in response to mitogens and growth factors, for example, which will result in an increased production of prostaglandins. The current theory is that the mechanism for the suppressor effect of NSAIDs on carcinogenesis is COX-2 inhibition. However, reliable data on the dose of aspirin or other NSAIDs for optimal benefit for tumor suppression are lacking, and it is still premature to give general recommendations on using NSAIDs for chemoprevention of gastrointestinal cancer.

A study undertaken in Newcastle, U.K., using FAP patients with colectomies, showed an effect of short-term aspirin and sulindac treatment. The normal rectal mucosa of these patients showed a significant reduction in crypt cell proliferation following treatment (John Burn, M.D., unpublished data, 2000), although other reports have failed to demonstrate any alterations.[41] Apoptosis appears to be reliably increased by NSAID treatment in the uninvolved rectal mucosa.

Two recent studies on aspirin and its role in colonic polyps have shown benefit.[42,43] Sandler et al.[42] performed a randomized, double-blind trial with the objective of determining the effect of aspirin on the incidence of colorectal adenomas. They randomly assigned 635 patients with a prior history of CRC to receive 325 mg of aspirin daily or placebo. Their results disclosed that

> one or more adenomas were found in 17 percent of patients in the aspirin group and 27 percent of patients in the placebo group (P = 0.004). The mean (±SD) number of adenomas was lower in the aspirin group than the placebo group (0.30 ± 0.87 vs. 0.49 ± 0.99, P = 0.003 by the Wilcoxon test). The adjusted relative risk of any recurrent adenoma in the aspirin group, as compared with the placebo group, was 0.65 (95 percent confidence interval, 0.46 to 0.91). The time to the detection of a first adenoma was longer in the aspirin group than in the placebo group (hazard ratio for the detection of a new polyp, 0.64; 95 percent confidence interval, 0.43 to 0.94; P = 0.022).

These authors concluded that the use of aspirin may give a significant reduction in the incidence of colorectal adenomas in patients with a prior history of CRC. Given the statistically significant findings favoring aspirin over placebo, the study was stopped.

Baron et al.[43] also performed a randomized, double-blind trial of aspirin as a chemoprevention agent against colorectal adenomas. They randomized 1121 patients with a recent history of histologically documented adenomas to receive placebo (372 patients), 81 mg of aspirin (377 patients), or 325 mg of aspirin (372 patients) daily. The findings disclosed that

> the incidence of one or more adenomas was 47 percent in the placebo group, 38 percent in the group given 81 mg of aspirin per day, and 45 percent in the group given 325 mg of aspirin per day (global P = 0.04). Unadjusted relative risks of any adenoma (as compared with the placebo group) were 0.81 in the 81-mg group (95 percent confidence interval, 0.69 to 0.96) and 0.96 in the 325-mg group (95 percent confidence interval, 0.81 to 1.13). For advanced neoplasms (adenomas measuring at least 1 cm in diameter or with tubulovillous or villous features, severe dysplasia, or invasive cancer), the respective relative risks were 0.59 (95 percent confidence interval, 0.38 to 0.92) and 0.83 (95 percent confidence interval, 0.55 to 1.23).

These authors concluded that aspirin provided a moderate chemopreventive effect on adenomas in the colon.

9.4.1 Possible Mechanisms of CRC Chemoprevention by Aspirin

The molecular targets for NSAIDs remain unclear. Prostaglandin inhibition has been implicated, given the pro-neoplastic effects of prostaglandins in colon carcinogenesis. Indeed, loss of prostaglandin receptors EP-2 was shown to retard neoplastic progression in the intestine. COX-2 is a logical target for prostaglandin production, given that it is selectively overexpressed early during colon carcinogenesis. Furthermore, elimination of COX-2 through transgenic technology markedly ameliorated the tumor number in an APC (adenomatous polyposis coli)-driven model of intestinal tumorigenesis.[44] However, the issue appears more complex. Eliminating COX-1 also decreased tumors in this model, although more modestly.[45] Furthermore, several lines of evidence suggest that non-COX mechanisms may be involved. For instance, NSAIDs without anti-COX activity (sulindac sulfone and R-fluoroibuprofen) have been shown to prevent experimental and/or clinical intestinal neoplasia. There are a myriad of putative molecular targets, including β-catenin, 15-lipoxygenase 1, PPAR δ, c-MYC, p21 cip/waf, BAX, AKT2, NFκb, and EGF signaling cascade.[46] Recent attention has focused on AKT as another critical NSAID target,[47] which we have recently demonstrated to be overexpressed in most sporadic colon cancers.[48]

Aspirin has anti-inflammatory properties due to its ability to decrease arachidonic acid metabolites, especially prostaglandins, by inhibiting cyclooxygenase enzymes COX-1 and COX-2. COX inhibition, especially of COX-2, is the key biochemical target of the chemopreventive activity of NSAIDs. Aspirin is a unique NSAID in that it irreversibly modifies the active site of COX. It is thought that prostaglandin E_2 plays a role in CRC because tumors produce increased amounts of this prostaglandin. The apparent lack of association between PGE_2 levels and Dukes staging, coupled with the increased levels in adenomas, suggests that PGE_2 is involved in the early stages of CRC development.[49]

Chan et al. have suggested that tumor suppression is due to increasing arachidonic acid within the same biochemical cascade[50]; this finding is of particular interest, as it demonstrates a potential role for aspirin in the stimulation of ceramide-mediated apoptosis. COX-1 and COX-2 inhibition may cause arachidonic acid to accumulate, leading, via sphingomyelinase, to an accumulation of ceramide transcription factors.[51,52] However, recent reports suggest that pathways independent of COX inhibition may play a role in carcinogenesis via effects on cell cycle arrest,[53] apoptosis,[54,55] necrosis,[56] inhibition of NF-κB,[57] cell interactions, activation of caspases,[56] long-chain fatty acid oxidation,[58] and activation of PPARγ.

The role of NSAIDs in MSI (microsatellite instability)-high colon cancers has been less clear. MSI-high cancers typically have less expression of putative NSAID targets such as COX-2[59] and AKT.[48] A number of studies have shown that NSAIDs can induce apoptosis in MSI-high colon cancer cell lines, potentially through induction of pro-apoptotic MMR (mismatch repair) proteins.[60]

Rüschoff et al.[61] showed that MSI in CRC cells deficient for a subset of the human MMR genes, namely *MLH1*, *MSH2*, and *MSH6*, was markedly reduced during exposure to aspirin or sulindac. The effect was reversible and time and concentration dependent. It appeared to be independent of proliferation rate and cyclooxygenase function. For example,

In contrast, the MSI phenotype of hPMS2-deficient endometrial cell line was unaffected by aspirin/ sulindac. We show that the MSI reduction in the susceptible MMR-deficient cells was confined to nonapoptotic cells, whereas apoptotic cells remained unstable and were eliminated from the growing population. These results suggest that aspirin/sulindac induces a genetic selection for microsatellite stability in a subset of MMR-deficient cells and may provide an effective prophylactic therapy for hereditary nonpolyposis colorectal cancer kindreds where alteration of the hMSH2 and hMLH1 genes are associated with the majority of cancer susceptibility cases.

Reddy and Rao[62] note that a variety of multidisciplinary epidemiology and molecular biology studies, in addition to preclinical investigations, have contributed immensely to approaches to chemoprevention, particularly in the case of NSAIDs. These authors found compelling evidence that several vital chemicals with anti-inflammatory properties and NSAIDs act to retard, block, or reverse colon carcinogenesis. This led to numerous opportunities for effective chemoprevention with selective cyclooxygenase-2 (COX-2) inhibitors, including celecoxib and rofecoxib, in a variety of preclinical models of colon cancer. They also found that naturally occurring COX-2 inhibitors such as curcumin and certain phytosterols were effective as chemopreventive agents against colon carcinogenesis and that there was only minimal gastrointestinal toxicity.

A reduction in CRC by NSAIDs is mediated through modulation of prostaglandin production by rate-limiting enzymes, namely cyclooxygenases (COXs). Side effects of COX-1 or COX-2 inhibitor included gastrointestinal ulceration and renal toxicity through the inhibition of constitutive COX-1. The chemopreventive effects of high-dose celecoxib in F344 rats indicated that

> administration of 500, 1000, or 1500 ppm celecoxib, during the initiation and post-initiation stages, significantly inhibited the incidence ($p < 0.01$ to $p < 0.0001$) as well as the multiplicity ($p < 0.01$ to $p < 0.0001$) of adenocarcinomas of the colon in a dose-dependant manner. Importantly, administration of 1500 ppm celecoxib during the promotion/progression stage also significantly suppressed the incidence and multiplicity of adenocarcinomas of the colon ($p < 0.01$). Also, administration of celecoxib to the rats during the initiation and post-initiation periods and throughout the promotion/progression stage strongly suppressed colon tumor volume ($p < 0.0002$ to $p < 0.001$).... Thus, in this model system, the chemopreventive efficacy of celecoxib is dose dependant when the COX-2 inhibitor is administered during the initiation and post-initiation periods. This study provides the first evidence that celecoxib is also very effective when it is given during the promotion/progression stage of colon carcinogenesis, indicating that the chemopreventive efficacy is achieved during the later stages of colon tumor development. This suggests that celecoxib may potentially be an effective chemopreventive agent for the secondary prevention of colon cancer in patients with familial adenomatous polyposis and sporadic polyps.

While the dose-dependence of aspirin and other NSAIDs in chemoprevention of experimental models of colon carcinogenesis is clear, the translation for clinical studies is much less clear. Even the most efficacious doses of aspirin in colon cancer prevention is also unclear. For instance, maximal inhibition of rectal prostaglandin is achieved by 81 mg of aspirin per day.[63] Epidemiological studies varied from a protective effect of aspirin with as few as two pills per month.[20] However, in the Nurses Health Study, there was some, but not compelling, evidence of a dose–response relationship, with maximal benefit being at 4 to 6 aspirin/week, with no greater protection in those who consumed either 7 to 14 or >14 aspirin per week.[26] This paradoxical relationship was also suggested by a recent polyp prevention study in which 81 mg (but not 325 mg) of aspirin a day led to a statistically significant polyp recurrence rate.[43]

Along with the dosage concerns, the best formulation of NSAIDs is equally unclear. Numerous NSAIDs (aspirin, ibuprofen, piroxicam, sulindac, nabumetone, celecoxib, etc.) have been shown to be effective in experimental models. However, the gastrointestinal (GI) toxicity of many of these agents has led to reluctance to implement this strategy. There has been great interest in using the new class of NSAIDs that are selective for COX-2, the isoform not found in normal gastrointestinal epithelium.

Another strategy has been to use nitric oxide(NO)-donating NSAIDs, which may be both safer on the gastrointestinal tract[64] and more efficacious[65] against CRC than conventional NSAIDs. Fiorucci et al.[64] studied NCX-4016, a nitric oxide-releasing derivative of aspirin with antiplatelet activity, in a double-blind, placebo-controlled study to investigate its effect on gastrointestinal mucosa and platelet functions in healthy human volunteers. Their findings disclosed that NCX-4016 is virtually devoid of gastric and duodenal toxicity,

> resulting in a total gastric and duodenal endoscopic score of 1.38 ± 0.3 and 1.25 ± 0.5 ($P < 0.0001$ vs. aspirin, not significant vs. placebo). NCX-4016 inhibited [arachidonic acid] AA-induced platelet aggregation as well as serum TXB_2 and platelet TXB_2 generation induced by AA to the same extent as aspirin (not significant vs. aspirin).

These authors concluded that addition of a NO-donating moiety to aspirin "results in a new chemical entity that maintains cyclooxygenase-1 and platelet inhibitory activity while nearly avoiding gastrointestinal damage."

9.5 COX INHIBITORS

COX-2, but not COX-1, is up-regulated early in most colon carcinogenesis. The selective COX-2 inhibitor celecoxib has been shown to suppress tumors in FAP patients, thus resulting in Food and Drug Administration (FDA) approval. However, several lines of evidence suggest that COX-2 inhibitors may be somewhat less efficacious than traditional, nonspecific inhibitors. This may be through activating the multitude of non-COX related pathways. While there have been no head-to-head studies, the approximately two-thirds suppression of colon polyp number and size with sulindac[66] appears superior to the one-third decrease seen with celecoxib.[67] This potential difference in efficacy may be overshadowed by the marked difference in safety. One factor that needs to be considered is potential cardiovascular benefit of aspirin and non-COX-2 selective NSAIDs.[68] The approximately 15% reduction in coronary events is of great significance given the prevalence of coronary artery disease in the population. Indeed, in the analysis by Ladabaum and colleagues,[69] while aspirin was not cost effective with respect to colon cancer prevention over a wide range of assumptions, it was estimated that if there was a minor decrease in cardiovascular mortality (0.1%), then aspirin would have a net beneficial effect on an average-risk population.

Martin et al.[70] evaluated 226 patients with colonic adenomas and 493 adenoma-free controls. They adjusted for sex, age, race, and body mass index. Their findings disclosed that those individuals in the highest tertile of regular NSAID use were substantially less likely to have adenomas (OR = 0.4; 95% CI: 0.2 to 0.7) compared with occasional or nonusers. When compared with the lowest tertile, individuals in the highest tertile of rectal mucosal apoptotic activity were less likely to have adenomas (OR = 0.12; 95% CI: 0.07 to 0.20). They concluded that "NSAID use and apoptotic activity were not correlated (r = 0.10). Mucosal proliferation was not related to adenomas or NSAID use … [and] NSAID use and higher levels of mucosal apoptosis are independently associated with a lower prevalence of adenomas." The study also showed a strong field effect for apoptosis. Many others have shown that NSAIDs increase apoptosis in the uninvolved mucosa.[71] However, another report in FAP patients suggested that rectal apoptosis rates may not correlate with adenoma regression.[72]

Clearly, knowledge of the natural history of CRC and its evolution from dysplastic aberrant crypts to adenomas and ultimately adenocarcinomas will likely provide numerous opportunities for chemopreventive interventions. In this context, Reddy and Rao[62] suggest that priority be given to the manner in which cells are modulated by NSAIDs inclusive of COX-2 inhibitors. In turn, there are many other chemopreventive agents that have undergone or are undergoing clinical trials in the interest of determining their preventive capabilities against CRC. For example, micronutrients and antioxidant vitamins have been employed as chemopreventive tools, but they have been steeped in controversy.[73]

In summary, the recognition of COX-2 and its upregulation role in colorectal adenomas has aided in the development of new chemopreventive drugs, namely COX-2 inhibitors. These drugs target the molecular mechanism of carcinogenesis and have been approved by the FDA for the chemoprevention of FAP. Indeed, these drugs will likely play an important role as chemopreventive agents for patients who are at increased risk for an increasing variety of familial CRC disorders.

9.5.1 CHEMISTRY AND COX ACTIVITY

Lal et al.[74] note that the DNA mismatch-repair-deficient Min mouse model (Apc+/–Msh2–/–) has genetic features of both FAP and HNPCC and, most importantly, rapidly develops numerous small and large bowel adenomas as well as colonic aberrant crypt foci. They investigated the effects of

COX inhibitors on intestinal adenomas and colonic aberrant crypt foci in this accelerated-polyposis, mismatch-repair-deficient Min mouse model, in addition to a standard Min mouse model. Their study demonstrated that a specific COX-2 inhibitor is effective in *preventing* small-bowel polyps in mismatch-repair-deficient Min mice and both small- and large-bowel polyps in standard Min mice.

Oshima and Taketo[75] note that the major target molecules of NSAIDs are cyclooxygenases (COX),

> which catalyze the rate-limiting step of prostaglandin biosynthesis. Whereas COX-1 is expressed constitutively in most tissues and is responsible for tissue homeostasis, COX-2 is inducible and plays an important role in inflammation and intestinal tumorigenesis. A genetic study using compound mutant mice of $COX-2^{(-)/(-)}$ and $Apc^{(\Delta 716)}$, which is a model for human familial adenomatous polyposis (FAP), directly demonstrated that induction of COX-2 is critical for intestinal polyp formation.

These authors call attention to findings from numerous investigations that have shown that COX-2 selective inhibitors suppress (a) intestinal polyp formation in Apc gene-mutant mice and (b) xenografted cancer cell growth. Furthermore, they note the important role of COX-2 in the stimulation of angiogenesis, which is induced in the polyp stromal cells. The location of COX-2 in colon cancers remains controversial, with Oshima and Taketo's data[75] suggesting stromal, while many human immunohistochemical analyses indicate epithelial cell staining.[76] In this regard, tumor transplantation studies show that host (stromal) COX-2 was more important than tumor COX-2 in establishment of metastatic focus.[77] However, COX-2 expression in human colon cancer cells has been shown to be predictive of prognosis.[78] In a recent report, Yao and colleagues[79] showed that COX-2 inhibitors, but not 5-fluorouracil, inhibited the development of liver metastasis in an experimental model. COX-2 is involved in the early stage of intestinal tumorigenesis. Such studies using animal models should aid in the comprehension of *in vivo* mechanism(s) of tumor suppression by NSAIDs or COX-2 inhibitors.

North[80] described the role of celecoxib as adjunctive treatment for FAP (MEDLINE search [1995 to March 2000] as well as the use of secondary sources). The FDA granted accelerated approval in December 1999 for the celecoxib for adjunctive therapy in patients with FAP, based on a 6-month, randomized control trial. This review found that aspirin and other NSAIDs cause a reduction in the incidence of CRC in the general population and that limited clinical trials in patients with FAP employing nonaspirin NSAIDs have revealed a reduction in polyp burden. However, the long-term impact of selective COX-2 inhibitors is not known, since celecoxib has not been investigated beyond 6 months in FAP affecteds.

Blanke[81] notes that expression of COX-2 is associated with a worse CRC prognosis and, in turn, inhibition of prostaglandin synthesis may prove beneficial in CRC. Furthermore, regular consumption of NSAIDs leads to a decrease in the incidence of and mortality rate from several different types of gastrointestinal cancers. Appropriate attention is called to findings from preclinical studies that show that NSAIDs block angiogenesis and suppress solid tumor metastasis. Furthermore, he notes that "the COX-2 inhibitors have safely and effectively been combined with chemotherapeutic agents in experimental studies. Ongoing clinical trials are currently assessing the potential therapeutic role of COX-2 inhibitors in both prevention and treatment of a diverse range of human cancers."

Marks and Fürstenberger[82] note that a dysregulation of arachidonic acid metabolism appears to play a disastrous role in tumor progression. However, the most potent, indeed promising, agents for chemoprevention are NSAIDs, due to their ability to inhibit prostaglandin biosynthesis and therein, "at the level of the 'pro-inflammatory' enzyme cyclooxygenase-2 (COX-2). A pathological overexpression of COX-2 resulting in excessive prostaglandin production has been found already in early stages of carcinogenesis and seems to be a consistent feature of neoplastic development in a wide variety of tissues."

9.6 SULINDAC

Giardiello et al.,[39] in reviewing the role of sulindac with respect to adenomatous polyps in patients with FAP, showed that regression of adenomatous polyps was evident in case reports dating back to 1983[83] and 1989.[84] These observations were also confirmed in randomized studies of sulindac.[66,85,86] This evidence prompted them to study the ability of sulindac to prevent adenomas in patients with FAP who were phenotypically normal.[39]

They performed a randomized, double-blind, placebo-controlled study involving 41 young individuals (ranging in age from 8 to 25 years) who were genotypically affected with FAP but phenotypically unaffected. They were given 75 or 150 mg of sulindac orally twice a day, or identically appearing placebo tablets, for 48 months.[39] Care was taken to evaluate the number and size of new colonic adenomas, as well as side effects of therapy in addition to levels of five major prostaglandins, which were serially measured in biopsy specimens of normal-appearing colorectal mucosa.

Findings disclosed that following four years of therapy,

> the average rate of compliance exceeded 76 percent in the sulindac group, and mucosal prostaglandin levels were lower in this group than in the placebo group. During the course of the study, adenomas developed in 9 of 21 subjects (43 percent) in the sulindac group and 11 of 20 subjects in the placebo group (55 percent) (P = 0.54). There were no significant differences in the mean number (P = 0.69) or size (P = 0.17) of polyps between the groups. Sulindac did not slow the development of adenomas, according to an evaluation involving linear longitudinal methods.

These authors concluded that there was no evidence showing that sulindac prevented the development of colonic adenomas in patients with FAP who were phenotypically normal. However, this is clearly a very select group, and the relatively small numbers leaves open the possibility of type II error. The polyps that occur in patients on sulindac appear molecularly distinct,[87] although the clinical implications of this finding remain unclear.

9.6.1 CRC Occurrence while Undergoing Sulindac Treatment

Cruz-Correa et al.[88] studied the effects of sulindac in 12 FAP patients with a mean age of 37.1 years with ileorectal anastomosis who received sulindac with a mean dosage of 158 mg/day for a mean period of 63.4 ± 31.3 months (range, 14 to 98 months). There was a significant decrease in polyp number in all 12 patients (p = 0.039) at a mean of 63.4 ± 31.3 months (p = 0.006). Recurrence of high-grade adenomas (tubulovillous, villous adenomas) was also prevented (p = 0.004). However, at 35 months of follow-up, a single patient developed Stage III cancer in the rectal stump.

Giardiello et al.[89] studied levels of five major prostaglandin metabolites in biopsy specimens of flat rectal mucosa from four patients with FAP before and after sulindac therapy, and from five healthy individuals. They found that the highest concentration of prostaglandin in rectal mucosa from FAP and control subjects was prostaglandin E_2. Furthermore, the concentration of thromboxane B2 alone was significantly elevated in FAP patients compared with controls (p = 0.016). When FAP patients were treated with sulindac, the prostaglandin metabolite levels were found to be significantly reduced when compared with pretreatment levels (p < 0.05), except for prostaglandin D2 (p = 0.07). In addition,

> prostaglandins D2, B2, F2alpha, and 6-keto-F1alpha levels also were significantly reduced in FAP patients on sulindac compared to healthy controls (P < 0.05). However, interpatient heterogeneity of response to sulindac was evident, with changes ranging from +19% to –89%, and the patient with the greatest reductions after sulindac developed colorectal cancer after 35 months of therapy. Sulindac treatment, at drug doses shown to regress colorectal adenomas in FAP patients, has heterogeneous effects on the level of major prostaglandins in their rectal mucosa and may not prevent colorectal cancer due to uncoupling of prostaglandin levels and carcinogenesis.

Lynch et al.[90] described an FAP patient who developed rectal carcinoma 15 months after beginning chemoprophylaxis with sulindac. There was metastatic adenocarcinoma in 6 of 20 perirectal lymph nodes. In addition to the carcinoma, her rectal mucosa contained two adenomas and multiple foci of adenomatous changes in flat mucosa. It was concluded that, while sulindac may alter the pathogenesis of FAP, those patients undergoing sulindac chemoprevention must be monitored closely by endoscopic examination. This surveillance should include an aggressive biopsy approach, since the absence of *polyps* does not prove the absence of malignant neoplastic changes.

Along this same vein, Matsuhashi et al.[91] described a rapidly growing invasive rectal cancer in 1 of 15 patients treated with sulindac for sporadic adenomatous colorectal polyps 16 months after sulindac treatment. They found that the adenomatous polyp had responded only partially to sulindac, and the rectal cancer developed following sulindac therapy and it showed immunostaining for COX-2. They concluded that while short-term sulindac therapy appears to cause some adenomatous colon polyps to *regress*, nevertheless it may *not* reliably *prevent* CRC in certain patients.

9.6.2 PERIAMPULLARY POLYPS

Richard et al.[92] showed that sulindac did not provide significant benefit for the control of periampullary polyps in FAP patients. However, a recent study suggested that celecoxib may decrease significant (>5% area covered) duodenal polyposis by 31%.[93] Whether this has clinical benefit with decreased duodenal carcinogenesis is unclear.

9.6.3 EXISULIND (SULINDAC SULFONE, FGN-1) IN FAP

van Stolk et al.[94] studied exisulind, which is the sulfone metabolite of sulindac without known effects on prostaglandin synthesis. Yet it promotes apoptosis and inhibits tumorigenesis in preclinical systems. A Phase I trial of this compound was performed in patients with FAP in order to determine tolerability and safety of the drug in a cancer chemoprevention setting. Findings disclosed the maximum safe dose of exisulind to be 300 mg p.o. twice a day in patients with subtotal colectomies. Dose escalation was limited by reversible hepatic dysfunction. Decrease in polyp numbers was not demonstrated. However, there was a trend toward increased apoptosis at the maximum tolerated dose. There was mucinous change histologically, which suggested that further investigation of drugs of this class might be warranted.

9.7 CHEMOPREVENTION AND CRC: GENETIC MODELS

Lynch[95] suggests that CRC is an excellent model for investigating cancer prevention through means of secondary approaches such as polypectomy for removal of precursor colonic adenomas as well as primary (chemoprevention) strategies. Attention was appropriately called to the manner in which the regular use of aspirin and other NSAIDs has led to a reduction in risk of CRC. Herein, a possible mechanism from this benefit relates to the mentioned decreased prostaglandin production,

> which is achieved through inhibition of cyclooxygenase (COX) activity, and possibly other pathways. Two isoforms of COX — COX-1 and COX-2 — have been identified. COX-2 is expressed in colorectal adenomas and carcinomas, both in humans and rodents. Inhibition of COX-2 has been shown to decrease the incidence of carcinogen-induced neoplasia in rats and to lower incidence of adenomas in murine models.

The subject of chemoprevention of CRC using COX-2 inhibitors in patients at increased risk for FAP, HNPCC, and sporadic adenomas was reviewed.[95] The role of the recently described COX-3[96] in carcinogenesis has not been assessed, although its role is unlikely given that acetaminophen inhibits this isoform but not many forms of experimental carcinogenesis.

9.7.1 ADENOMA REGRESSION IN FAP

Steinbach et al.[67] studied the effect of celecoxib on colorectal polyps in FAP-affected patients during a double-blind, placebo-control study. Seventy-seven patients were treated with celecoxib (100 or 400 mg twice daily) or a placebo for 6 months. Endoscopy was performed at the beginning and end of the study, and number and size of polyps were evaluated from photographs and videotapes. Response to therapy was expressed as the mean percent change from baseline. Findings disclosed that after 6 months, those patients who received 400 mg of celecoxib twice a day

> had a 28.0% reduction in the mean number of colorectal polyps (p = 0.003 for the comparison with placebo) and a 30.7% reduction in the polyp burden (the sum of polyp diameters) (p = 0.001), as compared with reductions of 4.5 and 4.9 percent, respectively, in the placebo group. The improvement in the extent of colorectal polyposis in the group receiving 400 mgs twice a day was confirmed by a panel of endoscopists who reviewed the videotapes. The reductions in the group receiving 100 mgs of celecoxib twice a day were 11.9% (p = 0.33 for the comparison with placebo) and 14.6% (p = 0.09), respectively. The incidence of adverse events were similar among the groups.

Thus, it was concluded that 6 months of twice-a-day treatment with 400 mg of celecoxib in FAP patients resulted in a significant reduction in the number of colorectal polyps.

9.7.2 ANGIOGENESIS AND COX-2

Masferrer[97] studied COX-2 and its putative role in effecting angiogenesis, which is a feature of both benign and malignant disease. In this review, Masferrer noted that selective COX-2 inhibitors may be able to modify the progression of polyps through the control of angiogenesis in patients with FAP.

Giles[98] has shown that the vascular endothelial growth factor (VEGF) has been implicated as the major pro-angiogenic factor that regulates multiple endothelial cell functions inclusive of mitogenesis. A direct relationship between VEGF and leukemic blasts as well as in malignant plasma cells has been identified. However, it is suggested that VEGF may have a function distinct from its role in angiogenesis. Furthermore, "current protocols with anti-VEGF agents in patients with hematologic malignancies involve the use of monoclonal antibody, blockers of the VEGF-receptor tyrosine kinase pathway, thalidomide (Thalomid) and its analogs, and cyclooxygenase inhibitors."

9.8 CURCUMIN

Perkins et al.[99] discuss the efficacy and pharmacokinetics of curcumin in the Min/+ mouse FAP model. Curcumin is the yellow pigment in turmeric, which "prevents the development of adenomas in the intestinal tract of the C57BI/6J Min/+ mouse, a model of human APC." They found that curcumin was not effective at 0.1% in the diet, but at 0.2% and 0.5%, it reduced adenoma multiplicity by 39 and 40% respectively, when compared with untreated mice. They concluded that curcumin may play an important role in the chemoprevention of human intestinal malignancies related to *APC* mutations. They suggest that a daily dose of 1.6 g of curcumin would be required for efficacy in humans. Furthermore, they call attention to the clear advantage of curcumin over NSAIDs in "its ability to decrease intestinal bleeding linked to adenoma maturation." Curcumin has been shown to decrease COX-2 expression and inhibit NF-κb,[100] suggesting similar molecular mechanisms as NSAIDs.

9.9 DISCUSSION

Peleg and Wilcox[101] review the historical background of salicylate-containing plant extracts that they note have been used since antiquity with respect to their analgesic, antipyretic, and anti-inflammatory effects.[102,103] Since that time, aspirin has become one of the most widely used medications worldwide in a variety of interventions, including those against cardiovascular disorders. It is now a prototype drug referred to generically as nonsteroidal anti-inflammatory drugs (NSAIDs), as discussed throughout this chapter.

Marnett and DuBois[104] provide an updated review of COX-2 and its role in CRC prevention. A key issue is that COX-2 is expressed at high levels in CRC as well as in a number of differing types of human tumors, but it is absent in the surrounding normal tissue. Herein, it appears to be a paradigm for CRC and can be extended to other anatomic sites. These authors credit Vane,[105] who demonstrated in 1971 that aspirin as well as other NSAIDs acted pharmacologically through inhibiting the production of prostaglandins. They then proposed that the cyclooxygenase enzyme was the molecular target. Experimental and clinical data, as discussed throughout this chapter, have since supported Vane's hypothesis. Importantly, with respect to CRC, epidemiologic studies have shown an approximate 40 to 50% reduction in mortality from CRC in individuals who have used aspirin or other NSAIDs on a regular basis. Herein, they suggest that there is strong evidence to suggest that inhibition of the COX enzymes by NSAIDs is important in this cancer risk reduction.[106]

Marnett and DuBois call attention to the Nurses Health Study,[26] which showed a protective effect on CRC through the use of aspirin that was only observed after 10 to 15 years of its use. They appropriately suggest that the risk–benefit ratio for chemoprevention of CRC would likely improve through its application among groups at high risk for CRC. In turn, they note that should selective COX-2 inhibitors prove to have fewer adverse effects than nonselective NSAIDs, then the risk–benefit ratio would improve. Meanwhile, human clinical trials for investigating the anticancer effects of selective COX-2 inhibitors should lead to a better understanding of those agents that will prove effective for cancer prevention.

Husain et al.[107] also note that there have been an increasing number of epidemiologic studies that suggest the role of aspirin as well as other NSAIDs contributes to the reduction of incidence and mortality from CRC, gastric, and esophageal cancers. However, the precise mechanisms by which NSAIDs lead to chemoprevention remain elusive. The best clues relate to cyclooxygenase, which converts arachidonic acid to prostaglandins. Therein, COX-1 and COX-2 appear to play an important role in that COX-2 is absent in normal mucosa but, conversely, is overexpressed in these GI cancers, including their precursor lesions. Furthermore, inhibition of COX-2 through selective NSAIDs or gene deletion contributes to colonic polyp regression.

Swamy et al.[108] used human HT-29 colon cancer cells to study the effectiveness of lovastatin and celecoxib, "individually or in combination on the induction of apoptosis in human HT-29 colon cancer cells." They also studied "the modulatory effect of lovastatin and celecoxib on lamin B levels, caspase-3 activity and expression in relationship to apoptosis in colon cancer cell lines." Importantly, their results showed that agents with different modes of action, when utilized in combinations, "will induce apoptosis synergistically by enhancing caspase-3 activities." These authors concluded that the regulation of apoptosis is enhanced by selective agents such as lovastatin and celecoxib combinations for colon cancer prevention.

The combinatorial approach to chemoprevention has great attraction in that, like chemotherapy, one could potentially use lower and therefore less toxic doses of each agent. For instance, Torrance and colleagues[109] showed that combining sulindac with an EGF inhibitor allowed lower doses and still led to an almost complete eradication of polyps in a mouse model. Jacoby et al.[110] showed in a mouse model that the NSAID piroxicam had synergistic effects with the polyamine inhibitor DMFO. Multiple agents may be important, since experimental data suggest that NSAIDs appear to preferentially target tumors in the distal colon.[111] Thus, a combinatorial approach may both reduce toxicity and target all regions of the colon.

9.8.1 Will NSAIDs Replace Screening and Prophylactic Colectomy in FAP?

Peleg and Wilcox[101] call attention to the mounting evidence from experimental cancer studies in rodents as well as interventional research on patients with FAP, which collectively suggest a chemopreventive effect of aspirin as well as nonaspirin NSAIDs against CRC. Needed, however, are a better understanding of the discrete chemopreventive mechanisms in concert with genetic susceptibility and risk factor profile features of those patients who will benefit from chemoprevention, verified as low-dose NSAIDs with differing targets for optimizing efficacy and reducing toxicity. Thus, in spite of these promising chemoprevention findings, the authors note that prophylactic colectomy is still mandated for patients with FAP, while early surveillance colonoscopies are necessary for HNPCC, and endoscopic screening will be essential for the prevention of sporadic CRC.

Marnett and DuBois[104] discuss how basic research in biochemistry and pharmacology can contribute to translational studies which, in this case, eventually led to approval of a drug by the FDA for use as a chemopreventive agent in humans. Furthermore, the authors believe that as the genome sequence is more clearly comprehended, additional targets will emerge that will become more effective as chemopreventive choices in the battle against CRC.

Suleiman et al.[112] employed a Markov process to follow a hypothetical cohort of 100,000 subjects aged 50 years until death. Comparing colonoscopy once per year and the use of aspirin to prevent CRC, they found that the use of aspirin saved fewer lives at higher cost. They found that the high complication cost, coupled with the lower efficacy of aspirin, rendered screening colonoscopy to be more cost effective for CRC prevention.

This study is based upon general-population expectations of CRC risk. The use of aspirin may prove more cost effective than colonoscopy when considering patients at inordinately high genetic risk for CRC, such as those patients from HNPCC families. Furthermore, since common polymorphisms in drug-metabolizing systems may affect the efficacy of aspirin in colorectal neoplasia prevention,[113] it is possible that these factors can be used to individualize chemopreventive strategies.

Ricciardiello et al.[73] support celecoxib as an agent for chemoprevention in FAP and suggest that these agents might be used in CRC prevention in patients with positive family histories of CRC.

9.10 CONCLUSION

The most cogent cancer prevention concern about the large number of chemopreventive agents reviewed in this chapter is the need to admonish patients who are taking these chemopreventive agents that they still need to adhere to highly targeted screening and, in disorders such as FAP, to prophylactic colectomy in order to prevent CRC. In other words, physicians must, at the outset of a chemoprevention trial, convince the patient that this is not a substitute for screening. Imperiale[114] emphasizes this point in his editorial on the papers by Sandler et al.[42] and Baron et al.,[43] saying that while aspirin may provide some degree of benefit in preventing CRC, "it cannot yet be recommended for this indication and is not a substitute for screening and surveillance." We fervently believe that this same admonition is well founded and should apply to every single chemoprevention agent that we have reviewed in this chapter.

ACKNOWLEDGMENT

This chapter was supported by revenue from Nebraska cigarette taxes awarded to Creighton University by the Nebraska Department of Health and Human Services. Its contents are solely the responsibility of the authors and do not necessarily represent the official views of the State of Nebraska or the Nebraska Department of Health and Human Services.

REFERENCES

1. Bentrem, D.J. and Jordan V.C., Tamoxifen, raloxifene and the prevention of breast cancer, *Minerva Endocrinol.*, 27, 139, 2002.
2. O'Regan, R.M. and Jordan, V.C., Tamoxifen to raloxifene and beyond, *Semin. Oncol.*, 28, 260–273, 2001.

3. Jordan, V.C., Progress in the prevention of breast cancer: concept to reality, *J. Steroid Boichem. Mol. Biol.*, 74, 269–277, 2000.
4. Jordan, V.C., Gapstur, S., and Morrow, M., Selective estrogen receptor modulation and reduction in risk of breast cancer, osteoporosis, and coronary heart disease, *J. Natl. Cancer Inst.*, 93, 1449–1457, 2001.
5. Narod, S.A., Dube, M.P., Klijn, J., Lubinski, J., Lynch, H.T., Ghadirian, P. et al., Oral contraceptives and the risk of breast cancer in BRCA1 and BRCA2 mutation carriers, *J. Natl. Cancer Inst.*, 94, 1773–1779, 2002.
6. Gwyn, K. and Sinicrope, F.A., Chemoprevention of colorectal cancer, *Am. J. Gastroenterol.*, 97, 13–21, 2002.
7. Hong, W.K. and Sporn, M.B., Recent advances in chemoprevention of cancer, *Science*, 278, 1073–1077, 1997.
8. Stratton, M.S., Current application of selective COX-2 inhibitors in cancer prevention and treatment, *Oncology*, 16 (Suppl. 4), 37–51, 2002.
9. Hawk, E., Lubet, R., and Limburg, P., Chemoprevention in hereditary colorectal cancer syndromes, *Cancer*, 86, 2551–2563, 1999.
10. King, J.E., Dozois, R.R., Lindor, N.M., and Ahlquist, D.A., Care of patients and their families with familial adenomatous polyposis, *Mayo Clin. Proc.*, 75, 57–67, 2000.
11. Moran, E.M., Epidemiological and clinical aspects of nonsteroidal anti-inflammatory drugs and cancer risks, *J. Environ. Pathol. Toxicol. Oncol.*, 21, 193–201, 2002.
12. Murphy, T.K., Calle, E.E., Rodriguez, C., Kahn, H.S., and Thun, M.J., Body mass index and colon cancer mortality in a large prospective study, *Am. J. Epidemiol.*, 152, 847–854, 2000.
13. Kune, G.A., Kune, S., and Watson, L.F., Colorectal cancer risk, chronic illnesses, operations, and medications: case control results from the Melbourne Colorectal Cancer Study, *Cancer Res.*, 48, 4399–4404, 1988.
14. Rosenberg, L., Louik, C., and Shapiro, S., Nonsteroidal antiinflammatory drug use and reduced risk of large bowel carcinoma, *Cancer*, 82, 2326–2333, 1998.
15. Logan, R.F.A., Little, J., Hawtin, P.G., and Hardcastle, J.D., Effect of aspirin and non-steroidal anti-inflammatory drugs on colorectal adenomas: case-control study of subjects participating in the Nottingham faecal occult blood screening programme, *Br. Med. J.*, 307, 285–289, 1993.
16. Suh, O., Mettlin, C., and Petrelli, N.J., Aspirin use, cancer, and polyps of the large bowel, *Cancer*, 72, 1171–1177, 1993.
17. Peleg, I.I., Maibach, H.T., Brown, S.H., and Wilcox, C.M., Aspirin and nonsteroidal anti-inflammatory drug use and the risk of subsequent colorectal cancer, *Arch. Intern. Med.*, 154, 394–399, 1994.
18. Muscat, J.E., Stellman, S.D., and Wynder, E.L., Nonsteroidal antiinflammatory drugs and colorectal cancer, *Cancer*, 74, 1847–1854, 1994.
19. Martinez, M.E., McPherson, R.S., Levin, B., and Annegers, J.F., Aspirin and other nonsteroidal anti-inflammatory drugs and risk of colorectal adenomatous polyps among endoscoped individuals, *Cancer Epidemiol. Biomarkers Prev.*, 4, 703–707, 1995.
20. Thun, M.J., Namboodiri, M.M., and Heath, C.J., Aspirin use and reduced risk of fatal colon cancer, *New England J. Med.*, 325, 1593–1596, 1991.
21. Thun, M.J., Namboodiri, M.M., Calle, E.E., Flanders, W.D., and Heath, C.W.J., Aspirin use and risk of fatal cancer, *Cancer Res.*, 53, 1322–1327, 1993.
22. Isomaki, H.A., Hakulinen, T., and Joutsenlahti, U., Excess risk of lymphomas, leukemia and myeloma in patients with rheumatoid arthritis, *J. Chron. Dis.*, 31, 691–696, 1978.
23. Gridley, G., McLaughlin, J.K., Ekbom, A., Klareskog, L., Adami, H.O., Hacker, D.G. et al., Incidence of cancer among patients with rheumatoid arthritis, *J. Natl. Cancer Inst.*, 85, 307–311, 1993.
24. Greenberg, E.R., Baron, J.A., Freeman, D.H.J., Mandel, J.S., and Haile, R., The Polyp Prevention Study Group: reduced risk of large-bowel adenomas among aspirin users, *J. Natl. Cancer Inst.*, 85, 912–916, 1993.
25. Schreinemachers, D.M. and Everson, R.B., Aspirin use and lung, colon, and breast cancer incidence in a prospective study, *Epidemiology*, 5, 138–146, 1994.
26. Giovannucci, E., Egan, K.M., Hunter, D.J., Stampfer, M.J., Colditz, G.A., Willett, W.C. et al., Aspirin and the risk of colorectal cancer in women, *New England J. Med.*, 333, 609–614, 1995.

27. Giovannucci, E., Rimm, E.B., Stampfer, M.J., Colditz, G.A., Ascherio, A., and Willett, W.C., Aspirin use and the risk for colorectal cancer and adenoma in male health professionals, *Ann. Intern. Med.*, 121, 241–246, 1994.
28. Gann, P.H., Manson, J.E., Glynn, J., Buring, J.E., and Hennekens, C.H., Low-dose aspirin and incidence of colorectal tumors in a randomized trial, *J. Natl. Cancer Inst.*, 85, 1220–1224, 1993.
29. Thun, M.J., NSAID use and decreased risk of gastrointestinal cancers, *Gastroenterol. Clin. North Am.*, 25, 333–348, 1996.
30. Thun, M.J., Calle, E.E., Namboodiri, M.M., Flanders, W.D., Coates, R.J., Byers, T. et al., Risk factors for fatal colon cancer in a large prospective study, *J. Natl. Cancer Inst.*, 84, 1491–1500, 1992.
31. Paganini-Hill, A., Aspirin and colorectal cancer: seems to reduce risk, *Br. Med. J.*, 307, 278–279, 1993.
32. Stürmer, T., Glynn, R.J., Lee, I.-M., Manson, J., Buring, J.E., and Hennekens, C.H., Aspirin use and colorectal cancer: post-trial follow-up data from the Physicians' Health Study, *Ann. Intern. Med.*, 128, 713–720, 1998.
33. Rossouw, J.E., Anderson, G.L., Prentice, R.L., LaCroix, A.Z., Kooperberg, C., Stefanick, M.L. et al., Risks and benefits of estrogen plus progestin in healthy postmenopausal women: principal results from the Women's Health Initiative randomized controlled trial, *JAMA*, 288, 321–333, 2002.
34. Davis, A.E. and Patterson, F., Aspirin reduces the incidence of colonic carcinoma in the dimethylhydrazine rat animal model, *Aust. N.Z. J. Med.*, 24, 301–303, 1994.
35. Oshima, M., Murai(Hata), N., Kargman, S., Arguello, M., Luk, P., Kwong, E. et al., Chemoprevention of intestinal polyposis in the $Apc^{\Delta716}$ mouse by rofecoxib, a specific cyclooxygenase-2 inhibitor, *Cancer Res.*, 61, 1733–1740, 2001.
36. Ahnen, D.J., Colon cancer prevention by NSAIDs: what is the mechanism of action? *Eur. J. Surg.*, 164, 111–114, 1998.
37. Piazza, G.A., Rahm, A.K., Finn, T.S., Fryer, B.H., Li, H., Stoumen, A.L. et al., Apoptosis primarily accounts for the growth-inhibitory properties of sulindac metabolites and involves a mechanism that is independent of cyclooxygenase inhibition, cell cycle arrest, and p53 induction, *Cancer Res.*, 57, 2452–2459, 1997.
38. Giardiello, F.M., Offerhaus, J.A., Tersmette, A.C., Hylind, L.M., Krush, A.J., Brensinger, J.D. et al., Sulindac induced regression of colorectal adenomas in familial adenomatous polyposis: evaluation of predictive factors, *Gut*, 38, 578–581, 1996.
39. Giardiello, F.M., Yang, V.W., Hylind, L.M., Krush, A.J., Petersen, G.M., Trimbath, J.D. et al., Primary chemoprevention of familial adenomatous polyposis with sulindac, *New England J. Med.*, 346, 1054–1059, 2002.
40. Sjödahl, R., Extent, mode, and dose dependence of anticancer effects, *Am. J. Med.*, 110 (1A), 66S–69S, 2001.
41. Pasricha, P.J., Bedi, A., O'Connor, K., Rashid, A., Akhtar, A.J., Zahurak, M.L. et al., The effects of sulindac on colorectal proliferation and apoptosis in familial adenomatous polyposis, *Gastroenterology*, 109, 994–998, 1995.
42. Sandler, R.S., Halabi, S., Baron, J.A., Budinger, S., Paskett, E., Keresztes, R. et al., A randomized trial of aspirin to prevent colorectal adenomas in patients with previous colorectal cancer, *New England J. Med.*, 348, 883–890, 2003.
43. Baron, J.A., Cole, B.F., Sandler, R.S., Haile, R.W., Ahnen, D., Bresalier, R. et al., A randomized trial of aspirin to prevent colorectal adenomas, *New England J. Med.*, 348, 891–899, 2003.
44. Oshima, M., Dinchuk, J.E., Kargman, S.L., Oshima, H., Hancock, B., Kwong, E. et al., Suppression of intestinal polyposis in $Apc^{\Delta716}$ knockout mice by inhibition of cyclooxygenase 2 (COX-2), *Cell*, 87, 803–809, 1996.
45. Chulada, P.C., Thompson, M.B., Mahler, J.F., Doyle, C.M., Gaul, B.W., Lee, C. et al., Genetic disruption of Ptgs-1, as well as Ptgs-2, reduces intestinal tumorigenesis in Min mice, *Cancer Res.*, 60, 4705–4708, 2000.
46. Chan, T.A., Nonsteroidal anti-inflammatory drugs, apoptosis, and colon-cancer chemoprevention, *Lancet Oncol.*, 3, 166–174, 2002.
47. Zhu, J., Song, X., Lin, H.P., Young, D.C., Yan, S., Marquez, V.E. et al., Using cyclooxygenase-2 inhibitors as molecular platforms to develop a new class of apoptosis-inducing agents, *J. Natl. Cancer Inst.*, 94, 1745–1757, 2002.

48. Roy, H.K., Olusola, B.F., Clemens, D.L., Karolski, W.J., Ratashak, A., Lynch, H.T. et al., AKT proto-oncogene overexpression is an early event during sporadic colon carcinogenesis, *Carcinogenesis*, 23, 201–205, 2002.

49. Gustafson-Svard, C., Lilja, I., Hallbook, O., and Sjodahl, R., Cyclo-oxygenase and colon cancer: clues to the aspirin effect? *Ann. Med.*, 29, 247–252, 1997.

50. Chan, T.A., Morin, P.J., Vogelstein, B., and Kinzler, K.W., Mechanisms underlying nonsteroidal antiinflammatory drug-mediated apoptosis, *PNAS*, 95, 681–686, 1998.

51. Ota, S., Tanaka, Y., Bamba, H., Kato, A., and Matsuzaki, F., Nonsteroidal anti-inflammatory drugs may prevent colon cancer through the suppression of hepatocyte growth factor expression, *Eur. J. Pharmacol.*, 367, 131–138, 1999.

52. Dong, Z., Huang, C., Brown, R.E., and Ma, W.-Y., Inhibition of activator protein 1 activity and neoplastic transformation by aspirin, *J. Biol. Chem.*, 272, 9962–9970, 1997.

53. Qiao, L., Hanif, R., Sphicas, E., Shiff, S.J., and Rigas, B., Effect of aspirin on induction of apoptosis in HT-29 human colon adenocarcinoma cells, *Biochem. Pharmacol.*, 55, 53–64, 1998.

54. Elder, D.J.E., Hague, A., Hicks, D.J., and Paraskeva, C., Differential growth inhibition by the aspirin metabolite salicylate in human colorectal tumor cell lines: enhanced apoptosis in carcinoma and *in vitro*-transformed adenoma relative to adenoma cell lines, *Cancer Res.*, 56, 2273–2276, 1996.

55. Schwenger, P., Bellosta, P., Vietor, I., Basilico, C., Skolnik, E.Y., and Vilcek, J., Sodium salicylate induces apoptosis via p38 mitogen-activated protein kinase but inhibits tumor necrosis factor-induced c-Jun N-terminal kinase/stress-activated protein kinase activation, *Proc. Natl. Acad. Sci. USA*, 94, 2869–2873, 1997.

56. Bellosillo, B., Pique, M., Barragán, M., Castaño, E., Villamor, N., Colomer, D. et al., Aspirin and salicylate induce apoptosis and activation of caspases in B-cell chronic lymphocytic leukemia cells, *Blood*, 92, 1406–1414, 1998.

57. Gupta, R.A. and DuBois, R.N., Aspirin, NSAIDs, and colon cancer prevention: mechanisms? *Gastroenterology*, 114, 1095–1098, 1998.

58. Petrescu, I. and Tarba, C., Uncoupling effects of diclofenac and aspirin in the perfused liver and isolated hepatic mitochondria of rat, *Biochim. Biophys. Acta*, 1318, 385–394, 1997.

59. Karnes, Jr., W.E., Shattuck-Brandt, R., Burgart, L.J., DuBois, R.N., Tester, D.J., Cunningham, J.M. et al., Reduced COX-2 protein in colorectal cancer with defective mismatch repair, *Cancer Res.*, 58, 5473–5477, 1998.

60. Goel, A., Chang, D.K., Ricciardiello, L., Gasche, C., and Boland, C.R., A novel mechanism for aspirin-mediated growth inhibition of human colon cancer cells, *Clin. Cancer Res.*, 9, 383–390, 2003.

61. Rüschoff, J., Wallinger, S., Dietmaier, W., Bocker, T., Brockhoff, G., Hofstädter, F. et al., Aspirin suppresses the mutator phenotype associated with hereditary nonpolyposis colorectal cancer by genetic selection, *Proc. Natl. Acad. Sci. USA*, 95, 11301–11306, 1998.

62. Reddy, B.S. and Rao, C.V., Novel approaches for colon cancer prevention by cyclooxygenase-2 inhibitors, *J. Environ. Pathol. Toxicol. Oncol.*, 21, 155–164, 2002.

63. Sample, D., Wargovich, M., Fischer, S.M., Inamdar, N., Schwartz, P., Wang, X. et al., A dose-finding study of aspirin for chemoprevention utilizing rectal mucosal prostaglandin E_2 levels as a biomarker, *Cancer Epidemiol. Biomarkers Prev.*, 11, 275–279, 2002.

64. Fiorucci, S., Santucci, L., Gresele, P., Faccino, R.M., Del Soldato, P., and Morelli, A., Gastrointestinal safety of NO-aspirin (NCX-4016) in healthy human volunteers: a proof of concept endoscopic study, *Gastroenterology*, 124, 600–607, 2003.

65. Rigas, B. and Williams, J.L., NO-releasing NSAIDs and colon cancer chemoprevention: a promising novel approach, *Int. J. Oncol.*, 20, 885–890, 2002.

66. Giardiello, F.M., Hamilton, S.R., Krush, A.J., Piantadosi, S., Hylind, L.M., Celano, P. et al., Treatment of colonic and rectal adenomas with sulindac in familial adenomatous polyposis, *New England J. Med.*, 328, 1313–1316, 1993.

67. Steinbach, G., Lynch, P.M., Phillips, R.K.S., Wallace, M.H., Hawk, E., Gordon, G.B. et al., The effect of celecoxib, a cyclooxygenase inhibitor, in familial adenomatous polyposis, *New England J. Med.*, 342, 1946–1952, 2000.

68. Lauer, M.S., Clinical practice: aspirin for primary prevention of coronary events, *New England J. Med.*, 346, 1468–1474, 2002.

69. Ladabaum, U., Chopra, C.L., Huang, G., Scheiman, J.M., Chernew, M.E., and Fendrick, A.M., Aspirin as an adjunct to screening for prevention of sporadic colorectal cancer: a cost-effectiveness analysis, *Ann. Intern. Med.*, 135, 769–781, 2001.

70. Martin, C., Connelly, A., Keku, T.O., Mountcastle, S.B., Galanko, J., Woosley, J.T. et al., Nonsteroidal anti-inflammatory drugs, apoptosis, and colorectal adenomas, *Gastroenterology*, 123, 1770–1777, 2002.

71. Keller, J., Offerhaus, G.J.A., Polak, M., Goodman, S.N., Zahurak, M.L., Hylind, L.M. et al., Rectal epithelial apoptosis in familial adenomatous polyposis patients treated with sulindac, *Gut*, 45, 822–828, 1999.

72. Keller, J.J., Offerhaus, G.J.A., Hylind, L.M., and Giardiello, F.M., Rectal epithelial apoptosis does not predict response to sulindac treatment or polyp development in presymptomatic familial adenomatous polyposis patients, *Cancer Epidemiol. Biomarkers Prev.*, 11, 670–671, 2002.

73. Ricciardiello, L., Roda, E., and Bazzoli, F., Chemoprevention in colorectal neoplasias: what is practical and feasible? *Dig. Dis.*, 20, 70–72, 2002.

74. Lal, G., Ash, C., Hay, K., Redston, M., Kwong, E., Hancock, B. et al., Suppression of intestinal polyps in Msh2-deficient and non-Msh2-deficient multiple intestinal neoplasia mice by a specific cyclooxygenase-2 inhibitor and by a dual cyclooxygenase-1/2 inhibitor, *Cancer Res.*, 61, 6131–6136, 2001.

75. Oshima, M. and Taketo, M.M., COX selectivity and animal models for colon cancer, *Curr. Pharm. Des.*, 8, 1021–1034, 2002.

76. Kargman, S.L., O'Neill, G.P., Vickers, P.J., Evans, J.F., Mancini, J.A., and Jothy, S., Expression of prostaglandin G/H synthase-1 and -2 protein in human colon cancer, *Cancer Res.*, 55, 2556–2559, 1995.

77. Williams, C.S., Tsujii, M., Reese, J., Dey, S.K., and DuBois, R.N., Host cyclooxygenase-2 modulates carcinoma growth, *J. Clin. Invest.*, 105, 1589–1594, 2000.

78. Sheehan, K.M., Sheahan, K., O'Donoghue, D.P., MacSweeney, F., Conroy, R.M., Fitzgerald, D.J. et al., The relationship between cyclooxygenase-2 expression and colorectal cancer, *JAMA*, 282, 1254–1257, 1999.

79. Yao, M., Kargman, S., Lam, E.C., Kelly, C.R., Zheng, Y., Luk, P. et al., Inhibition of cyclooxygenase-2 by rofecoxib attenuates the growth and metastatic potential of colorectal carcinoma in mice, *Cancer Res.*, 63, 586–592, 2003.

80. North, G.L., Celecoxib as adjunctive therapy for treatment of colorectal cancer, *Ann. Pharmacother.*, 35, 1638–1643, 2001.

81. Blanke, C.D., Celecoxib with chemotherapy in colorectal cancer, *Oncology*, 16 (Suppl. 3), 17–21, 2002.

82. Marks, F. and Fürstenberger, G., Cancer chemoprevention through interruption of multistage carcinogenesis: the lessons learnt by comparing mouse skin carcinogenesis and human large bowel cancer, *Eur. J. Cancer*, 36, 314–329, 2000.

83. Waddell, W.R. and Loughry, R.W., Sulindac for polyposis of the colon, *J. Surg. Oncol.*, 24, 83–87, 1983.

84. Waddell, W.R., Ganser, G.F., Cerise, E.J., and Loughry, R.W., Sulindac for polyposis of the colon, *Am. J. Surg.*, 157, 175–179, 1989.

85. Nugent, K.P., Farmer, K.C.R., Spigelman, A.D., Williams, C.B., and Phillips, R.K.S., Randomized controlled trial of the effect of sulindac on duodenal and rectal polyposis and cell proliferation in patients with familial adenomatous polyposis, *Br. J. Surg.*, 80, 1618–1619, 1993.

86. Labayle, D., Fischer, D., Vielh, P., Drouhin, F., Pariente, A., Bories, C. et al., Sulindac causes regression of rectal polyps in familial adenomatous polyposis, *Gastroenterology*, 101, 635–639, 1993.

87. Keller, J.J., Offerhaus, G.J.A., Drillenburg, P., Caspers, E., Musler, A., Ristimaki, A. et al., Molecular analysis of sulindac-resistant adenomas in familial adenomatous polyposis, *Clin. Cancer Res.*, 7, 4000–4007, 2001.

88. Cruz-Correa, M., Hylind, L.M., Romans, K.E., Booker, S.V., and Giardiello, F.M., Long-term treatment with sulindac in familial adenomatous polyposis: a prospective cohort study, *Gastorenterology*, 122, 641–645, 2002.

89. Giardiello, F.M., Spannhake, E.W., DuBois, R.N., Hylind, L.M., Robinson, C.R., Hubbard, W.C. et al., Prostaglandin levels in human colorectal mucosa: effects of sulindac in patients with familial adenomatous polyposis, *Dig. Dis. Sci.*, 43, 311–316, 1998.

90. Lynch, H.T., Thorson, A.G., and Smyrk, T., Rectal cancer after prolonged sulindac chemoprevention: a case report, *Cancer*, 75, 936–938, 1995.

91. Matsuhashi, N., Nakajima, A., Shinohara, K., Oka, T., and Yazaki, Y., Rectal cancer after sulindac therapy for a sporadic adenomatous colonic polyp, *Am. J. Gastroenterol.*, 93, 2261–2266, 1998.

92. Richard, C.S., Berk, T., Bapat, B.V., Haber, G., Cohen, Z., and Gallinger, S., Sulindac for periampullary polyps in FAP patients, *Int. J. Colorectal Dis.*, 12, 14–18, 1997.

93. Phillips, R.K.S., Wallace, M.H., Lynch, P.M., Hawk, E., Gordon, G.B., Saunders, B.P. et al., A randomised, double blind, placebo controlled study of celecoxib, a selective cyclooxygenase 2 inhibitor, on duodenal polyposis in familial adenomatous polyposis, *Gut*, 50, 857–860, 2002.

94. van Stolk, R., Stoner, G., Hayton, W.L., Chan, K., DeYoung, B., Kresty, L. et al., Phase I trial of exisulind (sulindac sulfone, FGN-1) as a chemopreventive agent in patients with familial adenomatous polyposis, *Clin. Cancer Res.*, 6, 78–89, 2000.

95. Lynch, P.M., COX-2 inhibition in clinical cancer prevention, *Oncology (Huntingt.)*, 15 (Suppl. 5), 21–26, 2001.

96. Chandrasekharan, N.V., Dai, H., Roos, K.L., Evanson, N.K., Tomsik, J., Elton, T.S. et al., COX-3, a cyclooxygenase-1 variant inhibited by acetaminophen and other analgesic/antipyretic drugs: cloning, structure, and expression, *Proc. Natl. Acad. Sci. USA*, 99, 13926–13931, 2002.

97. Masferrer, J., Approach to angiogenesis inhibition based on cyclooxygenase-2, *Cancer J.*, 7 (Suppl. 3), S144–S150, 2001.

98. Giles, F.J., The emerging role of angiogenesis inhibitors in hematologic malignancies, *Oncology (Huntingt.)*, 16 (Suppl. 4), 23–29, 2002.

99. Perkins, S., Verschoyle, R.D., Hill, K., Parveen, I., Threadgill, M.D., Sharma, R.A. et al., Chemopreventive efficacy and pharmacokinetics of curcumin in the Min/+ mouse, a model of familial adenomatous polyposis, *Cancer Epidemiol. Biomarkers Prev.*, 11, 535–540, 2002.

100. Plummer, S.M., Holloway, K.A., Manson, M.M., Munks, R.J., Kaptein, A., Farrow, S. et al., Inhibition of cyclo-oxygenase 2 expression in colon cells by the chemopreventive agent curcumin involves inhibition of NF-kappaB activation via the NIK/IKK signalling complex, *Oncogene*, 18, 6013–6020, 1999.

101. Peleg, I.I. and Wilcox, C.M., The role of eicosanoids, cyclooxygenases, and nonsteroidal anti-inflammatory drugs in colorectal tumorigenesis and chemoprevention, *J. Clin. Gastroenterol.*, 34, 117–125, 2002.

102. Gross, M. and Greenberg, L., *The Salicylates: a Critical Bibliographic Review*, Hillhouse Press, New Haven, CT, 1948.

103. Jack, D., One hundred years of aspirin, *Lancet*, 350, 437–439, 1997.

104. Marnett, L.J. and DuBois, R.N., COX-2: a target for colon cancer prevention, *Annu. Rev. Pharmacol. Toxicol.*, 42, 55–80, 2002.

105. Vane, J.R., Inhibition of prostaglandin synthesis as a mechanism of action for aspirin-like drugs, *Nat. New Biol.*, 231, 232–235, 1971.

106. Marx, J., Cancer research: anti-inflammatories inhibit cancer growth — but how? *Science* 291, 581–582, 2001.

107. Husain, S.S., Szabo, I.L., and Tarnawski, A.S., NSAID inhibition of GI cancer growth: clinical implications and molecular mechanisms of action, *Am. J. Gastroenterol.*, 97, 542–553, 2002.

108. Swamy, M.V., Cooma, I., Reddy, B.S., and Rao, C.V., Lamin B, caspase-3 activity, and apoptosis induction by a combination of HMG-CoA reductase inhibitor and COX-2 inhibitors: a novel approach in developing effective chemopreventive regimens, *Int. J. Oncol.*, 20, 753–759, 2002.

109. Torrance, C.J., Jackson, P.E., Montgomery, E., Kinzler, K.W., Vogelstein, B., Wissner, A. et al., Combinatorial chemoprevention of intestinal neoplasia, *Nat. Med.*, 6, 1024–1028, 2000.

110. Jacoby, R.F., Cole, C.E., Tutsch, K., Newton, M.A., Kelloff, G., Hawk, E.T. et al., Chemopreventive efficacy of combined piroxicam and difluoromethylornithine treatment of Apc mutant Min mouse adenomas, and selective toxicity against Apc mutant embryos, *Cancer Res.*, 60, 1864–1870, 2000.

111. Roy, H.K., Karolski, W.J., and Ratashak. A., Distal bowel selectivity in the chemoprevention of experimental colon carcinogenesis by the non-steroidal anti-inflammatory drug nabumetone, *Int. J. Cancer*, 92, 609–615, 2001.

112. Suleiman, S., Rex, D.K., and Sonnenberg, A., Chemoprevention of colorectal cancer by aspirin: a cost-effectiveness analysis, *Gastroenterology*, 122, 78–84, 2002.

113. Bigler, J., Whitton, J., Lampe, J.W., Fosdick, L., Bostick, R.M., and Potter, J.D., CYP2C9 and UGT1A6 genotypes modulate the protective effect of aspirin on colon adenoma risk, *Cancer Res.*, 61, 3566–3569, 2001.

114. Imperiale, T.F., Aspirin and the prevention of colorectal cancer, *New England J. Med.*, 348, 879–880, 2003.

10 Current Perspectives in Gastric Adenocarcinoma

Clinton Snedegar and Bhaskar Banerjee

CONTENTS

10.1 Introduction..143
10.2 Incidence and Epidemiology...143
10.3 Etiology...144
10.4 Pathology 145
10.5 Clinical Features..145
10.6 Diagnosis 146
10.7 Classification and Prognosis 147
10.8 Therapy 148
10.9 Summary 149
References 149

10.1 INTRODUCTION

Gastric adenocarcinoma is one of the leading causes of cancer morbidity and mortality worldwide. Gastric cancer comprises approximately 95% of all gastric tumors, and is the focus of this chapter. Other less common gastric tumors not discussed here include mucosa-associated lymphoid tissue (MALT) lymphoma, gastrointestinal stromal cell tumors, and carcinoid tumors.

10.2 INCIDENCE AND EPIDEMIOLOGY

Gastric adenocarcinoma is the world's second most common malignancy and the second leading cause of death from cancer. The highest rates of gastric adenocarcinoma are found in Japan and Eastern Asia, in many of the countries that once comprised the former Soviet Union, and in a few, more focal areas (Costa Rica, Portugal, for example). The prevalence of gastric cancer in the U.S. is relatively low, but not insignificant, at less than 10 new cases per 100,000 persons per year. In 2001, an estimated 22,800 gastric cancers were diagnosed in the U.S., with approximately 14,000 deaths.[1]

In the U.S., patients most commonly present after age 60. The age of presentation in many Asian countries, particularly Japan, is somewhat lower, likely due to earlier cancer detection through aggressive gastric cancer screening programs. Men are twice as likely as women to be affected. Death from cancer is more prevalent among African-Americans, Native and Hispanic Americans, and those of low socioeconomic status.[2]

The incidence of gastric adenocarcinoma has been declining worldwide over the last several decades, with most of that decline occurring in the developed Western world (Figure 10.1). The etiology for this decrease is unknown, although changes in patterns of environmental exposures, including *Helicobacter pylori* infection, have been postulated.

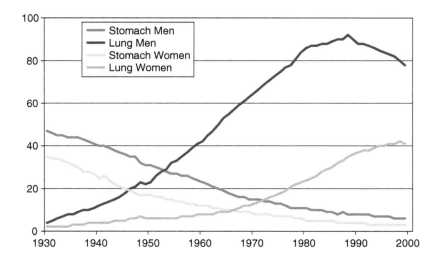

FIGURE 10.1 Gastric and lung cancer death rates by gender per 100,000, age-adjusted to the 2000 U.S. standard population, 1930–1999. (Adapted with permission from the American Cancer Society, Cancer Facts and Figures 2003.)

10.3 ETIOLOGY

Multiple risk factors have been identified throughout the years with regard to gastric cancer. Given the regional variations in disease prevalence, focus has generally been on environmental exposures and dietary concerns. However, with the relatively recent discovery of *Helicobacter pylori* as a pathogenic species, more attention is being given to it as a causative factor in gastric cancer.

There is a strong connection between gastric cancer and atrophic gastritis. The most common cause of atrophic gastritis is chronic infection with *H. pylori*, a Gram-negative curved bacillus that colonizes the mucin layer of the gastric lining. Its presence has been implicated in several upper gastrointestinal conditions, including nonulcer dyspepsia, acute and chronic gastritis, duodenal and gastric ulcer, mucosa-associated lymphoid tissue (MALT) lymphoma, and gastric adenocarcinoma.

The pattern of *H. pylori* infection in the stomach leads to either increased or decreased acid secretion. Increased acid secretion predisposes to duodenal ulcer disease, whereas reduced secretion yields pangastritis and eventually gastric atrophy, precursors of gastric cancer. However, most infected individuals do not have clinically significant disease. The basis for these disparities is not clear, though virulence factors of certain *H. pylori* strains and specific host genetic factors likely play a role.

H. pylori and the associated chronic inflammation are thought to initiate a cascade of events leading to atrophic gastritis, metaplasia, dysplasia, and cancer (Figure 10.2).[3] Once progression to dysplasia occurs, continued infection is probably unnecessary for continuation of the neoplastic sequence. In fact, the atrophic stomach may itself be unsuitable for continued *H. pylori* infection, and markers of infection may be negative at the time of diagnosis.[4]

Several nutritional factors have been suggested as risk factors for gastric cancer. Given all of the possible combinations of a given diet, it has been difficult to tease out which foods may be the most readily implicated, and certainly dietary habits differ greatly from region to region, even among high-risk regions. Briefly, foods with a high nitrate content (preserved foods) and high salt content (particularly pickled foods and salted fish and meat) may increase cancer risk.[5,6] It should be noted that the decrease in gastric cancer does tend to correlate with the advent of refrigeration and less need for traditional preserving methods. Fresh fruits and vegetables, on the other hand, particularly in the uncooked form, reduce cancer risk, perhaps through their antioxidant content.[7]

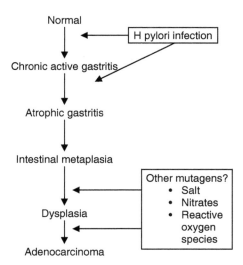

FIGURE 10.2 Proposed sequence of pathologic events in gastric cancer.

Cigarette smoking seems to increase gastric cancer risk approximately two-fold, though a dose–response relationship has not been identified.[8] Alcohol consumption is not by itself an independent risk factor.

10.4 PATHOLOGY

There are two histologic types of gastric adenocarcinoma as defined by Lauren in the 1960s: intestinal and diffuse.[9] The intestinal type is the predominant form, is characterized by glandular-type structures, and is usually well differentiated. It is more common in men than women, and the reason for this is unclear, although a hormonal basis has been postulated. This form is more frequently associated with identifiable environmental risk factors and chronic *H. pylori* infection and is thought to be preceded by the series of postulated precancerous steps as described in Figure 10.2. Overall, it is this form that is declining in frequency worldwide. The diffuse type of gastric cancer is declining at a slower rate than that of the intestinal type. This form is also more common in developed countries. It is less differentiated and is characterized by layers of cells without gland formation and occasional signet ring cells and mucin. There are no defined precancerous lesions of the diffuse type. It is found more frequently in younger patients and carries a worse prognosis.

10.5 CLINICAL FEATURES

The presenting symptoms of gastric adenocarcinoma are protean, nonspecific, and usually occur at a late stage of the disease, contributing to the generally poor prognosis at the time of diagnosis. The most common symptoms are weight loss and abdominal discomfort, though nausea, vomiting, dysphagia, early satiety, dyspepsia, and melena may also occur. Liver, lung, and bone are the most common sites for metastatic disease. Tumors may progress directly through the wall of the stomach, directly invading neighboring tissues, or lead to peritoneal carcinomatosis or drop metastases. Metastatic disease may yield jaundice in the case of liver metastasis, malignant ascites in the presence of carcinomatosis, or bone pain with bony metastasis, for example.

10.6 DIAGNOSIS

The physical examination in patients with gastric adenocarcinoma is usually normal. In advanced disease, a palpable abdominal mass, hepatomegaly, or ascites may be present. Lymph nodes may be palpated at the umbilicus (Sister Mary Joseph's node) or the supraclavicular region (Virchow's node). A perirectal mass may occur, with tumor deposition in the rectal pouch (Blumer's shelf), and palpable ovarian enlargement may occur with tumor seeding as well (Krukenberg tumor).

Laboratory testing is likewise frequently unrevealing until advanced disease occurs. Anemia and positive fecal occult blood testing may be present. Liver tests may be elevated in metastatic liver disease, while the alkaline phosphatase and calcium may be elevated with bony metastases. While certain tumor markers may be elevated (CEA, CA 19-9, CA 72-4), they are unreliable for diagnosis.

Upper gastrointestinal endoscopy is the gold standard in diagnosis. The gastric mucosa can be readily evaluated and tissue biopsies taken. The endoscopic appearance of gastric cancer can range from minor mucosal abnormalities, such as an abnormal fold or subtle ulceration, to a frank mass (Figure 10.3). Early gastric cancer may be more difficult to detect, and in Japan, dyes administered through the scope are sometimes used in an attempt to improve diagnostic accuracy. These techniques are not widely used in the U.S. Gastric cancer can often present with gastric ulceration, and sometimes distinguishing between peptic ulcer and cancer can be difficult. When a gastric ulcer is identified, repeat endoscopy after an appropriate course of acid suppression is recommended to document healing. Nonhealing ulcers should be extensively biopsied at their margins to assess for neoplasia. Linitis plastica (leather bottle stomach) is the result of an aggressively infiltrative tumor, with high-grade tumor cells spreading throughout most of the stomach yielding a coarse, noncompliant stomach. This form carries a poor prognosis, but is quite uncommon.

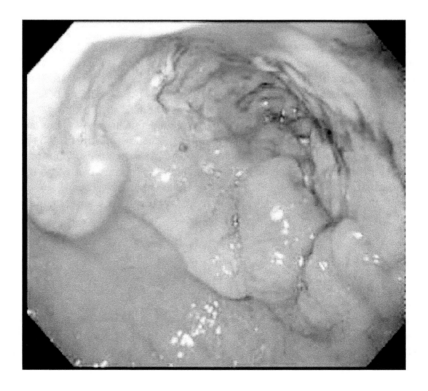

FIGURE 10.3 Endoscopic appearance of advanced gastric cancer showing a multilobed mass in the gastric antrum.

Radiography is primarily helpful for staging, rather than diagnostic purposes. Computed tomography [DB2](CT) is usually the initial study of choice for evaluating distant metastases, though accuracy of T and N staging (see Section 10.7) is limited. Magnetic resonance imaging (MRI) has no significant advantages over CT, and given its cost differential, is not routinely used in cancer staging.

There is little role in transabdominal ultrasound in detection or staging of gastric cancer. However, endoscopic ultrasound (EUS) allows close approximation of the sensor probe to the gastric wall and is able to stratify its major layers, providing for remarkable accuracy (80%) of T staging (depth of invasion).[10] Enlarged perigastric lymph nodes may also be detected. EUS may also help in differentiation of subtle gastric lesions where biopsy is unrevealing by determining layer of tumor origin and allowing for ultrasound-guided fine-needle aspiration or needle-core biopsy.

Upper gastrointestinal (GI) double-contrast radiography is used extensively in Japan as a screening method for detection of gastric cancer. In experienced hands, sensitivity of 70% and specificity of 90% for advanced gastric cancer have been reported, though this number is significantly lower in early gastric cancer.[11,12] This technique is not widely used for primary diagnosis in the U.S.

10.7 CLASSIFICATION AND PROGNOSIS

There are multiple classification systems for gastric adenocarcinoma. The histologic designation of intestinal- and diffuse-type cancers is discussed earlier in this chapter. Gastric cancer can also be divided into early versus advanced cancer, with early gastric cancer defined as lack of penetration past the submucosa, regardless of the presence of any lymph node involvement. Early gastric cancer carries with it a more favorable diagnosis after resection. These lesions are more frequently noted in Japan, where screening is directed at finding these tumors. Currently, however, the TNM staging system is the best for determining prognosis and therapy for advanced cancers. The factors that determine stage are depth of tumor penetration (T), amount of lymph node involvement (N), and the presence or absence of metastatic disease (M). The American Joint Committee on Cancer recently revised its TNM staging system for gastric cancer (Table 10.1).[13] This system allows

---

TABLE 10.1
American Joint Committee on Cancer's Staging for Gastric Carcinoma

Primary Tumor (T)

Tx	Primary cannot be assessed
T0	No evidence of primary tumor
Tis	Carcinoma *in situ*: intraepithelial tumor without invasion of the lamina propria
T1	Tumor invades lamina propria or submucosa
T2	Tumor invades muscularis propria or subserosa
T3	Tumor penetrates serosa (visceral peritoneum)
T4	Tumor invades adjacent structures

Regional Lymph Nodes

NX	Cannot be assessed
N0	No regional node mets
N1	Mets in 1–6 nodes
N2	Mets in 7–15 nodes
N3	Mets in >15 nodes

Distant Metastasis (M)

MX	Cannot be assessed
M0	No distant metastasis
M1	Distant metastasis

Note: T = tumor, N = lymph node involvement, M = metastasis.

TABLE 10.2
Four-Stage Grouping for Gastric
Carcinoma Using TNM Data

Stage	TNM-Stage
0	TisN0M0
IA	T1N0M0
IB	T1N1M0
	T2N0M0
II	T1N2M0
	T2N1M0
	T3N0M0
IIIA	T2N2M0
	T3N1M0
	T4N0M0
IIIB	T2N2M0
IV	T4N1M0
	Any N3
	Any M1

TABLE 10.3
Five- and Ten-Year Relative Survival for U.S. Cases Treated
by Gastrectomy, 1985–1996

AJCC Stage	5-Year Relative Survival (%)	10-Year Relative Survival (%)
IA	78	65
IB	58	42
II	34	26
IIIA	20	14
IIIB	8	3
IV	7	5

patients to be classified into one of four stages (Table 10.2).[13] Prognosis can be thus ascertained and an appropriate treatment course taken. Prognosis by stage of patients treated with gastrectomy can be found in Table 10.3.[14] Untreated patients with liver metastases have a life expectancy of 4 to 6 months, and patients with peritoneal carcinomatosis have a life expectancy of 4 to 6 weeks.[15]

10.8 THERAPY

Determination of suitability of surgical resection is the first therapeutic step, as resection remains the only potentially curative option. The type of resection is dependent upon the extent and location of the tumor within the stomach. A resection margin of at least 5 cm is commonly recommended. In most cases, total gastrectomy will be required to achieve this. Smaller distal (antral) cancers are the exception, and in this case, subtotal gastrectomy may be possible, with studies demonstrating similar 5-year survival to patients undergoing total gastrectomy for these lesions.[16,17] When total gastrectomy is performed, procedures are usually performed to limit the amount of postoperative alkaline reflux (i.e., Roux-en-Y).

What constitutes an appropriate extent of concomitant lymph node resection remains an area of debate. In the U.S., the majority of patients undergo resection of the perigastric nodes along the lesser and greater curvatures of the stomach without further regional lymph node dissection.[18] The Japanese literature has supported much more extensive lymphadenectomy, with increased 5-year survival rates in patients undergoing resection of regional lymph nodes up through the hepatoduode-nal ligament and root of the mesentery.[19,20] More recent studies conducted in Western populations suggest that more extensive lymphadenectomy substantially increases postoperative morbidity and mortality without conferring a significant survival benefit.[21–23] This disparity of results may arise out of the extensive Japanese screening effort and the ability to find cancers at an earlier stage and perhaps less understaging of disease.

In the U.S., given the late stage of most tumors at presentation, approximately 40% of patients will have recurrence of disease locally or distally within 5 years despite effort for curative resec-tion.[24] As many as 60% of patients who go to surgery will have unresectable disease at the time of initial operation and may undergo palliative resection or bypass rather than surgery for cure. Patients not undergoing resection or who have local disease recurrence are prone to tumor bleeding or gastric outlet obstruction. Endoscopy, angiography with embolization, or radiation therapy may be used to control bleeding. Expandable metal wall stents may be used in patients with pending obstruction.

Chemotherapy may be used as adjuvant therapy or for palliation, although gastric cancer is a relatively resistant tumor to this therapeutic modality. The currently available drugs yield disap-pointing results both alone and in combination, though clinical trials are ongoing with new agents. Radiation therapy may be used palliatively as noted above. Adjuvant chemoradiation therapy after surgical resection may yield a modest improvement in survival.[25]

Endoscopic mucosal resection is currently being used in Japan for cases of early gastric cancer of the intestinal type. In this method, saline is injected into the submucosa to lift the tumor away from the underlying layers, then ensnared and removed with electrocautery. This method can be attempted for selected early gastric cancers.[26]

10.9 SUMMARY

Gastric cancer is one of the world's most common and deadly malignancies, despite its recent decline in the Western industrialized world. *H. pylori* is now felt to be one of the major factors in disease pathogenesis, though dietary and other environmental factors may also play a role. The pattern of growth and spread of these cancers is such that they are often advanced at diagnosis. Upper GI endoscopy is the primary tool for diagnosis, while imaging modalities such as computed tomography and endoscopic ultrasound are used primarily for disease staging. Surgery remains the only potentially curative option, and adjuvant chemoradiation therapy may extend life expectancy. Nonetheless, 5-year mortality rates for postresection remain poor, and prognosis for unresectable disease is dismal.

REFERENCES

1. Greenlee, R., Hill-Harmon, M., Taylor Murray, T., and Thun, T., Cancer statistics, *CA Cancer J. Clin.*, 51, 15–36, 2001.
2. Howson, C., Hirayama, T., and Wylander, E., The decline in gastric cancer: epidemiology of an unplanned triumph, *Epidemiol. Rev.*, 8, 1, 1986.
3. Fox, J. and Wang, T., *Helicobacter pylori*: not a good bug after all, *New England J. Med.*, 345, 829–832, 2001.
4. Huang, J., Sridhar, S., Chen, Y., and Hunt, R., Meta-analysis of the relationship between *Helicobacter pylori* seropositivity and gastric cancer, *Gastroenterology*, 114, 1169–1179, 1998.
5. Ramon, J., Serra, L., Cerdo, C. et al., Dietary factors and gastric cancer risk: a case-control study in Spain, *Cancer*, 71, 1731, 1993.
6. Ji, B., Chow, W., Yang, G. et al., Dietary habits and stomach cancer in Shanghai, China, *Int. J. Cancer*, 76, 659, 1998.

7. Terry, P. and Yuen, O.N., Protective effect of fruits and vegetables on stomach cancer in a cohort of Swedish twins, *Int. J. Cancer*, 76, 35, 1998.

8. Ye, W., Ekstrom, A., Hansson, L. et al., Tobacco, alcohol and the risk of gastric cancer by sub-site and histologic type, *Int. J. Cancer*, 83, 223, 2000.

9. Lauren, P., The two histopathologic types of gastric carcinoma: diffuse and so-called intestinal type carcinoma: an attempt at a histo-clinical classification, *Acta Pathologica Microbiologica Scandinavica*, 64, 31–41, 1965.

10. Kelly, S., Harris, K., Berry, E. et al., A systematic review of the staging performance of endoscopic ultrasound in gastro-oesophageal carcinoma, *Gut*, 49, 534–539, 2001.

11. Archer, A. and Grant, D., Recent developments in diagnostic radiology of primary and recurrent gastric cancer, *Cancer Treat. Rev.*, 55, 106, 1991.

12. Montei, A., Graziana, L., Pesaresi, A. et al., Radiological diagnosis of early gastric cancer by routine double-contrast examination, *Gastrointest. Radiol.*, 7, 205, 1982.

13. American Joint Committee on Cancer, *AJCC Staging Manual*, 6th ed., Springer-Verlag, New York, 2002.

14. Hundahl, S., Phillips, J., and Menck, H., The national Cancer Data Base Report on poor survival of U.S. gastric carcinoma patients treated with gastrectomy: fifth edition, American Joint Committee on cancer staging, proximal disease, and the "different disease" hypothesis, *Cancer*, 88, 921, 2000.

15. Preusser, P., Achterrath, W., Wilke, H. et al., Chemotherapy of gastric cancer, *Cancer Treat. Rev.*, 15, 257, 1988.

16. Gouzi, J., Huguier, M., Fagniez, P. et al., Total versus subtotal gastrectomy for adenocarcinoma of the gastric antrum: a French prospective control study, *Ann. Surg.*, 209, 162–166, 1989.

17. Bozzetti, F., Marubini, E., Bonfanti, G., et al., Subtotal versus total gastrectomy for gastric cancer: five-year survival rates in a multicenter randomized Italian trial: Italian Gastrointestinal Tumor Study Group, *Ann. Surg.*, 230, 170–178, 1999.

18. Macdonald, J., Fleming, T., Peterson, R. et al., Adjuvant chemotherapy with 5-FU, adriamycin, and mitomycin-C (FAM) versus surgery alone for patients with locally advanced gastric adenocarcinoma: a Southwest Oncology Group study, *Ann. Surg. Oncol.*, 2, 488–494, 1995.

19. Kodama, Y., Sugimachi, K., Soejima, K. et al., Evaluation of extensive lymph node dissection for carcinoma of the stomach, *World J. Surg.*, 5, 241–248, 1981.

20. Maruyama, K., Sasako, M., Kinoshita, T. et al., Should systematic lymph node dissection be recommended for gastric cancer? *Eur. J. Cancer*, 34, 1480–1489, 1998.

21. Wanebo, H., Kennedy, B., Winchester, D. et al., Gastric carcinoma: does lymph node dissection alter survival? *J. Am. Coll. Surg.*, 183, 616–624, 1996.

22. Bonenkamp, J., Hermans, J., Sasako, M. et al., Patient survival after D1 and D2 resections for gastric cancer: Dutch Gastric Cancer Group, *New England J. Med.*, 340, 908–914, 1999.

23. Cuschieri, A., Weeden, S., Fielding, J. et al., Patient survival after D1 and D2 resections for gastric cancer: long-term results of the MRC randomized surgical trial: Surgical Co-operative Group, *Br. J. Cancer*, 79, 1522–1530, 1999.

24. Wanebo, H., Kennedy, B., Chmiel, J. et al., Cancer of the stomach: a patient care study by the American College of Surgeons, *Ann. Surg.*, 218, 583, 1993.

25. Macdonald, J., Smalley, S., Benedetti, J. et al., Chemoradiotherapy after surgery compared with surgery alone for adenocarcinoma of the stomach or gastroesophageal junction, *New England J. Med.*, 345, 725–730, 2001.

26. Watanabe, H., Mai, M., Shimoda, T. et al., Report of the meeting of the 72nd Japanese Gastric Cancer Congress, *Gastric Cancer*, 3, 1, 2000.

11 Adenocarcinoma of the Esophagus

Sandeep K. Tripathy and Bhaskar Banerjee

CONTENTS

11.1 Introduction...151
11.2 Epidemiology...151
11.3 Clinical Presentation...153
11.4 Risk Factors 153
11.5 Barrett's Esophagus and EAC..155
11.6 Staging and Preoperative Evaluation 157
11.7 Treatment and Prognosis 159
11.8 Palliation for Dysphagia...160
11.9 Conclusion 161
References 161

11.1 INTRODUCTION

The esophagus is a strong muscular tube that moves food from the oropharynx to the stomach. The lumen of the esophagus is lined by a thick protective stratified squamous epithelium. Underlying the epithelium is the lamina propria (containing scattered lymphoid aggregations) and the muscularis mucosa. The highly vascular submucosa lies below the muscularis mucosa and contains small mucous glands that aid in lubrication. These glands are located in the distal third of the esophagus. The muscularis propria is a thick inner circular and outer longitudinal layer of smooth muscle. In the proximal third of the esophagus, skeletal muscle can be seen in the muscularis mucosa (as the first part of swallowing is under voluntary control).

At the junction of the esophagus and stomach, the mucosa undergoes a transition from the protective stratified squamous epithelium to a tightly packed glandular secretory mucosa that contains columnar cells. The muscularis mucosa, the underlying submucosa, and the muscularis propria remain continuous between the mucosal junction.

Given the vast number of cell types present in the esophagus, a variety of tumors can arise. Spindle cell carcinoma, small-cell carcinoma, leiomyosarcoma, Kaposi's sarcoma, and malignant melanoma comprise rare malignant esophageal tumors. The vast majority of primary malignant tumors of the esophagus, however, include squamous cell carcinoma and adenocarcinomas.

11.2 EPIDEMIOLOGY

In the past, squamous cell cancer accounted for over 95% of malignancies in the esophagus. Over the past two decades, however, the incidence of squamous cell cancer has been decreasing, while that of adenocarcinoma has been increasing. Esophageal cancer ranks as the sixth most common malignancy in the world and constitutes seven percent of all gastrointestinal cancers.[1] Unfortunately, as the majority of patients with esophageal cancer present at an advanced stage, the prognosis is quite poor.

There is a striking geographic distribution of squamous cell carcinoma (SCC) of the esophagus. It is very prevalent in three regions: Asian belt (from Turkey extending through the southern states of the former Soviet Union, Iran, Iraq, and into northern China), parts of eastern and southern Africa, and northwestern France. Environmental factors such as diets deficient in certain vitamins, low soil levels of minerals, accumulation of carcinogenic agents in plants consumed, and the consumption of hot beverages that causes thermal injury to the esophagus are thought to play a role. SCC is rarely found in patients under the age of 40. SCC occurs two to three times more often in men and is five times more common in black men than in Caucasians.[2]

Adenocarcinomas of the esophagus and esophago-gastric junction have characteristics that distinguish them from squamous cell carcinoma. This includes a higher male to female ratio (7:1) and a higher incidence among whites than blacks. The disease usually affects patients over 50 years of age, with a peak around 55 to 65 years.[2]

In the last 20 years there has been a dramatic increase in the incidence of esophageal adeno-carcinoma (EAC) in the Western world. EAC is now more commonly encountered in the West than SCC of the esophagus (Figure 11.1). Several studies have documented this trend. Surgical series from 1926 to 1976 reported that EAC was uncommon (0.8 to 3.7% of esophageal cancers). Recent surgical series from major referral institutions showed 60 to 80% of patients with esophageal cancer had adenocarcinomas (compared with 10 to 15% about 10 years ago).[2]

Population studies in the U.S. and in Western Europe also confirm the rising rate of adenocarcinoma of the esophagus and esophago-gastric junction. Data from the Surveillance, Epidemiology, and End Results (SEER) program in the U.S indicated that the incidence of esophageal cancer in white males doubled from the 1970s to the 1980s.[2] Devesa et al. updated results of the SEER study and demonstrated that by 1994, the incidence of esophageal adenocarcinoma had surpassed that

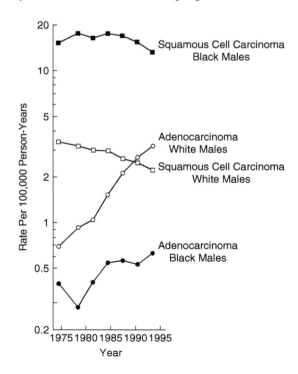

FIGURE 11.1 Trends in age-adjusted incidence for esophageal cancer.[3]

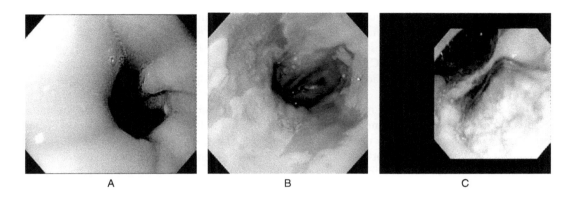

A   B   C

FIGURE 11.2 Endoscopic views of: (A) normal esophagus with pink mucosa; (B) Barrett's esophagus with the characteristic salmon red discoloration; (C) EAC at the gastro-esophageal junction seen as a mass in the distal esophagus.

of squamous cell carcinoma in white males.[3] The annual rates per 100,000 population rose from 0.7 during 1974 to 1976 to 3.2 during 1992 to 1994.[2] A recent study, using data accumulated in Europe, also found an increase in the incidence of adenocarcinoma of the esophagus in different regions of Europe during 1968 to 1995.[2]

11.3 CLINICAL PRESENTATION

Most patients with SCC of the esophagus present with dysphagia, indicating occlusion of 50 to 75% of the lumen. They can also present with weight loss, odynophagia, and persistent chest pain unrelated to swallowing. Patients with EAC also present with recent onset dysphagia and weight loss. The appearance of EAC (Figure 11.2) is indistinguishable from SCC of the esophagus on endoscopy. EACs are predominantly located in the distal esophagus and tend to invade the gastric cardia and fundus. Tumors can appear polypoid, ulcerated, and even varicoid.[1] The tumor can spread by direct extension to adjacent structures and via the lymph nodes to the neck, mediastinum, and upper abdomen. The tumor can also spread hematogenously to the liver, adrenal gland, and other organs. Radiographically, early cancer can appear as small sessile polyps, plaquelike lesions, or mucosal irregularities without a discrete mass. Advanced tumors may reveal infiltrating, ulcerative, or varicoid lesions on barium studies.[1]

11.4 RISK FACTORS

In the U.S. and other developed countries, the major cause for SCC of the esophagus is smoking and alcohol consumption. The risk of developing cancer was influenced most by the amount of alcohol consumed per day, the lifetime duration of cigarette smoking, the type of tobacco smoked (black tobacco had two-fold higher risk than blond or mixed tobacco), and time since quitting either habit. Recent data suggests that only after abstinence from drinking for 10 years does cancer risk decrease to levels of those who did not drink. After abstinence from smoking for five years, the risk of cancer is cut by 50%.[2]

Dietary factors likely play a major role in the pathogenesis of SCC of the esophagus. Inverse association of SCC with vitamins C and E, niacin, and β-carotene have been demonstrated.[2] Several studies of upper GI tract neoplasms in the U.S. and Europe suggest that fiber intake from fruits, vegetables, and grains may have a protective effect on the development of SCC of the esophagus. A recent Swedish study by Terry et al., however, found no association between the intake of cereal fiber and SCC risk.[4] Several studies have observed that the consumption of hot beverages and soups are positively associated with the risk of SCC.[2]

Endemic SCC of the esophagus in Southern Africa has been shown to have a positive association with the consumption of maize meal. This is believed to be due to the conversion of nonesterified linoleic acid in maize meal to prostaglandin E2, which affects pH and fluid content of the esophagus and could thus predispose to SCC of the esophagus. High levels of nonesterified linoleic acid were found in maize meal from Southern Africa as well as the foods prepared from it (beer, porridge, maize/pumpkin mash).[5]

In the Linxian province of China, where SCC is also endemic, riboflavin deficiency (manifest by cheilosis and glossitis) is common. In this region, the soil is low in molybdenum, a key factor in nitrate reductase (an enzyme necessary for converting nitrates from the soil to amino acids). When soil molybdenum content is low, plant conversion of nitrates to nitrosamines increases, resulting in increased nitrosamine exposure for those who eat the plants. Increased intake of nitrosamines, which are known carcinogens, may be one of a number of dietary and environmental factors that contributes to the development of gastroesophageal cancer in this population. It is not clear whether dietary molybdenum supplementation is beneficial in decreasing the risk of gastroesophageal cancer. In a large intervention trial, dietary supplementation of molybdenum (30 µg/day) and vitamin C (120 mg/day) did not decrease the incidence of gastroesophageal cancer or other cancers in residents of Linxian over a 5-year period.[6]

The cause for the increase in the incidence of EAC is unclear. Several risk factors for the development of adenocarcinoma of the esophagus have been proposed, including tobacco use, ethanol use, dietary factors, medications, *Helicobacter pylori* infection, Barrett's esophagus, and gastro-esophageal reflux disease (GERD). Further research into the causes of this malignancy as well as their preceding pathologic conditions will be needed in order to understand the changing epidemiologic conditions.

Tobacco and alcohol consumption have been found to be more prevalent in patients with EAC as compared with patients with uncomplicated Barrett's esophagus. Recent case-controlled studies suggest that the risk of esophageal and esophago-gastric junction adenocarcinoma is doubled among smokers.[2] Although the data suggest that tobacco is an etiologic factor for adenocarcinoma, it does not explain why the rate of EAC is increasing while squamous cell carcinoma of the esophagus (for which smoking is a major risk factor) is not increasing.

Obesity is a major risk factor for a number of chronic diseases and different types of cancers. A multicenter population-based case-control study revealed that excess weight was a strong risk factor for EAC. Risk rose with increasing BMI (body mass index). Interestingly, the greatest risk was seen in the youngest age group (<50 years old).[2] This suggests that obesity may be particularly important for early-onset tumors. It is unclear how obesity contributes to increased risk. One hypothesis suggests that obesity increases the risk of hiatal hernia and GERD, which in turn increases the risk of Barrett's esophagus (a precursor lesion to EAC). Several studies, however, have shown that obesity per se is a strong risk factor for EAC independent of reflux symptoms.[2]

Most dietary data pertain to squamous cell cancer of the esophagus. One multicenter case-controlled study identified a high intake of fat with increased risk for EAC. An inverse association between fruit and vegetable consumption and EAC has also been demonstrated.[2] Terry et al. found a strong inverse association between fiber intake and adenocarcinoma at the esophago-gastric junction. The association was with the consumption of cereal fiber (intake of fruits and vegetables were unrelated to the risk). A protective trend was seen with EAC, but the data were not statistically significant. It is believed that wheat fiber may act as a scavenger of nitrites (carcinogenic material) that are produced by swallowed air and saliva in the proximal portion of the stomach.[4]

Medications that relax the lower esophageal sphincter may contribute to increasing the subsequent risk for EAC by promoting GERD. Daily long-term users (>5 years) of any of these medications had an increased risk as compared with patients who had never used medications that relax the lower esophageal sphincter.[2] The association was particularly strong for anticholinergic medications.

H. pylori is an important risk factor for gastric cancer but not EAC.[2] It is of interest to note that there is an inverse relationship between *cagA*⁺ strain of *H. pylori* infection and risk of EAC. This is consistent with reports that the *cagA*⁻ strain is more commonly seen in patients with EAC, GERD, and Barrett's esophagus.[2]

Strong evidence supports the association of GERD and EAC. In a Swedish population-based case-control study, subjects with EAC were almost eight times as likely to report at least weekly symptoms of reflux or regurgitation than control subjects. A dose–response relationship was also noted to exist. If patient's symptoms were long-standing (>20 years) and severe, the odds ratio for EAC increased. No association was found between reflux symptoms and squamous cell carcinoma risk.[7] This suggests that recall bias did not explain the observed association. Similar studies in the U.S. have demonstrated a dose–response relationship between frequency of reflux symptoms and risk of EAC. Although there is a strong association of GERD with EAC, it should be mentioned that the annual incidence of cancer in patients with long-term GERD is 1 in 1000. More importantly, about 40% of patients who develop EAC do not even present with GERD symptoms.[8] For these reasons, the use of GERD symptoms to assess risk of the development of EAC is unlikely to be of much benefit.

A recent study by Freedman et al.[9] noted a moderately increased risk of EAC in patients who underwent cholecystectomy (CCK). There was no association of CCK with SCC of the esophagus. Patients with gallstone disease, who did not undergo CCK, did not have an increased incidence of EAC or SCC of the esophagus. It is postulated that after CCK, patients have increased reflux of bile and pancreatic juice from the duodenum to the stomach. This then reaches the esophageal mucosa, on which it has a toxic effect. Such toxic effects can predispose to the formation of EAC. Further studies are needed to identify a link between bile reflux and EAC.

11.5 BARRETT'S ESOPHAGUS AND EAC

In the 1950s, Norm Barrett noted a red, velvety glandular mucosa lining the lower part of the esophagus (Figure 11.2). He eventually recognized this glandular lining of the mucosa to be metaplastic. Barrett's esophagus (BE) is currently defined as a change in the esophageal epithelium of any length that can be recognized at endoscopy and is confirmed to have intestinal metaplasia by histology.[10] Intestinal metaplasia is confirmed by the presence of goblet cells that are best identified by Alcian blue stain at pH 2.5; the presence of columnar epithelium alone is not sufficient for diagnosis.[10] BE usually develops quickly to its fullest extent with little subsequent increase in length.[11] In addition, intestinal metaplasia can occur in extremely small segments of BE,[11] where sampling can be a problem. For patients with long-standing reflux symptoms (especially in patients older than 50), a single upper endoscopy is recommended to detect the presence of BE. In one study, 50% of cases with proven short-segment BE did not contain goblet cells on repeat biopsy, which emphasizes the need for taking multiple biopsies to diagnose the disorder.[10]

It is generally accepted that BE can lead to EAC. The mean annual incidence of EAC in BE is about 0.5%, but the risk of adenocarcinoma in patients with BE appears to be 30- to 125-fold greater than in the general population.[8] There is compelling evidence for a dysplasia-carcinoma sequence in BE, where low-grade dysplasia develops into high-grade dysplasia and finally carcinoma. However, the progression is not inevitable, and in many patients, dysplasia reverts back to normal.

Transformation of a normal cell into a malignant tumor cell involves a multistep process of genetic and epigenetic alterations. These changes allow the cell to proliferate, invade, and metastasize. In addition, the cells are released from normal regulatory and cell-death pathways. The progression from normal epithelium to BE to EAC is associated with a number of genetic changes. Loss of chromosomes (4q, 5 q, 16q, 18q) is frequently observed.[12] EGF/TGF-α (growth factors) and their receptor (EGFR) play an important role in the progression of normal esophageal epithelium to metaplasia, dysplasia, and finally carcinoma.[12] Abnormalities of the p16 and p53 tumor suppressor gene are among the most common somatic genetic lesions in the progression from BE to EAC.[12] These biomarkers have potential to become objective measures that can be used to stratify a patient's risk to progress to EAC.

A recent paper by Buskens et al.[13] showed that, in patients undergoing curative surgery for adenocarcinoma arising from BE, higher cyclooxygenase 2 (COX-2) expression was associated with a higher likelihood of distant metastases, local recurrence, and decreased survival. Five-year survival rates were 35 and 72% with high and low level of COX-2 expression, respectively. A causative role of COX-2 in EAC could not be determined from this study, but it does suggest that COX-2 levels may be useful as a prognostic indicator.[13] Another study demonstrated progressively greater COX-2 expression in Barrett's esophagus with dysplasia and adenocarcinoma compared with the normal squamous esophageal epithelium and normal columnar duodenal epithelium.[14] The potential role of COX-2 inhibitors in chemoprevention needs to be investigated.

Prospective studies show that dysplasia develops in 30% of patients with BE and is predominantly low grade. In low-grade dysplasia, the nuclei are cylindrical and located in the basal half of the cells. Two-thirds of the low-grade dysplasia are transient and tend to regress. A few patients progress to high-grade dysplasia or carcinoma,[11] which should be treated. High-grade dysplasia shows more architectural atypia, including marked crowding with back-to-back cribiform patterns. Loss of nuclear polarity and dilated irregular glands are also seen. Cases in which dysplasia is diagnosed should be reviewed by two pathologists experienced in gastrointestinal pathology. Because of the risk of synchronous adenocarcinoma in 25 to 50% of cases, esophagectomy or mucosectomy is recommended for patients with high-grade dysplasia. There is also a high frequency (25 to 50%) of developing adenocarcinoma in patients with high-grade dysplasia who are followed endoscopically.[11] High-grade dysplasia can be difficult to distinguish from adenocarcinoma. Invasion is diagnosed if there is single-cell invasion of the lamina propria, stromal reaction to glands, or if a solid mass of neoplastic cells is seen. Adenocarcinomas arising in BE are usually moderately or well differentiated. Detection of tumor when it is still confined to the mucosa and without lymph node metastasis is associated with a good prognosis at 5 years (about 80% survival). This changes to about 10% at 5 years when deep invasion or nodal metastases are found.[11]

Patients with BE should be enrolled in a surveillance program as long as they are candidates for esophageal resection. The aim of surveillance endoscopy is to detect invisible precursor lesions (dysplasia) in asymptomatic individuals and prevent the progression to invasive adenocarcinoma. Multiple blind biopsies are used to identify pathologic areas. Retrospective cohort studies have shown that surveillance leads to tumor detection at an earlier stage (with improved survival) than in patients not undergoing surveillance.[15] However, there has never been a prospective, long-term study comparing the effect of surveillance endoscopy with no surveillance in matched populations of BE patients. Such a study would require a large numbers of patients and raise ethical questions, and thus is unlikely to be done.

The American College of Gastroenterology has issued guidelines for the surveillance of BE.[10] In persons with established BE and no dysplasia, after a second confirming endoscopy, surveillance endoscopy should be performed every 2 to 3 years. It is likely that this interval will be further lengthened in the future. In patients with BE and low-grade dysplasia, endoscopy should be performed every 6 months on two occasions, then annually. At endoscopy, areas of ulceration, erosion, plaque, nodules, stricture, friability, or any other type of abnormality should be biopsied first. Four quadrant biopsies at every 2 cm are recommended.[10] Since dysplasia and cancer are focal, the chance of missing dysplasia or cancer on blind biopsies is reduced by taking a greater number of such biopsies. If BE with high-grade dysplasia is found, confirmation of histology by an expert pathologist is recommended; if confirmed, esophageal resection or aggressive endoscopic surveillance (four quadrant biopsies at 1-cm intervals, every 3 months) should be considered.[8,10,15] About 75% of practicing gastroenterologists recommend therapy when high-grade dysplasia is detected.[16]

Surveillance can only be practiced on patients once BE has been diagnosed. Currently, the detection of BE requires endoscopy with biopsy, which is both invasive and expensive. One-time screening endoscopy is recommended in patients over the age of 50 with long-term GERD; however, autopsy data have indicated that this fails to diagnose 19 out of 20 patients with BE[17]; surveillance is thus being performed on the tip of the "Barrett's iceberg." There are an estimated 20 million individuals with GERD in the U.S. It would be impractical and prohibitively expensive to endoscope this entire population. Unless a simple but inexpensive alternative to endoscopic diagnosis is found, the vast majority of patients with BE will remain undiagnosed, and the incidence of EAC is unlikely to decrease. The challenge in future may be to develop an inexpensive method of screening a large population, use markers to stratify the risk of cancer development, and then concentrate surveillance efforts on the high-risk group.

Areas of dysplasia or cancer in BE tend to be focal. The early lesions are usually very small in surface area (often <1 cm^2), making them easy to miss by blind biopsy. In addition, the process of multiple biopsies is time consuming, invasive, and expensive. There is now intense interest in the use of real-time optical techniques to identify early neoplasia in BE. A number of investigational techniques are being developed, including autofluorescence, elastic light scattering, Raman spectroscopy, and optical coherence tomography.[18] Elastic light scattering[19] (claims to measure the size of epithelial cell nuclei) and autofluorescence[18] (an optical assessment of chemical changes associated with tumors) have made the greatest advances. Autofluorescence is likely to be the first optical method used clinically. In this procedure, a fiber-optic probe passed through the biopsy channel of an endoscope will allow hundreds of areas of the esophagus to be sampled in real time. Such a process will eliminate "blind biopsies" and direct tissue biopsies to suspicious areas. Such a technique may reduce the number of tissue biopsies in a typical patient with BE from about 40 to as few as one or two. It should increase the yield of biopsies, reduce the time taken for the procedure, and significantly decrease the cost of histology. It is critical that such techniques be able to distinguish between low-grade dysplasia (no treatment needed) and high-grade dysplasia or cancer (surgical or other therapy indicated). Early data on autofluorescence indicate that this is indeed possible.[20]

Endoscopic ablation of Barrett's epithelium has been studied as a means of eradicating BE and hence reducing the risk of cancer. A variety of ablative techniques, including thermal lasers, argon beam coagulation, and photodynamic therapy, have been studied.[21] When combined with aggressive acid suppression therapy, the resulting epithelium is usually squamous after healing has occurred.[8] Reversion to squamous epithelium, however, may be incomplete, and patches of BE may be left behind. In addition, there have been reports of adenocarcinoma developing under squamous mucosa.[8] The risk of cancer in BE treated with ablative therapy remains unclear. At present, ablative techniques should be considered experimental. These procedures may be of benefit to patients who have dysplasia and are at highest risk for cancer but can not undergo surgical resection.

11.6 STAGING AND PREOPERATIVE EVALUATION

Because these tumors continue to be diagnosed at an advanced stage, overall prognosis of the patients has not improved. Systemic and local recurrences are common even after complete tumor resection and extensive lymphadenectomy. Patient evaluation and tumor staging are of utmost importance for selection of adequate treatment. The goal of preoperative evaluation is to determine if complete macroscopic and microscopic tumor resection can be achieved. In order to determine this, the following information must be obtained: (1) histologic tumor type, (2) presence of distant solid organ metastases, (3) localization of primary tumor in relation to the tracheo-bronchial tree, (4) T-staging, and (5) physiologic status of the patient.

Endoscopy with biopsy can usually differentiate between adenocarcinoma, squamous cell carcinoma, and other rare tumor entities. The differentiation of distal EAC (AEG type 1) from carcinoma of the gastrocardia (AEG type 2) and gastric cancer infiltrating the distal esophagus (AEG type 3) can be made through the evaluation of the esophago-gastric junction through retroflexion as well as contrast radiography. The presence of Barrett's esophagus indicates AEG type 1. It is important to differentiate AEG type 1 from types 2 and 3 because the latter two are treated on principles of gastric cancer.[22] It is also important to determine the length of Barrett's esophagus, as this entire section of esophagus would need to be removed when performing limited trans-hiatal resection.

In patients with distant solid organ metastases, a curative surgical approach is not possible. Organs most frequently affected include liver and lung. Evaluation of metastases in this area are performed by percutaneous ultrasound or CT (computed tomography) scan. Recent studies suggest that PET (positron-emission tomography) scanning may be of benefit in detecting metastatic disease to regional lymph nodes as well as distant organs.[22] Infiltration of the tracheo-bronchial tree or fistula formation excludes patients from primary resection. The relationship of the primary tumor to the tracheo-bronchial tree can best be determined by contrast radiography and high-resolution CT of the mediastinum. If close contact is seen between tumor and tracheo-bronchial tree, infiltration must be excluded by tracheal bronchoscopy.[22]

T staging mainly aims to assess tumor infiltration into mediastinal organs (Table 11.1). T_{is} denotes carcinoma *in situ*. T1 tumors invade the lamina propria or submucosa. With T1 tumors, it is important to distinguish between T1a (intramucosal) and T1b (submucosal) infiltration. T2 tumors invade the muscularis propria; T3 tumors invade the adventitia; and T4 tumors invade adjacent structures. Endoscopic ultrasound (EUS) is currently the most accurate technique in defining T parameters (Figure 11.3). The sensitivity of T staging for esophageal cancer by EUS is 85 to 95%.[23] The sensitivity for N staging with EUS is 70 to 80%. EUS fine-needle aspiration (FNA) can increase the accuracy of N staging to 90%.[23] A major problem with EUS, however, is the inability to advance the probe through stenotic lesions. Some endoscopists suggest dilation to allow for EUS evaluation. Although there have been increased reports of perforation, if general principles of dilation are followed, it should be safe. The use of blind probes, which can often be passed beyond the lesion

TABLE 11.1
TNM Staging System of Esophageal Carcinoma

	Primary tumor (T)		
Tx	Primary tumor cannot be assessed		
T0	No evidence of primary tumor		
T_{is}	Carcinoma *in situ*		
T1	Tumor invades lamina propria (T1a) or submucosa (T1b)		
T2	Tumor invades muscularis propria		
T3	Tumor invades adventitia		
T4	Tumor invades adjacent structures		
	Regional lymph nodes (N)		
Nx	Regional lymph nodes cannot be assessed		
N0	No regional lymph nodes involved		
N1	Distant metastasis		
	Staging (M)		
Stage 0	Tis	N0	M0
Stage I	T1	N0	M0
Stage IIa	T2	N0	M0
	T3	N0	M0
Stage IIb	T1	N1	M0
	T2	N1	M0
Stage III	T3	N1	M0
	T4	Any N	M0
Stage IV	Any T	Any N	M1

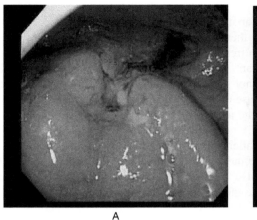 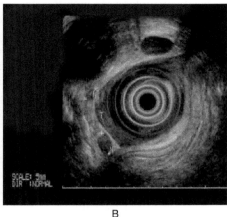

A B

FIGURE 11.3 (A) Endoscopic view of EAC in the thoracic esophagus seen as a multilobed ulcerated mass; (B) Endoscopic ultrasound showing the same lesion as a hypoechoic mass (labeled T). (Pictures courtesy of Dayna Early, MD, Washington University School of Medicine, St. Louis, MO.)

without dilation, may be an appropriate compromise.[23] Because appropriate treatment for esophageal carcinoma depends on accurate staging, EUS plays a pivotal role in the development of appropriate therapy.

It is also important that the patient have sufficient physiologic reserve and remain compliant in order to undergo such a major surgical procedure with substantial morbidity and mortality. Composite scores, which take into account the patient's general status, cardiac function, hepatic function, and respiratory function, have been used, and the result has been a reduction of postoperative mortality from 10% before 1990 to 2% since 1994.[22]

11.7 TREATMENT AND PROGNOSIS

Many surgeons consider lymphatic involvement in resectable esophageal adenocarcinoma equivalent to systemic disease; therefore, esophageal resection in these patients is essentially palliative. For this reason, wide peritumoral resection and aggressive lymph node dissection is not performed. Most surgeons would perform a trans-hiatal esophagectomy with esophageal stripping followed by gastric tube reconstruction via an esophago-gastric anastamosis through a cervical incision.[24] This procedure can be applied to most patients with carcinoma of the distal esophagus. It is contraindicated whenever tight adhesions or invasion of the tracheo-bronchial tree or the aorta are present. It is also contraindicated in patients with prior esophageal surgery.

Some surgeons believe that radical en bloc esophagectomy with extensive lymphadenectomy can have beneficial effects on cure rates even when there is lymph node invasion. Unlike surgery for gastric carcinoma, there are no standardized esophageal resections, and there is little consensus regarding the extent of lymph node dissection. Some surgeons recommend lymphadenectomy of upper abdominal and mediastinal compartments (two-field lymphadenectomy), while others recommend three-field lymphadenectomy (two-field lymphadenectomy with the additional resection of cervical lymph nodes).[25]

Resection rates have improved over the past decade, likely due to better preoperative staging and patient selection. From 1953 to 1978, 39% of patients were resectable. Recent data suggest that resection rates have increased to 62 to 97%.[24] Postoperative mortality rates also decreased from 14% in the 1980s to a current rate of 5%.[24] The cumulative 5-year survival of all series before the 1990s was 20%. More recent series reported overall 5-year survival rates from 26 to 45%.

Five-year survival rates vary with the stage of disease: 60 to 100% for stage 1, 24 to 82% for stage 2a, 38 to 71% for stage 2b, 12 to 56% for stage 3, and 0 to 31% for stage 4.[24]

Combinations of chemotherapy, radiation therapy, and surgery have been tried in order to improve the poor results from surgery alone in treatment of esophageal carcinoma. Most chemotherapy regimens include combinations of cisplatin, 5-fluorouracil, vindesine, and bleomycin. Most studies using chemotherapeutic agents have not shown improvement in long-term survival of patients.[24] Further work must be done to increase effectiveness of induction protocols, to develop new chemotherapeutic agents, and to choose the appropriate subgroups of patients to make chemotherapy a more viable option.

In view of the fact that a number of patients would not tolerate the extensive surgical procedures mentioned above, local endoscopic therapies including photodynamic therapy (PDT) and endoscopic mucosal resection (EMR) have been looked at for their possible role in the treatment of esophageal cancer. Photodynamic therapy using porphyrin compounds to sensitize the tumor to low-power laser light has been highly successful in ablating superficial esophageal cancer, and this is becoming a major indication for this novel treatment.[25] EMR involves the resection of the raised mucosa by snare cautery and is suitable for well-defined, small, superficial lesions. The previously identified mucosal lesion is raised by either the submucosal injection of epinephrine or the use of a "suck and ligate" technique with a ligator set to create a pseudopolyp. An endoscopic snare is then used to excise and remove the lesion. There is limited information on the role of endoscopic mucosal resection in the treatment of esophageal cancer. Ell et al.[26] performed endoscopic mucosal resections on 64 patients with BE and early cancer or high-grade dysplasia. EMR resulted in complete remission in 97% of patients with low-risk lesions (<20 mm, well or moderately differentiated histology, lesion limited to the mucosa, nonulcerated lesion). In contrast, remission was only achieved in 57% of patients with high-risk lesions (>20 mm, poorly differentiated histology, lesions extending into the submucosa, ulcerated lesion). In this study, follow-up was short, lesions were all visible by endoscopy, and recurrence in the short term was common. Comparisons with surgery will need to be done when long-term data from this procedure become available.[26]

11.8 PALLIATION FOR DYSPHAGIA

As the majority of patients are still not suitable for radical treatments due to age, physiologic status, or advanced disease, good palliation with minimal morbidity and mortality is important. Patients suffering dysphagia may benefit from intubation of strictures with insertion of rigid plastic tubes or self-expanding stents. Simple dilation offers short-term relief but does have significant risk of perforation. Some patients may benefit from chemotherapy or radiation therapy.

Rigid plastic tubes were initially used in the early 1990s. They are inserted under fluoroscopic control and in 90% of cases can be inserted with a procedure-related mortality of up to 15%. This mortality is usually due to perforation resulting from pre-insertion esophageal dilation. Today, self-expanding metal stents allow for insertion into narrower openings without predilation. Successful placement can occur in 95% of cases with mortality of less than 1.5%. Relief of dysphagia enabling a solid diet is achieved in the majority of cases. Although immediate results are good, follow-up complications can occur in up to 40% of cases.[27] This includes tumor in growth, bolus obstruction, migration of stents, and hemorrhage due to stent erosion into the esophageal wall. Covered stents are the treatment of choice for the palliation of perforated malignant strictures and malignant tracheo-esophageal fistulae.[27]

Recanalization of malignant strictures can also be achieved with the use of thermal laser, argon beam photocoagulation, and injection of absolute ethanol in attempts to treat dysphagia. The laser and argon beam photocoagulation are also good techniques for treating tumor migration and overgrowth into stents. These procedures have low mortality, and postprocedure complications tend to be minimal. Treatment usually needs to be repeated on a monthly basis.[27]

External beam radiation used as sole therapy or in combination with 5-fluorouracil (5 FU)* can be used to relieve dysphagia in over two-thirds of patients with SCC. Symptoms recur, however, due to recurrent cancer or fibrotic strictures. The most effective chemotherapeutic regimen in advanced esophageal cancer is epirubicin, cisplatin, and continuous infusion of 5 FU. Two-thirds of cases respond with improvement of dysphagia.[27]

For palliation of dysphagia in patients with advanced EAC, laser and argon beam are the first choice for friable intraluminal disease (with stent use in the majority of cases). Covered stents should be used to minimize ingrowth of tumor. Covered stents are also the treatment of choice for perforated cancers and malignant tracheo-esophageal cancer. Chemotherapy should only be used in the context of appropriate controlled clinical trials.

11.9 CONCLUSION

Despite advances in the staging and therapy for EAC, the overall prognosis for the patient has not improved markedly over the past decade. This is likely due to the advanced stage at the time of diagnosis. Multiple therapeutic regimens, including combinations of surgery, chemotherapy, and radiation therapy, have been used in an attempt to increase response rates in these patients. To date, the results have been less than optimal. Clearly, further research in better chemotherapy induction and more effective chemotherapeutic regimens is necessary to make this a viable treatment option in a subset of EAC patients.

An understanding of the molecular biology of EAC and BE, particularly the events that occur during the progression of BE to dysplasia to EAC, will help us to identify and treat patients at risk for developing the disease. More precise diagnosis and staging will result in better identification of potentially curative carcinoma, leading to better therapeutic strategies for each individual patient.

The risk factors for developing EAC still remain unclear. As we better understand the epidemiology of the disease, we will be better able to screen patients and prevent the progression to advanced EAC. Further research into better noninvasive screening methods for BE, less invasive methods to survey for dysplasia, validation of biological markers to identify those at risk to develop EAC, and use of chemotherapeutic and endoscopic ablation techniques will hopefully result in a better prognosis for patients presenting with EAC in the future.

REFERENCES

1. Kumbasar, B., Carcinoma of esophagus: radiologic diagnosis and staging, *Eur. J. Radiol.*, 42, 170–180, 2002.
2. Pera, M. and Pera, M., Recent changes in the epidemiology of esophageal cancer, *Surgical Oncol.*, 10, 81–90, 2001.
3. Devesa, S.S., Blot, W.J., and Fraumeni, J.F., Changing patterns in the incidence of esophageal and gastric carcinoma in the United States, *Cancer*, 83, 2049–2053, 1998.
4. Terry, P., Lagergren, J., Ye, W., Wolk, A., and Nyren, O., Inverse association between intake of cereal fiber and risk of gastric cardia cancer, *Gastroenterology*, 120, 387–391, 2001.
5. Sammon, A.M., Maize meal, non-esterified linoleic acid, and endemic cancer of the esophagus: preliminary findings, *Prostaglandins Other Lipid Mediators*, 57, 167–171, 1999.
6. Blot, W.J. et al., Nutrition intervention trials in Linxian, China: supplementation with specific vitamin/mineral combinations, cancer incidence, and disease specific mortality in the general population, *J. Natl. Cancer Inst.*, 85, 1483–1492, 1993.
7. Lagergren, J., Bergstrom, R., Lindgren, A., and Nyren, O., Symptomatic gastroesophageal reflux as a risk factor for esophageal adenocarcinoma, *New England J. Med.*, 340, 825–831, 1999.

---

* It is a fluorinated pyrimidine that is metabolized to an active form intracellularly. The active form inhibits DNA and RNA synthesis.

8. Shaheen, N. and Ransohoff, D.F., Gastroesophageal reflux, Barrett esophagus, and esophageal cancer: scientific review, *JAMA*, 287, 1972–1981, 2002.

9. Freedman, J. et al., Association between cholecystectomy and adenocarcinoma of the esophagus, *Gastroenterology*, 121, 548–553, 2001.

10. Sampliner, R.E. et al., Practice guidelines on the diagnosis, surveillance, and therapy of Barrett's esophagus, *Am. J. Gastro.*, 93, 1028–1032, 1998.

11. Clouston, A.D., Timely topic: premalignant lesions associated with adenocarcinoma of the upper gastrointestinal tract, *Pathology*, 33, 271–277, 2001.

12. Wijnhoven, B., Tilanus, H.W., and Dinjens, W., Molecular biology of Barrett's adenocarcinoma, *Ann. Surgery*, 33, 332–337, 2001.

13. Buskens, C.J., et al., Prognostic significance of elevated cyclooxygenase 2 expression in patients with adenocarcinoma of the esophagus, *Gastroenterology*, 122, 1800–1807, 2002.

14. Shirvani, V.N., Ouatu-Lascar, R., Kaur, B.S. et al., Cyclooxygenase-2 expression in Barrett's esophagus and adenocarcinoma: *ex-vivo* induction by bile salts and acid exposure, *Gastroenterology*, 118, 487–496, 2000.

15. Dulai, G.S., Surveying the case for surveillance, *Gastroenterology*, 122, 820–825, 2002.

16. Gross, C.P., Canto, M.I., Hixon, J., and Powe, N.R., Management of Barrett's esophagus: a national study of practice patterns and their cost implications, *Am. J. Gastroenterol.*, 94, 3440–3447, 1999.

17. Cameron, A.J., Zinmeister, A.R., Ballard, D.J., and Carney, J.A., Prevalence of columnar-lined Barrett's esophagus: comparison of population based and clinical findings, *Gastroenterology*, 99, 918–922, 1990.

18. Bohorfoush, A.G., Tissue spectroscopy for gastrointestinal diseases, *Endoscopy*, 28, 372– 380, 1996.

19. Wallace, B.M., Perelman, L.T., Backman, V. et al., Endoscopic detection of dysplasia in patients with Barrett's esophagus using light-scattering spectroscopy, *Gastroenterology*, 119, 677–682, 2000.

20. Banerjee, B., Agarwal, S., and Chandrasekhar, H.R., Autofluorescence spectroscopy to differentiate high grade dysplasia and cancer from low grade dysplasia in Barrett's esophagus, *Gastrointestinal Endoscopy*, 53, AB153, 2001.

21. Sharma, P., An update on strategies for eradication of Barrett's mucosa, *Am. J. Med.*, 111 (8A), 147S–152S, 2001.

22. Stein, H.J., Brucher, B.L.D.M., Sendler, A., and Siewert, J.R., Esophageal cancer: patient evaluation and pre-treatment staging, *Surgical Oncol.*, 10, 103–111, 2001.

23. Byrne, M.F. and Jowell, P.S., Gastrointestinal imaging: endoscopic ultrasound, *Gastroenterology*, 122, 1631–1648, 2002.

24. Lerut, T. et al., Cancer of the esophagus and gastro-esophageal junction: potentially curative therapies, *Surgical Oncol.*, 10, 113–122, 2001.

25. Sibille, A. et al. Long term survival after photodynamic therapy for esophageal cancer, *Gastroenterology*, 108, 337–344, 1995.

26. Ell, C. et al., Endoscopic mucosal resection of early cancer and high-grade dysplasia in Barrett's esophagus, *Gastroenterology*, 118, 670–677, 2000.

27. Mason, R., Palliation of oesophageal cancer, *Surgical Oncol.*, 10, 123–126, 2001.

12 Helicobacter pylori and Gastric Cancer

Shinya Toyokuni

CONTENTS

12.1 Summary 163
12.2 *Helicobacter pylori* as a Pathogen 163
    12.2.1 Discovery 163
    12.2.2 Characteristics of *Helicobacter Pylori* 164
    12.2.3 Cag A Protein and Cag Pathogenicity Island 165
12.3 Clinical Aspects of *Helicobacter pylori*-Induced Gastritis 166
    12.3.1 Historical Human Experiments 166
    12.3.2 Clinical Aspects: Diagnosis and Eradication 166
12.4 *Helicobacter pylori* and Gastric Cancer 167
    12.4.1 Epidemiology 167
    12.4.2 Pathology 168
    12.4.3 Experimental Evidence 168
12.5 *Helicobacter pylori* and Malignant Lymphoma 168
12.6 Conclusion 169
References 169

12.1 SUMMARY

A few years after the discovery of *Helicobacter pylori* (*H. pylori*), fundamental revision had to be made at the sections of gastritis and gastric cancer in the textbooks of pathology and gastroenterology. Nobody had imagined such a high incidence of *H. pylori* infection in humans, since it was thought that no bacteria could survive such a high pH in the gastric cavity. Almost 20 years have passed since the report of Marshall and Warren in 1984.[1] Now it is clear that there is a category of gastritis associated with *H. pylori* infection, leading to high incidence of gastric cancer and malignant lymphoma. There was a great impact in the report that removal of the bacteria by antibiotics cures longstanding uncontrollable gastric ulcer. The International Agency for Research on Cancer (IARC) has defined *H. pylori* as a Group 1 carcinogen.[2] Therefore, extensive studies are now in progress. Recent animal experiments revealed that *H. pylori* works mainly as a promoter. The question is still unanswered why only small fractions of people suffer from *H. pylori*-associated gastritis while so many people are infected.

12.2 *HELICOBACTER PYLORI* AS A PATHOGEN

12.2.1 DISCOVERY

In 1984, Warren and Marshall in Australia succeeded in the isolation and cultivation of a strain of Gram-negative bacteria, *Helicobacter pylori*.[1] This discovery has produced a great conceptual alteration not only in bacteriology, but also in pathology and gastroenterology. Now it is even difficult to discuss the etiology of gastritis, peptic ulcer, mucosa-associated lymphoid tissue lymphoma, and gastric cancer without considering *H. pylori*. Furthermore, diagnosis of *H. pylori* infection and treatment with antibiotics play an important role in the clinics of peptic ulcers and chronic gastritis.

12.2.2 CHARACTERISTICS OF *HELICOBACTER PYLORI*

Helicobacter pylori are noninvasive, non-spore forming, and spiral-shaped Gram-negative rod bacteria measuring approximately 3.5×0.5 μm. The main source of human *H. pylori* infection is suggested to be human feces. The stomach cavity presents with a strong acidic environment, and most of the microorganisms cannot survive in this acidic condition. How do *H. pylori* survive this environment? In the biopsied specimens of stomach, *H. pylori* are recognized in the mucus layer outside the gastric mucosa (Figure 12.1). According to the calculation of Falk et al.,[3] the number of this bacteria per 1 ml is 10^7 to 10^8, and 1% of them are attached to the gastric epithelia. It appears that these attached cells or cells near the epithelia can proliferate because of higher pH (>pH 4) and nutrients released from the epithelia. *Helicobacter pylori* cannot survive in anoxic condition, and high CO_2 presents a favorable environment.

Helicobacter pylori have several flagella at one end, and they move in the mucus. These bacteria are chemotactic toward urea and carbonic anion. *Helicobacter pylori* can produce energy for movement by converting urea to ammonia via urease. Ammonia helps also to neutralize the acidic condition nearby. Furthermore, chemotaxis to carbonic anions helps these bacteria to stick to the stomach.

In the normal stomach mucosa, almost no inflammatory cells are present (Figure 12.2). However, with the infection of *H. pylori*, inflammatory cells such as neutrophils and lymphocytes infiltrate the gastric mucosa (Figure 12.3). The host's mucosal immune response has been one of the hot subjects of recent investigation. Several factors have been proposed as possible virulence determinants of *H. pylori*. So far, the role of bacterial virulence factors such as vacuolating cytotoxin gene A (*vacA*), cytotoxin association gene A (*cag A*), and lipopolysaccharide (LPS) in the pathogenesis of *H. pylori* infection have been extensively studied.

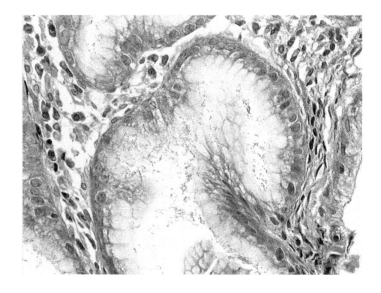

FIGURE 12.1 *Helicobacter pylori* present in the mucus layer of gastric mucosal surface (Giemsa staining).

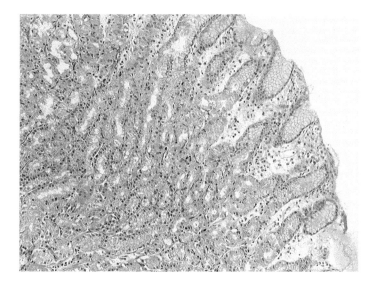

FIGURE 12.2 Normal gastric mucosa (hematoxylin and eosin staining). Note the surface epithelium and fundic glands in the deep portion.

12.2.3 CAG A PROTEIN AND CAG PATHOGENICITY ISLAND

The Cag A surface protein is one of the most investigated putative virulence factors, encoded by the *cag A* gene.[4,5] The gene is found in about 50 to 70% of *H. pylori* isolates in Western countries, and the production of the Cag A protein is reported to be associated with advanced gastrointestinal diseases, namely peptic ulcer disease, and gastric adenocarcinoma.[6–13] Cag A-positive *H. pylori* strains are known to induce secretion of interleukin-8 (IL-8), a chemokine, from gastric epithelial cells.[14] However, recent studies have indicated that it is not Cag A itself but the products of other genes in the cag pathogenicity island (PAI) that are responsible for IL-8 production.[15] PAI, an

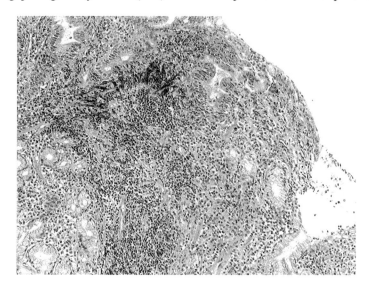

FIGURE 12.3 Chronic active gastritis (hematoxylin and eosin staining). Infiltration of lymphocytes and neutrophils is marked. Note the atrophic gastric glands.

approximately 40-kb region of genomic DNA in *H. pylori*, is observed in approximately 50 to 60% of *H. pylori* isolates in Western countries[16] and in more than 90% of such isolates in Japan.[17] In Western populations, patients carrying cag PAI-positive *H. pylori* have an increased risk for developing atrophic gastritis, noncardia gastric adenocarcinoma, and peptic ulcers. However, in East Asian populations, the relationship of cag PAI-positive *H. pylori* with disease is more difficult to establish, since practically all of the *H. pylori* have PAI in these countries.

An important breakthrough in this area occurred with the discovery that the type IV injection system of *H. pylori* crosses the inner and outer membrane of the bacteria, and, like a needle, inject Cag A into the cytosol of host gastric epithelial cells.[18] Recent reports revealed that Cag A protein would be phosphorylated on tyrosine residues by yet unidentified host cell kinases.[19,20] It was further shown that phosphorylated Cag A induces cytoskeletal changes that in turn induce cell elongation, cell spreading, and production of filopodia and lamellipodia. These data indicate that Cag A-mediated epithelial responses are at least partly independent of inflammation, although *H. pylori* virulence is associated with the presence of cag PAI.

The importance of cag PAI in inflammation was recently confirmed by a study using Mongolian gerbils. Ogura et al.[21] clearly demonstrated that the *vacA*-disrupted *H. pylori* strain induces inflammatory changes in the stomach of gerbils, whereas *cagE* (encoding a type IV secretion system) knock-out mutants did not cause severe gastritis or induce the metaplastic changes that often precede the development of cancer.

12.3 CLINICAL ASPECTS OF *HELICOBACTER PYLORI*-INDUCED GASTRITIS

12.3.1 HISTORICAL HUMAN EXPERIMENTS

Surprisingly, the pathogenicity of *H. pylori* to humans was confirmed by human experimentation. Barry Marshall, a resident doctor, was the first to do this.[1] After checking the absence of *H. pylori* in his stomach, he swallowed *H. pylori* isolated from a patient. After 8 days, he felt nausea, abdominal discomfort, headache, and foul breath. On the 10th day, acute gastritis with neutrophil infiltration was observed at the pyloric mucosa. Fortunately, on the 14th day, the bacteria spontaneously disappeared, and thereafter, the histology of the stomach was greatly improved. On the other hand, Morris swallowed 3×10^5 *H. pylori* after confirming the absence of this bacterium in his stomach.[22] On the third to sixth days, he felt severe abdominal pain and nausea. After 2 weeks, abdominal distention and discomfort appeared. He unfortunately failed to remove the bacteria after several trials with antibiotics. After 5 years he finally succeeded in their removal with simultaneous oral administration of bismuth colloid citrate, tetracycline, and metronidazol, with subsequent improvement of gastric histology. These human experiments revealed that symptoms of *H. pylori*-induced acute gastritis are quite different between each individual.

12.3.2 CLINICAL ASPECTS: DIAGNOSIS AND ERADICATION

With continued presence of *H. pylori* in the stomach mucosa, acute gastritis makes the transition to chronic gastritis, causing atrophy of the gastric mucosa. This means that glandular cells secreting enzymes and acids are lost, and that surface mucin-secreting foveolar cells proliferate. There is a proportional association between atrophy of the gastric mucosa and the presence of *H. pylori*.[23] Recently, it was clearly shown in a Japanese study (1526 patients with an average follow-up of 7.8 years) that gastric cancer develops in persons infected with *H. pylori* but not in uninfected persons. Those with histology of severe gastric atrophy, corpus-predominant gastritis, or intestinal metaplasia are at increased risk. Persons with *H. pylori* infection and nonulcer dyspepsia, gastric ulcers, or gastric hyperplastic polyps are also at risk.[24]

Diagnosis of *H. pylori* infection is made by the three independent methods: (1) histology of biopsied specimen,[25] (2) urea expiratory test, and (3) serum antibody titer. The noninvasive urea expiratory test and IgG serum test are accurate enough for predicting *H. pylori* status in untreated patients although biopsies are usually taken in Japan.[26] Since *H. pylori*-induced chronic gastritis leads to a high incidence of adenocarcinoma and lymphoma, eradication of *H. pylori* is now performed under defined standards. At the World Congress of Gastroenterology of 1990, it was agreed that eradication is necessary for duodenal ulcers with recurrence, incurability, or complication.[27] At the NIH conference of 1994, eradication was suggested for all *H. pylori*-positive peptic ulcers.[28] Further, at the conference held in the Netherlands in 1997, eradication was recommended for low-grade MALT (mucosa-associated lymphoid tissue) lymphoma, severe gastritis, and for stomach after endoscopic mucosal resection for early adenocarcinoma.[29]

12.4 *HELICOBACTER PYLORI* AND GASTRIC CANCER

12.4.1 EPIDEMIOLOGY

Until 10 years ago, it was believed that there is a continuous progression from surface gastritis to atrophic gastritis (over a 10-year duration) to metaplastic gastritis (with a further 10-year duration). However, the discovery of *H. pylori* has completely changed this notion. Kuipers et al.[10] performed an interesting prospective study and showed that atrophic gastritis with intestinal metaplasia (Figure 12.4) appeared in only 4% of cases in an *H. pylori*-negative group after 10 to 13 years of observation but in 28% in *H. pylori*-positive cases.[30]

Studies on the association of *H. pylori* with gastric cancer start with a report in 1991, which described the higher serum titer for *H. pylori* in patients who developed gastric adenocarcinoma in a prospective study.[31,32] There are basically two kinds of gastric adenocarcinoma: well-differentiated (intestinal type) and poorly differentiated (diffuse type). In a Japanese study that focused on early gastric cancer, the odds for serum positivity were higher in well-differentiated adenocarcinoma than in poorly differentiated adenocarcinoma.[33]

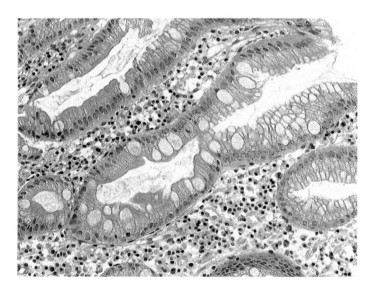

FIGURE 12.4 Chronic gastritis with intestinal metaplasia (hematoxylin and eosin staining). Note the ballooned cells filled with mucin. These cells are quite similar to intestinal cells.

12.4.2 PATHOLOGY

A number of studies have shown the association of gastric adenocarcinogenesis with intestinal metaplasia. Metaplasia means transformation of a mature form of cells into another type of mature cells in response to certain stimuli. In this case, gastric mucosal cells are converted to intestinal mucosal cells, especially goblet cells and Paneth cells. Indeed, it is believed that approximately one-third of all gastric adenocarcinoma is associated with intestinal metaplasia.[24,34]

However, based on the observation of *H. pylori*-associated gastritis, the number of *H. pylori* and the severity of inflammation are rather decreased at the stage of intestinal metaplasia. At present, the mechanism of the association of intestinal metaplasia and carcinogenesis is not well understood. This might be an adaptive change or a precancerous change.[35]

12.4.3 EXPERIMENTAL EVIDENCE

Recently, Hirayama et al. established an animal model for *H. pylori*-associated gastric disease by inoculating these bacteria into Mongolian gerbils.[36] Previous efforts to inoculate *H. pylori* into mice and rats were not successful. Various gastrointestinal diseases, such as gastritis, ulcers, intestinal metaplasia, and gastric cancer, can be developed in Mongolian gerbils. In other experiments, inoculation of *H. pylori* was used in combination with several kinds of carcinogens, including N-methyl-N-nitrosourea (MNU) (3 to 30 ppm) and N-methyl-N'-nitro-N-nitrosoguanidine (MNNG) (200 to 400 ppm).[37] In all of these experiments, high incidence (33 to 37%) of gastric adenocarcinoma was observed. Furthermore, this was reversed by eradicating the bacteria after inoculation. There are compelling reports on the carcinogenicity of *H. pylori* alone. Hirayama et al. reported single case of gastric adenocarcinoma out of 59 gerbils.[38] This incidence rate appears to be too low and might be a spontaneous incidence. Therefore, at present, *H. pylori* is considered to be a definite promoter in carcinogenesis, but it might not be a complete carcinogen.

Both intestinal- and diffuse-type gastric adenocarcinomas are associated with *H. pylori* infection. Since chronic inflammation has been associated with carcinogenesis,[39,40] the infection– inflammation pathway appears to be one of the major factors in carcinogenesis. Oxidative stress can be induced by infiltrating neutrophils and macrophages as well as the injected Cag proteins. This has been well summarized in a recent review.[41] Further studies are necessary to reveal the exact mechanism of *H. pylori*-associated adenocarcinogenesis.

12.5 *HELICOBACTER PYLORI* AND MALIGNANT LYMPHOMA

MALT is mucosa-associated lymphoid tissue that protects mucosa from infiltrating outside pathogens. The concept of MALT lymphoma (MALToma) was first presented by Isaacson and Wright in 1983.[42] This malignant lymphoma is classified into marginal-zone B-cell lymphoma with post-germinal center state in the revised WHO classification.[43] This kind of lymphoma appears not only in the stomach, but in the thyroid gland, salivary gland, orbit, lung, and mammary gland. However, the stomach is the most prevalent site. This is usually a low-grade lymphoma and might have been diagnosed as pseudolymphoma or reactive lymphoid hyperplasia. However, their monoclonality was established by clonal analysis with Southern blot of polymerase chain reaction.

Interestingly, this gastric lymphomagenesis is associated with *H. pylori* infection.[44] Surprisingly, eradication of *H. pylori* is the first choice of therapy.[45,46] Thus, a certain reactive process is involved in this lymphomagenesis. If not treated properly, this low-grade lymphoma has a potential to transform into a high-grade lymphoma.

12.6 CONCLUSION

Infection plays a role in carcinogenesis via mutation and modification of gene expression through inflammation. *Helicobacter pylori* infection in the stomach is prevalent in the world and constitutes a major risk factor for gastric carcinogenesis. *Helicobacter pylori* cause acute gastritis and then chronic gastritis, with associated mucosal atrophy and intestinal metaplasia. Approximately 10% of the people with *H. pylori* infection are reported to suffer from gastric adenocarcinoma eventually. *Helicobacter pylori* is a promoter in carcinogenesis, but it is not established yet whether it is an initiator. The International Agency for Research on Cancer declared *H. pylori* as a definite carcinogen based on the epidemiologic studies. Chemopreventive intervention as well as eradication of *H. pylori* is necessary to decrease the incidence of gastric cancer.

REFERENCES

1. Marshall, B. and Warren, J., Unidentified curved bacilli in the stomach of patients with gastritis and peptic ulceration, *Lancet*, 1, 1311, 1984.
2. Humans, I.W.G. o. t. E. o. C. R. t., Infection with *Helicobacter pylori*, in Schitosomes, Liver Flukes and Helicobacter pylori, *IARC Monographs on the Evaluation of Carcinogenic Risks to Humans*, WHO/IARC, Lyon, France, 1994, p. 177.
3. Falk, P. et al., Theoretical and experimental approaches for studying factors defining the *Helicobacter pylori*-host relationship, *Trends Microbiol.*, 8, 321, 2000.
4. Covacci, A. et al., Molecular characterization of the 128-kDa immunodominant antigen of *Helicobacter pylori* associated with cytotoxicity and duodenal ulcer, *Proc. Natl. Acad. Sci. USA*, 90, 5791, 1993.
5. Tummuru, M. et al., Cloning and expression of a high-molecular-mass major antigen of *Helicobacter pylori*: evidence of linkage to cytotoxin production, *Infect. Immun.*, 61, 1799, 1993.
6. Crabtree, J. et al., Mucosal IgA recognition of *Helicobacter pylori* 120 kDa protein, peptic ulceration, and gastric pathology, *Lancet*, 338, 332, 1991.
7. Ching, C. et al., Prevalence of Cag A-bearing *Helicobacter pylori* strains detected by the anti-Cag A assay in patients with peptic ulcer disease and in controls, *Am. J. Gastroenterol.*, 91, 949, 1996.
8. Cover, T. et al., Serologic detection of infection with cag A+ *Helicobacter pylori* strains, *J. Clin. Microbiol.*, 33, 1496, 1995.
9. Weel, J. et al., The interrelationship between cytotoxin-associated gene A, vacuolating cytotoxin, and *Helicobacter pylori*-related diseases, *J. Infect. Dis.*, 173, 1171, 1996.
10. Kuipers, E. et al., *Helicobacter pylori* and atrophic gastritis: importance of the cag A status, *J. Natl. Cancer Inst.*, 87, 1777, 1995.
11. Blaser, M. et al., Infection with *Helicobacter pylori* strains possessing cag A is associated with an increased risk of developing adenocarcinoma of the stomach, *Cancer Res.*, 55, 2111, 1995.
12. Xiang, Z. et al., Detection in an enzyme immunoassay of an immune response to a recombinant fragment of the 128 kilodalton protein (Cag A) of *Helicobacter pylori*, *Eur. J. Clin. Microbiol. Infect. Dis.*, 12, 739, 1993.
13. Parsonnet, J. et al., Risk for gastric cancer in people with Cag A positive or Cag A negative *Helicobacter pylori* infection, *Gut*, 40, 297, 1997.
14. Crabtree, J. et al., *Helicobacter pylori*-induced interleukin-8 expression in gastric epithelial cells is associated with Cag A positive phenotype, *J. Clin. Pathol.*, 48, 41, 1995.
15. Akopyants, N. et al., Analyses of the cag pathogenicity island of *Helicobacter pylori*, *Mol. Microbiol.*, 28, 37, 1998.
16. Censini, S. et al., Cag, a pathogenicity island of *Helicobacter pylori*, encodes type I-specific and disease-associated virulence factors, *Proc. Natl. Acad. Sci. USA*, 93, 14648, 1996.
17. Maeda, S. et al., Structure of cag pathogenicity island in Japanese *Helicobacter pylori isolates*, Gut, 44, 336, 1999.
18. Odenbreit, S. et al., Translocation of *Helicobacter pylori* Cag A into gastric epithelial cells by type IV secretion, Science, 287, 1497, 2000.

19. Stein, M. et al., Tyrosine phosphorylation of the *Helicobacter pylori* Cag A antigen after cag-driven host cell translocation, *Proc. Natl. Acad. Sci. USA*, 97, 1263, 2000.

20. Asahi, M. et al., *Helicobacter pylori* Cag A protein can be tyrosine phosphorylated in gastric epithelial cells, *J. Exp. Med.*, 191, 593, 2000.

21. Ogura, K. et al., Virulence factors of *Helicobacter pylori* responsible for gastric diseases in Mongolian gerbil, *J. Exp. Med.*, 192, 1601, 2000.

22. Morris, A. and Nicholson, G., Ingestion of *Campylobacter pyloridis* causes gastritis and raised fasting gastric pH, *Am. J. Gastroenterol.*, 82, 192, 1987.

23. Hui, P. et al., Pathologic changes of gastric mucosa colonized by *Helicobacter pylori, Hum. Pathol.*, 23, 548, 1992.

24. Uemura, N. et al., *Helicobacter pylori* infection and the development of gastric cancer, *New England J. Med.*, 345, 784, 2001.

25. Dixon, M. et al., Classification and grading of gastritis: the updated Sydney system, *Am. J. Surg. Pathol.*, 20, 1161, 1996.

26. Cutler, A. et al., Accuracy of invasive and noninvasive tests to diagnose *Helicobacter pylori* infection, *Gastroenterology*, 109, 136, 1995.

27. Hassall, E.J. and Dimmick, J.E., *Helicobacter pylori*: causal agent in peptic ulcer disease? in *Working Party Report of the World Congress of Gastroenterology*, Blackwell, Oxford, 1990, p. 36.

28. NIH Consensus Conference, *Helicobacter pylori* in peptic ulcer disease: NIH consensus development panel on *Helicobacter pylori* in peptic ulcer disease, *JAMA*, 272, 65, 1994.

29. Current European concepts in the management of *Helicobacter pylori* infection: the Maastricht Consensus Report, European *Helicobacter Pylori* Study Group, *Gut*, 41, 8, 1997.

30. Beales, I. et al., Long-term sequelae of *Helicobacter pylori* gastritis, *Lancet*, 346, 381, 1995.

31. Nomura, A. et al., *Helicobacter pylori* infection and gastric carcinoma among Japanese Americans in Hawaii, *New England J. Med.*, 325, 1132, 1991.

32. Parsonnet, J. et al., *Helicobacter pylori* infection and the risk of gastric carcinoma, *New England J. Med.*, 325, 1127, 1991.

33. Asaka, M. et al., What role does *Helicobacter pylori* play in gastric cancer?, *Gastroenterology*, 113, S56, 1997.

34. Hattori, T., Development of adenocarcinomas in the stomach, *Cancer*, 57, 1528, 1986.

35. Ota, H. et al., Intestinal metaplasia with adherent *Helicobacter pylori*: a hybrid epithelium with both gastric and intestinal features, *Hum. Pathol.*, 29, 846, 1998.

36. Hirayama, F. et al., Establishment of gastric *Helicobacter pylori* infection in Mongolian gerbils, *J. Gastroenterol.*, 31 (Suppl. 9), 24, 1996.

37. Tatematsu, M. et al., Induction of glandular stomach cancers in *Helicobacter pylori*–sensitive Mongolian gerbils treated with N-methyl-N-nitrosourea and N-methyl-N'-nitro-N-nitrosoguanidine in drinking water, *Jpn. J. Cancer Res.*, 89, 97, 1998.

38. Hirayama, F. et al., Development of poorly differentiated adenocarcinoma and carcinoid due to long-term *Helicobacter pylori* colonization in Mongolian gerbils, *J. Gastroenterol.*, 34, 450, 1999.

39. Farinati, F. et al., Oxidative DNA damage accumulation in gastric carcinogenesis, *Gut*, 42, 351, 1998.

40. Toyokuni, S., Reactive oxygen-species-induced molecular damage and its application in pathology, *Pathol. Int.*, 49, 91, 1999.

41. Naito, Y. and Yoshikawa, T., Molecular and cellular mechanisms involved in *Helicobacter pylori*–induced inflammation and oxidative stress, *Free Radical Biol. Med.*, 33, 323, 2002.

42. Isaacson, P. and Wright, D., Malignant lymphoma of mucosa-associated lymphoid tissue: a distinctive type of B-cell lymphoma, *Cancer*, 52, 1410, 1983.

43. Harris, N. et al., World Health Organization classification of neoplastic diseases of the hematopoietic and lymphoid tissues: report of the Clinical Advisory Committee meeting, *J. Clin. Oncol.*, 17, 3835, 1999.

44. Wotherspoon, A. et al., *Helicobacter pylori*–associated gastritis and primary B-cell gastric lymphoma, *Lancet*, 338, 1175, 1991.

45. Wotherspoon, A. et al., Regression of primary low-grade B-cell gastric lymphoma of mucosa-associated lymphoid tissue type after eradication of *Helicobacter pylori*, *Lancet*, 342, 575, 1993.

46. Thiede, C. et al., What role does *Helicobacter pylori* eradication play in gastric MALT and gastric MALT lymphoma? *Gastroenterology*, 113, S61, 1997.

13 Free Radicals, Oxidative Stress, and Cancer

Shampa Chatterjee and Aron B. Fisher

CONTENTS

13.1 Introduction..171
13.2 Free Radicals in Biological Systems and in the Environment..........................172
    13.2.1 Pathophysiological Conditions for Free-Radical
    Generation in Biological Systems..172
        13.2.1.1 Leakage from Damaged Mitochondrial Chain.....................173
        13.2.1.2 Reactions Involving Iron and Other Transition Metals.........173
        13.2.1.3 Ischemia/Reperfusion..173
        13.2.1.4 Inflammation..174
    13.2.2 Externally Generated Sources of Free Radicals..............................174
13.3 Role of Oxidative and Nitrosative Stress in Carcinogenesis...........................174
    13.3.1 Free-Radical-Induced Modifications of Biomolecules.....................174
        13.3.1.1 DNA Damage..174
        13.3.1.2 Lipid Peroxidation..175
        13.3.1.3 Protein Modifications..176
    13.3.2 Inflammation-Associated Carcinogenesis.......................................176
    13.3.3 Oxidative Stress, Signaling Pathways, and Transcriptional Factors
    Associated with Carcinogenesis..177
        13.3.3.1 MAP Kinases..177
        13.3.3.2 p53 Signaling..178
        13.3.3.3 NF-κB and AP-1 Activation and Induced
        Transformation Response...178
13.4 Free Radicals in Carcinogenesis and Apoptosis: The Critical Balance............179
13.5 Treatment and Protection Strategies..180
    13.5.1 Exposure to Free-Radical Generating Systems
    and Environmental Carcinogens...180
    13.5.2 Use of Nonsteroidal Anti-inflammatory Drugs...............................180
    13.5.3 Genes Involved in Chronic Inflammation......................................181
    13.5.4 Delivery of NO..181
13.6 Summary and Conclusions..181
References...181

13.1 INTRODUCTION

Oxygen free radicals and other reactive oxygen species (ROS) are generated in living cells as (a) by product of normal metabolism, (b) physiological signaling molecules, and (c) inflammatory responses to eliminate invading pathogenic microorganisms or by exogenous sources such as environmental agents and ionizing radiation.[1,2] It is now increasingly evident that free radicals are involved in the pathogenesis of several diseases, including cancer. A substantial body of work implicates free-radical-induced DNA damage as the cause of mutagenesis that drives carcinogenesis.[3–6]

In the development of cancer, an accepted paradigm is the multistep model of carcinogenesis in mouse skin. According to this model, three steps — initiation, promotion, and progression — are involved in the cancer development[7]. Free radicals play an important role in the induction of the first two steps. Initiation, which involves mutations and alterations in nucleic acids, is caused by free-radical attack. A majority of carcinogenic agents that are regarded as initiators produce free radicals that cause mutations, strand breaks, and sister chromatid exchange in DNA. Promotion involves the selective clonal proliferation of the initiated cells.[8] Promoters are compounds that stimulate formation of free radicals and modify the antioxidant defense system. Progression, the final stage, involves irreversible malignant transformation from preneoplastic lesions.[9]

Free radicals act in numerous other ways to promote carcinogenesis. Apart from DNA damage, radicals can oxidatively damage lipids and proteins.[10,11] While unrepaired DNA damage can be mutagenic and can cause uncontrolled cell growth, damage to lipids and proteins modifies membrane structure and function, cell–cell communication, and activity of enzymes, resulting in modulation of cell proliferation and growth.[12]

Free radicals are now known to regulate signal-transduction processes involved in cell differentiation, survival, and transformation.[13] These signaling processes can activate a number of early-response genes that are involved in tumor promotion. Thus at the cellular level, the free-radical-induced damage, termed as "oxidative stress," causes changes in gene expression, signal transduction, and posttranscriptional and posttranslational modifications that regulate cell growth and differentiation. The cumulative oxidative damage has thus been implicated in participation in cancer.[14,15]

In this chapter we discuss the free-radical-induced modification of the various cellular components and the mechanisms by which oxidative stress leads to carcinogenesis. The role of signaling pathways in regulating cell proliferation and in promoting carcinogenesis is also discussed.

13.2 FREE RADICALS IN BIOLOGICAL SYSTEMS AND IN THE ENVIRONMENT

A free radical is an atom or molecule that possesses one or more unpaired electrons. Since electrons are more stable when paired together in orbitals, radicals are generally unstable and are therefore highly reactive with a variety of substrates. Free radicals of importance in biological systems include reactive oxygen species (ROS) and reactive nitrogen species (RNS). ROS include superoxide anion ($O_2^{\cdot-}$), hydroxyl radical ($^{\cdot}OH$), lipid radical ($LOO^{\cdot}$), peroxy ($XOO^{\cdot}$) radicals, and singlet oxygen ($^1O_2$). Hydrogen peroxide (H_2O_2) and lipid hydroperoxides (LOOH) are not truly free radicals, but they are lumped with the ROS because of their increased reactivity with biomolecules. RNS include nitric oxide (NO), a diatomic free radical that is synthesized by a family of enzymes called NO synthases (NOS), and peroxynitrite ($ONOO^-$) formed by the reaction of NO with $O_2^{\cdot-}$. Nitrite radical ($NO_2^{\cdot}$) can arise from the decomposition of $ONOO^-$ (Figure 13.1). Other radical nitrogen species may form but are not well characterized in biological systems. Oxygen and nitrogen free radicals can be converted to other nonradical reactive species, such as hydrogen peroxide, hypochlorous acid (HOCl), hypobromous acid (HOBr), and peroxynitrite ($ONOO^-$). ROS, RNS, and reactive chlorine species are produced under physiologic and pathologic conditions. Thus ROS and RNS include radical and nonradical species.

13.2.1 PATHOPHYSIOLOGICAL CONDITIONS FOR FREE-RADICAL GENERATION IN BIOLOGICAL SYSTEMS

Some of the pathophysiological conditions for free-radical generation in biological systems are described in the following subsections.

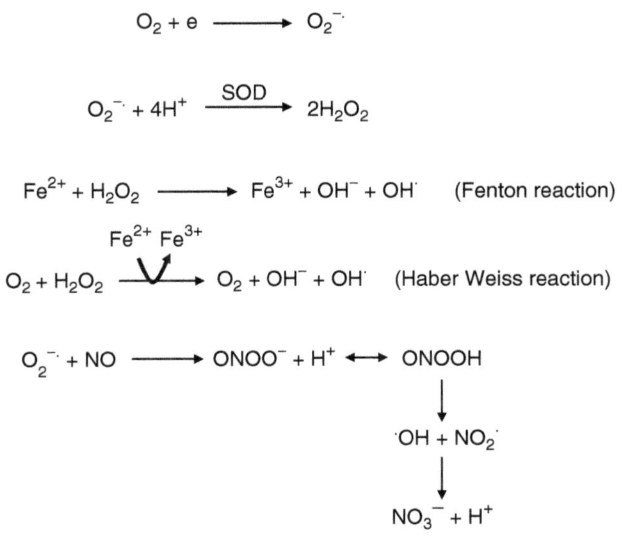

$$O_2 + e \longrightarrow O_2^{-\cdot}$$

$$O_2^{-\cdot} + 4H^+ \xrightarrow{\text{SOD}} 2H_2O_2$$

$$Fe^{2+} + H_2O_2 \longrightarrow Fe^{3+} + OH^- + OH^\cdot \quad \text{(Fenton reaction)}$$

$$O_2 + H_2O_2 \xrightarrow{Fe^{2+}\ Fe^{3+}} O_2 + OH^- + OH^\cdot \quad \text{(Haber Weiss reaction)}$$

$$O_2^{-\cdot} + NO \longrightarrow ONOO^- + H^+ \longleftrightarrow ONOOH$$

$$\downarrow$$

$$^\cdot OH + NO_2^\cdot$$

$$\downarrow$$

$$NO_3^- + H^+$$

FIGURE 13.1 Reaction mechanisms for generation of ROS and RNS in the cellular environment.

13.2.1.1 Leakage from Damaged Mitochondrial Chain

The mitochondrial electron transport chain reduces oxygen to water, where oxygen is reduced by accepting four electrons, leading to the formation of two water molecules. However, upon mitochondrial damage, a variable fraction of electrons can leak from the mitochondrial chain, leading to the univalent reduction of molecular oxygen, which generates $O_2^{\cdot-}$. The $O_2^{\cdot-}$ dismutates to H_2O_2 in a reaction catalyzed by superoxide dismutase. H_2O_2 can spontaneously but very slowly form OH^- and OH, a process that is greatly enhanced by the presence of transition metals such as Fe^{2+}.[16]

13.2.1.2 Reactions Involving Iron and Other Transition Metals

Under inflammation and other infectious conditions, the intracellular pool of active transition metal ions is released, which causes ROS generation by Haber—Weiss reaction. Apart from Fe, other metal ions such as Cu, Cr, Co, and Ni also catalyze this reaction, resulting in the formation of $^\cdot$OH radicals.[17]

13.2.1.3 Ischemia/Reperfusion

Ischemia reperfusion (I/R),[17–22] or the temporary interruption of blood supply to an organ followed by resumption of blood flow, causes ROS production primarily from the xanthine oxidase pathway. In tissues, hypoxanthine is metabolized by xanthine dehydrogenase, which uses NADH as the electron acceptor. However, xanthine oxidase is generated during anoxia from xanthine dehydrogenase by proteases activated due to elevated Ca^{2+} levels. Further, the decreased adenosine triphosphate (ATP) synthesis with ischemia is manifested by dephosphorylation of adenosine diphosphate (ADP) to adenosine monophosphate (AMP), which is degraded to adenosine, inosine, and hypoxanthine. With reperfusion, hypoxanthine provides the substrate for xanthine oxidase to use O_2 as an electron acceptor, leading to the formation of superoxide ($O_2^{\cdot-}$) and, following its dismutation, H_2O_2. These products can react, especially in the presence of transition metals such as Fe^{2+}, to produce the hydroxyl radical ($^\cdot$OH). Another route of ROS production during I/R is the NADPH oxidase pathway where, with the decreased shear shows with ischemia, membrane depolarization occurs, resulting in assembly of the NADPH oxidase subunit, thus leading to ROS generation.[20–22]

13.2.1.4 Inflammation

Polymorphonuclear leukocytes (PMN or neutrophils) participate in the systemic immune system. Neutrophils are an integral part of the body's defense mechanism against pathogens and, upon activation, are recruited to the site of injury, where they produce superoxide anions ($O_2^{\bullet-}$) via the NADPH oxidase pathway.[18,23] This enzyme is a multiprotein complex that is assembled upon PMN activation, and it reduces molecular oxygen (O_2) to the superoxide anion ($O_2^{\bullet-}$). Under inflammatory conditions, neutrophils also generate nitric oxide by both constitutive (nNOS) and inducible (iNOS) nitric oxide synthases.[24] At sites of inflammation, both NO and $O_2^{\bullet-}$ can be formed simultaneously, resulting in the generation of $ONOO^-$.[24,25]

13.2.2 EXTERNALLY GENERATED SOURCES OF FREE RADICALS

Externally generated sources of free radicals are radiation, cigarette smoke, ultraviolet (UV) light, and environmental pollutants.

13.3 ROLE OF OXIDATIVE AND NITROSATIVE STRESS IN CARCINOGENESIS

Oxidative and nitrosative stress occurs when the rate of production of ROS and RNS exceeds the antioxidant capability of the cell. In such cases, ROS/RNS interact with and modify the cellular proteins, lipids, and DNA, which results in altered target cell function. Excess ROS results in either alterations of DNA structure, such as point mutations or sister chromatid exchange, and chromosomal aberrations.[6,19,26] ROS and RNS cause alkylation, hydroxylation, and nitration of DNA bases. Unrepaired DNA damage results in the formation of new mutations and potentially new initiated cells that are promoted to neoplastic growth.[27] Mutations in nucleic acids also prevent binding of proteins at specific DNA sites, thus affecting transcription factor binding and signaling pathways and altering expression patterns of stress response genes. ROS- and RNS-initiated pathways of signal transduction, such as mitogen-activated protein kinases (MAPK) or the transcription factors, nuclear factor kappaB (NF-κB) and activator protein-1 (AP-1), eventually determine the course of cellular apoptosis and regeneration.[28,29]

Studies on some cell lines have shown that in tumor models such as mouse epidermal JB6 cells and MCF-7, ROS were observed to stimulate cell growth in monolayers.[28,30] In other cell lines, ROS can also be involved in the pathogenesis of cancer. By promoting cell proliferation in the transformed cancer cell lines MCF-7, HeLa, and Jurkat cells, reduced antioxidant levels were implicated in malignant transformation.[6,31,32] Overexpression of manganese superoxide dismutase (MnSOD), a normal cellular antioxidant, enzyme was reported to revert transformation or tumor-promotion response in these and other transformed cell lines, such as human melanoma (UACC-903) cells, human breast cancer (MCF-7) cells, and mouse epidermal JB6 cells.[33–35]

13.3.1 FREE-RADICAL-INDUCED MODIFICATIONS OF BIOMOLECULES

13.3.1.1 DNA Damage

Experimental and epidemiological evidence suggests that DNA oxidation is mutagenic and is a major contributor to human cancer.[1,2,36] ROS generate a variety of products of DNA bases such as 8-oxodG, a representative oxidative product of guanine and a highly mutagenic lesion that causes G-T transversion.[37,38] The OH radical causes oxidative damage to DNA, while $ONOO^-$ causes both oxidative damage and nitration of DNA bases. Hydroxyl radicals react with the double bond of the heterocyclic DNA bases to form an OH adduct. OH also adds to the C5 and C6 positions of thymine and cytosine, generating hydroxyl adducts that dehydrate or deaminate further. OH also abstracts an H atom from the sugar moiety of DNA, generating various sugar products. Further reactions of base and sugar radicals generate a variety of modified bases and sugars, base-free sites, strand breaks, and DNA–protein cross-links.[36,39,40]

RNS such as peroxynitrite and dinitrogen trioxide (N_2O_3) damage DNA and inhibit DNA repair mechanisms. Peroxynitrite reacts with DNA and RNA, yielding 8-oxyguanine and 8-nitroguanine (8-NO_2G).[41,42] In the presence of NADPH/NADH cytochrome reductases, 8-NO_2G further generates superoxide ($O_2^{\bullet-}$). The $O_2^{\bullet-}$ and OH radicals generated in this manner further increase DNA and RNA modifications. N_2O_3 forms nitrosoamines that are metabolized to alkylating agents that induce nitrosative deamination of DNA, where cytosine becomes uridine, adenine demethylates to hypoxanthine, and guanidine deaminates to xanthine. All of these cause DNA strand breaks that trigger secondary effects, such as up-regulation of tumor suppressor protein p53 and activation of nuclear enzyme poly (ADP-ribose) polymerase (PARP), which promotes further mutagenesis or apoptosis.[43,44] It has been generally accepted that DNA adduct formation participates in tumor initiation[45], while oxidative DNA damage would be involved in tumor promotion.[43,46]

Elevated DNA oxidation has been observed in several experimental models of cancer, including *Helicobacter pylori* infection-induced stomach cancer,[47] rodent renal cancer,[48] smoking- and diesel-exhaust-induced lung cancer,[49,50] and aflatoxin-induced liver damage.[51] Studies showing induction of DNA repair enzymes by oxidative carcinogens and suppression of carcinogenesis by administration of antioxidants have provided an association between oxidative DNA damage and carcinogenesis. In addition to DNA damage, ROS/RNS inhibit DNA repair by inhibiting nuclear enzymes associated with the repair of alkylated DNA. Besides DNA damage by direct free-radical attack, adducts formed by the reaction of DNA bases and the sugar backbone with products of lipid peroxidation and oxidized protein products also contribute to substantial DNA damage.

13.3.1.2 Lipid Peroxidation

ROS induce oxidative decomposition of omega-3 and omega-6 polyunsaturated fatty acids of membrane phospholipids (i.e., lipid peroxidation). This oxidative attack on polyunsaturated lipids initiates the lipid peroxidation (LPO) chain reaction that activates the arachidonic acid cascade. The role of lipid peroxidation in tumor promotion is mainly via the DNA adducts that are formed from LPO products.[52] The peroxidation products, including alkoxyl or peroxyl radicals, carbonyl containing compounds, alkenals, and hydroxyl-alkenals, form adducts with DNA and induce strand breaks and sister chromatid exchange.[53] Hydroxyl-alkenals such as *trans*-4-hydroxy-2-nonenal form DNA adducts, 1,N^6-ethenodeoxyadenosine (εdA), or 3,N^4-ethenodeoxycytidine (εdC) and 1,N^2 ethenodeoxyguanosine (εdG).[54] Etheno adducts are also formed after peroxidation of arachidonic acid.[55] High levels of such etheno products have been found in preneoplasia of rodents and patients, indicating that these promutagenic lesions can drive cells to uncontrolled differentiation.[56] Biomonitoring of etheno adduct (ε-DNA) levels showed that the increased amounts of etheno adduct from experimental animals paralleled the progression from benign to malignant disease stages.[56,57]

Apart from damaging DNA, products of lipid peroxidation such as malondialdehyde (MDA) and 4-hydroxynonenal (4-HNE) also stimulate leukocytes, plasma membrane guanylate cyclase, and phospholipase C, which mediate signaling cascades that trigger carcinogenesis or apoptosis. The presence of 4-HNE is reported to inhibit c-myc and c-fos, two early genes involved in cell differentiation.[58]

Lipid peroxidation can be used as a diagnostic marker in cancer. Studies with cancer patients have reported that the levels of malondialdehyde, the end product of lipid peroxidation,[59] are significantly higher in the serum of cancer patients as compared with healthy subjects. In another tissue study, malondialdehyde levels were reported to be significantly higher in specimens of human lung cancer tissue as compared with normal or noncancer lung tissue.[60] However some tumor cells are more resistant to lipid peroxidation than normal cells. Hepatoma cell membranes showed negligible lipid peroxide formation in the presence of O_2, generating systems as compared with normal rat liver membranes.[61] While many malignant cell lines show decreased levels of antioxidant enzymes such as MnSOD, glutathione peroxidase, and catalase,[3,62,63] making them more sensitive to lipid peroxidation, others have very low polyunsaturated fatty acid content, which accounts for the low concentrations of lipid peroxides with free-radical attack.[64]

13.3.1.3 Protein Modifications

Oxidative stress involves protein modification that occurs either by direct reaction with ROS or indirectly by reaction with secondary by-products of oxidative stress. Among the numerous agents that cause protein oxidation are ROS/RNS (H_2O_2, $O_2^{\bullet-}$, HOCl, ONOO$^-$), reduced transition metals (Cu, Fe), and the by-products of lipid and free amino acid oxidation. Oxidizing species directly attack the Cys and Met residues. Metals generally chelate with histidine, while Lys, Arg, Pro, and Thr form carbonyl groups on side chains. Indirect protein oxidation occurs with lipid peroxidation breakdown products such as 4-HNE and MDA, which add to Lys, His, and Cys residues.

As there are several mechanisms of oxidizing the various amino acyl side chains, there can be numerous different types of protein oxidative modifications. And because proteins have many different functions, oxidative modifications can lead to a range of functional consequences. Oxidation of the aminocarboxyl generates a carbonyl; other oxidative modifications include formation of protein HNE adducts, nitrated tyrosine residues and glycated end products. Modification of structural proteins or enzymes can lead to a loss of function. Modifications in proteins involved in signal pathways that regulate proliferation, angiogenesis, and apoptosis also occur and often correlate with progression of disease.[29,65] DNA-repair enzymes, when damaged, cause insufficient DNA repair. Apart from DNA-repair enzymes, p53 is reported to be posttranslationally modified after exposure to NO and its derivatives. The p53 protein mediates single-strand DNA annealing. The binding of p53 to DNA regulates cellular functions, such as gene transcription, DNA synthesis and repair, cell cycle arrest, apoptosis, and senescence.[66–68] Modifications in the p53 protein therefore affect these functions, contributing to the molecular pathogenesis of cancer.

13.3.2 Inflammation-Associated Carcinogenesis

Damage due to ROS-induced lipid peroxidation and DNA destruction is accompanied by an inflammatory response. Inflammation, under both infectious and noninfectious conditions, has been considered a major precursor for the development of cancer.[69] *Helicobacter pylori* infection is reported to increase the risk of gastric cancer.[70] Mice fed inflammatory agents such as dextran sodium sulfate develop inflammation and, eventually, colon cancer.[71,72] Inflammation alone can lead to cancer, even in the absence of a specific carcinogen. The surgical procedure of opening the duodenum to the gastroesophageal junction causes esophageal inflammation that can lead to cancer of the esophagus.[73,74]

The inducible cyclooxygenase (COX-2) increase that is reported with inflammation correlates well with increase in colorectal adenocarcinoma[75] in addition to breast, cervical, prostate, and lung tumors.[76] In general, increased levels of COX-2 increase expression of mitogenic metabolites such as prostaglandins. The COX-2 and prostaglandin overexpression stimulate cell proliferation, the inhibition of apoptosis, the induction of angiogenesis, and a direct mutagenic effect.[77,78] Chronic injury and irritation with or without infection is also associated with the release of cytokines, such as TNFα, interleukins IL-1β-6, and interferon γ, and with the eventual recruitment of leukocytes and mast cells to the damage site. Activated neutrophils undergo a "respiratory burst", i.e., increase in oxyzen nutilization to generate large amounts of free radicals by several oxidant-generating enzymes such as NADPH oxidase, inducible nitric oxide synthase, myeloperoxidase, and eosinophil peroxidase. These enzymes produce high concentrations of diverse free radicals such as $O_2^{\bullet-}$, NO,

H_2O_2, and HOCl, which also react with each other to generate the more potent free radicals such as $^\bullet OH$ and $ONOO^-$. These species can damage DNA, proteins, RNA, and lipids. Accumulation of mutations in DNA, posttranslational modification of proteins, and inadequate DNA repair can lead to increased local cell proliferation and shortened cell cycles in regions of chronic inflammation and can drive carcinogenesis.

Several infections play a role in the pathogenesis of cancer. Apart from *Helicobacter pylori* infection, which is a major environmental factor in the development of gastric adenocarcinoma, hepatitis B infection is reported to be associated with hepatocellular carcinoma. Cervical squamous cell cancer is associated with human papilloma virus.[79]

Besides ROS/RNS, alteration of host DNA can also be induced by integration of viral DNA into the genome. Such interactions often cause inactivation of the tumor suppressor genes such as p53. Moreover, viral infection causes immunosuppression, which can enhance malignancy. Several studies have targeted key molecules involved in inflammation such as interleukins, NO, and COX-2. Interleukin-10 is an anti-inflammatory cytokine that regulates inflammatory responses and free-radical release. IL-10$^{-/-}$ mice have a high incidence of colorectal adenocarcinomas, which can be reduced by exogenous IL-10 administration.[80] Administration of COX-2 inhibitor celecoxib caused tumor regression in patients with adenocarcinoma[81,82] and inhibited experimentally induced colon, breast, bladder, and skin carcinogenesis.[83,84]

Gene knock-out of COX-2 (an isoform of COX) or inhibition of COX has been shown to protect against intestinal polyposis in mice.[85] NO is also known to be involved in inflammation-mediated carcinogenesis. iNOS$^{-/-}$ mice treated with trinitrobenzene show significant resistance to colonic damage.[72] Also, when inoculated with B16-F1 melanoma cells, these mice develop fewer tumors.[86]

13.3.3 OXIDATIVE STRESS, SIGNALING PATHWAYS, AND TRANSCRIPTIONAL FACTORS ASSOCIATED WITH CARCINOGENESIS

13.3.3.1 MAP Kinases

Oxidative stress is known to mediate the phosphorylation or activation of protein kinases through cascades, involving the mitogen-activated protein kinase (MAPK). Three structurally related but functionally distinct MAPKs are extracellular signal-regulated kinase (ERK), c-Jun N-terminal kinase (JNK), and p38. These MAPKs trigger and coordinate gene responses that are responsible for proliferation, differentiation, inflammatory responses, and cell death.[87,88] ERKs can be stimulated by mitogens, while JNK and p38 can be activated by heat shock proteins and inflammatory cytokines. Activity of the transcription factor AP-1 is modulated by MAPKs. NF-κB activation is modulated by MAP kinase/ERK kinase kinase-1 (MEKK1), a kinase upstream of p38 and JNK. Antioxidants can attenuate MAPK activation, suggesting that the MAP kinase cascade, including activation of NF-κB and AP-1, is affected by ROS levels in cells.[89,90]

Activation of MAP kinases may to play a role in the pathogenesis of cancer. Application of 12-*O*-tetradecanoylphorbol-13-acetate (TPA), a tumor promoter, which acts through a free-radical-mediated mechanism, to the ears and dorsal skin of mice induced a rapid and sustained activation of ERK1/2 and p38 MAP kinase. NF-κB activation in this model was abolished by the ERK1/2 inhibitor U0126, while SB203580, a p38 inhibitor, had no effect.[91] In human hepatoma cells, ROS-induced in cell proliferation correlated well with ERK activation,[92] but not JNK or p38. Pathways recently, p38 MAP kinase has been shown to coprecipitate with p53 and phosphorylates p53.[93,94] Also, p38 was observed to phosphorylate p53 when exposed to DNA-damaging agents such as UV, etc.[95] Administration of U0126, a MAPK inhibitor, inhibited p53 phosphorylation. This suggests that the MAPK signaling pathway that leads to increased cell proliferation is triggered by ROS.

13.3.3.2 p53 Signaling

Gene p53, the tumor suppressor gene, is the most commonly altered gene in human cancer, with a mutation frequency of 50%.[96] Germline mutations of p53 have been found in Li–Fraumeni syndrome, an inherited syndrome with a high risk of developing a variety of cancers.[67] The p53 protein is rapidly induced by the stimulates of DNA damage. The currently accepted hypothesis is that p53 binds to altered DNA structures at specific sites and recruits DNA repair machinery to that site. This is strongly favored by the association of DNA repair proteins XPB and XPD with p53. In addition to binding specifically to DNA at the p53 consensus sites, p53 also binds nonspecifically to DNA. It is known that p53 binds to single-strand DNA, DNA duplex with free ends, and nicked DNA.[66]

The induced p53 functions as a transcription factor, and its modifications provide a means for p53 protein to transactivate several of the 20 downstream genes of p53 that control cell cycle regulation, apoptosis, and DNA repair. Among the genes downstream of p53 are *p21* in G1 growth arrest; *bax*, *mdm2*, and PIGs in apoptosis; and GADD45 and XPE in DNA repair. With DNA damage, the protein level and the transcriptional activation of p53 are rapidly induced, leading to expression of downstream effectors *p21, GADD45, mdm2, and bax.*[97]

The critical factor in the binding of p53 to DNA and the subsequent regulation of the downstream genes is the translocation of p53 to the nucleus. This process is mediated by ROS/RNS. Cells irradiated by UV showed p53 localization in the nucleus after 1 to 2 h and its emergence from the nucleus 12 to 24 h later.[98] Exposure often oligodendroglia-type cell line to H_2O_2 produced a translocation of p53 from cytosolic to nuclear compartments within 20 min of exposure. After 48 h of H_2O_2 treatment, nearly 60% of the cells exhibited p53 in the nuclei, at which time a large proportion of cells underwent apoptosis. These changes could be completely blocked by free-radical scavengers, indicating that ROS have the ability to signal p53 translocation to the nucleus. Moreover, p53 negatively regulates antioxidant levels. Transfection of HeLa cells with p53 led to a decrease in MnSOD levels. Based on this, it may be stated that a balance is normally maintained between p53 and SOD levels, and that overexpression of p53 can lead to an impairment of SOD activity. Thus when tumor cell are exposed to ROS, SOD in these cells is utilized to quench these free radicals, but this also leads to an increase in expression of p53. This in turn suppresses the expression of SOD, tilting the balance more toward a pro-oxidant state.

Exposure to ROS and RNS causes posttranslational modifications in p53 that inhibit cellular growth. Increase in iNOS expression causes mutations in p53 at codons 247 and 248. These have been observed in inflamed lesions of patients that have ulcerative colitis, associated with colon cancer. In colon tumors, there is a high correlation between iNOS activity and G:C to A:T mutations at methylcytosine sites in p53.[99] Similarly, increased iNOS expression was associated with mutations in p53 in stomach, brain, and breast cancers.[100–103] NO and its derivatives, which cause mutations in cancer-related genes such as p53 therefore act as endogenous initiators and promoters in carcinogenesis.

13.3.3.3 NF-κB and AP-1 Activation and Induced Transformation Response

NF-κB and AP-1 are two transcription factors that regulate genes implicated in ROS-induced oxidative stress. NF-κB is bound to the inhibitory protein IκBα in the cytoplasm. With a stimulus such as ROS, NF-κB is released from IκBα and translocates to the nucleus, where it can activate target gene transcription.[104] H_2O_2 and $O_2^{\bullet-}$ are known to activate NF-κB and blocking their production prevents NF-κB activation.[105] NF-κB is involved in the regulation of expression of cytokines, chemokines, growth factors, cell adhesion molecules, and cell cycle proteins, and it also regulates the expression of pro-apoptotic proteins such as Bcl-2 and anti-apoptotic proteins such as Bax. Therefore, NF-κB signaling may be a critical pathway for cancer progression and a potential therapeutic target.

Studies have shown the requirement of NF-κB for maintenance of tumor phenotype and inhibition of apoptosis.[106] Moreover, increased NF-κB activation has also been reported with tumor progression.[107] Inhibition of NF-κB by a nondegradable mutant IκBα, an antisense RNA, or a transgenic null mutation results in tumor regression.[106,108]

AP-1 activation also accompanies neoplastic transformation. Constitutive increase in AP-1 activity correlates with conversion of papillomas to carcinomas.[34,109–111] Tumor promoters can also increase AP-1 activity. In MCF-7 breast cancer cells, MnSOD suppressed tumor growth as well as the activities of AP-1 and NF-κB, thus providing a link between ROS generation, tumor formation, and transcription factors.[34] Tumor suppression by MnSOD also accompanied by inhibition of ROS. Exposure of HeLa cells to UV-C and hydrogen peroxide causes a rapid increase in AP-1 binding.[112]

In neoplastically transformed human keratinocytes, the inhibition of AP-1 causes suppression of tumor phenotype; however, this is accompanied by a concomitant inhibition of NF-κB activity, suggesting that both NF-κB- and AP-1-regulated gene expression may be required for neoplastic transformation.[113] In tumor-promotion-sensitive and tumor-promotion-resistant JB6 mouse epithelial cells, which are used as a model for studying tumor growth, activation of NF-κB by TNFα showed a concomitant increase in AP-1 activity. Overexpression of NF-κB protein in these cells significantly increases both NF-κB and AP-1 activities and confers transformation response to tumor promoters. Inhibition of NF-κB activation in P+JB36 cells using an antioxidant resulted in inhibition of AP-1 as well as suppression of growth.[28] Also, in human keratinocytes it was observed that inhibition of AP-1 by a dominant negative Jun resulted in inhibition of NF-κB activity, with suppression of tumor phenotype.[113] Thus, AP-1 and NFKB act concurrently and may contribute to neoplastic transformation in certain cell types.

13.4 FREE RADICALS IN CARCINOGENESIS AND APOPTOSIS: THE CRITICAL BALANCE

Under normal physiological conditions, ROS and RNS play a role in signal transduction and gene transcription in cells. Nitric oxide (NO) generated by activated macrophages is an important mediator of immune response. NO produced in the nervous system by neurons acts as a neurotransmitter. In addition, NO produced by the endothelial cells is essential for regulating vascular tone and vasodilation besides the relaxation and proliferation of vascular smooth muscle cells, leukocyte adhesion, platelet aggregation, and angiogenesis.

ROS trigger carcinogenesis by causing permanent DNA damage and also mutations in p53, as observed in skin, hepatocellular, and colon carcinoma. Insufficient repair of DNA causes a procarcinogenic response, triggering selection of cancerous cells and angiogenesis. ROS also modulate the activity of several transcription factors, such as the NF-κB and AP-1 pathways that regulate the *Jun* and c-*Fos* oncoproteins. In M14 melanoma cells, $O_2^{\bullet-}$ and H_2O_2 are reported to enhance tumor growth by inactivation of SH groups of caspases that regulate apoptosis.[114] All these are involved in initiation, promotion, and progression of carcinogenesis (Figure 13.2).[4] However, ROS can have a diametrically opposite effect and suppress or inhibit carcinogenesis (Figure 13.2). Generation of adequate amounts of ROS triggers tumor cell apoptosis by enhancing p53 expression and telomere shortening. Telomeres regulate cell division, and in human somatic cells, telomeres shorten with each cell division. However in tumors, telomeres are stabilized at constant length by the enzyme telomerase.[115] H_2O_2 treatment of nonproliferating human MRC-5 fibroblasts was reported to accelerate the shortening of telomeres and reduce cell growth and differentiation.[116]

Thus, low concentrations of ROS such as H_2O_2 promote cell proliferation, while intermediate doses result in growth arrest. However ROS generated in large amounts cause the oxidation of biomolecules such as amino acids, proteins, lipids, and DNA, which leads to cell injury and death. Aerobic organisms adapt to the oxidative shoes by increasing the production of glutathione,

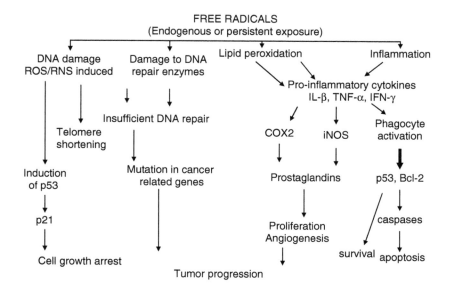

FIGURE 13.2 Free-radical-induced carcinogenesis and apoptosis. Free radicals can accelerate or inhibit carcinogenesis under different conditions. Free-radical-induced DNA damage and insufficient DNA repair can cause tumor progression, while inflammatory cytokines can stimulate p53, which causes growth arrest.

tocopherol, ascorbate, and β-carotene or by inducing endogenous antioxidant enzymes such as catalase, superoxide dismutase (SOD), thioredoxin, glutathione peroxidase (GPx), and reductase (GRS). Together, the antioxidant molecules and enzymes maintain an intracellular balance between reduction and oxidation.

13.5 TREATMENT AND PROTECTION STRATEGIES

13.5.1 EXPOSURE TO FREE-RADICAL GENERATING SYSTEMS AND ENVIRONMENTAL CARCINOGENS

Prevention of exposure to free-radical generating systems and decreasing the exposure to environmental carcinogens reduces the risk of cancer for radical-induced carcinogenesis. Scavenging free radicals and reducing free-radical synthesis are important approaches that can be achieved from natural or synthetic drugs or micronutrient supplementation. Although the specific role of dietary intake of antioxidants in cancer prevention remains to be elucidated, several cohort studies have shown that high intake of food rich in carotenoids and α tocopherol was associated significantly with reduced lung cancer risk.[117–119] However, in other similar studies, results have varied from low association, to no association, to inverse association between cancer and antioxidant intake.

13.5.2 USE OF NONSTEROIDAL ANTI-INFLAMMATORY DRUGS

Use of nonsteroidal anti-inflammatory drugs has been found to inhibit colorectal carcinogenesis, presumably by apoptosis of the neoplastic cells via mechanisms both dependent and independent of cyclooxygenase.[81–83,120,121] In randomized trials showed that drugs such as indomethacin, aspirin, etc. cause regression of colorectal tumors.[120,121]

13.5.3 Genes Involved in Chronic Inflammation

Targeting the genes that are involved in chronic inflammation with an increased cancer risk may provide an effective therapy. These may involve (a) targeting molecules that set off the inflammatory cascade, such as TNFα, TNFβ, and interleukins, (b) blocking protease activity with antitrypsin, or (c) reduction of free-radical production with catalase, SOD, and glutathione peroxidase.

13.5.4 Delivery of NO

Delivery of NO either by external administration or by gene therapy may prove beneficial, as NO-releasing agents have been reported to kill tumor cells.[122] Gene therapy to deliver iNOS cDNA into K-1735 melanoma cells suppressed tumorigenesis and abrogated metastasis in mice.[123] Transfection of human renal carcinoma cells with a retroviral iNOS cassette showed similar results.[124] The increase in iNOS activity in tumor cells increases NO concentration, resulting in higher NO-mediated DNA damage that triggers p53 accumulation and, eventually, p53-mediated growth arrest and apoptosis.[66]

13.6 SUMMARY AND CONCLUSIONS

Free radicals and oxidative stress encompass a range of changes that are associated with cancer risk. DNA damage, modification of proteins, changes in transcriptional activation, and gene expression are the factors that drive carcinogenesis. Activation of transcription factors that regulate cell proliferation and elevated signaling of MAPKs are involved in the ROS-mediated mitogenic signaling that leads to tumor promotion. Prospective chemoprevention strategies should therefore include the use of inhibitors of ROS and oxidant-generating enzymes as well as blockers of signaling molecules, inflammatory cascades, and gene targeting to increase the antioxidant defenses.

REFERENCES

1. Halliwell, B., Oxygen and nitrogen are pro-carcinogens — damage to DNA by reactive oxygen, chlorine and nitrogen species: measurement, mechanism and the effects of nutrition, *Mutat. Res.*, 443 (1–2), 37, 1999.
2. Halliwell, B.A., and Gutteridge, J.M., *Free Radicals in Biology and Medicine,* 3rd Ed., Oxford University Press, New York.
3. Oberley, L.W. and Oberley, T.D., The role of superoxide dismutase and gene amplification in carcinogenesis, *J. Theor. Biol.*, 106 (3), 403, 1984.
4. Loeb, L.A. et al., Mutagenesis by the autoxidation of iron with isolated DNA, *Proc. Natl. Acad. Sci. USA*, 85 (11), 3918, 1988.
5. Galaris, D. and Evangelou, A., The role of oxidative stress in mechanisms of metal-induced carcinogenesis, *Crit. Rev. Oncol. Hematol.*, 42 (1), 93, 2002.
6. Cerutti, P.A., Prooxidant states and tumor promotion, *Science*, 227 (4685), 375, 1985.
7. Hennings, H. et al., Malignant conversion of mouse skin tumours is increased by tumour initiators and unaffected by tumour promoters, *Nature*, 304 (5921), 67, 1983.
8. Finch, J.S. and Bowden, G.T., PCR/RFLP assay for copy number of mutant and wild-type alleles, *Biotechniques*, 21 (6), 1055, 1996.
9. DiGiovanni, J., Multistage carcinogenesis in mouse skin, *Pharmacol. Ther.*, 54 (1), 63, 1992.
10. Kasai, H., Analysis of a form of oxidative DNA damage, 8-hydroxy-2′-deoxyguanosine, as a marker of cellular oxidative stress during carcinogenesis, *Mutat. Res.*, 387 (3), 147, 1997.
11. Wiseman, H. and Halliwell, B., Damage to DNA by reactive oxygen and nitrogen species: role in inflammatory disease and progression to cancer, *Biochem. J.*, 313 (Pt. 1), 17, 1996.
12. Das, U.N., A radical approach to cancer, *Med. Sci. Monit.*, 8 (4), RA79, 2002.
13. Sauer, H., Wartenberg, M., and Hescheler, J., Reactive oxygen species as intracellular messengers during cell growth and differentiation, *Cell. Physiol. Biochem.*, 11 (4), 173, 2001.

14. Bauer, P.M. et al., Nitric oxide inhibits ornithine decarboxylase by S-nitrosylation, *Biochem. Biophys. Res. Commun.*, 262 (2), 355, 1999.

15. Dalton, T.P., Shertzer, H.G., and Puga, A., Regulation of gene expression by reactive oxygen, *Annu. Rev. Pharmacol. Toxicol.*, 39, 67, 1999.

16. Jeremy, J.Y. et al., Oxidative stress, nitric oxide, and vascular disease, *J. Card. Surg.*, 17 (4), 324, 2002.

17. Hernandez, L.A., Grisham, M.B., and Granger, D.N., A role for iron in oxidant-mediated ischemic injury to intestinal microvasculature, *Am. J. Physiol.*, 253 (1 Pt. 1), G49, 1987.

18. Fondevila, C., Busuttil, R.W., and Kupiec-Weglinski, J.W., Hepatic ischemia/reperfusion injury — a fresh look, *Exp. Mol. Pathol.*, 74 (2), 86, 2003.

19. Singh, N. and Aggarwal, S., The effect of active oxygen generated by xanthine/xanthine oxidase on genes and signal transduction in mouse epidermal JB6 cells, *Int. J. Cancer*, 62 (1), 107, 1995.

20. Fisher, A.B., Al-Mehdi, A.B., and Muzykantov, V., Activation of endothelial NADPH oxidase as the source of a reactive oxygen species in lung ischemia, *Chest*, 116 (Suppl. 1), 25S, 1999.

21. Fisher, A.B. et al., Lung ischemia: endothelial cell signaling by reactive oxygen species: a progress report, *Adv. Exp. Med. Biol.*, 510, 343, 2003.

22. Fisher, A.B. et al., Ischemia-reperfusion injury to the lung, *Ann. N.Y. Acad. Sci.*, 723, 197, 1994.

23. Wu, W. et al., 3-Bromotyrosine and 3,5-dibromotyrosine are major products of protein oxidation by eosinophil peroxidase: potential markers for eosinophil-dependent tissue injury *in vivo*, *Biochemistry*, 38 (12), 3538, 1999.

24. Ohshima, H., Tatemichi, M., and Sawa, T., Chemical basis of inflammation-induced carcinogenesis, *Arch. Biochem. Biophys.*, 417 (1), 3, 2003.

25. Beckman, J.S. and Koppenol, W.H., Nitric oxide, superoxide, and peroxynitrite: the good, the bad, and ugly, *Am. J. Physiol.*, 271 (5 Pt. 1), C1424, 1996.

26. Beckman, K.B. and Ames, B.N., Oxidative decay of DNA, *J. Biol. Chem.*, 272 (32), 19633, 1997.

27. Bertram, J.S., The molecular biology of cancer, *Mol. Aspects Med.*, 21 (6), 167, 2000.

28. Hsu, T.C. et al., Activator protein 1 (AP-1)- and nuclear factor kappaB (NF-kappaB)-dependent transcriptional events in carcinogenesis, *Free Radic. Biol. Med.*, 28 (9), 1338, 2000.

29. Marshall, H.E., Merchant, K., and Stamler, J.S., Nitrosation and oxidation in the regulation of gene expression, *FASEB J.*, 14 (13), 1889, 2000.

30. Jain, P.T. et al., Differential cytotoxicity in mouse epidermal JB6 cells: a potential mechanism for oxidant tumor promotion, *Mol. Carcinog.*, 11 (3), 164, 1994.

31. Warner, B.B. et al., Redox regulation of manganese superoxide dismutase, *Am. J. Physiol.*, 271 (1 Pt. 1), L150, 1996.

32. Burdon, R.H., Gill, V., and Rice-Evans, C., Oxidative stress and tumour cell proliferation, *Free Radic. Res. Commun.*, 11 (1–3), 65, 1990.

33. Porta, C. et al., Antioxidant enzymatic system and free radicals pathway in two different human cancer cell lines, *Anticancer Res.*, 16(5A), 2741, 1996.

34. Li, D.D. et al., The changes of AP-1 DNA binding activity and components in hippocampus of seizure-sensitive rat induced by kainate, *Sheng Li Xue Bao*, 50 (4), 385, 1998.

35. Amstad, P.A. et al., Manganese superoxide dismutase expression inhibits soft agar growth in JB6 clone41 mouse epidermal cells, *Carcinogenesis*, 18 (3), 479, 1997.

36. Dizdaroglu, M., Oxidative damage to DNA in mammalian chromatin, *Mutat. Res.*, 275 (3–6), 331, 1992.

37. Podmore, I.D. et al., Simultaneous measurement of 8-oxo-2′-deoxyguanosine and 8-oxo-2′-deoxyadenosine by HPLC-MS/MS, *Biochem. Biophys. Res. Commun.*, 277 (3), 764, 2000.

38. Bruner, S.D., Norman, D.P., and Verdine, G.L., Structural basis for recognition and repair of the endogenous mutagen 8-oxoguanine in DNA, *Nature*, 403 (6772), 859, 2000.

39. Pryor, W.A., Oxy-radicals and related species: their formation, lifetimes, and reactions, *Annu. Rev. Physiol.*, 48, 657, 1986.

40. Breen, A.P. and Murphy, J.A., Reactions of oxyl radicals with DNA, *Free Radic. Biol. Med.*, 18 (6), 1033, 1995.

41. Burney, S. et al., The chemistry of DNA damage from nitric oxide and peroxynitrite, *Mutat. Res.*, 424 (1–2), 37, 1999.

42. Mikkelsen, R.B. and Wardman, P., Biological chemistry of reactive oxygen and nitrogen and radiation-induced signal transduction mechanisms, *Oncogene*, 22 (37), 5734, 2003.

43. Naito, Y. and Yoshikawa, T., Molecular and cellular mechanisms involved in *Helicobacter pylori*-induced inflammation and oxidative stress, *Free Radic. Biol. Med.*, 33 (3), 323, 2002.

44. Szabo, E. et al., Peroxynitrite production, DNA breakage, and poly(ADP-ribose) polymerase activation in a mouse model of oxazolone-induced contact hypersensitivity, *J. Invest. Dermatol.*, 117 (1), 74, 2001.

45. Devanesan, P. et al., Catechol estrogen conjugates and DNA adducts in the kidney of male Syrian golden hamsters treated with 4-hydroxyestradiol: potential biomarkers for estrogen-initiated cancer, *Carcinogenesis*, 22 (3), 489, 2001.

46. Rusyn, I. et al., Expression of base excision repair enzymes in rat and mouse liver is induced by peroxisome proliferators and is dependent upon carcinogenic potency, *Carcinogenesis*, 21 (12), 2141, 2000.

47. Baik, S.C. et al., Increased oxidative DNA damage in *Helicobacter pylori*-infected human gastric mucosa, *Cancer Res.*, 56 (6), 1279, 1996.

48. Toyokuni, S. et al., Treatment of Wistar rats with a renal carcinogen, ferric nitrilotriacetate, causes DNA-protein cross-linking between thymine and tyrosine in their renal chromatin, *Int. J. Cancer*, 62 (3), 309, 1995.

49. Ballinger, S.W. et al., Mitochondrial genome damage associated with cigarette smoking, *Cancer Res.*, 56 (24), 5692, 1996.

50. Ichinose, T. et al., Lung carcinogenesis and formation of 8-hydroxy-deoxyguanosine in mice by diesel exhaust particles, *Carcinogenesis*, 18 (1), 185, 1997.

51. Shen, H.M. et al., Aflatoxin B1-induced 8-hydroxydeoxyguanosine formation in rat hepatic DNA, *Carcinogenesis*, 16 (2), 419, 1995.

52. Dianzani, M.U., Lipid peroxidation and cancer: a critical reconsideration, *Tumori*, 75 (4), 351, 1989.

53. Burcham, P.C., Genotoxic lipid peroxidation products: their DNA damaging properties and role in formation of endogenous DNA adducts, *Mutagenesis*, 13 (3), 287, 1998.

54. Chung, F.L., Chen, H.J., and Nath, R.G., Lipid peroxidation as a potential endogenous source for the formation of exocyclic DNA adducts, *Carcinogenesis*, 17 (10), 2105, 1996.

55. el Ghissassi, F. et al., Formation of 1,N6-ethenoadenine and 3,N4-ethenocytosine by lipid peroxidation products and nucleic acid bases, *Chem. Res. Toxicol.*, 8 (2), 278, 1995.

56. Bartsch, H. and Nair, J., New DNA-based biomarkers for oxidative stress and cancer chemoprevention studies, *Eur. J. Cancer*, 36 (10), 1229, 2000.

57. Bartsch, H., Keynote address: exocyclic adducts as new risk markers for DNA damage in man, *IARC Sci. Publ.*, 150, 1, 1999.

58. Barrera, G. et al., 4-Hydroxynonenal specifically inhibits c-myb but does not affect c-fos expressions in HL-60 cells, *Biochem. Biophys. Res. Commun.*, 227 (2), 589, 1996.

59. Uzun, K.V.H., Ozturk, T., Ozer, F., and Imecik, I., Diagnostic value of lipid peroxidation, *East. J. Med.*, 5 (2), 48, 2000.

60. Petruzzelli, S. et al., Pulmonary lipid peroxidation in cigarette smokers and lung cancer patients, *Chest*, 98 (4), 930, 1990.

61. Hostetler, K.Y., Zenner, B.D., and Morris, H.P., Phospholipid content of mitochondrial and microsomal membranes from Morris hepatomas of varying growth rates, *Cancer Res.*, 39 (8), 2978, 1979.

62. Tisdale, M.J. and Mahmoud, M.B., Activities of free radical metabolizing enzymes in tumours, *Br. J. Cancer*, 47 (6), 809, 1983.

63. Oberley, L.W. and Buettner, G.R., Role of superoxide dismutase in cancer: a review, *Cancer Res.*, 39 (4), 1141, 1979.

64. Cheeseman, K.H. et al., Lipid peroxidation and lipid antioxidants in normal and tumor cells, *Toxicol. Pathol.*, 12 (3), 235, 1984.

65. Stadtman, E.R. and Levine, R.L., Protein oxidation, *Ann. N.Y. Acad. Sci.*, 899, 191, 2000.

66. Liu, Y. and Kulesz-Martin, M., p53 protein at the hub of cellular DNA damage response pathways through sequence-specific and non-sequence-specific DNA binding, *Carcinogenesis*, 22 (6), 851, 2001.

67. Malkin, D. et al., Germ line p53 mutations in a familial syndrome of breast cancer, sarcomas, and other neoplasms, *Science*, 250 (4985), 1233, 1990.

68. Kastan, M.B. et al., Participation of p53 protein in the cellular response to DNA damage, *Cancer Res.*, 51 (23 Pt. 1), 6304, 1991.

69. Hussain, S.P., Hofseth, L.J., and Harris, C.C., Radical causes of cancer, *Natl. Rev. Cancer*, 3 (4), 276, 2003.

70. Lee, A., Animal models of *Helicobacter* infection, *Mol. Med. Today*, 5 (11), 500, 1999.

71. Cooper, H.S. et al., The role of mutant Apc in the development of dysplasia and cancer in the mouse model of dextran sulfate sodium-induced colitis, *Gastroenterology*, 121 (6), 1407, 2001.

72. Hokari, R. et al., Reduced sensitivity of inducible nitric oxide synthase-deficient mice to chronic colitis, *Free Radic. Biol. Med.*, 31 (2), 153, 2001.

73. Goldstein, S.R. et al., Development of esophageal metaplasia and adenocarcinoma in a rat surgical model without the use of a carcinogen, *Carcinogenesis*, 18 (11), 2265, 1997.

74. Chen, X. et al., Oxidative damage in an esophageal adenocarcinoma model with rats, *Carcinogenesis*, 21 (2), 257, 2000.

75. Sano, H. et al., Expression of cyclooxygenase-1 and -2 in human colorectal cancer, *Cancer Res.*, 55 (17), 3785, 1995.

76. Howe, L.R. et al., Cyclooxygenase-2: a target for the prevention and treatment of breast cancer, *Endocr. Relat. Cancer*, 8 (2), 97, 2001.

77. Goodwin, D.C., Landino, L.M., and Marnett, L.J., Effects of nitric oxide and nitric oxide-derived species on prostaglandin endoperoxide synthase and prostaglandin biosynthesis, *FASEB J.*, 13 (10), 1121, 1999.

78. Mei, J.M. et al., Expression of prostaglandin endoperoxide H synthase-2 induced by nitric oxide in conditionally immortalized murine colonic epithelial cells, *FASEB J.*, 14 (9), 1188, 2000.

79. Correa, P. and Miller, M.J., Carcinogenesis, apoptosis and cell proliferation, *Br. Med. Bull.*, 54 (1), 151, 1998.

80. Berg, D.J. et al., Enterocolitis and colon cancer in interleukin-10-deficient mice are associated with aberrant cytokine production and CD4(+) TH1-like responses, *J. Clin. Invest.*, 98 (4), 1010, 1996.

81. Tive, L., Celecoxib clinical profile, *Rheumatology (Oxford)*, 39 (Suppl. 2), 21, 2000.

82. Howe, L.R. et al., Celecoxib, a selective cyclooxygenase 2 inhibitor, protects against human epidermal growth factor receptor 2 (HER-2)/neu-induced breast cancer, *Cancer Res.*, 62 (19), 5405, 2002.

83. Kawamori, T. et al., Chemopreventive activity of celecoxib, a specific cyclooxygenase-2 inhibitor, against colon carcinogenesis, *Cancer Res.*, 58 (3), 409, 1998.

84. Harris, R.E. et al., Chemoprevention of breast cancer in rats by celecoxib, a cyclooxygenase 2 inhibitor, *Cancer Res.*, 60 (8), 2101, 2000.

85. Oshima, M. and Taketo, M.M., COX inhibitor, suppression of polyposis, and chemoprevention, *Nippon Yakurigaku Zasshi*, 120 (5), 276, 2002.

86. Konopka, T.E. et al., Nitric oxide synthase II gene disruption: implications for tumor growth and vascular endothelial growth factor production, *Cancer Res.*, 61 (7), 3182, 2001.

87. Dhar, A., Young, M.R. and Colburn, N.H., The role of AP-1, NF-kappaB and ROS/NOS in skin carcinogenesis: the JB6 model is predictive, *Mol. Cell. Biochem.*, 234–235 (1–2), 185, 2002.

88. Schramek, H., MAP kinases: from intracellular signals to physiology and disease, *News Physiol. Sci.*, 17, 62, 2002.

89. Xia, Z. et al., Opposing effects of ERK and JNK-p38 MAP kinases on apoptosis, *Science*, 270 (5240), 1326, 1995.

90. Wang, X. et al., The cellular response to oxidative stress: influences of mitogen-activated protein kinase signalling pathways on cell survival, *Biochem. J.*, 333 (Pt. 2), 291, 1998.

91. Chun, K.S. et al., Curcumin inhibits phorbol ester-induced expression of cyclooxygenase-2 in mouse skin through suppression of extracellular signal-regulated kinase activity and NF-kappaB activation, *Carcinogenesis*, 24 (9), 1515, 2003.

92. Liu, S.L. et al., Reactive oxygen species stimulated human hepatoma cell proliferation via cross-talk between PI3-K/PKB and JNK signaling pathways, *Arch. Biochem. Biophys.*, 406 (2), 173, 2002.

93. She, Q.B., Chen, N., and Dong, Z., ERKs and p38 kinase phosphorylate p53 protein at serine 15 in response to UV radiation, *J. Biol. Chem.*, 275 (27), 20444, 2000.

94. Bulavin, D.V. et al., Phosphorylation of human p53 by p38 kinase coordinates N-terminal phosphorylation and apoptosis in response to UV radiation, *EMBO J.*, 18 (23), 6845, 1999.

95. Oda, K. et al., p53AIP1, a potential mediator of p53-dependent apoptosis, and its regulation by Ser-46-phosphorylated p53, *Cell*, 102 (6), 849, 2000.

96. Harris, C.C. and Hollstein, M., Clinical implications of the p53 tumor-suppressor gene, *New England J. Med.*, 329 (18), 1318, 1993.

97. Levine, A.J., p53, the cellular gatekeeper for growth and division, *Cell*, 88 (3), 323, 1997.

98. Martinez, J.D. et al., Free radicals generated by ionizing radiation signal nuclear translocation of p53, *Cell. Growth Differ.*, 8 (9), 941, 1997.

99. Ambs, S. et al., Relationship between p53 mutations and inducible nitric oxide synthase expression in human colorectal cancer, *J. Natl. Cancer Inst.*, 91 (1), 86, 1999.

100. Hollstein, M. et al., p53 mutations in human cancers, *Science*, 253 (5015), 49, 1991.

101. Thomsen, L.L. et al., Nitric oxide synthase activity in human breast cancer, *Br. J. Cancer*, 72 (1), 41, 1995.

102. Ellie, E. et al., Differential expression of inducible nitric oxide synthase mRNA in human brain tumours, *Neuroreport*, 7 (1), 294, 1995.

103. Gallo, O. et al., Role of nitric oxide in angiogenesis and tumor progression in head and neck cancer, *J. Natl. Cancer Inst.*, 90 (8), 587, 1998.

104. Thanos, D. and Maniatis, T., NF-kappa B: a lesson in family values, *Cell*, 80 (4), 529, 1995.

105. Schulze-Osthoff, K. et al., Depletion of the mitochondrial electron transport abrogates the cytotoxic and gene-inductive effects of TNF, *EMBO J.*, 12 (8), 3095, 1993.

106. Finco, T.S. et al., Oncogenic Ha-Ras-induced signaling activates NF-kappaB transcriptional activity, which is required for cellular transformation, *J. Biol. Chem.*, 272 (39), 24113, 1997.

107. Beauparlant, P. et al., Disruption of I kappa B alpha regulation by antisense RNA expression leads to malignant transformation, *Oncogene*, 9 (11), 3189, 1994.

108. Latimer, M. et al., The N-terminal domain of IkappaB alpha masks the nuclear localization signal(s) of p50 and c-Rel homodimers, *Mol. Cell. Biol.*, 18 (5), 2640, 1998.

109. Dong, Z. et al., Blocking of tumor promoter-induced AP-1 activity inhibits induced transformation in JB6 mouse epidermal cells, *Proc. Natl. Acad. Sci. USA*, 91 (2), 609, 1994.

110. Domann, Jr., F.E. et al., Constitutive AP-1 DNA binding and transactivating ability of malignant but not benign mouse epidermal cells, *Mol. Carcinog.*, 9 (2), 61, 1994.

111. Dong, Z. et al., A dominant negative mutant of jun blocking 12-O-tetradecanoylphorbol-13-acetate-induced invasion in mouse keratinocytes, *Mol. Carcinog.*, 19 (3), 204, 1997.

112. Liu, Y. et al., Role of mitogen-activated protein kinase phosphatase during the cellular response to genotoxic stress: inhibition of c-Jun N-terminal kinase activity and AP-1-dependent gene activation, *J. Biol. Chem.*, 270 (15), 8377, 1995.

113. Li, J.J. et al., Induced expression of dominant-negative c-jun downregulates NFkappaB and AP-1 target genes and suppresses tumor phenotype in human keratinocytes, *Mol. Carcinog.*, 29 (3), 159, 2000.

114. Clement, M.V. and Pervaiz, S., Reactive oxygen intermediates regulate cellular response to apoptotic stimuli: an hypothesis, *Free Radic. Res.*, 30 (4), 247, 1999.

115. Saretzki, G. and von Zglinicki, T., Replicative senescence as a model of aging: the role of oxidative stress and telomere shortening — an overview, *Z. Gerontol. Geriatr.*, 32 (2), 69, 1999.

116. Oikawa, S. and Kawanishi, S., Site-specific DNA damage at GGG sequence by oxidative stress may accelerate telomere shortening, *FEBS Lett.*, 453 (3), 365, 1999.

117. Hirayama, T., Nutrition and cancer — a large scale cohort study, *Prog. Clin. Biol. Res.*, 206, 299, 1986.

118. Shekelle, R.B. et al., Serum cholesterol, beta-carotene, and risk of lung cancer, *Epidemiology*, 3 (4), 282, 1992.

119. Steinmetz, K.A. and Potter, J.D., Vegetables, fruit, and cancer prevention: a review, *J. Am. Diet. Assoc.*, 96 (10), 1027, 1996.

120. Hull, M.A., Gardner, S.H., and Hawcroft, G., Activity of the non-steroidal anti-inflammatory drug indomethacin against colorectal cancer, *Cancer Treat. Rev.*, 29 (4), 309, 2003.

121. Hussain, T., Gupta, S., and Mukhtar, H., Cyclooxygenase-2 and prostate carcinogenesis, *Cancer Lett.*, 191 (2), 125, 2003.

122. Hibbs, Jr., J.B., Taintor, R.R., and Vavrin, Z., Macrophage cytotoxicity: role for L-arginine deiminase and imino nitrogen oxidation to nitrite, *Science*, 235 (4787), 473, 1987.
123. Juang, S.H. et al., Use of retroviral vectors encoding murine inducible nitric oxide synthase gene to suppress tumorigenicity and cancer metastasis of murine melanoma, *Cancer Biother. Radiopharmacol.*, 12 (3), 167, 1997.
124. Juang, S.H. et al., Suppression of tumorigenicity and metastasis of human renal carcinoma cells by infection with retroviral vectors harboring the murine inducible nitric oxide synthase gene, *Hum. Gene Ther.*, 9 (6), 845, 1998.

14 Biotransformation and Mechanism of Action of Xenobiotics: What Lessons from the Past 40 Years?

Giovanni Pagano and Marco d'Ischia

CONTENTS

14.1 8-Methoxypsoralen (8-MOP): the "Obscure" Role of Oxygen 188
14.2 Xenobiotics with Redox-Dependent Toxicity in Werner
     Syndrome and Fanconi Anemia: the Oxidative Stress Link 189
14.3 P450 and Other Oxygen-Dependent Biotransformations: the Janus Facies
     of the Aerobic Lifestyle 192
14.4 Conclusions: What Lessons? 192
Acknowledgment 192
References 193

Early studies of the action mechanisms of carcinogens in the 1960s disclosed the critical roles of certain defined activities in determining carcinogen biotransformation from native and inactive forms (termed precarcinogens) into their active derivatives (the ultimate carcinogens).[1,2] Those studies, mostly the legacy of Eric Boyland,[3,4] opened the door to a broader approach to the action mechanisms of xenobiotics. This approach expanded the discussion from the fate of specific carcinogens to include the vast majority of xenobiotics, whose adverse effects are not due to their native molecules, but to their biotransformation products.[1,2,5]

The ensuing huge body of literature in this field has since pervaded disciplines such as pharmacology and toxicology and has become a widespread subject of predoctoral teaching for students in medicine, pharmacy, biology, and environmental sciences. Yet, as is often the case when a field has expanded at a rapid pace, some of the new knowledge appears not to have been integrated or properly assessed at a multidisciplinary level in the various areas. Thus, after some 40 years, it seems worthwhile to assess the state of the art, ascertain the general attitude in the biomedical circles, and learn what lessons can we take from the past.

A major focus for debate relies on the critical relationship between the detrimental effects (briefly termed "toxicity") of a given xenobiotic (e.g., DNA alterations) and their underlying mechanisms. Central to this issue is the distinction of whether the effects are due to the native molecules or to their biotransformation products. The distinction is not trivial, as it often underlies the very definition of the action mechanisms of those molecules and, consequently, of any proficiency or deficiency of the biological target in coping with the toxicity of the xenobiotic being considered.

In this chapter, we will provide some examples of potentially toxic or carcinogenic xenobiotics for which a biased or overly narrow perspective of their chemical properties, biotransformations, and/or the factors affecting their biotransformations has somewhat hindered a complete and balanced understanding of their mechanisms of action.[6–8]

14.1 8-METHOXYPSORALEN (8-MOP): THE "OBSCURE" ROLE OF OXYGEN

8-Methoxypsoralen (8-MOP) is a furocoumarin largely exploited in combination with ultraviolet (UV-A) radiation (PUVA phototherapy) for the treatment of vitiligo, psoriasis, and other skin disorders.[9] In addition to the pharmacological properties, 8-MOP and its congeners exert toxic and mutagenic effects that vary significantly, depending on their chemical structures.[9–11] According to the general view, the biological and pharmacological activity of 8-MOP, as well as its toxicity, are primarily the result of the covalent binding with DNA, to the point that psoralens were initially utilized as probes of nucleic acid structure and function.[9,10] Unlike other DNA modifying carcinogens, however, psoralens do not require biochemical modification (e.g., microsomal activation) in order to produce lethal or mutagenic effects. Indeed, metabolic products include chiefly ring-cleaved derivatives, which are not known to play a major role in 8-MOP bioactivity.[10,11] Yet, psoralens require photochemical activation by UV radiation in the UV-A range. The UV-A-induced DNA binding process involves: (a) noncovalent intercalative binding to the DNA helix, (b) formation of monoaddition products following irradiation at 365 nm,[10] and (c) absorption of a second photon by the monoadducts to form diadducts, which results in interstrand cross-linking.[10] This sequence of events has been generally accepted as the primary, if not the sole, determinant of psoralen bioactivity and mutagenic or carcinogenic effects.[12,13]

In fact, besides direct photoaddition with DNA (type I process), 8-MOP and related psoralens can also induce glutathione (GSH) depletion,[14,15] and in an aerobic environment, 8-MOP may be involved in oxygen-dependent type II photodynamic processes, leading to the production of reactive oxygen species[16–18] (Figure 14.1). These include singlet oxygen (energy transfer) and superoxide (electron transfer), which may target DNA, proteins, and membrane lipids, as well as,

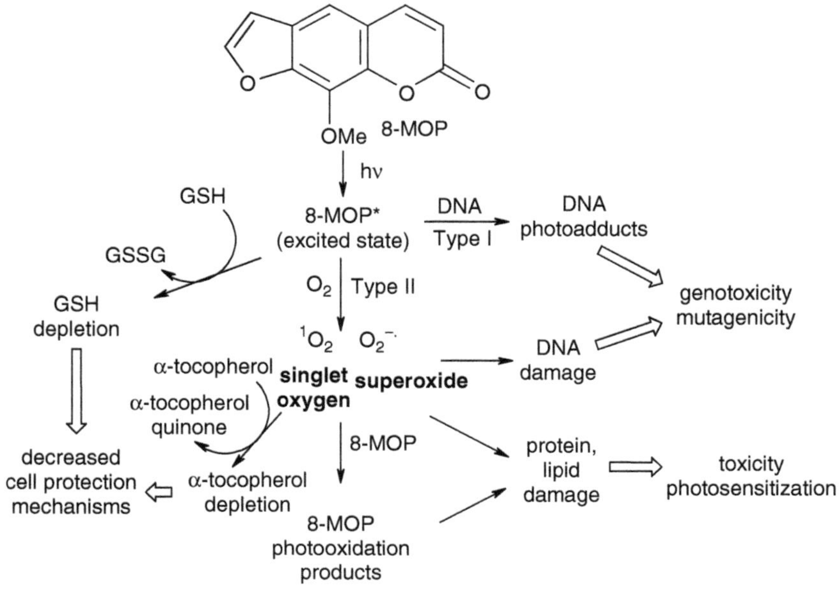

FIGURE 14.1 Simplified view of the mechanisms of 8-MOP phototoxicity.

notably, α-tocopherol and other antioxidants, causing extensive modifications of biomolecules.[19] In addition, and most relevant to the present discussion, these oxygen-derived species can modify 8-MOP and related furocoumarins, giving rise to cyclodimers, adducts with water, or products derived from photosplitting of the furan or pyrone rings. Oxidative products of furocoumarins can bind covalently to proteins and may be implicated in the hemolysis of red blood cells.[17] At the subcellular level, 8-MOP-associated oxidative damage has been related to mitochondrial alterations.[20] Oxygen-mediated damage induced by 8-MOP may be involved in a number of clinical side effects such as skin photosensitization, pigmentation, photoallergic reactions, and skin cancer.[16,17] Thus, in retrospect, it seems that the initial focus on direct DNA binding mechanisms has somewhat obscured a series of oxygen-dependent processes that should not be disregarded when looking at the mechanism of action of 8-MOP, including its toxicity and side effects.

14.2 XENOBIOTICS WITH REDOX-DEPENDENT TOXICITY IN WERNER SYNDROME AND FANCONI ANEMIA: THE OXIDATIVE STRESS LINK

Another case in point is represented by a set of xenobiotics whose toxicities have generally been defined without considering the biotransformations of these molecules and the underlying toxicity mechanisms.[6,7] Failure to keep these aspects in due consideration has resulted in widely held misconceptions as to the definition of certain peculiar disease states. Misuse of toxicity definitions has not been devoid of consequences in disease definitions, as has been the case for Werner syndrome[21] (WS) and for Fanconi anemia[22] (FA). The cells cultured from patients with these diseases display excessive sensitivity to a number of agents whose end products cause DNA damage[23–25] or inhibit activities, e.g., topoisomerase,[23,26] associated with the maintenance of genomic stability.[27] In the assumption that WS or FA cells were unable to cope with the toxicity of chemicals regarded as direct DNA-damaging agents, the phenotypes of these disorders have been related to deficiencies in DNA repair, with consequent implications on disease definition.[28–30] For example, excessive sensitivity to DEB and MMC is widely recognized as the most reliable diagnostic factor for FA, as FA cells show enhanced chromosomal breakage rates when cultured in the presence of DEB or MMC at concentrations that fail to affect non-FA ("normal") cells.[25] Thus, the sensitivity to DEB and MMC relates to the deeply rooted attribution of FA phenotype, referred to as "crosslinker sensitivity."[29–31] This term is due to the formation of DNA cross-links as an ultimate outcome of DEB and MMC toxicity; hence, the assumption that FA cells are unable to repair MMC- or DEB-induced cross-links has led to the notion of FA because of a DNA repair deficiency.[29,30] Evidence that has accrued over the past decades up to recent years, however, has gradually led to a reappraisal of the clinical history of FA as an oxidative stress-related disorder.[8]

MMC is an antitumor antibiotic known to alkylate DNA monofunctionally and bifunctionally, resulting in DNA crosslinking.[31,32] Chemically, it is characterized by a 5-amino-6-methyl-4,7-indolequinone moiety fused to a cyclopentane system bearing a highly tensioned three-member aziridine ring. It is therefore ranked among aziridinyl quinones, a class of bioactive agents that are known to require reductive activation by cellular reductases in order to bind to DNA.[33–37]

The requirement for a reductive activation,[36] in spite of an electrophilic quinonoid structure, is apparently paradoxical, yet it can be explained considering that the quinonoid system is fully substituted and cannot undergo nucleophilic substitution. Following reduction to the corresponding hydroquinone derivative, the electron-rich aromatic ring pushes electrons toward the aziridine ring, which is then opened to generate a highly electrophilic quinonemethidelike system.[37] In this latter reduced form, MMC becomes competent to first alkylate and then cross-link DNA at the 2-amino groups of guanine residues. It has been suggested that the cellular effect of MMC is related to the amount of alkylated sites (essentially 5'-CG sites) as well as to possible local DNA deformations (at minor alkylation sites).[38–41]

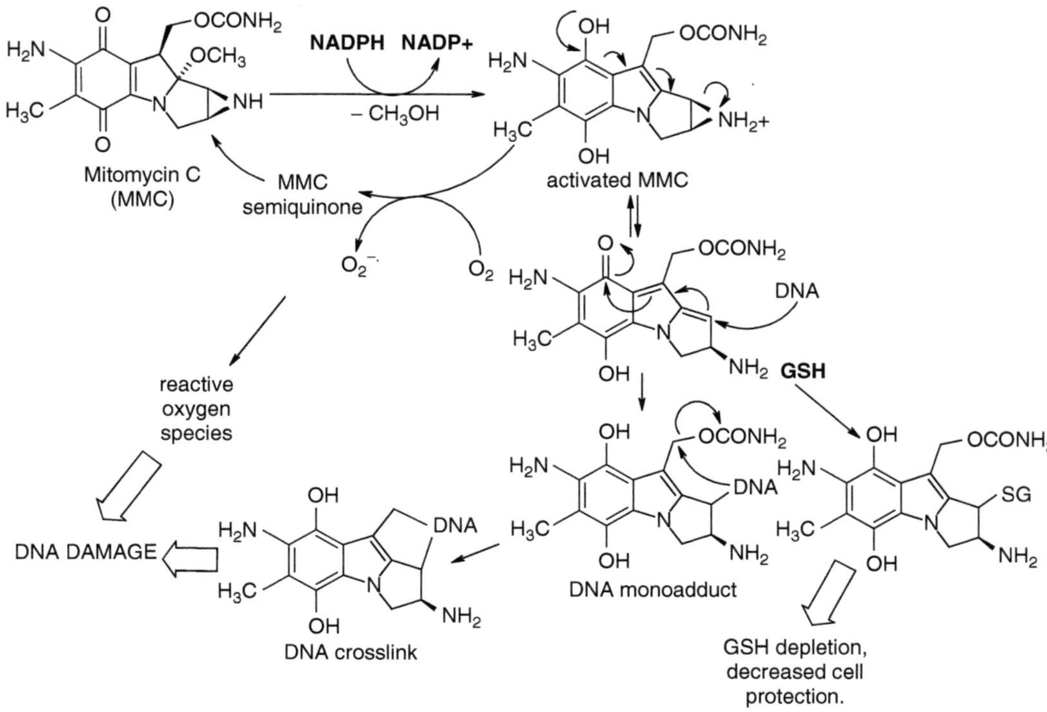

FIGURE 14.2 Simplified view of mechanisms involved in the bioactivity of MMC.

A simplified mechanism of action for MMC is shown in Figure 14.2. At least five different enzymes can catalyze MMC bioreduction *in vitro*, including a series of novel mitochondrial reductases such as a cytochrome P450 reductase.[42–44] Competition between reductases for MMC reduction appears to be based solely on protein levels rather than on enzyme kinetics.

The reductive activation steps required by this antibiotic make it likely that, under prooxidant conditions, alternative toxicity mechanisms may be involved. After reductive activation, e.g., by NADPH- and NADH-dependent enzymes, MMC can be partly diverted from DNA alkylation in an oxidative environment by redox interaction with oxygen, leading to formation of reactive oxygen species (ROS).[36,45] ROS, then, may contribute to DNA damage by oxidatively modifying bases in FA cells, whereby both antioxidant and repair mechanisms would eventually be overwhelmed. In normal (non-FA) cells, GSH levels are sufficient to scavenge reductively activated MMC.[35–37]

DEB is one of the key metabolites of the genotoxic carcinogen 1,3-butadiene, from which it is formed by a cytochrome P450-promoted dioxygenation.[46–49] The molecular mechanisms of DEB toxicity are not entirely elucidated, but there is sufficient evidence suggesting that its activity is due to its potent cross-linking effects and DNA alkylating properties. These stem from the peculiar 1,3-diepoxidation pattern of DEB,[46] in which the ring tension accounts for the marked reactivity toward nucleophiles, and the double epoxide ring allows for the formation of interstrand cross-links with many DNA sequences.

A substantial body of evidence suggests that individual sensitivity to the genotoxic and cytotoxic effects of DEB depends on oxygen levels and result in a dose-related GSH depletion.[50–52] The underlying implication is that DEB toxicity is modulated by GSH through a conjugation mechanism promoted by glutathione S-transferase (GST) (Figure 14.3).

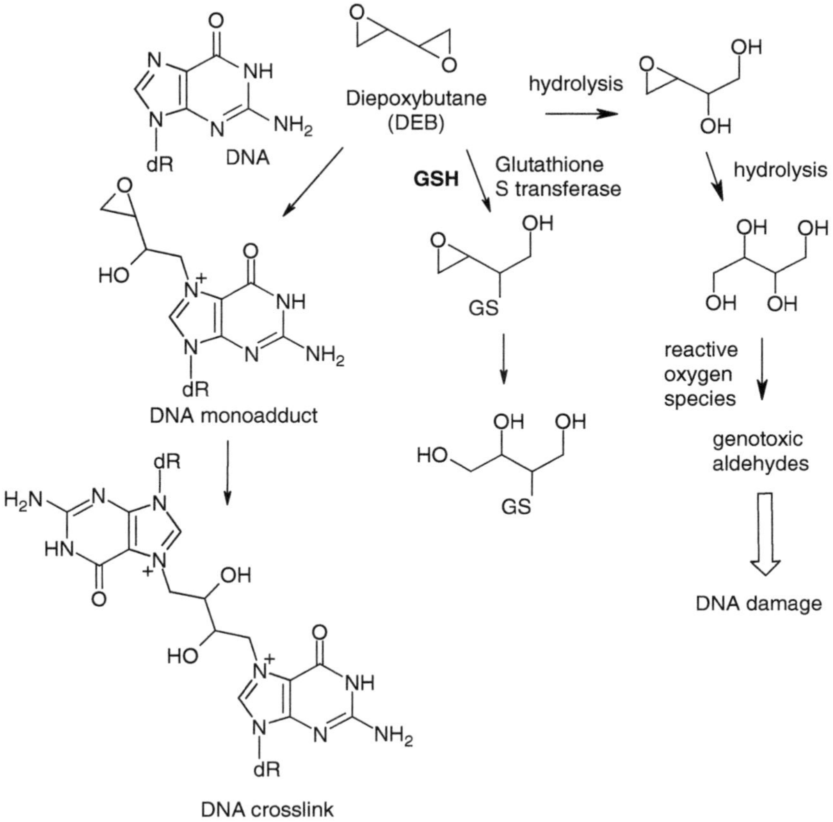

FIGURE 14.3

The role of GSH in DEB toxicity provides a plausible background to envisage oxidative stress-dependent GSH depletion and, especially, decreased GST activity as possible contributory factors affecting the detoxification potential of cells and individual susceptibility to DEB toxicity. This view was corroborated by a study in which pretreatment with phorone, a well-known intracellular glutathione depletor, was found to enhance DEB cytotoxicity, consistent with an active role played *in vivo* by the GSH-detoxifying system.[53]

Besides GST, another GSH-dependent enzyme has been implicated in the detoxification of DEB-induced DNA damage, i.e., GSH peroxidase (GSH-Px).[52] Studies of the effect of glutathione peroxidase (GSH-Px) in mammalian red blood cells (RBC) on spontaneous and DEB-induced sister chromatid exchange (SCE) in primary Big Blue mouse (BBM1) and Big Blue rat (BBR1) fibroblasts indicated that both GSH-Px and RBC alone or in combination were effective in significantly reducing DEB-induced SCE in both mouse and rat fibroblasts.[52] These results raise the intriguing possibility that H_2O_2 may be involved in DEB toxicity.[53]

In addition to conjugation with GSH, a possible transformation route of DEB is hydrolysis.[54] A comparative screening of the microsomal hydrolysis of 1,2:3,4-diepoxybutane (BDE) using tissues of rats, mice, and humans was found to give two main metabolites, erythritol and anhydro-erythritol. It is tempting to speculate that diol and polyol products of DEB hydrolysis, though expectedly inactive per se, may be sources of cytotoxic aldehydes (e.g., glyoxal) following oxidation and retroaldol reactions under oxidative stress conditions.

14.3 P450 AND OTHER OXYGEN-DEPENDENT BIOTRANSFORMATIONS: THE JANUS FACIES OF THE AEROBIC LIFESTYLE

The vast majority of metabolic transformations of xenobiotics stem from Phase I oxidative processes mediated by P450 and related activities leading to hydroxylation and other oxygenation reactions. Though they are primarily aimed at converting lipophilic xenobiotics into polar species more amenable to excretion and/or inactivation, Phase I biotransformations often generate highly reactive, harmful species that may elicit a variety of toxic responses.[5,6] The underlying implication is that even when an oxidative process like Phase I P450-mediated biooxidation is enzymatically controlled and is aimed at removing potentially toxic xenobiotics, it may eventually turn out to be more damaging than the parent compound. This is but one example of the dichotomy that is seen in various aspects of oxygen biology, e.g., the leakage of toxic and reactive oxygen species (like superoxide) from molecular systems that utilize oxygen for the essential physiological production of chemical energy (see for example the mitochondrial respiratory chain).

Stretching the analogy beyond the restricted context of biology, one could perhaps mention the most familiar case of oxygen-dependent reaction: combustion. Although useful and beneficial, improper or uncontrolled utilization of this process may be fatal and highly hazardous. Leonardo da Vinci's observation that "where the flame does not live, neither lives the breathing animal (c. 1493)" beautifully summarizes the complexity of phenomena underlying the dual roles of oxygen in nature.

The dichotomous behavior of oxygen is inherent to its peculiar free-radical nature, which accounts for its prominent role in biological processes. Oxygen is thermodynamically a good oxidizing agent, so that it can be efficiently utilized in metabolic reactions that produce energy, yet it is kinetically inert, implying that it requires catalytic activation. This may not always be under complete control, thus the evolution of systems that utilize the electron-accepting capabilities of oxygen implies the probability of undesired reactions that can damage cells. Very often, whether the oxidative biotransformation results in activation (toxicity) rather than inactivation (detoxification) depends on a very subtle balance of opposite factors subjected to genetically determined or circumstantial changes. These must be investigated and duly considered when looking at the toxicity mechanisms of xenobiotics or other organic substrates, otherwise the interpretation of biological phenomena may be incomplete or misleading.

14.4 CONCLUSIONS: WHAT LESSONS?

Lessons from the past 40 years indicate that the general involvement of redox processes must not be overlooked in future studies of xenobiotic metabolism. Researchers must avoid misleading opinions as to the toxicity of native species of xenobiotics vs. their derivatives and end products. Indeed, inquiries into the mode of action of xenobiotics should consider their aerobic environment from various perspectives as well as the possible environmental effects on their behavior.

ACKNOWLEDGMENT

The present paper relies on the activity of the authors in the European Research on Oxidative Stress — EUROS Project, supported by the European Commission DGXII (Contract BMH4-CT98-3107) and by the Italian Association for Fanconi Anemia Research (AIRFA). The precious help of Dr. Paola La Pietra and Dr. Elena De Nicola is gratefully acknowledged.

REFERENCES

1. Slaga, T.J. et al., Carcinogenicity and mutagenicity of benz(a)anthracene diols and diol-epoxides, *Cancer Res.*, 38, 1699, 1978.
2. Zajdela, F. et al., Carcinogenicity of chloroethylene oxide, an ultimate reactive metabolite of vinyl chloride, and bis(chloromethyl)ether after subcutaneous administration and in initiation-promotion experiments in mice, *Cancer Res.*, 40, 352, 1980.
3. Boyland, E. and Sims, P., Metabolism of polycyclic compounds: 24, the metabolism of benz[α]anthracene, *Biochem. J.*, 91, 493, 1964.
4. Boyland, E. and Chasseaud, L.F., Enzymes catalysing conjugations of glutathione with alpha-beta-unsaturated carbonyl compounds, *Biochem. J.*, 109, 651, 1968.
5. Marquardt, H., Microsomal metabolism of chemical carcinogens in animals and man, *IARC Sci. Publ.*, 16, 309, 1977.
6. Pagano, G., Redox-modulated xenobiotic action and ROS formation: a mirror or a window? *Hum. Exp. Toxicol.*, 21, 77, 2002.
7. Pagano, G., Manini, P., and Bagchi, D., Oxidative stress-related mechanisms in chromium(VI)-induced toxicity to Fanconi anemia cells, *Environ. Health Persp.*, 111, 1699, 2003.
8. Pagano, G. and Youssoufian, H., Fanconi's anaemia proteins: concurrent roles in cell protection against oxidative damage, *BioEssays*, 25, 589, 2003.
9. Averbeck, D., Recent advances in psoralen phototoxicity mechanism, *Photochem. Photobiol.*, 50, 859, 1989.
10. Song, P.S. and Tapley, Jr., K.J., Photochemistry and photobiology of psoralens, *Photochem. Photobiol.*, 29, 1177, 1979.
11. Parsons, B.J., Psoralen photochemistry, *Photochem. Photobiol.*, 32, 813, 1980.
12. Moustacchi, E. and Diatloff-Zito, C., DNA semi-conservative synthesis in normal and Fanconi anemia fibroblasts following treatment with 8-methoxypsoralen and near ultraviolet light or with X-rays, *Hum. Genet.*, 70, 236, 1985.
13. Averbeck, D., Papadopoulo, D., and Moustacchi E., Repair of 4,5′,8-trimethylpsoralen plus light-induced DNA damage in normal and Fanconi's anemia cell lines, *Cancer Res.*, 48, 2015, 1988.
14. d'Ischia, M., Napolitano, A., and Prota, G., Psoralens sensitize glutathione photooxidation *in vitro*, *Biochim. Biophys. Acta*, 993, 143, 1989.
15. Wheeler, L.A. et al., Depletion of cutaneous glutathione and the induction of inflammation by 8-methoxypsoralen plus UVA radiation, *J. Invest. Dermatol.*, 87, 658, 1986.
16. Punnonen, K. et al., Effects of *in vitro* UVA irradiation and PUVA treatment on membrane fatty acids and activities of antioxidant enzymes in human keratinocytes, *J. Invest. Dermatol.*, 96, 255, 1991.
17. Reinheckel, T. et al., Evaluation of UVA-mediated oxidative damage to proteins and lipids in extra-corporeal photoimmunotherapy, *Photochem. Photobiol.*, 69, 566, 1999.
18. Schmid, J. et al., The metabolism of 8-methoxypsoralen in man, *Eur. J. Drug Metabol. Pharmacokin.*, 5, 81, 1980.
19. Costantini, C. et al., Photodynamic degradation of vitamin E induced by psoralens, *Biochim. Biophys. Acta*, 1116, 291, 1992.
20. Rousset, S. et al., Mitochondrial alterations in Fanconi anemia fibroblasts following ultraviolet A or psoralen photoactivation, *Photochem. Photobiol.*, 75, 159, 2002.
21. Chen, L. and Oshima, J., Werner syndrome, *J. Biomed. Biotechnol.*, 2, 46, 2002.
22. Gordon-Smith, E.C. and Rutherford, T.R., Fanconi anemia: constitutional aplastic anemia, *Semin. Hematol.*, 28, 104, 1991.
23. Poot, M., Gollahon, K.A., and Rabinovitch P.S., Werner syndrome lymphoblastoid cells are sensitive to camptothecin-induced apoptosis in S-phase, *Hum. Genet.*, 104, 10, 1999.
24. Poot, M. et al., Werner syndrome cells are sensitive to DNA cross-linking drugs, *FASEB J.*, 15, 1224, 2001.
25. Auerbach, A.D. and Wolman, S.R., Susceptibility of Fanconi's anaemia fibroblasts to chromosome damage by carcinogens, *Nature*, 261, 494 1976.
26. Poot, M., Epe, B., and Hoehn, H., Cell cycle effects of the DNA topoisomerase inhibitors camptothecin and m-AMSA in lymphoblastoid cell lines from patients with Fanconi anemia, *Mutat. Res.*, 270, 185, 1992.

27. Poot, M. et al., Werner syndrome diploid fibroblasts are sensitive to 4-nitroquinoline-N-oxide and 8-methoxypsoralen: implications for the disease phenotype, *FASEB J.*, 7, 757, 2002.
28. Poon, P.K., O'Brien, R.L., and Parker, J.W., Defective DNA repair in Fanconi's anemia, *Nature*, 250, 223, 1974.
29. Auerbach, A.D. and Verlander, P.C., Disorders of DNA replication and repair, *Curr. Opin. Pediatr.*, 9, 600, 1997.
30. Grompe, M. and D'Andrea, A., Fanconi anemia and DNA repair, *Hum. Mol. Genet.*, 10, 2253, 2001.
31. Heinrich, M.C. et al., DNA cross-linker-induced G2/M arrest in group C Fanconi anemia lymphoblasts reflects normal checkpoint function, *Blood*, 91 275, 1998.
32. Tomasz, M. and Palom, Y., The mitomycin bioreductive antitumor agents: crosslinking and alkylation of DNA as the molecular basis of their activity, *Pharmacol. Ther.*, 76, 73, 1997.
33. Gargiulo, D. et al., Alkylation and crosslinking of DNA by the unnatural enantiomer of mitomycin C: mechanism of the DNA-sequence specificity of mitomycins, *J. Am. Chem. Soc.*, 117, 9388, 1995.
34. Iyer, V.N. and Szybalski, W., Mitomycins and porphyromycin: chemical mechanism of activation and crosslinking of DNA, *Science*, 145, 55, 1964.
35. Pritsos, C.A. and Sartorelli, A.C., Generation of reactive oxygen radicals through bioactivation of mitomycin antibiotics, *Cancer Res.*, 46, 3528, 1986.
36. Dusre, L. et al., DNA interstrand crosslink and free radical formation in a human multidrug resistant cell line from mitomycin C and its analogues, *Cancer Res.*, 50, 648, 1990.
37. Penketh, P.G. et al., Inhibition of DNA cross-linking by mitomycin C by peroxidase-mediated oxidation of mitomycin C hydroquinone, *J. Biol. Chem.*, 276, 34445, 2001.
38. Subramaniam, G. et al., Solution structure of a guanine-N7-linked complex of the mitomycin C metabolite 2,7-diaminomitosene and DNA: basis of sequence selectivity, *Biochemistry*, 40, 10473, 2001.
39. Paz, M.M., Das, A., and Tomasz, M., Mitomycin C linked to DNA minor groove binding agents: synthesis, reductive activation, DNA binding and cross-linking properties and *in vitro* antitumor activity, *Bioorg. Med. Chem.*, 7, 2713, 1999.
40. Cummings, J. et al., Enzymology of mitomycin C metabolic activation in tumor tissue: implications for enzyme-directed bioreductive drug development, *Biochem. Pharmacol.*, 56, 405, 1998.
41. Li, V.-S. et al., Concerning *in vitro* mitomycin-DNA alkylation, *J. Am. Chem. Soc.*, 118, 3765, 1996.
42. Joseph, P., Xu, Y., and Jaiswal, A.K., Non-enzymic and enzymic activation of mitomycin C: identification of a unique cytosolic activity, *Int. J. Cancer*, 65, 263, 1996.
43. Palom, Y. et al., Bioreductive metabolism of mitomycin C in EMT6 mouse mammary tumor cells: cytotoxic and non-cytotoxic pathways, leading to different types of DNA adducts: the effect of dicumarol, *Biochem. Pharmacol.*, 61, 1517, 2001.
44. Joseph, P. and Jaiswal, A.K., A unique cytosolic activity related but distinct from NQO1 catalyzes metabolic activation of mitomycin C, *Brit. J. Cancer*, 82, 1305, 2000.
45. Gutteridge, J.M.C., Quinlan, G.J., and Wilkins, S., Mitomycin C-induced deoxyribose degradation inhibited by superoxide dismutase: a reaction involving iron, hydroxyl and semiquinone radicals, *FEBS Lett.*, 167, 37, 1984.
46. Bartók, M. and Láng, K.L., Oxiranes, in *The Chemistry of Functional Groups, Suppl. E, Part 2 : the Chemistry of Ethers, Crown Ethers, Hydroxyl Groups and Their Sulphur Analogues*, Patai, S., Ed., John Wiley and Sons, Chichester, 1980, p. 609.
47. Nauhaus, S. et al., Characterization of urinary metabolites from Sprague-Dawley rats and B6C3F1 mice exposed to [1,2,3,4-13C]butadiene, *Chem. Res. Toxicol.*, 9, 764, 1996.
48. Kemper, R.A., Krause, R.J., and Elfarra, A.A., Metabolism of butadiene monoxide by freshly isolated hepatocytes from mice and rats: different partitioning between oxidative, hydrolytic, and conjugation pathways, *Drug Metabol. Dispos.*, 29, 830, 2001.
49. Krause, R.J. and Elfarra, A.A., Oxidation of butadiene monoxide to meso- and (±)-diepoxybutane by cDNA-expressed human cytochrome P450s and by mouse, rat, and human liver microsomes: evidence for preferential hydration of meso-diepoxybutane in rat and human liver microsomes, *Arch. Biochem. Biophys.*, 337, 176, 1997.
50. Vlachodimitropoulos, D. et al., GSTT1-dependent induction of centromere-negative and -positive micronuclei by 1,2:3,4-diepoxybutane in cultured human lymphocytes, *Mutagenesis*, 12, 397, 1997.

51. Boogaard, P.J., Sumner, S.C.J., and Bond, J.A., Glutathione conjugation of 1,2:3,4-diepoxybutane in human liver and rat and mouse liver and lung *in vitro*, *Toxicol. Appl. Pharmacol.*, 136, 307, 1996.

52. Erexson, G.L. and Tindall, K.R., Reduction of diepoxybutane-induced sister chromatid exchanges by glutathione peroxidase and erythrocytes in transgenic Big Blue mouse and rat fibroblasts, *Mutat. Res.*, 447, 267, 2000.

53. Spanò, M. et al., Diepoxybutane cytotoxicity on mouse germ cells is enhanced by *in vivo* glutathione depletion: a flow cytometric approach, *Mutat. Res.*, 397, 37, 1998.

54. Boogaard, P.J. and Bond, J.A., The role of hydrolysis in the detoxification of 1,2:3,4-diepoxybutane by human, rat, and mouse liver and lung *in vitro*, *Toxicol. Appl. Pharmacol.*, 141, 617, 1996.

15 Obesity and Cancer

Harry G. Preuss, Manashi Bagchi, Debasis Bagchi,
and G.R. Kaats

CONTENTS

15.1 Introduction...197
15.2 Etiology of Overweight and Obesity 198
15.3 Problems in Estimating Fat Gain and Loss 198
15.4 Association of Obesity with Cancer 199
15.5 Pathogenesis behind These Associations 200
15.6 Conclusions..201
References 202

15.1 INTRODUCTION

Many in the public are now realizing what should have been obvious over the last 10 to 20 years: there is an overweight–obesity epidemic in the U.S.[1–6] As of now, there are estimated to be more overweight Americans than nonoverweight Americans.[3] Over the last 10 to 15 years, obesity has increased in every state, in both genders, among smokers and nonsmokers, and across race/ ethnicity, age, and educational levels. The fact is that Americans have gained more weight in the past decade than in the previous four decades combined. Although it is generally accepted that overweight and obesity are highly prevalent in the American population, other countries, even so-called Third World countries, have similar experiences.[7] The World Health Organization (WHO) recently announced that obesity is also a global epidemic.[8]

A report published in *USA Today* in November 2003 examined obesity in Europe. The data indicate that the U.S. has more obese individuals than the European countries among adults. The U.K. has the next highest number of obese adults. Interestingly, the number of children in the U.S. listed as obese is considerably less by percentage than adults (15 vs. 31%). In contrast, countries such as Spain, Italy, Denmark, Sweden, and France have a proportionately higher number of children classified as obese than adults. This is most unfortunate, because the trend will be difficult to halt and reverse.[9] However, even more worrisome is that the overweight state, in addition to its negative effects on physical appearance, is associated with a number of chronic illnesses. The association of the obese state with various cardiovascular disorders, diabetes, arthritis, and gallstones are generally recognized, but the alarming association with certain cancers is just beginning to be noticed.[10–12] As a general truth, obesity is rapidly overtaking cigarette smoking as the leading cause of preventable death throughout the world.[13,14]

Obesity is a complex, multifactorial, chronic condition characterized by excess body fat result-ing from an imbalance between energy expenditure and caloric intake.[6,15] As mentioned above, more than one-half of U.S. adults are overweight, having a body mass index (BMI) greater than 25 kg/m^2, while more than a quarter (31%) of U.S. adults are obese, having a BMI of greater than 30 kg/m^2. A BMI of greater than 40 kg/m^2 is termed as "morbid obesity" or clinically severe obesity, a condition that affects more than 15 million Americans.[7] Low levels of physical activity, sedentary lifestyles, stress, depression, and consumption of high-fat and fast foods are largely responsible for unwanted weight gain.[16-20]

Recent studies have shown that approximately one-third of the variance in adult body weights is secondary to genetic influences.[1] Leptin, an adipocyte- and placenta-derived circulating protein, regulates, to some extent, the magnitude of fat stores in the body leading to obesity.[21] Gastrointestinal peptides, neurotransmitters, and adipose tissue may also have an etiologic role in obesity.[22] Low-caloric diets with/without exercise can help with temporary weight loss. Nevertheless, diet and exercise alone have not proven universally successful for long-term solutions in weight management. In addition, supplementation with drugs that suppress appetite, reduce food intake, increase energy expenditure, and/or influence nutrient partitioning or metabolism have potential efficacy, but unfortunately these are accompanied by adverse side effects — some life threatening.[23,24]

15.2 ETIOLOGY OF OVERWEIGHT AND OBESITY

For years, overeating and obesity were thought to result from psychological problems: an addictive personality, a lack of discipline, poor self-control, and/or traumatic childhood experiences. Many experts are now acknowledging that overeating and obesity are more likely a result of two concomitant factors not directly associated with psychological abnormalities: (a) a calorie-toxic environment that results in augmented intake of calories and refined carbohydrates along with a poor intake of fibers and (b) a sedentary lifestyle emanating from too much TV watching, computer surfing, and automobile driving in place of walking.[1,3,25-27]

15.3 PROBLEMS IN ESTIMATING FAT GAIN AND LOSS

One difficulty in managing the obesity epidemic is the inability to measure fat mass easily and economically with accuracy. The majority of research studies examining weight gain and loss characteristically use body weight measured on a weighing scale as the major end point. In practically all cases, the desire is to lose the real culprit in body weight gain and/or expansion — the "fat." Nevertheless, a weighing scale measures both fat and lean mass. The latter includes muscle, bone, internal organs, and fluids. Although the general assumption is that changes in body weight from scale measurements are due largely to fat loss, in most cases it is actually a combination of fat and lean body (muscle) mass loss.[28] In many articles discussing obesity and its role in many disease processes, the authors discuss fat loss or gains when they have really measured only body weight changes.

If one or both parents are obese, a child's chances of being overweight increase by 25 to 30%. Weight loss via diminution of muscle mass is unfortunate, because muscles determine, to some extent, the rate of metabolism. Gaining muscle can actually speed up metabolism and should be the goal of those trying to lose girth around the belly and hips, i.e., stay in shape. Men have more muscle than women. Thus, men expend up to 20% more calories than women at rest or after exercise. It has long been recognized that women have much more difficulty losing weight via increased exercise; this phenomenon may be explicable on the basis of lower muscle mass. During aging, muscle mass generally decreases (sarcopenia), while fat mass increases (obesity). Coupled with the lessened activity commonly seen in the elderly, fewer calories are needed. In most cases, the fat accumulation outpaces muscle loss, thus accounting for the usual increase in scale weight with aging.[29]

Logically, height should play a role in assessing the appropriateness of a given weight. Under ideal conditions, a taller person should weigh more than a shorter person. To take this into account, many, including governmental agencies, define the overweight/obese state based on the measurement of the body mass index (BMI). The BMI is based on both weight and height of the individual.

The BMI is the ratio of weight (in kilograms) to height (in meters) squared. In June 1998, a panel of experts was convened by the National Institutes of Health to provide guidelines for the measurement of overweight and obesity. Overweight was accepted as a BMI of 25 to 30, and obesity as BMI 30 or more. Those morbidly obese exceeded a BMI of 40. While using BMI may be an improvement over just examining body weight alone, this estimation still contains some flaws because it still uses body weight instead of fat mass. Muscle is more dense than fat. Accordingly, a muscular individual with a low percentage of body fat might be judged obese. Some of our best athletes become obese by using BMI criteria.

Unfortunately, interventions that may be the most effective in adding or preserving muscle are likely to be seen as ineffective by both scientific investigators and the public.[28] It is our view that measurement of body composition should be a requirement for any evaluation concerning overweight/obesity. The goal for those losing weight is to lose fat, not muscle. Individuals building up muscle mass want to know that their weight gain is truly muscle and not fat. Perhaps the best way to explain the preceding is via an example.

Two obese subjects are weighed after undergoing weight loss regimens: one has lost 7 lb on the scale, while the other has lost only 1 lb. Looking only at this single parameter, one would have to believe that the first subject is more successful than the second. Now, consider that further analyses show that the first subject lost 4 lb of fat and 3 lb of muscle to account for the 7-lb weight loss. The second subject also lost 4 lb of body fat but gained 3 lb of muscle, thus accounting for the small weight loss on the scale. We believe that the second subject shows better improvement despite the disappointing scale results caused by adding 3 lb of muscle. Through metabolic processes, this 3-lb gain in muscle mass on its own could theoretically account for a loss of 1 lb of fat in each ensuing month and prevent the return of the previously lost, unwanted fat — the so-called common yo-yo effect.[29]

In attempting to examine the correlation of increased weight with various cancers, a number of unanswered questions provide roadblocks. We will mention two of these problems. First, it is important to note that the exact amount of weight gain needed to augment the risk of certain cancers is unknown. It seems reasonable to assume that the higher the BMI, the greater is the risk for cancer.[30] Those with the highest girth show the most risk.[31] Second, it is apparent that very few articles examining weight loss actually follow fat loss. Nevertheless, in their discussions, the authors frequently equate their body weight measurements directly to fat mass. We bring this up primarily because the association of cancer will be made with overweight and obesity based upon scale weight. However, is it fat mass rather than the total mass, i.e., a combination of fat and lean body mass, that we believe is truly associated with certain cancers? We know that fat mass is very closely related to insulin resistance; and insulin resistance may play, at least to some extent, some role in the association between obesity and cancer.[32]

15.4 ASSOCIATION OF OBESITY WITH CANCER

While the association of cancer to obesity — defined as an abnormally high, unhealthy proportion of body fat — is generally accepted, results from studies attempting to link obesity to cancer have not always been positive. Therefore, it is important to realize that not all types of cancer may show such an association. Those with the highest certainty of a link include breast, colon. prostate, endometrium, cervix, ovary, kidney, gall bladder, liver, pancreas, rectum, and esophagus.[30,33,34]

The International Agency for Research in Lyon, France, noted the association between overweight, obesity, physical inactivity, and cancer.[33,34] Taken together, this working group proposed that excess body weight and physical inactivity account for approximately a quarter to one-third of cancers of the colon, endometrium, breast, kidney, and esophagus. They believed the enhanced risk could be attributed to alterations in the metabolism of endogenous hormones, including sex steroids, insulin, and insulin-like growth factors. In turn, these could lead to distortions in balance between cell proliferation, differentiation, and apoptosis.

In a recent prospective study of more than 900,000 U.S. adults (approximately one-half males and one-half females), 57,145 deaths occurred during a 16-year follow-up.[30] The chief end points were BMI and deaths from cancer. Compared with people of normal weight, those who were overweight or obese had a higher risk of death from cancers of the esophagus, colon, rectum, liver, gallbladder, pancreas, kidney, non-Hodgkin's lymphoma, and multiple myeloma — the higher the BMI, the greater the risk of cancer. The heaviest men were most likely to die from stomach or prostate cancer, while the heaviest women were most likely to die from breast, uterine, cervical, or ovarian cancer. The authors estimated that the current patterns of overweight and obesity in the U.S. could account for 14% of all deaths from cancer in men and 20% of those in women. The weight and height measurements were self-reported in this study.[30] The authors recognized that they had no direct measurements of adiposity, lean body mass, or even central obesity such as waist-to-hip ratios.

Borugian et al.[31] did examine waist-to-hip ratio (WHR) and compared it with breast cancer mortality in a study of 603 patients with incidental breast cancer in 1991–1992. The follow-up was 10 years. WHR was directly related to breast cancer mortality in postmenopausal women (for highest quartile vs. lowest) in ER-positive postmenopausal women. The menopausal state affects the role of obesity in breast cancer. Prior to menopause, obese women have a lower risk for breast cancer compared with normal-sized women.

Black women are typically diagnosed with breast cancer at a later stage than white women.[35] This may relate to the effects of the higher prevalence of severe obesity among black women compared with white women. While not proven, it could be hypothesized that the relatively greater increased body mass in black women could directly correlate with greater stimulation of cancer. However, obesity influences other aspects of breast cancer, especially the stage of diagnosis. The greater preponderance of excessive obesity among black women compared with white women has been given an important role in explaining their late diagnosis of breast cancer.[35] Severely overweight women may need special mammography procedures such as larger films and additional views in order to diagnose at an early stage.

In a study of approximately 1.1 million Norwegian women, 7882 histologically verified cases of ovarian cancer were noted.[36] Women who were overweight had an increased risk of ovarian cancer. This was predominant among women who had gained excess weight in their younger years and was not observed in older women (>60 years). Obesity and body fat distribution were associated more strongly with cervical adenocarcinoma rather than squamous cell carcinoma of the cervix.[37]

15.5 PATHOGENESIS BEHIND THESE ASSOCIATIONS

The exact mechanism(s) behind the association between obesity and some cancers is unknown. Obesity itself develops via complex mechanisms involving interactions between heredity and lifestyle changes that include nutritional and exercise considerations. Even the location of the fat deposits may provide clues as to the pathogenesis, as described below.[38] The general thought is that augmented fat deposits lead to hormonal imbalances. For example, high levels of estrogens contribute to the proclivity toward cancer. The increased risk of breast cancer after menopause in obese women is believed to be due to increased levels of estrogens.[39] Prior to menopause, the ovaries are the primary source of estrogen. After menopause, when ovaries stop producing estrogen, fat tissue becomes the primary source.[39] Logic dictates that more fat leads to more circulating estrogens and, hence, an increase in risk.

In an international study from three weight centers, blood samples were collected from almost 3000 women who were not using hormone-replacement therapy.[39] A total of 624 women developed breast cancer in the 2- to 12-year follow-up. Hormones in the afflicted were compared with the normals, and the subjects were categorized according to their BMI. In short, as BMI climbed, so did estrogen levels and breast cancer risk. Women with a BMI of 30 had an 18% higher risk of developing breast cancer than women with a BMI of 25.

Obesity not infrequently portends the development of insulin resistance, with high circulating levels of insulin and insulinlike hormones — very important growth factors.[40] Insulin is known to stimulate cell proliferation, suggesting that an overproduction or high concentrations may be responsible, at least to some extent, for cancer development. High insulin and insulinlike hormone levels have been associated with increased risk for breast cancer and poorer survival after breast cancer diagnosis.[31] An increased waist-to-hip ratio is a recognized marker of insulin resistance and hyperinsulinemia. The fact that an elevated ratio proved to be a predictor of breast cancer mortality supports the importance of some role for a perturbed insulin system in the pathogenesis of cancer, at least breast cancer.[31] Among premenopausal women, central obesity may be specifically associated with an increased risk of breast cancer.[41]

Noting that insulin and insulinlike growth factors such as IGF I and II and IGF-binding proteins such as IGFBP 1 and 3 have been implicated in breast cancer outcomes, 53 postmenopausal breast cancer survivors were randomly assigned to an exercise (n = 25) or control group (n = 28) for 15 weeks.[42] No significant intergroup differences were noted in fasting insulin, glucose, IGF II, or IGFBP 1. However, the changes in IGF I, IGFBP 3, and IGF 1/IGFBP 3 molar ratio between the exercise and control groups were significant. Accordingly, the possibility that exercise could be helpful in the treatment of breast cancer via its effects on the insulin system is strengthened.

An additional point to consider is that different forms of cancer may respond differently to obesity.

15.6 CONCLUSIONS

The importance of remaining in optimal physical shape becomes increasingly obvious with each passing year. Until recently, Western medicine was focused almost exclusively on treating acute disorders by the use of drugs to the neglect of preventive medicine. Nutrition and the use of dietary supplements are poorly taught in medical schools, often relegated as a component of "alternative medicine." However, this concept is changing, and even dietitians are now encouraging the use of multivitamins. The medical establishment has done little to promote knowledge of preventive approaches, such as glucosamine for osteoarthritis,[43] saw palmetto for symptoms of benign prostatic hypertrophy (BPH),[44] antioxidants for macular degeneration,[45] and fish oils for cardiovascular health.[46] The public is now beginning to hear about these preventive approaches, unfortunately long after they could have begun to help some.

It is worth repeating that the neglect of good nutritional and exercise advice must be remedied for a number of important reasons. While many have attempted to keep the "fat off" and the "muscles on" for appearance sake, we are now realizing that this approach also has many important health benefits. Obesity has long been associated with metabolic disorder such as cardiovascular perturbations, gall bladder disturbances, diabetes, and arthritis. Now it is even being associated with many severe, life-threatening cancers. Despite the obvious health benefits, keeping the fat off and the muscles on is not always easy to accomplish, especially in our calorie-toxic environment that discourages exercise. Keeping the fat off even becomes more of a difficulty as one ages due to shifts in metabolism and activity that favor the emergence of obesity.[47]

Accordingly, the major means to prevent and/or ameliorate obesity are to watch the calories and exercise. In years past, each medical specialty had its own preferred diets. Now it is becoming clear that one healthful dietary plan may suffice for all perturbations, including the metabolic syndrome, cardiovascular disorders, diabetes, cancer prevention, and even prostate disorders.[48] In fact, a good dietary program may help slow the aging process.[40] In general, the current advice is to keep the calories reasonable to reach and/or maintain ideal weight. While calories are the first dietary consideration, the form of those calories may also be important. Refined carbohydrates are absorbed rapidly (high glycemic index) and can stress the insulin system in such a way as to favor fat accumulation and muscle deterioration.[49] Thus, favoring more complex, slower absorbing carbohydrates in the diet, or the addition to the diet of more viscous soluble fiber that slows carbohydrate absorption, can be helpful in a weight-loss regimen.[50] In the case of the equally important second health component, i.e., exercise, it is worth emphasizing that aerobic exercising, like walking, can burn calories and exercise the heart. The addition of resistance training to maintain and even build muscle is also important, even for women and the elderly.[51]

REFERENCES

1. Brownell, K.D., Obesity: understanding and treating a serious, prevalent, and refractory disorder, *J. Consult. Clin. Psychol.,* 50, 820, 1992.
2. Bray, G.A., Obesity, in *Present Knowledge in Nutrition*, Ziegler, E.E. and Filer, Jr., L.J., Eds., ILSI Press, Washington, DC, 1996, pp. 19–32.
3. Guterman, L., Obesity problem swells worldwide, *Chron. Higher Educ.*, March 8, 2002, p. A18.
4. U.S. Department of Health and Human Services, *The Surgeon General's Call to Action To Prevent and Decrease Overweight and Obesity, 2001* , U.S. Government Printing Office, Washington, DC, 2001.
5. Critser, G., *Fat Land: How Americans Became the Fattest People in the World*, Houghton Mifflin, Boston, 2003; *New England J. Med.*, 348, 2161, 2003.
6. Jequier, E., Pathways to obesity, *Int. J. Obes. Relat. Metab. Disord.*, 260, S12, 2002.
7. Cole, T.J. et al., Establishing a standard definition for child overweight and obesity worldwide: international survey, *Br. Med. J.*, 320, 1240, 2000.
8. World Health Organization, Controlling the Global Obesity Epidemic; report available on-line at www.who.int/nut/obs.htm (updated Aug. 15, 2003).
9. Campbell, I., The obesity epidemic: can we turn the tide? *Heart*, 89, 35, 2003.
10. Manson, J.E. et al., Body weight and mortality among women, *New England J. Med.*, 333, 677, 1995.
11. Pi-Sunyer, F.X., Health implications of obesity, *Am. J. Clin. Nutr.*, 53, 1595S, 1991.
12. Pi-Sunyer, F.X., Medical hazards of obesity, *Ann. Intern. Med.*, 119, 655, 1993.
13. McGinnis, J.M. and Foege, W.H., Actual causes of death in the United States, *JAMA*, 270, 2207, 1993.
14. Sjostrom, L.V., Mortality of severely obese subjects, *Am. J. Clin. Nutr.*, 55, 5165, 1992.
15. Stunkard, A.J., Current views on obesity, *Am. J. Med.*, 100, 230, 1996.
16. Roberts, S.B., McCrory, M.A., and Saltzman, E., The influence of dietary composition on energy intake and body weight, *J. Am. Coll. Nutr.*, 21, 140S, 2002.
17. Popkin, B.M. et al., A review of dietary and environmental correlates of obesity with emphasis on developing countries, *Obes. Res.*, 3, 145S, 1995.
18. Morley, J.E., Neuropeptide regulation of appetite and weight, *Endocr. Rev.*, 8, 256, 1987.
19. Sobal, J. and Stunkard, A.J., Socioeconomic status and obesity: a review of the literature, *Psychol. Bull.*, 105, 260, 1989.
20. Namnoum, A.B., Obesity: a disease worth treating, *Female Patient*, 18, 33, 1993.
21. Frederich, R.C. et al., Leptin levels reflect body lipid content in mice: evidence for diet-induced resistance to leptin action, *Nat. Med.*, 1, 1311, 1995.
22. Bandini, L.G. et al., Validity of reported energy intake in obese and nonobese adolescents, *Am. J. Clin. Nutr.*, 52, 421, 1990.
23. Volmar, K.E. and Hutchins, G.M., Aortic and mitral fenfluramine-phenteramine valvulopathy in 64 patients treated with anorectic agents, *Arch. Pathol. Lab. Med.*, 125, 1555, 2001.
24. Cheng, T.O., Fen/Phen and valvular heart disease; the final link has now been established, *Circulation,* 102, E180, 2000.
25. Jeffery, R.W. and Utter, J., The changing environment and population obesity in the United States, *Obes. Res.*, 11, 12S, 2003.
26. Isomaa, B., A major health hazard: the metabolic syndrome, *Life Sci.*, 73, 2395, 2003.
27. Prentice, A.M. and Jeb, S.A., Obesity in Britain: gluttony or sloth? *Br. Med. J.*, 311, 437, 1995.
28. Crawford, V., Scheckenbach, R., and Preuss, H.G., Effects of niacin-bound chromium supplementation on body composition of overweight African-American women, *Diab. Obes. Metab.*, 1, 331, 1999.
29. Wescott, W.L. and La Rosa-Loud, R., Implementing a cellulite reduction exercise program, *Fitness Manage.*, 32, 2002.

30. Calle, E.E. et al., Overweight, obesity, and mortality from cancer in a prospectively studied cohort of U.S. adults, *New England J. Med.*, 348, 1625, 2003.

31. Borugian, M.J. et al., Waist-to-hip ratio and breast cancer mortality, *Am. J. Epidemiol.*, 158, 963, 2003.

32. Bertoli, A. et al., Lipid profile, BMI, body fat distribution, and aerobic fitness in men with metabolic syndrome, *Acta. Diabetol.*, 40 (Suppl. 1), S130, 2003.

33. Bianchini, F., Kaaks, R., and Vainio, H., Overweight, obesity, and cancer risk, *Lancet Oncol.*, 3, 565, 2002.

34. Vainio, H., Kaaks, R., and Bianchini, F., Weight control and physical activity in cancer prevention: international evaluation of the evidence, *Eur. J. Cancer. Prev.*, 119 (Suppl. 2), S94, 2002.

35. Jones, B.A. et al., Severe obesity as an explanatory factor for the black/white difference in stage at diagnosis of breast cancer, *Am. J. Epidemiol.*, 146, 394, 1997.

36. Engeland, A., Tretli, S., and Bjorge, T., Height, body mass index, and ovarian cancer: a follow-up of 1.1 million Norwegian women, *J. Natl. Cancer Inst.*, 95, 1244, 2003.

37. Lacey, Jr., J.V. et al., Obesity as a potential risk factor for adenocarcinomas and squamous cell carcinomas of the uterine cervix, *Cancer*, 98, 814, 2003.

38. Abu-Abid, S., Szold, A., and Klausner, J., Obesity and cancer, *J. Med.*, 33, 73, 2002.

39. Key, T.H. et al., Body mass index, serum sex hormones, and breast cancer risk in postmenopausal women, *J. Natl. Cancer Inst.*, 95, 1218, 2003.

40. Preuss, H.G., Bagchi, D., Clouatre, D., Insulin resistance: a factor in aging, in Ghen, M.J., Corso, N., Joiner-Bey, H., Klatz, R., and Dratz, A., Eds., *The Advanced Guide to Longevity Medicine*, Ghen, Landrum, SC, 2001.

41. Harvie, M. et al., Central obesity and breast cancer risk: a systematic review, *Obes. Rev.*, 4, 157, 2003.

42. Fairey, A.S. et al., Effects of exercise training on fasting insulin, insulin resistance, insulin-like growth factor binding proteins in postmenopausal breast cancer survivors: a randomized controlled trial, *Cancer Epidemiol. Biomarkers Prev.*, 12, 721, 2003.

43. Matheson, A.J. and Perry, C.M., Glucosamine: a review of its use in management of osteoarthritis, *Drugs Aging*, 20, 1041, 2003.

44. Preuss, H.G. et al., Randomized trial of a combination of natural products (cernitin, saw palmetto, β-sitosterol, vitamin E) on symptoms of benign prostatic hyperplasia (BPH), *Int. J. Urol.*, 33, 217, 2001.

45. Jampol, L.M., Antioxidants and zinc to prevent progression of age-related macular degeneration, *JAMA*, 286, 2466, 2001.

46. Kris-Etherton, P.M., Harris, W.S., and Appel, L.J., AHA Nutrition Committee, American Heart Association: omega-3 fatty acids and cardiovascular disease: new recommendations from the American Heart Association, *Arterioscler. Thromb. Vasc. Biol.*, 23, 151, 2003.

47. Inelman, E.M. et al., Can obesity be a risk factor in elderly people? *Obes. Rev.*, 4, 147, 2003.

48. Preuss, H.G. and Adderly, B., *The Prostate Cure*, Crown Publishers, New York, 1998.

49. Ludwig, D.S., Glycemic load comes of age, *J. Nutr.*, 133, 2695, 2003.

50. Zein, M. et al., Influence of oat bran on sucrose-induced blood pressure elevations in SHR, *Life Sci.*, 47, 1121, 1990.

51. de los Reyes, A.D., Bagchi, D., and Preuss, H.G., Overview of resistance training, diet, hormone replacement and nutritional supplements on age-related sarcopenia, *Res. Comm. Pharmacol. Toxicol.*, in press.

16 Hemangioendothelioma as a Model to Study the Antiangiogenic Effects of Dietary Chemopreventive Agents *In Vivo*

Gayle M. Gordillo and Chandan K. Sen

CONTENTS

16.1 Introduction...205
16.2 Endothelial Cell Neoplasms as an Experimental Model...206
    16.2.1 Endothelial Cell Neoplasms as a Model of Angiogenesis 206
    16.2.2 Murine Models of Endothelial Cell Neoplasms 206
    16.2.3 Clinical Relevance of the Endothelial Cell Neoplasm Model 206
16.3 The Rationale for Using Dietary Chemopreventive Strategies
    to Inhibit Angiogenesis 208
    16.3.1 Advantages of Using Dietary Chemopreventive Strategies.....................208
    16.3.2 *In Vivo* Evidence of Feasibility 208
    16.3.3 *In Vitro* Evidence to Support Inhibition of Angiogenesis
        by Dietary Chemoprevention 208
16.4 Summary..209
References 209

16.1 INTRODUCTION

Antiangiogenic strategies represent an approach with outstanding promise to cancer treatment. While many pharmaceutical antiangiogenic agents are being tested in clinical trials to determine their safety and efficacy,[1] diet-based approaches to inhibit angiogenesis have drawn major interest.[2–8] Dietary approaches are considered to be significantly safer because the antiangiogenic ingredient is present in foods with proven safety records. In the U.S., consumer demand for dietary supplements has risen steadily over the past few years. Over $12 billion worth of "nutraceuticals" were purchased in the U.S. in 1998 alone.[9] The U.S. government has responded to the increasing consumer demand for dietary supplements by specifying that a portion of the NIH (National Institutes of Health) budget be allocated toward research on alternative medicine.[10] Common features of dietary agents commonly used to treat cancer are strong antioxidant and free-radical-scavenging functions. One of the most prominent dietary agents being investigated for cancer treatment is green tea. Green tea contains an antioxidant called epigallocatechin gallate (EGCG) that is claimed to be substantially more potent than vitamins E and C and has antiangiogenic properties, which presumably contributes to the therapeutic benefits against cancer.[11–13] However, studies with green tea have had mixed results in clinical trials.[14] This may be attributable to a relative paucity of *in vivo* data on this topic, and, indeed, the need for more *in vivo* studies to refine the anticancer activity of dietary agents has been highlighted.[15]

16.2   ENDOTHELIAL CELL NEOPLASMS AS AN EXPERIMENTAL MODEL

16.2.1   ENDOTHELIAL CELL NEOPLASMS AS A MODEL OF ANGIOGENESIS

Vascular tumors, especially endothelial cell neoplasms, are an excellent model that can be used to study the effects of antiangiogenic agents *in vivo*. Endothelial cell neoplasms are highly angiogenic, as the growth of these lesions entails endothelial cell proliferation, formation of luminal channels communicating with host vessels, and creation of perfused vascular spaces. The fact that humans with proliferating hemangiomas, a benign endothelial cell neoplasm, can have urinary bFGF levels elevated 25- to 50-fold higher than normal individuals attests to the degree of angiogenic activity associated with endothelial cell neoplasms.[16] Halting the growth of these lesions requires the arrest of angiogenesis that is inherent in this process and is a direct indication of the antiangiogenic capabilities of the agent being tested. Several investigators have recognized this concept and used endothelial cell neoplasm models to evaluate the antiangiogenic effects of N-acetylcysteine, batimistat, IL-12, angiostatin, and AGM-1470.[17–21]

16.2.2   MURINE MODELS OF ENDOTHELIAL CELL NEOPLASMS

There are two murine models of endothelial cell neoplasms generated by subcutaneous injection of endothelial cells. One uses murine endothelial cells transformed with the middle T antigen of the murine polyoma virus, and the other uses endothelial cells derived from a spontaneously arising hemangioendothelioma (HE).[22–26] The endothelial cells that are virally transformed are on a mixed MHC background (H-2^d/H-2^b), making them suitable for use only in SCID mice.[26] The endothelial cells (EOMA) derived from the spontaneously arising HE are from the 129/J strain, which is commercially available (now called the 129P3/J) with a defined H-2^b MHC background. EOMA cells have also been well characterized with regard to endothelial cell phenotype, protein expression, response to inhibitors of angiogenesis, and even with regard to the development of Kassabach–Merritt syndrome.[19,20,22,27–31] The fact that the EOMA cell model is relatively well characterized and can be used with immunocompetent mice in a commercially available strain makes it the preferred model. The appearance of a 129P/3 mouse bearing an HE after EOMA injection is depicted in Figure 16.1.

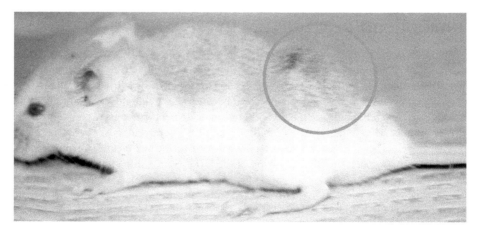

FIGURE 16.1 Murine hemangioendothelioma. A 129P/3 mouse is shown 7 days after subcutaneous injection with EOMA cells. The large subcutaneous mass is a hemangioendothelioma (HE). The violaceous hue is due to blood trapped within the lesion. HE lesions form with 100% efficacy after subcutaneous injection with EOMA cells.

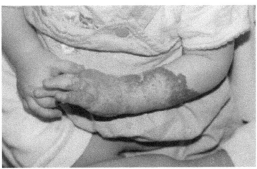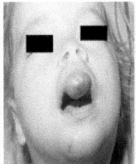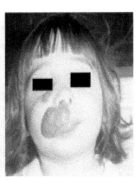

FIGURE 16.2 Children with endothelial cell neoplasms. The first (left) case is that of a 10-month-old girl with an endothelial cell neoplasm involving the upper extremity. Bleeding and ulceration frequently accompany these lesions. Her lesion shows some signs of involution, as evidenced by the gray, patulous skin in the central area of the lesion. Even with complete involution of the lesion, the site will never look like normal skin and will retain the gray color and texture difference seen in the central portion of this lesion. The other cases show facial lesions.

16.2.3 CLINICAL RELEVANCE OF THE ENDOTHELIAL CELL NEOPLASM MODEL

Criticisms related to the transformed status of EOMA cells are tempered by the fact that any results obtained with this model have direct clinical applicability. Hemangioendotheliomas (HE) are vascular neoplasms of borderline/intermediate malignancy.[32] They do not metastasize; however, in humans, the mortality rate ranges from 12 to 24%.[33] This lesion is also associated with the development of Kassabach–Merritt syndrome, a consumptive coagulopathy with a 20 to 30% mortality rate.[34,35] The fact that mice injected with EOMA cells develop Kassabach–Merritt syndrome is a good indicator of how closely this model mimics the human condition.[22] It may be argued that HE is not a common entity, but the true incidence cannot be determined because it is frequently misdiagnosed or mistakenly referred to as a hemangioma.[34,36–38] The HE lesion has several key characteristics in common with hemangiomas:[1] they proliferate rapidly during the first year of life;[2] they express cytokines and growth factors to facilitate proliferation; and[3] they have the potential to involute. The fact that HE and hemangiomas share these key characteristics significantly enhances the probability that results obtained with the EOMA model can be extrapolated to hemangiomas, which are the most common soft-tissue tumors in infants and children.

The overall incidence of endothelial cell neoplasms in children is approximately 10%, and although many of these lesions eventually involute, they have serious or life-threatening complications in 5 to 10% of affected children, and 50% of all lesions leave permanent residual deformity.[39] Because of the high-risk side-effect profiles of current treatment regimens — long-term high-dose steroids or interferon-alpha — many children and their parents are forced to accept a period of deformity, as shown in Figure 16.2, until involution occurs. These therapies are usually reserved for patients with life-, limb-, or vision-threatening complications, so a need for less toxic diet-based treatments is acute.

16.3 THE RATIONALE FOR USING DIETARY CHEMOPREVENTIVE STRATEGIES TO INHIBIT ANGIOGENESIS

16.3.1 ADVANTAGES OF USING DIETARY CHEMOPREVENTIVE STRATEGIES

Dietary chemoprevention represents an attractive approach to this problem, especially given the young age of the patient population. Dietary supplements could be taken orally, which is certainly advantageous compared with the daily injections required with interferon therapy or direct injections of the lesions with steroids. Presumably, the side-effect profile would be minimal and certainly much more favorable than the fevers, irritability, growth disturbance, immunosuppression, and spastic diplegia (a form of cerebral palsy) associated with the other treatment options discussed in the previous section. Indeed, the EOMA model has proven efficacy to demonstrate the antiangiogenic properties of edible berry extracts and to explore the feasibility of dietary chemoprevention for endothelial cell neoplasms.[40]

16.3.2 IN VIVO EVIDENCE OF FEASIBILITY

An *in vivo* model has been productively used[40] to document the antiangiogenic effects of berry extracts, which had been previously found to be antiangiogenic *in vitro*.[41] The incidence of HE formation was significantly reduced in mice injected with EOMA cells treated with blueberry and an optiberry mix. Mice that received EOMA cells treated with berry extracts and went on to develop HE had lesions that were significantly smaller than observed in control mice.[40] This study supports the feasibility to experimentally manipulate EOMA cells *in vitro* and then observe effects of that manipulation *in vivo*. This approach represents one of the few *in vivo* models that employ immunocompetent mice to examine the effects of dietary agents on angiogenesis-fed tumor growth. There are several studies that demonstrate the ability of dietary antioxidants to inhibit angiogenesis by limiting tumor growth or vessel development, but most use immunodeficient mice engrafted with human tumor cells.[21,42–44] The host immune response is a significant issue that cannot be overlooked as investigators accumulate the kind of animal data necessary to lead to successful outcomes in clinical trials.

16.3.3 IN VITRO EVIDENCE to SUPPORT INHIBITION OF ANGIOGENESIS BY DIETARY CHEMOPREVENTION

Despite the lack of *in vivo* data, there is a substantial body of *in vitro* data published on the topic of dietary antioxidant compounds and their antiangiogenic capabilities. This provides the scientific foundation to justify pursuing dietary chemoprevention strategies to inhibit angiogenesis in an *in vivo* setting. Anthocyanins are flavonoid glycosides found in berries with antioxidant properties.[45] The ability of endothelial cells to incorporate anthocyanins into the membrane and cytosol, where they maintain their antioxidant functions, has been documented *in vitro*.[46] Oral intake of anthocyanins has also been shown to achieve an increase in serum antioxidant functions *in vivo*.[47,48] Dietary antioxidants have been shown to inhibit a number of processes involved in tumor angiogenesis, many of which are known to be redox-sensitive processes. These include expression of growth factors, cytokines, and matrix metalloproteinases known to participate in angiogenesis, altering signal-transduction events, and modifying cell cycle control. For example, vascular endothelial growth factor (VEGF) is a critical growth required for tumor angiogenesis, and its expression is known to be redox sensitive.[49] Pure flavonoids, such as ferrulic acid, catechin, and rutin, have been shown to inhibit H_2O_2-inducible VEGF expression, and the combination of multiple flavonoids present in berry extracts seem to have a more potent effect on inhibiting both H_2O_2 as well as TNFα-inducible VEGF.[41] Other antioxidants such as tocopherol and EGCG have also been shown to have antiangiogenic effects by inhibiting VEGF induction or receptor activation.[12,44, 50]

Dietary antioxidants influence the expression of cytokines that participate in angiogenesis and whose expression is known to be redox sensitive. Platelet-derived growth factor BB (PDGF-BB) is a chemotactic stimulus for all cells of mesenchymal origin, including endothelial cells. Activation of the PDGF-BB receptor is an oxidant-dependent process.[51] EGCG inhibits PDGF-BB-induced intracellular signal transduction in vascular smooth muscle cells by inhibiting tyrosine phosphorylation of p42[MAPK] and p44[MAPK].[52] Monocyte chemoattractant protein-1 (MCP-1) is another chemotactic cytokine that is secreted by many neoplasms, including hemangiomas;[53] breast,[54,55] bladder,[56] and ovarian cancers;[57] and melanomas.[58] Expression of this cytokine has been shown to correlate with tumor proliferation for both hemangiomas[53] as well as breast cancer.[59,60] *In vitro* experiments have demonstrated that berry extracts inhibit inducible MCP-1 expression in EOMA cells and that this effect is mediated by inhibition of NF-κB, a redox-sensitive transcription factor.[40] In addition, EGCG and other plant-derived antioxidants inhibit the expression of matrix metalloproteinases 1 and 2 (MMP-1 and MMP-2).[61,62] The expression of these enzymes is redox sensitive.[63–65] MMP serves to remodel the extracellular matrix, supports endothelial migration and neovascularization, and recruits tumor-infiltrating leukocytes,[66] much like MCP-1.

Finally, dietary antioxidants may affect cell growth and possible tumorigenesis. When considering the treatment of endothelial cell neoplasms, limiting cell proliferation is one key aspect of an antiangiogenic strategy. Cyclins are molecules that directly stimulate cell division, and cyclin-dependent kinases (CDK) are needed to activate cyclin molecules. Thus the cyclin pathways represent an important target for tumor suppressors to inhibit. Two known tumor suppressor proteins that act in this manner are p21$^{WAF1/CIP1}$/p27^{KIP1} and p53. The tumor suppressor gene p53 is inactivated in half of all human cancers; it is needed to activate p21$^{WAF1/CIP1}$.[67,68] Antioxidants, including EGCG, can induce the transcription of p21$^{WAF1/CIP1}$ independent of p53, leading to the suppression of uncontrolled tumor cell proliferation.[69–71] In these cases, although tumor cell proliferation is inhibited, sensitivity to chemotherapy agents is not necessarily diminished. EGCG enhances sensitivity of ovarian cancer to doxoubicin and N-acetyl cysteine.[72,73] The mechanism for this increased sensitivity may be related to the fact that, in some cancers, c-Jun N terminal kinase-1 (JNK-1) is activated, leading to uncontrolled proliferation, and p21$^{WAF1/CIP1}$ can form a tight complex with this molecule to inactivate it.[74–76] While this has not been proven for endothelial cell neoplasms, constitutive activation of c-Jun has been shown to result in MCP-1 expression in endothelial cells.[77] Activated c-Jun may be present in EOMA cells or other neoplastic cells, with high levels of MCP-1 expression rendering them susceptible to this type of chemopreventive effect.

16.4 SUMMARY

Dietary chemopreventive strategies aimed at inhibiting angiogenesis are an alluring option to treat a variety of neoplasms. While it is unlikely that one approach will fit all cancers, it is imperative to develop models that permit experimental manipulation to identify optimal formulations of these naturally occurring compounds. As *in vivo* models become more standardized, the rate of data accumulation and the ability to define how dietary compounds affect signal transduction, gene expression, and cell behavior will be enhanced. Endothelial cell neoplasms are well suited for these purposes and should be utilized to collect *in vivo* data to increase the likelihood of success once clinical trials with human subjects are initiated.

REFERENCES

1. Marx, J., A boost for tumor starvation, *Science*, 301, 452–454, 2003.
2. Paper, D., Natural products as angiogenesis inhibitors, *Planta Med.*, 64, 686–695, 1998.
3. Kresty, L.A., Morse, M.A., Morgan, C., Carlton, P.S., Lu, J., Gupta, A., Blackwood, M., and Stoner, G.D., Chemoprevention of esophageal tumorigenesis by dietary administration of lyophilized black raspberries, *Cancer Res.*, 61, 6112–6119, 2001.
4. Jiang, C., Agarwal, R., and Lu, J., Anti-angiogenic potential of a cancer chemopreventive flavonoid antioxidant, silymarin: inhibition of key attributes of vascular endothelial cells and angiogenic cytokine secretion by cancer epithelial cells, *Biochemical Biophysical Rese. Commun.*, 276, 371–378, 2000.
5. Hisa, T., Kimura, Y., Takada, K., Suzuki, F., and Takigawa, M., Shikonin, an ingredient of Lithospermum erythrorhizon, inhibits angiogenesis *in vivo* and *in vitro*, *Anticancer Res.*, 18, 783–790, 1998.
6. Hayashi, A., Gillen, A.C., and Lott, J.R., Effects of daily oral administration of quercetin chalcone and modified citrus pectin on implanted colon-25 tumor growth in Balb-c mice, *Alternative Med. Rev.*, 5, 546–552, 2000.

7. Fotsis, T., Pepper, M.S., Montesano, R., Aktas, E., Breit, S., Schweigerer, L., Rasku, S., Wahala, K., and Adlercreutz, H., Phytoestrogens and inhibition of angiogenesis, *Baillieres Clinical Endocrinol. Metab.*, 12, 649–666, 1998.

8. Fotsis, T., Pepper, M.S., Aktas, E., Breit, S., Rasku, S., Adlercreutz, H., Wahala, K., Montesano, R., and Schweigerer, L., Flavonoids, dietary-derived inhibitors of cell proliferation and *in vitro* angiogenesis, *Cancer Res.*, 57, 2916–2921, 1997.

9. Zeisel, S., Health-regulation of "nutraceuticals," *Science*, 285, 1853–1855, 1999.

10. Couzin, J., Alternative medicine beefed-up NIH center probes unconventional therapies, *Science*, 282, 2175–2176, 1998.

11. Singh, A.K., Seth, P., Anthony, P., Husain, M.M., Madhavan, S., Mukhtar, H., and Maheshwari, R.K., Green tea constituent epigallocatechin-3-gallate inhibits angiogenic differentiation of human endothelial cells, *Arch. Biochem. Biophys.*, 401, 29–37, 2002.

12. Lamy, S., Gingras, D., and Beliveau, R., Green tea catechins inhibit vascular endothelial growth factor receptor phosphorylation, *Cancer Res.*, 62, 381–385, 2002.

13. Cao, Y. and Cao, R., Angiogenesis inhibited by drinking tea, *Nature*, 398, 381, 1999.

14. Webb, T., Green tea experiments in lab, clinic yield mixed results, *J. Natl. Cancer Inst.*, 92, 1038–1039, 2000.

15. Colic, M. and Pavelic, K., Molecular mechanisms of anticancer activity of natural dietetic products, *J. Molecular Med.*, 78, 333–336, 2000.

16. Takahashi, K., Mulliken, J., Kozakewich, H., Rogers, R., Folkman, J., and Ezekowitz, R., Cellular markers that distinguish the phases of hemangioma during infancy and childhood, *J. Clinical Invest.*, 93, 2357–2364, 1994.

17. Wang, C., Quevedo, M.E., Lannutti, B.J., Gordon, K.B., Guo, D., Sun, W., and Paller, A.S., In vivo gene therapy with interleukin-12 inhibits primary vascular tumor growth and induces apoptosis in a mouse model, *J. Investigative Dermatol.*, 112, 775–781, 1999.

18. Taraboletti, G., Garofalo, A., Belotti, D., Drudis, T., Borsotti, P., Scanziani, E., Brown, P.D., and Giavazzi, R., Inhibition of angiogenesis and murine hemangioma growth by batimastat, a synthetic inhibitor of matrix metalloproteinases, *J. Natl. Cancer Instit.*, 87, 293–298, 1995.

19. Lannutti, B., Gately, S., Quevedo, M., Soff, G., and Paller, A., Human angiostatin inhibits murine hemangioendothelioma tumor growth *in vivo*, *Cancer Res.*, 57, 5277–5280, 1997.

20. O'Reilly, M., Brem, H., and Folkman, J., Treatment of murine hemangioendotheliomas with the angiogenesis inhibitor AGM-1470, *J. Pediatric Surg.*, 30, 325–330, 1995.

21. Albini, A., Morini, M., D'Agostini, F., Ferrari, N., Campelli, F., Arena, G., Noonan, D.M., Pesce, C., and De Flora, S., Inhibition of angiogenesis-driven Kaposi's sarcoma tumor growth in nude mice by oral N-acetylcysteine, *Cancer Res.*, 61, 8171–8178, 2001.

22. Warner, E.D., Hoak, J.C., and Fry, G.L., Hemangioma, thrombocytopenia, and anemia: the Kassabach-Merritt syndrome in an animal model, *Arch. Pathol. Lab. Med.*, 91, 523–528, 1971.

23. Hoak, J.C., Warner, E.D., Cheng, H.F., Fry, G.L., and Hankenson, R.R., Hemangioma with thrombocytopenia and microangiopathic anemia (Kassabach-Merritt syndrome): an animal model, *J. Lab. Clinical Med.*, 77, 941–950, 1971.

24. Bautch, V.L., Toda, S., Hassell, J.A., and Hanahan, D., Endothelial cell tumors develop in transgenic mice carrying polyoma virus middle T oncogene, *Cell*, 51, 529–537, 1987.

25. Montesano, R., Pepper, M., Mohle-Steinlein, U., Risau, W., Wagner, E., and Orci, L., Increased proteolytic activity is responsible for the aberrant morphogenetic behavior of endothelial cells expressing middle T oncogene, *Cell*, 62, 435–445, 1990.

26. Taraboletti, G., Belotti, D., Dejana, E., Montovani, A., and Giavazzi, R., Endothelial cell migration and invasiveness are induced by a soluble factor produced by murine endothelioma cells transformed by polyoma virus middle T oncogene, *Cancer Res.*, 53, 3812–3816, 1993.

27. Obeso, J., Weber, J., and Auerbach, R., A hemangioendothelioma-derived cell line: its use as a model for the study of endothelial cell biology, *Lab. Invest.*, 63, 259–269, 1990.

28. O'Reilly, M., Boehm, T., Shing, Y., Fukai, N., Vasios, G., Lane, W., Flynn, E., Birkhead, J., Olsen, B., and Folkman, J., Endostatin: an endogenous inhibitor of angiogenesis and tumor growth, *Cell*, 88, 277–285, 1997.

29. Felbor, U., Dreier, L., Bryant, R., Ploegh, H., Olsen, B., and Mothes, W., Secreted cathepsin L generates endostatin from collagen XVIII, *EMBO J.*, 19, 1187–1194, 2000.

30. Sage, H. and Bornstein, P., Endothelial cells from umbilical vein and hemangioendothelioma secrete basement membrane largely to the exclusion of interstitial procollagens, *Arteriosclerosis*, 2, 27–36, 1982.
31. Wei, W., Moses, M., Wiederschain, D., Arbiser, J., and Folkman, J., The generation of endostatin is mediated by elastase, *Cancer Res.*, 59, 6052–6056, 1999.
32. Enzinger, F. and Weiss, S., *Soft Tissue Tumors*, C.V. Mosby, St. Louis, 1983.
33. Esterly, N., Cutaneous hemangiomas, vascular stains and malformations, and associated syndromes, *Curr. Prob. Dermatol.*, 7, 65–108, 1995.
34. Sarkar, M., Mulliken, J., Kozakewich, H., Robertson, R., and Burrows, P., Thrombocytopenic coagulopathy (Kassabach-Merritt phenomenon) is associated with kaposiform hemangioendothelioma and not common infantile hemangioma, *Plastic Reconstructive Surg.*, 100, 1377–1386, 1997.
35. Zuckerberg, L., Nickoloff, B., and Weiss, S., Kaposiform hemangioendothelioma of infancy and childhood: an aggressive neoplasm associated with Kasabach-Merritt syndrome and lymphangiomatosis, *Am. J. Surg. Pathol.*, 17, 321–328, 1993.
36. Vin-Christian, K., McCalmont, T., and Freiden, I., Kaposiform hemangioendothelioma, an aggressive, locally invasive vascular tumor that can mimic hemangioma of infancy, *Arch. Dermatol.*, 133, 1573–1578, 1997.
37. Hand, J. and Frieden, I., Vascular birthmarks of infancy: resolving nosologic confusion, *Am. J. Medical Genet.*, 108, 257–264, 2002.
38. Enjolras, O., Riche, M., Merland, J., and Escande, J., Management of alarming hemangiomas in infancy: a review of 25 cases, *Pediatrics*, 85, 491–498, 1990.
39. Paller, A.S., Responses to anti-angiogenic therapies, *J. Investigative Dermatol. Symp. Proc.*, 5, 83–86, 2000.
40. Atalay, M., Gordillo, G., Roy, S., Rovin, B., Bagchi, D., Bagchi, M., and Sen, C.K., Anti-angiogenic property of edible berry in a model of hemangioma, *FEBS Lett.*, 544, 252–257, 2003.
41. Roy, S., Khanna, S., Alessio, H.M., Vider, J., Bagchi, D., Bagchi, M., and Sen, C.K., Anti-angiogenic property of edible berries, *Free Radical Res.*, 36, 1023–1031, 2002.
42. Shklar, G. and Schwartz, J.L., Vitamin E inhibits experimental carcinogenesis and tumour angiogenesis, *Eur. J. Cancer, Part B, Oral Oncol.*, 32B, 114–119, 1996.
43. Marikovsky, M., Thiram inhibits angiogenesis and slows the development of experimental tumours in mice, *Br. J. Cancer*, 86, 779–787, 2002.
44. Malafa, M.P., Fokum, F.D., Smith, L., and Louis, A., Inhibition of angiogenesis and promotion of melanoma dormancy by vitamin E succinate, *Ann. Surgical Oncol.*, 9, 1023–1032, 2002.
45. Kahkonen, M.P., Hopia, A.I., and Heinonen, M., Berry phenolics and their antioxidant activity, *J. Agric. Food Chem.*, 49, 4076–4082, 2001.
46. Youdim, K.A., Martin, A., and Joseph, J.A., Incorporation of the elderberry anthocyanins by endothelial cells increases protection against oxidative stress, *Free Radical Biol. Med.*, 29, 51–60, 2000.
47. Cao, G. and Prior, R.L., Anthocyanins are detected in human plasma after oral administration of an elderberry extract, *Clinical Chem.*, 45, 574–576, 1999.
48. Cao, G., Russell, R.M., Lischner, N., and Prior, R.L., Serum antioxidant capacity is increased by consumption of strawberries, spinach, red wine or vitamin C in elderly women, *J. Nutr.*, 128, 2383–2390, 1998.
49. Sen, C.K., Khanna, S., Babior, B.M., Hunt, T.K., Ellison, E.C., and Roy, S., Oxidant-induced vascular endothelial growth factor expression in human keratinocytes and cutaneous wound healing, *J. Biological Chem.*, 277, 33284–33290, 2002.
50. Kondo, T., Ohta, T., Igura, K., Hara, Y., and Kaji, K., Tea catechins inhibit angiogenesis *in vitro*, measured by human endothelial cell growth, migration and tube formation, through inhibition of VEGF receptor binding, *Cancer Lett.*, 180, 139–144, 2002.
51. Sundaresan, M., Yu, Z.X., Ferrans, V.J., Irani, K., and Finkel, T., Requirement for generation of H_2O_2 for platelet-derived growth factor signal transduction, *Science*, 270, 296–299, 1995.
52. Ahn, H.Y., Hadizadeh, K.R., Seul, C., Yun, Y.P., Vetter, H., and Sachinidis, A., Epigallocathechin-3 gallate selectively inhibits the PDGF-BB-induced intracellular signaling transduction pathway in vascular smooth muscle cells and inhibits transformation of sis-transfected NIH 3T3 fibroblasts and human glioblastoma cells (A172), *Molecular Biol. Cell*, 10, 1093–1104, 1999.
53. Isik, F., Rand, R., Gruss, J., Benjamin, D., and Alpers, C., Monocyte chemoattractant protein-1 mRNA expression in hemangiomas and vascular malformations, *J. Surg. Res.*, 61, 71–76, 1996.

54. Saji, H., Koike, M., Yamori, T., Saji, S., Seiki, M., Matsushima, K., and Toi, M., Significant correlation of monocyte chemoattractant protein-1 expression with neovascularization and progression of breast carcinoma, *Cancer*, 92, 1085–1091, 2001.

55. Ueno, T., Toi, M., Saji, H., Muta, M., Bando, H., Kuroi, K., Koike, M., Inadera, H., and Matsushima, K., Significance of macrophage chemoattractant protein-1 in macrophage recruitment, angiogenesis, and survival in human breast cancer, *Clinical Cancer Res.*, 6, 3282–3289, 2000.

56. Amann, B., Pearbo, F., Wirger, A., Hugenschmidt, H., and Schultze-Seeman, W., Urinaryl levels of monocyte chemoattractant protein-1 correlate with tumor stage and grade in patients with bladder cancer, *Br. J. Urol.*, 82, 118–121, 1998.

57. Hefler, L., Tempfer, C., Heinze, G., Mayerhofer, K., Breitenecker, G., Leodolter, S., Reinthaller, A., and Kainz, C., Monocyte chemoattractant protein-1 serum levels in ovarian cancer patients, *Br. J. Cancer*, 81, 855–859, 1999.

58. Torisu, H., Ono, M., Kiryu, H., Furue, M., Ohmoto, Y., Nakayama, J., Nishioka, Y., Sone, S., and Kuwano, M., Macrophage infiltration correlates with tumor stage and angiogenesis in human malignant melanoma: possible involvement of TNFalpha and IL-1alpha, *Int. J. Cancer*, 85, 182–188, 2000.

59. Leek, R.D., Lewis, C.E., Whitehouse, R., Greenall, M., Clarke, J., and Harris, A.L., Association of macrophage infiltration with angiogenesis and prognosis in invasive breast carcinoma, *Cancer Res.*, 56, 4625–4629, 1996.

60. Bingle, L., Brown, N., and Lewis, C., The role of tumour-associated macrophages in tumor progression: implications for new anticancer therapies, *J. Pathol.*, 196, 254–265, 2002.

61. Wartenberg, M., Budde, P., De Marees, M., Grunheck, F., Tsang, S.Y., Huang, Y., Chen, Z.Y., Hescheler, J., and Sauer, H., Inhibition of tumor-induced angiogenesis and matrix-metalloproteinase expression in confrontation cultures of embryoid bodies and tumor spheroids by plant ingredients used in traditional Chinese medicine, *Lab. Invest.*, 83, 87–98, 2003.

62. Demeule, M., Brossard, M., Page, M., Gingras, D., and Beliveau, R., Matrix metalloproteinase inhibition by green tea catechins, *Biochimica Biophysica Acta*, 1478, 51–60, 2000.

63. Inoue, N., Takeshita, S., Gao, D., Ishida, T., Kawashima, S., Akita, H., Tawa, R., Sakurai, H., and Yokoyama, M., Lysophosphatidylcholine increases the secretion of matrix metalloproteinase 2 through the activation of NADH/NADPH oxidase in cultured aortic endothelial cells, *Atherosclerosis*, 155, 45–52, 2001.

64. Brenneisen, P., Briviba, K., Wlaschek, M., Wenk, J., and Scharffetter-Kochanek, K., Hydrogen peroxide H_2O_2 increases the steady-state mRNA levels of collagenase/MMP-1 in human dermal fibroblasts, *Free Radical Biol. Med.*, 22, 515–524, 1997.

65. Belkhiri, A., Richards, C., Whaley, M., McQueen, S.A., and Orr, F.W., Increased expression of activated matrix metalloproteinase-2 by human endothelial cells after sublethal H2O2 exposure, *Lab. Invest.*, 77, 533–539, 1997.

66. Shono, T., Ono, M., Izumi, H., Jimi, S.I., Matsushima, K., Okamoto, T., Kohno, K., and Kuwano, M., Involvement of the transcription factor NF-kappaB in tubular morphogenesis of human microvascular endothelial cells by oxidative stress, *Molecular Cell. Biol.*, 16, 4231–4239, 1996.

67. Hesketh, R., *The Oncogene and Tumor Suppressor Gene Facts Book*, Academic Press, New York, 1997.

68. Chinery, R., Brockman, J.A., Peeler, M.O., Shyr, Y., Beauchamp, R.D., and Coffey, R.J., Antioxidants enhance the cytotoxicity of chemotherapeutic agents in colorectal cancer: a p53-independent induction of p21WAF1/CIP1 via C/EBPbeta, *Nat. Med.*, 3, 1233–1241, 1997.

69. Liang, Y.C., Lin-Shiau, S.Y., Chen, C.F., and Lin, J.K., Inhibition of cyclin-dependent kinases 2 and 4 activities as well as induction of Cdk inhibitors p21 and p27 during growth arrest of human breast carcinoma cells by (–)-epigallocatechin-3-gallate, *J. Cell. Biochem.*, 75, 1–12, 1999.

70. Chinery, R., Brockman, J.A., Dransfield, D.T., and Coffey, R.J., Antioxidant-induced nuclear translocation of CCAAT/enhancer-binding protein beta; a critical role for protein kinase A-mediated phosphorylation of Ser299, *J. Biological Chem.*, 272, 30356–30361, 1997; [erratum] *J. Biological Chem.*, 273 (24), 15308, 1998.

71. Bai, F., Matsui, T., Ohtani-Fujita, N., Matsukawa, Y., Ding, Y., and Sakai, T., Promoter activation and following induction of the p21/WAF1 gene by flavone is involved in G1 phase arrest in A549 lung adenocarcinoma cells, *FEBS Lett.*, 437, 61–64, 1998.

72. Sadzuka, Y., Sugiyama, T., and Hirota, S., Modulation of cancer chemotherapy by green tea, *Clinical Cancer Res.*, 4, 153–156, 1998.

73. Adeyemo, D., Imtiaz, F., Toffa, S., Lowdell, M., Wickremasinghe, R.G., and Winslet, M., Antioxidants enhance the susceptibility of colon carcinoma cells to 5-fluorouracil by augmenting the induction of the bax protein, *Cancer Lett.*, 164, 77–84, 2001.
74. Patel, R., Bartosch, B., and Blank, J.L., p21WAF1 is dynamically associated with JNK in human T-lymphocytes during cell cycle progression, *J. Cell. Sci.*, 111, 2247–2255, 1998.
75. Shim, J., Lee, H., Park, J., Kim, H., and Choi, E.J., A non-enzymatic p21 protein inhibitor of stress-activated protein kinases, *Nature*, 381, 804–806, 1996.
76. Bost, F., McKay, R., Dean, N., and Mercola, D., The JUN kinase/stress-activated protein kinase pathway is required for epidermal growth factor stimulation of growth of human A549 lung carcinoma cells, *J. Biological Chem.*, 272, 33422–33429, 1997.
77. Wang, N., Verna, L., Hardy, S., Forsayeth, J., Zhu, Y., and Stemerman, M.B., Adenovirus-mediated overexpression of c-Jun and c-Fos induces intercellular adhesion molecule-1 and monocyte chemoat-tractant protein-1 in human endothelial cells, *Arteriosclerosis, Thrombosis Vascular Biol.*, 19, 2078–2084, 1999.

17 Modulation of Late Adverse Effects of Curative Radiation Therapy for Cancer

John Yarnold and Lone Gothard

CONTENTS

17.1 Introduction: Risk and Severity of Late Adverse Effects Impose Limit
on Total Dose of Curative Radiotherapy 215
17.2 Pathogenesis of Late Radiation-Induced
Adverse Effects..216
   17.2.1 Fibrosis Contributes to Late Radiation Adverse Events.................................216
   17.2.2 Pathogenesis of Fibrosis in Response to Radiation Injury
       May Be Related to Vascular Injury...216
17.3 Proof of Principle That Late Radiation Adverse Effects Can Be Modified 217
   17.3.1 Hyperbaric Oxygen (HBO) Therapy..217
       17.3.1.1 Rationale for Testing Hyperbaric Oxygen......................................217
       17.3.1.2 HBO for Mandibular Osteo-Radionecrosis and
              Soft Tissue Radionecrosis 217
       17.3.1.3 HBO for Radiation-Induced Hemorrhagic Cystitis.......................218
       17.3.1.4 HBO for Radiation-Induced Bowel Injury, Including Proctitis..............218
       17.3.1.5 HBO for Radiation-Induced Brachial Plexopathy (Nerve Damage) 218
       17.3.1.6 HBO for Radiation-Induced Arm Lymphedema (Arm Swelling) 219
   17.3.2 Antioxidants..220
       17.3.2.1 Rationale for Testing Antioxidants 220
       17.3.2.2 Superoxide Dismutase (SOD)...220
       17.3.2.3 Pentoxifylline and Tocopherol (Vitamin E) 220
       17.3.2.4 Grape Seed Proanthocyanidin Extract (GSPE) 221
17.4 Conclusions...222
References 222

17.1 INTRODUCTION: RISK AND SEVERITY OF LATE ADVERSE EFFECTS IMPOSE LIMIT ON TOTAL DOSE OF CURATIVE RADIOTHERAPY

Radiotherapy has a vital role in the curative treatment of many common malignancies, including cancers of the head and neck, prostate, bladder, rectum, lung, cervix, uterus, and breast.[1] There is a radiotherapy dose–response relationship for tumor control at most of these primary sites.[2,3] There is also a steep dose response for normal tissue injury late onset, progressive atrophy, and fibrosis.[4,5] For the individual patient, the prescribed dose aims to achieve the optimal balance between the risk of cancer recurrence and the risk of treatment-induced complications. Restrictions on the radiotherapy dose imposed by complication risk account for a high proportion of treatment failures and explain the irreducible risk of permanent morbidity in a proportion of survivors. An estimated 5% of patients treated with radiotherapy develop troublesome symptoms due to late radiation changes, and perhaps as many as 1% of patients have serious problems that are resistant to simple methods of treatment. Major surgery may be required as well as prolonged hospital care. Personal and social problems may be very distressing, and commonly those affected are unable to pursue gainful employment.

If the progression of late radiation-induced complications could be modified via a pharmaceutical approach, this would be of enormous benefit to patients, especially in terms of the quality of life of survivors. It would also raise the possibility of increasing radiotherapy dose and cancer cure prospects in selected patients who remain at high risk of local cancer recurrence after current standard treatment.

17.2 PATHOGENESIS OF LATE RADIATION-INDUCED ADVERSE EFFECTS

17.2.1 FIBROSIS CONTRIBUTES TO LATE RADIATION ADVERSE EVENTS

Tissue atrophy and fibrosis after radiotherapy are predominantly secondary responses to radiation-induced vascular injury.[6,7] Fibrosis is a pathological term describing increased interstitial collagen density. It explains tissue hardness (induration) and contraction after high-dose radio-therapy. The histological features of radiation fibrosis include zones of dense, irregular collagen fibers encasing and compressing adjacent structures.[8] Fibrosis is a prominent component of late radiotherapy change in many tissues, including skin, subcutaneous tissue, lung, muscle, heart, blood vessels, gastrointestinal tract, and urogenital system.[9] Tissue compliance and extensibility are reduced by a change in the proportion and organization of different collagen subtypes.[10] Fibrosis contributes to dose-limiting complications through tissue stiffness and contraction and is progressive with time. Stiffness reduces the compliance of organs with important elastic properties, including the lung, heart, and bladder.[11–13] Contraction of fibrous tissue causes strictures in hollow viscera such as the gastrointestinal tract and ureter, precipitating obstructive complications.

17.2.2 PATHOGENESIS OF FIBROSIS IN RESPONSE TO RADIATION INJURY MAY BE RELATED TO VASCULAR INJURY

Obliterative endarteritis is a classical pathological feature of late radiation-induced normal tissue injury.[14] The ischemia and hypoxia that accompany vascular atrophy are important causes of parenchymal atrophy and loss of tissue function.[15,16] Radiation fibrosis contributes to loss of function, and several models have been advanced to explain how this develops. A wound repair model describes chronic inflammation induced by chronic cell necrosis after radiation therapy as a potent source of fibrogenic cytokines and fibrogenesis, processes proposed to explain scarring in surgical wounds.[17] A rather different explanation is offered by a model that postulates a direct interaction between ionizing radiation and stromal fibroblasts, the latter undergoing differentiation after *in vitro* exposure to ionizing radiation to a postmitotic senescent phenotype characterized by up-regulated production of collagen, an important component of fibrotic tissue.[18,19]

A third mechanism focuses on tissue hypoxia secondary to vascular injury as a direct stimulus of fibrogenesis. It is hypothesized that a fibrogenic response to hypoxia, mediated via a range of hypoxia-regulated genes, contributes to tissue scarring, a mechanism increasingly recognized in other chronic fibrotic states. A close association has been noted between interstitial fibrosis and microvascular damage in progressive renal disease.[20] In this example, a sequence of events involving a heme-protein sensor is postulated, similar to the sensor involved in erythropoietin (*epo*) gene regulation. This activates signal pathways responsible for up-regulation of transcription factors, including hypoxia-inducible factor-1α (Hif-1α). [21,24] Several studies report up-regulated *coll-i* gene expression and increased production of collagen-α1(i) in response to low oxygen tension. [25,26] Down-regulation of collagenase (MMP-1) and up-regulation of metalloproteinase (TIMP-1) at mRNA and protein levels have also been reported. A role for these mechanisms in renal disease is supported by the interaction demonstrated between Hif-1α and hypoxia response elements in transient transfection experiments using a TIMP-1 promoter CAT reporter construct. [20]

The pathological implications for radiation-induced fibrosis are that progressive capillary injury may deregulate extracellular matrix metabolism via modulation of hypoxia-regulated genes. The clinical implications are that restoration of normal tissue oxygen tension via induction of angiogenesis may halt or reverse these processes. If temporal fluctuations in regional perfusion are a feature of irradiated vasculature, cyclical hypoxia may add reperfusion insult to ischemic injury via the generation of reactive oxygen species. [27] Therapeutically, reperfusion injury may be one of the processes targeted by antioxidant therapy that is reported to induce softening of subcutaneous fibrosis in patients with late radiation-induced injuries. [28] Regardless of the upstream pathways regulating extracellular matrix metabolism after radiotherapy, there is good evidence that the pluripotent cytokine TGF-β$_1$ plays a central role in the deposition of extracellular collagen. [18,19,29–32] The significance of postulating an active tissue response to radiotherapy injury is the potential it raises for modifying common and disabling symptoms.

17.3 PROOF OF PRINCIPLE THAT LATE RADIATION ADVERSE EFFECTS CAN BE MODIFIED

17.3.1 HYPERBARIC OXYGEN (HBO) THERAPY

17.3.1.1 Rationale for Testing Hyperbaric Oxygen

The rationale for testing hyperbaric oxygen (HBO) is that irradiated tissues are characterized by hypocellularity, fibrosis, low oxygen tensions, and shallow oxygen gradients. [33] At the pathological level, irradiated tissues have a capillary density 20 to 30% that of unirradiated areas. [33,34] During HBO exposure, there is a 7- to 10-fold rise in oxygen tension and a steep oxygen gradient across the damaged zone. Steep oxygen gradients appear to stimulate angiogenesis, fibroplasia, and other features of tissue restructuring. [35–37] In the early stage of HBO therapy, capillary budding and cellular collagen synthesis occur in animal systems. [34] This is followed by a rapid rise in the number of lumenized capillaries to 75 to 80% that of normal tissues. As new capillaries form, hypoxia is reversed and the oxygen gradient is lost, with down-regulation of angiogenesis. No increase in vascularity is seen in adjacent undamaged tissue. Further increases in oxygen tension abolish these effects and are toxic. How these physiological responses impact on functional disability is unclear. Take rectal injury, for example. It is not difficult to imagine how a renewed capillary network restores the integrity of the rectal epithelium and reduces rectal bleeding from telangiectatic vessels. It is also possible that reduced capillary leakage leads indirectly to improved rectal compliance via a reduction of edema in the rectal wall and ischiorectal fossa. It is also possible that restoration of tissue perfusion reverses a chronic stimulus of fibrogenesis perpetuated by radiation-induced vasculitis and hypoxia.

17.3.1.2 HBO for Mandibular Osteo-Radionecrosis and Soft Tissue Radionecrosis

Two comparative studies report benefit in patients with heavily damaged tissues. The first study tested preoperative hyperbaric oxygen against conventional penicillin cover in a randomized study of 74 patients requiring dental extraction following radical mandibular irradiation. [38] Only 5.4% of the patients in the hyperbaric oxygen group, compared with 30% of the patients in the penicillin-treated group, experienced failure of wound healing 6 months after surgery (P < 0.05). In 160 patients treated by hyperbaric oxygen or standard postoperative care following major soft-tissue surgery for radiotherapy injury, four-fold reductions in wound dehiscence, infection, and delayed wound healing were seen in the hyperbaric oxygen treated group (P < 0.01).[39] No adverse effects of hyperbaric oxygen were reported in either of these studies.

17.3.1.3 HBO for Radiation-Induced Hemorrhagic Cystitis

Level II (non-randomized) evidence of effect is available supporting the efficacy of HBO for the treatment of severe radiation-induced hemorrhagic cystitis refractory to conventional measures. The largest study describes 40 patients treated for 20 sessions of HBO.[40] A total of 37 of the 40 patients showed marked and durable (up to 5 years) reductions in bleeding frequency, including need for blood transfusions, and no adverse effects were reported. A consecutive series of 20 patients treated between 1989 and 1992 reported improvement of hematuria in 90% of cases.[41] Similar experiences based on smaller patient numbers are available.[41–45] No adverse effects of HBO were reported in any of these studies.

17.3.1.4 HBO for Radiation-Induced Bowel Injury, Including Proctitis

The efficacy of HBO in the management of radiation-induced bowel injury, including proctitis, is less well reported. There is a series of case studies and very small series reporting healing of pain, tenesmus, bleeding, diarrhea, and rectal ulceration.[46–51] More significantly, there are a number of retrospective studies describing complete or partial healing of rectal symptoms in patients resistant to conventional therapy for severe injuries. One of the largest retrospective studies describes 36 patients, including 9 with chronic necrotic wounds, 19 with chronic rectal bleeding, and 9 with chronic severe diarrhea.[52] After 67 (12-198) HBO sessions, the authors reported 9 complete responses, 12 partial responses, and 11 non-responses (in addition, one patient died from radiation injury, two from cancer, and one from liver cirrhosis). In a study of 38 patients treated with HBO for chronic, uncontrolled rectal bleeding, treatment resulted in an improvement in 61% of cases.[53] A third retrospective study of 18 patients reported complete or partial responses after 24 (12-40) sessions of HBO in 7 of 17 cases rectal bleeding, 2 of 4 pain syndromes, 3 of 4 cases of fecal incontinence, and 4 of 8 diarrheal syndromes.[54] These experiences are broadly repeated in other studies, including a cohort of 14 patients, 11 of whom presented with rectal bleeding, 5 with diarrhea, and 5 with tenesmus or colic.[55] After 40 (20-72) HBO sessions, 9 patients were completely healed, 3 patients improved substantially, and 2 were non-responders.

Although limited in many respects, these reports are encouraging, in that chronic hemorrhage from the bladder or rectum can be reliably scored in a semiquantitative manner, including the number of patients freed from the requirement for repeated blood transfusions over the months and years posttherapy. Taken together, the evidence lends strong support to the hypothesis that HBO is an effective treatment in a proportion of patients suffering complications following radiotherapy for pelvic cancer. However, the data are derived from retrospective studies and are characterized by a lack of consistency with respect to the scoring of symptoms and response criteria. Until well-designed randomized clinical trials are undertaken, patients will continue to be referred to hyperbaric facilities on an ad hoc basis, and effective treatment for a common and major morbidity may continue unrecognized and underutilized.

17.3.1.5 HBO for Radiation-Induced Brachial Plexopathy (Nerve Damage)

Our first randomized study testing hyperbaric oxygen therapy in patients with radiation-induced complication following treatment for cancer involved 34 volunteers with long-standing radiation-induced brachial plexopathy (RIBP).[56] RIBP is an untreatable complication of curative radiotherapy for early breast cancer, characterized by chronic neuropathic pain and limb paralysis. A group of 34 eligible research volunteers suffering from RIBP were randomized to HBO or control group. The HBO group breathed 100% oxygen for 100 min in a multiplace hyperbaric chamber on 30 occasions over a period of 6 weeks. The control group accompanied the HBO group and breathed a gas mixture equivalent to breathing 100% oxygen at surface pressure. All volunteers and investigators, except the operators of the hyperbaric chamber and the trial statistician, were blind to treatment assignments. The warm sensory threshold, which measures the function of small sensory fibers, was selected as the primary endpoint.

Pretreatment neurophysiological tests were grossly abnormal in the affected hand compared with the unaffected hand in both HBO and control groups, as expected, but no statistically significant differences were noted in either group at any time up to 12 months posttreatment. However, normalization of the warm sensory threshold in two of the HBO group was reliably recorded. Two cases with marked chronic arm lymphedema reported major and persistent improvements in arm volume for at least 12 months after treatment with HBO. These unexpected observations led to a study of HBO for arm lymphedema.

17.3.1.6 HBO for Radiation-Induced Arm Lymphedema (Arm Swelling)

Fibrosis is also believed to contribute to the development of lymphedema. The pathophysiology of lymphedema after radiotherapy and/or surgery involves obstruction of lymphatic flow, causing an imbalance between capillary filtration and lymph drainage.[57,58] Although physical removal of lymphatic vessels at surgery offers a partial explanation, the variable onset, progression rate, and ultimate severity indicate that this is not the only mechanism. Radiotherapy to the axilla is a potent cause of arm lymphedema in its own right, more so after any kind of surgical disturbance of the axilla.[59] The continuous accumulation and contraction of scar tissue over many years is considered to be a potent cause of progressive lymphatic obstruction in response to radiotherapy. The most vivid accounts of fibrosis are from the surgical records of affected patients describing scar tissue infiltrating and compressing axillary structures, especially the neurovascular sheath.

A total of 21 eligible research volunteers with a minimum 30% increase in arm volume in the years after axillary/supraclavicular radiotherapy (axillary surgery in 18 of 21 cases) were treated with HBO.[78] The volunteers breathed 100% oxygen at 2.4 ATA for 100 min in a multiplace hyperbaric chamber on 30 occasions over a period of 6 weeks. The volume of the ipsilateral limb, measured optoelectronically by a perometer and expressed as a percentage of contralateral limb volume, was selected as the primary endpoint. A secondary end point was local lymph drainage expressed as fractional removal rate of radioisotopic tracer, measured using lymphoscintigraphy. Three out of 19 evaluable patients experienced >20% reduction in arm volume at 12 months. A total of 6 out of 13 evaluable patients experienced a >25% improvement in ^{99}Tc-nanocolloid clearance rate from the ipsilateral forearm measured by quantitative lymphoscintigraphy at 12 months. Overall, there was a statistically significant, but clinically modest, reduction in ipsilateral arm volume at 12 months follow-up compared with baseline (P = 0.005). The mean percentage reduction in arm volume from baseline at 12 months was 7.51. Moderate or marked lessening of induration in the irradiated breast, pectoral fold, and/or supraclavicular fossa was recorded clinically in 8 of 15 evaluable patients. Of this group, 12 out of 19 evaluable patients volunteered that their arms felt softer, and 6 reported improvements in shoulder mobility at 12 months. Interpretation of these results is limited by the absence of a control group. However, measurement of limb volume by perometry[60,61] is reportedly reliable, and lymphoscintigraphy[62] is assumed to be operator independent. Taking all data into account, there is sufficient evidence to justify a double-blind randomized controlled trial of hyperbaric oxygen in this group of patients.

17.3.2 ANTIOXIDANTS

17.3.2.1 Rationale for Testing Antioxidants

Free radicals and their metabolites released by ionization of water mediate the classical cytotoxic effects of ionizing radiation via DNA strand breakage. The postulated antifibrotic effects of antioxidants prescribed years after exposure cannot be related to this phase of free-radical attack. Other spheres of free-radical effects are legion, in lipid membranes and aqueous environments, inside cells and outside. Although reactive oxygen species interact randomly with molecular targets, the effects are not merely stochastic in the sense that cell-signaling mediated by redox reactions have important roles in cellular homeostasis.[63] The processes modulated by antioxidants are not likely to be specific to the repair of radiotherapy injuries, since the fibrotic response to ionizing radiation appears to have much in common with responses to a range of chronic physical, chemical, or biological damaging agents, including UV radiation, bleomycin, ethanol, and viral infections. However, there may be mechanisms that relate more closely to tissue lesions caused by radiotherapy. For example, reactive oxygen species may drive the obliterative endarteritis, as suggested by elevated levels of oxidized tissue methionine and vascular endothelial growth factor (VEGF) in bronchial lavage fluid in patients given combined chemoradiotherapy for non-small-cell lung cancer.[64] Downstream in the fibrogenic pathway, oxidant-induced hydrolysis of matrix-bound TGF-β_1 latent complex with release of active TGF-β_1 may contribute to up-regulated collagen synthesis.[29]

17.3.2.2 Superoxide Dismutase (SOD)

The reversibility of radiation fibrosis in skin and subcutaneous tissues in response to antioxidants is supported by regression of induration reported in a French non-randomized pilot study involving intramuscular administration of bovine liposomal Cu/Zn superoxide dismutase (SOD), 5 mg twice weekly for 3 weeks, to 34 patients with 42 distinct zones of superficial fibrosis.[65] Softening of subcutaneous induration was noted in 86% of fibrotic zones, with an actuarial response rate of 70% by 5 years. Complete regressions were noted in 7 of 42 (17%) of the fibrotic zones. Supportive data are reported with topical applications of SOD over a period of several months in 40 patients with fibrosis after postmastectomy radiotherapy for early breast cancer.[66] These studies were not pursued after bovine spongiform encephalopathy (BSE) was recognized and bovine products withdrawn.

17.3.2.3 Pentoxifylline and Tocopherol (Vitamin E)

Experience with alpha-tocopherol, a synthetic derivative of vitamin E, in combination with pentoxifylline, an inhibitor of tumor necrosis factor alpha, is also consistent with a beneficial effect in patients with established fibrosis. A case report described marked regression of fibrosis 17 years after postmastectomy radiotherapy in a woman treated with alpha-tocopherol 400 mg once daily and pentoxifylline 400 mg three times daily for several months.[67] In a phase II trial testing alpha tocopherol 500 U twice daily and pentoxifylline 400 mg twice daily for 6 months in 43 patients with marked subcutaneous fibrosis after radiotherapy, 6 months regression and functional improvement were reported in all assessable sites of induration.[28,68] A case report of one woman treated with pentoxifylline 400 mg three times daily and vitamin E 200 mg twice daily for 18 months reported almost complete healing of an ulcerated radiation-induced fibrosis after therapy for breast cancer.[69] A pilot study in France evaluated the effect of pentoxifylline 800 mg once daily and tocopherol (vitamin E) 1000 IU once daily in women with a thin endometrium who were enrolled in an oocyte donation program. Medication was given for 6 months and appeared to improve the pregnancy rate in patients with a thin endometrium by increasing the endometrial thickness and improving ovarian function. According to the authors, this was especially noticeable in patients who had previously received total body irradiation.[70]

A pilot study testing a combination of pentoxifylline 800 mg o.d., tocopherol (vitamin E) 1000 IU daily, and clodonate reported clinical benefit, with more than 50% regression of progressive osteoradionecrosis observed at 6 months in 12 patients. This study included a woman who had developed severe exteriorized osteoradionecrosis following treatment for breast cancer 29 years previously. She had palpable breast fibrosis, including the sternum and a painful fistulous track in the upper part of the bone surrounded by local inflammatory signs, and chronic osteitis with sequestra extrusion. Magnetic resonance imaging showed deep radiation-induced fibrosis below this area without cancer recurrence, and complete bone destruction over an area of 7×4 cm. The above combination of pentoxifylline, vitamin E, and clodonate was administered for 3 years and resulted in completely reversing the progressive osteoradionecrosis and the associated radiation-induced fibrosis.[71]

Based on the empirical evidence of clinical regression of superficial radiation fibrosis after radiotherapy described above, our group conducted the first double-blind placebo-controlled trial of antioxidants in patients who had developed arm lymphedema and tissue fibrosis following radiotherapy treatment for breast cancer. A total of 68 eligible research volunteers with a minimum 20% increase in arm volume in the years after axillary/supraclavicular radiotherapy (axillary surgery in 49 of 68 cases) were randomized to active drugs or placebo. The volunteers were given dl-alpha tocopheryl acetate 500 mg twice daily and pentoxifylline 400 mg twice daily or corresponding placebo capsules over a period of 6 months. The volume of the ipsilateral limb, measured opto-electronically by a perometer and expressed as a percentage of contralateral limb volume, was selected as the primary end point. Secondary end points included clinician assessment of breast induration and validated patient self-assessment questionnaires.[72] No side effects have been reported. To date, the preliminary analysis fail to confirm the therapeutic effects reported in previous non-randomized studies.[78]

17.3.2.4 Grape Seed Proanthocyanidin Extract (GSPE)

Grape seed proanthocyanidin extract (GSPE, commercially available as ActiVin) is a rich source of polyphenolic flavonoids, a naturally occurring family of oligomeric compounds found in a wide range of fruit and vegetables.[73–75] The flavonoids in GSPE are thought to possess more powerful free-radical scavenging abilities than vitamin E (alpha-tocopherol) or vitamin C. For example, superoxide anions and hydroxyl radicals measured by chemiluminescence response and cytochrome c reduction were scavenged by GSPE (200 mg/l) in vitro with a maximum efficiency of 90% compared with vitamin C (19% at 100 mg/l, 500 μM) and vitamin E succinate (44% at 75 mg/l, 141.3 μM).[39] Potential problems affecting in vivo testing of naturally occurring extracts include uncertainties whether: (a) relevant ingredients vary in concentration in different batches of extract, (b) relevant ingredients are absorbed or modified after oral ingestion, (c) bioactive molecules reach target tissues in sufficient molar concentrations to exert an effect, and (d) observed clinical responses are related to the postulated ingredients (in this case proanthocyanidins).

Important uncertainties have been addressed with respect to GSPE, in that oral extract fed to adult male ICR(CD-1) mice at a dose of 100 mg/kg for 7 to 10 days offers a high degree of protection against acetaminophen-induced hepatic toxicity, amiodarone-induced lung toxicity, and doxorubicin-induced cardiac toxicity.[76] For example, four experimental groups of mice were fed saline alone, GSPE alone (100 mg/kg/day p.o. for 9 days prior to sacrifice), doxorubicin alone (20 mg/kg i.p. 48 h prior to sacrifice), and GSPE plus doxorubicin (the doxorubicin administered on day 7 of GSPE medication, 48 h prior to sacrifice on day 9). Doxorubicin alone caused a six-fold rise in serum creatine kinase, gross internucleosomal DNA fragmentation, and the appearance of apoptotic nuclei on tissue sections, changes not recorded in the animals fed GSPE plus doxorubicin.[76] From the in vivo effects noted in multiple organs, there seems little doubt that active ingredients are absorbed and distributed in a wide range of mouse tissues.

In humans, 40 research subjects with hypercholesterolemia (total cholesterol 210 to 300 mg/dl) have been randomized in a double-blind, placebo-controlled trial to placebo 100 mg BD, GSPE 100 mg BD, chromium polynicotinate 200 µg BD, or a combination of both.[74] One patient in the combination group experienced sinusitis, otitis media, and recurrent migraines following the ingestion of migraine-triggering foods, but no toxicity (predominantly gastrointestinal disturbance) was reported in the GSPE-alone group. With regard to efficacy, there was a significant reduction in total and LDL cholesterol in the combined group compared with placebo. The only other report of GSPE use in humans relates to three patients with advanced chronic pancreatitis, in whom GSPE 100 mg p.o. BD improved abdominal pain without any recorded toxicity.[77] In conclusion, the rodent data offer reliable data relating to the efficacy of oral GSPE in preventing oxidative damage in a wide range of tissues. Limited data raise no concerns about the safety of administering GSPE to humans.

Against this background, we are currently conducting a double-blind, placebo-controlled trial of GSPE in patients who have developed tissue induration after high-dose radiotherapy for early breast cancer. We aim to randomize 72 women with radiation-induced breast induration (hardness) to GSPE 100 mg TDS or placebo tablets in a ratio 2:1 (treatment:control). Volunteers take the trial medication for 6 months and are followed up for a further 6 months. The primary endpoint is breast induration, and secondary endpoints include patient self-assessment of breast hardness, pain, and tenderness together with urinary 8-OH-deoxyguanosine, plasma LDL, cholesterol, and lipid peroxides at 6 months after randomization. Accrual is well under way, and so far no side effects have been reported.

17.4 CONCLUSIONS

Moderate clinical evidence supports the notion that late adverse effects of radiation therapy can be modulated for the benefit of cancer survivors, encouraging safer escalation of radiation doses in patients who are not currently cured. The strongest evidence derives from an evaluation of hyperbaric oxygen therapy, although the dearth of well-designed randomized studies is a severe limitation. The same condition applies to the evaluation of antioxidants in the management of patients with radiation-induced complications. The scarcity of randomized controlled trials with blinding of investigator and research volunteer ensured by the use of placebo is particularly significant, given that many of the outcome measures are subjectively assessed. In this respect, arm lymphedema is a robust experimental system, since volume can be measured accurately using optical techniques in an operator-independent manner. In conclusion, much more work is needed to establish antioxidants as an effective treatment for radiation-induced late adverse effects, and much more needs to be understood concerning the molecular and cellular bases of any therapeutic effects observed.

REFERENCES

1. Souhami, R. and Tobias, J., *Cancer and Its Management*, Oxford, Blackwell, U.K., 1986.
2. Fletcher, G.H., Local results of irradiation in the primary management of localized breast cancer, *Cancer*, 29 (3), 545–551, 1972.
3. Fletcher, G.H., Clinical dose-response curves of human malignant epithelial tumours, *Br. J. Radiol.*, 46 (541), 1–12, 1973.
4. Turesson, I., The progression rate of late radiation effects in normal tissue and its impact on dose-response relationships, *Radiother. Oncol.*, 15 (3), 217–226, 1989.
5. Thames, H.D. et al., The high steepness of dose-response curves for late-responding normal tissues, *Radiother. Oncol.*, 15 (1), 49–53, 1989.
6. Hopewell, J. and Withers, H.R., Proposition: long-term changes in irradiated tissues are due principally to vascular damage in the tissues, *Med. Phys.*, 25 (12), 2265–2268, 1998.

7. Hopewell, J.W. et al., Microvasculature and radiation damage, *Recent Results Cancer Res.*, 130, 1–16, 1993.
8. Fajardo, L.F. and Berthrong, M., Radiation injury in surgical pathology, Part I, *Am. J. Surg. Pathol.*, 2 (2), 159–199, 1978.
9. Rubin, P. and Casarett, G.W., *Clinical Radiation Pathology*, Vols. I and II, WB Saunders Co., Philadelphia, 1968.
10. Lovell, C.R. et al., Type I and III collagen content and fibre distribution in normal human skin during ageing, *Br. J. Dermatol.*, 117 (4), 419–428. 1987.
11. Marks, L.B. et al., The response of the urinary bladder, urethra, and ureter to radiation and chemotherapy, *Int. J. Radiat. Oncol. Biol. Phys.*, 31 (5), 1257–1280, 1995.
12. Perez, C.A. et al., Impact of dose in outcome of irradiation alone in carcinoma of the uterine cervix: analysis of two different methods, *Int. J. Radiat Oncol. Biol. Phys.*, 21 (4), 885–898, 1991.
13. Coia, L.R., Myerson, R.J., and Tepper, J.E., Late effects of radiation therapy on the gastrointestinal tract, *Int. J. Radiat. Oncol. Biol. Phys.*, 31 (5), 1213–1236, 1995.
14. Hopewell, J.W. and Young, C.M., Changes in the microcirculation of normal tissues after irradiation, *Int. J. Radiat. Oncol. Biol. Phys.*, 4 (1–2), 53–58, 1978.
15. Grotendorst, G.R. et al., Chemoattractants in fibrotic disorders, *CIBA Found. Symp.*, 114, 150–163, 1985.
16. Herskind, C. et al., Differentiation state of skin fibroblast cultures versus risk of subcutaneous fibrosis after radiotherapy, *Radiother. Oncol.*, 47 (3), 263–269, 1998.
17. Clark, R.A.F., Wound repair: overview and general considerations, in *The Molecular and Cellular Biology of Wound Repair*, Clark, R.A.F., Ed., Plenum Press, New York, 1996.
18. Illsley, M.C. et al., Increased collagen production in fibroblasts cultured from irradiated skin and effect of TGF beta(1): clinical study, *Br. J. Cancer*, 83 (5), 650–654, 2000.
19. Rodemann, H.P. and Bamberg, M., Cellular basis of radiation-induced fibrosis, *Radiother. Oncol.*, 35 (2), 83–90, 1995.
20. Norman, J.T., Clark, I.M., and Garcia, P.L., Hypoxia promotes fibrogenesis in human renal fibroblasts, *Kidney Int.*, 58 (6), 2351–2366, 2000.
21. Ratcliffe, P.J., Molecular biology of erythropoietin, *Kidney Int.*, 44 (4): 887–904, 1993.
22. Morwenna, S. and Ratcliffe, W.P., Mammalian oxygen sensing and hypoxia inducible factor-1, *Int. J. Biochem. Cell. Biol.*, 29 (12), 1419–1432, 1997.
23. Semenza, G.L., Hypoxia-inducible facto 1 and the molecular physiology of oxygen homeostatis, *J. Lab. Clin. Med.*, 131 (3), 207–214, 1998.
24. Wenger, R.H. and Gassmann, M., Oxygen(es) and the hypoxia-inducible factor-1, *Biol. Chem.*, 378 (7), 609–616 1997.
25. Falanga, V. et al., Low oxygen tension increases mRNA levels of alpha 1 (I) procollagen in human dermal fibroblasts, *J. Cell. Physiol.*, 157 (2), 408–412, 1993.
26. Tamamori, M. et al., Stimulation of collagen synthesis in rat cardiac fibroblasts by exposure to hypoxic culture conditions and suppression of the effect by natriuretic peptides, *Cell. Biol. Int.*, 21 (3), 175–180, 1997.
27. Carden, D.L. and Granger, D.N., Pathophysiology of ischaemia-reperfusion injury, *J. Pathol.*, 190 (3), 255–266, 2000.
28. Delanian, S., Balla-Mekias, S., and Lefaix, J.L., Striking regression of chronic radiotherapy damage in a clinical trial of combined pentoxifylline and tocopherol, *J. Clin. Oncol.*, 17 (10), 3283–3290, 1999.
29. Barcellos-Hoff, M.H. and Dix, T.A., Redox-mediated activation of latent transforming growth factor-beta 1, *Mol. Endocrinol.*, 10 (9), 1077–1083, 1996.
30. Martin, M. et al., Temporal modulation of TGF-beta 1 and beta-actin gene expression in pig skin and muscular fibrosis after ionizing radiation, *Radiat. Res.*, 134 (1), 63–70, 1993.
31. Anscher, M.S. et al., Transforming growth factor beta as a predictor of liver and lung fibrosis after autologous bone marrow transplantation for advanced breast cancer, *N. Engl. J. Med.*, 328 (22), 1592–1598, 1993.
32. Autio, P. et al., Demonstration of increased collagen synthesis in irradiated human skin in vivo, *Br. J. Cancer*, 77 (12), 2331–2335, 1998.
33. Marx, R.E. et al., Relationship of oxygen dose to angiogenesis induction in irradiated tissue, *Am. J. Surg.*, 160 (5), 519–524, 1990.

34. Marx, R.E. and Johnson, R.P., Problem wounds in oral and maxillofacial surgery: the role of hyperbarix oxygen, in *Problem Wounds — the Role of Oxygen*, Davis, J.C. and Hunt, T.K., Eds., Elsevier, New York, 1988, pp. 65–123.

35. Knighton, D.R. et al., Oxygen tension regulates the expression of angiogenesis factor by macrophages, *Science*, 221 (4617), 1283–1285, 1983.

36. Knighton, D.R., Silver, I.A., and Hunt, T.K., Regulation of wound-healing angiogenesis-effect of oxygen gradients and inspired oxygen concentration, *Surgery*, 90 (2), 262–270, 1981.

37. Silver, I., Cellular microenvironment in healing and non-healing wounds, (4), 50–66.

38. Marx, R.E., Johnson, R.P., and Kline, S.N., Prevention of osteoradionecrosis: a randomized prospective clinical trial of hyperbaric oxygen versus penicillin, *J. Am. Dent. Assoc.*, 111 (1), 49–54, 1985.

39. Marx, R.E., Radiation injury to tissue, in *Hyperbaric Medicine Practice*, Best Publishing Co., 1994, chap. 23, pp. 448–501.

40. Bevers, R.F., Bakker, D.J., and Kurth, K.H., Hyperbaric oxygen treatment for haemorrhagic radiation cystitis, *Lancet,* 346 (8978), 803–805, 1995.

41. Lee, H.C. et al., Hyperbaric oxygen therapy in hemorrhagic radiation cystitis: a report of 20 cases, *Undersea Hyperb. Med.*, 21 (3), 321–327, 1994.

42. Schoenrock, G.J. and Cianci, P., Treatment of radiation cystitis with hyperbaric oxygen, *Urology*, 27 (3), 271–272, 1986.

43. Kindwall, E.P., Hyperbaric oxygen treatment of radiation cystitis, *Clin. Plast. Surg.*, 20 (3), 589–592, 1993.

44. Weiss, J.P. et al., Primary treatment of radiation-induced hemorrhagic cystitis with hyperbaric oxygen: 10-year experience, *J. Urol.*, 151 (6), 1514–1517, 1994.

45. Norkool, D.M. et al., Hyperbaric oxygen therapy for radiation-induced hemorrhagic cystitis, *J. Urol.*, 150 (2 Pt 1), 332–334, 1993.

46. Bem, J., Bem, S., and Singh, A., Use of hyperbaric oxygen chamber in the management of radiation-related complications of the anorectal region: report of two cases and review of the literature, *Dis. Colon Rectum*, 43 (10), 1435–1438, 2000.

47. Neurath, M.F. et al., A new treatment for severe malabsorption due to radiation enteritis, *Lancet,* 347 (9011), 1302, 1996.

48. Hamour, A.A. and Denning, D.W., Hyperbaric oxygen therapy in a woman who declined colostomy, *Lancet*, 348 (9021), 197, 1996.

49. Moulin, C. et al., Value of hyperbaric oxygen in the hemostatic treatment of chronic radiation-induced recto-sigmoiditis, *Gastroenterol. Clin. Biol.*, 17 (6–7), 520–521, 1993.

50. Nakada, T. et al., Therapeutic experience of hyperbaric oxygenation in radiation colitis: report of a case, *Dis. Colon Rectum*, 36 (10), 962–965, 1993.

51. Kitta, T. et al., The treatment of chronic radiation proctitis with hyperbaric oxygen in patients with prostate cancer, *BJU Int.*, 85 (3), 372–374, 2000.

52. Gouello, J.P. et al., The role of hyperbaric oxygen therapy in radiation-induced digestive disorders. 36 cases, *Presse Med.*, 28 (20), 1053–1057, 1999.

53. Aanderud, L. et al., Hyperbaric oxygen treatment for radiation reactions, *Tidsskr. Nor Laegeforen*, 120 (9), 1020–1022, 2000.

54. Woo, T.C., Joseph, D., and Oxer, H., Hyperbaric oxygen treatment for radiation proctitis, *Int. J. Radiat. Oncol. Biol. Phys.*, 38 (3), 619–622, 1997.

55. Warren, D.C. et al., Chronic radiation proctitis treated with hyperbaric oxygen, *Undersea Hyperb. Med.*, 24 (3), 181–184, 1997.

56. Pritchard, J. et al., Double-blind randomized phase II study of hyperbaric oxygen in patients with radiation-induced brachial plexopathy, *Radiother. Oncol.*, 58 (3), 279–286, 2001.

57. Stanton, A.W., Levick, J.R., and Mortimer, P.S., Current puzzles presented by postmastectomy oedema (breast cancer related lymphoedema), *Vasc. Med.*, 1 (3), 213–225, 1996.

58. Mortimer, P.S., Managing arm oedema in women with breast cancer, *Drug Ther. Bull.*, 1998.

59. Kissin, M.W. et al., Risk of lymphoedema following the treatment of breast cancer, *Br. J. Surg.*, 73 (7), 580–584, 1986.

60. Stanton, A.W. et al., Validation of an optoelectronic limb volumeter (Perometer), *Lymphology*, 30 (2), 77–97, 1997.

61. Goltner, E. et al., Objective measurement of lymphedema after mastectomy, *Dtsch. Med. Wochenschr.*, 110 (24), 949–952, 1985.
62. Stanton, A.W. et al., Differences in lymph drainage between swollen and non-swollen regions in arms with breast-cancer-related lymphoedema, *Clin. Sci. (Lond.)*, 101 (2), 131–140, 2001.
63. Finkel, T. and Holbrook, N.J., Oxidants, oxidative stress and the biology of ageing, *Nature*, 408 (6809), 239–247, 2000.
64. Beinert, T. et al., Further evidence for oxidant-induced vascular endothelial growth factor up-regulation in the bronchoalveolar lavage fluid of lung cancer patients undergoing radio-chemotherapy, *J. Cancer Res. Clin. Oncol.*, 126 (6), 352–356, 2000.
65. Delanian, S. et al., Successful treatment of radiation-induced fibrosis using liposomal Cu/Zn superoxide dismutase: clinical trial, *Radiother. Oncol.*, 32 (1), 12–20, 1994.
66. Perdereau, B. et al., Superoxide dismutase (Cu/Zn) in cutaneous application in the treatment of radiation-induced fibrosis, *Bull. Cancer*, 81 (8), 659–669, 1994.
67. Gottlober, P. et al., The treatment of cutaneous radiation-induced fibrosis with pentoxifylline and vitamin E: an empirical report. *Strahlenther Onkol.*, 172 (1), 34–38, 1996.
68. Delanian, S. and Lefaix, J.L., The association of pentoxifylline and tocopherol is useful in the treatment of human superficial radiation-induced fibrosis: preliminary results, 1996.
69. Fischer, M., Wohlrab, J., and Marsch, W., Crux medicorum ulcerated radiation-induced fibrosis: successful therapy with pentoxifylline and vitamin E, *Eur. J. Dermatol.*, 11 (1), 38–40, 2001.
70. Ledee-Bataille, N. et al., Combined treatment by pentoxifylline and tocopherol for recipient women with a thin endometrium enrolled in an oocyte donation programme, *Hum. Reprod.*, 17 (5), 1249–1253, 2002.
71. Delanian, S. and Lefaix, J.L., Complete healing of severe osteoradionecrosis with treatment combining pentoxifylline, tocopherol and clodronate, *Br. J. Radiol.*, 75 (893), 467–469, 2002.
72. Fayers, P. et al., *EORTC QLQ-C30 Scoring Manual*, 2nd ed., EORTC Quality of Life Study Group, 1999, pp. 1–77.
73. Rice-Evans, C.A., Miller, N.J., and Paganga, G., Structure-antioxidant activity relationships of flavonoids and phenolic acids, *Free Radic. Biol. Med.*, 20 (7), 933–956, 1996.
74. Bagchi, D. et al., Free radicals and grape seed proanthocyanidin extract: importance in human health and disease prevention, *Toxicology*, 148 (2–3), 187–197, 2000.
75. Rapport, L. and Lockwood, B., Proanthocyanidins and grape seed extract, *Pharm. J.*, 266, 581–584, 2001.
76. Bagchi, D. et al., Protection against drug- and chemical-induced multiorgan toxicity by a novel IH636 grape seed proanthocyanidin extract, *Drugs Exp. Clin. Res.*, 27 (1), 3–15, 2001.
77. Banerjee, B. and Bagchi, D., Beneficial effects of a novel IH636 grape seed proanthocyanidin extract in the treatment of chronic pancreatitis, *Digestion*, 63 (3), 203–206, 2001.
78. Gothard, L. et al., Non-randomised phase II trial of hyperbaric oxgen therapy in patients with chronic are lymphoedema and tissue fibrosis after radiotherapy for early breast cancer, *Radiother. oncol.*, 70(3), 217–224, 2004.

Section III

Phytopharmaceuticals and Chemoprevention

18 Development of Selected Phytochemicals for Cancer Chemoprevention

Vernon E. Steele

CONTENTS

18.1 Drug Development Overview...229
18.2 Genistein...231
18.3 Resveratrol..232
18.4 Tea Polyphenols..232
18.5 Lycopene...234
18.6 Summary..234
References 234

18.1 DRUG DEVELOPMENT OVERVIEW

One of the most important public health and medical practices of the new century is the prevention of cancer. Much progress has been made in this new field, but much work remains before widespread use becomes practice. Cancer chemoprevention is defined as the inhibition, reversal, and retardation of the cancer process. The cancer process, called carcinogenesis, requires two to four decades to reach the end point called malignant cancer. The process follows multiple, diverse, and complex pathways in a stochastic process of clonal evolution. Many have shown that these pathways are responsive to inhibition, reversal, or retardation at various points. Much basic research is devoted to identifying the many genetic lesions and epigenetic processes associated with the progression of cancer. Normal regulatory genes are mutated, making them insensitive to normal regulatory signals. Tumor suppressor genes are deleted or mutated, rendering them inactive. Thus there is a wide range of defects in cellular machinery that can lead to evolution of the cancer phenotype. To conquer this diverse disease, we must attack multiple key pathways at once for a predetermined period of time.

Strategies to develop chemopreventive drugs for use have been evolving for many years.[1-3] This program has developed as a linear-array strategy by which agents enter and flow through the system in stepwise pattern. In this context, we have begun to examine the chemopreventive potential of a wide variety of synthetic and natural products. Epidemiological evidence has related decreased cancer risk to increased consumption of phytoestrogens and lignans in a vegetarian diet.[4] The scientific rationale for conducting chemoprevention trials in humans is based largely on epidemiological evidence suggesting that dietary components may be inhibitors of carcinogenesis.[5] Examples of these include vitamins and minerals such as β-carotene,[6,7] vitamin E,[8-10] calcium,[11] and selenium.[12,13] Other important sources for identifying potential chemopreventives include cancer research literature, particularly experimental carcinogenesis data from animal

TABLE 18.1
Mechanisms and Targets for Selected Dietary Chemopreventive Agents

Agent	Mechanisms of Action	Molecular Targets
Genistein	Inhibits activation of carcinogen	Cytochrome P450
	Modulates hormone activity	estrogen receptor
	Modulates growth-factor receptors	IGF-I receptor
	Induces terminal differentiation/apoptosis	TGF-β
	Inhibits angiogenesis	FGF receptor
Resveratrol	Modulates hormone activity	aromatase enzymes
	Antioxidation	LDL cholesterol
	Anti-inflammatory	COX enzymes
	Signal transduction	NF-κB and AP1
Tea polyphenols	Inhibits activation of carcinogen	Cytochrome P450
	Modulates hormone activity	5α reductase enzyme
	Inhibits polyamine synthesis	ornithine decarboxylase
	Induces apoptosis	caspase enzyme
	Antioxidation	DNA, lipids
Lycopene	Enhances intercellular junctions	Connexin 43
	Antioxidation	DNA, lipids
	Deactivation of carcinogen	GST, GSH

studies, which have conclusively shown that specific chemical substances can, in fact, prevent the entire cancer process or prevent the progression of premalignant lesions to invasive cancers.[14,15] Thus far, several thousand chemical agents have been identified, some of which exist as distinct chemical entities, and others consisting of defined and undefined mixtures. However, only through prospective, doubled-blinded, placebo-controlled, clinical chemoprevention trials can we determine whether these agents are safe and work effectively as cancer preventive agents in humans. Key strategies to streamline the selection and prioritization process from preliminary *in vitro* screening to definitive human clinical trials are of vast importance to the National Cancer Institute's (NCI) drug discovery and development program, and these strategies represent an ongoing applied science problem of major significance.

The four natural products presented in Table 18.1 have the potential to attack multiple pathways in the cancer process. Generally, natural products have little known toxicity, which is a highly desirable characteristic for cancer-preventing agents. Since food-derived products are perceived as safe and not perceived as "medicine," natural products may be acceptable for long-term use in healthy people. Extracts make up most of the food-derived natural products. These extracts usually contain multiple classes of compounds or multiple compounds within one class. For example, green tea extracts may contain primarily polyphenols of multiple types. Typically, a single compound is isolated as the active fraction or one having the most activity. Then this compound is synthesized and tested in parallel with the parent extract. If the purified compound and the parent extract have similar efficacy, then the parent extract is further developed as a cost-effective agent for clinical trials. The pure compound is then dropped from further testing, since it might be more toxic by itself and usually is more expensive to produce.

The second important strategy for development of natural products is the careful characterization of the extract. To achieve reproducible mixtures, much effort needs to be put into a defined extraction protocol, which is checked against standard biochemical analysis such as high-pressure liquid chromatography. Extracts must meet stringent peak analysis to ensure that the major peaks of the extract analysis are quantitatively and qualitatively present and with only a small acceptable amount of batch-to-batch variance. Reproducible preparations are necessary to ensure comparison of test results using multiple batches. Contradictory results may arise if batch quality is not properly controlled. Definition of growth conditions may be critically important and may vastly affect the nutrient content of the final extract. The amount of nutrients in the soil, the growth temperature, the amount of sunlight, and the moisture content of the soil harboring the vegetable or berry could have an orders-of-magnitude effect on the final composition of the extract.

Many of the food-derived extracts found to prevent cancer also typically have benefits in reducing morbidity from other diseases. For example, grape juice extract, which is a promising lead cancer preventive agent, has also been documented to reduce markers of coronary heart disease.[16] Such extracts inhibited the oxidation of LDL *in vitro* and could have profound effects in reducing the risk of coronary heart disease, since the oxidation of LDL has been implicated in the pathogenesis of arteriosclerosis.[17]

18.2 GENISTEIN

Genistein is a naturally occurring isoflavone and phytoestrogen that has been identified in various human dietary sources, notably soybeans. Epidemiological evidence shows an inverse correlation of phytoestrogens and lignans in a vegetarian diet and cancer risk.[18] There have been a large number of mechanisms proposed for the anticancer activity of genistein, including: inhibition of tyrosine kinase activity,[19] inhibition of ornithine decarboxylase,[20] antioxidant and anti-inflammatory activity.[21] inhibition of prostaglandin synthetase,[22] and inhibition of benzo[a]pyrene metabolism.[23]

The ability of genistein to inhibit the formation of foci of rat colon aberrant crypts was seen at dietary concentrations of 75 and 150 ppm.[24] The addition of genistein at 75 mg/kg diet significantly reduced the frequency of aberrant crypts by 29.3%, while in the group receiving 150 mg genistein/kg in its diet, the frequency of aberrant crypts was reduced by 34.1%. The marginal difference between these two groups would not justify stating that a dose response was observed. While the number of aberrant crypt foci (ACF) per colon was reduced with administration of genistein, the number of aberrant crypts per focus was not reduced. However, later studies using a rat azoxymethane-induced tumor model showed that high concentrations actually slightly enhanced tumor incidence.[25]

More recently, an isoflavone-enriched extract of soy, designated PTI G-2535, provided by Protein Technologies, Inc., has been tested. When added to the diet, this extract showed significant efficacy in preventing MNU-induced rat prostate carcinoma (unpublished results). This extract has undergone preclinical toxicology testing.[26] A 90-day oral toxicity study with PTI G-2535 in rats and dogs, as well as teratology studies, indicated no signs of toxicity. However there remained a genotoxicity concern associated with the ability of isoflavones to inhibit topoisomerase, which could lead to DNA strand breaks. Misra's report[26] describes results from two *in vitro* genotoxicity studies, one *in vivo* genotoxicity study, and a single carcinogenicity study conducted in p53 knock-out mice. Bacterial mutagenesis experiments using six tester strains were basically negative. Mouse lymphoma cell mutagenesis revealed statistically significant increases in mutation frequency, and the increases were dose related. In contrast, dietary genistein had no effect on survival, weight gain, or the incidence or types of tumors that developed in cancer-prone rodents lacking the p53 tumor suppressor gene.

PTI G-2535 has completed Phase I human trials.[27] These studies showed a lack of significant genotoxicity of purified soy isoflavones (genistein, diademing, and glycogen) in 20 patients with prostate cancer. Further clinical studies investigated the pharmacokinetic profiles of single doses of this isoflavone mixture in women and men.[28,29] Phase II clinical trials are under way for the prevention of prostate cancer with PTI-2535.

18.3 RESVERATROL

Various biological activities relevant to chemoprevention have been demonstrated by chemicals found in grapes, including antioxidant, antimutagenic, signal transduction, and anti-inflammatory activity.[30–32] Jang et al.[33] showed that resveratrol inhibited the hydroperoxidase activity of cyclooxygenase 1 and 2 (COX-1 and COX-2), demonstrated anti-inflammatory activity, inhibited free-radical formation, and inhibited DMBA (dimethylbenz[a]anthracene)-induced mouse mammary lesions *in vitro*. Based on this large number of relevant chemopreventive mechanisms and *in vitro* studies, resveratrol was entered into the chemoprevention development process in the two cell culture assays and in the aberrant colon crypt study. Resveratrol is a naturally occurring phytoalexin that has been identified in various human dietary sources and is concentrated in grapes.

Resveratrol, 3,4′,5-trihydroxystilbene, is a phytoalexin that is found in abundant quantities in grape skins and stems. It is now thought to be nature's fungicide against molds that attack grapes. However it has a number of properties that make it an attractive cancer-preventive agent. Among these properties are inhibition of cyclooxygenase, aromatase, and free radicals. In assays measuring induction of DMBA hyperplastic alveolar nodules in organ cultures of mouse mammary gland, resveratrol greatly reduced the percentage of glands that were affected by the carcinogen.[33] In that same report, the authors present evidence that DMBA/12-O-tetradecanoylphorbol-13-acetate-induced mouse skin cancers were inhibited by this natural phytoalexin. Resveratrol also inhibited benzo[a]pyrene-induced morphologically transformed rat tracheal epithelial cells in culture.[34] In the same paper, it is reported that resveratrol inhibited anchorage-independent growth of human A427 lung cancer cells by over 60%. In further studies using azoxymethane-treated rat colons, resveratrol significantly inhibited aberrant crypt formation.[34] Further studies are planned to examine inhibition of colon and prostate tumors in long-term carcinogenesis inhibition studies.

Recently it has been shown that resveratrol induces the cAMP/kinase-A system, which makes it a pro-apoptotic and cell-cycle-repressing agent in breast cancer cells.[35] Having such mechanistic activity would suggest that resveratrol might be an effective agent against ER-negative breast cancer cells. A review of resveratrol's activities has been recently published by Aziz et al.[36]

Preclinical toxicology and pharmacology studies using resveratrol are underway in rats and dogs. In addition, a 6-month carcinogenicity study using p53+/– mice is in process. Future Phase I clinical trials are planned and will provide safety information and guide in selecting nontoxic doses for Phase II trials. In one published clinical study, a group of healthy subjects consumed 500 ml daily of grape juice that had been enriched with 4 mg of trans-resveratrol/L.[37] No adverse effects were noted in this study.

18.4 TEA POLYPHENOLS

Tea (*Camellia sinensis*) is one of the most popular beverages consumed worldwide. Tea leaves are primarily manufactured as green, black, or oolong, with black tea representing approximately 80% of tea products consumed. Green tea is the nonoxidized, nonfermented product, and it contains several polyphenolic components such as epicatechin, epicatechin gallate (ECG), epigallocatechin (EGC), and epigallocatechin gallate (EGCG). EGCG is the major green tea polyphenol (GTP) (>40% dry weight). The major components of black tea (the fermented product) are theaflavins (1–2% dry weight) and thearubigins (10–20% dry weight). Theaflavins, which determine the quality and flavor of the tea, are formed by oxidation of quinones derived from the epicatechins. Thearubigins are highly colored flavonol oxidation products, which are often bound to peptides or proteins. Oolong tea is the partially oxidized/fermented product, which retains a considerable amount of the original polyphenolic material.[38]

The NCI testing effort used tea extracts, tea polyphenol-enriched fractions, isolated catechin fractions, and a theaflavin-rich fraction. The black and green tea extracts were tested in both caffeinated and decaffeinated forms. The percentage of polyphenols in both green and black teas were approximately the same. However, levels of EGCG and caffeine in green tea are four- to five-fold that of black tea. Theaflavins are unique to black tea due to the oxidation process.

In an effort to determine the cancer chemopreventive activity of tea compounds, we conducted several *in vitro* mechanistic assays and *in vitro* cell-transformation assays to screen for efficacy of these compounds.[39] The mechanistic assays measure (a) inhibition of DNA adduct formation, (b) inhibition of free-radical formation, and (c) enhancement of glutathione (GSH), glutathione-*S*-transferase (GST), ornithine decarboxylase (ODC), and NAD(P)H:quinone reductase (QR) activity. No extracts were positive in all six assays, but several were positive in four to five assays.

Several tea compounds, including the catechins, were assayed for activity in three *in vitro* cell-transformation assays: the rat tracheal epithelial (RTE), the human lung cell (A427), and the mouse mammary organ culture (MMOC) assays. In the RTE assay, black tea extract (BTE) and black tea polyphenols (BTP) inhibited transformation by 22 to 37%. On the other hand, the decaffeinated BTE inhibited RTE transformation by 86 to 100%. The theaflavin mixture was also very effective. Green tea extract (GTE) inhibited transformation from 70 to 83%. The decaffeinated GTE had a somewhat lower (47 to 53%) inhibition at comparable doses. Green tea polyphenols also had a high percentage of inhibition (74 to 100%) at similar doses. Two of the four tea catechins, ECG and EGCG, were also effective. Therefore, both black and green tea extracts or polyphenols were very effective at inhibiting B[*a*]P-induced RTE transformation. The inhibition of anchorage independence in human lung tumor (A427) cells was measured following treatment with the tea extracts. BTE inhibited anchorage-independent growth by 50 to 90%, while BTP inhibited anchorage-independent growth by 90 to 100% at similar concentrations. Decaffeinated BTE was more inhibitory (82 to 93%) than BTE. The theaflavins, GTE, and decaffeinated GTE showed little inhibition, while GTP was quite effective (74 to 92% inhibition). All the green tea catechins were effective in inhibiting anchorage-independent growth of A427 cells. In the MMOC assay, the two black tea extracts (BTE and decaffeinated BTE) were quite effective. Surprisingly, the polyphenol and theaflavin mixtures were ineffective in inhibiting DMBA-induced mammary gland transformation. Green tea extracts were largely ineffective, although decaffeinated GTE, GTP, and EGCG were effective.

Following the observation of significant *in vitro* activity, animal efficacy studies in the rat azoxymethane (AOM)-induced colon aberrant crypts and *N*-nitrosomethylbenzylamine (NMBA)-induced esophageal carcinogenesis models were tested with these compounds.[40–42] The chemopreventive activity of green and black tea compounds was evaluated to reduce the occurrence of premalignant lesions (ACF). In this model, low-dose BTP (360 ppm in drinking water) and high-dose EGCG (1200 ppm in drinking water) inhibited the formation of ACF by 26% and 35%, respectively.[40] In the rat NMBA esophageal carcinogenesis model, the tea compounds were administered in the drinking water at concentrations of 360 and 1200 ppm for 2 weeks before administration of the carcinogen (subcutaneously, three times weekly for 5 weeks) for 25 weeks.[42] At the end of the study (week 25), tumor multiplicity was significantly reduced ($p < 0.05$) in the group receiving 1200 ppm theaflavins. At 20 weeks, similar results were obtained in the high-dose (1200 ppm) EGCG group. High-dose (1200 ppm) BTP, GTP, and EGCG also reduced tumor multiplicity significantly ($p < 0.05$) at 15 weeks. Tumor incidence was not affected by any of the tea compounds at any time point during the study.

In the rat AOM-induced colon carcinogenesis model, none of the tea compounds was effective.[41] Recently it was found that both green and black tea prevented the formation of basal cell carcinoma in a transgenic mouse model.[43] Preclinical efficacy studies of tea extracts in other labs have shown positive effects in the lung and liver,[44] prostate,[45] and stomach.[46]

Polyphenon E is a well-standardized decaffeinated green tea catechin mixture containing about one-half EGCG, with the other half containing mostly other catechin polyphenols. Recently Chang et al.[47] examined the genotoxicity of Polyphenon E and saw no mutagenicity in the *Salmonella typhimurium*-E assay. Also, no significant increases in revertant colonies were found in the Big Blue transgenic mutation assay at high doses of 2 g/kg/day. In a human Phase I study, 20 healthy subjects were given 200 to 800 mg of EGCG or Polyphenon E in a single-dose study.[48] The investigators concluded that similar EGCG levels were found in the plasma regardless of whether the subjects received EGCG or Polyphenon E.

Randomized placebo-controlled Phase II clinical trials are under way for the prevention of skin, oral, and lung cancers by topical or oral administration of Polyphenon E or other tea polyphenol extracts.

18.5 LYCOPENE

Lycopene is one of the most abundant carotenoids in human serum. Lycopene has exhibited a wide range of anticancer activities, including antioxidant and free-radical scavenging activities, induction of differentiation and gap junctional communication, and inhibition of transformation of mouse mammary organ cultures.[49,50] A recent review of animal studies on lycopene has been published.[51] Azoxymethane-induced aberrant crypt foci in rats were strongly inhibited by lycopene given in the diet.[52] These data were followed up by studies by Kim et al.,[53] where lycopene was shown to inhibit dimethylhydrazine (DMH)-induced cancer in mouse colon. Mouse lung adenomas that were induced by multiple carcinogens were inhibited by lycopene given in the drinking water.[54,55] Lycopene also inhibited hamster cheek-pouch carcinogenesis when given lycopene by gavage.[56]

Phase II clinical trials are under way for the prevention of prostate cancer with lycopene administration. A study by Bowen et al.[57] showed that lycopene was accumulated in the human prostate following administration of tomato paste (about 30 mg lycopene/day). A second study in patients with localized prostate cancer[58] showed that lycopene, given as a tomato oleoresin extract (about 30 mg/day), resulted in lower PSA (prostate specific antigen) values and a less aggressive prostate disease at the time of surgery. Further studies with larger Phase II trials are planned to evaluate lycopene to prevent cancer of the prostate.

18.6 SUMMARY

The use of food, food extracts, and dietary supplements present a safe chemopreventive strategy. Four such agents that were presented in this article have unique properties and mechanisms of action that potentially would make them effective cancer preventive agents. At the same time, these agents are extremely nontoxic and could be given to people who are at low risk for cancer. These agents could also be taken in high doses for long periods of time, which may slow the progression of a wide variety of cancers. Strategies involving combinations of these agents should seriously be considered, since a much wider range of mechanisms possibly could result in higher efficacy than single agents alone and may preclude cancer in several organ sites. Randomized placebo-controlled clinical trials are underway for many of these agents and should provide the data necessary to make future public health recommendations.

REFERENCES

1. Boone, C.W., Kelloff, G.J., and Malone, W.E., Identification of candidate cancer chemopreventive agents and their evaluation in animal models and human clinical trials: a review, *Cancer Res.*, 50, 2, 1990.
2. Kelloff, G.J. et al., Recent results in preclinical and clinical drug development of chemopreventive agents at the National Cancer Institute, in *Cancer Chemoprevention*, Wattenberg, L., Lipkin, M., Boone, C., and Kelloff, G., Eds., CRC Press, Boca Raton, FL, 1992, p. 41.
3. Kelloff, G.J. et al., Progress in cancer chemoprevention: perspectives on agent selection and short-term clinical trials, *Cancer Res.*, 54, 2051, 1994.

4. Adlercreutz, H., Does fiber-rich food containing animal lignan precursors protect against both colon and breast cancer? An extension of the "fiber hypothesis," *Gastroenterology*, 86, 761, 1984.
5. Bertram, J.S., Kolonel, L.N., and Meyskens, F.L., Rationale and strategies for chemoprevention of cancer in humans, *Cancer Res.*, 47, 3012, 1987.
6. Bertram, J.S., Cancer prevention by carotenoids: mechanistic studies in cultured cells, *Ann. N.Y. Acad. Sci.*, 691, 177, 1993.
7. Moon, R.C. et al., Retinoid chemoprevention of lung cancer, in *Cancer Chemoprevention*, Wattenberg, L., Lipkin, M., Boone, C.W., and Kelloff, G.J. Eds., Boca Raton, FL, CRC Press, 1992, p. 83.
8. Blot, W.J. et al., Nutrition intervention trials in Linxian, China: supplementation with specific vitamin/mineral combinations, cancer incidence, and disease-specific mortality in the general population, *J. Natl. Cancer Inst.*, 8, 1483, 1993.
9. Ip, C. and White, G., Mammary cancer chemoprevention by inorganic and organic selenium: single agent treatment or in combination with vitamin E and their effects on *in vitro* immune functions, *Carcinogenesis*, 8, 1763, 1987.
10. Wald, N.J. et al., Serum vitamin E and subsequent risk of cancer, *Br. J. Cancer*, 56, 69, 1987.
11. Whitfield, J.M. et al., Calcium–cell cycle regulator, differentiator, killer, chemopreventor, and maybe, tumor promoter, *J. Cell. Biochem.*, 22, 74, 1995.
12. El-Bayoumy, K. et al., Chemoprevention of cancer by organoselenium compounds, *J. Cell. Biochem.*, 22, 92, 1995.
13. Ip, C. et al., Comparative effect of inorganic and organic selenocyanate derivatives in mammary cancer chemoprevention, *Carcinogenesis*, 15, 187, 1994.
14. Moon, R.C. and Mehta, R.G., Chemoprevention of experimental carcinogenesis in animals, *Prev. Med.*, 18, 576, 1989.
15. Wattenberg, L.W., Chemoprevention of cancer, *Cancer Res.*, 45, 1, 1985.
16. Frankel, E.N. and Meyer, A.S., Antioxidants in grapes and grape juices and their potential health effects, *Pharmaceut. Biol.*, 36, 14, 1998.
17. Esterbauer, H. et al., The role of lipid peroxidation and antioxidants in oxidative modification of LDL, *Free Radical Biol. Med.*, 13, 341, 1992.
18. Adlercreutz, H. et al., Urinary excretion of lignans and isoflavonoid phytoestrogens in Japanese men and women consuming a traditional Japanese diet, *Am. J. Clin. Nutr.*, 54, 1093, 1991.
19. Akiyama, T. et al., Genistein, a specific inhibitor of tyrosine-specific protein kinases, *J. Biol. Chem.*, 262, 5592, 1987.
20. Majumdar, A.P.N., Role of tyrosine kinases in gastrin induction of ornithine decarboxylase in colonic mucosa, *Am. J. Physiol.*, 259, G626, 1990.
21. Wei, H. et al., Inhibition of tumor promoter-induced hydrogen peroxide formation *in vitro* and *in vivo* by genistein, *Nutr. Cancer*, 20, 1, 1993.
22. Coyne, D.W. and Morrison, A.R., Effect of the tyrosine kinase inhibitor genistein, on interleukin-1 stimulated PGE_2 production in mesangial cells, *Biochem. Biophys. Res. Commun.*, 173, 718, 1990.
23. Chae, Y.H. et al., Effects of synthetic and naturally occurring flavonoids on benzo[a]pyrene metabolism by hepatic microsomes prepared from rats treated with cytochrome P-450 inducers, *Cancer Lett.*, 60, 15, 1991.
24. Steele, V.E. et al., Cancer chemoprevention agent development strategies for genistein, *J. Nutr.*, 125, 713S, 1995.
25. Rao, C.V. et al., Enhancement of experimental colon cancer by genistein, *Cancer Res.*, 57, 3717, 1997.
26. Misra, R.R. et al., Genotoxicity and carcinogenicity studies of soy isoflavones, *Int. J. Toxicol.*, 21, 277, 2002.
27. Miltyk, W. et al., Lack of significant genotoxicity of purified soy isoflavones (genistein, daidzein, and glycitein) in 20 patients with prostate cancer, *Am. J. Clin. Nutr.*, 77, 875, 2003.
28. Bloedon, L.T. et al., Safety and pharmacokinetics of purified soy isoflavones: single-dose administration to postmenopausal women, *Am. J. Clin. Nutr.*, 76, 1126, 2002.
29. Busby, M.G. et al., Clinical characteristics and pharmacokinetics of purified soy isoflavones: single-dose administration to healthy men, *Am. J. Clin. Nutr.*, 75, 126, 2002.
30. Frankel, E.N. and Meyer, A.S., Antioxidants in grapes and grape juices and their potential health effects, *Pharm. Biol.*, 36, 14, 1998.
31. Subbaramaiah, K. et al., Resveratrol inhibits the expression of cyclooxygenase-2 in human mammary and oral epithelial cells, *Pharm. Biol.*, 36, 35, 1998.

32. Jang, M. and Pezzuto, J.M., Resveratrol blocks eicosanoid production and chemically induced cellular transformation: implications for cancer chemoprevention, *Pharm. Biol.*, 36, 28, 1998.
33. Jang, M. et al., Cancer chemopreventive activity of resveratrol, a natural product derived from grapes, *Science*, 275, 218, 1997.
34. Steele, V.E. et al., Cancer chemoprevention drug development strategies for resveratrol, *Pharm. Biol.*, 36, 1, 1998.
35. El-Mowafy, A.M. and Alkjalaf, M., Resveratrol activates adenylcyclase in human breast cancer cells: a novel, estrogen receptor-independent cytostatic mechanism, *Carcinogenesis*, 24, 869, 2003.
36. Aziz, M.H., Kumar, R., and Ahmad, N., Cancer chemoprevention by resveratrol: *in vitro* and *in vivo* studies and the underlying mechanisms (review), *Int. J. Oncol.*, 23 (1), 17, 2003.
37. Pace-Asciak, C.R. et al., Wines and grape juices as modulators of platelet aggregation in healthy human subjects, *Clin. Chim. Acta*, 246 (1–2), 163, 1996.
38. IARC Working Group, Tea, *IARC Monogr.*, 51, 207, 1991.
39. Steele, V.E. et al., Chemopreventive properties of green and black tea and selected extracts in *in vitro* bioassays, *Carcinogenesis*, 21 (1), 63, 2000.
40. Steele, V.E. et al., Preclinical efficacy studies of green and black tea extracts, *Fed. Proc. Exp. Biol.*, 220, 210, 1999.
41. Weisburger, J.H. et al., Effect of tea extracts, polyphenols, and epigallocatechin gallate on azoxymethane-induced colon cancer, *Proc. Soc. Exp. Biol. Med.*, 217, 104, 1998.
42. Morse, M.A. et al., Effects of theaflavins on *N*-nitrosomethylbenzylamine-induced esophageal tumorigenesis, *Nutr. Cancer*, 29, 7, 1997.
43. Hebert, J.L. et al., Chemoprevention of basal cell carcinomas in the ptc1 +/− mouse-green and black tea, *Skin Pharmacol Appl. Skin Physiol.*, 14, 358, 2001.
44. Cao, J. et al., Chemopreventive effects of green and black tea on pulmonary and hepatic carcinogenesis, *Fundam. Appl. Toxicol.*, 29 (2), 244, 1996.
45. Gupta, S. et al., Inhibition of prostate carcinogenesis in TRAMP mice by oral infusion of green tea polyphenols, *Proc. Natl. Acad. Sci. USA*, 98 (18), 10350, 2001.
46. Katiyar, S.K., Agarwal, R., and Mukhtar, H., Protective effects of green tea polyphenols administered by oral intubation against chemical carcinogen-induced forestomach and pulmonary neoplasia in A/J mice, *Cancer Lett.*, 73 (2–3), 167, 1993.
47. Chang, P.Y. et al., Genotoxicity and toxicity of the potential cancer-preventive agent Polyphenon E, *Environ. Mol. Mutagen.*, 41 (1), 43, 2003.
48. Chow, H.H. et al., Phase I pharmacokinetic study of tea polyphenols following single-dose administration of epigallocatechin gallate and Polyphenon E, *Cancer Epidemiol. Biomarkers Prev.*, 10, 53, 2001.
49. Kelloff, G.J. et al., Strategy and planning for chemopreventive drug development: clinical development plans: beta-carotene and other carotenoids, *J. Cell. Biochem.*, 20 (Suppl.), 110, 1994.
50. Heber, D. and Lu, Q.Y., Overview of mechanisms of action of lycopene, *Exp. Biol. Med.*, 227, 920, 2002.
51. Cohen, L.A., A review of animal model studies of tomato carotenoids, lycopene, and cancer prevention, *Exp. Biol. Med.*, 227, 864, 2002.
52. Wargovich, M.J. et al., Efficacy of potential chemopreventive agents on rat colon aberrant crypt formation and progression, *Carcinogenesis*, 21 (6), 1149, 2000.
53. Kim, J.M. et al., Chemopreventive effects of carotenoids and curcumins on mouse colon carcinogenesis after 1,2-dimethylhydrazine initiation, *Carcinogenesis*, 19 (1), 81, 1998.
54. Kim, D.J. et al., Chemoprevention by lycopene of mouse lung neoplasia after combined initiation treatment with DEN, MNU and DMH, *Cancer Lett.*, 120 (1), 15, 1997.
55. Kim, D.J. et al., Chemoprevention of lung cancer by lycopene, *Biofactors*, 13 (1–4), 95, 2000.
56. Bhuvaneswari, V. et al., Chemopreventive efficacy of lycopene on 7,12-dimethylbenz[a]anthracene-induced hamster buccal pouch carcinogenesis, *Fitoterapia*, 72 (8), 865, 2001.
57. Bowen, P. et al., Tomato sauce supplementation and prostate cancer: lycopene accumulation and modulation of biomarkers of carcinogenesis, *Exp. Biol. Med.*, 227, 886, 2002.
58. Kucuk, O. et al., Effects of lycopene supplementation in patients with localized prostate cancer, *Exp. Biol. Med.*, 227, 881, 2002.

19 Phytochemicals as Potential Cancer Chemopreventive Agents

Rajendra G. Mehta and John M. Pezzuto

CONTENTS

19.1 Introduction...237
19.2 Chemoprevention by Phytochemicals Present in Foods....................................238
19.3 Discovery of New Phytochemicals as Chemopreventive Agents.......................239
    19.3.1 Selection of Plants...239
    19.3.2 Selection of Bioassays..240
    19.3.3 Selection of Secondary Screening Systems..240
        19.3.3.1 Mouse Mammary Gland Organ Culture (MMOC) Assay 240
        19.3.3.2 Aberrant Crypt Foci (ACF) Assay...241
    19.3.4 Structural Characterization of Potential Chemopreventive Agents 241
    19.3.5 Efficacy in Experimental Carcinogenesis Models 241
19.4 From Candidate to Clinical Trial 243
    19.4.1 Preclinical Toxicity..243
    19.4.2 Clinical Trials 244
        19.4.2.1 Trial Design..244
        19.4.2.2 Group Size and Power Calculations...244
        19.4.2.3 Compliance...244
        19.4.2.4 Molecular Targets and Surrogate Endpoints as Biomarkers.........244
19.5 Summary...245
Acknowledgment..245
References 245

19.1 INTRODUCTION

As with most human disease states, there is general agreement that prevention of cancer development is preferable to the treatment of cancer with chemotherapeutic agents. In general, chemotherapy requires treatment of patients with toxic drugs that often result in untoward side effects. Since the beneficial effects of these agents may outweigh adverse reactions, therapeutic use is often tolerated. On the other hand, chemopreventive agents are any and all natural or synthetic compounds that can inhibit, suppress, or reverse the development and progression of cancer.[1,2] Obviously, for the treatment of ostensibly healthy people, these compounds must be absolutely nontoxic. For individuals at high risk for developing cancer, there must be a solid risk/benefit ratio. Chemopreventive agents can be administered as dietary components in food, dietary supplements, or pharmaceutical preparations. Phytochemicals present in edible or nonedible plant materials may function as chemopreventive or chemotherapeutic agents. Considering the sequence of carcinogenesis — including initiation, promotion, and progression — in the stage of progression, chemoprevention ends and chemotherapy begins. However, it is not very clear when promotion ends and progression begins (Figure 19.1).

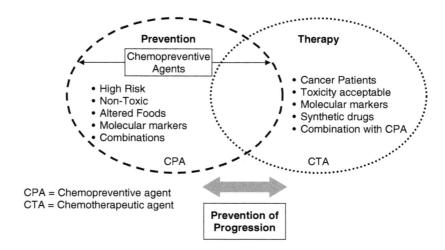

FIGURE 19.1 Schematic diagram showing the range of efficacy of chemopreventive agents.

Therefore, at least on a conceptual level, there is a time when chemopreventive agents can be successfully utilized for blocking the progression of cancer.[3] Furthermore, it is possible to use nontoxic chemopreventive agents in combination with chemotherapeutic agents in order to suppress toxicity.

The field of chemoprevention has evolved over the years from the concept and classification of chemopreventive agents as selective for initiation, promotion, or progression phases, to the current concept of molecular chemoprevention. It has become clear that in addition to mediating a phenotypic response to suppress transformation, inhibit the growth of transformed cells, or reverse the transformed phenotype to a normal cell type, a chemopreventive agent must modulate a genomic response.[4,5] This would, in turn, modify signal-transduction processes within the cell to acquire or maintain normal cell function. Thus, chemopreventive agents can be defined as chemicals or a combination of chemicals that can alter or reverse the expression or function of a molecular target responsible for transforming normal cells or supporting the proliferation of transformed cells. In turn, the agent prevents or delays carcinogenesis. However, it is important to realize that molecular targets for drug discovery, although extremely relevant, cannot replace conventional procedures for assaying efficacy. The potential of newly discussed agents to function in a chemopreventive setting must be established using experimental methods for biological interventions.

19.2 CHEMOPREVENTION BY PHYTOCHEMICALS PRESENT IN FOODS

For the past two decades, there have been extensive efforts to evaluate the chemopreventive role of micronutrients that are present in edible natural products. Largely, foods selected for evaluation of chemopreventive properties are identified on the basis of epidemiological studies guided by suggestions that food habits in various regions in the world are associated with reduced cancer rates. A good example is soybean consumption and reduced incidence of breast cancer in Japanese women.[6] Epidemiological evidence suggests that the increased consumption of soybean products by Japanese women is directly related to lower breast cancer incidence in these women. Moreover, migration to Western countries, followed by reduction in soy food, is associated with an elevation of cancer incidence. Based on these observations, the efficacy of genistein and daidzein in experimental mammary carcinogenesis models has been determined. The phytoestrogen genistein, which is also a cell-differentiating agent, provides protection against breast cancer.[7,8] However, genistein enhances carcinogenesis of other target organs.[9] On the other hand, soy food has been positively correlated with a reduction of breast cancer. Interestingly, recent studies with modified soy devoid of genistein showed chemopreventive activity similar to that of soy containing genistein.[10] Clearly, attributing effects of a whole food to one or two components present in the food may be misleading. At the same time, determining the mechanism of action of a whole food with numerous components within the food is very challenging.

There are many well-established chemopreventive agents either derived from foods or synthesized as pure chemopreventive agents. These include polyphenols such as epigallocatechin gallate and theaflavin in green tea[11]; selenium in foods grown in selenium-rich soil or in Brazil nuts[12]; carotenoids in vegetables[13,14]; vitamin D in dairy products[15]; resveratrol in grapes[16]; indole-3-carbinol, isothiocyanates, sulforaphane, etc., in cruciferous vegetables[17]; and limonene in fruits.[18] Since many of these chemopreventive agents and their mechanisms in cancer prevention are described in this volume, constituents of common foods will not be described here.

19.3 DISCOVERY OF NEW PHYTOCHEMICALS AS CHEMOPREVENTIVE AGENTS

Many chemotherapeutic agents used in the clinic are derived from natural products or designed on the basis of original compounds present in natural products.[19,20] Examples include taxol, camptothecin, and vincristine. However, there are about 250,000 plants known. Thus, it seems likely that the medicinal value of many plants is largely unknown, since it would be very difficult and cost prohibitive to evaluate plants on a random basis for their possible medicinal value. Based on these considerations, there must be a well-defined rational approach to select natural products for identification of chemopreventive or chemotherapeutic agents. It is a monumental task to identify a potent chemical from the complex mixture of substances present in a plant, establish its efficacy, understand possible mechanisms of action, and perform the steps required for a possible clinical trial. In recent years, several groups of investigators have developed focused approaches for drug discovery. For example, chemopreventive agents derived from Asian plants have been evaluated using a mouse lung model.[21] Many of these chemopreventive agents are known chemicals present in common foods, including ascorbic acid, capsaicin, lecithin from soybean, and caffeic acid. Some novel compounds were also reported. These include 2-(allylthio)pyrazine and eupatilin (5,7-dihydroxy-3,4,6-trimethoxyflavone) in *Artemisia asiatica*.[22] Many other reports often focus on selected agents rather than *de novo* drug discovery.

Our approach for discovering natural product-based chemopreventive agents involves a multiple-step design. The overall progression from bench to clinic is streamlined and divided into nine steps. These include: selection of plants, evaluation of extracts using selective bioassays for possible efficacy, selection of active fractions, determination of novel chemical structures from the active fractions, synthesis of active chemopreventive agents, examination of novel chemopreventive agent with *in vivo* experimental carcinogenesis models, synthesis of analogs with altered functional groups to reduce toxicity and increase efficacy, determination of mechanism of action of new chemopreventive agents, and development of new agents for clinical trials.

19.3.1 SELECTION OF PLANTS

Judicious selection of plants for further evaluation and discovery of new chemopreventive agents is a challenge. The process of selecting and evaluating plants is based on available literature, epidemiological evidence, and traditional medicinal use by people in the region where the plant is available. In addition, wherever possible, priority is given to edible plants. Along these lines, we also utilize the database of plants maintained at the University of Illinois. This valuable resource, Natural Product Alert (NAPRALERT), constitutes a collection of more than 150,000 scientific articles. Thus far, more than 3000 plants have been collected. The various plant parts, such as leaves, roots, stems, or fruit, are extracted with ethyl acetate and evaluated for potential chemopreventive properties using several mechanistically based bioassays.[23,24]

TABLE 19.1
Summary of Evaluation of Plant Extracts and Chemopreventive Agents in Primary Screens and Their Efficacy

Primary Screen	Active Extracts or Agents/Total	Percent Effective
Quinone reductase	64/2061	3.1
COX-1 and COX-2 inhibition	57/1448	2.3
Ornithine decarboxylase	89/1608	5.5
JB-6 cell transformation	32/124	25.8
HL-60 cell differentiation	110/1761	6.1
Antiestrogen (Ishikawa cells)	202/1845	10.9
ER competition	7/390	1.8
Antiproliferation (Col-2 colon cancer cells)	296/1915	15.5
Total primary screen	1311/17871	8.5
Secondary screen (MMOC)	59/225	26.2
Secondary screen (ACF)	03/04	75
DMBA-TPA skin *in vivo*	03/04	75
Mammary carcinogenesis	5/5	100

19.3.2 SELECTION OF BIOASSAYS

Based on molecular targets and known signaling pathways for active chemopreventive agents, we selected a number of assays predicted to identify efficacious plant extracts. The assays are subsequently used to direct the isolation of new chemopreventive agents. Selection of bioassays is subjective and varies among laboratories. We have adopted a strategy that encompasses a variety of pathways and is capable of detecting the major classes of chemopreventive agents.[23,24] Additional assays introduced as new mechanistic pathways are identified and reported in the literature. At the present time, we evaluate the efficacy of plant extracts to induce quinone reductase with Hepa 1c1c7 cells, inhibit phorbol ester-induced ornithine decarboxylase activity with mouse 308 cells, displace estrogen binding with estrogen receptors α and β, inhibit cyclooxygenases 1 and 2, induce terminal differentiation with HL-60 cells, inhibit transformation with JB6 cells, and inhibit aromatase activity. As shown in Table 19.1, there are differential responses with each of these assays. Each procedure yields unique information. Single bioassays or molecular targets cannot enable the selection or detection of all efficacious chemopreventive agents.

19.3.3 SELECTION OF SECONDARY SCREENING SYSTEMS

In addition to primary *in vitro* bioassays, an additional discriminatory assay is of great value. As shown in Table 19.1, several hundred active extracts have been identified. In order to further evaluate and rank order these active extracts, the MMOC system was introduced.

19.3.3.1 Mouse Mammary Gland Organ Culture (MMOC) Assay

This assay is a good predictor of inhibitors of mammary carcinogenesis. Mammary glands of immature mice respond to hormones and 7,12-dimethylbenz(*a*)anthracene (DMBA) in whole gland culture through development of mammary alveolar lesions (MAL).[25] Inhibition of MAL by extracts or chemopreventive agents is used as a parameter for assessing the efficacy of test agents or extracts.[26] Only active plant extracts are subjected to this secondary screening protocol. To date, approximately 18,000 biochemical assays have been completed on plant extracts and isolates, with approximately

1300 yielding activity in at least one of the assays. In MMOC, 225 assays were performed, and 26% (59 samples) showed inhibition of MAL development (Table 19.1). Extracts significantly inhibiting development of MAL are further fractionated to obtain pure active isolates. This eliminates a large number of lead plant extracts.

19.3.3.2 Aberrant Crypt Foci (ACF) Assay

The MMOC assay is a good predictor of mammary carcinogenesis, but it has been reported that many compounds that are active in mammary carcinogenesis models are not able to prevent carcinogenesis of the gastrointestinal (GI) tract. There appears to be target-organ specificity for chemopreventive agents.[27] For example, retinoids are generally ineffective against carcinogenesis of the gastrointestinal tract, whereas they are active inhibitors of mammary carcinogenesis. On the other hand, certain COX-2 inhibitors, such as piroxicam or aspirin, are effective inhibitors of colon carcinogenesis but are ineffective against breast cancer development. Thus, using only MMOC may result in identification of false negative for colon or other GI cancers. Therefore, we recently introduced azoxymethane-induced aberrant crypt foci in CF1 mice as another secondary model. The model has been well established for both rats and mice.[28,29] Treatment of animals with azoxymethane induces development of ACF in the colon. Simultaneous treatment of the animals with chemopreventive agents suppresses the number and severity of the ACF. Using this assay, we showed that piroxicam and deguelin are very effective against ACF formation,[30] whereas piroxicam was ineffective in MMOC. Table 19.1 summarizes the efficacy of some of the new chemopreventive agents in this model.

19.3.4 Structural Characterization of Potential Chemopreventive Agents

Typically, isolates are obtained by fractionation of the active extract. A plant extract showing activity in a bioassay and secondary model (MMOC or ACF) is selected and fractionated using high-pressure liquid chromatography (HPLC) or other chromatographic procedures. All fractions are subjected to the extract-specific *in vitro* bioassay that was initially responsible for the selection of the extract for further evaluation. Structural elucidation of the compounds is carried out using physical and spectroscopic methods.[24] Once the compound is identified and is available in sufficient quantities for bioassays, the activity of the chemopreventive agent is confirmed both with *in vitro* bioassays and the secondary discriminator. Further studies are designed based on the efficacy of the isolates. Often, the amount of the chemopreventive agent available from the natural source is not sufficient for determining efficacy with *in vivo* experimental carcinogenesis models. If possible, chemical synthesis of active newly discovered chemopreventive agent(s) is performed. As a part of our overall program to discover novel chemopreventive agents, we have successfully implemented this strategy and synthesized several potential chemopreventive leads. These include brassinin,[31] resveratrol,[32] deguelin,[33] and 4'-bromoflavone.[34]

19.3.5 Efficacy in Experimental Carcinogenesis Models

There are numerous experimental carcinogenesis models reported for evaluating chemopreventive agents.[35] The ideal model requires that the cancers developing as a result of carcinogen insult be target-organ specific, but otherwise, the carcinogen does not have adverse effects in the animals. Descriptions of the most commonly used models are summarized elsewhere.[27] In our experimental program for the discovery of chemopreventive agents from natural products, three models of carcinogenesis are used in the primary approach. These include the DMBA-induced and 12-*O*-tetradecanoylphorbol-13-acetate (TPA)-promoted (two-stage) skin carcinogenesis model in mice[33,35] DMBA- or *N*-nitroso-*N*-methylurea (NMU)-induced rat mammary carcinogenesis model,[34] and benzo(a)pyrene-induced lung tumor model in mice.[27] These models have been used rigorously by numerous investigators and have resulted in consistent and reproducible results.

Prior to the commencement of a carcinogenesis inhibition experiment, a dose-selection study is conducted. It is of prime importance for chemoprevention experiments to select a dose that is efficacious and nontoxic. This is imperative, since the agent would be administered to people at a high risk of developing tumors who are otherwise disease free, but more directly, the results of the carcinogenesis studies are confounded by toxic responses. The dose-selection study is often a 6-week experiment in which a relatively small group of animals receive increasing concentrations of the chemopreventive agent.[27] If possible, the median dose is selected from the literature based on a compound with a similar chemical structure; otherwise, the process is semiempirical. Laboratory animals are weighed twice weekly and observed daily for any signs of toxicity. The maximum tolerated dose (MTD) (highest dose with no apparent toxicity) is identified, and 80% of this value is normally used as a high dose for carcinogenesis experiments.

Among the newly identified potential chemopreventive agents, we evaluated the efficacy of brassinin, deguelin, 4'-bromoflavone, and resveratrol in the two-stage skin and NMU-induced mammary carcinogenesis models. These agents (except 4'-bromoflavone) were selected from the secondary screen using the MMOC assay. All of these agents (except 4'-bromoflavone) were chemopreventive in both these models. We reported that brassinin was effective when treated in the two-stage skin carcinogenesis model.[31] The suppression of papilloma formation was observed during both initiation and promotion phases. Numerous analogs of brassinin and flavones have been synthesized as modulators of quinone reductase activity. Among these agents, 4'-bromoflavone exhibited extremely potent activity. All these agents were evaluated in the MMOC assay. However, as expected based on results obtained with other halogenated compounds, 4'-bromoflavone did not show any activity in MMOC. Both brassinin and 4'-bromoflavone were evaluated in the DMBA-induced mammary carcinogenesis model, as anti-initiators. Brassinin was given from –3 weeks to +1 week, and 4'-bromoflavone was given from –1 to +1 week, in relation to DMBA treatment. Results showed remarkable suppression of tumor incidence by both these agents. Brassinin suppressed tumor incidence from 87% in control rats to 52% in the treatment group, whereas 4'-bromoflavone reduced the incidence from 94 to 20% in the control and treatment groups, respectively. These agents are now being evaluated for their role as antipromotional agents in the DMBA-induced mammary carcinogenesis model.

We also determined the efficacy of deguelin, a rotenone found in an African plant, *Mundulea sericea*,[33,36] in skin and mammary carcinogenesis models. Deguelin, a potent inhibitor of ornithine decarboxylase (ODC) activity, inhibited carcinogen-induced mammary lesions in the MMOC assay. In the two-stage skin carcinogenesis model, deguelin suppressed induction of papillomas by nearly 100% by topical application.[33] However, recent experiments in ultraviolet (UV)-induced skin carcinogenesis studies indicated that deguelin was not effective. Deguelin was originally selected with ODC, and this was considered a molecular target for deguelin activity. Further, tumor suppression by deguelin in a two-stage skin carcinogenesis model correlated with ODC activity. However, in the UV-induced model, ODC does not seem to be affected. Therefore, the lack of the deguelin effect in this model may support the previous suggestion that deguelin activity is mediated by suppression of ODC activity. The efficacy of deguelin was also determined in the NMU-induced rat mammary carcinogenesis model. Although there was no effect on tumor incidence, multiplicity was reduced by 50% (from 6.8 tumors per rat to 3.2 tumors per rat in the high-deguelin group). We also showed that deguelin (5 mg/kg body weight, daily) inhibited development of azoxymethane-induced aberrant crypt formation from 29 ± 4.3 in control mice to 7 ± 1.5 in deguelin-treated CF1 mice. Promising activity has also been shown against lung cancer.[37] Thus, deguelin may be developed as a promising new chemopreventive agent.[36]

We originally isolated resveratrol from the roots of *Cassia quinquangulata,* a plant collected in Peru, and reported its possible effects as a chemopreventive agent. In recent years, considerable effort has been directed toward understanding the protective effects of resveratrol. Resveratrol is present in grapes, red wine, and peanuts. Cyclooxygenase was identified as one molecular target.

TABLE 19.2
Selective New Chemopreventive Agents Derived from Natural Products

Plant	Chemopreventive Agent	Primary Activity	Secondary Screen	*In Vivo* Efficacy
Brassica spp.	brassinin, sulforaphane	quinone reductase	MMOC	skin, mammary gland
Cassia quinquangulata	resveratrol	COX inhibition	MMOC	skin, mammary, prostate,
Mundulea sericea	deguelin	ODC	MMOC, ACF	skin, mammary, melanoma, lung
Casimirora edulis	zapotin	differentiation	MOC, ACF	N.D.
Brucea javanica	brusatol	differentiation	MMOC	leukemia

Resveratrol inhibited cyclooxygenase activity and suppressed development of carcinogen-induced mammary lesion formation in MMOC significantly, indicating its potential to function as a chemopreventive agent. As recently reviewed,[16] a large number of chemopreventive studies have now been performed with resveratrol, and the compound demonstrates a plethora of relevant activities. Recent reports have indicated that resveratrol could be efficacious against mammary and prostate carcinogenesis. There is very little known about the metabolism of resveratrol. In mice and rats, resveratrol is converted to resveratrol sulfate and resveratrol glucuronide.[38] Whether the sulfates or glucuronides of resveratrol exhibit chemopreventive activity is under investigation.

Over the past several years, we have identified a variety of chemopreventive agents derived from edible and nonedible plants, and synthetic analogs of some of these potentially valuable chemopreventive agents have been produced. These include brassinin analogs, 4′-bromoflavone, and sulforaphane (present in broccoli) analogs such as sulforamate[39] and oxomate. Other agents of interest include zapotin from sapote (fruits commonly consumed in Asia and Mexico) and chemopreventive agents present in tomatillo. A summary of the effectiveness of some newly identified chemopreventive agents is given in Table 19.2.

Selection of chemopreventive agents for further development is not a trivial task. Factors such as structure, efficacy, availability, mechanism, and relationship to known active agents need to be taken into account. Only a few selective agents can advance to preclinical toxicity and clinical trials.

19.4 FROM CANDIDATE TO CLINICAL TRIAL

19.4.1 PRECLINICAL TOXICITY

Prior to clinical trials, it is mandatory to determine toxicity profiles for chemopreventive agents in at least two mammalian species. The requirements for chemopreventive agents, for obvious reasons, are more stringent than those of chemotherapeutic agents. The general approach with chemopreventive agents is to design intervention trials for people who are either at higher risk of developing cancer or those who should be protected from developing second primary tumors. Since the benefit-to-risk ratio must be very large for this group of patients, it is unacceptable to have any significant toxic side effects.

Preclinical toxicity experiments are usually carried out as 28-day gavage studies using rats and dogs as models. The experiments are carried out under good laboratory practice (GLP) guidelines. Animals are treated with four increasing dose levels for 28 days. The rats and dogs are observed at least once a day. Complete clinical pathology and necropsy are conducted at the end of the study. The tissue sections are examined for possible toxicity by board certified veterinary pathologists. These studies provide a complete toxicity profile for the compound at four dose levels in two species. For chemopreventive trials, genetic toxicity and reproductive toxicities may be required. If the agent appears to be safe at all dose levels with no toxicity, then it moves forward to FDA approval for a Phase I clinical trial.

19.4.2 CLINICAL TRIALS

Phase I clinical trials for chemopreventive agents are carried out with healthy volunteers. The Phase I trial characterizes the acute and chronic tolerance of the selected compound as well as pharmacokinetics, absorption, metabolism, and retention. The initial doses are selected from experimental carcinogenesis studies or from previous clinical studies with comparable agents. Phase II clinical trials are performed to determine the biological effects of the chemopreventive agent. Here, the focus is on the evaluation of intermediate end points.[39] Effects of chemopreventive agents can subsequently be correlated with the duration and dose level of the test agent. The design of clinical trials for chemopreventive agents must take into consideration many selective issues.

19.4.2.1 Trial Design

The clinical trial can be initiated with one agent. These studies are relatively straightforward. The chemopreventive agent is tested at various doses, and biomarkers are evaluated during the course of the trial or at the end of the treatment period. However, this intervention does not take into consideration other modulating and compounding factors such as nutrition or secondary interactions of the biomarkers. Therefore, in recent years, a factorial design has received a great deal of attention. In a factorial design, more than one chemopreventive agent can be studied. It also provides an opportunity to identify involvement of other factors responsible for the preventive effects of the agent(s). Another advantage of factorial designs is that one arm of the study can be terminated without affecting the outcome of the other arm.

19.4.2.2 Group Size and Power Calculations

The number of participants to obtain statistically significant results based on hypothetical results is very crucial in the study design. This is the single most important factor responsible for a successful chemoprevention trial. It is generally believed that volunteers at high risk of developing cancer tend to commit to the trial for a longer period of time. Unlike chemotherapeutic agents, the trial with chemopreventive agents requires inclusion of a relatively large number of participants. This often is one of the major limiting factors for chemoprevention trials.

19.4.2.3 Compliance

The compliance and commitment of taking the treatment agent is also very crucial for a successful clinical trial. Although pill counting has been successfully used as one criterion for compliance, serum levels of the test agent in the volunteers is obviously the most desired parameter to verify compliance. Low compliance compromises the clinical trials by reducing the power of the study and may offset the grouping errors.

19.4.2.4 Molecular Targets and Surrogate Endpoints as Biomarkers

The most accurate end point for any chemotherapeutic or chemoprevention trial is mortality at the end of the study. However, this would require a very large volunteer population and is neither feasible nor cost effective. Therefore, selective early end-point biomarkers are used, which may be agent selective.[40] Cell proliferation, micronuclei, and early precancerous lesions have all been used to various degrees as markers. However, in recent years, considerable attention has been given to identifying molecular targets for a given chemopreventive agent. Modulation of these biological components — which can be enzyme activity, RNA expression, receptor binding, protein expression or a receptor-mediated specific functional response — would be extremely useful in predicting responses for a given chemopreventive agent. This set of parameters must be custom designed and linked specifically to the chemopreventive agent.

19.5 SUMMARY

Cancer chemoprevention has received major attention in recent years.[41] The focus largely has been on determining the efficacy of well-established chemopreventive agents derived from food or assessing analogs in a variety of experimental models. Yet, there appears to be a critical need to identify new efficacious and nontoxic chemopreventive agents. In this chapter, we have described a bioassay-guided fractionation approach for the discovery of new chemopreventive agents from natural products. Using a rational plant selection process and evaluating plant extracts by mechanism-based bioassays followed by fractionation and structural elucidation, we have identified a variety of new chemopreventive agents from approximately 3000 plants. We also have in place a secondary screening system to select agents that show the greatest promise and are target-organ specific in their actions. Lastly, a stepwise protocol for advancing newly discovered chemopreventive agents to clinical trials is described. Clearly, this approach offers great hope for the eventual control of cancer in human beings.

ACKNOWLEDGMENT

"Natural Inhibitors of Carcinogenesis" is supported by the National Cancer Institute under the auspices of program project grant P01 CA48112. The authors gratefully acknowledge this support and the contributions of the numerous investigators and collaborators who participated in the project.

REFERENCES

1. Sporn, M.B. and Hong, K.W., Recent advances in chemoprevention of cancer, *Science,* 278, 1073, 1997.
2. Kelloff, G.J., Sigman. C.C., and Greenwald, P., Cancer chemoprevention: progress and promise, *Eur. J. Cancer,* 35, 2031, 1999.
3. Mehta R.G. and Pezzuto J.M., Discovery of cancer preventive agents from natural products: from plants to prevention, *Curr. Oncol. Rep.*, 4, 478, 2002.
4. Einspahr, J.G. et al., Chemoprevention of human skin cancer, *Crit. Rev. Oncol. Hematol.*, 41, 269, 2002.
5. Milner, J.A. et al., Molecular targets for nutrients involved in cancer prevention, *Nutr. Cancer*, 41, 1, 2001.
6. Messina, M.J. and Loprinzi, C.L., Soy for breast cancer survivors: a critical review of the literature, *J. Nutr.,* 131, 3095S, 2001.
7. Lamariniere, C.A., Timing and exposure and mammary cancer risk, *J. Mammary Gland Biol. Neoplasia,* 7, 67, 2002.
8. Constantinou, A.I. et al., Chemopreventive effects of soy protein and purified isoflavones on DMBA-induced mammary carcinogenesis in Sprague-Dawley rats, *Nutr. Cancer,* 41, 75, 2001.
9. Rao, C.V. et al., Enhancement of experimental colon cancer by genistein, *Cancer Res.*, 57, 3717, 1997.
10. Gupta S. et al., Inhibition of prostate carcinogenesis in TRAMP mice by oral infusion of green tea polyphenols, *Proc. Natl. Acad. Sci. USA*, 98, 10350, 2001.
11. Demeule M. et al., Green tea catechins as novel antitumor and antiangiogenic compounds, *Curr. Med. Chem. Anti-Cancer Agents*, 4, 441, 2002.
12. Ip, C., Dong, Y., and Ganther, H.E., New concepts in selenium chemoprevention, *Cancer Metastasis Rev.*, 21, 28128, 2002.
13. Nishino, H. et al., Carotenoids in cancer chemoprevention, *Cancer Metastasis Rev.*, 21, 257, 2002.
14. Bowen, P. et al., Tomato sauce supplementation and prostate cancer: lycopene accumulation and modulation of biomarkers of carcinogenesis, *Exp. Biol. Med.*, 227, 886, 2002.
15. Mehta, R.G. and Mehta, R.R., Vitamin D and cancer, *J. Nutr. Biochem.*, 13, 252, 2002.
16. Bhat, K.P. and Pezzuto, J.M., Cancer chemopreventive activity of resveratrol, *Ann. N.Y. Acad. Sci.*, 957, 210, 2002.

17. Murillo, G. and Mehta, R.G. Cruciferous vegetables and cancer chemoprevention, *Nutr. Cancer*, 41, 17, 2001.

18. Gould, M.N., Cancer chemoprevention and therapy by monoterpenes, *Environ. Health Perspect.*, 4, 977, 1997.

19. Eckardt, S., Recent progress in development of the anticancer agents, *Curr. Med. Chem. Anti-Cancer Agents*, 3, 419, 2002.

20. Lacombe, D. et al., Cancer drug development in Europe: a selection of new agents under development at the European Drug Development Network, *Cancer Invest.*, 21, 137, 2003.

21. Yun, T.K., Update from Asia: Asian studies on cancer chemoprevention, *Ann. N.Y. Acad. Sci.*, 889, 157, 1999.

22. Seo, H.J. and Surh, Y.J., Eupatilin, a pharmacologically active flavone derived from *Artemisia* plants induces apoptosis in human promyelocytic leukemia cells, *Mutat. Res.*, 49, 6191, 2001.

23. Pezzuto, J.M., Plant derived anticancer agents, *Biochem. Pharmacol.*, 53, 121, 1997.

24. Kinghorn, A.D. et al., Cancer chemopreventive agents discovered by activity guided fractionation: a review, *Curr. Org. Chem.*, 2, 597, 1998.

25. Steele, V.E. et al., Use of *in vitro* assays to predict the efficacy of chemopreventive animals in whole animals, *J. Cell. Biochem.*, 265, 29, 1997.

26. Su, B.-N. et al., Constituents of the bark and twigs of *Artocarpus dadah* with cyclooxygenase inhibitory activity, *J. Nat. Prod.*, 65, 163, 2002.

27. Moon, R.C., Mehta, R.G., and Rao, K.V.N., Retinoids and cancer, in *The Retinoids*, 2nd ed., Sporn, M.B., Roberts, A.B., and Goodman, D.S., Eds., Raven Press, New York, 1994, 273.

28. Wargovich, M.J. et al., Efficacy of potential chemopreventive agents on rat colon aberrant crypt formation and progression, *Carcinogenesis*, 6, 1149, 2000.

29. Yang, K. et al., Chemoprevention studies of the flavonoids quercetin and rutin in normal and azoxymethane-treated mouse colon, *Carcinogenesis*, 21, 1655, 2000.

30. Murillo, G. et al., Deguelin suppresses the formation of carcinogen-induced aberrant crypt foci in the colons of CF-1 mice, *Int. J. Cancer*, 104, 7, 2003.

31. Mehta, R.G. et al., Cancer chemopreventive activity of bassinin, a phytoalexin from cabbage, *Carcinogenesis*, 16, 399, 1995.

32. Geshäuler C. et al., Rotenoids mediate potent chemopreventive activity through transcriptional regulation of ornithine decarboxylase, *Nature Med.*, 1, 260, 1995.

33. Jang, M. et al., Cancer chemopreventive activity of resveratrol, a natural product derived from grapes, *Science*, 275, 218, 1997.

34. Song, L.L. et al., Cancer chemopreventive activity mediated by 4'-bromoflavone, a potent inducer of phase II detoxification enzymes, *Cancer Res.*, 59, 578, 1999.

35. Mehta, R.G., Experimental basis for breast cancer, *Eur. J. Cancer*, 36, 1275, 2000.

36. Udeani, G.O. et al., Cancer chemopreventive activity mediated by deguelin, a naturally occurring rotenoid, *Cancer Res.*, 57, 3424, 1997.

37. Mehta, R.G. et al., Induction of atypical ductal hyperplasia in mouse mammary gland organ culture, *J. Natl. Cancer Inst.*, 93, 1103, 2001.

38. Yu, C. et al., Human, rat and mouse metabolism of resveratrol, *Pharm. Res.*, 19, 1907, 2002.

39. Gerh user, C. et al., Cancer chemopreventive potential of sulforamate, a novel analog of sulforaphane that induces phase 2 drug-metabolizing enzymes, *Cancer Res.*, 57, 272, 1997.

40. Kelloff, G.J. et al., Surrogate end-point biomarkers in chemopreventive drug development, *IARC Sci. Publ.*, 154, 13, 2001.

41. Alberts, D.S. et al., Prevention of cancer in the next millennium: report of the Chemoprevention Working Group to the American Association for Cancer Research, *Cancer Res.*, 59, 4743, 1999.

20 History of Natural Supplements in Cancer Therapy and Prevention

Gottumukkala V. Subbaraju and Golakoti Trimurtulu

CONTENTS

20.1 Introduction..247
20.2 Cancer and Ayurveda: an Overview...248
20.3 Role of Natural Products in Cancer Cure...249
    20.3.1 Mistletoe 250
    20.3.2 Hoxsey 250
    20.3.3 Laetrile 251
    20.3.4 Essiac 252
20.4 Natural Supplements in the Prevention of Cancers...253
    20.4.1 Garlic 253
    20.4.2 Coenzyme Q10 254
    20.4.3 Ginseng 254
    20.4.4 Green Tea..255
    20.4.5 Resveratrol 256
    20.4.6 Turmeric...257
    20.4.7 Soy Isoflavones 258
20.5 Antioxidants Useful as Adjuncts to Radiation Therapy 258
    20.5.1 Aloe..259
    20.5.2 Tulsi 259
20.6 Natural Products That Were Developed into Anticancer Drugs.......................260
    20.6.1 Yew 260
    20.6.2 *Vinca Rosea* 261
    20.6.3 Mayapple 261
    20.6.4 Camptotheca 262
20.7 Conclusions...262
References 264

20.1 INTRODUCTION

Cancer is the most dreaded of all diseases in terms of the great devastation it causes in the life of patients and family. The American Cancer Society estimates that about 1,284,900 new cases of cancer were diagnosed in the U.S. in the year 2002, and about 555,500 Americans are expected to die of the disease; that is, more than 1500 people a day. Cancer is the second leading cause of death in the U.S., exceeded only by heart diseases. In the U.S., men have a little less than a one in two lifetime risk of developing cancer. For women, the risk is a little more than one in three.[1] According to the World Health Organization, the global cancer rates could increase by 50% to 15 million by 2020.[2] With such a high probability that one can contract this terrible menace, coupled with rapidly rising health care costs, the emphasis has now shifted to prevention of the disease rather than cure through radiation treatment or chemotherapy.

20.2 CANCER AND AYURVEDA: AN OVERVIEW

Humans have traditionally been dependent on nature for the healing and prevention of afflictions and revitalizing body systems for longevity. Natural supplements have been offering relief for nearly every known illness since the dawn of mankind, long before today's conventional medical practices. *Swabhavoparana*, Sanskrit for "remission by nature," was clearly illustrated through the use of herbs and other natural secrets in Ayurveda, the oldest healing science known in India.[3] Ayurveda is considered a branch of *Atharva Veda*, one of the four main texts of Hindu spirituality known as *Vedas*. Fairly comprehensive information about herbs has been recorded in *Charaka Samhita* and *Shustruta Samhita*, the two most important Vedic books. The description of cancer can be dated to earlier than 2000 B.C. by its description given in Atharva Veda under the name *apachi*. Sushruta, one of the reorganizers of Ayurveda (400 B.C.), described the superficial tumors under the name *arbuda*. The malignancies of internal organs were described as *gulma*.[4]

Ayurveda is the complete health care system embracing the whole domain of medical science. The Vedic texts consider cancer as a life within a body created as a result of emotional imbalance.[5] Illness is a sign that life is not being lived in balance. The writers of Ayurveda laid down the rules for maintaining perfect health and suggested remedies for restoring health after the onset of a disease. The holistic approach in Ayurveda seeks the root cause of an illness and restores balance rather than just treating the symptoms alone.

The Ayurvedic concept of cancer is based on three *doshas* (body constitutions). The imbalance in doshas (*vayu*, *pitta*, and *kapha*) is the beginning of pathogenesis in the development of cancer.[4] The disease is classified according to the dosha vitiated, and the treatment is chosen appropriately to augment the dosha using various procedures, including herbal medications.[6] The herbs are classified into groups, depending upon their enhancing or decreasing effect on each dosha. The general cancer healing process include cleansing therapies and treatment with immune-boosting herbs such as *Satavari* (*Asparagus racemosus*), *Ashwagandha* (*Withania somnifera*), *Vidari kand* (*Puereria tuberosa*), etc.[6] Ayurveda offers unique therapeutic procedures like *pancha karma* (five actions) to detoxify the body using several processes, including treatment with herbal concoctions.[7] *Poorva karma* is a pretreatment procedure performed to enhance the immune system.[8]

The treatment of cancer and other diseases using the Ayurvedic system of health care is still a common practice in India. The first recorded cancer clinical study based on the Ayurvedic concept was undertaken in India under the sponsorship of the Central Council for Research in Ayurveda and Sidda, and the results were compiled as a publication entitled *Ayurvedic Drugs in the Management of Cancer*.[9] The study involves the classification and analysis of the histopathology of the disease and treatment modalities on patients having various types of tumors. In one study involving 400 cancer patients, the Ayurvedic formulation containing herbal drugs bhallatak (*Semecarpus anacardium*), rohitak (*Amoora rohitaka*), madhuyasti (*Glycyrrhiza glabra*), and tamra bhasma was evaluated alone or in combination with other treatment modalities, chemotherapy, and radiotherapy. The patients were monitored for 10 years. Significant improvement and increase in survival rate was observed in patients treated with the Ayurvedic drug alone. The combination of the Ayurvedic drug along with modern chemotherapy or radiotherapy showed much better response than the Ayurvedic drug alone.[10] It was also observed that the patients who received *poorva karma* prior to chemotherapy showed longer survival times. This might be the first recorded cancer study based on Ayurveda concepts.

The current popular trends for healthy living and disease prevention have their roots in Ayurveda, which emphasizes health promotion and disease prevention through proper diets and rejuvenating herbs. Other cultures across the globe also have their own contemporary traditional health care systems.

20.3 ROLE OF NATURAL PRODUCTS IN CANCER CURE

The search for natural products as potential anticancer agents dates back to Papyrus Ebers, the Egyptian, who listed more than 700 drugs, mostly from plants, during the middle of the second millennium B.C.[11] It is thus ironic that at the beginning of the 21st century, humankind is still ill-equipped to wage a war on its meanest enemy, cancer, despite the advent of modern allopathic medicine and the research that has been documented since the turn of 20th century. For centuries, our greatest health concern was quickly responding to swiftly spreading overt infectious diseases. The use of plants in the treatment of cancer has not been widely embraced by traditional medicine, has not been supported by documented cures, and has been limited mostly to treatment of easily detectable skin cancers. However, the popularity of herbal medicine as a natural supplement, particularly for the cure and prevention of cancer, has gained momentum since the beginning of the 20th century.

The conception and development of cancer is a slow process that progresses over many years and stages. Despite the complexity of cancer progression, there are many opportunities to intervene during its path of development and to wage battle with tumors. Fortunately, nature provides us with a wide variety of beneficial metabolites in the form of dietary and herbal supplements that can interact at one or many stages of the cancer progression, starting from the initial exposure to carcinogens through outright malignancy in the final step. These beneficial compounds include vitamins, essential fatty acids, amino acids, and a wide array of nonessential nutrients, called phytochemicals. According to dietary guidelines advocated by the American Cancer Society and the U.S. National Cancer Institute, eating plenty of fruits and vegetables provides the first line of defense against disease initiation in general and cancer in particular.[12] There is also extensive literature covering the importance of nutritional supplements such as vitamins, minerals, essential fatty acids, and amino acids in the prevention and cure of cancer. The other substances in fruits and vegetables are nonnutritional phytochemicals. Some of these phytochemicals, like flavonoids, offer several-fold better protection than the vitamins. However, it is difficult to provide all of the friendly chemical agents through dietary sources alone. Herbal supplements are therefore vital to fill this void and provide optimum levels of a broad range of beneficial ingredients.

These phytochemical anticancer supplements can be categorized broadly into three groups, based on where they interrupt the cancer formation process:

Supplements that nourish and build up the immune system, enhancing its anticancer surveillance function to prevent cancers in the first place and to help the body fight existing cancers.

Supplements that are antioxidants, scavenging free radicals before they reach target sites and initiate carcinogenesis. These compounds also activate enzymes that break down carcinogens, reducing them to harmless compounds.

Supplements that directly attack the cancer and fight for remission and cure.

In reality, however, the boundaries between the anticancer supplements in these three groups are not distinct. Rather, these phytochemical supplements show a great deal of overlap in their beneficial effects, and most often they constitute a three-pronged approach.

Folklore in different cultures around the world provides a valuable source of information regarding the use of herbal supplements in cancer treatment. Many of these "traditional" cancer treatments have come into practice since the beginning of the 20th century, even though there was no absolute evidence that these herbal supplements could cure cancer. When some of these unconventional natural treatments became popular, they were challenged by the medical establishment, which vigorously condemned them as ineffectual and unproven quackery. Some of the therapies, like Hoxsey and essiac (see Sections 20.3.2 and 20.3.4), ignited national movements, with supporters of herbal medicine on one side, demanding the right to choose in health care decisions, and agencies of organized medicine on the other side preventing unconventional medicine for lack of documented evidence for their efficacy. The U.S. bans the use of some of these therapies as cancer treatments, even though they are legal in some other countries. Nevertheless, these unproven treatments are still being used by desperate cancer patients everywhere. A brief history of some of the popular unconventional herbal medicines is given in the following subsections.

20.3.1 Mistletoe

The Latin name for mistletoe is *Viscum album*. It is a semiparasitic plant that grows on old apple, ash, and hawthorn trees. It is one of very few cancer medications with a rich traditional history.[13] Ancient Greeks used it as an anticancer medication, and it was a popular herbal medicine of the Druids, who used it for treating epilepsy and nervous system disorders. Dr. Rudolf Steiner, founder of the Society for Cancer Research, advocated its use in 1920 for the treatment of human cancer.[14] Presently, it is widely used in Europe and North America to treat arthritis and rheumatism and for palliative treatment of malignant tumors. *Viscum album* is listed in the *Homeopathic Pharmacopoeia of the United States*.[15] Mistletoe is marketed under the brand names Iscador, Isorel, Helixor, Eurixor, Vysorel, etc.

The extracts of mistletoe have been shown to inhibit the growth of a variety of animal and human cancer cells *in vitro* and tumor growth *in vivo* in a number of animal models.[13] Iscador was found to be selectively cytotoxic to tumor cells while causing no damage to cells of the immune system.[16] The antitumor effects of mistletoe reportedly depend on several factors, like the type of host tree, the time of collection during the year, and the manufacturing process. The active principles are found to be lectins, viscotoxins, and alkaloids.

The mistletoe lectins inhibited the growth of multidrug-resistant human colon cancer line HT 29 (mdr) *in vitro*.[17] Numerous studies have reported immunostimulatory and proapoptotic effects for mistletoe extracts, preparations, and components.[13] The mistletoe extracts and preparations also significantly inhibited carcinogenesis and metastatis.[18] Animal studies have shown that mistletoe preparations suppressed side effects when used as an adjunct to chemotherapy and radiation therapy.[19]

Numerous human clinical studies have been reported on mistletoe or its preparations as a treatment for cancer. In most of the studies, the survival and tumor recurrence were the end points. Many of these studies attributed beneficial effects to mistletoe products in terms of improvements in survival, improvements in quality of life, and/or stimulation of the immune system.[20] However, most of the studies have serious flaws in study design, and they have often yielded inconsistent outcomes.[13] The variation in the type of mistletoe products used from one study to the other makes the analogy difficult. Although mistletoe is one of the most widely studied herbal medicines for cancer, well-designed clinical trials are still needed to decipher its absolute potential in cancer treatment. Mistletoe is one of the best among the unconventional therapies with moderate evidence of efficacy, and the reported side effects have generally been minimal.[21]

20.3.2 Hoxsey

Hoxsey is an herbal formula named after the folk healer Harry Hoxsey, who inherited the formula from his father. According to the story, Hoxsey's grandfather, John Hoxsey, discovered this cancer therapy accidentally in 1840s when his stallion was cured of cancer after grazing on some unusual plants in the pasture. He conducted animal experiments using the herbs in the pasture and came up with a formula that he passed down to his family. Harry Hoxsey started his first clinic in 1924 and then became famous. By the 1950s, Hoxsey had established clinics in 17 states, and the clinic in Dallas, Texas, was known as the largest privately owned cancer medical facility in the world.[22]

Similar to the advocates of other unconventional drugs, Hoxsey endured the wrath of the organized medical profession, the news media, court actions, and the Federal Drug Administration until his last operation in the U.S. was closed down during the 1960s.[23] The reports claiming miraculous cures from the Hoxsey treatment were anecdotal. Some of the patients who claimed cure from the treatment had not been diagnosed with biopsy-proven cancer at the beginning of the treatment, and many of the patients were simultaneously undergoing conventional therapies.

Hoxsey kept the ingredients of the formula secret for a long time, but eventually he had to reveal the composition in pursuit of an FDA (U.S. Food and Drug Administration) litigation. He indicated that the formula contained a basic solution of cascara (*Rhamnus purshiana*) and potassium iodide, along with one or more of several plant substances — poke root (*Phytolacca americana*), burdock root (*Arctium lappa*), berberis root (*Berberis vulgaris*), buckthorn bark (*Rhamnus frangula*), Stilling root (*Stillingia sylvatica*), and prickly ash bark (*Zanthoxylum americanum*) — depending on the patient's condition.[24]

Both the American Medical Association (AMA) and the National Cancer Institute (NCI) reviewed the case histories and dismissed the claims attributed to Hoxsey.[25] The FDA subsequently issued a public warning against Hoxsey in 1956, stating that the FDA had not found a single verified cure of internal cancer effected by the Hoxsey treatment. The warning also stated that NCI's review of case histories submitted by Hoxsey failed to provide conclusive evidence in support of his formula's therapeutic effect.[26]

The therapeutic efficacy of Hoxsey's formula is still a big question, but many of the individual herbs in the formula have a history of folk use in many cultures around the world. Many of the herbs also have scientific evidence in support of their anticancer properties. The activity of poke root in stabilizing the growth of epitheliomas and increasing patient's survival time was well known long before Harry Hoxsey started offering his formulation as a cancer treatment.[27] Pokeweed is also known to possess mitogenic activity and to serve a role in modulating the immune system.[28] Burdock, which is known as a folk remedy for new growths and ulcerations, was found to be cytostatic and antimutagenic.[29,30] Berberrubine is an antitumor alkaloid isolated from berberis root.[31] The University of Virginia study found an antileukemic principle from buckthorn.[32]

In spite of these positive attributes, Hoxsey's formulation could not pass the evaluation of the medical establishment. However, it did survive longer than any of the other unconventional therapies of the last century.

20.3.3 LAETRILE

Laetrile is a levorotatory mandelonitrile beta-diglucoside that has often been used as a synonym for chemically related amygdalin. Amygdalin was first isolated from bitter almonds by two French scientists in 1830.[33] It is also a constituent of lentils, lima beans, cashews, brown rice, and millet. The kernels of the apricot, peach, or bitter almond are the commercial source for its production. It was known as a cancer medicine in Russia as early as 1845, and its first reported use in the U.S. was during the 1920s.[34]

Laetrile was very popular as a cancer medicine in 1970s.[35,43] The evaluation of laetrile by NCI and other investigators on various solid tumors and leukemia implanted in mice and rats found no anticancer activity, although some reports claim positive results.[36,37] The reports on the use of laetrile as an anticancer treatment are mostly anecdotal and case reports.[38] The published data from clinical studies or case series provided no conclusive evidence in support of its beneficial effects.[39,40]

In the early 1960s, the Cancer Advisory Council of the California Department of Health studied laetrile, and a U.S. Senate subcommittee held hearings on laetrile in 1977. They both found no value for laetrile in the treatment, alleviation, or cure of cancer.

In the early 1970s, the FDA approved an IND application on laetrile study, but later rejected the same due to questions regarding its efficacy.[38,41] The FDA's role in health care matters was challenged in courts in several states by laetrile supporters. As a consequence, laetrile use was legalized in several states, but in 1980, the U.S. Supreme Court reaffirmed the FDA's authority by overturning lower-court decisions.[42]

In response to mounting pressure from laetrile supporters, NCI undertook a retrospective analysis of 67 patients who had undergone laetrile therapy. The review could not, however, muster conclusive evidence supporting its use as a cancer treatment.[43] In continuation of this study, NCI sponsored a phase I clinical study in 1980 and a phase II clinical study in 1982 with 178 patients. The study again offered no cures, and the survival times were comparable with those who received no treatment. In addition, several patients experienced symptoms of cyanide toxicity due to elevated blood cyanide levels.[39,44] It was proposed that the metabolic actions of beta-glucosidase enzyme caused the breakdown of laetrile in the body into glucose, benzaldehyde, and hydrogen cyanide.[45]

There was always an abyss of disagreement between the proponents and critics of laetrile regarding its usefulness, origin, chemical structure, mechanism of action, and therapeutic effects.[37,46] It was a source of intense national debate regarding freedom of choice in health care decisions, and the debate was guided more by sociological and political factors than medical. In the end, laetrile could not live up to the scientific scrutiny, and it was not approved for use in the U.S.

20.3.4 ESSIAC

Essiac was an herbal treatment for cancer developed in the 1920s by Caisse, a Canadian nurse. She learned of this remedy through a patient who had been cured of cancer from a formula offered by a Native American Ojibwa herbalist. Caisse started using the medicine in the mid-1920s and became so highly popular that her story was the subject of two biographies.[47]

The name *essiac* was actually derived by spelling her name backwards. Her practice was in technical violation of Canadian law, so essiac supporters mustered 50,000 signatures on a petition to the government to legalize her practice. The bill to authorize her practice, however, fell short by three legislative votes. Caisse continued treating patients for a long time. In 1959, Caisse collaborated with Charles Brusch at the Brusch Clinic in Cambridge for the clinical evaluation of essiac. The study found that essiac was beneficial.[48]

However, Memorial Sloan-Kettering Cancer Center (MSKCC) tested the samples submitted by Caisse in mouse sarcoma S-180 and found no activity.[49] Later, Caisse came to a collaborative agreement with Resperin Corp. to develop essiac, and the latter filed a preclinical new drug submission with Health and Welfare Canada in 1978. The submission was later suspended in 1982 due to Resperin's failure to fulfill the agreement.[50]

At the request of Health and Welfare Canada, the U.S. National Cancer Institute conducted animal tests on a sample submitted by Resperin Corp. In a mouse lymphocytic leukemia system, essiac showed no activity, but it did show toxicity at the highest levels tested.[51]

In 1980, Health and Welfare Canada undertook a retrospective review of case summaries of patients who had received essiac treatment. The study suggested that a majority of the patients showed no response to essiac. The small number of patients who had their disease stabilized were known to have also received some form of conventional therapy.[52] The essiac supporters were highly critical of this study.

In conclusion, essiac was a tremendously popular cancer therapy during the early half of the 20th century. There is no documented evidence to support its effectiveness in humans. Like other cancer medications of traditional origin, it could not withstand scientific validation.

20.4 NATURAL SUPPLEMENTS IN THE PREVENTION OF CANCERS

Although mainstream medical researchers continue to argue against unconventional cancer therapies, they have a different perspective regarding the role of herbal supplements in prevention of cancer. There is mounting scientific evidence that vitamins and herbal supplements play a key role in prevention of cancer and enhancing the quality of life of cancer victims. Indeed, epidemiological and clinical evidence support the chemopreventive efficacy of supplements such as green tea, resveratrol, etc.

Nearly 80 to 90% of all cancers are known to be caused directly or indirectly by environmental factors.[53] The modification of macromolecules such as DNA, proteins, lipids, etc. by reactive oxygen species (ROS) is the first step in the initiation of several diseases, including cancer as well as cardiovascular, neurodegenerative, and inflammatory diseases. The human body has an immune system that is designed to counteract disease. Unfortunately, conventional cancer therapies compromise the body's immune system, and this compromised immunity may promote tumorigenesis.

Oxidation plays a key role in the initiation of carcinogenesis in biological systems, so it should not be surprising that antioxidants play an important role in the inhibition of carcinogenesis.[54] A broad variety of antioxidants like flavonoids, isothiocyanates, anthocyanadines, polyphenols, CoQ10, etc. have demonstrated significant activity in the prevention of cancer induction. Some of these antioxidants are more potent than vitamin supplements. For example, phytochemicals such as flavonoids effect several-fold better protection than vitamin C against oxidative DNA. Natural herbal supplements also exhibit beneficial synergistic effects when used along with vitamins and minerals.[55]

The recent literature has indicated that several natural products of plant origin with a broad spectrum of structural types not only effect chemoprevention of carcinogenesis but also inhibit the growth of established tumors.[56] The molecular basis for the interaction of natural supplements with a number of cellular events has been explored. Some of the modes of interaction include antioxidation; angiogenesis; prevention of DNA damage; COX-2 (cyclooxygenase 2) inhibition; oncogene activation; p53 mutation; induction of p57 cell cycle regulator; modulation of cytochrome P450; stimulation of natural killer cell cytotoxicity as well as T-helper cell proliferation; and inhibition of an array of enzymes such as ornithine decarboxylase, topoisomerase II, protein kinase C, and tyrosine kinase C.[57,107,119,127,135] Protein kinase C and tyrosine kinase C, two key enzymes responsible for carcinogenesis and tumor spread, respectively, were strongly inhibited by a number of phytochemicals.[58,59] The following subsections briefly review some of the important antioxidants and biological response modifiers that play a key role in cancer prevention.

20.4.1 GARLIC

Garlic (*Allium sativum*) has a history as a medicinal agent going back thousands of years. Hippocrates, Dioscorides, and other prominent medical experts of ancient and medieval eras deciphered its medicinal uses.[60] Garlic is mentioned in the Bible, and it may be one of the most widely used natural supplements. Its first known origin may be Central Asia (3000 B.C.), later spreading to other parts of the world.[60] It is a regular part of the diet for most people in Asian and Middle Eastern countries. Its potential therapeutic uses include antimicrobial, antithrombotic, hypolipidemic, antiarthritic, hypoglycemic, and antitumor activity.[61]

Studies demonstrated that garlic and its constituents effectively inhibit experimentally induced carcinogen formation, carcinogen bioactivation, and tumor proliferation at a number of sites including skin, mamma, and colon.[62] A population-based study conducted by researchers at the National Cancer Institute showed that people with daily consumption of vegetables from the allium group had a statistically significantly lower risk of developing prostate cancer.[63] Other epidemiological studies suggest that regular intake of garlic is also associated with reduced risk of esophageal, stomach, and colon cancers.[64–66] Garlic also showed significant reductions in both tumor volume and mortality against transitional cell carcinoma in MBT2 murine bladder carcinoma model.[67] In addition to its chemopreventive effects, some of the sulfur compounds in garlic also inhibit the growth of induced and transplantable tumors in animal models.[68]

The beneficial chemopreventive properties of garlic were attributed to organosulfur compounds in garlic. S-Allylcysteine and S-allylmercapto-L-cysteine, the major compounds in aged garlic, showed the highest antioxidant activity.[68] Crushing or aging of garlic causes unstable molecules such as allicin to convert into stable molecules such as diallyl disulfide (DADS), S-allylcysteine and S-allylmercaptocysteine.[69] Crushing garlic releases an enzyme, allinase, that triggers a series of chemical reactions. Garlic also contains vitamins A and C, potassium, phosphorus, sulfur (including 75 different sulfur compounds), selenium, and a number of amino acids. The highly complex mixture of phytochemicals in garlic makes the standardization of garlic extract difficult. This might be one of the reasons for inconsistencies in some of the experimental studies. Further studies are required to form a definitive conclusion regarding its beneficial effects toward chemoprevention and cancer cure.

One of the important virtues of garlic is its strengthening of the immune system, essential for fighting cancer.[70] Garlic also showed protective effects on the cardiovascular system, and it is a potent natural antibiotic.[71–73]

20.4.2 COENZYME Q10

The coenzyme Q10 is a natural substance belonging to the family of 2,3-dimethoxy-5-methyl-6-polyprenyl-1,4-benzoquinone, widely known as ubiquinone for its ubiquitous occurrence in animal and plant tissues. This coenzyme plays a pivotal biological role in the electron transport chain in both photosynthetic and respiratory processes.[74] CoQ10 has a clinical effect in human congestive heart failure and hypertension.[75,76] It also augments the compromised immune system and prevents cardiotoxicity of some of the cancer drugs.[77]

In countries like Japan, it is a drug prescribed for those having suffered from heart disease. However, in the U.S., it is a dietary supplement available from health food stores or mail-order businesses. Coenzyme Q10 has been shown to have both anticancer and immune-system enhancing properties in animals, and the results suggest that it might inhibit tumor-associated cytokines.[78] In one study, 6 out of 32 high-risk breast cancer patients obtained partial to complete regression when supplemented with 90- to 390-mg daily doses of CoQ10.[79] A patient treated with a daily dose of 390 mg for 1 month showed no sign of palpable tumor mass. Another patient with intraductal cancer showed complete remission following treatment for 3 months with a daily dose of 300 mg, and there was also no trace of distant metastases.[79]

One study involving breast cancer patients showed that CoQ10 concentrations in tumor tissues were significantly depleted compared with the surrounding normal tissues. Administration of coenzyme Q10 by dietary supplementation may induce a protective effect on breast tissue.[80]

CoQ10 is an immunomodulating agent and is essential for the optimal function of the immune system.[81] Its cellular deficiency can result from several reasons, including aging and cancer. Correction of deficiency requires supplementation of CoQ10 at concentrations higher than those available in the regular diet.

20.4.3 GINSENG

Ginseng is the most thoroughly studied plant with the longest history of traditional use for cancer and other ailments.[82] The Asian variety called *Panax ginseng* is the center of ginseng research and the most sought after herb in Chinese medicine. In Chinese tradition, ginseng is considered as a tonic that gives vital energy to the body. Its usage in Western medicine was picked up after scientists described ginseng as an adaptogen, something that normalizes bodily functions.[83] Ginseng sales peaked during the 1990s.

Ginsenosides are a major group of saponins in ginseng; other constituents include eleuthero-sides, phenolics, flavonoids, polyacetylenic compounds, alkaloids and polysaccharides, etc. The proponents of ginseng considered it to be one of the promising agents for cancer treatment.[84] It exhibited anticarcinogenesis in animal models.[85] In a case control study of 1987 pairs of human patients from Korea Cancer Center Hospital, scientists established that the use of *Panax ginseng* reduced the risk of cancer by 50%.[86] The research results accumulated to date suggest that ginseng can inhibit cancer formation by bolstering the immune system. This enhanced immune stimulation also helps in augmenting the immune suppression that follows chemotherapy. Expression of stimulated macrophage phagocytosis, increased antibody production, and enhanced natural killer (NK) cell activity upon exposure to ginseng or its constituents in cultures and animal models corroborated its anticarcinogenesis and immunomodulatory properties.[87] Some investigations have manifested that ginseng can potentiate the effects of chemotherapeutic agents.[88] Another research study indicated the reversal of p-glycoprotein-mediated multidrug resistance by ginsenoside Rg(3).[89]

Contradictory to popular belief, some of the recent literature indicates that ginseng may not have any effect on immune function and carcinogenesis.[90] Ginseng users were led to panic following a media stint in the wake of Dr. Ronald K. Siegel's article in the *Journal of the American Medical Association* stating that long-term use of ginseng may cause insomnia, high blood pressure, skin eruptions, and nervousness.[91]

Even after 3000 years of history and more than 3000 research publications, there are more questions than answers regarding the medical use of ginseng. The scientific data pouring in to date has been a source of further confusion for both consumers and health care professionals alike.

20.4.4 Green Tea

Tea is the most widely consumed beverage among all cultures worldwide. Though all tea preparations originate from the same plant source, namely *Camellia sinensis*, a majority of studies aimed at exploring the usefulness of tea are focused on green tea. The epidemiological data demonstrate that people who consume large amounts of green tea have a lower risk of developing various cancers. The data on animal and human studies manifested that green tea has preventive effects on both chronic inflammatory diseases and lifestyle-related diseases such as cancer and cardiovascular diseases.[92] The beneficial effects of green tea have been attributed to the polyphenol, epigallocatechin gallate (EGCG). EGCG and other green tea polyphenols are powerful antioxidants.

The cancers that exhibited a significant inverse relationship between green tea consumption and disease incidence are colon, rectal, pancreatic, stomach, ovarian, and gastrointestinal tract cancers.[93] In most of the cases, the epidemiological data were corroborated by animal studies and mechanistic findings of the actions. For example, at least two studies indicate that people who consume tea regularly have a reduced risk of prostate cancer.[94] In a laboratory study, the main constituent of green tea, EGCG, inhibited growth and regressed human prostate and breast cancers in athymic nude mice.[95] Further studies have shown that EGCG inhibited ornithine decarboxylase, an important contributor in the development of prostate cancer.[96]

A prospective cohort study with over 8000 patients showed that daily consumption of at least 10 cups of green tea delayed onset of cancer, whereas daily consumption of over 5 cups per day by stage I and stage II breast cancer patients in a follow-up study resulted in lower recurrence rate and longer disease-free period than those consuming fewer than four cups per day.[97] Hence, green tea may hold potential as an agent for cancer prevention before cancer onset and prevention following cancer treatment.

Green tea and EGCG exhibited chemopreventive effects against the adverse effects of ultraviolet (UV) irradiation and cigarette smoking.[98] They have also shown immune modulatory effects and ameliorated immune dysfunction in mice bearing Lewis lung carcinoma.[99]

Green tea has been shown to have a significant chemopreventive effect against the initiation, promotion, and progression stages of carcinogenesis, and reports of animal models show that it provides protection against the development and progression of skin, lung, mammary gland, and gastrointestinal tumors.[100,102] Green tea is relatively inexpensive and associated with few adverse effects, and its major constituents inhibit tumor invasion and angiogenesis, the two crucial steps for the growth and metastasis of all solid tumors.[101,102] The cancer preventive and inhibitory effects of EGCG were related to induction of apoptosis and cell cycle arrest, inhibition of carcinogenesis, regulation of transcription factors (AP-1 and NF-κB), inhibition of gene expression (TNFα, VEGF, and NOS), modulation of enzyme activities (ornithine decarboxylase, matrix metalloproteinase, urokinase plasminogen activator, protein kinase C and protein phosphatase 2A, cyclooxygenase and lipooxygenase, protein tyrosine kinase), inhibition of angiogenesis, inhibition of metastasis, and antioxidation.[102]

Green tea was found to enhance the inhibitory effects of doxorubicin by 2.5-fold in mice bearing ehrlich ascites carcinoma.[103] In addition to potentiating the activity of cancer drugs, green tea polyphenols and EGCG inhibit the binding and efflux of cancer drugs by P-gp, a protein involved in the multidrug-resistance phenotype of cancer cells.[104] EGCG has also been shown to inhibit metastasis of B16 melanoma cell lines in both experimental and spontaneous systems.[105] The inhibitory effects of EGCG were found to be cancer-cell specific, with little or no effect on the normal counterparts.[106] The differential responses to green tea between normal and malignant cells were ascribed to induction of p57, a cell cycle regulator.[107]

Despite the foregoing evidence, there are a few epidemiological studies that show no correlation between green tea consumption and cancer incidence.[108] Further human studies are needed to clearly establish the potential of green tea as a food supplement for prevention of cancer and other diseases.

20.4.5 RESVERATROL

Resveratrol is a natural phytoalexin found in grapes, *Polygonum cuspidatum*, and a number of other medicinal plants. Red wine is the main source of resveratrol in the human diet. It has been implicated as the responsible factor behind the so-called French paradox, an epidemiological phenomenon of low rate of heart disease among French and Greek populations despite their high fatty diet.[109] The cardioprotective properties of resveratrol are corroborated by its antioxidant properties, antiplatelet aggregation properties, and its ability to elevate production of nitric oxide and high-density lipoprotein.[110] Based on epidemiological data, carcinogenesis and coronary heart disease are linked to dietary lifestyle and share common pathways at the molecular level.[111] Resveratrol has gained popularity in recent years following numerous reports indicating its cancer chemopreventive activity.[112]

Resveratrol was reported to be a potent cancer chemopreventive agent in assays representing all three major stages of carcinogenesis.[113] The anti-initiation effects were supported by its antioxidant and antimutagenic activities and by its induction of phase II drug metabolizing enzymes. The anti-inflammatory and inhibition of cyclooxygenase (COX-1) potentials of resveratrol correlate with its antitumor promotion effect. The induction of human promyelocytic cell differentiation in HL-60 cells corresponds to its ability to inhibit the antiprogression stage of carcinogenesis.[114] In a dose-dependent manner, resveratrol inhibited preneoplastic lesions in a mouse mammary gland culture and also carcinogen-induced tumorigenesis in a two-stage mouse skin model.[113]

Resveratrol is a selective inhibitor of human cytochrome P450 1A1, a mixed-functional oxidase needed to metabolically activate carcinogens for tumor initiation.[115] Resveratrol prevented lung carcinogenesis in human bronchial epithelial cells exposed to carcinogenic polycyclic aromatic hydrocarbons through the modulation of P450, and this may have contributed to the lower risk of lung cancer among consumers of red wine compared with those on other beverages.[116] A significant rise in serum antioxidant capacity was observed in a study on normal individuals who were ingesting red wine.[117] Resveratrol significantly reduced tumor volume (42%), tumor weight (44%), and metastasis to the lung (56%) in mice bearing highly metastatic Lewis lung carcinoma (LLC) tumors.[118] Resveratrol was shown to be a potent inhibitor of activation of NF-κB, a transcription factor associated with inflammatory diseases and oncogenesis.[119] This might be a probable molecular basis for its chemopreventive effect. It was reported to be a potent inducer of apoptotic cell death in human melanoma, HL-60 leukemia, and T47D breast carcinoma cancer cell lines, and it was shown to arrest cell division at S2/G2 phase of cell cycle.[120]

Resveratrol prevented the formation of colon tumors and reduced the formation of small-intestinal tumors by 70% in mice genetically predisposed to develop intestinal tumors.[121] It was shown to be a remarkable inhibitor of ribonucleotide reductase and DNA synthesis in mammalian cells, and it offered protection against hydrogen peroxide-induced DNA damage in human peripheral blood lymphocytes.[122] The selective inhibition of cell proliferation in mice (L1210) and human (HL-60) leukemia cell lines by resveratrol may be utilized for *ex vivo* purging of leukemia cells from bone marrow autograft without significant loss in the hematopoietic activity of progenitor cells.[123] Resveratrol is a phytoestrogen and exhibits various degrees of estrogen-receptor agonism in different test systems, which can further broaden its spectrum of biological functions.[124]

The foregoing data suggest that resveratrol may be of great value in preventing cancer and other degenerative diseases. But conclusive evidence of its potential is still lacking. The National Cancer Institute is currently funding clinical studies on resveratrol in healthy humans and patients with early-stage cancers. It has jointly awarded a research grant totaling $1.7 million to the University of Lichester and the University of Michigan to carry out a research study for this purpose.[125]

20.4.6 TURMERIC

Turmeric (*Curcuma longa*) is a famous herb known in Ayurveda and Chinese medicine for thousands of years for its amazing healing properties. The fleshy underground rhizomes are the medicinally important part of this plant, since these contain curcuminoids, which are biologically active yellow pigments. Turmeric is a regular condiment in food items like curries in Indian and other Asian cuisine. The yellow powder made from the root has great spiritual significance in Hindu culture, and the dry roots are exchanged as a gesture of goodwill, particularly during marriage ceremonies.

In folk traditions, turmeric has been used to treat diarrhea, fever, parasitic worms, leprosy, bladder and kidney inflammations, etc. Turmeric has been reported to protect humans against the development and growth of various cancers.[126] The potential health benefits of turmeric were attributed to curcumins. Natural curcuminoids were found to be powerful antioxidants.[127] Curcumin was shown to inhibit tumorigenesis during both initiation and promotion stages in several experimental models.[128,134] Turmeric/curcumin has shown chemoprevention against carcinogen-induced oral, forestomach, skin, mammary, liver, and colon tumors in mice and rats.[129,136] Curcumin inhibited proliferation of prostate cancer cells *in vitro* and inhibited the growth of implanted LNCaP tumor *in vivo*.[130] It is currently being evaluated in phase I clinical study for the treatment of colon cancer.[131] It has also been found to be highly effective at inhibiting recurring melanoma in people at high risk.[132] Turmeric and curcumin exhibited remarkable beneficial symptomatic relief effects in patients with external cancerous lesions.[132] They also demonstrated, when administered in diet, chemopreventive effect on oral precancerous lesions in hamsters.[133]

Epidemiological data suggest that curcumin may be responsible for the lower rate of colorectal cancer in Asian countries. The anticancer activity of curcumin may partly be due to the preferential arrest of cancer cells in G2/M phase of the cell cycle and the induction of apoptosis.[134] Curcumin is a specific inhibitor of cyclooxygenase-2 expression and inhibits lipooxygenase activity, the enzymes known to have a role in carcinogenesis.[135] Curcumin exhibits chemoprevention of carcinogenesis by inhibiting cytochrome P450 activity and enhancing glutathione-S-transferase activity.[136] The inhibitory effects of curcumin on cancer growth might be due to its inhibitory effects on ornithine decarboxylase, an enzyme that makes tumors grow rapidly.[137] It was shown that curcumin inhibits the activation of NF-κB and suppresses tumor promotion by blocking signal-transduction pathways in the target cells. Recent studies also indicate that curcumin can significantly reduce adriamycin-induced cardiotoxicity in rats.[138]

20.4.7 SOY ISOFLAVONES

Soy-rich diets have been credited with the low incidence of breast, stomach, and prostate cancers in Asian populations compared with their much higher prevalence in the Western hemisphere, where soy consumption is low.[139] There has been a multifold increase in research studies correlating consumption of soybean products to prevention of breast, colon, uterine, and prostate cancers. Soy was found to be one of the potent natural supplements with cancer-prevention properties.[140] It is a host of two important isoflavones, namely genistein and daidzein. These substances are shown to have various anticarcinogenic properties, including prevention of oxidation, inhibition of protein tyrosine phosphorylation, triggering of apoptosis, regulation of gene transcription, modulation of transcription factors, antiangiogenesis, and inhibition of DNA topoisomerage.[141] Daidzein and genistein down-regulate the expression of prostate androgen-regulated transcript-1 (PART-1) gene, a possible prostate cancer tumor marker.[142] These isoflavones are known as phytoestrogens because of their estrogenic property, and they are being used as a natural alternative to hormone replacement therapy (HRT). Recent studies indicate that soy isoflavones are cardioprotective and help to reduce menopausal hot flashes and increase bone density in women.[143] The U.S. Food and Drug Administration recently approved a health claim that soy protein as part of a diet low in saturated fats and cholesterol may reduce the risk of heart disease.[144]

A community-based prospective cohort study of Japanese men and women in Takayama, Japan, established the inverse relationship between risk of death from stomach cancer and soy intake.[145] Epidemiological studies indicated that women who consume a traditional Japanese diet excrete 1000 times more genistein than those from societies with high breast cancer rates, supporting the role of genistein in lowering the risk factor. These isoflavones inhibit the growth of estrogen-dependant tumor cells by blocking the cell uptake of this hormone. However, recent *in vivo* studies on animals indicate that genistein and soy diets containing varying amounts of genistein can stimulate the growth of estrogen-dependent (MCF-7) tumors in a dose-dependent manner.[146] The effect of phytoestrogens on breast cancer risk thus seems to be complex, and the outcome may depend on the dose and time of administration. A review of epidemiological studies and studies involving immigrant women in Western countries demonstrates that a soy-containing diet may be beneficial against breast cancer if consumed before puberty and during adolescence.[147,148] These studies found no negative effects of soy on breast cancer and suggested that consumption of soy only during the adult life had little or no effect on the risk of breast cancer. Thus postmenopausal women who have breast cancer or those at high risk of developing breast cancer should be extra cautious until the results from well-documented clinical trials are available.

20.5 ANTIOXIDANTS USEFUL AS ADJUNCTS TO RADIATION THERAPY

With the advent of modern radiotherapy, survival rates of people with certain types of cancers have increased significantly. Unfortunately, the side effects of radiation impose devastating effects on the quality of life of cancer patients. Moreover, radiation therapy often leads to secondary malignancies such as melanoma, medullablastoma, soft-tissue sarcoma, leukemia, carcinomas of the pancreas, etc. Because of the potential danger associated with radiation therapy, necessary steps should be taken to protect normal tissues during the treatment. The *in vitro* and *in vivo* studies on herbal supplements have indicated the possibility that some of them might have potential as adjuncts to radiation therapy. Aloe and tulsi, for example, besides having significant chemopreventive properties, exhibit beneficial effects needed for their use as adjuncts to radiation therapy.

20.5.1 ALOE

Aloe (*Aloe vera*) is a cactuslike plant originally indigenous to Africa. *Aloe vera* is the most prominent among the more than 300 species belonging to this family. It was widely regarded as a master healing plant in folkloric medicine.[149] In the last few decades, aloe has gained popularity as an adjunct treatment to heal burns and side effects following radiation therapy. This important application, which correlates with its traditional usage as a remedy to cuts and burns, has gained momentum following reports of a successful treatment of X-ray and radium burns in the 1930s. A large number of published reports support aloe's protective effects against radiation-induced skin injury and radiation ulcers.[150,165] A randomized, blinded clinical study with *Aloe vera* gel exhibited a protective effect in preventing skin reactions in patients undergoing radiation therapy,[151] whereas two other phase III clinical studies found no significant protective effect with *Aloe vera* treatment.[152] Aloe was found to be antibacterial and antifungal. It increases blood flow to wounded areas and stimulates fibroblasts, the skin cells responsible for wound healing.

Aloe exhibited antitumor,[153] anticarcinogenic,[154] and cancer chemopreventive[155] properties in animal models. According to a Russian report, treatment of experimental tumors in mice and rats with aloe juice contributed to reduction of tumor mass, metastatic foci, and metastasis frequency.[156] Bioassay-guided purification of an aloe species afforded a hydroxyanthraquinone compound called aloe-emodine, which significantly inhibited P-388 lymphocytic leukemia and human neuroectodermal tumors implanted in mice.[157,158] It also inhibited human hepatoma cell lines,[159] Hep G2 and Hep 3B, human Merkel cell carcinoma,[160] and human lung squamous cell carcinoma[161] cell lines in cultures. The tumor-prevention and anticancer activities of aloe are due mostly to the immune stimulation and immune modulation through activation of macrophages (antigens), causing the release of substances like interferons, interleukines, and tumor necrosis factor.[162] *Aloe vera* gel inhibited angiogenesis, a process of forming new blood vessels to feed the tumor.[163] It also potentiated the effects of 5-fluorouracil and cyclophosphamide in combination therapy.[156]

The only clinical study on record on the antitumor activity of *Aloe vera* demonstrated that an aloe and melatonin combination might delay the onset of metastasis in patients with advanced-stage solid tumors, including breast cancer.[164] Aloe is very safe when applied either orally and topically, and it also appears to be safe when injected in high doses. The FDA's approval of *Aloe vera* as a flavoring agent is a further indication of its safety.[165] The benefits associated with the use of aloe may provide great impetus to the research community to design and implement further controlled studies on this important plant.

20.5.2 TULSI

Tulsi (*Ocimum sanctum*) is the most revered of all sacred medicinal plants in India. In traditional medicine, it is cherished for its holistic healing properties to restore healthy living by warding off imbalances in the body and mind. In Indian herbal medicine, it is used as a remedy for bronchitis, gastric disorders, hepatic disorders, and skin diseases. Tulsi is also considered to be a diaphoretic, an antiperiodic, an anthelmintic, a cardiotonic, and an antipyretic as well as a blood purifier and an anti-inflammatory.[166]

In the only human clinical study reported to date, tulsi significantly reduced the fasting and postprondrial sugar levels during treatment.[167] Modern scientific research on tulsi has demonstrated some impressive findings that could have potential application for use in cancer prevention and cure. *Ocimum sanctum* was reported to be adaptogenic in rats and mice. Oral administration of *Ocimum sanctum* extract significantly prevented chemically induced carcinogenesis in Swiss albino mice and also in hamster buccal pouch model.[168,169] The chemopreventive efficacy of *Ocimum sanctum* oil at 100 µg/kg against 20-methylcholanthrene-induced fibrosarcoma in mice was comparable with that of 80 mg/kg of vitamin E.[170]

Ocimum sanctum has also been shown to play a potential role in the management of immuno-logical disorders, including allergies and asthma.[171] Its ethanolic extract induced cytotoxicity at 50 µg/ml against fibrosarcoma cells in cultures and mediated a significant shrinkage of tumor volume in mice bearing Sarcoma 180 solid tumors, extending their life span.[172] Recent literature also attributes potential for the use of tulsi in combination with radiation therapy. The leaf extract of *Ocimum sanctum* affords *in vivo* protection against cytogenetic damage by restoring the radiation-induced depletion of glutathione transferase and other enzymes to their normal levels.[173] Further studies have indicated that two flavonoids, orientin and vicenin isolated from the extract, demon-strated significant protection to the human lymphocytes in cultures against radiation lethality and also provided *in vivo* protection against death from radiation-induced gastrointestinal syndrome.[174] The *Ocimum sanctum* extract and the pure flavonoids were also found to be excellent protectors of bone marrow against radiation-induced clastogenesis and stem cell lethality in the mouse at a low nontoxic dose.[175] The radiation protection, which was comparable or even better than the known synthetic radioprotectors like WR-2721, was further enhanced in combination therapy.[176]

Ursolic acid, a triterpene constituent of *Ocimum sanctum*, protected the rat liver microsomes *in vitro* against free-radical-induced lipid peroxidation.[177] In animal studies, ursolic acid exhibited remarkable inhibitory activity against tumor promotion.[178] Topical cosmetic preparations containing ursolic acid and its isomer oleanolic acid are proprietary products in Japan, and both products are reported to be useful in skin cancer therapy.[179] The foregoing data open a tremendous possibility that *Ocimum sanctum* or its constituents may hold promise as adjuncts to human radiation therapy.

The items covered in the foregoing discussion are only a few of the widely known natural supplements with potential anticancer and chemopreventive properties. There are a large number of other phytochemicals and plant extracts reported to possess potential activities. Maitake mushroom (*Grifola frondosa*), horse chestnut (*Aesculus hippocastanum*), and shark cartilage are some examples suggested for cancer cure; and licorice (*Glycyrrhiza glabra*), ginkgo (*Ginkgo biloba*), and silymarin from milk thistle (*Silybum marianum*) are some examples known for cancer prevention properties.

20.6 NATURAL PRODUCTS THAT WERE DEVELOPED INTO ANTICANCER DRUGS

Nine of the current top-20 drugs, with annual sales of around $16.5 billion, are derived from or based on natural products. Many of the anticancer drugs currently being used or currently being developed were originally derived from nature, mostly from plants. For some other anticancer drugs, nature provided the initial lead from which the analogs were prepared and developed into new drugs.

Knowledge of traditional medicine affords a valuable approach to the development of new anticancer drugs. The search for plant species related to those used in folk medicine using chemo-taxonomic principles is another successful approach for developing new drugs. A fair number of anticancer drugs currently on the market were developed based upon their use in traditional medicine as a folk remedy, although not all of them were known to be active against cancer. This section of the chapter covers the natural supplements that are currently being used as treatments against cancers or other diseases. Note, however, that these products are also the original source for the development of important anticancer drugs currently in the market.

20.6.1 YEW

The yew tree has been associated very closely with cancer treatment during the last 15 years. Native Americans used Pacific yew (*Taxus brevifolia*) as a disinfectant, an abortifacant, and a cure for skin cancer. The yew tree has traditionally been used by many different cultures across the globe for thousands of years as a treatment for a variety of ailments. The yew tree was considered sacred by many civilizations, and its strong red wood was used to make household tools and implements of war. Chinese healers used yew to treat arthritis, and Native Americans used Pacific yew to treat skin diseases, bronchitis, headaches, dizziness, rheumatism, scurvy, arthritis, and stomach and lung problems. The salve made of crushed needles mixed with butter was used in India for the treatment of skin cancers.[180] Yew has been known to be poisonous since ancient times, as revealed by Latin and Greek literature.[181] Pacific yew tea, tincture, and salve are being used as dietary supplements in the northwestern U.S. In India, the tincture made from the young shoots of *Taxus baccata* has been used for giddiness, diarrhea, and severe biliousness, etc.[182]

The real miracle in the medical history of yew was the discovery of paclitaxel as a powerful anticancer agent from the Pacific yew, *Taxus brevifolia*. A massive research effort by the NCI revealed the cancer-killing properties of the Pacific yew. Its development into a viable cancer medicine in the form of Taxol© took almost three decades. Taxol and its related congener, cephalomannine, are the only natural products that have showed potential antitumor activity, although more than 100 analogs have been isolated and characterized. Taxol has exhibited a mode of action different from that of any previously known antitumor agents, and it has become the prototype for a new class of chemotherapeutic agents. Taxol kinetically stabilizes microtubule dynamics, thereby preventing depolymerization of the microtubule network essential for cell replication. Today, Taxol is the most promising and widely used drug for the treatment of breast and ovarian cancers.

Although Pacific yew is the most promising source of Taxol, the main source — the bark of the tree — offers a meager 0.004%. To date, nature alone has the efficient mechanism to concoct such a complex structure; a viable synthetic process for large-scale production has yet to be accomplished. However, the major impediment to commercial supply has been mitigated by a semisynthetic process involving the use of a chemical that is naturally more abundant, 10-deacetyl-baccatin, which can be extracted from the needles of the English yew (*Taxus baccata*). This process is environmentally benign as the needles, unlike the bark, are a renewable resource.

20.6.2 VINCA ROSEA

Another plant that changed the course of anticancer research was rosy periwinkle (*Catharanthus roseus*). This common garden plant, also known as *Vinca rosea*, is native to Madagascar. It has long been prescribed in folk medicine for treating dysentery, menstrual disorders, and diabetes. The discovery of its cancer-fighting properties was serendipitous, and it happened when the plant was being investigated for treatment of diabetes. Dr. Robert L. Noble at the University of Western Ontario received an unforeseen envelope containing leaves from Madagascar periwinkle with information that the herbal tea from the leaves was being used in Jamaica to treat diabetes.[183] The leaves did not produce considerable effect on blood glucose levels, but surprisingly it reduced the white blood cell count. This result led to the speculation that periwinkle might hold a cure for cancers like leukemia. In 1957, Dr. Gordon Svoboda of Eli Lilly Corp., who was working in collaboration with Dr. Nobel, tested crude extract of the whole plant on mice infected with P-1534 leukemia. Interestingly, 60 to 80% of the mice experienced prolonged life. In 1958, after several years of hard work, Dr. Nobel and Dr. Charles T. Beer successfully isolated a complex alkaloid constituent, vinblastine, that was responsible for inhibition of white blood cell count. Eli Lilly undertook the clinical trials, ultimately developing and marketing vinblastine as Velban©. Later, a scarce chemical cousin of vinblastine, namely vincristine, was isolated and developed as Oncovin©. These two compounds act by binding to tubulin protein, thus preventing the cell from undergoing necessary changes for cell division. Presently, these two compounds are routinely used to treat Hodgkin's disease and childhood leukemia. This is another example where new cancer drugs were discovered because of scientific investigation of a lead from a traditional folk remedy.

20.6.3 MAYAPPLE

The mayapple is Minnesota's native plant and ranges from Quebec to Texas and Florida. *Podophyllum peltatum* is its Latin name. The plant and its constituents have been in the medical literature for more than 250 years. The folkloric use of mayapple derives from various Native American

Phytopharmaceuticals in Cancer Chemoprevention

groups, who used the plant both as medicine and poison. During late 19th century, the Penobscot Indians of Maine used underground parts of the mayapple for the treatment of cancer, and it was also used by physicians in Louisiana and Mississippi for venereal warts.[184] Podophyllin, the resin prepared from an alcoholic extract of the rhizomes and roots, was very popular as a cathartic, vermifuge, and emetic. It was produced commercially in the U.S. and exported to Europe. Podophyllin was included in the *U.S. Pharmacopoeia* as an official drug from 1820 until 1942.

Inspired by mayapple's traditional application and its action against warts, Hartwell and coworkers at the U.S. National Cancer Institute explored its potential as a possible cure in cancer treatment. They identified the correct structure for the major toxin, podophyllotoxin. Podophyllotoxin exhibited anti-cancer activity in animal studies and human studies, but clinical trials were abandoned when the substance was found to be too toxic. Later, Sandoz Ltd. in Basel, Switzerland, picked up the podophyllotoxin research in the 1960s and developed several new analogs by incorporating subtle changes in the general structure. Two of the semisynthetic analogs, teniposide (VM-26) and etoposide (VP-16), were found to be very active yet relatively safe to human consumption and were approved by the FDA. These two compounds have been widely used to treat non-Hodgkin's lymphomas, leukemias, small-cell lung cancer, and testicular cancer. Teniposide is also being prescribed to treat brain tumors and childhood leukemias. Podophyllin is an example of a product where a new pharmaceutical has been evolved based on a generous lead provided by traditional medicine.

20.6.4 CAMPTOTHECA

Camptotheca acuminata is native to southern China and Tibet, where it is popularly called *xi shu*, which means "happy tree." Preparations made from camptotheca have been used in China as a cancer treatment, especially for cancers of the stomach and liver and leukemia.[185] The National Cancer Institute evaluated camptotheca in 1958 and confirmed its anticancer properties. A quinoline alkaloid, camptothecin, was isolated in 1966, and it was found to be responsible for the anticancer activity. Studies aimed at its mode of action revealed that camptothecin inhibited DNA replication by inhibiting topoisomerase I, an enzyme that mediates the winding/unwinding action of DNA strands during cell proliferation. The first human studies were conducted using sodium salt of camptothecin, but they were abandoned later due to severe side effects. Further studies revealed that the sodium salt was several times less potent than camptothecin, yet the salt was responsible for all the toxicity. Camptothecin was chemically modified to generate an array of potent anticancer drugs, such as 9-aminocamptothecan, CPT-11, irinotecan, and topotecan. Topotecan (Hycamtin®) and irinotecan HCl (Camptosar®) are the only compounds approved by the FDA for treatment against ovarian cancer and metastatic colorectal cancers, respectively.

Without the effort of scientists who believed in the efficacy of traditional systems of natural medicine, the second-most-important source of anticancer drugs would never have been known in modern allopathic medicine.

20.7 CONCLUSIONS

Plants have catered to the medical needs of human civilization since prehistoric times, but it is only during the last 100 years that mankind has begun to scientifically characterize their active principles and put them to use in modern medicine. According to the World Health Organization, up to 80% of the world's population relies mainly on herbal medicine for primary health care. In countries like India, the practitioners of "traditional" medicine outnumber registered doctors.

The World Wildlife Fund estimates that plants have contributed more than 7000 compounds produced by the pharmaceutical industry in industrialized nations.[186] The folk medicine and indigenous medicine developed by different native cultures across the world is based mainly on herbs. Indeed, more than 200 of the herbal drugs listed in the *U.S. Pharmacopoeia* were used by Indians in Mexico, the West Indies, and Central and South America.[187] The popular traditional medical systems — Chinese medicine, Ayurveda, and Unani — have suggested cancer cures involving herbal medicine. Dr. J. Hartwell and other researchers compiled a long list of plants that have historically been used around the planet for treating cancer.[188] During the last two decades, cancer patients, driven by the well-founded desire for improved quality of life coupled with the freedom to choose, have moved toward natural herbal therapies. Studies indicate that up to 60% of cancer patients use unconventional medical therapies, and nearly half of them do so without informing their physicians.[189] Some of the herbs have also been used for purposes other than cancer cure, especially for alleviating symptoms arising from mainstream cancer therapies.

Despite the popularity of some of the herbal cancer therapies known around the world, there is no absolute evidence that any herbal supplement will actually cure cancer. Unlike modern cancer medicines, which are thoroughly studied and developed toward a chosen goal, herbal medicines have been adapted from empirical evidence based on their long record of use through primitive trial-and-error methods. Because of the absence of controlled studies and lack of technical monitoring of extraction and preparation processes, the general scientific acceptance of herbal medications among the established medical community remains very low. The meager number of human studies that did exist were mostly undertaken retrospectively, leaving scope for potential errors. Unconventional cancer therapies like essiac and laetrile caused a war of attrition between people on both sides of the issue of unconventional medical treatments. Many of these battles were rather political, and some grew to the status of a national movement. The important step to bring an upward swing in the popularity for herbal supplements would be initiating double-blind, placebo-controlled clinical studies on human subjects. Unfortunately, no pharmaceutical giants will ever be inclined to invest millions of dollars to conduct such studies on well-known substances that are already being used for the same purpose. The onus is now in the hands of the government agencies, like the U.S. NCI and its counterparts in other countries, to take up the clinical trials on important natural herbal therapies for the benefit of the general public.

Even though herbal supplements have yet to muster strong support for their positive role in cancer cure, they have already established a niche as cancer chemoprevention agents. Cancer chemoprevention is a relatively new but important medical science. Oxidation within biological systems plays a crucial role in carcinogenesis. Chemoprevention by naturally occurring antioxidant herbal supplements appears to be a practical approach to fight cancer. There is overwhelming scientific evidence from epidemiological studies, *in vitro* studies, and *in vivo* studies on animals that a large number of herbal supplements have a profound influence on cancer prevention before cancer onset and following cancer treatment. Preliminary studies demonstrated that antioxidant supplements could work as complementary treatments to reduce the side effects of chemotherapy and radiotherapy and also to enhance the efficacy of some cancer drugs.[190,191] However, the combination of antioxidants and chemotherapeutic drugs may not be completely immune to potential problems for patients due to possible interference with the biological functioning of the primary drug. Antioxidants like soy isoflavones exhibit paradoxical effects with respect to breast cancer. Such ambiguities can be kept in check by considering the dosage and timing of administration. The medical community should therefore consider the research generated during the last few decades on the benefits of natural antioxidants and work with the patients to judiciously choose an effective agent at the proper dosage.

The current pace of technological advancement has given us a great deal of room to maneuver in identifying anticancer herbs and active ingredients. Scientists all over the world are focusing on locating biological response modifiers that inhibit cancer growth by immune system stimulation and by modulating the activity of hormones, enzymes, and biological factors. Manufacturers of herbal products, for their part, are concentrating more on standardization of the herbs with respect to active principle or biological activity for greater consistency in activity profile. The standardized herbals are now seen as a compromise between crude plant medicine and phytopharmaceutical drugs. Governmental and nonprofit organizations are making an effort to improve the standards and legal setup to ensure the quality and safety of the products. Overall, scientists see a greater advantage by integrating the knowledge of traditional herbal medicine with mainstream medical research for the advancement of effective cancer therapies.

Estimates suggest that a mere 1% of plant species have been screened for their potential medicinal value. Unfortunately, more than 50 million acres of tropical forest are lost every year due to human activity. Unless we expedite the bioprospecting efforts, thousands of untapped species and potential molecules will be lost forever. Finding a new cancer cure from nature is like finding a needle in a haystack. Consequently, scientists are embarking upon a new strategy involving the knowledge of indigenous people, which should have a better potential than randomly collected samples. For example, mayapple is an example where two cancer drugs, etoposide and teniposide, were developed based on the traditional knowledge of indigenous people. Mayapple was used by the Penobscot Indians of Maine to treat certain tumors like venereal warts. Vincristine (Oncovine®) and vinblastine (Velban®) are cancer drugs based on folk knowledge available on Madagascar periwinkle, although for conditions other than cancer. Similarly, Pacific yew (*Taxus brevifolia*), widely used by native tribes in western North America, became the most popular anticancer drug of the last century, namely Paclitaxel®. Ethanobotany is a rapidly expanding discipline in both academia and industrial arenas, particularly for exploring indigenous knowledge on the medicinal value of plants.

Cancer is an emerging health concern in many countries, and its incidence in some Western countries has reached epidemic proportions. With many cancers showing resistance to existing chemotherapeutic drugs, there is a clear need to develop new cancer medications. Many researchers believe that nature is still the best source to look for new therapies for cancer cure and prevention. After all, nature is the ultimate chemist.

REFERENCES

1. U.S. cancer statistics, 1999 incidence.
2. World Cancer Report, World Health Organization, Geneva, April 3, 2003.
3. Swami, S.S.T., *The Ayurveda Encyclopedia: Natural Secrets to Healing, Prevention and Longevity*, Sri Satguru Publications, New Delhi, India, 1998, p. 39.
4. Prasad, G.C., *Ayurvedic Drugs in the Management of Cancer*, Council for Research in Ayurveda and Sidda, Janakpuri, New Delhi, 1999, p. 5.
5. Swami, S.S.T., *The Ayurveda Encyclopedia, Natural Secrets to Healing, Prevention and Longevity*, Sri Satguru Publications, New Delhi, India, 1998, p. 9.
6. Swami, S.S.T., *The Ayurveda Encyclopedia, Natural Secrets to Healing, Prevention and Longevity*, Sri Satguru Publications, New Delhi, India, 1998, pp. 494–501.
7. Swami, S.S.T., *The Ayurveda Encyclopedia, Natural Secrets to Healing, Prevention and Longevity*, Sri Satguru Publications, New Delhi, India, 1998, pp. 169–247.
8. Prasad, G.C., *Ayurvedic Drugs in the Management of Cancer*, Council for Research in Ayurveda and Sidda, Janakpuri, New Delhi, 1999, pp. 78–93.
9. Prasad, G.C., *Ayurvedic Drugs in the Management of Cancer*, Council for Research in Ayurveda and Sidda, Janakpuri, New Delhi, 1999.
10. Prasad, G.C., *Ayurvedic Drugs in the Management of Cancer*, Council for Research in Ayurveda and Sidda, Janakpuri, New Delhi, 1999, pp. 103–122.
11. Hartwell, J.L., *Lloydia*, 30, 379–436, 1967; Suffness, M., Pettit, G.R., Fornsworth, N.R., and Wall, M.E., *HerbalGram*, 26, 38, 1992.
12. Steinmetz, K.A. and Potter, J.D., Vegetables, fruits and cancer prevention: a review, *J. Am. Diet. Assoc.*, 96, 1027–1039, 1996.
13. Cancer Net, Complementary/Alternative Medicine Information; available on-line at http://cancernet.nci.nih.gov/cam/mistletoe.html.
14. U.S. Congress, Office of Technology Assessment, *Unconventional Cancer Treatments*, U.S. Government Printing Office, Washington, DC, 1990, pp. 81–86.
15. Monograph 9444 Visc., Homoeopathic Pharmacopoeia Convention of the U.S., Washington, DC, 2002.
16. Kuttan, G., Vasudevan, D.M., and Kuttan, R., *J. Ethnopharmacol.*, 29 (1), 35–41, 1990; Kuttan, G., Vasudevan, D.M., and Kuttan, R., *Cancer Lett.*, 41 (3), 307–314, 1988.
17. Valentiner, U., Pfuller, U., Baum, C., and Schumacher, U., Toxicology, 171(2–3), 187–199, 2002.

18. Braun, J.M., Ko, H.L., Schierholz, J.M., Weir, D., Blackwell, C.C., and Beuth, J., *Cancer Lett.*, 170 (1), 25–31, 2001; Kuttan, G., Menon, L.G., Antony, S., and Kuttan, R., *Anticancer Drugs*, 8 (Suppl. 1), S15–16, 1997.
19. Beuth, J., Ko, H.L., Tunggal, L. et al., *In Vivo*, 8 (6), 989–992, 1994; Rentea, R., Lyon, E., and Hunter, R., *Lab. Invest.*, 44 (1), 43–48, 1981.
20. Kleijnen, J. and Knipschild, P., *Phytomedicine*, 1, 255–260, 1994.
21. Hall, A.H., Spoerke, D.G., Rumack, B.H. et al., *Ann. Emergency Med.*, 15 (11), 3120–1323, 1986.
22. Hoxsey, H.M., *You Do Not Have To Die*, Milestone Books, New York, 1956, pp. 62–64.
23. *The Medical Messiahs: a Social History of Health Quackery in Twentieth-Century America*, Young J. Harvey, Princeton University Press, Princeton, NJ, p. 387, 1967.
24. Comment on court opinion, *JAMA*, 145, 252–253, 1951; *The Medical Messiahs: a Social History of Health Quackery in Twentieth-Century America*, Young J. Harvey, Princeton University Press, Princeton, NJ, p. 375, 1967.
25. Cough medicine for cancer? *JAMA*, 155, 667–668, 1954.
26. Larrack, G.P., Public warning against Hoxsey cancer treatment, *Consumer Rep.*, 21, 303, 1956.
27. Milliard, F.R., *Medical Surgical Rep. (Philadelphia)*, 75, 420–422, 1986.
28. Farnes, P., Barker, B.E., Brownhill, L.E. et al., *Lancet*, 2, 1100–1101, 1964; Barker, B.E., Farnes, P., and LaMarche, P.H., *Pediatrics*, 38, 490–493, 1966; Downing, H.J., Kemp, G.C., and Denborough, M.A., *Nature*, 217, 654–655, 1968.
29. Dombradi, C.A. and Foldeak, S., *Tumori*, 52, 173–175, 1966; Koshimizu, K., Ohigashi, H., Tokuda, H., Kondo, A., and Yamagachi, K., *Cancer Lett.*, 39, 247–257, 1988.
30. Morita, K., Kada, T., and Namiki, M., *Mutat. Res.*, 129, 25–31, 1984.
31. Hoshi, A., Ikekawa, T., and Ikeda, Y., *Gann*, 67, 321–325, 1976.
32. Kupchan, S.M. and Karim, A., *Lloydia*, 39, 223–224, 1976.
33. Viehoever, A. and Mack, H., Bio-chemistry of amygdalin, *Am. J. Pharm.*, 107, 397–450, 1935.
34. National Cancer Institute, Web site for complementary and alternative medicine; available on-line at http://www.cancer.gov/cancerinfo.
35. Laetrile at Sloan-kettering: acase study, in Moss, R.W., *The Cancer Industry: The Classic Expose on Cancer Establishment*, First Equinox Press, New York, 1996, pp. 153–183.
36. Wodinsky, I. and Swiniarski, J.K., *Cancer Chemother. Rep.*, 59 (5), 939–950, 1975; Laster, Jr., W.R. and Schabel, Jr., F.M., *Cancer Chemother. Rep.*, 59 (5), 951–965, 1975; Hill, G.J. et al., *Cancer Res.*, 36, 2102–2107, 1976.
37. Laetrile at Sloan-Kettering; a case study, in Moss R.W., *The Cancer Industry: The Classic Expose on Cancer Establishment*, First Equinox Press, New York, 1996, pp. 131–152.
38. Lewis, J.P., *West. J. Med.*, 127 (1), 55–62, 1977.
39. Moertel, C.G., Fleming, T.R., Rubin, J. et al., *New England J. Med.*, 306 (4), 201–206, 1982; Moertel, C.G., Ames, M.M., Kovach, J.S. et al., *J. Am. Medical Assoc.*, 245 (6), 591–594, 1981.
40. Cancer Commission of the California Medical Association, The treatment of cancer with "laetriles," *Calif. Med.*, 78 (4), 320–326, 1953; Navarro, M.D., Five years experience with laetrile therapy in advanced cancer, *Acta Unio Internationalis Contra Cancrum*, 15 (Suppl. 1), 209–221, 1959.
41. Dorr, R.T. and Paxinos, J., The current status of laetrile, *Ann. Intern. Med.*, 89 (3), 389–397, 1978;
42. Curt, G.A., *Principles Practice Oncol. Updates*, 4 (12), 1–10, 1990; Curran, W.J., *New England J. Med.*, 302 (11), 619–621, 1980.
43. Ellison, N.M., Byar, D.P., and Newell, G.R., Special report on laetrile: the NCI laetrile review — results of the National Cancer Institute's retrospective laetrile analysis, *New England J. Med.*, 299 (10), 549–552, 1978.
44. Sun M., *Science*, 212 (4496), 758–759, 1981.
45. Lewis, W.H. and Elvin-Lewis, M.P.F., *Medical Botany: Plants Affecting Man's Health*, John Wiley and Sons, New York, 1977; Ames, M.M., Moyer, T.P., Kovach, J.S. et al., *Cancer Chemother. Pharmacol.*, 6 (1), 51–57, 1981.
46. Young, J.H., Laetrile in historical perspective, in Merkle, G.E., Petersen, J.C., editors. *Politics, Science, and Cancer: The Laetrile Phenomenon*, Westview Press, Boulder, CO, 1980; American Cancer Society. Laetrile Background Information. American Cancer Society, New York, 1977.
47. Gary, G. L., *The Calling of An Angel: Nature's Care for Cancer*, Silent Walker Publishing, Los Angeles, 1988; Thomas, R., *The Essiac Report: Canada's Remarkable Unknown Cancer Remedy*, Alternative Treatment Information Network, Los Angeles, 1993.

48. Thomas, R., *The Essiac Report: Canada's Remarkable Unknown Cancer Remedy*, Alternative Treatment Information Network, Los Angeles, 1993, p. 37.

49. Hutchinson, D.J., Memorial Sloan-Kettering Cancer Center, Rye, NY, personal communication, Sep. 1988 and Mar. 1989; in Office of Technology Assessment, Unconventional Cancer Treatments, 73–74.

50. Henderson, I.W.D., director, Bureau of Human Prescription Drugs, Health Protection Branch, Health and Welfare Canada, Vanier, ON, letter to J.W. Meakin, executive director, Ontario Cancer Treatment and Research Foundation, Toronto, ON, Nov. 19, 1982; in Office of Technology Assessment, Unconventional Cancer Treatments, 74.

51. Greenberg, N.H., Developmental Therapeutics Program, Division of Cancer Treatment, National Cancer Institute, National Institutes of Health, U.S. Department of Health and Human Services, memo to J.A.R. Mead, acting associate director, Developmental Therapeutic Program, Division of Cancer Treatment, National Cancer Institute, Nov. 1, 1983.

52. Walters, R., *Options: The Alternative Cancer Therapy Book*, Avery Publishing Group, Garden City, NY, 1993, p. 113.

53. Jaggi, O.P., *Cancer, Causes, Prevention and Treatment*, Orient Paperbacks, Delhi, 1990, p. 20.

54. Swanson, C.A., Fruits, vegetables and cancer risk: the role of phytochemicals, in *Phytochemicals, a New Paradigm*, Bidlack, W.R., Omaye, S.T. et al., Eds., Technomic Pub., Basel, 1998, pp. 1–12.

55. Noroozi, M., Anderson, W.J. et al., *Am. J. Clin. Nutr.*, 67, 1210–1218, 1998.

56. Blaylock, R.L., New Developments in phytoprevention and treatment of cancer, *J. Am. Neutr. Accoc.*, 2, 19–29, 1999.

57. Cao, Y., and Cao, R., Nature, 398, 381, 1999; Liu, G.A. and Zheng, R.L., Pharmazie, 57(12), 852–854, 2002; Mikstacka, R., Gnojkowski, J., and Baer-Dubowska, W., *Acta Biochim. Pol.*, 49(4) 917–925, 2002; See, D.M., Broumand, N., Sahl, L., and Tilles, J.G., *Immunopharmacology*, 35, 229–235, 1997; Weisburger, J.H., Hara, Y., Dolan, L., Luo, F.Q., Pittman, B., and Zang, E., Mutat. Res. 371, 57–63, 1999; Lin, J.K., Chen, Y.C., Huang, Y.T., and Lin-Shiau, S.Y., *J. Cell. Biochem. Suppl.*, 28/29, 39–48, 1997; Bachrach, U. and Wang, Y.C., Amino Acids. 22(1), 1–13, 2002.

58. Ferriola, P.C., Cody, V., and Middleton, E. Jr., *Biochem. Pharmacol.*, 38, 1617–1624, 1989.

59. Agullo G., Gamet-Payrastre, L., Marteni, S., Viala, C., Remesy, C., Chap, H., and Payrastre, B. *Biochem. Pharmacol.*, 53, 1649–1657, 1997.

60. Bakhru, H.K., *Herbs That Heal: Natural Remedies for Good Health*, Orient Paperbacks, New Delhi, 1990,1992.

61. Bakhru, H.K., *Herbs That Heal: Natural Remedies for Good Health*, Orient Paperbacks, New Delhi, 1990, pp. 93–96.

62. Milner, J.A., *J. Nutr.*, 131 (Suppl. 3), 1027S–1031S, 2001.

63. Hsing, A.W., Chokkalingam, A.P., Gao, Y.T., Madigan, M.P., Deng, J., Gridley, G., and Fraumeni, Jr., J.F., *J. Natl. Cancer Inst.*, 94 (21), 1648–1651, 2002.

64. Dorant, E., van den Brandt, P.A., Goldbohm, R.A. et al., *Br. J. Cancer*, 67, 424–429, 1993.

65. Fleischauer, A.T., Poole, C., and Arab, L., *Am. J. Clin. Nutr.*, 72, 1047–1052, 2000.

66. Fleischauer, A.T. and Arab, L., *J. Nutr.*, 131(35), 10325–10405, 2001.

67. Riggs, D.R., DeHaven, J.I., and Lamm, D.L., *Cancer*, 79, 1987–1994, 1997.

68. Thomson, M. and Ali, M., *Curr. Cancer Drug Targets*, 3 (1), 67–81, 2003.

69. Borek, C., *J. Nutr.*, 131 (3s), 1010S–1015S, 2001.

70. Kyo, E., Uda, N., Kasuga, S., and Itakura, Y., *J. Nutr.*, 131 (3s), 1075S–1079S, 2001.

71. Koscielny, J., Klussendorf, D., Latza, R. et al., *Atherosclerosis*, 144, 237–249, 1999.

72. Warshafsky, S., Kamer, R., and Sivak, S., *Ann. Int. Med.*, 119, 599–605, 1993.

73. Hughes, B.G. and Lawson, L.D., *Phytother. Res.*, 5, 154–158, 1991.

74. Bentley, R.M. and Campbell, I.M., in *The Chemistry of Quinoid Compounds*, Part 2, Patai, S., Ed., Wiley, New York, 1974, pp. 683–736.

75. Littaru, G.P., Ho, L., and Folkers, K., *Int. J. Vitam. Nitr. Res.*, 42, 413–434, 1972.

76. Yamagami, T., Shibata, N., and Folkers, K., *Res. Commun. Chem. Pathoal. Pharmacol.*, 14, 721–727, 1976.

77. Combs, A.B., Acosta, D., and Folkers, K., IRCS, *Med. Sci. Biochem.*, 4, 403, 1976.

78. Hodges, S., Hertz, N., Lockwood, K., and Lister, R., *Biofactors*, 9 (2–4), 365–370, 1999.

79. Lockwood, K., Moesgaard, S., and Folkers, K., *Biochem. Biophys. Res. Commun.*, 199 (3), 1504–1508, 1994; Lockwood, K., Moesgaard, S., Hanioka, T., and Folkers, K., Mol. Aspects Med., 15 (Suppl.), 5231–5240, 1994.

80. Portakal, O., Ozkaya, O., Erden Inal, M., Bozan, B., Kosan, M., and Sayek, I., *Clin. Biochem.*, 33 (4), 279–284, 2000.
81. Folkers, K. and Wolaniuk, A., *Drugs Exp. Clin. Res.*, 11 (8), 539–545, 1985.
82. Bergner, P., *The Healing Power of Ginseng and the Tonic Herbs*, Prima Publishing, Rocklin, CA, 1996; Moramaco, J., *The Complete Ginseng Hand Book: a Practical Guide for Energy, Health and Longevity*, Contemporary Books, IL, 1998, pp. 1–40.
83. Bergner, P., *The Healing Power of Ginseng and the Tonic Herbs*, Prima Publishing, Rocklin, CA, 1996, p. 104.
84. Boik, J., *Cancer and Natural Medicine: a Textbook of Basic Science and Clinical Research*, Oregon Medical Press, Princeton, MN, 1996, pp. 180–183.
85. Rhee, R.H. et al., *Planta Medica*, 57, 125–128, 1991; Yun, T.K. and Lee, Y.S., *Korean J. Ginseng Sci.*, 18, 89–94, 1994; Yun, T.K. and Lee, Y.S., *Korean J. Ginseng Sci.*, 18, 160–164, 1994.
86. Yun, T-K. and Choi, S-Y., *Cancer Epidemiol. Biomarkers Prev.*, 4, 401–408, 1995; Yun, T-K. and Choi, S-Y., *Int. J. Epidemiol.*, 19, 871–876, 1990.
87. Wang, B., Cui, J., and Lui, A., The effect of ginseng on immune responses, in *Advances in Chinese Medicinal Materials Research*, Chang, H.M., Yeung, H.W. and Koo, A., Eds., World Scientific, Singapore, 1985; Yun, Y.S. et al., Effect of red ginseng on natural killer cell activity in mice with lung adenoma induced by urethan and benzo(a)pyrene, *Cancer Detection Prev.*, (Suppl. 1), 301–309, 1987; Kim, J.Y., Germolec, D.R., and Luster, M.I., *Immunopharmacol. Immunotoxicol.*, 12 (2), 257–276, 1990.
88. Hasegawa, H. et al., *Planta Medica*, 61 (5), 409–413, 1995; Tong, C.N., Matsuda, H., and Kubo, M., *J. Pharmaceut. Soc. Jpn.*, 112 (11), 856–865, 1992.
89. Kim, S.W., Kwon, H.Y., Chi, D.W., Shim, J.H., Park, J.D., Lee, Y.H., Pyo, S., and Rhee, D.K., *Biochem. Pharmacol.*, 65 (1), 75–82, 2003.
90. Wang, H., Actor, J.K., Indrigo, J., Olsen, M., and Dasgupta, A., *Clin. Chim. Acta*, 327 (1–2), 123–128, 2003; Vogler, B.K., Pittler, M.H., and Ernst, E., *Eur. J. Clin. Pharmacol.*, 55, 567–575, 1999.
91. Siegel, R.K., Ginseng abuse syndrome, *JAMA*, 241, 1614–1615, 1979.
92. Sueoka, N., Suganuma, M., Sueoka, E., Okabe, S., Matsuyama, S., Imai, K., Nakachi, K., and Fujiki, H., *Ann. N.Y. Acad. Sci.*, 928, 274–280, 2001.
93. Ji, B.T., Chow, W.H., Hsing, A.W., McLaughlin, J.K., Dai, Q., Gao, Y.T., Blot, W.J., and Fraumeni, Jr., J.F., *Int. J. Cancer*, 70 (3), 255–258, 1997; Yu, G.P., Hsieh, C.C., Wang, L.Y., Yu, S.Z., Li, X.L., and Jin, T.H., *Cancer Causes Control*, 6 (6), 532–538, 1995; Zhang, M., Binns, C.W., and Lee, A.H., *Cancer Epidemiol. Biomarkers Prev.*, 11 (8), 713–718, 2002.
94. Kinlen, L.J., Willows, A.N., Goldblatt, P., and Yudkin, J., *Br. J. Cancer*, 58: 397–401, 1988; Heilburn, L.K., Nomura, A., and Stemmermann, G.N., *Br. J. Cancer*, 54, 677–683, 1986.
95. Liao, S., Umekita, Y., Guo, J., Kokontis, J.M., and Hiipakka, R.A., *Cancer Lett.*, 96, 239–243, 1995.
96. Gupta, S., Ahmad, N., Mohan, R.R., Husain, M.M., and Mukhtar, H., *Cancer Res.*, 59, 2115–2120, 1999.
97. Fujiki, H., *J. Cancer Res. Clinical Oncol.*, 125 (11), 589–597, 1999; Fujiki, H., Suganuma, M., Okabe, S., Sueoka, N., Komori, A., Sueoka, E., Kozu, T., Tada, Y., Suga, K., Imai, K., and Nakachi, K., *Mutat. Res.*, 402 (1–2), 307–310, 1998.
98. Afaq, F., Adhami, V.M., Ahmad, N., and Mukhtar, H., *Oncogene*, 22 (7), 1035–1044, 2003; Conney, A.H., Wang, Z.Y., Huang, M.T., Ho, C.T., and Yang, C.S., *Prev. Med.*, 21 (3), 361–369, 1992; Shim, J.S., Kang, M.H., Kim, Y.H., Roh, J.K., Roberts, C., and Lee, I.P., *Cancer Epidemiol. Biomarkers Prev.*, 4 (4), 387–391, 1995.
99. Zhu, M., Gong, Y., and Yang, Z., *Zhonghua Yu Fang Yi Xue Za Zhi*, 32 (5), 270–274, 1998; Zhu, M., Gong, Y., Yang, Z., Ge, G., Han, C., and Chen, J., *Nutr. Cancer*, 35 (1), 64–72, 1999.
100. Fujiki, H., Yoshizawa, S., Horiuchi, T., Suganuma, M., Yatsunami, J., Nishiwaki, S., Okabe, S., Nishiwaki-Matsushima, R., Okuda, T., and Sugimura, T., *Prev. Med.*, 21 (4), 503–509, 1992; Huang, M.T., Ho, C.T., Wang, Z.Y., Ferraro, T., Finnegan-Olive, T., Lou, Y.R., Mitchell, J.M., Laskin, J.D., Newmark, H., Yang, C.S. et al., *Carcinogenesis*, 13 (6), 947–954, 1992; Mukhtar, H., Katiyar, S.K., and Agarwal, R., *J. Invest. Dermatol.*, 102 (1), 3–7, 1994.
101. Takada, M., Nakamura, Y., Koizumi, T., Toyama, H., Kamigaki, T., Suzuki, Y., Takeyama, Y., and Kuroda, Y., *Pancreas*, 25 (1), 45–48, 2002; Cao, Y. and Cao, R., *Nature*, 398, 381.
102. Jung, Y.D. and Ellis, L.M., *Int. J. Exp. Pathol.*, 82 (6), 309–316, 2001.
103. Sadzuka, Y., Sugiyama, T., and Hirota, S., *Clin. Cancer Res.*, 4 (1), 153–156, 1998.
104. Jodoin, J., Demeule, M., and Beliveau, R., *Biochim. Biophys. Acta*, 1542 (1–3), 149–159, 2002.

105. Taniguchi, S., Fujiki, H., Kobayashi, H., Go, H., Miyado, K., Sadano, H., and Shimokawa, R., *Cancer Lett.*, 65 (1), 51–54, 1992.
106. Wang, Y.C. and Bachrach, U., *Amino Acids*, 22 (2), 131–143, 2002.
107. Hsu, S.D., Singh, B.B., Lewis, J.B., Borke, J.L., Dickinson, D.P., Drake, L., Caughman, G.B., and Schuster, G.S., *Gen. Dent.*, 50 (2), 140–146, 2002.
108. Tsubono, Y., Nishino, Y., Komatsu, S., Hsieh, C.C., Kanemura, S., Tsuji, I., Nakatsuka, H., Fukao, A., Satoh, H., and Hisamichi, S., *New England J. Med.*, 344 (9), 632–636, 2001; Nagano, J., Kono, S., Preston, D.L., and Mabuchi, K., *Cancer Causes Control*, 12 (6), 501–508, 2001.
109. Kopp, P., *Eur. J. Endocrinol.*, 138, 619–620, 1998; Constant, J., *Coron. Artery Dis.*, 8 (10), 645–649, 1997.
110. Lin, J-K. and Tsai, S-H., *Proc. Natl. Sci. Coun. ROC(B)*, 23 (3), 99–106, 1999.
111. Bhat, K.P.L., Kosmeder, II, J.W., and Pezzuto, J.M., *Antioxid. Redox. Signal*, 3 (6), 1041–1064, 2001.
112. Savouret, J.F. and Quesne, M., *Biomed. Pharmacother.*, 56 (2), 84–87, 2002.
113. Jang, M., Cai, L., Udeani, G.O., Slowing, K.V., Thomas, C.F., Beecher, C.W., Fong, H.H., Farnsworth, N.R., Kinghorn, A.D., Mehta, R.G., Moon, R.C., and Pezzuto, J.M., *Science*, 275 (5297), 218–220, 1997.
114. Jang, M. and Pezzuto, J.M., *Drugs Exp. Clin. Res.*, 25 (2–3); 65–77, 1999.
115. Chun, Y.J., Kim, M.Y., and Guengerich, F.P., *Biochem. Biophys. Res. Commun.*, 262 (1), 20–24, 1999.
116. Mollerup, S., Ovrebo, S., and Haugen, A., *Int. J. Cancer*, 92 (1), 18–25, 2001.
117. Whitehead, T.P., Robinson, D., Allaway, S., Syms, J., and Hale, A., *Clin. Chem.*, 41 (1), 32–35, 1995.
118. Kimura, Y. and Okuda, H., *J. Nutr.*, 131 (6), 1844–1849, 2001.
119. Manna, S.K., Mukhopadhyay, A., and Aggarwal, B.B., *J. Immunol.*, 164 (12), 6509–6519, 2000.
120. Niles, R.M., McFarland, M., Weimer, M.B., Redkar, A., Fu, Y.M., and Meadows, G.G., *Cancer Lett.*, 190 (2), 157–163, 2003; Ragione, F.D., Cucciolla, V., Borriello, A., Pietra, V.D., Racioppi, L., Soldati, G., Manna, C., Galletti, P., and Zappia V., *Biochem. Biophys. Res. Commun.*, 250 (1), 53–58, 1998.
121. Schneider, Y., Duranton, B., Gosse, F., Schleiffer, R., Seiler, N., and Raul, F., *Nutr. Cancer*, 39 (1), 102–107, 2001.
122. Fontecave, M., Lepoivre, M., Elleingand, E., Gerez, C., Guittet, O., *FEBS Lett.*, 421 (3), 277–279, 1998.
123. Gautam, S.C., Xu, Y.X., Dumaguin, M., Janakiraman, N., and Chapman, R.A., *Bone Marrow Transplant*, 25 (6), 639–645, 2000.
124. Gehm, B.D., McAndrews, J.M., Chien, P.Y., and Jameson, J.L., *Proc. Natl. Acad. Sci. USA*, 94 (25), 14138–14143, 1997.
125. Press release, University of Leicester, Leicester, U.K., Nov. 2002.
126. Kuttan, R. et al., *Cancer Lett.*, 29, 197–202, 1985; Nagabhushan, N. and Bhide, S.V., *J. Am. Coll. Nutr.*, 11, 192–198, 1992.
127. Ruby, A.J., Kuttan, G., Babu, K.D., Rajasekharan, K.N., and Kuttan, R., *Cancer Lett.*, 94 (1), 79–83, 1995.
128. Huang, M.T., Newmark, H.L., and Frenkel, K., *J. Cell. Biochem. Suppl.*, 27, 26–34, 1997.
129. Thapliyal, R., Naresh, K.N., Rao, K.V., and Maru, G.B., *Toxicol. Lett.*, 139 (1), 45–54, 2003; Singh, S.V., Hu, X., Srivastava, S.K., Singh, M., Xia, H., Orchard, J.L., and Zaren, H.A., *Carcinogenesis*, 19 (8), 1357–1360, 1998; Azuine, M.A. and Bhide, S.V., *J. Ethnopharmacol.*, 44 (3), 211–217, 1994; Singletary, K., MacDonald, C., Wallig, M., and Fisher, C., *Cancer Lett.*, 103 (2), 137–141, 1996; Rao, C.V., Rivenson, A., Simi, B., and Reddy, B.S., *Cancer Res.*, 55 (2), 259–266, 1995.
130. Dorai, T., Cao, Y.C., Dorai, B., Buttyan, R., and Katz, A.E., *Prostate*, 47 (4), 293–303, 2001.
131. Dalton, L., How curry combats cancer, C&E News, 81, 8, 2003.
132. Kuttan, R., Sudheeran, P.C., and Joseph, C.D., *Tumori*, 73 (1), 29–31, 1987.
133. Krishnaswamy, K., Goud, V.K., Sesikeran, B., Mukundan, M.A., and Krishna, T.P., *Nutr. Cancer*, 30 (2), 163–166, 1998.
134. Chauhan, D.P., *Curr. Pharm. Des.*, 8 (19), 1695–1706, 2002.
135. Goel, A., Boland, C.R., and Chauhan, D.P., *Cancer Lett.*, 172 (2), 111–118, 2001; Plummer, S.M., Holloway, K.A., Manson, M.M., Munks, R.J., Kaptein, A., Farrow, S., Howells, L., *Oncogene*, 18 (44), 6013–6020, 1999.

136. Azuine, M.A. and Bhide, S.V., *Nutr. Cancer*, 17 (1), 77–83, 1992.
137. Huang, M.T., Newmark, H.L., and Frenkel, K., *J. Cell. Biochem. Suppl.*, 27, 26–34, 1997.
138. Venkatesan, N., *Br. J. Pharm.*, 124, 425–427, 1998.
139. Pathak, S.K., Sharma, R.A., and Mellon, J.K., *Int. J. Oncol.*, 22 (1), 5–13, 2003; Sarkar, F.H. and Li, Y., *Cancer Metastasis Rev.*, 21 (3–4), 265–280, 2002; Ganry, O., *Eur. J. Cancer Prev.*, 11 (6), 519–522, 2002.
140. Birt, D.F., Hendrich, S., and Wang, W., *Pharmacol. Ther.*, 90 (2–3), 157–177, 2001.
141. Toyomura, K. and Kono, S., *Asian Pac. J. Cancer Prev.*, 3 (2), 125–132, 2002; Sarkar, F.H. and Li, Y., *Cancer Metastasis Rev.*, 21 (3–4), 265–280, 2002; Ren, M.Q., Kuhn, G., Wegner, J., and Chen, J., *Eur. J. Nutr.*, 40 (4), 135–146, 2001.
142. Yu, L., Blackburn, G.L., and Zhou, J.R., *J. Nutr.*, 133 (2), 389–392, 2003.
143. This, P., De La Rochefordiere, A., Clough, K., Fourquet, A., and Magdelenat, H., *Endocr. Relat. Cancer*, 8 (2), 129–134, 2001.
144. U.S Federal Register, 64, 57700–57733, 1999.
145. Nagata, C., Takatsuka, N., Kawakami, N., and Shimizu, H., *Br. J. Cancer*, 87 (1), 31–36, 2002.
146. Hsieh, C.Y., Santell, R.C., Haslam, S.Z., and Helferich, W.G., *Cancer Res.*, 58, 3833–3838, 1998; Allred, C.D., Allred, K.F., Ju, Y.H., Virant, S.M., and Helferich, W.G., *Cancer Res.*, 61 (13), 5045–5050, 2001.
147. Wu, A.H., Wan, P., Hankin, J., Tseng, C.C., Yu, M.C., and Pike, M.C., *Carcinogenesis*, 23 (9), 1491–1496, 2002.
148. Adlercreutz, H., *J. Steroid Biochem. Mol. Biol.*, 83 (1–5), 113–118, 2002.
149. Moss, R.W., *Cancer Therapy: the Independent Consumer's Guide to Non-Toxic Treatment and Prevention*, Equinox Press, New York, 1992, pp. 126–127.
150. Lee, C.K., Han, S.S., Shin, Y.K., Chung, M.H., Park, Y.I., Lee, S.K., and Kim, Y.S., *Int. J. Immunopharmacol.*, 21 (5), 303–310, 1999; Sato, Y., Ohta, S., Sakurai, N., and Shinoda, M., *Yakugaku Zasshi*, 109 (2), 113–118, 1989.
151. Olsen, D.L., Raub, Jr., W., Bradley, C., Johnson, M., Macias, J.L., Love, V., and Markoe, A., *Oncol. Nurs. Forum*, 28 (3), 543–547, 2001.
152. Heggie, S., Bryant, G.P., Tripcony, L., Keller, J., Rose, P., Glendenning, M., and Heath, J., *Cancer Nurs.*, 25 (6), 442–451, 2002; Williams, M.S., Burk, M., Loprinzi, C.L., Hill, M., Schomberg, P.J., Nearhood, K., O'Fallon, J.R., Laurie, J.A., Shanahan, T.G., Moore, R.L., Urias, R.E., Kuske, R.R., Engel, R.E., and Eggleston, W.D., *Int. J. Radiat. Oncol. Biol. Phys.*, 36 (2), 345–349, 1996.
153. Reynolds, T. and Dweck, A.C., *J. Ethnopharmacol.*, 68 (1–3), 3–37, 1999; Corsi, M.M., Bertelli, A.A., Gaja, G., Fulgenzi, A., and Ferrero, M.E., *Int. J. Tissue React.*, 20 (4), 115–118, 1998.
154. Singh, R.P., Dhanalakshmi, S., and Rao, A.R., *Phytomedicine*, 7 (3), 209–219, 2000.
155. Kim, H.S., Kacew, S., and Lee, B.M., *Carcinogenesis*, 20 (8), 1637–1640, 1999.
156. Gribel, N.V. and Pashinskii, V.G., *Vopr. Onkol.*, 32 (12), 38–40, 1986.
157. Kupchan, S.M. and Karim, A., *Lloydia*, 39 (4), 223–224, 1976.
158. Pecere, T., Gazzola, M.V., Mucignat, C., Parolin, C., Vecchia, F.D., Cavaggioni, A., Basso, G., Diaspro, A., Salvato, B., Carli, M., and Palu, G., *Cancer Res.*, 60 (11), 2800–2804, 2000.
159. Kuo, P.L., Lin, T.C., and Lin, C.C., *Life Sci.*, 71 (16), 1879–1892, 2002.
160. Wasserman, L., Avigad, S., Beery, E., Nordenberg, J., and Fenig, E., *Am. J. Dermatopathol.*, 24 (1), 17–22, 2002.
161. Lee, H.Z., Hsu, S.L., Liu, M.C., and Wu, C.H., *Eur. J. Pharmacol.*, 431 (3), 287–295, 2001.
162. Zhang, L. and Tizard, I.R., *Immunopharmacology*, 35 (2), 119–128, 1996; Ralamboranto, L. et al., *Archives de l Institut Pasteur de Madagascar*, 50 (1), 227–256, 1982; Harris, C. et al., *Biotherapy*, 3 (4), 207–213, 1991.
163. Davis, R.H., Stewart, G.J., and Bregman, P.J., *J. Am. Podiatric Medical Assoc.*, 82 (3), 140–148, 1992.
164. Lissoni, P., Giani, L., Zerbini, S., Trabattoni, P., and Rovelli, F., *Nat. Immune.*, 16 (1), 27–33, 1998.
165. Klein, A.D. and Penneys, N.S., Aloe vera, *J. Am. Acad. Dermatol.*, 18 (4 pt1), 714–720, 1988, and erratum in: *J. Am. Acad. Dermatol.*, 19(1pt 1), 82, 1988.
166. Cevallier, A., *Encyclopedia of Medicinal Plants*, Darling Kindersley Ltd., London, 1996, p. 118; Satyavati, G.V., Gupta, A.K., and Tandon, N., *Medicinal Plants of India*, Vol. 2, Indian Council of Medical Research, New Delhi, 1987, p. 355.
167. Agrawal, P., Rai, V., and Singh, R.B., *Int. J. Clin. Pharmacol. Ther.*, 34 (9), 406–409, 1996.

168. Prashar, R., Kumar, A., Banerjee, S., and Rao, A.R. *Anticancer Drugs*, 5 (5), 567–572, 1994; Aruna, K. and Sivaramakrishnan, V.M., *Food Chem. Toxicol.*, 30 (11), 953–956, 1992.
169. Karthikeyan, K., Ravichandran, P., and Govindaswamy, S. *Oral. Oncol.*, 35 (1), 112–119, 1999.
170. Prakash, J. and Gupta, S.K., *J. Ethnopharmacol.*, 72 (1–2), 29–34, 2000.
171. Rajasekaran, M. et al., *J. Drug Dev.*, 2 (3), 179–182, 1989.
172. Karthikeyan, K., Gunasekaran, P., Ramamurthy, N., and Govindaswamy, S., *Pharm. Biol.*, 37 (4), 285–290, 1999.
173. Devi, P.U. and Ganasoundari, A., *Indian J. Exp. Biol.*, 37 (3), 262–268, 1999; Ganasoundari, A., Devi, P.U., and Rao, M.N., *Mutat. Res.*, 373 (2), 271–276, 1997.
174. Vrinda, B. and Devi, P.U. *Mutat. Res.*, 498 (1–2), 39–46, 2001; Devi, P.U. Ganasoundari, A., Rao, B.S., and Srinivasan, K.K., *Radiat. Res.*, 151 (1), 74–78, 1999.
175. Ganasoundari, A., Devi, P.U., and Rao, M.N., *Mutat. Res.*, 373 (2), 271–276, 1997; Ganasoundari, A., Zare, S.M., and Devi, P.U. *Br. J. Radiol.*, 70, 599–602, 1997; Devi, P.U. Bisht, B.S., and Vanitha, M., *Br. J. Radiol.*, 71, 782–784, 1998.
176. Devi, P.U., Bisht, K.S., and Vinitha, M., *Br. J. Radiol.*, 71, 782–784, 1998; Ganasoundari, A., Devi, P.U., Rao, B.S., *Mutat. Res.*, 397 (2), 303–312, 1998.
177. Balanehru, S. and Nagarajan, B., *Biochem. Int.*, 24 (5), 981–990, 1991.
178. Tokuda, H., Ohigashi, H., Koshimizu, K., and Ito, Y., *Cancer Lett.*, 33, 279–285, 1986.
179. Ishida, M., Okubo, T., Koshimizu, K., Daito, H., Tokuda, H., Kin, T., Yamamoto, T., and Yamazaki, N., *Chemical Abstr.*, 113, 12173y, 1990; Muto, Y., Ninomiya, M., and Fujiki, H., *Jpn. J. Clinical Oncol.*, 20, 219–224, 1990.
180. Christy, M.M., The Pacific Yew Story, monograph, Wishland Publishing, Scottsdale, AZ, 1999.
181. Bryan-Brown, T., The pharmacological actions of toxine, *J. Pharm. Pharmacol.*, 5, 205–219, 1932.
182. National Institute of Science Communication, *The Wealth of India*, Vol. X, CSIR, New Delhi, 1998, pp. 132–143.
183. Canadian Hall of Fame, Biography of Dr. Robert L. Noble, 1999–2002; Anon., 100 Plant facts for campaigning conservationists, Fact no. 84, *Plant Talk, Bulletin of the National Tropical Botanical Garden*, Jan. 1998.
184. Kroll, D.J., *HerbalGram, J. Am. Botanical Counc.*, 56, 64, 2002; Persinos, G.J., *World I*, 3, 189, 1988; Lewis, W.H. and Elvis-Lewis, M.R.F., *Medical Botany: Plants Affecting Human Health*, John Wiley & Sons, New York, 1977.
185. Duke, J.A. and Ayenshu, E.S., *Medicinal Plants of China*, Reference Publications, Algonac, MI, 1985.
186. World Wildlife Fund, Fact Sheet 7; available on-line at www.wwf.org.uk/researcher/programmethemes/plants/ 0000000181.asp#med.
187. Gillespie, S.G., Herbal Drugs and Phytomedicinal Agents, 2001; available on-line at http://www.arches.uga. edu/~cmccall/5920_PHRM/Keeping_up/Herbs_and_Phytos.pdf.
188. Hartwell, J.L., *Plants Used against Cancer*, Quarterman Publications, Lawrence, MA, 1982; Graham, J.G., Quinn, M.L., Fabricant, D.S., and Farnsworth, N.R., *J. Ethnopharmacol.*, 73 (3), 347–377, 2000.
189. Jones, H.A., Metz, J.M., Devine, P., Hahn, S.M., and Whittington, R., *Urology*, 59 (2), 272–276, 2002.
190. Kuhlmann, M.K. and Horsch, E. et al., *Arch. Toxicol.*, 72, 536–540, 1998; Venkatesan, N., *Br. J. Pharmacol.*, 124, 425–427, 1998.
191. Hofmann, J., Fiebig, H.H. et al., *Int. J. Cancer* 45, 536–539, 1990; Schmbia, G., Ranelletti, F.O., and Panici, P.B., et al., *Cancer Chemother. Pharmacol.*, 34, 459–464, 1994.

21 Vitamin C, Vitamin E, and β-Carotene in Cancer Chemoprevention

Jane Higdon and Balz Frei

CONTENTS

21.1 Antioxidant and Other Biological Functions 272
    21.1.1 Vitamin C..272
    21.1.2 Vitamin E..273
    21.1.3 β-Carotene 274
21.2 Potential Mechanisms for Cancer Chemoprevention 275
    21.2.1 Prevention of DNA Damage 275
        21.2.1.1 Biomarkers of DNA Damage 276
        21.2.1.2 Observational Studies...276
        21.2.1.3 Supplementation Studies..277
    21.2.2 Inhibition of Nitrosation..281
    21.2.3 Modulation of Cell Signaling Resulting in Antiproliferative
        and Pro-Apoptotic Effects 281
        21.2.3.1 Vitamin C 282
        21.2.3.2 Vitamin E 282
        21.2.3.3 β-Carotene 282
    21.2.4 Enhancement of Immune System Function..283
        21.2.4.1 Vitamin C 283
        21.2.4.2 Vitamin E 283
        21.2.4.3 β-Carotene 283
21.3 Evidence for Cancer Chemoprevention 284
    21.3.1 Inhibition of Carcinogenesis in Animal Models..284
    21.3.2 Prospective Cohort Studies 284
        21.3.2.1 Total Cancer Incidence 284
        21.3.2.2 Lung Cancer..285
        21.3.2.3 Gastric Cancer..287
        21.3.2.4 Colorectal Cancer...287
        21.3.2.5 Bladder Cancer...288
        21.3.2.6 Prostate Cancer 289
        21.3.2.7 Breast Cancer 289
    21.3.3 Intervention Trials: Cancer...290
        21.3.3.1 Esophageal and Gastric Cancer 290
        21.3.3.2 Lung Cancer..290
        21.3.3.3 Why Did β-Carotene Supplementation Increase Lung
                Cancer Risk in the ATBC and CARET Studies? 290

         21.3.3.4 Prostate Cancer 291
         21.3.3.5 Other Cancers...291
     21.3.4 Intervention Trials: Precancerous Lesions and Surrogate
             Endpoint Biomarkers...292
         21.3.4.1 Oral Leukoplakia..292
         21.3.4.2 Cervical Intraepithelial Neoplasia 292
         21.3.4.3 Colorectal Adenomas and Colorectal Epithelial
                   Cell Proliferation...293
         21.3.4.4 *Helicobacter pylori* Infection and Precancerous Lesions
                   of Gastric Mucosa...294
21.4 Summary and Conclusions...294
References 296

Diets rich in fruits and vegetables have been consistently associated with decreased risk of cancer and other chronic diseases. Such diets are also relatively rich in antioxidants, including vitamin C, vitamin E, β-carotene, and other carotenoids. These antioxidant nutrients have been hypothesized to contribute substantially to the protective effects of fruits and vegetables against cancer.[1,2] Strong and consistent inverse associations between dietary β-carotene intakes or serum retinol concentrations and cancer risk led Peto and colleagues[3] to hypothesize that β-carotene might be the operative cancer chemopreventive factor. Strong and consistent inverse associations between dietary or plasma levels of antioxidant nutrients and cancer risk were an important part of the rationale for several long-term cancer chemoprevention trials of antioxidant supplements in high-risk populations.[4–6] In the past decade, the results of mechanistic studies, prospective cohort studies, and intervention trials have raised a number of questions regarding the role and effectiveness of antioxidant nutrients in cancer chemoprevention. This chapter reviews the evidence that has accumulated over the past decade of active research regarding the potential of vitamin C, vitamin E, and β-carotene as cancer chemopreventive agents.

21.1 ANTIOXIDANT AND OTHER BIOLOGICAL FUNCTIONS

21.1.1 VITAMIN C

The physiological functions of vitamin C are related to its efficacy as a reducing agent or electron donor. Vitamin C is known to be a specific electron donor for eight human enzymes.[7] Three of those enzymes participate in the posttranslational hydroxylation of collagen, which is essential for the formation of stable collagen helices.[8] Many of the manifestations of the vitamin C-deficiency disease, scurvy, are related to defective collagen synthesis. Vitamin C is also necessary for the maximal activity of two dioxygenase enzymes required for L-carnitine biosynthesis, and it serves as a cosubstrate for dopamine-β-monooxygenase, the enzyme that catalyzes the conversion of the neurotransmitter dopamine to norepinephrine.[9] To prevent scurvy, an adult must consume about 10 mg/d of vitamin C, an amount easily obtained by as little as one serving per day of most fruits and vegetables. The current recommended dietary allowance (RDA) for vitamin C is 90 mg/d for men and 75 mg/d for women.[10]

Several properties make vitamin C an ideal antioxidant in biological systems. First, the low one-electron reduction potentials of ascorbate and the ascorbyl radical (Figure 21.1) enable these compounds to react with and reduce virtually all physiologically relevant reactive oxygen species (ROS) and reactive nitrogen species (RNS), including superoxide, hydroperoxyl radicals, aqueous peroxyl radicals, singlet oxygen, ozone, nitrogen dioxide, nitroxide radicals, and hypochlorous acid. When ascorbate scavenges ROS or RNS, the resulting ascorbyl radical is neither strongly oxidizing nor strongly reducing, and it reacts poorly with oxygen. The ascorbyl radical may scavenge another

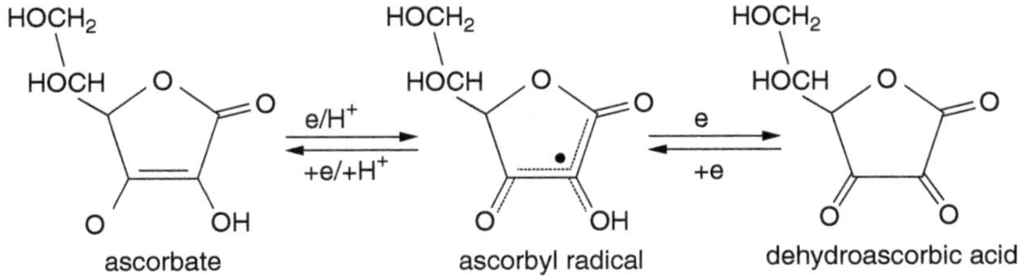

FIGURE 21.1 Structure of vitamin C. Oxidation of ascorbate by two successive one-electron oxidation steps results in the formation of the ascorbyl radical and dehydroascorbic acid, respectively.

radical or rapidly dismutate to form ascorbate and dehydroascorbic acid.[9] Vitamin C also acts as a co-antioxidant by regenerating α-tocopherol from the α-tocopheroxyl radical. This may be an important function, since α-tocopherol can act as a pro-oxidant *in vitro* in the absence of co-antioxidants such as vitamin C.[11]

21.1.2 Vitamin E

The term *vitamin E* describes a family of eight antioxidants, α-, β-, γ-, and δ-tocopherol and α-, β-, γ-, and δ-tocotrienol, which are synthesized only by plants. Tocopherols and tocotrienols are characterized by a substituted hydroxylated ring system (chromanol ring). However, tocotrienols have an unsaturated side chain, while tocopherols have a saturated side chain with three asymmetric carbon atoms (Figure 21.2). Out of eight possible tocopherol stereoisomers, only the *RRR*-stereoisomer occurs naturally.[12]

Unless otherwise noted, the vitamin E supplements discussed in this chapter contain either natural *RRR*-α-tocopherol (also known as *d*-α-tocopherol) or synthetic α-tocopherol (*all rac*-α-tocopherol or *dl*-α-tocopherol), which contains equal amounts of all eight stereoisomers. Only *RRR*-α-tocopherol and the other 2*R*-stereoisomers of α-tocopherol are selectively maintained in plasma,[13] suggesting that *all rac*-α-tocopherol is only half as potent as *RRR*-α-tocopherol. For this reason, the only forms of vitamin E that meet the most recent RDA of 15 mg/d for adults are the naturally occurring *RRR*-α-tocopherol and the other three synthetic 2*R*-stereoisomers of α-tocopherol.[14]

Vitamin E is a lipid-soluble antioxidant present in cell membranes and lipoproteins. All forms of vitamin E (tocopherols and tocotrienols) are capable of inhibiting lipid peroxidation because they react with lipid peroxyl radicals much faster than these radicals can react with polyunsaturated fatty acids (PUFA) to propagate the chain reaction of lipid peroxidation.[15] For example, lipid peroxidation is inhibited when α-tocopherol donates a phenolic hydrogen (electron) to the lipid peroxyl radical, resulting in the formation of a lipid hydroperoxide and an α-tocopheroxyl radical.[16] The potentially dangerous lipid hydroperoxide can be enzymatically reduced to an alcohol. The α-tocopheroxyl radical is relatively stable and may undergo one of several reactions: (1) reduction back to α-tocopherol by a co-antioxidant such as vitamin C; (2) reaction with another radical to form nonradical products; (3) further oxidation to form α-tocopherol quinone; or (4) in the absence of co-antioxidants, during *in vitro* lipoprotein oxidation, the α-tocopheroxyl radical may oxidize PUFA through a process termed tocopherol-mediated peroxidation (TMP).[11] However, TMP has not been conclusively demonstrated *in vivo*. While most research has focused on its role as an antioxidant, other biological activities of vitamin E have been identified that are not necessarily related to its antioxidant activity, including protein kinase C inhibition by α-tocopherol and regulation of gene expression by several forms of vitamin E.[17]

The function of γ-tocopherol in humans is presently unclear. Because γ-tocopherol is initially absorbed in the same manner as α-tocopherol, small amounts are detectable in blood and tissue. However, α-tocopherol is preferentially incorporated into VLDL (very low density lipoprotein) by the hepatic α-tocopherol transfer protein (α-TTP), while γ-tocopherol and other forms of vitamin E

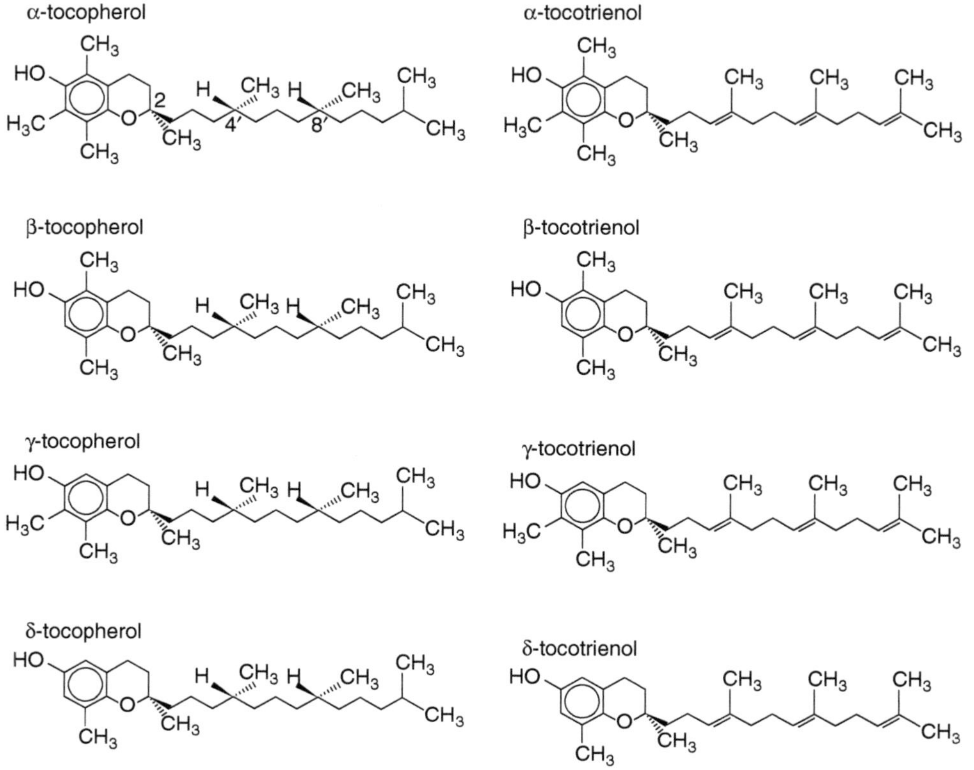

FIGURE 21.2 Structures of the eight naturally occurring forms of vitamin E. Tocopherols are in the *RRR*-configuration, and tocotrienols are in the *R*-configuration.

are rapidly degraded and eliminated through biliary and urinary excretion.[12] The selective incorporation of α-tocopherol into lipoproteins by α-TTP explains why human plasma concentrations of γ-tocopherol are generally 4 to 10 times lower than those of α-tocopherol despite the fact that γ-tocopherol is the most abundant form of vitamin E in the U.S. diet.[18] Due to a lack of one of the electron-donating methyl groups on its chromanol ring, γ-tocopherol is a slightly less potent electron donor (antioxidant) than α-tocopherol and, therefore, a slightly less potent inhibitor of lipid peroxidation. However, this structural difference makes γ-tocopherol more efficient at trapping electrophiles, such as RNS associated with inflammation.[19] Recent evidence that γ-tocopherol or its metabolites may have important physiological functions, in addition to observations that high doses of α-tocopherol decrease plasma and tissue levels of γ-tocopherol, have led some scientists to call for additional research on the effects of γ-tocopherol on human health.[20]

21.1.3 β-Carotene

Carotenoids are a group of more than 600 naturally occurring pigments, of which only about 50 can be bioconverted to vitamin A. Because they are highly insoluble in water, carotenoids circulate in the blood with lipids in lipoproteins. In cells, they are present in cell membranes, while liver and adipose tissue are the major tissues where intact carotenoids accumulate. Carotenoids have a 40-carbon skeleton with a series of conjugated double bonds that can be cyclized on each end (Figure 21.3). β-Carotene is the primary provitamin A carotenoid in the human diet. Oxidative cleavage of β-carotene by the 15-15′-dioxygenase enzyme at the central carbon-carbon double bond yields retinal, which can be oxidized to the retinoid (retinoic acid) or reduced to retinol.[21] The only

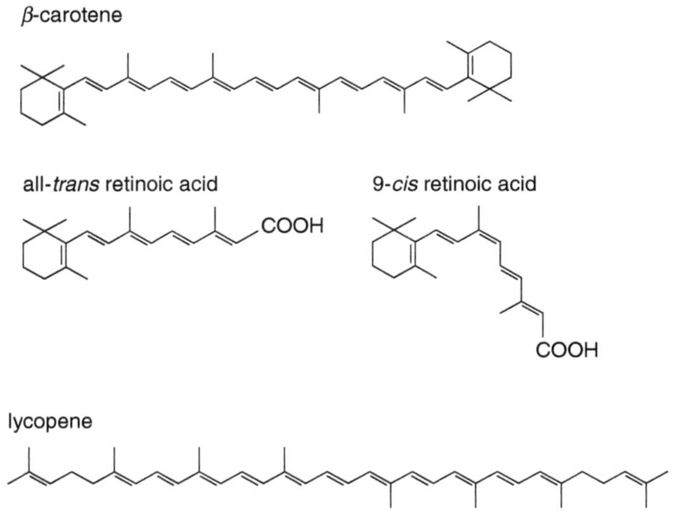

β-carotene

all-*trans* retinoic acid

9-*cis* retinoic acid

lycopene

FIGURE 21.3 Structures of β-carotene, all-*trans* retinoic acid, 9-*cis* retinoic acid, and lycopene. β-Carotene is a provitamin A carotenoid that can undergo central cleavage to form two molecules of retinal. Retinal can undergo further metabolism to form the retinoids, all-*trans* retinoic acid and 9-*cis* retinoic acid, which regulate gene expression via nuclear retinoic acid receptors. Lycopene is a nonprovitamin A carotenoid.

function of β-carotene currently recognized in humans by the Food and Nutrition Board of the Institute of Medicine is to act as a source of vitamin A in the diet.[22] Vitamin A plays essential roles in visual function, immune system function, and cell growth and differentiation. The retinoic acid isomers, 9-*cis*- and all-*trans*-retinoic acid are retinoids that regulate gene expression through nuclear retinoic acid receptors (RAR and RXR).

In plants, carotenoids have an important antioxidant function as effective quenchers of singlet oxygen formed during photosynthesis. Singlet oxygen is an oxidant most often generated by photosensitization reactions, although it can also be generated during lipid peroxidation.[15] Although important for plants, the importance of singlet oxygen quenching to human health is less clear. The ability of β-carotene to inhibit lipid peroxidation is dependent on the system studied. In organic solvents, β-carotene inhibited lipid peroxidation at low oxygen (O_2) concentrations comparable with those in most tissues, but it acted as a pro-oxidant at very high O_2 concentrations.[23] The pro-oxidant effect was seen at 100% oxygen, but not at ambient conditions (21% O_2) or at tissue concentrations, which are much lower.[24] β-Carotene did not protect lipids in LDL[25] or liver microsomes[26] from oxidation *in vitro* at low or high O_2 concentrations, and evidence that β-carotene is an effective antioxidant in humans is limited.[27,28] At present, it is unclear whether β-carotene plays an important antioxidant role in humans *in vivo*, and it is possible that its biological effects are exerted through mechanisms unrelated to its antioxidant activity.[29]

21.2 POTENTIAL MECHANISMS FOR CANCER CHEMOPREVENTION

21.2.1 PREVENTION OF DNA DAMAGE

DNA contains reactive groups in its bases that are highly susceptible to attack by ROS and RNS. It has been proposed that oxidative damage to DNA occurs *in vivo* at a rate of 10^4 oxidative hits per cell per day.[30] Some carcinogens generate free radicals during their metabolism, which may damage DNA and predispose cells to malignant changes. Inflammatory cells also generate ROS and RNS, and chronic inflammation has been associated with the development of cancer in a number of tissues.[31] Most oxidative lesions are repaired, e.g., by specific DNA glycosylases, but repair is not 100% efficient, and oxidative lesions to DNA accumulate with age.[32] If not repaired, oxidative DNA damage may lead to mutations. Such mutations increase the risk of cancer if they occur in critical genes, such as those encoding tumor suppressor proteins or growth factors. A direct link between cancer and oxidative DNA damage is still lacking.[33] However, increased levels of oxidized DNA bases have been observed at sites of chronic inflammation and in preneoplastic lesions,[34] while cellular and urinary levels of the oxidized DNA base, 8-oxoguanine, are elevated in smokers.[35,36]

21.2.1.1 Biomarkers of DNA Damage

More than 20 different oxidative lesions of DNA have been identified, of which 8-oxo-2'-deoxyguanosine (8-oxodG) is one of the most frequent and best studied.[37,38] Attack of guanine in the C-8 position by a hydroxyl radical, peroxynitrite, or singlet oxygen results in the formation of 8-oxoguanine (8-oxogua), a mutagenic base damage product. Guanine normally forms a base pair with cytosine, but 8-oxogua may also pair with adenine. After replication, what was formerly a G:C base pair is replaced by a T:A base pair, resulting in a transversion mutation. In tissues such as lymphocytes, the measurement of damaged DNA bases represents the steady-state balance between damage and repair in that particular tissue at the time of sampling.[39] The derivatization required for gas chromatography/mass spectrometry (GC/MS) assays of 8-oxogua, 8-oxodG, and other oxidized base products increases the likelihood of artifactual oxidation compared with HPLC-EC (high-pressure liquid chromatography–EC) methods for measuring these products, explaining the higher values of oxidized DNA bases reported with GC/MS.[38] The significance of urinary measurements of 8-oxodG is less clear, since it may result from degradation of DNA from necrotic or apoptotic cells rather than representing DNA repair in viable cells.[40]

Single-cell alkaline gel electrophoresis (the Comet assay) is a sensitive assay of DNA strand breaks. The Comet assay can be adapted to measure oxidative damage to DNA bases by measuring strand breaks induced by the treatment of DNA with relevant repair enzymes, e.g., endonuclease III (EndoIII) for oxidized pyrimidines and formamidopyrimidine glycosylase (Fpg) for 8-oxogua and ring-opened pyrimidines, such as those resulting from the breakdown of alkylated bases.[33]

21.2.1.2 Observational Studies

If the prevention of oxidative DNA damage plays an important role in the cancer chemopreventive effects of dietary antioxidants, then one would expect to see an inverse relationship between antioxidant intake and biomarkers of oxidative DNA damage. Lymphocyte ascorbate concentrations were inversely associated with lymphocyte 8-oxodG concentrations in 105 men and women ($r = -0.28$).[41] However, plasma concentrations of α-tocopherol and carotenoids were positively correlated with lymphocyte concentrations of 8-oxodG in a study of 52 healthy women ($r = 0.29$ to 0.41).[42] Increased oxidative damage to sperm DNA could result in increased frequencies of heritable mutations. In 24 healthy men, seminal fluid ascorbate concentrations were inversely related to 8-oxodG levels in sperm DNA.[43] Prenatal development may represent a period of increased vulnerability to the effects of oxidative DNA damage. Plasma α-tocopherol levels in 30 pregnant women were inversely correlated to placental tissue levels of 8-oxodG at birth.[44]

A few studies have used the Comet assay to examine the association between plasma antioxidant concentrations and oxidative DNA damage. One study of healthy, nonsmoking men and women from Madrid found oxidized pyrimidines, but not altered purines, in lymphocyte DNA to be inversely associated with plasma β-carotene, lutein, and total carotenoid concentrations ($r = -0.4$).[45] Neither plasma vitamin C nor vitamin E concentrations were significantly associated with oxidized pyrimidines or altered purines in lymphocyte DNA. A smaller study of 11 healthy Slovakian men followed monthly for one year found only plasma lycopene concentrations to be inversely associated with oxidized pyrimidines in lymphocyte DNA (r = –0.24), while plasma ascorbate concentrations were inversely associated with DNA strand breaks and altered purines (r = –0.23).

Although urinary 8-oxodG excretion was not associated with dietary intake of vitamin C, vitamin E, or β-carotene in 83 healthy men and women,[46] plasma ascorbate concentrations were inversely associated with creatinine-indexed urinary 8-oxodG in 184 nonsmoking men and women (r = –0.11).[47] Furthermore, plasma α-carotene and β-carotene concentrations were inversely associated with urinary 8-oxodG levels in a sample of 210 middle-aged men.[48] Overall, the findings of several small, cross-sectional studies indicate inverse associations between measures of antioxidant and carotenoid intake and biomarkers of oxidative DNA damage and repair.

21.2.1.3 Supplementation Studies

There is some evidence that antioxidant supplementation decreases oxidative DNA damage *in vivo* (Table 21.1). Sperm concentrations of 8-oxodG were significantly increased in men who were depleted of vitamin C, while repletion with 60 mg/d brought sperm 8-oxodG concentrations back to baseline.[43] A small placebo-controlled intervention trial in 12 male smokers examined the effect of supplementation with 500 mg/d of vitamin C, 200 IU/d of vitamin E, or 9 mg/d of β-carotene for 4 weeks.[49] Only vitamin E supplementation significantly decreased lymphocyte 8-oxodG levels compared with the baseline. In contrast, a crossover trial in 9 male smokers and 12 male nonsmokers supplemented with 250 mg/d of vitamin C, 250 mg/d of α-tocopherol, 60 mg/d of β-carotene, or a combination of vitamin C and α-tocopherol (250 mg/d of each) for 4-week periods found that β-carotene supplementation decreased leukocyte 8-oxodG levels in nonsmokers compared with placebo, but increased them in smokers.[50] In a small uncontrolled study in 11 workers exposed to environmental tobacco smoke, daily supplementation with a combination of 3 mg of β-carotene, 60 mg of vitamin C, 30 IU of α-tocopherol, 40 mg of zinc, and 2 mg of copper for two months resulted in leukocyte 8-oxodG levels that were significantly lower than baseline levels.[51] However, in a larger placebo-controlled study of 111 heavy smokers, supplementation with a combination of 250 mg/d of vitamin C, 200 IU/d of α-tocopherol, and 6 mg/d of β-carotene for 6 months did not result in significant differences in 8-oxodG levels or polyaromatic hydrocarbon DNA adducts in lymphocytes or oral mucosal cells.[52]

Supplementing 20 healthy volunteers with 500, 1000, or 5000 mg/d of vitamin C did not change leukocyte 8-oxodG concentrations compared with placebo.[53] In another supplementation trial, 500 mg/d of vitamin C for 6 weeks was found to decrease lymphocyte 8-oxodG levels as measured by HPLC and to decrease lymphocyte 8-oxogua levels as measured by GC/MS.[54,55] However, levels of another oxidized DNA base measured by GC/MS, 8-oxoadenine (8-oxoade), were increased. Because serum and urinary 8-oxodG concentrations measured by ELISA (enzyme-linked immunosorbent assay) increased after vitamin C supplementation was discontinued, the investigators hypothesized that vitamin C supplementation may have increased the activity of DNA repair enzymes.[54] Two other supplementation trials evaluated changes in 13 oxidized DNA bases measured by GC-MS, including 8-oxogua and 8-oxoade, after cosupplementation with vitamin C and iron. In individuals with relatively high plasma vitamin C levels at the start of the study, some oxidized bases, including 8-oxoade, were decreased after cosupplementation, while others were increased.[56] Lack of a placebo control made this study's results difficult to interpret. In a subsequent placebo-controlled crossover trial conducted by the same group of investigators, 6 weeks of supplementation with either 260 mg/d of vitamin C or 260 mg/d of vitamin C plus 14 mg/d of iron resulted in lymphocyte 8-oxoade and 8-oxogua levels that were not significantly different from placebo.[57]

A large number of studies have consistently shown that urinary 8-oxodG excretion is not significantly altered by supplementation with vitamin C,[47,58,59] α-tocopherol,[47,58,59] β-carotene,[60,61] various combinations of vitamin C and α-tocopherol,[47,58,59] or a combination of vitamin C, α-tocopherol, and β-carotene.[62] However, the relevance of urinary 8-oxodG levels as a marker of DNA damage and repair in viable cells is unclear.

TABLE 21.1
Oxidative DNA Damage *in vivo*: Intervention Trials Supplementing Vitamin C, Vitamin E, or β-Carotene

Reference	Subjects	Treatment	Duration	Biomarker (Method)	Results
		Cell Measurements			
Fraga et al., 1991[43] Jacob et al., 1991[102]	10 men	Sequential: 1. vitamin C, 5 mg/d 2. vitamin C, 10 mg/d or 20 mg/d 3. vitamin C, 60 mg/d or 250 mg/d	32 days 28 days 28 days	sperm, lymphocyte, urine 8-oxodG (HPLC-EC)	sperm 8-oxodG: 1. ↑ vs. baseline 2. ↑ vs. baseline 3. ↓ vs. phase 2 lymphocyte and urine 8-oxodG: ↔ for all phases
Podmore et al., 1998[55] Cooke et al., 1998[54]	30 nonsmokers	Sequential: 1. calcium carbonate, 500 mg/d 2. vitamin C, 500 mg/d 3. washout	6 weeks 6 weeks 12 weeks	lymphocyte 8-oxoade, 8-oxogua (GC-MS) 8-oxodG (HPLC-EC) serum 8-oxodG (ELISA) urine 8-oxodG (ELISA)	vitamin C vs. calcium: lymphocyte 8-oxoade ↑ 8-oxogua → 8-oxodG ↑ serum 8-oxodG ↑ washout vs. calcium: urine 8-oxodG ↑
Rehman et al., 1998[56]	38 adults (20 with high plasma ascorbate)	1. vitamin C, 60 mg/d + Fe, 14 mg/d 2. vitamin C, 260 mg/d + Fe, 14 mg/d	12 weeks	leukocyte DNA base damage products (GC/MS)	vs. baseline; both treatments high plasma ascorbate: ↑ week 6, ↔ week 12 low plasma ascorbate: ↔ week 6, ↔ week 12
Protegegente et al., 2000[57]	20 nonsmokers	1. vitamin C, 260 mg/d 2. vitamin C, 260 mg/d + Fe, 14 mg/d 3. placebo	6 weeks	leukocyte DNA base damage products (GC/MS)	1) ↔ vs. placebo 2) ↔ vs. placebo

Study	Subjects	Supplements	Duration	Assay	Results
Vojdani et al., 2000[53]	20 healthy volunteers	1. vitamin C, 500 mg/d 2. vitamin C, 1000 mg/d 3. vitamin C, 5000 mg/d 4. placebo	14 days	leukocyte 8-oxodG (HPLC-EC)	↔ vs. placebo for all vitamin C supplements
Lee et al., 1998[49]	12 male smokers	1. vitamin E, 200 IU/d 2. β-carotene, 9 mg/d 3. vitamin C, 500 mg/d 4. placebo	4 weeks	lymphocyte 8-oxodG (HPLC-EC)	vitamin E: ↓ vs. baseline all other groups: ↔ vs. baseline
Howard et al., 1998[51]	11 adults exposed to environmental tobacco smoke	β-carotene, 3 mg/d + vitamin C, 60 mg/d + α-tocopherol, 30 IU/d + zinc, 40 mg/d + copper, 2 mg/d	56 days	leukocyte 8-oxodG (HPLC-EC)	↔ vs. baseline
Welch et al., 1999[50]	9 smokers, 12 nonsmokers	1. vitamin C, 250 mg/d 2. α-tocopherol, 250 mg/d 3. vitamin C, 250 mg/d + α-tocopherol, 250 mg/d 4. β-carotene, 60 mg/d 5. placebo	4 weeks	leukocyte 8-oxodG (HPLC-EC)	β-carotene: ↑ in smokers, ↓ in nonsmokers all other supplements: ↔ in smokers and nonsmokers
Jacobson et al., 2000[52]	111 heavy smokers	1. vitamin C, 250 mg/d + α-tocopherol, 200 IU/d + β-carotene, 6 mg/d 2. placebo	6 months	mononuclear and oral cells 8-oxodG (ELISA)	↔ vs. placebo at 1, 3, and 6 months
Duthie et al., 1996[66]	37 male smokers, 53 male nonsmokers	1. vitamin C, 100 mg/d + α-tocopherol, 280 mg/d + β-carotene, 25 mg/d 2. placebo	20 weeks	lymphocyte-oxidized pyrimidines (Comet assay + endonuclease III-sensitive sites)	↓ vs. placebo
Jenkinson et al., 1999[70]	21 male nonsmokers	1. 5% PUFA[a] diet + α-tocopherol, 80 mg/d 2. 15% PUFA diet + α-tocopherol, 80 mg/d 3. 5% PUFA diet 4. 15% PUFA diet	4 weeks	lymphocyte-oxidized pyrimidines (Comet assay + endonuclease III-sensitive sites)	α-tocopherol-inhibited ↑ associated with 15% PUFA diet

(Continued)

TABLE 21.1 (CONTINUED)
Oxidative DNA Damage *in vivo*: Intervention Trials Supplementing Vitamin C, Vitamin E, or β-Carotene

Reference	Subjects	Treatment	Duration	Biomarker (Method)	Results
		Urinary Measurement			
Witt et al., 1992[62]	11 healthy men	vitamin C, 1000 mg/d + α-tocopherol, 533 mg/d + β-carotene, 10 mg/d 90 min submaximal exercise for 3 days	30 days	urine 8-oxodG (HPLC-EC)	↔ pre- vs. post-supplementation ↔ pre- vs. post-exercise
van Poppel et al., 1995[60]	122 male smokers	1. β-carotene, 20 mg/d 2. placebo	14 weeks	urine 8-oxodG (HPLC-EC)	↔ vs. placebo
Prieme et al., 1997[58]	142 male smokers	1. α-tocopherol, 200 mg/d + vitamin C, 500 mg/d 2. α-tocopherol, 200 mg/d 3. vitamin C, 500 mg/d 4. slow-release vitamin C, 500 mg/d 5. placebo	60 days	urine 8-oxodG (HPLC-EC)	↔ vs. placebo for all supplement groups
Sumida et al., 1997[61]	14 healthy males	1. β-carotene, 30 mg/d 2. placebo, maximal exercise bout	33 days	urine 8-oxodG (HPLC-EC)	↔ β-carotene vs. placebo ↔ with exercise
Huang et al., 2000[47]	184 nonsmokers	1. vitamin C, 500 mg/d 2. α-tocopherol, 400 IU/d 3. vitamin C, 500 mg/d + α-tocopherol, 400 IU/d 4. placebo	2 months	urine 8-oxodG (ELISA)	↔ vs. placebo for all supplement groups
Porkkala-Sarataho et al., 2000[59]	22 male smokers, 26 male nonsmokers	1. α-tocopherol, 182 mg/d 2. vitamin C, 500 mg/d 3. α-tocopherol, 182 mg/d + vitamin C, 500 mg/d 4. placebo	1 year	urine 8-oxodG (HPLC-EC)	↔ vs. placebo for all supplement groups

a PUFA = polyunsaturated fatty acids.

A number of antioxidant supplementation trials have used the Comet assay to assess DNA damage, but relatively few have employed DNA repair enzymes to assess oxidative DNA damage. Although antioxidant supplementation has resulted in decreased leukocyte DNA strand breaks after exposure to *ex vivo* oxidative stress, such as hydrogen peroxide or radiation,[63–66] the majority of intervention trials have not demonstrated that supplementation with vitamin C, vitamin E, or β-carotene can decrease endogenous DNA strand breaks in leukocytes[50,65,67,68] or gastric mucosa.[69] Several studies have employed the Comet assay with the DNA repair enzyme, endonuclease III, to examine the effect of antioxidant supplementation on oxidized pyrimidine levels in lymphocytes. In a placebo-controlled trial, supplementation with a combination of 100 mg/d of vitamin C, 280 mg/d of α-tocopherol, and 25 mg/d of β-carotene for 20 weeks significantly decreased lymphocyte levels of oxidized pyrimidines in smokers and nonsmokers compared with a placebo.[66] Increasing the dietary polyunsaturated fatty acid (PUFA) content from 5 to 15% resulted in a significant increase in lymphocyte levels of oxidized pyrimidines in 21 healthy males, which was inhibited by supplementing 80 mg/d of α-tocopherol for a total intake of 90 mg/d.[70] Compared with a diet depleted of carotenoids, the addition of carrot juice providing 22 mg/d of β-carotene and 16 mg/d of α-carotene for 2 weeks significantly decreased lymphocyte levels of oxidized pyrimidines in 23 healthy men.[71] However, it is not possible to conclude that the carotenoids in the carrot juice were responsible for the antioxidant effect.

In summary, trials of antioxidant supplementation of smokers and nonsmokers with vitamin C, α-tocopherol, or β-carotene did not generally find significant decreases in biomarkers of oxidative DNA damage, in contrast to observational studies, which suggest an inverse association between dietary antioxidant intake and oxidative DNA damage.

21.2.2 Inhibition of Nitrosation

Mutagenic *N*-nitroso compounds such as nitrosamines have been implicated in the etiology of cancers of the nasopharynx, esophagus, stomach, colon, urinary bladder, and lung.[72,73] Vitamin C and vitamin E may provide protection from these cancers by inhibiting the formation of *N*-nitroso compounds.[74] *N*-nitroso compounds are formed by reaction of nitrite, common in cured foods and cigarette smoke, with amines and amides.[72] Nitrosating compounds can also be formed from nitric oxide generated by inflammatory cells expressing inducible nitric oxide synthase.[75] Increased vitamin C intake has been demonstrated to decrease nitrosation in humans.[76,77] Although somewhat less effective than vitamin C, α-tocopherol has also been found to decrease nitrosation in humans.[78] Interestingly, γ-tocopherol, the predominant form of dietary vitamin E, has recently been shown to be a more effective inhibitor of nitrosation than α-tocopherol *in vitro* and *in vivo*.[19,20]

21.2.3 Modulation of Cell Signaling Resulting in Antiproliferative and Pro-Apoptotic Effects

Cell cycle arrest and apoptosis are important mechanisms for limiting the adverse effects of DNA damage in living organisms. Following DNA damage, cells can be transiently arrested at the G_1/S, S, or G_2/M damage checkpoints, where DNA is either repaired or the cell undergoes apoptosis. Inhibition of normal apoptosis pathways contributes to carcinogenesis by promoting genetic instability and the accumulation of mutations as well as the proliferation of transformed cells.[79] ROS can modulate the expression of genes involved in the regulation of cell cycle, apoptosis, and growth through activation or inactivation of redox-sensitive transcription factors, such as p53, nuclear factor-kappa B (NF-κB), activator protein 1 (AP-1), and specificity protein 1 (SP-1). The maintenance of cellular redox homeostasis represents one mechanism by which antioxidants may limit the pathogenic expression of anti-apoptotic and proliferative genes.[80]

21.2.3.1   Vitamin C

Pretreatment of hamster ovary cells with physiological concentrations of vitamin C (50 µM) enhanced cell cycle arrest at the G$_2$/M checkpoint when the cells were exposed to oxidative stress.[81] In the absence of oxidative stress, vitamin C did not affect cell cycle progression. The p53 tumor-suppressor protein is known to play a critical role in inducing cell cycle arrest at the G$_1$/S DNA damage checkpoint, and mutations of the *p53* gene have been associated with numerous cancers.[15] Cervical cells infected with cancer-producing strains of human papilloma virus (HPV) synthesize oncoproteins that target p53 for destruction. Addition of vitamin C to HPV-18 positive HeLa cells resulted in stabilization of p53 and increased sensitivity to apoptosis induced by chemotherapeutic agents.[82] MLH-1 is a protein that plays a critical role in DNA mismatch repair, and mutations in the *MLH-1* gene have been found to increase the risk of hereditary nonpolyposis colorectal cancer.[83] In response to DNA damage, activation of MLH-1 can cooperate with p53 to induce cell cycle arrest, or MLH-1 activation can lead to apoptosis through an alternative pathway induced by a p53 homologue known as p73. Recent studies in cultured HaCaT keratinocytes showed that 10 to 100 µM ascorbate significantly up-regulates the expression of *MLH-1* and *p73*.[84]

21.2.3.2   Vitamin E

Protein kinase C (PKC) is a family of enzymes that plays a major role in cell signaling and the modulation of gene expression related to cell growth, proliferation, and differentiation. PKC activation may also play a critical role in tumor promotion in some types of cancer.[85] α-Tocopherol has been found to inhibit PKCα activity in a number of cell types and in animal models of atherosclerosis.[86] PKCα inhibition by α-tocopherol appears unrelated to its antioxidant function[87]; rather, α-tocopherol may inhibit PKCα activity by promoting its dephosphorylation or by modulating diacylglycerol kinase activity.[88,89]

The α-tocopherol homologue, α-tocopherol succinate (α-TOS) but not α-tocopherol has been found to be a potent inducer of apoptosis in a number of cancer cell lines, though not in primary cell lines.[90] Additionally, Intraperitoneal administration of α-TOS to nude mice bearing human colorectal cancer xenografts limited tumor growth more effectively than α-tocopherol. In cell culture, maximal apoptotic activity required the inhibition of PKCα, which may be caused by α-tocopherol, and PKC-independent mechanisms mediated by the succinyl moiety. The requirement for esterification of the succinate moiety indicates that the pro-apoptotic effect of α-TOS is not likely to be an antioxidant effect. Because α-tocopherol esters are hydrolyzed to α-tocopherol during intestinal uptake, parenteral administration would likely be required to affect cells other than those lining the gastrointestinal tract.[17]

21.2.3.3   β-Carotene

The retinoic acid isomers 9-*cis*-retinoic acid and all-*trans*-retinoic acid induce cell growth and differentiation by binding to nuclear retinoic acid receptors (RAR and RXR) and regulating gene transcription.[21] However, β-carotene may also promote differentiation and inhibit proliferation through mechanisms unrelated to its provitamin A activity. One form of communication between neighboring cells results from the exchange of small molecules through gap junctions formed by proteins known as connexins. Gap junctional intercellular communication is important for cell differentiation and control of the cell cycle, and it is often deficient in tumor cells. Up-regulation of gap junctional communication in tumor cells through increased *connexin* expression is associated with decreased proliferation.[91] Carotenoids, including β-carotene, can induce gap junction communication between fibroblasts in culture by increasing the expression of *connexin 43*.[92] This effect was not related to the ability of carotenoids to quench singlet oxygen, but it is possibly related to the ability of carotenoids or their metabolites to induce gene expression.[93] In contrast to its effects on fibroblasts, β-carotene did not increase *connexin 43* expression in cultured lung cells,[94] and supplementing rats with high doses of β-carotene decreased hepatocyte intracellular communication, although low-dose supplementation enhanced it.[95]

21.2.4 ENHANCEMENT OF IMMUNE SYSTEM FUNCTION

Viral and bacterial infections are associated with increased risk of certain cancers.[96] Moreover, immunocompetent hosts may be better able to protect themselves from cancer by eliminating tumors and modulating the immunogenic phenotypes of tumors.[97] In either case, a competent immune system may decrease the risk of several types of cancer. Thus, preventing or alleviating immune system dysfunction may be another mechanism by which vitamin C, vitamin E, and β-carotene contribute to cancer prevention.

21.2.4.1 Vitamin C

Phagocytes and lymphocytes take up vitamin C, resulting in intracellular concentrations that are 100-fold higher than in plasma.[8] *In vitro*, vitamin C has been found to stimulate several indices of phagocytic function in macrophages[98] and to inhibit T-lymphocyte apoptosis.[99] This may be important because some tumors appear to be able to increase lymphocyte apoptosis and, thus, escape immune-mediated killing.[100,101] In humans, studies of the effect of vitamin C on immune function have been somewhat contradictory. Vitamin C depletion reduced the delayed-type hypersensitivity (DTH) responses in otherwise healthy men, but it had no effect on lymphocyte proliferation.[102] However, two other studies found that vitamin C supplementation with at least 1000 mg/d increased *ex vivo* measures of lymphocyte proliferation.[103,104] Vitamin C supplementation decreased leukocyte apoptosis at doses of 500 and 1000 mg/d for 2 weeks, and supplementation of at least 1000 mg/d increased natural killer cell activity in healthy individuals[53] and individuals exposed to immunosuppressing toxic chemicals.[105]

21.2.4.2 Vitamin E

Decreased DTH response and lymphocyte proliferation, and altered cytokine patterns are among the changes in the immune response associated with aging.[106] Vitamin E supplementation has been found to reverse age-associated declines in immune function in elderly individuals. Supplementation of elderly adults with 800 mg/d of α-tocopherol for 1 month increased the production of interleukin-2 (IL-2), an important factor in cell-mediated immunity.[107] Furthermore, supplementation of elderly adults with 200 mg/d of α-tocopherol for 4 months resulted in significantly higher antibody responses to vaccines for hepatitis B and tetanus, as well as increased DTH responses.[108] In aged rats, α-tocopherol supplementation enhances T-lymphocyte function through at least two mechanisms: (1) directly, by increasing the proliferation and IL-2 production of naïve T lymphocytes and (2) indirectly, by preventing age-associated increases in prostaglandin E_2 (PGE_2) production by macrophages. PGE_2 has a direct inhibitory effect on T-lymphocyte activation, resulting in decreased IL-2 production.[109] Stimulated macrophages from aged mice produce significantly more PGE_2 and exert higher cyclooxygenase (COX) activity than those from young mice, but α-tocopherol supplementation of aged mice results in macrophage PGE_2 production similar to that of young mice. Recent data from *in vitro* studies in murine macrophages suggests that vitamin E may inhibit COX activity by preventing the formation of peroxynitrite, which enhances COX activity by serving as a substrate for the enzyme's peroxidase activity.[110] Recently, γ-tocopherol and its major metabolite 2,7,8-trimethyl-2-(β-carboxyethyl)-6-hydroxychroman (γ-CEHC) were found to inhibit COX-2 activity in cultured macrophages and endothelial cells more effectively than α-tocopherol.[111]

21.2.4.3 β-Carotene

Although vitamin A is required for normal immune function,[112] a role for β-carotene in enhancing immune system function other than as a source of vitamin A has not been identified.[22]

21.3 EVIDENCE FOR CANCER CHEMOPREVENTION

21.3.1 INHIBITION OF CARCINOGENESIS IN ANIMAL MODELS

Experiments in animal models provide some support for the notion that increased consumption of vitamin C, vitamin E, or β-carotene inhibits tumorigenesis induced by chemical carcinogens and ultraviolet (UV) radiation. β-Carotene supplementation has been found to inhibit carcinogenesis in the skin of mice induced by a two-stage tumor model using the mutagen, 7,12-dimethylbenz[a]anthracene (DMBA), and the tumor promoter, 12-O-tetradecanoylphorbol 13-acetate.[113,114] In contrast, the effects of β-carotene supplementation on UV-induced skin carcinogenesis in animal models have been mixed.[115,116] Topical vitamin E has been found to decrease UV-induced skin carcinogenesis in mice,[117] but it may promote carcinogenesis in mice initiated with DMBA.[118] Vitamin C, vitamin E, and β-carotene supplementation each inhibited DMBA-induced carcinogenesis in the buccal pouch of hamsters, a model for oral carcinogenesis.[119–121] Supplementation with β-carotene and vitamin C has been found to inhibit chemically induced pancreatic carcinogenesis in rats and hamsters.[122–124] Recently, α-tocopherol supplementation was found to inhibit hepatic carcinogenesis in transgenic mice that overexpress transforming growth factor-α (TGF-α).[125,126]

21.3.2 PROSPECTIVE COHORT STUDIES

Numerous retrospective case-control studies have identified inverse associations between dietary and serum levels of vitamin C, vitamin E, and β-carotene and cancer risk at many different sites.[96] Although case-control studies are useful in identifying differences in nutrient intake between individuals who develop cancer and matched cancer-free controls, they may be biased if those with cancer differ from controls in the accuracy of their dietary recall or if the disease affects serum micronutrient levels. Prospective cohort and nested case-control studies avoid many of the limitations of retrospective case-control studies and provide stronger support for relationships between nutrient intake and disease. The following discussion will review the results of prospective cohort and case-control studies examining the association of measures of vitamin C, vitamin E, and β-carotene intake with cancer incidence published since 1990. Some earlier studies may not have effectively controlled for smoking exposures; since smokers tend to eat fewer fruits and vegetables, this could bias findings with respect to micronutrient intake and cancer risk.[127,128] The reader is referred to several excellent reviews for information on epidemiological studies assessing the relationships between these nutrients and cancer risk published prior to 1990.[96,129–131]

21.3.2.1 Total Cancer Incidence

Five out of eight prospective cohort studies observed a significant inverse relationship between some measure of vitamin C intake and total cancer incidence or mortality.[132–139] In a 7-year study of more than 11,000 men and women in the U.S., the risk of death from cancer was 24% lower in women with a dietary vitamin C intake of more than 225 mg/d than in women whose dietary intake was less than 155 mg/d.[133] No significant relationship between vitamin C intake and cancer mortality was observed in men. In a study of 1556 men followed for 24 years, men with dietary vitamin C intakes higher than 113 mg/d had a risk of death from cancer that was 39% lower than men with dietary vitamin C intakes lower than 82 mg/d.[134] None of the four studies that directly assessed supplement use found significant associations between vitamin C supplements and overall cancer mortality.[132,133,135,136]

Serum or plasma ascorbate levels reflect total (dietary and supplemental) vitamin C intake as well as cigarette smoking, which is consistently associated with decreased plasma ascorbate levels.[10] Three out of the four prospective studies that assessed serum or plasma ascorbate concentrations found them to be inversely associated with overall cancer mortality in men. In a study of 2974 Swiss men followed for 17 years, plasma ascorbate levels consistent with vitamin C intakes higher than 55 mg/d were associated with a risk of death from cancer that was 25% lower than plasma concentrations reflecting vitamin C intakes lower than 55 mg/d.[137] A study in the U.S. that followed more than 7000 men and women for 12 to 16 years found that men with serum ascorbate concentrations in the highest quartile were 62% less likely to die of cancer than men with serum concentrations in the lowest quartile, which was consistent with intakes of less than 60 mg/d.[138] No significant association between serum ascorbate concentrations and cancer mortality was observed in women. In a study of more than 19,000 men and women followed for 4 years, men in the highest quintile of plasma ascorbate, corresponding to a mean intake of 109 mg/d, were 53% less likely to die of cancer than men in the lowest plasma ascorbate quintile, corresponding to a mean intake of 51 mg/d.[139] Again, no significant association between plasma ascorbate concentration and cancer mortality was observed in women. Taken together, the data from prospective studies suggest that low dietary vitamin C intakes (less than 60 mg/d) increases cancer risk, especially in men.[9]

Over the past decade, prospective cohort studies have not found vitamin E intake from diet[136] and supplements,[133,135] or plasma vitamin E concentrations,[136,137,140] to be associated with total cancer incidence or mortality. Dietary β-carotene intake was not significantly associated with cancer mortality in middle-aged men followed for 24 years[134] or in a large cohort of elderly men and women followed for 8 years.[133] Lipid-adjusted plasma carotene levels (20% α-carotene and 80% β-carotene) were inversely associated with total cancer mortality over 17 years in a cohort of Swiss men,[137] but plasma β-carotene levels were not associated with total cancer mortality in a cohort of Swedish men followed for 25 years. In evaluating prospective studies utilizing plasma β-carotene levels, it is important to keep in mind that they are highly correlated with vegetable intake, which has been consistently inversely associated with cancer risk.[141,142] In order to examine the effect of multiple dietary micronutrients on total cancer mortality, Van Hoydonck and colleagues[143] constructed an oxidative balance score, in which those with low scores had high dietary intakes of vitamin C and β-carotene and low dietary iron intakes. In 2814 male smokers followed over 10 years, those with high oxidative balance scores (mean intake: vitamin C, 60 mg/d; β-carotene, 1.3 mg/d; iron, 17 mg/d) had total cancer mortality rates that were 62% higher than those with low oxidative balance scores (mean intake: vitamin C, 134 mg/d; β-carotene, 5 mg/d; iron, 16 mg/d).

21.3.2.2 Lung Cancer

Five out of 10 prospective cohort studies have observed significant inverse associations between vitamin C intake and lung cancer risk.[133,137,144–151] In a cohort of 872 Dutch men followed over a period of 25 years, those with dietary vitamin C intakes greater than 100 mg/d had a risk of lung cancer that was 60% lower than those with intakes less than 60 mg/d.[148] In a much larger cohort of 58,279 Dutch men followed over 6 years, those with dietary vitamin C intakes higher than 138 mg/d had a risk of lung cancer that was 23% lower than those with intakes less than 82 mg/d.[151] However, dietary vitamin C intake was inversely related to lung cancer risk only in nonsmokers in a cohort of 4538 Finnish men.[144] In a large U.S. cohort of more than 10,000 men and women followed over 19 years, men and women with dietary vitamin C intakes greater than 113 mg/d had a risk of lung cancer that was 34% lower than those with intakes less than 23 mg/d,[149] while in a larger cohort of more than 47,000 U.S. men and women, total vitamin C intake (diet and supplements) was inversely related to lung cancer risk in men, but not women.[150] In those studies that evaluated supplement use, no associations between vitamin C supplement use and lung cancer risk were observed.[133,146,151] Plasma ascorbate concentrations were not associated with lung cancer risk in two prospective studies.[137,147] Although smoking remains the most significant risk factor for lung cancer, prospective studies suggest that dietary vitamin C intakes of greater than 100 mg/d are associated with a decreased risk of lung cancer in men compared with dietary intakes of less than 60 mg/d. Although this effect is likely to hold true for women, fewer prospective studies have examined the relationship between vitamin C intake and lung cancer in women, despite the fact that it is the most common cause of cancer mortality in women.

Only two out of nine prospective studies observed an inverse association between measures of vitamin E intake and lung cancer risk.[133,137,144,147–152] None of the prospective studies that examined vitamin E supplement use found it to be inversely associated with lung cancer risk.[133,150,151] In a cohort of more than 10,000 men and women followed for 19 years, dietary vitamin E intake was inversely associated with lung cancer risk only in smokers in the lowest tertile of cumulative smoking exposure (≤33 pack years).[149] Analysis of baseline dietary vitamin E intake and serum α-tocopherol levels in the 29,133 Finnish male smokers who participated in the Alpha-Tocopherol Beta-Carotene (ATBC) study revealed that those with dietary vitamin E intakes averaging 15.7 mg/d had a risk of lung cancer that was 23% lower than those with intakes averaging 7.6 mg/d.[152] Cholesterol-adjusted baseline serum α-tocopherol concentrations were also inversely associated with lung cancer risk, with a stronger inverse relationship observed in those men who had smoked for less than 40 years. These subgroup analyses suggest that dietary vitamin E intake may be protective in smokers below a certain threshold of cumulative tobacco exposure.

Dietary intakes of β-carotene or provitamin A carotenoids were inversely associated with lung cancer risk in only 4 out of 11 prospective studies.[133,144–146,148–151,153–155] In a study of 4538 Finnish men followed for 20 years, dietary carotenoid (estimated at 45% β-carotene) intake was not significantly associated with lung cancer risk in smokers, but the risk of lung cancer in nonsmokers in the highest tertile of carotenoid intake was 2.5 times lower than that of nonsmokers in the lowest tertile.[144] In 561 Dutch men followed for 20 years, those with low stable β-carotene intakes (i.e., intakes in the lowest tertile assessed by diet history in 1960, 1965, and 1970) had more than twice the risk of lung cancer than those with stable β-carotene intakes in the highest tertile.[148] Dietary intake of provitamin A carotenoids (calculated by summing the vitamin A intake from fruits and vegetables) was inversely associated with lung cancer risk in smokers, but not nonsmokers, in a cohort of more than 10,000 men and women followed for 19 years.[149] Smokers with provitamin A carotenoid intakes in the highest quartile had a risk of lung cancer that was 51% lower than smokers with intakes in the lowest quartile. In a U.S. cohort of 27,456 men and 20,456 women followed for 7 years, men in the highest tertile of carotenoid intake (determined by a questionnaire designed to assess vitamin A intake) had a risk of lung cancer 27% lower than that of men in the lowest tertile of intake.[150] Carotenoid intake was not significantly associated with lung cancer in women; however, only 130 cases of lung cancer were identified in women compared with 395 cases in men.

In contrast, several of the largest prospective studies using the best assessments of dietary content of total and individual carotenoids observed significant inverse associations between the dietary intake of some carotenoids and lung cancer risk, but not for β-carotene.[151,153] Analysis of dietary carotenoid intake and lung cancer risk in the Nurses' Health Study (77,283 women followed for 12 years) and the Health Professionals Follow-up Study (46,924 men followed for 10 years) revealed no significant association between dietary β-carotene intake and lung cancer risk, even when the two cohorts were pooled.[153] However, total carotenoid intake, α-carotene, and lycopene intake were significantly inversely associated with lung cancer risk in the pooled sample, suggesting that dietary intake of some carotenoids decreases lung cancer risk. In a nested case-control study of a Dutch cohort comparing dietary carotenoid intakes in 939 men who developed lung cancer over a 6-year follow-up period with 1525 cancer-free men, only β-cryptoxanthin and lutein plus zeaxanthin intakes were inversely associated with lung cancer risk.[151] However, adjustment for dietary vitamin C intake, which was also inversely associated with lung cancer risk, eliminated all significant associations between individual carotenoids and lung cancer risk. In general, results of recent prospective studies using more accurate estimates of dietary carotenoid content suggest that diets rich in a number of carotenoids, rather than only β-carotene, may be inversely associated with lung cancer risk.

In contrast, serum β-carotene was inversely associated with lung cancer risk in three out of four prospective studies. Lipid-adjusted plasma carotene levels (80% β-carotene and 20% α-carotene) were inversely associated with lung cancer risk in 2974 Swiss men followed for 17 years.[137] Serum β-carotene concentrations as well as serum concentrations of other carotenoids were inversely associated with lung cancer in a nested case-control study that followed men and women in the U.S. for up to 15 years.[147] A recent analysis of baseline dietary and serum carotenoid levels in the 27,084 male smokers who participated in the ATBC intervention trial found that baseline serum β-carotene concentrations were inversely associated with lung cancer risk, but dietary β-carotene intake was not.[154] Interestingly, baseline lycopene, β-cryptoxanthin, lutein plus zeaxanthin, and total carotenoid intakes were significantly inversely associated with lung cancer risk, suggesting that the intake of carotenoids other than β-carotene may partially explain the inverse association of serum β-carotene concentrations with lung cancer risk. It is important to note that serum β-carotene levels are affected by factors other than dietary intake, leaving open the possibility that low serum β-carotene levels reflect other physiological processes related to cancer risk. For example, cigarette smoking is known to decrease serum β-carotene concentrations. Moreover, serum β-carotene concentrations are inversely associated with markers of systemic inflammation, such as serum C-reactive protein (CRP) levels, in both smokers and nonsmokers.[156,157] Thus, it is possible that systemic inflammation, rather than low β-carotene intake, could explain inverse associations between serum β-carotene concentrations and lung cancer risk, even after adjusting the data for smoking.

21.3.2.3 Gastric Cancer

Case-control studies have consistently demonstrated strong inverse associations between gastric cancer risk and vitamin C intake from fruits and vegetables.[96,158–161] In a large ecologic study that followed more than 12,000 men in seven countries for 25 years, dietary vitamin C intake was also inversely associated with gastric cancer mortality. However, only one out of three prospective cohort studies that specifically reported stomach cancer risk observed an inverse association with dietary vitamin C intake that bordered on significance.[162–164] After 6 years of follow-up of more than 120,000 Dutch men and women, the risk of stomach cancer in those consuming less than 55 mg/d of vitamin C was 30% higher than that of the two-thirds consuming more than 55 mg/d (p = 0.06).[162] A prospective study of 2974 Swiss men found low plasma vitamin C levels consistent with intakes of less than 55 mg/d to be associated with increased stomach cancer mortality after 12 years of follow-up,[165] but the inverse relationship was not significant at 17 years of follow-up.[137] A very large prospective study found that those who used vitamin C supplements for less than 10 years at enrollment had a risk of stomach cancer that was 32% lower than those who did not use vitamin C supplements.[166] However, the risk of stomach cancer in those who had been using vitamin C supplements for 10 years or longer was not different than the risk in nonusers. Although prospective studies offer limited evidence that low vitamin C intakes increase the risk of gastric cancer, the relationship may be complicated by *Helicobacter pylori* (*H. pylori*) infection, a known risk factor for gastric cancer that can decrease the vitamin C content of gastric juice (see Section 21.3.4.4).[167]

In general, prospective cohort studies have not observed dietary or supplemental vitamin E intake to be significantly associated with gastric cancer risk.[137,162,164,166] Despite the fact that numerous case-control studies have found increased β-carotene intake from fruits and vegetables to be inversely associated with gastric cancer risk,[96,168] only one out of four prospective studies found a significant inverse association between β-carotene intake and gastric cancer risk.[137,162–164] In that study of 34,691 postmenopausal women, those whose intake of dietary α-carotene plus β-carotene was in the highest tertile had a risk of gastric cancer 70% lower than those whose intake was in the lowest tertile.[164] However, only 26 cases of gastric cancer were identified during the 7-year follow-up period. Although the evidence for a protective effect of fruit and vegetable intake against gastric cancer is strong, the evidence that α-tocopherol or β-carotene play a role in that protective effect is weak.

21.3.2.4 Colorectal Cancer

Although fruit and, more consistently, vegetable intakes are inversely associated with colorectal cancer risk, the epidemiological evidence for vitamin C as a specific protective factor is less consistent.[169] Since 1990, two out of four prospective studies have observed a significant inverse association between vitamin C intake and the risk of colon or rectal cancer.[133,137,170,171] In a prospective cohort study that followed 4277 elderly men and 7300 elderly women over 8 years, women who consumed more than 225 mg/d of vitamin C had a 39% lower risk of colon cancer than women who consumed less than 155 mg/d, but no association between vitamin C intake and colon cancer risk was observed for men.[133] A similar decrease in colon cancer risk was observed in women who took vitamin C supplements. A recent study of colon cancer mortality in more than 700,000 U.S. men and women found that the use of vitamin C supplements for at least 10 years was associated with a 60% reduction in rectal cancer risk, but was not significantly associated with colon cancer risk.[171]

Neither dietary nor serum β-carotene levels have been significantly inversely associated with colorectal cancer risk,[133,137,170,172] and only one out of seven prospective studies observed a significant inverse association between measures of vitamin E intake and colorectal cancer risk.[133,137,170–174] In that study, total vitamin E intake (diet plus supplements) but not dietary vitamin E intake was inversely associated with colon cancer risk in 35,215 postmenopausal women followed for 5 years.[170] In fact, women who used a daily supplement containing ≥30 IU of vitamin E had a risk of colon cancer that was 50% lower than that of women who did not use a vitamin E supplement. Further subgroup analysis revealed that total vitamin E intake was inversely associated with colon cancer risk only in those women who had no family history of colon cancer.[175] However, the use of vitamin E supplements was not associated with significantly lower risks of colorectal cancer in several other large cohorts, including a cohort of 711,891 men and women followed for 14 years[171] and 87,998 women followed for 16 years through the Nurses' Health Study cohort.[174] In 47,344 men followed for 10 years through the Health Professionals Follow-Up Study, a 20 to 27% decrease in the risk of colon cancer for those who took at least 300 IU/d of supplemental vitamin E compared with those who had never used vitamin E supplements was not statistically significant.[174]

21.3.2.5 Bladder Cancer

Dietary vitamin C intake was not significantly associated with bladder cancer in four prospective studies.[133,176–178] However, vitamin C supplement use was significantly associated with a reduction in bladder cancer risk in a cohort of elderly men followed for 8 years,[133] and a trend toward an inverse relationship (p = 0.08) was observed for at least 10 years of vitamin C supplement use in the Health Professionals Follow-Up Study.[176] However, a prospective study of supplement use in 991,522 men and women did not observe a significant association between vitamin C supplement use and bladder cancer mortality.[179] Overall, the evidence for a protective effect of dietary vitamin C intake against bladder cancer is weak, although there is some evidence for a protective effect of supplemental vitamin C intake, which is more likely to increase urinary vitamin C concentrations than dietary intake.

Recent prospective studies have not observed significant inverse associations between dietary intake or serum levels of β-carotene and bladder cancer risk.[133,176–178,180] Similarly, significant associations between dietary vitamin E or plasma α-tocopherol levels and bladder cancer risk have not been observed in recent prospective studies.[177,178,180] However, two out of three prospective studies found significant inverse associations between measures of vitamin E supplement use and bladder cancer risk. In 12 years of follow-up, men in the Health Professionals Follow-Up Study with vitamin E intakes averaging 415 mg/d had a risk of bladder cancer that was 36% lower than that of men whose vitamin E intakes averaged 5 mg/d, and men who took vitamin E supplements for at least 10 years had a risk of bladder cancer that was 32% lower than that of men who had never used vitamin E supplements.[153] Additionally, in a cohort of 991,522 men and women followed over 16 years, those who had used vitamin E supplements for at least 10 years had a risk of death from bladder cancer that was 40% lower than the risk of those who had never used vitamin E supplements.[179] Although residual confounding of supplement use by other healthy behaviors cannot be ruled out in these studies, they support the idea that the use of vitamin E supplements over relatively long periods of time may decrease the incidence of bladder cancer and highlight the need for further research into the role of the urinary metabolites of vitamin E in bladder carcinogenesis.

21.3.2.6 Prostate Cancer

The results of recent prospective studies do not support a relationship between vitamin C intake and prostate cancer risk.[133,137,181,182] Only two out of eight prospective studies found a measure of vitamin E intake to be inversely associated with prostate cancer risk.[133,137,182–187] Serum vitamin E levels were inversely associated with prostate cancer mortality in smokers but not nonsmokers in a cohort of Swiss men followed over 17 years.[137] In another cohort, a nested case-control study that compared baseline plasma vitamin E concentrations in 117 men who developed prostate cancer and 233 men who remained cancer free during 17 years of follow-up found that plasma γ-tocopherol but not α-tocopherol was inversely associated with prostate cancer risk.[187] Moreover, significant inverse associations for plasma α-tocopherol and toenail selenium concentrations with prostate cancer risk were observed only when γ-tocopherol levels were also high. However, two other prospective studies did not observe significant inverse associations between plasma γ-tocopherol concentrations and prostate cancer risk.[184,185] Despite recent findings that lycopene intake is inversely associated with prostate cancer risk,[185,188] recent prospective studies do not support a relationship between β-carotene intake and prostate cancer risk.[133,137,181–184,188]

21.3.2.7 Breast Cancer

Although the etiology of breast cancer is multifactorial with genetic and hormonal influences, increased oxidative stress may also contribute to breast cancer risk in susceptible individuals.[189] However, none of the seven prospective studies that assessed dietary vitamin C intake,[133,190–195] nor the single prospective study that measured plasma vitamin C levels,[196] observed a significant association with breast cancer risk in women. Two recent prospective studies found dietary vitamin C intake to be inversely associated with breast cancer risk in subgroups. In the Nurses' Health Study cohort, premenopausal women with a family history of breast cancer who consumed an average of 205 mg/d of vitamin C from foods had a risk of breast cancer that was 63% lower than that of women who consumed an average of 70 mg/d.[194] In the Swedish Mammography Cohort, women with a body mass index (BMI) greater than 25 kg/m^2 who consumed an average of 110 mg/d had a 39% lower risk of breast cancer than those who consumed an average of 31 mg/d.[195] Still, there is little evidence from prospective studies that vitamin C intake is associated with breast cancer risk in the majority of cases.

None of the recent prospective studies that examined serum tocopherol levels[197–199] or vitamin E intake from diet or supplements[133,190–195] observed significant inverse associations with breast cancer risk. Likewise, seven out of eight recent prospective studies found no significant association between dietary β-carotene intake and breast cancer risk.[133,190–195,200] Although β-carotene intake was not associated with breast cancer risk in the Nurses' Health Study cohort, subgroup analysis revealed that dietary β-carotene intake was inversely associated with the risk of premenopausal breast cancer, as was lutein plus zeaxanthin intake.[194] For women with a family history of breast cancer, the risk of premenopausal breast cancer was 62% lower for those who consumed an average of 7694 μg/d of β-carotene compared with those women consuming 1609 μg/d. Two out of four recent prospective studies found serum β-carotene levels to be inversely associated with breast cancer risk.[197–199,201] In both studies that observed a significant inverse association, women with serum β-carotene concentrations in the highest quartile had approximately one-half the breast cancer risk of those women with serum β-carotene concentrations in the lowest quartile.[199,201] Interestingly, total serum carotenoid concentrations were also inversely associated with breast cancer risk in those studies, and serum lycopene levels were inversely associated with breast cancer risk in all four of those studies, although not significantly in one of them.[201] These findings suggest that increasing consumption of a number of dietary carotenoids instead of β-carotene in isolation may offer some protection from breast cancer.

21.3.3 INTERVENTION TRIALS: CANCER

21.3.3.1 Esophageal and Gastric Cancer

A randomized controlled trial in Linxian, China, assessed the effects of 5 years of supplementation with four different vitamin/mineral combinations in 26,584 undernourished men and women at very high risk for esophageal and gastric cancer.[4] Esophageal cancer incidence and mortality were not significantly altered by a combination of β-carotene (15 mg/d), α-tocopherol (30 mg/d), and selenium (50 μg/d) or a combination of vitamin C (120 mg/d) and molybdenum (30 μg/d), and neither vitamin/mineral combination affected the prevalence of esophageal dysplasia or early esophageal cancer in a subgroup that underwent endoscopy at the end of the 5-year intervention.[202] In contrast, supplementation with the β-carotene, α-tocopherol, and selenium combination was associated with a significant 21% decrease in gastric cancer mortality, although supplementation with the vitamin C and molybdenum combination did not significantly alter the risk of gastric cancer.[4] Unfortunately, no attempt was made to eradicate *H. pylori*, which likely limited increases in gastric juice vitamin C levels. A randomized trial that will assess the efficacy of *H. pylori* treatment and/or antioxidant supplementation in reducing the prevalence of gastric dysplasia and gastric cancer in a high-risk area of China is currently underway.[203]

21.3.3.2 Lung Cancer

Epidemiological data suggesting an inverse relationship between serum levels and dietary intake of β-carotene with lung cancer led to two randomized controlled intervention trials aimed at determining whether supplementation of β-carotene with other micronutrients could decrease the risk of lung cancer in high-risk populations.[3] Using a 2 × 2 factorial design, the ATBC cancer prevention trial studied the effects of supplementing 29,133 Finnish male smokers with 50 mg/d of α-tocopherol and 20 mg/d of β-carotene for 6 years.[5] The β-Carotene and Retinol Efficacy Trial (CARET) examined the effect of supplementing 18,314 men and women in the U.S. who were smokers or former smokers, or who had a history of occupational asbestos exposure, with 30 mg/d of β-carotene and 25,000 IU/d of retinol over 4 years.[6] Unexpectedly, the risk of lung cancer was significantly increased by 16% in those supplemented with β-carotene in the ATBC trial and by 28% in those supplemented with β-carotene and retinol in the CARET trial. α-Tocopherol supplementation was not significantly associated with lung cancer risk in the ATBC trial. The Physicians' Health Study (PHS) examined the effect of β-carotene supplementation (50 mg on alternate days) on cancer risk in 22,071 male physicians in the U.S., of whom only 11% were current smokers.[204] In that lower risk population, β-carotene supplementation for more than 12 years was not associated with an increased risk of lung cancer. In fact, β-carotene supplementation did not increase or decrease the risk of any type of cancer in the group as a whole.[205] An analysis of the effect of 2 years of β-carotene supplementation (50 mg on alternate days) in 39,876 women who took part in the Women's Health Study (13% current smokers) also found no association between β-carotene supplementation and lung cancer.[206]

21.3.3.3 Why Did β-Carotene Supplementation Increase Lung Cancer Risk in the ATBC and CARET Studies?

It is not yet known why high-dose β-carotene supplementation significantly increased the risk of lung cancer in smokers and other high-risk individuals, but a number of *in vitro* and *in vivo* studies published since those surprising results were made public offer some insight into possible mechanisms. In humans, β-carotene from supplements is much more readily absorbed than β-carotene from fruits and vegetables, resulting in plasma and tissue β-carotene levels many times higher than could be achieved by a carotenoid-rich diet.[207] The ferret, unlike most rodents, is an appropriate species for modeling the effects of β-carotene supplementation and cigarette smoke exposure because tissue accumulation of β-carotene and lung pathology in response to tobacco exposure in ferrets is similar to that of humans.[208] In ferrets, the presence of high concentrations of β-carotene in lung tissue results in increased formation of oxidative metabolites of β-carotene.[209] These oxidative metabolites could promote carcinogenesis in the lung by inducing cytochrome P450 enzymes that bioactivate procarcinogens in cigarette smoke,[210,211] enhancing DNA binding of benzo[a]pyrene metabolites,[212] or altering retinoid signaling.[209] In the CARET study, the increased risk of lung cancer was limited to current smokers.[213] All of the participants were smokers in the ATBC study, but the effect of β-carotene was stronger in those who smoked at least 20 cigarettes daily.[214] Alcohol use is also known to increase cytochrome P450 activity, and heavier alcohol use appeared to increase the effect of β-carotene supplementation on lung cancer risk in the ATBC and the CARET studies.[213,214]

The lung-cancer-promoting effects associated with β-carotene supplementation in the ATBC and CARET studies may also have been dose related, which would explain the discrepancy between these intervention trials and the consistent inverse associations between dietary β-carotene consumption and lung cancer risk. When ferrets were given a dose of β-carotene equivalent to a human dose of 30 mg/d, retinoid signaling in lung tissue was diminished and lung cell proliferation was increased.[215] These potentially procarcinogenic changes were enhanced by cigarette smoke exposure. In contrast, these effects were not observed when ferrets were given a dose of β-carotene equivalent to a human dose of 6 mg/d (the amount supplied by five to nine servings of fruits and vegetables). In fact, cigarette smoke-induced lung damage (squamous metaplasia) was slightly decreased. The lungs of cigarette smokers have been called a "free radical-rich, antioxidant-poor environment."[208] Cigarette smokers have significantly lower plasma vitamin C concentrations than nonsmokers with equal vitamin C intakes,[216] as do nonsmokers exposed to environmental tobacco smoke.[217] Studies *in vitro* suggest that vitamin C can reduce the β-carotene radical back to β-carotene,[218] potentially decreasing the formation of its oxidative metabolites. Vitamin E may also stabilize β-carotene and prevent the formation of excentric cleavage products known as apo-carotenals.[219] Apo-carotenals can be further oxidized to their corresponding retinoic acids. The biological activities of apo-carotenal metabolites have not been well characterized, but there is some evidence that they differ from those of retinoic acid.[24,209] Thus, high concentrations of β-carotene along with relative deficiencies of vitamin C and possibly vitamin E in an oxidizing environment may also have contributed to an increased risk of lung cancer.

21.3.3.4 Prostate Cancer

Although supplementation with β-carotene (50 mg on alternate days) for more than 12 years did not significantly alter prostate cancer incidence in the entire PHS sample, β-carotene supplementation significantly decreased prostate cancer incidence by 32% in those with baseline serum β-carotene concentrations in the lowest quartile.[220] In the ATBC trial, supplementation with 50 mg/d of α-tocopherol for 6 years resulted in a significant 34% decrease in prostate cancer incidence, while supplementation with β-carotene did not significantly alter prostate cancer incidence.[5] The mechanism for the protective effect of this relatively low dose of vitamin E (equivalent to 25 mg *RRR*-α-tocopherol) is not clear. Inhibition of tumor angiogenesis by α-tocopherol represents one potential mechanism that has been demonstrated in animal studies.[221] Vascular endothelial growth factor (VEGF) is a cytokine that plays a critical role in tumor angiogenesis. When serum VEGF concentrations were compared in a subset of ATBC study participants over the intervention period, a significant 11% decrease was observed in those taking α-tocopherol supplements, while serum VEGF concentrations increased by 10% in those taking a placebo.[222]

21.3.3.5 Other Cancers

A placebo-controlled trial involving 1805 patients with a previous nonmelanoma skin cancer found supplementation with 50 mg/d of β-carotene to be of no benefit in preventing the occurrence of new nonmelanoma skin cancers over a 5-year period.[223] Likewise, supplementation of men in the PHS with 50 mg of β-carotene on alternate days for more than 12 years did not significantly decrease the incidence of nonmelanoma skin cancer.[224] Second cancers of the head and neck are the leading cause of death in patients diagnosed with early-stage head and neck cancers. Although there is some evidence that β-carotene supplementation can increase the regression of oral precancerous lesions, a placebo-controlled trial in 264 patients who had been curatively treated for an early-stage oral, pharyngeal, or laryngeal squamous cell carcinoma found supplementation with 50 mg/d of β-carotene for more than 4 years to be of no benefit in preventing second primary tumors.[225] The effect of α-tocopherol and β-carotene supplementation on a number of cancers has been assessed in the ATBC study cohort. Neither supplement significantly decreased the incidence of colorectal,[226] urinary tract,[227] gastric,[228] or pancreatic cancer.[229] Recently, a large randomized placebo-controlled trial found that supplementation with a combination of antioxidants (250 mg/d of vitamin C, 600 mg/d of vitamin E, 20 mg/d of β-carotene) for more than 5 years did not significantly affect total cancer incidence or total cancer mortality in 20,536 adults at high risk for cardiovascular diseases.

21.3.4 Intervention Trials: Precancerous Lesions and Surrogate Endpoint Biomarkers

21.3.4.1 Oral Leukoplakia

Oral leukoplakia is a white lesion of the mouth that has the potential to develop into squamous cell carcinoma. The rate of malignant transformation is highly variable in different populations and ranges from less than 1% to approximately 20% in 1 to 30 years.[230] A number of small uncontrolled trials have reported some degree of clinical improvement in 44 to 71% of those treated after supplementation with β-carotene (30 to 90 mg/d)[231,232] or combinations of β-carotene and other antioxidants[233,234] for 3 to 6 months. However, these trials may be too short to provide information about the actual benefit of β-carotene therapy, since less than half of all leukoplakias undergo malignant transformation within 2 years of diagnosis. In a longer trial, 50 men and women with oral leukoplakia were treated with 60 mg/d of β-carotene for 6 months.[235] In the 23 participants who consented to biopsies, 39% improved by at least one grade of dysplasia. After 6 months, those who were judged to have had a clinical response to the treatment (52%) were randomized to continued β-carotene treatment at 60 mg/d or a placebo for an additional 12 months. The rate of relapse was similar between the groups treated with β-carotene (18%) and placebo (17%), suggesting that remissions induced by high-dose β-carotene supplementation are sustainable for at least one year in most oral leukoplakia patients. Two randomized controlled trials have examined the effect of β-carotene alone or in combination with other nutrients on oral leukoplakia in high-risk populations. In Uzbekistan, a combination of 100,000 IU/week of retinol, 40 mg/d of β-carotene, and 80 mg/week of vitamin E for 20 months significantly decreased the prevalence of oral leukoplakia by 35% in a study of 532 men with oral leukoplakia or chronic esophagitis.[236a] Supplementation of 131 Indian oral leukoplakia patients with 360 mg/week of β-carotene significantly improved the regression rate to 33% compared with 10% for placebo.[236b] However, supplementation with 300 IU/week of retinol resulted in a regression rate of 52%. Smokers and tobacco chewers are at increased risk of oral leukoplakia. Given the potential for long-term oral β-carotene supplementation to increase lung cancer risk, its utility as a long-term treatment for oral leukoplakia must be questioned.

21.3.4.2 Cervical Intraepithelial Neoplasia

Invasive cervical cancer ultimately develops from the progression of precancerous lesions of the cervix called cervical intraepithelial neoplasia (CIN). Mild dysplasia (CIN I) regresses spontaneously in approximately 60% of cases and has a low rate of progression to invasive carcinoma. However, 16 to 36% of lesions classified as moderate to severe dysplasia (CIN II or CIN III) will eventually progress to invasive cervical cancer.[237] Infection with an oncogenic strain of human papilloma virus (HPV) is now considered to be a causal factor for cervical cancer.[238] Although one small uncontrolled trial reported that 6 months of supplementation with 30 mg/d of β-carotene resulted in a 70% regression of CIN I-II lesions, four larger randomized placebo-controlled trials did not find β-carotene supplementation (10 to 30 mg/d) for 3 to 24 months to increase regression or decrease progression of CIN lesions.[239–241] A 2-year randomized controlled trial in 141 women with the low-grade cervical lesions, squamous atypia or CIN I did not find β-carotene at 30 mg/d, vitamin C at 500 mg/d, or a combination of the two to improve regression to normal or decrease progression to CIN II.[242] Findings from intervention trials in precancerous cervical lesions do not support observations from retrospective case-control studies of inverse associations between serum and dietary β-carotene levels and cervical cancer risk.

21.3.4.3   Colorectal Adenomas and Colorectal Epithelial Cell Proliferation

Several intervention trials have examined the effect of antioxidant supplementation on colorectal adenomas (precancerous polyps) and colorectal epithelial cell proliferation as biomarkers of colorectal cancer risk. Although four out of five familial adenomatous polyposis (FAP) patients who took 3 g/d of vitamin C for 4 months experienced regression of rectal adenomas,[243] a placebo-controlled trial of 3 g/d of vitamin C for 18 months in 36 FAP patients did not find a significant difference in the regression of rectal adenomas between vitamin C and placebo treatment.[244] Other intervention trials aimed at decreasing the growth or recurrence of colorectal adenomas have employed combinations of antioxidant supplements. While one preliminary trial suggested marked beneficial effects of a combination of vitamins C, E, and A given for 18 months,[245] a large randomized controlled trial in 751 patients with previous colorectal adenomas did not find 30 mg/d of β-carotene or a combination of 400 mg/d of α-tocopherol and 1000 mg/d of vitamin C to decrease adenoma recurrence over a 4-year period.[246] Similarly, several smaller randomized controlled trials did not find supplementation with β-carotene[247] or combinations of α-tocopherol and vitamin C[248,249] to decrease colorectal adenoma recurrence over periods of 2 to 4 years. Analysis of the occurrence of colorectal adenomas (146 cases) in 15,538 male smokers who participated in the ATBC cancer prevention trial revealed that α-tocopherol supplementation (50 mg/d) increased the risk of being diagnosed with a colorectal adenoma over 6 years, but that β-carotene supplementation (20 mg/d) did not significantly affect that risk.[250] Men taking α-tocopherol supplements reported slightly more prediagnostic bleeding, which may have led to more colonoscopies and an increase in the detection of colorectal adenomas in this group. α-Tocopherol supplementation did not affect the risk of colorectal cancer in the ATBC cancer prevention study.[172]

   Several small intervention trials have used indices of colorectal epithelial cell proliferation as intermediate endpoints. Supplementation with 3 g/d of vitamin C for at least 18 months resulted in a significantly decreased proliferation index in rectal epithelial cells compared with placebo in patients with FAP.[244] In another small trial, supplementation with 750 mg/d of vitamin C or 9 mg/d of β-carotene but not 160 mg/d of vitamin E for one month decreased cell proliferation in colonic crypts of adenoma patients compared with the baseline.[251] However, supplementation of colorectal adenoma or colorectal cancer patients with 30 mg/d of β-carotene for 3 months decreased indices of colonic cell proliferation when compared with a placebo.[252] A study of colon cancer patients randomized to receive a combination of 1000 mg/d of vitamin C, 70 mg/d of α-tocopherol, 30,000 IU/d of vitamin A, and 2000 mg/d of calcium or placebo for 6 months found that colonic epithelial cell proliferation rates decreased from baseline in both groups, but they did not differ significantly between the treatment and placebo groups.[253] In general, controlled trials have not found vitamin C, vitamin E, or β-carotene supplementation to decrease the recurrence of colorectal adenomas, and there is only limited evidence that vitamin C or β-carotene supplementation decreases indices of colonic cell proliferation.

21.3.4.4   *Helicobacter pylori* Infection and Precancerous Lesions of Gastric Mucosa

Chronic infection with *H. pylori* is associated with an increased risk of gastric cancer, and it also appears to inhibit secretion of vitamin C into the gastric juice. Consequently, less vitamin C is available to scavenge RNS and prevent the formation of potentially carcinogenic *N*-nitroso compounds.[254] *Helicobacter pylori* infection and low serum levels of vitamin C have been associated with increased risk of progression of precancerous conditions of the gastric mucosa, such as atrophic gastritis or intestinal metaplasia and dysplasia, to gastric cancer in high-risk populations.[255,256] Additionally, gastric mucosal concentrations of α-tocopherol and β-carotene are decreased in the presence of gastric atrophy and intestinal metaplasia.[257] In order to assess the effect of α-tocopherol (50 mg/d) and β-carotene (20 mg/d) supplementation on the occurrence of precancerous or cancerous lesions in the ATBC cancer prevention study, serum levels of pepsinogen I were determined at baseline.[258] The 2132 male smokers that had low serum pepsinogen I levels, indicative of atrophic gastritis, were invited to have gastroscopy after a median of 5 years of supplementation. Of the 1344 men who underwent gastroscopy, 4.7% had neoplastic changes, including moderate to severe dysplasia and carcinoma. Neither α-tocopherol nor β-carotene supplementation was significantly associated with the occurrence of gastric neoplasms in smokers with atrophic gastritis.

In a placebo-controlled study designed to evaluate the effect of *H. pylori* eradication with triple antibiotic therapy or supplementation with β-carotene (30 mg/d) or vitamin C (2 g/d) on the course of gastric atrophy and metaplasia, 856 high-risk patients were followed by gastric biopsy over a 6-year period.[259] Each treatment resulted in significant five-fold increase in the regression rate for gastric atrophy and a three-fold increase in the regression rate for intestinal metaplasia. Successful eradication of *H. pylori* resulted in eight-fold and five-fold increases in regression rates for gastric atrophy and intestinal metaplasia, respectively. A randomized controlled trial of 500 mg/d of vitamin C after *H. pylori* eradication therapy in patients with intestinal metaplasia resulted in complete resolution of intestinal metaplasia in 9 out of 29 patients who received vitamin C for 6 months compared with 1 out of 29 who received placebo.[260] The use of vitamin C or β-carotene supplementation in combination with *H. pylori* eradication therapy shows promise in decreasing the progression of precancerous gastric lesions. Additional studies in high-risk populations are already ongoing, and these should provide additional information on the utility of antioxidant therapy for the prevention of gastric cancer.[167]

21.4   SUMMARY AND CONCLUSIONS

Epidemiological data provide strong and consistent evidence of an inverse association between fruit and vegetable intake and the incidence of a variety of cancers, especially vitamin C-rich citrus fruits and carotenoid-rich fruits and vegetables.[261] The evidence that vitamin C, vitamin E, and β-carotene play a substantial role in the anticancer effects of high fruit and vegetable intakes is much less convincing. The science of cancer chemoprevention would be limited if it approached cancer as only one disease, rather than many diseases with a variety of etiologies in varying target tissues. Similarly, to approach vitamin C, vitamin E, and β-carotene simply as antioxidants without considering their varied chemistry, metabolism, and biological activities would limit scientific investigation into their utility as cancer chemopreventive agents.

Numerous controlled supplementation trials have failed to demonstrate that vitamin C, vitamin E, or β-carotene decrease *in vivo* oxidative DNA damage in humans (Table 21.1). Yet, the findings of prospective cohort studies and controlled intervention trials suggest that these compounds may be beneficial in preventing specific types of cancer and in specific populations (Table 21.2).

TABLE 21.2
Vitamin C, Vitamin E, and β-Carotene: Summary of Recent Findings from Prospective Cohort Studies and Intervention Trials of Cancer Prevention

Cancer Site	Type of Studies	Antioxidant Nutrient	Summary of Findings
All sites	prospective cohort studies	vitamin C	Plasma ascorbate was inversely associated with total cancer mortality in men.[137–139]
	intervention trials	vitamin C	Supplementation with vitamin C and molybdenum for 5.3 years did not change total cancer mortality in an undernourished, high-risk population.[4]
		β-carotene	Supplementation of men for 12 years[205] and women for 2 years[206] with β-carotene did not change total cancer incidence in well-nourished populations.
		combinations	Supplementation with α-tocopherol, β-carotene, and selenium for 5.3 years decreased total cancer mortality by 13% in an undernourished, high-risk population.[4]
			Supplementation with vitamin C, α-tocopherol, and β-carotene for 5 years did not change total cancer mortality in a well-nourished population at high risk for cardiovascular diseases.[262]
Lung	prospective cohort studies	vitamin C	Dietary vitamin C intakes >100 mg/d were associated with decreased lung cancer risk compared with <60 mg/d in men.[148–151]
			Further research on women is needed.
	intervention trials	β-carotene	Supplementation of smokers with β-carotene for 6 years and of smokers and former asbestos workers with β-carotene + retinol for 4 years increased lung cancer risk by 16% and 36%, respectively.[213,214]
			Supplementation of a lower risk population (11% smokers) with β-carotene for 12 years did not affect lung cancer incidence.[205]
Stomach	prospective cohort studies	vitamin C	Limited evidence suggests vitamin C supplement use is inversely associated with gastric cancer risk.[166]
	intervention trials	combinations	Supplementation with α-tocopherol, β-carotene, and selenium for 5.3 years decreased gastric cancer mortality by 21% in an undernourished, high-risk population.[4]
Urinary bladder	prospective cohort studies	vitamin E	Vitamin E supplement use was inversely associated with bladder cancer incidence or mortality in two large cohorts.[176,179]
Prostate	intervention trials	vitamin E	Supplementation with α-tocopherol for 6 years decreased prostate cancer incidence in smokers by 34%.[5]
Head and neck	intervention trials	β-carotene	Supplementation with β-carotene for 6 months improved the regression of oral leukoplakia.[235,236]
			Supplementation with β-carotene for >4 years did not affect the recurrence of oral, pharyngeal, or laryngeal squamous cell carcinoma.[225]

Supplementation with β-carotene, α-tocopherol, and selenium in a malnourished population at high risk of gastric cancer in Linxian, China, was associated with a 21% decrease in mortality from gastric cancer.[4] In contrast, supplementation of well-nourished populations with vitamin C, α-tocopherol, and β-carotene, individually or in combinations, has not been found to decrease overall cancer incidence or mortality.[5,205,262] Two notable exceptions to these findings involve prostate cancer. Supplementation of Finnish smokers with 50 mg/d of synthetic α-tocopherol for 6 years was associated with a 34% decrease in prostate cancer incidence,[5] and β-carotene (50 mg on alternate days) supplementation of male physicians with low baseline serum β-carotene concentrations for 12 years was associated with a 32% decrease in prostate cancer incidence.[220] These data may indicate that men with low intake or increased destruction of these compounds, e.g., through oxidative stress, represent subgroups that could benefit from supplementation.

The investigation of interactions between genetic polymorphisms and cancer risk may result in the identification of individuals or groups that could benefit from supplementation. Glutathione S-transferase (GST) is a phase II xenobiotic metabolizing enzyme that plays an important role in the detoxification of a number of carcinogens. Genetic polymorphisms in the gene for GST μ-1 (*GSTM1*) may influence individual susceptibility to certain carcinogens. Findings from the ATBC cancer prevention trial indicate that individuals with the *GSTM1* null genotype may be more susceptible to the adverse effects of cumulative tobacco exposure on lung cancer risk, and that α-tocopherol supplementation may attenuate that susceptibility.[263]

Although further confirmation is needed, findings in two large cohorts that long-term vitamin E supplement use is associated with decreased incidence of and mortality from bladder cancer highlight the need for more research on the potential for vitamin E and its urinary metabolites to inhibit urinary bladder carcinogenesis.[176,179]

The unexpected finding that high-dose β-carotene supplementation (five times that found in five to nine servings of carotenoid-rich fruits and vegetables) increased lung cancer risk in two high-risk groups[213,214] highlights the need for thorough mechanistic research on carotenoids and their metabolites with respect to retinoid signaling in carcinogenesis, in addition to their antioxidant or pro-oxidant effects. Recent epidemiological studies indicating that diets rich in a variety of carotenoids are inversely associated with the risk of certain types of cancer should not be used to justify high-dose supplementation of one carotenoid in high-risk groups without supportive mechanistic research in appropriate models. Although the past decade of research has provided considerable insight into mechanisms by which antioxidant vitamins and carotenoids can influence carcinogenesis, it has not provided evidence that supplementation with these compounds can replace the need for a diet that is rich in fruits and vegetables with respect to cancer chemoprevention.

REFERENCES

1. Block, G., Patterson, B., and Subar, A., Fruit, vegetables, and cancer prevention: a review of the epidemiological evidence, *Nutr. Cancer*, 18, 1, 1992.
2. Steinmetz, K.A. and Potter, J.D., Vegetables, fruit, and cancer prevention: a review, *J. Am. Diet. Assoc.*, 96, 1027, 1996.
3. Peto, R., Doll, R., Buckley, J.D., and Sporn, M.B., Can dietary beta-carotene materially reduce human cancer rates? *Nature*, 290, 201, 1981.
4. Blot, W.J., Li, J.Y., Taylor, P.R., Guo, W., Dawsey, S., Wang, G.Q., Yang, C.S., Zheng, S.F., Gail, M., Li, G.Y. et al., Nutrition intervention trials in Linxian, China: supplementation with specific vitamin/mineral combinations, cancer incidence, and disease-specific mortality in the general population, *J. Natl. Cancer Inst.*, 85, 1483, 1993.
5. The Alpha-Tocopherol, Beta Carotene Cancer Prevention Study Group, The effect of vitamin E and beta carotene on the incidence of lung cancer and other cancers in male smokers, *New England J. Med.*, 330, 1029, 1994.
6. Omenn, G.S., Goodman, G.E., Thornquist, M.D., Balmes, J., Cullen, M.R., Glass, A., Keogh, J.P., Meyskens, F.L., Valanis, B., Williams, J.H., Barnhart, S., and Hammar, S., Effects of a combination

of beta carotene and vitamin A on lung cancer and cardiovascular disease, *New England J. Med.*, 334, 1150, 1996.

7. Levine, M., Rumsey, S.C., Daruwala, R., Park, J.B., and Wang, Y., Criteria and recommendations for vitamin C intake, *JAMA*, 281, 1415, 1999.

8. Levine, M., Conry-Cantilena, C., Wang, Y., Welch, R.W., Washko, P.W., Dhariwal, K.R., Park, J.B., Lazarev, A., Graumlich, J.F., King, J., and Cantilena, L.R., Vitamin C pharmacokinetics in healthy volunteers: evidence for a recommended dietary allowance, *Proc. Natl. Acad. Sci. USA*, 93, 3704, 1996.

9. Carr, A.C. and Frei, B., Toward a new recommended dietary allowance for vitamin C based on antioxidant and health effects in humans, *Am. J. Clin. Nutr.*, 69, 1086, 1999.

10. Food and Nutrition Board, Institute of Medicine, Vitamin C, in *Dietary Reference Intakes for Vitamin C, Vitamin E, Selenium, and Carotenoids*, National Academy Press, Washington, DC, 2000, p. 95.

11. Upston, J.M., Terentis, A.C., and Stocker, R., Tocopherol-mediated peroxidation of lipoproteins: implications for vitamin E as a potential antiatherogenic supplement, *FASEB J.*, 13, 977, 1999.

12. Traber, M.G., Vitamin E, in *Nutrition in Health and Disease*, 9th ed., Shils, M., Olson, J.A., Shike, M., and Ross, A.C., Eds., Williams & Wilkins, Baltimore, 1999, p. 347.

13. Kiyose, C., Muramatsu, R., Kameyama, Y., Ueda, T., and Igarashi, O., Biodiscrimination of alpha-tocopherol stereoisomers in humans after oral administration, *Am. J. Clin. Nutr.*, 65, 785, 1997.

14. Food and Nutrition Board, Institute of Medicine, Vitamin E, in *Dietary Reference Intakes for Vitamin C, Vitamin E, Selenium, and Carotenoids*, National Academy Press, Washington, DC, 2000, p. 95.

15. Halliwell, B. and Gutteridge, J.M.C., *Free Radicals in Biology and Medicine*, 3rd ed., Oxford University Press, New York, 1999.

16. Buettner, G.R., The pecking order of free radicals and antioxidants: lipid peroxidation, alpha-tocopherol, and ascorbate, *Arch. Biochem. Biophys.*, 300, 535, 1993.

17. Brigelius-Flohe, R., Kelly, F.J., Salonen, J.T., Neuzil, J., Zingg, J.M., and Azzi, A., The European perspective on vitamin E: current knowledge and future research, *Am. J. Clin. Nutr.*, 76, 703, 2002.

18. Behrens, W.A. and Madere, R., Alpha- and gamma-tocopherol concentrations in human serum, *J. Am. Coll. Nutr.*, 5, 91, 1986.

19. Jiang, Q., Lykkesfeldt, J., Shigenaga, M.K., Shigeno, E.T., Christen, S., and Ames, B.N., Gamma-tocopherol supplementation inhibits protein nitration and ascorbate oxidation in rats with inflammation, *Free Radic. Biol. Med.*, 33, 1534, 2002.

20. Jiang, Q., Christen, S., Shigenaga, M.K., and Ames, B.N., Gamma-tocopherol, the major form of vitamin E in the US diet, deserves more attention, *Am. J. Clin. Nutr.*, 74, 714, 2001.

21. Solomons, N.W., Vitamin A and carotenoids, in *Present Knowledge in Nutrition*, 8th ed., Bowman, B.A. and Russell, R.M., Eds., ILSI Press, Washington, DC, 2001, p. 127.

22. Food and Nutrition Board, Institute of Medicine, Beta-carotene and other carotenoids, in *Dietary Reference Intakes for Vitamin C, Vitamin E, Selenium, and Carotenoids*, National Academy Press, Washington, DC, 2000, p. 325.

23. Burton, G.W. and Ingold, K.U., Beta-carotene: an unusual type of lipid antioxidant, *Science*, 224, 569, 1984.

24. Krinsky, N.I., The antioxidant and biological properties of the carotenoids, *Ann. N.Y. Acad. Sci.*, 854, 443, 1998.

25. Hatta, A. and Frei, B., Oxidative modification and antioxidant protection of human low density lipoprotein at high and low oxygen partial pressures, *J. Lipid Res.*, 36, 2383, 1995.

26. Liebler, D.C., Stratton, S.P., and Kaysen, K.L., Antioxidant actions of beta-carotene in liposomal and microsomal membranes: role of carotenoid-membrane incorporation and alpha-tocopherol, *Arch. Biochem. Biophys.*, 338, 244, 1997.

27. Allard, J.P., Royall, D., Kurian, R., Muggli, R., and Jeejeebhoy, K.N., Effects of beta-carotene supplementation on lipid peroxidation in humans, *Am. J. Clin. Nutr.*, 59, 884, 1994.

28. Winklhofer-Roob, B.M., Puhl, H., Khoschsorur, G., van't Hof, M.A., Esterbauer, H., and Shmerling, D.H., Enhanced resistance to oxidation of low density lipoproteins and decreased lipid peroxide formation during beta-carotene supplementation in cystic fibrosis, *Free Radic. Biol. Med.*, 18, 849, 1995.

29. Pryor, W.A., Stahl, W., and Rock, C.L., Beta carotene: from biochemistry to clinical trials, *Nutr. Rev.*, 58, 39, 2000.

30. Woodall, A.A. and Ames, B.N., Diet and oxidative damage to DNA: the importance of ascorbate as an antioxidant, in *Vitamin C in Health and Disease*, Packer, L. and Fuchs, J., Eds., Marcel Dekker, New York, 1997, p. 193.

31. Wiseman, H. and Halliwell, B., Damage to DNA by reactive oxygen and nitrogen species: role in inflammatory disease and progression to cancer, *Biochem. J.*, 313 (Pt. 1), 17, 1996.

32. Ames, B.N., Shigenaga, M.K., and Hagen, T.M., Oxidants, antioxidants, and the degenerative diseases of aging, *Proc. Natl. Acad. Sci. USA*, 90, 7915, 1993.

33. Collins, A.R., Oxidative DNA damage, antioxidants, and cancer, *Bioessays*, 21, 238, 1999.

34. Halliwell, B., Vitamin C and genomic stability, *Mutat. Res.*, 475, 29, 2001.

35. Asami, S., Manabe, H., Miyake, J., Tsurudome, Y., Hirano, T., Yamaguchi, R., Itoh, H., and Kasai, H., Cigarette smoking induces an increase in oxidative DNA damage, 8-hydroxydeoxyguanosine, in a central site of the human lung, *Carcinogenesis*, 18, 1763, 1997.

36. Poulsen, H.E., Loft, S., Prieme, H., Vistisen, K., Lykkesfeldt, J., Nyyssonen, K., and Salonen, J.T., Oxidative DNA damage *in vivo*: relationship to age, plasma antioxidants, drug metabolism, glutathione-S-transferase activity and urinary creatinine excretion, *Free Radic. Res.*, 29, 565, 1998.

37. Halliwell, B., Can oxidative DNA damage be used as a biomarker of cancer risk in humans? Problems, resolutions and preliminary results from nutritional supplementation studies, *Free Radic. Res.*, 29, 469, 1998.

38. Loft, S. and Poulsen, H.E., Antioxidant intervention studies related to DNA damage, DNA repair and gene expression, *Free Radic. Res.*, 33 Suppl., S67, 2000.

39. Loft, S., Deng, X.S., Tuo, J., Wellejus, A., Sorensen, M., and Poulsen, H.E., Experimental study of oxidative DNA damage, *Free Radic. Res.*, 29, 525, 1998.

40. Cooke, M.S., Lunec, J., and Evans, M.D., Progress in the analysis of urinary oxidative DNA damage, *Free Radic. Biol. Med.*, 33, 1601, 2002.

41. Lenton, K.J., Therriault, H., Fulop, T., Payette, H., and Wagner, J.R., Glutathione and ascorbate are negatively correlated with oxidative DNA damage in human lymphocytes, *Carcinogenesis*, 20, 607, 1999.

42. Bianchini, F., Elmstahl, S., Martinez-Garcia, C., van Kappel, A.L., Douki, T., Cadet, J., Ohshima, H., Riboli, E., and Kaaks, R., Oxidative DNA damage in human lymphocytes: correlations with plasma levels of alpha-tocopherol and carotenoids, *Carcinogenesis*, 21, 321, 2000.

43. Fraga, C.G., Motchnik, P.A., Shigenaga, M.K., Helbock, H.J., Jacob, R.A., and Ames, B.N., Ascorbic acid protects against endogenous oxidative DNA damage in human sperm, *Proc. Natl. Acad. Sci. USA*, 88, 11003, 1991.

44. Daube, H., Scherer, G., Riedel, K., Ruppert, T., Tricker, A.R., Rosenbaum, P., and Adlkofer, F., DNA adducts in human placenta in relation to tobacco smoke exposure and plasma antioxidant status, *J. Cancer Res. Clin. Oncol.*, 123, 141, 1997.

45. Collins, A.R., Olmedilla, B., Southon, S., Granado, F., and Duthie, S.J., Serum carotenoids and oxidative DNA damage in human lymphocytes, *Carcinogenesis*, 19, 2159, 1998.

46. Loft, S., Vistisen, K., Ewertz, M., Tjonneland, A., Overvad, K., and Poulsen, H.E., Oxidative DNA damage estimated by 8-hydroxydeoxyguanosine excretion in humans: influence of smoking, gender and body mass index, *Carcinogenesis*, 13, 2241, 1992.

47. Huang, H.Y., Helzlsouer, K.J., and Appel, L.J., The effects of vitamin C and vitamin E on oxidative DNA damage: results from a randomized controlled trial, *Cancer Epidemiol. Biomarkers Prev.*, 9, 647, 2000.

48. Kristenson, M., Kucinskiene, Z., Schafer-Elinder, L., Leanderson, P., and Tagesson, C., The LiVicordia study: lower serum levels of beta-carotene in Lithuanian men are accompanied by higher urinary excretion of the oxidative DNA adduct, 8-hydroxydeoxyguanosine, *Nutrition*, 19, 11, 2003.

49. Lee, B.M., Lee, S.K., and Kim, H.S., Inhibition of oxidative DNA damage, 8-OHdG, and carbonyl contents in smokers treated with antioxidants (vitamin E, vitamin C, beta-carotene and red ginseng), *Cancer Lett.*, 132, 219, 1998.

50. Welch, R.W., Turley, E., Sweetman, S.F., Kennedy, G., Collins, A.R., Dunne, A., Livingstone, M.B., McKenna, P.G., McKelvey-Martin, V.J., and Strain, J.J., Dietary antioxidant supplementation and DNA damage in smokers and nonsmokers, *Nutr. Cancer*, 34, 167, 1999.

51. Howard, D.J., Ota, R.B., Briggs, L.A., Hampton, M., and Pritsos, C.A., Oxidative stress induced by environmental tobacco smoke in the workplace is mitigated by antioxidant supplementation, *Cancer Epidemiol. Biomarkers Prev.*, 7, 981, 1998.

52. Jacobson, J.S., Begg, M.D., Wang, L.W., Wang, Q., Agarwal, M., Norkus, E., Singh, V.N., Young, T.L., Yang, D., and Santella, R.M., Effects of a 6-month vitamin intervention on DNA damage in heavy smokers, *Cancer Epidemiol. Biomarkers Prev.*, 9, 1303, 2000.

53. Vojdani, A., Bazargan, M., Vojdani, E., and Wright, J., New evidence for antioxidant properties of vitamin C, *Cancer Detect. Prev.*, 24, 508, 2000.
54. Cooke, M.S., Evans, M.D., Podmore, I.D., Herbert, K.E., Mistry, N., Mistry, P., Hickenbotham, P.T., Hussieni, A., Griffiths, H.R., and Lunec, J., Novel repair action of vitamin C upon *in vivo* oxidative DNA damage, *FEBS Lett.*, 439, 363, 1998.
55. Podmore, I.D., Griffiths, H.R., Herbert, K.E., Mistry, N., Mistry, P., and Lunec, J., Vitamin C exhibits pro-oxidant properties, *Nature*, 392, 559, 1998.
56. Rehman, A., Collis, C.S., Yang, M., Kelly, M., Diplock, A.T., Halliwell, B., and Rice-Evans, C., The effects of iron and vitamin C co-supplementation on oxidative damage to DNA in healthy volunteers, *Biochem. Biophys. Res. Commun.*, 246, 293, 1998.
57. Proteggente, A.R., Rehman, A., Halliwell, B., and Rice-Evans, C.A., Potential problems of ascorbate and iron supplementation: pro-oxidant effect *in vivo*? *Biochem. Biophys. Res. Commun.*, 277, 535, 2000.
58. Prieme, H., Loft, S., Nyyssonen, K., Salonen, J.T., and Poulsen, H.E., No effect of supplementation with vitamin E, ascorbic acid, or coenzyme Q10 on oxidative DNA damage estimated by 8-oxo-7,8-dihydro-2′-deoxyguanosine excretion in smokers, *Am. J. Clin. Nutr.*, 65, 503, 1997.
59. Porkkala-Sarataho, E., Salonen, J.T., Nyyssonen, K., Kaikkonen, J., Salonen, R., Ristonmaa, U., Diczfalusy, U., Brigelius-Flohe, R., Loft, S., and Poulsen, H.E., Long-term effects of vitamin E, vitamin C, and combined supplementation on urinary 7-hydro-8-oxo-2′-deoxyguanosine, serum cholesterol oxidation products, and oxidation resistance of lipids in nondepleted men, *Arterioscler. Thromb. Vasc. Biol.*, 20, 2087, 2000.
60. van Poppel, G., Poulsen, H., Loft, S., and Verhagen, H., No influence of beta carotene on oxidative DNA damage in male smokers, *J. Natl. Cancer Inst.*, 87, 310, 1995.
61. Sumida, S., Doi, T., Sakurai, M., Yoshioka, Y., and Okamura, K., Effect of a single bout of exercise and beta-carotene supplementation on the urinary excretion of 8-hydroxy-deoxyguanosine in humans, *Free Radic. Res.*, 27, 607, 1997.
62. Witt, E.H., Reznick, A.Z., Viguie, C.A., Starke-Reed, P., and Packer, L., Exercise, oxidative damage and effects of antioxidant manipulation, *J. Nutr.*, 122, 766, 1992.
63. Green, M.H., Lowe, J.E., Waugh, A.P., Aldridge, K.E., Cole, J., and Arlett, C.F., Effect of diet and vitamin C on DNA strand breakage in freshly isolated human white blood cells, *Mutat. Res.*, 316, 91, 1994.
64. Panayiotidis, M. and Collins, A.R., Ex vivo assessment of lymphocyte antioxidant status using the Comet assay, *Free Radic. Res.*, 27, 533, 1997.
65. Brennan, L.A., Morris, G.M., Wasson, G.R., Hannigan, B.M., and Barnett, Y.A., The effect of vitamin C or vitamin E supplementation on basal and H_2O_2-induced DNA damage in human lymphocytes, *Br. J. Nutr.*, 84, 195, 2000.
66. Duthie, S.J., Ma, A., Ross, M.A., and Collins, A.R., Antioxidant supplementation decreases oxidative DNA damage in human lymphocytes, *Cancer Res.*, 56, 1291, 1996.
67. Anderson, D., Phillips, B.J., Yu, T.W., Edwards, A.J., Ayesh, R., and Butterworth, K.R., The effects of vitamin C supplementation on biomarkers of oxygen radical generated damage in human volunteers with "low" or "high" cholesterol levels, *Environ. Mol. Mutagen.*, 30, 161, 1997.
68. Astley, S., Langrish-Smith, A., Southon, S., and Sampson, M., Vitamin E supplementation and oxidative damage to DNA and plasma LDL in type 1 diabetes, *Diabetes Care*, 22, 1626, 1999.
69. White, K.L., Chalmers, D.M., Martin, I.G., Everett, S.M., Neville, P.M., Naylor, G., Sutcliffe, A.E., Dixon, M.F., Turner, P.C., and Schorah, C.J., Dietary antioxidants and DNA damage in patients on long-term acid-suppression therapy: a randomized controlled study, *Br. J. Nutr.*, 88, 265, 2002.
70. Jenkinson, A.M., Collins, A.R., Duthie, S.J., Wahle, K.W., and Duthie, G.G., The effect of increased intakes of polyunsaturated fatty acids and vitamin E on DNA damage in human lymphocytes, *FASEB J.*, 13, 2138, 1999.
71. Pool-Zobel, B.L., Bub, A., Muller, H., Wollowski, I., and Rechkemmer, G., Consumption of vegetables reduces genetic damage in humans: first results of a human intervention trial with carotenoid-rich foods, *Carcinogenesis*, 18, 1847, 1997.
72. Hecht, S.S., Approaches to cancer prevention based on an understanding of N-nitrosamine carcinogenesis, *Proc. Soc. Exp. Biol. Med.*, 216, 181, 1997.
73. Mirvish, S.S., Role of N-nitroso compounds (NOC) and N-nitrosation in etiology of gastric, esophageal, nasopharyngeal and bladder cancer and contribution to cancer of known exposures to NOC, *Cancer Lett.*, 93, 17, 1995.

74. Tannenbaum, S.R., Wishnok, J.S., and Leaf, C.D., Inhibition of nitrosamine formation by ascorbic acid, *Am. J. Clin. Nutr.*, 53, 247S, 1991.

75. Mannick, E.E., Bravo, L.E., Zarama, G., Realpe, J.L., Zhang, X.J., Ruiz, B., Fontham, E.T., Mera, R., Miller, M.J., and Correa, P., Inducible nitric oxide synthase, nitrotyrosine, and apoptosis in *Helicobacter pylori* gastritis: effect of antibiotics and antioxidants, *Cancer Res.*, 56, 3238, 1996.

76. Leaf, C.D., Vecchio, A.J., Roe, D.A., and Hotchkiss, J.H., Influence of ascorbic acid dose on N-nitrosoproline formation in humans, *Carcinogenesis*, 8, 791, 1987.

77. Mirvish, S.S., Grandjean, A.C., Reimers, K.J., Connelly, B.J., Chen, S.C., Morris, C.R., Wang, X., Haorah, J., and Lyden, E.R., Effect of ascorbic acid dose taken with a meal on nitrosoproline excretion in subjects ingesting nitrate and proline, *Nutr. Cancer*, 31, 106, 1998.

78. Mirvish, S.S., Effects of vitamins C and E on N-nitroso compound formation, carcinogenesis, and cancer, *Cancer*, 58, 1842, 1986.

79. Reed, J.C., Dysregulation of apoptosis in cancer, *J. Clin. Oncol.*, 17, 2941, 1999.

80. Lavrovsky, Y., Chatterjee, B., Clark, R.A., and Roy, A.K., Role of redox-regulated transcription factors in inflammation, aging and age-related diseases, *Exp. Gerontol.*, 35, 521, 2000.

81. Bijur, G.N., Briggs, B., Hitchcock, C.L., and Williams, M.V., Ascorbic acid-dehydroascorbate induces cell cycle arrest at G2/M DNA damage checkpoint during oxidative stress, *Environ. Mol. Mutagen.*, 33, 144, 1999.

82. Reddy, V.G., Khanna, N., and Singh, N., Vitamin C augments chemotherapeutic response of cervical carcinoma HeLa cells by stabilizing P53, *Biochem. Biophys. Res. Commun.*, 282, 409, 2001.

83. Peltomaki, P., Deficient DNA mismatch repair: a common etiologic factor for colon cancer, *Hum. Mol. Genet.*, 10, 735, 2001.

84. Catani, M.V., Costanzo, A., Savini, I., Levrero, M., de Laurenzi, V., Wang, J.Y., Melino, G., and Avigliano, L., Ascorbate up-regulates MLH1 (Mut L homologue-1) and p73: implications for the cellular response to DNA damage, *Biochem. J.*, 364, 441, 2002.

85. Gopalakrishna, R. and Jaken, S., Protein kinase C signaling and oxidative stress, *Free Radic. Biol. Med.*, 28, 1349, 2000.

86. Azzi, A., Breyer, I., Feher, M., Ricciarelli, R., Stocker, A., Zimmer, S., and Zingg, J., Nonantioxidant functions of alpha-tocopherol in smooth muscle cells, *J. Nutr.*, 131, 378S, 2001.

87. Tasinato, A., Boscoboinik, D., Bartoli, G.M., Maroni, P., and Azzi, A., d-Alpha-tocopherol inhibition of vascular smooth muscle cell proliferation occurs at physiological concentrations, correlates with protein kinase C inhibition, and is independent of its antioxidant properties, *Proc. Natl. Acad. Sci. USA*, 92, 12190, 1995.

88. Clement, S., Tasinato, A., Boscoboinik, D., and Azzi, A., The effect of alpha-tocopherol on the synthesis, phosphorylation and activity of protein kinase C in smooth muscle cells after phorbol 12-myristate 13-acetate down-regulation, *Eur. J. Biochem.*, 246, 745, 1997.

89. Ricciarelli, R., Tasinato, A., Clement, S., Ozer, N.K., Boscoboinik, D., and Azzi, A., alpha-Tocopherol specifically inactivates cellular protein kinase C alpha by changing its phosphorylation state, *Biochem. J.*, 334 (Pt. 1), 243, 1998.

90. Neuzil, J., Weber, T., Schroder, A., Lu, M., Ostermann, G., Gellert, N., Mayne, G.C., Olejnicka, B., Negre-Salvayre, A., Sticha, M., Coffey, R.J., and Weber, C., Induction of cancer cell apoptosis by alpha-tocopheryl succinate: molecular pathways and structural requirements, *FASEB J.*, 15, 403, 2001.

91. Bertram, J.S., Carotenoids and gene regulation, *Nutr. Rev.*, 57, 182, 1999.

92. Cooper, D.A., Eldridge, A.L., and Peters, J.C., Dietary carotenoids and lung cancer: a review of recent research, *Nutr. Rev.*, 57, 133, 1999.

93. Stahl, W., Nicolai, S., Briviba, K., Hanusch, M., Broszeit, G., Peters, M., Martin, H.D., and Sies, H., Biological activities of natural and synthetic carotenoids: induction of gap junctional communication and singlet oxygen quenching, *Carcinogenesis*, 18, 89, 1997.

94. Banoub, R.W., Fernstrom, M., and Ruch, R.J., Lack of growth inhibition or enhancement of gap junctional intercellular communication and connexin43 expression by beta-carotene in murine lung epithelial cells *in vitro*, *Cancer Lett.*, 108, 35, 1996.

95. Krutovskikh, V., Asamoto, M., Takasuka, N., Murakoshi, M., Nishino, H., and Tsuda, H., Differential dose-dependent effects of alpha-, beta-carotenes and lycopene on gap-junctional intercellular communication in rat liver *in vivo*, *Jpn. J. Cancer Res.*, 88, 1121, 1997.

96. World Cancer Research Fund, *Food, Nutrition, and the Prevention of Cancer: a global perspective*, American Institute for Cancer Research, Washington, DC, 1997.

97. Dunn, G.P., Bruce, A.T., Ikeda, H., Old, L.J., and Schreiber, R.D., Cancer immunoediting: from immunosurveillance to tumor escape, *Nat. Immunol.*, 3, 991, 2002.

98. Del Rio, M., Ruedas, G., Medina, S., Victor, V.M., and De la Fuente, M., Improvement by several antioxidants of macrophage function *in vitro*, *Life Sci.*, 63, 871, 1998.

99. Campbell, J.D., Cole, M., Bunditrutavorn, B., and Vella, A.T., Ascorbic acid is a potent inhibitor of various forms of T cell apoptosis, *Cell. Immunol.*, 194, 1, 1999.

100. Radoja, S. and Frey, A.B., Cancer-induced defective cytotoxic T lymphocyte effector function: another mechanism how antigenic tumors escape immune-mediated killing, *Mol. Med.*, 6, 465, 2000.

101. O'Connell, J., Bennett, M.W., Nally, K., Houston, A., O'Sullivan, G.C., and Shanahan, F., Altered mechanisms of apoptosis in colon cancer: Fas resistance and counterattack in the tumor-immune conflict, *Ann. N. Y. Acad. Sci.*, 910, 178, 2000.

102. Jacob, R.A., Kelley, D.S., Pianalto, F.S., Swendseid, M.E., Henning, S.M., Zhang, J.Z., Ames, B.N., Fraga, C.G., and Peters, J.H., Immunocompetence and oxidant defense during ascorbate depletion of healthy men, *Am. J. Clin. Nutr.*, 54, 1302S, 1991.

103. Anderson, R., Oosthuizen, R., Maritz, R., Theron, A., and Van Rensburg, A.J., The effects of increasing weekly doses of ascorbate on certain cellular and humoral immune functions in normal volunteers, *Am. J. Clin. Nutr.*, 33, 71, 1980.

104. Panush, R.S., Delafuente, J.C., Katz, P., and Johnson, J., Modulation of certain immunologic responses by vitamin C, III: potentiation of *in vitro* and *in vivo* lymphocyte responses, *Int. J. Vitam. Nutr. Res. Suppl.*, 23, 35, 1982.

105. Heuser, G. and Vojdani, A., Enhancement of natural killer cell activity and T and B cell function by buffered vitamin C in patients exposed to toxic chemicals: the role of protein kinase-C, *Immunopharmacol. Immunotoxicol.*, 19, 291, 1997.

106. Han, S.N. and Meydani, S.N., Antioxidants, cytokines, and influenza infection in aged mice and elderly humans, *J. Infect. Dis.*, 182 (Suppl. 1), S74, 2000.

107. Meydani, S.N., Barklund, M.P., Liu, S., Meydani, M., Miller, R.A., Cannon, J.G., Morrow, F.D., Rocklin, R., and Blumberg, J.B., Vitamin E supplementation enhances cell-mediated immunity in healthy elderly subjects, *Am. J. Clin. Nutr.*, 52, 557, 1990.

108. Meydani, S.N., Meydani, M., Blumberg, J.B., Leka, L.S., Siber, G., Loszewski, R., Thompson, C., Pedrosa, M.C., Diamond, R.D., and Stollar, B.D., Vitamin E supplementation and *in vivo* immune response in healthy elderly subjects: a randomized controlled trial, *JAMA*, 277, 1380, 1997.

109. Han, S.N., Adolfsson, O., and Meydani, S.N., Vitamin E and enhancement of the immune response in the aged: cellular and molecular mechanisms, in *The Antioxidant Vitamins C and E*, Packer, L., Traber, M.G., Kramer, K., and Frei, B., Eds., AOCS Press, Champaign, IL, 2002, p. 216.

110. Beharka, A.A., Wu, D., Serafini, M., and Meydani, S.N., Mechanism of vitamin E inhibition of cyclooxygenase activity in macrophages from old mice: role of peroxynitrite, *Free Radic. Biol. Med.*, 32, 503, 2002.

111. Jiang, Q., Elson-Schwab, I., Courtemanche, C., and Ames, B.N., Gamma-tocopherol and its major metabolite, in contrast to alpha-tocopherol, inhibit cyclooxygenase activity in macrophages and epithelial cells, *Proc. Natl. Acad. Sci. USA*, 97, 11494, 2000.

112. Ross, A.C., Vitamin A and retinoids, in *Nutrition in Health and Disease*, 9th ed., Shils, M., Olson, J.A., Shike, M., and Ross, A.C., Eds., Williams & Wilkins, Baltimore, 1999, p. 305.

113. Ponnamperuma, R.M., Shimizu, Y., Kirchhof, S.M., and De Luca, L.M., Beta-carotene fails to act as a tumor promoter, induces RAR expression, and prevents carcinoma formation in a two-stage model of skin carcinogenesis in male Sencar mice, *Nutr. Cancer*, 37, 82, 2000.

114. Chen, L.C., Sly, L., Jones, C.S., Tarone, R., and De Luca, L.M., Differential effects of dietary beta-carotene on papilloma and carcinoma formation induced by an initiation-promotion protocol in SENCAR mouse skin, *Carcinogenesis*, 14, 713, 1993.

115. Mathews-Roth, M.M. and Krinsky, N.I., Carotenoids affect development of UV-B induced skin cancer, *Photochem. Photobiol.*, 46, 507, 1987.

116. Black, H.S., Okotie-Eboh, G., and Gerguis, J., Diet potentiates the UV-carcinogenic response to beta-carotene, *Nutr. Cancer*, 37, 173, 2000.

117. Krol, E.S., Kramer-Stickland, K.A., and Liebler, D.C., Photoprotective actions of topically applied vitamin E, *Drug Metab. Rev.*, 32, 413, 2000.

118. Mitchel, R.E. and McCann, R., Vitamin E is a complete tumor promoter in mouse skin, *Carcinogenesis*, 14, 659, 1993.

119. Sawant, S.S. and Kandarkar, S.V., Role of vitamins C and E as chemopreventive agents in the hamster cheek pouch treated with the oral carcinogen-DMBA, *Oral. Dis.*, 6, 241, 2000.

120. Schwartz, J.L., Sloane, D., and Shklar, G., Prevention and inhibition of oral cancer in the hamster buccal pouch model associated with carotenoid immune enhancement, *Tumour Biol.*, 10, 297, 1989.

121. Shklar, G., Schwartz, J., Trickler, D., and Reid, S., Regression of experimental cancer by oral administration of combined alpha-tocopherol and beta-carotene, *Nutr. Cancer*, 12, 321, 1989.

122. Woutersen, R.A., Appel, M.J., and Van Garderen-Hoetmer, A., Modulation of pancreatic carcinogenesis by antioxidants, *Food Chem. Toxicol.*, 37, 981, 1999.

123. Majima, T., Tsutsumi, M., Nishino, H., Tsunoda, T., and Konishi, Y., Inhibitory effects of beta-carotene, palm carotene, and green tea polyphenols on pancreatic carcinogenesis initiated by N-nitorsobis (2-oxopropyl)amine in Syrian golden hamsters, *Pancreas*, 16, 13, 1998.

124. Appel, M.J. and Woutersen, R.A., Effects of dietary beta-carotene and selenium on initiation and promotion of pancreatic carcinogenesis in azaserine-treated rats, *Carcinogenesis*, 17, 1411, 1996.

125. Factor, V.M., Laskowska, D., Jensen, M.R., Woitach, J.T., Popescu, N.C., and Thorgeirsson, S.S., Vitamin E reduces chromosomal damage and inhibits hepatic tumor formation in a transgenic mouse model, *Proc. Natl. Acad. Sci. USA*, 97, 2196, 2000.

126. Kakizaki, S., Takagi, H., Fukusato, T., Toyoda, M., Horiguchi, N., Sato, K., Takayama, H., Nagamine, T., and Mori, M., Effect of alpha-tocopherol on hepatocarcinogenesis in transforming growth factor-alpha (TGF-alpha) transgenic mice treated with diethylnitrosamine, *Int. J. Vitam. Nutr. Res.*, 71, 261, 2001.

127. Ziegler, R.G., Mayne, S.T., and Swanson, C.A., Nutrition and lung cancer, *Cancer Causes Control*, 7, 157, 1996.

128. Stram, D.O., Huberman, M., and Wu, A.H., Is residual confounding a reasonable explanation for the apparent protective effects of beta-carotene found in epidemiologic studies of lung cancer in smokers? *Am. J. Epidemiol.*, 155, 622, 2002.

129. Knekt, P., Role of vitamin E in the prophylaxis of cancer, *Ann. Med.*, 23, 3, 1991.

130. Block, G., Epidemiologic evidence regarding vitamin C and cancer, *Am. J. Clin. Nutr.*, 54, 1310S, 1991.

131. Ziegler, R.G., Vegetables, fruits, and carotenoids and the risk of cancer, *Am. J. Clin. Nutr.*, 53, 251S, 1991.

132. Enstrom, J.E., Kanim, L.E., and Klein, M.A., Vitamin C intake and mortality among a sample of the United States population, *Epidemiology*, 3, 194, 1992.

133. Shibata, A., Paganini-Hill, A., Ross, R.K., and Henderson, B.E., Intake of vegetables, fruits, beta-carotene, vitamin C and vitamin supplements and cancer incidence among the elderly: a prospective study, *Br. J. Cancer*, 66, 673, 1992.

134. Pandey, D.K., Shekelle, R., Selwyn, B.J., Tangney, C., and Stamler, J., The Western Electric Study: dietary vitamin C and beta-carotene and risk of death in middle-aged men, *Am. J. Epidemiol.*, 142, 1269, 1995.

135. Losonczy, K.G., Harris, T.B., and Havlik, R.J., Vitamin E and vitamin C supplement use and risk of all-cause and coronary heart disease mortality in older persons: the Established Populations for Epidemiologic Studies of the Elderly, *Am. J. Clin. Nutr.*, 64, 190, 1996.

136. Sahyoun, N.R., Jacques, P.F., and Russell, R.M., Carotenoids, vitamins C and E, and mortality in an elderly population, *Am. J. Epidemiol.*, 144, 501, 1996.

137. Eichholzer, M., Stahelin, H.B., Gey, K.F., Ludin, E., and Bernasconi, F., Prediction of male cancer mortality by plasma levels of interacting vitamins: 17-year follow-up of the prospective Basel study, *Int. J. Cancer*, 66, 145, 1996.

138. Loria, C.M., Klag, M.J., Caulfield, L.E., and Whelton, P.K., Vitamin C status and mortality in US adults, *Am. J. Clin. Nutr.*, 72, 139, 2000.

139. Khaw, K.T., Bingham, S., Welch, A., Luben, R., Wareham, N., Oakes, S., and Day, N., European Prospective Investigation into Cancer and Nutrition: relation between plasma ascorbic acid and mortality in men and women in EPIC-Norfolk prospective study: a prospective population study, *Lancet*, 357, 657, 2001.

140. Kilander, L., Berglund, L., Boberg, M., Vessby, B., and Lithell, H., Education, lifestyle factors and mortality from cardiovascular disease and cancer: a 25-year follow-up of Swedish 50-year-old men, *Int. J. Epidemiol.*, 30, 1119, 2001.

141. Martini, M.C., Campbell, D.R., Gross, M.D., Grandits, G.A., Potter, J.D., and Slavin, J.L., Plasma carotenoids as biomarkers of vegetable intake: the University of Minnesota Cancer Prevention Research Unit Feeding Studies, *Cancer Epidemiol. Biomarkers Prev.*, 4, 491, 1995.

142. McEligot, A.J., Rock, C.L., Flatt, S.W., Newman, V., Faerber, S., and Pierce, J.P., Plasma carotenoids are biomarkers of long-term high vegetable intake in women with breast cancer, *J. Nutr.*, 129, 2258, 1999.

143. Van Hoydonck, P.G., Temme, E.H., and Schouten, E.G., A dietary oxidative balance score of vitamin C, beta-carotene and iron intakes and mortality risk in male smoking Belgians, *J. Nutr.*, 132, 756, 2002.

144. Knekt, P., Jarvinen, R., Seppanen, R., Rissanen, A., Aromaa, A., Heinonen, O.P., Albanes, D., Heinonen, M., Pukkala, E., and Teppo, L., Dietary antioxidants and the risk of lung cancer, *Am. J. Epidemiol.*, 134, 471, 1991.

145. Chow, W.H., Schuman, L.M., McLaughlin, J.K., Bjelke, E., Gridley, G., Wacholder, S., Chien, H.T., and Blot, W.J., A cohort study of tobacco use, diet, occupation, and lung cancer mortality, *Cancer Causes Control*, 3, 247, 1992.

146. Steinmetz, K.A., Potter, J.D., and Folsom, A.R., Vegetables, fruit, and lung cancer in the Iowa Women's Health Study, *Cancer Res.*, 53, 536, 1993.

147. Comstock, G.W., Alberg, A.J., Huang, H.Y., Wu, K., Burke, A.E., Hoffman, S.C., Norkus, E.P., Gross, M., Cutler, R.G., Morris, J.S., Spate, V.L., and Helzlsouer, K.J., The risk of developing lung cancer associated with antioxidants in the blood: ascorbic acid, carotenoids, alpha-tocopherol, selenium, and total peroxyl radical absorbing capacity, *Cancer Epidemiol. Biomarkers Prev.*, 6, 907, 1997.

148. Ocke, M.C., Bueno-de-Mesquita, H.B., Feskens, E.J., van Staveren, W.A., and Kromhout, D., The Zutphen study: repeated measurements of vegetables, fruits, beta-carotene, and vitamins C and E in relation to lung cancer, *Am. J. Epidemiol.*, 145, 358, 1997.

149. Yong, L.C., Brown, C.C., Schatzkin, A., Dresser, C.M., Slesinski, M.J., Cox, C.S., and Taylor, P.R., First National Health and Nutrition Examination Survey, NHANES I epidemiologic followup study: intake of vitamins E, C, and A and risk of lung cancer, *Am. J. Epidemiol.*, 146, 231, 1997.

150. Bandera, E.V., Freudenheim, J.L., Marshall, J.R., Zielezny, M., Priore, R.L., Brasure, J., Baptiste, M., and Graham, S., Diet and alcohol consumption and lung cancer risk in the New York State Cohort (United States), *Cancer Causes Control*, 8, 828, 1997.

151. Voorrips, L.E., Goldbohm, R.A., Brants, H.A., van Poppel, G.A., Sturmans, F., Hermus, R.J., and van den Brandt, P.A., A prospective cohort study on antioxidant and folate intake and male lung cancer risk, *Cancer Epidemiol. Biomarkers Prev.*, 9, 357, 2000.

152. Woodson, K., Tangrea, J.A., Barrett, M.J., Virtamo, J., Taylor, P.R., and Albanes, D., Serum alpha-tocopherol and subsequent risk of lung cancer among male smokers, *J. Natl. Cancer Inst.*, 91, 1738, 1999.

153. Michaud, D.S., Feskanich, D., Rimm, E.B., Colditz, G.A., Speizer, F.E., Willett, W.C., and Giovannucci, E., Intake of specific carotenoids and risk of lung cancer in two prospective US cohorts, *Am. J. Clin. Nutr.*, 72, 990, 2000.

154. Holick, C.N., Michaud, D.S., Stolzenberg-Solomon, R., Mayne, S.T., Pietinen, P., Taylor, P.R., Virtamo, J., and Albanes, D., Dietary carotenoids, serum beta-carotene, and retinol and risk of lung cancer in the alpha-tocopherol, beta-carotene cohort study, *Am. J. Epidemiol.*, 156, 536, 2002.

155. Rohan, T.E., Jain, M., Howe, G.R., and Miller, A.B., A cohort study of dietary carotenoids and lung cancer risk in women (Canada), *Cancer Causes Control*, 13, 231, 2002.

156. Kritchevsky, S.B., Bush, A.J., Pahor, M., and Gross, M.D., Serum carotenoids and markers of inflammation in nonsmokers, *Am. J. Epidemiol.*, 152, 1065, 2000.

157. Erlinger, T.P., Guallar, E., Miller, III, E.R., Stolzenberg-Solomon, R., and Appel, L.J., Relationship between systemic markers of inflammation and serum beta-carotene levels, *Arch. Intern. Med.*, 161, 1903, 2001.

158. Ekstrom, A.M., Serafini, M., Nyren, O., Hansson, L.E., Ye, W., and Wolk, A., Dietary antioxidant intake and the risk of cardia cancer and noncardia cancer of the intestinal and diffuse types: a population-based case-control study in Sweden, *Int. J. Cancer*, 87, 133, 2000.

159. Kaaks, R., Tuyns, A.J., Haelterman, M., and Riboli, E., Nutrient intake patterns and gastric cancer risk: a case-control study in Belgium, *Int. J. Cancer*, 78, 415, 1998.

160. Mayne, S.T., Risch, H.A., Dubrow, R., Chow, W.H., Gammon, M.D., Vaughan, T.L., Farrow, D.C., Schoenberg, J.B., Stanford, J.L., Ahsan, H., West, A.B., Rotterdam, H., Blot, W.J., and Fraumeni, Jr., J.F., Nutrient intake and risk of subtypes of esophageal and gastric cancer, *Cancer Epidemiol. Biomarkers Prev.*, 10, 1055, 2001.

161. Palli, D., Russo, A., and Decarli, A., Dietary patterns, nutrient intake and gastric cancer in a high-risk area of Italy, *Cancer Causes Control*, 12, 163, 2001.

162. Botterweck, A.A., van den Brandt, P.A., and Goldbohm, R.A., Vitamins, carotenoids, dietary fiber, and the risk of gastric carcinoma: results from a prospective study after 6.3 years of follow-up, *Cancer*, 88, 737, 2000.

163. Chyou, P.H., Nomura, A.M., Hankin, J.H., and Stemmermann, G.N., A case-cohort study of diet and stomach cancer, *Cancer Res.*, 50, 7501, 1990.

164. Zheng, W., Sellers, T.A., Doyle, T.J., Kushi, L.H., Potter, J.D., and Folsom, A.R., Retinol, antioxidant vitamins, and cancers of the upper digestive tract in a prospective cohort study of postmenopausal women, *Am. J. Epidemiol.*, 142, 955, 1995.

165. Stahelin, H.B., Gey, K.F., Eichholzer, M., Ludin, E., Bernasconi, F., Thurneysen, J., and Brubacher, G., Plasma antioxidant vitamins and subsequent cancer mortality in the 12-year follow-up of the prospective Basel Study, *Am. J. Epidemiol.*, 133, 766, 1991.

166. Jacobs, E.J., Connell, C.J., McCullough, M.L., Chao, A., Jonas, C.R., Rodriguez, C., Calle, E.E., and Thun, M.J., Vitamin C, vitamin E, and multivitamin supplement use and stomach cancer mortality in the Cancer Prevention Study II cohort, *Cancer Epidemiol. Biomarkers Prev.*, 11, 35, 2002.

167. Reed, P.I., Vitamin C, *Helicobacter pylori* infection and gastric carcinogenesis, *Int. J. Vitam. Nutr. Res.*, 69, 220, 1999.

168. De Stefani, E., Boffetta, P., Brennan, P., Deneo-Pellegrini, H., Carzoglio, J.C., Ronco, A., and Mendilaharsu, M., Dietary carotenoids and risk of gastric cancer: a case-control study in Uruguay, *Eur. J. Cancer Prev.*, 9, 329, 2000.

169. Bostick, R.M., Diet and colon cancer, in *Preventive Nutrition: the Comprehensive Guide for Health Professionals*, 2nd ed., Bendich, A. and Deckelbaum, R.J., Eds., Humana Press, Totowa, NJ, 2001, p. 47.

170. Bostick, R.M., Potter, J.D., McKenzie, D.R., Sellers, T.A., Kushi, L.H., Steinmetz, K.A., and Folsom, A.R., Reduced risk of colon cancer with high intake of vitamin E: the Iowa Women's Health Study, *Cancer Res.*, 53, 4230, 1993.

171. Jacobs, E.J., Connell, C.J., Patel, A.V., Chao, A., Rodriguez, C., Seymour, J., McCullough, M.L., Calle, E.E., and Thun, M.J., Vitamin C and vitamin E supplement use and colorectal cancer mortality in a large American Cancer Society cohort, *Cancer Epidemiol. Biomarkers Prev.*, 10, 17, 2001.

172. Malila, N., Virtamo, J., Virtanen, M., Pietinen, P., Albanes, D., and Teppo, L., Dietary and serum alpha-tocopherol, beta-carotene and retinol, and risk for colorectal cancer in male smokers, *Eur. J. Clin. Nutr.*, 56, 615, 2002.

173. Longnecker, M.P., Martin-Moreno, J.M., Knekt, P., Nomura, A.M., Schober, S.E., Stahelin, H.B., Wald, N.J., Gey, K.F., and Willett, W.C., Serum alpha-tocopherol concentration in relation to subsequent colorectal cancer: pooled data from five cohorts, *J. Natl. Cancer Inst.*, 84, 430, 1992.

174. Wu, K., Willett, W.C., Chan, J.M., Fuchs, C.S., Colditz, G.A., Rimm, E.B., and Giovannucci, E.L., A prospective study on supplemental vitamin E intake and risk of colon cancer in women and men, *Cancer Epidemiol. Biomarkers Prev.*, 11, 1298, 2002.

175. Sellers, T.A., Bazyk, A.E., Bostick, R.M., Kushi, L.H., Olson, J.E., Anderson, K.E., Lazovich, D., and Folsom, A.R., Diet and risk of colon cancer in a large prospective study of older women: an analysis stratified on family history (Iowa, United States), *Cancer Causes Control*, 9, 357, 1998.

176. Michaud, D.S., Spiegelman, D., Clinton, S.K., Rimm, E.B., Willett, W.C., and Giovannucci, E., Prospective study of dietary supplements, macronutrients, micronutrients, and risk of bladder cancer in US men, *Am. J. Epidemiol.*, 152, 1145, 2000.

177. Zeegers, M.P., Goldbohm, R.A., and Brandt, P.A., Are retinol, vitamin C, vitamin E, folate and carotenoids intake associated with bladder cancer risk? Results from the Netherlands Cohort Study, *Br. J. Cancer*, 85, 977, 2001.

178. Michaud, D.S., Pietinen, P., Taylor, P.R., Virtanen, M., Virtamo, J., and Albanes, D., Intakes of fruits and vegetables, carotenoids and vitamins A, E, C in relation to the risk of bladder cancer in the ATBC cohort study, *Br. J. Cancer*, 87, 960, 2002.

179. Jacobs, E.J., Henion, A.K., Briggs, P.J., Connell, C.J., McCullough, M.L., Jonas, C.R., Rodriguez, C., Calle, E.E., and Thun, M.J., Vitamin C and vitamin E supplement use and bladder cancer mortality in a large cohort of US men and women, *Am. J. Epidemiol.*, 156, 1002, 2002.

180. Knekt, P., Aromaa, A., Maatela, J., Alfthan, G., Aaran, R.K., Nikkari, T., Hakama, M., Hakulinen, T., and Teppo, L., Serum micronutrients and risk of cancers of low incidence in Finland, *Am. J. Epidemiol.*, 134, 356, 1991.

181. Daviglus, M.L., Dyer, A.R., Persky, V., Chavez, N., Drum, M., Goldberg, J., Liu, K., Morris, D.K., Shekelle, R.B., and Stamler, J., Dietary beta-carotene, vitamin C, and risk of prostate cancer: results from the Western Electric Study, *Epidemiology*, 7, 472, 1996.

182. Schuurman, A.G., Goldbohm, R.A., Brants, H.A., and van den Brandt, P.A., A prospective cohort study on intake of retinol, vitamins C and E, and carotenoids and prostate cancer risk (Netherlands), *Cancer Causes Control*, 13, 573, 2002.

183. Hsing, A.W., Comstock, G.W., Abbey, H., and Polk, B.F., Serologic precursors of cancer: retinol, carotenoids, and tocopherol and risk of prostate cancer, *J. Natl. Cancer Inst.*, 82, 941, 1990.

184. Nomura, A.M., Stemmermann, G.N., Lee, J., and Craft, N.E., Serum micronutrients and prostate cancer in Japanese Americans in Hawaii, *Cancer Epidemiol. Biomarkers Prev.*, 6, 487, 1997.

185. Gann, P.H., Ma, J., Giovannucci, E., Willett, W., Sacks, F.M., Hennekens, C.H., and Stampfer, M.J., Lower prostate cancer risk in men with elevated plasma lycopene levels: results of a prospective analysis, *Cancer Res.*, 59, 1225, 1999.

186. Chan, J.M., Stampfer, M.J., Ma, J., Rimm, E.B., Willett, W.C., and Giovannucci, E.L., Supplemental vitamin E intake and prostate cancer risk in a large cohort of men in the United States, *Cancer Epidemiol. Biomarkers Prev.*, 8, 893, 1999.

187. Helzlsouer, K.J., Huang, H.Y., Alberg, A.J., Hoffman, S., Burke, A., Norkus, E.P., Morris, J.S., and Comstock, G.W., Association between alpha-tocopherol, gamma-tocopherol, selenium, and subsequent prostate cancer, *J. Natl. Cancer Inst.*, 92, 2018, 2000.

188. Giovannucci, E., Ascherio, A., Rimm, E.B., Stampfer, M.J., Colditz, G.A., and Willett, W.C., Intake of carotenoids and retinol in relation to risk of prostate cancer, *J. Natl. Cancer Inst.*, 87, 1767, 1995.

189. Ambrosone, C.B., Oxidants and antioxidants in breast cancer, *Antioxid. Redox Signal.*, 2, 903, 2000.

190. Graham, S., Zielezny, M., Marshall, J., Priore, R., Freudenheim, J., Brasure, J., Haughey, B., Nasca, P., and Zdeb, M., Diet in the epidemiology of postmenopausal breast cancer in the New York State Cohort, *Am. J. Epidemiol.*, 136, 1327, 1992.

191. Rohan, T.E., Howe, G.R., Friedenreich, C.M., Jain, M., and Miller, A.B., Dietary fiber, vitamins A, C, and E, and risk of breast cancer: a cohort study, *Cancer Causes Control*, 4, 29, 1993.

192. Kushi, L.H., Fee, R.M., Sellers, T.A., Zheng, W., and Folsom, A.R., Iowa Women's Health Study: intake of vitamins A, C, and E and postmenopausal breast cancer, *Am. J. Epidemiol.*, 144, 165, 1996.

193. Verhoeven, D.T., Assen, N., Goldbohm, R.A., Dorant, E., van't Veer, P., Sturmans, F., Hermus, R.J., and van den Brandt, P.A., Vitamins C and E, retinol, beta-carotene and dietary fibre in relation to breast cancer risk: a prospective cohort study, *Br. J. Cancer*, 75, 149, 1997.

194. Zhang, S., Hunter, D.J., Forman, M.R., Rosner, B.A., Speizer, F.E., Colditz, G.A., Manson, J.E., Hankinson, S.E., and Willett, W.C., Dietary carotenoids and vitamins A, C, and E and risk of breast cancer, *J. Natl. Cancer Inst.*, 91, 547, 1999.

195. Michels, K.B., Holmberg, L., Bergkvist, L., Ljung, H., Bruce, A., and Wolk, A., Dietary antioxidant vitamins, retinol, and breast cancer incidence in a cohort of Swedish women, *Int. J. Cancer*, 91, 563, 2001.

196. Wu, K., Helzlsouer, K.J., Alberg, A.J., Comstock, G.W., Norkus, E.P., and Hoffman, S.C., A prospective study of plasma ascorbic acid concentrations and breast cancer (United States), *Cancer Causes Control*, 11, 279, 2000.

197. Dorgan, J.F., Sowell, A., Swanson, C.A., Potischman, N., Miller, R., Schussler, N., and Stephenson, Jr., H.E., Relationships of serum carotenoids, retinol, alpha-tocopherol, and selenium with breast cancer risk: results from a prospective study in Columbia, Missouri (United States), *Cancer Causes Control*, 9, 89, 1998.

198. Hulten, K., Van Kappel, A.L., Winkvist, A., Kaaks, R., Hallmans, G., Lenner, P., and Riboli, E., Carotenoids, alpha-tocopherols, and retinol in plasma and breast cancer risk in northern Sweden, *Cancer Causes Control*, 12, 529, 2001.

199. Sato, R., Helzlsouer, K.J., Alberg, A.J., Hoffman, S.C., Norkus, E.P., and Comstock, G.W., Prospective study of carotenoids, tocopherols, and retinoid concentrations and the risk of breast cancer, *Cancer Epidemiol. Biomarkers Prev.*, 11, 451, 2002.

200. Jumaan, A.O., Holmberg, L., Zack, M., Mokdad, A.H., Ohlander, E.M., Wolk, A., and Byers, T., Beta-carotene intake and risk of postmenopausal breast cancer, *Epidemiology*, 10, 49, 1999.

201. Toniolo, P., Van Kappel, A.L., Akhmedkhanov, A., Ferrari, P., Kato, I., Shore, R.E., and Riboli, E., Serum carotenoids and breast cancer, *Am. J. Epidemiol.*, 153, 1142, 2001.

202. Taylor, P.R., Li, B., Dawsey, S.M., Li, J.Y., Yang, C.S., Guo, W., and Blot, W.J., Linxian Nutrition Intervention Trials Study Group: Prevention of esophageal cancer: the nutrition intervention trials in Linxian, China, *Cancer Res.* 54, 2029s, 1994.
203. Gail, M.H., You, W.C., Chang, Y.S., Zhang, L., Blot, W.J., Brown, L.M., Groves, F.D., Heinrich, J.P., Hu, J., Jin, M.L., Li, J.Y., Liu, W.D., Ma, J.L., Mark, S.D., Rabkin, C.S., Fraumeni, Jr., J.F., and Xu, G.W., Factorial trial of three interventions to reduce the progression of precancerous gastric lesions in Shandong, China: design issues and initial data, *Control. Clin. Trials*, 19, 352, 1998.
204. Hennekens, C.H., Buring, J.E., Manson, J.E., Stampfer, M., Rosner, B., Cook, N.R., Belanger, C., LaMotte, F., Gaziano, J.M., Ridker, P.M., Willett, W., and Peto, R., Lack of effect of long-term supplementation with beta carotene on the incidence of malignant neoplasms and cardiovascular disease, *New England J. Med.*, 334, 1145, 1996.
205. Cook, N.R., Lee, I.M., Manson, J.E., Buring, J.E., and Hennekens, C.H., Effects of beta-carotene supplementation on cancer incidence by baseline characteristics in the Physicians' Health Study (United States), *Cancer Causes Control*, 11, 617, 2000.
206. Lee, I.M., Cook, N.R., Manson, J.E., Buring, J.E., and Hennekens, C.H., Beta-carotene supplementation and incidence of cancer and cardiovascular disease: the Women's Health Study, *J. Natl. Cancer Inst.*, 91, 2102, 1999.
207. Yeum, K.J. and Russell, R.M., Carotenoid bioavailability and bioconversion, *Annu. Rev. Nutr.*, 22, 483, 2002.
208. Wang, X.D. and Russell, R.M., Procarcinogenic and anticarcinogenic effects of beta-carotene, *Nutr. Rev.*, 57, 263, 1999.
209. Wang, X.D., Liu, C., Bronson, R.T., Smith, D.E., Krinsky, N.I., and Russell, M., Retinoid signaling and activator protein-1 expression in ferrets given beta-carotene supplements and exposed to tobacco smoke, *J. Natl. Cancer Inst.*, 91, 60, 1999.
210. Gradelet, S., Leclerc, J., Siess, M.H., and Astorg, P.O., Beta-Apo-8'-carotenal, but not beta-carotene, is a strong inducer of liver cytochromes P4501A1 and 1A2 in rat, *Xenobiotica*, 26, 909, 1996.
211. Paolini, M., Antelli, A., Pozzetti, L., Spetlova, D., Perocco, P., Valgimigli, L., Pedulli, G.F., and Cantelli-Forti, G., Induction of cytochrome P450 enzymes and over-generation of oxygen radicals in beta-carotene supplemented rats, *Carcinogenesis*, 22, 1483, 2001.
212. Salgo, M.G., Cueto, R., Winston, G.W., and Pryor, W.A., Beta carotene and its oxidation products have different effects on microsome mediated binding of benzo[a]pyrene to DNA, *Free Radic. Biol. Med.*, 26, 162, 1999.
213. Omenn, G.S., Goodman, G.E., Thornquist, M.D., Balmes, J., Cullen, M.R., Glass, A., Keogh, J.P., Meyskens, Jr., F.L., Valanis, B., Williams, Jr., J.H., Barnhart, S., Cherniack, M.G., Brodkin, C.A., and Hammar, S., Risk factors for lung cancer and for intervention effects in CARET, the Beta-Carotene and Retinol Efficacy Trial, *J. Natl. Cancer Inst.*, 88, 1550, 1996.
214. Albanes, D., Heinonen, O.P., Taylor, P.R., Virtamo, J., Edwards, B.K., Rautalahti, M., Hartman, A.M., Palmgren, J., Freedman, L.S., Haapakoski, J., Barrett, M.J., Pietinen, P., Malila, N., Tala, E., Liippo, K., Salomaa, E.R., Tangrea, J.A., Teppo, L., Askin, F.B., Taskinen, E., Erozan, Y., Greenwald, P., and Huttunen, J.K., Alpha-tocopherol and beta-carotene supplements and lung cancer incidence in the alpha-tocopherol, beta-carotene cancer prevention study: effects of base-line characteristics and study compliance, *J. Natl. Cancer Inst.*, 88, 1560, 1996.
215. Liu, C., Wang, X.D., Bronson, R.T., Smith, D.E., Krinsky, N.I., and Russell, R.M., Effects of physiological versus pharmacological beta-carotene supplementation on cell proliferation and his-topathological changes in the lungs of cigarette smoke-exposed ferrets, *Carcinogenesis*, 21, 2245, 2000.
216. Lykkesfeldt, J., Christen, S., Wallock, L.M., Chang, H.H., Jacob, R.A., and Ames, B.N., Ascorbate is depleted by smoking and repleted by moderate supplementation: a study in male smokers and nonsmokers with matched dietary antioxidant intakes, *Am. J. Clin. Nutr.*, 71, 530, 2000.
217. Tribble, D.L., Giuliano, L.J., and Fortmann, S.P., Reduced plasma ascorbic acid concentrations in nonsmokers regularly exposed to environmental tobacco smoke, *Am. J. Clin. Nutr.*, 58, 886, 1993.
218. Burke, M., Edge, R., Land, E.J., and Truscott, T.G., Characterisation of carotenoid radical cations in liposomal environments: interaction with vitamin C, *J. Photochem. Photobiol. B*, 60, 1, 2001.
219. Yeum, K.J., dos Anjos Ferreira, A.L., Smith, D., Krinsky, N.I., and Russell, R.M., The effect of alpha-tocopherol on the oxidative cleavage of beta-carotene, *Free Radical Biol. Med.*, 29, 105, 2000.

220. Cook, N.R., Stampfer, M.J., Ma, J., Manson, J.E., Sacks, F.M., Buring, J.E., and Hennekens, C.H., Beta-carotene supplementation for patients with low baseline levels and decreased risks of total and prostate carcinoma, *Cancer*, 86, 1783, 1999.

221. Shklar, G. and Schwartz, J.L., Vitamin E inhibits experimental carcinogenesis and tumour angiogenesis, *Eur. J. Cancer B Oral Oncol.*, 32B, 114, 1996.

222. Woodson, K., Triantos, S., Hartman, T., Taylor, P.R., Virtamo, J., and Albanes, D., Long-term alpha-tocopherol supplementation is associated with lower serum vascular endothelial growth factor levels, *Anticancer Res.*, 22, 375, 2002.

223. Greenberg, E.R., Baron, J.A., Stukel, T.A., Stevens, M.M., Mandel, J.S., Spencer, S.K., Elias, P.M., Lowe, N., Nierenberg, D.W., Bayrd, G. et al., The Skin Cancer Prevention Study Group: a clinical trial of beta carotene to prevent basal-cell and squamous-cell cancers of the skin, *New England J. Med.*, 323, 789, 1990.

224. Frieling, U.M., Schaumberg, D.A., Kupper, T.S., Muntwyler, J., and Hennekens, C.H., A randomized, 12-year primary-prevention trial of beta carotene supplementation for nonmelanoma skin cancer in the physician's health study, *Arch. Dermatol.*, 136, 179, 2000.

225. Mayne, S.T., Cartmel, B., Baum, M., Shor-Posner, G., Fallon, B.G., Briskin, K., Bean, J., Zheng, T., Cooper, D., Friedman, C., and Goodwin, Jr., W.J., Randomized trial of supplemental beta-carotene to prevent second head and neck cancer, *Cancer Res.*, 61, 1457, 2001.

226. Albanes, D., Malila, N., Taylor, P.R., Huttunen, J.K., Virtamo, J., Edwards, B.K., Rautalahti, M., Hartman, A.M., Barrett, M.J., Pietinen, P., Hartman, T.J., Sipponen, P., Lewin, K., Teerenhovi, L., Hietanen, P., Tangrea, J.A., Virtanen, M., and Heinonen, O.P., Effects of supplemental alpha-tocopherol and beta-carotene on colorectal cancer: results from a controlled trial (Finland), *Cancer Causes Control*, 11, 197, 2000.

227. Virtamo, J., Edwards, B.K., Virtanen, M., Taylor, P.R., Malila, N., Albanes, D., Huttunen, J.K., Hartman, A.M., Hietanen, P., Maenpaa, H., Koss, L., Nordling, S., and Heinonen, O.P., Effects of supplemental alpha-tocopherol and beta-carotene on urinary tract cancer: incidence and mortality in a controlled trial (Finland), *Cancer Causes Control*, 11, 933, 2000.

228. Malila, N., Taylor, P.R., Virtanen, M.J., Korhonen, P., Huttunen, J.K., Albanes, D., and Virtamo, J., Effects of alpha-tocopherol and beta-carotene supplementation on gastric cancer incidence in male smokers (ATBC Study, Finland), *Cancer Causes Control*, 13, 617, 2002.

229. Rautalahti, M.T., Virtamo, J.R., Taylor, P.R., Heinonen, O.P., Albanes, D., Haukka, J.K., Edwards, B.K., Karkkainen, P.A., Stolzenberg-Solomon, R.Z., and Huttunen, J., The effects of supplementation with alpha-tocopherol and beta-carotene on the incidence and mortality of carcinoma of the pancreas in a randomized, controlled trial, *Cancer*, 86, 37, 1999.

230. Lodi, G., Sardella, A., Bez, C., Demarosi, F., and Carrassi, A., Systematic review of randomized trials for the treatment of oral leukoplakia, *J. Dent. Educ.*, 66, 896, 2002.

231. Garewal, H.S., Meyskens, Jr., F.L., Killen, D., Reeves, D., Kiersch, T.A., Elletson, H., Strosberg, A., King, D., and Steinbronn, K., Response of oral leukoplakia to beta-carotene, *J. Clin. Oncol.*, 8, 1715, 1990.

232. Toma, S., Benso, S., Albanese, E., Palumbo, R., Cantoni, E., Nicolo, G., and Mangiante, P., Treatment of oral leukoplakia with beta-carotene, *Oncology*, 49, 77, 1992.

233. Kaugars, G.E., Silverman, Jr., S., Lovas, J.G., Brandt, R.B., Riley, W.T., Dao, Q., Singh, V.N., and Gallo, J., A clinical trial of antioxidant supplements in the treatment of oral leukoplakia, *Oral Surg. Oral Med. Oral Pathol.*, 78, 462, 1994.

234. Barth, T.J., Zoller, J., Kubler, A., Born, I.A., and Osswald, H., Redifferentiation of oral dysplastic mucosa by the application of the antioxidants beta-carotene, alpha-tocopherol and vitamin C, *Int. J. Vitam. Nutr. Res.*, 67, 368, 1997.

235. Garewal, H.S., Katz, R.V., Meyskens, F., Pitcock, J., Morse, D., Friedman, S., Peng, Y., Pendrys, D.G., Mayne, S., Alberts, D., Kiersch, T., and Graver, E., Beta-carotene produces sustained remissions in patients with oral leukoplakia: results of a multicenter prospective trial, *Arch. Otolaryngol. Head Neck Surg.*, 125, 1305, 1999.

236a. Zaride, D., Evstifeeva, T., and Boyle, P., Chemoprevention of oral leukoplakia and chromic esophagitis in an area of high incidence of oral and esophagent cancer, *Ann. Epidemiol.* 3, 225, 1993

236b. Sankaranarayanan, R., Mathew, B., Varghese, C., Sudhakaran, P.R., Menon, V., Jayadeep, A., Nair, M.K., Mathews, C., Mahalingam, T.R., Balaram, P., and Nair, P.P., Chemoprevention of oral leukoplakia with vitamin A and beta carotene: an assessment, *Oral Oncol.*, 33, 231, 1997.

237. Giuliano, A.R. and Gapstur, S., Can cervical dysplasia and cancer be prevented with nutrients? *Nutr. Rev.*, 56, 9, 1998.

238. Einstein, M.H. and Goldberg, G.L., Human papilloma virus and cervical neoplasia, *Cancer Invest.*, 20, 1080, 2002.

239. de Vet, H.C., Knipschild, P.G., Willebrand, D., Schouten, H.J., and Sturmans, F., The effect of beta-carotene on the regression and progression of cervical dysplasia: a clinical experiment, *J. Clin. Epidemiol.*, 44, 273, 1991.

240. Romney, S.L., Ho, G.Y., Palan, P.R., Basu, J., Kadish, A.S., Klein, S., Mikhail, M., Hagan, R.J., Chang, C.J., and Burk, R.D., Effects of beta-carotene and other factors on outcome of cervical dysplasia and human papilloma virus infection, *Gynecol. Oncol.*, 65, 483, 1997.

241. Keefe, K.A., Schell, M.J., Brewer, C., McHale, M., Brewster, W., Chapman, J.A., Rose, G.S., McMeeken, D.S., Lagerberg, W., Peng, Y.M., Wilczynski, S.P., Anton-Culver, H., Meyskens, F.L., and Berman, M.L., A randomized, double blind, Phase III trial using oral beta-carotene supplementation for women with high-grade cervical intraepithelial neoplasia, *Cancer Epidemiol. Biomarkers Prev.*, 10, 1029, 2001.

242. Mackerras, D., Irwig, L., Simpson, J.M., Weisberg, E., Cardona, M., Webster, F., Walton, L., and Ghersi, D., Randomized double-blind trial of beta-carotene and vitamin C in women with minor cervical abnormalities, *Br. J. Cancer*, 79, 1448, 1999.

243. DeCosse, J.J., Adams, M.B., Kuzma, J.F., LoGerfo, P., and Condon, R.E., Effect of ascorbic acid on rectal polyps of patients with familial polyposis, *Surgery*, 78, 608, 1975.

244. Bussey, H.J., DeCosse, J.J., Deschner, E.E., Eyers, A.A., Lesser, M.L., Morson, B.C., Ritchie, S.M., Thomson, J.P., and Wadsworth, J., A randomized trial of ascorbic acid in polyposis coli, *Cancer*, 50, 1434, 1982.

245. Roncucci, L., Di Donato, P., Carati, L., Ferrari, A., Perini, M., Bertoni, G., Bedogni, G., Paris, B., Svanoni, F., Girola, M. et al., Colorectal Cancer Study Group of the University of Modena and the Health Care District 16: Antioxidant vitamins or lactulose for the prevention of the recurrence of colorectal adenomas, *Dis. Colon Rectum*, 36, 227, 1993.

246. Greenberg, E.R., Baron, J.A., Tosteson, T.D., Freeman, Jr., D.H., Beck, G.J., Bond, J.H., Colacchio, T.A., Coller, J.A., Frankl, H.D., Haile, R.W. et al., Polyp Prevention Study Group: a clinical trial of antioxidant vitamins to prevent colorectal adenoma, *New England J. Med.*, 331, 141, 1994.

247. MacLennan, R., Macrae, F., Bain, C., Battistutta, D., Chapuis, P., Gratten, H., Lambert, J., Newland, R.C., Ngu, M., Russell, A. et al., The Australian Polyp Prevention Project: randomized trial of intake of fat, fiber, and beta carotene to prevent colorectal adenomas, *J. Natl. Cancer Inst.*, 87, 1760, 1995.

248. McKeown-Eyssen, G., Holloway, C., Jazmaji, V., Bright-See, E., Dion, P., and Bruce, W.R., A randomized trial of vitamins C and E in the prevention of recurrence of colorectal polyps, *Cancer Res.*, 48, 4701, 1988.

249. DeCosse, J.J., Miller, H.H., and Lesser, M.L., Effect of wheat fiber and vitamins C and E on rectal polyps in patients with familial adenomatous polyposis, *J. Natl. Cancer Inst.*, 81, 1290, 1989.

250. Malila, N., Virtamo, J., Virtanen, M., Albanes, D., Tangrea, J.A., and Huttunen, J.K., The effect of alpha-tocopherol and beta-carotene supplementation on colorectal adenomas in middle-aged male smokers, *Cancer Epidemiol. Biomarkers Prev.*, 8, 489, 1999.

251. Paganelli, G.M., Biasco, G., Brandi, G., Santucci, R., Gizzi, G., Villani, V., Cianci, M., Miglioli, M., and Barbara, L., Effect of vitamin A, C, and E supplementation on rectal cell proliferation in patients with colorectal adenomas, *J. Natl. Cancer Inst.*, 84, 47, 1992.

252. Frommel, T.O., Mobarhan, S., Doria, M., Halline, A.G., Luk, G.D., Bowen, P.E., Candel, A., and Liao, Y., Effect of beta-carotene supplementation on indices of colonic cell proliferation, *J. Natl. Cancer Inst.*, 87, 1781, 1995.

253. Cascinu, S., Ligi, M., Del Ferro, E., Foglietti, G., Cioccolini, P., Staccioli, M.P., Carnevali, A., Luigi Rocchi, M.B., Alessandroni, P., Giordani, P., Catalano, V., Polizzi, V., Agostinelli, R., Muretto, P., and Catalano, G., Effects of calcium and vitamin supplementation on colon cell proliferation in colorectal cancer, *Cancer Invest.*, 18, 411, 2000.

254. Schorah, C.J., Vitamin C and gastric cancer prevention, in *Vitamin C: the State of the Art in Disease Prevention Sixty Years after the Nobel Prize*, Paoletti, R., Sies, H., Bug, J., Grossi, E., and Poli, A., Eds., Springer-Verlag, Milan, 1998, p. 41.

255. Zhang, L., Blot, W.J., You, W.C., Chang, Y.S., Liu, X.Q., Kneller, R.W., Zhao, L., Liu, W.D., Li, J.Y., Jin, M.L. et al., Serum micronutrients in relation to pre-cancerous gastric lesions, *Int. J. Cancer*, 56, 650, 1994.

256. You, W.C., Zhang, L., Gail, M.H., Chang, Y.S., Liu, W.D., Ma, J.L., Li, J.Y., Jin, M.L., Hu, Y.R., Yang, C.S., Blaser, M.J., Correa, P., Blot, W.J., Fraumeni, Jr., J.F., and Xu, G.W., Gastric dysplasia and gastric cancer: *Helicobacter pylori*, serum vitamin C, and other risk factors, *J. Natl. Cancer Inst.*, 92, 1607, 2000.

257. Zhang, Z.W., Patchett, S.E., Perrett, D., Domizio, P., and Farthing, M.J., Gastric alpha-tocopherol and beta-carotene concentrations in association with *Helicobacter pylori* infection, *Eur. J. Gastroenterol. Hepatol.*, 12, 497, 2000.

258. Varis, K., Taylor, P.R., Sipponen, P., Samloff, I.M., Heinonen, O.P., Albanes, D., Harkonen, M., Huttunen, J.K., Laxen, F., and Virtamo, J., Helsinki Gastritis Study Group: gastric cancer and premalignant lesions in atrophic gastritis: a controlled trial on the effect of supplementation with alpha-tocopherol and beta-carotene, *Scand. J. Gastroenterol.*, 33, 294, 1998.

259. Correa, P., Fontham, E.T., Bravo, J.C., Bravo, L.E., Ruiz, B., Zarama, G., Realpe, J.L., Malcom, G.T., Li, D., Johnson, W.D., and Mera, R., Chemoprevention of gastric dysplasia: randomized trial of antioxidant supplements and anti-*Helicobacter pylori* therapy, *J. Natl. Cancer Inst.*, 92, 1881, 2000.

260. Zullo, A., Rinaldi, V., Hassan, C., Diana, F., Winn, S., Castagna, G., and Attili, A.F., Ascorbic acid and intestinal metaplasia in the stomach: a prospective, randomized study, *Aliment. Pharmacol. Ther.*, 14, 1303, 2000.

261. Smith-Werner, S.A. and Giovannucci, E., Fruit and vegetable intake and cancer, in *Nutritional Oncology*, Heber, D., Blackburn, G.L., and Go, V.L., Eds., Academic Press, San Diego, 1999, p. 153.

262. Anon., MRC/BHF Heart Protection Study of antioxidant vitamin supplementation in 20,536 high-risk individuals: a randomised placebo-controlled trial, *Lancet*, 360, 23, 2002.

263. Woodson, K., Stewart, C., Barrett, M., Bhat, N.K., Virtamo, J., Taylor, P.R., and Albanes, D., Effect of vitamin intervention on the relationship between GSTM1, smoking, and lung cancer risk among male smokers, *Cancer Epidemiol. Biomarkers Prev.*, 8, 965, 1999.

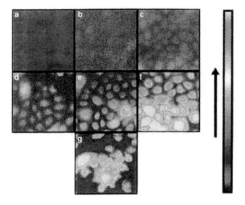

Figure 7.1

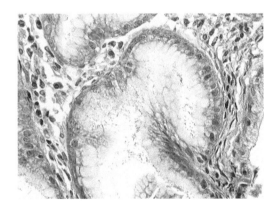

Figure 12.1

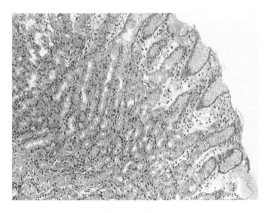

Figure 12.2

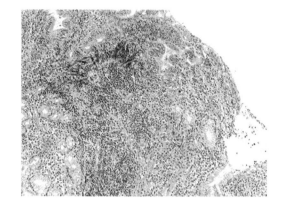

Figure 12.3

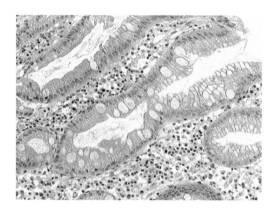

Figure 12.4

DMN   DMN+GSPE

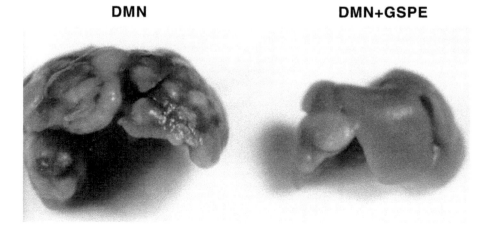

Figure 22.1

DMN **DMN+GSPE**

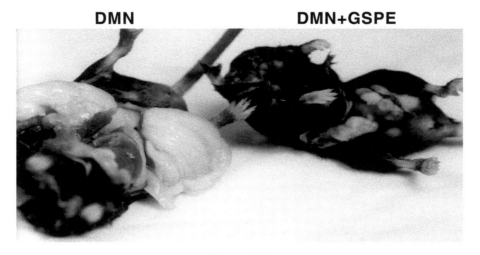

Figure 22.2

DMN **DMN+GSPE**

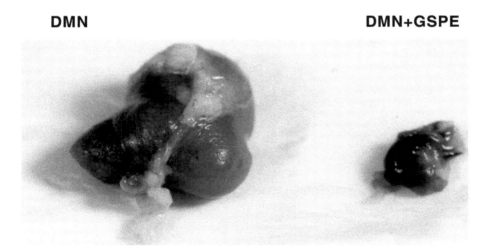

Figure 22.3

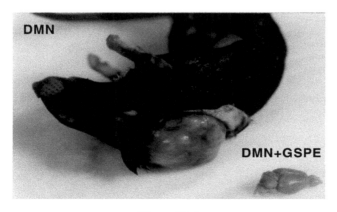

Figure 22.4

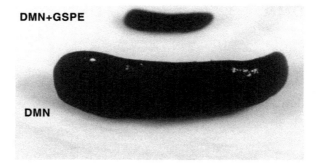

Figure 22.5

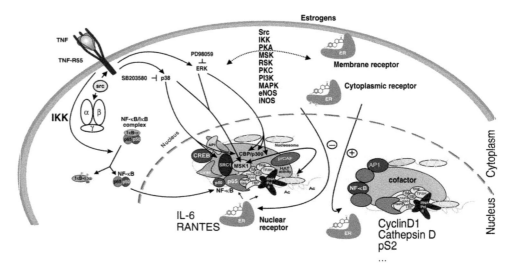

Figure 25.3

A) Control

B) D-fraction-treated

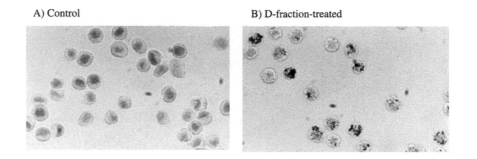

Figure 30.3

22 Roles of Polyphenols, Flavonoids, and Oligomeric Proanthocyanidins in Cancer Chemoprevention

Sidhartha D. Ray and Debasis Bagchi

CONTENTS

22.1 Introduction 311
22.2 Phytochemicals in Cancer Chemoprevention 312
22.3 OPCs and Their Role *In Vivo* 320
22.4 OPC and Cytochrome P450..322
22.5 OPC and Cell-Cycle Analysis 325
22.6 OPC and Metalloproteinases 327
22.7 OPC's Role in DNA Damage and DNA Repair 330
22.8 OPCs and Programmed Cell Death..332
22.9 Miscellaneous Mechanisms 337
22.10 Conclusions 339
References 344

22.1 INTRODUCTION

Chemoprevention is a relatively new term compared with *chemotherapy*. In fact, cancer chemoprevention is the newest branch of oncology research, even though the scientific community has used the idea for decades. Chemoprevention literally means a process of repeated exposure to an agent that completely prevents a lesion from occurring or overturns morphogenetically altered tissues and restores them to a normal state. This definition can have several interpretations: (1) selected removal of altered or mutated cells or tissues; (2) selected blockage of further propagation of morphogenetically altered cells without harming preexisting cells; (3) specialized interference with intracellular metabolic events or pathways that temporarily or permanently alter cellular functions in abnormal cells; (4) specific influence on the genomic machinery in favor of the host; (5) selective enhancement of carcinogen clearance by altering absorption or metabolism; and lastly, (6) preparation and protection of normal cells from future episodes of such noxious attacks. These propositions are based on the concepts that cell proliferation in premalignant conditions can be slowed down or arrested and that new tissue can grow beneath lesions and displace initiated cells or tissues. Many investigators now believe that all common cancers in viscera may develop through a state of dysplasia. Dysplasia is the antecedent change in tissues that occurs during the development of devastating malignancy. Considerable resources have been invested in efforts to identify and develop compounds that can stop, suppress, delay, or eliminate the progression of any type of cancer. Surprisingly, the use of phytochemicals as anticancer agents has captured the most attention.[1]

We now know that cancer cells are developed in three phases:

Phase I (initiation): When something (such as a free radical, carcinogen, or radiation) irreversibly alters a cell's genetic makeup, it causes the cell to divide more frequently than it should.

Phase II (promotion): When the damaged or genetically altered cell fails to fix its own damage, it goes into uncontrolled growth.

Phase III (progression): The growing tumor itself builds a blood supply network through angiogenesis and invades the surrounding tissue.

The last decade has witnessed an unprecedented interest in research of bioflavonoids or natural products. Given their safety and the fact that they are not perceived as "conventional medicine," a lengthy list of phytochemicals and plant-derived natural products is being considered for development as anticancer agents that may find widespread, long-term use in populations at normal risk. There is currently an explosion in the development of novel, naturally derived anticancer therapeutics, and from a developmental standpoint, nutraceutical therapy is at a critical impasse in which the abundance of molecular biological information about malignancy ascertained over the last couple of decades is being rationally used to develop anticancer therapeutics. From a commercial standpoint, pharmaceutical companies that never dared to venture toward anticancer naturopathy are clearly branching out to explore this profitable avenue. Several companies have launched exploratory clinical trials in combination therapy, which involves the use of conventional medications along with selected nutraceutical supplements.

22.2 PHYTOCHEMICALS IN CANCER CHEMOPREVENTION

Among many phytochemicals, pycnogenol, quercetin, resveratrol, curcumin, green tea extract, soy proteins, isoflavones, and particularly grape seed proanthocyanidins have incurred the most attention. In addition, a switch in the general interest from conventional medications to herbal supplements is evident by the fact that in the U.S. alone, nutraceutical sales have exceeded the multibillion dollar mark in the last few years. Due to this slowly evolving intense interest in these naturally derived products, the role of some of the classical vitamins (such as vitamin D, vitamin E, and vitamin C) with unsettled beneficial roles has come under microscopic scrutiny. With the hope of breaking new ground, several investigators are attempting to define beneficial roles for these historical moieties (vitamins) in the presence of proanthocyanidins. Of course, consumers are ever hopeful of conquering the painful sequel of this dreadful disease with a noninvasive miracle product. Coincidentally, nutraceuticals provide the best hope.

Proanthocyanidins are a subgroup of flavonoids that, in general, are nontoxic and, if absorbed, are very active *in vivo*. Oligomeric proanthocyanidins (OPCs) are present, particularly in high concentrations, in the seeds of red grapes and the bark of maritime pine trees. The use of OPCs as a dietary supplement to increase the body's antioxidant power is becoming increasingly popular. Grape seed extract is classified as a classic proanthocyanidin as well as an OPC. Pycnogenol, a pine bark extract, belongs to the same class of compounds. Grape seed proanthocyanidin extract (GSPE), which is predominantly an admixture of OPCs commonly found in red grape seeds, has been the object of intense investigation in recent years. Grape seed extracts are a group of natural antioxidants that are known to possess a broad spectrum of pharmacological activity and protect cells from a variety of toxic insults. Some studies claim a superior ability of GSPE to scavenge free radicals compared with vitamin C, vitamin E, and β-carotene.[2–6]

These OPCs share similar properties, and in general function mainly as free-radical scavengers. Some of the OPCs show an anti-inflammatory property and serve as enhancers of the strength of connective tissue and blood vessels. In addition to serving as a powerful antioxidant, GSPE has been used to improve edema (swelling) in individuals with chronic venous insufficiency. No side effects have been reported with this compound. Our laboratories have reported numerous antitoxic effects of GSPE *in vivo*, and those are: (a) counteraction of acetaminophen-induced hepatotoxicity,[7] (b) protection from acetaminophen-induced nephrotoxicity,[8,9] (c) reduction of $CdCl_2$-induced nephrotoxicity,[10] (d) reversal of acute high-dose dimethylnitrosamine (DMN)-induced splenotoxicity,[10] (e) inhibition of amiodarone-induced pulmonary toxicity,[8,9] (f) substantial reduction of pesticide (MOCAP)-induced neurotoxicity and mortality,[10] (g) reduction of doxorubicin-induced cardiotoxicity,[8,9] (h) protection against myocardial ischemia-reperfusion injury and oxidative stress,[11,12] (i) protection against atherosclerosis,[13] and (j) reduction of acute and chronic stress-induced gastrotoxicity.[14] Furthermore, GSPE demonstrated protection against (a) smokeless-tobacco-induced oxidative stress and apoptotic cell death in cultured human keratinocyte cells[6,15] and (b) cancer-chemotherapeutic-drug-induced cytotoxicity toward human liver cells.[16–18] It was also demonstrated that GSPE induces selective cytotoxicity toward human lung cancer, breast cancer, and gastric adenocarcinoma cells in cultures, while enhancing the growth and viability of human normal gastric mucosal cells and J774A.1 murine macrophage cells.[19–21] The most remarkable aspect of these studies was the ability of GSPE to selectively modulate and influence two cell-death processes, apoptosis (programmed) and necrosis (unprogrammed), in a variety of target organs *in vivo*. Between these two cell-death processes, the 21st-century biomedical research community (including the cancer biologists) is pursuing apoptosis research very aggressively. Several laboratories pioneering in OPC research have already shown beneficial effects of GSPE at the cellular, subcellular, and molecular levels (Table 22.1 and Table 22.2). Scientists are very hopeful that apoptosis research may yield several cancer-cure-related clues during the first quarter of this century.

Despite the numerous well-known complications of naturopathy and antioxidant therapy (antioxidants could blunt the effect of standard therapies, particularly alkylating, platinum, and tumor antibiotic agents, which are oxidative in nature), the biomedical research community has witnessed a remarkable optimism in the attitude of many conventional practitioners toward naturopathy and antioxidant therapy in recent years. Although this may appear as a theoretical concern, a plethora of published papers shows that this proposed interaction of anti- and pro-oxidant therapies has tremendous implications, and it is time to put this controversy in the proper perspective.[22,23] Approaches to cancer therapy have been remarkably consistent for the last several decades. Surgery, radiation, and chemotherapy alone — or their reasonable combinations — have been the cornerstones of conventional treatment. However, the clinical success of these therapies has reached a plateau. Indeed, numerous cancer-treatment specialists even question the validity of chemotherapy as a treatment for most cancers. Clearly, there is a need for new therapies that can increase the efficacy of cancer treatment. Careful application of antioxidants may be a means of helping to raise cancer therapy to a new level of success.

Every plant species has a unique combination of phenolic/polyphenolic constituents, which is why different herbs, all rich in these substances, have very different effects on the body. For example, pycnogenol is a complex combination of a particular group of proanthocyanidins, and its distinctive chemical "blueprint" accounts for its properties.[24] The National Cancer Institute has arbitrarily classified these chemopreventive agents into first, second, and third generations of compounds or agents of interest. Retinoids, vitamin A, β-cis-retinoic acid, tamoxifen, and finesteride are considered to be the first-generation chemopreventive agents, whereas green tea and several other polyphenolic compounds, including OPCs, have been categorized into the third generation of putative chemopreventive agents.

The discipline of chemoprevention is rather broad and needs focus. Initial efforts of chemoprevention primarily included precise identification and removal of offending carcinogenic substances from the diet or the environment, with the hypothesis that "cure is better than prevention." This approach was considered viable in the past because plotting of the cancer-etiology map was incomplete.

TABLE 22.1
Examples of Execution of Anticarcinogenic Action via Modulation of Intracellular Molecular Targets by Phenols, Polyphenols, Flavonoids, and OPCs

Name of Compound	Intracellular Molecular Target	Model System	Reference
Curcumin	Inhibits oxidative stress and DNA damage	Jurkat T-cells	Kelly, M.R. et al., *Mut. Res.*, 485, 309, 2001
Tannic acid	Inhibits protease activity; increases p27/Bax expression; induces G1 arrest	Jurkat cells	Nam, S. et al., *Can. Epidem. Biomar. Prev.*, 10, 1083, 2001
Anethole, eugenol, and isoeugenol	Promotes TNF-signaling, lipid peroxidation, NF-κβ activation	—	Chainy, G.B. et al., *Oncogene*, 19, 2943, 2000
Epicatechin	Promotes oxidative stress and inhibition of caspase-3-activation	human fibroblasts	Spencer, J.P. et al., *Biochem. J.*, 354, 493, 2001
Red clover isoflavone biochanin A (trifolium pratense)	Inhibits DMBA-induced DNA damage; reduces CYP1A1 and -1B1 mRNA and interference of XRE-dependent transactivation	MCF7 cells	Chan, H.Y. et al., *Br. J. Nutr.*, 90, 87, 2003
GSPE	Increases bcl-2 expression and decreases p53 and c-myc expression	normal human Chang liver cells	Joshi, S.S. et al., *Curr. Pharm. Biotechnol.*, 2, 187, 2001
Viscum album (preparation of mistletoe lectin)	Decreases bcl-2 and bcl-X expression; inhibits cell proliferation	lymphoblastoid cells	Huyen, D.V. et al., *Chemotherapy*, 47, 366, 2001
Eupatilin extract	Promotes caspase-9, -3, -7 activation, cytochrome-c release, PARP cleavage	HL-60 cells	Seo, H.J. and Surh, Y.J., *Mut. Res.*, 496, 191, 2001
Gallic acid derivative	Inhibits DNA replication and S-G2 progression	Jurkat-T cells, K-562 cells	Smith, D.M. and Dou, Q.P., *Int. J. Mol. Med.*, 7, 645, 2001
Ellagic acid	Down-regulates IGF-II; activates p21	SW-480 colon cancer cells	Narayanan, B.A. et al., *Anticancer Res.*, 21, 359, 2001
Grape seed extract	Inhibits constitutive activation of MAPK/p38	breast carcinoma MDA-MB468 cells	Agarwal, C. et al., *Clin. Cancer Res.*, 6, 2921, 2000
Baicalein and baicalin	Functions as an anti-angiogenic	chorioallantoic membrane and human umbilical vein cells	Liu, J.J. et al., *Int. J. Cancer*, 106, 559, 2003
Sophoranone	Inhibits mitochondrial complex III and IV, and causes cytochrome-c release	U-937 cells	Kajimoto, S.S. et al., *Int. J. Cancer*, 99, 879, 2002

TABLE 22.1 (CONTINUED)
Examples of Execution of Anticarcinogenic Action via Modulation of Intracellular Molecular Targets by Phenols, Polyphenols, Flavonoids, and OPCs

Name of Compound	Intracellular Molecular Target	Model System	Reference
Aucubin, chlorogenic acid, ferulic acid, p-coumaric acid, and vanillic acid	Enhances the activity of human lymphocyte proliferation and secretion of IFN-gamma	human peripheral blood mononuclear cells	Liang, L.C. et al., *Planta Medica*, 69, 600, 2003
Quercetin and flavopiridol	Inhibits CD34+ cell proliferation and activation of caspase-3 activity	acute myelogenous leukemia (AML) cell lines	Liesveld, J.L. et al., *Leuk. Res.*, 27, 517, 2003
Biochanin A, morin, phloretin, and silymarin	Interacts with Pgp and modulates drug efflux from cells	breast cancer cell lines MCF7 and MDA435/LCC6	Zhang, S. and Morris, M.E., *J. Pharmacol. Exp. Ther.*, 304, 1258, 2003
Morin	Promotes growth arrest at G2/M; induction of GADD45; phosphorylation and inactivation of the cell-cycle kinase, cdc2	squamous carcinoma cells	Brown, J. et al., *Carcinogenesis*, 24, 171, 2003
Quercetin (QU), galangin (GA), and chrysin	Promotes DNA fragmentation and apoptosis	human promyelocytic leukemia HL-60 cells	Cipak, L. et al., *Leuk. Res.*, 27, 65, 2003
Daidzein and genistein	Functions as Akt/protein kinase B and focal adhesion kinase (FAK)	metastatic ER-negative MDA-MB-231 cells	Brownson, M.D. et al., *J. Nutr.*, 132 (11 Suppl.), 3482S, 2002
Myricetin and myricetin-3-galactoside	Functions as poison topoisomerase-I	cancer cell lines	Lopez-Lazaro, M.J., *Enzyme Inhib. Med. Chem.*, 17, 25, 2002
Baicalein	Functions as potent inhibitor of CYP1A1 and CYP1B1 gene expressions; reduces DMBA-DNA adduct formation	MCF7 breast cancer cells	Chan, H.Y. et al., *Biomed. Pharmacother.*, 56, 269, 2002
Quercetin and luteolin	Modulates protein tyrosine kinase and EGFR-signaling pathway	MiaPaCa-2 cancer cells	Lee, L.T. et al., *Anticancer Res.*, 22, 1615, 2002
Quercetin and amentoflavone	Down-regulates COX-2 and iNOS enzyme activities	A549 human lung adenocarcinoma cell line	Banerjee, T. et al., *Prostaglandins Leuko. Essen. Fatty Acids*, 66, 485, 2002
Anthocyanidins, dihydrochalcones, dihydroflavonols, and flavonolignans	Inhibits some matrix-proteases instrumental to inflammation and cancer invasion (leukocyte elastase and gelatinases)	*in vitro*	Sartor, L. et al., *Biochem. Pharmacol.*, 64, 229, 2002

TABLE 22.2
Examples of Anticancer Effects of Phenols, Polyphenols, Flavonoids, and OPCs on Pro- and Anti-Apoptosis-Based Mechanisms

Name of the Compound	Molecular Mechanism	Model System	Reference
Flax-seed, isoflavones, or rutin	Increases *Brca1* and *Brca2* mRNA	ovariectomized Wistar rats	Vissac-Sabatier, C. et al., *Cancer Res.*, 63, 6607, 2003
Catechins, flavones, and antocyanines	Inhibits the cancer-associated enzyme telomerase	*in vivo* and *in vitro*	Naasani, I. et al., *Cancer Res.*, 63, 824, 2003
Curcumin	Inhibits camptothecin-, mechlorethamine-, and doxorubicin-induced apoptosis	MCF7, MDA-MB-231, BT-474 human breast cancer cells	Somasundaram, S. et al., *Cancer Res.*, 62, 3868, 2002
Salvianolic acid (*Salvia miltiorrhiza*)	Induces apoptosis	*in vivo* rabbit model: neointima	Hung, H.H et al., *Histol. Histopathol.*, 16, 175, 2001
Apocynin and spinach leaf antioxidant extract	Promotes antioxidative stress; functions as free-radical scavenger	rat hepatoma and murine B-16 melanoma cells	Zhang, G. et al., *Nutr. Cancer*, 38, 265, 2000
Grape seed proanthocyanidin extract	Modulates dimethylnitrosamine-induced apoptosis and necrosis	male B6C3F1 mice	Ray S.D. et al., *Am. Assoc. Cancer Res. Meeting*, 2000
Withania somnifera root extract	Modulates DMBA-induced skin cancer and 20-methylcholanthrene-induced fibrosarcoma tumors	Swiss albino mice	Prakash, J. et al., *Nutr. Cancer*, 42, 91, 2002; *Phytother. Res.*, 15, 240, 2001
Alpinia oxyphylla Miquel (Zingiberaceae)	Suppresses skin tumor promotion and induction of apoptosis in HL-60 cells	female ICR mice	Lee, S. et al., *Carcinogenesis*, 19, 1377, 1998
Green tea/caffeine	Inhibits UV-induced carcinogenesis, p53 + ve cells, p21(WAF1/CIP1) + ve cells, apoptotic cells	SKH-1 mice	Lu, Y.P. et al., *Cancer Res.*, 60, 4785, 2000
Gingko biloba extract	Inhibits benzo(a)pyrene-induced forestomach carcinogenesis and doxorubicin cardiotoxicity	female Swiss albino mice	Agha, A.M. et al., *J. Exp. Clin. Cancer Res.*, 20, 39, 2001
Polyphenolic fraction from grape seeds	Inhibits DMBA-initiated and 12-O-tetradecanoylphorbol 13-acetate-promoted	SENCAR mouse skin two-stage carcinogenesis protocol	Zhao, J. et al., *Carcinogenesis*, 20, 1737, 1999

TABLE 22.2 (CONTINUED)
Examples of Anticancer Effects of Phenols, Polyphenols, Flavonoids, and OPCs
on Pro- and Anti-Apoptosis-Based Mechanisms

Name of the Compound	Molecular Mechanism	Model System	Model System
Curcumin (diferuloyl methane), the major pigment from the *Curcuma longa* L.	Modulates integrin receptors, collagenase activity, and expression of Nm23 (enhances the expression of antimetastatic proteins, tissue-inhibitor metalloproteinase) and E-cadherin; considerably lowered in lung metastases	B16F10 melanoma cells implanted in C57BL6 mice	Ray S. et al., *J. Environ. Pathol. Toxicol. Oncol.*, 22, 49, 2003
Hesperidin and beta-cryptoxanthin	Inhibits lung cancer induced by 4-(methylnitrosamino)-1-(3-pyridyl)-1-butanone (NNK)	male A/J mice	Kohno, H. et al., *Cancer Lett.*, 174, 141, 2001
Tannic acid	Modulates anticholangiocarcinoma (malignancy of the biliary tract)	Balb/c athymic mice	Marienfeld, C. et al., *Hepatology*, 37, 1097, 2003
Polyphenon E and EGCG	Modulates cervical tissue proliferation and associated changes	human papilloma virus-induced chronic cervicitis, dysplasia	Ahn, W.S. et al., *Eur. J. Cancer Prev.*, 12, 383, 2003
Apigenin and quercetin	Inhibits melanoma growth and metastatic potential	syngeneic mice	Caltagirone, S. et al., *Int. J. Cancer*, 87, 595, 2000
Green tea polyphenol EGCG	Reduces signaling via the phosphatidylinositol-3-kinase, Akt kinase to NF-κB pathway	mammary tumor virus-Her-2/neu NF639 cell	Pianetti, S. et al., *Cancer Res.*, 62, 652, 2002
Quercetin and naringin	Reduces lung cancer	humans	Le Marchand, L. et al., *J. Natl. Cancer Inst.*, 92, 154, 2000
Nobiletin (*Citrus depressa*)	Suppresses pro-matrix metalloproteinase-9, progelatinase-B, interstitial procollagenase	TPA-stimulated HT-1080 cells	Sato, T. et al., *Cancer Res.*, 62, 1025, 2002
Catechin, trans-resveratrol, quercetin, and gallic acid	Inhibits mouse skin cancer	DMBA+TPA-induced mouse skin cancer	Soleas, G.J. et al., *Clin. Biochem.*, 35, 119, 2002
Quercetin and luteolin	Modulates protein tyrosine kinase and EGFR-signaling pathway	MiaPaCa-2 cancer cells	Lee, L.T. et al., *Anticancer Res.*, 22, 1615, 2002
Green tea powder	Suppresses tumor growth and hyperlipidemia	rats implanted with ascites hepatoma cells (AH109A)	Zhang, G. et al., *Biosci. Biotechnol. Biochem.*, 66, 711, 2002
Silibinin	Inhibits advanced human prostate carcinoma	athymic nude mouse	Anon., *Cancer Res.*, 62, 3063, 2002

Secondary chemoprevention strategies predominantly dictate the following: (a) early detection of premalignant disease, (b) early treatment with a drug (or a team of drugs), and (c) early treatment with a dietary supplement (or a team of dietary supplements). It also recommends intervention at a stage that can blunt any stage (initiation, promotion, or progression) of the cancer development. Among the various protocols that are in use, early diagnosis of cancer with secondary prevention has gained maximum popularity. Significant progress has been made in early detection tools. Routine physical examinations now routinely screen for prostate cancer (by detecting blood PSA changes), cervical cancer (by cervical smears), colon cancer (by occult blood testing and/or proctosigmoidoscopy), and breast cancer (by mammography). The lesion that needs to be detected in secondary preventive efforts is a tissue change that is described as premalignant, preinvasive, or dysplastic.

Types and concentrations of OPCs vary within plants and among species due to variables involved during plant growth, soil, nutrition, weather conditions, its own microflora, and the age of the plant. Flavonoids, the largest group of plant polyphenols and the most studied, include more than 4000 different compounds. The basic flavonoid skeleton of OPCs allows a multitude of variations in chemical structure, giving rise to flavonols (quercetin, kempferol, myricetin), flavones (apigenin, luteolin), flavanones (catechin, epicatechin), anthocyanidins, and isoflavonoids (genistein, daidzein). The flavonoids occur in fruits, vegetables, nuts, seeds, flowers, leaves, and bark. Besides serving as antioxidants, OPCs follow unique mechanisms to protect plants from ultraviolet (UV) light and attacks from insects, fungi, viruses, and bacteria.[25] These pigments provide a wide assortment of attractive colors to pollinators, and their chemical diversity is known to regulate plant growth by modulating hormones.

Other significant allied members include the *Allium* genus (garlic, onions, and chives), members of the Labiatae family (mint family: basil, mints, oregano, rosemary, sage, and thyme), members of the Zingiberaceae family (turmeric, ginger), licorice root, green tea, flax, members of the Umbelliferae family (carrot family: anise, caraway, celery, chervil, cilantro, coriander, cumin, dill, fennel, and parsley), and tarragon.[26] Researchers have identified a host of cancer chemoprotective phytochemicals in these herbs. In addition, many herbs contain a variety of phytosterols, triterpenes, saponins, and carotenoids, which have been shown in studies of legumes, fruits, and vegetables to be cancer-chemoprotective. These beneficial substances dispense their role as antioxidants, electrophile scavengers, and immune system boosters; they inhibit nitrosation and the formation of DNA adducts with carcinogens, inhibit hormonal actions and metabolic pathways associated with the development of cancer, and induce phase I or II detoxication enzymes. Several phytochemicals inhibit tumor formation by stimulating the protective phase II enzyme, glutathione transferase. (EC 2.5.1.18: GT is a detoxifying enzyme that catalyzes the reaction of glutathione with electrophiles to form compounds that are less toxic, more water-soluble, and can be excreted easily.)

There is overwhelming epidemiological evidence for the protective effect of plant-derived OPCs against cancer and heart disease (Table 22.2, Figure 22.1 through Figure 22.5). It is not known which dietary components are responsible for this protective effect, but it is often assumed that antioxidants such as carotenoids, vitamin E, and vitamin C are responsible for the protection. Although a coherent picture is slowly emerging, several large chemoprevention trials involving interventions with these antioxidant micronutrients have been carried out with conflicting results. Since most plants are abundantly loaded with carotenoids, vitamin E, vitamin C, and several other antioxidant micronutrients, custom studies are needed to dissect pinpointed roles for OPCs for these well-characterized vitamins. It is also not known clearly whether the ultimate outcomes are due to a synergistic, additive, or potentiating cooperation between these arrays of substances. More-stringent experiments are needed in this direction to resolve the actual role of OPCs.

DMN   DMN+GSPE

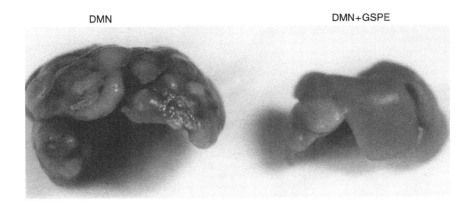

FIGURE 22.1 Protection of dimethylnitrosamine-induced liver tumor by grape seed extract exposure.

DMN   DMN+GSPE

FIGURE 22.2 Protection of dimethylnitrosamine-induced urinary bladder cancer by grape seed extract exposure.

DMN+GSPE   DMN

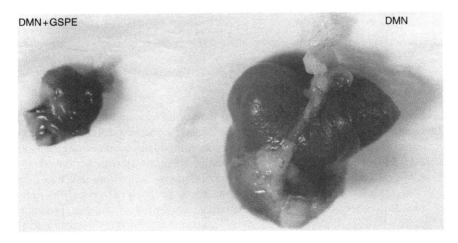

FIGURE 22.3 Protection of dimethylnitrosamine-induced stomach cancer by grape seed extract exposure.

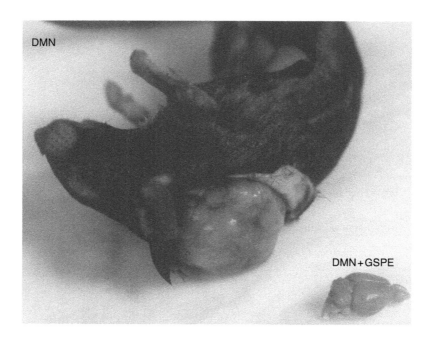

FIGURE 22.4 Protection of dimethylnitrosamine-induced brain cancer by grape seed extract exposure.

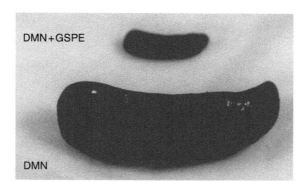

FIGURE 22.5 Protection of dimethylnitrosamine-induced spleen cancer by grape seed extract exposure.

22.3 OPCs AND THEIR ROLE *IN VIVO*

Flavonoids have been studied since the 1940s, and their antioxidant activity is undisputed to date. With the immense volume of research being released every year regarding the effects of radical oxygen species on human health, the role of flavonoid antioxidants cannot be ignored (Table 22.1 and Table 22.2). Monomeric and oligomeric proanthocyanidins (flavonols, flavones, catechins, flavanones, anthocyanidins, isoflavonoids, and their oligomers) are mostly attached to gallic acid. These molecules are plentiful in the bark of pine trees, grape seeds and skins, peanut skins, cranberries, tea, and other sources. Recent literature recognizes oligomeric proanthocyanidins (OPCs) as "nature's biological response modifiers," which modify the body's reaction to compounds such as allergens, microorganisms, and carcinogens. The biological fate (absorption, biodistribution, metabolism, degradation, and elimination) of OPCs, nonnutritive plant factors including dietary flavonoids, is of great importance relative to their reported health-protective effects. An estimate of the absolute bioavailability (expressed as a percentage) of OPCs from a food source is derived from knowledge of the proportion of molecules that enters the blood circulation intact after consumption and following their passage across the gut wall and other vital organs. Therefore, the extent of absorption does not entirely indicate actual amounts of bioavailability. A multitude of factors such as absorption, distribution, metabolism, and elimination also contribute to the bioavailability and subsequent effectiveness of OPCs following ingestion. Bioavailability thus estimates the exposure of the body to the OPCs in question. Absolute bioavailability can only be determined experimentally by administering the flavonoid, for example, both orally and intravenously, and then calculating the ratio of plasma levels with time.[27] However, because of rapid alteration in the structures of OPCs *in vivo*, estimates can be inaccurate. Nevertheless, several investigators have attempted to determine OPC concentrations *in vivo* in a range of animal models.

The extent of absorption of dietary flavonoids is an important unsolved problem in judging their potential health effects. Only the colonic members of the gut microflora hydrolyze the glycosides, which degrade flavonoids at the same time.[28] It has often been stated that flavonoids present in foods cannot be absorbed from the intestine if they are sugar-bound as glycosides. Most of the information on subsequent metabolism is derived from animal studies, but there is little available data in humans. The limited source of knowledge on absorption and metabolism of the flavonoids has been generated by studying isolated flavonoids or single foods. There is a very large volume of *in vitro* data on the effects of flavonoids, but very few *in vivo* studies. Whether the *in vitro* data can predict *in vivo* effects (including in humans) remains an intriguing question. More studies are needed to investigate the absorption and metabolism of various flavonoids in individual foods and combinations of foods.[29]

Many OPCs exhibit kaleidoscopic property; they are very similar in chemical structure, but slight variations yield different biological effects. For example, rutin is the same as quercetin except for having a sugar molecule; hesperidin is very similar to rutin; and diosmin is almost identical to hesperidin. Similarly, most OPCs are present in specific proportions in different genus and species of plants, and these flavonoids are ubiquitous in the plant kingdom. Among numerous OPCs, quercetin and its glycosides are the most commonly encountered dietary flavonoids in the plant kingdom. In several investigations, laboratory animals and humans were exposed to pure quercetin (or its glycoside derivatives) directly or included in a diet.[30–34] It was difficult to determine which form was absorbed, whether the glycoside, aglycone, or both forms. The analysis by Hollman et al.[35] concluded that human absorption of the quercetin glycosides from onions (52%) was far better than that of the pure aglycone (24%).

The sugar moiety is an important determinant of the absorption and bioavailability. The absorption of quercetin rutinoside (absorbed from colon) was far less than glucoside (absorbed from intestine) derivative in humans. However, Walle et al.[36] did not detect any quercetin glucosides in the ileostomy fluid, whereas a substantial amount of quercetin aglycone was identified, suggesting that quercetin glucosides were hydrolyzed to quercetin in the small intestine and then absorbed. Hydrophilic glycosides (e.g., rutin, naringin, hesperidin, baicalin, daidzein, and phellamurin) generally cannot be transported across membranes by passive diffusion.[29,37] The absorption of quercetin glucoside's parent form led to a speculation that it was transported across the gut wall by the intestinal sodium-glucose transporter.[30–33] Upon hydrolysis by the enzymes released by enteric microflora, the sugar moiety of flavonoid glycosides were cleaved and resulted in more lipophilic aglycones, which became permeable through the gut wall. Morand et al.[38] reported that the nature of glycosylation greatly influences the efficiency of quercetin absorption in rats. Quercetin 3-glucoside is absorbed in the small intestine, and the plasma level of its glucuronides/sulfates is three times higher than quercetin itself. This fact can be explained by the higher water solubility of quercetin 3-glucoside than quercetin. Olthof et al.[39] suggested that the quercetin glucosides were rapidly absorbed in humans irrespective of the position of the glucose moiety. Numerous laboratories in the world have made extensive efforts to resolve absorption and bioavailability issues of diverse OPCs.[40–63]

Since the standards of flavonoid-metabolite conjugates are unavailable, they are typically determined as their aglycones after hydrolysis with β-glucuronidase/sulfatase enzymes. Moreover, flavonoid aglycones are also vulnerable to oxidation; therefore, either β-glucuronidase and sulfatase alone or their mixture are used for the hydrolysis of glucuronides/sulfates of flavonoids in serum. Likewise, flavonoid distribution was found to be markedly different if they were administered through different routes (orally vs. i.v., orally vs. i.p., etc.). It has been reported that some OPC metabolites can cross the blood–brain barrier. Cell-free extracts of human small intestine and liver were used to investigate whether there is glucosidase activity toward flavonoid glycosides. The study revealed that small intestine and liver were not capable of hydrolyzing all forms of flavonoid glycosides. In fact, flavonoids are capable of undergoing hydroxylation, methylation, and reduction reactions, although conjugation reactions with sulfate and/or glucuronic acid were found to be the most common pathway for flavonoid metabolism. Some studies found sulfation metabolism to be more prominent than the glucuronidation pathway for quercetin, morin, naringenin, hesperidin, daidzein, and baicalein.[29,37] The conjugated metabolites of flavonoids are still believed to possess antioxidation ability *in vivo*, although it may be weaker than the aglycone parent forms.

Gut microflora is known to play a prominent role in the gastrointestinal metabolism of flavonoids.[28] Enzymes released by the resident microbes have the ability to gradually hydrolyze the glycosides into aglycones, which are absorbed by the intestine. The unabsorbed aglycones that pass through the small intestine are degraded by colonic microflora into phenolic acids. Microbiological techniques have isolated two distinct species:

1. *Enterococcus casseliflavus*, capable of utilizing the sugar moiety of the glycoside but does not further degrade the aglycone
2. *Eubacterium ramulus*, capable of degrading the aromatic ring system by detachment of the A ring from the residual flavonoid molecules and opening of the heterocyclic C ring

The sulfates and glucuronides of flavonoids are ionized under physiological pH and are water soluble; therefore, they are readily excreted into bile and urine. When excreted into bile, the conjugated metabolites are passed into the duodenum and metabolized by enterobacteria, which hydrolyze the sulfates/glucuronides and further fragment the flavonoid aglycones into aromatic acids. The resulting metabolites can be reabsorbed and enter an enterohepatic circulation to result in a second peak of serum profile. The structure of flavonoid conjugates determines the extent of biliary excretion and enterohepatic circulation. The half-life of elimination also can be prolonged, and the plasma levels of quercetin metabolites can be detected up to 24 h after flavonoid consumption, indicating a possible buildup of quercetin metabolites in plasma after repeated intake of onion. In contrast to enterohepatic circulation, the variation of urinary excretion of flavonoid metabolites among individuals is very large. Previous studies have shown that the urinary excretion of flavonoid-metabolite conjugates, based on the fraction of intake, can vary from trace to 24.6%, depending on the source and structure of flavonoids.

22.4 OPC AND CYTOCHROME P450

The anticarcinogenicity of some flavonoids has been attributed to modulation of the cytochrome P450 enzymes, which metabolize procarcinogens to their activated forms. However, the mechanism by which flavonoids inhibit some P450-mediated activities while activating others is an imposing question. One of the oldest prototype flavonoids, α-naphthoflavone, is known to exert a dual role on two forms of human cytochrome P450 isozymes (1A1 and 3A4), whose benzo[*a*]pyrene hydroxylation activities are respectively inhibited and stimulated by this compound. This flavonoid inhibited P450 1A1 binding to benzo[*a*]pyrene via a classical competitive mechanism. In contrast, α-naphthoflavone stimulated P450 3A4 by selectively binding and activating an otherwise inactive subpopulation of this P450 and promoting benzo[*a*]pyrene binding to the latter. These observations suggest that flavonoids enhance activity by increasing the pool of active P450 molecules within this P450 macrosystem. Activators in other biological systems may similarly exert their effect by expanding the population of active receptor molecules.[64] Pharmacological and toxicological effects of flavonoids arise from a dual mode of action, because P450-mediated drug and carcinogen metabolism can be inhibited via classical competitive inhibition or enhanced by conformational induction of selected P450 molecules to an active form. The latter mechanism is similar to the recently described activation of receptors by agonists[65,66] and may prove to be generally applicable in protein interactions with small molecules. Through understanding the diverse mechanisms of flavonoid action, the role of dietary flavonoids in modulating P450-mediated reactions is clearer, and the identification and design of flavonoids as chemopreventive agents becomes easier.

Using *in vitro* and *in vivo* approaches, researchers have identified many herbs and natural compounds isolated from herbs as substrates, inhibitors, and/or inducers of various CYP enzymes. For example, St.-John's-wort is a potent inducer of CYP3A4, which is mediated by activating the orphan pregnane X receptor. It also contains ingredients that inhibit CYP1A2, CYP2C9, CYP2C19, CYP2D6, and CYP3A4.[67] Many other common medicinal herbs also exhibited inducing or inhibiting effects on the CYP system, with the latter being competitive, noncompetitive, or mechanism based. It appears that the regulation of CYPs by herbal products is complex, depending on the herb type, their administration dose and route, the target organ, and species.[68] Due to the difficulties in identifying the active constituents responsible for the modulation of CYP enzymes, prediction of herb–drug metabolic interactions is difficult. However, herb–CYP interactions may have important clinical and toxicological consequences. For example, induction of CYP3A4 by St.-John's-wort may partly provide an explanation for the enhanced plasma clearance of a number of drugs, such as cyclosporine and indinavir, which are known substrates of CYP3A4, although other mechanisms including modulation of gastric absorption and drug transporters are possibly involved. In contrast, many organosulfur compounds, such as diallyl sulfide from garlic, are potent inhibitors of CYP2E1; this may provide an explanation for garlic's chemopreventive effects, as many mutagens require activation by CYP2E1. Therefore, known or potential herb–CYP interactions exist, and further studies on their clinical and toxicological roles are warranted.

A marked effect on the P450 isozymes may dramatically affect the plasma concentration of drugs, resulting in either overdose or rapid loss of their therapeutic effects. The other important issue is the involvement of flavonoids in the process of carcinogenesis. Increased activation of carcinogens via induction of specific CYPs could be detrimental. On the other hand, down-regulation or inhibition of CYPs involved in carcinogen activation and the scavenging of biological reactive intermediates formed from carcinogens by CYP-mediated reactions can be a beneficial properties of various flavonoids. Induction of CYP activity by flavonoids proceeds via various mechanisms, including direct stimulation of gene expression through a specific receptor or CYP protein or through mRNA stabilization. Certain flavonoids induce CYPs via binding to aryl hydrocarbon receptor (AhR), a ligand-activated transcription factor. This mechanism is associated with the enhanced activity of CYP1-family enzymes, including CYP1A1, 1A2, and 1B1, that are responsible for the activation of carcinogens such as benzo[*a*]pyrene, 7,12-dimethyl benzo[*a*]anthracene, and aflatoxin B1.

Interaction of grapefruit juice flavonoids with several drugs is becoming an important area of focus in clinical practice. The mechanism of this drug–nutrient interaction is inhibition of the cytochrome P450 3A4 isoenzyme (CYP3A4), resulting in increased bioavailability of the drug. Grapefruit juice has been shown to inhibit this enzyme in the wall of the small intestine.[69] The component of grapefruit juice that causes this inhibition has not been elucidated, but it is believed to be a bioflavonoid. Grapefruit juice contains a psoralen, which may serve as the inhibitor. This effect reduces the metabolism of cyclosporin, terfenadine, and calcium channel blockers (other than amlodipine and diltiazem), resulting in increased plasma concentrations, which could be clinically important.[70,71]

A number of other flavonoids, such as fisetin, galangin, myricetin, kempferol, chrysin, and apigenin, are also potent inhibitors of P-form PST-mediated sulfation, with IC_{50} values <1 μM. Studies of structural analogs indicated the flavonoid 7-hydroxyl group as particularly important for potent inhibition. Potential human metabolites of quercetin were poor inhibitors. Curcumin, genistein, and ellagic acid (other polyphenolic natural products) were also inhibitors of P-form PST, with IC_{50} values of 0.38 to 34.8 μM. Quercetin was also shown to inhibit sulfoconjugation by the human hepatoma cell line Hep G2. Although less potent in this intact cell system (IC_{50} 2 to 5 μM), quercetin was still more potent than 2,6-dichloro-4-nitrophenol, the classical P-form PST inhibitor that has also been shown to be an inhibitor *in vivo*. These observations suggest the potential for clinically important drug interactions as well as a possible role for flavonoids as chemopreventive agents in sulfation-induced carcinogenesis.[72]

The inhibition of gene expression of CYP1-family enzymes through blocking AhR plays an important role in the chemopreventive properties of flavonoids. For instance, quercetin, one of the most popular natural flavonoids, binds as an antagonist to AhR and inhibits the CYP1A1 mRNA transcription and protein expression, thus reducing benzo[a]pyrene-DNA adduct formation. The metabolism of many carcinogens involves activation of the compound by phase I enzymes and detoxification by phase II enzymes. Although cytochrome P450s (phase I enzymes) catalyze many essential reactions, an imbalance between phase I and phase II enzymes leads to an increase in DNA adducts, which is one of the biomarkers for susceptibility to carcinogenesis. Flavonoids are potent inhibitors of many enzymes, including phase I enzymes, but they are very weak phase II enzyme inducers.[73] It seems that food rich in certain flavonoids may protect against certain forms of lung cancer and that decreased bioactivation of carcinogens by inhibition of CYP1A1 and CYP3A4 could be important mechanisms by which these foods may protect against lung cancer.[74,75] Although flavone supplementation can result in increased liver weight and P450 cytochrome content, feeding 1% quercetin to rats has no effect on hepatic drug-metabolizing enzymes. Flavonoids, especially quercetin, are found to strongly inhibit PhIP (2-amino-1-methyl-6-phenylimidazo-[4,5-b]pyridine) N-hydroxylation in rat liver microsomes, resulting in a nearly 85 to 90% decrease in activity at 100 μM, or 0.2%. In the same way, flavonoids are found to be inhibitory in human liver microsomes (approximately 85 to 90% at 100 μM, or 0.02%) of PhIP and 4-aminobiphenyl.[76] In addition to preventing the activation of carcinogens, flavonoids are also shown to increase the cellular detoxification capacity. *In vivo* studies have shown that some of them were able to enhance phase I and phase II enzymes when they were administered by oral route, and this might contribute to their anticarcinogenic activity. Several flavonoid members are capable of modulating phase I and phase II enzymes in preparations obtained from experimental animals or humans.

Reports by several researches have clearly shown that several OPCs produce a more reducing intracellular environment and outperform vitamin C, vitamin E, and other classical antioxidants by more than an order of magnitude in their ability to scavenge free radicals.[21] Scientists have only begun to understand how antioxidants influence gene expression regulation, cell proliferation, differentiation, survival, and death, although underlying mechanisms remain incomprehensible. It was recently reported that strong antioxidants, such as pyrrolidine dithiocarbamate and N-acetyl cysteine, cause partial growth inhibition *in vitro* and *in vivo* when added to human colorectal adenocarcinoma cells grown in cell culture, and when fed to mice with implanted tumors.[70] Moreover, when antioxidants were used in conjunction with chemotherapy agents such as 5-fluorouracil and doxorubicin, they potentiate or synergize the cytotoxicity of chemotherapy agents, resulting in a complete remission. In such circumstances, chemotherapeutic agents alone afforded only partial remission.

Recent studies indicate that some of the most potent molecules controlling cell growth and possible tumorigenesis are tumor suppressor molecules.[77] Such molecules influence gene expression, design signals to execute a certain function, and regulate critical proteins involved in the initiation of cell division. Interestingly, the biological effects of flavonoids seem to occur mainly through their interaction with protein tyrosine kinases[78] and cyclooxygenases.[79] Polymethoxylated flavonoids, such as tangeretin and nobiletin, are more potent inhibitors of tumor cell growth than free hydroxylated flavonoids. They also possess potent anti-invasive and antimetastatic activities.[80,81] Tangeretin is a polymethoxylated flavone, 5,6,7,8,4′-pentamethoxyflavone, that is concentrated in the peel of citrus fruits and probably acts as a natural resistance factor against pathogenic fungi. Grapefruit and orange juice have been shown to inhibit human breast cancer cell proliferation[82,83] and to interact with drug administration. Several biological activities have been shown for tangeretin itself, including the ability to enhance gap junctional intercellular communication, to counteract tumor promoter-induced inhibition of intercellular communication, and to inhibit cancer cell proliferation.[84]

22.5 OPC AND CELL-CYCLE ANALYSIS

One of the key tasks that a cell needs to accomplish on a second-to-second basis is the integration of a number of distinct signals into an appropriate biological response. Very frequently, the response is proliferation, differentiation, or quiescence. Tight control of cellular growth is essential to ensure normal tissue patterning and to prevent pathological responses associated with excessive proliferation. The core components controlling cellular proliferation and other growth-related decisions of the cell are the cyclin-dependent kinases (CDKs) and the kinase regulatory subunits, the cyclins.

There are a number of different cyclins and CDK proteins with unique and overlapping functions that have been described to date (Table 22.1 and Table 22.2). Different CDK/cyclin complexes are activated in an organized manner at specific stages of the cell cycle. While the levels of cyclins drastically fluctuate (hence the name "cyclins"), CDKs are generally expressed at relatively constant amounts. Interestingly, the dynamics of changes in the levels of cyclins may be achieved by influencing the transcriptional activity of the genes, by modulation of post-translational modifications, and by alterations of the overall stability of the cyclin proteins.

The role of each cyclin is to achieve a specific CDK and direct the activated kinase complex to the appropriate substrate. The activated phosphorylated substrates then execute the key steps in cell-cycle progression. Active CDK/cyclin holoenzymes are presumed to hyperphosphorylate pRb and the related pocket proteins p107 and p130 from mid G1 to mitosis. The interaction among members of the E2F family of transcription factors and individual pocket proteins is a complex regulatory event that determines whether E2F proteins function as transcriptional activators or repressors.[85–88] Phosphorylation of pocket proteins is involved in the transactivation of genes with functional E2F-binding sites, including several growth and cell-cycle regulators (i.e., c-myc, pRb, p34^{cdc2}, cyclin E, cyclin A) and genes encoding proteins that are required for nucleotide and DNA biosynthesis (i.e., DNA polymerase α, histone H2A, proliferating cell nuclear antigen, thymidine kinase).[89] For example, the B-type cyclins partner with CDK1 kinase to control the passage through M-phase, whereas D-type cyclins cooperate with CDK4 and CDK6 kinases to assist transit of the cells through mid/late G1, ultimately activating cyclin E/CDK2 complexes that move the cells through the end of the G1 phase. Activity of the cyclin-A/CDK2 complex increases in parallel with the entrance into the S phase. It is presumed that the cyclin-A/CDK2 complex also propels the G1 through the S phase. As the cells progress through S phase, CDK2 gets dissociated from the complex and becomes associated with CDK1. The various cyclin/CDK complexes are in turn regulated at a number of different levels by several distinct mechanisms.

Vascular proliferative diseases (i.e., atherosclerosis and restenosis) and cancer are associated with excessive cell growth. The crucial role of CDK/cyclin holoenzymes in the control of cell proliferation has prompted great interest in the development of chemical inhibitors of CDK activity that would be expected to inhibit cellular proliferation and, therefore, may attenuate the development of vascular occlusive lesions and tumor progression.

A common signal-transduction strategy that cells usually follow is to activate cyclin-dependent kinases (CDKs) to activate cyclin molecules by phosphorylation in order to execute specific functions. Flavopiridol and its dichloro derivative L86-8276 are relatively potent CDK inhibitors (IC_{50} values in the nanomolar range). The crystal structure of a complex between CDK2 and L86-8276 shows that the aromatic portion of the inhibitor binds to the adenine-binding pocket of the CDK.[90] Moreover, the position of the phenyl group enables the inhibitor to make contacts with the enzyme that are not observed in the ATP (adenosine triphosphate) complex structure. Flavopiridol shows higher specificity toward CDK4 ($IC_{50} = 65$ nM) than toward CDK1 and 2 (IC_{50} values of 500 nM and 100 nM, respectively); the IC_{50} values of flavopiridol for other protein kinases are all in the millimolar range.[91,92] Flavopiridol has been used in different human xenografted tumors, including head and neck squamous cell carcinoma (HNSCC),[93] colon carcinoma,[94] prostate cancer,[95] and leukemia and lymphoma xenografts.[96] Treatment of mice bearing HNSCC xenografts with flavopiridol, given as a daily intraperitoneal (i.p.) injection for five consecutive days at milligram per kilogram levels, resulted in a 23% reduction in tumor growth, reaching a 60% reduction 10 weeks after the end of the treatment. In this study, it was shown that flavopiridol served as a tumor-cell apoptosis inducer.[93] Human tumors of lymphohematopoietic origin, including HL-60 and SUDHL-4 subcutaneous xenografts and Nalm/6 and AS283 disseminated disease models, also showed regression when flavopiridol was injected intravenously (i.v.) into mice at a concentration of 7.5 mg/kg. In addition to inducing apoptosis in these tumor models, flavopiridol also had a marked proapoptotic effect on normal lymphoid organs, such as spleen, thymus, and intestinal lymphoid tissues, when administered into normal animals.[96] Tangeretin (5,6,7,8,4'-pentamethoxyflavone) is concentrated in the peel of citrus fruits. DNA flow cytometric analysis indicated that tangeretin blocked cell-cycle progression at G1 phase in colorectal carcinoma COLO 205 cells.[97]

Natural flavonoids like quercetin and genistein are known as inhibitors of several protein kinases. Quercetin arrests human leukemic T cells in late G1 phase, and genistein arrests cell-cycle progression in G2-M. Similarly, apigenin has the ability to put a G2-M block. Other natural flavonoids were reported to posses antimitotic activity, and synthetic flavonoids like L86-8275, a flavonoid containing chlorophenol and hydroxymethyl piperidinyl substituents, have previously been shown to inhibit the growth of several breast cancer cell lines, to block the cell-cycle progression from G1 to M, and to inhibit the kinase activity of CDC2. Some of the most potent tumor suppressor molecules are actually inhibitors of CDK2 and CDK4. Two such molecules are known as p21WAF1/CIP1 and p27KIP1. Another common tumor suppressor molecule, p53, is needed to activate p21WAF1/CIP1. It has also been shown that antioxidants induce transcription of p21WAF1/CIP1 without a need for p53, which is inactivated in almost one-half of human tumors. Further research indicated that the transcription factor, which activates transcription of the p21WAF1/CIP1 gene, is C/EBPβ (also known as NF-IL6). The researchers went even further and showed that C/EBPβ in its activated form moves from cytoplasm to the nucleus, where it stimulates transcription of p21WAF1/CIP1 by binding to the CCAAT enhancer sequence of DNA. It was also shown that the possible first step in the activation of p21WAF1/CIP1 is antioxidant-induced activation of protein kinase A. The reduced form of protein kinase A binds to plasma membrane, becomes activated, and then phosphorylates C/EBPβ. This causes its translocation to the nucleus and induction of transcription of p21WAF1/CIP1.

Importantly, a series of recent papers on the *in vitro* and *in vivo* anticancer activities of dietetic products report a similar mechanism of action. For instance, it has been shown that plant flavonoids induce p21WAF1/CIP1 in A549 human lung adenocarcinoma cells. This results in growth arrest and apoptosis. The growth arrest is independent of p53. It has also been found that genistein from soybeans induces p21WAF1/CIP1 and blocks the G1- to S-phase transition in mouse fibroblast and melanoma cells. Other researchers have shown that green tea extract enhances the chemotherapy activity of doxorubicin (*in vitro* and *in vivo*) toward human ovarian cancer cells with low sensitivity to doxorubicin. Also, (−)-epigallocatechin-3-gallate (EGCG) is reported to induce the CDK inhibitors p21WAF1/CIP1 and p27KIP1 in a p53-independent manner and to inhibit CDK2 and CDK4, with subsequent induction of cell-cycle arrest at G1 phase in MCF7 breast cancer cells. A flavonoid has been identified that binds directly to CDK2 and CDK4 and inhibits both of these CDKs directly. This flavonoid, termed flavopiridol, has been shown to cause growth arrest in a human breast carcinoma cell line. The three-dimensional structure of the complex between CDK2 and this particular flavonoid has been determined. These data will be used to design more potent CDK inhibitors in the future.

Finely ground natural clinoptilolite-zeolite (a naturally occurring mineral powder) has also been shown to induce activation of p21WAF1/CIP1. However, *in vitro* experiments disclosed that activated zeolite particles inhibit protein kinase B/akt, another kinase involved in antiapoptotic processes and cancer promotion.[98] This happens only when growth of cells in tissue cultures is stimulated by the addition of growth factors. Zeolite particles might adsorb growth factors or prevent interaction of protein kinase B with membranes, where it is phosphorylated by phosphatidylinositol-3-kinase. It has recently been shown that inactivation of protein kinase B by the novel tumor suppressor molecule PTEN, for example, also results in induction of the tumor suppressor–CDK inhibitor p27KIP1. There also seems to be a relationship between p21WAF1/CIP1 and another protein kinase involved in cell "decisions" about proliferation, arrest, or apoptosis, namely c-Jun N-terminal kinase (JNK)1. JNK1 is a member of the recently discovered stress-activated protein kinases. Interestingly, while such kinase activation results in apoptosis in reaction to stress, its activation in some cancer cells actually promotes uncontrolled proliferation. This is particularly obvious in the human-lung adenocarcinoma cell line A549. Research has shown that p21WAF1/CIP1 inhibits JNK1, and recent work indicates that those two molecules form a tight complex. Inactivation of JNK1 may be part of the reason why antioxidants enhance cytotoxicity of chemotherapeutic agents toward cancer cells while, on the other hand, they also protect neurons from apoptosis caused by free-radical damage. Our studies have demonstrated that GSPE can reduce cardiomyocyte apoptosis through the inhibition of ischemia-reperfusion-induced activation of proapoptotic genes JNK1 and c-Jun.[12]

Accumulation of multiple genetic defects by cells ultimately leads to the development of a malignant phenotype. A crucial event occurs when hyperproliferation progresses to tumor spread by means of invasion of surrounding tissues and metastasis to remote sites. It is extremely difficult to treat after the spread. The majority of normal cell types are constrained by adhesion to each other and to the extracellular matrix (ECM). Normal cells are not static: they are able to move in a regulated fashion by modulating their adhesion in a controlled way. In fact, some normal events involve controlled invasion, for example, neural crest migration and angiogenesis. In this context, angiogenesis is of particular relevance to cancer. Tumors cannot grow extensively unless they develop a blood supply. In contrast, malignant invasion and metastasis are deregulated processes in which excessive growth is accompanied by abnormal adhesive relationships. In particular, the relationship between malignant cells and the ECM becomes unbalanced. As a result, excessive degradation of matrix structures, such as basement membranes, leads to abnormal and misdirected motility and the spread of the malignant cells.

22.6 OPC AND METALLOPROTEINASES

Key players in the spread of cancer are three different proteolytic enzymes that degrade the ECM. These proteolytic enzymes are: matrix metalloproteinases (MMPs), plasminogen activators, and cathepsins. In normal noncancer systems, these enzymes make important contributions and are vital to normal cellular function, tissue homeostasis, and repair (including renovation). In cancer, their dysregulation produces inappropriate matrix degradation, thus promoting invasion and metastasis. Because of their intimate association with cancer, there is significant interest in these enzymes as targets for antimetastatic drug therapy.

The MMP family includes over 20 Zn-dependent endoproteinases (MMP1 through MMP24), and each one of them is a distinct gene product. They are broadly classified into four different categories: (1) collagenases: MMP1, 8 and, 13; (2) gelatinases: MMP2 and 8; (3) stromelysins: MMP3, 7, 10, 11, 12; and (4) the membrane-associated MMPs. These enzymes are secreted as zymogens (inactive precursors), and they have three principal domains that are homologous throughout the family. Typically, transcription, activation, and inhibition regulate MMPs. Transcriptional regulation of MMP expression has been shown to be increased by a number of mechanisms, including treatment of cultured cells with concanavalin-A, PMA, TNFα (tumor necrosis factor α), and TGFβ (transforming growth factor β). Decreased expression has been observed with calmodulin antagonists.

MMP7, MMP9, and metalloelastase (MMP12) sometimes block angiogenesis by converting plasminogen to angiotensin, a powerful angiogenesis antagonist. In human cancers, it appears that MMPs are predominantly expressed in stromal fibroblasts, e.g., in breast cancer, where high levels of mRNAs have been found in fibroblasts adjacent to the carcinoma cells. However, the tumor cell surface is also occasionally associated with MMP proteins. Cooperative interaction between stromal and tumor cells may lead to up-regulation of the MMP activity, leading to invasion. An intriguing observation is that elevated MT1–MMP expression has been found in fibroblasts treated with conditioned medium from invasive breast carcinomas, suggesting the release of a soluble potentiator of MMP expression by cancer cells. Numerous cases of abnormal expression of MMPs in cancer have been reported. Poor prognosis has been associated with elevated levels of MMP1 expression in colorectal cancer and with elevated MMP2 expression in breast cancer, malignant astrocytomas, and glioblastomas. Elevated levels have also been reported in cancers of the uterine cervix, stomach, lung, brain, and head and neck. Although direct evidence is reported for MMPs and serine proteinases to support their role in glioma invasion, much of the evidence for the involvement of cysteine proteinases remains circumstantial. In kidney cancers, the ratio of MMP2 and 9 to TIMP1 and 2 has been found to increase. (TIMPs, tissue inhibitors of metalloproteinases, have the ability to form complexes with MMPs). More interestingly, changes of MMP2 and 9 have been shown to be predictors of whether a tumor is organ-confined or metastatic. Regulation of IGF (insulinlike growth factor) activity by MMP9 through degradation of IGF BP3 has been demonstrated in a human prostate adenocarcinoma cell line. Recently, it has been shown that cleavage of cell-associated FasL by MMP7 yields soluble FasL, which induces apoptosis and involution of the mouse prostate following castration. The importance of MMPs in disease is evidenced by the fact that the synthetic MMP inhibitors are already undergoing clinical trials.[99]

Kinases such as protein kinase A, protein kinase B, protein kinase C, JNK1, CDK2, and CDK4 are either activated or deactivated by these antioxidants. This can happen directly or indirectly through activation of some transcription factors, such as NF-IL6, or tumor suppressor genes, such as p21WAF1/CIP1 and p27KIP1. Similarly PCNA (proliferating cell nuclear antigen) has no known enzymatic activity; it can, however, interact with and regulate the activities of a variety of proteins. PCNA and its adapter protein, replication factor C (RFC), cooperate to form a moving clamp that is an attachment point for DNA polymerases δ (delta) and ε (epsilon). PCNA is central to a decision pathway between a diverse array of important cellular processes, including cell-cycle control, DNA replication, nucleotide excision repair, postreplication mismatch repair, and at least one apoptotic pathway. Even though PCNA has been shown to be relatively stable during the cell cycle, PCNA is present at very low levels in nonproliferating transformed cells, provided that the cells remain quiescent for prolonged periods. Measurement of this protein is a reliable and sensitive marker during the transition between the nonproliferating to proliferating stage.

Cell proliferation plays an important role in multistage carcinogenesis with multiple genetic changes. Several cellular components, such as polyamines and polyamine synthetic enzyme activities, have been associated with cell proliferation. A decrease in the numbers of PCNA-positive cells reflects a decrease in S-phase cells and, thus, reduced proliferative activity. Recent advances in molecular analysis of the cell cycle have revealed how fidelity is normally achieved by the coordinated activity of cyclin-dependent kinase, checkpoint controls, and repair pathways, and how this fidelity can be abrogated by specific genetic changes. In one of the studies, morin served as a chemopreventive agent on large-bowel tumorigenesis induced by azoxymethane when fed during the postinitiation phase.[100]

What we now need is a better understanding of the way in which active substances from orally administered dietetic products come to regulate the activity of enzymes and regulatory molecules, such as the transcription factors that regulate gene expression, which in turn affects cell proliferation, survival, and death. Investigators should also try to isolate active components from dietetic supplements in their pure form and use them at a range of higher concentrations. This also may improve therapy by removing constituents with opposing activities that may be in the parent extract. We should also study the bioavailability of such substances and try to improve their adsorption into the body and their residence time in tumor tissues. More *in vivo* studies should also be performed. Synergistic interactions of such substances with chemotherapy agents should be studied. Synergistic effects should be used in better treatment of resistant tumors. This will hopefully lead to well-controlled studies with human volunteers. Fortunately, such efforts are currently underway. Better cancer treatment therapies with milder side effects are desperately needed.

Finally, we should also mention that modulation of activity of protein kinases by antioxidants is a provocative way to improve current cancer therapies and prevention strategies. Studies on the role of diet in preventing colon cancer have recently shown that antioxidants are less beneficial than had been expected. Instead, it seems that agents that modulate acetylation of histones, such as butyrate, are more important in colon cancer prevention. Such agents are also capable of inducing p21WAF1/CIP1 and p27KIP1. Modulation of the activity of some other nuclear-regulatory proteins, such as heat-shock-related proteins, and growth-arrest proteins, such as gadd45, mdm2, and p53, can also be used to improve current cancer therapies. An information-intensive approach to the molecular pharmacology of cancer is currently being used to select natural and synthetic compounds with anticancer activity.

Cancer cell resistance to chemotherapy is often mediated by overexpression of P-glycoprotein (Pgp), a plasma membrane ABC (ATP-binding cassette) transporter that extrudes cytotoxic drugs at the expense of ATP hydrolysis. On the other hand, Pgp in normal tissues may serve as a cellular defense mechanism against naturally occurring xenobiotics. Structurally different flavonoids were tested for their ability to modulate adriamycin accumulation and efflux in Pgp-expressing HCT-15 colon cells. Micromolar concentrations of quercetin, kempferol, and galangin inhibited the accumulation of adriamycin. Flavonoid-mediated stimulation of efflux was rapid and was blocked by verapamil, vinblastine, and quinidine (multidrug-resistant reversal agents). Since quercetin showed partial protection against adriamycin-induced cell growth inhibition, it was concluded that certain flavonoids might acutely up-regulate the apparent activity of Pgp. In another study, the effects of flavonols on Pgp activity were studied in rat liver cells by assessing the transmembrane transport of Pgp substrates like Rhodamine-123 and doxorubicin. The results indicated that flavonols, reported to strongly up-regulate the apparent activity of Pgp in cancer cell lines, might modulate the transport of putative Pgp substrates differently in normal rat hepatocytes. Moreover, quercetin and genistein potentiated the effects of adriamycin and daunorubicin, respectively, in a multidrug-resistant MCF7 human breast cancer cell line. Another *in vitro* study used mouse brain capillary endothelial cells (MBEC4 cells) to investigate the effect of quercetin, and the results showed that quercetin imparts a dual role at low concentrations of stimulated activity of Pgp, whereas the activity was inhibited at higher concentrations.

Flavonoids have the potential to cause interactions with xenobiotics. To understand the pharmacokinetics of a "precipitant drug" is very important for elucidating the mechanism of interaction with an "object drug." The significant role of the conjugated metabolites of flavonoids for interaction with xenobiotics is speculated. The direct *in vitro* effects of the sulfates/glucuronides on the modulation of CYPs, Pgp, and MRPs are worthy of investigation. For the visualization of multidrug resistance *in vivo*, the use of single-photon-emission tomography (SPET) and positron-emission tomography (PET) are feasible to study the functionality of Pgp and MRP transporters. Modulation of CYPs, Pgp, and MRP by flavonoids may be beneficial in detoxication, in chemoprevention, or in suppression of drug resistance. Because flavonoids are very common constituents in various herbs, it is proposed that herbs represent a potential and possibly an overlooked cause for drug interaction. Medications whose absorption and metabolism are mediated by CYPs, Pgp, or MRP should require close monitoring when coadministered with flavonoid-containing herbs or dietary supplements.

Since the first report on grapefruit juice–drug interaction was published in 1989, flavonoid-drug interactions have captured serious attention. After additional research demonstrated that the citrus flavanone naringin, present in grapefruit juice, was not a causative agent in initiating adverse reactions, the role of other flavonoid constituents, such as naringenin and quercetin, were investigated for their effects on CYP3A4 isozyme, Pgp, and MRP, and they were found to be potent inhibitors of CYP3A4 and modulators of Pgp. In the meantime, effects of St.-Johns-wort (SJW) were undergoing serious scrutiny due to their antidepressant effect. However, important interactions between SJW and many drugs were published during the same time frame. Simultaneous use of SJW with CYP3A enhanced unexpected excretions of therapeutic drugs, e.g., cyclosporin, resulting in a subtherapeutic level that could result in acute heart-transplant rejections. Although not all investigations yielded the same results, most agreed that extracts of *Hypericum perforatum* activated CYP3A4. Although induction of CYP3A4 isozymes could explain a majority of interactions, such an effect may not explain all the drug interactions. Induction of intestinal Pgp is known to lower plasma digoxin levels, and there is evidence that SJW could induce intestinal Pgp in laboratory rodents and humans. SJW is derived from the leaves and flowering tops of *Hypericum perforatum*. The alcoholic extracts contain 0.1 to 0.3% hypericin and a very low amount of flavonoids (e.g., quercetin, rutin, kempferol, luteolin, apigenin, and quercitrin) and up to 6% hyperforin. From the results of *in vivo* interaction studies using flavonoids (e.g., phellamurin, quercetin) as the precipitant drug, the content of flavonoids in SJW might be responsible for the interaction with CYP 3A4/Pgp substrates, e.g., cyclosporin. Although the evidence of these results is appealing, how such interactions influence the overall outcome of a carcinogenic initiation needs further elucidation.

D-glucaric acid (calcium glucarate) is a nontoxic garlic/onion derivative and also a potent β-glucuronidase inhibitor (1,4-GL; active ingredient allyl sulfides). The role of β-glucuronidase in detoxification pathways is very well characterized. 1,4-GL increases detoxification of carcinogens and tumor promoters by inhibiting β-glucuronidase and preventing hydrolysis of their glucuronides. 1,4-GL and its precursors, such as calcium D-glucarate, may exert their anticancer action, in part, through alterations in steroidogenesis accompanied by changes in the hormonal environment and the proliferative status of the target organ. Glucarates may directly detoxify any environmental agents responsible for cancer formation. It has been postulated that D-glucarate exerts some of its effects by equilibrium conversion to D-glucarolactone, a potent β-glucuronidase inhibitor. Laboratory studies comparing calcium glucarate (CGT) with a known chemopreventive agent, 4-HPR, during initiation phase (I), promotion phase (P), and initiation plus promotion phase (I+P) together, showed that CGT reduced tumor multiplicity by 28, 42, and 63% for the various stages, respectively, compared with 4-HPR, which reduces tumor multiplicity by 63, 34, and 63%, respectively. The maximum effect occurred during the P and I+P phases. In particular, studies show that the chemopreventive effect was synergistic when CGT was used together with 4-HPR.

22.7 OPC'S ROLE IN DNA DAMAGE AND DNA REPAIR

DNA is the essential carrier of genetic information in all living cells. How is the huge amount of DNA in organisms — from bacteria to humans — maintained and protected from the ravages of noxious agents in the environment? The chemical stability of the DNA molecule is not unusually great; DNA undergoes several types of spontaneous modifications, and it can also react with many physical and chemical agents, some of which are endogenous products of the cellular metabolism (e.g., reactive oxygen species or biological reactive intermediates of drug metabolism), while others, including ionizing radiation and ultraviolet light, are threats from the external environment. The resulting alterations of DNA structure are generally incompatible with its essential role in the preservation and transmission of genetic information. Damage to DNA can cause genetic alterations, and if genes that control cell growth are involved, these mutations can lead to the development of cancer. Of course, DNA damage may also result in cell death, which can have serious consequences for the organism of which the cell is a part, e.g., loss of irreplaceable neurons in the brain or cardiac myocytes in the heart. Accumulation of damaged DNA may also contribute to some of the features of aging.

It is not surprising that a complex set of cellular surveillance and repair mechanisms has evolved to reverse the potentially deleterious damage that would otherwise destroy the precious blueprint for life. Some of these DNA repair systems are so important that life cannot be sustained without them. An increasing number of human hereditary diseases that are characterized by severe developmental problems or a predisposition to cancer are linked to deficiencies in DNA repair. DNA serves as a critical target in apoptosis, necrosis, and cancer. At the molecular level, damage induced in DNA directly or indirectly through epigenetic processes stimulates repair of lesions in DNA. Such cells survive and continue to divide, usually after a short delay in the cell cycle, to allow for repair. Some cells are killed either directly from DNA damage or secondarily from apoptosis. Other cells become resistant to killing, and in some cases this resistance becomes associated with the increased ability to repair DNA damage.[101] Some cells, however, apparently become permanently arrested by mechanisms that are separable from the cyclin-dependent kinase activity of CDKN1A.[102] This permanent arrest appears to be secondary to DNA damage and can involve increased expression of p53 and p21.[103]

Numerous *in vitro* and *in vivo* studies have shown protection of genomic integrity by flavonoids.[104] The effects of the flavonoids quercetin, apigenin, and chrysin on the genetic toxicity of 2-amino-3-methylimidazo[4,5-f]quinoline (IQ) and benzo[a]pyrene (BaP) were investigated at subcytotoxic concentrations in Chinese hamster V79 cells expressing human or rat cytochromes P450. In V79 r1A2-NH and V79 h1A1-MZ cells, none of the flavonoids increased DNA strand breaks (SB) or produced detectable DNA adducts. Neither IQ nor BaP produced DNA damage in the absence of expressed CYP1A2 or CYP1A1, respectively. DNA damage measured as SB and DNA adducts was detectable in V79 r1A2-NH cells expressing rat CYP1A2 when treated with IQ, and this was inhibited by quercetin. Likewise, DNA damage was elevated in V79 h1A1-MZ cells expressing human CYP1A1 when treated with BaP, and this was inhibited by chrysin and apigenin, but not by quercetin. The specificity of CYP1A1 inhibition by chrysin and apigenin and CYP1A2 inhibition by quercetin was confirmed by ethylresorufin-*O*-deethylase assay.[105]

Similarly, several flavonoids have shown their anticarcinogenic effects in various models. The soybean isoflavone genistein was demonstrated to be an effective inhibitor of dimethyl-benz[a]anthracene (DMBA)-induced DNA damage in MCF7 cells by curbing cytochrome P450 CYP1 enzymes. The red clover (*Trifolium pratense*) isoflavone biochanin-A is a methylated derivative of genistein, and its antimutagenic effect in bacterial cells has been shown previously. Because of its protection against chemical carcinogenesis in an animal model, using a semiquantitative reverse-transcription polymerase chain reaction and a xenobiotic response-element (XRE)-luciferase reporter assay, biochanin A could reduce xenobiotic-induced CYP1A1 and CYP1B1 mRNA abundances through the interference of XRE-dependent transactivation. Since the biotransformation of DMBA was dependent on CYP1 enzyme activities, biochanin A was able to decrease the DMBA-induced DNA lesions. One study illustrated that the red clover isoflavone could protect against polycyclic aromatic hydrocarbon-induced DNA damage.[106]

Flavonoids of grape juice may increase plasma antioxidant capacity, resulting in reduced DNA damage in peripheral lymphocytes by neutralizing release of reactive oxygen species

(ROS).[107] The anthracycline antibiotic adriamycin (doxorubicin) is one of the most effective anticancer chemotherapeutic agents available today. However, its use is seriously limited due to its cardiotoxicity. Among the standard strategies to attenuate adriamycin toxicity are dosage optimization, synthesis, and use of analogs or combined therapy with antioxidants. The most promising results come from the combination of the drug delivery together with an antioxidant in order to reduce oxidative stress.[108] Several OPCs have shown promise in this direction.[109–111]

The impact of oxidative stress and the initiative it takes to promote cancer have been described from a number of perspectives in the foregoing paragraphs. Besides causing macromolecular damage, oxidative stress can be devastating to the vital architecture of the cell. DNA frequently becomes a prime target of this event, ultimately perturbing the cell-cycle balance and apoptosis. It is interesting that agents that induce DNA damage serve as double-edged swords. These agents need to work in a concerted manner to culminate DNA damage in cancerous cells, sparing the genomic integrity of normal cells. Similarly, agents that facilitate DNA repair or block DNA damage probably take precautionary steps prior to exerting their influence in a complex environment. In this context, the mechanism of apoptosis is behind the idea that chemotherapy must create a lethal injury to DNA to facilitate the death of malignant cells; however, occurrence of apoptosis is predominantly dependant on DNA integrity. A dose of chemotherapy that does not produce necrosis can trigger apoptosis, either immediate or delayed. Additionally, antiapoptotic mutations can result in drug resistance in human tumors. At least one antioxidant (quercetin) has been demonstrated to overcome such an antiapoptotic blockage.

A wide variety of unique DNA-damaging natural products are in clinical development, including ET-743 derived from a marine tunicate, HMAF (MGI-114) derived from mushrooms, and the bacterium-derived rebeccamycin analog, all of which are nonclassical inhibitors of topoisomerase II. How natural products achieve this task is unknown. From numerous previous studies, it is very clear that grape seed OPCs show prolific anti-DNA damaging activity in normal cells and promote DNA-damaging activity in cancerous cells.[2–10,19,21] Future studies will unravel the underlying mechanisms.

If the conditions of carcinogen exposure generate abnormalities in cells, they may undergo limited proliferation to form preneoplastic lesions. During these processes, further alterations of DNA — the result of transpositions and other error-prone processes — are possible, leading to the formation of a neoplastic cell. These neoplastic cells may remain dormant or undergo progressive growth to form neoplasms. For suppressed neoplastic cells, promoters facilitate their proliferation possibly by the interruption of tissue growth control, leading to neoplastic development. Some neoplasms undergo qualitative changes in their phenotypic properties, possibly including transition from benign to malignant behavior. Benign neoplasms are progressively growing and expansive, but they do not invade or metastasize to remote sites, as do malignancies. Progression may reflect the selection during growth of a population with a genotype coding for advantageous phenotypic properties. New genotypes could arise in neoplasms through additional changes in DNA produced by reactive carcinogens, errors in DNA replication, alterations in chromosome constitution, or hybridization of different cell types. The neoplasm that ultimately emerges is in most cases the progeny of a single cell, i.e., it is a monoclonal population. Nevertheless, there is cellular heterogeneity in neoplasms, perhaps owing to lower fidelity in DNA replication, and these neoplasms display abnormalities in the expression of numerous gene products. This cellular diversity in solid neoplasms is a major obstacle in developing effective methods of chemotherapy.

22.8 OPCs AND PROGRAMMED CELL DEATH

A successful therapy for established cancers requires readjusting the quantitative cell kinetic relationship so that cell death exceeds cancer cell proliferation. Maintenance of this negative relationship could eradicate cancer and cure the patient. However, no therapeutic regimens are presently available to accomplish this goal repeatedly when used against solid cancers such as breast, prostate, lung, and colon. An understanding of what normally regulates the balance between cell proliferation and cell death in these tissues is critical to maintain this negative relationship. The growth of any tissue, whether normal or malignant, is determined by the quantitative relationship between the rate of cell proliferation and the rate of cell death. In the adult, these cell kinetic rates are balanced such that neither involution nor overgrowth of tissues normally occurs. Because a cell must undergo a series of molecular changes to acquire the malignant phenotype, and because these changes are often induced by agents or treatments that damage the cell over an extended period of time, any events that derange this kinetic balance in favor of an enhanced survival of initiated or damaged cells promote the carcinogenic process. In this regard, certain promoting agents of carcinogenesis function not to enhance proliferation but, rather, to decrease the death of neoplastic-initiated cells. Thus, understanding what normally regulates cell death and how this regulation is deranged in carcinogenesis has important ramifications for chemoprevention. Also important for therapy of established cancers is the realization that a fundamental requirement for full transformation of premalignant cells into cancer cells is that they must undergo a series of genetic changes that disrupt this homeostatic balance. This disruption allows the rate of proliferation of cancer cells to exceed cell death, resulting in the continuous net accumulation of malignant cells. It is this continuous net accumulation of malignant cells that produces the lethality of cancer.

The overall cell cycle that controls cell number is, thus, composed of a multicompartment system in which the cell has multiple capabilities: the cell can be (1) metabolically active but not undergoing either proliferation or death (G_0 cell), (2) undergoing cell proliferation ($G_0 \rightarrow G_1 \rightarrow S \rightarrow G_2 \rightarrow$ mitosis), or (3) undergoing cell death by either the programmed cell death (PCD) pathway ($D_1 \rightarrow F \rightarrow D_2$ apoptotic cellular fragmentation) or the nonprogrammed (necrotic) pathway. Cells in any phase (G_0, G_1, S, G_2, or mitosis) can be induced to enter the PCD pathway. Indeed, proliferating cells characteristically undergo PCD if their progression through the cell cycle is sufficiently perturbed. Various independent investigators have demonstrated that proliferating cells can be induced to undergo PCD at any stage of the proliferative cell cycle (G_1, S, G_2, M). Entrance into and progression through the proliferative cell cycle is not absolutely required for PCD. For example, immature thymocytes can undergo PCD when they are proliferatively quiescent (i.e., out of cycle in G_0) without any attempt to enter even the earliest part of G_1. Likewise, androgen-dependent prostatic glandular cells undergo PCD following androgen ablation in G_0 without entrance into the cell cycle. Similar to normal prostate glandular cells, androgen-dependent prostatic cancer cells are induced to undergo PCD by androgen ablation, and this PCD also does not require progression into the S phase. The apoptotic cell death kinetics reported in several studies strongly suggest participation of GSPE in such mechanisms.[6–10,21]

Previous studies have demonstrated that when the DNA of a proliferating cell is sufficiently damaged with a variety of chemicals, such as alkylating agents or mitogenic agents, the cell arrests in G_2 and undergoes PCD. Likewise, agents such as 5-fluorodeoxyuridine that inhibit the progression through the S phase of proliferative cell cycle induce PCD. Mimosine, an agent that arrests cells in the G_1 phase of the proliferative cell cycle, also can induce PCD. The mechanism that enables the cell to "sense" such cellular arrest during the proliferative cell cycle and activate PCD is not understood. Flavonoids and their impact on various stages of the cell cycle have been described elsewhere in this chapter. One possible explanation for the cytotoxic effects of cell-cycle arrest is that these agents dissociate the normally integrated cell-cycle events that lead to "unbalanced growth" and, eventually, to programmed death of the cells. Regardless of the mechanism, these results clearly demonstrate that PCD can be activated not only in cells in G, but also in cells in the various parts of the proliferative cell cycle.

Indeed, several well-designed studies have already shown the precise efficacies of polyphenols, flavonoids, and OPCs on specific cell-cycle-related events. The ability of these naturally derived agents to interfere with cell-cycle-related events is no longer a matter of speculation. In one study, it was shown that GSPE negatively influenced normal growth of a cancer cell line, but it very effectively potentiated cancer cell death in the presence of daunorubicin.[16–18] Based on the results of such experiments, it can be argued that GSPE has the potential to alter the cell cycle and PCD-related events. The cell suicide kinetics in this study followed a biphasic pattern. The first cycle of massive apoptotic death was observed at 3 months and the second at 9 months. Both cycles were characterized by genomic DNA fragmentation. The timing of occurrence of DNA fragmentation and cell suicide was found to be overlapping in groups 3 and 5. Based on this data, it can be logically concluded that massive apoptosis led to hyperproliferation related to compensatory mechanisms at 3 months, whereas the second cycle of death observed at 6 months may be more related to cell-cycle-related events. However, further studies are needed to ascertain these complex *in vivo* speculations.

How do cells sense damage? How do they activate the PCD pathway, and how do various agents such as oncoproteins modify this sensing mechanism? Clear answers to these questions may resolve many issues. Modulation of responsiveness to PCD has emerged as a new type of treatment resistance, and one that is shared by a variety of tumor types. Although the role of OPCs in reducing chemoresistance remains a mystery, it can be counted as the most viable alternative at the present time. Numerous chemotherapeutic agents selectively choose to kill cancer cells via PCD pathway. However, in clinical practice, the majority of human tumors, especially solid tumors, remain resistant to most therapeutic agents, even when used in combination. It is becoming increasingly clear that the most important determinant of tumor resistance may be a generalized resistance to induction of PCD, rather than resistance based on specific alterations in the drug/target interaction. Alterations in a variety of oncogenes and tumor suppressor genes have been shown to modulate responsiveness to PCD induction by chemotherapeutic agents and radiation. Additional work is needed to delineate the function of these modulatory proteins and to define the cellular machinery involved in the PCD pathway itself. Further understanding of these cellular pathways should help in developing new strategies for overcoming cancer resistance and may yield more effective therapies in the future. Future investigations will unravel such important uses for OPC.

Some of the most interesting new therapies in development are those that specifically target tumors with specific oncogene abnormalities. At this time, the most intense activity in this regard is being directed against malignant cells with mutations or deficiencies of the p53 tumor suppressor oncogene. An example of p53-targeted therapeutics is the Onyx-015 attenuated adenoviral vector, which specifically incorporates into the genome and divides in cells with p53 mutations or deficiencies. Other viral vectors that restore p53 suppressor function are also being evaluated. Therapeutics are also being directed at inhibiting abnormally active proliferative oncogenes. Abnormalities in the *ras* oncogene are associated with malignant proliferation. One of several chemical steps involved in the post-transcriptional processing of the Ras protein is farnesylation, a process that is being currently targeted. Several highly specific inhibitors of farnesyl transferase are currently entering clinical development. In the whole scenario of oncogene-activation-dependent cancer, besides p53, several others have gained considerable attention. Bcl-2 family members (Bad, Bax, Bid, Bcl-XL, Bcl-Xs), in particular, have been the center of the debate. There are several reports describing how OPCs modulate p53 and an array of pro- and antiapoptotic genes.[6,7,15–18,21]

Dietary and endogenous antioxidants prevent cellular damage by reacting with and eliminating oxidizing free radicals. However, in cancer treatment, a mode of action of certain chemotherapeutic agents involves the generation of free radicals to cause cellular damage and necrosis of malignant cells. Such scenarios logically develop concern as to whether exogenous antioxidant compounds taken concurrently during chemotherapy could reduce the beneficial effect of chemotherapy on malignant cells. The importance of this concern was published in a recent study, which estimated that 23% of cancer patients take antioxidants. How antioxidant therapies protect normal cells against damage from cancer therapies, while often increasing their cytotoxic effect against malignant cells is an open-ended question. While the answer to this is not entirely mapped out, there are numerous concepts that help us understand such convoluted issues. Recent evidence suggests that radiation and chemotherapy often harm DNA to a relatively minor extent, which causes the cells to undergo apoptosis, rather than necrosis. Since many antioxidant treatments stimulate apoptotic pathways, the potential exists for a synergistic effect with radiation or chemotherapy with antioxidants. A second possibility is that the defensive mechanisms of many cancer cells are known to be impaired. This presumably makes tumor cells unable to use the extra antioxidants in a repair capacity. An experimental murine ascites tumor cell line was found to have negligible level of catalase compared with normal cells. This led to a buildup of hydrogen peroxide in the cells upon treatment with vitamin C, in turn leading to cell death. The cytotoxic effects of vitamin C were completely eliminated by the addition of catalase to the cell culture. Since publication of these findings, most human tumor cell lines studied have proved to be similarly low in catalase.

The attitude of many conventional practitioners toward antioxidant therapy for cancer has been hostile. This low level of confidence is due to the lack of adequate data from tightly controlled human trials. Others have raised the argument that antioxidants could blunt the effect of standard therapies, particularly alkylating, platinum, and tumor antibiotic agents, which are oxidative in nature. While this appears to be a valid theoretical concern, the evidence presented here shows that this proposed interaction of anti- and pro-oxidant therapies has tremendous implications in *in vivo* models.[22,23] Modes of cancer therapy have been remarkably consistent for the last several decades. Surgery, radiation, and chemotherapy alone, or their reasonable combinations, have been the cornerstones of conventional treatment. Although surprising, the clinical success of these therapies has reached a plateau. Numerous experts in the field even questioned the validity of chemotherapy as a treatment for most cancers. Clearly, there is a need for new therapies that can increase the efficacy of cancer treatment. Careful application of antioxidants may be a means of helping to raise cancer therapy to a new level of success. Based on the available evidence, GSPE exposure with precautions may ameliorate events and conditions associated with cancer. Since GSPE is a relatively new player in this field, careful clinical and human trials are needed before this can be prescribed on a regular basis. Moreover, the daily human dose has to be determined prior to the execution of such strategies. OPC of choice along with a team of other antioxidants may be another novel way to approach anticancer therapy.[19–21]

Numerous animal studies have been published demonstrating decreased tumor size and/or increased longevity with the combination of chemotherapy and antioxidants. A recent study was conducted on small-cell lung cancer in humans using combination chemotherapy of cyclophosphamide, Adriamycin (doxorubicin), and vincristine with radiation and a combination of antioxidants, vitamins, trace elements, and fatty acids. The conclusion was "antioxidant treatment, in combination with chemotherapy and irradiation, prolonged the survival time of patients" compared with expected outcome without the composite oral therapy. Two human studies found melatonin plus chemotherapy to induce greater tumor response than chemotherapy alone. The treatments producing these positive results would have been advised against by those advocating no antioxidant use during chemotherapy.

Based on the use of diverse molecular probes and protein synthesis inhibitors, it is obvious that the machinery for executing PCD is already present within cells. However, numerous studies have demonstrated that the signal transduction needed to activate this preexisting death machinery sometimes requires new protein synthesis. This suggests that there are three types of gene products involved in the process of PCD.

The first are those involved in generating the signal transduction for activation of the death process in healthy, undamaged cells. These type-I gene products often are highly contextual with regard to the particular cell type (i.e., during differentiation of the particular cell type, specific gene products are selected for such regulatory roles). Thus, the same gene product can have both the ability to stimulate cell proliferation or PCD, depending on the differentiation status of the cell. For example, TGFβ is a cell-type-specific type-I gene that can stimulate proliferation of mesenchymal cells while stimulating the death of certain epithelial cells. Genetic changes that induce loss of function of these type-I genes allow the affected cells to be resistant to specific induction of physiological cell death.

The second type of PCD genes encodes the proteins involved in determining the sensitivity to activation of PCD that is initiated by pathological damage to the cell (e.g., from radiation, viral infection, chemotherapy). This second type of PCD gene includes p53 and *bcl*-2. The functional expression of the gene p53 increases the sensitivity of the expressing cells to activation of programmed death induced by a wide variety of damaging agents. Genetic alteration leading to the loss of function of this type-II gene decreases the sensitivity of the affected cells to pathological PCD (e.g., from radiation, chemotherapy, viral infection). In contrast, *bcl*-2 is a gene whose functional expression decreases the sensitivity of the expressing cells to activation of PCD induced by the same types of damaging agents. Thus, genetic alterations leading to the loss of function of this type of gene increases the sensitivity of the affected cells to pathological PCD.

Unlike the first two types of genes that primarily encode proteins involved in the signal-transduction-induced activation of PCD, a third group of genes encodes proteins that are the actual machinery needed to propel the cell death itself. For instance, loss of function of genes such as p53 or overexpression of other genes such as *bcl*-2 and *bcl*-XL would increase the threshold for activation of PCD (Table 22.1 and Table 22.2). Thus, cancer cells harboring these changes would become more resistant to cell-damaging agents and also would be more likely to survive removal of growth/survival factors. New therapeutic strategies aimed at decreasing the threshold for activation of PCD (e.g., by inhibiting the *bcl*-2 protein function) are currently under development, as well as agents capable of directly triggering the PCD cascade.

The precise molecular modes of action of quercetin and rutin in altering the spatial and temporal death program of colonic cells remain to be elucidated. The apoptosis-inducing action of quercetin has also been observed *in vivo* in regenerating liver after partial hepatectomy, in cells of disparate lineages, including colonic adenocarcinoma cell lines. Wei et al.[112] and Nagasaka and Nakamura[113] have observed that the apoptotic action of the flavonoid on tumor cells is related to the inhibition of the synthesis of heat-shock proteins, a class of bioactive molecules known to play an important role in cell survival. Since quercetin was shown to be an effective protein kinase inhibitor, an alternative possibility is that the death-inducing action of this compound resides, at least partly, in promoting dephosphorylation of key proteins involved in the control of apoptotic events. Consistent with this notion is the recurring observation that extensive tyrosine dephosphorylation is a contributory determinant in the decision of cells to die an apoptotic death, including normal murine cells of the large intestine. Likewise, nonsteroidal anti-inflammatory drugs (NSAIDs) have been reported to increase apoptosis in intestinal cancer cells of mice. Since quercetin inhibits cyclooxygenase activity, a tenable hypothesis is that NSAIDs and the flavonoid share a common path of action in promoting the apoptotic process. Similarly, animals treated with isorel (an ingredient found in apple, pine and fir trees; *Viscum album* sp.) had, on the average, three times fewer lung metastases than the controls.[114] Thus, it was concluded that both local and systemic effects of the application of isorel could be of benefit for the tumor-bearing organism, resulting in immunomodulation combined with tumor growth inhibition and reduction of metastases. Antitumor effects observed based on *in vitro* experiments could be the result not only of the mistletoe lectins (derived from *Viscum album* plants) and the other high-MW factors, but also of the very low-MW (<500 Da) substances, both of which deserve further analysis. Along these lines, it is worth mentioning that continuous exposure to a GSPE-mixed diet significantly attenuated liver, bladder, stomach, spleen, and brain cancers induced by DMN (Ray et al.[115]; see Figure 22.1 through Figure 22.5).[115]

Mounting evidence indicates that the acquired ability to resist apoptosis is a hallmark of most, and perhaps all, types of cancer. As scientists learn more about apoptosis failure mechanisms in cancer, they also are gaining a greater understanding of why many tumors are resistant to the lethal effects of radiation and chemotherapy or even their combination, which both act by inducing suicide machinery. Insights from these studies will shed light on and assist in constructing strategies to overcome treatment resistance and offer important clues about new drugs that target genes and protein products in the apoptotic pathways to encourage selective cell death. Researchers are still vigorously exploring apoptosis regulatory mechanisms, how these might be repaired through genetic therapies, and how apoptosis can be selectively triggered, through tailored treatments, to induce suicide pathways in cancer cells while leaving healthy cells alone. Experiments are in progress in our laboratories to examine whether grape seed extract has the ability to lessen the radiation and chemotherapy resistance of tumor cells.

22.9 MISCELLANEOUS MECHANISMS

In contrast to observations in mice, dietary treatment with quercetin resulted in a dose-dependent increase in the yield of colonic tumors in azoxymethane (AOM)-treated rats. Recently, Barotto et al.[116] have shown that quercetin enhances pretumorous lesions in the rat NMU (N-nitroso-N-methylurea) model of pancreatic carcinogenesis. Sugimura[117] recently reviewed the conflicting evidence for a genotoxic activity of quercetin and concluded that the flavonoid may be mutagenic, but it is not a carcinogenic agent. Quercetin has also been used in clinical trials in cancer patients. Since cancer is a major public health issue and flavonoids are consumed daily, further clarification of the full biological and biochemical effects of quercetin and rutin in humans are in progress in several laboratories.

The modulating effects of dietary feeding of two flavonoids, diosmin and hesperidin, both alone and in combination, during the initiation and postinitiation phases on colon carcinogenesis initiated with AOM were investigated in male F344 rats. Animals were initiated with AOM by weekly s.c. injections of 15 mg/kg body weight for 3 weeks to induce colon neoplasms. Rats were fed diets containing diosmin (1000 ppm), hesperidin (1000 ppm), or diosmin (900 ppm) + hesperidin (100 ppm) for 5 weeks (initiation treatment) or 28 weeks (postinitiation treatment). The others contained the groups of rats treated with diosmin, hesperidin alone or in combination, and untreated. At the end of the study (32 weeks), the incidence and multiplicity of neoplasms (adenoma and adenocarcinoma) in the large intestine of rats initiated with AOM together with, or followed by, a diet containing diosmin or hesperidin were significantly smaller than those of rats given AOM alone (p < 0.001). The combination regimen during the initiation and postinitiation stages also inhibited the development of colonic neoplasms, but the tumor data did not indicate any beneficial effect of diosmin and hesperidin administered together as compared with when these agents were given individually. In addition, feeding of diosmin and hesperidin, both alone and in combination, significantly inhibited the development of aberrant crypt foci.[118] As for cell proliferation biomarkers, dietary exposure of diosmin and hesperidin significantly decreased the 5'-bromodeoxyuridine-labeling index and argyrophilic nuclear organizer region's number in crypt cells, colonic mucosal ornithine decarboxylase activity, and polyamine levels in the blood. These results indicate that diosmin and hesperidin, both alone and in combination, act as a chemopreventive agent against colon carcinogenesis, and such effects may be partly due to suppression of cell proliferation in the colonic crypts, although precise mechanisms should be clarified.[118]

Pycnogenol, derived from the bark of the French maritime pine tree (Pinus maritima), is a complex of more than 40 antioxidant compounds.[24,63] Many of these compounds, described by chemists as flavonoids and polyphenolic organic acids, are found in fruits and vegetables, although not in the potency or specific ratio achieved through supplements. Pycnogenol is one of those substances that is synergistic with other antioxidants, such as vitamins C and E. Flavonoids, such as those found in Pycnogenol, are a big part of the reason why fruits and vegetables are good for people and reduce the risk of disease.[21,24,63]

The membrane protein mediating the ATP-dependent transport of lipophilic substances conjugated to glutathione, glucuronate, or sulfate, have been identified as members of the multidrug resistance protein (MRP) family. A soybean isoflavone, genistein, was found as an inhibitor on the basis of its effect on drug accumulation in MRP1-overexpressing cells. More studies concerning the effects of flavonoids on MRP are needed in the future.

As one of the most promising cancer chemopreventive agents, β-carotene has been studied extensively. However, other natural carotenoids have also suppressed tumorigenesis, and some are more potent than β-carotene. For example, α-carotene shows higher potency than β-carotene in suppressing tumorigenesis in mouse skin and lung models. In the two-stage mouse skin carcinogenesis model (initiator, 7,12-dimethylbenz[*a*]anthracene [DMBA]; promoter, 12-*O*-tetradecanoylphorbol-13-acetate [TPA]), topical application of α-carotene at a dose of 200 nmol per painting twice a week significantly decreased the mean number of skin tumors per mouse, and its potency was higher than that of β-carotene. The greater potency of α-carotene over β-carotene in the suppression of tumor promotion was confirmed in the two-stage mouse lung carcinogenesis model (initiator, 4-nitroquinoline-1-oxide; promoter, glycerol). Oral administration of α-carotene (0.05% in drinking water) significantly decreased the mean number of lung tumors per mouse. In contrast, β-carotene did not show any suppression of lung tumor formation under the same experimental conditions. These results suggest that future investigations should target antitumorigenic activity of natural carotenoids other than β-carotene.

Similarly, another study examined the effect of a natural carotenoid, fucoxanthin, on tumorigenesis in mice. Fucoxanthin, like β-carotene, is one of the most abundant carotenoids in nature; it is widely distributed in marine organisms, including edible seaweeds. Fucoxanthin (0.005% in drinking water) suppressed N-ethyl-N′-nitro-N-nitroso-guanidine-induced mouse duodenal carcinogenesis. The mean number of duodenal tumors per mouse was significantly decreased by fucoxanthin; control mice developed 1.28 tumors/mouse, whereas fucoxanthin-treated mice had 0.55 tumors/mouse ($p < 0.05$). Fucoxanthin also suppressed tumor promotion in two-stage carcinogenesis of mouse skin (initiator, DMBA; promoter, TPA); it completely suppressed skin tumor formation during the whole period of the experiment, up to 20 weeks of promotion, at a dose of 0.6 μmol per painting.

The genus *Lilium*, with more than 90 species, is distributed throughout the northern temperate zone. Bulbs of some *Lilium* plants have long been used as food and/or medicinal material. Recently, new derivatives of steroidal saponins isolated from these bulbs were found to show potent inhibitory activity against TPA-stimulated 32P incorporation into phospholipids of cultured cells. For example, methyl esters of (25R)-27-O-[(S)-3-hydroxy-3-methylglutaryl]-spirost-5-ene-3β,27-diol-3-O-[O-α-L-rhamnopyranosyl-(1-2)-O-[β-D-glucopyranosyl-(1-4)]-β-D-glucopyranoside] and (25R)-27-O-[(S)-3-hydroxy-3-methylglutaryl]-spirost-5-ene-3β,27-diol3-O-[O-α-L-rhamnopyranosyl-(1-2)-O-[α-L-arabinopyranosyl-(1-3)]-β-D-glucopyranoside] significantly inhibited antitumor-promoter activity (inhibition of 40.3% at the dose of 5 μg/ml and 50.8% at the dose of 50 μg/ml, respectively). Thus, these agents might be useful for cancer chemoprevention.

Biochemoprevention, a biotechnology-based concept for cancer control, is cancer chemoprevention.[119] For example, intracellular prevention of carcinogenesis by the endogenous production of plant substances in animal cells appears to be a potentially viable mode. Development of a vaccine containing cloned human cells with the ability to produce anticancer phytochemicals may be another approach. Phytoene, a plant substance produced in animal cells by introducing a phytoene synthesis enzyme gene (*crtB*), resulted in the reduction of the rate of cell transformation. Biochemoprevention might become one of the fundamental technologies for human cancer control for the future.

Green tea and, to a lesser extent, black tea are rich sources of still another group of flavonoids called catechins. These catechins, including the closely related epigallocatechin-3-gallate (EGCG), epigallocatechin, and epicatechin-3-gallate, form about 30% of the dry weight of tea leaves. Literally dozens of studies by Japanese researchers — avid drinkers of green tea — have shown that catechins can prevent free-radical damage to cholesterol and lower the risk of heart disease and cancer. Population-based studies have shown that tea drinkers have a lower than average risk of heart disease, stroke, and several cancers, including esophageal, stomach, and lung cancers. A study by Jankun et al.[120] described one of the ways green tea prevents cancer: cancer cells use the enzyme urokinase to invade cells and metastasize. Omnipresent in green tea, EGCG is a potent inhibitor of urokinase activity. A single cup of green tea contains enough EGCG to temporarily inhibit urokinase activity, and it is safer than the synthetic drugs that block urokinase activity.

22.10 CONCLUSIONS

Trillions of dollars have been spent on cancer research over the past few decades, and a review of the research data makes it clear that there is no single magic bullet for cancer prevention. However, a healthy lifestyle, appropriate stress reduction, and a nutritionally balanced diet coupled with flavonoid-rich produce provide an excellent start. Cancer prevention strategy should be started early on, using lifestyle choices that minimize the body's exposure to noxious toxicants and maximize the body's exposure to circulating antioxidants. Since each individual has his or her own distinct physiology with its own inimitable requirements, additional antioxidant supplementation may be extremely beneficial. Many researchers agree that in order to maintain optimum health, the levels of intake of antioxidants should be toward the upper limit. While the definitions of *safe* and *unsafe* still need to be established, naturally derived anticancer agents are undergoing intense scrutiny. Therefore, chemopreventive strategies must be custom-designed for those with a high risk of cancer due to family history, or for those who have existing cancer or who live in high-cancer-risk areas. The fact is, no one knows which of the myriad of chemicals in a carrot or broccoli does the most to keep our cells healthy. To address these issues scientifically will take decades of clinical study and research. Even then, age, heredity, and other unknown risks will still make prevention an inexact science. Despite all these unresolved issues, there is reason to be optimistic. Each scientific study provides another clue to the evolving mystery of how diet affects cancer. But as with any good mystery, some clues hold more weight than others, and no single piece of evidence can give us all the answers. The "body of evidence" formed by many studies must be considered as a whole as we investigate the mystery of the flavonoid–cancer connection.

Flavonoids are a gold mine for human health. The biological fates of flavonoids are still in question. In light of the considerable effort that has been expended by scientists from the academic, governmental, and private sectors in identifying, characterizing, and utilizing potential cancer chemopreventive agents, it is reasonable to inquire about the progress that has been made to date and the promise that this field holds in the fight against cancer. Adequate nutrition, specifically including sufficient fruits and vegetables, is the key in the fight against cancer, whether as a preemptive strike or as a therapy applied at a stage where the cancer needs to be wiped out or killed. Given the effects of solar radiation, smoking, and pesticides, very few people, perhaps as low as 0.1%, die because of old age. The vast majority succumb to one disease or another and die prematurely. What is absolutely necessary is a medicine for the new millennium, and with all this evidence at hand, it is just a matter of time before our diets become adjusted and food manufacturers begin to produce foods that are more functionally geared toward the fight against cancer. The common fatal cancers occur, in large part, as a result of lifestyle and other environmental factors and are, in principle, preventable, as recognized by an expert World Health Organization committee in 1964. The lion's share of available data suggests that over 85% of the diagnosed cancers are potentially preventable, if not curable. There is substantial evidence that cancer is largely an avoidable disease provided some preventive measures along with lifestyle modification are applied. Some of the avoidable causes include excessive tobacco use, alcohol consumption, unwarranted use of dietary supplements, use of additives, uncontrolled use of endocrine-active products, reproductive and sexual behavior, occupation, pollution, industrial products, medicines and medical procedures, geophysical factors, infection, and other unknown causes. These environmental factors (extrinsic factors) play a dominant role in the incidence of cancer in humans. Epidemiological studies estimate that a significant percentage of all cancers are chiefly associated with the environment in which we conduct our day-to-day activities.

Nutrition and dietary carcinogens together constitute one of the three major causes of carcinogenesis. The knowledge of nutrition and carcinogenic factors can contribute to suppression of carcinogenesis and promotion of anticarcinogenic actions. Diet very much depends on locality, history, race, and religion, but at the same time, the food industry can be an effective partner in cancer prevention. Development of a malignancy is the result of a series of cellular events, progressing step by step. If dietary improvements delay any of these steps, the final outcome is considerably shifted to an older age, resulting in end-of-life cancer. Nutrition and food carcinogens continue to be a most challenging subject for research in cancer control.[121] Some of the extensively debated hypothetical or actual means whereby diet may affect the incidence of cancer are: (a) ingestion of powerful, direct-acting carcinogens or their precursors affecting the formation of carcinogens in the body; (b) agents affecting transport and/or activation and deactivation of carcinogens; (c) agents affecting "promotion-stage" of cells that are already initiated; (d) vulnerability due to malnutrition; and (e) susceptibility due to excessive nutrition. Experimental findings and human observation provide many indications that dietary factors are of a major importance in determining the risk of cancers of the gastrointestinal tract, some cancers of the female sex organs, and epithelial cancers in general. Pharmaceutical research in this decade has already witnessed an intense interest in the field of discovery of dietary supplements derived from natural sources, mostly from herbs, that could be of prognostic importance in a variety of ailments, including cancer.

The protective effect of vegetables also was observed for hormone-related cancers. In one of the published studies of 2400 Greek women, it was noted that vegetable and fruit consumption were independently associated with significant reductions in the incidence of breast cancer.[122] In recent times, there has been a worldwide campaign to promote increased use of fruits and vegetables. The Five-a-Day for Better Health program was designed to encourage the consumption of at least five servings of fruit per day. In fact, some of the recent surveys show that one in every nine Americans had eaten no fruit and no vegetable on the day of the survey.

Two types of animal studies have been conducted to demonstrate the role of flavonoids as anticarcinogenic agents: studies using dietary components that are known to have high concentrations of flavonoids and studies using pure flavonoids. Flavonoids may exhibit anticarcinogenicity by blocking formation of endogenous or exogenous carcinogens, thereby preventing initiation of carcinogenesis.[73] Some studies have demonstrated that soybean products in the diet inhibited the development of liver, stomach, colon, skin, mammary, and prostate cancers. These plant-derived estrogen analogs (phytoestrogens) may confer significant health advantages, including cholesterol reduction, antioxidant activity, and possibly a reduced cancer risk. However, the concern has also been raised that phytoestrogens may be endocrine disrupters and major health hazards. Tea extracts (green, black, or decaffeinated) and its polyphenolic components have been shown to decrease skin cancer incidence in mice. Flavonoids have been used to prevent the development of cancers of the gastrointestinal tract. Diosmin and hesperidin were found to reduce the incidence of oral cancer and were equally effective when used before and during the chemical exposure or after the chemical exposure. Dietary quercetin treatment has shown a decrease in the incidence of small-intestinal cancers after multicarcinogen treatment in rats. Dietary catechin and its analog, epigallocatechin gallate, were found to decrease the incidence of 1,2-dimethylhydrazine-induced colon cancer in rats. Flavonoids have been shown to reduce the cancers of the reproductive system. Quercetin, genistein, and catechin have shown a reduction in the incidence of mammary cancer and tumors in rats. Studies have demonstrated the effectiveness of flavonoids in the prevention of skin cancer. Flavonoids, if ingested in adequate amounts, are associated with significant reductions in the risk of lung cancer.[123]

Flavonoids are also known to act as hormones.[124] Although the estrogenic property of flavonoids has been proposed to play a role in cancer prevention, the molecular link between the estrogenic property of flavonoids and their ability to prevent cancer has not been defined clearly. Flavonoids have been found to interfere with the activity of enzymes involved in the metabolism of steroid hormones. However, it seems that their ability to prevent cancer could be especially important if the exposure starts at the neonatal stage, as neonatal exposure has been shown to lead to long-term changes in the reproductive organs. As described above, there are multiple mechanisms that have been proposed to mediate the anticancer effect of flavonoids. Conceivably, some mechanisms may be more important in regulating normal tissue development (e.g., enhancing cell differentiation) and protecting against the irreversible transformation of normal cells (e.g., antioxidant), while other mechanisms may be more important for the control of existing cancer (e.g., kinase inhibition, type

II estrogen binding sites). Nevertheless, the possibility that these mechanisms are working collectively, perhaps in a hierarchical structure, cannot be ruled out.

The lifetime chance of getting cancer of the colon is 6%, prostate 17%, breast 14%, and lung 7%. At the same time, proper diet, supplementation, and lifestyle changes can reduce cancer of the colon by up to 50%, prostrate by 15%, and lung by 90%. Dr. Gabriel Deldman, Director of the American Cancer Society, sums it up well by saying, "We don't need years of research. If people implement what we know today, cancer rates would drop. It's that simple" (LamMD.com).

The critical genetic alterations underlying neoplasia involve at least two types of genes, oncogenes and oncosuppressor (tumor suppressor) genes or antioncogenes. More than 50 different oncogenes have been found to have transforming capabilities. Among these are ras, myc, fos, and src. They code for oncoproteins located in the nucleus or cytoplasm that are involved in growth control, as growth factors, factor recognition complexes, signal transduction elements, protein kinases, and DNA-binding proteins (transcriptional activators). Oncogenes in their normal state, referred to as proto-oncogenes, are present in all cells. Activated oncogenes are defined by their ability to act dominantly to transform cells. Activation can occur through point mutations, chromosomal translocation, amplification, or retroviral insertion (insertional mutagenesis). Activated oncogenes are dominant over the corresponding proto-oncogene allele, suggesting that expression of at least two oncogenes may be required for full neoplastic conversion. Oncosuppressor genes are capable of suppressing neoplastic behavior in tumor cells in which they are present. Thus, they are recessive genes. The first oncosuppressor gene identified was the human retinoblastoma (Rb-1) gene. Recently gene p53, which encodes the cellular protein p53 and various chromosomal deletions, has been identified as a candidate oncosuppressor gene (interestingly, the mutated p53 gene is an oncogene). The lack of expression of oncosuppressor genes or inactivation of their products may lead to neoplastic conversion.

In rodents, under some conditions, the symmetric dialkyl compounds exhibit delicate and yet specific organotropism, i.e., they preferentially cause cancer in a given organ. This occurs more so in rats than in mice, with the latter showing mostly liver neoplasms. For example, diethyl- and dimethylnitrosamine usually cause liver cancer in rats, whereas the dibutyl derivative causes cancer of the urinary bladder and the diamyl compound cancer of the lung. The dose rate also plays a role. Dimethylnitrosamine administered to rats in moderate doses for a long time leads to cancer of the liver, whereas fewer high doses or, indeed, a single large dose results in renal carcinomas. Because it is characterized by a deficiency in liver activity of the alkyl acceptor protein involved in repair of O^6-alkylation of guanine, hamsters are usually extremely sensitive to liver carcinogenesis by dimethylnitrosamine. Although clear-cut species differences in carcinogen metabolism have been reported, systematic studies unraveling such differences in flavonoid effects are lacking. Along these lines (a first approximation to species and strain-related differences in metabolic activation), the large differences found in rodents appear to be linked to the Ah locus, which in turn, reflects the presence and/or inducibility of distinct forms of CYP 450. Under well-defined *in vitro* conditions, the extent of metabolic activation can be quantified by measuring the released mutagenic metabolites with cell-mediated mutation assays. The subsequent DNA damage by the active metabolites can be determined directly and compared in the cultured cells in order to examine the relationships between species and tissue specificity in chemical carcinogenesis and the genotoxic effects of carcinogens.[125] Regarding organotropism of carcinogens in specific hosts, some proximate carcinogens have transport forms such as glucuronic acid conjugates, often generated in the liver, which release a reactive proximate carcinogen in a specific target organ such as urinary bladder or large bowel, where it is further converted to the reactive electrophile. Both species differences and organotropism are also functions of the DNA-repair systems. For example, the hamster is more susceptible than the rat to liver cancer induced by dimethylnitrosamine because of its deficiency in repair of O^6 alkylation of guanine in DNA. In the rat, a single large dose of dimethylnitrosamine induces kidney cancer, but not liver cancer, because of the ability of liver repair enzymes to eliminate alkylated forms of DNA from the liver.

Age is an important factor that does influence chemical carcinogenesis. Newborn animals exhibit higher sensitivity to certain carcinogens than older animals. Thus, injection of newborn mice with only a few doses of any one of a number of genotoxic chemical carcinogens results in tumor, primarily of the liver and lung, approximately a year after the administration of the carcinogen. Carcinogens may be effective when administered to young animals but ineffective when administered to older animals. It has also been shown that DMN induces significant dose-dependent increases in DNA repair in the young and middle-aged animals fed *ad libitum*, while significantly lower levels of DNA repair were observed in cultures from old animals fed *ad libitum*. These results indicate an age-related decline in the specific hepatic microsomal isozymes required to metabolically activate DMN to a DNA-reactive species and suggest a decline in cytochrome P450 IIB1. It is known that cytochrome P450 IIB1 is responsible for the increase in DMN demethylation and the subsequent DNA damage and repair, and not cytochrome P450 IIE1.[126]

Epidemiological data show that some types of cancer occur more frequently in men than in women, and vice versa. In experimental animals, likewise, some chemical carcinogens induce cancer more frequently at specific target organs in one than the other sex, even when a nonendocrine organ is involved. Diethyl- or dimethylnitrosamine produces liver cancer often, but not always, with similar efficiency in males and females. In nonendocrine target organs, the sex-linked effectiveness of a given carcinogen stems mostly from a sex-dependent activation of enzyme systems necessary for the conversion of a procarcinogen to the reactive ultimate carcinogen. Immunological factors have been found in some instances to alter the rate and extent of tumor development. Immunological mechanisms have been thought to play a role in the greater sensitivity of newborn animals to chemical carcinogens. For instance, in some species and strains, the immunological competence in the newborn is either totally lacking or considerably less than that present in the adult animal. However, immunological status is more relevant in the development of metastases than in the early effects in the carcinogenic process.

Nitrosamines have been suspected in the etiology of esophageal and gastric cardia cancer in the high-incidence area of Linxian in the Henan province in northern China, but marginal deficiencies in riboflavin, vitamins A and C, and other micronutrients may also be involved.[127] A joint U.S.–China nutritional intervention study with investigators from the Cancer Institute of the Chinese Academy of Medical Sciences and the U.S. National Cancer Institute tested the effects of the following four combinations of nutrients on 29,584 subjects in an eight-group design: (1) retinol and zinc, (2) riboflavin and niacin, (3) vitamin C and molybdenum, and (4) vitamin E, β-carotene, and selenium. Supplementation with Group 4 nutrients significantly decreased mortality rate from stomach cancer, primarily due to the decrease in deaths resulting from adenocarcinomas of the gastric cardia; it also lowered the total mortality rate and showed signs of other beneficial effects.[127] Another study of nutrition and gastric cancer in the high-incidence area of Linqu in the Shandong province in northern China (in collaboration with the Beijing Institute for Cancer Research and the U.S. National Institutes of Health) found significantly lower serum concentrations of vitamin C and β-carotene among individuals with intestinal metaplasia; an intervention trial with vitamins C and E and selenium (combined) is ongoing in Linqu. Other studies are also elucidating the mechanisms for the pathogenesis of adenocarcinoma at the gastroesophageal junction with the use of a rat model. Such studies are expected to shed light on the etiology and prevention of gastroesophageal cancers in humans.

Clinical trials are the essential gateways to development of new cancer therapies. Several years ago, the Food and Drug Administration (FDA) approved topotecan, a new class of drugs, to treat patients whose advanced ovarian cancers have not responded favorably to standard treatments. Topotecan works by interfering with the DNA-uncoiling enzyme, topoisomerase I, that creates and reseals nicks in the DNA before cell division. A drastic change in the function of this key enzyme ultimately causes tumor cell lethality. Approval of topotecan is significant because it is the first of a series of drugs with a new mechanism of action, which opens up a new dimension of research into the development of unique methods of treatment. These approvals signify the importance of natural products and their derivatives in the search for a better cancer cure. Topotecan, which is marketed as Hycamtin by SmithKline Beecham Pharmaceuticals, is derived from the bark of a Chinese tree known scientifically as *Camptotheca acuminata*. Efforts in this direction have led to the production of a semisynthetic topotecan by SmithKline Beecham, in partnership with NCDDG. In addition, research teams at NCI and other institutions had synthesized several other modified camptothecins, such as irinotecan. Topoisomerase I, the target enzyme of these new agents, exists in both normal and tumor cells and appears to take advantage of the growth rate difference between them. Experts in this field are not sure why, but there are two possible explanations. Topoisomerase I is most vulnerable to the "S" phase of the cell cycle. In this phase, cells copy their DNA while preparing to divide; therefore, the target enzyme is present more frequently and at a higher concentration. Since rapidly proliferating tumors have a greater percent of cells in the "S" phase than do normal, slow-growing cells, tumors are more susceptible to the action of topotecan. A second explanation might be that rapid metabolic processes of tumor cells enable greater influx of the drug.

Whatever may be the mechanism, the outcome of the trial results were positive after topotecan treatment. In a series of NCI and SmithKline Beecham joint clinical trials, the compound showed activity against a variety of solid tumors, including ovarian cancer, small-cell lung cancer, and several others. In the European trials conducted by the European Organization for the Research and Treatment of Cancer, topotecan exhibited very promising results and powerful antitumor activity against a variety of tumor types. In later phase III trials, examining topotecan as second-line therapy against ovarian cancers, response rates were 10 to 15% among patients who had not previously responded to standard treatment and 25 to 30% among patients who had responded to first-line therapy.[128] In one study comparing topotecan and taxol for the treatment of advanced, recurrent ovarian cancer, topotecan showed a 20% response rate, whereas taxol had a 12% response rate. According to some researchers, topotecan is as unique as taxol in terms of a unique mechanism of action and novelty of approach. Response rates as high as 39% also have been seen in phase III trials using the drug as a first-line treatment for small-cell lung cancer. Future studies will evaluate the efficacy of topotecan when used in combination with other drugs, such as cisplatin.

Over the past three years, more than 1000 scientific studies have been published on soy flavonoids, technically known as isoflavones. These isoflavones function as antioxidants, and they seem to possess properties that distinguish them from other types of flavonoids. The principal soy isoflavones are genistein, daidzein, and glycitein. They have very mild estrogenic effects without being true estrogen hormones and, as such, play paradoxical but beneficial roles. Several human studies have found that soy isoflavones mimic estrogen and promote bone density, likely reducing the risk of osteoporosis in postmenopausal women. But the isoflavones also seem to prevent estrogen from attaching to and stimulating cells. It is for this reason, researchers believe, that soy isoflavones ease the intensity of menopausal hot flashes and may lower the long-term risk of breast cancer. Laboratory experiments also indicate that isoflavones can significantly influence the activity of tyrosine kinase, an enzyme that promotes cancer growth. Genistein has been demonstrated to block angiogenesis and trigger the destruction of cancer cells based on similar mechanisms.[129–130]

Cells in our body are dynamic, very active 24 hours a day, and many of them are constantly dividing. Incorporations of bad mutations during a vulnerable phase are few, but they do happen. When enough mutations have occurred, the result can be cancer. In fact, we are all walking around with millions of premalignant cells. If we live long enough, we will come down with one form of cancer or another. Advancements in technology have increased life expectancy up to platinum standards in industrialized nations. Cancer is fast replacing cardiovascular disease as the number one cause of death in the U.S. Unfortunately, cancer chemoprevention research is two decades behind cancer treatment research. It makes more sense to treat precancerous lesions than to wait for people to develop fully blown cancer. For example, over 1 trillion molecules of oxygen go through each cell every day, inflicting about 100,000 free-radical wounds on the DNA. By age 30, a few million free-radical lesions per cell exist in each of our body's cells. By age 50, about 30% of our cellular protein has been damaged by free radicals. The solution is to fight the free radicals with antioxidants natural or synthetic, whichever it may be. The cornerstone of cancer prevention research is to design drugs that thwart the carcinogenic process at points where genetic lesions may occur. The current trend in research is to focus on drugs that limit the damage caused by substances that cause cell mutation. Another goal is to use gene therapy to stop the random genetic mutation that results in cancer. A third goal is to intercept "free radicals" and errant oxygen molecules that are released during normal cellular metabolism in order to prevent the damage they can do to cells and to prevent them from triggering genetic mutations. Finally, another goal is to design a mechanism-based protective infrastructure inside the cell to safeguard vital intracellular targets long before a carcinogenic exposure.

REFERENCES

1. Surh, Y.J., Cancer chemoprevention with dietary phytochemicals, *Nat. Rev. Cancer*, 3, 768, 2003.
2. Bagchi, D. et al., Free radicals and grape seed proanthocyanidin extract: importance in human health and disease prevention, *Toxicology*, 148, 187, 2000.
3. Bagchi, D. et al., Protective effects of grape seed proanthocyanidins and selected antioxidants against TPA-induced hepatic and brain lipid peroxidation and DNA fragmentation, and peritoneal macrophage activation in mice, *Gen. Pharmacol.*, 30, 771, 1998.
4. Bagchi, D. et al., Comparative *in vitro* and *in vivo* free radical scavenging abilities of a novel grape seed proanthocyanidin extract and selected antioxidants, in *Natural Antioxidants and Anticarcinogens in Nutrition, Health and Disease*, Kumpulainen, J.T. and Salonen, J.T., Eds., Royal Society of Chemistry, Cambridge, U.K., 1999, p. 178.
5. Bagchi, D. et al., Hydrogen peroxide-induced modulation of intracellular oxidized states in cultured macrophage J774A.1 and neuroactive PC-12 cells, and protection by a novel grape seed proanthocyanidin extract, *Phytotherapy Res.*, 12, 568, 1998.
6. Bagchi, M. et al., Smokeless tobacco, oxidative stress, apoptosis and antioxidants in human oral keratinocytes, *Free Radical Biol. Med.*, 26, 992, 1999.
7. Ray, S.D., Kumar, M.A., and Bagchi, D., A novel proanthocyanidin IH636 grape seed extract increases *in vivo* Bcl-X$_L$ expression and prevents acetaminophen-induced programmed and unprogrammed cell death in mouse liver, *Arch. Biochem. Biophys.*, 369, 42, 1999.
8. Ray, S.D. et al., *In vivo* protection of DNA damage associated apoptotic and necrotic cell deaths during acetaminophen-induced nephrotoxicity, amiodarone-induced lung toxicity and doxorubicin-induced cardiotoxicity by a novel IH636 grape seed proanthocyanidin extract, *Res. Comm. Mol. Pathol. Pharmacol.*, 107, 137, 2000.
9. Bagchi, D. et al., Protection against drug- and chemical-induced multiorgan toxicity by a novel IH636 grape seed proanthocyanidin extract, *Drugs Exptl. Clin. Res.*, 27, 3, 2001.
10. Ray, S.D. et al., Unique organo-protective properties of a novel IH636 grape seed proanthocyanidin extract on cadmium chloride-induced nephrotoxicity, dimethylnitrosamine (DMN)-induced spleenotoxicity and MOCAP-induced neurotoxicity in mice, *Res. Comm. Mol. Pathol. Pharmacol.*, 107, 105, 2000.
11. Sato, M. et al., Cardioprotective effects of grape seed proanthocyanidin against ischemic reperfusion injury, *J. Mol. Cell. Cardiol.*, 31, 1289, 1999.
12. Bagchi, D. et al., Molecular mechanisms of cardioprotection by a novel grape seed proanthocyanidin extract, *Mutation Res.*, 523/524, 87, 2003.
13. Vinson, J.A. et al., Beneficial effects of a novel IH636 grape seed proanthocyanidin extract and a niacin-bound chromium in a hamster atherosclerosis model, *Mol. Cell. Biochem.*, 240, 99, 2002.
14. Bagchi, M. et al., Acute and chronic stress-induced oxidative gastrointestinal injury in rats, and the protective ability of a novel grape seed proanthocyanidin extract, *Nutrition Res.*, 19, 1189, 1999.
15. Bagchi, M. et al., Protective effects of antioxidants against smokeless tobacco-induced oxidative stress and modulation of Bcl-2 and p53 genes in human oral keratinocytes, *Free Radical Res.*, 35, 181, 2001.
16. Joshi, S.S. et al., Amelioration of the cytotoxic effects of chemotherapeutic agents by grape seed proanthocyanidin extract, *Antiox. Redox. Signal.*, 1, 563, 1999.
17. Joshi, S.S. et al., Chemopreventive effects of grape seed proanthocyanidin extract on Chang liver cells, *Toxicology*, 155, 83, 2000.

18. Joshi, S.S., Kuszynski, C.A., and Bagchi, D., The cellular and molecular basis of health benefits of grape seed proanthocyanidin extract, *Curr. Pharm. Biotech.*, 2, 187, 2001.

19. Ye, X. et al., The cytotoxic effects of a novel IH636 grape seed proanthocyanidin extract on cultured human cancer cells, *Mol. Cell. Biochem.*, 196, 99, 1999.

20. Krohn, R.L. et al., Differential effect of a novel grape seed proanthocyanidin extract on cultured human normal and malignant cells, in *Natural Antioxidants and Anticarcinogens in Nutrition, Health and Disease*, Kumpulainen, J.T. and Salonen, J.T., Eds., Royal Society of Chemistry, Cambridge, U.K., 1999, p. 443.

21. Bagchi, D. and Preuss, H.G., Oligomeric proanthocyanidins in human health and disease prevention, in *Critical Reviews in Oxidative Stress and Aging: Advances in Basic Science, Diagnostics and Intervention*, Cutler, R.G. and Rodriguez, H., Eds., World Scientific Publishing, NJ, 2002, p. 640.

22. Gressier, B. et al., Pro-oxidant properties of methotrexate: evaluation and prevention by an anti-oxidant drug, *Pharmazie*, 49, 679, 1994.

23. Mayne, S.T., Graham, S., and Zheng, T.Z., Dietary retinol: prevention or promotion of carcinogenesis in humans? *Cancer Causes Control*, 2, 443, 1991.

24. Packer, L., Rimbach, G., and Virgili, F., Antioxidant activity and biologic properties of a procyanidin-rich extract from pine (*Pinus maritima*) bark, pycnogenol, *Free Radical Biol. Med.*, 27, 704, 1999.

25. Mittal, A., Elmets, C.A., and Katiyar, S.K., Dietary feeding of proanthocyanidins from grape seeds prevents photocarcinogenesis in SKH-1 hairless mice: relationship to decreased fat and lipid peroxidation, *Carcinogenesis*, 24, 1379, 2003.

26. Craig, W.J., Health-promoting properties of common herbs, *Amer. J. Clin. Nutr.*, 70 (3 Suppl.), 491S, 1999.

27. Wiseman, S., Mulder, T., and Rietveld, A., Tea flavonoids: bioavailability *in vivo* and effects on cell signaling pathway *in vitro*, *Antioxid. Redox Signal.*, 3, 1009, 2001.

28. Turner, N.J., Thomson, B.M., and Shaw, I.C., Bioactive isoflavones in functional foods: the importance of gut microflora on bioavailability, *Nutr. Rev.*, 61 (6, Pt. 1), 204, 2003.

29. Chao, P.D. et al., Flavonoids in herbs: biological fates and potential interactions with xenobiotics, *J. Food Drug Anal.*, 10, 219, 2002.

30. Hollman, P.C. et al., Absorption of dietary quercetin glycosides and quercetin in healthy ileostomy volunteers, *Am. J. Clin. Nutr.*, 62, 1276, 1995.

31. Hollman, P.C., Hertog M.G., and Katan M.B., Role of dietary flavonoids in protection against cancer and coronary heart disease, *Biochem. Soc. Trans.*, 3, 785, 1996.

32. Hollman, P.C. et al., Bioavailability of the dietary antioxidant flavonol quercetin in man, *Cancer Lett.*, 114, 139, 1997.

33. Hollman, P.C., Bioavailability of flavonoids, *Eur. J. Clin. Nutr.*, 51 (Suppl. 1), S66, 1997.

34. Manach, C. et al., Bioavailability of rutin and quercetin in rats, *FEBS Lett.*, 409, 12, 1997.

35. Hollman, P.C. and Katan M.B., Health effects and bioavailability of dietary flavonols, *Free Radic. Res.*, 31, S75, 1999.

36. Walle, T. et al., Quercetin glucosides are completely hydrolyzed in ileostomy patients before absorption, *J. Nutr.*, 130, 2658, 2000.

37. Chao, P.D. et al., Quercetin glucuronides inhibited 2-aminofluorene acetylation in human acute mycloid HL-60 leukemia cells, *Phytomedicine*, 9, 625, 2002.

38. Morand, C. et al., Respective bioavailability of quercetin aglycone and its glycosides in a rat model, *Biofactors*, 12, 169, 2000.

39. Olthof, M.R. et al., Bioavailabilities of quercetin-3-glucoside and quercetin-4'-glucoside do not differ in humans, *J. Nutr.*, 130, 1200, 2000.

40. Ameer, B. et al., Flavanone absorption after naringin, hesperidin, and citrus administration, *Clin. Pharmacol. Ther.*, 60, 34, 1996.

41. Choudhury, R. et al., Interactions of the flavonoid naringenin in the gastrointestinal tract and the influence of glycosylation, *Biochem. Biophys. Res. Commun.*, 265, 410, 1999.

42. Felgines, C. et al., Bioavailability of the flavanone naringenin and its glycosides in rats, *Am. J. Physiol. Gastrointest. Liver Physiol.*, 279, G1148, 2000.

43. Erlund, I. et al., Plasma kinetics and urinary excretion of the flavanones naringenin and hesperetin in humans after ingestion of orange juice and grapefruit juice, *J. Nutr.*, 131, 235, 2001.

44. Erlund, I. et al., Pharmacokinetics of quercetin from quercetin aglycone and rutin in healthy volunteers, *Eur. J. Clin. Pharmacol.*, 56, 545, 2000.

45. Donovan, J.L. et al., Procyanidins are not bioavailable in rats fed a single meal containing a grapeseed extract or the procyanidin dimer B3, *Br. J. Nutr.*, 87, 299, 2002.

46. Baba, S. et al., Absorption and urinary excretion of procyanidin B2 [epicatechin-(4beta-8)-epicatechin] in rats, *Free Radic. Biol. Med.*, 33, 142, 2002.

47. Baba, S. et al., Absorption and urinary excretion of (−)-epicatechin after administration of different levels of cocoa powder or (−)-epicatechin in rats, *J. Agric. Food Chem.*, 49, 6050, 2001.

48. Baba, S. et al., *In vivo* comparison of the bioavailability of (+)-catechin, (+)-epicatechin and their mixture in orally administered rats, *J. Nutr.*, 131, 2885, 2001.

49. Vaidyanathan, J.B. and Walle, T., Glucuronidation and sulfation of the tea flavonoid (−)-epicatechin by the human and rat enzymes, *Drug Metab. Dispos.*, 30, 897, 2002.

50. Schramm, D.D. et al., Food effects on the absorption and pharmacokinetics of cocoa flavonols, *Life Sci.*, 73, 857, 2003.

51. Crespy, V. et al., The splanchnic metabolism of flavonoids highly differed according to the nature of the compound, *Am. J. Physiol. Gastrointest. Liver Physiol.*, 284, G980, 2003.

52. Carbonaro, M., Grant, G., and Pusztai, A., Evaluation of polyphenol bioavailability in rat small intestine, *Eur. J. Nutr.*, 40, 84, 2001.

53. Huynh, H.T. and Teel, R.W., Selective induction of apoptosis in human mammary cancer cells (MCF-7) by pycnogenol, *Anticancer Res.*, 20, 2417, 2000.

54. Virgili, F. et al., Ferulic acid excretion as a marker of consumption of a French maritime pine (*Pinus maritima*) bark extract, *Free Radical Biol. Med.*, 28, 1249, 2000.

55. Piskula, M.K., Factors affecting flavonoids absorption, *Biofactors*, 12, 175, 2000.

56. Wang, M.Y. and Su, C., Cancer preventive effect of *Morinda citrifolia* (Noni), *Ann. N.Y. Acad. Sci.*, 952, 161, 2001.

57. Boveris, A., Valdez, L., and Alvarez, S., Inhibition by wine polyphenols of peroxynitrite-initiated chemiluminescence and NADH oxidation, *Ann. N.Y. Acad. Sci.*, 957, 90, 2002.

58. Yamashita, S. et al., Absorption and metabolism of antioxidative polyphenolic compounds in red wine, *Ann. N.Y. Acad. Sci.*, 957, 325, 2002.

59. Lin, Y.T. et al., Degradation of flavonoid aglycones by rabbit, rat and human fecal flora, *Biol. Pharm. Bull.*, 26, 747, 2003.

60. Boue, S.M. et al., Evaluation of the estrogenic effects of legume extracts containing phytoestrogens, *J. Agric. Food. Chem.*, 51, 2193, 2003.

61. Ross, J.A. and Kasum, C.M., Dietary flavonoids: bioavailability, metabolic effects, and safety, *Ann. Rev. Nutr.*, 22, 19, 2002.

62. Rechner, A.R. et al., The metabolism of dietary polyphenols and the relevance to circulating levels of conjugated metabolites, *Free Radical Res.*, 36, 1229, 2002.

63. Rohdewald, P., A review of the French maritime pine bark extract (pycnogenol), a herbal medication with a diverse clinical pharmacology, *Int. J. Clin. Pharmacol. Ther.*, 40, 158, 2002.

64. Koley, A.P., Differential mechanisms of cytochrome P450 inhibition and activation by alpha-naphthoflavone, *J. Biol. Chem .*, 272, 3149, 1997.

65. Kenakin, T., Agonist-receptor efficacy, II: agonist trafficking of receptor signals, *Trends Pharmacol. Sci.*, 16, 232, 1995.

66. Bruns, R.F., Conformational induction versus conformational selection: evidence from allosteric enhancers, *Trends Pharmacol. Sci.*, 17, 189, 1996.

67. Obach, R.S., Inhibition of human cytochrome P450 enzymes by constituents of St. John's wort, an herbal preparation used in the treatment of depression, *J. Pharmacol. Exp. Ther.*, 294, 88, 2000.

68. Ray, S.D. et al., Differential effects of IH636 grape seed proanthocyanidin extract and a DNA repair modulator 4-aminobenzamide on liver microsomal cytochrome P450 2E1-dependent aniline hydroxylation, *Mol. Cell. Biochem.*, 218, 27, 2001.

69. Bailey, D.G. et al., Grapefruit juice--drug interactions, *Br. J. Clin. Pharmacol.*, 46, 101, 1998.

70. Fuhr, U. and Kummert, A.L., The fate of naringin in humans: a key to grapefruit juice–drug interactions? *Clin. Pharmacol. Ther .*, 58, 365, 1995.

71. Honig, A. et al., Effect of arterial chemoreceptor stimulation with almitrine bismesylate on plasma renin activity, aldosterone, ACTH and cortisol in anaesthetized, artificially ventilated cats, *Clin. Exp. Pharmacol. Physiol.*, 23, 106, 1996.

72. Eaton, E.A. et al., Flavonoids, potent inhibitors of the human P-form phenolsulfotransferase: potential role in drug metabolism and chemoprevention, *Drug Metab. Dispos.*, 24, 232, 1996.

73. Williamson, G., Faulkner K., and Plumb, G.W., Glucosinolates and phenolics as antioxidants from plant foods, *Eur. J. Cancer Prev.*, 7, 17, 1998.
74. de Groot, H. and Rauen, U., Tissue injury by reactive oxygen species and the protective effects of flavonoids, *Fundam. Clin. Pharmacol.*, 12, 249, 1998.
75. Le Marchand, L., Intake of flavonoids and lung cancer, *J. Natl. Cancer Inst.*, 92, 154, 2000.
76. Hammons, G.J., Effects of chemoprotective agents on the metabolic activation of the carcinogenic arylamines PhIP and 4-aminobiphenyl in human and rat liver microsomes, *Nutr. Cancer*, 33, 46, 1999.
77. Chinery, R. et al., Antioxidants enhance the cytotoxicity of chemotherapeutic agents in colorectal cancer: a p53-independent induction of p21WAF1/CIP1 via C/EBPbeta, *Nat. Med.*, 3, 1233, 1997.
78. Abou-Shoer, M. et al., Flavonoids from *Koelreuteria henryi* and other sources as protein-tyrosine kinase inhibitors, *J. Nat. Prod.*, 56, 967, 1993.
79. Laughton, M.J, Inhibition of mammalian 5-lipoxygenase and cyclo-oxygenase by flavonoids and phenolic dietary additives: relationship to antioxidant activity and to iron ion-reducing ability, *Biochem. Pharmacol.*, 42, 1673, 1991.
80. Bracke, M.E. et al., Anti-invasive activity of 3,7-dimethoxyflavone *in vitro*, *J. Pharm. Sci.*, 83, 1217, 1994.
81. Hirano, T. et al., Citrus flavone tangeretin inhibits leukaemic HL-60 cell growth partially through induction of apoptosis with less cytotoxicity on normal lymphocytes, *Br. J. Cancer*, 72, 1380, 1995.
82. So, F.V. et al., Inhibition of human breast cancer cell proliferation and delay of mammary tumorigenesis by flavonoids and citrus juices, *Nutr. Cancer*, 26, 167, 1996.
83. Fuhr, U., Drug interactions with grapefruit juice: extent, probable mechanism and clinical relevance, *Drug Saf.*, 18, 251, 1998.
84. Chaumontet, C. et al., Lack of tumor promoting effects of flavonoids: studies on rat liver preneoplastic foci and on *in vivo* and *in vitro* gap junctional intercellular communication, *Nutr. Cancer*, 26, 251, 1996.
85. Helin, K., Harlow, E., and Fattaey, A., Inhibition of E2F-1 transactivation by direct binding of the retinoblastoma protein, *Mol. Cell. Biol.*, 13, 6501, 1993.
86. Weinberg, R.A., The retinoblastoma protein and cell cycle control, *Cell*, 81, 323, 1995.
87. Mayol, X. and Grana, X., The p130 pocket protein: keeping order at cell cycle exit/re-entrance transitions, *Front Biosci.*, 3, D11, 1998.
88. Dyson, N., The regulation of E2F by pRB-family proteins, *Genes Dev.*, 12, 2245, 1998.
89. Lavia, P. and Jansen-Durr, P., E2F target genes and cell-cycle checkpoint control, *Bioessays*, 21, 221, 1999.
90. De Azevedo, Jr., W.F. et al., Structural basis for specificity and potency of a flavonoid inhibitor of human CDK2, a cell cycle kinase, *Proc. Natl. Acad. Sci. USA*, 93, 2735, 1996.
91. Losiewicz, M.D. et al., Potent inhibition of CDC2 kinase activity by the flavonoid L86-8275, *Biochem. Biophys. Res. Commun.*, 201, 589, 1994.
92. Carlson, B.A. et al., Flavopiridol induces G1 arrest with inhibition of cyclin-dependent kinase (CDK) 2 and CDK4 in human breast carcinoma cells, *Cancer Res.*, 56, 2973, 1996.
93. Patel, V., Flavopiridol, a novel cyclin-dependent kinase inhibitor, suppresses the growth of head and neck squamous cell carcinomas by inducing apoptosis, *J. Clin. Invest.*, 102, 1674, 1998.
94. Kerr, J.S. et al., Novel small molecule alpha vs. integrin antagonists: comparative anti-cancer efficacy with known angiogenesis inhibitors, *Anticancer Res.*, 19, 959, 1999.
95. Drees, M. et al., Flavopiridol (L86-8275): selective antitumor activity *in vitro* and activity *in vivo* for prostate carcinoma cells, *Clin. Cancer Res.*, 3, 273, 1997.
96. Arguello, F. et al., Flavopiridol induces apoptosis of normal lymphoid cells, causes immunosuppression, and has potent antitumor activity *in vivo* against human leukemia and lymphoma xenografts, *Blood*, 91, 2482, 1998.
97. Pan, M.H. et al., Tangeretin induces cell-cycle G1 arrest through inhibiting cyclin-dependent kinases 2 and 4 activities as well as elevating Cdk inhibitors p21 and p27 in human colorectal carcinoma cells, *Carcinogenesis*, 23, 1677, 2002.
98. Pavelic, K. et al., Recent advances in molecular genetics of breast cancer, *J. Mol. Med.*, 79, 566, 2001.
99. Rooprai, H.K., Evaluation of the effects of swainsonine, captopril, tangeretin and nobiletin on the biological behaviour of brain tumour cells *in vitro*, *Neuropathol. Appl. Neurobiol.*, 27, 29, 2001.
100. Tanaka, T. et al., Modifying effects of a flavonoid morin on azoxymethane-induced large bowel tumorigenesis in rats, *Carcinogenesis*, 20, 1477, 1999.
101. Joiner, M.C. et al., Hypersensitivity to very-low single radiation doses: its relationship to the adaptive response and induced radioresistance, *Mutat. Res.*, 358, 171, 1996.
102. Savell, J. et al., Permanent growth arrest in irradiated human fibroblasts, *Radiat. Res.*, 155, 554, 2001.

103. Te Poele, R.H. et al., DNA damage is able to induce senescence in tumor cells *in vitro* and *in vivo*, *Cancer Res.*, 62, 1876, 2002.
104. Bagchi, D., Ray, S.D., Bagchi, M., Preuss, H.G., and Stohs, S.J., Mechanistic pathways of antioxidant cytoprotection by a novel IH636 grape seed proanthocyanidin extract, *Indian J. Exp. Biol.*, 40, 717, 2002.
105. Lautraite, S. et al., Flavonoids inhibit genetic toxicity produced by carcinogens in cells expressing CYP1A2 and CYP1A1, *Mutagenesis*, 17, 45, 2002.
106. Chan, H.Y., Wang, H., and Leung, L.K., The red clover (*Trifolium pratense*) isoflavone biochanin A modulates the biotransformation pathways of 7,12-dimethylbenz[a]anthracene, *Br. J. Nutr.*, 90, 87, 2003.
107. Park, Y.K. et al., Daily grape juice consumption reduces oxidative DNA damage and plasma free radical levels in healthy Koreans, *Mutat. Res.*, 529, 77, 2003.
108. Quiles, J.L. et al., Antioxidants, nutrients and adriamycin toxicity, *Toxicology*, 180, 79, 2002.
109. Van Acker, S.A., Intracerebroventricular administration of a glucocorticoid receptor antagonist enhances the cardiovascular responses to brief restraint stress, *Eur. J. Pharmacol.*, 430, 87, 2001.
110. Snyder, R.D. and Gillies, P.J., Evaluation of the clastogenic, DNA intercalative, and topoisomerase II-interactive properties of bioflavonoids in Chinese hamster V79 cells, *Environ. Mol. Mutagen.*, 40, 266, 2002.
111. Borek, C., Antioxidant health effects of aged garlic extract, *J. Nutr.*, 131, 1010S, 2001.
112. Wei, Y.Q. et al., Induction of apoptosis by quercetin: involvement of heat shock protein, *Cancer Res.*, 54, 4952, 1994.
113. Nagasaka, Y. and Nakamura, K., Modulation of heat-induced activation of mitogen-activated protein (MAP) kinase by quercetin, *Biochem. Pharmacol.*, 56, 1151, 1998.
114. Zarkovic, N. et al., An overview on anticancer activities of the *Viscum album* extract isorel, *Cancer Biother. Radiopharm.*, 16, 55, 2001.
115. Ray, S.D. et al., IH636 Grape seed proanthocyanidin extract (GSPE) exposure significantly attenuates dimethylnitrosamine (DMN)-induced liver cancer and mortality in B6C3F1 mice, *Proc. Am. Assoc. Cancer Res.*, 41, 460, 2000.
116. Barotto, N.N., Quercetin enhances pretumorous lesions in the NMU model of rat pancreatic carcinogenesis, *Cancer Lett.*, 129, 1, 1998.
117. Sugimura, T., Nutrition and dietary carcinogens, *Carcinogenesis,* 21, 387, 2000.
118. Tanaka, T., Modulation of N-methyl-N-amylnitrosamine-induced rat oesophageal tumourigenesis by dietary feeding of diosmin and hesperidin, both alone and in combination, *Carcinogenesis*, 18, 761, 1997.
119. Papadimitrakopoulou, V.A. et al., Biologic correlates of a biochemoprevention trial in advanced upper aerodigestive tract premalignant lesions, *Cancer Epidemiol. Biomarkers Prev.*, 11, 1605, 2002.
120. Jankun, J. et al., Why drinking green tea could prevent cancer, *Nature*, 387, 561, 1997.
121. Sugimura, T., Nutrition and dietary carcinogens, *Carcinogenesis*, 21, 387, 2000.
122. Katsouyanni, K. et al., Diet and breast cancer: a case-control study in Greece, *Int. J. Cancer*, 38, 815, 1986.
123. Stefani, E.D. et al., Dietary antioxidants and lung cancer risk: a case-control study in Uruguay, *Nutr. Cancer*, 34, 100, 1999.
124. Baker, M.E., Flavonoids as hormones: a perspective from an analysis of molecular fossils, *Adv. Exp. Med. Biol.*, 439, 249, 1998.
125. Hsu, I.C. et al., Cell and species differences in metabolic activation of chemical carcinogens, *Mutat. Res.*, 177, 1, 1987.
126. Shaddock, J.G., Snawder, J.E., and Casciano, D.A., Cryopreservation and long-term storage of primary rat hepatocytes: effects on substrate-specific cytochrome P450-dependent activities and unscheduled DNA synthesis, *Cell. Biol. Toxicol.*, 9, 345, 1993.
127. Zou, X.N. et al., Seasonal variation of food consumption and selected nutrient intake in Linxian, a high risk area for esophageal cancer in China, *Int. J. Vitam. Nutr. Res.*, 72, 375, 2002.
128. Poveda, A., Ovarian cancer treatment: what is new, *Int. J. Gynecol. Cancer*, 13 (Suppl. 2), 241, 2003.
129. Fotsis, T. et al., Genistein, a dietary-derived inhibitor of *in vitro* angiogenesis, *Proc. Natl. Acad. Sci. USA*, 90, 2690, 1993.
130. Fotsis, T. et al., Genistein, a dietary ingested isoflavonoid, inhibits cell proliferation and *in vitro* angiogenesis, *J. Nutr.*, 125 (3 Suppl.), 790S, 1995.

23 Curcumin Derived from Turmeric (*Curcuma longa*): a Spice for All Seasons

Bharat B. Aggarwal, Anushree Kumar, Manoj S. Aggarwal, and Shishir Shishodia

CONTENTS

23.1 Introduction..350
23.2 Anticancer Properties of Curcumin 351
    23.2.1 Curcumin Inhibits Tumorigenesis...351
    23.2.2 Curcumin Exhibits Antiproliferative Effects against Cancer Cells....................352
    23.2.3 Curcumin Down-Regulates the Activity of Epidermal Growth Factor Receptor (EGFR) and Expression of HER2/neu 353
    23.2.4 Curcumin Down-Regulates the Activation of Nuclear Factor-κB (Nf-κB) 354
    23.2.5 Curcumin Down-Regulates the Activation of STAT3 Pathway 354
    23.2.6 Curcumin Activates Peroxisome Proliferator-Activated Receptor-γ (PPAR-γ)...355
    23.2.7 Curcumin Down-Regulates the Activation of Activator Protein-1 (AP-1) and C-Jun N-Terminal Kinase (JNK) 355
    23.2.8 Curcumin Suppresses the Induction of Adhesion Molecules 355
    23.2.9 Curcumin Down-Regulates Cyclooxygenase-2 (COX-2) Expression 355
    23.2.10 Curcumin Inhibits Angiogenesis..356
    23.2.11 Curcumin Suppresses the Expression of MMP9 and Inducible Nitric Oxide Synthase (iNOS)....................................356
    23.2.12 Curcumin Down-Regulates Cyclin D1 Expression 356
    23.2.13 Curcumin Is Chemopreventive 356
    23.2.14 Curcumin Inhibits Tumor Growth and Metastasis in Animals 357
    23.2.15 Curcumin Inhibits Androgen Receptors and AR-Related Cofactors 358
23.3 Effect of Curcumin on Atherosclerosis and Myocardial Infarction....................359
    23.3.1 Curcumin Inhibits the Proliferation of Vascular Smooth Muscle Cells..359
    23.3.2 Curcumin Lowers Serum Cholesterol Levels............................360
    23.3.3 Curcumin Inhibits LDL Oxidation 361
    23.3.4 Curcumin Inhibits Platelet Aggregation 362
    23.3.5 Curcumin Inhibits Myocardial Infarction 362
23.4 Other Effects of Curcumin..363
    23.4.1 Curcumin Suppresses Diabetes...363
    23.4.2 Curcumin Stimulates Muscle Regeneration 364

23.4.3   Curcumin Enhances Wound Healing 365
23.4.4   Curcumin Suppresses Symptoms Associated with Arthritis 365
23.4.5   Curcumin Reduces the Incidence of Cholesterol
          Gallstone Formation 366
23.4.6   Curcumin Modulates Multiple Sclerosis 366
23.4.7   Curcumin Blocks the Replication of HIV 367
23.4.8   Curcumin Affects Alzheimer's Disease 367
23.4.9   Curcumin Protects against Cataract Formation 368
23.4.10   Curcumin Protects from Drug-Induced Myocardial Toxicity 368
23.4.11   Curcumin Protects from Alcohol-Induced Liver Injury 368
23.4.12   Curcumin Protects from Drug-Induced Lung Injury 369
23.4.13   Curcumin Prevents Adriamycin-Induced Nephrotoxicity 371
23.4.14   Curcumin Protects from Scarring 371
23.4.15   Curcumin Protects from Inflammatory Bowel Disease..371
23.4.16   Curcumin Enhances the Immunosuppressive Activity
          of Cyclosporine 372
23.4.17   Curcumin Protects against Various Forms of Stress 372
23.4.18   Curcumin Protects against Endotoxin Shock 372
23.4.19   Curcumin Protects against Pancreatitis 372
23.4.20   Curcumin Inhibits Multidrug Resistance (MDR)..372
23.5 Curcumin Metabolism 373
23.6 Clinical Experience with Curcumin 374
23.7 Curcumin Analogs...376
23.8 Sources of Curcumin 377
23.9 Conclusion 378
Acknowledgment 379
References 379

23.1 INTRODUCTION

Curcuma longa or turmeric is a tropical plant native to southern and southeastern tropical Asia. A perennial herb belonging to the ginger family, turmeric measures up to 1 m high with a short stem and tufted leaves (Figure 23.1A). The parts used are the rhizomes. Perhaps the most active component in turmeric is curcumin, which may make up 2 to 5% of the total spice in turmeric (Figure 23.1B). Curcumin is a diferuloylmethane present in extracts of the plant. Curcuminoids are responsible for the yellow color of turmeric and curry powder. They are derived from turmeric by ethanol extraction. The pure orange-yellow, crystalline powder is insoluble in water. The structure of curcumin ($C_{21}H_{20}O_6$) was first described in 1815 by Vogel and Pellatier and in 1910 was shown to be diferuloylmethane by Lampe et al.[1] Chemical synthesis in 1913 confirmed its identity.[2]

Turmeric is widely consumed in the countries of its origin for a variety of uses, including as a dietary spice, a dietary pigment, and an Indian folk medicine for the treatment of various illnesses. It is used in the textile and pharmaceutical industries[3] and in Hindu religious ceremonies in one form or another. Current traditional Indian medicine uses it for biliary disorders, anorexia, cough, diabetic wounds, hepatic disorders, rheumatism, and sinusitis.[4] The old Hindu texts have described it as an aromatic stimulant and carminative.[5] Powder of turmeric mixed with slaked lime is a household remedy for the treatment of sprains and swelling caused by injury, applied locally over the affected area. In some parts of India, the powder is taken orally for the treatment of sore throat. This nonnutritive phytochemical is pharmacologically safe, considering that it has been consumed as a dietary spice, at doses up to 100 mg/day, for centuries.[6] Recent phase I clinical trials indicate

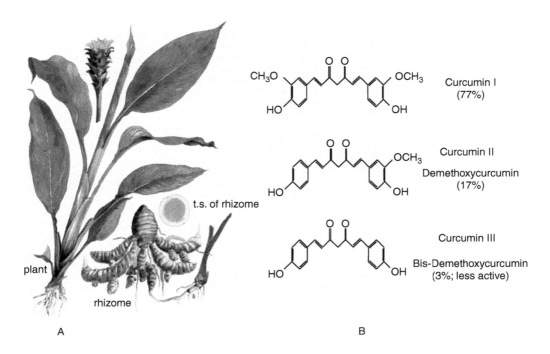

FIGURE 23.1 The plant *Curcuma longa* (panel A), from which curcumin is derived, and its structure (panel B).

that people can tolerate a dose as high as 8 g/day.[7] In the U.S., curcumin is used as a coloring agent in cheese, spices, mustard, cereals, pickles, potato flakes, soups, ice-creams, and yogurts (www.kalsec.com).

Curcumin is not water-soluble, but it is soluble in ethanol or in dimethylsulfoxide. The degradation kinetics of curcumin under various pH conditions and the stability of curcumin in physiological matrices have been established.[8] When curcumin was incubated in $0.1M$ phosphate buffer and serum-free medium (pH 7.2 at 37°C), about 90% decomposed within 30 min. A series of pH conditions ranging from 3 to 10 were tested, and the results showed that decomposition was pH-dependent and occurred faster at neutral-basic conditions. It is more stable in cell culture medium containing 10% fetal calf serum and in human blood. Less than 20% of curcumin decomposed within 1 h, and after incubation for 8 h, about 50% of curcumin still remained. Trans-6-(4'-hydroxy-3'-methoxyphenyl)-2,4-dioxo-5-hexenal was predicted to be the major degradation product, and vanillin, ferulic acid, and feruloyl methane were identified as minor degradation products. The amount of vanillin increased with incubation time.

Numerous studies have indicated that curcumin has antioxidant and anti-inflammatory properties. A Medline search revealed over 1000 publications describing various activities of this polyphenol. The following sections describe some of its major biological and clinical effects.

23.2 ANTICANCER PROPERTIES OF CURCUMIN

23.2.1 CURCUMIN INHIBITS TUMORIGENESIS

Numerous reports suggest that curcumin has chemopreventive and chemotherapeutic effects (Figure 23.2). Its anticancer potential in various systems was recently reviewed by our laboratory.[9] Curcumin blocks tumor initiation induced by benzo[a]pyrene and 7,12dimethylbenz[a]anthracene,[10]

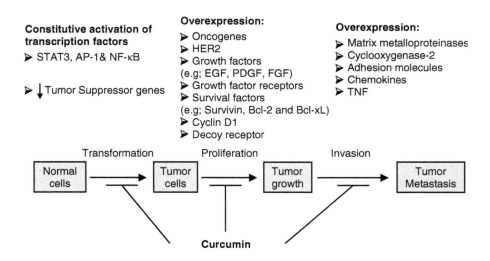

FIGURE 23.2 Various steps involved in tumorigenesis and metastasis and their suppression by curcumin.

and it suppresses phorbol ester-induced tumor promotion.[11,12] *In vivo,* curcumin was found to suppress carcinogenesis of the skin,[12–15] the forestomach,[16,17] the colon,[18–20] and the liver[21] in mice. Curcumin also suppresses mammary carcinogenesis.[22–24]

23.2.2 CURCUMIN EXHIBITS ANTIPROLIFERATIVE EFFECTS AGAINST CANCER CELLS

Compounds that block or suppress the proliferation of tumor cells have potential as anticancer agents. Curcumin has been shown to inhibit the proliferation of a wide variety of tumor cells, including B-cell and T-cell leukemia,[25–28] colon carcinoma,[29] and epidermoid carcinoma cells.[30] It has also been shown to suppress the proliferation of various breast carcinoma cell lines in culture.[31–33] We showed that the growth of the breast tumor cell lines BT20, SKBR3, MCF-7, T47D, and ZR75-1 is completely inhibited by curcumin, as indicated by MTT dye uptake, [³H] thymidine incorporation, and clonogenic assay.[31] We also showed that curcumin can overcome Adriamycin resistance in MCF-7 cells.[31] Recently, we have shown that curcumin can activate caspase-8, which leads to cleavage of Bid, thus resulting in sequential release of mitochondrial cytochrome C and activation of caspase-9 and caspase-3.[34] More recently, we have demonstrated that curcumin can suppress the proliferation of multiple myeloma cells.[35] Woo et al.[36] have demonstrated that curcumin can cause cell damage by inactivating the Akt-related cell survival pathway and release of cytochrome c, providing a new mechanism for curcumin-induced cytotoxicity.

Zheng et al.[37] explored the apoptosis-inducing effects of curcumin in human ovarian tumor A2780 cells. They found that curcumin could significantly inhibit the growth of ovarian cancer cells by inducing apoptosis through up-regulation of caspase-3 and down-regulation of expression of NF-κB. Studies have also been performed to examine the synergy of curcumin with other antiproliferative agents. Deeb et al.[38] investigated whether curcumin and TNF-related apoptosis-inducing ligand (TRAIL) cooperatively interact to promote death of LNCaP cells. At concentrations at which neither of the two agents alone produced significant cytotoxicity in LNCaP cells, cell death was markedly enhanced (two- to three-fold) if tumor cells were treated with curcumin and TRAIL together. The combined curcumin and TRAIL treatment increased the number of hypodiploid cells and induced DNA fragmentation in LNCaP cells. The combined treatment induced cleavage of procaspase-3, procaspase-8, and procaspase-9, truncation of BID, and release

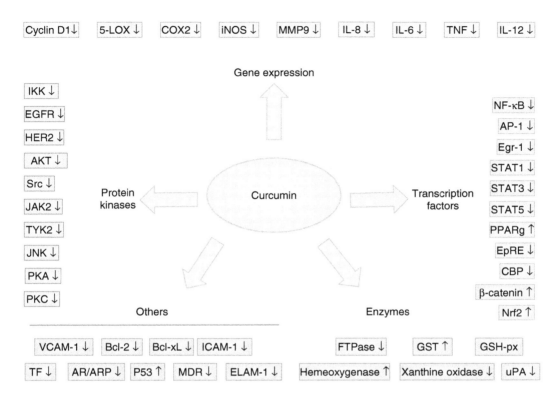

FIGURE 23.3 Molecular targets shown to be regulated by curcumin.

of cytochrome c from the mitochondria, indicating that both the extrinsic (receptor mediated) and intrinsic (chemical induced) pathways of apoptosis are triggered in prostate cancer cells treated with a combination of curcumin and TRAIL. These results define a potential use of curcumin to sensitize prostate cancer cells for TRAIL-mediated immunotherapy.

Chan et al.[39] demonstrated that curcumin increased the sensitivity of ovarian cancer cells (CAOV3 and SKOV3) to cisplatin. The effect was obtained both when the compound was added simultaneously with cisplatin and when it was added 24 h before. Curcumin inhibited the production of interleukin 6 (IL-6) in these cell lines (Figure 23.3), suggesting that one of the mechanisms for synergy between cisplatin and curcumin involved reducing the autologous production of IL-6. However, the synergy was also observed in the low IL-6 producer, SKOV3, indicating that additional targets were responsible. The down-regulation of IL-6 by curcumin was also noted in multiple myeloma cells.[35]

23.2.3 CURCUMIN DOWN-REGULATES THE ACTIVITY OF EPIDERMAL GROWTH FACTOR RECEPTOR (EGFR) AND EXPRESSION OF HER2/NEU

HER2/neu and epithelial growth factor receptor (EGFR) activity represent one possible mechanism by which curcumin suppresses the growth of breast cancer cells. Almost 30% of the breast cancer cases have been shown to overexpress the HER2/neu protooncogene,[40] and both HER2 and EGF receptors stimulate proliferation of breast cancer cells. Overexpression of these two proteins correlates with progression of human breast cancer and poor patient prognosis.[40] Curcumin has been shown to down-regulate the activity of EGFR and HER2/neu[30,41] and to deplete the cells of HER2/neu protein.[42] Additionally, we have recently found that curcumin can down-regulate bcl-2 expression, which may contribute to its antiproliferative activity.[43]

Like geldanamycin, curcumin has been shown to provoke the intracellular degradation of HER2.[44] HER2 mutations, however, limit the capacity of geldanamycin to disrupt the tyrosine kinase activity of HER2. Thus these HER2 mutants are resistant to geldanamycin-induced degradation, but they maintain their sensitivity to curcumin through ErbB-2 degradation.

23.2.4 CURCUMIN DOWN-REGULATES THE ACTIVATION OF NUCLEAR FACTOR-κB (NF-κB)

Curcumin may also operate through suppression of NF-κB activation (Figure 23.3). NF-κB is a nuclear transcription factor required for the expression of genes involved in cell proliferation, cell invasion, metastasis, angiogenesis, and resistance to chemotherapy.[45] This factor is activated in response to inflammatory stimuli, carcinogens, tumor promoters, and hypoxia, which is frequently encountered in tumor tissues.[46] Several groups, including ours, have shown that activated NF-κB suppresses apoptosis in a wide variety of tumor cells,[47–49] and it has been implicated in chemoresistance.[47] We have shown that cells that overexpress NF-κB are resistant to paclitaxel-induced apoptosis.[50] Furthermore, the constitutively active form of NF-κB has been reported in human breast cancer cell lines in culture,[51] carcinogen-induced mouse mammary tumors,[52] and biopsies from patients with breast cancer.[53] Our laboratory has shown that various tumor promoters, including phorbol ester, TNF, and H_2O_2, activate NF-κB and that curcumin down-regulates the activation.[54] Subsequently, others showed that curcumin-induced down-regulation of NF-κB is mediated through suppression of IκBα kinase activation.[55,56] Recently, Shishodia et al.[57] have shown that curcumin down-regulated cigarette smoke-induced NF-κB activation through inhibition of IκBα kinase in human lung epithelial cells. This led to the down-regulation of cyclin D1, cyclooxygenase 1 (COX-2), and matrix metalloproteinase 9 (MMP9) by curcumin. Philip et al.[58] have recently reported that curcumin down-regulates osteopontin (OPN)-induced NF-κB-mediated promatrix metalloproteinase-2 activation through IκBα/IKK signaling.

23.2.5 CURCUMIN DOWN-REGULATES THE ACTIVATION OF STAT3 PATHWAY

Numerous reports suggest that IL-6 promotes survival and proliferation of various tumors, including multiple myeloma (MM) cells, through the phosphorylation of a cell signaling protein, signal transducers, and activators of transcription (STAT3). Thus agents that suppress STAT3 phosphorylation have potential for the treatment of MM. Bharti et al.[59] demonstrated that curcumin inhibited IL-6-induced STAT3 phosphorylation and consequent STAT3 nuclear translocation. Curcumin had no effect on STAT5 phosphorylation but inhibited the interferon α-induced STAT1 phosphorylation. The constitutive phosphorylation of STAT3 found in certain MM cells was also abrogated by treatment with curcumin. Curcumin-induced inhibition of STAT3 phosphorylation was reversible. Compared with AG490, a well-characterized JAK2 inhibitor, curcumin was a more rapid (30 min vs. 8 h) and more potent (10 μM vs. 100 μM) inhibitor of STAT3 phosphorylation. Similarly, at the dose of curcumin that completely suppressed proliferation of MM cells, AG490 had no effect. In contrast, the STAT3 inhibitor peptide that can inhibit the STAT3 phosphorylation mediated by Src blocked the constitutive phosphorylation of STAT3 and also suppressed the growth of myeloma cells. TNF-α and lymphotoxin (LT) also induced the proliferation of MM cells, but through a mechanism independent of STAT3 phosphorylation. In addition, dexamethasone-resistant MM cells were found to be sensitive to curcumin. Overall, these results demonstrated that curcumin was a potent inhibitor of STAT3 phosphorylation, and this plays a role in curcumin's suppression of proliferation of MM.

Li et al.[60] showed that curcumin suppressed oncostatin-M-stimulated STAT1 phosphorylation, DNA-binding activity of STAT1, and c-Jun N-terminal kinase activation without affecting Janus kinase 1 (JAK1), JAK2, JAK3, ERK1/2, and p38 phosphorylation. Curcumin also inhibited OSM-induced MMP1, MMP3, MMP13, and TIMP3 gene expression.

Natarajan et al.[61] showed that treatment of activated T cells with curcumin inhibited IL-12-induced tyrosine phosphorylation of Janus kinase 2, tyrosine kinase 2, and STAT3 and STAT4 transcription factors. The inhibition of the Janus kinase-STAT pathway by curcumin resulted in a decrease in IL-12-induced T-cell proliferation and Th1 differentiation.

23.2.6 CURCUMIN ACTIVATES PEROXISOME PROLIFERATOR-ACTIVATED RECEPTOR-γ (PPAR-γ)

Activation of PPAR-γ inhibits the proliferation of nonadipocytes. The level of PPAR-γ is dramatically diminished along with activation of hepatic stellate cells (HSC). Xu et al.[62] demonstrated that curcumin dramatically induced the gene expression of PPAR-γ and activated PPAR-γ in activated HSC (Figure 23.3). Blocking its trans-activating activity by a PPAR-γ antagonist markedly decreased the effects of curcumin on inhibition of cell proliferation.

23.2.7 CURCUMIN DOWN-REGULATES THE ACTIVATION OF ACTIVATOR PROTEIN-1 (AP-1) AND C-JUN N-TERMINAL KINASE (JNK)

AP-1 is another transcription factor that has been closely linked with proliferation and transformation of tumor cells.[63] The activation of AP-1 requires the phosphorylation of c-jun through activation of stress-activated kinase JNK.[64] The activation of JNK is also involved in cellular transformation.[65] Curcumin has been shown to inhibit the activation of AP-1 induced by tumor promoters[66] and JNK activation induced by carcinogens.[67]

Dickinson et al.[68] have demonstrated that the beneficial effects elicited by curcumin appear to be due to changes in the pool of transcription factors that compose EpRE and AP-1 complexes, affecting gene expression of glutamate-cysteine ligase and other phase II enzymes. Squires et al.[69] have demonstrated that curcumin suppresses the proliferation of tumor cells through inhibition of Akt/PKB activation.

23.2.8 CURCUMIN SUPPRESSES THE INDUCTION OF ADHESION MOLECULES

The expression of various cell surface adhesion molecules such as intercellular cell adhesion molecule-1, vascular cell adhesion molecule-1, and endothelial leukocyte adhesion molecule-1 on endothelial cells is absolutely critical for tumor metastasis.[70] The expression of these molecules is in part regulated by nuclear factor NF-κB.[71] We have shown that treatment of endothelial cells with curcumin blocks the cell surface expression of adhesion molecules, and this accompanies the suppression of tumor-cell adhesion to endothelial cells.[72] We also have demonstrated that down-regulation of these adhesion molecules is mediated through the down-regulation of NF-κB activation.[72] Jaiswal et al.[73] showed that curcumin treatment causes p53- and p21-independent G(2)/M phase arrest and apoptosis in colon cancer cell lines. Their results suggest that curcumin treatment impairs both Wnt signaling and cell-cell adhesion pathways, resulting in G(2)/M-phase arrest and apoptosis in HCT-116 cells.

23.2.9 CURCUMIN DOWN-REGULATES CYCLOOXYGENASE-2 (COX-2) EXPRESSION

Overexpression of COX-2 has been shown to be associated with a wide variety of cancers, including colon,[74] lung,[75] and breast[76] cancers. The role of COX-2 in suppression of apoptosis and tumor cell proliferation has been demonstrated.[77] Furthermore, Celebrex, a specific inhibitor of COX-2, has been shown to suppress mammary carcinogenesis in animals.[78] Several groups have shown that curcumin down-regulates the expression of COX-2 protein in different tumor cells,[29,56] most likely through the down-regulation of NF-κB activation,[56] which is needed for COX-2 expression.

23.2.10 CURCUMIN INHIBITS ANGIOGENESIS

For most solid tumors, including breast cancer, angiogenesis (blood vessel formation) is essential for tumor growth and metastasis.[79] The precise mechanism that leads to angiogenesis is not fully understood, but growth factors that cause proliferation of endothelial cells have been shown to play a critical role in this process. Curcumin has been shown to suppress the proliferation of human vascular endothelial cells *in vitro*[80] and abrogate the fibroblast growth-factor-2-induced angiogenic response *in vivo*,[81] thus suggesting that curcumin is also an antiangiogenic factor. Indeed curcumin has been shown to suppress angiogenesis *in vivo*.[82]

To elucidate possible mechanisms of antiangiogenic activity by curcumin, Park et al.[83] performed cDNA microarray analysis and found that curcumin modulated cell-cycle-related gene expression. Specifically, curcumin induced G0/G1- and G2/M-phase cell-cycle arrest; up-regulated CDKIs, p21WAF1/CIP1, p27KIP1, and p53; and slightly down-regulated cyclin D1 and cdc2 in ECV304 cells. The up-regulation of CDKIs by curcumin played a critical role in the regulation of cell-cycle distribution in these cells, which may underlie the antiangiogenic activity of curcumin.

23.2.11 CURCUMIN SUPPRESSES THE EXPRESSION OF MMP9 AND INDUCIBLE NITRIC OXIDE SYNTHASE (iNOS)

The MMPs make up a family of proteases that play a critical role in tumor metastasis.[84] One of them, MMP9, has been shown to be regulated by NF-κB activation, and curcumin has been shown to suppress its expression.[85] Curcumin has also been demonstrated to down-regulate iNOS expression, also regulated by NF-κB and involved in tumor metastasis.[86] These observations suggest that curcumin must have antimetastatic activity. Indeed, there is a report suggesting that curcumin inhibits tumor metastasis.[87]

23.2.12 CURCUMIN DOWN-REGULATES CYCLIN D1 EXPRESSION

Cyclin D1, a component subunit of cyclin-dependent kinase Cdk4 and Cdk6, is a rate-limiting factor in progression of cells through the first gap (G1) phase of the cell cycle.[88] Cyclin D1 has been shown to be overexpressed in many cancers including breast, esophagus, head and neck, and prostate.[89–94] It is possible that the antiproliferative effects of curcumin are due to inhibition of cyclin D1 expression. We found that curcumin can indeed down-regulate cyclin D1 expression,[35,43,95] and this down-regulation occurred at the transcriptional and posttranscriptional level.

23.2.13 CURCUMIN IS CHEMOPREVENTIVE

Several studies suggest that curcumin has chemopreventive potential. Huang et al.[96] found that topical application of curcumin inhibits tumor initiation by benzo[a]pyrine (BaP) and tumor promotion by TPA in mouse skin. Dietary curcumin (commercial grade) inhibits BaP-induced forestomach carcinogenesis, N-ethyl-N'-nitro-N-nitrosoguanidine (ENNG)-induced duodenal carcinogenesis, and azoxymethane (AOM)-induced colon carcinogenesis. Dietary curcumin had little or no effect on 4-(methylnitrosamino)-1-(3-pyridyl)-1-butanone (NNK)-induced lung carcinogenesis and 7,12-dimethylbenz[a]anthracene (DMBA)-induced breast carcinogenesis in mice. Poor circulating bioavailability of curcumin may account for the lack of lung and breast carcinogenesis inhibition.

Perkins et al.[97] showed that curcumin prevents the development of adenomas in the intestinal tract of the C57Bl/6J Min/+ mouse, a model of human familial adenomatous polyposis coli (APC). To aid in the rational development of curcumin as a colorectal cancer-preventive agent, the group explored the link between its chemopreventive potency in the Min/+ mouse and levels of drug and metabolites in target tissue and plasma. Mice received dietary curcumin for 15 weeks, after which adenomas were enumerated. Levels of curcumin and metabolites were determined by high-performance liquid chromatography (HPLC) in plasma, tissues, and feces of mice after either long-term ingestion of dietary curcumin or a single dose of [^{14}C] curcumin (100 mg/kg) via the intraperitoneal (i.p.) route. Whereas curcumin at 0.1% in the diet was without effect, at 0.2 and 0.5%, it reduced adenoma multiplicity by 39 and 40%, respectively, compared with untreated mice. Hematocrit values in untreated Min/+ mice were drastically reduced compared with those in wild-type C57Bl/6J mice. Dietary curcumin partially restored the suppressed hematocrit. Traces of curcumin were detected in the plasma. Its concentration in the mucosa of the small intestine, between 39 and 240 nmol/g of tissue, reflected differences in dietary concentration. [^{14}C] curcumin disappeared rapidly from tissues and plasma within 2 to 8 h after dosing. Curcumin may be useful in the chemoprevention of human intestinal malignancies related to Apc mutations. The comparison of dose, resulting curcumin levels in the intestinal tract, and chemopreventive potency suggested tentatively that a daily dose of 1.6 g of curcumin is required for efficacy in humans. A clear advantage of curcumin over nonsteroidal anti-inflammatory drugs is its ability to decrease intestinal bleeding linked to adenoma maturation.

Helicobacter pylori is a Group 1 carcinogen and is associated with the development of gastric and colon cancer. Mahady et al.[98] have demonstrated that curcumin inhibits the growth of *H. pylori* cagA+ mouse strains *in vitro*, and this may be one of the mechanisms by which curcumin exerts its chemopreventive effects.

In another study, Perkins and coworkers[97] found that the nonsteroidal anti-inflammatory drug aspirin and curcumin retard adenoma formation when administered long-term to Apc(Min/+) mice, a model of human familial APC. Aspirin administered to Apc (Min/+) mice postweaning was not effective, though curcumin given postweaning was active. Here the hypothesis was tested that dietary aspirin (0.05%) or curcumin (0.2%) prevents or delays adenoma formation in offspring when administered to Apc (Min/+) mothers and up to the end of weaning. Whereas curcumin was without effect when administered afterward, aspirin reduced the numbers of intestinal adenomas by 21%. When aspirin given up to the end of weaning was combined with curcumin administered from the end of weaning for the rest of the animals' lifetime, intestinal adenoma numbers were reduced by 38%. The combination was not superior to intervention postweaning with curcumin alone. These results show that aspirin exerts chemopreventive activity in the Apc(Min/+) mouse during tumor initiation/early promotion, while curcumin is efficacious when given at a later stage of carcinogenic progression. Thus, the results suggest that in this mouse model, aspirin and curcumin act during different "windows" of neoplastic development.

Recently, Van der Logt et al.[99] demonstrated that curcumin exerted its anticarcinogenic effects in gastrointestinal cancers through the induction of UDP-glucuronosyltransferase enzymes.

23.2.14 CURCUMIN INHIBITS TUMOR GROWTH AND METASTASIS IN ANIMALS

Kuttan et al.[100] examined the anticancer potential of curcumin *in vitro* using tissue culture methods and *in vivo* in mice using Dalton's lymphoma cells grown as ascites. Initial experiments indicated that curcumin reduced the development of animal tumors. They encapsulated curcumin (5 mg/ml) into neutral and unilamellar liposomes prepared by sonication of phosphatidylcholine and cholesterol. An aliquot of liposomes (50 mg/kg) was given i.p. to mice the day after giving the Dalton's lymphoma cells and continued for 10 days. After 30 days and 60 days, surviving animals were counted. When curcumin was used in liposomal formulations at concentration of 1 mg/animal, all animals survived 30 days, and only two of the animals developed tumors and died before 60 days.

Busquets[101] showed that systemic administration of curcumin (20 μg/kg body weight) for 6 consecutive days to rats bearing the highly cachectic Yoshida AH-130 ascites hepatoma resulted in an important inhibition of tumor growth (31% of total cell number). Interestingly, curcumin was also able to reduce by 24% *in vitro* tumor cell content at concentrations as low as 0.5 μ*M* without promoting any apoptotic events. Although systemic administration of curcumin has previously been shown to facilitate muscle regeneration, administration of the compound to tumor-bearing rats did not result in any changes in muscle wasting, when compared with the untreated tumor-bearing animals. Indeed, both the weight and protein content of the gastrocnemius muscle significantly decreased as a result of tumor growth, and curcumin was unable to reverse this tendency. It was concluded that curcumin, in spite of having clear antitumoral effects, has little potential as an anticachectic drug in the tumor model used in the study.

Menon et al.[102] reported curcumin-induced inhibition of B16F-10 melanoma lung metastasis in mice. Oral administration of curcumin at concentrations of 200 nmol/kg body weight reduced the number of lung tumor nodules by 80%. The life span of the animals treated with curcumin was increased by 143.85%.[102] Moreover, lung collagen hydroxyproline and serum sialic acid levels were significantly lower in treated animals than in the untreated controls. Curcumin treatment (10 μg/ml) significantly inhibited the invasion of B16F-10 melanoma cells across the collagen matrix of a Boyden chamber. Gelatin zymographic analysis of the trypsin-activated B16F-10 melanoma cells' sonicate revealed no metalloproteinase activity. Curcumin treatment did not inhibit the motility of B16F-10 melanoma cells across a polycarbonate filter *in vitro*. These findings suggest that curcumin inhibits the invasion of B16F-10 melanoma cells by inhibition of MMPs, thereby inhibiting lung metastasis.

Curcumin decreases the proliferative potential and increases apoptotic potential of both androgen-dependent and androgen-independent prostate cancer cells *in vitro*, largely by modulating the apoptosis-suppressor proteins and by interfering with the growth factor receptor signaling pathways as exemplified by the EGF receptor. To extend these observations, Dorai et al.[103] investigated the anticancer potential of curcumin in a nude mouse prostate cancer model. The androgen-dependent LNCaP prostate cancer cells were grown, mixed with Matrigel, and injected subcutaneously. The experimental group received a synthetic diet containing 2% curcumin for up to 6 weeks. At the end point, mice were killed, and sections taken from the excised tumors were evaluated for pathology, cell proliferation, apoptosis, and vascularity. Results showed that curcumin induced a marked decrease in the extent of cell proliferation, as measured by the Brdu (bromodeoxyuridine) incorporation assay, and a significant increase in the extent of apoptosis, as measured by an *in situ* cell death assay. Moreover, a significant decrease in the microvessel density, as measured by CD31 antigen staining, was also seen. It was concluded that curcumin was a potentially therapeutic anticancer agent, as it significantly inhibited prostate cancer growth, as exemplified by LNCaP *in vivo*, and it had the potential to prevent the progression of this cancer to its hormone refractory state.

23.2.15 CURCUMIN INHIBITS ANDROGEN RECEPTORS AND AR-RELATED COFACTORS

Nakamura et al.[104] have evaluated the effects of curcumin in cell growth, activation of signal transduction, and transforming activities of both androgen-dependent and -independent cell lines. The prostate cancer cell lines LNCaP and PC-3 were treated with curcumin, and its effects on signal transduction and expression of androgen receptor (AR) and AR-related cofactors were analyzed. Their results showed that curcumin down-regulates transactivation and expression of AR, AP-1, NF-κB, and CREB (cAMP response element-binding protein)-binding protein (CBP). It also inhibited the transforming activities of both cell lines, as evidenced by reduced colony forming ability in soft agar. These studies suggest that curcumin has a potential therapeutic effect on prostate cancer cells through down-regulation of AR and AR-related cofactors, AP-1, NF-κB, and CBP.

Overall, numerous mechanisms as indicated above could account for the tumor-suppressive effects of curcumin (Figure 23.3). Curcumin also has modulatory effects in diseases besides cancer (Figure 23.4). These effects are described in Section 23.3 and Section 23.4.

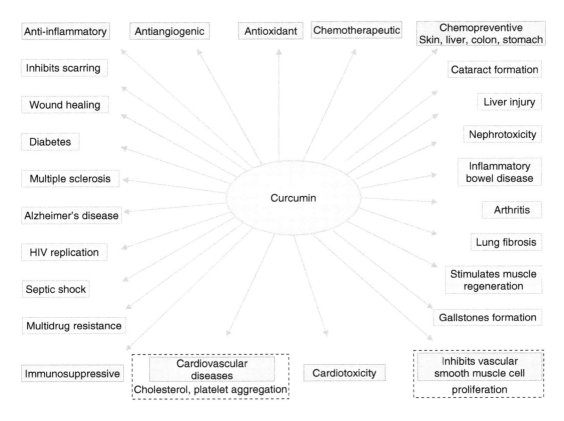

FIGURE 23.4 Effect of curcumin on various diseases.

23.3 EFFECT OF CURCUMIN ON ATHEROSCLEROSIS AND MYOCARDIAL INFARCTION

23.3.1 CURCUMIN INHIBITS THE PROLIFERATION OF VASCULAR SMOOTH MUSCLE CELLS

The proliferation of peripheral blood mononuclear cells (PBMC) and vascular smooth muscle cells (VSMC) is a hallmark of atherosclerosis. Huang et al.[10] investigated the effects of curcumin on the proliferation of PBMC and VSMC from the uptake of [^{3}H]thymidine. Curcumin dose dependently inhibited the response to phytohemaglutinin and the mixed lymphocyte reaction in human PBMC at dose ranges of 1 to 30 μM and 3 to 30 μM, respectively. Curcumin (1 to 100 μM) dose-dependently inhibited the proliferation of rabbit VSMC stimulated by fetal calf serum. Curcumin had a greater inhibitory effect on platelet-derived growth factor (PDGF)-stimulated proliferation than on serum-stimulated proliferation. Analogs of curcumin (cinnamic acid, coumaric acid, and ferulic acid) were much less effective than curcumin as inhibitors of serum-induced smooth muscle cell proliferation. This suggested that curcumin may be useful for the prevention of the pathological changes associated with atherosclerosis and restenosis.

Chen and Huang[67] examined the possible mechanisms underlying curcumin's antiproliferative and apoptotic effects using the rat VSMC cell line A7r5. Curcumin (1 to 100 μM) inhibited serum-stimulated [^{3}H]thymidine incorporation of both A7r5 cells and rabbit VSMC. Cell viability, as determined by the trypan blue dye exclusion method, was unaffected by curcumin at the concentration range 1 to 10 μM in A7r5 cells. However, the number of viable cells after 100-μM curcumin treatment was less than the basal value. Following curcumin (1 to 100 μM) treatment, cell cycle analysis revealed a G0/G1 arrest and a reduction in the percentage of cells in S phase. Curcumin at 100 μM also induced cell apoptosis, as demonstrated by hematoxylin-eosin staining, TdT-mediated dUTP nick end labeling, DNA laddering, cell shrinkage, chromatin condensation, and DNA fragmentation. The membranous protein tyrosine kinase activity stimulated by serum in A7r5 cells was significantly reduced by curcumin (10 to 100 μM). On the other hand, phorbol myristate acetate-stimulated cytosolic protein kinase C (PKC) activity was reduced by 100 μM curcumin. The levels of c-myc mRNA and bcl-2 mRNA were significantly reduced by curcumin but had little effect on the p53 mRNA level. These results demonstrate that curcumin inhibited cell proliferation, arrested cell cycle progression, and induced cell apoptosis in VSMC. These results may explain how curcumin prevents the pathological changes of atherosclerosis and postangioplasty restenosis.

23.3.2 CURCUMIN LOWERS SERUM CHOLESTEROL LEVELS

Numerous studies suggest that curcumin lowers serum cholesterol levels.[105–111] Soudamini et al. [108] investigated the effect of oral administration of curcumin on serum cholesterol levels and on lipid peroxidation in the liver, lung, kidney, and brain of mice treated with carbon tetrachloride, paraquat, and cyclophosphamide. Oral administration of curcumin significantly lowered the increased peroxidation of lipids in these tissues produced by these chemicals. Administration of curcumin also significantly lowered the serum and tissue cholesterol levels in these animals, indicating that the use of curcumin helps in conditions associated with peroxide-induced injury such as liver damage and arterial diseases. Soni and Kuttan examined the effect of curcumin administration in reducing the serum levels of cholesterol and lipid peroxides in 10 healthy human volunteers receiving 500 mg of curcumin per day for 7 days.[112] A significant decrease in the level of serum lipid peroxides (33%), an increase in high-density lipoproteins (HDL) cholesterol (29%), and a decrease in total serum cholesterol (12%) were noted. Because curcumin reduced serum lipid peroxides and serum cholesterol, the study of curcumin as a chemopreventive substance against arterial diseases was suggested.

Curcuma xanthorrhiza Roxb., a medicinal plant used in Indonesia (known as "temu lawak" or "Javanese turmeric"), has been shown to exert diverse physiological effect. However, little attention has been paid to its effect on lipid metabolism. Yasni et al.[113] investigated the effects of *C. xanthorrhiza* on serum and liver lipids, serum HDL cholesterol, apolipoprotein, and liver lipogenic enzymes. In rats given a cholesterol-free diet, *C. xanthorrhiza* decreased the concentrations of serum triglycerides, phospholipids, and liver cholesterol and increased the concentrations of serum HDL cholesterol and apolipoproteins. The activity of liver fatty acid synthase, but not glycerophosphate dehydrogenase, was decreased by the medicinal plant. In rats on a high-cholesterol diet, *C. xanthorrhiza* did not suppress the elevation of serum cholesterol, although it did decrease liver cholesterol. Curcuminoids prepared from *C. xanthorrhiza* had no significant effects on the serum and liver lipids. These studies, therefore, indicate that *C. xanthorrhiza* contains an active principle other than the curcuminoids that can modify the metabolism of lipids and lipoproteins.

In later studies, Yasni et al.[114] identified the major component (approx. 65%) of the essential oil as alpha-curcumene. Addition of essential oils (0.02%), prepared by steam distillation, to a purified diet lowered hepatic triglyceride concentration without influencing serum triglyceride levels, whereas addition of the hexane-soluble fraction (0.5%) lowered the concentration of serum and hepatic triglycerides. Rats fed the essential oil and hexane-soluble fraction had lower hepatic fatty acid synthase activity. The fraction containing α-curcumene, prepared from the hexane-soluble fraction by silica gel column chromatography, suppressed the synthesis of fatty acids from [^{14}C] acetate in primary cultured rat hepatocytes.

Skrzypczak-Jankun et al.[115] showed the three-dimensional structural data and explained how curcumin interacts with the fatty-acid-metabolizing enzyme soybean lipoxygenase. Curcumin binds to lipoxygenase in a noncompetitive manner. Trapped in that complex, it undergoes photodegradation in response to x-rays, but utilizes enzyme catalytic ability to form the peroxy complex Enz-Fe-O-O-R as 4-hydroperoxy-2-methoxy-phenol, which later is transformed into 2-methoxycyclohexa-2,5-diene-1,4-dione. However, when Rukkumani et al.[116] compared the effects of curcumin and photo-irradiated curcumin on alcohol- and polyunsaturated fatty acid-induced hyperlipidemia, they found that photo-irradiated curcumin was more effective than curcumin in treating the above pathological conditions.

23.3.3 Curcumin Inhibits LDL Oxidation

The oxidation of low-density lipoproteins (LDL) plays an important role in the development of atherosclerosis. Atherosclerosis is characterized by oxidative damage, which affects lipoproteins, the walls of blood vessels, and subcellular membranes. Several studies suggest that curcumin inhibits oxidation of LDL.[117–120] Naidu and Thippeswamy[120] examined the effect of curcumin on copper-ion-induced lipid peroxidation of human LDL by measuring the formation of thiobarbituric acid reactive substance (TBARS) and relative electrophoretic mobility of LDL on agarose gel. Curcumin inhibited the formation of TBARS effectively throughout the incubation period of 12 h and decreased the relative electrophoretic mobility of LDL. Curcumin at 10 μM produced 40 to 85% inhibition of LDL oxidation. The inhibitory effect of curcumin was comparable with that of BHA but more potent than ascorbic acid. Further, curcumin significantly inhibited both initiation and propagation phases of LDL oxidation.

Ramirez-Tortosa et al.[118] evaluated the effect of curcumin on LDL oxidation susceptibility and plasma lipids in atherosclerotic rabbits. A total of 18 rabbits were fed for 7 weeks on a diet containing 95.7% standard chow, 3% lard, and 1.3% cholesterol to induce atherosclerosis. The rabbits were divided into groups, two of which were also orally treated with turmeric extract at doses of 1.66 (group A) and 3.2 (group B) mg/kg body weight. A third group (group C) acted as an untreated control. Plasma and LDL lipid composition, plasma alpha-tocopherol, plasma retinol, LDL TBARS, and LDL lipid hydroperoxides were assayed, and aortic atherosclerotic lesions were evaluated. The low but not the high dosage of turmeric extracts decreased the susceptibility of rabbit LDL to lipid peroxidation. Both doses produced lower levels of total plasma cholesterol than the control group. Moreover, the lower-dosage group had lower levels of cholesterol, phospholipids, and triglycerides than the group treated with the 3.2-mg dosage.

Quiles et al.[117] evaluated the antioxidant capacity of a *C. longa* extract on the lipid peroxidation of liver mitochondria and microsome membranes in atherosclerotic rabbits. Male rabbits fed a 3% (w/w) lard and 1.3% (w/w) cholesterol diet were randomly assigned to three groups. Two groups were treated with different dosages of a turmeric extract (A and B), and the third group (control) was treated with a curcumin-free solution. Basal and *in vitro* 2,2′-azobis(2-amidinopropane)dihydrochloride-induced hydroperoxide and TBARS production in liver mitochondria and microsomes were analyzed. Group A had the lowest concentration of mitochondrial hydroperoxides. In microsomes, the basal hydroperoxide levels were similar in all groups, but after the induction of oxidation, group C registered the highest value; TBARS production followed the same trend in mitochondria. These findings suggest that active compounds in curcuma extract may be protective against lipoperoxidation of subcellular membranes in a dosage-dependent manner.

Asai and Miyazawa[119] examined the effect of curcumin on lipid metabolism in rats fed a control, moderately high-fat diet (15 g soybean oil/100 g diet) and those given supplements of 0.2 g curcuminoids/100 g diet. Liver triacylglycerol and cholesterol concentrations were significantly lower in rats fed curcumin than in control rats. Plasma triacylglycerols in the very-low-density lipoproteins fraction were also lower in curcumin-fed rats than in control (P < 0.05). Hepatic acyl-CoA oxidase activity of the curcumin group was significantly higher than that of the control. Furthermore, epididymal adipose tissue weight was significantly reduced with curcuminoid intake in a dose-dependent manner. These results indicated that dietary curcuminoids have lipid-lowering potency *in vivo*, probably due to alterations in fatty acid metabolism.

23.3.4 CURCUMIN INHIBITS PLATELET AGGREGATION

Platelet aggregation contributes to the pathway resulting in atherosclerosis. There are reports suggesting that curcumin can inhibit platelet aggregation.[121–123] Srivastava et al.[122] examined the effect of curcumin on platelet aggregation and vascular prostacyclin synthesis. *In vitro* and *ex vivo* effects of curcumin and acetylsalicylic acid (ASA) on the synthesis of prostacyclin (PGI_2) and on platelet aggregation have been studied in rats. Both drugs inhibited adenosine diphosphate and epinephrine (adrenaline)- and collagen-induced platelet aggregation in monkey plasma. Pretreatment with ASA (25 to 100 mg/kg), but not curcumin (100 to 300 mg/kg), inhibited PGI_2 synthesis in rat aorta. In the *in vitro* system, curcumin also caused a slight increase in the synthesis of PGI_2, while ASA inhibited it. Curcumin may, therefore, be preferable in patients prone to vascular thrombosis and requiring antiarthritic therapy.

Srivastava et al. showed that curcumin inhibited platelet aggregation induced by arachidonate, adrenaline, and collagen.[123] This compound inhibited thromboxane B2 production from exogenous [^{14}C] arachidonate (AA) in washed platelets and concomitantly increased the formation of 12 lipoxygenase products. Moreover, curcumin inhibited the incorporation of [^{14}C] AA into platelet phospholipids and inhibited the deacylation of AA-labeled phospholipids (liberation of free AA) on stimulation with calcium ionophore A23187. Curcumin's anti-inflammatory properties may, in part, be explained by the compound's effects on eicosanoid biosynthesis.

23.3.5 CURCUMIN INHIBITS MYOCARDIAL INFARCTION

The effect of curcumin on myocardial infarction (MI) in the cat and the rat has been investigated.[124–126] Dikshit et al.[124] examined the prevention of ischemia-induced biochemical changes by curcumin in the cat heart. Myocardial ischemia was induced by the ligation of the left descending coronary artery. Curcumin (100 mg/kg, i.p.) was given 30 min before ligation. Cats were killed and hearts were removed 4 h after coronary artery ligation. Levels of glutathione (GSH), malonaldehyde (MDA), myeloperoxidase (MPO), superoxide dismutase (SOD), catalase, and lactate dehydrogenase (LDH) were estimated in the ischemic and nonischemic zones. Curcumin protected the animals against decrease in the heart rate and blood pressure following ischemia. In the ischemic zone, after 4 h of ligation, an increase in the level of MDA and activities of MPO and SOD (cytosolic fraction) were observed. Curcumin pretreatment prevented the ischemia-induced elevation in MDA contents and LDH release, but it did not affect the increase in MPO activity. Thus curcumin prevented ischemia-induced changes in the cat heart.

Nirmala and Puvanakrishnan[125] investigated the effect of curcumin on lysosomal hydrolases (β-glucuronidase, β-N-acetylglucosaminidase, cathepsin B, cathepsin D, and acid phosphatase) in serum and heart after isoproterenol (ISO)-induced MI. Rats treated with ISO (30 mg/100 g body weight) showed a significant increase in serum lysosomal hydrolase activities, which were found to decrease after curcumin treatment. ISO administration to rats resulted in decreased stability of the membranes, which was reflected by the lowered activity of cathepsin D in mitochondrial, lysosomal, and microsomal fractions. Curcumin treatment returned the activity levels almost to normal, showing that curcumin restored the normal function of the membrane. Histopathological studies of the infarcted rat heart also showed a decreased degree of necrosis after curcumin treatment. Nirmala and Puvanakrishnan[125] also examined the effect of curcumin on the biochemical changes induced by ISO administration in rats. ISO caused a decrease in body weight and an increase in heart weight, water content, and levels of serum marker enzymes, namely creatine kinase (CK), LDH, and LDH1 isozyme. It also produced electrocardiographic changes such as an increased heart rate, reduced R amplitude, and elevated ST. Curcumin at a concentration of 200 mg/kg, when administered orally, decreased serum enzyme levels, and the electrocardiographic changes were restored toward normalcy. MI was accompanied by the disintegration of membrane polyunsaturated fatty acids expressed by an increase in TBARS, a measure of lipid peroxides, and by the impairment of natural scavenging, characterized by a decrease in the levels of SOD, catalase, glutathione peroxidase, ceruloplasmin, α-tocopherol, GSH, and ascorbic acid. Oral pretreatment with curcumin 2 days before and during ISO administration decreased the effect of lipid peroxidation. It has a membrane-stabilizing action by inhibiting the release of β-glucuronidase from nuclei, mitochondria, lysosomes, and microsomes. Curcumin given before and during treatment decreased the severity of pathological changes and thus could have a protective effect against the damage caused by MI.

Nirmala et al.[126] showed that curcumin treatment modulates collagen metabolism in ISO-induced myocardial necrosis in rats. This study evaluated whether curcumin had any specific role in the synthesis and degradation of collagen in rat heart with myocardial necrosis induced by ISO. The effect of curcumin (200 mg/kg) was examined on ISO-induced myocardial necrosis and collagen metabolism. The incorporation of [^{14}C] proline into collagen was studied as an index of collagen synthesis. The heart-weight/body-weight ratio, heart RNA/DNA ratio, and protein increased significantly in ISO-treated animals. Curcumin given before and during treatment with ISO reversed these changes and attenuated the development of cardiac hypertrophy 2 weeks after the second dose of ISO. Increased fractional synthesis rate and enhanced degradation of newly synthesized collagen were observed in ISO-treated animals. Curcumin before and during treatment with ISO decreased the degree of degradation of the existing collagen matrix and collagen synthesis 2 weeks after the second dose of ISO. The observed effects could have been due to free-radical scavenging capacity and inhibition of lysosomal enzyme release by curcumin.

Enzymes of the SOD family are key regulators of cellular oxidant stress caused by ischemia-reperfusion. In particular, the mitochondrial-associated MnSOD enzyme has been implicated in protection from ischemia-reperfusion injury. Shahed et al.[127] investigated the effect of curcumin compounds on expression of antioxidant enzymes mRNAs *in vivo* in rat kidney after ureteral obstruction or ischemia-reperfusion injury. Curcumin exhibited renoprotective properties by modulating the expression of MnSOD.

23.4 OTHER EFFECTS OF CURCUMIN

23.4.1 CURCUMIN SUPPRESSES DIABETES

Arun and Nalini[128] investigated the efficacy of turmeric and curcumin on blood sugar and polyol pathway in diabetic albino rats. Alloxan was used to induce diabetes. Administration of turmeric or curcumin reduced the blood sugar, hemoglobin, and glycosylated hemoglobin levels significantly.[129] Turmeric and curcumin supplementation also reduced the oxidative stress encountered by the diabetic rats, as demonstrated by lower levels of TBARS, which may have been due to the decreased influx of glucose into the polyol pathway, leading to an increased NADPH/NADP ratio and elevated activity of the potent antioxidant enzyme GPx. Moreover, the activity of sorbitol dehydrogenase, which catalyzes the conversion of sorbitol to fructose, was lowered significantly by treatment with turmeric or curcumin. These results also appeared to reveal that curcumin was more effective in attenuating diabetes mellitus-related changes than turmeric.

Babu et al.[130] also examined the influence of dietary curcumin on the progression of experimentally induced diabetes induced by cholesterol feeding in the albino rat. Albino rats fed 0.5% curcumin diet or 1% cholesterol diet were rendered diabetic with streptozotocin injection. Diabetic rats maintained on curcumin diet for 8 weeks excreted less albumin, urea, creatinine, and inorganic phosphorus. Urinary excretion of the electrolytes sodium and potassium were also significantly lowered under curcumin treatment. Dietary curcumin also partially reversed the abnormalities in plasma albumin, urea, creatine, and inorganic phosphorus in diabetic animals. On the other hand, glucose excretion or the fasting sugar level was unaffected by dietary curcumin, and so also the body weights were not improved to any significant extent. The curcumin diet lowered liver weight and lowered lipid peroxidation in plasma and urine at the end of the study compared with controls. The extent of lipid peroxidation was still higher in cholesterol-fed diabetic groups compared with diabetic rats fed with control diet. Thus, the study reveals that curcumin feeding improves the metabolic status in diabetic conditions, despite no effect on hyperglycemic status or body weight. The mechanism by which curcumin improves this situation is probably by virtue of its hypocholesterolemic influence and its antioxidant and free-radical-scavenging properties.

In another study, Babu et al.[131] showed the hypolipidemic action of curcumin in rats with streptozotocin-induced diabetes. Rats were maintained on 0.5% curcumin-containing diet for 8 weeks. The diet lowered blood cholesterol significantly exclusively by decreasing the LDL-VLDL fraction. A significant decrease in blood triglyceride and phospholipids was also brought about by dietary curcumin. In a parallel study, wherein diabetic animals were maintained on a high-cholesterol diet, the extents of hypercholesterolemia and phospholipidemia were higher than those maintained on the control diet. Curcumin lowered cholesterol and phospholipid levels in these animals also. Liver cholesterol and triglyceride and phospholipid contents were elevated under diabetic conditions. Dietary curcumin showed a distinct tendency to counter these changes in lipid fractions of liver. This effect of curcumin was also seen in diabetic animals maintained on a high-cholesterol diet. Dietary curcumin significantly countered renal cholesterol and triglyceride elevation in diabetic rats. In order to understand the mechanism of hypocholesterolemic action of dietary curcumin, activities of hepatic cholesterol-7a-hydroxylase and HMG-CoA reductase were measured. Hepatic cholesterol-7α-hydroxylase activity was markedly higher in curcumin-fed diabetic animals, suggesting a higher rate of cholesterol catabolism.

Suresh and Srinivasan[132] showed amelioration of renal lesions associated with diabetes by dietary curcumin in Wistar rats with streptozotocin-induced diabetes. For these studies, curcumin was fed at 0.5% in the diet for 8 weeks. Renal damage was assessed by the amount of proteins excreted in the urine and the extent of leaching of the renal tubular enzymes NAG, LDH, AsAT, AlAT, and alkaline and acid phosphatases. The integrity of the kidney was assessed by measuring the activities of several key enzymes of the renal tissue: glucose-6-phosphate dehydrogenase, glucose-6-phosphatase, and LDH (carbohydrate metabolism); aldose reductase and sorbitol dehydrogenase (polyol pathway); and transaminases, ATPases, and membrane polyunsaturated/polysaturated fatty acid ratio (membrane integrity). Data on enzymuria, albuminuria, activity of kidney ATPases, and fatty acid composition of renal membranes suggested that dietary curcumin significantly inhibited the progression of renal lesions in diabetes. These findings were corroborated by histological examination of kidney sections. This beneficial influence was possibly mediated through curcumin's ability to lower blood cholesterol levels.

23.4.2 CURCUMIN STIMULATES MUSCLE REGENERATION

Skeletal muscle is often the site of tissue injury due to trauma, disease, developmental defects, or surgery. Yet to date no effective treatment is available to stimulate the repair of skeletal muscle. Thaloor et al.[133] investigated the kinetics and extent of muscle regeneration *in vivo* after trauma following systemic administration of curcumin to mice. Biochemical and histological analyses indicated faster restoration of normal tissue architecture in mice treated with curcumin after only 4 days of daily intraperitoneal injection, whereas controls required over 2 weeks to restore normal tissue architecture. Curcumin acted directly on cultured muscle precursor cells to stimulate both cell proliferation and differentiation under appropriate conditions. The authors suggested that this effect of curcumin was mediated through suppression of NF-κB; inhibition of NF-κB-mediated transcription was confirmed using reporter gene assays. They concluded that NF-κB exerts a role in regulating myogenesis and that modulation of NF-κB activity within muscle tissue is beneficial for muscle repair. The striking effects of curcumin on myogenesis suggest therapeutic applications for treating muscle injuries.

23.4.3 Curcumin Enhances Wound Healing

Tissue repair and wound healing are complex processes that involve inflammation, granulation, and remodeling of the tissue. Perhaps the earliest report that curcumin has wound-healing activity was reported by Gujral and coworkers.[3] Sidhu et al.[134] examined the wound-healing capacity of curcumin in rats and guinea pigs. Punch wounds in curcumin-treated animals closed faster in treated than in untreated animals. Biopsies of the wound showed reepithelialization of the epidermis and increased migration of various cells, including myofibroblasts, fibroblasts, and macrophages in the wound bed. Multiple areas within the dermis showed extensive neovascularization, and Masson's trichrome staining showed greater collagen deposition in curcumin-treated wounds. Immunohistochemical localization showed an increase of transforming growth factor beta 1 (TGF-β1) in curcumin-treated wounds as compared with untreated wounds. *In situ* hybridization and polymerase chain reaction analysis also showed an increase in the mRNA transcripts of TGF-β1 and fibronectin in curcumin-treated wounds. Because TGF-β1 is known to enhance wound healing, it possible that curcumin modulates TGF-β1 activity.

To further understand its therapeutic effect on wound healing, the antioxidant effects of curcumin on H_2O_2-induced and hypoxanthine-xanthine oxidase-induced damage to cultured human keratinocytes and fibroblasts were investigated by Phan et al.[135] Cell viability was assessed by colorimetric assay and quantification of LDH release. Exposure of human keratinocytes to curcumin at 10 μg/ml significantly protected against the keratinocytes from H_2O_2-induced oxidative damage. Interestingly, exposure of human dermal fibroblasts to curcumin at 2.5 μg/ml showed significant protective effects against H_2O_2. No protective effects of curcumin on either fibroblasts or keratinocytes against hypoxanthine-xanthine oxidase-induced damage were found. These investigators thus concluded that curcumin indeed possessed powerful inhibitory capacity against H_2O_2-induced damage in human keratinocytes and fibroblasts and that this protection may contribute to wound healing.

Mani et al.[136] investigated the effect of curcumin treatment by topical application in dexamethasone-impaired cutaneous healing in a full-thickness punch-wound model in rats. They assessed healing in terms of histology, morphometry, and collagenization on the fourth and seventh days postwounding and analyzed the regulation of TGF-β1, its receptors type I (tIrc) and type II (tIIrc), and iNOS. Curcumin significantly accelerated healing of wounds with or without dexamethasone treatment, as revealed by a reduction in the wound width and gap length compared with controls. Curcumin treatment enhanced expression of TGF-β1 and TGF-β tIIrc in both normal and impaired healing wounds. Macrophages in the wound bed showed an enhanced expression of TGF-β1 mRNA in curcumin-treated wounds, as evidenced by *in situ* hybridization. iNOS levels were increased following curcumin treatment in unimpaired wounds, but not so in the dexamethasone-impaired wounds. Their study indicated an enhancement in dexamethasone-impaired wound repair by topical curcumin and its differential regulatory effect on TGF-β1, its receptors, and iNOS in this cutaneous wound-healing model.

23.4.4 Curcumin Suppresses Symptoms Associated with Arthritis

Deodhar et al.[137] were the first to report on the antirheumatic activity of curcumin in human subjects. They performed a short-term double-blind crossover study in 18 patients with "definite" rheumatoid arthritis to compare the antirheumatic activity of curcumin (1200 mg/day) with phenylbutazone (300 mg/day). Subjective and objective assessment in patients who were taking corticosteroids just prior to the study showed significant ($P < 0.05$) improvements in morning stiffness, walking time, and joint swelling following 2 weeks of curcumin therapy.

Liacini et al.[138] examined the effect of curcumin in articular chondrocytes. Interleukin-1 (IL-1), the main cytokine instigator of cartilage degeneration in arthritis, induces matrix metalloproteinase-3 (MMP3) and MMP13 RNA and protein in chondrocytes through activation of mitogen-activated protein kinase (MAPK), AP-1, and NF-κB transcription factors. Curcumin achieved 48 to 99% suppression of MMP3 and 45 to 97% of MMP13 in human chondrocytes and 8 to 100% (MMP3) and 32 to 100% (MMP13) in bovine chondrocytes. Inhibition of IL-1 signal transduction by these agents could be useful for reducing cartilage resorption by MMPs in arthritis.

23.4.5 CURCUMIN REDUCES THE INCIDENCE OF CHOLESTEROL GALLSTONE FORMATION

Hussain and Chandrasekhara[110] studied the efficacy of curcumin in reducing the incidence of cholesterol gallstones induced by feeding a lithogenic diet in young male mice. Feeding a lithogenic diet supplemented with 0.5% curcumin for 10 weeks reduced the incidence of gallstone formation to 26%, as compared with 100% incidence in the group fed with the lithogenic diet alone. Biliary cholesterol concentration was also significantly reduced by curcumin feeding. The lithogenic index, which was 1.09 in the cholesterol-fed group, was reduced to 0.43 in the 0.5% curcumin supplemented group. Further, the cholesterol:phospholipid ratio of bile was also reduced significantly when 0.5% curcumin-supplemented diet was fed. A dose-response study with 0.2, 0.5, and 1% curcumin-supplemented lithogenic diets showed that 0.5% curcumin was more effective than a diet with 0.2 or 1% curcumin. How curcumin mediates antilithogenic effects in mice was further investigated by this group.[111] For this purpose, the hepatic bile of rats was fractionated by gel filtration chromatography, and the low molecular weight (LMW) protein fractions were tested for their ability to influence cholesterol crystal growth in model bile. The LMW protein fraction from the lithogenic-agent-fed control group's bile shortened the nucleation time and increased the crystal growth rate and final crystal concentration. But with the LMW protein fractions from the bile of rats given curcumin, the nucleation times were prolonged, and the crystal growth rates and final crystal concentrations were decreased. The LMW fractions were further purified into three different sugar-specific proteins by affinity chromatography. A higher proportion of LMW proteins from the control group bile was bound to Con-A, whereas higher proportions of LMW proteins from the groups fed with curcumin were bound to wheat germ agglutinin (WGA) and helix pomatia lectin. The Con-A-bound fraction obtained from the control group showed a pronucleating effect. In contrast, the WGA-bound fraction obtained from the curcumin group showed a potent antinucleating activity.

23.4.6 CURCUMIN MODULATES MULTIPLE SCLEROSIS

Multiple sclerosis (MS) is an inflammatory disease of the central nervous system (CNS) that afflicts more than 1 million people worldwide. The destruction of oligodendrocytes and myelin sheath in the CNS is the pathological hallmark of MS. MS is an inflammatory autoimmune disease of the CNS resulting from myelin antigen-sensitized T cells in the CNS. Experimental allergic encephalomyelitis (EAE), a CD4+ Th1 cell-mediated inflammatory demyelinating autoimmune disease of the CNS, serves as an animal model for MS. IL-12 plays a crucial proinflammatory role in the induction of neural antigen-specific Th1 differentiation and pathogenesis of CNS demyelination in EAE and MS.

Natarajan and Bright[61] investigated the effect of curcumin on the pathogenesis of CNS demyelination in EAE. *In vivo* treatment of SJL/J mice with curcumin significantly reduced the duration and clinical severity of active immunization and adoptive transfer of EAE.[61] Curcumin inhibited EAE in association with a decrease in IL-12 production from macrophage/microglial cells and differentiation of neural antigen-specific Th1 cells. *In vitro* treatment of activated T cells with curcumin inhibited IL-12-induced tyrosine phosphorylation of Janus kinase 2, tyrosine kinase 2, and STAT3 and STAT4 transcription factors. The inhibition of Janus kinase-STAT pathway by curcumin resulted in a decrease in IL-12-induced T-cell proliferation and Th1 differentiation. These findings show that curcumin inhibits EAE by blocking IL-12 signaling in T cells and suggest its use in the treatment of MS and other Th1-cell-mediated inflammatory diseases.

23.4.7 CURCUMIN BLOCKS THE REPLICATION OF HIV

Transcription of type 1 human immunodeficiency virus (HIV-1) provirus is governed by the viral long-terminal repeat (LTR). Drugs can block HIV-1 replication by inhibiting the activity of its LTR. Li et al.[139] examined the effect of curcumin on HIV-1 LTR-directed gene expression and virus replication. Curcumin was found to be a potent and selective inhibitor of HIV-1 LTR-directed gene expression, at concentrations that have minor effects on cells. Curcumin inhibited p24 antigen production in cells either acutely or chronically infected with HIV-1 through transcriptional repression of the LTR. Sui et al.[140] examined the effect on the HIV-1 and HIV-2 proteases by curcumin and curcumin boron complexes. Curcumin was a modest inhibitor of HIV-1 (IC_{50} = 100 μM) and HIV-2 (IC_{50} = 250 μM) proteases. Simple modifications of the curcumin structure raised the IC_{50} value, but complexes of the central dihydroxy groups of curcumin with boron lowered the IC_{50} to a value as low as 6 μM. The boron complexes were also time-dependent inactivators of the HIV proteases. The increased affinity of the boron complexes may reflect binding of the orthogonal domains of the inhibitor in intersecting sites within the substrate-binding cavity of the enzyme, while activation of the α, β-unsaturated carbonyl group of curcumin by chelation to boron probably accounts for time-dependent inhibition of the enzyme.

Mazumder et al.[141] examined the effect of curcumin analogs with altered potencies against HIV-1 integrase. They reported that curcumin inhibited HIV-1 integrase activity. They also synthesized and tested analogs of curcumin to explore the structure-activity relationships and mechanism of action of this family of compounds in more detail. They found that two curcumin analogs, dicaffeoylmethane and rosmarinic acid, inhibited both activities of integrase for IC_{50} values below 10 μM. They demonstrated that lysine 136 may play a role in viral DNA binding and that two curcumin analogs had equivalent potencies against both an integrase mutant and wild-type integrase, suggesting that the curcumin-binding site and the substrate-binding site may not overlap. Combining one curcumin analog with the recently described integrase inhibitor NSC 158393 resulted in integrase inhibition that was synergistic, again suggesting that drug-binding sites may not overlap. They also determined that these analogs could inhibit binding of the enzyme to the viral DNA, but that this inhibition is independent of divalent metal ion. Furthermore, kinetic studies of these analogs suggest that they bind to the enzyme at a slow rate. These studies can provide mechanistic and structural information to guide the future design of integrase inhibitors.

The transcription of HIV-1 provirus is regulated by both cellular and viral factors. Various pieces of evidence suggest that Tat protein secreted by HIV1-infected cells may have additional activity in the pathogenesis of AIDS because of its ability to also be taken up by noninfected cells. Barthelemy et al.[142] showed that curcumin used at 10 to 100 nM inhibited Tat transactivation of HIV1-LTR lacZ by 70 to 80% in HeLa cells. To develop more efficient curcumin derivatives, the researchers synthesized and tested in the same experimental system the inhibitory activity of reduced curcumin (C1), which lacks the spatial structure of curcumin; allyl-curcumin (C2), which possesses a condensed allyl derivative on curcumin that plays the role of metal chelator; and tocopheryl-curcumin (C3), whose structural alterations enhance the antioxidant activity of the molecule. Results obtained with the C1, C2, and C3 curcumin derivatives showed a significant inhibition (70 to 85%) of Tat transactivation. Despite the fact that tocopheryl-curcumin (C3) failed to scavenge O^{2-}, this curcumin derivative exhibited the most activity; 70% inhibition was obtained at 1 nM, while only 35% inhibition was obtained with the curcumin.

23.4.8 CURCUMIN AFFECTS ALZHEIMER'S DISEASE

Brain inflammation in Alzheimer's disease (AD) patients is characterized by increased cytokines and activated microglia. Epidemiological studies suggest reduced AD risk is associated with long-term use of nonsteroidal anti-inflammatory drugs (NSAIDs). Whereas chronic ibuprofen suppressed inflammation and plaque-related pathology in an Alzheimer transgenic APPSw mouse model

(Tg2576), excessive use of NSAIDs targeting cyclooxygenase I can cause gastrointestinal, liver, and renal toxicity. One alternative NSAID is curcumin. Lim et al.[143] found that curcumin reduced oxidative damage and amyloid pathology in an Alzheimer transgenic mouse model. To evaluate whether it could affect Alzheimer-like pathology in the APPSw mice, they tested the effect of a low (160 ppm) and a high (5000 ppm) dose of dietary curcumin on inflammation, oxidative damage, and plaque pathology. Low and high doses significantly lowered oxidized proteins and IL-1β, a proinflammatory cytokine usually elevated in the brains of these mice. With low-dose, but not high-dose, curcumin treatment, the astrocytic marker glial fibrillary acidic protein was reduced, and insoluble beta-amyloid (Aβ), soluble Aβ, and plaque burden were significantly decreased, by 43 to 50%. However, levels of amyloid precursor in the membrane fraction were not reduced. Microgliosis was also suppressed in neuronal layers but not adjacent to plaques. In view of its efficacy and apparent low toxicity, this Indian spice component has promise for the prevention of Alzheimer's disease.

23.4.9 CURCUMIN PROTECTS AGAINST CATARACT FORMATION

Age-related cataractogenesis is a significant health problem worldwide. Oxidative stress has been suggested to be a common underlying mechanism of cataractogenesis, and augmentation of the antioxidant defenses of the ocular lens has been shown to prevent or delay cataractogenesis. Awasthi et al.[144] tested the efficacy of curcumin in preventing cataractogenesis in an *in vitro* rat model. Rats were maintained on an AIN-76 diet for 2 weeks, after which they were given a daily dose of corn oil alone or 75 mg curcumin/kg in corn oil for 14 days. Their lenses were removed and cultured for 72 h *in vitro* in the presence or absence of 100 μmol 4-hydroxy-2-nonenal (4-HNE)/L, a highly electrophilic product of lipid peroxidation. The results of these studies showed that 4-HNE led to opacification of cultured lenses, as indicated by the measurements of transmitted light intensity using digital image analysis. However, the lenses from curcumin-treated rats were resistant to 4-HNE-induced opacification. Curcumin treatment significantly induced the glutathione-*S*-transferase (GST) isozyme rGST8-8 in rat lens epithelium. Because rGST8-8 utilizes 4-HNE as a preferred substrate, we suggest that the protective effect of curcumin may be mediated through the induction of this GST isozyme. These studies suggest that curcumin may be an effective protective agent against cataractogenesis induced by lipid peroxidation.

23.4.10 CURCUMIN PROTECTS FROM DRUG-INDUCED MYOCARDIAL TOXICITY

Cardiotoxicity is one of the major problems associated with administration of many chemotherapeutic agents. Venkatesan[145] examined the protective effect of curcumin on acute adriamycin (ADR) myocardial toxicity in rats. ADR toxicity, induced by a single intraperitoneal injection (30 mg/kg), was revealed by elevated serum creatine kinase (CK) and LDH. The level of the lipid peroxidation products, conjugated dienes, and malondialdehyde were markedly elevated by ADR. ADR also caused a decrease in myocardial glutathione content and glutathione peroxidase activity and an increase in cardiac catalase activity. Curcumin treatment (200 mg/kg) 7 days before and 2 days following ADR significantly ameliorated the early manifestation of cardiotoxicity (ST segment elevation and an increase in heart rate) and prevented the rise in serum CK and LDH exerted by ADR. ADR-treated rats that received curcumin displayed a significant inhibition of lipid peroxidation and augmentation of endogenous antioxidants. These results suggest that curcumin inhibits ADR cardiotoxicity and might serve as a novel combination chemotherapeutic agent with ADR to limit free-radical-mediated organ injury.

23.4.11 CURCUMIN PROTECTS FROM ALCOHOL-INDUCED LIVER INJURY

Because induction of NF-κB-mediated gene expression has been implicated in the pathogenesis of alcoholic liver disease (ALD) and curcumin inhibits the activation of NF-κB, Nanji et al.[146]

determined whether treatment with curcumin would prevent experimental ALD and elucidated the underlying mechanism. Four groups of rats (six rats/group) were treated by intragastric infusion for 4 weeks. One group received fish oil plus ethanol (FE); a second group received fish oil plus dextrose (FD). The third and fourth groups received FE or FD supplemented with 75 mg/kg/day of curcumin. Liver samples were analyzed for histopathology, lipid peroxidation, NF-κB binding, TNFα, IL-12, monocyte chemotactic protein-1, macrophage inflammatory protein-2, COX-2, iNOS, and nitrotyrosine. Rats fed FE developed fatty liver, necrosis, and inflammation, which was accompanied by activation of NF-κB and the induction of cytokines, chemokines, COX-2, iNOS, and nitrotyrosine formation. Treatment with curcumin prevented both the pathological and biochemical changes induced by alcohol. Because endotoxin and the Kupffer cell are implicated in the pathogenesis of ALD, they also investigated whether curcumin suppressed the stimulatory effects of endotoxin in isolated Kupffer cells. Curcumin blocked endotoxin-mediated activation of NF-κB and suppressed the expression of cytokines, chemokines, COX-2, and iNOS in Kupffer cells. Thus curcumin prevented experimental ALD, in part by suppressing induction of NF-κB-dependent genes.

Hepatic fibrogenesis occurs as a wound-healing process after many forms of chronic liver injury. Hepatic fibrosis ultimately leads to cirrhosis if not treated effectively. During liver injury, quiescent hepatic stellate cells (HSC), the most relevant cell type, become active and proliferative. Oxidative stress is a major and critical factor for HSC activation. Activation of peroxisome proliferator-activated receptor-γ (PPAR-γ) inhibits the proliferation of nonadipocytes. The level of PPAR-γ is dramatically diminished along with activation of HSC during liver injury. Xu et al.[62] examined the effect of curcumin on HSC proliferation. They hypothesized that curcumin inhibits the proliferation of activated HSC by inducing PPAR-γ gene expression and reviving PPAR-γ activation. Their results indicated that curcumin significantly inhibited the proliferation of activated HSC and induced apoptosis *in vitro*. They also demonstrated, for the first time, that curcumin dramatically induced the expression of the PPAR-γ gene and activated PPAR-γ in activated HSC. Blocking its transactivating activity by a PPAR-γ antagonist markedly abrogated the effects of curcumin on inhibition of cell proliferation. These results provided a novel insight into mechanisms underlying the inhibition of activated HSC growth by curcumin. The characteristics of curcumin, including antioxidant potential, reduction of activated HSC growth, and no adverse health effects, make it a potential candidate for prevention and treatment of hepatic fibrosis.

23.4.12 CURCUMIN PROTECTS FROM DRUG-INDUCED LUNG INJURY

Cyclophosphamide causes lung injury in rats through its ability to generate free radicals, with subsequent endothelial and epithelial cell damage. Venkatesan and Chandrakasan[147] examined the effect of curcumin on cyclophosphamide-induced early lung injury. In order to observe the protective effects of curcumin on cyclophosphamide-induced early lung injury, healthy, pathogen-free male Wistar rats were exposed to 20 mg/100 g body weight of cyclophosphamide, given intraperitoneally as a single injection. Prior to cyclophosphamide intoxication, curcumin was administered orally daily for 7 days. At various times (2, 3, 5, and 7 days after insult), serum and lung samples were analyzed for angiotensin-converting enzyme (ACE), lipid peroxidation, reduced glutathione, and ascorbic acid. Bronchoalveolar lavage fluid was analyzed for biochemical constituents. The lavage cells were examined for lipid peroxidation and glutathione content. Excised lungs were analyzed for antioxidant enzyme levels. Biochemical analyses revealed increased lavage fluid total protein, albumin, ACE, LDH, N-acetyl-beta-D-glucosaminidase, alkaline phosphatase, acid phosphatase, lipid peroxide, GSH, and ascorbic acid levels 2, 3, 5, and 7 days after cyclophosphamide intoxication. Increased levels of lipid peroxidation and decreased levels of GSH and ascorbic acid were seen in serum, lung tissue, and lavage cells of cyclophosphamide-treated groups. Serum ACE activity increased, which coincided with the decrease in lung tissue levels. Activities of antioxidant enzymes were reduced with time in the lungs of cyclophosphamide-treated groups. A significant reduction in the lavage fluid biochemical constituents and in lipid peroxidation products in the serum, lung, and lavage cells occurred concomitantly with an increase in antioxidant defense mechanisms in curcumin-fed cyclophosphamide rats. Therefore, the study indicated that curcumin is effective in moderating the cyclophosphamide-induced early lung injury.

In another study, Venkatesan et al.[148] investigated the effect of curcumin on bleomycin (BLM)-induced lung injury. The data indicated that BLM-mediated lung injury resulted in increases in lung lavage fluid biomarkers such as total protein, ACE, LDH, N-acetyl-beta-D-glucosaminidase, lipid peroxidation (LPO) products, SOD, and catalase. BLM administration also increased the levels of malondialdehyde in bronchoalveolar lavage fluid and bronchoalveolar lavage (BAL) cells and led to greater amounts of alveolar macrophage (AM) SOD activity. In contrast, lower levels of reduced GSH were observed in lung lavage fluid, BAL cells, and AM. Stimulated superoxide anion and H_2O_2 release by AM from BLM-treated rats were higher. Curcumin treatment significantly reduced lavage fluid biomarkers. In addition, it restored the antioxidant status in BLM rats. These data suggested that curcumin treatment reduces the development of BLM-induced inflammatory and oxidant activity. Therefore, curcumin offers the potential for a novel pharmacological approach in the suppression of drug- or chemical-induced lung injury.

Punithavathi et al.[149] also evaluated the ability of curcumin to suppress BLM-induced pulmonary fibrosis in rats. A single intratracheal instillation of BLM (0.75 U/100 g, sacrificed 3, 5, 7, 14, and 28 days post-BLM) resulted in significant increases in total cell numbers, total protein, and ACE and in alkaline phosphatase activities in bronchoalveolar lavage fluid. Animals with fibrosis had a significant increase in lung hydroxyproline content. AM from BLM-treated rats elaborated significant increases in TNF-α and in superoxide and nitric oxide production in culture medium. Interestingly, oral administration of curcumin (300 mg/kg) 10 days before and daily thereafter throughout the experimental time period inhibited BLM-induced increases in total cell counts and biomarkers of inflammatory responses in BALF. In addition, curcumin significantly reduced the total lung hydroxyproline in BLM-treated rats. Furthermore, curcumin remarkably suppressed the BLM-induced AM production of TNF-α, SOD, and nitric oxide. These findings suggest that curcumin is a potent anti-inflammatory and antifibrotic agent against BLM-induced pulmonary fibrosis in rats. Punithavathi et al.[150] also examined whether curcumin prevented amiodarone-induced lung fibrosis in rats. They found that curcumin had a protective effect on amiodarone-induced pulmonary fibrosis. Curcumin inhibited the increases in lung myeloperoxidase activity, TGF-β1 expression, lung hydroxyproline content, and expression of type I collagen and c-Jun protein in amiodarone-treated rats.

Paraquat (PQ), a broad-spectrum herbicide, can cause lung injury in humans and animals. An early feature of PQ toxicity is the influx of inflammatory cells, releasing proteolytic enzymes and oxygen free radicals that can destroy the lung epithelium and cause pulmonary fibrosis. Suppressing early lung injury before the development of irreversible fibrosis is critical to effective therapy. Venkatesan[151] showed that curcumin confers remarkable protection against PQ-induced lung injury. A single intraperitoneal injection of PQ (50 mg/kg) significantly increased the levels of protein, angiotensin-converting enzyme (ACE), alkaline phosphatase, N-acetyl-beta-D-glucosaminidase (NAG), thiobarbituric acid reactive substances (TBARS), and neutrophils in the bronchoalveolar lavage fluid (BALF), while it decreased GSH levels. In PQ-treated rat BAL cells, TBARS concentration was increased at the same time as glutathione content was decreased. In addition, PQ caused a decrease in ACE and glutathione levels and an increase in levels of TBARS and myeloperoxidase activity in the lung. Interestingly, curcumin prevented the general toxicity and mortality induced by PQ and blocked the rise in BALF protein, ACE, alkaline phosphatase, NAG, TBARS, and neutrophils. Likewise, it prevented the rise in TBARS content in both BAL cell and lung tissue and MPO activity of the lung, reduced lung ACE, and abolished BAL cell and lung glutathione levels. These findings indicate that curcumin has important therapeutic potential in suppressing PQ lung injury.

23.4.13 Curcumin Prevents Adriamycin-Induced Nephrotoxicity

Nephrotoxicity is another problem observed in patients given chemotherapeutic agents. Venkatesan et al.[145,152] showed that curcumin prevents adriamycin (ADR)-induced nephrotoxicity in rats. Treatment with curcumin markedly protected against ADR-induced proteinuria, albuminuria, hypoalbuminemia, and hyperlipidemia. Similarly, curcumin inhibited ADR-induced increase in urinary excretion of N-acetyl-β-D-glucosaminidase (a marker of renal tubular injury), fibronectin, glycosaminoglycan, and plasma cholesterol. It restored renal function in ADR-treated rats, as judged by the increase in glomerular filtration rate (GFR). The data also demonstrated that curcumin protected against ADR-induced renal injury by suppressing oxidative stress and increasing kidney glutathione content and glutathione peroxidase activity. In like manner, curcumin abolished ADR-stimulated kidney microsomal and mitochondrial lipid peroxidation. These data suggest that administration of curcumin is a promising approach in the treatment of nephrosis caused by ADR.

23.4.14 Curcumin Protects from Scarring

Keloid and hypertrophic scars commonly occur after injuries. Overproliferation of fibroblasts, overproduction of collagen, and contraction characterize these pathological scars. Current treatment of excessive scars with intralesional corticosteroid injections used individually or in combination with other methods often have unsatisfactory outcomes, frustrating both the patient and the clinician. Phan et al.[153] investigated the inhibitory effects of curcumin on keloid fibroblasts (KF) and hypertrophic scar-derived fibroblasts (HSF) by proliferation assays, fibroblast-populated collagen lattice contraction, and electron microscopy. Curcumin significantly inhibited KF and HSF proliferation in a dose- and time-dependent manner. Curcumin seemed to have potent effects in inhibiting proliferation and contraction of excessive scar-derived fibroblasts.

23.4.15 Curcumin Protects from Inflammatory Bowel Disease

Inflammatory bowel disease (IBD) is characterized by oxidative and nitrosative stress, leukocyte infiltration, and up-regulation of proinflammatory cytokines. Ukil et al.[154] recently investigated the protective effects of curcumin on 2,4,6-trinitrobenzene sulfonic acid-induced colitis in mice, a model for IBD. Intestinal lesions were associated with neutrophil infiltration, increased serine protease activity (may be involved in the degradation of colonic tissue), and high levels of malondialdehyde. Dose-response studies revealed that pretreatment of mice with curcumin at 50 mg/kg daily i.g. for 10 days significantly ameliorated diarrhea and the disruption of colonic architecture. Higher doses (100 and 300 mg/kg) had comparable effects. In curcumin-pretreated mice, there was a significant reduction in the degree of both neutrophil infiltration and lipid peroxidation in the inflamed colon as well as decreased serine protease activity. Curcumin also reduced the levels of NO and O_2^- associated with the favorable expression of Th1 and Th2 cytokines and inducible NO synthase. Consistent with these observations, NF-κB activation in colonic mucosa was suppressed in the curcumin-treated mice. These findings suggested that curcumin exerts beneficial effects in experimental colitis and may, therefore, be useful in the treatment of IBD.

Salh et al.[155] also showed that curcumin is able to attenuate colitis in the dinitrobenzene (DNB) sulfonic acid-induced murine model of colitis. When given before the induction of colitis, it reduced macroscopic damage scores, NF-κB activation, and myeloperoxidase activity, and it attenuated the DNB-induced message for IL-1β. Western blotting analysis revealed a reproducible DNB-induced activation of p38 MAPK in intestinal lysates detected by a phosphospecific antibody. This signal was significantly attenuated by curcumin. Furthermore, Salh's group showed that the immunohistochemical signal is dramatically attenuated at the level of the mucosa by curcumin. Thus they concluded that curcumin attenuates experimental colitis through a mechanism that also inhibits the activation of NF-κB and effects a reduction in the activity of p38 MAPK. They proposed that this agent may have therapeutic implications for human IBD.

23.4.16 CURCUMIN ENHANCES THE IMMUNOSUPPRESSIVE ACTIVITY
OF CYCLOSPORINE

Chueh et al.[156] have demonstrated that curcumin enhances the immunosuppressive activity of cyclosporine in rat cardiac allografts and in mixed lymphocyte reactions. Their study demonstrated for the first time the effectiveness of curcumin as a novel adjuvant immunosuppressant with cyclosporine both *in vivo* and *in vitro*. The immunosuppressive effects of curcumin were studied in rat heterotrophic cardiac transplant models, using Brown-Norway hearts transplanted to WKY hosts. In the Brown-Norway-to-WKY model, curcumin alone significantly increased the mean survival time, to 20.5 to 24.5 days as compared with 9.1 days in nontreated controls. The combination of curcumin and subtherapeutic doses of cyclosporine further prolonged the mean survival time to 28.5 to 35.6 days, better than that of curcumin or cyclosporine alone. Cytokine analysis revealed significantly reduced expression of interleukin-2, IFNγ, and granzyme B in the day-3 specimens of the curcumin and curcumin plus cyclosporine-treated allografts compared with the nontreated allograft controls.

23.4.17 CURCUMIN PROTECTS AGAINST VARIOUS FORMS OF STRESS

Curcumin has been identified as a potent inducer of hemoxygenase-1 (HO-1), a redox-sensitive inducible protein that provides protection against various forms of stress. Curcumin stimulated the expression of Nrf2, an increase associated with a significant increase in HO-1 protein expression and HO-1 activity.[157]

23.4.18 CURCUMIN PROTECTS AGAINST ENDOTOXIN SHOCK

Madan and Ghosh[158] have demonstrated that curcumin exerts protective effects in high-dose endotoxin shock by improving survival and reducing the severity of endotoxin shock symptoms such as lethargy, diarrhea, and watery eyes following a challenge with lipopolysaccharide. They demonstrated that curcumin inhibits the transmigration and infiltration of neutrophils from blood vessels to the underlying liver tissue and, hence, inhibits the damage to the tissue. Curcumin blocks the induced expression of ICAM-1 and VCAM-1 in liver and lungs.

23.4.19 CURCUMIN PROTECTS AGAINST PANCREATITIS

Gukovsky et al.[159] reported that curcumin ameliorates pancreatitis in two rat models. In both cerulein pancreatitis and pancreatitis induced by a combination of ethanol diet and low-dose curcumin, curcumin decreased the severity of the disease. Curcumin markedly inhibited NF-κB and AP-1, IL-6, TNFα, and iNOS in the pancreas. Based on these studies, Gukovsky et al. suggested that curcumin may be useful for treatment of pancreatitis.

23.4.20 CURCUMIN INHIBITS MULTIDRUG RESISTANCE (MDR)

The effect of curcumin on apoptosis in multidrug-resistant cell lines has been reported. Piwocka et al.[160] demonstrated that curcumin induced cell death in multidrug-resistant CEM(P-gp4) and LoVo(P-gp4) cells in a caspase-3-independent manner. Mehta et al.[31] also examined the antiproliferative effects of curcumin against multidrug-resistant (MDR) lines, which were found to be highly sensitive to curcumin. The growth-inhibitory effect of curcumin was time- and dose-dependent and was correlated with its inhibition of ornithine decarboxylase activity. Curcumin preferentially arrested cells in the G2/S phase of the cell cycle.

23.5 CURCUMIN METABOLISM

Numerous studies have been performed on the biotransformation of curcumin. Lin et al.[161] showed that curcumin was first biotransformed to dihydrocurcumin and tetrahydrocurcumin and that these compounds subsequently were converted to monoglucuronide conjugates. Thus, curcumin-glucuronide, dihydro-curcumin-glucuronide, tetrahydrocurcumin-glucuronide, and tetrahydrocurcumin are major metabolites of curcumin in mice.

Since the systemic bioavailability of curcumin is low, its pharmacological activity may be mediated, in part, by its metabolites. To investigate this possibility, Ireson et al.[162] compared curcumin metabolism in human and rat hepatocytes in suspension with that in rats *in vivo*. Analysis by high-performance liquid chromatography with detection at 420 and 280 nm permitted characterization of metabolites with both intact diferoylmethane structure and increased saturation of the heptatrienone chain. Chromatographic inferences were corroborated by mass spectrometry. The major metabolites in suspensions of human or rat hepatocytes were identified as hexahydrocurcumin and hexahydrocurcuminol. In rats, *in vivo*, curcumin administered i.v. (40 mg/kg) disappeared from the plasma within 1 h of dosing. After p.o. administration (500 mg/kg), parent drug was present in plasma at levels near the detection limit. The major products of curcumin biotransformation identified in rat plasma were curcumin glucuronide and curcumin sulfate, whereas hexahydrocurcumin, hexahydrocurcuminol, and hexahydrocurcumin glucuronide were present in small amounts. To test the hypothesis that curcumin metabolites resemble their progenitor in that they can inhibit COX-2 expression, curcumin and four of its metabolites at a concentration of 20 μM were compared in terms of their ability to inhibit phorbol ester-induced prostaglandin E2 (PGE2) production in human colonic epithelial cells. Curcumin reduced PGE2 levels to preinduction levels, whereas tetrahydrocurcumin, hexahydrocurcumin, and curcumin sulfate had only weak PGE2 inhibitory activity, and hexahydrocurcuminol was inactive. The results suggested that (a) the major products of curcumin biotransformation by hepatocytes occurred only at low abundance in rat plasma after curcumin administration and (b) metabolism of curcumin by reduction or conjugation generates species with reduced ability to inhibit COX-2 expression. Because the gastrointestinal tract seems to be exposed more prominently to unmetabolized curcumin than any other tissue, the results support the clinical evaluation of curcumin as a colorectal cancer chemopreventive agent.

Curcumin has very poor bioavailability. In Ayurveda, black pepper (*Piper nigrum*), long pepper (*Piper longum*), and ginger (*Zingiber officinalis*) are collectively termed Trikatu, and are essential ingredients of numerous prescriptions that are used for a wide range of disorders. Numerous studies suggest that Trikatu has a bioavailability-enhancing effect.[163] Since curcumin belongs to the same family as ginger, it has similar enhancer activity.[164]

Shoba et al.[164] found that curcumin has poor bioavailability due to its rapid metabolism in the liver and intestinal wall. In this study, the effect of combining piperine, a known inhibitor of hepatic and intestinal glucuronidation, was evaluated on the bioavailability of curcumin in rats and healthy human volunteers. When curcumin was given alone, in the dose 2 g/kg to rats, moderate serum concentrations were achieved over a period of 4 h. Concomitant administration of piperine at 20 mg/kg increased the serum concentration of curcumin for a short period of 1 to 2 h. Time to maximum was significantly increased, while elimination half-life and clearance significantly decreased, and the bioavailability was increased by 154%. On the other hand, after a dose of 2 g curcumin alone in humans, serum levels were either undetectable or very low. Concomitant administration of piperine at 20 mg/kg produced much higher concentrations from 15 min to 1 h later, and the increase in bioavailability was 2000%. The study shows that in the dosages used, piperine enhances the serum concentration, extent of absorption, and bioavailability of curcumin in both rats and humans, with no adverse effects.

In the studies by Kumar et al.[165] natural biodegradable polymers, namely bovine serum albumin and chitosan, were used to encapsulate curcumin to form a depot drug-delivery system. Microspheres were prepared by emulsion-solvent evaporation method coupled with chemical cross-linking of the natural polymers. As much as 79.49 and 39.66% of curcumin could be encapsulated into the biodegradable carriers with albumin and chitosan respectively. *In vitro* release studies indicated a biphasic drug-release pattern, characterized by a typical burst-effect followed by a slow release that continued for several days. It was evident from Kumar's study that the curcumin-biodegradable microspheres could be successfully employed as a prolonged-release drug-delivery system for better therapeutic management of inflammation as compared with oral or subcutaneous administration of curcumin. Kumar et al.[166] synthesized bioconjugates of curcumin to improve its systemic delivery. Di-*O*-glycinoyl curcumin (I) and 2′-deoxy-2′-curcuminyl uridine (2′-cur-U) (IV) were quite potent against multiresistant microorganisms. These bioconjugates served the dual purpose of facilitating systemic delivery as well as providing therapeutic agents against viral diseases.

23.6　CLINICAL EXPERIENCE WITH CURCUMIN

Nine different studies of the safety and efficacy of curcumin in humans have been reported (Table 23.1). For example Deodar et al.[137] performed a short-term, double-blind, crossover study in 18 patients (age 22 to 48 years) to compare the antirheumatic activity of curcumin and phenylbutazone. They administered 1200 mg curcumin/day or 300 mg phenylbutazone/day for 2 weeks. These investigators reported that curcumin was well tolerated, had no side effects, and showed comparable antirheumatic activity.

Lal et al.[167] administered curcumin orally to patients suffering from chronic anterior uveitis (CAU) at a dose of 375 mg three times per day for 12 weeks. Of 53 patients enrolled, 32 completed the 12-week study. They were divided into two groups: one group of 18 patients received curcumin alone, whereas the other group of 14 patients, who had a strong PPD reaction, in addition received antitubercular treatment. The patients in both the groups started improving after 2 weeks of treatment. All the patients who received curcumin alone improved, whereas the group receiving antitubercular therapy along with curcumin had a response rate of 86%. Follow-up of all the patients for the next 3 years indicated a recurrence rate of 55% in the first group and of 36% in the second

TABLE 23.1
Clinical Studies with Curcumin in Human Subjects

Study	Patients	Dose	Comments	Ref.
Double blind, crossover study	18 pts. (22–48 yrs)	1200 mg/d ×2 wks	Anirheumatic	137
	46 male pts. (15–68)	400 mg; 3 × /d × 5 d	Inguinal hernia	169
	111 pts (40–85 yrs)	Topical	HNSCC, breast, vulva, skin	170
	10 volun.	500 mg/d × 7 d	Serum cholesterol and LPO	112
	40 pts.	625 mg; 4 × /d × 8 wks	Well-tolerated	171
	53 pts.	375 mg; 3 × /d × 12 wks	Chronic anterior uveitis	167
	8 pts.	375 mg; 3 × /d 6–22 mo	Idiopathic inflammation, orbital pseudotumors	168
Prospective Phase I	25 pts.	500 mg–12,000 mg/d × 3 mo	H&N cancers	7
	15 pts.	36–180 mg 4 mo	Colorectal, serum GST-down	172

group. Four of 18 (22%) patients in the first group and 3 of 14 patients (21%) in the second group lost their vision in the follow-up period because of various complications, e.g., vitritis, macular edema, central venous block, cataract formation, glaucomatous optic nerve damage, etc. None of the patients reported any side effects. The efficacy of curcumin and recurrences following treatment are comparable with corticosteroid therapy, which is at present considered the only available standard treatment for this disease. The lack of side effects with curcumin is its greatest advantage compared with corticosteroids. A double-blind multicenter clinical trial of this drug for CAU would be highly desirable to further validate the results of the study.

Satoskar et al.[169] evaluated the anti-inflammatory properties of curcumin in patients with postoperative inflammation. They studied 46 male patients (between the ages of 15 and 68 years) having inguinal hernia and/or hydrocoele. After the hernia operation, spermatic cord edema, and tenderness were evaluated. Either curcumin (400 mg) or placebo (250 mg lactose) or phenylbutazone (100 mg) was administered three times a day for a period of 5 days from the first postoperative day. Curcumin was found to be quite safe, and phenylbutazone and curcumin produced a better anti-inflammatory response than placebo.[169]

Kuttan et al.[170] used an ethanol extract of turmeric as well as an ointment of curcumin and showed that it produces remarkable symptomatic relief in patients with external cancerous lesions. Reduction in smell were noted in 90% of the cases and reduction in itching in almost all cases. Dry lesions were observed in 70% of the cases, and a small number of patients (10%) had a reduction in lesion size and pain. In many patients, the effect continued for several months. An adverse reaction was noticed in only 1 of the 62 patients evaluated.

Soni et al.[109] examined the effect of curcumin on serum levels of cholesterol and lipid peroxides in 10 healthy human volunteers. A dose of 500 mg of curcumin per day for 7 days significantly decreased the level of serum lipid peroxides (33%), increased HDL cholesterol (29%), and decreased total serum cholesterol (11.63%). The results suggest curcumin as a chemopreventive substance against arterial diseases.

James[171] led a New England clinical trial of curcumin's effectiveness as an antiviral agent in 40 participants. Two dropped out; 23 were randomized to a high-dose group (four capsules, four times a day) and 15 to a low-dose group (three capsules, three times a day) for 8 weeks. Though it had no antiviral effects, curcumin was well tolerated, and most participants liked taking curcumin and felt better.

Lal et al.[168] described for the first time the clinical efficacy of curcumin in the treatment of patients suffering from idiopathic inflammatory orbital pseudotumors. Curcumin was administered orally at a dose of 375 mg, three times per day, for a period of 6 to 22 months in eight patients. They were followed up for a period of 2 years at 3-month intervals. Five patients completed the study, of which four recovered completely. In the remaining patient, the swelling regressed completely, but some limitation of movement persisted. No side effect was noted in any patient, and there was no recurrence. Thus curcumin could be used as a safe and effective drug in the treatment of idiopathic inflammatory orbital pseudotumors.

Cheng et al.[7] examined the toxicology, pharmacokinetics, and biologically effective dose of curcumin in humans. This prospective phase I study evaluated curcumin in patients with one of the following five high-risk conditions: (1) recently resected urinary bladder cancer; (2) arsenic Bowen's disease of the skin; (3) uterine cervical intraepithelial neoplasm (CIN); (4) oral leukoplakia; and (5) intestinal metaplasia of the stomach. Curcumin was taken orally for 3 months. Biopsy of the lesion sites was done immediately before and 3 months after starting curcumin treatment. The starting dose was 500 mg/day. If no toxicity of grade II or higher was noted in at least three successive patients, the dose was escalated to 1,000, 2,000, 4,000, 8,000, or 12,000 mg/day in order. The concentration of curcumin in serum and urine was determined by high-pressure liquid chromatography (HPLC). A total of 25 patients were enrolled in this study. There was no treatment-related toxicity for doses up to 8000 mg/day. Beyond 8000 mg/day, the bulky volume of the drug was unacceptable to the patients. The serum concentration of curcumin usually peaked at 1 to 2 h after oral intake of curcumin and gradually declined within 12 h. The average peak serum concentrations after taking 4000 mg, 6000 mg, and 8000 mg of curcumin were 0.51 ± 0.11 μM, 0.63 ± 0.06 μM, and 1.77 ± 1.87 μM, respectively. Urinary excretion of curcumin was undetectable. One of four patients with CIN and one of seven patients with oral leukoplakia developed frank malignancies in spite of curcumin treatment. In contrast, histological improvement of precancerous lesions was seen in one of two patients with recently resected bladder cancer, two of seven patients with oral leukoplakia, one of six patients with intestinal metaplasia of the stomach, one of four patients with CIN, and two of six patients with Bowen's disease. In conclusion, this study demonstrated that curcumin is not toxic to humans at doses up to 8000 mg/day when taken by mouth for 3 months. These results also suggested a biological effect of curcumin in the chemoprevention of cancer.

Sharma et al.[172] examined the pharmacodynamics and pharmacokinetics of curcumin in humans in a dose-escalation pilot study. A novel standardized Curcuma extract in proprietary capsule form was given at doses between 440 and 2200 mg/day, containing 36 to 180 mg of curcumin. Fifteen patients with advanced colorectal cancer refractory to standard chemotherapies received Curcuma extract daily for up to 4 months. The activity of glutathione-S-transferase and levels of a DNA adduct (M(1)G) formed by malondialdehyde, a product of lipid peroxidation and prostaglandin biosynthesis, were measured in patients' blood cells. Oral Curcuma extract was well tolerated, and dose-limiting toxicity was not observed. Neither curcumin nor its metabolites were detected in blood or urine, but curcumin was recovered from feces. Curcumin sulfate was identified in the feces of one patient. Ingestion of 440 mg of Curcuma extract for 29 days was accompanied by a 59% decrease in lymphocytic glutathione-S-transferase activity. At higher dose levels, this effect was not observed. Leukocytic M(1)G levels were constant within each patient and unaffected by treatment. Radiologically stable disease was demonstrated in five patients for 2 to 4 months of treatment. The results suggested that (a) Curcuma extract can be administered safely to patients at doses of up to 2.2 g daily, equivalent to 180 mg of curcumin; (b) curcumin has low oral bioavailability in humans and may undergo intestinal metabolism; and (c) larger clinical trials of Curcuma extract are merited.

23.7 CURCUMIN ANALOGS

Commercial curcumin isolated from the rhizome of *Curcuma longa* Linn. contains three major curcuminoids (approximately 77% curcumin, 17% demethoxycurcumin, and 3% bisdemethoxycurcumin) (Figure 23.1). Commercial curcumin, pure curcumin, and demethoxycurcumin are approximately equipotent as inhibitors of TPA-induced tumor promotion in mouse skin, whereas bisdemethoxycurcumin is somewhat less active.[15] Besides natural curcumin, several analogs of curcumin have been synthesized and tested.[173,174] Tetrahydrocurcumin, an antioxidative substance that is derived from curcumin by hydrogenation, has been shown to have a protective effect on oxidative stress in cholesterol-fed rabbits.[175] Kumar et al.[176] have developed an analog of curcumin, 4-hydroxy-3-methoxybenzoic acid methyl ester (HMBME), that targets the Akt/NF-κB signaling pathway. They demonstrated the ability of this novel compound to inhibit the proliferation of human and mouse PCA cells. Overexpression of constitutively active Akt reversed the HMBME-induced growth inhibition and apoptosis, illustrating the direct role of Akt signaling in HMBME-mediated growth inhibition and apoptosis. HMBME-mediated inhibition of Akt kinase activity may have a potential in suppressing/decreasing the activity of major survival/antiapoptotic pathways.

Using an *in vitro* SVR assay, Robinson et al.[177] have demonstrated potent antiangiogenic properties in aromatic enone and dienone analogs of curcumin. Based on a simple pharmacophore model, the aromatic enone and aromatic dienone analogs of curcumin were prepared using standard drug design concepts.

Devasena et al.[178] examined the protective effect of a curcumin analog [bis-1,7-(2-hydroxyphenyl)-hepta-1,6-diene-3,5-dione] on hepatic lipid peroxidation and antioxidant status during 1,2-dimethylhydrazine-induced colon carcinogenesis in male Wistar rats. They observed that the curcumin analog exerted chemopreventive effects against cancer development at extrahepatic sites by modulating hepatic biotransformation enzymes and antioxidant status. The effect was comparable with that of curcumin. They proposed that the hydroxyl group in the aromatic ring is responsible for the protective effect rather than the methoxy group. Mishra et al.[179] synthesized a novel curcumin conjugate, namely 1,7-bis(4-O-glycinoyl-3-methoxyphenyl)-1,6-heptadiene-3,5,dione (I) that was attached to the deoxy-11 mer, 5′-GTT AGG GTT AG-3′, a complementary sequence of telomerase RNA template. This novel anticancer prodrug has the potential to target the telomerase sequence.

The antitumor properties of metal chelates of synthetic curcuminoids have also been investigated. John et al.[180] examined four synthetic curcuminoids, 1,7-bis(4-hydroxy-3-methoxyphenyl)-1, 6-heptadiene-3,5-dione (curcumin 1); 1,7-bis(piperonyl)-1,6-heptadiene-3,5-dione (piperonyl curcumin); 1,7-bis(2-hydroxynaphthyl)-1,6-heptadiene-2,5-dione (2-hydroxynaphthyl curcumin); and 1,1-bis(phenyl)-1,3,8,10-undecatetraene-5,7-dione (cinnamyl curcumin) and their copper(II) complexes for their possible cytotoxic and antitumor activities. Copper chelates of synthetic curcuminoids showed enhanced antitumor activity. In addition to these, a novel curcumin derivative, hydrazinocurcumin (HC), was synthesized and examined for its biological activities by Shim et al.[181] HC potently inhibited the proliferation of bovine aortic endothelial cells at nanomolar concentrations (IC_{50} = 520 nM) without cytotoxicity. Snyder et al.[182] reported the synthesis of several different structural analogs of curcumin and examined their antitumor and antiangiogenic properties. They found analogs that are more potent than native curcumin.

23.8 SOURCES OF CURCUMIN

There are several good sources of curcumin with purity ranging from 60 to 98% (Table 23.2). Some companies supply curcumin in a powder form; other companies supply it in capsules. Curcumin supplied by Life Extension is in 900-mg capsules; turmeric curcumin is in 500-mg capsules. Likewise, Iherb, Club Natural, Immune Support, Nature's, Big Fitness, Powerhouse Gym, MMS

TABLE 23.2
Sources of Curcumin

Human use

Sabinsa (www.sabinsa.com/products/circumin_book.htm), Piscataway, NJ

Kalsec (ww.kalsec.com/products/turmeric_over.cfm), Kalamazoo, MI

Life Extension (www.lef.org/hewshop/items/item00552.html?source=WebProtProd)

Turmeric Curcumin (www.turmeric-curcumin.com/)

Iherb (www.iherb.com/curcumin1.html)

Club Natural (www.clubnatural.com/curex9550180.html), Irvine, CA

American Nutrition (www.AmericanNutrition.com)

Amerifit (www.amerifit.geomerx.com/items/categories.cfm?category=2), Bloomfiled, CT

XKMS (www.xkms.org/WebVitamins-32/Curcumin-Power-60C.htm)

Immune Support (www.Immunesupport.com/shop/prodlisting.cfm?NOTE=NOC)

Nature's (ww.naturesnutrition.com/SKU/55114.htm)

Big Fitness (www.bfwse.com/jr-021.html)

Powerhoue Gym (store.yahoo.com/musclespot/curcumin95.html)

MMS Pro (www.mmspro.com)

Herbal Fields (www.herbalfields.com/curcumin.html)

Research use

Sigma Aldrich (www.sigmaaldrich.com/cgibin/hsrun/Distributed/HahtShop/HAHTpage/HS_CatalogSearch)

Calbiochem (www.calbiochem.com/Products/ProductDetail_CBCB.asp?catNO=239802)

LKT Laboratories (www.lktlabs.com)

Pro, and American Nutrition supply 500-mg capsules; Amerifit supplies 1700-mg capsules; and XKMS supplies 300-mg capsules. Curcumin combined with piperine (also bioperine derived from black pepper), which has a higher bioavailability than curcumin alone, is available from Life Extension in a formulation referred to as "super curcumin."[164]

23.9 CONCLUSION

From all these studies, it is clear that curcumin exhibits activities against cancer, cardiovascular diseases, and diabetes, the major ailments in the U.S. This drug has also shown therapeutic effects against Alzheimer's disease, multiple sclerosis, cataract formation, HIV, and drug-induced nonspecific toxicity in the heart, lung, and kidney. Several of the studies establishing curcumin's potential were carried out in animals. Further testing of curcumin in humans is required to confirm these observations. A clinical development plan for using curcumin to treat cancer was recently described by the NCI. Studies also show that in countries such as India where curcumin is consumed, the profile of cancer incidence is very different than those that do not (such as the U.S.; see Table 23.3).

TABLE 23.3
Comparison of Cancer Incidence in U.S. (Curcumin Non-Users) and India (Curcumin Users)

Cancer	U.S. Cases	U.S. Deaths	India Cases	India Deaths
Breast	660	160	79	41
Prostate	690	130	20	9
Colon/rectum	530	220	30	18
Lung	660	580	38	37
Head and neck SCC	140	44	153	103
Liver	41	44	12	13
Pancreas	108	103	8	8
Stomach	81	50	33	30
Melanoma	145	27	1.8	1
Testis	21	1	3	1
Bladder	202	43	15	11
Kidney	115	44	6	4
Brain, nervous system	65	47	19	14
Thyroid	55	5	12	3
Endometrial cancers	163	41	132	72
Ovary	76	50	20	12
Multiple myeloma	50	40	6	5
Leukemia	100	70	19	17
Non-Hodgkin's lymphoma	180	90	17	15
Hodgkin's disease	20	5	7	4

Showing cases per 1 million persons calculated on the basis of current consensus:
Endometrial cancers include Cervix uteri and Corpus uteri.

GLOBOCAN 2000: Cancer Incidence, Mortality and Prevalence Worldwide, Version 1.0. IARC Cancer Base No. 5. Lyon, IARC Press, 2001.

How curcumin produces its therapeutic effects is not fully understood, but they are probably mediated in part through the antioxidant and anti-inflammatory action of curcumin. It is quite likely that curcumin mediates its effects through other mechanisms as well. Over a dozen different cellular proteins and enzymes have been identified to which curcumin binds. High-throughput ligand-interacting technology can reveal more molecular targets of curcumin. Microarray gene chip technology may in the future indicate which genes are regulated by curcumin.

ACKNOWLEDGMENT

This research was supported by The Clayton Foundation for Research (to BBA), by the Department of Defense U.S. Army Breast Cancer Research Program BC010610 (to BBA), and by a P50 Head and Neck SPORE grant from the National Institutes of Health (to BBA). We would like to thank Walter Pagel for a careful review of the manuscript.

REFERENCES

1. Lampe, V., Milobedeska, J., and Kostanecki, V., Ber. *Dtsch. Chem. Ges.*, 43, 2163, 1910.
2. Lampe, V. and Milobedeska, J., *Ber. Dtsch. Chem. Ges.*, 46, 2235, 1913.
3. Srimal, R.C. and Dhawan, B.N., Pharmacology of diferuloyl methane (curcumin), a non-steroidal anti-inflammatory agent, *J. Pharm. Pharmacol.*, 25 (6), 447–452, 1973.
4. Jain, S.K. and DeFilipps, R.A., *Medicinal Plants of India*, Reference Publications, Algonac, MI, 1991, p. 120.
5. Nadkarni, A.K., *Indian Materia Medica*, Vol. 1, Popular Book Depot, Bombay, 1954.
6. Ammon, H.P. and Wahl, M.A., Pharmacology of Curcuma longa, *Planta Med.*, 57 (1), 1–7, 1991.
7. Cheng, A.L., Hsu, C.H., Lin, J.K., Hsu, M.M., Ho, Y.F., Shen, T.S., Ko, J.Y., Lin, J.T., Lin, B.R., Ming-Shiang, W., Yu, H.S., Jee, S.H., Chen, G.S., Chen, T.M., Chen, C.A., Lai, M.K., Pu, Y.S., Pan, M.H., Wang, Y.J., Tsai, C.C., and Hsieh, C.Y., Phase I clinical trial of curcumin, a chemopreventive agent, in patients with high-risk or pre-malignant lesions, *Anticancer Res.*, 21 (4B), 2895–2900, 2001.
8. Wang, Y.J., Pan, M.H., Cheng, A.L., Lin, L.I., Ho, Y.S., Hsieh, C.Y., and Lin, J.K., Stability of curcumin in buffer solutions and characterization of its degradation products, *J. Pharm. Biomed. Anal.*, 15 (12), 1867–1876, 1997.
9. Aggarwal, B.B., Kumar, A., and Bharti, A.C., Anticancer potential of curcumin: preclinical and clinical studies, *Anticancer Res.*, 23 (1A), 363–398, 2003.
10. Huang, H.C., Jan, T.R., and Yeh, S.F., Inhibitory effect of curcumin, an anti-inflammatory agent, on vascular smooth muscle cell proliferation, *Eur. J. Pharmacol.*, 221 (2–3), 381–384, 1992.
11. Huang, M.T., Lou, Y.R., Xie, J.G., Ma, W., Lu, Y.P., Yen, P., Zhu, B.T., Newmark, H., and Ho, C.T., Effect of dietary curcumin and dibenzoylmethane on formation of 7,12-dimethylbenz[a]anthracene-induced mammary tumors and lymphomas/leukemias in Sencar mice, *Carcinogenesis*, 19 (9), 1697–1700, 1998.
12. Conney, A.H., Lysz, T., Ferraro, T., Abidi, T.F., Manchand, P.S., Laskin, J.D., and Huang, M.T., Inhibitory effect of curcumin and some related dietary compounds on tumor promotion and arachidonic acid metabolism in mouse skin, *Adv. Enzyme Regul.*, 31, 385–396, 1991.
13. Lu, Y.P., Chang, R.L., Lou, Y.R., Huang, M.T., Newmark, H.L., Reuhl, K.R., and Conney, A.H., Effect of curcumin on 12-O-tetradecanoylphorbol-13-acetate- and ultraviolet B light-induced expression of c-Jun and c-Fos in JB6 cells and in mouse epidermis, *Carcinogenesis*, 15 (10), 2363–2370, 1994.
14. Limtrakul, P., Lipigorngoson, S., Namwong, O., Apisariyakul, A., and Dunn, F.W., Inhibitory effect of dietary curcumin on skin carcinogenesis in mice, *Cancer Lett.*, 116 (2), 197–203, 1997.
15. Huang, M.T., Newmark, H.L., and Frenkel, K., Inhibitory effects of curcumin on tumorigenesis in mice, *J. Cell. Biochem. Suppl.*, 27, 26–34, 1997.
16. Huang, M.T., Lou, Y.R., Ma, W., Newmark, H.L., Reuhl, K.R., and Conney, A.H., Inhibitory effects of dietary curcumin on forestomach, duodenal, and colon carcinogenesis in mice, *Cancer Res.*, 54 (22), 5841–5847, 1994.

17. Piper, J.T., Singhal, S.S., Salameh, M.S., Torman, R.T., Awasthi, Y.C., and Awasthi, S., Mechanisms of anticarcinogenic properties of curcumin: the effect of curcumin on glutathione linked detoxification enzymes in rat liver, *Int. J. Biochem. Cell. Biol.*, 30 (4), 445–456, 1998.

18. Rao, C.V., Rivenson, A., Simi, B., and Reddy, B.S., Chemoprevention of colon carcinogenesis by dietary curcumin, a naturally occurring plant phenolic compound, *Cancer Res.*, 55(2); 259–266, 1995.

19. Kim, J.M., Araki, S., Kim, D.J., Park, C.B., Takasuka, N., Baba-Toriyama, H., Ota, T., Nir, Z., Khachik, F., Shimidzu, N., Tanaka, Y., Osawa. T., Uraji, T., Murakoshi, M., Nishino, H., and Tsuda, H., Chemopreventive effects of carotenoids and curcumins on mouse colon carcinogenesis after 1,2-dimethylhydrazine initiation, *Carcinogenesis*, 19 (1), 81–85, 1998.

20. Kawamori, T., Lubet, R., Steele, V.E., Kelloff, G.J., Kaskey, R.B., Rao, C.V., and Reddy, B.S., Chemopreventive effect of curcumin, a naturally occurring anti-inflammatory agent, during the promotion/progression stages of colon cancer, *Cancer Res.*, 59 (3), 597–601, 1999.

21. Chuang, S.E., Cheng, A.L., Lin, J.K., and Kuo, M.L., Inhibition by curcumin of diethylnitrosamine-induced hepatic hyperplasia, inflammation, cellular gene products and cell-cycle-related proteins in rats, *Food Chem. Toxicol.*, 38 (11); 991–995, 2000.

22. Singletary, K., MacDonald, C., Wallig, M., and Fisher, C., Inhibition of 7,12-dimethylbenz[a]anthracene (DMBA)-induced mammary tumorigenesis and DMBA-DNA adduct formation by curcumin, *Cancer Lett.*, 103 (2), 137–141, 1996.

23. Chan, M.M., Huang, H.I., Fenton, M.R., and Fong, D., In vivo inhibition of nitric oxide synthase gene expression by curcumin, a cancer preventive natural product with anti-inflammatory properties, *Biochem. Pharmacol.*, 55 (12), 1955–1962, 1998.

24. Inano, H., Onoda, M., Inafuku, N., Kubota, M., Kamada, Y., Osawa, T., Kobayashi, H., and Wakabayashi, K., Chemoprevention by curcumin during the promotion stage of tumorigenesis of mammary gland in rats irradiated with gamma-rays, *Carcinogenesis*, 20 (6), 1011–1018, 1999.

25. Kuo, M.L., Huang, T.S., and Lin, J.K., Curcumin, an antioxidant and anti-tumor promoter, induces apoptosis in human leukemia cells, *Biochim. Biophys. Acta*, 1317 (2), 95–100, 1996.

26. Han, S.S., Chung, S.T., Robertson, D.A., Ranjan, D., and Bondada, S., Curcumin causes the growth arrest and apoptosis of B cell lymphoma by downregulation of egr-1, c-myc, bcl-XL, NF-kappa B, and p53, *Clin. Immunol.*, 93 (2), 152–161, 1999.

27. Piwocka, K., Zablocki, K., Wieckowski, M.R., Skierski, J., Feiga, I., Szopa, J., Drela, N., Wojtczak, L., and Sikora, E., A novel apoptosis-like pathway, independent of mitochondria and caspases, induced by curcumin in human lymphoblastoid T (Jurkat) cells, *Exp. Cell. Res.*, 249 (2), 299–307, 1999.

28. Abe, Y., Hashimoto, S., and Horie, T., Curcumin inhibition of inflammatory cytokine production by human peripheral blood monocytes and alveolar macrophages, *Pharmacol. Res.*, 39 (1), 41–47, 1999.

29. Chen, H., Zhang, Z.S., Zhang, Y.L., and Zhou, D.Y., Curcumin inhibits cell proliferation by interfering with the cell cycle and inducing apoptosis in colon carcinoma cells, *Anticancer Res.*, 19 (5A), 3675–3680, 1999.

30. Korutla, L. and Kumar, R., Inhibitory effect of curcumin on epidermal growth factor receptor kinase activity in A431 cells, *Biochim. Biophys. Acta*, 1224 (3), 597–600, 1994.

31. Mehta, K., Pantazis, P., McQueen, T., and Aggarwal, B.B., Antiproliferative effect of curcumin (diferuloylmethane) against human breast tumor cell lines, *Anticancer Drugs*, 8 (5), 470–481, 1997.

32. Ramachandran, C. and You, W., Differential sensitivity of human mammary epithelial and breast carcinoma cell lines to curcumin, *Breast Cancer Res. Treat.*, 54 (3), 269–278, 1999.

33. Simon, A., Allais, D.P., Duroux, J.L., Basly, J.P., Durand-Fontanier, S., and Delage, C., Inhibitory effect of curcuminoids on MCF-7 cell proliferation and structure-activity relationships, *Cancer Lett.*, 129 (1), 111–116, 1998.

34. Anto, R.J., Mukhopadhyay, A., Denning, K., and Aggarwal, B.B., Curcumin (diferuloylmethane) induces apoptosis through activation of caspase-8, BID cleavage and cytochrome c release: its suppression by ectopic expression of Bcl-2 and Bcl-xl, *Carcinogenesis*, 23 (1), 143–150, 2002.

35. Bharti, A.C., Donato, N., Singh, S., and Aggarwal, B.B., Curcumin (diferuloylmethane) down-regulates the constitutive activation of nuclear factor-kappa B and IkappaBalpha kinase in human multiple myeloma cells, leading to suppression of proliferation and induction of apoptosis, *Blood*, 101 (3), 1053–1062, 2003.

36. Woo, J.H., Kim, Y.H., Choi, Y.J., Kim, D.G., Lee, K.S., Bae, J.H., Min, D.S., Chang, J.S., Jeong, Y.J., Lee, Y.H., Park, J.W., and Kwon, T.K., Molecular mechanisms of curcumin-induced cytotoxicity: induction of apoptosis through generation of reactive oxygen species, down-regulation of Bcl-XL and IAP, the release of cytochrome c and inhibition of Akt, *Carcinogenesis*, 24 (7), 1199–1208, 2003.
37. Zheng, L.D., Tong, Q.S., and Wu, C.H., Inhibitory effects of curcumin on apoptosis of human ovary cancer cell line A2780 and its molecular mechanism, *Ai Zheng*, 21 (12), 1296–1300, 2002.
38. Deeb, D., Xu, Y.X., Jiang, H., Gao, X., Janakiraman, N., Chapman, R.A., and Gautam, S.C., Curcumin (diferuloyl-methane) enhances tumor necrosis factor-related apoptosis-inducing ligand-induced apoptosis in LNCaP prostate cancer cells, *Mol. Cancer Ther.*, 2 (1), 95–103, 2003.
39. Chan, M.M., Fong, D., Soprano, K.J., Holmes, W.F., and Heverling, H., Inhibition of growth and sensitization to cisplatin-mediated killing of ovarian cancer cells by polyphenolic chemopreventive agents, *J. Cell. Physiol.*, 194 (1), 63–70, 2003.
40. Slamon, D.J., Clark, G.M., Wong, S.G., Levin, W.J., Ullrich, A., and McGuire, W.L., Human breast cancer: correlation of relapse and survival with amplification of the HER-2/neu oncogene, *Science*, 235 (4785), 177–182, 1987.
41. Korutla, L., Cheung, J.Y., Mendelsohn, J., and Kumar, R., Inhibition of ligand-induced activation of epidermal growth factor receptor tyrosine phosphorylation by curcumin, *Carcinogenesis*, 16 (8), 1741–1745, 1995.
42. Hong, R.L., Spohn, W.H., and Hung, M.C., Curcumin inhibits tyrosine kinase activity of p185neu and also depletes p185neu, *Clin. Cancer Res.*, 5 (7), 1884–1891, 1999.
43. Mukhopadhyay, A., Bueso-Ramos, C., Chatterjee, D., Pantazis, P., and Aggarwal, B.B., Curcumin downregulates cell survival mechanisms in human prostate cancer cell lines, *Oncogene*, 20 (52), 7597–7609, 2001.
44. Tikhomirov, O. and Carpenter, G., Identification of ErbB-2 kinase domain motifs required for geldanamycin-induced degradation, *Cancer Res.*, 63 (1), 39–43, 2003.
45. Baldwin, A.S., Control of oncogenesis and cancer therapy resistance by the transcription factor NF-kappaB, *J. Clin. Invest.*, 107 (3), 241–246, 2001.
46. Pahl, H.L., Activators and target genes of Rel/NF-kappaB transcription factors, *Oncogene*, 18 (49), 6853–6866, 1999.
47. Wang, C.Y., Mayo, M.W., and Baldwin, Jr., A.S., TNF- and cancer therapy-induced apoptosis: potentiation by inhibition of NF-kappaB, *Science*, 274 (5288), 784–787, 1996.
48. Lee, H., Arsura, M., Wu, M., Duyao, M., Buckler, A.J., and Sonenshein, G.E., Role of Rel-related factors in control of c-myc gene transcription in receptor-mediated apoptosis of the murine B cell WEHI 231 line, *J. Exp. Med.*, 181 (3), 1169–1177, 1995.
49. Giri, D.K. and Aggarwal, B.B., Constitutive activation of NF-kappaB causes resistance to apoptosis in human cutaneous T cell lymphoma HuT-78 cells. Autocrine role of tumor necrosis factor and reactive oxygen intermediates, *J. Biol. Chem.*, 273 (22), 14008–14014, 1998.
50. Manna, S.K. and Aggarwal, B.B., Lipopolysaccharide inhibits TNF-induced apoptosis: role of nuclear factor-kappaB activation and reactive oxygen intermediates, *J. Immunol.*, 162 (3), 1510–1518, 1999.
51. Nakshatri, H., Bhat-Nakshatri, P., Martin, D.A., Goulet, Jr., R.J., and Sledge, Jr., G.W., Constitutive activation of NF-kappaB during progression of breast cancer to hormone-independent growth, *Mol. Cell. Biol.*, 17 (7), 3629–3639, 1997.
52. Kim, D.W., Sovak, M.A., Zanieski, G., Nonet, G., Romieu-Mourez, R., Lau, A.W., Hafer, L.J., Yaswen, P., Stampfer, M., Rogers, A.E., Russo, J., and Sonenshein, G.E., Activation of NF-kappaB/Rel occurs early during neoplastic transformation of mammary cells, *Carcinogenesis*, 21 (5), 871–879, 2000.
53. Sovak, M.A., Bellas, R.E., Kim, D.W., Zanieski, G.J., Rogers, A.E., Traish, A.M., and Sonenshein, G.E., Aberrant nuclear factor-kappaB/Rel expression and the pathogenesis of breast cancer, *J. Clin. Invest.*, 100 (12), 2952–2960, 1997.
54. Singh, S. and Aggarwal, B.B., Activation of transcription factor NF-kappa B is suppressed by curcumin (diferuloylmethane) [corrected], *J. Biol. Chem.*, 270 (42), 24995–5000, 1995.
55. Jobin, C., Bradham, C.A., Russo, M.P., Juma, B., Narula, A.S., Brenner, D.A., and Sartor, R.B., Curcumin blocks cytokine-mediated NF-kappa B activation and proinflammatory gene expression by inhibiting inhibitory factor I-kappa B kinase activity, *J. Immunol.*, 163 (6), 3474–3483, 1999.

56. Plummer, S.M., Holloway, K.A., Manson, M.M., Munks, R.J., Kaptein, A., Farrow, S., and Howells, L., Inhibition of cyclo-oxygenase 2 expression in colon cells by the chemopreventive agent curcumin involves inhibition of NF-kappaB activation via the NIK/IKK signalling complex, *Oncogene*, 18 (44), 6013–6020, 1999.

57. Shishodia, S., Potdar, P., Gairola, C.G., and Aggarwal, B.B., Curcumin (diferuloylmethane) down-regulates cigarette smoke-induced NF-kappaB activation through inhibition of IkappaBalpha kinase in human lung epithelial cells: correlation with suppression of COX-2, MMP-9 and cyclin D1, *Carcinogenesis*, 24 (7), 1269–1279, 2003.

58. Philip, S. and Kundu, G.C., Osteopontin induces nuclear factor kappa B-mediated promatrix metalloproteinase-2 activation through I kappa B alpha/IKK signaling pathways, and curcumin (diferuloylmethane) down-regulates these pathways, *J. Biol. Chem.*, 278 (16), 14487–14497, 2003.

59. Bharti, A.C., Donato, N., and Aggarwal, B.B., Curcumin (diferuloylmethane) inhibits constitutive and interleukin-6-inducible STAT3 phosphorylation in human multiple myeloma cells, *J. Immunol.*, in press, 2003.

60. Li, W.Q., Dehnade, F., and Zafarullah, M., Oncostatin M-induced matrix metalloproteinase and tissue inhibitor of metalloproteinase-3 genes expression in chondrocytes requires Janus kinase/STAT signaling pathway, *J. Immunol.*, 166 (5), 3491–3498, 2001.

61. Natarajan, C. and Bright, J.J., Curcumin inhibits experimental allergic encephalomyelitis by blocking IL-12 signaling through Janus kinase-STAT pathway in T lymphocytes, *J. Immunol.*, 168 (12), 6506–6513, 2002.

62. Xu, J., Fu, Y., and Chen, A., Activation of peroxisome proliferator-activated receptor-gamma contributes to the inhibitory effects of curcumin on rat hepatic stellate cell growth, *Am. J. Physiol. Gastrointest. Liver Physiol.*, 285 (1), G20–30, 2003.

63. Karin, M., Liu, Z., and Zandi, E., AP-1 function and regulation, *Curr. Opin. Cell. Biol.*, 9 (2), 240–246, 1997.

64. Xia, Y., Makris, C., Su, B., Li, E., Yang, J., Nemerow, G.R., and Karin, M., MEK kinase 1 is critically required for c-Jun N-terminal kinase activation by proinflammatory stimuli and growth factor-induced cell migration, *Proc. Natl. Acad. Sci. USA*, 97 (10), 5243–5248, 2000.

65. Huang, C., Li, J., Ma, W.Y., and Dong, Z., JNK activation is required for JB6 cell transformation induced by tumor necrosis factor-alpha but not by 12-O-tetradecanoylphorbol-13-acetate, *J. Biol. Chem.*, 274 (42), 29672–29676, 1999.

66. Huang, M.T., Lysz, T., Ferraro, T., Abidi, T.F., Laskin, J.D., and Conney, A.H., Inhibitory effects of curcumin on *in vitro* lipoxygenase and cyclooxygenase activities in mouse epidermis, *Cancer Res.*, 51 (3) 813–819, 1991.

67. Chen, Y.R. and Tan, T.H., Inhibition of the c-Jun N-terminal kinase (JNK) signaling pathway by curcumin, *Oncogene*, 17 (2), 173–178, 1998.

68. Dickinson, D.A., Iles, K.E., Zhang, H., Blank, V., and Forman, H.J., Curcumin alters EpRE and AP-1 binding complexes and elevates glutamate-cysteine ligase gene expression, *FASEB J.*, 17 (3), 473–475, 2003.

69. Squires, M.S., Hudson, E.A., Howells, L., Sale, S., Houghton, C.E., Jones, J.L., Fox, L.H., Dickens, M., Prigent, S.A., and Manson, M.M., Relevance of mitogen activated protein kinase (MAPK) and phosphotidylinositol-3-kinase/protein kinase B (PI3K/PKB) pathways to induction of apoptosis by curcumin in breast cells, *Biochem. Pharmacol.*, 65 (3), 361–376, 2003.

70. Ohene-Abuakwa, Y. and Pignatelli, M., Adhesion molecules in cancer biology, *Adv. Exp. Med. Biol.*, 465, 115–126, 2000.

71. Iademarco, M.F., Barks, J.L., and Dean, D.C., Regulation of vascular cell adhesion molecule-1 expression by IL-4 and TNF-alpha in cultured endothelial cells, *J. Clin. Invest.*, 95 (1), 264–271, 1995.

72. Kumar, A., Dhawan, S., Hardegen, N.J., and Aggarwal, B.B., Curcumin (diferuloylmethane) inhibition of tumor necrosis factor (TNF)-mediated adhesion of monocytes to endothelial cells by suppression of cell surface expression of adhesion molecules and of nuclear factor-kappaB activation, *Biochem. Pharmacol.*, 55 (6), 775–783, 1998.

73. Jaiswal, A.S., Marlow, B.P., Gupta, N., and Narayan, S., Beta-catenin-mediated transactivation and cell-cell adhesion pathways are important in curcumin (diferuylmethane)-induced growth arrest and apoptosis in colon cancer cells, *Oncogene*, 21 (55), 8414–8427, 2002.

74. Fournier, D.B. and Gordon, G.B., COX-2 and colon cancer: Potential targets for chemoprevention, *J. Cell. Biochem.*, 77 (S34), 97–102, 2000.

75. Hida, T., Yatabe, Y., Achiwa, H., Muramatsu, H., Kozaki, K., Nakamura, S., Ogawa, M., Mitsudomi, T., Sugiura, T., and Takahashi, T., Increased expression of cyclooxygenase 2 occurs frequently in human lung cancers, specifically in adenocarcinomas, *Cancer Res.*, 58 (17), 3761–3764, 1998.

76. Harris, R.E., Alshafie, G.A., Abou-Issa, H., and Seibert, K., Chemoprevention of breast cancer in rats by celecoxib, a cyclooxygenase 2 inhibitor, *Cancer Res.*, 60 (8) 2101–2103, 2000.

77. Williams, C.S., Mann, M., and DuBois, R.N., The role of cyclooxygenases in inflammation, cancer, and development, *Oncogene*, 18 (55), 7908–7916, 1999.

78. Reddy, B.S., Hirose, Y., Lubet, R., Steele, V., Kelloff, G., Paulson, S., Seibert, K., and Rao, C.V., Chemoprevention of colon cancer by specific cyclooxygenase-2 inhibitor, celecoxib, administered during different stages of carcinogenesis, *Cancer Res.*, 60 (2), 293–297, 2000.

79. Folkman, J., Can mosaic tumor vessels facilitate molecular diagnosis of cancer? *Proc. Natl. Acad. Sci. USA*, 98 (2), 398–400, 2001.

80. Singh, A.K., Sidhu, G.S., Deepa, T., and Maheshwari, R.K., Curcumin inhibits the proliferation and cell cycle progression of human umbilical vein endothelial cell, *Cancer Lett.*, 107 (1), 109–115, 1996.

81. Mohan, R., Sivak, J., Ashton, P., Russo, L.A., Pham, B.Q., Kasahara, N., Raizman, M.B., and Fini, M.E., Curcuminoids inhibit the angiogenic response stimulated by fibroblast growth factor-2, including expression of matrix metalloproteinase gelatinase B, *J. Biol. Chem.*, 275 (14), 10405–10412, 2000.

82. Arbiser, J.L., Klauber, N., Rohan, R., van Leeuwen, R., Huang, M.T., Fisher, C., Flynn, E., and Byers, H.R., Curcumin is an *in vivo* inhibitor of angiogenesis, *Mol. Med.*, 4 (6) 376–383, 1998.

83. Park, M.J., Kim, E.H., Park, I.C., Lee, H.C., Woo, S.H., Lee, J.Y., Hong, Y.J., Rhee, C.H., Choi, S.H., Shim, B.S., Lee, S.H., and Hong, S.I., Curcumin inhibits cell cycle progression of immortalized human umbilical vein endothelial (ECV304) cells by up-regulating cyclin-dependent kinase inhibitor, p21WAF1/CIP1, p27KIP1 and p53, *Int. J. Oncol.*, 21 (2) 379–383, 2002.

84. Kumar, A., Dhawan, S., Mukhopadhyay, A., and Aggarwal, B.B., Human immunodeficiency virus-1-tat induces matrix metalloproteinase-9 in monocytes through protein tyrosine phosphatase-mediated activation of nuclear transcription factor NF-kappaB, *FEBS Lett.*, 462 (1–2), 140–144, 1999.

85. Lin, L.I., Ke, Y.F., Ko, Y.C., and Lin, J.K., Curcumin inhibits SK-Hep-1 hepatocellular carcinoma cell invasion *in vitro* and suppresses matrix metalloproteinase-9 secretion, *Oncology*, 55 (4) 349–353, 1998.

86. Pan, M.H., Lin-Shiau, S.Y., and Lin, J.K., Comparative studies on the suppression of nitric oxide synthase by curcumin and its hydrogenated metabolites through down-regulation of IkappaB kinase and NFkappaB activation in macrophages, *Biochem. Pharmacol.*, 60 (11), 1665–1676, 2000.

87. Menon, L.G., Kuttan, R., and Kuttan, G., Anti-metastatic activity of curcumin and catechin, *Cancer Lett.*, 141 (1–2), 159–165, 1999.

88. Baldin, V., Lukas, J., Marcote, M.J ., Pagano, M., and Draetta, G., Cyclin D1 is a nuclear protein required for cell cycle progression in G1, *Genes Dev.*, 7 (5), 812–821, 1993.

89. Bartkova, J., Lukas, J., Muller, H., Lutzhoft, D., Strauss, M., and Bartek, J., Cyclin D1 protein expression and function in human breast cancer, *Int. J. Cancer*, 57 (3), 353–361, 1994.

90. Adelaide, J., Monges, G., Derderian, C., Seitz, J.F., and Birnbaum, D., Oesophageal cancer and amplification of the human cyclin D gene CCND1/PRAD1, *Br. J. Cancer*, 71 (1), 64–68, 1995.

91. Caputi, M., Groeger, A.M., Esposito, V., Dean, C., De Luca, A., Pacilio, C., Muller, M.R., Giordano, G.G., Baldi, F., Wolner, E., and Giordano, A., Prognostic role of cyclin D1 in lung cancer: relationship to proliferating cell nuclear antigen, *Am. J. Respir. Cell. Mol. Biol.*, 20 (4), 746–750, 1999.

92. Nishida, N., Fukuda, Y., Komeda, T., Kita, R., Sando, T., Furukawa, M., Amenomori, M., Shibagaki, I., Nakao, K., Ikenaga, M. et al., Amplification and overexpression of the cyclin D1 gene in aggressive human hepatocellular carcinoma, *Cancer Res.*, 54 (12), 3107–3110, 1994.

93. Gumbiner, L.M., Gumerlock, P.H., Mack, P.C., Chi, S.G., deVere White, R.W., Mohler, J.L., Pretlow, T.G., and Tricoli, J.V., Overexpression of cyclin D1 is rare in human prostate carcinoma, *Prostate*, 38 (1), 40–45, 1999.

94. Drobnjak, M., Osman, I., Scher, H.I., Fazzari, M., and Cordon-Cardo, C., Overexpression of cyclin D1 is associated with metastatic prostate cancer to bone, *Clin. Cancer Res.*, 6 (5), 1891–1895, 2000.

95. Mukhopadhyay, A., Banerjee, S., Stafford, L.J., Xia, C.X., Liu, M., and Aggarwal, B.B., Curcumin-induced suppression of cell proliferation correlates with downregulation of cyclin D1 expression and CDK4-mediated retinoblastoma protein phosphorylation, *Oncogene*, 21 (57), 8852–8862, 2002.

96. Huang, M.T., Ma, W., Yen, P., Xie, J.G., Han, J., Frenkel, K., Grunberger, D., and Conney, A.H., Inhibitory effects of topical application of low doses of curcumin on 12-O-tetradecanoylphorbol-13-acetate-induced tumor promotion and oxidized DNA bases in mouse epidermis, *Carcinogenesis*, 18 (1), 83–88, 1997.

97. Perkins, S., Clarke, A.R., Steward, W., and Gescher, A., Age-related difference in susceptibility of Apc(Min/+) mice towards the chemopreventive efficacy of dietary aspirin and curcumin, *Br. J. Cancer*, 88 (9), 1480–1483, 2003.

98. Mahady, G.B., Pendland, S.L., Yun, G., and Lu, Z.Z., Turmeric (Curcuma longa) and curcumin inhibit the growth of *Helicobacter pylori*, a group 1 carcinogen, *Anticancer Res.*, 22 (6C), 4179–4181, 2002.

99. Van der Logt, E.M., Roelofs, H.M., Nagengast, F.M., and Peters, W.H., Induction of rat hepatic and intestinal UDP-glucuronosyltransferases by naturally occurring dietary anticarcinogens, *Carcinogenesis*, 2003.

100. Kuttan, R., Bhanumathy, P., Nirmala, K., and George, M.C., Potential anticancer activity of turmeric (Curcuma longa), *Cancer Lett.*, 29 (2), 197–202, 1985.

101. Busquets, S., Carbo, N., Almendro, V., Quiles, M.T., Lopez-Soriano, F.J., and Argiles, J.M., Curcumin, a natural product present in turmeric, decreases tumor growth but does not behave as an anticachectic compound in a rat model, *Cancer Lett.*, 167 (1), 33–38, 2001.

102. Menon, L.G., Kuttan, R., and Kuttan, G., Inhibition of lung metastasis in mice induced by B16F10 melanoma cells by polyphenolic compounds, *Cancer Lett.*, 95 (1–2), 221–225, 1995.

103. Dorai, T., Cao, Y.C., Dorai, B., Buttyan, R., and Katz, A.E., Therapeutic potential of curcumin in human prostate cancer, III: Curcumin inhibits proliferation, induces apoptosis, and inhibits angiogenesis of LNCaP prostate cancer cells *in vivo*, *Prostate*, 47 (4), 293–303, 2001.

104. Nakamura, K., Yasunaga, Y., Segawa, T., Ko, D., Moul, J.W., Srivastava, S., and Rhim, J.S., Curcumin down-regulates AR gene expression and activation in prostate cancer cell lines, *Int. J. Oncol.*, 21 (4), 825–830, 2002.

105. Rao, D.S., Sekhara, N.C., Satyanarayana, M.N., and Srinivasan, M., Effect of curcumin on serum and liver cholesterol levels in the rat, *J. Nutr.*, 100 (11), 1307–1315, 1970.

106. Patil, T.N. and Srinivasan, M., Hypocholesteremic effect of curcumin in induced hypercholesteremic rats, *Indian J. Exp. Biol.*, 9 (2), 167–169, 1971.

107. Keshavarz, K., The influence of turmeric and curcumin on cholesterol concentration of eggs and tissues, *Poult. Sci.*, 55 (3), 1077–1083, 1976.

108. Soudamini, K.K., Unnikrishnan, M.C., Soni, K.B., and Kuttan, R., Inhibition of lipid peroxidation and cholesterol levels in mice by curcumin, *Indian J. Physiol. Pharmacol.*, 36 (4), 239–243, 1992.

109. Soni, K.B., Rajan, A., and Kuttan, R., Reversal of aflatoxin induced liver damage by turmeric and curcumin, *Cancer Lett.*, 66 (2); 115–121, 1992.

110. Hussain, M.S. and Chandrasekhara, N., Effect of curcumin on cholesterol gall-stone induction in mice, *Indian J. Med. Res.*, 96; 288–291, 1992.

111. Hussain, M.S. and Chandrasekhara, N., Biliary proteins from hepatic bile of rats fed curcumin or capsaicin inhibit cholesterol crystal nucleation in supersaturated model bile, *Indian J. Biochem. Biophys.*, 31 (5), 407–412, 1994.

112. Soni, K.B. and Kuttan, R., Effect of oral curcumin administration on serum peroxides and cholesterol levels in human volunteers, *Indian J. Physiol. Pharmacol.*, 36 (4), 273–275, 1992.

113. Yasni, S., Imaizumi, K., Nakamura, M., Aimoto, J., and Sugano, M., Effects of Curcuma xanthorrhiza Roxb. and curcuminoids on the level of serum and liver lipids, serum apolipoprotein A-I and lipogenic enzymes in rats, *Food Chem. Toxicol.*, 31 (3), 213–218, 1993.

114. Yasni, S., Imaizumi, K., Sin, K., Sugano, M., Nonaka, G., and Sidik, Identification of an active principle in essential oils and hexane-soluble fractions of Curcuma xanthorrhiza Roxb. showing triglyceride-lowering action in rats, *Food Chem. Toxicol.*, 32 (3), 273–278, 1994.

115. Skrzypczak-Jankun, E., Zhou, K., McCabe, N.P., Selman, S.H., and Jankun, J., Structure of curcumin in complex with lipoxygenase and its significance in cancer, *Int. J. Mol. Med.*, 12 (1), 17–24, 2003.

116. Rukkumani, R., Sri Balasubashini, M., Vishwanathan, P., and Menon, V.P., Comparative effects of curcumin and photo-irradiated curcumin on alcohol- and polyunsaturated fatty acid-induced hyperlipidemia, *Pharmacol. Res.*, 46 (3), 257–264, 2002.

117. Quiles, J.L., Aguilera, C., Mesa, M.D., Ramirez-Tortosa, M.C., Baro, L., and Gil, A., An ethanolic-aqueous extract of Curcuma longa decreases the susceptibility of liver microsomes and mitochondria to lipid peroxidation in atherosclerotic rabbits, *Biofactors*, 8 (1–2), 51–57, 1998.
118. Ramirez-Tortosa, M.C., Mesa, M.D., Aguilera, M.C., Quiles, J.L., Baro, L., Ramirez-Tortosa, C.L., Martinez-Victoria, E., and Gil, A., Oral administration of a turmeric extract inhibits LDL oxidation and has hypocholesterolemic effects in rabbits with experimental atherosclerosis, *Atherosclerosis*, 147 (2), 371–378, 1999.
119. Asai, A. and Miyazawa, T., Dietary curcuminoids prevent high-fat diet-induced lipid accumulation in rat liver and epididymal adipose tissue, *J. Nutr.*, 131 (11), 2932–2935, 2001.
120. Naidu, K.A. and Thippeswamy, N.B., Inhibition of human low density lipoprotein oxidation by active principles from spices, *Mol. Cell. Biochem.*, 229 (1–2), 19–23, 2002.
121. Srivastava, R., Dikshit, M., Srimal, R.C., and Dhawan, B.N., Anti-thrombotic effect of curcumin, *Thromb. Res.*, 40 (3), 413–417, 1985.
122. Srivastava, R., Puri, V., Srimal, R.C., and Dhawan, B.N., Effect of curcumin on platelet aggregation and vascular prostacyclin synthesis, *Arzneimittelforschung*, 36 (4), 715–717, 1986.
123. Srivastava, K.C., Bordia, A., and Verma, S.K., Curcumin, a major component of food spice turmeric (Curcuma longa), inhibits aggregation and alters eicosanoid metabolism in human blood platelets, *Prostaglandins Leukot. Essent. Fatty Acids*, 52 (4), 223–227, 1995.
124. Dikshit, M., Rastogi, L., Shukla, R., and Srimal, R.C., Prevention of ischaemia-induced biochemical changes by curcumin and quinidine in the cat heart, *Indian J. Med. Res.*, 101, 31–35, 1995.
125. Nirmala, C. and Puvanakrishnan, R., Effect of curcumin on certain lysosomal hydrolases in isoproterenol-induced myocardial infarction in rats, *Biochem. Pharmacol.*, 51 (1), 47–51, 1996.
126. Nirmala, C., Anand, S., and Puvanakrishnan, R., Curcumin treatment modulates collagen metabolism in isoproterenol induced myocardial necrosis in rats, *Mol. Cell. Biochem.*, 197 (1–2), 31–37, 1999.
127. Shahed, A.R., Jones, E., and Shoskes, D., Quercetin and curcumin up-regulate antioxidant gene expression in rat kidney after ureteral obstruction or ischemia/reperfusion injury, *Transplant Proc.*, 33 (6), 2988, 2001.
128. Arun, N. and Nalini, N., Efficacy of turmeric on blood sugar and polyol pathway in diabetic albino rats, *Plant Foods Hum. Nutr.*, 57 (1), 41–52, 2002.
129. Srinivasan, M., Effect of curcumin on blood sugar as seen in a diabetic subject, *Indian J. Med. Sci.*, 26 (4), 269–270, 1972.
130. Babu, P.S. and Srinivasan, K., Influence of dietary curcumin and cholesterol on the progression of experimentally induced diabetes in albino rat, *Mol. Cell. Biochem.*, 152 (1), 13–21, 1995.
131. Babu, P.S. and Srinivasan, K., Hypolipidemic action of curcumin, the active principle of turmeric (Curcuma longa) in streptozotocin induced diabetic rats, *Mol. Cell. Biochem.*, 166 (1–2), 169–175, 1997.
132. Suresh Babu, P. and Srinivasan, K., Amelioration of renal lesions associated with diabetes by dietary curcumin in streptozotocin diabetic rats, *Mol. Cell. Biochem.*, 181 (1–2), 87–96, 1998.
133. Thaloor, D., Miller, K.J., Gephart, J., Mitchell, P.O., and Pavlath, G.K., Systemic administration of the NF-kappaB inhibitor curcumin stimulates muscle regeneration after traumatic injury, *Am. J. Physiol.*, 277 (2 Pt. 1); C320–329, 1999.
134. Sidhu, G.S., Singh, A.K., Thaloor, D., Banaudha, K.K., Patnaik, G.K., Srimal, R.C., and Maheshwari, R.K., Enhancement of wound healing by curcumin in animals, *Wound Repair Regen.*, 6 (2), 167–177, 1998.
135. Phan, T.T., See, P., Lee, S.T., and Chan, S.Y., Protective effects of curcumin against oxidative damage on skin cells *in vitro*: its implication for wound healing, *J. Trauma*, 51 (5), 927–931, 2001.
136. Mani, H., Sidhu, G.S., Kumari, R., Gaddipati, J.P., Seth, P., and Maheshwari, R.K., Curcumin differentially regulates TGF-beta1, its receptors and nitric oxide synthase during impaired wound healing, *Biofactors*, 16 (1–2), 29–43, 2002.
137. Deodhar, S.D., Sethi, R., and Srimal, R.C., Preliminary study on antirheumatic activity of curcumin (diferuloyl methane), *Indian J. Med. Res.*, 71, 632–634, 1980.
138. Liacini, A., Sylvester, J., Li, W.Q., and Zafarullah, M., Inhibition of interleukin-1-stimulated MAP kinases, activating protein-1 (AP-1) and nuclear factor kappa B (NF-kappaB) transcription factors down-regulates matrix metalloproteinase gene expression in articular chondrocytes, *Matrix Biol.*, 21 (3), 251–262, 2002.

139. Li, C.J., Zhang L.J., Dezube, B.J., Crumpaker, C.S., Pardee, A.B., Three inhibitors of type 1 human immunodeficiency virus long terminal repeat-directed gene expression and virus replication. *Proc. Natl. Acad. Sci. USA,* 90 (5), 1839–1842, 1993.

140. Sui, Z., Salto, R., Li, J., Craik, C., and Ortiz de Montellano, P.R., Inhibition of the HIV-1 and HIV-2 proteases by curcumin and curcumin boron complexes, *Bioorg. Med. Chem.,* 1 (6), 415–422, 1993.

141. Mazumder, A., Neamati, N., Sunder, S., Schulz, J., Pertz, H., Eich, E., and Pommier, Y., Curcumin analogs with altered potencies against HIV-1 integrase as probes for biochemical mechanisms of drug action, *J. Med. Chem.,* 40 (19), 3057–3063, 1997.

142. Barthelemy, S., Vergnes, L., Moynier, M., Guyot, D., Labidalle, S., and Bahraoui, E., Curcumin and curcumin derivatives inhibit Tat-mediated transactivation of type 1 human immunodeficiency virus long terminal repeat, *Res. Virol.,* 149 (1), 43–52, 1998.

143. Lim, G.P., Chu, T., Yang, F., Beech, W., Frautschy, S.A., and Cole, G.M., The curry spice curcumin reduces oxidative damage and amyloid pathology in an Alzheimer transgenic mouse, *J. Neurosci.,* 21 (21), 8370–8377, 2001.

144. Awasthi, S., Srivatava, S.K., Piper, J.T., Singhal, S.S., Chaubey, M., and Awasthi, Y.C., Curcumin protects against 4-hydroxy-2-trans-nonenal-induced cataract formation in rat lenses, *Am. J. Clin. Nutr.,* 64 (5), 761–766, 1996.

145. Venkatesan, N., Curcumin attenuation of acute adriamycin myocardial toxicity in rats, *Br. J. Pharmacol.,* 124 (3), 425–427, 1998.

146. Nanji, A.A., Jokelainen, K., Tipoe, G.L., Rahemtulla, A., Thomas, P., and Dannenberg, A.J., Curcumin prevents alcohol-induced liver disease in rats by inhibiting the expression of NF-kappa B-dependent genes, *Am. J. Physiol. Gastrointest. Liver Physiol.,* 284 (2), G321–327, 2003.

147. Venkatesan, N. and Chandrakasan, G., Modulation of cyclophosphamide-induced early lung injury by curcumin, an anti-inflammatory antioxidant, *Mol. Cell. Biochem.,* 142 (1), 79–87, 1995.

148. Venkatesan, N., Punithavathi, V., and Chandrakasan, G., Curcumin protects bleomycin-induced lung injury in rats, *Life Sci.,* 61 (6); PL51–58, 1997.

149. Punithavathi, D., Venkatesan, N., and Babu, M., Curcumin inhibition of bleomycin-induced pulmonary fibrosis in rats, *Br. J. Pharmacol.,* 131 (2), 169–172, 2000.

150. Punithavathi, D., Venkatesan, N., and Babu, M., Protective effects of curcumin against amiodarone-induced pulmonary fibrosis in rats, *Br. J. Pharmacol.,* 139 (7), 1342–1350, 2003.

151. Venkatesan, N., Pulmonary protective effects of curcumin against paraquat toxicity, *Life Sci.,* 66 (2), PL21–28, 2000.

152. Venkatesan, N., Punithavathi, D., and Arumugam, V., Curcumin prevents adriamycin nephrotoxicity in rats, *Br. J. Pharmacol.,* 129 (2), 231–234, 2000.

153. Phan, T.T., Sun, L., Bay, B.H., Chan, S.Y., and Lee, S.T., Dietary compounds inhibit proliferation and contraction of keloid and hypertrophic scar-derived fibroblasts *in vitro*: therapeutic implication for excessive scarring, *J. Trauma,* 54 (6), 1212–1224, 2003.

154. Ukil, A., Maity, S., Karmakar, S., Datta, N., Vedasiromoni, J.R., and Das, P.K., Curcumin, the major component of food flavour turmeric, reduces mucosal injury in trinitrobenzene sulphonic acid-induced colitis, *Br. J. Pharmacol.,* 139 (2), 209–218, 2003.

155. Salh, B., Assi, K., Templeman, V., Parhar, K., Owen, D., Gomez-Munoz, A., and Jacobson, K., Curcumin attenuates DNB-induced murine colitis, *Am. J. Physiol. Gastrointest. Liver Physiol.,* 285 (1), G235–243, 2003.

156. Chueh, S.C., Lai, M.K., Liu, I.S., Teng, F.C., and Chen, J., Curcumin enhances the immunosuppressive activity of cyclosporine in rat cardiac allografts and in mixed lymphocyte reactions, *Transplant Proc.,* 35 (4), 1603–1605, 2003.

157. Balogun, E., Hoque, M., Gong, P., Killeen, E., Green, C.J., Foresti, R., Alam, J., and Motterlini, R., Curcumin activates the haem oxygenase-1 gene via regulation of Nrf2 and the antioxidant-responsive element, *Biochem. J.,* 371 (Pt. 3); 887–895, 2003.

158. Madan, B. and Ghosh, B., Diferuloylmethane inhibits neutrophil infiltration and improves survival of mice in high-dose endotoxin shock, *Shock,* 19 (1), 91–96, 2003.

159. Gukovsky, I., Reyes, C.N., Vaquero, E.C., Gukovskaya, A.S., and Pandol, S.J., Curcumin ameliorates ethanol and nonethanol experimental pancreatitis, *Am. J. Physiol. Gastrointest. Liver Physiol.,* 284 (1), G85–95, 2003.

160. Piwocka, K., Bielak-Mijewska, A., and Sikora, E., Curcumin induces caspase-3-independent apoptosis in human multidrug-resistant cells, *Ann. N.Y. Acad. Sci.*, 973, 250–254, 2002.

161. Lin, J.K., Pan, M.H., and Lin-Shiau, S.Y., Recent studies on the biofunctions and biotransformations of curcumin, *Biofactors*, 13 (1–4), 153–158, 2000.

162. Ireson, C.R., Jones, D.J., Orr, S., Coughtrie, M.W., Boocock, D.J., Williams, M.L., Farmer, P.B., Steward, W.P., and Gescher, A.J., Metabolism of the cancer chemopreventive agent curcumin in human and rat intestine, *Cancer Epidemiol. Biomarkers Prev.*, 11 (1), 105–111, 2002.

163. Johri, R.K. and Zutshi, U., An Ayurvedic formulation "Trikatu" and its constituents, *J. Ethnopharmacol.*, 37 (2), 85–91, 1992.

164. Shoba, G., Joy, D., Joseph, T., Majeed, M., Rajendran, R., and Srinivas, P.S., Influence of piperine on the pharmacokinetics of curcumin in animals and human volunteers, *Planta Med.*, 64 (4), 353–356, 1998.

165. Kumar, V., Lewis, S.A., Mutalik, S., Shenoy, D.B., Venkatesh, and Udupa, N., Biodegradable microspheres of curcumin for treatment of inflammation, *Indian J. Physiol. Pharmacol.*, 46 (2), 209–217, 2002.

166. Kumar, S., Dubey, K.K., Tripathi, S., Fujii, M., and Misra, K., Design and synthesis of curcumin-bioconjugates to improve systemic delivery, *Nucleic Acids Symp. Ser.*, (44), 75–76, 2000.

167. Lal, B., Kapoor, A.K., Asthana, O.P., Agrawal, P.K., Prasad, R., Kumar, P., and Srimal, R.C., Efficacy of curcumin in the management of chronic anterior uveitis, *Phytother. Res.*, 13 (4), 318–322, 1999.

168. Lal, B., Kapoor, A.K., Agrawal, P.K., Asthana, O.P., and Srimal, R.C., Role of curcumin in idiopathic inflammatory orbital pseudotumours, *Phytother. Res.*, 14 (6), 443–447, 2000.

169. Satoskar, R.R., Shah, S.J., and Shenoy, S.G., Evaluation of anti-inflammatory property of curcumin (diferuloyl methane) in patients with postoperative inflammation, *Int. J. Clin. Pharmacol. Ther. Toxicol.*, 24 (12), 651–654, 1986.

170. Kuttan, R., Sudheeran, P.C., and Joseph, C.D., Turmeric and curcumin as topical agents in cancer therapy, *Tumori*, 73 (1), 29–31, 1987.

171. James, J.S., Curcumin: clinical trial finds no antiviral effect, *AIDS Treat. News*, (242), 1–2, 1996.

172. Sharma, R.A., McLelland, H.R., Hill, K.A., Ireson, C.R., Euden, S.A., Manson, M.M., Pirmohamed, M., Marnett, L.J., Gescher, A.J., and Steward, W.P., Pharmacodynamic and pharmacokinetic study of oral curcuma extract in patients with colorectal cancer, *Clin. Cancer Res.*, 7 (7), 1894–1900, 2001.

173. Ishida, J., Ohtsu, H., Tachibana, Y., Nakanishi, Y., Bastow, K.F., Nagai, M., Wang, H.K., Itokawa, H., and Lee, K.H., Antitumor agents, part 214: synthesis and evaluation of curcumin analogues as cytotoxic agents, *Bioorg. Med. Chem.*, 10 (11), 3481–3487, 2002.

174. Dinkova-Kostova, A.T. and Talalay, P., Relation of structure of curcumin analogs to their potencies as inducers of phase-2 detoxification enzymes, *Carcinogenesis*, 20 (5), 911–914, 1999.

175. Naito, M., Wu, X., Nomura, H., Kodama, M., Kato, Y., and Osawa, T., The protective effects of tetrahydrocurcumin on oxidative stress in cholesterol-fed rabbits, *J. Atheroscler. Thromb.*, 9 (5), 243–250, 2002.

176. Kumar, A.P., Garcia, G.E., Ghosh, R., Rajnarayanan, R.V., Alworth, W.L., and Slaga, T.J., 4-Hydroxy-3-methoxybenzoic acid methyl ester: a curcumin derivative targets Akt/NF kappa B cell survival signaling pathway: potential for prostate cancer management, *Neoplasia*, 5 (3), 255–266, 2003.

177. Robinson, T.P., Ehlers, T., Hubbard, I.R., Bai, X., Arbiser, J.L., Goldsmith, D.J., and Bowen, J.P., Design, synthesis, and biological evaluation of angiogenesis inhibitors: aromatic enone and dienone analogues of curcumin, *Bioorg. Med. Chem. Lett.*, 13 (1), 115–117, 2003.

178. Devasena, T., Rajasekaran, K.N., and Menon, V.P., Bis-1,7-(2-hydroxyphenyl)-hepta-1,6-diene-3,5-dione (a curcumin analog) ameliorates DMH-induced hepatic oxidative stress during colon carcinogenesis, *Pharmacol. Res.*, 46 (1), 39–45, 2002.

179. Mishra, S., Tripathi, S., and Misra, K., Synthesis of a novel anticancer prodrug designed to target telomerase sequence, *Nucleic Acids Res. Suppl.*, (2), 277–278, 2002.

180. John, V.D., Kuttan, G., and Krishnankutty, K., Anti-tumour studies of metal chelates of synthetic curcuminoids, *J. Exp. Clin. Cancer Res.*, 21 (2), 219–224, 2002.

181. Shim, J.S., Kim, D.H., Jung, H.J., Kim, J.H., Lim, D., Lee, S.K., Kim, K.W., Ahn, J.W., Yoo, J.S., Rho, J.R., Shin, J., and Kwon, H.J., Hydrazinocurcumin, a novel synthetic curcumin derivative, is a potent inhibitor of endothelial cell proliferation, *Bioorg. Med. Chem.*, 10 (9), 2987–2992, 2002.

182. Snyder, J.P., Davis, M.C., and Adams, B., Curcumin analogs with anti-tumor and anti-angiogenic properties, U.S. Patent application, 2002/0019382, 2002.

24 Tea in Chemoprevention of Cancer

Vaqar Mustafa Adhami, Moammir Hasan Aziz, Nihal Ahmad, and Hasan Mukhtar

CONTENTS

24.1 Tea and Cancer: an Introduction 390
24.2 Consumption, Composition, and Chemistry of Tea 391
    24.2.1 Consumption 391
    24.2.2 Composition 392
        24.2.2.1 Green Tea 392
        24.2.2.2 Black Tea and Oolong Tea 392
    24.2.3 Chemistry 393
24.3 Anticarcinogenic Effects of Tea: Experimental Studies 394
    24.3.1 Prevention against Skin Tumorigenesis 394
        24.3.1.1 Prevention against Skin Tumor Initiation 394
        24.3.1.2 Prevention against Skin Tumor Promotion 394
        24.3.1.3 Prevention against Stage I and Stage II Skin Tumor Promotion 395
        24.3.1.4 Prevention against Malignant Conversion of Chemically Induced Benign Skin Papillomas to Carcinomas 395
        24.3.1.5 Prevention against UVB Radiation-Induced Photocarcinogenesis 396
        24.3.1.6 Effect on the Growth of Established Skin Tumors 397
    24.3.2 Prevention against Forestomach and Lung Tumorigenesis 398
    24.3.3 Prevention against Esophageal Tumorigenesis 398
    24.3.4 Prevention against Duodenum and Small Intestine Tumorigenesis 399
    24.3.5 Prevention against Colon Tumorigenesis 399
    24.3.6 Prevention against Liver Tumorigenesis 399
    24.3.7 Prevention against Mammary Carcinogenesis 400
    24.3.8 Prevention against Pancreatic Carcinogenesis 400
    24.3.9 Prevention against Prostate Carcinogenesis 401
    24.3.10 Modulatory Effect of Green Tea on Cancer Chemotherapy 402
    24.3.11 Cell-Specific Biological Effects of Tea 402
24.4 Anti-Inflammatory Effects of Tea 403
    24.4.1 Prevention against Tpa-Caused Inflammatory Responses 403
    24.4.2 Prevention against UVB Radiation-Caused Inflammatory Responses 403
    24.4.3 Prevention against UVB Radiation-Induced Immunosuppression 404
    24.4.4 Inhibition of Tumor Promoter-Caused Induction of Cytokines 404
24.5 Mechanisms of Biological Effects of Tea 404
    24.5.1 Prevention against Mutagenicity and Genotoxicity 404
    24.5.2 Inhibition of Biochemical Markers of Tumor Initiation: Cytochrome P450-Dependent Metabolism 405

24.5.3 Inhibition of Biochemical Markers of Tumor Promotion 405
24.5.4 Effects on Detoxification Enzymes 406
24.5.5 Trapping of Activated Metabolites of Carcinogens 407
24.5.6 Antioxidant and Free-Radical Scavenging Activity 407
24.5.7 Inhibition of Tumor Angiogenesis by Tea..408
24.5.8 Recent Advances in the Mechanisms of Biological Effects of Tea......................409
       24.5.8.1 Green Tea Modulates Mitogen-Activated Protein Kinases...................409
       24.5.8.2 EGCG Inhibits Urokinase Activity 409
       24.5.8.3 Green Tea Induces Apoptosis and Cell Cycle Arrest 410
       24.5.8.4 EGCG Suppresses Extracellular Signals and Cell
               Proliferation through EGF Receptor Binding..410
       24.5.8.5 EGCG Down-Regulates Nuclear Transcription Factor-κB....................410
       24.5.8.6 EGCG and Theaflavins Inhibit Tumor Promoter-Induced
               Activator Protein 1 Activation and Cell Transformation.......................411
24.6 Tea and Cancer: Epidemiological Studies 411
     24.6.1 Cancer of the Esophagus and Nasopharynx..411
     24.6.2 Cancer of the Stomach..412
     24.6.3 Cancer of the Bladder, Kidney, and Urinary Tract 413
     24.6.4 Cancer of the Colon, Rectum, and Uterus 413
     24.6.5 Cancer of the Prostate...414
     24.6.6 Cancer of the Liver, Lung, Breast, and Pancreas..415
24.7 Conclusion and Future Directions 415
References 417

24.1 TEA AND CANCER: AN INTRODUCTION

Tea is a popular beverage made from the leaves of *Camellia sinensis,* an evergreen shrub of the Theaceae family. According to popular Chinese belief, tea consumption can be traced to 2737 B.C., when Shen Nung, then emperor of China, discovered and used tea for the first time. Tea consumption rapidly spread across the globe, and the tea plant is currently cultivated in approximately 30 countries. According to an estimate of the Food and Agriculture Organization of the United Nations, world tea production in 2002 reached 3 million tons, and the demand for tea continues to grow. There are three basic types of tea — black, green, and oolong — based on the method of brewing and processing. About 20% of the total tea manufactured is green tea, which is consumed mainly in Asian countries such as Japan and China; 78% is black tea, which is predominantly consumed in Western and some Asian countries; the other 2% is oolong tea, which is produced and consumed in southeastern China.

Green tea is produced by steaming or drying fresh leaves at temperatures where the chemical composition of the extract remains unchanged. Green tea contains polyphenols, including flavonols, flavondiols, flavonoids, and phenolic acids. Most of the green tea polyphenols are flavonols, known as catechins. Some major green tea catechins are (+)-catechin, (+)-gallocatechin, (–)-epicatechin, (–)-epicatechin-3-gallate, (–)-epigallocatechin, and (–)-epigallocatechin-3-gallate.

The picture-perfect image of serenity can be viewed as relaxing with an interesting book and sipping a cup of hot tea. For many thousands of years, the harvesting and processing of the leaves of the tea plant *Camellia sinensis* has become as an integral part of human society and traditional culture. Because of its characteristic flavor and pharmacological properties, tea — next to water — is the most popular beverage consumed worldwide (Harbowy and Balentine, 1997). The per capita consumption of tea in the U.S. is approximately 340 g. Although the largest total consumption of tea is registered in India (540,000 metric tons, 620 g per capita), Ireland has the largest per capita consumption of tea (3220 g). Tea contains several polyphenolic components that are antioxidant in nature, and many studies have shown that tea polyphenols possess the ability to prevent oxidant-induced cellular damage (Harbowy and Balentine, 1997; Katiyar and Mukhtar, 1996). In recent years, studies from many laboratories around the world, conducted in various organ-specific animal bioassay systems, have shown that tea and its polyphenolic constituents are capable of affording protection against a variety of cancer types. Most of the studies conducted have used green tea, although a limited number of studies have also shown the anticancer efficacy of black tea.

The magnitude of the problem of cancer and the failure of conventional strategies to effect a marked diminution in the total number of deaths from this disease indicate that other preventive measures must be seriously considered. There are many possible strategies to reduce cancer-related deaths: (1) prevention, (2) early diagnosis and intervention, (3) successful treatment of localized cancer, and (4) improved management of nonlocalized cancer. Among these, prevention appears to be the most practical approach for dealing with the cancer problem.

Chemoprevention, by definition, is a means of cancer control in which the occurrence of the disease can be entirely prevented, slowed, or reversed by the administration of one or more naturally occurring and/or synthetic compounds (Ames, 1983; Ames and Gold, 1990; Kohlmeier et al., 1997; Morse and Stoner, 1993; Mukhtar and Ahmad, 1999; Sporn, 1991; Wattenberg, 1990). The expanded definition of cancer chemoprevention also includes the chemotherapy of precancerous lesions, which are called preinvasive neoplasia, dysplasia, or intraepithelial neoplasia, depending on the organ system (Boone et al., 1990, Boone et al., 1992). Such chemopreventive compounds are known as anticarcinogens, and ideally they should have (a) little or no toxic effects, (b) high efficacy in multiple sites, (c) capability of oral consumption, (d) a known mechanism of action, (e) low cost, and (f) human acceptance.

Chemoprevention of cancer thus differs from conventional cancer treatment in that the goal of this approach is to lower the rate of cancer incidence. This approach is promising because conventional therapies and surgery have not been fully effective against the high incidence and low survival rate of most types of cancer. Furthermore, the chemopreventive approach appears to have practical implications in reducing cancer risk. Individuals cannot easily control environmental factors that may contribute to carcinogenesis, but they can choose the food and beverages they consume.

In recent years, the naturally occurring compounds, especially the antioxidants, present in the common diet and beverages consumed by human population have gained considerable attention as chemopreventive agents for potential human benefit (Ames, 1983; Ames and Gold, 1990; Kohlmeier et al., 1997; Morse and Stoner, 1993; Mukhtar and Ahmad, 1999; Sporn, 1991; Wattenberg, 1990). Abundant epidemiological, experimental, and metabolic studies have provided convincing evidence that nutrition plays an important causative role in the initiation, promotion, and progression stages of several types of human cancers (Ames, 1983; Boone et al., 1992; Wattenberg, 1990). It is becoming clear that, in addition to substances that pose a cancer risk, the human diet also contains agents that are capable of affording protection against some forms of cancer (Ames, 1983; Ames and Gold, 1990; Kohlmeier et al., 1997). This collective information strongly suggests that the occurrence of cancer can be prevented or slowed by dietary intake of substances that have the capacity to afford protection against the occurrence of cancer. This chapter presents a critical evaluation of this topic.

24.2 CONSUMPTION, COMPOSITION, AND CHEMISTRY OF TEA

24.2.1 CONSUMPTION

The tea plant originated in Southeast Asia and is presently cultivated in over 30 countries around the globe. Tea as a beverage is consumed worldwide, although at greatly varying levels of consumption. A large segment of the world's population consumes virtually no tea. Not only does tea consumption vary from country to country, but there is also enormous variation within a given population, ranging from no tea to as many as 20 or more cups per day. Although definite data are not available, it is generally accepted that next to water, tea is the most consumed beverage in the world, with a per capita worldwide consumption of approximately 120 ml per day (Katiyar and Mukhtar, 1996). Of the total tea produced in the world, 78% is black tea, 20% is green tea, and less than 2% is oolong tea. Green tea is produced in relatively few countries and is mainly consumed

TABLE 24.1
Polyphenolic Composition of Green and Black Tea (% w/w)

Constituent	Green Tea	Black Tea
Catechins	30–42	3–10
Flavonols	5–10	6–8
Other flavonoids	2–4	—
Theogallin	2–3	—
Gallic acid	0.5	—
Quinic acid	2.0	—
Theanine	4–6	—
Methylxanthines	7–9	8–11
Theaflavins	—	3–6
Thearubigins	—	12–18

in China, Japan, India, and a few countries in North Africa and the Middle East. Of the total tea consumed in the world, black tea accounts for about 78%, which is mainly consumed in the Western countries and some Asian countries. Oolong tea production and consumption is confined to south-eastern China and Taiwan (Katiyar and Mukhtar, 1996).

24.2.2 COMPOSITION

The composition of tea-leaf varies with climate, season, horticultural practices, variety of the plant, and age of the leaf, i.e., the position of the leaf on the harvested shoot. Three main varieties of the commercial tea are available: green (unfermented), Oolong (partially fermented), and black (fully fermented). Their composition varies according to the manufacturing process, which differs in the degree of "enzymatic oxidation" or fermentation. In Table 24.1, principal polyphenolic components present in a typical green and black tea beverage are shown, but variations may be considerable. Oolong tea composition in general falls between that of green and black teas.

24.2.2.1 Green Tea

The manufacturing process for green tea involves rapid steaming or pan frying of freshly harvested leaves to inactivate enzymes, preventing fermentation and thereby producing a dry stable product. Green teas are generally produced in two different varieties, white tea and yellow tea, the latter being less fermented because of a process known as wilting. There is also some difference in manufacture between Chinese green tea and Japanese green tea. Because of the increasing popularity of green tea, a wide variety of products introduced into the market contain green tea. Epicatechins are the main constituent compounds in green tea, giving it the characteristic color and flavor.

24.2.2.2 Black Tea and Oolong Tea

In the production of black and oolong tea, fresh leaves are allowed to wither until the moisture content of the leaves is reduced to about 55% of the original leaf weight, resulting in the concentration of polyphenols and the deterioration of leaf structural integrity. This step gives the typical aroma to the tea. The withered leaves are rolled and crushed, initiating fermentation of the polyphenols. This process is known as maceration, and the fermenting mass is known as "dhool." The process used to crush the leaf plays an important role in the final grading of tea. During these processes, the catechins are converted to theaflavins and thearubigins. Theaflavins are astringent compounds contributing importantly to the color and taste of the black tea. The thearubigin fraction is a mixture of substances, with a molecular-weight distribution of 1,000 to 40,000, and accounts for 15% of dry-weight solids of black tea.

Oolong teas are prepared by firing the leaves shortly after rolling to terminate the oxidation and dry the leaves. Normal oolong tea is considered to be about half fermented compared with black tea. Oolong tea extracts contain catechins at a level of 8 to 20% of the total dry matter. The fermentation process results in the oxidation of simple polyphenols to more complex condensed polyphenols, giving black and oolong teas their characteristic colors and flavors (Harbowy and Balentine, 1997).

24.2.3 CHEMISTRY

The chemical composition of green tea is more or less similar to that of the fresh leaf with regard to the major components. Green tea contains polyphenolic compounds, including flavonols, flavon-diols, flavonoids, and phenolic acids. These compounds account for up to 30% of the dry weight of green tea leaves. Most of the polyphenols present in green tea are flavonols, commonly known as catechins. Some major catechins present in green tea are (–)-epicatechin (EC), (–)-epicatechin-3-gallate (ECG), (–)-epigallocatechin (EGC), and (–)-epigallocatechin-3-gallate (EGCG). The chemical structures of these compounds are given in Figure 24.1. In addition, caffeine, theobromine, theophylline, and phenolic acids such as gallic acids are also present in green tea (Table 24.1).

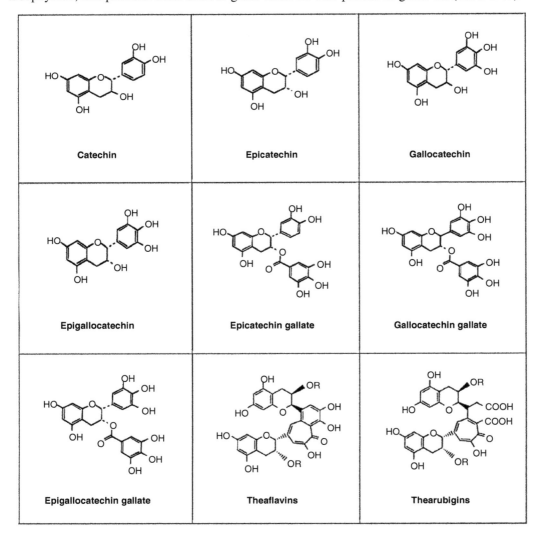

FIGURE 24.1 Structures of major polyphenolic constituents present in green and black tea.

During the fermentation process involved in the manufacture of black tea, the monomeric flavon-3-ols undergo polyphenol oxidase-dependent oxidative polymerization, leading to the formation of bisflavonols, theaflavins, thearubigins, and some other oligomers. Theaflavins (1 to 2%, on dry-weight basis) contain benzotropolone rings with dihydroxy or trihydroxy substitution systems. About 10 to 20% of the dry weight of black tea is due to thearubigins, which are even more extensively oxidized and polymerized. The structures of theaflavins and thearubigins are shown in Figure 24.1. Oolong tea contains monomeric catechins, theaflavins, and thearubigins. The other characteristic components of oolong tea include epigallocatechin esters, theasinensins, dimeric catechins, and dimeric proanthocyanidins.

24.3 ANTICARCINOGENIC EFFECTS OF TEA: EXPERIMENTAL STUDIES

24.3.1 PREVENTION AGAINST SKIN TUMORIGENESIS

24.3.1.1 Prevention against Skin Tumor Initiation

Utilizing several tumor bioassay protocols, studies from laboratories all over the world (Katiyar et al., 1992a; Katiyar and Mukhtar, 1996; Mukhtar et al., 1992; Mukhtar et al., 1994; Shanmugaratnam et al., 1978) have shown that the topical application as well as the oral feeding of a polyphenolic fraction isolated from green tea (hereafter referred to as GTP) to SENCAR, CD-1, and BALB/C mice results in significant prevention against the occurrence of skin tumorigenesis. In a complete carcinogenesis protocol, the topical application of GTP on the backs of BALB/C mice for 7 days prior to application of 3-methylcholanthrene was found to result in significant prevention against the development of skin tumors (Wang et al., 1989b).

Studies were also conducted to assess whether GTP possesses antitumor-initiating effects. In these studies, a two-stage skin carcinogenesis protocol in SENCAR mice was employed. Topical application of GTP for 7 days prior to the single application of 7,12-dimethylbenz(a)anthracene (DMBA) as the initiating agent, followed by twice weekly applications of the tumor promoter 12-O-tetradecanoyl-phorbol-13-acetate (TPA), resulted in significant prevention in the occurrence of tumorigenesis (Wang et al., 1989b). That study showed a considerable delay in the latency period for the appearance of the first tumor and subsequent tumor growth in GTP-treated animals. Oral feeding of GTP (0.05%, w/v) in drinking water for 50 days prior to the DMBA-TPA treatment or its continuous feeding during the entire period of the tumor protocol was also found to result in significant prevention both in terms of the tumor incidence and tumor multiplicity. EGCG is the major constituent present in green tea (Katiyar et al., 1992b), and a single cup contains up to 200 mg, and therefore the intake of EGCG in individuals who habitually consume green tea is substantial. For this reason, it was of interest to examine the utility of this tea component as a cancer chemopreventive agent. The topical application of EGCG prior to DMBA/TPA protocol resulted in significant prevention against skin tumor initiation in SENCAR mouse skin. In these experiments, skin application of EGCG to the SENCAR mouse prior to that of carcinogen treatment was found to result in 30% inhibition in carcinogen metabolite binding to epidermal DNA, suggesting that EGCG may be inhibiting the metabolism of the precarcinogen.

24.3.1.2 Prevention against Skin Tumor Promotion

The phorbol-type tumor promoter TPA is most widely employed in two-stage skin tumorigenesis protocol (Agarwal and Mukhtar, 1991; DiGiovanni, 1992; Katiyar and Mukhtar, 1996). Studies from the laboratory of the authors of this chapter investigated the effect of GTP on TPA-induced skin tumor promotion in DMBA-initiated SENCAR mouse. Topical application of varying doses of GTP prior to that of TPA resulted in significant prevention against skin tumor promotion in a dose-dependent manner (Katiyar et al., 1992c). The pretreatment of the animals with GTP showed substantially lower tumor body burden, such as a decrease in total number of tumors per group, number of tumors per animal, tumor volume per mouse, and average tumor size, as compared with the animals that did not receive GTP. Another study has shown that the topical application of GTP to CD-1 mice also inhibited TPA-induced tumor promotion in DMBA-initiated skin (Huang et al., 1992). Topical application of GTP or its major component, EGCG, has been shown to inhibit tumor promotion mediated by TPA as well as by other skin tumor promoters such as teleocidin and okadaic acid (Huang et al., 1992; Yoshizawa et al., 1987). Another study (Fujiki et al., 1990) showed that the topical application of EGCG to mouse skin also inhibited teleocidin-promoted tumorigenesis in mouse skin.

It was shown that EGCG inhibited a tumor-promoting activity of okadaic acid in a two-stage carcinogenesis experiment on mouse skin. EGCG treatment, prior to okadaic acid, completely inhibited the tumor formation in mice up to 20 weeks of the treatment (Yoshizawa, 1996). This inhibitory effect of EGCG was dose dependent. This study further showed that a topical application of 5 mg EGCG immediately reduced the specific binding of ^{3}H-okadaic acid to a particulate fraction of mouse skin to as low as 30% of control. According to the Scatchard analysis, the reduction of specific ^{3}H-okadaic acid binding was mainly due to the reduction of the binding sites, not due to the change of the affinity. The reduction of the specific binding was closely related to the inhibitory effect of EGCG on tumor promotion by okadaic acid.

24.3.1.3 Prevention against Stage I and Stage II Skin Tumor Promotion

Skin tumor promotion is often divided into two operational stages known as stage I and stage II. Experiments were performed to determine which of these two stages of tumor promotion is inhibited by GTP. Topical application of GTP concurrently with each application of either TPA (stage I) or mezerein (stage II) resulted in significant prevention against DMBA-initiated tumor formation in SENCAR mouse in terms of tumor multiplicity (42 and 50%, respectively) and tumor growth (54 and 43%, respectively) in stage I and stage II (Katiyar et al., 1993a). Furthermore, this study showed that the sustained inhibition of tumor promotion by GTP required an uninterrupted application of GTP in conjunction with each promotional treatment of either of the tumor promoters. Under this treatment regimen, compared with non-GTP treated positive controls, the GTP treatment afforded a significant inhibition (74%) in the growth/development of the tumor. It can therefore be concluded that GTP is capable of inhibiting stage I as well as stage II skin tumor promotion, and that the inhibition of tumor promotion depends on the duration of GTP treatment.

24.3.1.4 Prevention against Malignant Conversion of Chemically Induced Benign Skin Papillomas to Carcinomas

The progression of benign tumors to malignant cancer is the most critical step in carcinogenesis, since malignant lesions are capable of metastatic spread that eventually results in fatal consequences (Athar et al., 1991). Studies were conducted to assess the possible role of GTP in prevention against the conversion of chemically induced benign skin papillomas to squamous cell carcinomas in SENCAR mice. An enhanced rate of malignant conversion was achieved in the stabilized papilloma yield by twice-weekly topical applications of either a free-radical-generating compound, benzoyl-peroxide, or a genotoxic agent, 4-nitroquinoline-N-oxide, whereas spontaneous malignant conversion was associated with topical application of acetone. In these protocols, the pre-application of GTP 30 min prior to skin application of acetone, benzoyl peroxide, or 4-nitroquinoline-N-oxide resulted in significant prevention when assessed for the conversion of papillomas to carcinomas. This study suggested a significant preventive effect of green tea against tumor progression induced by free-radical generating compounds and genotoxic agents (Katiyar et al., 1993a).

In a study from our laboratory (Katiyar et al., 1997), we evaluated the protective effect of GTP against the induction and subsequent progression of papillomas to squamous cell carcinomas (SCCs) in experimental protocols where papillomas were developed with a low or high probability of their malignant conversion. Topical application of GTP (6 mg/animal) 30 min prior to that of TPA either once a week for 5 weeks (high-risk TPA protocol) or once a week for 20 weeks (low-risk TPA protocol) or mezerein (MEZ) twice a week for 20 weeks (high-risk MEZ protocol) in DMBA-initiated mouse skin resulted in significant protection against skin tumor promotion in terms of tumor incidence (32 to 60%), multiplicity (49 to 63%), and tumor volume per mouse (73 to 90%) at the termination of the experiment at 20 weeks. In three separate malignancy-progression experiments when papilloma yields in DMBA-initiated and TPA- or MEZ-promoted low- and high-risk protocols were stabilized at 20 weeks, animals were divided into two subgroups. These animals were either topically treated twice weekly with acetone (0.2 ml/animal, spontaneous malignant-conversion group) or with GTP (6 mg/animal in 0.2 ml acetone) for an additional period of 31 weeks. During these treatment regimens, all suspected carcinomas were recorded, and each one was verified histopathologically either at the time when the tumor-bearing mouse died or became moribund, or at the termination of the experiment at 51 weeks. GTP resulted in significant protection against the malignant conversion of papillomas to SCC in all the protocols employed. At the termination of the experiment at 51 weeks, these protective effects were evident in terms of mice with carcinomas (35 to 41%), carcinomas per mouse (47 to 55%), and percent malignant conversion of papillomas to carcinomas (47 to 58%).

The kinetics of malignant conversion suggest that a subset of the papillomas formed during the early phase of tumor promotion in all the protocols had a higher probability of malignant conversion into SCCs, because all of the positive control groups (acetone treated) produced nearly the same number of carcinomas (33 to 38 in a group of 20 animals) at the end of the progression period. In the GTP-treated group of animals, fewer carcinomas were formed (14 to 20 in a group of 20 animals), which shows the ability of GTP to protect against the malignant conversion of papillomas with a higher probability of malignant conversion to SCCs. The results of this study suggested that irrespective of the risk involved, GTP might be highly useful in affording protection against skin cancer risk.

24.3.1.5 Prevention against UVB Radiation-Induced Photocarcinogenesis

Ultraviolet B (UVB) radiation (280 to 320 nm) present in the solar spectrum is the major risk factor for skin cancer in humans (Elmets, 1991). Studies in our laboratory assessed the effect of topical application as well as oral feeding (through drinking water) of GTP to SKH-1 hairless mice on UVB radiation-induced photocarcinogenesis (Wang et al., 1991). The chronic oral feeding of GTP (0.1%, w/v) in drinking water to mice during the entire period of UVB exposure was found to result in significantly lower tumor body burden as compared with non-GTP fed animals. The topical application of GTP before UVB radiation exposure also afforded some prevention against photocarcinogenesis. However, the observed protection was lower compared with the oral feeding of GTP in drinking water. This observation was validated in terms of percentage of mice with tumors and number of tumors per mouse.

Another study (Wang et al., 1992c) showed that the infusion of green tea extracts (1.25%, w/v) as a sole source of drinking water to mice afforded substantial prevention against UVB radiation-induced intensity of red color and area of skin lesions, as well as UVB radiation-induced tumor initiation and tumor promotion. In a subsequent study, Wang et al. (Wang et al., 1994) showed that black tea consumption by SKH-1 hairless mice markedly reduced tumor formation when the animals were initiated with DMBA followed by multiple UVB exposures. The oral administration of black tea, green tea, and decaffeinated black or green tea (0.63% or 1.25%, w/v) as the sole source of drinking water 2 weeks prior to and during 31 weeks of UVB treatment was found to reduce tumor risk in terms of number of tumors per animal as well as tumor size. These studies suggest that consumption of tea as a sole source of drinking water may reduce the risk of some forms of human cancers induced by solar UV radiation.

Liu et al. (1995) analyzed p53 and H-ras mutations in UV-induced skin tumors in SKH-1 mice with and without treatment with green tea. They found that mutations in exon 6 of the p53 gene are unique for tumors from the UV/green tea group. They suggested that green tea might, somehow, select certain p53 mutations during the chemoprevention process. However, there is a need for additional studies to validate the suggestion that tea consumption offers selective advantage during clonal expansion by the tumor-promoting agents. The effect of topically or orally administered pure EGCG on photocarcinogenesis and immunosuppression induced by UVB radiation in BALB/cAnNHsd mice was also assessed (Gensler et al., 1996). This study showed that the induction of skin tumors by UV radiation was significantly reduced by topical but not by oral administration of EGCG through a mechanism distinct from inhibition of photoimmunosuppression.

In a study from our laboratory (Chatterjee et al., 1996), a novel ^{32}P postlabeling method was employed to detect UVB-induced DNA lesions in the epidermis and its prevention by GTP. This study showed that epidermal DNA from UVB-exposed mice at 24 h contain up to five DNA lesions. Quantitation of these lesions showed that their formation increased in a UVB dose-dependent manner. Treatment of DNA samples with the bacteriophage DNA repair enzyme T4 endonuclease V confirmed that four of these lesions are pyrimidine dimers. While some of these lesions were repaired 18 h after UVB irradiation, 30% of them persisted even 48 h postirradiation. Topical application of GTP to the skin of the mice prior to UVB exposure was found to prevent the formation of pyrimidine dimers.

Huang et al. (1997) showed an inhibitory effect of p.o.-administered green and black tea on UVB-induced complete carcinogenesis, but the decaffeinated teas were either inactive at moderate dose levels or they enhanced the tumorigenic effect of UVB at high dose levels. The oral administration of caffeine was found to have an inhibitory effect on UVB-induced complete carcinogenesis. This study concluded that caffeine was a biologically important constituent of tea.

In a study by Lu et al. (2002), SKH-1 hairless mice were irradiated with ultraviolet B (UVB) twice weekly for 20 weeks. These tumor-free mice, which had a high risk of developing skin tumors during the next several months, were then treated topically with caffeine (6.2 μmol) or (−)-epigallocatechin gallate (EGCG, 6.5 μmol) once a day 5 days a week for 18 weeks in the absence of further treatment with UVB. Topical applications of caffeine to these mice decreased the number of nonmalignant and malignant skin tumors per mouse by 44 and 72%, respectively. Topical applications of EGCG decreased the number of nonmalignant and malignant tumors per mouse by 55 and 66%, respectively. Immunohistochemical analysis showed that topical applications of caffeine or EGCG increased apoptosis, as measured by the number of caspase 3-positive cells in nonmalignant skin tumors, by 87 or 72%, respectively, and in squamous cell carcinomas by 92 or 56%, respectively. However, there was no effect on apoptosis in nontumor areas of the epidermis. Topical applications of caffeine or EGCG had a small inhibitory effect on proliferation in nonmalignant tumors, as measured by bromodeoxyuridine (BrdUrd) labeling (16 to 22%), and there was also a similar, but nonsignificant, inhibitory effect on proliferation in malignant tumors. The results suggested a need for further studies to determine whether topical applications of caffeine or EGCG can inhibit sunlight-induced skin cancer in humans.

24.3.1.6 Effect on the Growth of Established Skin Tumors

In multistage skin tumorigenesis studies in SENCAR mice, we observed that GTP treatment results in a reduction of tumor size in terms of total tumor volume per group and total tumor volume per animal when compared with the non-GTP treated animals (Katiyar et al., 1992c). In another study (Wang et al., 1992d), it was demonstrated that the consumption of water extracts of green tea (WEGT) or GTP not only resulted in a decrease in tumor formation and multiplicity, but also markedly reduced the tumor size. In this study, it was also shown that feeding WEGT or GTP or giving EGCG intraperitoneally inhibited tumor growth and caused partial regression of established skin papillomas in female CD-1 mice. These observations suggest that green tea may also possess chemotherapeutic effects. Lu et al. (1997) evaluated the effect of oral administration of black tea on the growth of chemically or UV radiation-induced established skin papillomas, keratoacanthomas, and squamous cell carcinoma in CD-1 mice. They also evaluated the effect of oral administration of black tea on apoptosis, mitotic index, and incorporation of bromodeoxy-uridine into DNA of these tumors. The results suggested that black tea inhibited the growth of papillomas in mice. The treatment also inhibited proliferation and enhanced apoptosis in both nonmalignant as well as malignant skin tumors.

24.3.2 PREVENTION AGAINST FORESTOMACH AND LUNG TUMORIGENESIS

In a study from our laboratory, WEGT (2.5%, w/v) was given to female A/J mice as the sole source of drinking water, and diethylnitrosamine (DEN) (20 mg/kg body weight of animal) and BαP (2 mg/animal) were employed as test carcinogens (Katiyar et al., 1993a). It was observed that compared with only DEN-treated mice, those fed with WEGT and DEN showed significantly lower tumor yield both in lungs and forestomach. In terms of appearance of total number of tumors per mouse, a significant prevention in the occurrence of tumorigenesis in forestomach (80 to 85%) and lung (41 to 61%) at all the stages of carcinogenesis (initiation, promotion, and complete carcinogenesis protocols) was observed. In the case of BαP-induced tumorigenicity, WEGT also resulted in 60 to 75% and 25 to 35% prevention in the total number of tumors per mouse in the forestomach and lung, respectively, at all stages of carcinogenesis.

The same protocols (Katiyar et al., 1993b) were used to assess the protective effects of GTP against DEN- and BαP-induced tumorigenesis in forestomach and lung of female A/J mice. The data showed that oral feeding of GTP (0.2%, w/v) significantly affords prevention against DEN-induced total number of tumors per mouse in forestomach (68 to 82%) and lung tumorigenesis (37 to 45%) during initiation, promotion, and complete carcinogenesis protocols. GTP also showed comparable chemopreventive effects when BαP was used as a carcinogen. In a further study (Katiyar et al., 1993c), we also demonstrated that the administration of GTP (5 mg/animal) by gavage 30 min prior to the challenge with carcinogen afforded significant prevention against both DEN- and BαP-induced forestomach and lung tumorigenesis in A/J mice.

Studies by Wang et al. (Wang et al., 1992a, Wang et al., 1992b) have shown that WEGT (0.63, 1.2, or 1.25%, w/v) treatment to A/J mice prevents against DEN- and BαP- induced forestomach and lung tumorigenesis. In one study (Xu et al., 1992), it was shown that 2% green tea infusion or 560 ppm EGCG feeding in drinking water for 13 weeks affords prevention against 4-(methylnitrosamino)-1-(3-pyridyl)-1-butanone (NNK)-induced lung tumorigenesis in A/J mice. In a subsequent study, Shi et al. (Shi et al., 1994) showed that when decaffeinated green or black tea extracts were given to female A/J mice as the sole source of drinking water before an i.p. injection of NNK (100 mg/kg body weight), a significant reduction in lung tumor multiplicity was observed. Shim et al. (Shim et al., 1995) reported the chemopreventive effects of daily green tea consumption among cigarette smokers. In this study, the sister-chromatid exchange frequency in peripheral lymphocytes was used as a marker for mutagenic response, and green tea was shown to block the cigarette-induced increase in sister chromatid exchange frequency.

24.3.3 PREVENTION AGAINST ESOPHAGEAL TUMORIGENESIS

The chemopreventive effects of five different varieties of Chinese teas, including green tea, against N-nitroso-methylbenzylamine (NMBzA)-induced esophageal tumorigenicity in Wistar rats were assessed (Chen, 1992). The oral administration of 2% tea infusion as the sole source of drinking fluid to rats during the entire experimental period resulted in the inhibition of esophageal tumorigenesis induced by NMBzA. All five types of green and black teas tested were found to be effective and resulted in a reduction of tumor incidence by 26 to 53%, and of tumor multiplicity by 58 to 75%. The oral feeding of green tea also resulted in inhibition of esophageal tumor formation induced by precursors of NMBzA (Gao et al., 1990) or by nitrososarcosine in mice (Oguni et al., 1992).

24.3.4 PREVENTION AGAINST DUODENUM AND SMALL INTESTINE TUMORIGENESIS

Some studies (Fujiki et al., 1992; Fujita et al., 1989) assessed the chemopreventive effects of EGCG against N-ethyl-N′-nitro-N-nitrosoguanidine (ENNG)-induced duodenum tumorigenicity in C57BL/6 mice. Compared with non-EGCG-fed animals, the oral feeding of EGCG (after treatment with carcinogen) in drinking water resulted in significant prevention against ENNG-caused tumor promotion in duodenum as observed by a decrease in tumor incidence and number of tumors per mouse. In another study (Hirose et al., 1993), the authors used multiorgan carcinogenesis model to assess the preventive effect of GTP. In this study, F344 rats were pretreated with a combination of five carcinogens for 4 weeks, and their diet was supplemented with 1% GTP given during or after the carcinogen exposure period. The GTP treatment resulted in an inhibition of adenoma and adenocarcinoma formation in the small intestine.

24.3.5 PREVENTION AGAINST COLON TUMORIGENESIS

Using a rat colon tumorigenicity model, it was shown (Yamane et al., 1991) that one week after subcutaneous administration of azoxymethane to fisher rats, oral feeding of GTP (0.01 or 0.1%, w/v) in drinking water for an additional 10 weeks resulted in the inhibition of azoxymethane-induced colon tumorigenesis. However, in another study (Hirose et al., 1993) with a multiorgan tumorigenesis model using F344 rats, the administration of GTP (1%) in the diet did not inhibit tumorigenesis in the colon. In another study (Narisawa and Fukaura, 1993), a very low dose of GTP against colon carcinogenesis in F344 rats was tested. In this study, a total of 129 female F344 rats were given an intrarectal instillation of 2 mg of N-methyl-N-nitrosourea three times a week for 2 consecutive weeks, and drinking water was replaced with WEGT (0.05, 0.01, or 0.002%) throughout the experiment. Autopsies performed at the 35th week revealed significant lowering in the incidence of colon carcinomas in rats ingested with WEGT as compared with control animals.

A pyrolysis product in cooked foods, 2-amino-1-methyl-6-phenylimidazo[4,5-b]pyridine (PhIP), is a known colon carcinogen and has been implicated in the etiology of human colon cancer. In order to identify chemoprotection strategies that could be carried out in humans, a pilot study (Huber et al., 1997) was conducted in which PhIP-DNA adduct levels were quantified in the colons of male F344 rats that had been subjected to 16 different putative chemoprotection regimens, followed by a gavage of PhIP (50 mg/kg) and sacrifice 24 h later. Out of 16 chemopreventive agents used, the strongest inhibition (67%) of PhIP-DNA adduct formation in the colon was observed upon pretreatment with black tea.

The effect of black tea and tea with milk was evaluated on colon cancer induced by azoxymethane in F344 rat models (Weisburger et al., 1997). Foci of aberrant crypts in the colon were decreased in both tea-fed and tea-and-milk-fed animals. The study showed that tea consumption reduced the production of foci of aberrant crypts in the colon and that milk potentiates these inhibitory effects.

24.3.6 PREVENTION AGAINST LIVER TUMORIGENESIS

Administration of 5% green tea leaf in a diet given to rats from 10 days prior to treatment with the carcinogenic compound aflatoxin B1 (AFB1) until 3 days after the treatment resulted in a significant inhibition of AFB1-induced γ-glutamyl transpeptidase-positive foci in the rat liver (Chen et al., 1987). In another study, it was also shown that 2.5% green tea leaf in the diet given to rats produced significant inhibition of DEN-induced hepatocarcinogenesis (Li, 1991). Mao showed the inhibitory effect of epicatechin complex on DEN-induced liver precancerous lesions, variant-cell foci, and nodule formation in rats (Mao, 1993). Epicatechin complex treatment resulted in a marked decrease in the number of N-ras overexpressed lesions. In another study, it was shown that the administration of decaffeinated black tea extract by oral gavage to male Swiss mice resulted in a decrease of tobacco-induced liver tumors (Nagabhushan et al., 1991).

The pathogenesis of pulmonary tumors induced by a tobacco carcinogen, 4-(methylnitrosamino)-1-(3-pyridyl)-1-butanone (NNK), and its inhibition by black tea was characterized in female A/J mice (Yang et al., 1997). In this short-term model, the administration of black tea polyphenols (0.3%) through the drinking water significantly inhibited NNK-induced early bronchiolar cell proliferation, as measured immunohistochemically by the incorporation of bromodeoxyuridine (BrdUrd). Administration of black tea also inhibited the progression of adenoma to adenocarcinoma, as determined by both malignant tumor incidence and multiplicity. The cell proliferation rate in adenomas was also suppressed by black tea treatment. This study demonstrates the antiproliferative activities of black tea and its polyphenols. The authors suggested that such activities, at the early and late stages of lung tumorigenesis, might be important for the cancer-chemopreventive activities of black tea.

In another study (Qin et al., 1997), in a Fischer rat model, the effect of green tea (given through diet) was examined on the initiation of aflatoxin B1 (AFB1)-induced hepatocarcinogenesis, as assessed by hepatic AFB1-DNA binding *in vivo*, AFB1 metabolism *in vitro*, and by the appearance of AFB1-induced glutathione S-transferase placental form (GST-P)-positive hepatocytes detected by immunohistochemical methods. Green tea feeding did not affect the microsome-mediated AFB1 binding to exogenous DNA, but it enhanced microsome-mediated formation of nontoxic hydroxylated metabolites of AFB1. Hepatic nuclear AFB1-DNA binding *in vivo* was significantly inhibited by green tea. AFB1-induced GST-P positive single hepatocytes were also inhibited significantly by green tea treatment. The authors concluded that green tea inhibits initiation of AFB1-induced hepatocarcinogenesis in the rat by modulation of AFB1 metabolism, thereby inhibiting AFB1-DNA binding and AFB1-induced GST-P-positive hepatocytes.

24.3.7 Prevention against Mammary Carcinogenesis

In a study by Hirose et al. (1994), the effect of 1% green tea catechins on mammary gland carcinogenesis in female Sprague–Dawley rats pretreated with DMBA was evaluated. This study showed that although the final incidences and multiplicities of mammary tumors were not significantly affected by administration of green tea catechins, the number of survivors in the green-tea-fed group at the end of the experiment (36 weeks) was significantly higher than in the basal diet group. The average size of the palpable mammary tumors was significantly smaller in the group fed with green tea catechins. This study also indicated that green tea catechins inhibit rat mammary gland carcinogenesis after DMBA initiation. Sakamoto et al. (1995) have shown that GTP and BTP (black tea phenols) suppressed the growth of canine mammary tumor cells CMT-13 in culture when added at 25 ppm or more for at least 24 hours. This study also showed that BTP is a more effective inhibitor of mammary tumor cells than GTP. Komori et al. (1993) reported that EGCG and WEGT treatment inhibited the growth of lung and mammary cancer cell lines with almost identical potencies. The authors postulated that EGCG and the components present in WEGT would block the interaction of tumor promoters, hormones, and growth factors with their receptors.

In another study (Weisburger et al., 1997), the effect of tea or tea and milk instead of drinking water was evaluated in rat models of cancer in the mammary gland. Solutions of 1.25% (w/v) black tea or 1.85% (v/v) milk in tea were prepared three times per week. SD rats were given tea beginning at 42 days of age; one group was gavaged with 5 μg DMBA at 49 days of age; another group received 8.4 mg 2-amino-3-methylimidazo[4,5-f]quinoline (IQ) twice per week beginning at 49 days of age and then 14 mg twice a week for 4 weeks more. The groups on DMBA were killed 33 weeks later, and those on IQ 39 weeks later. Tea decreased the mammary gland tumor multiplicity and volume, and milk potentiated these inhibitory effects.

24.3.8 Prevention against Pancreatic Carcinogenesis

In a study by Harada et al. (1991), Syrian golden hamsters were used as a carcinogenesis model. The hamsters were treated with N-nitroso-bis(2-oxopropyl)amine and fed with a protein-deficient diet consisting of DL-ethionine and L-methionine for tumor promotion. In this study, the dietary supplementation with GTP (500 mg/kg per day) during the promotion stage was found to reduce pancreatic tumorigenesis as compared with non-GTP supplemented animals (Harada et al., 1991).

24.3.9 PREVENTION AGAINST PROSTATE CARCINOGENESIS

Prostate cancer is considered an ideal candidate disease for chemoprevention because of its high latency period and because it is commonly diagnosed in men over the age of 50. Thus, even a moderate delay in the progression of this type of cancer by chemopreventive agents could result in a substantial reduction of incidence of this disease and, more importantly, improve the quality of life of patients with the disease (Liberman, 2001). Lio et al. (1995) showed that intraperitoneal administration of EGCG rapidly reduced the size of human prostate tumor growth in nude mice. These authors further suggested that there might be a possible relationship between the higher consumption of green tea and the lower incidence of prostate cancer in some Asian countries. In our ongoing research program on green tea and cancer chemoprevention, we attempted to systematically evaluate the effect of green tea consumption on prostate carcinogenesis (Gupta et al., 1999). We initially showed that green tea constituent EGCG induces apoptosis in human prostate cancer cell line DU145 (Ahmad et al., 1997). In a subsequent study, we showed that ornithine decarboxylase (ODC), a rate-controlling enzyme in the polyamine biosynthesis pathway, is overexpressed in prostate cancer and prostate fluid in humans (Mohan et al., 1999). The induction of ODC activity is known to be mediated by high testosterone level, and exposure of prostate cancer cells to EGCG, and infusion of green tea to Cpb:WU rats, caused a down-regulation of ODC activity in prostate cancer cells as well as in rats (Gupta et al., 1999). Because androgen action is intimately associated with proliferation and differentiation of prostate cancers, and given the fact that prostate cancer progresses from an androgen-sensitive to an androgen-insensitive stage in humans, we reasoned that agents capable of inhibiting growth and/or inducing apoptosis in both androgen-sensitive and androgen-insensitive cells would be useful for the management of prostate cancer.

We showed that EGCG, the major polyphenolic constituent present in green tea, induced antiproliferative effects against both androgen-sensitive and androgen-insensitive human prostate cancer cells, and this effect was mediated by deregulation in cell cycle and induction of apoptosis (Gupta et al., 2000). EGCG treatment was found to result in a dose-dependent inhibition of cell growth in both androgen-insensitive DU145 and androgen-sensitive LNCaP cells. In both the cell types, EGCG treatment also resulted in a dose-dependent G_0/G_1-phase arrest of the cell cycle. We also observed that EGCG causes an induction of G_1-phase cyclin kinase inhibitors, which inhibit the cyclin–cyclin-dependent kinase complexes operative in the G_0/G_1 phase of the cell cycle, thereby causing an arrest, which may be an irreversible process ultimately leading to apoptotic cell death (Gupta et al., 2003). This was the first systematic study showing the involvement of each component of cyclin kinase inhibitor- cyclin–cyclin-dependent kinase machinery during cell cycle arrest and apoptosis of human prostate carcinoma cells by EGCG. These results suggested that EGCG negatively modulates prostate cancer cell growth by affecting mitogenesis as well as inducing apoptosis irrespective of the androgen status (Gupta et al., 2000).

We also utilized a cDNA microarray technique to elucidate how EGCG alters the program of gene expression in prostate carcinoma LNCaP cells (Wang and Mukhtar, 2002). Fluorophore-labeled cDNA probes synthesized from the untreated LNCaP cells or the cells treated for 12 h with EGCG (12 μM), a physiologically achievable dose, were competitively hybridized to the microarray that contained a total of 250 kinases and phosphatase genes. Such high-throughput screening identified a number of EGCG-responsive gene candidates. Of these, we found that EGCG induced a subset of genes that functionally could exhibit inhibitory effects on cell growth. The genes repressed by EGCG mostly belonged to the G-protein signaling network. Interestingly, the protein kinase C-α (PKC-α) form, whose inhibition of expression has been shown to inhibit cell growth in some cancer cells, was selectively repressed by EGCG, while the expression of six other PKC isoforms (β, δ, ϵ, μ, η and ζ) was unaffected. These EGCG-responsive genes may provide key insights from which to understand mechanisms of action of other polyphenolic compounds in prostate cancer chemoprevention.

We then reasoned that these preclinical studies must be conducted with human-achievable doses of green tea in a model system that mimics human disease. Transgenic adenocarcinoma of the mouse prostate (TRAMP) is one such model for prostate cancer that closely mimics progressive forms of human disease (Gupta et al., 2001). Using this model for chemoprevention of prostate cancer, we provided convincing evidence that oral infusion of green tea polyphenols (equivalent to human consumption of 6 cups of green tea) to TRAMP mice inhibits prostate carcinogenesis (Gupta et al., 2001). It is known that the development of prostate cancer depends upon a variety of factors as it progresses from a small latent carcinoma to a large metastatic lesion. Therefore, it will be important to establish the stage of prostate carcinogenesis that is most responsive to green tea. We have developed plans to investigate this in a systematic fashion.

24.3.10 MODULATORY EFFECT OF GREEN TEA ON CANCER CHEMOTHERAPY

The biochemical modulations in cancer chemotherapy have been studied extensively (Sadzuka et al., 1998, and references therein). The use of modulators increases the number of medications, which adds to the patient's burden. Moreover, the side effects due to modulatory drugs can pose a threat to a patient's life. Therefore, the substances used in foods and beverages can serve as excellent biochemical modulators to increase the efficacy of therapy. In this regard, Sadzuka et al. (1998) recently showed that the oral administration of green tea enhanced the tumor-inhibitory effects of doxorubicin on Ehrlich ascites carcinomas implanted in CDF_1 and BDF_1 mice. The study showed that green tea treatment increases the concentration of doxorubicin in tumor, but not in normal tissue. The study also demonstrated significant enhancement of antitumor activity of doxorubicin in implanted M5076 ovarian sarcoma, which normally possesses very low sensitivity toward this drug.

24.3.11 CELL-SPECIFIC BIOLOGICAL EFFECTS OF TEA

Many tumor promoters are known to inhibit gap-junctional intercellular communication that is regarded as an important mechanism of promotion. Sigler and Ruch (1993) assessed WEGT and the individual constituents of green tea for their effects on gap-junctional intercellular communication in p,p'-dichlorodiphenyltrichloroethane-, TPA-, and dieldrin-treated WB-F344 rat liver epithelial cells. All three of these tumor promoters showed inhibition of gap-junctional intercellular communication in a dose-responsive manner at noncytolethal concentrations. WEGT enhanced gap-junctional intercellular communication (20 to 80%) in promoter-treated cells. EGCG and ECG also enhanced gap-junctional intercellular communication in p,p'-dichlorodiphenyltrichloroethane-treated cells. These data suggest that WEGT may inhibit tumor promotion by enhancing gap-junctional intercellular communication. Studies by Mitsui et al. (1995) showed that tea polyphenols, specifically EGCG, killed 3Y1 cells transformed by E1A gene of human adenovirus type 12 (E1A-3Y1 cells) at 100 times lower concentration than the parental 3Y1 cells. EGCG was also found to exert a strong E1A-3Y1 cell-specific toxicity, while EC and ECG did not. EGCG (0.05 to 0.1%, w/w), when given in drinking water to C3H/HeNCrj mice for 65 weeks, reduced the incidence of hepatoma-bearing mice from 83.3% (non-EGCG) to 56.0–52.2% (EGCG), and also reduced the average number of hepatoma per mouse. Nishida et al. (1994) showed that EGCG inhibited the growth and secretion of α-fetoprotein by human hepatoma-derived PLC/PRF/5 cells without decreasing their viability. In another study (Lea et al., 1993), the polyphenols extracted from green or black tea were found to be strong inhibitors for DNA synthesis in HTC rat hepatoma cells as well as in DS19 mouse erythroleukemia cells.

Intraperitoneal administration of EGCG was found to significantly inhibit the metastasis of lung caused by intravenous injection of lung carcinoma cells (Taniguchi et al., 1992). WEGT has also been shown to prevent the induction of hepatocyte, derived from male B6C3F1 mice, cytolethality by glucose oxidase, xanthine oxidase, and paraquat in a dose-dependent manner. WEGT also prevented gap-junctional-mediated intercellular communication by phenobarbital, lindane, and paraquat (Klaunig, 1992).

The influence of EGCG and EGC on doxorubicin-resistant murine sarcoma (S180-dox) and human colon carcinoma (SW620-dox) cell lines was studied (Stammler and Volm, 1997). Both polyphenols showed a sensitizing effect on the cell lines by treatment with doxorubicin. These results suggested that protein kinase C might be inhibited by EGCG and EGC, thereby reducing the expression of some drug-resistance-related proteins.

Another study (Valcic et al., 1996) showed the inhibitory effect of six green tea catechins on the growth of four selected human tumor cell lines (MCF-7 breast carcinoma, HT-29 colon carcinoma, A-427 lung carcinoma, and UACC-375 melanoma). EGCG was the most potent of the seven green tea components against three out of the four cell lines (i.e., MCF-7 breast cancer, HT-29 colon cancer, and UACC-375 melanoma).

Grinberg et al. (1997) addressed the question of whether tea polyphenols in green and black tea at concentrations close to plasma levels would impart antioxidant effects on red blood cells (RBCs), subjected to oxidative stress. This study showed a substantial effect of tea polyphenols against oxidant-mediated biochemical and morphological alterations in RBCs. In another study, Zhang et al. (1997) showed that the four polyphenols purified from jasmine green tea — EC, EGC, ECG, and EGCG, — exhibited strong protection for RBC membrane to hemolysis induced by 2,2'-azo-bis(2-amidinopropane)dihydrochloride, an azo-free radical initiator.

24.4 ANTI-INFLAMMATORY EFFECTS OF TEA

24.4.1 PREVENTION AGAINST TPA-CAUSED INFLAMMATORY RESPONSES

In view of the fact that edema and hyperplasia are often used as an early marker of skin tumor promotion, we assessed the effect of preapplication of GTP on these parameters (Katiyar et al., 1992c). In this study, we found that the topical application of GTP to SENCAR mouse skin results in significant prevention against TPA-caused effects on the enzyme activities of cyclooxygenase and lipoxygenase, which play a role in inflammatory responses. The prior application of GTP to the mouse dorsal skin was found to result in significant inhibition of TPA-induced epidermal edema and hyperplasia. In further studies, we showed that single or multiple application of GTP to SENCAR mouse ear skin prior to or after the application of TPA affords significant prevention against TPA-induced edema (Katiyar et al., 1993a). This study also showed that the preapplication of GTP affords significant prevention against TPA-induced hyperplasia in the ear skin. The occurrence of prevention by GTP was 75 and 90% in terms of epidermal thickness and vertical cell layers, respectively. GTP was also found to afford prevention against TPA-caused infiltration of polymorphonuclear leukocytes in the dermis.

24.4.2 PREVENTION AGAINST UVB RADIATION-CAUSED INFLAMMATORY RESPONSES

In our laboratories, experiments were performed to assess whether GTP possesses chemopreventive effects against UVB-radiation-induced inflammatory changes in murine skin (Agarwal et al., 1993). We showed that the oral feeding of GTP (0.2%, w/v) through drinking water to SKH-1 hairless mice for 30 days followed by irradiation with UVB (900 mJ/cm^2) resulted in significant protection against UVB radiation-caused cutaneous edema and depletion of the antioxidant-defense system in epidermis. GTP also afforded prevention against UVB radiation-caused induction of epidermal ODC and cyclooxygenase activities in a time-dependent manner.

24.4.3 PREVENTION AGAINST UVB RADIATION-INDUCED IMMUNOSUPPRESSION

Chronic UV exposure to the skin is known to cause diverse biological effects, including induction of inflammation, alteration in cutaneous immune cells, and impairment of contact hypersensitivity responses (Elmets, 1991; Kripke, 1984). There is strong evidence that UVB radiation can cause skin cancer in humans and laboratory animals (DeFabo and Noonan, 1983; Noonan et al., 1981). In experimental studies on mice, studies from our laboratory (Katiyar et al., 1995a) have shown that topical application of GTP (1 to 6 mg/animal) for 30 min or more, prior to or after exposure to a single dose of UVB (2 kJ/m^2), resulted in significant prevention against local and systemic suppression (25 to 90% and 23 to 95%, respectively) of contact hypersensitivity and inflammation (70 to 80%) in C3H/HeN mice. The preventive effects of GTP was found to have an inverse correlation with the doses of UVB (2 to 32 kJ/m^2). Among the four main epicatechin derivatives present in GTP, EGCG was found to be the most effective in affording prevention against UVB-caused suppression of contact hypersensitivity.

24.4.4 INHIBITION OF TUMOR PROMOTER-CAUSED INDUCTION OF CYTOKINES

Cytokines are known to play an important role in a variety of physiological and pathological processes, including inflammation, wound healing, immunity, and hematopoiesis. Interleukin-1α (IL-1α) plays an important role in both immune and inflammatory reactions, and it has been shown to be induced in response to various skin tumor promoters (Katiyar et al., 1995b). We conducted studies to assess whether pretreatment of the animal with GTP can afford preventive effects against tumor-promoter-caused induction of IL-1α expression in the murine skin model system. Northern blot analysis of IL-1α gene expression in mouse skin revealed that the topical application of GTP or BTP prior to treatment with TPA results in significant inhibition of TPA-induced expression of epidermal IL-1α mRNA. These inhibitory effects were found to be dependent on the dose of GTP or BTP used. GTP also inhibited IL-1α mRNA and protein expression induced by skin tumor promoters like mezerein, benzoyl peroxide, and anthralin. In this study, EGCG and ECG showed maximum inhibitory effects at equimolar dose as compared with other epicatechin derivatives.

Studies from our laboratory (Challa et al., 1998) have recently shown that GTP treatment ameliorated UVB-induced oxidative burst as measured by H$_2$O$_2$ and myeloperoxidase production. GTP treatment to mice also partially blocked UVB-induced infiltration of leukocytes and appeared to inhibit IL-10 production in skin, as shown by immunohistochemistry. In this study, GTP application to mice prior to UVB irradiation was found to result in complete reversal of UVB-induced inhibition of contact hypersensitivity, but showed only partial reversal of induction of tolerance to 2,4-dinitrofluorobenzene. These data suggested that green tea, and the polyphenols present therein, might be useful against inflammatory dermatoses and immunosuppression caused by solar radiation in humans. The validation of these studies to human population exposed to low levels of UV radiation chronically through solar radiation is an area for further study.

24.5 MECHANISMS OF BIOLOGICAL EFFECTS OF TEA

24.5.1 PREVENTION AGAINST MUTAGENICITY AND GENOTOXICITY

Tea has been shown to suppress the mutagenicity of products formed in a model nitrosation reaction system (Mukhtar et al., 1992). GTP and WEGT were found to significantly inhibit the mutagenicity induced by BαP, aflatoxin B1, 2-aminofluorene, and methanol extract of coal tar pitch in bacterial or mammalian cell test systems (Wang et al., 1989a). In a study, Jain et al. (1989) showed that tea extracts inhibited N-methyl-N'-nitro-N-nitrosoguanidine (MNNG)-induced mutagenicity in vitro as well as in intragastric tract of rats. It was also observed that the galloyl-group-containing compounds in green tea, such as EGCG, ECG, and EGC, were antimutagenic in nature in the Escherichia coli B/r WP$_2$ assay system (Kada et al., 1985; Shimoi et al., 1986). The oral feeding of green tea or black tea extracts to rats was shown to inhibit chromosomal aberrations in rat bone marrow cells if the extracts were given 24 hours prior to aflatoxin B1 treatment (Ito et al., 1989). Cheng et al. (1991) showed that compared with other chemopreventive agents such as ellagic acid, ascorbic acid, α–tocopherol, and β-carotene, GTP afforded stronger inhibitory effects against mutagenicity of cigarette smoke condensate as assessed by Ames test.

The heterocyclic amines (HCAs), formed during the cooking of meats and fish are thought to be the genotoxic carcinogens associated with important types of human cancer such as cancer of the breast, colon, or pancreas in meat-eating populations (Weisburger et al., 1994). Weisburger et al. (1994) studied the effect of black tea, green tea, theaflavin gallate (polyphenol in black tea), and EGCG (polyphenol in green tea) on the formation of typical HCAs — 2-amino-3,8-dimethylimidazo(4,5-f)quinoxaline and 2-amino-1-methyl-6-phenylimidazo(4,5-b)pyridine — using the model *in vitro* systems of Jagerstad. This study revealed that although the teas as such were either less effective or not effective, the polyphenols were inhibitory in the production of the HCAs 2-amino-3,8-dimethylimidazo(4,5-f)quinoxaline or 2-amino-1-methyl-6-phenylimidazo(4,5-b)pyridine. In this study, it was suggested that the tea polyphenols represent another approach to lower the formation of HCAs and its associated cancer risk.

24.5.2 INHIBITION OF BIOCHEMICAL MARKERS OF TUMOR INITIATION: CYTOCHROME P450-DEPENDENT METABOLISM

The major enzyme system that is responsible for the metabolism of procarcinogens to their DNA-binding metabolites is cytochrome P450 (P450) (Conney, 1982; Mukhtar et al., 1991). This binding to DNA is considered essential for tumor initiation. Wang et al. (1988) studied the interaction of GTP and its constituent polyphenols EC, EGC, ECG, and EGCG with P450 and associated monooxygenase activities. The addition of EC, EGC, ECG, EGCG, and GTP to microsomes prepared from rat liver resulted in a dose-dependent inhibition of P450-dependent arylhydrocarbon hydroxylase, 7-ethoxycoumarin-O-deethylase, and 7-ethoxyresorufin-O-deethylase activities. That same study also showed that epidermal arylhydrocarbon hydroxylase activity and epidermal enzyme-mediated binding of BaP and DMBA to DNA were inhibited by these polyphenols. Sohn et al. (1994) found a significant increase in hepatic P450 1A1, 1A2, and 2B1 activities in rats consuming green tea or black tea. Tea consumption, however, did not affect the P450 2E1 and 3A4 activities. Tea consumption also resulted in an increase in enzyme activity of the phase II enzyme UDP-glucuronyltransferase but not in glutathione S-transferase. In another study, Shi et al. (1994) showed that EGCG inhibited the catalytic activities of several P450 enzymes and was more potent against P450 1A and 2B1 than 2E1. Bu-Abbas et al. (1994) used aqueous extracts of tea (2.5%) as the sole source of drinking water for rats for 4 weeks and determined hepatic cytochrome P450 activity by using chemical probes with selectivity for particular isoforms. Feeding of green tea resulted in an increase in the O-demethylation of methoxyresorufin and pentoxyresorufin. Immunoblot analysis revealed increases in the apo-protein levels of CYP1A2 and CYP4A1 following treatment with green tea.

Recently, Chen et al. (1996) reported a comparative study on the induction of CYP1A2 with different teas in male Fischer 344 rats. In this study, CYP1A2-dependent O-methoxyresorufin demethylase (MROD) activity in liver microsomes was observed. The induction was also shown by intragastric administration of caffeine (100 mg/kg). The CYP1A2 protein, as determined by immunoblot analysis, and the concentrations of tea polyphenols and caffeine in plasma were also measured. This study demonstrates that caffeine, not tea polyphenols, is the component in tea responsible for the induction of this enzyme.

24.5.3 INHIBITION OF BIOCHEMICAL MARKERS OF TUMOR PROMOTION

A number of studies report that the consumption of green tea inhibits tumor promotion as assessed by inhibition of the biochemical markers of tumor promotion (Katiyar and Mukhtar, 1996).

Topical application of TPA on mouse skin results in induction of ODC activity followed by an increase in the levels of polyamines, epidermal hyperplasia, and inflammation and in the number of dark basal keratinocytes (Agarwal and Mukhtar, 1991; DiGiovanni, 1992; Katiyar et al., 1992c). However, it is not yet clear which of these parameters or many others are obligatory or sufficient for the process of tumor promotion. ODC plays an essential role in cell proliferation and differentiation, and its induction is considered to be closely associated with, though not sufficient for, the tumor-promoting activity of a variety of tumor promoters (Agarwal and Mukhtar, 1991; DiGiovanni, 1992). TPA-mediated inflammation in skin is believed to be governed by cyclooxygenase- and lipoxygenase-catalyzed metabolites of arachidonic acid, specifically prostaglandins and hydroxyeicosatetraenoic acids, respectively (Katiyar and Mukhtar, 1996, and references therein). The importance of induction of epidermal ODC, cyclooxygenase, and lipoxygenase activities in skin tumor promotion is evident from the fact that several inhibitors of these enzymes inhibit tumor promotion in murine skin (Katiyar and Mukhtar, 1996, and references therein).

Skin application of GTP in mouse was shown to inhibit TPA-mediated induction of epidermal ODC activity in a dose-dependent manner (Agarwal et al., 1992). The inhibitory effect of GTP was also dependent on the time of its application relative to TPA treatment. Topical application of GTP to SENCAR mouse skin was also found to inhibit the induction of epidermal ODC activity caused by several structurally different mouse skin tumor promoters. Also, the prior application of GTP to mouse skin was found to result in significant inhibition of TPA-induced epidermal edema and hyperplasia (Katiyar et al., 1992c; Katiyar et al., 1993a). As quantitated by the formation of prostaglandins and hydroxyeicosatetraenoic acid metabolites from, respectively, cyclooxygenase- and lipoxygenase-catalyzed metabolism of arachidonic acid, skin application of GTP to SENCAR mice was also found to result in significant inhibition of TPA-caused effects on these two enzymes (Katiyar et al., 1992c). Inhibition of all of these pathways alone or in combination may contribute to overall antitumor-promoting effects of green tea (Katiyar and Mukhtar, 1996).

24.5.4 EFFECTS ON DETOXIFICATION ENZYMES

Studies from our laboratory (Wang et al., 1992b) have shown that the topical application or oral administration of GTP to SENCAR mice inhibited carcinogen-DNA adduct formation in epidermis after topical application of ^{3}H-BP or ^{3}H-DMBA. In another study from this laboratory (Khan et al., 1992), it was also shown that the chronic oral administration of 0.2% GTP in drinking water to mice for 4 weeks resulted in significant enhancement in glutathione peroxidase (86 to 129%), catalase (59 to 92%), NADPH-quinone oxidoreductase (53 to 71%), and glutathione S-transferase (GST) (28 to 30%) activities in the small bowel, the lung, and the liver. Modulations by green tea may be expected to have cancer-chemopreventive functions on the enzymatic pathways that (a) play a role in detoxification of carcinogenic metabolites formation by P450 and other enzymes and (b) are key determinants for cancer initiation (Katiyar and Mukhtar, 1996, and references therein). The phase II enzyme GST not only catalyzes the conjugation of hydroquinones and epoxides of PAH with reduced glutathione for their excretion, but also shows low activity toward organic hydroperoxides for their detoxification from cells/tissue (Laskin et al., 1992). Sohn et al. (1994) reported that both green and black tea increased the activities of cytochrome p450 1A1, 1A2, and 2B1 in the liver of male F344 rats receiving 2% (w/v) solutions of either tea for 6 weeks. In another study, Bu-Abbas et al. (1994) found that in rats given 2.5% green tea through drinking water, the apoprotein levels of CYP4A1 and CYP1A2 were significantly increased. Green tea was also found to potently inhibit the activity of ethoxyresorufin-O-deethylase (EROD) in induced rat liver microsomes (Obermeir et al., 1995), increase the activity of glutathione-s-transferase (Chou et al., 2000), and strongly inhibit the benzo[a]pyrene adduct formation with human DNA (Steele et al., 2000). These pathways alone or in combination may contribute to overall chemopreventive effects of green tea against cancer.

The effect of green tea intake on the metabolism of 2-amino-3-methylimidazo[4,5-f]quinoline (IQ) was studied by Embola et al. (2001). IQ belongs to a new class of heterocyclic arylamines that are formed during cooking through browning of meats and fish. Rats fed a 2% solution of green tea showed significant differences in the recovery of the three major metabolites, namely, IQ-sulfamate, IQ-5-O-sulfate, and IQ-5-O-glucuronide. Green tea, therefore, influences the manner in which the food carcinogen IQ is metabolized and excreted in urine. The formation of glucuronides, increased by green tea, was found to be a key means of detoxification of the heterocyclic amine, IQ (Embola et al., 2001). In another study, Maliakal et al. (2001) reported that tea consumption modulates the activities of hepatic drug-metabolizing enzymes in rats. CYP1A2 and cytosolic glutathione-S-transferase activities were markedly increased by tea treatment.

Muto et al. (2001) studied the effect of four epicatechin derivatives — EGCG, ECG, EGC, and EC — on the metabolic activation of benzo[*a*]pyrene (BaP), 2-amino-1-methyl-6-phenylimidazo-[4,5-b]pyridine (PhIP), and aflatoxin B1 (AFB1) by human cytochrome P450. BaP, PhIP, and AFB1 were activated by respective human CYP1A1, CYP1A2, and CYP3A4 expressed in the membrane fraction of genetically engineered *Salmonella typhimurium* TA1538 cells harboring the human CYP and human NADPH-CYP reductase (OR) when the membrane fraction was added to *S. typhimurium* TA98. The galloylated catechins ECG and EGCG inhibited the mutagenic activation potently, while EGC and EC showed relatively weak inhibitory effects. Catechins also inhibited the oxidations of typical substrates catalyzed by human CYPs, namely ethoxycoumarin O-deethylation by CYP1A1, ethoxyresorufin O-deethylation by CYP1A2, and midazolam 1′-hydroxylation by CYP3A4.

24.5.5 Trapping of Activated Metabolites of Carcinogens

Another important mechanism by which green tea prevents against cancer is by trapping the active metabolites of carcinogens (Katiyar and Mukhtar, 1996). Flavonols are a group of chemicals that possess strong nucleophilic centers at two positions. This property provides an opportunity for the flavonols to react with electrophilic carcinogenic species to form flavonol-carcinogen adducts that may result in prevention of tumorigenesis. In general, the initial step in carcinogenesis is the metabolic activation of chemical carcinogens by the P450-dependent biotransformation reaction. For example, the ubiquitous environmental pollutant BαP is known to cause cancer in experimental animals only after its metabolic activation to highly reactive molecules (Agarwal and Mukhtar, 1993; Katiyar and Mukhtar, 1996). The ultimate carcinogenic metabolite of BαP is BαP diolepoxide-2 (BPDE-2), the formation of which is catalyzed by successive enzymatic steps catalyzed by P450 and epoxide hydrolase (Conney, 1982; Mukhtar et al., 1991). Our studies have shown that tea polyphenols interact with BPDE-2 and that topical application of GTP prior to BPDE-2 treatment resulted in inhibition of skin tumor initiation (Khan et al., 1988).

24.5.6 Antioxidant and Free-Radical Scavenging Activity

The generation of reactive oxygen species (ROS) in biological systems, either by normal metabolic pathways or as a consequence of exposure to chemical carcinogens, has been extensively studied (Wattenberg, 1990). It is now universally agreed that ROS generation contributes to the multistage process of carcinogenesis (Agarwal and Mukhtar, 1993; Wattenberg, 1990). It has been suggested that peroxides and superoxide anion (O_2^-) produce cytotoxicity/genotoxicity in the cellular systems (Perchellet and Perchellet, 1989). The source of hydrogen peroxide (H_2O_2) in cells/tissues is mainly through superoxide dismutase-mediated dismutation of O_2^-, which is generated in the cells/tissues by endogenous enzyme systems as well as by the nonenzymatic pathways (Perchellet and Perchellet, 1989). Additionally, the highly reactive hydroxide radical (^-OH) generated from H_2O_2 is known to damage DNA and produce pathological alterations. The two-electron reduction of the metabolic products of PAH such as quinones, catalyzed by NADPH quinone reductase (QR), has been considered to be a detoxification pathway, since the resulting hydroquinones may be conjugated and excreted through mercapturic acid pathway (Katiyar and Mukhtar, 1996).

Our studies have shown that the administration of chemopreventive agents results in an increase in the levels of antioxidant enzymes in various organs of the test animals (Khan et al., 1992).

In view of this fact, we assessed the effect of oral feeding of GTP (0.2%, w/v) to SKH-1 hairless mice (via drinking water for 30 days) on the activities of antioxidant enzymes, namely glutathione peroxidase and catalase, and the phase II detoxifying enzymes GST and QR (Khan et al., 1992). We found that GTP consumption resulted in a significant increase in glutathione peroxidase, catalase, and QR activities in small bowel, liver and lungs, and an increase in GST activity in small bowel and liver. In another study comparing DEN- or BαP-fed animals, feeding of WEGT (2.5%) or GTP (0.2%) in drinking water with DEN or BαP to female A/J mice resulted in a significant increase in GST activity in liver and small bowel and QR activity in small bowel, lung, and stomach (Katiyar et al., 1993b). These studies suggest that these may be contributing factors for the cancer chemopreventive effects of green tea.

It is becoming clear that the anticarcinogenic properties of tea are due to the antioxidant effect of the epicatechins present therein (Katiyar and Mukhtar, 1996). In a study by Katiyar et al. (1994), it was found that EGCG, EGC, and ECG from green tea significantly inhibit Fe^{3+}/ADP-supported spontaneous lipid peroxidation in mouse epidermal microsomes. Interestingly, each of these epicatechins was also effective in inhibiting photoenhanced lipid peroxidation generated by incubating epidermal microsomes in the presence of silicon phthalocyanine and 650 nm irradiation. EGCG, which is also the major constituent in GTP, showed maximum inhibitory effects compared with the other epicatechins. This study provides the evidence for the antioxidant property of epicatechins. In another study, Terao et al. (1994) also showed the antioxidant property of EC and ECG by measuring the inhibition of lipid peroxidation in large unilamellar liposomes composed of egg yolk phosphatidylcholine. This study provided the evidence that EC and ECG serve as powerful antioxidants against lipid peroxidation when phospholipid bilayers are exposed to aqueous oxygen radicals (Terao et al., 1994). The concept of antioxidant activities is also supported by the findings that EGCG inhibited the formation of 8-hydroxy-deoxyguanosine in HeLa cells (Bhimani et al., 1993). The oral administration of green tea inhibited the formation of 8-hydroxy-deoxyguanosine in mice (Xu et al., 1992), and topical treatment with GTP inhibited TPA-induced hydrogen peroxide formation (Katiyar and Mukhtar, 1996).

The mechanism of antioxidant property of these epicatechins can be explained as follows. Tea polyphenols such as EGCG, ECG, and EGC are strong scavengers against superoxide anion radicals and hydroxyl radicals, which are the two major reactive oxygen species considered to be responsible for the damage of DNA and other cellular molecules and can initiate lipid peroxidation. Tea flavonols can react with and trap the peroxy radicals, thereby terminating lipid peroxidation chain reactions. The reactive oxygen species play important roles in carcinogenesis through damaging DNA, altering gene expression, or affecting cell growth and differentiation (Cerruti, 1989; Namiki, 1990).

24.5.7 INHIBITION OF TUMOR ANGIOGENESIS BY TEA

The ability of cancer cells to move from their original sites and invade surrounding tissues is a fact that makes cancer a deadly and life-threatening disease. Once tumors become aggressive and metastasize to other organs, even systemic chemotherapy may be in vain. Jankun et al. (1997) showed that the main flavonol of green tea, EGCG, inhibits urokinase, one of the hydrolases implicated in tumor invasion. EGCG was found to inhibit tumor cell invasion and directly suppress the activity of matrix metalloproteases (MMPs) MMP2 and MMP9, two of the proteases most frequently overexpressed in cancer and angiogenesis, and essential in cutting through the basement membrane barriers (Jung and Ellis, 2001). Oral feeding of green tea polyphenols to animals that spontaneously develop prostate cancer resulted in inhibition of tumor growth, metastasis, and angiogenesis. This was found to be due to inhibition of vascular endothelial growth factor, a marker of angiogenesis, and also due to suppression of the activities of MMP2 and MMP9.

In a study by Cao and Cao (1999), it was proposed that the anticancer activity of EGCG is associated with the inhibition of urokinase, which is one of the most frequently expressed enzymes in human cancers. Employing computer-based molecular modeling, it was demonstrated that EGCG binds to urokinase, blocking the histidine 57 and serine 195 of the urokinase catalytic triad and extending toward arginine 35 from a positively charged loop of urokinase. These calculations were verified by assessing the inhibition of urokinase activity by the spectrophotometric amidolytic assay.

In a recent study, we investigated the effect of EGCG on the tube formation of human umbilical vein endothelial cells (HUVEC) on matrigel. Tube formation was inhibited by treatment both prior to plating and after plating endothelial cells on matrigel with EGCG. EGCG treatment also was found to reduce the migration of endothelial cells in matrigel plug model. Zymography was performed to determine if EGCG had any effect on MMPs. Zymographs of EGCG-treated culture supernatants modulated the gelatinolytic activities of secreted proteinases, indicating that EGCG may be exerting its inhibitory effect by regulating proteinases. These findings suggest that EGCG also acts as an angiogenesis inhibitor by modulating protease activity during endothelial morphogenesis (Singh et al., 2002).

24.5.8 Recent Advances in the Mechanisms of Biological Effects of Tea

24.5.8.1 Green Tea Modulates Mitogen-Activated Protein Kinases

It has been shown that the activation of mitogen-activated protein kinases (MAPK) by GTP might be a potential signaling pathway in the regulation of antioxidant-responsive element-mediated phase II enzyme gene expression (Yu et al., 1997). In this study, it was shown that GTP induces chloramphenicol acetyltransferase (CAT) activity in human hepatoma HepG2 cells transfected with a plasmid construct containing an antioxidant-responsive element (ARE) and a minimal glutathione S-transferase Ya promoter linked to the CAT reporter gene. This indicates that GTP stimulates the transcription of phase II detoxifying enzymes through the ARE. The authors studied the involvement of MAPKs' extracellular signal-regulated kinase 2 (ERK2) and c-Jun N-terminal kinase 1 (JNK1). Potent activation of ERK2 was seen following treatment of HepG2 cells with GTP. Similar to ERK2, JNK1 was also activated by treatment with GTP. GTP treatment also increased mRNA levels of the immediate-early genes c-jun and c-fos, as determined by reverse transcriptase-coupled polymerase chain reaction.

Exposure of normal human epidermal keratinocytes (NHEK) to UVB radiation induces intracellular release of hydrogen peroxide (oxidative stress) and phosphorylation of mitogen-activated protein kinase cell signaling pathways. In a recent study (Katiyar et al., 2001), we demonstrated that pretreatment of normal human epidermal keratinocytes (NHEK) with EGCG inhibits UVB-induced H_2O_2 production and H_2O_2-mediated phosphorylation of MAPK signaling pathways. We found that treatment of EGCG (20 µg/ml of medium) to NHEK before UVB exposure (30 mJ/cm^2) inhibited UVB-induced H_2O_2 production concomitant with the inhibition of UVB-induced phosphorylation of ERK1/2, JNK, and p38 proteins. We also observed that when these cells were pretreated with EGCG or with the known antioxidant ascorbic acid (as positive control), H_2O_2-induced phosphorylation of ERK1/2, JNK, and p38 was found to be significantly inhibited. These findings demonstrated that EGCG has the potential to inhibit UVB-induced oxidative stress-mediated phosphorylation of MAPK signaling pathways, suggesting that EGCG could be useful in attenuation of oxidative stress-mediated and MAPK-caused skin disorders in humans.

24.5.8.2 EGCG Inhibits Urokinase Activity

In a publicized study, it was shown that the anticancer activity of EGCG in green tea might be due to the inhibition of the enzyme urokinase, one of the most frequently expressed enzymes in human cancers (Jankun et al., 1997). Using molecular modeling, the authors demonstrated that EGCG binds to urokinase, blocking His 57 and Ser 195 of the urokinase catalytic triad and extending toward Arg 35 from a positively charged loop of urokinase. They verified this computer-based calculation by assessing the inhibition of urokinase activity using an amidolytic assay in which the release of a chromogen, on specific cleavage by urokinase, was spectrophotometrically quantified.

24.5.8.3   Green Tea Induces Apoptosis and Cell Cycle Arrest

In recent years, apoptosis has become a challenging issue in biomedical research, and the life span of both normal and cancer cells within a living system is regarded to be significantly affected by the rate of apoptosis. In addition, apoptosis is a discrete way of cell death different from necrotic cell death and regarded to be an ideal way of cell elimination (Fesus et al., 1995, and references therein). Thus, the chemopreventive agents that can modulate apoptosis may be able to affect the steady-state cell population, which can be useful in the management and therapy of cancer. In recent years, many cancer chemopreventive agents have been shown to induce apoptosis, and conversely, several tumor promoters have also been shown to inhibit apoptosis (Boolbol et al., 1996; Mills et al., 1995; Wright et al., 1994). Therefore, it is reasonable to assume that there may be wider implications for cancer control for chemopreventive agents (a) with proven effects in animal tumor bioassay systems and/or human epidemiology and (b) that cause induction of apoptosis of cancer cells.

Only a limited number of chemopreventive agents are known to induce apoptosis (Jiang et al., 1996). In a study (Ahmad et al., 1997), we found that green tea constituent EGCG induced apoptosis and cell cycle arrest in human epidermoid carcinoma cells A431. A promising observation of this study was that the EGCG-mediated apoptotic response was specific to cancer cells, as the induction of apoptosis was also observed in human carcinoma keratinocytes HaCaT, human prostate carcinoma cells DU145, and mouse lymphoma cells LY-R, but not in normal human epidermal keratinocytes. In another study (Ahmad et al., 2002), EGCG treatment of human epidermoid carcinoma A431 cells resulted in a dose- as well as time-dependent decrease in the total pRb as well as other members of the pRb family, namely p130 and p107. This response was accompanied by down-regulation in the protein expression of the E2F (1 through 5) family of transcription factors and their heterodimeric partners DP1 and DP2. This study demonstrated that EGCG causes a down-regulation of hyperphosphorylated pRb protein with a relative increase in hypophosphorylated pRb that, in turn, compromises the availability of "free" E2F. This series of events leads to stoppage of cell cycle progression at the G1 to S phase transition, thereby causing G0/G1 arrest and subsequent apoptotic cell death. Following this, we conducted a systematic study and showed the involvement of each component of cyclin-dependent kinase inhibitor cyclin–cyclin-dependent kinase machinery during cell cycle arrest and apoptosis of human prostate carcinoma cells by EGCG (Gupta et al., 2003).

24.5.8.4   EGCG Suppresses Extracellular Signals and Cell Proliferation through EGF Receptor Binding

Liang et al. (1997), using thymidine incorporation assay, demonstrated that EGCG could significantly inhibit DNA synthesis in A431 cells. EGCG also inhibited the protein tyrosine kinase activities of EGF-R, PDGF-R (platelet-derived growth factor R), and FGF-R, but not of pp60[v-src], PKC, and PKA. EGCG also inhibited the phosphorylation of EGF-R by EGF and blocked the binding of EGF to its receptor. These finding suggested that EGCG might inhibit the process of tumor formation via blocking cellular signal transduction pathways.

24.5.8.5   EGCG Down-Regulates Nuclear Transcription Factor -κB

Lin and Lin (1997) examined the effects of EGCG on nitric oxide production from murine peritoneal macrophages and sought the possible mechanisms. Their data suggested that EGCG blocked early-event nitric oxide synthase induction via the inhibition of binding of transcription factor NF-κB to the iNOS promoter, thereby inhibiting the induction of iNOS transcription. In a recent study, we demonstrated that EGCG treatment of normal human epidermal keratinocytes (NHEK) cells resulted in a significant dose- and time-dependent inhibition of UVB-mediated activation and nuclear translocation of a NF-κB/p65. These data suggest that EGCG protects against the adverse effects of UV radiation via modulations in NF-κB pathway and provide a molecular basis for the photo-chemopreventive effect of EGCG (Afaq et al., 2003).

24.5.8.6 EGCG and Theaflavins Inhibit Tumor Promoter-Induced Activator Protein 1 Activation and Cell Transformation

Dong et al. (1997) used JB6 mouse epidermal cell line, a system that has extensively been used as an *in vitro* model for tumor promotion studies, to examine antitumor promotion effects of EGCG and theaflavins at the molecular levels. EGCG and theaflavins inhibited EGF- or TPA- induced cell transformation in a dose-dependent manner. EGCG and theaflavins also inhibited AP-1-dependent transcriptional activity and DNA binding activity. This study further showed that the inhibition of AP-1 activation occurs through the inhibition of a c-Jun NH_2-terminal kinase dependent pathway.

24.6 TEA AND CANCER: EPIDEMIOLOGICAL STUDIES

In recent years, the availability of increasingly sophisticated biochemical approaches to molecular epidemiology has made the process of cancer formation easier to understand and comprehend (Wright, 1991). Such understanding has offered opportunities for defining the risk to individuals and for modulating this risk by means of agents that can alter critical steps in the multistage process of carcinogenesis. The epidemiological studies, published in 1991 by the International Agency for Research on Cancer (IARC), on tea consumption and its effects on various types of cancer did not find conclusive results. The study concluded that there is inadequate evidence for the carcinogenicity of tea in humans and in experimental animals (IARC, 1991). It was also concluded that the available epidemiological data do not provide sufficient indication that tea consumption has a statistically significant causative effect on human cancers. However, the possible harmful effects of excessive tea consumption, tea consumption at very high temperature, or the consumption of salted tea were not ruled out, and this is an area in which virtually no study has been conducted as yet. Since the publication of this position paper by IARC, many laboratory and epidemiological studies defining an association between tea consumption and cancer risk and prevention have been published. This section provides a general overview of the pertinent epidemiological studies on tea consumption and cancer prevention of different sites. It is important to mention that a few studies have shown either no effect or enhanced effect of tea consumption and cancer risk.

24.6.1 CANCER OF THE ESOPHAGUS AND NASOPHARYNX

The epidemiological studies conducted in China revealed that most of the areas with higher esophageal cancer mortality rates are in the northern provinces, where tea is either not produced or is infrequently consumed. In a case-control study, Gao et al. (1994) showed that the consumption of green tea reduces the risk of esophageal cancer. This population-based case-control study of esophageal cancer in urban Shanghai, People's Republic of China, suggested a protective effect of green tea consumption. The authors concluded, "Although these findings are consistent with studies in laboratory animals, indicating that green tea can inhibit esophageal carcinogenesis, further investigations are definitely needed." Zheng et al. (1995) have conducted a cohort study in post-menopausal women in Iowa that revealed inverse associations between tea consumption and cancer risk for oropharyngeal and esophageal cancers. The daily consumption of tea was found to be associated with an over 50% lower risk of these cancers.

In Linzhou (formerly Linxian), which is a high-incidence area for esophageal cancer in northern China, the consumption of tea is rare and is not believed to be a contributing factor (Yang, 1980). In a follow-up study, decaffeinated green tea was not shown to have beneficial effects in alleviating esophageal precancerous lesions and abnormal cell proliferation patterns (Wang et al., 2002). Two case-control studies in southern Brazil (Victora et al., 1987) and northern Italy (La Vecchia et al., 1992) also indicated that there was no relationship between esophageal cancer risk and the frequency of tea consumption. Case-control studies conducted by several investigators (Bashirov et al., 1968; Cook-Mozaffari et al., 1979; De Jong et al., 1974; Kaufman et al., 1965) showed no association between drinking of tea at normal temperature (35 to 47°C) and esophageal cancer, but ingestion of tea at hot temperature (55 to 67°C) was associated with an increased risk of esophageal cancer (Dhar et al., 1993). A large hospital-based case-control for 1248 cases with esophageal cancer and the same number of controls in South China showed that Congou, a grade of Chinese black tea, may protect against cancers of the esophagus and reduce the risk of a combination of alcohol drinking and smoking (especially smoking), regardless of temperature when drinking (Ke et al., 2002). In another study conducted in Singapore, no association between tea drinking and esophageal cancer was found (De Jong et al., 1974). Yang and Wang (1993) have suggested that the high temperature of tea or hot tea itself, rather than the components present in tea, may be an important factor in human esophageal cancer.

In three case-control studies (Henderson et al., 1976; Lin et al., 1973; Shanmugaratnam et al., 1978), no correlation between tea consumption and nasopharyngeal cancer was observed.

A limited number of studies suggested a positive association between tea consumption and the occurrence of esophageal cancer. Morton (1986) and Kapadia et al. (1983) suggested that the excessive consumption of tea in the geographical zone between Iran to northern China may be a causative factor for the high incidence of esophageal cancer. A geographical correlation study in the Caspian littoral of Iran (Lubin et al., 1985) indicated that individuals in the high-incidence area consumed more tea than in low-incidence areas. The case-control studies conducted in Kazakstan (Bashirov et al., 1968; Kaufman et al., 1965) revealed that only the consumption of very hot tea was associated with higher risk of esophageal cancer. Another case-control study by Cook-Mozaffari et al. (1979) indicated that only the ingestion of very hot tea had a statistically significant association with esophageal cancer.

A recent case control study was designed to evaluate interactions between lifestyle factors and the development of esophageal cancers in Huaian City of Jiangsu Province, China. A total of 141 cases of esophageal cancer patients and 223 population-based controls were evaluated, and tea drinking was found to be a protective factor, with tea consumption reducing the risk of esophageal cancers (Gao et al., 2002).

24.6.2 CANCER OF THE STOMACH

An epidemiological study (Oguni et al., 1992) conducted in Shizuoka Prefecture, Japan, indicated that the cancer death rate in this tea-producing area, especially from stomach cancer, was lower than the national average. A case-control study in Kyushu, Japan, showed that the individuals consuming green tea frequently or in larger quantities have a lower risk of gastric cancer (Kono et al., 1988). A cohort study conducted in Iowa (Zheng et al., 1995) showed an inverse relationship between tea consumption and stomach cancer. A population-based case-control study to evaluate risk factors of gastric cancer conducted in areas with contrasting incidence rates in Sweden confirmed that tea had a statistically significant preventive effect when it was consumed during adolescence (Hansson et al., 1993).

To establish the correlation of tea consumption with stomach cancer, a number of case-control studies were conducted in Buffalo (Graham et al., 1967), Kansas City (Higginson, 1966), Nagoya, Japan (Tazima and Tominaga, 1985), Piraeus, Greece (Trichopoulos, 1985), Milan, Italy (La Vecchia et al., 1992), Spain (Agudo et al., 1992), and Turkey (Demirer et al., 1990). These studies indicated that there was no statistically significant association between tea consumption and cancer of the stomach. A cohort study conducted in London (Kinlen et al., 1988) showed a positive association between black tea consumption and stomach cancer. Another case control study in Taipei, Taiwan, also suggested that green tea consumption is a risk factor for gastric cancer (Lee et al., 1990).

In one study (Goldbohm et al., 1996), an association was sought between black tea consumption and the subsequent risk of stomach, colorectal, lung, and breast cancers in The Netherlands Cohort

Study on Diet and Cancer among 58,279 men and 62,573 women aged 55 to 69 years. The consumption of tea showed an inverse association with cancer. However, tea drinkers appeared to smoke less and to eat more vegetables and fruits than nondrinkers do. When smoking and dietary factors were taken into account, tea in itself did not appear to protect against stomach cancer. Gao et al. (2002) conducted a case control study designed to evaluate interactions between lifestyle factors and the development of stomach cancers in Huaian City of Jiangsu Province, China. A total of 153 cases of stomach cancer patients and 223 population-based controls were evaluated, and tea drinking was found to be a protective factor, with tea consumption reducing the risk of stomach cancers.

24.6.3 CANCER OF THE BLADDER, KIDNEY, AND URINARY TRACT

In a cohort study, conducted in Iowa it was found that daily tea consumption reduced the risk of kidney cancers in postmenopausal women (Zheng et al., 1995). Several other case-control studies were conducted to determine whether there was an association between tea consumption and cancers of the bladder and urinary tract (Armstrong et al., 1976; Claude et al., 1986; D'Avanzo et al., 1992; Hartge et al., 1983; Heilbrun et al., 1986; Howe et al., 1980; Iscovich et al., 1987; Jensen et al., 1986; Kinlen et al., 1988; Kunze et al., 1992; La Vecchia et al., 1989; La Vecchia et al., 1992; Miller et al., 1978; Morgan and Jain, 1974; Nomura et al., 1991; Ohno et al., 1985; Risch et al., 1988; Simon et al., 1975; Slattery et al., 1988; Stocks, 1970). No positive relationship was observed in any of these studies. Similarly, in five other case-control studies, no association was found between green tea consumption and renal cell cancer (Armstrong et al., 1976; Goodman et al., 1986; La Vecchia et al., 1992; McCredie et al., 1988; Yu et al., 1986). In a cohort study (Kinlen et al., 1988), a positive correlation was observed between tea consumption and cancer of the kidney.

Bianchi et al. (2000) examined the association between usual adult tea consumption and risk of bladder and kidney cancers in a population-based case-control study that included 1452 bladder cancer cases, 406 kidney cancer cases, and 2434 controls. For bladder cancer, the age- and sex-adjusted odds ratios were 0.9 for <1.0 cup/day, 1.0 for 1.0–2.6 cups/day, and 0.9 for >2.6 cups/day. When more extreme cut points were used, persons who consumed >5 cups/day (>90th percentile) had a suggestive decreased risk (odds ratio = 0.7), but there was no evidence of a dose–response relationship. This study offered only minimal support for an inverse association between tea consumption and bladder or kidney cancer risk.

24.6.4 CANCER OF THE COLON, RECTUM, AND UTERUS

Several studies (Armstrong et al., 1976; Claude et al., 1986; D'Avanzo et al., 1992; Goodman et al., 1986; Hartge et al., 1983; Heilbrun et al., 1986; Howe et al., 1980; Iscovich et al., 1987; Jensen et al., 1986; Kinlen et al., 1988; Kunze et al., 1992; La Vecchia et al., 1989; La Vecchia et al., 1992; McCredie et al., 1988; Miller et al., 1978; Morgan and Jain, 1974; Nomura et al., 1991; Ohno et al., 1985; Risch et al., 1988; Simon et al., 1975; Slattery et al., 1988; Yu et al., 1986) indicated that the consumption of black tea decreased the risk for rectal cancer, while several other studies (Dales et al., 1979; Higginson, 1966; Miller et al., 1983; Morgan and Jain, 1974; Phillips and Snowdon, 1985; Tazima and Tominaga, 1985) showed no correlation. Two more studies (Oguni et al., 1992; Stocks, 1970) concluded that there was a possible negative association between tea consumption and cancer of the uterus. These studies also indicated that more studies are required to draw a final conclusion. Three other studies (Heilbrun et al., 1986; La Vecchia et al., 1992; Stocks, 1970) indicated a positive association between tea consumption and colon and rectal cancer. Baron et al. (1997) used data from an adenoma-prevention trial to investigate these associations. Patients with at least one recent large bowel adenoma were followed with colonoscopy 1 and 4 years after their qualifying examinations. Adenomas detected at the year-4 colonoscopy were used as end points. The study showed that there was no apparent association between the intake of tea and the risk of recurrent colorectal adenomas. In a large population-based case-control study

(Ji et al., 1997) conducted in Shanghai, China, newly diagnosed cancer cases (931 colon and 884 rectum) during 1990–1993 among residents 30 to 74 years of age were included. An inverse association with each cancer was observed with increasing amount of green tea consumption.

An association between black tea consumption and the subsequent risk of colorectal cancer was sought in The Netherlands Cohort Study on Diet and Cancer (Goldbohm et al., 1996). No association was observed between tea consumption and risk of colorectal cancer: The risk among tea drinkers in each consumption category was similar to that among nondrinkers. The authors concluded that this study did not support the hypothesis that consumption of black tea protects against four of the major cancers in humans, and there was no evidence that black tea has a cancer-enhancing effect.

A prospective study in Japan (Nagano et al., 2001) examined the association between green tea consumption and cancer incidence. A total of 38,540 subjects (14,873 men, mean age 52.8 years; 23,667 women, mean age 56.8 years) responded to a mail survey carried out between 1979 and 1981. A self-administered questionnaire ascertained consumption frequency of green tea using precoded answers (never, once per day, twice to four times per day, and five or more times per day). Follow-up continued until 31 December 1994. The study analyzed solid cancers (n = 3881); hematopoietic cancers (188); cancers of all sites combined (4069); and cancer of specific sites with more than 100 cases, i.e., stomach (901), colon (432), rectum (193), liver (418), gallbladder (122), pancreas (122), lung (436), breast (281), and bladder (122). Poisson regression was used to allow for city, gender, age, radiation exposure, smoking status, alcohol drinking, body-mass index, education level, and calendar time. Green tea consumption was virtually unrelated to incidence of cancers under study. The relative risks of all cancers for those consuming green tea twice to four times per day and five or more times per day were 1.0 and 0.98 respectively, as compared with those consuming green tea once per day or less. The study concluded that there was no evidence to suggest that regular green tea consumption is related to reduced cancer risks.

24.6.5 CANCER OF THE PROSTATE

As yet, no detailed case-control study has been conducted to assess the effect of consumption of green tea for human prostate cancer. Published data seeking an association between tea consumption and prostate cancer risk have considered undefined tea preparations, mostly black tea. At least two epidemiological studies have shown that people who regularly consume tea have lower incidence of prostate cancer (Heilbrun et al., 1986; Jain et al., 1998). Heilbrun et al. (1986), in a prospective cohort study employing 7833 men of Japanese ancestry living in Hawaii, observed a weaker but significant negative association between black tea intake (more than one cup per day) and prostate cancer incidence (p = 0.02). In a case-control study conducted in three geographical areas of Canada, Jain et al. (1998) observed a decrease in prostate cancer risk with tea intake of more than 2 cups per day. Other epidemiological studies conducted in Italy (La Vecchia et al., 1992), Utah (Slattery and West, 1993), and Canada (Ellison, 2000) did not find any difference of risk for prostate cancer between tea drinkers and nondrinkers. However, most of these studies include populations that were predominantly black tea drinkers. It should be noted that most of these studies lacked parallel for comparison in categorization of tea consumption, the type of tea consumed, and the ethnicity of the subjects, which weakens the overall impact of the study.

Epidemiological investigations seeking an association between green tea and prostate cancer should be undertaken to establish the validity of cell culture and animal data to human prostate cancer patients. A phase I trial of green tea extract in patients with solid tumors concluded that 1.0 g/m² of green tea extract three times a day (equivalent to 7 to 8 Japanese cups [120 ml] of green tea three times daily) is well tolerated and is not associated with any side effects (Pisters, 2001). On the basis of this study, a recent phase II clinical trial explored the antineoplastic effects of green tea in patients with metastatic androgen-independent prostate cancer (Jatoi et al., 2003). In this study, 48 patients were instructed to take 6 g of green tea per day orally in six divided doses.

Patients were monitored monthly for response and toxicity. The study concluded that green tea carries limited antineoplastic activity among patients with androgen-independent prostate cancer. However, it should be noted that this study was conducted in patients with metastatic androgen-independent prostate cancer and, therefore, in principle does not mimic to assess the chemopreventive effects of green tea. For an ideal chemoprevention study, a population with high risk for prostate cancer development should be considered.

24.6.6 CANCER OF THE LIVER, LUNG, BREAST, AND PANCREAS

In Shizuoka Prefecture of Japan, a negative association between green tea consumption and liver cancer incidence was observed (Oguni et al., 1992), while no relationship was observed in three other studies (Heilbrun et al., 1986; La Vecchia et al., 1992; Stocks, 1970). A number of studies have shown positive as well as negative associations between tea consumption and lung cancer (Morgan and Jain, 1974; Oguni et al., 1992; Stocks, 1970; Tewes et al., 1990). Several other case-control studies indicated no association between tea consumption and breast cancer (La Vecchia et al., 1986; Lubin et al., 1985; Mabuchi et al., 1985; Rosenberg et al., 1985; Schairer et al., 1987). No association between pancreatic cancer and tea consumption was observed in a number of individual studies (Bueno de Mesquita et al., 1992; Heilbrun et al., 1986; Hiatt et al., 1988; Jain et al., 1991; La Vecchia et al., 1992; Mabuchi et al., 1985; Mack et al., 1986; MacMahon et al., 1981; Morgan and Jain, 1974; Schairer et al., 1987), whereas one case-control study (Kinlen and McPherson, 1984) showed a positive correlation. In a recent large population-based case-control study (Ji et al., 1997) conducted in Shanghai, China, 451 newly diagnosed pancreas cancer cases among residents 30 to 74 years of age were included. This study showed an inverse association with cancer with increasing amount of green tea consumption.

Goldbohm et al. (1996) conducted an epidemiological study to define an association between black tea consumption and the subsequent risk of stomach, colorectal, lung, and breast cancers in The Netherlands Cohort Study on Diet and Cancer. They found that the risk among tea drinkers in each consumption category was similar to that among nondrinkers. They further concluded that this investigation did not support the hypothesis that black tea consumption protects against four of the major cancers in humans. However, in this study, a cancer-enhancing effect was also not evident. In another prospective study, Harnack et al. (1997) tested the hypothesis that tea intake increases the risk of cancer of the exocrine pancreas. The results from a cohort of 33,976 white American women with 9 years of follow-up showed that tea consumption is not related to the incidences of cancer. A Swedish mammography screening cohort, a large population-based prospective cohort study comprising 59,036 women aged 40 to 76 years, was conducted to examine relation between tea drinking and breast cancer incidence. During 508,267 person-years of follow-up, 1271 cases of invasive breast cancers were diagnosed. Women who reported drinking 4 or more cups of coffee/tea per day had a covariate-adjusted hazard ratio of breast cancer of 0.94 compared with women who reported drinking 1 cup a week or less. In this large cohort of Swedish women, consumption of coffee, tea, and caffeine was not associated with breast cancer incidence (Michels, 2002).

On the basis of available epidemiological studies and careful observations made in laboratory animals, it is reasonable to conclude that tea consumption is unlikely to have preventive effects against cancer risk at body sites in all populations. However, it is becoming clear that it may have preventive effects against cancer risk of some body sites in some populations. Collectively, these data suggest that more case-controlled studies should be undertaken.

24.7 CONCLUSION AND FUTURE DIRECTIONS

Although not sufficiently explained, it is estimated that almost one-third of all cancers are caused by dietary substances. Thus, dietary habits are an important factor in the development of human cancers. The best example of this association is the fact that changes in dietary habits modulate the incidence of cancer in experimental animals, and epidemiological studies suggest a similar association in the human population. The usefulness of dietary substances for prevention against the occurrence of cancer is increasingly recognized as one practical approach in this direction (Challa et al., 1997; Katiyar and Mukhtar, 1996). Changing lifestyle, as reflected in dietary habits and culinary practices, has been recognized as a major factor for human cancer risk. There is suggestive evidence in humans that diets rich in fruits, or containing reduced levels of fats, particularly those derived from animal sources, are less likely to lead to cancer. Because tea is a popular beverage consumed worldwide, the relationship between tea consumption and human cancer incidence is an important concern, and epidemiological studies have indicated that tea consumption does not pose a causative effect on human cancers. Indeed, epidemiological observations as well as laboratory studies have indicated that tea consumption is likely to have beneficial effects in reducing certain cancer types in some populations. However, possible harmful effects of the consumption of excessive amounts of tea, tea at very high temperature, or salted tea cannot be ruled out.

Although a considerable body of information provides compelling evidence for the preventive potential of tea against cancer, a clear understanding of the mechanisms by which tea polyphenols retard the induction, growth, and subsequent progression of cancer is necessary in order to examine the effect of tea polyphenols on health and to devise better strategies against cancer. Black tea, the major form of tea, is consumed mostly in Western countries. However, the chemistry, biological activities, and chemopreventive properties of black tea, especially the polyphenols present therein, are not well understood. Research in this area is needed.

In view of the available experimental data on the protection of mouse skin carcinogenesis and inflammation by tea polyphenols, an intervention study on human skin carcinogenesis and inflammatory responses including a dose–response effect could be of great importance. Similarly, the data collected on the prevention of prostate cancer in transgenic mice by green tea polyphenols provides convincing evidence for a chemopreventive trial in a high-risk human population. Since only limited data are available on the bioavailability of tea polyphenols following tea consumption by the human population, studies on the absorption, distribution, and metabolism of green and black tea polyphenols in animals and humans are of great importance. In this regard a method has been developed for the analysis of plasma and urinary tea polyphenols in human subjects (Lee et al., 1995). This methodology may prove valuable for studying the effects of tea consumption, bioavailability of polyphenols, and an association of the two with human cancer.

Because the causative factors are different for different populations, tea consumption may affect carcinogenesis only in selected situations rather than having a general effect on all cancers. Thus, there is a need to define the population that could benefit from tea consumption. Such intervention studies in various populations may provide useful information on the protective effects of tea polyphenols on cancer of specific organs or in specific populations. After careful evaluation of additional studies, it may be possible to recommend consumption of tea polyphenols by humans. Such agents do not necessarily have to be consumed by tea drinking. They can be supplemented in other food items, for example in cosmetic products, consumer items, and in vitamin supplements and other products. This approach involves the production of "designer items" for consumption by human population.

Because research findings in laboratory animals clearly point to a role of tea polyphenols in cancer chemoprevention, the well-defined and naturally occurring polyphenols found in tea should be evaluated in clinical intervention in human trials. In this regard, a phase I clinical trial has suggested that green tea is well tolerated and can be safely administered without any side effects over a period of 6 months (Pisters et al., 2001). A recent phase II clinical trial, however, concluded that green tea carries only a limited antineoplastic activity in patients with androgen-dependent prostate cancer (Jatoi et al., 2003). It is worth mentioning here that this trial in principle does not mimic a true chemopreventive trail. An ideal chemopreventive study should be conducted in a population with high risk for prostate cancer development.

REFERENCES

Afaq, F., Adhami, V.M., Ahmad, N., and Mukhtar, H., Inhibition of ultraviolet B-mediated activation of nuclear factor kappaB in normal human epidermal keratinocytes by green tea constituent (–)-epigallocatechin-3-gallate, *Oncogene*, 22, 1035–1044, 2003.

Agarwal, R. and Mukhtar, H., Cutaneous chemical carcinogenesis, in *Pharmacology of the Skin*, Mukhtar, H., Ed., CRC Press, Boca Raton, FL, 1991, pp. 371–387.

Agarwal, R., Katiyar, S.K., Zaidi, S.I.A., and Mukhtar, H., Inhibition of tumor promoter-caused induction of ornithine decarboxylase activity in SENCAR mice by polyphenolic fraction isolated from green tea and its individual epicatechin derivatives, *Cancer Res.*, 52, 3582–3588, 1992.

Agarwal, R. and Mukhtar, H., Oxidative stress in skin chemical carcinogenesis, in *Oxidative Stress in Dermatology*, Fuchs, J. and Packer, L., Eds., Marcel Dekker, New York, 1993, pp. 207–241.

Agarwal, R., Katiyar, S.K., Khan, S.G., and Mukhtar, H., Protection against ultraviolet B radiation-induced effects in the skin of SKH-1 hairless mice by a polyphenolic fraction isolated from green tea, *Photochem. Photobiol.*, 58, 695–700, 1993.

Agudo, A., Gonzalez, C.A., Marcos, G., Sanz, M., Saigi, E., Verge, J., Boleda, M., and Ortego, J., Consumption of alcohol, coffee, and tobacco, and gastric cancer in Spain, *Cancer Causes Control*, 3, 137–143, 1992.

Ahmad, N., Feyes, D.K., Nieminen, A.-L., Agarwal, R., and Mukhtar, H., Green tea constituent epigallocatechin-3-gallate and induction of apoptosis and cell cycle arrest in human carcinoma cells, *J. Natl. Cancer Inst.*, 89, 1881–1886, 1997.

Ahmad, N., Adhami, V.M., Gupta, S., Cheng, P., and Mukhtar, H., Role of the retinoblastoma (pRb)-E2F/DP pathway in cancer chemopreventive effects of green tea polyphenol epigallocatechin-3-gallate, *Arch. Biochem. Biophys.*, 398, 125–131, 2002.

Ames, B.N., Dietary carcinogens and anticarcinogens, *Science*, 221, 1256–1262, 1983.

Ames, B.N. and Gold, L.S., Too many rodent carcinogens: mitogenesis increases mutagenesis, *Science*, 5249, 970–971, 1990.

Armstrong, B., Garrod, A., and Doll, R.A., Retrospective study of renal cancer with special reference to coffee and animal protein consumption, *Br. J. Cancer*, 33, 127–136, 1976.

Athar, M., Agarwal, R., Wang, Z.Y., Lloyd, J.R., Bickers, D.R., and Mukhtar, H., All trans retinoic acid protects against free radical generating compounds-mediated conversion of chemically and ultraviolet B radiation-induced skin papillomas to carcinomas, *Carcinogenesis*, 12, 2325–2329, 1991.

Baron, J.A., Greenberg, E.R., Haile, R., Mandel, J., Sandler, R.S., and Mott, L., Coffee and tea and the risk of recurrent colorectal adenomas, *Cancer Epidemiol. Biomarkers Prev.*, 6, 7–10, 1997.

Bashirov, M.S., Nugmanov, S.N., and Kolycheva, N.I., On the epidemiology of cancer of the esophagus in the Aktiubinsk region of the Kazakhstan SSR, *Voprosy Onkologii*, 14, 3–7, 1968.

Bhimani, R.S., Troll, W., Grunberger, D., and Frenkel, K., Inhibition of oxidative stress in HeLa cells by chemopreventive agents, *Cancer Res.*, 53, 4528–4533, 1993.

Bianchi, G.D., Cerhan, J.R., Parker, A.S., Putnam, S.D., See, W.A., Lynch, C.F., and Cantor, K.P., Tea consumption and risk of bladder and kidney cancers in a population-based case-control study, *Am. J. Epidemiol.*, 151, 377–383, 2000.

Boolbol, S.K., Dannenberg, A.J., Chadburn, A., Martucci, C., Guo, X.J., Ramonetti, J.T., Abreu-Goris, M., Newmark, H., Lipkin, M.L., DeCosse, J.J., and Bertgnolli, M.M., Cyclooxygenase-2 overexpression and tumor formation are blocked by sulindac in murine model of familial polyposis, *Cancer Res.*, 56, 2556–2560, 1996.

Boone, C.W., Kelloff, G.J., and Malone, W.E., Identification of candidate cancer chemopreventive agents and their evaluation in animal models and human clinical trials: a review, *Cancer Res.*, 50, 2–9, 1990.

Boone, C.W., Kelloff, G.J., and Steele, V.E., Natural history of intraepithelial neoplasia in humans with implications for cancer chemoprevention strategy, *Cancer Res.*, 52, 1651–1659, 1992.

Bu-Abbas, A., Clifford, M.N., Walker, R., and Ioannides, C., Selective induction of rat hepatic CYP1 and CYP4 proteins and of peroxisomal proliferation by green tea, *Carcinogenesis*, 15, 2575–2579, 1994.

Bueno de Mesquita, H.B., Maisonneuve, P., Moerman, C.J., Runia, S., and Boyle, P., Lifetime consumption of alcoholic beverages, tea and coffee and exocrine carcinoma of the pancreas: a population-based case-control study in The Netherlands. *Int. J. Cancer*, 50, 514–522, 1992.

Cao, Y. and Cao, R., Angiogenesis inhibited by drinking tea, *Nature*, 398, 381, 1999.

Cerruti, P.A., Mechanisms of action of oxidant carcinogens, *Cancer Detection Prev.*, 14, 281–284, 1989.

Challa, A., Ahmad, N., and Mukhtar, H., Cancer prevention through sensible nutrition, *Int. J. Oncol.*, 11, 1387–1392, 1997.

Challa, A., Katiyar, S.K., Cooper, K.D., and Mukhtar, H., Inhibition of UV-radiation-caused induction of oxidative stress and immunosuppression in C3H/HeN mice by polyphenols from green tea, *J. Investigative Dermatol.*, accepted, 1998.

Chatterjee, M.L., Agarwal, R., and Mukhtar, H., Ultraviolet B radiation-induced DNA lesions in mouse epidermis: an assessment using a novel ^{32}P-postlabelling technique, *Biochemical Biophysical Res. Commn.*, 229, 590–595, 1996.

Chen, J.S., The effect of Chinese tea on the occurrence of esophageal tumors induced by N-nitrosomethyl-benzylamine in rats, *Preventive Med.*, 21, 385–391, 1992.

Chen, L., Bondoc, F.Y., Lee, M.J., Hussin, A.H., Thomas, P.E., and Yang, C.S., Caffeine induces cytochrome P4501A2: induction of CYP1A2 by tea in rats, *Drug Metab. Disposal*, 24, 529–533, 1996.

Chen, Z.Y., Yan, R.Q., Qin, G.Z., and Chia, K.B., Effect of six edible plants on the development of aflatoxin B_1-induced γ-glutamyltranspeptidase positive hepatocyte foci in rats, *Chung Hua liu Tsa Chih*, 9, 109–111, 1987.

Cheng, S., Lin, P., Ding, L., Hu, X., Oguni, I., and Hara, Y., Inhibition of green tea extract on mutagenicity and carcinogenicity, *Proc. Int. Symp. Tea Sci., Jpn.*, 195-199, 1991.

Chou, F.P., Chu, Y.C., Hsu, J.D., Chiang, H.C., and Wang, C.J., Specific induction of glutathione S-transferase GSTM2 subunit expression by epigallocatechin gallate in rat liver, *Biochem. Pharmacol.*, 60, 643–650, 2000.

Claude, J., Kunze, E., Frentzel-Beyme, R., Paczkowski, K., Schneider, J., and Schubert, H., Life-style and occupational risk factors in cancer of the lower urinary tract, *Am. J. Epidemiol.*, 124, 578–589, 1986.

Conney, A.H., Induction of microsomal enzymes by foreign chemicals and carcinogenesis by polycyclic aromatic hydrocarbons, *Cancer Res.*, 42, 4875–4917, 1982.

Cook-Mozaffari, P.J., Azordegan, F., Day, N.E., Ressicand, A., Sabai, C., and Aramesh, B., Oesophageal cancer studies in the Caspian Littoral of Iran: results of a case-control study, *Br. J. Cancer*, 39, 293–309, 1979.

Dales, L.G., Friedman, G.D., Ury, H.K., Grossman, S., and Williams, S.R., A case-control study of relationships of diet and other traits to colorectal cancer in American blacks, *Am. J. Epidemiol.*, 109, 132–144, 1979.

D'Avanzo, B., La Vecchia, C., Franceschi, S., Negri, E., Talamani, R., and Buttino, I., Coffee consumption and bladder cancer risk, *Eur. J. Cancer*, 28A, 1480–1484, 1992.

DeFabo, E.C. and Noonan, F.P., Mechanism of immune suppression by ultraviolet radiation *in vivo*, I: evidence for the existence of a unique photoreceptor in skin and its role in photoimmunology, *J. Experimental Med.*, 157, 84–98, 1983.

De Jong, U.W., Breslow, N., Hong, J.G.E, Sridharan, M., and Shanmugaratnam, K., Aetiological factors in oesophageal cancer in Singapore Chinese, *Int. J. Cancer*, 13, 291–303, 1974.

Demirer, T., Icli, F., Uzunalimoglu, O., and Kucuk, O., Diet and stomach cancer incidence: a case-control study in Turkey, *Cancer*, 65, 2344–2348, 1990.

Dhar, G.M., Shah, G.N., and Hafiza, N.B., Epidemiological trend in the distribution of cancer in Kashmir Valley, *J. Epidemiol. Community Health*, 47, 290–292, 1993.

DiGiovanni, J., Multistage carcinogenesis in mouse skin, *Pharmacol. Ther.*, 54, 63–128, 1992.

Dong, Z., Ma, W.-Y., Huang, C., and Yang, C.S., Inhibition of tumor promoter-induced activator protein 1 activation and cell transformation by tea polyphenols, (–)-epigallocatechin gallate and theaflavins, *Cancer Res.*, 57, 4414–4419, 1997.

Ellison, L.F., Tea and other beverage consumption and prostate cancer risk: a Canadian retrospective cohort study, *Eur. J. Cancer Prev.*, 9, 125–130, 2000.

Elmets, C.A., Cutaneous photocarcinogenesis, in *Pharmacology of the Skin*, Mukhtar, H., Ed., CRC Press, Boca Raton, FL, 1991, pp. 389–416.

Embola, C.W., Weisburger, M.C., and Weisburger, J.H., Green tea and the metabolism of 2-amino-3-meth-ylimidazo, *Food Chem. Toxicol.*, 39, 629–633, 2001.

Fesus, L., Szondy, Z., and Uray, I., Probing the molecular program of apoptosis by cancer chemopreventive agents, *J. Cell. Biochem.*, 22 (Suppl.), 151–161, 1995.

Fujiki, H., Suganuma, M., Suguri, H., Takagi, K., Yoshizawa, S., Ootsuyama, A., Tanooka, H., Okuda, T., Kobayashi, M., and Sugimura, T., New antitumor promoters: (–)-epigallocatechin gallate and sarco-phytols A and B, *Basic Life Sci.*, 52, 205–212, 1990.

Fujiki, H., Yoshizawa, S., Horiuchi, T., Suganuma, M., Yatsunami, S., Nishiwaki, S., Okabe, S., Nishiwaki, M.R., Okuda, T., and Sugimura, T., Anticarcinogenic effect of (–)-epigallocatechin gallate, *Preventive Med.*, 21, 503–509, 1992.

Fujita, Y., Yamane, T., Tanaka, M., Kuwata, K., Okuzumi, J., Takahashi, T., Fujiki, H., and Okuda, T., Inhibitory effect of (–)-epigallocatechin gallate on carcinogenesis with N-ethyl-N′-nitro-N-nitrosoguanidine in mouse duodenum, *Jpn. J. Cancer Res.*, 80, 503–505, 1989.

Gao, G.D., Zhou, L.F., and Qi, G., Initial study of antitumorigenesis of green tea: animal test and flow cytometry, *Tumor*, 10, 42–44, 1990.

Gao, Y.T., McLaughlin, J.K., Blot, W.J., Ji, B.T., Dai, Q., and Fraumeni, Jr., J.F., Reduced risk of esophageal cancer associated with green tea consumption, *J. Natl. Cancer Inst.*, 86, 855–858, 1994.

Gao, C.M., Takezaki, T., Wu, J.Z., Li, Z.Y., Liu, Y.T., Li, S.P., Ding, J.H., Su, P., Hu, X., Xu, T.L., Sugimura, H., and Tajima, K., Glutathione-S-transferases M1 (GSTM1) and GSTT1 genotype, smoking, consumption of alcohol and tea and risk of esophageal and stomach cancers: a case-control study of a high-incidence area in Jiangsu Province, China, *Cancer Lett.*, 188, 95–102, 2002.

Gensler, H.L., Timmermann, B.N., Valcic, S., Wachter, G.A., Dorr, R., Dvorakova, K., and Alberts, D.S., Prevention of photocarcinogenesis by topical administration of pure epigallocatechin gallate isolated from green tea, *Nutr. Cancer*, 26, 325–335, 1996.

Goldbohm, R.A., Hertog, M.G., Brants, H.A., van Poppel, G., and van den Brandt, P.A., Consumption of black tea and cancer risk: a prospective cohort study, *J. Natl. Cancer Inst.*, 88, 93–100, 1996.

Goodman, M.T., Morgenstern, H., and Wynder, E.L., A case-control study of factors affecting the development of renal cell cancer, *Am. J. Epidemiol.*, 124, 926–941, 1986.

Graham, S., Lilienfeld, A.M., and Tidings, J.E., Dietary and purgation factors in the epidemiology of gastric cancer, *Cancer*, 20, 2224–2234, 1967.

Grinberg, L.N., Newmark, H., Kitrossky, N., Rahamim, E., Chevion, M., and Rachmilewitz, E.A., Protective effects of tea polyphenols against oxidative damage to red blood cells, *Biochem. Pharmacol.*, 54, 973–978, 1997.

Gupta, S., Ahmad, N., Mohan, R.R., Husain, M.M., and Mukhtar, H., Prostate cancer chemoprevention by green tea: *in vitro* and *in vivo* inhibition of testosterone-mediated induction of ornithine decarboxylase, *Cancer Res.*, 59, 2115–2120, 1999.

Gupta, S., Ahmad, N., Nieminen, A.L., and Mukhtar, H., Growth inhibition, cell-cycle dysregulation, and induction of apoptosis by green tea constituent (–)-epigallocatechin-3-gallate in androgen-sensitive and androgen-insensitive human prostate carcinoma cells, *Toxicol. Appl. Pharmacol.*, 164, 82–90, 2000.

Gupta, S., Hastak, K., Ahmad, N., Lewin, J.S., and Mukhtar, H., Inhibition of prostate carcinogenesis in TRAMP mice by oral infusion of green tea polyphenols, *Proc. Natl. Acad. Sci. USA*, 98, 10350–10355, 2001.

Gupta, S., Hussain, T., and Mukhtar, H., Molecular pathway for (–)-epigallocatechin-3-gallate-induced cell cycle arrest and apoptosis of human prostate carcinoma cells, *Arch. Biochem. Biophys.*, 410, 177–185, 2003.

Hansson, L.-E., Nyren, O., Bergstrom, R., Wolk, A., Lindgren, A., Baron, J., and Adami, H.O., Diet and risk of gastric cancer: a population-based case-control study in Sweden, *Int. J. Cancer*, 55, 181–189, 1993.

Harada, N., Takabayashi, F., Oguni, I., and Hara, Y., Anti-promotion effect of green tea extracts on pancreatic cancer in golden hamster induced by N-nitroso-bis(2-oxopropyl)amine, *Int. Symp. Tea Sci., Japan*, 200–204, 1991.

Harbowy, M.E. and Balentine, D.A., Tea chemistry, *Critical Rev. Plant Sci.*, 16, 415–480, 1997.

Harnack, L.J., Anderson, K.E., Zheng, W., Folsom, A.R., Sellers, T.A., and Kushi, L.H., Smoking, alcohol, coffee, and tea intake and incidence of cancer of the exocrine pancreas: the Iowa women's health study, *Cancer Epidemiol., Biomarkers Prev.*, 6, 1081–1086, 1997.

Hartge, P., Hoover, R., West, D.W., and Lyon, J.L., Coffee drinking and risk of bladder cancer, *J. Natl. Cancer Inst.*, 70, 1021–1026, 1983.

Heilbrun, L.K., Nomura, A., and Stemmermann, G.N., Black tea consumption and cancer risk: a prospective study, *Br. J. Cancer*, 54, 677–683, 1986.

Henderson, B.E., Louie, E., Soo Hoo Jing, J., Buell, P., and Gardner, M.B., Risk factors associated with nasopharyngeal carcinoma, *New England J. Med.*, 295, 1101–1106, 1976.

Hiatt, R.A., Klatsky, A.L., and Armstrong, M.A., Pancreatic cancer, blood glucose and beverage consumption, *Int. J. Cancer*, 41, 794–797, 1988.

Higginson, J., Etiological factors in gastrointestinal cancer in man, *J. Natl. Cancer Inst.*, 37, 527–545, 1966.

Hirose, M., Hoshiya, T., Akagi, K., Takahashi, S., Hara, Y., and Ito, N., Effects of green tea catechins in a rat multi-organ carcinogenesis model, *Carcinogenesis*, 14, 1549–1553, 1993.

Hirose, M., Hoshiya, T., Akagi, K., Futakuchi, M., and Ito, N., Inhibition of mammary gland carcinogenesis by green tea catechins and other naturally occurring antioxidants in female Sprague-Dawley rats pretreated with 7,12, dimethylbenz(a)anthracene, *Cancer Lett.*, 83, 149–156, 1994.

Howe, G.R., Burch, J.D., Miller, A.B., Cook, G.M., Esteve, J., Morrison, B., Gordon, P., Chambers, L.W., Fodor, G., and Winsor, G.M., Tobacco use, occupation, coffee, various nutrients, and bladder cancer, *J. Natl. Cancer Inst.*, 64, 701–713, 1980.

Huang, M.-T., Ho, C.-T., Wang, Z.Y., Ferraro, T., Finnegan-Olive, T., Lou, Y.-R., Mitchell, J.M., Laskin, J.D., Newmark, H., Yang, C.S., and Conney, A.H., Inhibitory effect of topical application of a green tea polyphenol fraction on tumor initiation and promotion in mouse skin, *Carcinogenesis*, 13, 947–954, 1992.

Huang, M.-T., Xie, J.-G., Wang, Z.Y., Ho, C.-T., Lou, Y.-R., Wang, C.-X., Hard, G.C., and Conney, A.H., Effects of tea and caffeine on UVB light-induced complete carcinogenesis in SKH-1 mice: demonstration of caffeine as a biologically important constituent of tea, *Cancer Res.*, 57, 2623–2629, 1997.

Huber, W.W., McDaniel, L.P., Kaderlik, K.R., Teitel, C.H., Lang, N.P., and Kadlubar, F.F., Chemoprotection against the formation of colon DNA adducts from the food-borne carcinogen 2-amino-1-methyl-6-phenylimidazo[4,5-b]pyridine (PhIP) in the rat, *Mutation Res.*, 376, 115–122, 1997.

IARC monographs, The evaluation of the carcinogenic risk to humans: coffee, tea, mate, methylxanthines and methylglyoxal, *Int. Agency Res. Cancer Working Group Lyon*, 51, 1–513, 1991.

Iscovich, J., Castelletto, R., Esteve, J., Munoz, N., Colanzi, R., Coronel, A., Deamezola, I., Tassi, V., and Arslan, A., Tobacco smoking, occupational exposure and bladder cancer in Argentina, *Int. J. Cancer*, 40, 734–740, 1987.

Ito Y, Ohnishi, S., and Fujie, K., Chromosome aberrations induced by aflatoxin B1 in rat bone marrow cells *in vivo* and their suppression by green tea, *Mutation Res.*, 222, 253–261, 1989.

Jain, A.K., Shimoi, K., Nakamura, Y., Kada, T., Hara, Y., and Tomita, I., Crude tea extracts decrease the mutagenic activity of N-methyl-N′-nitro-N-nitrosoguanidine *in vitro* and in intragastric tract of rats, *Mutation Res.*, 210, 1–8, 1989.

Jain, M., Howe, G.R., and St. Louis, P., Coffee and alcohol as determinants of risk of pancreas cancer: a case-control study from Toronto, *Int. J. Cancer*, 47, 384–389, 1991.

Jain, M.G., Hislop, G.T., Howe, G.R., Burch, J.D., and Ghadirian, P., Alcohol and other beverage use and prostate cancer risk among Canadian men, *Int. J. Cancer*, 78, 707–711, 1998.

Jankun, J., Selman, S.H., Swiercz, R., and Skrzypczak-Jankun, E., Why drinking green tea could prevent cancer, *Nature*, 387, 561, 1997.

Jatoi, A., Ellison, N., Burch, A.P., Sloan, A.J., Dakhil, R.S., Novotny, P., Tan, W., Fitch, T.R., Rowland, K.M., Young, C.Y., and Flynn, P.J., A phase II trail of green tea in the treatment of patients with androgen independent metastatic prostate carcinoma, *Cancer*, 97, 1441–1446, 2003.

Jensen, O.M., Wahrendorf, J., Knudsen, J.B., and Sorensen, B.L., The Copenhagen case-control study of bladder cancer, II: Effect of coffee and other beverages, *Int. J. Cancer*, 37, 651–657, 1986.

Ji, B.T., Chow, W.H., Hsing, A.W., McLaughlin, J.K., Dai, Q., Gao, Y.T., Blot, W.J., and Fraumeni, Jr., J.F., Green tea consumption and the risk of pancreatic and colorectal cancers, *Int. J. Cancer*, 70, 255–258, 1997.

Jiang, M.C., Yang-Yen, H.F., Yen, J.J.Y., and Lin, J.K., Curcumin induces apoptosis in immortalized NIH 3T3 and malignant cancer cell lines, *Nutr. Cancer*, 26, 111–120, 1996.

Jung, Y.D. and Ellis, L.M., Inhibition of tumour invasion and angiogenesis by epigallocatechin gallate (EGCG), a major component of green tea, *Int. J. Exp. Pathol.*, 82, 309–316, 2001.

Kada, T., Kaneko, K., Matsuzaki, S., Matsuzaki, T. and Hara, Y., Detection and chemical identification of natural bio-antimutagens, a case of the green tea factor, *Mutation Res.*, 150, 127–132, 1985.

Kapadia, G.J., Rao, S., and Morton, J.F., Herbal tea consumption and esophageal cancer, in *Carcinogens and Mutagens in the Environment*, Stich, H.F., Ed., CRC Press, Boca Raton, FL, 1983, pp. 3–12.

Katiyar, S.K., Agarwal, R., and Mukhtar, H., Green tea in chemoprevention of cancer, *Comprehensive Therapy*, 18, 3–8, 1992a.

Katiyar, S.K., Agarwal, R., Wang, Z.Y., Bhatia, A.K., and Mukhtar, H., (–)-Epigallocatechin-3-gallate in *Camellia sinensis* leaves from Himalayan region of Sikkim: inhibitory effects against biochemical events and tumor initiation in SENCAR mouse skin, *Nutr. Cancer*, 18, 73–83, 1992b.

Katiyar, S.K., Agarwal, R., Wood, G.S., and Mukhtar, H., Inhibition of 12-O-tetradecanoylphorbol-13-acetate-caused tumor promotion in 7,12-dimethylbenz[a]anthracene-initiated SENCAR mouse skin by a polyphenolic fraction isolated from green tea, *Cancer Res.*, 52, 6890–6897, 1992c.

Katiyar, S.K., Agarwal, R., Ekker, S., Wood, G.S., and Mukhtar, H., Protection against 12-O-tetradecanoylphorbol-13-acetate-caused inflammation in SENCAR mouse ear skin by polyphenolic fraction isolated from green tea, *Carcinogenesis*, 14, 361–365, 1993a.

Katiyar, S.K., Agarwal, R., Zaim, M.T., and Mukhtar, H., Protection against N-nitrosodiethylamine and benzo(a)pyrene-induced forestomach and lung tumorigenesis in A/J mice by green tea, *Carcinogenesis*, 14, 849–855, 1993b.

Katiyar, S.K., Agarwal, R., and Mukhtar, H., Protective effects of green tea polyphenols administered by oral intubation against chemical carcinogen-induced forestomach and pulmonary neoplasia in A/J mice, *Cancer Lett,*, 73, 167–172, 1993c.

Katiyar, S.K., Agarwal, R., and Mukhtar, H., Inhibition of spontaneous and photo-enhanced lipid peroxidation in mouse epidermal microsomes by epicatechin derivatives from green tea, *Cancer Lett.*, 79, 61–66, 1994.

Katiyar, S.K., Elmets, C.A., Agarwal, R., and Mukhtar, H., Protection against ultraviolet B radiation-induced local and systemic suppression of contact hypersensitivity in mice by green tea polyphenols, *J. Investigative Dermatol.*, 104, 690, 1995a.

Katiyar, S.K., Agarwal, R., Korman, N.J., Rupp, C.O., and Mukhtar, H., Inhibition of tumor promoter-caused induction of interleukin-1 (IL-1)-alpha, tumor necrosis factor (TNF)-alpha, and ornithine decarboxylase (ODC) gene expression in SENCAR mouse skin by tea polyphenols, *Proc. Am. Assoc. Cancer Res.*, 36, 594, 1995b.

Katiyar, S.K. and Mukhtar, H., Tea in chemoprevention of cancer: epidemiologic and experimental studies (review), *Int. J. Oncol.*, 8, 221–238, 1996.

Katiyar, S.K., Mohan, R.R., Agarwal, R., and Mukhtar, H., Protection against induction of mouse skin papillomas with low and high risk of conversion to malignancy by green tea polyphenols, *Carcinogenesis*, 18, 497–502, 1997.

Katiyar, S.K., Afaq, F., Azizuddin, K., and Mukhtar, H., Inhibition of UVB-induced oxidative stress-mediated phosphorylation of mitogen-activated protein kinase signaling pathways in cultured human epidermal keratinocytes by green tea polyphenol (–)-epigallocatechin-3-gallate, *Toxicol. Appl. Pharmacol.*, 176, 110–117, 2001.

Kaufman, B.D., Liberman, I.S., and Tyshetskii, V.I., Data concerning the incidence of oesophageal cancer in the Gurjev region of the Kazakh SSR (in Russian), *Voprosy Onkologii*, 11, 78–85, 1965.

Ke, L., Yu, P., Zhang, Z.X., Huang, S.S., Huang, G., and Ma, X.H., Congou tea drinking and oesophageal cancer in South China, *Br. J. Cancer*, 86, 346–347, 2002.

Khan, S.G., Katiyar, S.K., Agarwal, R., and Mukhtar, H., Enhancement of antioxidant and phase II enzymes by oral feeding of green tea polyphenols in drinking water to SKH-1 hairless mice: possible role in cancer chemoprevention, *Cancer Res.*, 52, 4050–4052, 1992.

Khan, W.A., Wang, Z.Y., Athar, M., Bickers, D.R., and Mukhtar, H., Inhibition of the skin tumorigenicity of (+)-7β, 8α-dihydroxy-9α, 10α-epoxy-7,8,9,10-tetrahydrobenzo(a)pyrene by tannic acid, green tea polyphenols and quercetin in SENCAR mice, *Cancer Lett.*, 42, 7–12, 1988.

Kinlen, L.J. and McPherson, K., Pancreas cancer and coffee and tea consumption: a case-control study, *Br. J. Cancer*, 49, 93–96, 1984.

Kinlen, L.J., Willows, A.N., Goldblatt, P., and Yudkin, J., Tea consumption and cancer, *Br. J. Cancer*, 58, 397–401, 1988.

Klaunig, J.E., Chemopreventive effects of green tea components on hepatic carcinogenesis, *Preventive Med.*, 21, 510–519, 1992.

Kohlmeier, L., Weterings, K.G.C., Steck, S., and Kok, F.J., Tea and cancer prevention: an evaluation of the epidemiologic literature, *Nutr. Cancer*, 27, 1–13, 1997.

Komori, A., Yatsunami, J., Okabe, S., Abe, S., Hara, K., Suganuma, M., Kim, S.J., and Fujiki, H., Anticarcinogenic activity of green tea polyphenols, *Jpn. J. Clinical Oncol.*, 23, 186–190, 1993.

Kono, S., Ikeda, M., Tokudome, S., and Kuratsune, M., A case-control study of gastric cancer and diet in northern Kyushu, Japan, *Jpn. J. Cancer Res.*, 79, 1067–1074, 1988.

Kripke, M.L., Immunological unresponsiveness induced by ultraviolet radiation, *Immunol. Rev.*, 80, 87–102, 1984.

Kunze, E., Chang-Claude, J., and Frentzel-Beyme, R., Life style and occupational risk factors for bladder cancer in Germany: a case-control study, *Cancer*, 69, 1776–1790, 1992.

Laskin, J.D., Heck, D., and Laskin, D.L., Inhibitory effects of a green tea polyphenol fraction on 12-O-tetradecanoylphorbol-13-acetate-induced hydrogen peroxide formation in mouse epidermis, in *Phenolic Compounds in Foods and Health II: Antioxidant and Cancer Prevention*, Huang, M.T., Ho, C.T., and Lee, C.Y., Eds., Washington, DC, 1992, pp. 308–314.

La Vecchia, C., Talamini, R., Decarli, A., Franceschi, S., Parazzini, F., and Tognoni, G., Coffee consumption and the risk of breast cancer, *Surgery*, 100, 477–481, 1986.

La Vecchia, C., Negri, E., Decarli, A., D'Avanzo, B., Liberati, C., and Franceschi, S., Dietary factors in the risk of bladder cancer, *Nutr. Cancer*, 12, 93–101, 1989.

La Vecchia, C., Negri, E., Franceschi, S., D'Avanzo, B., and Boyle, P., Tea consumption and cancer risk, *Nutr. Cancer*, 17, 27–31, 1992.

Lea, M.A., Xiao, Q., Sadhukhan, A.K., Cottle, S., Wang, Z.Y., and Yang, C.S., Inhibitory effects of tea extracts and -(–)-epigallocatechin gallate on DNA synthesis and proliferation of hepatoma and erythroleukemia cells, *Cancer Lett.*, 68, 231–236, 1993.

Lee, H.H., Wu, H.Y., Chuang, Y.C., Chang, A.S., Chao, H.H., Chen, K.Y., Chen, H.K., Lai, G.M., Huang, H.H., and Chen, C.J., Epidemiological characteristics and multiple risk factors of stomach cancer in Taiwan, *Anticancer Res.*, 10, 875–881, 1990.

Lee, J.S., Phase I trial of oral green tea extract in adult patients with solid tumors, *J. Clin. Oncol.*, 19, 1830–1838, 2001.

Lee, M.J., Wang, Z.Y., Li, H., Chen, L., Sun, Y., Gobbo, S., Balentine, D.A., and Yang, C.S., Analysis of plasma and urinary tea polyphenols in human subjects, *Cancer Epidemiol. Biomarkers Prev.*, 4, 393–399, 1995.

Li, Y., Comparative study on the inhibitory effect of green tea, coffee and levamisole on the hepatocarcinogenic action of diethylnitrosamine, *Chung Hua Chung liu Tsa Chih (Chin. J. Cancer)*, 13, 193–195, 1991.

Liang, Y.-C., Lin-shiau, S.-Y., Chen, C.-F., and Lin, J.-K., Suppression of extracellular signals and cell proliferation through EGF receptor binding by (–)-epigallocatechin gallate in human A431 epidermoid carcinoma cells, *J. Cell. Biochem.*, 67, 55–65, 1997.

Liberman, R., Prostate cancer chemoprevention: strategies for designing efficient clinical trials, *Urology*, 57, 224–229, 2001.

Lin, T.M., Chen, K.P., Lin, C.C., Hsu, M.M., Tu, S.M., Chiang, T.C., Jung, P.F., and Hirayama, T., Retrospective study on nasopharyngeal carcinomas, *J. Natl. Cancer Inst.*, 51, 1403–1408, 1973.

Lin, Y.L. and Lin, J.K., (–)-Epigallocatechin-3-gallate blocks the induction of nitric oxide synthase by down-regulating lipopolysaccharide-induced activity of transcription factor nuclear factor-kappaB, *Molecular Pharmacol.*, 52, 465–472, 1997.

Lio, S., Umekita, Y., and Guo, J., Growth inhibition and regression of human prostate and breast tumors in athymic mice by tea epigallocatechin gallate, *Cancer Lett.*, 96, 239–243, 1995.

Liu, Q., Wang, Y., Crist, K.A., Huang, M.-T., Conney, A.H., and You, M., Analysis of p^{53} and H-*ras* mutations in UV- and UV/green tea-induced tumorigenesis in the skin of SKH-1 mice, *Proc. Am. Assoc. Cancer Res.*, 36, 591, 1995.

Lu, Y.P., Lou, Y.R., Xie, J.G., Yen, P., Huang, M.-T., and Conney, A.H., Inhibitory effect of black tea on the growth of established skin tumors in mice: effects on tumor size, apoptosis, mitosis and bromodeoxyuridine incorporation into DNA, *Carcinogenesis*, 18, 2163–2169, 1997.

Lu, Y.P., Lou, Y.R., Xie, J.G., Peng, Q.Y., Liao, J., Yang, C.S., Huang, M.T., and Conney, A.H., Topical applications of caffeine or (–)-epigallocatechin gallate (EGCG) inhibit carcinogenesis and selectively increase apoptosis in UVB-induced skin tumors in mice, *Proc. Natl. Acad. Sci. USA*, 99, 12455–12460, 2002.

Lubin, F., Ron, E., Wax, Y., and Modan, B., Coffee and methylxanthines and breast cancer: a case-control study, *J. Natl. Cancer Inst.*, 74, 569–573, 1985.

Mabuchi, K., Bross, D.S., and Kessler, I.I., Epidemiology of cancer of the vulva: a case-control study, *Cancer*, 55, 1843–1848, 1985a.

Mabuchi, K., Bross, D.S., and Kessler, I.I., Risk factors for male breast cancer, *J. Natl. Cancer Inst.*, 74, 371–375, 1985b.

Mack, T.M., Yu, M.C., Hanisch, R., and Henderson, B.E., Pancreas cancer and smoking, beverage consumption, and past medical history, *J. Natl. Cancer Inst.*, 76, 49–60, 1986.

MacMahon, B., Yen, S., Trichopoulos, D., Warren, K., and Nardi, G., Coffee and cancer of the pancreas, *New England J. Med.*, 304, 630–633, 1981.

Maliakal, P.P., Coville, P.F., and Wanwimolruk, S., Tea consumption modulates hepatic drug metabolizing enzymes in Wistar rats, *J. Pharm. Pharmacol.*, 53, 569–577, 2001.

Mao, R., The inhibitory effects of epicatechin complex on diethylnitrosamine induced initiation of hepatocarcinogenesis in rats, *Chung Hua Yu Fang I Hsueh Tsa Chih*, 27, 201–204, 1993.

McCredie, M., Ford, J.M., and Stewart, J.H., Risk factors for cancer of the renal parenchyma, *Int. J. Cancer*, 42, 13–16, 1988.

Michels, K.B., Holmberg, L., Bergkvist, L., Wolk, A., Coffee, tea, and caffeine consumption and breast cancer incidence in a cohort of Swedish women, *Ann. Epidemiol.*, 12, 21–26, 2002.

Miller, A.B., Howe, G.R., Jain, M., Craib, K.J., and Harrison, L., Food items and food groups as risk factors in a case-control study of diet and colo-rectal cancer, *Int. J. Cancer*, 32, 155–161, 1983.

Miller, C.T., Neutel, C.I., Nair, R.C., Marrett, L.D., Last, J.M., and Collins, W.E., Relative importance of risk factors in bladder carcinogenesis, *J. Chronic Dis.*, 31, 51–56, 1978.

Mills, J.J., Chari, R.S., Boyer, I.J., Gould, M.N., and Jirtle, R.L., Induction of apoptosis in liver tumors by the monoterpene perillyl alcohol, *Cancer Res.*, 55, 979–983, 1995.

Mitsui, T., Yamada, K., Yamashita, K., Matsuo, N., Okuda, A., Kimura, G., and Sugano, M., E1A-3Y1 cell-specific toxicity of tea polyphenols and their killing mechanism, *Int. J. Oncol.*, 6, 377–383, 1995.

Mohan, R.R., Challa, A., Gupta, S., Bostwick, D.G., Ahmad, N., and Mukhtar, H., Overexpression of ornithine decarboxylase in prostate cancer and prostatic fluid in humans, *Clin. Cancer Res.*, 5, 143–147, 1999.

Morgan, R.W. and Jain, M.G., Bladder cancer: smoking, beverages and artificial sweeteners, *Can. Medical Assoc. J.*, 111, 1067–1070, 1974.

Morse, M.A. and Stoner, G.D., Cancer chemoprevention: principles and prospects, *Carcinogenesis*, 14, 1737–1746, 1993.

Morton, J.F., The potential carcinogenicity of herbal tea, *Environ. Carcinogenesis Rev. (J. Environ. Sci. Health)*, C4, 203–223, 1986.

Mukhtar, H., Agarwal, R., and Bickers, D.R., Cutaneous metabolism of xenobiotics and steroid hormones, in *Pharmacology of the Skin*, Mukhtar, H., Ed., CRC Press, Boca Raton, FL, 1991, pp. 89–110.

Mukhtar, H., Wang, Z.Y., Katiyar, S.K., and Agarwal, R., Tea components: antimutagenic and anticarcinogenic effects, *Preventive Med.*, 21, 351–360, 1992.

Mukhtar, H., Katiyar, S.K., and Agarwal, R., Green tea and skin: anticarcinogenic effects, *J. Investigative Dermatol.*, 102, 3–7, 1994.

Mukhtar, H. and Ahmad, N., Cancer chemoprevention: future holds in multiple agents, *Toxicol. Appl. Pharmacol.*, 158, 207–210, 1999.

Muto, S., Fujita, K., Yamazaki, Y., and Kamataki, T., Inhibition by green tea catechins of metabolic activation of procarcinogens by human cytochrome P450, *Mutation Res.*, 479, 197–206, 2001.

Nagabhushan, M., Sarode, A.V., Nair, J., Amonkar, A.J., D'Souza, A.V., and Bhide, S.V., Mutagenicity and carcinogenicity of tea, *Camellia sinensis*, *Indian J. Experimental Biol.*, 29, 401–406, 1991.

Nagano, J., Kono, S., Preston, D.L., and Mabuchi, K., A prospective study of green tea consumption and cancer incidence, Hiroshima and Nagasaki (Japan), *Cancer Causes Control*, 12, 501–508, 2001.

Namiki, M., Antioxidants/antimutagens in foods, *CRC Crit. Rev. Food Sci. Nutr.*, 29, 273–300, 1990.

Narisawa, T. and Fukaura, Y., A very low dose of green tea polyphenols in drinking water prevents N-methyl-N-nitrosourea-induced colon carcinogenesis in F344 rats, *Jpn. J. Cancer Res.*, 84, 1007–1009, 1993.

Nishida, H., Omori, M., Fukutomi, Y., Ninomiya, M., Nishiwaki, S., Suganuma, M., Moriwaki, H., and Muto, Y., Inhibitory effects of (–)-epigallocatechin gallate on spontaneous hepatoma in C3H/HeNCrj mice and human-derived PLC/PRF/5 cells, *Jpn. J. Cancer Res.*, 85, 221–225, 1994.

Nomura, A.M., Kolonel, L.N., Hankin, J.H., and Yoshizawa, C.N., Dietary factors in cancer of the lower urinary tract, *Int. J. Cancer*, 48, 199–205, 1991.

Noonan, F.P., DeFabo, E.C., and Kripke, M.L., Suppression of contact hypersensitivity by UV radiation and its relationship to UV-induced suppression of tumor immunity, *Photochem. Photobiol.*, 34, 683–689, 1981.

Obermeier, M.T., White, R.E., and Yang, C.S., Effects of bioflavonoids on hepatic P450 activities, *Xenobiotica*, 25, 575–584, 1995.

Oguni, I., Chen, S.J., Lin, P.Z., and Hara, Y., Protection against cancer risk by Japanese green tea, *Preventive Med.*, 21, 332, 1992.

Ohno, Y., Aoki, K., Obata, K., and Morrison, A.S., Case-control study of urinary bladder cancer in metropolitan Nagoya, *Monogr. Natl. Cancer Inst.*, 69, 229–234, 1985.

Perchellet, J. and Perchellet, E.M., Antioxidants and multistage carcinogenesis in mouse skin, *Free Radicals Biol. Med.*, 7, 377–408, 1989.

Phillips, R.L. and Snowdon, D.A., Dietary relationships with fatal colorectal cancer among Seventh-Day Adventists, *J. Natl. Cancer Inst.*, 74, 307–317, 1985.

Pisters, K.M., Newman, R.A., Coldman, B., Shin, D.M., Khuri, F.R., Hong, W.K., Glisson, B.S., 2001.

Qin, G., Gopalan-Kriczky, P., Su, J., Ning, Y., and Lotlikar, P.D., Inhibition of aflatoxin B1-induced initiation of hepatocarcinogenesis in the rat by green tea, *Cancer Lett.*, 112, 149–154, 1997.

Risch, H.A., Burch, J.D., Miller, A.B., Hill, G.B., Steele, R., and Howe, G.R., Dietary factors and the incidence of cancer of the urinary bladder, *Am. J. Epidemiol.*, 127, 1179–1191, 1988.

Rosenberg, L., Miller, D.R., Helmrich, S.P., Kaufman, D.W., Schottenfeld, D., Stolley, P.D., and Shapiro, S., Breast cancer and the consumption of coffee, *Am. J. Epidemiol.*, 122, 391–399, 1985.

Sadzuka, Y., Sugiyama, T., and Hirota, S., Modulation of cancer chemotherapy by green tea, *Clinical Cancer Res.*, 4, 153–156, 1998.

Sakamoto, K., Reddy, D., Hara, Y., and Milner, J.A., Impact of green or black tea polyphenols on canine mammary tumor cells in culture, *Proc. Am. Assoc. Cancer Res.*, 36, 595, 1995.

Schairer, C., Brinton, L.A., and Hoover, R.N., Methylxanthines and breast cancer, *Int. J. Cancer*, 40, 469–473, 1987.

Shanmugaratnam, K., Tye, C.Y., Goh, E.H., and Chia, K.B., Etiological factors on naso-pharyngeal carcinoma: a hospital-based, retrospective, case-control, questionnaire study, in *Nasopharyngeal Carcinoma, Etiology and Control*, Ito, Y. Ed., IARC, Lyon, 1978, pp. 199–212.

Shi, S.T., Wang, Z.Y., Theresa, J.S., Hong, J.Y., Chen, W.F., Ho, C.T., and Yang, C.S., Effect of green tea and black tea on 4-(methylnitrosamino)-1-(3-pyridyl)-1-butanone bioactivation, DNA methylation and lung tumorigenesis in A/J mice, *Cancer Res.*, 54, 4641–4647, 1994.

Shim, J.S., Kang, M.H., Kim, Y.H., Roh, J.K., Toberts, C., and Lee, I.P., Chemopreventive effects of green tea (*Camellia sinensis*) among cigarette smokers, *Cancer Epidemiol. Biomarkers Prev.*, 4, 387–391, 1995.

Shimoi, K., Nakamura, Y., Tomita, I., Hara, Y., and Kada, T., The pyrogallol related compounds reduce UV-induced mutations in *Escherichia coli* B/r WP2, *Mutation Res.*, 173, 239–244, 1986.

Sigler, K. and Ruch, R.J., Enhancement of gap junctional intercellular communication in tumor promoter-treated cells by components of green tea, *Cancer Lett.*, 69, 15–19, 1993.

Simon, D., Yen, S., and Cole, P., Coffee drinking and cancer of the lower urinary tract, *J. Natl. Cancer Inst.*, 54, 587–591, 1975.

Singh, A.K., Seth, P., Anthony, P., Husain, M.M., Madhavan, S., Mukhtar, H., Green tea constituent epigallocatechin-3-gallate inhibits angiogenic differentiation of human endothelial cells, *Arch. Biochem. Biophys.*, 401, 29–37, 2002.

Slattery, M.L., West, D.W., and Robison, L.M., Fluid intake and bladder cancer in Utah, *Int. J. Cancer*, 42, 17–22, 1988.

Slattery, M.L. and West, D.W., Smoking, alcohol, coffee, tea, caffeine, and theobromine: risk of prostate cancer in Utah (United States), *Cancer Causes Control*, 4, 559–563, 1993.

Sohn, O.S., Surace, A., Fiala, E.S., Richie, Jr., J.P., Colosimo, S., Zang. E., and Weisburger, J.H., Effects of green and black tea on hepatic xenobiotic metabolizing systems in the male F344 rat, *Xenobiotica*, 24, 119–127, 1994.

Sporn, M.B., Carcinogenesis and cancer: different perspectives on the same disease, *Cancer Res.*, 51, 6215–6218, 1991.

Stammler, G. and Volm, M., Green tea catechins (EGCG and EGC) have modulating effects on the activity of doxorubicin in drug-resistant cell lines, *Anti-Cancer Drugs*, 8, 265–268, 1997.

Steele, V.E., Kelloff, G.J., Balentine, D., Boone, C.W., Mehta, R., Bagheri, D., Sigman, C.C., Zhu, S., and Sharma, S., Comparative chemopreventive mechanisms of green tea, black tea and selected polyphenol extracts measured by *in vitro* bioassays, *Carcinogenesis*, 21, 63–67, 2000.

Stocks, P., Cancer mortality in relation to national consumption of cigarettes, solid fuel, tea and coffee, *Br. J. Cancer*, 24, 215–225, 1970.

Taniguchi, S., Fujiki, H., Kobayashi, H., Go, H., Miyado, K., Sadano, H., and Shimokawa, R., Effect of (−)-epigallocatechin gallate, the main constituent of green tea, on lung metastasis with mouse B16 melanoma cell lines, *Cancer Lett.*, 65, 51–54, 1992.

Tazima, K. and Tominaga, S., Dietary habits and gastro-intestinal cancers: a comparative case-control study of stomach and large intestinal cancers in Nagoya, Japan, *Jpn. J. Cancer Res.*, 76, 705–716, 1985.

Terao, J., Piskula, M., and Yao, Q., Protective effect of epicatechin, epicatechin gallate, and quercetin on lipid peroxidation in phospholipid bilayers, *Arch. Biochem. Biophys.*, 308, 278–284, 1994.

Tewes, F.J., Koo, L.C., Meisgen, T.J., and Rylander, R., Lung cancer risk and mutagenicity of tea, *Environ. Res.*, 52, 23–33, 1990.

Trichopoulos, D., Ouranos, G., Day, N.E., Tzonou, A., Manousos, O., Papadimitriou, C., and Trichopoulos, A., Diet and cancer of the stomach: a case-control study in Greece, *Int. J. Cancer*, 36, 291–297, 1985.

Valcic, S., Timmermann, B.N., Alberts, D.S., Wachter, G.A., Krutzsch, M., Wymer, J., and Guillen, J.M., Inhibitory effect of six green tea catechins and caffeine on the growth of four selected human tumor cell lines, *Anticancer-Drugs*, 7, 461–468, 1996.

Victora, C.G., Munoz, N., Day, N.E., Barcelos, L.B., Peccin, D.A., and Braga, N.M., Hot beverages and oesophageal cancer in southern Brazil: a case-control study, *Int. J. Cancer*, 39, 710–716, 1987.

Wang, L.D., Zhou, Q., Feng, C.W., Liu, B., Qi, Y.J., Zhang, Y.R., Gao, S.S., Fan, Z.M., Zhou, Y., Yang, C.S., Wei, J.P., and Zheng, S., Intervention and follow-up on human esophageal precancerous lesions in Henan, northern China, a high-incidence area for esophageal cancer, *Gan To Kagaku Ryoho*, 29 (Suppl. 1), 159–172, 2002.

Wang, S.I. and Mukhtar, H., Gene expression profile in human prostate LNCaP cancer cells by (–)epigallo-catechin-3-gallate, *Cancer Lett.*, 182, 43–51, 2002.

Wang, Z.Y., Das, M., Bickers, D.R., and Mukhtar, H., Interaction of epicatechins derived from green tea with rat hepatic cytochrome P-450, *Drug Metab. Disposal*, 16, 98–103, 1988.

Wang, Z.Y., Cheng, S.J., Zhou, Z.C., Athar, M., Khan, W.A., Bickers, D.R., and Mukhtar, H., Antimutagenic activity of green tea polyphenols, *Mutation Res.*, 223, 273–285, 1989a.

Wang, Z.Y., Khan, W.A., Bickers, D.R., and Mukhtar, H., Protection against polycyclic aromatic hydrocarbon-induced skin tumor initiation in mice by green tea polyphenols, *Carcinogenesis*, 10, 411–415, 1989b.

Wang, Z.Y., Agarwal, R., Bickers, D.R., and Mukhtar, H., Protection against ultraviolet B radiation-induced photocarcinogenesis in hairless mice by green tea polyphenols, *Carcinogenesis*, 12, 1527–1530, 1991.

Wang, Z.Y., Agarwal, R., Khan, W.A., and Mukhtar, H., Protection against benzo(a)pyrene and N-nitrosodi-ethylamine-induced lung and forestomach tumorigenesis in A/J mice by water extracts of green tea and licorice, *Carcinogenesis*, 13, 1491–1494, 1992a.

Wang, Z.Y., Hong, J.Y., Huang, M.-T,, Reuhl, K.R., Conney, A.H., and Yang, C.S., Inhibition of N-nitrosodi-ethylamine- and 4-(methylnitrosamino)-1-(3-pyridyl)-1-butanone-induced tumorigenesis in A/J mice by green tea and black tea, *Cancer Res.*, 52, 1943–1947, 1992b.

Wang, Z.Y., Huang, M.-T., Ferraro, T., Wong, C.-Q., Lou, Y.-R., Reuhl, K., Iatropoulos, M., Yang, C.S., and Conney, A.H., Inhibitory effect of green tea in the drinking water on tumorigenesis by ultraviolet light and 12-O-tetradecanoylphorbol-13-acetate in the skin of SKH-1 mice, *Cancer Res.*, 52, 1162–1171, 1992c.

Wang, Z.Y., Huang, M.-T., Ho, C.-T., Chang, R., Ma, W., Ferraro, T., Reuhl, K.R., Yang, C.S., and Conney, A.H., Inhibitory effect of green tea on the growth of established skin papillomas in mice, *Cancer Res.*, 52, 6657–6665, 1992d.

Wang, Z.Y., Huang, M.-T., Lou, Y.-R., Xie, J.-G., Reuhl, K.R., Newmark, H.L., Ho, C.-T., Yang, C.S., and Conney, A.H., Inhibitory effects of black tea, green tea, decaffeinated black tea, and decaffeinated green tea on ultraviolet B light-induced skin carcinogenesis in 7,12-dimethylbenz(a)anthracene-initiated SKH-1 mice, *Cancer Res.*, 54, 3428–3435, 1994.

Wattenberg, L.W., Inhibition of carcinogenesis by naturally occurring and synthetic compounds, in *Antimu-tagenesis and Anticarcinogenesis, Mechanisms II*, Kuroda, Y., Shankel, D.M., and Waters, M.D., Eds., Plenum Publishing, New York, 1990, pp. 155–166.

Weisburger, J.H., Nagao, M., Wakabayashi, K., and Oguri, A., Prevention of heterocyclic amine formation by tea and tea polyphenols, *Cancer Lett.*, 83, 143–147, 1994.

Weisburger, J.H., Rivenson, A., Garr, K., and Aliaga, C., Tea, or tea and milk, inhibit mammary gland and colon carcinogenesis in rats, *Cancer Lett.*, 114, 323–327, 1997.

Wright, A.S., Emerging strategies for the determination of human carcinogens: detection, identification, exposure monitoring, and risk evaluation, in *Human Carcinogen Exposure, Biomonitoring and Risk Assessment*, Garner, R.C., Farmer, B., Steel, G.T., and Wright, A.S., Eds., IRL Press, Oxford, U.K., 1991, pp. 3–23.

Wright, S.C., Zhong, J., and Larrick, J.W., Inhibition of apoptosis as a mechanism of tumor promotion, *FASEB J.*, 8, 654–660, 1994.

Xu, Y., Ho, C.-T., Amin, S.G., Han, C., and Chung, F.L., Inhibition of tobacco-specific nitrosamine-induced lung tumorigenesis in A/J mice by green tea and its major polyphenol as antioxidants, *Cancer Res.*, 52, 3875–3879, 1992.

Yamane, T., Hagiwara, N., Tateishi, M., Akachi, S., Kim, M., Okuzumi, J., Kitao, Y., Inagake, M., Kuwata, K., and Takahashi, T., Inhibition of azoxymethane-induced colon carcinogenesis in rat by green tea polyphenol fraction, *Jpn. J. Cancer Res.*, 82, 1336–1339, 1991.

Yang, C.S., Research on esophageal cancer in China: a review, *Cancer Res.*, 40, 2633–2644, 1980.

Yang, C.S. and Wang, Z.Y., Tea and cancer, *J. Natl. Cancer Inst.*, 85, 1038–1049, 1993.

Yang, G., Wang, Z.Y., Kim, S., Liao, J., Seril, D.N., Chen, X., Smith, T.J., and Yang, C.S., Characterization of early pulmonary hyperproliferation and tumor progression and their inhibition by black tea in a 4-(methylnitrosamino)-1-(3-pyridyl)-1-butanone-induced lung tumorigenesis model with A/J mice, *Cancer Res.* 57, 1889–1894, 1997.

Yoshizawa, S., Horiuchi, T., Fujiki, H., Yoshida, T., Okuda, T., and Sugimura, T., Antitumor promoting activity of (–)-epigallocatechin gallate, the main constituent of "tannin" in green tea, *Phytother. Res.*, 1, 44–47, 1987.

Yoshizawa, S., (–)-Epigallocatechin gallate, the main constituent of Japanese green tea, inhibits tumor promotion of okadaic acid, *Fukuoka Igaku Zasshi*, 87, 215–221, 1996.

Yu, M.C., Mack, T.M., Hanisch, R., Cicioni, C., and Henderson, B.E., Cigarette smoking, obesity, diuretic use, and coffee consumption as risk factors for renal cell carcinoma, *J. Natl. Cancer Inst.*, 77, 351–356, 1986.

Yu, R., Jiao, J.J., Duh, J.L., Gudehithlu, K., Tan, T.H., and Kong, A.N., Activation of mitogen-activated protein kinases by green tea polyphenols: potential signaling pathways in the regulation of antioxidant-responsive element-mediated phase II enzyme gene expression, *Carcinogenesis*, 18, 451–456, 1997.

Zhang, A., Zhu, Q.Y., Luk, Y.S., Ho, K.Y., Fung, K.P., and Chen, Z.Y., Inhibitory effects of jasmine green tea epicatechin isomers on free radical-induced lysis of red blood cells, *Life Sci.*, 61, 383–394, 1997.

Zheng, W., Doyle, T.J., Hong, C.P., Kushi, L.H., Sellers, T.A., and Folsom, A.R., Tea consumption and cancer incidence in a prospective cohort study of postmenopausal women, *Proc. Am. Assoc. Cancer Res.*, 36, 278, 1995.

25 Phytoestrogens in Cancer Prevention: Characterization and Beneficial Effects of Kurarinone, a New Flavanone and a Major Phytoestrogen Constituent of *Sophora flavescens* Ait

A. De Naeyer, W. Vanden Berghe, D. De Keukeleire, and G. Haegeman

CONTENTS

25.1 Introduction..428
    25.1.1 Phytoestrogens...428
    25.1.2 Estrogens...428
    25.1.3 The Estrogen Receptor 429
    25.1.4 Classification..429
25.2 Biological Effects of Phytoestrogens 430
    25.2.1 Estrogenic and Antiestrogenic Activities of Phytoestrogens.............430
    25.2.2 Effects on Enzymes...431
    25.2.3 Effects on Sex Hormone-Binding Globulin (SHBG) 432
    25.2.4 Effects on Menstrual Cycle Length and Endogenous Hormones 432
    25.2.5 Angiogenesis and Endothelial Cell Proliferation 432
    25.2.6 Antioxidant Efficacy of Phytoestrogens.......................................432
    25.2.7 Anti-inflammatory Effects 433
25.3 Potential Health Benefits of Phytoestrogens...433
    25.3.1 Cancer Chemopreventive...433
    25.3.2 Relief of Menopausal Ailments 435
        25.3.2.1 Osteoporosis Amelioration...436
        25.3.2.2 Prevention of Cardiovascular Diseases.........................436
        25.3.2.3 Prevention of Neurodegenerative Disorders 436
25.4 Isolation and Characterization of Kurarinone..436
    25.4.1 Material and Methods 436
        25.4.1.1 Plant Materials 436
        25.4.1.2 Extraction of Polyphenols...437
        25.4.1.3 Estrogen-Dependent *In Vitro* Bioassays 438
        25.4.1.4 Cell Proliferation *In Vitro* Bioassay..............................438

    25.4.2 Results...439
            25.4.2.1 Bioassay-Guided Fractionation of the Polyphenolic
                     Extract of *Sophora flavescens* Ait. 439
            25.4.2.2 Isolation and Identification of an Estrogenically Active
                     Component from *Sophora flavescens* Ait..............................439
            25.4.2.3 Screening for Anticancer Activity 442
25.5 Conclusion 442
References 444

25.1 INTRODUCTION

25.1.1 Phytoestrogens

In the early 1940s, Australian sheep ranchers began to notice a peculiar and frightening trend, as the ewes became sterile and fewer lambs were counted each spring. By the mid-1940s, the ranching industry in Australia was in a state of crisis and faced certain financial ruin unless the cause of the mysterious infertility in the sheep could be found. What could be causing this disastrous sterility? Genetic mutation? Radiation? Poisonous chemicals? The Australian department of agriculture was called in. A cadre of veterinary doctors and scientists identified clover as the source of the sterility. The ranchers were unaware that the innocuous-looking clover (*Trifolium subterraneum*), which they had recently begun planting in their fields to feed the sheep, had been producing large quantities of estrogen-mimicking compounds. It took another decade before scientists finally pinned down the exact compounds that were causing the sterility, namely the isoflavones genistein and formononetin.[1] Although these compounds are, in fact, rather weak estrogens, plants make up for that fact by producing them in comparatively large quantities, up to 5% of the dry weight in the clover fodder.

These plant-derived nonsteroidal compounds with estrogen-like biological activity have been defined as phytoestrogens. The estrogen-mimicking nature of plant compounds is not restricted to clover. By 1975, several hundred plants had been found to exhibit estrogenic activity or to contain estrogenically active compounds.[2] As the number of plant species demonstrated to contain phytoestrogens expanded, interest increased in the potential use of these compounds for medicinal purposes. At the same time, concerns arose regarding the presence of nonendogenous estrogens in the diet for human and wildlife populations.

25.1.2 Estrogens

Nonsteroidal compounds with estrogenic or antiestrogenic activities mediate their effects by either mimicking or inhibiting the action of endogenous estrogens. Therefore, an understanding of the action of the body's endogenous estrogens is essential. Estrogen is a generic term for estrous-producing compounds, synthesized mainly by the ovaries of female mammals. Both the ovaries and testes produce testosterones and estrogens, with the quantities varying according to gender. Estrogens influence growth, differentiation, and the function of many target tissues in the body, including those of the male and the female reproductive system (mammary glands, uterus, ovary, testes, and prostate). Estrogens also play an important role in bone maintenance, and they have a protective effect in the cardiovascular system. In the central nervous system, estrogens influence both sexual/mating behavior and nonproductive events, including memory and learning.[3] Natural and synthetic estrogen antagonists are used in a number of clinical applications, for example in breast cancer therapy. The estrogen dependency of breast tumors is the key for intervention with antiestrogens such as tamoxifen. Use of estrogens in the treatment and prevention of osteoporosis and in estrogen-replacement therapy of postmenopausal women is also common.

25.1.3 THE ESTROGEN RECEPTOR

The pathway of receptor-mediated estrogenic effects has been fairly well elucidated. Produced in the ovaries and testes, estrogens diffuse in and out of all cells, but they are retained with high affinity and specificity in target cells by an intranuclear binding protein, termed the estrogen receptor (ER). Once bound by estrogens, the ER undergoes a conformational change, allowing the receptor to regulate the expression of specific target genes by binding to target sequences called estrogen response elements (ERE) in the promoter region of the respective target genes.[4,5] Recognition of the complexity of estrogen and estrogen receptor (ER) signaling has substantially increased in the last several years.[6,7] In their genomic role, estrogens enter the cell and bind to ERs, which are members of a superfamily of ligand-regulated transcription factors. However, estrogens also exert nongenomic effects that occur independently of gene transcription. Typically, these relatively rapid events are initiated at the plasma membrane and result in the activation of intracellular signaling pathways.

Biochemical analysis of the estrogen receptor has indicated that it consists of several functional domains. There is a hypervariable N-terminal domain that contributes to the transactivation function, a highly conserved central domain responsible for specific DNA binding, dimerization, and nuclear localization, and a C-terminal domain involved in ligand binding and ligand-dependent transactivation.[4,8–10] Our understanding of estrogen signaling has undergone a true paradigm shift over recent years, following the discovery in 1995 of a second estrogen receptor, estrogen receptor β (ERβ), which differs from ERα in the C-terminal ligand-binding domain and in the N-terminal transactivation domain.[9] The mode according to which phytoestrogenic compounds mediate their effects is still unclear. Assumptions range from mimicking normal estrogenic action to competitive inhibitory effects that may block normal estrogenic action. Structural studies clearly reveal changes in ER conformation upon binding of classic estrogens as compared with phytoestrogens, which may already suggest ligand-selective activities.[8,11–16] Interestingly, phytoestrogens seem to preferentially mediate their effects via ERβ, whereas classical estrogen acts via both receptors ERα and ERβ.[17,18] As such, phytoestrogens may act as natural selective estrogen-receptor modulators that elicit distinct clinical effects from estrogens used for hormone replacement by selectively recruiting coregulatory proteins to ERβ that trigger transcriptional pathways.[17]

Another possibility deserving consideration is that some of the phytoestrogen effects may be attributable to properties that do not involve estrogen receptors, such as effects on enzymes, protein synthesis, cell proliferation, angiogenesis, calcium transport, Na^+/K^+ adenosine triphosphatase, growth factor action, vascular smooth muscle cells, lipid oxidation, and cell differentiation.[19] Some of these properties will be further discussed.

25.1.4 CLASSIFICATION

Isoflavones are widespread in the plant kingdom, as they are predominantly found in leguminous plants and are especially abundant in soy.[20] Lignans exist as minor constituents of building blocks in the formation of lignin in plant cell walls. They are found widely in cereals, legumes, fruit, and vegetables, with exceptionally high concentrations in flaxseed.[19,21]

The structural similarities between these substances, endogenous mammalian estrogens, and potent synthetic estrogens (Figure 25.1 and Figure 25.2) have attracted a lot of attention.[22] Numerous studies have revealed that the biological activity of 17β-estradiol (Figure 25.2) greatly depends on the presence of at least two hydroxyl groups, one located in the A-ring of the steroid nucleus and the other at C_{17}. Unlike 17β-estradiol, phytoestrogens are not steroids, but they possess hydroxyl groups (present or introduced by hydroxylation) that can be positioned in a stereochemical alignment resembling that of 17β-estradiol. The distances between two hydroxyl groups in isoflavones, mammalian lignans, and 17β-estradiol are comparable. It appears that this feature is an essential factor for strong binding to the estrogen receptor.[23] Phytoestrogens are also structurally related to the antiestrogen tamoxifen, which is widely used in the treatment of breast cancer.

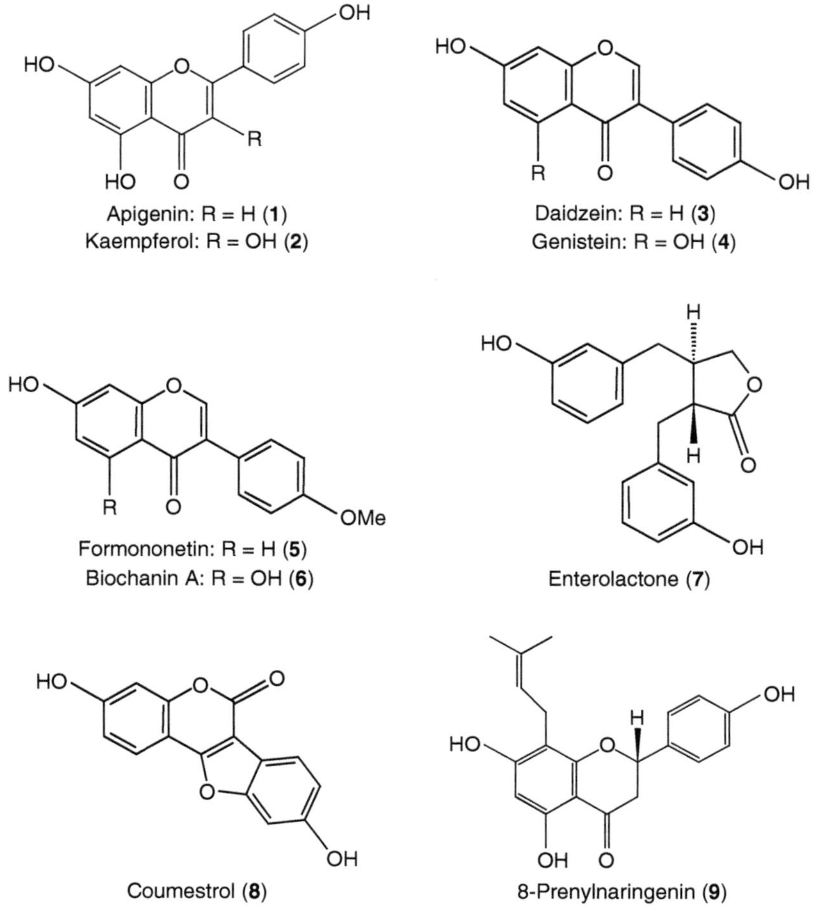

FIGURE 25.1 Structures of nonsteroidal phytoestrogens. The most prevalent classes of phytoestrogens are: (1) flavones: apigenin; (2) kaempferol; (3) isoflavones: daidzein; (4) genistein; (5) formononetin; (6) biochanin A; (7) lignans: enterolactone; (8) coumestans: coumestrol; (9) flavanones: 8-prenylnaringenin.

25.2 BIOLOGICAL EFFECTS OF PHYTOESTROGENS

25.2.1 ESTROGENIC AND ANTIESTROGENIC ACTIVITIES OF PHYTOESTROGENS

Phytoestrogens are of biological interest because they exhibit weak estrogenic and antiestrogenic actions both *in vitro* and *in vivo*. Many studies have shown that phytoestrogens bind to the estrogen receptor and show significant estrogenic effects in cell cultures, animals, and humans. The relative estrogenic activity of various phytoestrogens has been assessed using human cell culture bioassays. Relative potencies of the common phytoestrogens compared with 17β-estradiol (value 100) were established as follows: coumestrol, 0.202; genistein, 0.084; equol, 0.061; daidzein, 0.013; biochanin A, <0.006; and formononetin, <0.0006.[24] Comparable levels of bioactivity were produced by 17β-estradiol and isoflavones, indicating that the estrogen receptor complexes, formed by 17β-estradiol and isoflavones, are functionally equivalent. On the other hand, antiestrogenic effects of phytoestrogens have also been observed *in vivo*. When administered in combination with 17β-estradiol, genistein functions as an antiestrogen, thereby decreasing uterine 17β-estradiol uptake in an animal model.[25]

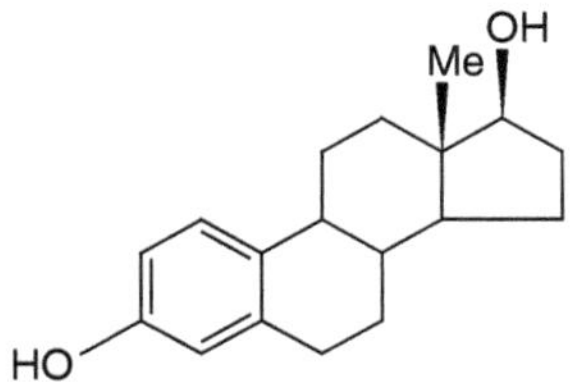

17β-estradiol

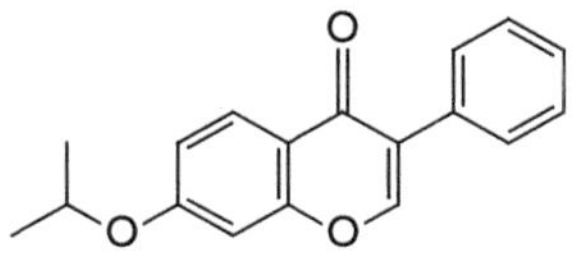

Ipriflavone

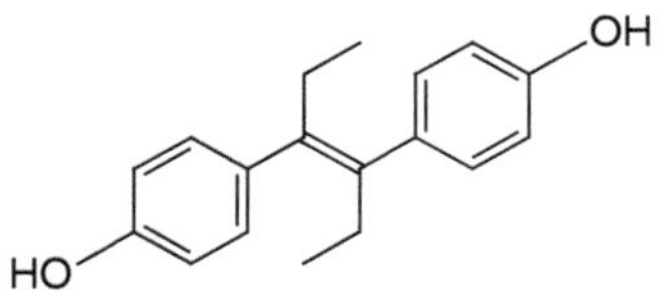

Diethylstilbestrol

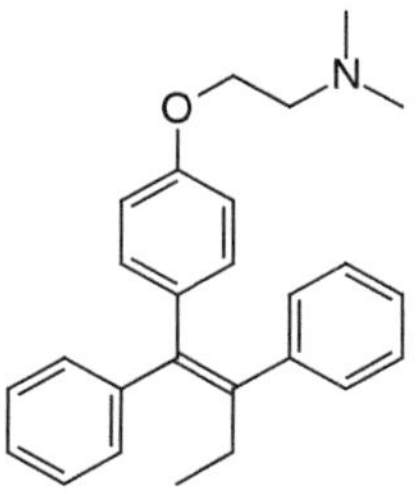

Tamoxifen

FIGURE 25.2 Structures of 17β-estradiol and synthetic antiestrogens.

Phytoestrogens might exert both estrogenic and antiestrogenic effects on metabolism, depending on several factors, including their concentrations, the concentrations of endogenous estrogens, and individual characteristics such as gender and menopausal status.[26] Phytoestrogens exhibit 100 to 1000 times less estrogenic activity than 17β-estradiol, but they may be present in the body in concentrations 100-fold higher than endogenous estrogens.[14] The antiestrogenic activity of phytoestrogens could be partially explained by their competition with endogenous 17β-estradiol for the estrogen receptor. This partial estrogenic/antiestrogenic behavior is a common feature of many weak estrogens.[27] Understanding the estrogenic activity of phytoestrogens is complicated by the existence of two estrogen receptor subtypes, ERα and ERβ. Importantly, comparison of the relative molar binding affinities shows that the most common phytoestrogens have a stronger binding affinity for ERβ than ERα, suggesting that ERβ may be an important mediator of the action of phytoestrogens.[15,17,18,28]

25.2.2 Effects on Enzymes

Phytoestrogens inhibit, to a greater or lesser extent, many of the most important steroid biosynthetic and metabolic enzymes. These activities form an essential part of their possible disease-preventing effects. Most lignans are weak inhibitors of the steroid aromatase and, therefore, reduce conversion of androstenedione to estrone, thereby lowering the risk of estrogen-dependent cancers.[29] Genistein, equol, biochanin A, formononetin, daidzein, and enterolactone, as well as, to a lesser degree, coumestrol and enterodiol were found to inhibit 5α-reductase, the key prostate-specific enzyme that converts testosterone to the more potent metabolite, dihydrotestosterone, which is important in promotion and progression of prostate diseases.[30] Genistein, the most extensively studied isoflavone, also has inhibitory effects on a number of tyrosine-specific protein kinases, cell-cycle kinases, DNA topoisomerases I and II, and protein histidine kinase, which play an important role in cell proliferation and transformation.[31–37]

25.2.3 EFFECTS ON SEX HORMONE-BINDING GLOBULIN (SHBG)

Estrogens and androgens are relatively insoluble in aqueous solutions and, therefore, are bound to transport proteins in the circulation, primarily SHBG and albumin. Only a small portion of steroids (<2%) is transported in the free form. It is assumed that free or unbound steroids are biologically active. Changes in total hormone concentrations result in relatively small alterations in the size of the free-hormone fraction, whereas changes in SHBG concentrations result in relatively large changes in the amount of free and bound hormones. Both lignans and isoflavones have been reported to stimulate the synthesis of SHBG in the liver. This is consistent with the observation that the proportion and concentration of free 17β-estradiol is significantly and directly correlated with the urinary lignan concentrations in body fluids. SHBG is associated with breast cancer risk, and it is likely that the effect of phytoestrogens on plasma SHBG production may also be involved in the disease-preventive effects.[38,39]

25.2.4 EFFECTS ON MENSTRUAL CYCLE LENGTH AND ENDOGENOUS HORMONES

Cassidy et al. established that isoflavone-rich diets can exert hormonal effects and interfere with the regulation of the menstrual cycle in premenopausal women (increased length of the menstrual cycle or delayed menstruation in premenopausal women, and reduced levels of luteinizing hormone [LH], follicle-stimulating hormone [FSH], and progesterone).[40,41] Recently, dietary phytoestrogens have been shown to decrease serum 17β-estradiol, a finding that could also be related to changes in the length of the menstrual cycle.[14] Some of the reported biological effects of a diet of soy products containing isoflavones are similar to those induced by the potent synthetic antiestrogen tamoxifen. Tamoxifen, when used therapeutically in breast cancer patients, led to a decrease in circulating concentrations of LH and FSH, which were consistently associated with a longer follicular phase.[39] At this time, it would appear to be premature to specifically attribute the longer menstrual cycle of Asian women to their high intake of phytoestrogens, although breast cancer patients have been shown to have a significantly shorter menstrual cycle than control subjects.[14]

25.2.5 ANGIOGENESIS AND ENDOTHELIAL CELL PROLIFERATION

In vitro and *in vivo* studies have both suggested that isoflavonoids and flavonoids exert multiple suppressive effects on angiogenesis, including the inhibition of several kinases (tyrosine kinase, protein kinase C, 1-phosphatidylinositol kinase, cdc2, and cyclin-dependent kinases) and the downregulation of matrix metalloprotease 2 (MMP-2). However, the mechanisms whereby soy products alter the angiogenic switch and the balance between angiogenic growth factors and inhibitors *in vivo* are still under investigation. Experiments with transplantable cancers placed orthopically within the same normal tissue may provide the preclinical data required to extend these observations to human trials[14,42]

25.2.6 ANTIOXIDANT EFFICACY OF PHYTOESTROGENS

Antioxidant activity is a fundamental property underlying a wide range of biological effects, including antibacterial, anti-inflammatory, antiallergic, antithrombotic, antimutagenic, anticarcinogenic, antiaging, and vasodilatory effects.

The polyphenolic nature of phytoestrogens gives them the ability to act as antioxidants, and thus to inhibit or delay the oxidation of other molecules by preventing the initiation or propagation of oxidizing chain reactions. The antioxidant activity of phytoestrogens has been reported both *in vitro* and *in vivo*, and it may help to lower the risk of cancer by protecting cells, for example, against lipid peroxidation and by modulating the production of the prostaglandines and leukotrienes involved in carcinogenesis.[43,44] Genistein is reported to be a potent inhibitor of hydrogen peroxide production in HL60 cells and an inhibitor of the generation of superoxide anion by xanthine/xanthine oxidase. Genistein has also been shown *in vivo* (mice) to increase the activities of antioxidant enzymes like catalase, superoxide dismutase, glutathione peroxidase, and glutathione reductase.[14] The antioxidant properties of phytoestrogens might also be beneficial in preventing the development of atherosclerosis by reducing lipid peroxidation and increasing the resistance of low-density lipoprotein (LDL) to oxidation.[38]

25.2.7 ANTI-INFLAMMATORY EFFECTS

The inflammatory response is a highly regulated physiological process that is critically important to homeostasis. Interleukin-6 (IL-6) is a proinflammatory cytokine that is normally tightly regulated and expressed at low levels, except during infection, trauma, aging, or other stress.[45] Among several factors that can down-regulate IL-6 gene expression are estrogen and testosterone hormones. After menopause or andropause, IL-6 levels are elevated, even in the absence of infection, trauma, or stress.[46] It has been proposed that the age-associated increase in IL-6 accounts for certain of the phenotypic changes of advanced age, including cardiovascular disease, osteoporosis, arthritis, type 2 diabetes, certain cancers, lymphoproliferative disorders, multiple myeloma, periodontal disease, frailty, chronic inflammatory disease, Alzheimer's disease, and functional decline.[47] The functional interaction, or "cross talk," between estrogen receptor (ER) and the proinflammatory transcription factor, nuclear factor (NF)-κB, demonstrated *in vitro*, has been suggested to play a key role in estrogen prevention of those age-related conditions *in vivo*.[48–50] In this respect, restoring hormonal imbalance by hormone replacement therapy (HRT) (either by synthetic or plant-derived hormone preparations) in the aging population is an attractive therapeutic option, although the risks/benefits of HRT are still an area of hot debate. Many hope that phytoestrogens can exert the cardioprotective, antiosteoporotic, and other beneficial effects of the estrogens used in hormone replacement therapy in postmenopausal women without adversely affecting the risk of thrombosis and the incidence of breast and uterine cancers. Although there are many positive indications that phytoestrogens can fulfill this role, it remains to be proven. Controlled interventional studies are lacking, and many questions remain unanswered[38, 51–56]

The onset of inflammatory gene expression is driven by the transcription factor NF-κB, the transcriptional activity of which is regulated at multiple levels[57,58] First, NF-κB activity is regulated by cytoplasmic degradation of the IκB inhibitor and nuclear translocation. Second, the nuclear p65 transactivation potential can be further influenced by posttranslational modifications, such as phosphorylation or acetylation. Ser276 phosphorylation seems to be highly important, considering its crucial role in interaction with and engagement of the cofactor acetylase CBP/p300. Recently, we have identified MSK1 as an essential kinase in the NF-κB pathway, where it is responsible for p65 phosphorylation at Ser276, as well as for H3 phosphorylation of Ser10 in IL-6 promoter-associated chromatin (Figure 25.3).[59–62] This allows formation of a transcription-competent IL-6 enhanceosome in which NF-κB is a central switch modulating IL-6 gene transcription rates. To evaluate anti-inflammatory potencies of phytoestrogens, further investigations are needed to determine the extent to which various classes of (phyto)estrogens selectively modulate the multiple regulatory levels of NF-κB-driven gene expression.[63,64]

25.3 POTENTIAL HEALTH BENEFITS OF PHYTOESTROGENS

25.3.1 CANCER CHEMOPREVENTIVE

It has been well established that cancer rates differ strikingly in various populations. Hormone-related cancers of breast, ovary, endometrium, and prostate have been reported to vary by as much as 5- to 20-fold between populations, and migrant studies indicate that the difference is largely attributed to environmental factors rather than genetics.[65–68] The highest incidences of these cancers are typically observed in populations with Western lifestyles that include relatively

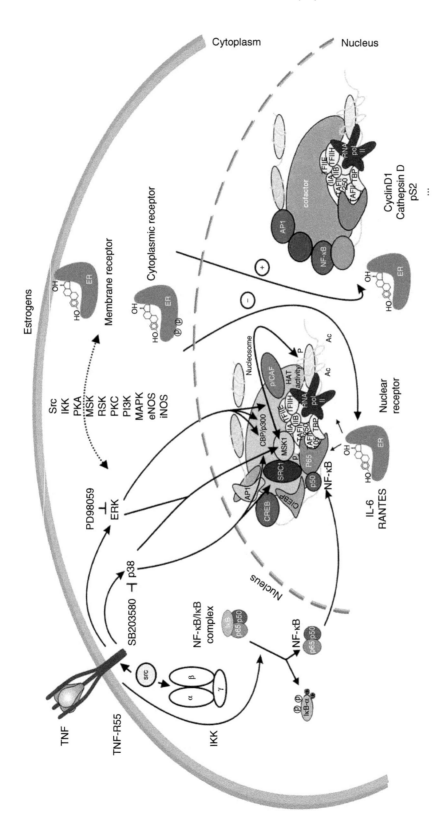

FIGURE 25.3 TNF-mediated IL-6 gene induction involves a dual signaling pathway: (1) release of the transcription factor NF-κB from the cytoplasm, its migration to the nucleus, and binding onto NF-κB-responsive sequences in cellular gene promoters; and (2) concomitant activation of MAPK and, subsequently, of nuclear MSK1 that phosphorylates NF-κB p65 (now acting as a platform for CBP/p300 interaction) and the surrounding histone tails as a first step for chromatin relaxation and gene activation. ERα/β activation can modulate gene expression at multiple levels (as indicated by arrows), resulting in either gene repression (i.e., IL-6, RANTES, etc.) or gene activation (i.e., cyclinD1, pS2, CathepsinD, etc.).

high-fat, meat-based, and low-fiber diets, whereas the lowest rates are typically observed in Asian populations with Eastern lifestyles that include plant-based diets with a high content of phytoestrogens.[69–75] Migrants from Asian to Western countries, who maintain their traditional diet, do not show increased risk for these diseases, whereas an increased risk accompanies a change toward a Western diet.[76,77] Much of the evidence is based on the differences in consumption of soy products as the major source of isoflavones in different areas of the world. The consumption of soy products is estimated to be highest in particular Japanese populations, with isoflavone levels in the diet up to 200 mg/day. Throughout Asia, the consumption of legumes is estimated to supply 25 to 45 mg total isoflavones in the diet each day, compared with Western countries with a consumption of less than 5 mg/day.[78]

Isoflavones and other phytoestrogens have been considered to exert anticarcinogenic actions, mainly through antiestrogenic, antiaromatase, or antiproliferative mechanisms.[68,79,80] Soy seems to protect against breast cancer if consumed throughout life, particularly before and during adolescence. Whether the phytoestrogens are responsible for the protection is not known; it is more likely that the soybean products or grain–fiber complexes are protective in their entirety. Soy and isoflavones seem to protect against prostate cancer, but more studies with humans are needed before any definite conclusions can be drawn. With regard to colon cancer, the situation is even less clear. Soy does not seem to be protective, and there is no good evidence indicating that lignans themselves have a direct effect on the development of colon cancer.[72] The current evidence suggests that phytoestrogens may play a role in the prevention and treatment of several types of cancer, although there are some studies suggesting that they may actually promote carcinogenesis. A report by McMichael-Phillips et al. showed an increase in proliferation of breast lobular epithelium,[81] whereas another study[82] documented an increase in estradiol levels and increased frequency of hyperplastic cells in aspirated breast fluid after supplementation with soybeans. Allred et al.[83–86] reported that soy protein diets containing varying concentrations of genistein stimulated the growth of estrogen-dependent breast cancer cells *in vivo* in a dose-dependent manner. Of note is that many of the effects, positive and negative, have been shown with very high concentrations and not at levels likely to be achieved by eating foods containing phytoestrogens. Additionally, the role of such factors as bioavailability, phytoestrogen absorption, duration of exposure, and the potential influence of other dietary components remains uncertain at this time.[71]

25.3.2 Relief of Menopausal Ailments

Natural menopause is associated with a rapid decline in circulating estrogen and, apart from the loss of reproductive function, the decline in sex hormones has many implications for nonreproductive tissues. Estrogen deficiency in postmenopausal women can lead to unpleasant symptoms such as hot flashes and vaginal dryness, with a long-term increased risk of bone loss, cardiovascular disease, and neurological disorders (dementia). An isoflavone-rich diet may help approximately two-thirds of postmenopausal women to better cope with these ailments.[38,51–53,87] Because the average life expectancy for woman in Western countries exceeds 80 yr and women spend more than a third of their lifetime in postmenopause, the possible implications of estrogen deficiency on the rates of cardiovascular disease and osteoporosis are of enormous public health importance.

There is no unifying mechanism that would be able to explain all consequences of menopause on the metabolism of organs as diverse as bone, blood vessels, or adipose tissue. However, menopause-triggered changes in the levels of proinflammatory cytokines are beginning to emerge as a common theme that may have a significant impact on the function of all of these tissues.[46,47] The exact mechanisms by which estrogen interferes with cytokine activity are still incompletely known but may potentially include interactions of the ER with transcription factors (like NF-κB, AP1), modulation of nitric oxide antioxidative effects, plasma membrane actions, and changes in immune cell functions. Despite recent doubts arising from short-term placebo-controlled HRT studies in postmenopausal women, there is still substantial reason to believe that estrogen therapy may have beneficial effects in the long run on selected patients.[38,51–56]

25.3.2.1 Osteoporosis Amelioration

Postmenopausal osteoporosis is a major public health problem. An estimated 30 to 50% of Western women will, in the course of their lifetime, suffer a clinical fracture and the associated morbidity. Estrogen deficiency is a key factor in the pathogenesis of postmenopausal osteoporosis. The prevalence of vertebral and hip fractures increases with advanced age. As a larger proportion of women survive into old age, osteoporotic fractures constitute a progressively greater burden of both illness and demand on resources. The majority of evidence for the antiosteoporotic efficacy of HRT is based on results of case-control and cohort studies, but its effectiveness for therapy would be strengthened greatly if this could be confirmed for randomized trials.[54,55]

Soy isoflavones positively help maintenance of bone mass in ovariectomized rodents, although daidzein may be more efficient than genistein in this respect. One study has indicated that isoflavone-rich soy protein may attenuate bone loss in the lumbar spine of postmenopausal women, and that this effect is due to isoflavones rather than to soy protein.[88] Indirect evidence for the potential benefits of phytoestrogens with regard to bone metabolism comes from a growing number of studies of ipriflavone (7-isopropoxyisoflavone), a synthetic isoflavone. In doses ranging from 200 to 600 mg/day, this drug has proved to be effective in promoting bone mass and preventing bone loss. Pharmacological studies revealed that a main metabolite of ipriflavone in humans is daidzein, which constitutes around 10% of the breakdown products of ipriflavone.[89]

25.3.2.2 Prevention of Cardiovascular Diseases

Epidemiological data highlight a lower incidence of coronary heart disease in Asian countries than in Western countries, as well as in vegetarians relative to omnivores. Although this may be related to factors such as lower cholesterol intake and higher dietary antioxidant and fiber consumption, a protective role attributable to dietary soy products in Asian populations has also been hypothesized.[71] Many experimental, epidemiological, and interventional studies have attempted to identify a role for phytoestrogens in cardioprotection and to evaluate the underlying mechanism(s). Data analysis revealed that phytoestrogen intake appears to be associated with favorable effects on lipid and lipoprotein profiles and on vascular function both in primates and in humans.[38,90,91]

25.3.2.3 Prevention of Neurodegenerative Disorders

Estrogen replacement therapy has furthermore been shown to improve episodic and semantic memory in postmenopausal women, and, remarkably, a high soy diet improves memory within weeks in both young male and female volunteers.[92] Understanding how phytoestrogens affect cognitive function will be an exciting goal for the future.

25.4 ISOLATION AND CHARACTERIZATION OF KURARINONE

25.4.1 Material and Methods

25.4.1.1 Plant Materials

Sophora flavescens Aiton (Leguminosae or *kurara* in Japanese) is a perennial from the leguminous plant family. It is an evergreen shrub growing to 1.5 m by 1 m at a slow rate (Figure 25.4). It is in leaf all year, in flower from July to August, and the seeds ripen in September. The flowers are hermaphrodite (have both male and female organs) and are pollinated by insects. The plant can survive in light (sandy), medium (loamy), or heavy (clay) soils but requires well-drained soil. It can be cultivated on acidic, neutral, and basic (alkaline) soils. It cannot grow in the shade.

INFLORESCENCE   FLOWERS

FIGURE 25.4 Views of Sophora flavescens Ait.

It is found in Japan, the Korean peninsula, and China. Another name for this plant is *kujin*, which literally means "the bitter root." The Chinese characters for this plant also point to the fact that although this root may taste bitter, it has superior medicinal qualities. The roots of this species are commercially available as the generic Kosam in Korea, and it has been applied frequently in folk medicine as an antipyretic, analgesic, anthelmintic, and a stomachic drug. Externally the root has been applied to treat rashes and a variety of other ailments. Although hitherto it has not found widespread application in cosmetic products, it is used to a certain extent for its moisture-retention and disinfectant properties.[93–95]

25.4.1.2 Extraction of Polyphenols

Sophora flavescens roots were ground thoroughly to obtain a smooth powder. The grinding procedure was necessary to maximize the extraction yield. (Due to cell rupturing, the solvent can extract the secondary metabolites stored in cell vacuoles more efficiently.) The plant material was extracted three times with isooctane by refluxing during 1 h under nitrogen. These initial extraction steps were intended to defatten the material and to remove lipophilic constituents. The isooctane extracts were discarded, and the residue was then extracted with methanol/water 3:1 (v/v) (three times; refluxing during 1 h under nitrogen). The combined methanol/water extracts were filtered and concentrated (30°C), leaving a residue to be considered as a polyphenolic fraction of *Sophora flavescens* Ait. Stock solutions were prepared by dissolving 20-mg powderous polyphenolic fractions in 1 ml methanol.

An aliquot of the polyphenolic extract was qualitatively analyzed using reversed-phase HPLC (high-performance liquid chromatography). HPLC is currently the most widely applied separation technique for the analysis of complex mixtures of polyphenols. Successful analytical separations were obtained on an octadecylsilica (ODS) column (Alltech, Alltima C_{18} 5 μ, 250 × 6 mm) using the following gradient system: methanol/acetonitrile 50:50 (v/v) as organic solvent (solvent B) and 0.05% formic acid in water as aqueous solvent (solvent A). As basic gradient for the separation of polyphenolic compounds in extracts, fractions, and subfractions, we used the following profile: 0 to 3 min: 15% B in A (isocratic); 3 to 45 min: 15% B in A to 95% B in A; 45 to 50 min: 95% B in A (isocratic); 50 to 57 min: 95% B in A to 15% B in A; 57 to 60 min: 15% B in A (isocratic).

According to the obtained chromatographic profiles, this gradient was then further optimized for more efficient separation. HPLC analysis of the polyphenolic extract of *Sophora flavescens* Ait.

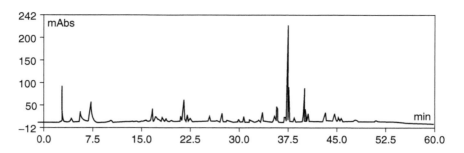

FIGURE 25.5 HPLC analysis of a polyphenolic extract of *Sophora flavescens* Ait.

showed that the components of interest are mainly present in the relatively nonpolar region of the chromatogram (Figure 25.5).

25.4.1.3 Estrogen-Dependent *In Vitro* Bioassays

The first bioassay relies on an estrogen-inducible yeast (*Saccharomyces cerevisiae*) screen, expressing the estrogen receptor. Yeast cells normally do not contain estrogen receptors. Therefore, the DNA sequence of the human estrogen receptor (hER) was stably integrated into the main chromosome of the yeast. The yeast cells also contain expression plasmids carrying the reporter gene *lac*-Z, encoding the enzyme β-galactosidase. This enzyme is secreted into the medium and metabolizes the yellow chromogenic substrate, chlorophenol red-β-D-galactopyranoside (CPRG), to chlorophenol red, which can be measured by its absorbance at 540 nm. The absorbance reflects the activity of the receptor.[96]

Similar results can be obtained by the Ishikawa-Var-I assay. This second bioassay, developed by Littlefield et al.,[97] is based on the ability of compounds with estrogenic activity to stimulate alkaline phosphatase activity in a human endometrial adenocarcinoma cell line (Ishikawa-Var-I).[97] The human Ishikawa cell line was established by Nishida et al.[98] from a well-differentiated endometrial adenocarcinoma. A variant of this cell line (Ishikawa-Var-I) is unresponsive to estrogens with respect to proliferation, but it is sensitive to the stimulation of alkaline phosphatase activity. The estrogenic activity is assayed by monitoring the formation of p-nitrophenol at 405 nm following enzymatic hydrolysis of p-nitrophenyl phosphate. The stimulation of alkaline phosphatase activity has been shown to be specific to estrogens, and no other type of steroids, including androgens, progestins, mineralocorticoids, or glucocorticoids, produce the effect. The cells also respond to nonsteroidal estrogens, including phytoestrogens. These responses can be blocked by pure antiestrogens, indicating that they are mediated by the estrogen receptor.[24]

25.4.1.4 Cell Proliferation *In Vitro* Bioassay

Human MCF-7/6 breast cancer cells, obtained from Dr. H. Rochefort (Unité d' Endocrinologie Cellulaire et Moléculaire, Montpellier, France) were investigated using the sulforhodamine-B (SRB) bioassay, a colorimetric *in vitro* anticancer drug screening test. The cells, originally derived from a pleural effusion of a metastasized mammary adenocarcinoma, possess fully characterized estrogen receptors, and their proliferation *in vitro* is modulated by 17β-estradiol and by other ligands of the steroid/thyroid receptor superfamily.[99] The bioassay estimates the inhibition rate of cell proliferation after continuous exposure of the MCF-7/6 cells, seeded at 5×10^3 cells/well in a 96-well microtiter plate, to varying concentrations of kurarinone (1/500 to $1/10^5$ dilutions were prepared in medium (1/2 DMEM, F12 HAM) from a stock solution (40 mM) that was obtained by dissolving lyophilized compound in methanol (control on solvent effects). After

72 h of incubation at 37°C, the MCF-7/6 cells were fixed to the bottom of the wells by adding 50% trichloroacetic acid. After incubation for 1 h at 2°C, the wells were washed (5×) with water and allowed to dry. SRB was added to the wells, and after 30 min, the wells were washed (4×) with 1% glacial acetic acid to remove unbound dye. Bound SRB was redissolved in Tris buffer (pH 10.8), and the optical densities were measured at 490 nm using a V_{max} plate reader (Molecular Devices, Palo Alto, CA).[100,101]

25.4.2 Results

25.4.2.1 Bioassay-Guided Fractionation of the Polyphenolic Extract of *Sophora flavescens* Ait

Upon investigation of a polyphenolic extract of *S. flavescens* by means of an estrogen-inducible yeast screen and the Ishikawa-Var-I bioassay, both assays clearly revealed significant estrogenic activity at concentrations above 0.02% with respect to dry polyphenolic extract (Figure 25.6A and Figure 25.6B). The polyphenolic extract of *S. flavescens* (1 ml diluted 10-fold in water) was further subfractionated (SF-1 to SF-5) according to polarity (SF-1 being most polar) using solid-phase extraction (Bond Elut 200 mg C_{18} cartridge, Varian) by successive elutions (2 ml) of water (SF-1), methanol/water 1:3 (v/v) (SF-2), methanol/water 1:1 (v/v) (SF-3), methanol/water 3:1 (v/v) (SF-4), and methanol (SF-5) (Figure 25.7). Estrogenic activity of each fraction was defined using the above-mentioned *in vitro* bioassays (Figure 25.6A and Figure 25.6C).

25.4.2.2 Isolation and Identification of an Estrogenically Active Component from *Sophora flavescens* Ait

Only one major compound (SF-x) was present in the estrogenically active fractions, and this was isolated using semipreparative HPLC. The principles of separation are the same as for qualitative HPLC (see Section 25.4.1.2), with the difference that higher amounts of material can be loaded onto the ODS-column (Alltech, Econosil, C_{18} 10 μ, 250 × 22 mm). For the identification of SF-x, a combination of spectroscopic techniques was used. Electrospray ionization in the mass spectrometer (HP1100 LC/MSD, Hewlett-Packard) with positive ionization mode gave a pseudo-molecular ion with m/z = 439. ^{1}H-NMR, ^{13}C-NMR, DEPT, HMQC, and COSY spectra were recorded on a Varian-300 (300 MHz) spectrometer. Analysis of the COSY spectrum showed the presence of a lavandulyl (5-methyl-2-isopropenyl-hex-4-enyl) side chain. From the HMQC spectrum and a DEPT experiment, it appeared that SF-x possesses a disubstituted flavanone skeleton.

> ^{1}H-NMR (500 MHz, CD$_3$OD): δ 1.46 (3H, s), 1.55 (3H, s), 1.66 (3H, s), 2.0 (2H, m), 2.48 (1H, m), 2.61 (2H, m), 2.67 (1H, dd, J = 3.5, 17.1 Hz), 2.85 (1H, dd, J = 13.5, 16.5 Hz), 3.79 (3H, s), 4.50 (1H, m), 4.56 (1H, m), 4.96 (1H, m), 5.35 (1H, m), 6.09 (1H, s), 6.33 (1H, dd, J = 2.5, 4.1 Hz), 6.34 (1H, d, J = 2.5 Hz), 7.30 (1H, d, J = 2.5 Hz)
> ^{13}C-NMR (75 MHz, CD$_3$OD): δ 16.7, 18.0, 24.7, 27.0, 31.2, 44.4, 47.8, 54.8, 74.3, 92.1, 102.1, 104.5, 106.4, 108.3, 110.1, 117.2, 123.6, 127.3, 130.9, 148.5, 155.5, 158.3, 160.6, 163.6, 163.7, 192.8.

By comparing the spectroscopic data with that reported in the literature, SF-x was identified as kurarinone, a lavandulyl flavanone (Figure 25.8).[95,102] The estrogenic activity of pure kurarinone was assayed again by the yeast screen (Figure 25.9). Established EC_{50} values were 4.6 μ*M* and 0.37 n*M* for kurarinone and 17β-estradiol, respectively. These values indicate that kurarinone is a rather weak estrogenic compound compared with the endogenous female hormone 17β-estradiol and has similar estrogenic potency as other classes of phytoestrogens.[18,103]

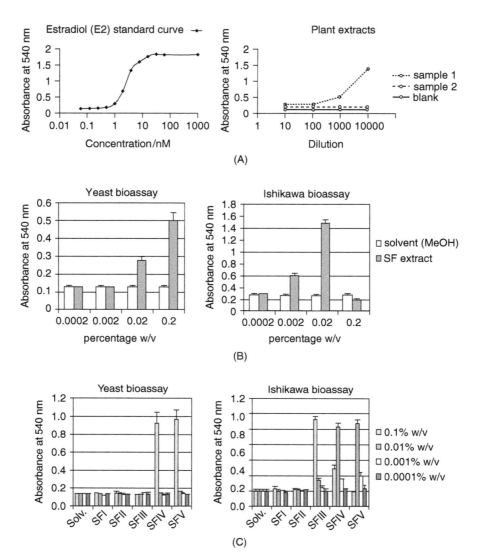

FIGURE 25.6 A: Dose–response curve for 17β-estradiol (E_2) (left panel) and results of the recombinant yeast bioassay for two representative polyphenolic extracts (right panel). The blank corresponds to the absorbance of the reaction medium. The observed values are the sum of the absorbances of the blank and the real absorbances of the investigated samples. Estrogenic activity can be derived from the dose–response curve of 17β-estradiol by extrapolation. Upon comparing figures from the left and right panels, it can be concluded that sample 1 possesses a relatively high estrogenic activity and that sample 2 is absolutely inactive. Dilution 1 of 10 in sample 1 gives an absorbance of 1.3905. By extrapolation, this corresponds to a maximal activity equivalent to ±16 nM 17β-estradiol. B: Dose–response data relating to the estrogenicity of a polyphenolic extract of *Sophora flavescens* Ait. according to an estrogen-inducible yeast screen (left panel) and the Ishikawa Var-I assay (right panel). Reduced values observed in the Ishikawa Var-I assay at the highest concentration can be attributed to cellular toxicity. C: Dose–response data relating to the estrogenicity of five subfractions of a polyphenolic extract of *S. flavescens* (concentrations expressed as percentages of w/v) according to an estrogen-inducible yeast screen (left panel) and the Ishikawa Var-I assay (right panel). From this figure, it appears that estrogenic activity is present only in the less polar fractions SF-IV and SF-V.

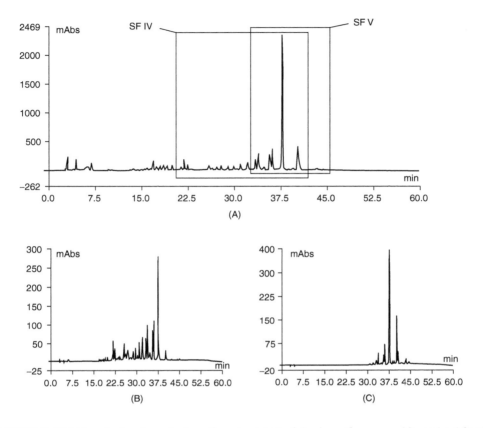

FIGURE 25.7 HPLC analysis of a polyphenolic extract (A) of *Sophora flavescens* Ait. and subfractions SF-IV (B) and SF-V (C).

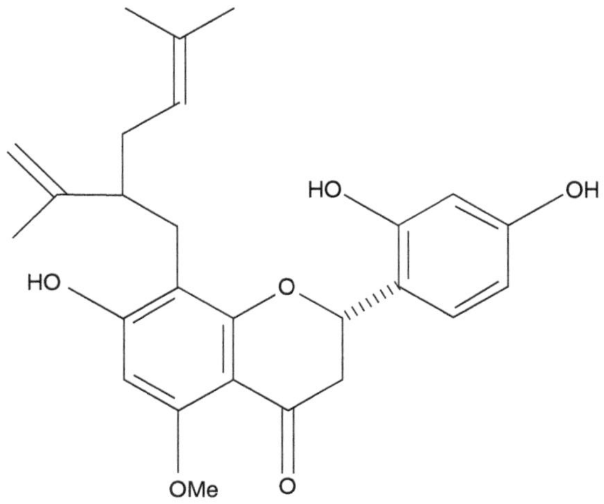

FIGURE 25.8 Chemical structure of kurarinone.

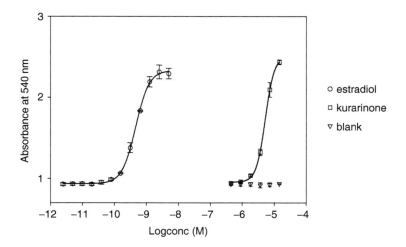

FIGURE 25.9 Dose–response curves relating to the estrogenicity of 17β-estradiol (E$_2$) and of kurarinone according to an estrogen-inducible yeast screen.

25.4.2.3 Screening for Anticancer Activity

In line with observations that suggest a lower incidence of breast, endometrial, and prostate cancers in populations that are daily exposed to phytoestrogens, the *in vitro* antitumor activity of kurarinone has been demonstrated in various cell lines, including A549 (non-small-cell lung), SK-OV-3 (ovary), SK-MEL-2 (skin), XF498 (central nerve system), HCT-15 (colon), and HL-60 (human myeloid leukemia).[95] We complemented these data with a study on the influence of a polyphenolic extract of *S. flavescens*, the estrogenically active subfractions SF-IV and SF-V (data not shown), as well as kurarinone on the growth of MCF-7/6 breast cancer cells using the sulforhodamine-B (SRB) assay, a colorimetric *in vitro* anticancer drug screening test (see Section 25.4.1.4). Figure 25.10 illustrates a dose-dependent growth inhibition upon incubation of the MCF-7/6 cells with 80- and 40-μ*M* concentrations of kurarinone (gives, respectively, 80 and 60% growth inhibition). This initial screening gives a clear indication that kurarinone has a growth-inhibitory effect with regard to MCF-7/6 breast cancer cells. Further investigation is necessary to fully unravel the anticarcinogenic effects of kurarinone by which it modulates genes that are related to the homeostatic control of cell cycle and apoptosis. As it has been found that phytoestrogens inhibit the activation of the nuclear transcription factor, NF-κB, and of the Akt signaling pathway, both of which are also known to maintain a balance between cell survival and programmed cell death (apoptosis),[58,104] it will be interesting to explore cross talk of kurarinone with the NF-κB signaling pathway.

25.5 CONCLUSION

In the past decade, nutrition research has undergone an important shift in focus from epidemiology and physiology to molecular biology and genetics. This is a result of a growing realization that the effects of nutrition on health and disease cannot be understood without a profound understanding of how nutrients act at the molecular level. Micronutrients and macronutrients can be potent dietary signals that influence the metabolic programming of cells and have an important role in the control of homeostasis.[105]

Phytoestrogens are plant-derived molecules with estrogen-like action. There are numerous reports on the potential beneficial role of nutritional phytoestrogens in human health (cancer chemoprevention, relief of postmenopausal symptoms, osteoporosis amelioration). Herbal extracts

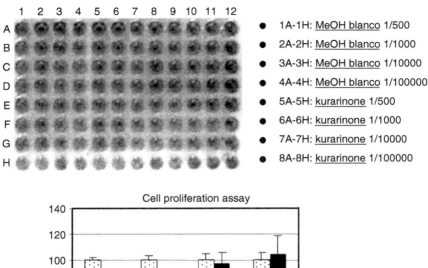

1 2 3 4 5 6 7 8 9 10 11 12

A
B
C
D
E
F
G
H

● 1A-1H: MeOH blanco 1/500
● 2A-2H: MeOH blanco 1/1000
● 3A-3H: MeOH blanco 1/10000
● 4A-4H: MeOH blanco 1/100000
● 5A-5H: kurarinone 1/500
● 6A-6H: kurarinone 1/1000
● 7A-7H: kurarinone 1/10000
● 8A-8H: kurarinone 1/100000

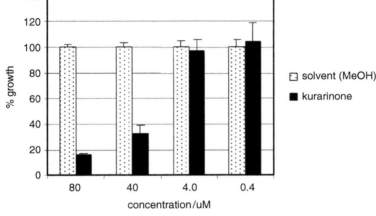

Cell proliferation assay

□ solvent (MeOH)
■ kurarinone

FIGURE 25.10 Upper panel: Visualization of the influence of methanol (solvent) and kurarinone on the growth of MCF-7/6 breast cancer cells after staining with sulforhodamine-B; samples were tested in octoplate. Lower panel: Dose–response relationships of the cytotoxic effects of kurarinone on MCF-7/6 breast cancer cells according to the sulforhodamine-B assay presented in the upper panel.

are frequently used in traditional Chinese medicine for the treatment of hormonal disturbances. Often, these preparations are formulated as complex mixtures, without any knowledge about the nature of the active components. Therefore, not all of these herbs and dietary supplements are free of risk. The number of newly formulated herbal products is growing rapidly, and because herbs, vitamins, and other natural products cannot be patented, industry-sponsored research remains limited.[56,87,106]

In this chapter, we investigated *Sophora flavescens* roots, applied frequently in folk medicine, for its active components. Estrogenicity-guided fractionation of the extract was conducted and led to the isolation of kurarinone, a prenylated flavonoid. Estrogenic activity of kurarinone was evaluated using the recombinant yeast screen and the Ishikawa bioassays. Compared with the endogenous female hormone 17β-estradiol, kurarinone exhibited a 10,000-fold weaker agonistic activity, which is in a similar range as other phytoestrogens.[14] Furthermore we demonstrated that kurarinone dose-dependently inhibits the growth of MCF-7/6 breast cancer cells, reducing growth by 20 and 40% at concentrations of 80 and 40 μ*M*, respectively.

Currently we are further evaluating the anti-inflammatory and antiproliferative potencies of kurarinone and exploring in more detail the molecular mechanisms involved. Altogether, we believe kurarinone is a promising molecule, demonstrating attractive properties for therapeutic applications, and may contribute to the development of alternative treatments for postmenopausal disturbances (osteoporosis, hot flashes, mood fluctuations).[107]

Furthermore, the identification of compounds in medicinal plant extracts with potential NF-κB inhibitory activity and/or inhibitory effects on targets of the inflammatory/apoptotic cascade might lead to the development of new anti-inflammatory, immunomodulatory, or apoptotic drugs for clinical and endemic use. These nonsteroidal modulators may serve as lead compounds for developing new pharmaceuticals to be used in the treatment of inflammatory diseases, hormonal disturbances, cancer, or conditions associated with an increased activity of the transcription factor NF-κB. In conclusion, kurarinone holds promise as a prophylactic compound that may have beneficial effects in the treatment of various postmenopausal disturbances and neoplastic disorders.

REFERENCES

1. Bennetts, H.W., Underwood, E.J., and Shier, F.L., A specific breeding problem of sheep on subterranean clover pastures in Western Australia, *Aust. Vet. J.*, 22, 2, 1946.
2. Farnsworth, N.R., Bingel, A.S., Cordell, G.A., Crane, F.A., and Fong, H.H.S., Potential value of plants as sources of new antifertility agents, II, *J. Pharm. Sci.*, 64, 717, 1975.
3. Clark, J.H., Schrader, W.T., and O'Malley, B.W., Mechanism of action of steroid hormones, in *Textbook of Endocrinology,* Wilson, J. and Foster, D.W., Eds., Saunders, Philadelphia, 1992.
4. Moggs, J.G. and Orphanides, G., Estrogen receptors: orchestrators of pleiotropic cellular responses, *EMBO Rep.*, 2, 775, 2001.
5. Murdoch, F.E. and Gorski, J., The role of ligand in estrogen receptor regulation of gene expression, *Mol. Cell. Endocrinol.*, 78, C103, 1991.
6. Segars, J.H. and Driggers, P.H., Estrogen action and cytoplasmic signaling cascades, part I: membrane-associated signaling complexes, *Trends Endocrinol. Metab.*, 13, 349, 2002.
7. Driggers, P.H. and Segars, J.H., Estrogen action and cytoplasmic signaling pathways, part II: the role of growth factors and phosphorylation in estrogen signaling, *Trends Endocrinol. Metab.*, 13, 422, 2002.
8. Métivier, R., Stark, A., Flouriot, G., Hubner, M.R., Brand, H., Penot, G., Manu, D., Denger, S., Reid, G., Kos, M., Russell, R.B., Kah, O., Pakdel, F., and Gannon, F., A dynamic structural model for estrogen receptor-α activation by ligands, emphasizing the role of interactions between distant A and E domains, *Mol. Cell.*, 10, 1019, 2002.
9. Weihua, Z., Andersson, S., Cheng, G., Simpson, E.R., Warner, M., and Gustafsson, J.A., Update on estrogen signaling, *FEBS Lett.*, 546, 17, 2003.
10. Pike, A.C., Brzozowski, A.M., and Hubbard, R.E., A structural biologist's view of the estrogen receptor, *J. Steroid. Biochem. Mol. Biol.*, 74, 261, 2000.
11. Levenson, A.S., Svoboda, K.M., Pease, K.M., Kaiser, S.A., Chen, B., Simons, L.A., Jovanovic, B.D., Dyck, P.A., and Jordan, V.C., Gene expression profiles with activation of the estrogen receptor α-selective estrogen receptor modulator complex in breast cancer cells expressing wild-type estrogen receptor, *Cancer Res.*, 62, 4419, 2002.
12. Jordan, V.C., The secrets of selective estrogen receptor modulation: cell-specific coregulation, *Cancer Cell.*, 1, 215, 2002.
13. Gustafsson, J.A., Therapeutic potential of selective estrogen receptor modulators, *Curr. Opin. Chem. Biol.*, 2, 508, 1998.
14. Benassayag, C., Perrot-Applanat M., and Ferre, F., Phytoestrogens as modulators of steroid action in target cells, *J. Chromatogr. B Analyt. Technol. Biomed. Life Sci.*, 777, 233, 2002.
15. Jordan, V.C., Schafer J.M., Levenson A.S., Liu H., Pease, K.M., Simons, L.A., and Zapf, J.W., Molecular classification of estrogens, *Cancer Res.*, 61, 6619, 2001.
16. Pike, A.C., Brzozowski, A.M., Hubbard, R.E., Bonn, T., Thorsel, A.G., Engström, O., Ljunggren, J., Gustafsson, J.A., and Carlquist, M., Structure of the ligand-binding domain of estrogen receptor β in the presence of a partial agonist and a full antagonist, *EMBO J.*, 18, 4608, 1999.
17. An, J., Tzagarakis-Foster, C., Scharschmidt, T.C., Lomri, N., and Leitman, D.C., Estrogen receptor β-selective transcriptional activity and recruitment of coregulators by phytoestrogens, *J. Biol. Chem.*, 276, 17808, 2001.

18. Kuiper, G.G., Lemmen, J.G., Carlsson, B., Corton, J.C., Safe, S.H., van der Saag, P.T., van der Burg, B., and Gustafsson, J.A., Interaction of estrogenic chemicals and phytoestrogens with estrogen receptor β, *Endocrinology*, 139, 4252, 1998.

19. Knight, D.C. and Eden, J.A., A review of the clinical effects of phytoestrogens, *Obstet. Gynecol.*, 87, 897, 1996.

20. Wang, H. and Murphy, P.A., Isoflavone composition of American and Japanese soybeans in Iowa: effects of variety, crop year and location, *J. Agric. Food Chem.*, 42, 1674, 1994.

21. Thompson, L.U., Robb, P., Serraino, M., and Cheung, F., Mammalian lignan production from various foods, *Nutr. Cancer*, 16, 43, 1991.

22. Havsteen, B.H., The biochemistry and medical significance of the flavonoids, *Pharmacol. Ther.*, 96, 67, 2002.

23. Miksicek, R.J., Commonly occurring plant flavonoids have estrogenic activity, *Mol. Pharmacol.*, 44, 37, 1993.

24. Markiewicz, L., Garey, J., Adlercreutz, H., and Gurpide, E., *In vitro* bioassays of non-steroidal phytoestrogens, *J. Steroid. Biochem. Mol. Biol.*, 45, 399, 1993.

25. Kitts, W.D., Newsome, F.E., and Runeckles, V.C., The estrogenic and antiestrogenic effects of coumestrol and zearalenol on the immature rat uterus, *Can. J. Anim. Sci.*, 63, 823, 1983.

26. Adlercreutz, H. and Mazur, W., Phyto-oestrogens and Western diseases, *Ann. Med.*, 29, 95, 1997.

27. Tham, D.M., Gardner, C.D., and Haskell, W.L., Clinical review 97: potential health benefits of dietary phytoestrogens: a review of the clinical, epidemiological, and mechanistic evidence, *J. Clin. Endocrinol. Metab.*, 83, 2223, 1998.

28. Sun, J., Meyers, M.J., Fink, B.E., Rajendran, R., Katzenellenbogen, J.A., and Katzenellenbogen, B.S., Novel ligands that function as selective estrogens or antiestrogens for estrogen receptor-α or estrogen receptor-β, *Endocrinology*, 140, 800, 1999.

29. Adlercreutz, H., Bannwart, C., Wahala, K., Makela, T., Brunow, G., Hase, T., Arosemena, P.J., Kellis, Jr., J.T., and Vickery, L.E., Inhibition of human aromatase by mammalian lignans and isoflavonoid phytoestrogens, *J. Steroid Biochem. Mol. Biol.*, 44, 147, 1993.

30. Evans, B.A., Griffiths, K., and Morton, M.S., Inhibition of 5 a-reductase in genital skin fibroblasts and prostate tissue by dietary lignans and isoflavonoids, *J. Endocrinol.*, 147, 295, 1995.

31. Markovits, J., Linassier, C., Fosse, P., Couprie, J., Pierre, J., Jacquemin-Sablon, A., Saucier, J.M., Le Pecq, J.B., and Larsen, A.K., Inhibitory effects of the tyrosine kinase inhibitor genistein on mammalian DNA topoisomerase, II, *Cancer Res.*, 49, 5111, 1989.

32. Matsukawa, Y., Marui, N., Sakai, T., Satomi, Y., Yoshida, M., Matsumoto, K., Nishino, H., and Aoike, A., Genistein arrests cell cycle progression at G2-M, *Cancer Res.*, 53, 1328, 1993.

33. Agarwal, R., Cell signaling and regulators of cell cycle as molecular targets for prostate cancer prevention by dietary agents, *Biochem. Pharmacol.*, 60, 1051, 2000.

34. Casagrande, F. and Darbon, J.M., Effects of structurally related flavonoids on cell cycle progression of human melanoma cells: regulation of cyclin-dependent kinases CDK2 and CDK1, *Biochem. Pharmacol.*, 61, 1205, 2001.

35. Kim, M.H., Gutierrez, A.M., and Goldfarb, R.H., Different mechanisms of soy isoflavones in cell cycle regulation and inhibition of invasion, *Anticancer Res.*, 22, 3811, 2002.

36. Akiyama, T., Ishida, J., Nakagawa, S., Ogawara, H., Watanabe, S., Itoh, N., Shibuya, M., and Fukami, Y., Genistein, a specific inhibitor of tyrosine-specific protein kinases, *J. Biol. Chem.*, 262, 5592, 1987.

37. Dixon, R.A. and Ferreira, D., Genistein, *Phytochemistry*, 60, 205, 2002.

38. Bolego, C., Poli, A., Cignarella, A., and Paoletti, R., Phytoestrogens: pharmacological and therapeutic perspectives, *Curr. Drug Targets*, 4, 77, 2003.

39. Adlercreutz, H., Hockerstedt, K., Bannwart, C., Bloigu, S., Hamalainen, E., Fotsis, T., and Ollus, A., Effect of dietary components, including lignans and phytoestrogens, on enterohepatic circulation and liver metabolism of estrogens and on sex hormone binding globulin (SHBG), *J. Steroid Biochem.*, 27, 1135, 1987.

40. Cassidy, A., Bingham, S., and Setchell, K.D., Biological effects of a diet of soy protein rich in isoflavones on the menstrual cycle of premenopausal women, *Am. J. Clin. Nutr.*, 60, 333, 1994.

41. Cassidy, A., Bingham, S., and Setchell, K., Biological effects of isoflavones in young women: importance of the chemical composition of soyabean products, *Br. J. Nutr.*, 74, 587, 1995.

42. Fotsis, T., Pepper, M.S., Aktas, E., Breit, S., Rasku, S., Adlercreutz, H., Wahala, K., Montesano, R., and Schweigerer, L., Flavonoids, dietary-derived inhibitors of cell proliferation and *in vitro* angiogenesis, *Cancer Res.*, 57, 2916, 1997.

43. Cotelle, N., Role of flavonoids in oxidative stress, *Curr. Top. Med. Chem.*, 1, 569, 2001.

44. Kulling, S.E., Lehmann, L., and Metzler, M., Oxidative metabolism and genotoxic potential of major isoflavone phytoestrogens, *J. Chromatogr. B Anal. Technol. Biomed. Life Sci.*, 777, 211, 2002.

45. Kiecolt-Glaser, J.K., Preacher, K.J., MacCallum, R.C., Atkinson, C., Malarkey, W.B., and Glaser, R., Chronic stress and age-related increases in the proinflammatory cytokine IL-6, *Proc. Natl. Acad. Sci. USA*, 100, 9090, 2003.

46. Pfeilschifter, J., Koditz, R., Pfohl, M., and Schatz, H., Changes in proinflammatory cytokine activity after menopause, *Endocr. Rev.*, 23, 90, 2002.

47. Ershler, W.B. and Keller, E.T., Age-associated increased interleukin-6 gene expression, late-life diseases, and frailty, *Annu. Rev. Med.*, 51, 245, 2000.

48. Galien, R. and Garcia, T., Estrogen receptor impairs interleukin-6 expression by preventing protein binding on the NF-κB site, *Nucleic Acids Res.*, 25, 2424, 1997.

49. Ray, P., Ghosh, S.K., Zhang, D.H., and Ray, A., Repression of interleukin-6 gene expression by 17β-estradiol: inhibition of the DNA-binding activity of the transcription factors NF-IL6 and NF-κB by the estrogen receptor, *FEBS Lett.*, 409, 79, 1997.

50. Evans, M.J., Eckert, A., Lai, K., Adelman, S.J., and Harnish, D.C., Reciprocal antagonism between estrogen receptor and NF-κB activity *in vivo*, *Circ. Res.*, 89, 823, 2001.

51. Wuttke, W., Jarry, H., Westphalen, S., Christoffel, V., and Seidlova-Wuttke, D., Phytoestrogens for hormone replacement therapy? *J. Steroid Biochem. Mol. Biol.*, 83, 133, 2002.

52. Wells, G., Tugwell, P., Shea, B., Guyatt, G., Peterson, J., Zytaruk, N., Robinson, V., Henry, D., O'Connell, D., and Cranney, A., Meta-analyses of therapies for postmenopausal osteoporosis, V: meta-analysis of the efficacy of hormone replacement therapy in treating and preventing osteoporosis in postmenopausal women, *Endocr. Rev.*, 23, 529, 2002.

53. Marsden, J., The menopause, hormone replacement therapy and breast cancer, *J. Steroid Biochem. Mol. Biol.*, 83, 123, 2002.

54. Cranney, A., Guyatt, G., Griffith, L., Wells, G., Tugwell, P., and Rosen, C., Meta-analyses of therapies for postmenopausal osteoporosis, IX: summary of meta-analyses of therapies for postmenopausal osteoporosis, *Endocrine Rev.*, 23, 570, 2002.

55. Guyatt, G.H., Cranney, A., Griffith, L., Walter, S., Krolicki, N., Favus, M., and Rosen, C., Summary of meta-analyses of therapies for postmenopausal osteoporosis and the relationship between bone density and fractures, *Endocrinol. Metab. Clin. North Am.*, 31, 659, 2002.

56. Kronenberg, F. and Fugh-Berman, A., Complementary and alternative medicine for menopausal symptoms: a review of randomized, controlled trials, *Ann. Intern. Med.*, 137, 805, 2002.

57. Li, Q. and Verma, I.M., NF-κB regulation in the immune system, *Nat. Rev. Immunol.*, 2, 725, 2002.

58. Richmond, A., NF-κB, chemokine gene transcription and tumour growth, *Nat. Rev. Immunol.*, 2, 664, 2002.

59. Vermeulen, L., De Wilde, G., Damme, P., Vanden Berghe, W., and Haegeman, G., Transcriptional activation of the NF-κB p65 subunit by mitogen- and stress-activated protein kinase-1 (MSK1), *EMBO J.*, 22, 1313, 2003.

60. Vanden Berghe, W., De Bosscher, K., Vermeulen, L., De Wilde, G., and Haegeman, G., Induction and repression of NF-κB-driven inflammatory genes, *Ernst Schering Res. Found. Workshop*, 40, 233, 2002.

61. Vermeulen, L., De Wilde, G., Notebaert, S., Vanden Berghe, W., and Haegeman, G., Regulation of the transcriptional activity of the nuclear factor-κB p65 subunit, *Biochem. Pharmacol.*, 64, 963, 2002.

62. Vanden Berghe, W., Vermeulen, L., De Wilde, G., De Bosscher, K., Boone, E., and Haegeman, G., Signal transduction by tumor necrosis factor and gene regulation of the inflammatory cytokine interleukin-6, *Biochem. Pharmacol.*, 60, 1185, 2000.

63. Holmes-McNary, M. and Baldwin, Jr., A.S., Chemopreventive properties of trans-resveratrol are associated with inhibition of activation of the IκB kinase, *Cancer Res.*, 60, 3477, 2000.

64. Quaedackers, M.E., Van Den Brink, C.E., Wissink, S., Schreurs, R.H., Gustafsson, J.A., Van Der Saag, P.T., and Van Der Burg, B.B., 4-Hydroxytamoxifen transrepresses nuclear factor-κB activity in human osteoblastic U2-OS cells through estrogen receptor (ER) α, and not through ERβ, *Endocrinology*, 142, 1156, 2001.

65. Ziegler, R.G., Hoover, R.N., Pike, M.C., Hildesheim, A., Nomura, A.M., West, D.W., Wu-Williams, A.H., Kolonel, L.N., Horn-Ross, P.L., Rosenthal, J.F. et al., Migration patterns and breast cancer risk in Asian-American women, *J. Natl. Cancer Inst.*, 85, 1819, 1993.

66. Bange, J., Zwick, E., and Ullrich, A., Molecular targets for breast cancer therapy and prevention, *Nat. Med.*, 7, 548, 2001.

67. Nathanson, K.L., Wooster, R., Weber, B.L., and Nathanson, K.N., Breast cancer genetics: what we know and what we need, *Nat. Med.*, 7, 552, 2001.

68. Ali, S. and Coombes, C., Endocrine-responsive breast cancer and strategies for combating resistance, *Nat. Rev. Mol. Cell. Biol.*, 2, 101, 2002.

69. Parkin, D.M., Cancers of the breast, endometrium and ovary: geographic correlations, *Eur. J. Cancer Clin. Oncol.*, 25, 1917, 1989.

70. Birt, D.F., Hendrich, S., and Wang, W., Dietary agents in cancer prevention: flavonoids and isoflavonoids, *Pharmacol. Ther.*, 90, 157, 2001.

71. Kris-Etherton, P.M., Hecker, K.D., Bonanome, A., Coval, S.M., Binkoski, A.E., Hilpert, K.F., Griel, A.E., and Etherton, T.D., Bioactive compounds in foods: their role in the prevention of cardiovascular disease and cancer, *Am. J. Med.*, 113 (Suppl. 9B), 71S, 2002.

72. Adlercreutz, H., Phytoestrogens and cancer, *Lancet Oncol.*, 3, 364, 2002.

73. Adlercreutz, H., Mazur, W., Bartels, P., Elomaa, V., Watanabe, S., Wahala, K., Landstrom, M., Lundin, E., Bergh, A., Damber, J.E., Aman, P., Widmark, A., Johansson, A., Zhang, J.X., and Hallmans, G., Phytoestrogens and prostate disease, *J. Nutr.*, 130, 658S, 2000.

74. Valsta, L.M., Kilkkinen, A., Mazur, W., Nurmi, T., Lampi, A.M., Ovaskainen, M.L., Korhonen, T., Adlercreutz, H., and Pietinen, P., Phytoestrogen database of foods and average intake in Finland, *Br. J. Nutr.*, 89, 822, 2003.

75. Brownson, D.M., Azios, N.G., Fuqua, B.K., Dharmawardhane, S.F., and Mabry, T.J., Flavonoid effects relevant to cancer, *J. Nutr.*, 132, 3482S, 2002.

76. Kolonel, L.N., Variability in diet and its relation to risk in ethnic and migrant groups, *Basic Life Sci.*, 43, 129, 1988.

77. Knight, D.C. and Eden, J.A., Phytoestrogens — a short review, *Maturitas*, 22, 167, 1995.

78. Coward, L., Barnes, N.C., Setchell, K.D., and Barnes, S., The isoflavones genistein and daidzein in soybean foods from American and Asian diets, *J. Agric. Food Chem.*, 41, 1961, 1993.

79. Rosenberg, Z.R.S., Jenkins, D.J., and Diamandis, E.P., Flavonoids and steroid hormone-dependent cancers, *J. Chromatogr. B Anal. Technol. Biomed. Life Sci.*, 777, 219, 2002.

80. Powles, T.J., Antiestrogenic chemoprevention of breast cancer-the need to progress, *Eur. J. Cancer*, 39, 572, 2003.

81. McMichael-Phillips, D.F., Harding, C., Morton, M., Roberts, S.A., Howell, A., Potten, C.S., and Bundred, N.J., Effects of soy-protein supplementation on epithelial proliferation in the histologically normal human breast, *Am. J. Clin. Nutr.*, 68, 1431S, 1998.

82. Petrakis, N.L., Barnes, S., King, E.B., Löwenstein, J., Wiencke, J., Lee, M.M., Miike, R., Kirk, M., and Coward, L., Stimulatory influence of soy protein isolate on breast secretion in pre- and postmenopausal women, *Cancer Epidemiol. Biomarkers Prev.*, 5, 785, 1996.

83. Ju, Y.H., Allred, C.D., Allred, K.F., Karko, K.L., Doerge, D.R., and Helferich, W.G., Physiological concentrations of dietary genistein dose dependently stimulate growth of estrogen-dependent human breast cancer (MCF-7) tumors implanted in athymic nude mice, *J. Nutr.*, 131, 2957, 2001.

84. Allred, C.D., Ju, Y.H., Allred, K.F., Chang, J., and Helferich, W.G., Dietary genistin stimulates growth of estrogen-dependent breast cancer tumors similar to that observed with genistein, *Carcinogenesis*, 22, 1667, 2001.

85. Allred, C.D., Allred, K.F., Ju, Y.H., Virant, S.M., and Helferich, W.G., Soy diets containing varying amounts of genistein stimulate growth of estrogen-dependent (MCF-7) tumors in a dose-dependent manner, *Cancer Res.*, 61, 5045, 2001.

86. Ju, Y.H., Doerge, D.R., Allred, K.F., Allred, C.D., and Helferich, W.G., Dietary genistein negates the inhibitory effect of tamoxifen on growth of estrogen-dependent human breast cancer (MCF-7) cells implanted in athymic mice, *Cancer Res.*, 62, 2474, 2002.

87. Albertazzi, P. and Purdie, D., The nature and utility of the phytoestrogens: a review of the evidence, *Maturitas*, 42, 173, 2002.

88. Alekel, D.L., Germain, A.S., Peterson, C.T., Hanson, K.B., Stewart, J.W., and Toda, T., Isoflavone-rich soy protein isolate attenuates bone loss in the lumbar spine of perimenopausal women, *Am. J. Clin. Nutr.*, 72, 844, 2000.

89. Gambacciani, M., Spinetti, A., Cappagli, B., Taponeco, F., Felipetto, R., Parrini, D., Cappelli, N., and Fioretti, P., Effects of ipriflavone administration on bone mass and metabolism in ovariectomized women, *J. Endocrinol. Invest.*, 16, 333, 1993.

90. Carroll, K.K., Review of clinical studies on cholesterol-lowering response to soy protein, *J. Am. Diet. Assoc.*, 91, 820, 1991.

91. Setchell, K.D. and Cassidy, A., Dietary isoflavones: biological effects and relevance to human health, *J. Nutr.*, 129, 758S, 1999.

92. File, S.E., Jarrett, N., Fluck, E., Duffy, R., Casey, K., and Wiseman, H., Eating soya improves human memory, *Psychopharmacology (Berl.)*, 157, 430, 2001.

93. Ryu, S.Y., Kim, S.-K., No, Z., and Ahn, J.W., A novel flavonoid from *Sophora flavescens*, *Planta Medica*, 62, 361, 1996.

94. Woo, E.R., Kwak, J.H., Kim, H.J., and Park, H., A new prenylated flavonol from the roots of *Sophora flavescens*, *J. Natl. Prod.*, 61, 1552, 1998.

95. Kang, T.H., Jeong, S.J., Ko, W.G., Kim, N.Y., Lee, B.H., Inagaki, M., Miyamoto, T., Higuchi, R., and Kim, Y.C., Cytotoxic lavandulyl flavanones from *Sophora flavescens*, *J. Nat. Prod.*, 63, 680, 2000.

96. Routledge, E.J. and Sumpter, J.P., Estrogenic activity of surfactants and some of their degradation products assessed using a recombinant yeast screen, *Environ. Toxicol. Chem.*, 15, 241, 1996.

97. Littlefield, B.A., Gurpide, E., Markiewicz, L., McKinley, B., and Hochberg, R.B., A simple and sensitive microtiter plate estrogen bioassay based on stimulation of alkaline phosphatase in Ishikawa cells: estrogenic action of delta 5 adrenal steroids, *Endocrinology*, 127, 2757, 1990.

98. Nishida, M., Kasahara, K., Kaneko, M., Iwasaki, H., and Hayashi, K., Establishment of a new human endometrial adenocarcinoma cell line, Ishikawa cells, containing estrogen and progesterone receptors, *Nippon Sanka Fujinka Gakkai Zasshi*, 37, 1103, 1985.

99. Bracke, M.E., Van Larebeke, N.A., Vyncke B.M., and Mareel, M.M., Retinoic acid modulates both invasion and plasma membrane ruffling of MCF-7 human mammary carcinoma cells *in vitro*, *Br. J. Cancer*, 63, 867, 1991.

100. Skehan, P., Storeng, R., Scudiero, D., Monks, A., McMahon, J., Vistica, D., Warren, J.T., Bokesch, H., Kenney, S., and Boyd, M.R., New colorimetric cytotoxicity assay for anticancer-drug screening, *J. Natl. Cancer Inst.*, 82, 1107, 1990.

101. Loberg, L.I., Engdahl, W.R., Gauger, J.R., and McCormick, D.L., Cell viability and growth in a battery of human breast cancer cell lines exposed to 60 Hz magnetic fields, *Radiat. Res.*, 153, 725, 2000.

102. Kuroyanagi, M., Arakawa, T., Hirayama, Y., and Hayashi, T., Antibacterial and antiandrogen flavonoids from *Sophora flavescens*, *J. Nat. Prod.*, 62, 1595, 1999.

103. Coldham, N.G., Dave, M., Sivapathasundaram, S., McDonnell, D.P., Connor, C., and Sauer, M.J., Evaluation of a recombinant yeast cell estrogen screening assay, *Environ. Health Perspect.*, 105, 734, 1997.

104. Karin, M. and Lin, A., NF-κB at the crossroads of life and death, *Nat. Immunol.*, 3, 221, 2002.

105. Müller, M. and Kersten, S., Nutrigenomics: goals and strategies, *Natl. Rev. Genet.*, 4, 315, 2003.

106. Ewies, A.A., Phytoestrogens in the management of the menopause: up-to-date, *Obstet. Gynecol. Surv.*, 57, 306, 2002.

107. Cos, P., De Bruyne, T., Apers, S., Berghe, D.V., Pieters, L., and Vlietinck, A.J., Phytoestrogens: recent developments, *Planta Medica*, 69, 589, 2003.

26 Resveratrol in the Chemoprevention and Chemotherapy of Breast Cancer

Wisit Tangkeangsirisin and Ginette Serrero

CONTENTS

26.1 Scope of this Review 449
26.2 Introduction 450
26.3 Chemical Structure of Resveratrol 450
26.4 Resveratrol: Estrogen or Antiestrogen in Mammary Epithelial Cells? 451
26.5 Effect of Resveratrol on Breast Cancer Cell Proliferation *In Vitro* 452
26.6 Mechanism of Inhibition of Breast Cancer 453
    26.6.1 Effect of Resveratrol on Cell Cycle and Apoptosis 453
    26.6.2 Inhibition of Autocrine Growth-Factor Loops by Resveratrol...........................454
    26.6.3 Stimulation of Autocrine Growth Inhibitors 454
    26.6.4 Antioxidant Activity...454
    26.6.5 Induction of Cell Differentiation 455
    26.6.6 Cyclooxygenase Inhibition 455
    26.6.7 Anti-initiation Activity 455
26.7 Resveratrol in Other Cancers..455
26.8 Effect of Resveratrol *In Vivo* 456
26.9 Bioavailability of *Trans*-Resveratrol..456
26.10 Benefits of Using Resveratrol..456
26.11 Conclusion..457
References 457

26.1 SCOPE OF THIS REVIEW

Chemoprevention is an emerging and promising approach to control human cancer. Several natural products have been proven to have potentials as anticancer agents. Resveratrol is one such compound. Recently, increasing interest has focused on examining the effect of resveratrol in breast cancer prevention and treatment. The present review summarizes the recent studies on the role of resveratrol in cancer prevention and therapy, especially in human breast cancer, that provide new insights into the molecular mechanisms of resveratrol. The estrogenic activity of resveratrol, as well as its *in vitro* and *in vivo* antiproliferative and chemoprotective activities, will be discussed here.

0-8493-1560-3/05/$0.00+$1.50
© 2005 by CRC Press LLC

TABLE 26.1
Pharmacological Activities of Resveratrol

Pharmacological Activity	References
Vasorelaxation	95, 96
Antioxidant	97, 98
Antiplatelet aggregation	99–103
Cardioprotective activity	104–109
Cyclooxygenase inhibition	74, 110, 111
Antiangiogenesis	112–115
Protein kinase inhibition	116–119
Reduction of oxidative stress	120, 121
Estrogenic activity	5, 19, 122
Immunomodulatory activity	123
Antibacterial	124, 125
Inhibition of nitric oxide synthase	126, 127
Tyrosine kinase inhibition	128, 129
Anticancer	See Table 26.2

26.2 INTRODUCTION

The search for phytoestrogens to use as a natural source of estrogen for hormone replacement therapy has been ongoing for many years. The interest in this topic is based on the fact that populations in geographical areas having a diet high in phytoestrogens have a lower incidence of breast cancer.[1] An ideal compound for this application would be one able to supplement the beneficial effects of estrogen in bone, liver, cardiovascular system, and central nervous system without showing growth-stimulatory effects in breast and uterus. Among the various compounds in this category, resveratrol and quercetin have been particularly investigated.

Resveratrol (*trans*-3,4′,5-trihydroxystilbene) is a phytoalexin enriched in grape seed and skin.[2] Its structure is similar to the synthetic estrogen diethylstilbestrol. As a result, resveratrol is also categorized as a phytoestrogen. Resveratrol has received a lot of interest in the past several years, as it has been reported to have many diverse beneficial pharmacological effects (Table 26.1). In particular, resveratrol has been implicated as the major contributor to the cardioprotective effect of red wine, occasionally referred as the "French paradox."[3] Resveratrol has also been found in Japanese and Chinese traditional medicine, in the form of extracts from *Polygonum cuspidatum*, for curing headaches, inflammation, amenorrhea, and cancers.[4]

26.3 CHEMICAL STRUCTURE OF RESVERATROL

Resveratrol occurs naturally in different chemical forms: *cis*-resveratrol, *trans*-resveratrol, and their 3-*O*-β-glucosides (piceid) (Figure 26.1). Each shows differences in estrogenic and antiproliferative activities.[5-8] Among these compounds, *trans*-resveratrol is the most widely studied. It is believed that *trans*-resveratrol and its glycoside are the active forms.[9] *trans*-Resveratrol is found in red grape juice at a concentration of 0.2 to 3 mg/l.[10,11] *cis*-Resveratrol is also found in red grape wine, but at lower concentrations. The amount of piceid isomers is slightly higher than aglycones in red wine. On the other hand, in another source of resveratrol, *Polygonum cuspidatum*, resveratrol was found in the form of *trans*-resveratrol glucoside, whereas *cis*-resveratrol was undetectable in this plant.[2] Conversion from *trans*-resveratrol to *cis*-resveratrol is detectable but not significant in normal volunteers.[12]

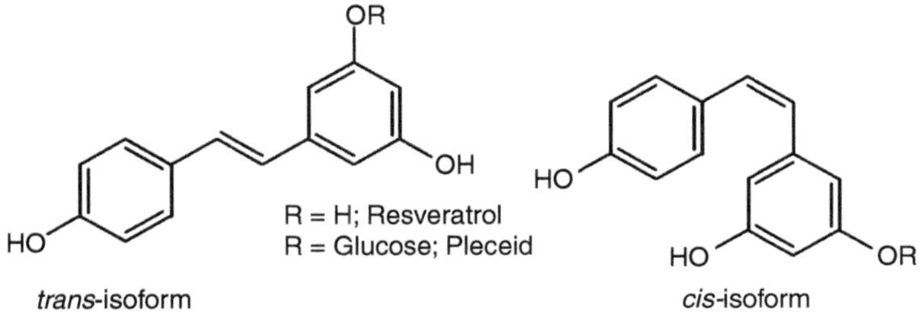

R = H; Resveratrol
R = Glucose; Pleceid

trans-isoform   *cis*-isoform

FIGURE 26.1 Chemical structures of resveratrol isoforms found in natural products.

26.4 RESVERATROL: ESTROGEN OR ANTIESTROGEN IN MAMMARY EPITHELIAL CELLS?

The investigation of the effect of resveratrol in inhibiting cell proliferation in breast cancer cells has unveiled many possible mechanisms of action that will be discussed here. Concerning breast cancer cells, extensive literature has examined whether resveratrol effect on breast cancer cells was due to the structural similarity between resveratrol and the pure estrogen agonist diethylstilbestrol (Figure 26.2). Based on these similarities, it has been hypothesized that resveratrol would be able to present estrogenic activity or modulate response to estrogen, which would explain its *in vitro* and *in vivo* biological properties.

The relationship between resveratrol and estradiol has been investigated using a variety of bioassays from receptor binding studies to transcriptional activation of estrogen-responsive genes. Based on these studies, resveratrol classification ranges from superagonist to antagonist with respect to estrogen, depending on the experimental conditions used and the assay systems examined.[13-20]

Concerning the ability of resveratrol to bind to the estrogen receptors, several reports suggest that resveratrol binds to estrogen receptors with a lower affinity (IC_{50} for estrogen receptor α [ERα] = 58.5 μM; IC_{50} for estrogen receptor β [ERβ] = 130 μM) compared with 17β-estradiol (K_D = 0.5 to 2.0 nM).[13,15,19,21] Since the discovery of the ERβ isoforms, studies have focused on comparing the ability of resveratrol to interact with both estrogen receptors isoforms (ERα and ERβ). Resveratrol binds to both estrogen receptor isoforms with similar affinity,[19] whereas other phytoestrogens,

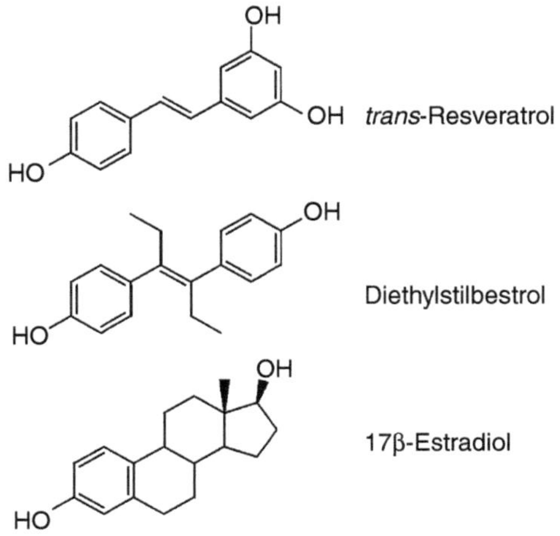

trans-Resveratrol

Diethylstilbestrol

17β-Estradiol

FIGURE 26.2 Structural similarity between *trans*-resveratrol and other estrogen agonists.

such as genistein, preferentially bind to ERβ over ERα.[20] These data indicate that resveratrol may compete with estrogen and thereby interfere with estrogen activity mediated by either one of the estrogen receptor isoforms.

Although resveratrol and diethylstilbestrol display similarities in their chemical structure and ability to bind to estrogen receptors, a molecular-modeling study suggests that the interactions between key moieties on ERα are different for resveratrol and the pure agonists estradiol and diethylstilbestrol.[14] These distinct interactions may explain the unique biological activity of resveratrol in estrogen responsiveness. The molecular interaction of resveratrol with ERβ has not yet been fully investigated.

Controversial reports have been presented in the past few years about the hormonal activity and antiproliferative effect of resveratrol in estrogen-dependent human breast cancer. Estrogen activates gene transcription by binding to estrogen receptors. The ligand—receptor complex trans-activates specific gene expression by binding to a unique DNA sequence called an estrogen-responsive element (ERE) contained in the promoter of many estrogen-responsive genes, thereby recruiting cellular transcriptional machinery to the sites.[22] Using a reporter gene under the control of an estrogen-responsive element within the promoter region, transcriptional activity can be quantitated in the investigation of the estrogenic or antiestrogenic activity of specific compounds such as resveratrol. In CHO-K1 cells transfected with ERα or ERβ, resveratrol at the concentration of 10 and 50 μM increased the activity of luciferase reporter gene under the control of consensus estrogen responsive element (ERE) or ERE from c-fos, pS2, or progesterone receptor gene promoter.[19] In MCF-7 cells, Gehm et al.[15] also showed that resveratrol transactivated ERE-tk109-luc and ERE-tk81-luc reporter gene constructs. These results indicate that resveratrol acted as an estrogen receptor agonist in these conditions.

Similar results were obtained by examining the effect of resveratrol on the transcriptional control of known estrogen-responsive genes. In the estrogen-receptor-positive human breast cancer cell line MCF-7, cultivated in the absence of estrogen, resveratrol stimulated transcription of known estrogen-responsive genes such as pS2 and progesterone receptor (PR).[17,19,23] Resveratrol added alone also induced TGFα expression, another estrogen-inducible gene, in a dose-dependent fashion. This stimulation was inhibited by the pure antiestrogen ICI 182,780.[16] However, using the same bioassays in the human endometrial adenocarcinoma Ishikawa cells, resveratrol did not show any estrogenic effect but instead showed antiestrogenic activity.[13] As a single agent, resveratrol acted as an ERα agonist by inducing PR mRNA expression in MCF-7 cells maintained in the absence of estradiol.[17] However, when resveratrol was added in the presence of estradiol in MCF-7 cells, then it acted as an estrogen antagonist and inhibited estrogen effect on cell proliferation and on PR expression in MCF-7 cells and T47D cells.[17,23]

These data would indicate that resveratrol acts as an agonist when added as a single agent in the absence of estrogen. However, it acts as an antagonist in the presence of E2.[17] In summary, evidence reported in the literature would suggest that the activity of resveratrol as an estrogen agonist or antagonist depends on the cell type,[24] the ER isoforms, and the promoter context.[19]

26.5 EFFECT OF RESVERATROL ON BREAST CANCER CELL PROLIFERATION *IN VITRO*

Similar estrogen agonistic or antagonistic effects of resveratrol have been observed when the effect of resveratrol on cell proliferation was examined. Resveratrol has been shown to have growth-stimulatory effect when added alone but to inhibit cell proliferation when added in the presence of estradiol in several ER-positive breast cancer cell lines.[17,18,23,25-30] In these cells, it has been established that the growth-inhibitory effect of resveratrol was ER-mediated and that resveratrol acted as an estradiol antagonist. On the contrary, resveratrol appeared to be an agonist by mediating ER-dependent gene transcription but inhibited growth and proliferation. In some reports, a superagonistic effect was observed.[5,15] The difference can be explained by the fact that these various studies used different resveratrol concentrations.

In MCF-7 cells, resveratrol at low dose (0.1 to 10 μM) induced proliferation when added in the absence of estrogen. At higher doses (25 to 100 μM), a growth-inhibitory effect was observed.[17,21] In contrast, when added with estrogen, resveratrol only displayed a dose-dependent growth-inhibitory effect, even at doses that were found to be stimulatory in the absence of estrogen. Moreover, in ER-negative human breast cancer cells, resveratrol also inhibited cell proliferation at all concentrations tested.[18,21,30] Taken together, it is proposed that resveratrol mediated its effect on cell proliferation either by ER-dependent pathway in ER-positive cells by antagonizing estrogen mitogenic activity or by ER-independent pathways that would be important in ER-positive as well as ER-negative cells. On the other hand, in breast cancer cell lines, several reports have found no correlation between ER expression and sensitivity to resveratrol, suggesting that the antiproliferative effect of resveratrol is independent of ER status.[18,21,28]

Of interest are the studies by Levenson et al.,[16] who investigated the estrogenic activity of resveratrol in MDA-MB-231, an estrogen-receptor-negative human breast cancer cell line, by transfecting either wild-type or mutant ERα, mutD351Y ER, which has a low binding affinity for ligands. The results showed that resveratrol acted as an ER agonist by increasing the expression of the estrogen-responsive gene TGFα in only wild-type ERα cells. However, resveratrol inhibited cell proliferation in both mutant and wild-type ER transfected MDA-MB-231 cells.[16]

Based on these data, it is mainly essential to focus on the various pathways by which resveratrol would inhibit cell proliferation, whether these pathways are ER-mediated or not.

26.6 MECHANISM OF INHIBITION OF BREAST CANCER

In addition to directly inhibiting estradiol mitogenic activity by competing with estradiol for binding to ER, there are several other possible pathways by which resveratrol would inhibit cell proliferation. These pathways have been examined mainly by *in vitro* studies and are summarized in Table 26.2.

26.6.1 EFFECT OF RESVERATROL ON CELL CYCLE AND APOPTOSIS

Several components of the cell-cycle machinery are regulated by estrogen receptors.[31] Estrogen regulates expression and function of cyclin D1 and c-Myc and activates cyclin E-Cdk2 complexes by promoting dissociation of p21, a cyclin-dependent kinase inhibitor, from the complexes. Pozo-Guisado et al.[30] reported that resveratrol transiently induced p21 expression in MCF-7 cells. Similar results were reported in MDA-MB-231 cells stably transfected with wild-type or mutant ER,[16] human epidermoid carcinoma,[32] human hepatoma HepG2,[33] human prostate cancer LNCaP,[34] and human erythroleukemic K562.[35] The p21 induction resulted in the inhibition of G1/S transition. In contrast, in MDA-MB-231 cells, resveratrol did not affect p21 expression, but it did decrease cyclin

TABLE 26.2
Anticancer Activity of Resveratrol

Anticancer Activity	References
Antimutagenesis	61, 130
Antioxidant, free radical scavenger	84, 131, 132
Antipromotion	36, 133
Antiprogression	61, 134
Apoptosis induction	32, 37, 135, 136
Anti-initiation	80
Inhibition of aryl hydrocarbon receptor	137, 138
Inhibition of COX-2 expression	74, 139
Inhibition of autocrine growth-factor expression	17
Stimulation of TGF expression	18

D1/CDK4 and cyclin B1/cdc2 kinase activity.[30] These results suggest once again that antiproliferative mechanisms of resveratrol are diverse and depend on cell context.

In several cancer cell lines, resveratrol is capable of inducing apoptosis. In breast cancer cells, Pozo-Guisado et al.[30] showed that resveratrol induced apoptosis in MCF-7 but not in MDA-MB-231 cells. Nevertheless, it inhibited cell proliferation in both cell lines. This result suggests that the mechanisms of antiproliferative effect of resveratrol are both apoptotic and nonapoptotic cell death.

Resveratrol induces apoptosis by a p53-dependent pathway in some cell lines such as leukemic cells,[36] hepatocellular carcinoma,[33] prostate cancer cell,[37] epidermal cell,[38] and kidney cells.[39] However, others reported the induction of apoptosis by resveratrol in a p53-independent manner.[40,41]

26.6.2 INHIBITION OF AUTOCRINE GROWTH-FACTOR LOOPS BY RESVERATROL

Autocrine growth factors have been shown to play an important role in the regulation of breast cancer cell growth.[42,42] Moreover, several autocrine growth factors produced by breast cancer cells are also known to be the targets of estrogen regulation. Transforming growth factor α (TGFα) has a growth-promoting effect on breast cancer cells. Its expression is induced by estrogen and inhibited by antiestrogen treatment.[44] Insulin-like growth factor I (IGF-I) and its receptor (IGF-IR) have also been described as targets of estrogen regulation, and their induction or increase of expression are important for E2-mediated proliferative effect in MCF-7 cells.[45] PC-cell-derived growth factor (PCDGF) is a recently characterized autocrine growth factor that mediates the mitogenic activity of estrogen.[46] Its expression is stimulated by estradiol and inhibited by tamoxifen.[47] Reports by Lu and Serrero[17,18] indicated that the expression of TGFα, IGF-IR, and PCDGF were inhibited by resveratrol in MCF-7 cells cultivated in the presence of E2. It is possible that the inhibition of one or several of these autocrine loops represents one mechanism by which resveratrol would inhibit proliferation of breast cancer cells.

26.6.3 STIMULATION OF AUTOCRINE GROWTH INHIBITORS

TGFβs have been shown to act as negative autocrine growth regulators that inhibit the proliferation of breast cancer cells.[48] The TGFβ family includes three isoforms: TGFβ1, TGFβ2, and TGFβ3. Growth stimulation of E2-dependent breast cancer cells is accompanied by the down-regulation of TGFβ2 and TGFβ3 mRNA expression.[49] It has been demonstrated that resveratrol treatment of MCF-7 cells strongly stimulated TGFβ2 mRNA expression in a dose-dependent fashion without any change in the expression of TGFβ1 and TGFβ3.[17] TGFβ2 has been regarded as a negative growth regulator for breast cancer cells[48,50] and a marker for measuring antiestrogenic activity, since its expression is stimulated by antiestrogen. This result would provide an additional mode of action for resveratrol to inhibit cell proliferation by stimulating the expression of breast cancer cells growth inhibitor.

26.6.4 ANTIOXIDANT ACTIVITY

Environmental pollutants (including pesticides, toxic chemical wastes, and cigarette smoke) produce oxygen free radicals, resulting in oxidative damage with lipid peroxidation. Free radicals also activate procarcinogenesis and inhibit cellular and antioxidant defense systems, causing tumor progression and carcinogenesis. A large number of naturally occurring chemicals have been shown to have antioxidant and free-radical scavenging effect. Basly et al.[5] showed that both *cis*- and *trans*-resveratrol are free-radicals scavengers and pro-oxidant compounds.

Polyphenols in red wine inhibit cell proliferation in both hormone-sensitive (MCF-7 and T47D) and hormone-resistant breast cancer cell lines (MDA-MB-231).[26] The author suggested that inhibitory effect was partly due to the antioxidant action of resveratrol.

26.6.5 INDUCTION OF CELL DIFFERENTIATION

Promotion of cellular differentiation is a promising postulated mechanism for chemoprevention. Several compounds have been reported to induce cell differentiation, including retinoids,[51] vitamin D3,[52,53] and genistein.[54,55] Resveratrol is capable of inducting differentiation in some hematological malignant cell lines,[56-58] colon cancers,[59] and neuroblastoma.[60] Jang et al.[61] reported that incubation of the promyelocytic leukemia HL-60 cells with resveratrol induces cell differentiation and reduction of DNA synthesis. However, the mechanism by which resveratrol induces cell differentiation is not yet defined. It has not yet been determined whether resveratrol induces differentiation of breast cancer cells, although a report by Lu and Serrero[17] indicates that this may not be the case for MCF-7 cells.

26.6.6 CYCLOOXYGENASE INHIBITION

Considerable interest has been shown in recent years about the use of cyclooxygenase as a molecular target for chemotherapy and chemoprevention. In fact, a current strategy for chemoprevention is to investigate the potential of agents that inhibit cyclooxygenase (COX). This enzyme is present in two isoforms: COX-1 and COX-2. COX-1 is the constitutive and physiological form, while COX-2 is the inducible form.[62] The significance of COX enzymes comes from the fact that COX-2 is overexpressed in various cancerous tumors.[63-71] Several reports have indicated that inhibition of COX-2 activity *in vivo* resulted in a reduction of both tumor incidence and tumor progression in adenomatous and intestinal polyp model.[72,73] Interestingly, it has been shown that resveratrol inhibited COX expression in several cell types. In murine resident peritoneal macrophages, resveratrol exerted an antioxidant activity and inhibited COX-2 induction stimulated by lipopolysaccharide, PMA, reactive oxygen, or hydrogen peroxide at the transcriptional level.[74] In mammary epithelial cells and oral epithelial cells, resveratrol inhibited PMA-mediated COX-2 induction.[75] This evidence suggests that inhibition of COX-2 activity may be one of the anticancer outcomes of resveratrol treatment.

26.6.7 ANTI-INITIATION ACTIVITY

The chemopreventive effect of resveratrol results from inhibition of the expression of several cytochromes that prevent the tumor initiation process. 4-Hydroxyestradiol (4-OHE2) is known as a more potent carcinogen than its parent compound estradiol (E2).[76] Conversion from E2 to 4-OHE2 in several tissues is catalyzed by a specific enzyme called cytochrome P450 1B1 (CYP1B1). The increased expression of this enzyme was found in benign and malignant breast cancer, whereas it remained low in normal breast cells.[77] Resveratrol has been described as an inhibitor of CYP1B1.[78] The inhibition of the catalytic activity of CYP1B1 by resveratrol would then confer protection against the carcinogenic activity of estrogen metabolites in mammary epithelial cells. Moreover, resveratrol inhibits other cytochrome P450 isoforms such as CYP1A1, CYP2B1, and CYP2E1.[79-81] These enzymes catalyze the metabolic activation of xenobiotics.

26.7 RESVERATROL IN OTHER CANCERS

Resveratrol has been reported to inhibit all three stages of carcinogenesis: initiation, progression, and promotion in several types of cancer. Several reports suggested that resveratrol inhibits cancer initiation in various model systems. Sharma et al. found that resveratrol reduced 12-*O*-tetradecanoylphorbol-13-acetate (TPA)-induced free-radical formation in human promyelocytic leukemia cells HL-60.[82] In another model, resveratrol also inhibited 7,12-dimethylbenz(*a*)anthracene (DMBA)-induced preneoplastic lesion in a mouse mammary gland organ culture system, and it inhibited DMBA-induced mutation in *Salmonella typhimurium* strain TM677, a bacterial cancer initiation model.[83] In another distinct study, resveratrol was suggested to possess anti-initiation activity by inhibiting free-radical formation.[84]

Resveratrol also showed preferential biological activity by inhibiting progression in human promyelocytic leukemia cells. Furthermore, in the same study, resveratrol was able to reduce the incidence of skin tumor formation induced by TPA, a known cancer promoter, in nude mice.[61] Taken together, the evidence suggests that resveratrol has an advantage over other candidate chemopreventive agents by acting at several stages of cancer development.

26.8 EFFECT OF RESVERATROL *IN VIVO*

A limited number of studies have been carried out to examine the efficacy of resveratrol *in vivo*. Daily oral administration of 100-μM resveratrol in drinking water in female rats resulted in ovarian hypertrophy and disruption of the estrous cycle, whereas no behavioral response was observed.[85] Others also confirmed that resveratrol treatment has no effect on increasing uterine weight. Resveratrol has no effect on body weight, serum cholesterol, and bone growth in rat, but it was able to antagonize the effect of estradiol on lowering serum cholesterol.[86] Taken together, resveratrol is a weak estrogen receptor agonist and may act as a mixed agonist/antagonist in rat model *in vivo*.

Concerning the *in vivo* effect of resveratrol on tumor growth, it was shown that resveratrol reduced the incidence and number of tumor foci of *N*-methyl-*N*-nitrosourea-induced tumors in rat model.[22] However, using 4T1 breast cancer cells, Bove et al.[25] reported that although resveratrol inhibited cell proliferation *in vitro*, inhibition of tumor growth was not observed *in vivo*.

26.9 BIOAVAILABILITY OF *TRANS*-RESVERATROL

Resveratrol is mainly absorbed through the small intestine and eliminated from the body in urine as native and conjugated forms, resveratrol 3-*O*-β-glucuronide and resveratrol-3-sulfate.[87] When administered at 25 mg/70 kg body weight in normal male volunteers, resveratrol serum concentration did not reach the therapeutic concentration considered as being active *in vitro* (5 to 50 m*M*).[12] However, it has been reported that the tissue concentration of resveratrol could reach the therapeutic range when regular or moderate amounts of red wine were consumed. This may be due to the fact that the hydrolysis of β-glucuronide-conjugated resveratrol in liver and small intestine may enhance the concentration of tissue resveratrol.[88]

26.10 BENEFITS OF USING RESVERATROL

Menopause is a condition of estrogen deficiency that triggers several symptoms such as hot flash, anxiety, irritability, and bone loss. Synthetic hormones have been extensively used as estrogen replacement therapy. However, many recent studies show the possible adverse effects of estrogen replacement therapy in increasing breast cancer risk, leading to the arrest of current clinical studies.[89] It was suggested that resveratrol could present similar beneficial effects and also offer the advantages of being used as a chemoprevention and estrogen supplement in postmenopausal women.

In support of this possibility, resveratrol shows estrogenlike effect by stimulating proliferation and differentiation of osteoblasts. Mizutani et al.[90] reported that resveratrol increased alkaline phosphatase activity, an osteoblast-specific marker, and DNA synthesis in mouse osteoblastic MC3T3-C1 cells in a dose-dependent fashion. Instead of accelerating bone loss like tamoxifen, resveratrol not only prevented bone loss but also stimulated osteoblast formation. Thus, resveratrol may serve as a beneficial agent in the prevention of and therapy for osteoporosis in postmenopausal women. It is also suggested that resveratrol has cardioprotective effect in isolated ischemic/reperfused rat heart model by its free-radical-scavenging effect.[91,92]

Resveratrol may be beneficial particularly in postmenopausal women for cardioprotection and osteoporosis prevention.[93,94] This effect may be due to the estrogenic effect of resveratrol. Moreover, as shown above (Section 26.6.7) by its ability to modulate activity of cytochrome P450 enzymes, resveratrol would have positive effect as a chemopreventive agent. However, no studies have yet been conducted to examine the possible clinical application of resveratrol for cardioprotection, osteoporosis prevention, or chemoprevention.

26.11 CONCLUSION

The natural product resveratrol has been investigated for its extensive biological activities. Although resveratrol is an ER agonist, its binding activity to ER can protect from more potent forms of estrogen that could have long-term adversarial effects and therefore makes resveratrol an attractive product to use in chemoprevention. In breast cancer model, resveratrol demonstrates antiproliferative effect both *in vitro* and *in vivo* via several pathways. Profiles of overall effects on the growth of individual cell types as well as specific gene activation/inhibition should be further examined to provide definite conclusions about resveratrol mode of action.

It has been reported that resveratrol presented other benefits in other organs. These pharmacological and therapeutic properties strengthen the concept that resveratrol supplements may be considered in future strategies of using natural products in cancer prevention and therapy. However, appropriate long-term clinical trials will have to be conducted before any final conclusions can be made about the potential benefit of resveratrol as a chemopreventive agent.

REFERENCES

1. Kris-Etherton, P.M., Hecker, K.D., Bonanome, A., Coval, S.M., Binkoski, A.E., Hilpert,. F., Griel, A.E., and Etherton, T.D., Bioactive compounds in foods: their role in the prevention of cardiovascular disease and cancer, *Am. J. Med.*, 113 (Suppl. 9B), 71S–88S, 2002.
2. Burns, J., Yokota, T., Ashihara, H., Lean, M.E., and Crozier, A., Plant foods and herbal sources of resveratrol, *J. Agric. Food Chem.*, 50 (11), 3337–3340, 2002.
3. Constant, J., Alcohol, ischemic heart disease, and the French paradox, *Coron. Artery Dis.*, 8 (10), 645–649, 1997.
4. Kimura, Y., Okuda, H., and Arichi, S., Effects of stilbenes on arachidonate metabolism in leukocytes, *Biochim. Biophys. Acta*, 834 (2), 275–278, 1985.
5. Basly, J.P., Marre-Fournier, F., Le Bail, J.C., Habrioux, G., and Chulia, A.J., Estrogenic/antiestrogenic and scavenging properties of (E)- and (Z)-resveratrol, *Life Sci.*, 66 (9), 769–777, 2000.
6. Waffo-Teguo, P., Hawthorne, M.E., Cuendet, M., Merillon, J.M., Kinghorn, A.D., Pezzuto, J.M., and Mehta, R.G., Potential cancer-chemopreventive activities of wine stilbenoids and flavans extracted from grape (*Vitis vinifera*) cell cultures, *Nutr. Cancer*, 40 (2), 173–179, 2001.
7. Waffo-Teguo, P., Lee, D., Cuendet, M., Merillon, J., Pezzuto, J.M., and Kinghorn, A.D., Two new stilbene dimer glucosides from grape (*Vitis vinifera*) cell cultures, *J. Nat. Prod.*, 64 (1), 136–138, 2001.
8. Waffo Teguo, P., Fauconneau, B., Deffieux, G., Huguet, F., Vercauteren, J., and Merillon, J.M., Isolation, identification, and antioxidant activity of three stilbene glucosides newly extracted from *Vitis vinifera* cell cultures, *J. Nat. Prod.*, 61 (5), 655–657, 1998.
9. Soleas, G.J., Diamandis, E.P., and Goldberg, D.M., Resveratrol: a molecule whose time has come? And gone?, *Clin. Biochem.*, 30 (2), 91–113, 1997.
10. Romero-Perez, A.I., Lamuela-Raventos, R.M., Andres-Lacueva, C., and de La Torre-Boronat, M.C., Method for the quantitative extraction of resveratrol and piceid isomers in grape berry skins: effect of powdery mildew on the stilbene content, *J. Agric. Food Chem.*, 49 (1), 210–215, 2001.
11. Wang, Y., Catana, F., Yang, Y., Roderick, R., and van Breemen, R.B., An LC-MS method for analyzing total resveratrol in grape juice, cranberry juice, and in wine, *J. Agric. Food Chem.*, 50 (3), 431–435, 2002.
12. Goldberg, D.M., Yan, J., and Soleas, G.J., Absorption of three wine-related polyphenols in three different matrices by healthy subjects, *Clin. Biochem.*, 36 (1), 79-87, 2003.

13. Bhat, K.P. and Pezzuto, J.M., Resveratrol exhibits cytostatic and antiestrogenic properties with human endometrial adenocarcinoma (Ishikawa) cells, *Cancer Res.*, 61 (16), 6137–6144, 2001.

14. el-Mowafy, A.M., Abou-Zeid, L.A., and Edafiogho, I., Recognition of resveratrol by the human estrogen receptor-alpha: a molecular modeling approach to understand its biological actions, *Med. Princ. Pract.*, 11 (2), 86–92, 2002.

15. Gehm, B.D., McAndrews, J.M., Chien, P.Y., and Jameson, J.L., Resveratrol, a polyphenolic compound found in grapes and wine, is an agonist for the estrogen receptor, *Proc. Natl. Acad. Sci., USA*, 94 (25), 14138–14143, 1997.

16. Levenson, A.S., Gehm, B.D., Pearce, S.T., Horiguchi, J., Simons, L.A., Ward, III J.E., Jameson, J.L., and Jordan, V.C., Resveratrol acts as an estrogen receptor (ER) agonist in breast cancer cells stably transfected with ER alpha, *Int. J. Cancer*, 104 (5), 587–596, 2003.

17. Lu, R. and Serrero, G., Resveratrol, a natural product derived from grape, exhibits antiestrogenic activity and inhibits the growth of human breast cancer cells, *J. Cell. Physiol.*, 179 (3), 297–304, 1999.

18. Serrero, G. and Lu, R., Effect of resveratrol on the expression of autocrine growth modulators in human breast cancer cells, *Antioxid. Redox Signal*, 3 (6), 969–979, 2001.

19. Bowers, J.L., Tyulmenkov, V.V., Jernigan, S.C., and Klinge, C.M., Resveratrol acts as a mixed agonist/antagonist for estrogen receptors alpha and beta, *Endocrinology*, 141 (10), 3657–3667, 2000.

20. Morito, K., Hirose, T., Kinjo, J., Hirakawa, T., Okawa, M., Nohara, T., Ogawa, S., Inoue, S., Muramatsu, M., and Masamune, Y., Interaction of phytoestrogens with estrogen receptors alpha and beta, *Biol. Pharm. Bull.*, 24 (4), 351–356, 2001.

21. Schmitt, E., Lehmann, L., Metzler, M., and Stopper, H., Hormonal and genotoxic activity of resveratrol, *Toxicol. Lett.*, 136 (2), 133–142, 2002.

22. Klinge, C.M., Estrogen receptor interaction with estrogen response elements, *Nucleic Acids Res.*, 29 (14), 2905–2919, 2001.

23. Bhat, K.P., Lantvit, D., Christov, K., Mehta, R.G., Moon, R.C., and Pezzuto, J.M., Estrogenic and antiestrogenic properties of resveratrol in mammary tumor models, *Cancer Res.*, 61 (20), 7456–7463, 2001.

24. Yoon, K., Pallaroni, L., Stoner, M., Gaido, K., and Safe, S., Differential activation of wild-type and variant forms of estrogen receptor alpha by synthetic and natural estrogenic compounds using a promoter containing three estrogen-responsive elements, *J. Steroid Biochem. Mol. Biol.*, 78 (1), 25–32, 2001.

25. Bove, K., Lincoln, D.W., and Tsan, M.F., Effect of resveratrol on growth of 4T1 breast cancer cells *in vitro* and *in vivo*, *Biochem. Biophys. Res. Commun.*, 291 (4), 1001–1005, 2002.

26. Damianaki, A., Bakogeorgou, E., Kampa, M., Notas, G., Hatzoglou, A., Panagiotou, S., Gemetzi, C., Kouroumalis, E., Martin, P.M., and Castanas, E., Potent inhibitory action of red wine polyphenols on human breast cancer cells, *J. Cell. Biochem.*, 78 (3), 429–441, 2000.

27. Hsieh, T.C., Burfeind, P., Laud, K., Backer, J.M., Traganos, F., Darzynkiewicz, Z., and Wu, J.M., Cell cycle effects and control of gene expression by resveratrol in human breast carcinoma cell lines with different metastatic potentials, *Int. J. Oncol.*, 15 (2), 245–252, 1999.

28. Mgbonyebi, O.P., Russo, J., and Russo, I.H., Antiproliferative effect of synthetic resveratrol on human breast epithelial cells, *Int. J. Oncol.*, 12 (4), 865–869, 1998.

29. Nakagawa, H., Kiyozuka, Y., Uemura, Y., Senzaki, H., Shikata, N., Hioki, K., and Tsubura, A., Resveratrol inhibits human breast cancer cell growth and may mitigate the effect of linoleic acid, a potent breast cancer cell stimulator, *J. Cancer Res. Clin. Oncol.*, 127 (4), 258–264, 2001.

30. Pozo-Guisado, E., Alvarez-Barrientos, A., Mulero-Navarro, S., Santiago-Josefat, B., and Fernandez-Salguero, P.M., The antiproliferative activity of resveratrol results in apoptosis in MCF-7 but not in MDA-MB-231 human breast cancer cells: cell-specific alteration of the cell cycle, *Biochem. Pharmacol.*, 64 (9), 1375–1386, 2002.

31. Prall, O.W., Sarcevic, B., Musgrove, E.A., Watts, C.K., and Sutherland, R.L., Estrogen-induced activation of Cdk4 and Cdk2 during G1-S phase progression is accompanied by increased cyclin D1 expression and decreased cyclin-dependent kinase inhibitor association with cyclin E-Cdk2, *J. Biol. Chem.*, 272 (16), 10882–10894, 1997.

32. Ahmad, N., Adhami, V.M., Afaq, F., Feyes, D.K., and Mukhtar, H., Resveratrol causes WAF-1/p21-mediated G(1)-phase arrest of cell cycle and induction of apoptosis in human epidermoid carcinoma A431 cells, *Clin. Cancer Res.*, 7 (5), 1466–1473, 2001.

33. Kuo, P.L., Chiang, L.C., and Lin, C.C., Resveratrol-induced apoptosis is mediated by p53-dependent pathway in Hep G2 cells, *Life Sci.*, 72 (1), 23–34, 2002.

34. Mitchell, S.H., Zhu, W., and Young, C.Y., Resveratrol inhibits the expression and function of the androgen receptor in LNCaP prostate cancer cells, *Cancer Res.*, 59 (23), 5892–5895, 1999.

35. Rodrigue, C.M., Arous, N., Bachir, D., Smith-Ravin, J., Romeo, P.H., Galacteros, F., and Garel, M.C., Resveratrol, a natural dietary phytoalexin, possesses similar properties to hydroxyurea towards erythroid differentiation, *Br. J. Haematol.*, 113 (2), 500–507, 2001.

36. Huang, C., Ma, W.Y., Goranson, A., and Dong, Z., Resveratrol suppresses cell transformation and induces apoptosis through a p53-dependent pathway, *Carcinogenesis*, 20 (2), 237–242, 1999.

37. Lin, H.Y., Shih, A., Davis, F.B., Tang, H.Y., Martino, L.J., Bennett, J.A., and Davis, P.J., Resveratrol-induced serine phosphorylation of p53 causes apoptosis in a mutant p53 prostate cancer cell line, *J. Urol.*, 168 (2), 748–755, 2002.

38. She, Q.B., Bode, A.M., Ma, W.Y., Chen, N.Y., and Dong, Z., Resveratrol-induced activation of p53 and apoptosis is mediated by extracellular-signal-regulated protein kinases and p38 kinase, *Cancer Res.*, 61 (4), 1604–1610, 2001.

39. Shih, A., Lin, H.Y., Davis, F.B., and Davis, P.J., Thyroid hormone promotes serine phosphorylation of p53 by mitogen-activated protein kinase, *Biochemistry*, 40 (9), 2870–2878, 2001.

40. Mahyar-Roemer, M., Katsen, A., Mestres, P., and Roemer, K., Resveratrol induces colon tumor cell apoptosis independently of p53 and precede by epithelial differentiation, mitochondrial proliferation and membrane potential collapse, *Int. J. Cancer*, 94 (5), 615–622, 2001.

41. Soleas, G.J., Goldberg, D.M., Grass, L., Levesque, M., and Diamandis, E.P., Do wine polyphenols modulate p53 gene expression in human cancer cell lines? *Clin. Biochem.*, 34 (5), 415–420, 2001.

42. Yee, D., The insulin-like growth factors and breast cancer ?*? revisited, *Breast Cancer Res. Treat.*, 47 (3), 197–199, 1998.

43. Dickson, R.B. and Lippman, M.E., Growth factors in breast cancer, *Endocr. Rev.*, 16 (5), 559–589, 1995.

44. Dickson, R.B., Thompson, E.W., and Lippman, M.E., Regulation of proliferation, invasion and growth factor synthesis in breast cancer by steroids, *J. Steroid Biochem. Mol. Biol.*, 37 (3), 305–316, 1990.

45. Stewart, A.J., Johnson, M.D., May, F.E., and Westley, B.R., Role of insulin-like growth factors and the type I insulin-like growth factor receptor in the estrogen-stimulated proliferation of human breast cancer cells, *J. Biol. Chem.*, 265 (34), 21172–21178, 1990.

46. Lu, R. and Serrero, G., Inhibition of PC cell-derived growth factor (PCDGF, epithelin/granulin precursor) expression by antisense PCDGF cDNA transfection inhibits tumorigenicity of the human breast carcinoma cell line MDA-MB-468, *Proc. Natl. Acad. Sci. USA* , 97 (8), 3993–3998, 2000.

47. Lu, R. and Serrero, G., Stimulation of PC cell-derived growth factor (epithelin/granulin precursor) expression by estradiol in human breast cancer cells, *Biochem. Biophys. Res. Commun.*, 256 (1), 204–207, 1999.

48. Arteaga, C.L. and Moses, H.L., TGF-beta in mammary development and neoplasia, *J. Mammary Gland Biol. Neoplasia*, 1 (4), 327–329, 1996.

49. Jeng, M.H., ten Dijke, P., Iwata, K.K., and Jordan, V.C., Regulation of the levels of three transforming growth factor beta mRNAs by estrogen and their effects on the proliferation of human breast cancer cells, *Mol. Cell. Endocrinol.*, 97 (1–2), 115–123, 1993.

50. Knabbe, C., Lippman, M.E., Wakefield, L.M., Flanders, K.C., Kasid, A., Derynck, R., and Dickson, R.B., Evidence that transforming growth factor-beta is a hormonally regulated negative growth factor in human breast cancer cells, *Cell*, 48 (3), 417–428, 1987.

51. Paik, J., Blaner, W.S., Sommer, K.M., Moe, R., and Swisshlem, K., Retinoids, retinoic acid receptors, and breast cancer, *Cancer Invest.*, 21 (2), 304–312, 2003.

52. Grunberg, E., Eckert, K., Karsten, U., and Maurer, H.R., Effects of differentiation inducers on cell phenotypes of cultured nontransformed and immortalized mammary epithelial cells: a comparative immunocytochemical analysis, *Tumour Biol.*, 21 (4), 211–223, 2000.

53. Lazzaro, G., Agadir, A., Qing, W., Poria, M., Mehta, R.R., Moriarty, R.M., Das Gupta, T.K., Zhang, X.K., and Mehta, R.G., Induction of differentiation by 1alpha-hydroxyvitamin D(5) in T47D human breast cancer cells and its interaction with vitamin D receptors, *Eur. J. Cancer*, 36 (6), 780–786, 2000.

54. Lamartiniere, C.A., Cotroneo, M.S., Fritz, W.A., Wang, J., Mentor-Marcel, R., and Elgavish, A., Genistein chemoprevention: timing and mechanisms of action in murine mammary and prostate, *J. Nutr.*, 132 (3), 552S–558S, 2002.

55. Cotroneo, M.S., Wang, J., Fritz, W.A., Eltoum, I.E., and Lamartiniere, C.A., Genistein action in the prepubertal mammary gland in a chemoprevention model, *Carcinogenesis*, 23 (9), 1467–1474, 2002.

56. Asou, H., Koshizuka, K., Kyo, T., Takata, N., Kamada, N., and Koeffier, H.P., Resveratrol, a natural product derived from grapes, is a new inducer of differentiation in human myeloid leukemias, *Int. J. Hematol.*, 75 (5), 528–533, 2002.

57. Jang, M. and Pezzuto, J.M., Cancer chemopreventive activity of resveratrol, *Drugs Exp. Clin. Res.*, 25 (2–3), 65–77, 1999.

58. Ragione, F.D., Cucciolla, V., Borriello, A., Pietra, V.D., Racioppi, L., Soldati, G., Manna, C., Galletti, P., and Zappia, V., Resveratrol arrests the cell division cycle at S/G2 phase transition, *Biochem. Biophys. Res. Commun.*, 250 (1), 53–58, 1998.

59. Wolter, F. and Stein, J., Resveratrol enhances the differentiation induced by butyrate in caco-2 colon cancer cells, *J. Nutr.*, 132 (7), 2082–2086, 2002.

60. Melzig, M.F. and Escher, F., Induction of neutral endopeptidase and angiotensin-converting enzyme activity of SK-N-SH cells *in vitro* by quercetin and resveratrol, *Pharmazie*, 57 (8), 556–558, 2002.

61. Jang, M., Cai, L., Udeani, G.O., Slowing, K.V., Thomas, C.F., Beecher, C.W., Fong, H.H., Farnsworth, N.R., Kinghorn, A.D., Mehta, R.G., Moon, R.C., and Pezzuto, J.M., Cancer chemopreventive activity of resveratrol, a natural product derived from grapes, *Science*, 275 (5297), 218–220, 1997.

62. Kiefer, W. and Dannhardt, G., COX-2 inhibition and the control of pain, *Curr. Opin. Investig. Drugs*, 3 (9), 1348–1358, 2002.

63. Karim, M.M., Hayashi, Y., Inoue, M., Imai, Y., Ito, H., and Yamamoto, M., Cox-2 expression in retinoblastoma, *Am. J. Ophthalmol.*, 129 (3), 398–401, 2000.

64. Shono, T., Tofilon, P.J., Bruner, J.M., Owolabi, O., and Lang, F.F., Cyclooxygenase-2 expression in human gliomas: prognostic significance and molecular correlations, *Cancer Res.*, 61 (11), 4375–4381, 2001.

65. Wardlaw, S.A., March, T.H., and Belinsky, S.A., Cyclooxygenase-2 expression is abundant in alveolar type II cells in lung cancer-sensitive mouse strains and in premalignant lesions, *Carcinogenesis*, 21 (7), 1371–1377, 2000.

66. Schlosser, W., Schlosser, S., Ramadani, M., Gansauge, F., Gansauge, S., and Beger, H.G., Cyclooxygenase-2 is overexpressed in chronic pancreatitis, *Pancreas*, 25 (1), 26–30, 2002.

67. Tomozawa, S., Tsuno, N.H., Sunami, E., Hatano, K., Kitayama, J., Osada, T., Saito, S., Tsuruo, T., Shibata, Y., and Nagawa, H., Cyclooxygenase-2 overexpression correlates with tumour recurrence, especially haematogenous metastasis, of colorectal cancer, *Br. J. Cancer*, 83 (3), 324–328, 2000.

68. Khuri, F.R., Wu, H., Lee, J.J., Kemp, B.L., Lotan, R., Lippman, S.M., Feng, L., Hong, W.K., and Xu, X.C., Cyclooxygenase-2 overexpression is a marker of poor prognosis in stage I non-small cell lung cancer, *Clin. Cancer Res.*, 7 (4), 861–867, 2001.

69. Kagoura, M., Toyoda, M., Matsui, C., and Morohashi, M., Immunohistochemical expression of cyclooxygenase-2 in skin cancers, *J. Cutan. Pathol.*, 28 (6), 298–302, 2001.

70. Hosomi, Y., Yokose, T., Hirose, Y., Nakajima, R., Nagai, K., Nishiwaki, Y., and Ochiai, A., Increased cyclooxygenase 2 (COX-2) expression occurs frequently in precursor lesions of human adenocarcinoma of the lung, *Lung Cancer*, 30 (2), 73–81, 2000.

71. Zhang, H. and Sun, X.F., Overexpression of cyclooxygenase-2 correlates with advanced stages of colorectal cancer, *Am. J. Gastroenterol.*, 97 (4), 1037–1041, 2002.

72. Jacoby, R.F., Seibert, K., Cole, C.E., Kelloff, G., and Lubet, R.A., The cyclooxygenase-2 inhibitor celecoxib is a potent preventive and therapeutic agent in the min mouse model of adenomatous polyposis, *Cancer Res.*, 60 (18), 5040–5044, 2000.

73. Oshima, M., Murai, N., Kargman, S., Arguello, M., Luk, P., Kwong, E., Taketo, M.M., and Evans, J.F., Chemoprevention of intestinal polyposis in the Apcdelta716 mouse by rofecoxib, a specific cyclooxygenase-2 inhibitor, *Cancer Res.*, 61 (4), 1733–1740, 2001.

74. Subbaramaiah, K., Chung, W.J., Michaluart, P., Telang, N., Tanabe, T., Inoue, H., Jang, M., Pezzuto, J.M., and Dannenberg, A.J., Resveratrol inhibits cyclooxygenase-2 transcription and activity in phorbol ester-treated human mammary epithelial cells, *J. Biol. Chem.*, 273 (34), 21875–21882, 1998.

75. Martinez, J. and Moreno, J.J., Effect of resveratrol, a natural polyphenolic compound, on reactive oxygen species and prostaglandin production, *Biochem. Pharmacol.*, 59 (7), 865–870, 2000.

76. Liehr, J.G. and Roy, D., Free radical generation by redox cycling of estrogens, *Free Radic. Biol. Med.*, 8 (4), 415–423, 1990.

77. Liehr, J.G. and Ricci, M.J., 4-Hydroxylation of estrogens as marker of human mammary tumors, *Proc. Natl. Acad. Sci. USA*, 93 (8), 3294–3296, 1996.

78. Chang, T.K., Lee, W.B., and Ko, H.H., *trans*-Resveratrol modulates the catalytic activity and mRNA expression of the procarcinogen-activating human cytochrome P450 1B1, *Can. J. Physiol. Pharmacol.*, 78 (11), 874–881, 2000.

79. Ciolino, H.P. and Yeh, G.C., Inhibition of aryl hydrocarbon-induced cytochrome P-450 1A1 enzyme activity and CYP1A1 expression by resveratrol, *Mol. Pharmacol.*, 56 (4), 760–767, 1999.

80. Huynh, H.T. and Teel, R.W., Effects of plant-derived phenols on rat liver cytochrome P450 2B1 activity, *Anticancer Res.*, 22 (3), 1699–1703, 2002.

81. Mikstacka, R., Gnojkowski, J., and Baer-Dubowska, W., Effect of natural phenols on the catalytic activity of cytochrome P450 2E1, *Acta Biochim. Pol.*, 49 (4), 917–925, 2002.

82. Sharma, S., Stutzman, J.D., Kelloff, G.J., and Steele, V.E., Screening of potential chemopreventive agents using biochemical markers of carcinogenesis, *Cancer Res.*, 54 (22), 5848–5855, 1994.

83. Shamon, L.A., Chen, C., Mehta, R.G., Steele, V., Moon, R.C., and Pezzuto, J.M., A correlative approach for the identification of antimutagens that demonstrate chemopreventive activity, *Anticancer Res.*, 14 (5A), 1775–1778, 1994.

84. Fauconneau, B., Waffo-Teguo, P., Huguet, F., Barrier, L., Decendit, A., and Merillon, J.M., Comparative study of radical scavenger and antioxidant properties of phenolic compounds from *Vitis vinifera* cell cultures using *in vitro* tests, *Life Sci.*, 61 (21), 2103–2110, 1997.

85. Henry, L.A. and Witt, D.M., Resveratrol: phytoestrogen effects on reproductive physiology and behavior in female rats, *Horm. Behav.*, 41 (2), 220–228, 2002.

86. Turner, R.T., Evans, G.L., Zhang, M., Maran, A., and Sibonga, J.D., Is resveratrol an estrogen agonist in growing rats? *Endocrinology*, 140 (1), 50–54, 1999.

87. Yu, C., Shin, Y.G., Chow, A., Li, Y., Kosmeder, J.W., Lee, Y.S., Hirschelman, W.H., Pezzuto, J.M., Mehta, R.G., and van Breemen, R.B., Human, rat, and mouse metabolism of resveratrol, *Pharm. Res.*, 19 (12), 1907–1914, 2002.

88. Day, A.J., DuPont, M.S., Ridley, S., Rhodes, M., Rhodes, M.J., Morgan, M.R., and Williamson, G., Deglycosylation of flavonoid and isoflavonoid glycosides by human small intestine and liver beta-glucosidase activity, *FEBS Lett.*, 436 (1), 71–75, 1998.

89. Solomon, C.G. and Dluhy, R.G., Rethinking postmenopausal hormone therapy, *New England J. Med.*, 348 (7), 579–580, 2003.

90. Mizutani, K., Ikeda, K., Kawai, Y., and Yamori, Y., Resveratrol stimulates the proliferation and differentiation of osteoblastic MC3T3-E1 cells, *Biochem. Biophys. Res. Commun.* , 253 (3), 859–863, 1998.

91. Hung, L.M., Chen, J.K., Huang, S.S., Lee, R.S., and Su, M.J., Cardioprotective effect of resveratrol, a natural antioxidant derived from grapes, *Cardiovasc. Res.*, 47 (3), 549–555, 2000.

92. Sato, M., Maulik, G., Bagchi, D., and Das, D.K., Myocardial protection by protykin, a novel extract of *trans*-resveratrol and emodin, *Free Radic. Res.*, 32 (2), 135–144, 2000.

93. Bagchi, D., Das, D.K., Tosaki, A., Bagchi, M., and Kothari, S.C., Benefits of resveratrol in women's health, *Drugs Exp. Clin. Res.*, 27 (5–6), 233–248, 2001.

94. Mizutani, K., Ikeda, K., Kawai, Y., and Yamori, Y., Protective effect of resveratrol on oxidative damage in male and female stroke-prone spontaneously hypertensive rats, *Clin. Exp. Pharmacol. Physiol.*, 28 (1–2), 55–59, 2001.

95. Fitzpatrick, D.F., Hirschfield, S.L., and Coffey, R.G., Endothelium-dependent vasorelaxing activity of wine and other grape products, *Am. J. Physiol.*, 265 (2, Pt. 2), H774–778, 1993.

96. Chen, C.K. and Pace-Asciak, C.R., Vasorelaxing activity of resveratrol and quercetin in isolated rat aorta, *Gen. Pharmacol.*, 27 (2), 363–366, 1996.

97. Bastianetto, S. and Quirion, R., Natural extracts as possible protective agents of brain aging, *Neurobiol. Aging*, 23 (5), 891–897, 2002.

98. Soleas, G.J., Grass, L., Josephy, P.D., Goldberg, D.M., and Diamandis, E.P., A comparison of the anticarcinogenic properties of four red wine polyphenols, *Clin. Biochem.*, 35 (2), 119-14, 2002.

99. Olas, B., Zbikowska, H.M., Wachowicz, B., Krajewski, T., Buczynski, A., and Magnuszewska, A., Inhibitory effect of resveratrol on free radical generation in blood platelets, *Acta Biochim. Pol.*, 46 (4), 961–966, 1999.

100. Olas, B., Wachowicz, B., Szewczuk, J., Saluk-Juszczak, J., and Kaca, W., The effect of resveratrol on the platelet secretory process induced by endotoxin and thrombin, *Microbios*, 105 (410), 7–13, 2001.

101. Orsini, F., Pelizzoni, F., Verotta, L., Aburjai, T., and Rogers, C.B., Isolation, synthesis, and antiplatelet aggregation activity of resveratrol 3-O-beta-D-glucopyranoside and related compounds, *J. Nat. Prod.*, 60 (11), 1082–1087, 1997.

102. Dobrydneva, Y., Williams, R.L., and Blackmore, P.F., *trans*-Resveratrol inhibits calcium influx in thrombin-stimulated human platelets, *Br. J. Pharmacol.*, 128 (1), 149–157, 1999.

103. Pace-Asciak, C.R., Rounova, O., Hahn, S.E., Diamandis, E.P., and Goldberg, D.M., Wines and grape juices as modulators of platelet aggregation in healthy human subjects, *Clin. Chim. Acta*, 246 (1–2), 163–182, 1996.

104. Park, J.W., Choi, Y.J., Jang, M.A., Lee, Y.S., Jun, D.Y., Suh, S.I., Baek, W.K., Suh, M.H., Jin, I.N., and Kwon, T.K., Chemopreventive agent resveratrol, a natural product derived from grapes, reversibly inhibits progression through S and G2 phases of the cell cycle in U937 cells, *Cancer Lett.*, 163 (1), 43–49., 2001.

105. Orallo, F., Alvarez, E., Camina, M., Leiro, J.M., Gomez, E., and Fernandez, P., The possible implication of *trans*-Resveratrol in the cardioprotective effects of long-term moderate wine consumption, *Mol. Pharmacol.*, 61 (2), 294–302, 2002.

106. Hung, L.M., Su, M.J., Chu, W.K., Chiao, C.W., Chan, W.F., and Chen, J.K., The protective effect of resveratrols on ischaemia-reperfusion injuries of rat hearts is correlated with antioxidant efficacy, *Br. J. Pharmacol.*, 135 (7), 1627–1633, 2002.

107. Imamura, G., Bertelli, A.A., Bertelli, A., Otani, H., Maulik, N., and Das, D.K., Pharmacological preconditioning with resveratrol: an insight with iNOS knockout mice, *Am. J. Physiol. Heart Circ. Physiol.*, 282 (6), H1996–2003, 2002.

108. Wu, J.M., Wang, Z.R., Hsieh, T.C., Bruder, J.L., Zou, J.G., and Huang, Y.Z., Mechanism of cardio-protection by resveratrol, a phenolic antioxidant present in red wine (review), *Int. J. Mol. Med.*, 8 (1), 3–17, 2001.

109. Zou, J.G., Wang, Z.R., Huang, Y.Z., Cao, K.J., and Wu, J.M., Effect of red wine and wine polyphenol resveratrol on endothelial function in hypercholesterolemic rabbits, *Int. J. Mol. Med.*, 11 (3), 317–320, 2003.

110. Moreno, J.J., Resveratrol modulates arachidonic acid release, prostaglandin synthesis, and 3T6 fibroblast growth, *J. Pharmacol. Exp. Ther.*, 294 (1), 333–338, 2000.

111. MacCarrone, M., Lorenzon, T., Guerrieri, P., and Agro, A.F., Resveratrol prevents apoptosis in K562 cells by inhibiting lipoxygenase and cyclooxygenase activity, *Eur. J. Biochem.*, 265 (1), 27–34, 1999.

112. Brakenhielm, E., Cao, R., and Cao, Y., Suppression of angiogenesis, tumor growth, and wound healing by resveratrol, a natural compound in red wine and grapes, *FASEB J.*, 15 (10), 1798–1800, 2001.

113. Cao, Y., Cao, R., and Brakenhielm, E., Antiangiogenic mechanisms of diet-derived polyphenols, *J. Nutr. Biochem.*, 13 (7), 380–390, 2002.

114. Igura, K., Ohta, T., Kuroda, Y., and Kaji, K., Resveratrol and quercetin inhibit angiogenesis *in vitro*, *Cancer Lett.*, 171 (1), 11–16, 2001.

115. Kimura, Y. and Okuda, H., Effects of naturally occurring stilbene glucosides from medicinal plants and wine on tumour growth and lung metastasis in Lewis lung carcinoma-bearing mice, *J. Pharm. Pharmacol.*, 52 (10), 1287–1295, 2000.

116. Atten, M.J., Attar, B.M., Milson, T., and Holian, O., Resveratrol-induced inactivation of human gastric adenocarcinoma cells through a protein kinase C-mediated mechanism, *Biochem. Pharmacol.*, 62 (10), 1423–1432, 2001.

117. Garcia-Garcia, J., Micol, V., de Godos, A., and Gomez-Fernandez, J.C., The cancer chemopreventive agent resveratrol is incorporated into model membranes and inhibits protein kinase C alpha activity, *Arch. Biochem. Biophys.*, 372 (2), 382–388, 1999.

118. Fremont, L., Belguendouz, L., and Delpal, S., Antioxidant activity of resveratrol and alcohol-free wine polyphenols related to LDL oxidation and polyunsaturated fatty acids, *Life Sci.*, 64 (26), 2511–2521, 1999.

119. Stewart, J.R., Ward, N.E., Ioannides, C.G., and O'Brian, C.A., Resveratrol preferentially inhibits protein kinase C-catalyzed phosphorylation of a cofactor-independent, arginine-rich protein substrate by a novel mechanism, *Biochemistry*, 38 (40), 13244–13251, 1999.

120. Karlsson, J., Emgard, M., Brundin, P., and Burkitt, M.J., *trans*-Resveratrol protects embryonic mesencephalic cells from tert-butyl hydroperoxide: electron paramagnetic resonance spin trapping evidence for a radical scavenging mechanism, *J. Neurochem.*, 75 (1), 141–150, 2000.

121. Wang, M., Jin, Y., and Ho, C.T., Evaluation of resveratrol derivatives as potential antioxidants and identification of a reaction product of resveratrol and 2, 2-diphenyl-1-picryhydrazyl radical, *J. Agric. Food Chem.*, 47 (10), 3974–3977, 1999.

122. Ashby, J., Tinwell, H., Pennie, W., Brooks, A.N., Lefevre, P.A., Beresford, N., and Sumpter, J.P., Partial and weak oestrogenicity of the red wine constituent resveratrol: consideration of its super-agonist activity in MCF-7 cells and its suggested cardiovascular protective effects, *J. Appl. Toxicol.*, 19 (1), 39–45, 1999.

123. Gao, X., Xu, Y.X., Janakiraman, N., Chapman, R.A., and Gautam, S.C., Immunomodulatory activity of resveratrol: suppression of lymphocyte proliferation, development of cell-mediated cytotoxicity, and cytokine production, *Biochem. Pharmacol.*, 62 (9), 1299–1308, 2001.

124. Chan, M.M., Antimicrobial effect of resveratrol on dermatophytes and bacterial pathogens of the skin, *Biochem. Pharmacol.*, 63 (2), 99–104, 2002.

125. Daroch, F., Hoeneisen, M., Gonzalez, C.L., Kawaguchi, F., Salgado, F., Solar, H., and Garcia, A., *In vitro* antibacterial activity of Chilean red wines against *Helicobacter pylori*, *Microbios*, 104 (408), 79–85, 2001.

126. Matsuda, H., Kageura, T., Morikawa, T., Toguchida, I., Harima, S., and Yoshikawa, M., Effects of stilbene constituents from rhubarb on nitric oxide production in lipopolysaccharide-activated macrophages, *Bioorg. Med. Chem. Lett.*, 10 (4), 323–327, 2000.

127. Tsai, S.H., Lin-Shiau, S.Y., and Lin, J.K., Suppression of nitric oxide synthase and the down-regulation of the activation of NFkappaB in macrophages by resveratrol, *Br. J. Pharmacol.*, 126 (3), 673–680, 1999.

128. Palmieri, L., Mameli, M., and Ronca, G., Effect of resveratrol and some other natural compounds on tyrosine kinase activity and on cytolysis, *Drugs Exp. Clin. Res.*, 25 (2–3), 79–85, 1999.

129. Jayatilake, G.S., Jayasuriya, H., Lee, E.S., Koonchanok, N.M., Geahlen, R.L., Ashendel, C.L., McLaughlin, J.L., and Chang, C.J., Kinase inhibitors from *Polygonum cuspidatum*, *J. Nat. Prod.*, 56 (10), 1805–1810, 1993.

130. Uenobe, F., Nakamura, S., and Miyazawa, M., Antimutagenic effect of resveratrol against Trp-P-1, *Mutat. Res.*, 373 (2), 197–200, 1997.

131. De Salvia, R., Festa, F., Ricordy, R., Perticone, P., and Cozzi, R., Resveratrol affects in a different way primary versus fixed DNA damage induced by H(2)O(2) in mammalian cells *in vitro*, *Toxicol. Lett.*, 135 (1–2), 1–9, 2002.

132. Ferry-Dumazet, H., Garnier, O., Mamani-Matsuda, M., Vercauteren, J., Belloc, F., Billiard, C., Dupouy, M., Thiolat, D., Kolb, J.P., Marit, G., Reiffers, J., and Mossalayi, M.D., Resveratrol inhibits the growth and induces the apoptosis of both normal and leukemic hematopoietic cells, *Carcinogenesis*, 23 (8), 1327–1333, 2002.

133. Nielsen, M., Ruch, R.J., and Vang, O., Resveratrol reverses tumor-promoter-induced inhibition of gap-junctional intercellular communication, *Biochem. Biophys. Res. Commun.*, 275 (3), 804–809, 2000.

134. Schneider, Y., Vincent, F., Duranton, B., Badolo, L., Gosse, F., Bergmann, C., Seiler, N., and Raul, F., Anti-proliferative effect of resveratrol, a natural component of grapes and wine, on human colonic cancer cells, *Cancer Lett.*, 158 (1), 85–91, 2000.

135. Clement, M.V., Hirpara, J.L., Chawdhury, S.H., and Pervaiz, S., Chemopreventive agent resveratrol, a natural product derived from grapes, triggers CD95 signaling-dependent apoptosis in human tumor cells, *Blood*, 92 (3), 996–1002, 1998.

136. Dorrie, J., Gerauer, H., Wachter, Y., and Zunino, S.J., Resveratrol induces extensive apoptosis by depolarizing mitochondrial membranes and activating caspase-9 in acute lymphoblastic leukemia cells, *Cancer Res.*, 61 (12), 4731–4739, 2001.

137. Ciolino, H.P., Daschner, P.J., and Yeh, G.C., Resveratrol inhibits transcription of CYP1A1 *in vitro* by preventing activation of the aryl hydrocarbon receptor, *Cancer Res.*, 58 (24), 5707–5712, 1998.

138. Revel, A., Raanani, H., Younglai, E., Xu, J., Han, R., Savouret, J.F., and Casper, R.F., Resveratrol, a natural aryl hydrocarbon receptor antagonist, protects sperm from DNA damage and apoptosis caused by benzo(a)pyrene, *Reprod. Toxicol.*, 15 (5), 479–486, 2001.

139. Mutoh, M., Takahashi, M., Fukuda, K., Matsushima-Hibiya, Y., Mutoh, H., Sugimura, T., and Wakabayashi, K., Suppression of cyclooxygenase-2 promoter-dependent transcriptional activity in colon cancer cells by chemopreventive agents with a resorcin-type structure, *Carcinogenesis*, 21 (5), 959–963, 2000.

27 Berries and Fruits in Cancer Chemoprevention*

Ronald L. Prior and Jim Joseph

CONTENTS

27.1 Introduction..465
27.2 Epidemiology of Fruit and Vegetable Consumption and Cancer......................466
27.3 Major Phytochemicals in Berries..467
27.4 Studies of Fruit Phytochemicals on Carcinoma Cell Lines
     or Other *In Vitro* Systems...469
     27.4.1 Berry Extracts..469
     27.4.2 Proanthocyanidins..470
     27.4.3 Anthocyanins 471
     27.4.4 Protocatechuic Acid...472
     27.4.5 Quercetin..472
     27.4.6 Resveratrol...472
     27.4.7 Salicylates 473
27.5 Fruits and Berries and Cancer Prevention in Animal Models 473
27.6 Summary and Conclusions...474
References 475

27.1 INTRODUCTION

There is an abundance of data showing that there are significant associations among oxidative stress, inflammation, and cancer. It appears, for example, that oxidative stress-induced DNA damage may be a major contributor to colorectal[1] and other gastric cancers.[2] Additional evidence suggests that prostatic carcinogenesis involves reductions in glutathione S-transferase, which increases the vulnerability of the prostatic cells to carcinogens.[3] In a similar vein, Halliwell[4] has postulated that agents that increase DNA damage increase the risk of cancer development. He reviews agents that can contribute to DNA damage, and all can increase oxidative stress and or inflammation (e.g., cigarette smoke). Additional reviews have implicated the involvement of inflammation in ovarian cancer, since inflammation is induced during the process of ovulation (Ness et al., 1999).

Plants, including food plants (fruits and vegetables), synthesize a vast array of chemical compounds that are not involved in their primary metabolism. These "secondary compounds" instead serve a variety of ecological functions, ultimately, to enhance the plant's survivability. Interestingly, these compounds also may be responsible for the multitude of beneficial effects of fruits and vegetables on an array of health-related bioactivities, two of the most important of which

---

* Mention of a trade name, proprietary product or specific equipment does not constitute a guarantee by the U.S. Department of Agriculture and does not imply its approval to the exclusion of other products that may be suitable.

0-8493-1560-3/05/$0.00+$1.50
© 2005 by CRC Press LLC

may be their antioxidant and anti-inflammatory properties. Recent evidence also indicates that fruit- and vegetable-derived flavonoids may alter additional factors known to be important in carcino-genesis, including signaling, cell-cycle regulation, and angiogenesis.[5]

If this is the case, then it could be postulated that the inclusion of an abundance of fruits and vegetables in the diet may significantly reduce the incidence of several forms of cancer. The purpose of the present chapter is to briefly review the epidemiology of diet and cancer occurrence and to discuss the mechanisms involved in the beneficial effects.

27.2 EPIDEMIOLOGY OF FRUIT AND VEGETABLE CONSUMPTION AND CANCER

Fruits and vegetables are associated with lower incidence and lower mortality rates of cancer in several human cohort and case-control studies for all common cancer sites.[6–13] Individuals with low fruit and vegetable intake experienced about twice the risk of cancer compared with those with high intake. A statistically significant protective effect of fruit and vegetable consumption has been found in 128 of 156 dietary studies in which results were expressed in terms of relative risk. More recent studies continue to confirm these earlier findings, although there are some studies not supportive of this conclusion.

Although some previous studies suggested that colorectal cancer risk decreases with higher intake of vegetables, fruits, and grains, few studies have examined these factors in relation to occurrence of colorectal polyps. In case-control data from 488 matched pairs in which consumption of vegetables, fruits, and grains in the diet was measured with a food frequency questionnaire in the year before sigmoidoscopy, subjects (aged 50 to 74 years) with frequent consumption of vegetables, fruits, and grains had a decreased polyp prevalence. Specifically, the adjusted odds ratio comparing the highest with the lowest quintile of intake for vegetables was 0.47 (95% confidence interval [CI], 0.29 to 0.76), for fruits was 0.65 (95% CI, 0.40 to 1.05), and for grains was 0.55 (95% CI, 0.33 to 0.91). The authors also found inverse associations for high-carotenoid vegetables, cruciferae, high vitamin C fruits, garlic, and tofu (or soybeans). After further adjusting for potentially anticarcinogenic constituents of these foods, high-carotenoid vegetables, cruciferous vegetables, garlic, and tofu (or soybeans) remained inversely associated with polyps. These findings support the hypothesis that high intake of vegetables, fruits, or grains decreases the risk of polyps and suggest that any protective effects might reflect unmeasured constituents in these foods.[14]

A group of 119 patients with ulcerative colitis, colorectal polyps or previous carcinoma, and age- and sex-matched controls were assessed for fruit and vegetable intake.[15] Patients in all groups consumed 12.8% less energy than the controls (P < 0.02) and 28.9% less fruit and vegetables (P < 0.0001). Patients with neoplastic disease consumed 21% less fruit and vegetables than controls (n = 60; P < 0.01). In this group of patients at increased risk of colorectal cancer, subjects selected diets that contained significantly less fruit and vegetables than control subjects with no symptoms of neoplastic disease.[15] In contrast to the preponderance of observations indicating that fruits and vegetables may confer protection against some chronic diseases, Michels et al.[16] found that their frequent consumption did not appear to confer protection from colon or rectal cancer.

Franceschi et al.[17] studied the effects of 26 types or groups of vegetables and fruit on the risk of cancer using data from two case-control studies that included 1225 cases of cancer of the colon, 728 cases of cancer of the rectum, 2569 cases of cancer of the breast, and 5155 hospital controls interviewed between 1991 and 1996 in six Italian areas. Consumption of most vegetables was inversely associated with cancer of the colon and rectum. High fruit intake was associated only with a reduction of rectal cancer.[17]

In the prospective Kuopio ischemic heart disease risk factor (KIHD) study, the risk of all-cause and non-CVD-related deaths was studied in 2641 men.[18] During a mean follow-up time of 12.8 yr, noncardiovascular and all-cause mortality were lower among men with the highest consumption of fruits, berries, and vegetables. The relative risk for men in the highest fifth of fruit, berry, and vegetable intake for all-cause death and non-CVD-related death was 0.66 and 0.68, respectively, compared with men in the lowest fifth. These data indicate that a high fruit, berry, and vegetable intake is associated with reduced risk of mortality in middle-aged Finnish men.[18]

Serafini et al.[19] investigated whether the total dietary antioxidant potential of fruit and vegetables was an appropriate means of estimating the antioxidant impact on gastric cancer risk in a large population-based study. Intake of antioxidant equivalents was inversely associated with the risk of both cardia and distal gastric cancer (odds ratio [OR], 0.65; 95% confidence interval [CI], 0.48 to 0.89 for the highest quartile of TRAP). Controlling for smoking, the inverse relationship between TRAP values and cancer displayed a clearer dose–response pattern. Individuals who had never smoked with the highest antioxidant intake had the lowest risk of cancer, 0.44 (95% CI, 0.27 to 0.71). The results suggested that dietary intake of antioxidants measured as total antioxidant potential is inversely associated with risk of both cardia and distal gastric cancer.[19]

Other studies have demonstrated reduced risk of cancer that is specifically related to dietary fruit intake. In a case-controlled study in Uruguay, fruits were associated with strong reductions in the risk for esophageal cancer.[20] Twelve of 15 dietary antioxidants displayed significant inverse associations with esophageal cancer risk. The strongest effect was observed for high intake of β-cryptoxanthin, but α-carotene, lycopene, and β-sitosterol were also associated with significant reductions in risk.[20] In the Welsh Caerphilly study,[21] vegetable and fruit consumption was, independently from other risk factors, inversely related to mortality from cancer of the digestive tract, mainly due to an inverse association with fruit consumption (RR for the highest quartile versus the lowest was 0.3; 95% CI, 0.1 to 0.8). Vegetable and fruit consumption were also inversely related to all-cause cancer mortality, and the strongest association was observed for fruit consumption (RR in the highest versus lowest quartile was 0.5; 95% CI, 0.3 to 1.0). These investigators concluded that consumption of vegetables and particularly the consumption of fruit could lower the risk of dying from cancer in middle-aged men.[21]

Increased childhood fruit intake was associated with reduced risk of incident cancer. In fully adjusted logistic regression models, odds ratios (95% confidence intervals) with increasing quartiles of fruit consumption were 1.0 (reference), 0.66 (0.48 to 0.90), 0.70 (0.51 to 0.97), 0.62 (0.43 to 0.90); p value for linear trend = 0.02. The association was weaker for cancer mortality. There was no clear pattern of association between the other dietary factors and total cancer risk. Childhood fruit consumption may have a long-term protective effect on cancer risk in adults.[22] Increasing evidence has shown that exposure to these nutrients in early infancy may have long-terms effects.[23]

Thus, the scientific evidence regarding a role for vegetable and fruit consumption in cancer prevention is generally consistent and is supportive of current dietary recommendations of increased consumption of fruit and vegetables. However, what is not clear from the available literature is what phytochemicals are responsible for the anticarcinogenicity and whether specific fruits, vegetables, or cereal grains might be more effective than others in preventing age-related diseases.

27.3 MAJOR PHYTOCHEMICALS IN BERRIES

Fruits, and in particular berries, tend to contain the highest antioxidant capacity (Table 27.1).[24] A typical serving of berries might contain 4,000 to 13,000 μmol Trolox equivalents/100 g fresh weight, whereas many of the other fruits and vegetables may contain in the range of 100 to 2000 μmol Trolox equivalents/100 g fresh weight, as determined by the ORAC (oxygen radical absorbance capacity) procedure. Total antioxidant capacity tends to relate to total phenolics, at least within the fruits and berries (Table 27.1). The two major classes of polyphenolics or flavonoids that contribute in a major way to the measured total antioxidant capacity include proanthocyanidins (Table 27.2) and anthocyanins.[25] Anthocyanins are common colored plant flavonoids, occurring as glycosides of the respective anthocyanidin chromophores, and may

TABLE 27.1
Total Antioxidant Capacity of Selected Fruits and Berries

Fruit/Berry	TAC[a] (μ mol TE/g)	TP[b] (mg GAE/g)	Serving Size[c] (g)	TAC per serving (μ mol TE)
Blueberry, wild	92.6	8.0	145 (1c)	13427
Blueberry, cultivated	62.2	5.3	145 (1c)	9019
Cranberries	94.6	7.1	95 (1c whole)	8983
Blackberries	53.5	6.6	144 (1c)	7703
Raspberries	49.3	5.0	123 (1c)	6058
Strawberries	35.8	3.7	149 (1c)	5330
Cherries	33.6	3.4	145 (1c)	4873
Black plums	73.4	4.8	66 (1 fruit)	4844
Plums	62.4	3.7	66 (1 fruit)	4118

[a] Total antioxidant capacity (TAC): sum of hydrophilic and lipophilic ORAC expressed as μ mol Trolox equivalents per gram of fresh weight.

[b] Total phenolics (TP) expressed as milligrams of gallic acid equivalents per gram of fresh weight.

[c] Serving size comes from USDA handbook *Composition in Foods*.

Source: Wu, X. et al., *J. Agric. Food Chem.* 52, 4026–4037, 2004.

represent substantial constituents of the human diet. Anthocyanins may be in a similar range of about 20 to 700+ mg/100 g fresh weight,[25,26] with blueberry and black raspberry being particularly high.

Based upon our survey of common foods,[27] 39 foods were found to contain proanthocyanidins. Of these foods, 46% were berries or fruits. Total proanthocyanidins may range from about 20 to over 400 mg/100 g fresh weight in fruits and vegetables. The composition of the proanthocyanidins is quite diverse in that fruits and berries contain both oligomeric (DP † 10) and polymeric compounds (DP > 10), and their structures contain B-type as well as A-type linkages (Table 27.1). Berries like blueberry, black currant, green grapes, and red grapes contain greater than 70% of proanthocyanidins as polymers (DP > 10), while others have little or no polymers (raspberries, cherries, marionberries, blackberries). Thus, it is not clear from the data currently available from *in vivo* or *in vitro* studies whether these diverse proanthocyanidins will have similar biological effects. Most likely, this will not be the case. Because good analytical methods have not been available until recently,[28] it is not clear in most studies exactly what proanthocyanidins are being studied. Because the polymers and probably the higher oligomers will not be absorbed intact, *in vitro* cell culture studies of these compounds are suspect unless the studies involve a cell associated with the gastrointestinal tract, where cells will be exposed to these high-molecular-weight compounds.

Although these are the major polyphenolics in berries, it may or may not be these particular components that are responsible for some of the health effects attributed to berries. In our present scientific environment, it is common to take the reductionist point of view, where researchers study single compounds as new therapeutic agents, but in reality, it may be that the end result is based on hundreds of distinct molecules. Although this idea presents a huge challenge scientifically, it may not be as far-fetched today as it once seemed.[29] In the following section, we will discuss some of the literature that indicates possible involvement of at least the anthocyanins and pro-anthocyanidins in prevention of cancers.

TABLE 27.2
Composition of Proanthocyanidins in Various Fruits and Berries

Fruit/Berry	1-3 mers (%)	4-10 mers (%)	>10 mers (%)	>10 mers (mg/100 g)	Total (mg/100 g)
Grape seed (dried)	38.7	30.1	31.1	1100.1 – 86.3	3532.3 – 106
Cranberries	12.4	31.8	55.8	233.5 – 49.1	418.8 – 75.3
Plums, Black Diamond®	21.9	41.2	36.9	94.6 – 8.7	256.6 – 18.7
Plums, black	15.8	35.6	48.5	115.3 – 2.0	237.9 – 3.1
Plums	30.9	42.5	26.5	57.3 – 24.4	215.9 – 50.7
Blueberries	9.2	19.0	71.7	129.0 – 47.3	179.8 – 50.8
Black currants	4.6	13.9	82.8	122.4 – 28.0	147.8 – 33.0
Strawberries	11.9	35.9	52.3	75.8 – 13.4	145.0 – 24.9
Apples, five varieties	23.3	45.0	29.9	31.1 – 11.9	104.0 – 28.8
Apple sauce	23.9	40.3	35.8	16.9 – 0.4	47.2 – 0.6
Raspberries	71.5	28.5	0	n.d.	30.2 – 23.4
Grapes, green	6.4	21.6	72.2	58.9 – 14.2	81.5 – 15.0
Grapes, red	7.0	20.2	73.1	44.6 – 9.9	61.0 – 12.3
Peaches	24.8	42.5	32.7	22.0 – 7.7	67.3 – 20.9
Pears, green cultivars	4.7	27.0	57.2	24.2 – 15.3	42.3 – 18.6
Pears	24.5	34.8	41.1	13.1 – 11.3	31.9 – 7.8
Blackberries	51.9	42.5	5.6	1.5 – 0.0	27.0 – 17.5
Marion berries	75.3	24.7	0	n.d.	8.9 – 0.1
Nectarines	25.8	42.1	32.0	7.3 – 6.5	22.8 – 14.6
Cherries	53.8	46.2	0	n.d.	18.2 – 3.2

Note: The number of mers is defined as the number of monomeric units in a particular oligomer.

Source: Gu, L. et al., *J. Nutr.*, 2003 (submitted). Gu, L. et al., *J. Nutr.* 134, 613–617, 2004.

27.4 STUDIES OF FRUIT PHYTOCHEMICALS ON CARCINOMA CELL LINES OR OTHER *IN VITRO* SYSTEMS

Cell culture systems have been utilized to study the effects of fruit phytochemicals on the cancer process. Although useful as a potential screening tool, care must be taken in interpreting the data. Some of the potential drawbacks to this approach are that concentrations of materials above that which might be observed *in vitro* must be used, and in many cases one cannot be certain that the form of the compound applied to the cells *in vitro* is the same as what might be delivered to the cells *in vivo*. The exceptions may be the tissues of the gastrointestinal tract, where concentration of phytochemicals to which the cells lining the gut might be exposed may be several orders of magnitude higher than what might be observed in the blood or other body tissues. Both berry extracts as well as purified compounds found in berries have been studied in cell culture and other *in vitro* systems.

27.4.1 BERRY EXTRACTS

Raspberries are particularly rich in phenolic phytochemicals. Proliferation of HepG(2) human liver cancer cells was significantly inhibited in a dose-dependent manner after exposure to raspberry extracts.[30] Yoshizawa et al.[31] screened 43 small-fruit juices and found that four of them (*Actinidia polygama* Maxim., *Rosa rugosa* Thunb., *Vaccinium smallii* A. Gray, and *Sorbus sambucifolia* Roem) strongly inhibited the proliferation of all cancer cell lines examined and yet were substantially less cytotoxic toward normal human cell lines. Bilberry extract was found to be the most effective at inhibiting the growth of HL60 human leukemia cells and HCT116 human colon carcinoma cells *in vitro*.[32] Bilberry extract induced apoptotic cell bodies and nucleosomal DNA fragmentation in HL60 cells. The proportion of apoptotic cells induced by bilberry extract in HCT116 was much lower than that in HL60 cells, and DNA fragmentation was not induced in the former. Of the several berry extracts tested, that from bilberry contained the largest amounts of phenolic compounds, including anthocyanins, and showed the greatest radical scavenging activity using 1,1-diphenyl-2-picrylhydrazyl (DPPH).[32] Pure delphinidin and malvidin, like the glycosides isolated from the bilberry extract, induced apoptosis in HL60 cells. These results indicate that the bilberry extract and the anthocyanins, bearing delphinidin or malvidin as the aglycone, inhibit the growth of HL60 cells through the induction of apoptosis. Only pure delphinidin and the delphinidin glycoside isolated from the bilberry extract, but not malvidin and its glycoside, inhibited the growth of HCT116 cells.[32]

Crude extracts of anthocyanin and proanthocyanidin fractions from fruit extracts of four Vaccinium species (low-bush blueberry, bilberry, cranberry, and lingonberry) were not highly active in quinone reductase (QR) induction, whereas the ethyl acetate extracts were active QR inducers. Further fractionation of the bilberry ethyl acetate extract revealed that the majority of inducer potency was contained in a hexane/chloroform subfraction. In contrast to their effects on QR, crude extracts of low-bush blueberry, cranberry, and lingonberry were active inhibitors of ornithine decarboxylase (ODC) activity. The greatest activity in these extracts appeared to be contained in the polymeric proanthocyanidin fractions of the low-bush blueberry, cranberry, and lingonberry fruits. Thus, components in the hexane/chloroform fraction of bilberry and of the proanthocyanidin fraction of low-bush blueberry, cranberry, and lingonberry exhibit potential anticarcinogenic activity, as evaluated by *in vitro* screening tests.[33]

Extracts from freeze-dried strawberries and black raspberries, along with ellagic acid, were analyzed for antitransformation activity in a Syrian hamster embryo (SHE) cell transformation model. None of the extracts or ellagic acid by themselves produced an increase in morphological transformation. For assessment of chemopreventive activity, SHE cells were treated with each agent and benzo[*a*]pyrene (BaP) for 7 days. Ellagic acid and two methanol fractions produced a dose-dependent decrease in transformation compared with BaP treatment only, suggesting that a methanol extract from strawberries and black raspberries may display chemopreventive activity. The possible mechanism by which these methanol fractions inhibited cell transformation appeared to involve interference of uptake, activation, detoxification of B[a]P, and intervention of DNA binding and DNA repair.[34]

Roy et al.[35] tested the effects of multiple berry extracts on angiogenesis. Vascular endothelial growth factor (VEGF) is known to play a crucial role in the vascularization of tumors, which in adult skin remains normally quiescent. However, skin retains the capacity for brisk initiation of angiogenesis during skin cancers. Six berry extracts (wild blueberry, bilberry, cranberry, elderberry, raspberry seed, and strawberry), but not GSPE (grape seed proanthocyanidin extract), significantly inhibited both H_2O_2 as well as TNFα-induced VEGF expression by the human HaCaT keratinocytes.[35] In a hemangioma mouse model of *in vivo* angiogenesis, wild blueberry and a berry mix significantly inhibited inducible monocyte chemotactic protein 1 (MCP1) expression in endothelioma cells.[36] MCP1 is known to be responsible for recruiting macrophages to sites of infection or inflammation and facilitate angiogenesis. Endothelioma cells pretreated with berry powders showed diminished ability to form hemangioma. Histological analysis demonstrated markedly decreased infiltration of macrophages in hemangioma of treated mice compared with placebo-treated controls.[36]

27.4.2 Proanthocyanidins

Proanthocyanidin fractions from low-bush blueberry, cranberry, and lingonberry were effective in inhibiting the tumor promoter 12-*O*-tetradecanoylphorbol-13-acetate (TPA) induction of ODC activity.[33] However, differences likely exist between the individual proanthocyanidins as compounds, with doubly linked units showing negligible cytotoxic effects. Ye et al.[37] demonstrated that GSPE exhibited cytotoxicity toward some cancer cells, while enhancing the growth and viability of the normal cells that were examined. Cytotoxic effects of GSPE were observed on the MCF-7 breast cancer, A-427 lung cancer, and gastric adenocarcinoma cells, while no cytotoxicity toward the neoplastic K562 myelogenous leukemic cells was observed. However, GSPE enhanced the growth and viability of the normal human gastric mucosal cells and J774A.1 murine macrophage cells.[37]

Treatment of Caco-2 cells with 50 µg/ml of procyanidin-enriched (PE) extracts from cocoa caused a 70% growth inhibition with a blockade of the cell cycle at the G2/M phase. PE extracts caused a significant decrease of ODC and S-adenosylmethionine decarboxylase activities, two key enzymes of polyamine biosynthesis, which might be an important target in the antiproliferative effects of cocoa polyphenols.[38]

In cell culture experiments, gallotannin treatment for 3 days inhibited the growth of T-84 cells (a human colon cancer line), with a concentration resulting in half-maximal inhibition estimated to be 20 µg/ml. The treatment was not cytotoxic to cells at 1 to 40 µg/ml. Interestingly, at 10 µg/ml, gallotannin induced apoptosis in T-84 cells. Collectively, these findings support a potential role for tannins as chemopreventive agents against colon cancer.[39]

Ruan et al.[40] isolated three compounds, identified as β-sitosterol-[stigmasta-5,22-dien-3-ol] and β-sitosterol-3-β-D-glucose, from freeze-dried powder of blackberries, that demonstrated anticancer activity. Using the Ames assay, Tate et al.[41] found eight different varieties of blackberries that strongly suppressed 2-amino anthracene mutagenesis but had minimal effect on methyl methanesulfonate mutagenesis. However, the varieties seem to vary greatly in their ability to inhibit mutagenesis induced by ultraviolet radiation.[41]

27.4.3 ANTHOCYANINS

Like other flavonoids, anthocyanins are also expected to have antioxidant and antimutagenic properties *in vivo*, although only limited data are available. From the published data, it appears that anthocyanins can be effective *in vivo* as antioxidants when included in the diet at 1 or 2 g/kg diet in animal experiments. These levels in the diet provide 20 to 40 mg per day, levels that are much higher on a body-weight basis than found in the typical diet of humans.[25,42]

Using purified or synthesized flavonoids, Kamei[43] studied the effect of the chemical structure of 2-benzoic flavone on the growth of human tumor cells (HCT-15) *in vitro*. The type of sugar combined with the A phenolic ring of flavonoids played an important role in the tumor suppression; i.e., glucose attachment at the A phenolic ring caused suppression of tumor cell growth, but other sugars at that position, such as rhamnose and rutinoside, did not suppress the growth of the cells. Flavonoids with OH groups conjugated to the 3′, 4′, and 5′ of the B phenolic ring had stronger antitumor effects than those with the OH groups attached at the 3′ and 4′ or 4′ only, although anthocyanins were generally more effective than the other flavonoids.

Pool-Zobel et al.[44] studied anthocyanin fractions or concentrates from *Aronia melanocarpa* Elliot, from elderberry, macqui, and tintorera fruits, as well as pure compounds in human colon tumor cells. H_2O_2-induced DNA strand breaks were reduced in cells treated with the complex plant extracts. In contrast, endogenous generation of oxidized DNA bases was not prevented. The intracellular steady state of oxidized DNA bases was not altered by anthocyanins or anthocyanidins. These authors questioned the anticancer potential of anthocyanidins within specific tissues, such as the colon.[44] However, Katsube et al.[32] showed that not only the aglycones but also the glycosides of anthocyanins inhibited cancer cell growth and induced apoptosis in cancer cells. The strongest inhibitory effects were observed with delphinidin- and malvidin-glycosides on HL-60 human leukemia cells. HCT116 human colon carcinoma cells were less sensitive to induction of apoptosis by anthocyanins than HL60 cells. However, the concentrations of the anthocyanins in the gut will be much greater than in the plasma or other tissues. Maximum plasma concentrations of anthocyanins in human subjects have been reported to be in the 1- to 120-nM range,[42] whereas concentrations of 20 to 200 μM were used in the studies by Katsube et al.[32]

27.4.4 PROTOCATECHUIC ACID

Protocatechuic acid (PCA, 3,4-dihydroxybenzoic acid) is a simple phenolic acid with strong antioxidant capacity that is present in some fruits, vegetables, and nuts. In the rat, PCA has been detected as a metabolite of dietary anthocyanins[45]; however, we have not detected PCA in human plasma or urine samples following anthocyanin consumption.[46,47] PCA has been shown to be efficacious in reducing the carcinogenicity of 4-nitroquinoline-1-oxide in the oral cavity, N-methyl-N-nitrosourea in the glandular stomach, azoxymethane (AOM) in colon, and diethylnitrosamine in liver.[48–50] Dietary PCA administration at 0.5 and 1.0 g/kg diet during the initiation or postinitiation phase significantly inhibited intestinal carcinogenesis induced by AOM, as revealed by the reduction of tumor incidence and multiplicity. PCA also inhibited bromodeoxyuridine labeling index and exerted a pronounced inhibitory effect on the colonic ODC levels.[48] Feeding of PCA during initiation or postinitiation phases significantly decreased the development of tongue neoplasms (squamous cell papilloma and carcinoma) and preneoplasia (hyperplasia and dysplasia), and decreased the labeling index of bromodeoxyuridine and the number and area of silver-stained nucleolar organizer regions per cell nucleus, known as cell proliferation indices, of the tongue squamous epithelium. In addition, PCA exposure during either initiation or postinitiation phase decreased polyamine levels in the oral mucosa. These results clearly indicated that PCA inhibited rat oral carcinogenesis in both initiation and postinitiation phases when administered together with or following treatment with 4-NQO.[50]

Tseng et al.[51] isolated PCA from *Hibiscus sabdariffa* L. and evaluated it for its ability to inhibit the 12-O-tetradecanoylphorbol-13-acetate (TPA)-induced promotion of skin tumors in female CD-1 mice. Topical application of PCA to mice 5 min prior to TPA treatment twice weekly for 20 weeks, which were initiated with benzo[a]pyrene (BaP), inhibited the incidence of tumors in mice by 56 to 81%. The number of tumors in mice pretreated with PCA was two to four, and that of mice treated only with TPA was 6.6. Protective effects of PCA were also evident by its significant suppression of the TPA-induced hyperplasia in the skin and edema of mouse ears by 65 and 73% at doses of 10 and 20 μmol, respectively. When applied to the dorsal surface of CD-1 mice before TPA application, PCA (5, 10, or 20 μmol) inhibited the induction of epidermal ODC and myeloperoxidase activity by 6.5 nmol TPA. The same doses of PCA also reduced the formation of hydrogen peroxide in mouse skin when compared with that of the TPA-treated group. These results indicate that PCA possesses potential as a cancer chemopreventive agent against tumor promotion.[51]

27.4.5 QUERCETIN

Quercetin is a flavonoid that is found in a large number of fruits and vegetables, and it may be the most abundant flavonoid in the human diet.[52] Quercetin has been found to inhibit growth in the human breast carcinoma cell line MCF-7 by inhibiting cell-cycle progression through transient M-phase accumulation and subsequent G2 arrest, and by inducing apoptosis.[53] Duthie and Dobson[54] found that quercetin and related flavonoids, myricetin, kaempferol, and rutin, protected against oxidative DNA damage in human colonocytes *in vitro*. Quercetin and myricetin, but not kaempferol and rutin, protected Caco-2 cells against oxidative attack from hydrogen peroxide, which causes an increased DNA strand breakage. In addition, quercetin decreased hydrogen peroxide-mediated inhibition of growth.[54]

27.4.6 RESVERATROL

Resveratrol is a natural polyphenolic compound produced by a number of plants and is found in high concentrations in peanuts, seeds, grapes, and some berries. Epidemiological studies strongly suggest that resveratrol may act as a cancer chemopreventive compound. Resveratrol strongly inhibits cell proliferation at the µmolar range in a time- and dose-dependent manner. Resveratrol appears to block the cell cycle at the G2/M transition, while there is an increase in the cell number in the S phase. During this inhibition process, resveratrol increases the content of cyclins A and B1 as well as cyclin-dependent kinases CDK1 and CDK2. Moreover, resveratrol promotes CDK1 phosphorylation. Resveratrol appears to exert a strong inhibition of SW480 human colorectal tumor cell proliferation by modulating cyclin and cyclin-dependent kinase activities.[55] In other studies, resveratrol at 10 and 100 μM induced significant dose-dependent inhibition in cell growth as well as in DNA synthesis in human oral squamous carcinoma cells (SCC-25). Quercetin exhibited a biphasic effect, with stimulation at 1 and 10 μM and a slight inhibition at 100 μM in cell growth and DNA synthesis. Combining 50 μM of resveratrol with 10, 25, and 50 μM of quercetin resulted in a gradual and significant increase in the inhibitory effect of quercetin on cell growth and DNA synthesis. These investigators have suggested that resveratrol or a combination of resveratrol and quercetin can act as cancer chemopreventive agents in the oral squamous carcinoma cell model.[56]

27.4.7 SALICYLATES

Paterson[57] has proposed that natural salicylates contribute to the recognized benefits of a healthy fruit and vegetable diet. Salicylic acid is responsible for the anti-inflammatory action of aspirin, and it may cause the reduced risk of colorectal cancer observed in those who take aspirin. Salicylic acid and other salicylates occur naturally in fruits and plants, and diets rich in these are believed to reduce the risk of colorectal cancer. Serum salicylic acid concentrations are greater in vegetarians than nonvegetarians, and there is overlap between concentrations in vegetarians and those taking low-dose aspirin. However, Janssen et al.[58] concluded that the content of (acetyl)salicylic acid in diets may be too low to affect disease risk. Additional work is needed to confirm this concept.

27.5 FRUITS AND BERRIES AND CANCER PREVENTION IN ANIMAL MODELS

Fruits or berries have been studied in five animal cancer models. Black raspberries and strawberries are the primary whole foods tested for anticancer effects, although there are other studies ongoing that have not been published. Grape seed extract has also been studied. Protective effects have been observed in the colon, esophagus, and skin models, but not with mammary or lung cancer models. There was no effect of feeding grape seed proanthocyanidins on 7,12-dimethylbenz[a]anthracene-induced rat mammary tumorigenesis nor on the liver carcinogen-metabolizing enzymes cytochrome P450 1A and glutathione S-transferase.[59] The lack of action of grape seed proanthocyanidins on mammary tumorigenesis in part may be due to lack of effect of the liver carcinogen-metabolizing enzymes CYP450 1A and glutathione S-transferase. Also, lyophilized strawberries in the diet (10%) failed to inhibit 4-(methylnitrosamino)-1-(3-pyridyl)-1-butanone (NNK)- and benzo[a]pyrene (BaP)-induced lung tumorigenesis in the A/J mouse.[60]

Lyophilized black raspberries (BR) at 2.5, 5, or 10% (w/w) of the diet have been shown to inhibit several measures of AOM-induced colon carcinogenesis and modulate an important marker of oxidative stress in the Fischer 344 rat.[61] Aberrant crypt foci (ACF) multiplicity decreased 36, 24, and 21% in the 2.5, 5, and 10% BR groups, respectively, relative to the AOM-only group after 9 weeks on the diet. Total tumor multiplicity declined 42, 45, and 71%, respectively, after 33 weeks on the diet. Although not significant, a decrease in tumor burden (28, 42, and 75%) was observed in all BR groups. Adenocarcinoma multiplicity decreased by 28, 35, and 80% in the same treatment groups. Urinary 8-OHdG levels were reduced by 73, 81, and 83%. In studies of esophageal tumors in the Fisher 344 rat, dietary BR (5% of diet) significantly reduced tumor incidence and multiplicity, proliferation, and preneoplastic lesion development.[62] The hamster cheek pouch was used by Casto et al.[63] to evaluate the ability of black raspberries to inhibit oral cavity tumors. Male Syrian golden hamsters, 3 to 4 weeks of age, were fed 5 and 10% lyophilized BR in the diet for two weeks prior to treatment with 0.2% 7,12-dimethylbenz(*a*)anthracene in dimethylsulfoxide and for 10 weeks thereafter. BR had chemopreventive activity in inhibiting tumor formation in the oral cavity.[63] Strawberries in the diet have also been shown to effectively inhibit NMBA-induced tumorigenesis in the rat esophagus.[64] However, the inhibitory effect of the berries could not be attributed solely to the content of ellagic acid in the berries.[65]

Feeding female rats diets containing 0.1 to 1.0% grape seed proanthocyanidins was associated with a significant 72 to 88% inhibition of AOM-induced aberrant crypt foci formation and a 20 to 56% inhibition of ODC activity in the distal third of the colon. Feeding the grape proanthocyanidins resulted in no significant effect on the activity of liver cytochrome P450 2E1.[59] Gali-Muhtasib et al.[39] have shown that naturally occurring tannins possess antitumor promotion activity in a mouse skin and, more recently, that they protect against colon tumors. Mice were given tannins by intraperitoneal injections, by gavage, or in drinking water before treatment with 1,2-dimethylhydrazine (DMH) for 24 weeks. Alternatively, mice were given tannins by intraperitoneal injection or gavage for only 2 weeks before DMH administration, then tannin administration was discontinued, and mice were treated with DMH for 24 weeks. The multiplicity, size, and distribution of ACF and tumors were significantly inhibited by gallotannin and red alder bark in the above treatment regimens. The most effective treatments included gallotannin given by gavage, red alder bark extract by intraperitoneal injection, and either tannin dissolved in drinking water. Extent of inhibition of ACF and tumors was gender independent. The fact that both methods of administering the treatments were effective adds credence to the data, as we do not know what components are absorbed with many of the tannins. Further, these are generally quite large molecules, and they likely are not absorbed intact, so the material that is injected likely is not the same as what might be absorbed following ingestion.

Zhao et al.[66] assessed the antitumor-promoting effect of a polyphenolic fraction isolated from grape seeds (GSE) employing the 7,12-dimethylbenz[*a*]anthracene (DMBA)-initiated and 12-*O*-tetradecanoylphorbol-13-acetate (TPA)-promoted SENCAR mouse skin two-stage carcinogenesis protocol as a model system. Following tumor initiation with DMBA, topical application of GSE at doses of 0.5 and 1.5 mg/mouse/application to the dorsal-initiated mouse skin resulted in a highly significant inhibition of TPA tumor promotion. The observed antitumor-promoting effects of GSE were dose dependent and were evident in terms of a reduction in tumor incidence (35 and 60% inhibition), tumor multiplicity (61 and 83% inhibition), and tumor volume (67 and 87% inhibition) at both 0.5 and 1.5 mg GSE, respectively. Of the compounds studied in the GSE, procyanidin B5-3'-gallate showed the most potent antioxidant activity in an epidermal lipid peroxidation assay and may be one of the active anticancer components.[66]

Carlsen et al.[67] suggested that antioxidant-rich berries were able to modulate oxidative stress-related gene expression, and this might be important in the chemopreventive effects of fruits and vegetables. Berry feeding (blueberries or blackberries) was found to induce μ-glutamylcysteine synthetase heavy subunit promoter activity in brain and skeletal muscle and also to increase total glutathione in some, but not all, muscle tissues of mice. However, liver activity was decreased. μ-Glutamylcysteine synthetase is the rate-limiting enzyme in glutathione production, which is a primary antioxidant. Ellagic acid, a phenolic found in many berries, produced a similar pattern of response.

27.6 SUMMARY AND CONCLUSIONS

Increased consumption of fruits and vegetables clearly is related to decreased cancer risk. However, whether the protection can be related to particular fruits or vegetables or whether it is related to overall intake of antioxidants is not clear. Diets rich in fruits and vegetables contain a complex mixture of antioxidants (including ascorbate, carotenoids, vitamin E, and numerous phenolics), but the diet may also contain pro-oxidants, including iron, copper, H_2O_2, heme, lipid peroxides, and aldehydes. Nitrite is frequently present in the diet, leading to generation of reactive nitrogen species in the stomach. In considering the biological importance of dietary antioxidants, attention has usually focused on those that are absorbed through the gastrointestinal tract into the rest of the body. However, the high levels of antioxidants present in certain foods may play an important role in protecting the gastrointestinal tract itself from oxidative damage, and in delaying the development of stomach, colon, and rectal cancer. Since many of the antioxidant phytochemicals may not be as readily absorbed as other nutrients, their concentrations can be much higher in the lumen of the gastrointestinal (GI) tract than are ever achieved in plasma or other body tissues, making antioxidant action in the GI tract the likely mechanism of action.[68]

Indeed, the protective effects that have been observed with berries (black raspberries and strawberries, primarily) and grape seed extracts have been with animal models of various cancers of the GI tract. Studies of cancers of the breast and lung have not been protective with berries or grape seed extracts. However, one must realize that there are relatively few studies with fruits and berries in the numerous cancer models.

The particular components in berries that may have anticancer effects are numerous, but positive results have been observed with anthocyanins, proanthocyanidins, protocatechuic acid, quercetin, resveratrol, ellagic acid, salicylates, and others. Again, the protective effects are likely not due to one particular compound but to the joint effects of several.

Finally, it should also be mentioned that since cancer incidence rises exponentially with age in humans and other mammals (e.g.,[69–71]), and since cancer is intimately associated with oxidative stress and inflammation (see above), then in addition to the above factors, fruits and berries may be able to reduce the incidence of cancer by reducing age-related increases in vulnerability to these insults. It is well known, for example, that CNS (central nervous system) vulnerability to oxidative stress (for review, see[72,73]) and inflammation[74] increases during aging, and dietary antioxidant/anti-inflammatory agents may reduce these increases in sensitivity. In this regard, the literature is replete with studies in which a large variety of dietary agents have been employed to alter behavioral and neuronal deficits with aging. These studies have included such nutritional supplements as vitamins C or E, garlic,[75] herbals (e.g., ginseng, *Ginkgo biloba*, Ding lang; see Cantuti-Castelvetri[76]), and dietary fatty acids (reviewed in Youdim and Joseph[75]). Additional studies have suggested that berry fruit both retards and reverses[77,78] both behavioral and CNS changes that occur during aging. If this is the case, then it may be that dietary fruits and berries may also reduce cancer incidence by altering the deleterious effects of aging. Although the data are limited, fruits and berries may be protective through antioxidant mechanisms in preventing DNA damage, but they may also affect cell division, apoptosis, and angiogenesis.

REFERENCES

1. Seril, D.N., Liao, J., Yang, G.Y., and Yang, C.S., Oxidative stress and ulcerative colitis-associated carcinogenesis: studies in humans and animal models, *Carcinogenesis*, 24, 353–362, 2003.
2. Ernst, P., Review article: the role of inflammation in the pathogenesis of gastric cancer, *Aliment Pharmacol. Ther.*, 1, 13–18, 1999.
3. Nelson, W.G., DeWeese, T.L., and DeMarzo, A.M., The diet, prostate inflammation, and the development of prostate cancer, *Cancer Metastasis Rev.*, 21, 3–16, 2002.
4. Halliwell, B., Effect of diet on cancer development: is oxidative DNA damage a biomarker? *Free Radical Biol. Med.*, 32, 968–974, 2002.
5. Le Marchand, L., Cancer preventive effects of flavonoids—a review, *Biomed. Pharmacother.*, 56, 296–301, 2002.
6. Ames, B.M., Shigena, M.K., Hagen, T.M., Oxidants, antioxidants and the degenerative diseases of aging, *Proc. Natl. Acad. Sci.*, 7915–7922, 1993.
7. Doll, R., An overview of the epidemiological evidence linking diet and cancer, *Proc. Nutr.*, 49, 119–131, 1990.

8. Dragsted, L.O., Strube, M., and Larsen, J.C., Cancer-protective factors in fruits and vegetables: biochemical and biological background, *Pharmacol. Toxicol.*, 72, 116–135, 1993.

9. Willett, W.C., Micronutrients and cancer risk, *Am. J. Clin. Nutr.*, 59, 162S–165S, 1994.

10. Steinmetz, K.A. and Potter, J.D., Vegetables, fruit, and cancer, I: epidemiology, *Cancer Causes Control*, 2, 325–357, 1991.

11. Steinmetz, K.A. and Potter, J.D., Vegetables, fruit, and cancer, II: mechanisms, *Cancer Causes Control*, 2, 427–442, 1991.

12. Steinmetz, K.A. and Potter, J.D., Vegetables, fruit, and cancer prevention: a review, *J. Am. Dietetic Assoc.*, 96, 1027–1039, 1996.

13. Block, G., Patterson, B., and Subar, A., Fruit, vegetables, and cancer prevention: a review of the epidemiological evidence, *Nutr. Cancer*, 18, 1–29, 1992.

14. Witte, J.S., Longnecker, M.P., Bird, C.L., Lee, E.R., Frankl, H.D., and Haile, R.W., Relation of vegetable, fruit, and grain consumption to colorectal adenomatous polyps, *Am. J. Epidemiol.*, 144, 1015–1025, 1996.

15. Matthew, J.A., Fellows, I.W., Prior, A., Kennedy, H.J., Bobbin, R., and Johnson, I.T., Habitual intake of fruits and vegetables amongst patients at increased risk of colorectal neoplasia, *Cancer Lett.*, 114, 255–258, 1997.

16. Michels, K.B., Edward, G., Joshipura, K.J., Rosner, B.A. et al., Prospective study of fruit and vegetable consumption and incidence of colon and rectal cancers, *J. Natl. Cancer Inst.*, 92, 1740–1752, 2000; [erratum], *J. Natl. Cancer Inst.*, 93 (11), 879, 2001.

17. Franceschi, S., Parpinel, M., La Vecchia, C., Favero, A., Talamini, R., and Negri, E., Role of different types of vegetables and fruit in the prevention of cancer of the colon, rectum, and breast, *Epidemiology*, 9, 338–341, 1998.

18. Rissanen, T.H., Voutilainen, S., Virtanen, J.K., Venho, B. et al., Low intake of fruits, berries and vegetables is associated with excess mortality in men: the Kuopio ischaemic heart disease risk factor (KIHD) study, *J. Nutr.*, 133, 199–204, 2003.

19. Serafini, M., Bellocco, R., Wolk, A., and Ekstrom, A.M., Total antioxidant potential of fruit and vegetables and risk of gastric cancer, *Gastroenterology*, 123, 985–991, 2002.

20. De Stefani, E., Brennan, P., Boffetta, P., Ronco, A.L., Mendilaharsu, M., and Deneo-Pellegrini, H., Vegetables, fruits, related dietary antioxidants, and risk of squamous cell carcinoma of the esophagus: a case-control study in Uruguay, *Nutr. Cancer*, 38, 23–29, 2000.

21. Hertog, M.G., Bueno-de-Mesquita, H.B., Fehily, A.M., Sweetnam, P.M., Elwood, P.C., and Kromhout, D., Fruit and vegetable consumption and cancer mortality in the Caerphilly study, *Cancer Epidemiol., Biomarkers Prev.*, 5, 673–677, 1996.

22. Maynard, M., Gunnell, D., Emmett, P.M., Frankel, S., and Davey Smith, G., Fruit, vegetables, and antioxidants in childhood and risk of adult cancer: the Boyd Orr cohort, *J. Epidemiol. Community Health*, 57, 218–225, 2003.

23. Lucas, A., Programming by early nutrition: an experimental approach, *J. Nutr.*, 128, 401S–406S, 1998.

24. Wu, X., Beecher, G., Holden, J., Haytowitz, D., Gebhardt, S.E., and Prior, R.L., Lipophilic and hydrophilic antioxidant capacities of common foods in the U.S., *J. Agric. Food Chem.*, 52, 4026–4037, 2004.

25. Prior, R.L., Absorption and metabolism of anthocyanins: potential health effects, in *Phytochemicals: Necgabusns to Mechanisms*, Meskin, M., Ed., CRC Press, Boca Raton, FL, 2003.

26. Macheix, J., Fleuriet, A., and Billot, J., *Fruit Phenolics*, CRC Press, Boca Raton, FL, 1990.

27. Gu, L., Kelm, M.A., Hammerstone, J.F., Holden, J. et al., Development of a Database of Procyanidin Profile and Content in Foods, presented at 5th International Food Data Conference and 27th U.S. National Nutrient Databank Conference, Washington, DC, 2003.

28. Gu, L., Kelm, M., Hammerstone, J.F., Beecher, G.R. et al., Fractionation of polymeric procyanidins from lowbush blueberry and quantification of procyanidins in selected foods with an optimized normal-phase HPLC-MS fluorescent detection method, *J. Agric. Food Chem.*, 50, 4852–4860, 2002.

29. Longtin, R., The pomegranate: nature's power fruit? *JNCI Cancer Spectrum*, 95, 346–348, 2003.

30. Liu, M., Li, X.Q., Weber, C., Lee, C.Y., Brown, J., and Liu, R.H., Antioxidant and antiproliferative activities of raspberries, *J. Agricultural Food Chem.*, 50, 2926–2930, 2002.

31. Yoshizawa, Y., Kawaii, S., Urashima, M., Fukase, T. et al., Antiproliferative effects of small fruit juices on several cancer cell lines, *Anticancer Res.*, 20, 4285–4289, 2000.

32. Katsube, N., Iwashita, K., Tsushida, T., Yamaki, K., and Kobori, M., Induction of apoptosis in cancer cells by bilberry (*Vaccinium myrtillus*) and the anthocyanins, *J. Agricultural Food Chem.*, 51, 68–75, 2003.

33. Bomser, J., *In vitro* anticancer activity of fruit extracts from vaccinium species, *Planta Medica*, 62, 212–216, 1996.

34. Xue, H., Aziz, R.M., Sun, N., Cassady, J.M. et al., Inhibition of cellular transformation by berry extracts, *Carcinogenesis*, 22, 351–356, 2001; [erratum], *Carcinogenesis*, 22 (5), 831–833, 2001.

35. Roy, S., Khanna, S., Alessio, H.M., Vider, J. et al., Anti-angiogenic property of edible berries, *Free Radical Res.*, 36, 1023–1031.

36. Atalay, M., Gordillo, G., Roy, S., Rovin, B. et al., Anti-angiogenic property of edible berry in a model of hemangioma, *FEBS Lett.*, 544, 252–257, 2003.

37. Ye, X., Krohn, R.L., Liu, W., Joshi, S.S. et al., The cytotoxic effects of a novel IH636 grape seed proanthocyanidin extract on cultured human cancer cells, *Molecular Cell. Biochem.*, 196, 99–108, 1999.

38. Carnesecchi, S., Schneider, Y., Lazarus, S.A., Coehlo, D., Gosse, F., and Raul, F., Flavanols and procyanidins of cocoa and chocolate inhibit growth and polyamine biosynthesis of human colonic cancer cells, *Cancer Lett.*, 175, 147–155, 2002.

39. Gali-Muhtasib, H.U., Younes, I.H., Karchesy, J.J., and El-Sabban, M.E., Plant tannins inhibit the induction of aberrant crypt foci and colonic tumors by 1,2-dimethylhydrazine in mice, *Nutr. Cancer*, 39, 108–116, 2001.

40. Ruan, J., Zhao, X., Cassady, J.M., and Stoner, G.D., Study on the constituents from freeze-dried power of blackberries (*Rubus ursinus*), *Zhong Yao Cai*, 24, 645–647, 2001.

41. Tate, P., Kuzmar, A., Smith, S.W., Wedge, D.E., and Larcom, L.L., Comparative effects of eight varieties of blackberry on mutagenesis, *Nutr. Res.*, 2003, (in press).

42. Prior, R.L., Fruits and vegetables in the prevention of cellular oxidative damage, *Am. J. Clin. Nutr.*, 78, 5705–5785, 2003.

43. Kamei, H., Influence of OH group and sugar bonded to flavonoids on flavonoid-mediated suppression of tumor growth *in vitro*, *Cancer Biother. Radiopharm.*, 11, 247–249, 1996.

44. Pool-Zobel, B.L., Bub, A., Schroder, N., and Rechkemmer, G., Anthocyanins are potent antioxidants in model systems but do not reduce endogenous oxidative DNA damage in human colon cells, *Eur. J. Nutr.*, 38, 227–234, 1999.

45. Tsuda, T., Horio, F., and Osawa, T., The role of anthocyanins as an antioxidant under oxidative stress in rats, *Biofactors*, 13, 133–139, 2000.

46. Cao, G., Muccitelli, H.U., Sanchez-Moreno, C., and Prior, R.L., Anthocyanins are absorbed in glycated forms in elderly women: a pharmacokinetic study, *Am. J. Clin. Nutr.*, 73, 920–926, 2001.

47. Wu, X., Cao, G., and Prior, R.L., Absorption and metabolism of anthocyanins in human subjects following consumption of elderberry or blueberry, *J. Nutr.*, 132, 1865–1871, 2002.

48. Tanaka, T., Kojima, T., Suzui, M., and Mori, H., Chemoprevention of colon carcinogenesis by the natural product of a simple phenolic compound protocatechuic acid: suppressing effects on tumor development and biomarkers expression of colon tumorigenesis, *Cancer Res.*, 53, 3908–3913, 1993.

49. Tanaka, T., Kojima, T., Kawamori, T., and Mori, H., Chemoprevention of digestive organs carcinogenesis by natural product protocatechuic acid, *Cancer*, 75, 1433–1439, 1995.

50. Tanaka, T., Kawamori, T., Ohnishi, M., Okamoto, K., Mori, H., and Hara, A., Chemoprevention of 4-nitroquinoline 1-oxide-induced oral carcinogenesis by dietary protocatechuic acid during initiation and postinitiation phases, *Cancer Res.*, 54, 2359–2365, 1994.

51. Tseng, T.H., Hsu, J.D., Lo, M.H., Chu, C.Y. et al., Inhibitory effect of hibiscus protocatechuic acid on tumor promotion in mouse skin, *Cancer Lett.*, 126, 199–207, 1998.

52. Hollman, P.C. and Katan, M.B., Absorption, metabolism and health effects of dietary flavonoids in man, *Biomed. Pharmacother.*, 51, 305–310, 1997.

53. Choi, J.A., Kim, J.Y., Lee, J.Y., Kang, C.M. et al., Induction of cell cycle arrest and apoptosis in human breast cancer cells by quercetin, *Int. J. Oncol.*, 19, 837–844, 2001.

54. Duthie, S.J. and Dobson, V.L. Dietary flavonoids protect human colonocyte DNA from oxidative attack *in vitro*, *Eur. J. Nutr.*, 38, 28–34, 1999.

55. Delmas, D., Passilly-Degrace, P., Jannin, B., Malki, M.C., and Latruffe, N., Resveratrol, a chemopreventive agent, disrupts the cell cycle control of human SW480 colorectal tumor cells, *Int. J. Molecular Med.*, 10, 193–199, 2002.

56. Elattar, T.M. and Virji, A.S., Modulating effect of resveratrol and quercetin on oral cancer cell growth and proliferation, *Anti-Cancer Drugs*, 10, 187–193, 1999.

57. Paterson, J.R. and Lawrence, J.R., Salicylic acid: a link between aspirin, diet and the prevention of colorectal cancer, *Quart. J. Med.*, 94, 445–448, 2001.

58. Janssen, P.L., Katan, M.B., van Staveren, W.A., Hollman, P.C., and Venema, D.P., Acetylsalicylate and salicylates in foods, *Cancer Lett.*, 114, 163–164, 1997.

59. Singletary, K.W. and Meline, B., Effect of grape seed proanthocyanidins on colon aberrant crypts and breast tumors in a rat dual-organ tumor model, *Nutr. Cancer*, 39, 252–258, 2001.

60. Carlton, P.S., Kresty, L.A., and Stoner, G.D., Failure of dietary lyophilized strawberries to inhibit 4-(methylnitrosamino)-1-(3-pyridyl)-1-butanone- and benzo[a]pyrene-induced lung tumorigenesis in strain A/J mice, *Cancer Lett.*, 159, 113–117, 2000.

61. Harris, G.K., Gupta, A., Nines, R.G., Kresty, L.A. et al., Effects of lyophilized black raspberries on azoxymethane-induced colon cancer and 8-hydroxy-2′-deoxyguanosine levels in the Fischer 344 rat, *Nutr. Cancer*, 40, 125–133, 2001.

62. Kresty, L.A., Morse, M.A., Morgan, C., Carlton, P.S. et al., Chemoprevention of esophageal tumorigenesis by dietary administration of lyophilized black raspberries, *Cancer Res.*, 61, 6112–6119, 2001.

63. Casto, B.C., Kresty, L.A., Kraly, C.L., Pearl, D.K. et al., Chemoprevention of oral cancer by black raspberries, *Anticancer Res.*, 22, 4005–4015, 2002.

64. Carlton, P.S., Kresty, L.A., Siglin, J.C., Morse, M.A. et al., Inhibition of N-nitrosomethylbenzylamine-induced tumorigenesis in the rat esophagus by dietary freeze-dried strawberries, *Carcinogenesis*, 22, 441–446, 2001.

65. Stoner, G.D., Kresty, L.A., Carlton, P.S., Siglin, J.C., and Morse, M.A., Isothiocyanates and freeze-dried strawberries as inhibitors of esophageal cancer, *Toxicological Sci.*, 52, 95–100, 1999.

66. Zhao, J., Wang, J., Chen, Y., and Agarwal, R., Anti-tumor-promoting activity of a polyphenolic fraction isolated from grape seeds in the mouse skin two-stage initiation-promotion protocol and identification of procyanidin B5-3′-gallate as the most effective antioxidant constituent, *Carcinogenesis*, 20, 1737–1745, 1999.

67. Carlsen, H., Myhrstad, M.C.W., Thoresen, M., Moskaug, J.O., and Blomhoff, R., Berry intake increases the activity of the glutamylcysteine synthetase promoter in transgenic reporter mice, *J. Nutr.*, 133, 2137–2140, 2003.

68. Halliwell, B., Zhao, K., and Whiteman, M., The gastrointestinal tract: a major site of antioxidant action? *Free Radical Res.*, 33, 819–830, 2000.

69. Krtolica, A. and Campisi, J., Integrating epithelial cancer, aging stroma and cellular senescence, *Adv. Gerontol.*, 11, 109–116, 2003.

70. Rabeneck, L., Davila, J.A., and El-Serag, H.B., Is there a true "shift" to the right colon in the incidence of colorectal cancer? *Am. J. Gastroenterol.*, 98 (6), 1400–1409.

71. Benz, C.C. and Thor, A.D., Understanding the age dependency of breast cancer biomarkers, *Adv. Gerontol.*, 11, 117–120, 2003.

72. Joseph, J.A., Denisova, N., Fisher, D., Bickford, P., Prior, R., and Cao, G., Age-related neurodegeneration and oxidative stress: putative nutritional intervention, *Neurologic Clinics*, 16, 747–755, 1998.

73. Joseph, J.A., Denisova, N., Fisher, D., Shukitt-Hale, B. et al., Membrane and receptor modifications of oxidative stress vulnerability in aging: nutritional considerations, *Ann. N.Y. Acad. Sci.*, 854, 268–276, 1998.

74. Chang, H.N., Wang, S.R., Chiang, S.C., Teng, W.J. et al., The relationship of aging to endotoxin shock and to production of TNFα, *J. Gerontol.*, 51, M220-M222, 1996.

75. Youdim, K.A. and Joseph, J.A., A possible emerging role of phytochemicals in improving age-related neurological dysfunctions: a multiplicity of effects, *Free Radical Biol. Med.*, 30, 583–594, 2001.

76. Cantuti-Castelvetri, I., Neurobehavioral aspects of antioxidants in aging, *Int. J. Developmental Neurosci.*, 18, 367–381, 2000.

77. Joseph, J.A., Shukitt-Hale, B., Denisova, N.A., Prior, R.L. et al., Long-term dietary strawberry, spinach, or vitamin E supplementation retards the onset of age-related neuronal signal-transduction and cognitive behavioral deficits, *J. Neurosci.*, 18, 8047–8055, 1998.

78. Joseph, J.A., Shukitt-Hale, B., Denisova, N.A., Bielinski, D. et al., Reversals of age-related declines in neuronal signal transduction, cognitive and motor behavior deficits with blueberry, spinach, or strawberry dietary supplementation, *J. Neurosci.*, 19, 8114–8121, 1999.
79. Gu, L., Kelm, M.A., Hammerstone, J.F., Beecher, G. et al., Concentrations of oligomeric and polymeric of flavan-3-ols (proanthocyanidins) in common and infant foods and estimation of normal consumption, *J. Nutr.*, 134, 613–617, 2004.

28 Palm Tocotrienols and Cancer

Kalanithi Nesaretnam

CONTENTS

28.1 Introduction..481
28.2 Vitamin E Content of Palm Oil 482
28.3 Extraction of Palm Vitamin E 482
28.4 Vitamin E Activity of Tocotrienols 483
28.5 Metabolic Fate of Tocotrienols 484
28.6 Antioxidant Properties of Tocotrienols 485
28.7 Anticancer Properties of Tocotrienols...486
28.8 Conclusion 489
References 489

28.1 INTRODUCTION

Vitamin E is a generic term that refers to an entire class of compounds that is further divided into two subgroups called tocopherols and tocotrienols.[1] Tocopherols and tocotrienols have the same basic chemical structure characterized by a long phytyl chain attached at the 1-position of a chromane ring structure (Figure 28.1). However, tocopherols have a saturated phytyl chain, whereas that of tocotrienols is unsaturated. Each subgroup of vitamin E contains several isoforms, and individual tocopherol and tocotrienol isoforms differ from each other on the basis of the number of methyl groups bound to their chromane ring. Specifically, α-tocopherol and α-tocotrienol contain methyl groups at the 5-, 7-, and 8-positions, whereas their δ-isoforms lack the methyl group at the 5-position, and their δ-isomers lack methyl groups at the 5- and 7-positions on the chromane ring. This difference in the phytyl chain saturation or chromane ring methylation may be critical in determining the physiological activity of individual tocopherol and tocotrienol isoforms.

Vitamin E in palm oil occurs as a complex mixture of tocopherols and tocotrienols. Palm oils containing the various tocotrienol isomers of vitamin E have been suggested to have interesting biological and physiological properties that are not generally evident with tocopherol-rich vitamin E preparations. These include potential blood cholesterol lowering effects, more efficient antioxidant activity in biological systems, and possible anticancer properties, all of which have been reported.[2]

Vegetable oils, especially the seed oils, are rich sources of tocopherols. Refining waste from the edible oil industry has emerged as an important raw material for the extraction of vitamin E, and the residues of the soybean refining industry are one such source. Tocotrienols, on the other hand, are found predominantly in palm oil and in cereals such as barley and rice bran oils. With the emergence of palm oil as the second largest edible oil in world markets, technological advances have been made to similarly extract tocotrienol-rich palm vitamin E from the refinery wastes of the industry. The resultant product, often termed palm tocotrienol-rich fraction (TRF), is currently available on a commercial scale from Malaysia. A similar tocotrienol-rich vitamin E preparation derived from rice bran oil is also available in the U.S.

α-tocopherol

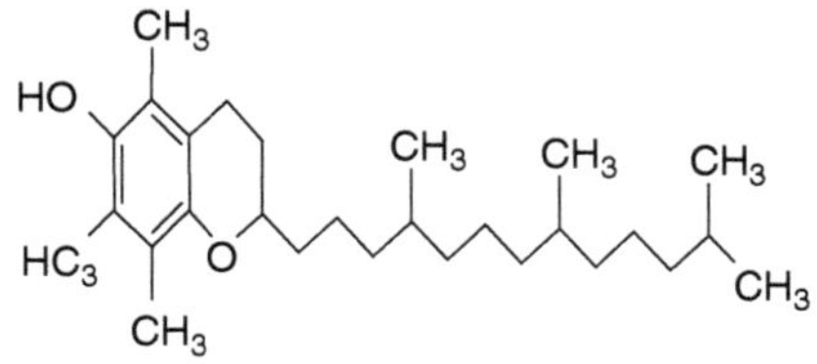

Tocotrienols (T3)

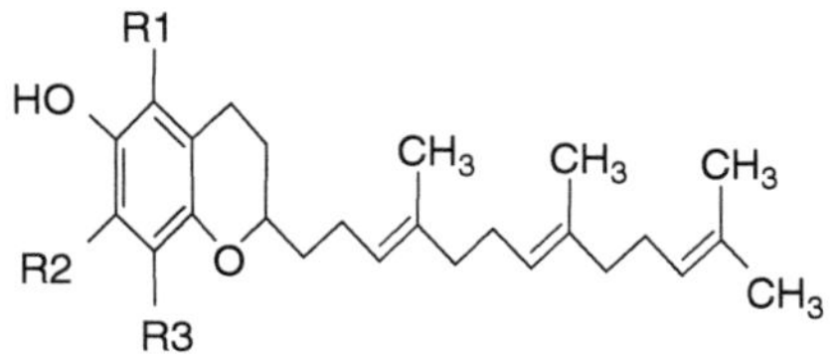

	R1	R2	R3
α-T3	CH3	CH3	CH3
γ-T3	H	CH3	CH3
δ-T3	H	H	CH3

FIGURE 28.1 Structures of various homologs of tocopherol and tocotrienol.

28.2 VITAMIN E CONTENT OF PALM OIL

The vitamin E content of crude palm oil ranges between 600 and 1000 parts per million (ppm) and is a mixture of tocopherols (21 to 34%) and tocotrienols (66 to 79%). The major tocotrienol isomers occurring in palm oil are α (28%), γ (37%), and δ (14%). Only a negligible amount of β-tocotrienol (1 to 2%) occurs in palm oil. In comparison, rice bran oil, the only current comparable commercial source of tocotrienols, differs from TRF in its composition and normally consists of the tocopherols (α, 5%; β, 1%; and δ, 15%) and tocotrienols (α, 2%; β, 7%; γ, 65%; and δ, 3%). The analysis of palm vitamin E usually requires a high-performance liquid chromatography (HPLC) system fitted with a fluorescence detector and an integrator. Figure 28.2 represents a typical HPLC chromatogram of the vitamin E present in palm oil. In biological tissue, the levels of tocotrienols found are extremely low, even when the diet is supplemented. In view of this, the HPLC analytical system used must be properly optimized. Relatively pure tocotrienol standards are not readily available. In their absence, it is recommended that the quantification of tocotrienols be done using the four common tocopherol isomers as external standards.

Samples isolated by the HPLC may be further characterized by either electron impact or tandem mass spectroscopy. High-resolution mass spectroscopic analysis of α-tocotrienol shows a molecular ion peak M+ at *m/z* 424, which corresponds to the molecular formula $C_{29}H_{44}O_2$. Fragmentation peaks, *m/z* 205 ($C_{13}H_{17}O_2$) and *m/z* 203 ($C_{13}H_{15}O_2$) are formed after the loss of the phytyl chain. The β-, γ-, and δ-tocotrienols show corresponding M+ peaks at 410, 410, and 396, respectively.

28.3 EXTRACTION OF PALM VITAMIN E

The Malaysian Palm Oil Board (MPOB), in association with Japanese researchers, developed a patented process (Australian patent PI7565/88) for the extraction of palm vitamin E, and this process has been commercialized. It can produce vitamin E concentrates ranging from 80 to 95% pure. Palm fatty acid distillate (PFAD) containing about 0.4% vitamin E (tocopherols and tocotrienols) is used as the starting raw material in the process. Apart from vitamin E, PFAD contains almost

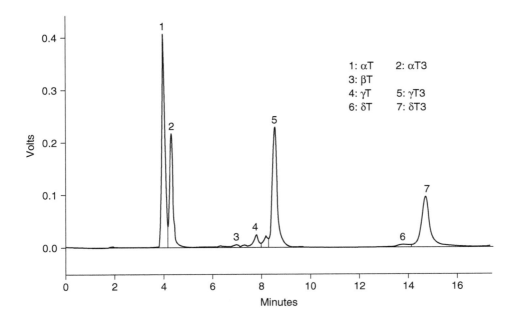

FIGURE 28.2 Palm vitamin E tocol composition.

80% FFA, 14.5% acylglycerols (mono-, di-, and triacylglycerols), 0.4% phytosterols (campesterol, β-sitosterol, stigmasterol, and cholesterol), and 1.5% hydrocarbons (mostly squalene).

The principle used to isolate the tocopherols and tocotrienols from PFAD first involves the conversion of FFA and triacylglycerols into methyl esters by an esterification process.[3] These esters are removed by distillation, leaving behind a vitamin E concentrate that is about 5 to 10% pure. Concentration of this crude mixture is a two-stage process: First, the mixture is subjected to crystallization, and then the slurry is passed through an ion-exchange column, resulting in a vitamin E concentrate that is 60 to 70% pure. Washing and then drying the concentrate and passing it through a second molecular distillation stage achieves further purification of this preparation. The product is then deodorized under vacuum to remove any remaining odoriferous materials and finally dissolved in a precise quantity of palm super olein prior to encapsulation. The flow diagram of the MPOB process is given in Figure 28.3. The overall yield in this process is about 65%, and a final purity of the vitamin E up to 80 to 95% is achievable.

28.4 VITAMIN E ACTIVITY OF TOCOTRIENOLS

Refined palm oil contains approximately 500 to 700 ppm vitamin E, which is present as the RRR-α-tocopherol (30%) and tocotrienol (70%) isomers. In contrast, oils such as corn, soybean, and sunflower are good sources of the tocopherols but contain no tocotrienols. Historically, vitamin E activity (one international unit, IU) has been defined as 1 mg of all rac-α-tocopheryl acetate, whereas 1 mg of all RRR-α-tocopherol equals 1.49 IU. In addition, vitamin E activity in foods is expressed as the α-tocopherol equivalent (α-TE), which is the activity of 1 mg of RRR-α-tocopherol. On this basis, conversion factors for each milligram of the different tocopherols and tocotrienols present in palm oil to α-TE have been calculated as follows: α-tocopherol, 1.0; γ-tocopherol, 0.5; δ-tocopherol, 0.1; α-tocotrienol, 0.3; and β-tocotrienol, 0.05. The factors for γ- and δ-tocotrienol are presently unknown.[4] The conversion factors are based on the ability of each isomer to overcome specific vitamin E deficiency symptoms such as fetal resorption, muscular dystrophy, and encephalomalacia. Since these factors are based on rat fetal resorption assays, their relevance to humans

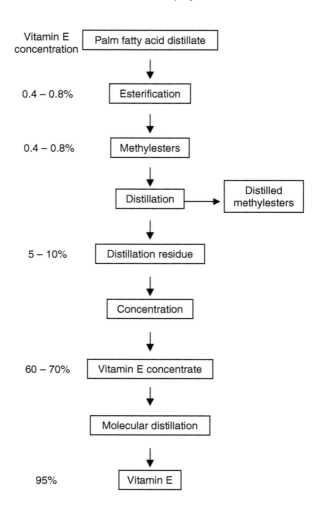

FIGURE 28.3 Process flowchart for palm vitamin E production.

is often questioned. In addition, their biological activity is also based on their antioxidant activities, but this appears misleading. For example, α-tocotrienol has only one-third the biological activity of α-tocopherol, yet it has been reported to be a better antioxidant than α-tocopherol.[5,6]

28.5 METABOLIC FATE OF TOCOTRIENOLS

Together with dietary fat, both tocopherols and tocotrienols are absorbed in the digestive tract, incorporated into chylomicrons, and transported in the lymphatic system.[7] However, following chylomicron clearance, tocotrienols disappear from circulating plasma, whereas tocopherol, especially α-tocopherol, is preferentially secreted into plasma. α-Tocopherol transfer protein, which controls plasma α-tocopherol levels, has an affinity of only 12% for α-tocotrienol (compared with 100% for α-tocopherol). The reason for the low plasma tocotrienol concentrations compared with tocopherols is the very fast clearance, as demonstrated by Yap et al.[8]

The mode of tocotrienol elimination from the body is not known. However, the rapid disappearance of the tocotrienols could be a result of metabolism. Carboxyethyl-hydroxychromans

(α- and γ-CEHC) are human urinary metabolites of α- and γ-tocopherols, respectively.[9] Lodge et al.[10] demonstrated that α- or γ-tocotrienols are metabolized to the same carboxyethyl-hydroxychromans (α- and γ-CEHC) as are α- and γ-tocopherols, respectively, and that these CEHCs are detectable in urine. The subjects (n = 5 men, 1 woman) were healthy nonsmokers, who did not take antioxidant supplements. A single supplement was consumed with breakfast each week for 4 weeks in the following order: 125 mg γ-tocotrienyl acetate, 500 mg γ-tocotrienyl acetate, 125 mg α-tocotrienyl acetate, then 500 mg α-tocotrienyl acetate. Complete urine collections were obtained 24 h before as well as on the day of supplementation, and on the subsequent two days. The fraction of the dose excreted as CEHC was approximately 1 and 5% for α- and γ-tocotrienols, respectively. These relatively low percentages suggest that tocotrienols that are rapidly cleared from plasma are eliminated apparently by a mechanism other than metabolism, or that the metabolites are excreted by other routes.

Hayes et al.[11] reported the presence of tocotrienols in all tissues except the brain of hamsters fed palm tocotrienols, with adipose tissue being especially rich in its ability to accumulate appreciable amounts of tocotrienols. The presence of α- and γ-tocotrienols has been reported, even when the diet was not especially enriched with tocotrienols. Skin has been suggested to be an important storage and excretory site for vitamin E, and the accumulation of tocotrienols could be especially beneficial in protecting the lipid barrier of the stratum corneum.[12]

28.6 ANTIOXIDANT PROPERTIES OF TOCOTRIENOLS

Vitamin E compounds are well known for their antioxidant property.[13] This property depends primarily on the phenolic group in the chromanol ring, rather than the side chain.[14,15] Tocotrienols, like tocopherols, are capable of scavenging and quenching free radicals. Their antioxidative activity, however, resides mainly with its "chain breaking" property, which neutralizes peroxyl and alkoxyl radicals generated during lipid peroxidation.[13,16] Peroxidation of membrane lipids is known to modify and inactivate cellular components, which can have damaging effects on crucial cellular factors, leading to disease. In the case of low-density lipids (LDL), peroxidation has emerged as the initiating step in the pathogenesis of atherosclerosis. Such modification of LDL causes recognition of the LDL particle by the scavenger cell receptor present on macrophages. This pathway is unregulated and nonsaturable, which consequently leads to massive deposition of cholesterol into these cells. This is typically associated with the formation of foam cells in the atherosclerotic lesions.[17]

For many years, α-tocopherol was generally considered as the most potent antioxidant against lipid peroxidation in the vitamin E group.[18] However, considerable discrepancy has recently been found in its relative antioxidant effectiveness when compared with tocotrienols and the individual isomers. We have studied the *in vitro* effect of tocotrienol-rich fraction (TRF) on the peroxidation potential in rat liver mitochondria and microsomes, using ascorbate and NADPH-induced systems.[19] Our results showed that TRF was able to inhibit lipid peroxidation in both mitochondria and microsomes in a dose-dependent manner. Further experiments carried out showed that when compared with α-tocopherol, TRF showed better lipid peroxidation potential[20] (Figure 28.4).

When the individual isomers were tested, α-tocotrienol was found to be a better antioxidant than α-tocopherol.[5,6] Serbinova et al.[5] observed a remarkably higher antioxidant activity with tocotrienols against lipid peroxidation in rat liver liver microsomes than with γ-tocopherol. Kamat and Devasagayam[21] observed similar results in rat brain mitochondria and noted a stronger effect with γ-tocotrienol. In an elaborate study, a number of mechanisms were shown to contribute to its higher antioxidant activity compared with α-tocopherol. They were:[1] a more uniform distribution in the membrane lipid bilayer,[2] a more efficient interaction of the chromanol ring with lipid radicals, and[3] a higher recycling efficiency from chromanoxyl radicals.[22]

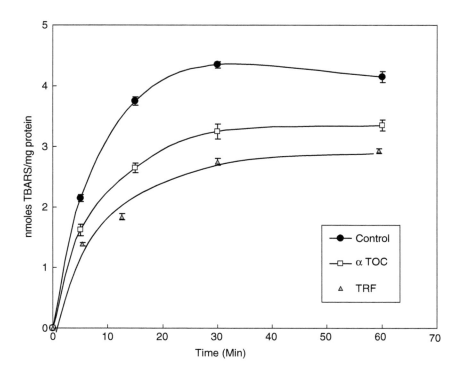

FIGURE 28.4 Ascorbate-Fe^{2+}-induced lipid peroxidation in rat liver microsomes and its prevention by TRF and α-tocopherol as a function of time. Lipid peroxidation was measured as thiobarbituric acid reactive substances (TBARS). The concentration of α-tocopherol or TRF was 5 μM. Values are mean ±SE from five experiments.

28.7 ANTICANCER PROPERTIES OF TOCOTRIENOLS

Evidence suggests that cancer development occurs in two stages, initiation and promotion. Initiation involves a permanent, irreversible genetic change in the cell's DNA. This generally causes DNA strand breaks, which can lie dormant in the cell and do not alone cause cancer. Promotion elicits the second stage of carcinogenesis, which stimulates the initiated cell, transforming it to a cancerous cell. This process is reversible and may require continued stimulation to promote mutations. Many components have both initiating and promoting activities. However, free radicals are thought to act principally as promoting agents. Superoxide radicals and many peroxides are tumor promoters. Free radicals have been associated with gross chromosomal damage and inhibition of the biological natural repair systems. Free radicals can alter gene expression by the mobilization of calcium stores, which activates a variety of cellular kinases, phosphatases, and transcription factors. Lipid peroxidation by free radicals has also been implicated as a causative factor in cancer development.[23]

Increased intake of antioxidants, either through the diet or as dietary supplements, has been associated with a reduced incidence of cancer. Long-term scientific studies have shown that low antioxidant levels are associated with increased incidence of certain cancers. Anticarcinogenic and antipromoter activity of some antioxidants is thought to offer some protection against cancer.

Our studies on breast cancer looked at the protective role palm oil played in mammary carcinogenesis, compared the positive effect conferred by palm oil, and linked it to the presence of vitamin E.[24] Further work was then carried out *in vitro* on the growth of estrogen-receptor-positive and receptor-negative human breast cancer cells using both TRF as well as the individual tocotrienols.

Breast cancer was chosen as a subject for investigation as it is the leading cause of death among women worldwide. An estimated 570,000 cases occurred in 1999, accounting for 9% of all new cases of cancer.[25] In Malaysia, breast cancer represented 9% of total cancer incidence in 1993. While it is estimated that 1800 women developed breast cancer annually, the number had increased substantially by 1996. In that year, 3500 new breast cancer cases were reported in the whole country, and 220 women who sought treatment at government breast cancer clinics died in the same year. More attention should be focused on measures that can be taken to lower the risk of women developing breast cancer. It is therefore reassuring to know that the development of more effective biological therapies for breast cancer is under intense investigation. Investigation of fat-soluble vitamins (A,D, E, and K) for anticancer properties have shown them to possess antiproliferative effects.

The antiproliferative effect of tocotrienols has also been studied *in vitro* on human breast cancer cell lines. The first set of experiments showed that TRF inhibited the incorporation of [^{3}H] thymidine into these cells by 50% at a concentration of 180 μg/ml of TRF (Figure 28.5). In comparison, α-tocopherol did not significantly inhibit growth at concentrations up to 500 μg/ml. Subsequent experiments showed growth of the cells was inhibited by 50% in the presence of 180 μg/ml, whereas no growth inhibition was observed with a concentration of 500 μg/ml of α-tocopherol.[26]

These results indicated that the inhibition of proliferation and growth of the cells by TRF was probably due to tocotrienols. To confirm this, estrogen-receptor-negative human breast cancer cells were incubated with varying concentrations of individual tocotrienols. Experiments on the incorporation of [^{3}H] thymidine gave an IC$_{50}$ of 28 μg/ml for γ-tocotrienol and 90 μg/ml for both α- and δ-tocotrienols.[27] Further experiments carried out on estrogen-receptor-positive human breast cancer cells showed that TRF inhibited proliferation of the cells with maximal effect at 8 μg/ml. In contrast, α-tocopherol had no effect (Figure 28.5) The individual tocotrienol fractions (γ and δ) showed even

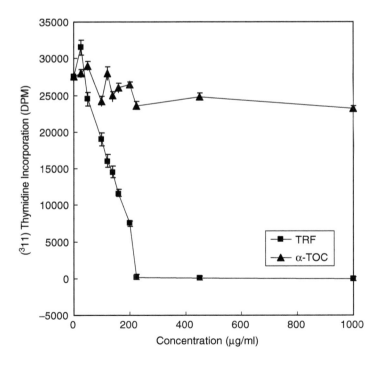

FIGURE 28.5 Effect of the tocotrienol-rich fraction (TRF) and α-tocopherol on the proliferation of MDA-MB-435 cells. The cells were incubated with various concentrations of TRF or α-T for 48 h. [^{3}H] thymidine (0.5 μCi/well) was then added, and the cells were harvested after 4 h to evaluate the incorporation of thymidine into DNA. Points are the average of mean values from three experiments.

greater inhibitory effects at lower concentrations from 1 to 6 μg/ml, with γ-tocotrienol having the greatest effect. In addition to their antiproliferative effect, γ- and δ-tocotrienols also resulted in decreased expression of insulinlike growth factor binding proteins (IGFBP)-2 and -4. Recent evidence suggests IGFBPs can influence interactions between IGFs and their receptors. As such, IGFBPs play an important role in mediating IGF-induced growth of breast cancer. The expression of pS2, a well-established estrogen-regulated gene, was also determined. Northern blotting of pS2 mRNA from the breast cancer cells demonstrated that TRF did not inhibit expression of the gene, indicating that the mechanism of inhibition by tocotrienols was different from that of an estrogen antagonist such as tamoxifen, which is widely used in breast cancer therapy.[28] Similar results were obtained by Guthrie et al.[29] on estrogen-receptor-positive human breast cancer cells. They showed that tocotrienols inhibited proliferation of the cells, as measured by ³[H] thymidine incorporation. The IC_{50}s for TRF, α-tocopherol, α-, γ-, and δ-tocotrienols were 4, 125, 6, 2, and 2 μg/ml, respectively.

A recent study showed that tocotrienols were significantly more potent than tocopherols in suppressing epidermal growth factor (EGF)-dependent normal mammary epithelial cell growth. EGF is a potent mitogen for normal mammary epithelial cells, and an initial event in EGF-receptor mitogenic signaling is activation of protein kinase C (PKC). Studies were conducted to determine whether the antiproliferative effects of specific tocopherol and tocotrienol isoforms were associated with a reduction in EGF-receptor mitogenic signaling and/or PKC activation.[30] Normal mammary epithelial cells isolated from midpregnant BALB/c mice were grown in primary culture, maintained on serum-free media containing 10 ng/mL EGF as a mitogen, and treated with various doses (0 to 250 μm) of α-, γ-, or δ-tocopherol or α-, γ-, or δ-tocotrienol. Treatment with growth-inhibitory doses of δ-tocopherol (100 μm), α-tocotrienol (50 μm), or γ- or δ-tocotrienol (10 μm) did not affect EGF receptor levels, EGF-induced EGF-receptor tyrosine kinase activity, or total intracellular levels of PKC (alpha). However, these treatments were found to inhibit EGF-induced PKC (alpha) activation, as determined by its translocation from the cytosolic to the membrane fraction. Treatment with 250 μm of α- or γ-tocopherol had no effect on EGF-receptor mitogenic signaling or cell growth. These findings demonstrate that the inhibitory effects of specific tocopherol and tocotrienol isoforms on EGF-dependent normal mammary epithelial cell mitogenesis occurs downstream from the EGF receptor and appears to be mediated, at least in part, by a reduction in PKC (alpha) activation.

Apart from the ability of TRF and individual tocotrienols to inhibit human breast cancer cell proliferation, they have been tested and show promising results in other cancer model studies. Palm TRF reduced the severity of hepatocarcinogenesis in rats treated with 2-acetyl-aminofluorene. Hepatic cell damage was reduced by TRF along with decreased activities of plasma glutaryl-transpeptidase. The degree of severity of hepatocarcinoma was also reduced in rats supplemented with TRF compared with α-tocopherol. Palm tocotrienols also inhibited the proliferation of Caski epithelial cancer cells, whereas α-tocopherol had no effect. The antiproliferative effect of palm tocotrienols was attributed to an increase in apoptosis, as measured by increased DNA fragmentation.[31]

Other evidence that vitamin E from palm oil may possess anticancer properties has come from experiments with transplantable tumors in mice.[32] When injected intraperitoneally into mice, α- and γ-tocotrienols prolonged the life of mice bearing Ehrlich sarcoma, sarcoma 180, or IMC carcinoma. γ-Tocotrienol was more effective than α-tocotrienol in each case, and α-tocopherol was least effective. In addition to studies on experimental animals, Komiyama and Yamaoka[32] showed the growth inhibition of human and mouse tumor cells (H69, HeLa, and P388) when the cells were exposed to tocotrienols for 72 h *in vitro*. Survival of mice receiving transplantable tumor cells was also increased if they were fed tocotrienols in a dose-responsive way.[33] Tocotrienols have also caused a delay in the onset of subcutaneous lymphoma by 2 to 4 weeks in HRS/J hairless mice, a strain genetically susceptible to subcutaneous lymphoma.[34]

The precise mechanism for the antiproliferative property of tocotrienols is uncertain, but it may lie in its prenylated side chain involved in the production of isoprenoid intermediates from the mevalonate biosynthetic pathway.[35,36] These intermediates are thought to be involved in the prenylation of several signal-transduction proteins, including the Ras protein, essential for normal cell growth.

28.8 CONCLUSION

Collectively known as vitamin E, tocotrienols are identical in structure to tocopherols except for the degree of saturation in their side chain. The prenyl side chain of tocotrienol has been postulated to be responsible for the differential membrane distribution and metabolism of tocotrienols when compared with tocopherols. Future investigations need to examine the molecular mechanism of action of tocotrienols in order to achieve a more comprehensive understanding of their complex but beneficial effects on cancer.

REFERENCES

1. IUPAC-IUB Joint Commission on Biochemical Nomenclature (JCBN), Nomenclature of tocopherols and related compounds: recommendations 1981, *Eur. J. Biochem.*, 123, 473–475, 1982.
2. Therialut, A., Chao, Q., Wang, A., Gapor, A., and Adeli, K., Tocotrienol: a review of its therapeutic potential, *Clin. Biochem.*, 32, 309–319, 1999.
3. Sundram, K. and Gapor, A., Vitamin E from palm oil: its extraction and nutritional properties, *Lipid Technol.*, 4, 137–141, 1996.
4. Sokal, R.R., Vitamin E deficiency and neurological disorders, in *Vitamin E in Health and Disease*, Packer, L. and Fuchs, J., Eds., Marcel Dekker, New York, 1993, pp. 815–849.
5. Serbinova, E., Kagan, V., Han, D., and Packer, L., Free radical recycling and intramembrane mobility in the antioxidant properties of α-tocopherol and α-tocotrienol, *Free Radical Biol. Med.*, 10, 263–275, 1991.
6. Suzuki, Y.J., Tsuchiya, M., and Wassall, S.R., Structural and dynamic membrane properties of α-tocopherol and α-tocotrienol: implications to the molecular mechanism of their antioxidant potency, *Biochemistry*, 32, 10692–10696, 1993.
7. Kayden, H.J. and Traber, M.G., Absorption, lipoprotein transport and regulation of plasma concentrations of vitamin E in humans, *J. Lipid Res.*, 34, 343–358, 1993.
8. Yap, S.P., Kuen, K.H., and Wong, J.W., Pharmacokinetics and bioavailability of alpha-, gamma- and delta-tocotrienols under different food status, *J. Pharm. Pharmacol.*, 1, 67–71, 2001.
9. Brigelius-Flohe, R. and Traber, M.G., Vitamin E: function and metabolism, *FASEB J.*, 13, 1145–1155, 1999.
10. Lodge, J.K., Ridlington, J., Leonard, S., Vaule, H., and Traber, M.G., α- and γ-tocotrienols are metabolized to carboxyethyl-hydroxychroman derivatives and excreted in human urine, *Lipids*, 36, 43–48, 2001.
11. Hayes, K.C., Pronczuk, A., and Liang, J.S., Differences in the plasma transport and tissue concentrations of tocopherols and tocotrienols: observations in humans and hamsters, *Proc. Soc. Exp. Biol. Med.*, 202, 353–359, 1993.
12. Traber, M.G., Podda, M., Weber, C., Thiele, J., Rallis, M., and Packer, L., Diet derived and topically applied tocotrienols accumulate in skin and protect the tissue against ultraviolet light-induced oxidative stress, *Asia Pac. J. Clin. Nutr.*, 6, 63–67, 1997.
13. Kamal-eldin, A. and Appelqvist, L.A., The chemistry and antioxidant properties of tocopherols and tocotreinols, *Lipids*, 31, 671–701, 1996.
14. Niki, E., Kawakami, A., Satio, M., Yamamoto, Y., Tucshiya, J., and Kamiya, Y., Effect of phytyl side chain of vitamin E on its antioxidant activity, *J. Biol. Chem.*, 260, 2191–2196, 1985.
15. Burton, G.W. and Ingold, K.U., Vitamin E as *in vitro* and *in vivo* antioxidant, *Ann. N.Y. Acad. Sci.*, 570, 7–22, 1989.
16. Burton, G. and Traber, M.G., Vitamin E: antioxidant activity, biokinetics and bioavailability, *Ann. Rev. Nut.*, 10, 857–882, 1990.
17. Steinberg, D., Parthasarathy, S., Carew, T.E., Khoo, J.C., and Witztum, J.L., Beyond cholesterol: modification of low density lipoprotein that increases its atherogenicity, *New England J. Med.*, 320, 915–924, 1989.

18. Burton, G.W., Joyce, A., and Ingold, K.U., Is vitamin E the only lipid soluble, chain breaking, antioxidant combination in human blood plasma and erythrocyte membranes? *Arch. Biochem. Biophys.*, 221, 281–290, 1983.

19. Nesaretnam, K., Devasagayam, T.P.A., Singh, B.B., and Basiron, Y., Influence of palm oil or its tocotrienol-rich fraction on the lipid peroxidation potential of rat liver mitochondria and microsomes, *Biochem. Mol. Biol. Inter.*, 30, 159–167, 1993.

20. Kamat, J.P., Sarma, H.D., Devasagayam, T.P.A., Nesaretnam, K., and Basiron, Y., Tocotrienols from palm oil as effective inhibitors of protein oxidation and lipid peroxidation in rat liver microsomes, *Mol. Cell. Biochem.*, 170, 131–138, 1998.

21. Kamat, J.P. and Devasagayam, T.P.A., Tocotrienols from palm oil as potent inhibitors of lipid peroxidation and protein oxidation in rat brain mitochondria, *Neuro. Lett.*, 195, 179–182, 1995.

22. Serbinova, E., Khwaja, S., and Catudioc, J., Palm oil vitamin E protects against ischaemia/reperfusion injury in the isolated perfused Langendorff heart, *Nutr. Res.*, 12, S205–S215, 1996.

23. Wang, M., Dhingra, K., Hittleman, W.N., Liehr, J.G., De Andrade, M., and Li, D., Lipid peroxidation-induced putative malondialdehyde-DNA adducts in human breast tissues, *Cancer Epidemiol. Biomarkers Prev.*, 5, 705–710, 1996.

24. Nesaretnam, K., Khor, H.T., Ganeson, J., Chong, Y.H., Sundram, K., and Gapor, A., The effect of vitamin E tocotrienols from palm oil on chemically induced mammary carcinogenesis in female rats, *Nutr. Res.*, 12, 879–892, 1992.

25. Bundred, N.J., Prognostic and predictive factors in breast cancer, *Cancer Treat. Rev.*, 27, 137–142, 2001.

26. Nesaretnam, K., Guthrie, N., Chambers, A.F., and Carroll, K.K., Effect of tocotrienols on the growth of human breast cancer cell line in culture, *Lipids*, 30, 1139–1143, 1995.

27. Carroll, K.K., Guthrie, N., Nesaretnam, K., Gapor, A., and Chambers, A.F., Anticancer properties of tocotrienols from palm oil, in *Nutrition, Lipids, Health and Disease*, Ong, A.S.H., Niki, E., and Packer, L., Eds., AOCS Press, pp. 117–121.

28. Nesaretnam, K., Stephen, R., Dils, R., and Darbre, P., Tocotrienols inhibit the growth of human breast cancer cells irrespectively of estrogen receptor status, *Lipids*, 33, 461–469, 1998.

29. Guthrie, N., Gapor, A., Chambers, A., and Caroll, K.K., Inhibition of proliferation of estrogen receptor-negative MDA-MB-435 and -positive MCF-7 human breast cancer cells by palm oil tocotrienols and tamoxifen, alone and in combination, *J. Nutr.*, 127, 544S–548S, 1997.

30. Sylvester, K., McIntyre, B.S., Gapor, A., and Brsiki, K.P., Vitamin E inhibition of normal mammary epithelial cell growth is associated with a reduction in protein kinase C(alpha) activation, *Cell. Prolif.*, 34, 347–357, 2001.

31. Wan Zurinah, W.N., Zanariah, J., San, M.M., Marzuki, A., Gapor, A., Shamaan, N.A., and Khalid, A.K., Effect of tocotrienols on hepatocarcinogenesis induced by 2-acetylaminofluorene in rats, *Am. J. Clin. Nutr.*, 53, 1076S–1081S, 1991.

32. Komiyama, K. and Yamaoka, Studies on the biological activity of tocotrienols, *Chem. Pharm. Bull.*, 37, 1369–1371, 1991.

33. Kato, A., Yamaoka, M., Tanaka, A., Komiyama, K., and Umezawa, I., Physiological effect of tocotrienols, *Yakugaku Zasshi*, 34, 375–376, 1985.

34. Tan, B., Antitumour effects of palm carotenes and tocotrienols in HRS/J hairless female mice, *Nutr. Res.*, 12, S163–S173, 1992.

35. Crowell ,P.L., Chang, R.R., Ren, Z.B., Elson, C.E., and Gould, M.N., Selective inhibition of isoprenylation of 21-26 kDA proteins by the anticarcinogen d-limonene and its metabolites, *J. Biol. Chem.*, 266, 1769–1785, 1991.

36. Elson, C.E. and Yu, S.G., The chemoprevention of cancer by mevanolate-derived constituents of fruits and vegetables, *J. Nutr.*, 124, 607–614, 1994.

29 Pycnogenol® in Cancer Chemoprevention

Fabio Virgili, R. Ambra, R. Canali, and O. Gulati

CONTENTS

29.1   Introduction 491
29.2   Historical Landmarks..492
29.3   Bioavailability 493
29.4   Cancer as a Disease Characterized by Unbalanced Redox Conditions...........493
29.5   Phytochemical Antioxidants as Preventive Factors
        in Cancer Development...495
29.6   Procyanidins and Cancer 495
29.7   Antioxidant Capacity of Pycnogenol...496
29.8   Specific Binding to Proteins and Enzyme Inhibition.......................................498
29.9   New Avenues To Explore by Using New Approaches.......................................500
29.10   Conclusion...502
References 503

29.1   INTRODUCTION

Polyphenol-rich foods and extracts are commonly referred to as "nutraceuticals," a category consisting of food items that significantly and positively affect human health. There is growing interest in the use of polyphenol-rich plant extracts as dietary food supplements because they have a wide spectrum of beneficial activity for human health. The main established feature of polyphenols is their strong antioxidant activity. Recent *in vitro* studies have provided evidence that polyphenols can also affect enzyme activity, cell response, and gene expression, providing a novel mechanistic perspective underlying biological activity of polyphenols.

Cancer is the second leading cause of death in the U.S. and, perhaps, in the world, among the most productive age group. Cancer is a complex multistep and multifactorial disease. The combination of multiple genetic defects occurs, at least in part, because of exposure to environmental pollution, unhealthy diets, and lifestyle factors. In fact, the importance of diet is being recognized, in particular the role of antioxidant nutrients in modulating the oxidative damage leading to genomic alterations.[1] Recent research has demonstrated that molecules of nutritional interest are able to directly affect gene expression and cell response to mutagenic stimuli, opening new avenues in understanding of the molecular background of cancer development as well as its prevention and treatment. Supplementation with nutrients having a strong antioxidant capacity is therefore considered a possible expedient strategy not only to reduce the risk of cancer, but also as a possible adjuvant to anticancer therapy.

The standardized extract from the bark of the French maritime pine is one of the most utilized herbal-sourced food supplements in the world. It is patented under the brand name of Pycnogenol®, and its history and biological profile have been reviewed in the literature.[2,3] Pycnogenol is obtained by water extraction of the bark of the *Pinus pinaster,* using a validated standardized manufacturing procedure.

Even though the unique and very complex chemical composition of Pycnogenol has not been completely elucidated, its main constituents are known to be phenolic compounds, broadly divided into monomers (catechin, epicatechin, and taxifolin) and condensed flavonoids classified as procyanidins/proanthocyanidins. Pycnogenol also contains phenolic acids (such as caffeic, ferulic, and *p*-hydroxybenzoic acids) as important constituents and glycosylation products, e.g., glucopyranosyl derivatives of either flavonols or phenolic acids, as minute constituents.[2]

Pycnogenol has been reported to have a strong antioxidant activity either *in vitro* utilizing cultured cells[4,5] or *in vivo* in experimental animals[6] and in humans,[7] and to participate to the cellular antioxidant network. Further beneficial effects such as vasorelaxation,[8] immunomodulatory function,[9] and anti-inflammatory activities[6,10] have been reported, confirming the potential of this extract as an effective phytochemical.

Recently, some reports have led to the hypothesis that Pycnogenol could be beneficial in not only mitigating the adverse effects of anticancer chemotherapy,[11] but possibly also inhibiting the development, growth, and progression of cancer.[12–14] This activity is associated not only with the strong antioxidant capacity of the polyphenols contained in Pycnogenol, but also with its ability to modulate cell response to different stimuli.

This chapter discusses the antioxidant capacity of Pycnogenol and reviews some of the recent advances and hypotheses regarding its ability to significantly affect cell signaling and cell response in carcinogenesis.

29.2 HISTORICAL LANDMARKS

The use of pine bark extracts has its roots in ancient traditional medicine in different countries of the world.[15] In general, the utilization of Pycnogenol in the past was considered suitable for many of those conditions where vitamin C deficiency is now known to be involved, such as scurvy, skin disorders, and wound healing. This utilization anticipated the understanding of the complex nature of the interaction between different antioxidants and the important role that polyphenols may play in the recycling and interplay with other antioxidants and, finally, in human health and disease.

The bark of the pine has been utilized medicinally for more than 2000 years, with the earliest indications of its therapeutic use dating to the 4th century B.C., when Hippocrates mentioned its use in treating inflammatory diseases.[16] Its utilization continued even as an "emergency food" until the last world war in the northern Europe.[17]

As an early background of its capacity to interplay with ascorbic acid, it is interesting to mention that pine bark was utilized against inflammation and also to overcome the symptoms of scurvy.[18] Other uses of pine bark extract were suggested by the naturalist Hieronymus Bosch and included topical application on skin ulcers and other skin disorders.[19] In the so-called New World, the bark of the pine was consumed by native Americans either as food or to obtain infusions to be utilized as a remedy for different diseases that are now recognized to have some free-radical involvement in inflammatory conditions.[20–22]

29.3 BIOAVAILABILITY

The extent and the kinetics of absorption, metabolism, and pathways of elimination of the compounds contained in Pycnogenol are still not fully understood. Pycnogenol is a complex mixture of different molecular entities, and it is extremely difficult to define and assess the real bioavailability of all components constituting the extract, since different components are likely to follow a different pattern of absorption, biotransformation, and excretion.[23]

Up to few years ago, the bioavailability of Pycnogenol could be only indirectly inferred from the evidence of the biological effects detectable after its administration. We have made an experimental approach to evaluate the bioavailability of this complex mixture by considering one of its components as a "tracking molecule" and following its route after ingestion. Ferulic acid is a useful and expedient biomarker for the study of the metabolic fate of plant extracts, since its corresponding metabolites are easily detected into both plasma and urine.[24,25] Utilizing this approach, we have demonstrated in a controlled study in humans that, after oral supplementation with Pycnogenol, all subjects displayed a significant, though variable, level of excretion of ferulic acid, providing evidence that at least a part of the phenolic components of Pycnogenol are absorbed and metabolized by humans.[26] Specific molecules derived from the biotransformation of the flavonoid component of Pycnogenol have been now identified in plasma and urines after oral supplementation in humans.[27]

More recently, it has been shown that supplementation with Pycnogenol in human subjects is associated with a significant increase of total plasma antioxidant capacity.[7] This observation confirms that the antioxidant capacity of the mixture is bioavailable, indirectly supporting the bioavailability of Pycnogenol components.

Pycnogenol has also recently been reported to be absorbed percutaneously by humans and may therefore exert its anti-inflammatory and antioxidant activity from this route, protecting the skin from ultraviolet (UV)-radiation-induced damage.[28]

29.4 CANCER AS A DISEASE CHARACTERIZED BY UNBALANCED REDOX CONDITIONS

In order to understand the molecular basis for possible beneficial effects of Pycnogenol as a preventive agent in mutagenesis and carcinogenesis, it is important to address some specific aspects of this complex phenomenon. Many recent epidemiological studies have established a close link between antioxidant defense and lowered risk of morbidity and mortality from cancer. Oxidative stress has now been recognized as one of the major factors contributing to the increased risk of several diseases, including cancer.[29–31] The term "reactive oxygen and nitrogen species" (RONS) collectively denotes both oxygen- and nitrogen-centered radicals such as superoxide ($O_2^{\cdot-}$), hydroxyl ($^-OH^{\cdot}$), and nitric oxide (NO^-) as well as nonradical species derived from oxygen, such as hydrogen peroxide (H_2O_2), hypochlorous acid (HOCl), and peroxynitrite ($ONOO^-$). RONS are generated during normal metabolism, are an integral part of normal cellular function, and usually do little harm because of intracellular mechanisms that reduce their damaging effects. However, increased or prolonged production of these species can overwhelm the defense mechanisms and is thought to accompany most forms of tissue injury associated with chronic inflammation or immune injuries, which can lead to degenerative diseases such as cancer.[32,33]

The overproduction of RONS may damage biological systems through peroxidation of membrane lipids, oxidative damage of nucleic acids and carbohydrates, and oxidation of susceptible groups in proteins.[32,34] It is now well documented that oxidative stress plays an important role in the induction of many different types of cancers such as skin cancer,[35] colorectal cancer,[36,37] and breast carcinoma.[38]

The involvement of RONS in carcinogenesis is a part of a very complex network, including the individual genetic characteristics and specific events such as the activation of chemical carcinogens, pathological ionic fluxes across destabilized cell membrane, DNA mutation, and altered cyclic nucleotide levels.[39] Figure 29.1 shows a simplified picture of this network starting from the generation of RONS overwhelming the cellular capacity to quench their reactivity and finally leading to the

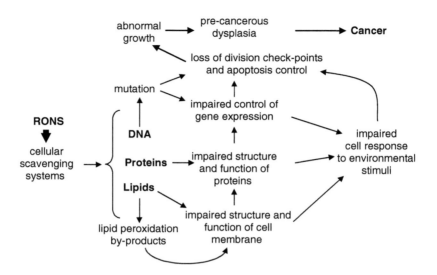

FIGURE 29.1 From reactive oxygen and nitrogen species (RONS) to cancer: RONS escaping from antioxidant defense systems can target lipids, proteins, and DNA, in turn inducing a cascade of events that can either be repaired or result in the impairment of cell response to stimuli and growth control. These impairments initially lead to tissue dysplasia and finally to cancer.

loss of the control of cell growth. Premutagenic lesions of DNA, believed to be important in the development of human cancers, can arise through covalent modification of bases by reaction with oxidants and products of lipid peroxidation derived through normal aerobic metabolism.[29] Mutations in cancer-related genes may lead to a cycle of death and regeneration.[29]

Besides direct oxidative DNA damage, reactive oxygen and nitrogen species can also induce etheno (epsilon)–DNA adducts mainly via *trans*-4-hydroxy-2-nonenal, generated as the major aldehyde by lipid peroxidation of omega-6 polyunsaturated fatty acids (PUFA).[40] In patients affected by familial adenomatous polyposis, an increased epsilon–DNA adduct level has been observed.[40] These adducts, caused by oxidative stress, could be premutagenic, causing genomic instability that drives colorectal adenomas to malignancy.[36] It has also been suggested that etheno–DNA adducts could be promising biomarkers for quantifying increased DNA damage in early stages of colon carcinogenesis.[41] A persistent oxidative stress in colorectal carcinoma is also suggested by Rozalski et al.[42] who found an increased level of 8-oxodGuanosine in lymphocyte DNA of cancer patients.

It has been suggested that pleiotropic genetic disorders such as ataxia telangiectasia (AT) — characterized by progressive neurodegeneration, immunodeficiency, and increased incidence of cancer — can be linked to a continuous state of oxidative stress. The phenotype of AT appears to be the functional inactivation of the AT-mutated gene product, which is thought to act as a sensor of RONS and oxidative damage of cellular macromolecules and DNA.[43]

Carcinoma of the breast is the third most common cancer worldwide and accounts for the highest morbidity and mortality. An increasing number of studies emphasize the need to evaluate the use of lipid peroxidation and antioxidant level as biomarkers in human mammary tumors.[38]

Oxidative stress can lead not only to direct damage of DNA, but can also induce signal-transduction events that lead to unwanted gene expression profiles[33,44] Generation of ROS, acting through the MAP (mitogen-activated protein) kinase pathway, stimulates transcription by activating transcription factors involved in transformation response, such as activator protein 1 (AP-1) and nuclear factor kappaB (NF-κB). It has been reported that antioxidants not only inhibit the activation of AP-1 and NF-κB, but can also prevent the transformation of epidermal JB6 cells, a cell lineage particularly responsive to transformation.[45]

29.5 PHYTOCHEMICAL ANTIOXIDANTS AS PREVENTIVE FACTORS IN CANCER DEVELOPMENT

Antioxidant micronutrients are one of the body's primary defenses against free radicals and ROS. There is considerable support from animal studies for a protective effect of antioxidant micronutrients against cancer. The relationship between fruit and vegetable intake and cancer prevention has been well documented in epidemiological studies, which suggest that the protective effects of these foods against the risk of cancer work at several sites.[46,47] However, it remains unknown which of their chemical moieties reacts at specific steps or sites in carcinogenesis. As flavonoids comprise the most common group of plant polyphenols,[48] many studies have focused on their possible chemopreventive role.[49] Human intervention studies have suggested that foods rich in certain flavonoids may protect against certain forms of lung cancer.[50] There is also evidence that flavonoids have antimutagenic activity[51] and can inhibit nuclear damage.[52] It has also been demonstrated that quercetin may counteract the tumor activity of phorbol ester tumor promoter (TPA) on mouse skin.[53] Furthermore, when certain flavonoids were applied topically to mouse skin in conjunction with TPA, there was a preventive inhibition of skin papilloma formation.[54]

The biological activities of flavonoids are structure dependent, and their role in human cancer prevention, supported by epidemiological studies, may be due to their antioxidant properties and inhibitory role in various stages of tumor development.[55] Many mechanisms of action have been identified for isoflavone/flavone prevention of cancer, including antiproliferation, induction of cell-cycle arrest and apoptosis, prevention of oxidation, induction of detoxification enzymes, regulation of host immune system, and changes in cellular signaling. Some combination of these mechanisms will be found to be responsible for cancer prevention by these compounds.[56]

Plant extracts with a chemical composition comparable with that of Pycnogenol have been reported to beneficially affect different aspects of tumor development. The general features of flavonoids within this topic have been reviewed by Bracke et al.,[57] who considered this class of plant-derived molecules as possible anti-invasive agents and an important tool for the study of the mechanisms of tumor development. In the 1960s, Ulubelen and coworkers[58] reported that the extract of the bark of *Abies concolor* has antitumor activity against colon adenocarcinoma, according to the U.S. Cancer Chemiotherapy National Service Center, and attributed this activity to the tannin component of the mixture. Similarly, an antitumor-promoting activity has been reported in mouse epidermis *in vivo* for the purified flavonoid catechin and oligomeric proanthocyanidins from different sources.[59] The same paper reports that the extract from an American variety of pine is the most effective among other bark moieties in reducing the incidence and yield of skin tumor induced by TPA.

29.6 PROCYANIDINS AND CANCER

The procyanidin-rich extract from the bark of *Hammamelis virginiana* has been reported as a powerful inhibitor of mutagenesis induced by nitro-aromatic compounds utilizing the Ames test. This inhibitory activity was not observed when the mixture's tannin component was removed, and the oligomeric proanthocyanidin fraction, obtained by chromatographic separation, showed the most evident antimutagenic activity.[60] In another study utilizing the Ames test and the somatic mutation and recombination in *D. melanogaster*, Grimmer and Parbhoo[61] observed that the procyanidin-rich fraction from an extract from *Sorghum bicolor* had the most significant antimutagenic effect with respect to other fractions composed of smaller molecules. The procyanidins from *Vitis vinifera*, which have a molecular profile similar to that of Pycnogenol, have been reported for their antimutagenic activity by Liviero and coworkers[62] in a model based on the rate of spontaneous mutation in *S. cerevisiae* at both the mitochondrial and nuclear level. The authors attributed this effect mainly to the antioxidant capacity of the mixture. An attempt to classify the antitumor activity of different tannins on the basis of a structure–activity relationship has been tried by Miyamoto and coworkers,[63] who tested over 60 tannins and related polyphenols in animals inoculated with sarcoma cells. These authors concluded that the presence of a dimeric structure seems to be essential for marked antitumor activity. All these reports, taken together, corroborate the hypothesis that oligomeric procyanidins obtained from different sources have the capacity to inhibit both cancer promotion and development. It is clearly easy to speculate that the majority of these observations may apply also for the maritime pine bark extract Pycnogenol.

Thus, it is possible to speculate about two different mechanisms for the activity of antioxidants in inhibiting both cancer promotion and development. The first is to be considered a nonspecific effect, somehow shared by any bioavailable molecules having an antioxidant capacity. In this context, molecules significantly different from one another, such as for instance tocopherols and polyphenols, may have the same final effect on the biological system. In fact, their activity results from their capacity of buffering within a physiologically appropriate window the cellular redox status in the presence of a pro-oxidant stimulus or challenge.

The second is more specific to the subject of this chapter, i.e., the procyanidins from pine bark. In this case, the inhibiting effect may be due to their capacity to modulate cell response to specific stimuli by interacting with proteins that affect enzyme activity and, possibly, the protein–DNA interaction. This ability, in turn, may affect different pathways in cell signaling and gene expression. In contrast to the nonspecific antioxidant capacity, this activity is considered to be quite specific for any given bioavailable compound, and small differences between the chemical structures of similar molecules are likely to result in great differences in their biological activity. This simply means that it is extremely unlikely that all polyphenols will display the same ability in modulating enzyme activity or gene expression, even though they share the same chemical structure (the phenylbenzo-pyrone, three-ring structure), in much the same way that your house key, though very similar to other house keys, is not able to unlock the door of someone else's house.

29.7 ANTIOXIDANT CAPACITY OF PYCNOGENOL

The most obvious feature of Pycnogenol, owing to the basic chemical structure of its components, is its strong antioxidant activity. Phenolic acids, polyphenols, and in particular flavonoids are composed of one (or more) aromatic rings bearing one or more hydroxyl groups and are therefore potentially able to quench free radicals by forming resonance-stabilized phenoxyl radicals.[64] The ability to quench the reactivity of superoxide radical anion ($O_2^{\bullet-}$), hydroxyl radical (HO$^{\bullet}$), lipid peroxyl radical, and the reactive nitrogen species, such as nitric oxide (NO$^{\bullet}$) and peroxynitrite (ONOO$^-$), which are among the most important free radicals in the biological environment and in human health and disease, have been investigated either *in vitro* or *in vivo*. Moreover Pycnogenol flavonoids may play a fundamental role in the antioxidant network by recycling other free-radical molecules, thus significantly contributing to cellular and extracellular defenses against oxidative stress, as originally suggested by Szent-Gyorgyi.[65]

Various studies have addressed the antioxidant capacity of Pycnogenol in simplified assay systems *in vitro,* cultured cell models, perfused organs, and *in vivo*. Blazso and coworkers[10] tested the *in vitro* $O_2^{\bullet-}$ scavenging activity of Pycnogenol as well as three different chromatographic fractions separated according to their molecular dimension, reporting that the most active fraction was the one containing oligomeric procyanidins. Pycnogenol has also been reported to be an efficient scavenger of both $O_2^{\bullet-}$ [10,66,67] and OH$^{\bullet}$.[66] Quite interestingly, when compared with other phytochemicals and plant extracts, Pycnogenol has been demonstrated to be the highest-ranking *in vitro* oxygen free-radical scavenger.[68] Macrides and coworkers[69] have also reported that Pycnogenol was the most potent procyanidin-containing plant extract in inhibiting the degradation of deoxyribose challenged by iron ascorbate. Most interestingly, within the context of this chapter, it has been reported that Pycnogenol is able to protect DNA from iron/ascorbate-induced damage, as indicated by a significant decrease in single- and double-strand DNA breaks.[70]

Both NO• and ONOO⁻ are attracting increasing interest as oxidizing molecules. NO• is a relatively unreactive molecule, although it displays a high affinity for the heme group of some proteins.[71] It may become potentially harmful once its concentration overwhelms its neurotransmitter and second-messenger functions, and in particular when, in activated macrophages, NO• reacts with $O_2^{•-}$ to generate ONOO⁻. Pycnogenol has been reported to significantly decrease, in a dose-dependent fashion, the accumulation of nitrite after spontaneous decomposition of sodium nitroprusside, thus acting as nitric oxide radical scavenger.[66]

The capacity of Pycnogenol to protect cellular systems has also been investigated, in particular in cultured endothelial cells and macrophages. In a bovine line of cultured normal endothelial cells (pulmonary artery endothelial cells, PAEC), preincubation with 20 to 80 µg/ml Pycnogenol is associated with significant protection from both lipid peroxidation and cell damage induced by *tert*-butylhydroperoxide (t-BHP).[72] In the same cellular system, Pycnogenol has been reported to be associated with a dose-dependent decrease in the steady-state production of both $O_2^{•-}$ and hydrogen peroxide (H_2O_2), and decreased the rate of H_2O_2 accumulation following treatment with xanthine/xanthine oxidase (X/XO) as a superoxide generating system.[73] The authors report that a significant increase in GSH levels, an increased activity of the GSH redox enzymes (GSH reductase and GSH peroxidases), and an increase in the enzymatic activity of both superoxide dismutase (SOD) and catalase occurs after incubation with Pycnogenol, suggesting an induction of protein synthesis, since pretreatment with cycloheximide (an inhibitor of protein synthesis) abolished its effect.[73] These hypotheses have never been confirmed more directly to provide a possible molecular explanation for the observed increase of GSH-related metabolism. In fact, a capacity of Pycnogenol to specifically enhance cellular antioxidant capacity would have a tremendous potential impact on a wide spectrum of pathophysiological conditions, because induction of the synthesis of antioxidant enzymes would amplify the organism's antioxidant defenses and modulate all cellular responses that are regulated by the cellular redox status.[33]

The preincubation with Pycnogenol has been reported to significantly decrease the extent of the oxidative burst induced by zymosan treatment in a dose-dependent manner in the murine macrophage cell line J774.[70] This observation corroborates a possible beneficial role for Pycnogenol on the modulation of macrophage activity and on chronic inflammatory conditions.[6,66]

A protective activity has also been reported for the eye, which is constantly exposed to photooxidative stress. Pycnogenol was found to protect both purified rod outer segments and the pigment epithelium of the retina from lipid peroxidation induced *in vitro* by ferric iron.[74] Not surprisingly, ascorbate did not protect against such iron-driven lipid peroxidation, whereas flavonoids related to Pycnogenol components, such as the gallic ester of catechin, enhanced the protective effect of α-tocopherol.[74,75] The nature of the inhibitory activity reported in this study was not investigated, and therefore it is not clear if Pycnogenol and other tested compounds reduced lipid peroxidation because of their free-radical-quenching ability or by chelating ferric ions, or both.

In hearts isolated from rats after dietary supplementation of Pycnogenol (50 mg/day for 4 weeks and then 30 mg/day for 3 weeks) subjected to ischemia/reperfusion injury, neither dietary Pycnogenol nor its addition to the perfusate significantly decreased tissue damage, assessed both as release of low-molecular-weight iron and mitochondrial oxidative phosphorylation.[76] However, in the same study, catechin, which is the major component of procyanidins, as well as one of the main polyphenol monomers contained in Pycnogenol, was found to affect all of the above parameters, significantly protecting the heart from ischemia/reperfusion injury.

It is now clear that the various antioxidants, either in a single cell or in a complete tissue or organ, do not work in an isolated manner. Rather, their activities interact, and complex pathways of reciprocal oxidation–reduction and recycling occur, assisting overall system homeostasis.[77] Polyphenols, and in particular flavonoids, may play an important role in such interplay. Due to their average redox potential of 700 to 540 mV,[78] they are capable of reducing many of the oxygen free-radical species having redox potentials in the range of 2300 to 1000 mV.[79] The resulting radical can be reduced back to the parental nonradical species by either ascorbic acid, uric acid, or tocopherol (280, 590, and 500 mV redox potential, respectively[79]).

Among a wide number of natural plant extracts, Pycnogenol displayed the strongest effect on the lifetime of the ascorbyl radical and the ascorbate radical.[80] It was therefore proposed that Pycnogenol and polyphenols in general may regenerate ascorbic acid from its corresponding radical. In the endothelial-like cell line ECV 304 challenged either by cocultured activated macrophages or with $ONOO^-$, we have demonstrated that the preincubation with Pycnogenol is associated with an increase of the endogenous steady-state levels of α-tocopherol and with protection of α-tocopherol and GSH in conditions of oxidative stress.[4,5] All these data, taken together, suggest that polyphenols, and in particular Pycnogenol, may play a central and pivotal role in the network of antioxidants inside the cell as well as in the body as a whole.

Recently, Saliou and coworkers[81] have reported that oral supplementation of Pycnogenol reduces UV-induced erythema in the human skin and also significant inhibits the activation, the binding to DNA, and the NF-κB-dependent gene expression in the immortalized human keratinocyte cell line HaCaT. In the same context, Pycnogenol has been observed to protect GSH levels and to partially suppress the cytotoxicity induced by UV treatment, both in human primary keratinocytes and in HaCaT cells.[82]

Besides the "quenching" effects described above, phenolics, and in particular Pycnogenol, may act as preventive antioxidants by chelating transition metals, thus preventing the formation of hypervalent metal forms involved in the initiation of the peroxidation process.[83] The ability of phenolic molecules to act as metal chelators appears obvious owing to the presence of adjacent hydroxyl groups in the same aromatic ring. Despite this, relatively few studies have been performed to characterize such activity.[84,85] The interaction between either iron or copper and procyanidins extracted from *Vitis vinifera*, which have a composition quite similar to that of Pycnogenol, has been studied by Maffei Facino et al.[86] The authors observed the formation of a distinct new peak of absorption at 557 nm resulting from the combination of procyanidins and ferric iron and indicating the formation of a complex characterized by a procyanidin/iron ratio of 1:2. The stability constant of the procyanidin/iron complex was reported to be approximately the same of another strong iron chelating agent, nitrilotriacetate (NTA). In the same paper, the authors also reported observing the formation of a procyanidin/ copper complex in a ratio of 1:4, the stability of which was reported to be of the same rank of magnitude of allopurinol and glycine.[86] In a preliminary unpublished study, we have observed that Pycnogenol is able to fully protect bovine serum albumin from fragmentation induced by free radicals generated by either iron or copper ions and hydrogen peroxide (Figure 29.2). This effect is the result of the combination of chelating and "shielding" capacity due to the affinity of procyanidins to proteins (see Section 29.8). The biological consequences of this protective effect are quite obviously in the maintenance of both the structural and functional integrity of proteins in the presence of a noncontrolled flux of free radicals, such as those occurring in chronic inflammations.

29.8 SPECIFIC BINDING TO PROTEINS AND ENZYME INHIBITION

The biological effects of polyphenols, and in particular of flavonoids, are thought to result from their most widely studied property, their antioxidant activity. Little is known about their ability to bind to proteins. Interaction between small molecules such as phenolic acids and polyphenols and

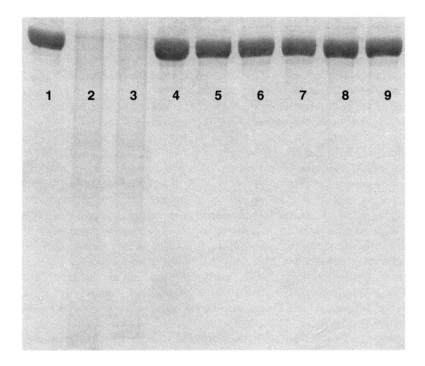

FIGURE 29.2 Pycnogenol acts as a transition metal chelator and free-radical scavenger, protecting bovine serum albumin from fragmentation induced by incubation in the presence of either Cu (as $CuSO_4$) or Fe (as FeNTA) ions plus hydrogen peroxide. DETAPAC and Desferal have been used as reference metal chelators. Lane 1 = BSA; Lane 2 = BSA + $CuSO_4$ + H_2O_2; Lane 3 = BSA + FeNTA + H_2O_2; Lane 4 = same as 2 + DETAPAC; Lane 5 = same as 3 + Desferal; Lane 6 = same as 2 + 1 µg/ml Pycnogenol; Lane 7 = same as 2 + 5 µg/ml Pycnogenol; Lane 8 = same as 3 + 1 µg/ml Pycnogenol; Lane 9 = same as 3 + 5 µg/ml Pycnogenol.

macromolecules in biological systems may result in a modification of their physicochemical properties, therefore affecting the characteristics of the entire biological system.

Pycnogenol has been reported to significantly inhibit the activity of enzymes generating free radicals in biological systems, such as horseradish peroxidase (HRP), lipoxygenase, and xanthine oxidase.[87,88] The inhibitory effect of procyanidins on enzyme activities has also been reported for other plant extracts with characteristics similar to those of Pycnogenol. Maffei Facino and coworkers[89] reported that procyanidins from *Vitis vinifera* inhibit xanthine oxidase, elastase, collagenase, β-glucuronidase, and hyaluronidase, which are considered key enzymes in the microvascular endothelium and extravascular matrix.

We have demonstrated that Pycnogenol significantly affects iNOS (inducible nitric oxide synthase) activity both *in vitro*[90] and in murine macrophages (RAW 264.7 cell line) activated by LPS and IFN-γ.[66] *In vitro*, we found that Pycnogenol had a remarkable modulatory effect on iNOS enzyme activity, producing a slight stimulatory effect at low concentrations, while acting as a powerful inhibitor of iNOS activity at higher, though still physiologically achievable, concentrations. As with many polyphenols, the inhibitory activity of Pycnogenol on iNOS is likely to be nonspecific, or partially specific, and due to the high affinity of phenolics for proteins. This is supported by the noncompetitive type of inhibitory activity that was observed. Interestingly, purified catechin and epicatechin, which are the basic blocks of procyanidin structure, were found to inhibit iNOS activity to a significantly lesser extent.[90]

29.9   NEW AVENUES TO EXPLORE BY USING NEW APPROACHES

The determination of the global picture of the effect of Pycnogenol on gene expression through genomic techniques is fundamental for a better understanding of its activity at the molecular level. cDNA arrays probably represent the best tool to open new avenues to explore in an effort to provide more insight into gene regulation and signal-transduction pathways. This technique has been recently utilized to study different molecular aspects of cancer biology such as in nasopharyngeal carcinoma,[91] renal[92] and colorectal cancer,[93] and tumor-specific T cell.[94]

Very briefly, cDNA arrays are based on the hybridization of labeled cDNA, obtained from the reverse transcription of RNA extracted from cells or tissues with a large number of specific genomic DNA fragments spotted and immobilized onto a nylon membrane, plastic, or glass. The intensity of the hybridization is proportional to the level of expression of specific mRNA. As a result, the differential expression of specific genes, either in response to a stimulus or in the presence of a pathological condition, can be assessed. The search for differentially expressed genes provides a better insight into the molecular mechanisms of nutrients, enabling the design of related "hypothesis driven" experiments.

In order to emphasize the importance and the possibility of this kind of approach, we present here some preliminary information obtained in a study of the effect of Pycnogenol on gene expression in a human tumor cell line, U937. This cell line has been obtained from a pleural effusion of a human Caucasian histiocytic lymphoma with a lymphoblast morphology. The cell line U937 is one of only a few human lines still expressing many of the monocytic like characteristics exhibited by cells of histiocytic origin.

Cells were incubated with 5 µg/ml Pycnogenol for 16 h, and gene expression was evaluated by Atlas cDNA array (Clontech), enabling the simultaneous detection of the differential expression level of 1200 genes related to cancer induction, development, and growth. We observed a very interesting effect of Pycnogenol that can be summarized as a significant down-regulation of genes encoding for oncogenic hyperproliferative proteins and an up-regulation of genes encoding for proteins controlling cell division. This effect is not present in normal human monocytes. Figure 29.3 shows an overview of the effect of Pycnogenol on the pattern of gene expression in U937 human lymphoma. As mentioned before, these data must be considered preliminary and need further confirmation. However, as we believe that these data can be of interest and provide background for more investigation, we present some of the genes that displayed the most significant regulation with respect to untreated cells and a brief description of their known functions, leaving it to the reader to develop any possible conclusions.

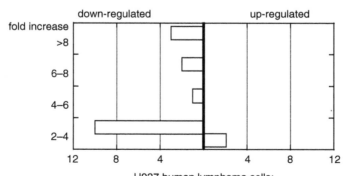

FIGURE 29.3 Summary of the changes in the pattern of mRNA expression induced by 16-h treatment with 5 µg/ml Pycnogenol in U937 human tumor cells. Only changes greater than 2-fold are considered significant and reported.

UP-REGULATED GENES

p73 (Monoallelically Expressed p53-Related Protein)

This gene was originally identified by Kaghad and coworkers.[95] It shares the sequence and some functional similarities with p53, and it can therefore be expected to act as a tumor suppressor. In fact, p53 gene is the most frequently mutated tumor suppressor identified in human cancers.[96] The ability of p53 to inhibit cell growth is due, at least in part, to its ability to bind to specific DNA sequences and activate transcription of target genes, such as that encoding the cell-cycle inhibitor p21 (Waf1/Cip1). The p73 gene maps to a region that is frequently deleted in neuroblastoma, indicating that the loss of p73 function might have a role in the development of this tumor. It has been shown that p73 is able to activate the transcription of p53-responsive genes and inhibit cell growth in a p53-like manner by inducing apoptosis.[97]

Vimentin

Vimentins are class-III intermediate filaments (IF) found in various nonepithelial cells, in particular in mesenchymal cells. These genes are highly expressed in fibroblasts, but also, though in a lesser extent, in T and B lymphocytes, while little or no expression occurs in Burkitt's lymphoma cell lines.

IF proteins form the largest family of cytoskeletal proteins in mammalian cells. The function of these proteins has long been thought to be only structural. However, this specific function is not sufficient to explain their diverse tissue- and differentiation-specific expression patterns. It is now emerging that IFs are a central part of a framework controlling essential cell processes, and in particular in signal transduction.[98]

Vimentin is one of the most prominent phosphoproteins in various cells of mesenchymal origin. The phosphorylation process is enhanced in the course of cell division, at which time vimentin filaments are significantly reorganized. In fact, it has been shown that the differentiation and decreased proliferation of U937 human promonocytic cells, induced by several drugs, is associated by surface accumulation of integrins and by a progressive increase of vimentin and lamin A and C mRNAs,[99,100] suggesting that vimentin IFs could play an important role in the differentiation process.

DOWN-REGULATED GENES

Leukocyte IgG Receptor FC-gammaRIII (CD16a)

FC-gammaRIII is a receptor for the FC region of monomeric or complexed IGG, mediating the antibody-dependent cellular cytotoxicity and other antibody-dependent responses, such as phagocytosis and release of inflammatory mediators. Monocytes expressing CD16a and the LPS receptor CD14 are characterized by a proinflammatory pattern expression of cytokines, i.e., a lower expression of the anti-inflammatory interleukin-10 and an unchanged expression of tumor necrosis factor α (TNFα)[101] and by an high expression of β1- and β2-integrins, which are involved in their attachment at sites of inflammation.[102]

The down-regulation of CD16a pairs with the down-regulation of the HLA-DR antigen-associated invariant subunit (CD74). CD74 is a transmembrane glycoprotein that plays an important role in the immune response, as it influences the expression and peptide loading of the MHC class II molecules on antigen-presenting cells. Tumor cells overexpressing CD74 are thought to escape attacking cytotoxic lymphocytes by suppressing the host immune response.

CDC10

CDC10 is a protein belonging to the septins family. Septins are well-conserved GTPases found in animals and fungi. In yeast, they are required for the formation of 10-nm filaments and for the maintenance of cell polarity. The septins also appear to be involved in various other aspects of the organization of the cell surface. Nakatsuru et al.[103] cloned a human cDNA homologous to the gene of *Saccharomyces cerevisiae*. Using differential display and RT-PCR, Nagata et al.[104] found an involvement of this gene in human neuroblastomas.

29.10 CONCLUSION

Pycnogenol is a plant extract gaining worldwide popularity as a dietary supplement because of its diversified profile of beneficial effects in human health. Its buffering mechanism may help in reducing the toxic effects of anticancer drugs having a pro-oxidant nature.

The use of Pycnogenol as a dietary supplement may help to reduce the risk of cancer. There is ample clinical and experimental evidence indicating that Pycnogenol's strong and bioavailable antioxidant capacity places it at the top rank of biological substances to be considered beneficial for human redox balance.

On the basis of the data presented, we can draw a rough map addressing the cellular compartments where Pycnogenol could produce beneficial effects to counteract cancer promotion and development:

As an effective *bona fide* antioxidant in both plasma and intracellular membrane, Pycnogenol can significantly contribute to the maintenance of cellular redox homeostasis, in particular in the course of events that are likely to overwhelm the capacity to cope with an increased production of RONS, such as in chronic inflammations, thereby reducing the possibility of cellular damage at different targets. The ability of Pycnogenol to act as a lipid peroxidation chain breaker is also likely to reduce the toxic consequences of a free-radical-induced cellular stress.

In both membrane and cytosol, Pycnogenol binds to proteins, affecting their activity and also protecting them from oxidative damage and inactivation. The protection of proteins acting as transcription factors also results in a functional control of gene transcription and expression. Its ability to enhance the cellular antioxidant network and the expression of antioxidant enzymatic machinery may play an important role in protection from cancer promotion.

In the nucleus, the antioxidant capacity results in the protection of DNA and, therefore, of genetic information.

To summarize, Pycnogenol may contribute to the protection from cancer at three different levels: at a preventive level, as a specific modulator of cellular functions, and as an adjuvant in anticancer therapy. Beyond these interrelated activities, Pycnogenol also possibly directly affects gene expression, thus opening new hypothetical pathways as shown in Figure 29.4. Recent observations addressing its ability to directly affect transduction pathways and gene expression confirm that these are probably the avenues to explore in the future, and surely they warrant further investigation. The search for such an effect would also provide a background to understand whether there is a specific molecular mechanism underlying the extremely wide spectrum of biological activities of this procyanidin-rich extract.

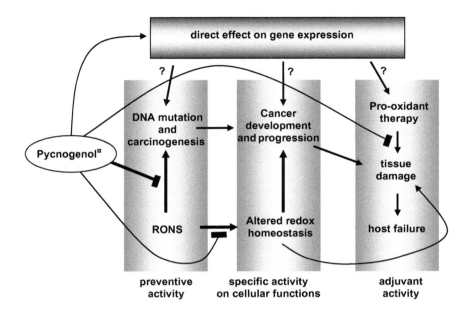

FIGURE 29.4 Pycnogenol is in theory able to contribute to the protection from cancer at three different levels: at a preventive level, as a specific modulator of cellular functions, and as adjuvant in anticancer therapy. On the top of these activities interplaying one another, Pycnogenol also possibly directly affects gene expression, opening new hypothetic pathways for its beneficial activity.

REFERENCES

1. Go, V.L.W., Wong, D.A., and Butrum, R., Diet, nutrition and cancer prevention: where are we going from here? *J. Nutrition*, 131, 3121S–3126S, 2001.
2. Packer, L., Rimbach, G., and Virgili, F., Antioxidant activity and biologic properties of a procyanidin-rich extract from pine (*Pinus maritima*) bark, pycnogenol, *Free Radical Biol. Med.*, 27, 704–724, 1999.
3. Rohdewald, P., Pycnogenol, in *Flavonoids in Health and Disease*, Rice-Evans, C. and Packer, L., Eds., Marcel Dekker, New York, 1997, pp. 405–419.
4. Virgili, F., Kim, D., and Packer, L., Procyanidins extracted from pine bark protect a-tocopherol in ECV 304 endothelial cells challenged by activated RAW 264.7 macrophages: role of nitric oxide and peroxynitrite, *FEBS Lett.*, 431, 315–318, 1998.
5. Rimbach, G., Virgili, F., Park, Y.C., and Packer, L., Effect of procyanidins from *Pinus maritima* on glutathione levels in endothelial cells challenged by 3-morpholinosyndonimine or activated macrophages, *Redox Rep.*, 4, 171–177, 1999.
6. Blazso, G., Gabor, M., and Rohdewald, P., Antiinflammatory activities of procyanidin-containing extract from *Pinus pinaster* Ait. after oral and cutaneous application, *Pharmazie*, 52, 380–382, 1997.
7. Devaraj, S., Vega-Lopez, S., Kaul, N., Schoenlau, F., Rohdewald, P., and Jialal, I., Supplementation with a pine bark extract rich in polyphenols increases plasma antioxidant capacity and alters lipoprotein profile, *Lipids*, 37, 931–934, 2002.
8. Fitzpatrick, D.F., Bing, B., and Rohdewald, P., Endothelium-dependent vascular effects of Pycnogenol, *J. Cardiovascular Pharmacol.*, 32, 509–515, 1998.
9. Liu, F.J., Zhang, Y.X., and Lau, B.H.S., Pycnogenol enhances immune and haemopoietic functions in senescence-accelerated mice, *Cell. Molecular Life Sci.*, 54, 1998.

10. Blazso, G., Gabor, M., Sibbel, R., and Rohdewald, P., Antiinflammatory and superoxide radical scavenging activities of procyanidins containing extract from the bark of *Pinus pinaster* Sol. and its fractions, *Pharm. Pharmacol.*, 3, 217–220, 1994.

11. Feng, W.H., Wei, H.L., and Liu, G.T., Effect of Pycnogenol on the toxicity of heart, bone marrow and immune organs as induced by antitumor drugs, *Phytomed.*, 9, 414–418, 2002.

12. Huynh, H.T. and Teel, R.W., Selective induction of apoptosis in human mammary cancer cells (MCF-7) by Pycnogenol, *Anticancer Res.*, 20, 2417–2420, 2000.

13. Huynh, H.T. and Teel, R.W., Effects of intragastrically administered Pycnogenol on NNK metabolism in F344 rats, *Anticancer Res.*, 19, 2095–2099, 1999.

14. Peng, Q., Wei, Z., and Lau, B.H., Pycnogenol inhibits tumor necrosis factor-alpha-induced nuclear factor kappa B activation and adhesion molecule expression in human vascular endothelial cells, *Cell. Molecular Life Sci.*, 57, 834–841, 2000.

15. Drehsen, G., From ancient pine bark uses to Pycnogenol, in *Antioxidant Food Supplements in Human Health*, Hiramatzu, M., Packer, L., and Yoshikawa, T., Eds., Academic Press, San Diego, 1998.

16. Kollesch, J. and Nickel, D., *Antike Heilkunst, Ausgewählte Texte aus den medizinischen Schriften der Griechen und Römer*, Reclam, Stuttgart, 1994.

17. Defence SCCoN, On the Possibility of Relaying on Wild Plants and Animals as a Source of Nutrition, Helsinki, 1979.

18. Dragendorff, G., *Die Heilpfanzen der verschiedenen Völker und Zeiten, Ihre Anwendung, wesentlichen Bestandteile und Geschichte*, Ferd Enke, Stuttgart, 1898.

19. Hoppe, B., *Das Kräuterbuch des Hieronimyus Bosch, Wissenschafts-historische Untersuchung*, Stuttgart, 1969.

20. Chandler, F.R., Freeman, L., and Hooper, S.N., Herbal remedies of the maritime Indians, *J. Ethnopharmacol.*, 1, 49–68, 1979.

21. Fielder, M., *Plant Medicine and Folklore*, Winchester Press, New York, 1975.

22. Youngken, H.W., The drugs of the North American Indians, *Am. J. Pharmacol.*, 97, 251–271, 1924.

23. Hollman, P.C., Bioavailability of flavonoids, *Eur. J. Clinical Nutr.*, 51 (Suppl. 1), S66–S69, 1997.

24. Choudhury, R., Chowrimootoo, G., Srai, K., Debnam, E., Rice-Evans, C.A., Interactions of the flavonoid naringenin in the gastrointestinal tract and the influence of glycosylation, *Biochemical Biophysical Res. Commn.*, 265, 410–415, 1999.

25. Bourne, L.C. and Rice-Evans, C.A., Detecting and measuring bioavailability of phenolics and flavonoids in humans: pharmacokinetics of urinary excretion of dietary ferulic acid, *Methods Enzymol.*, 299, 91–106, 1998.

26. Virgili, F., Pagana, G., Bourne, L. et al., Ferulic acid excretion as a marker of consumption of a French maritime pine (*Pinus maritima*) bark extract, *Free Radical Biol. Med.*, 28, 1249–1256, 2000.

27. Grosse-Duweler, K. and Rohdewald, P., Urinary metabolites of French maritime pine bark extract in humans, *Pharmazie*, 55, 364–368, 2000.

28. Sarikaki, V., Rallis, M., Tanojo, H., Weber, S., and Packer, L., *In Vitro* Percutaneous Absorption of Pine Bark Extract in Human's Skin, paper presented at 4th Teupitzer Colloquium: Free Radicals and Skin, Teupitz, Germany, 2001.

29. Emerit, I., Reactive oxygen species, chromosome mutation, and cancer: possible role of clastogenic factors in carcinogenesis, *Free Radical Biol. Med.*, 16, 99–109, 1994.

30. Halliwell, B., Oxygen and nitrogen are pro-carcinogens: damage to DNA by reactive oxygen, chlorine and nitrogen species: measurement, mechanism and the effects of nutrition, *Mutation Res.*, 443, 37–52, 1999.

31. Shackelford, R.E., Kaufmann, W.K., and Paules, R.S., Oxidative stress and cell cycle checkpoint function, *Free Radical Biol. Med.*, 28, 1387–1404, 2000.

32. Halliwell, B., Antioxidants in human health and disease, *Ann. Rev. Nutr.*, 16, 33–50, 1996.

33. Suzuki, Y.J., Forman, H.J., and Sevanian, A., Oxidants as stimulators of signal transduction, *Free Radical Biol. Med.*, 22, 269–285, 1997.

34. Kehrer, J.P., Free radicals as mediators of tissue injury and disease, *Crit. Rev. Toxicol.*, 23, 21–48, 1993.

35. Trouba, K.J., Hamadeh, H.K., Amin, R.P., and Germolec, D.R., Oxidative stress and its role in skin disease, *Antioxid Redox Signal*, 4, 665–673, 2002.

36. Bartsch, H. and Nair, J., Potential role of lipid peroxidation derived DNA damage in human colon carcinogenesis: studies on exocyclic base adducts as stable oxidative stress markers, *Cancer Detection Prev.*, 26, 308–312, 2002.

37. Gackowski, D., Banaszkiewicz, Z., Rozalski, R., Jawien, A., and Olinski, R., Persistent oxidative stress in colorectal carcinoma patients, *Int. J. Cancer*, 101, 395–397, 2002.
38. Kumaraguruparan, R., Subapriya, R., Viswanathan, P., and Nagini, S., Tissue lipid peroxidation and antioxidant status in patients with adenocarcinoma of the breast, *Clinica Chimica Acta; Int. J. Clinical Chem.*, 325, 165, 2002.
39. Demopoulos, H.B., Pietronigro, D.D., Flamm, E.S., and Seligman, M.L., The possible role of free radical reactions in carcinogenesis, *J. Environmental Pathol. Toxicol.*, 3, 273–303, 1980.
40. Bartsch, H., Nair, J., and Owen, R.W., Exocyclic DNA adducts as oxidative stress markers in colon carcinogenesis: potential role of lipid peroxidation, dietary fat and antioxidants, *Biol. Chem.*, 383, 915–921, 2002.
41. Nair, J., Barbin, A., Velic, I., and Bartsch, H., Etheno DNA-base adducts from endogenous reactive species, *Mutation Res.*, 424, 59–69, 1999.
42. Rozalski, R., Gackowski, D., Roszkowski, K., Foksinski, M., and Olinski, R., The level of 8-hydroxyguanine, a possible repair product of oxidative DNA damage, is higher in urine of cancer patients than in control subjects, *Cancer Epidemiol. Biomarkers Prev.*, 11, 1072–1075, 2002.
43. Reichenbach, J., Schubert, R., Schindler, D., Muller, K., Bohles, H., and Zielen, S., Elevated oxidative stress in patients with ataxia telangiectasia, *Antioxid Redox Signal*, 4, 465–469, 2002.
44. Sen, C.K. and Packer, L., Antioxidant and redox regulation of gene expression, *FASEB J.*, 10, 709–720, 1996.
45. Dhar, A., Young, M.R., and Colburn, N.H., The role of AP-1, NF-kappaB and ROS/NOS in skin carcinogenesis: the JB6 model is predictive, *Molecular Cell. Biochem.*, 234–235, 185–193.
46. Block, G., Patterson, B., and Subar, A., Fruit, vegetables and cancer prevention: a review of the epidemiological evidence, *Nutr. Cancer*, 18, 1–30, 1992.
47. Steinmetz, K.A. and Potter, J.D., Vegetable, fruit and cancer prevention: a review, *J. Am. Diet Assoc.*, 96, 1027–1039, 1996.
48. Middleton, E.J. and Kandashwami, C., The impact of plant flavonoids on mammalian biology: implications for immunity, inflammation and cancer, in *The Flavonoids: Advances in Research since 1986*, Harborne, J.H. and Liss, A.R., Eds., New York, 1993, 619–652.
49. Rosenberg-Zand, R., Jenkins, D., and Diamandis, E., Flavonoids and steroid hormone-dependent cancers, *J. Chromatogr B Amelyt Techmoe*, 777, 219, 2002.
50. Le Marchand, L., Murphy, S.P., Hankin, J.H., Wilkens, L.R., and Kolonel, L.N., Intake of flavonoids and lung cancer, *J. Natl. Cancer Inst.*, 92, 154–160, 2000.
51. Ogawa, S., Hirayama, T., Nohara, M., Tokuda, M., Hirai, K., and Fukui, S., The effect of quercetin on the mutagenicity of 2-acetylaminofluorene and benzo[alpha]pyrene in *Salmonella typhimurium* strains, *Mutation Res.*, 142, 103–107, 1985.
52. Wargovich, M.J., Eng, V.W., and Newmark, H.L., Inhibition by plant phenols of benzo[a]pyrene-induced nuclear aberrations in mammalian intestinal cells: a rapid *in vivo* assessment method, *Food Chemical Toxicol.: Int. J. Br. Industrial Biological Res. Assoc.*, 23, 47–49, 1985.
53. Kato, R., Nakadate, T., Yamamoto, S., and Sugimura, T., Inhibition of 12-O-tetradecanoylphorbol-13-acetate-induced tumor promotion and ornithine decarboxylase activity by quercetin: possible involvement of lipoxygenase inhibition, *Carcinogenesis*, 4, 1301–1305, 1983.
54. Nakadate, T., Yamamoto, S., Aizu, E., and Kato, R., Effects of flavonoids and antioxidants on 12-O-tetradecanoyl-phorbol-13-acetate-caused epidermal ornithine decarboxylase induction and tumor promotion in relation to lipoxygenase inhibition by these compounds, *Gann*, 75, 214–222, 1984.
55. Kuo, S.M., Dietary flavonoid and cancer prevention: evidence and potential mechanism, *Crit. Rev. Oncogenesis*, 8, 47–69, 1997.
56. Birt, D.F., Hendrich, S., and Wang, W., Dietary agents in cancer prevention: flavonoids and isoflavonoids, *Pharmacol. Ther.*, 90, 157–177.
57. Bracke, M.E., Van Canwenberg, R.M.-L., Mareeel, M.M., Castronovo, V., and Foidart, J.-M., Flavonoids: tools for the study of tumor invasion *in vitro*, in *Plant Flavonoids in Biology and Medicine: Biochemical, Pharmacological and Structure-Activity Relationship*, Alan R. Liss, New York, 1986, pp. 441–444.
58. Ulubelen, A., Caldwell, M.E., and Cole, J.R., Phytochemical investigation of Abies concolor, *J. Pharmaceutical Sci.*, 55, 1308–1310, 1966.
59. Chen, G., Perchellet, E.M., Gao, X.M. et al., Ability of m-chloroperoxybenzoic acid to induce the ornithine decarboxylase marker of skin tumor promotion and inhibition of this response by gallotannins,

oligomeric proanthocyanidins, and their monomeric units in mouse epidermis *in vivo*, *Anticancer Res.*, 15, 1183–1189.

60. Dauer, A., Metzner, P., and Schimmer, O., Proanthocyanidins from the bark of *Hamamelis virginiana* exhibit antimutagenic properties against nitroaromatic compounds, *Planta Medica*, 64, 324–327, 1998.

61. Grimmer, H.R., Parbhoo, V., and R.M.M., Antimutagenicity of polyphenol-rich fractions from sorghum bicolor grain, *J. Sci. Food Agric.*, 59, 251–256, 1992.

62. Liviero, L., Puglisi, P.P., Morazzoni, P., and Bombardelli, F., Antimutagenic activity of procyanidins from *Vitis vinifera*, *Fitoterapia*, 65, 203–210, 1994.

63. Miyamoto, K., Kishi, N., Koshiura, R., Yoshida, T., Hatano, T., and Okuda, T., Relationship between the structure and the antitumor activities of tannins, *Chem. Pharm. Bull.*, 35, 814–822, 1987.

64. Rice-Evans, C.A. and Miller, N.J., Antioxidant activities of flavonoids as bioactive components of food, *Biochem. Soc. Trans.*, 24, 790–795, 1996.

65. Benthsath, A., Rusznyak, S., and Szent-Gyorgyi, A., Vitamin nature of flavones, *Nature*, 138, 798, 1936.

66. Virgili, F., Kobuchi, H., and Packer, L., Procyanidins extracted from *Pinus maritima* (Pycnogenol): scavengers of free radical species and modulators of nitrogen monoxide metabolism in activated murine RAW 264.7 macrophages, *Free Radical Biol. Med.*, 24, 1120–1129, 1998.

67. Elstner, E.F. and Kleber, E., Radical scavenger properties of leucocyanidine, in *Flavonoids in Biology and Medicine, III: Current Issues in Flavonoids Research*, Das, N.P., Ed., National University of Singapore Press, Singapore, 1990, 227–235.

68. Noda, Y., Anzai, K., Mori, A., Kohno, M., Shinmei, M., and Packer, L., Hydroxyl and superoxide anion radical scavenging activities of natural source of antioxidants using the computerized JES-FR30 ESR spectrometer system, *Biochem. Mol. Biol. Int.*, 42, 35–44, 1997.

69. Macrides, T.A., Shihata, A., Kalafatis, N., and Wright, P.F.A., A comparison of the hydroxyl radical scavenging properties of the shark bile steroid 5-b-scymnol and plant pycnogenols, *Biochem. Mol. Biol. Int.*, 42, 1249–1260, 1997.

70. Nelson, A.B., Lau, B.H.S., Ide, N., and Rong, Y., Pycnogenol inhibits macrophage oxidative burst, lipoprotein oxidation and hydroxyl radical induced DNA damage, *Drug Develop. Indust. Med.*, 24, 1–6, 1998.

71. Staedler, J., Schmalix, W.A., and Doehmer, J., Inhibition of biotransformation by nitric oxide (NO) overproduction and toxic consequences, *Toxicol. Lett.*, 82, 215–219, 1995.

72. Rong, Y., Li, L., and Lau, B.H., Pycnogenol protects vascular endothelial cells from t-butyl hydroperoxide induced oxidant injury, *Biotechnol. Ther.*, 5, 117–126, 1994–1995.

73. Wei, Z., Peng, Q., and Lau, B.H.S., Pycnogenol enhances endothelial cell antioxidant defences, *Redox. Rep.*, 3, 147–155, 1997.

74. Ueda, T., Ueda, T., and Armstrong, D., Preventive effect of natural and synthetic antioxidants on lipid peroxidation in the mammalian eye, *Ophthalmic Res.*, 28, 184–192, 1996.

75. Chida, M., Suziki, K., Nokanishi-Ueda, T. et al., *In vitro* testing of antioxidants and biochemical endpoints in bovine retinal tissue, *Ophtalmic Res.*, 31, 407–415, 1999.

76. van Jaarsveld, H., Kuyl, J.M., Schulemburg, D.H., and Wiid, N.M., Effect of flavonoids in the outcome of myocardial mitochondrial ischemia/reperfusion injury, *Res. Comm. Mol. Pathol. Pharmacol.*, 91, 65–75, 1996.

77. Packer, L., Interactions among antioxidants in health and disease: vitamin E and its redox cycle, *Proc. Soc. Exp. Biol. Med.*, 200, 271–276, 1991.

78. Jovanovic, S.V., Steeken, S., Tosic, M., Marjanovic, B., and Simic, M.G., Flavonoids as antioxidants, *J. Am. Chem. Soc.*, 116, 4846–4851, 1994.

79. Buettner, G.R., The pecking order of free radicals and antioxidants lipid peroxidation alpha tocopherol and ascorbate, *Arch. Biochem. Biophys.*, 300, 535–543, 1993.

80. Cossins, E., Lee, R., and Packer, L., ESR studies of vitamin C regeneration, order of reactivity of natural source phytochemical preparations, *Biochem. Mol. Biol. Int.*, 45, 583–598, 1998.

81. Saliou, C., Rimbach, G., Moini, H. et al., Solar ultraviolet-induced erythema in human skin and nuclear factor-kappa-B-dependent gene expression in keratinocytes are modulated by a French maritime pine bark extract, *Free Radical Biol. Med.*, 30, 154–160, 2001.

82. Bito, T., Roy, S., Sen, C.K., and Packer, L., Pine bark extract pycnogenol downregulates IFN-gamma-induced adhesion of T cells to human keratinocytes by inhibiting inducible ICAM-1 expression, *Free Radical Biol. Med.*, 28, 219–227, 2000.

83. Cao, G., Sofic, E., and Prior, R.L., Antioxidant and prooxidant behavior of flavonoids: structure-activity relationships, *Free Radical Biol. Med.*, 22, 749–760, 1997.

84. Nardini, M., D'Aquino, M., Tomassi, G., Gentili, V., Di Felice, M., and Scaccini, C., Inhibition of human low-density lipoprotein oxidation by caffeic acid and other hydroxycinnamic acid derivatives, *Free Radical Biol. Med.*, 19, 541–552, 1995.

85. Decker, D.A., Phenolics: prooxidant or antioxidants, *Nutr. Rev.*, 55, 396–398, 1997.

86. Maffei Facino, R., Carini, M., Aldini, G. et al., Procyanidines from *Vitis vinifera* seeds protect rabbit hearth from ischemia/reperfusion injury: antioxidant intervention and/or iron and copper sequestering ability, *Planta Medica*, 62, 495–502, 1996.

87. Moini, H., Guo, Q., and Packer, L., Enzyme inhibition and protein-binding action of the procyanidin-rich French maritime pine bark extract, pycnogenol: effect on xanthine oxidase, *J. Agricultural Food Chem.*, 48, 5630–5639, 2000.

88. Moini, H., Guo, Q., and Packer, L., Protein binding of procyanidins: studies using polyacrylamide gel electrophoresis and French maritime pine bark extract, *Methods Enzymol.*, 335, 333–337.

89. Maffei Facino, R., Carini, M., Aldini, G., Bombardelli, E., Morazzoni, P., and Morelli, R., Free radical scavenging action and anti-enzyme activities of procyanidins from *Vitis vinifera*, *Drug Res.*, 44, 592–601, 1994.

90. Kobuchi, H., Virgili, F., and Packer, L., Assay of the inducible form of nitric oxide synthase activity: effect of flavonoids and plant extracts, *Methods Enzymol.*, 301, 504–513, 1999.

91. Guo, X., Lui, W.O., Qian, C.N. et al., Identifying cancer-related genes in nasopharyngeal carcinoma cell lines using DNA and mRNA expression profiling analyses, *Int. J. Oncol.*, 21, 1197–1204, 2002.

92. Wilhelm, M., Veltman, J.A., Olshen, A.B. et al., Array-based comparative genomic hybridization for the differential diagnosis of renal cell cancer, *Cancer Res.*, 62, 957–960, 2002.

93. Zimmer, R. and Thomas, P., Expression profiling and interferon-beta regulation of liver metastases in colorectal cancer cells, *Clinical Experimental Metastasis*, 19, 541–550, 2002.

94. Zhang, X., Chen, Z., Huang, H., Gordon, J., and Xiang, J., DNA microarray analysis of the gene expression profiles of naive versus activated tumor-specific T cells, *Life Sci.*, 71, 3005, 2002.

95. Kaghad, M., Bonnet, H., Yang, A. et al., Monoallelically expressed gene related to p53 at 1p36, a region frequently deleted in neuroblastoma and other human cancers, *Cell*, 90, 809–819, 1997.

96. Hill, K.A. and Sommer, S.S., p53 as a mutagen test in breast cancer, *Environmental Molecular Mutagenesis*, 39, 216–227, 2002.

97. Jost, C.A., Marin, M.C., and Kaelin, W.G., p73 is a simian [correction of human] p53-related protein that can induce apoptosis, *Nature*, 389, 191–194, 1997.

98. Paramio, J.M. and Jorcano, J.L., Beyond structure: do intermediate filaments modulate cell signalling? *Bioessays: News Rev. Molecular, Cell. Developmental Biol.*, 24, 836–844, 2002.

99. Perez, C., Vilaboa, N.E., and Aller, P., Etoposide-induced differentiation of U937 promonocytic cells: AP-1-dependent gene expression and protein kinase C activation, *Cell Growth Differentiation: Molecular Biol. J. Am. Assoc. Cancer Res.*, 5, 949–955, 1994.

100. Hass, R., Giese, G., Meyer, G. et al., Differentiation and retrodifferentiation of U937 cells: reversible induction and suppression of intermediate filament protein synthesis, *Eur. J. Cell Biol.*, 51, 265–271, 1990.

101. Frankenberger, M., Sternsdorf, T., Pechumer, H., Pforte, A., and Ziegler-Heitbrock, H.W., Differential cytokine expression in human blood monocyte subpopulations: a polymerase chain reaction analysis, *Blood*, 87, 373–377, 1996.

102. Schmitz, G., Herr, A.S., and Rothe, G., T-lymphocytes and monocytes in atherogenesis, *Herz*, 23, 168–177, 1998.

103. Nakatsuru, S., Sudo, K., and Nakamura, Y., Molecular cloning of a novel human cDNA homologous to CDC10 in *Saccharomyces cerevisiae*, *Biochem. Biophys. Res. Commn.*, 202, 82–87, 1994.

104. Nagata, T., Takahashi, Y., Asai, S. et al., The high level of hCDC10 gene expression in neuroblastoma may be associated with favorable characteristics of the tumor, *J. Surgical Res.*, 92, 267–275, 2000.

30 Overview of the Use of Maitake Mushroom and Fraction D in Cancer

Harry Preuss, Sensuke Konno, and Debasis Bagchi

CONTENTS

30.1 Background...509
30.2 Maitake Fraction D..510
30.3 Beta-Glucans: an Introduction 510
30.4 Immune-Enhancing Functions of Beta-Glucans.......................................510
30.5 Therapeutic Action of Beta-Glucans...511
30.6 Apoptosis..512
30.7 Maitake and Beta-Glucans: Research on Health Benefits.........................513
30.8 Effects of Maitake on Side Effects of Chemotherapy...............................514
30.9 Summary and Conclusions...514
References 515

30.1 BACKGROUND

Traditional healers in the Far East have long recognized the health-enhancing potential of mushrooms. In fact, mushrooms are considered by many in this same region to be mythic foods of immortality, i.e., some mushrooms are regarded as longevity herbs that preserve youth and maintain health. Although this chapter will discuss several mushrooms with therapeutic potential, we will especially focus on one mushroom — maitake. Common lore states that maitake is so named because people who found it in the deep mountains, knowing of its luscious flavor and health benefits, began dancing with joy. In feudal Japan, maitake mushroom was exchanged for the same weight of silver by local lords who, in turn, offered it to the shogun. *Grifola frondosa* is the botanical name of maitake, a designation that refers to a mythical beast — half-lion and half-eagle.

In the present era, extracts derived from maitake mushroom and other mushrooms are considered by some to be frontline medical "drugs," even though the scientific study of mushrooms only began during the last two decades.[1-6] A number of mushrooms are rich in minerals (potassium, calcium, and magnesium), various vitamins (D_2, B_2, niacin, and C), and amino acids. However, some believe the most important ingredients found in maitake are polysaccharide compounds called beta-glucans that exhibit strong immune-support activity.[7-25] Maitake mushroom, indigenous to the northeastern part of Japan, is an especially rich source of beta-glucans.[7,13,17] A major attribute of beta-glucans is their effects on the immune system.

Several natural dietary supplements are commonly accepted to benefit the immune system, i.e., act as biological response modifiers that stimulate nonspecific immunity. Among these are various synthetic and biological agents, each with completely different chemical structures. Herbs such as astragalus, echinacea, and others have been documented to possess a nonspecific immunomodulatory effect.[27,28] In contrast, most of the medicinal mushrooms such as reishi, shiitake, cordyceps, and maitake show a common ability to enhance immune function by stimulating cell-mediated immunity.[29] Quite simply, mushrooms seem to activate cells in the immune system, including macrophages and T cells, that appear to have significant cancer and infection-fighting properties. It is important to note that while many fungi can be used as important sources of beta-glucans, maitake is substantially different from the others because it appears to retain efficacy when administered orally. To use an example, lentinan derived from shiitake must be administered intravenously to be most effective.[29] To sum up, maitake is a very effective mushroom when compared with others, can be taken orally, and is completely safe. Also worth noting is that different factors in the crude maitake powder offer many additional health benefits, including blood-sugar control, cholesterol lowering, high blood-pressure reduction, and even weight loss.[30-36]

30.2 MAITAKE FRACTION D

The majority of studies on the immunological capabilities and cancer effects of maitake have been performed mainly with one of its bioactive extracts, referred to as maitake Fraction D. Fraction D, weighing approximately 1×10^6 Da, is a hot-water-extractable fraction prepared by standardized procedures.[1,11] A protein-bound polysaccharide, consisting of either B-1,6 linked glucan with B-1,3 branches or of B-1,3 glucan branched with B-1,6 glucosides, is the major component of this fraction. Although the major beneficial effects of Fraction D seem to derive from its immune-enhancing potential, other very different physiological mechanisms may contribute to the overall therapeutic effect, e.g., antiangiogenesis and apoptosis. These mechanisms will be described more fully below.

30.3 BETA-GLUCANS: AN INTRODUCTION

Today, many mushrooms, in addition to maitake, are being recognized for their important health benefits. This is primarily because they are an excellent source of beta-glucan polysaccharide compounds that promote well-being.[3] Beta-glucans (also known as beta-1,3/1,6-D-glucan or beta-1,3/1,6-glucan) are nondigestible polysaccharides (long-chain carbohydrate) found in the cell walls of maitake mushroom. As with any beta-glucan (or polyglucose), its beta-1,3/1,6-D-glucan consists of multiple glucose units linked together. Different from common table sugar, a dimer composed of glucose and fructose that can be deleterious to health when consumed in excess, beta-glucans are healthful. Indeed, we need to take in, at the least, small amounts of beta-glucans for our bodies to function properly. However, these are no longer readily available in the average diet to any great extent, and they are scarcely present in our modern, overprocessed diets.

30.4 IMMUNE-ENHANCING FUNCTIONS OF BETA-GLUCANS

These complex sugars are a foundation for multicellular immune intelligence — the ability of immune cells to communicate, cohere, and work together to keep and maintain an individual in a healthy state. Prudent consumption of beta-glucans benefits the immune system significantly. Consuming individuals generally reactivate their immune system and enhance their ability to fight a variety of diseases. Further, this allows them to more easily ward off a multitude of infections.

How do beta-glucans operate? Beta-1,3/1,6-D-glucan molecules readily bind to macrophages containing specific protein-based receptor complexes on their cell walls. Certain beta-glucans are known to dock onto these receptors. This activates the macrophages so that they can aid an individual in the fight against cancer and infectious diseases,[26] i.e., the binding of beta-1,3/1,6-D-glucan to the macrophage enhances its ability to detect and scavenge many different health threats. An overall impression can be made that the body's macrophages and beta-glucans were meant for each other. Bacterial infections respond remarkably to these polysaccharides, as do many viral infections, from the common cold and flu to herpes and HIV. Beta-glucans are reported also to mitigate the toxic effects of radiation and chemotherapy while augmenting their cancer-killing effects. This, in turn, can lead to prolonged survival and improved quality of life for cancer patients.[1,2]

Up to now, copious studies have been conducted on various beta-glucans, and much information can be derived from such studies. Beta-glucans are classified by scientists as semiessential nonvitamin factors and are believed to protect against certain diseases. These polysaccharides are found in a variety of foods, including oats, nutritional yeasts, and other medicinal mushrooms, but the chemical structures of maitake's beta-glucans, especially those found in the D fraction, are unique due to their greater degree of molecular branching.[6,7] Observers have known for a very long time that the Monkey's Bench mushroom family, of which maitake is a member, possesses significant anticancer effects. Therefore, they have used the ingredients extracted by boiling these substances down as a medicinal hot-water treatment for certain cancers. Interesting to note is that beta-glucans in maitake are water soluble. It is now generally accepted that the antitumor effects of maitake are the result of the activity of various beta-glucans found in the fruit body and mycelium, which is the mass of interwoven filaments that form the vegetative portion of the maitake mushroom and are submerged in soil or organic matter.[14]

Researchers obtained various fractions by continually refining down the elements in the fruiting body of maitake. The results of this research were first published in the 1980s.[6–22] It was the D fraction of maitake extract, the fraction obtained last, that was found to possess the most potent antitumor activity, leading to the highest reduction rate in cancer proliferation.[1,2,37–41] Focused research on the beta-glucans found in the D fraction of maitake demonstrates positive effects not only on macrophages but also natural killer cells and various T cells. Additional research further demonstrates high cancer inhibition with oral administration of any source of beta-glucan.[8,10,20–22]

30.5 THERAPEUTIC ACTION OF BETA-GLUCANS

No matter whether an infection is bacterial, viral, fungal, or via parasite infestation — maitake's beta-glucans have been found to effectively activate the immune system and enhance the body's healing response.[1] Complementing antibiotics and chemotherapeutic agents, beta-glucans influence generalized immune enhancement by supporting a fundamental aspect of immune function — the workings of the primitive macrophage. Indeed, beta-glucans may be essential to optimal immune function for virtually all life forms. Meanwhile, more studies on various beta-glucans, all with similar configurations and derived from a variety of sources, continue to demonstrate an even broader array of therapeutic benefits. A literature review on maitake confirms that this natural medicine is an important adjuvant in cancer therapeutics, both for mitigating the damaging side effects of chemotherapy and radiation, as well as for improving the body's innate immune defenses and improving the main outcome — living cancer free. In addition to prevention and treatment of a wide variety of cancers, beta-glucans have been shown to give some protection against surgical-, chemotherapeutic-, and radiation-induced side effects.

Another way that maitake's beta-glucans may aid in the battle against major disease processes is by stimulating production of tumor inhibitors. In certain experimental models, systemic macrophage activation and certain cytokine releases seem to be critical for clearing tissues of tumor cells and inhibiting metastasis. In 1995, Dr. Mitsuhiro Okazaki and coresearchers[43] reported that maitake mushroom stimulates release, or, at least, "primes" the body to release tumor necrosis factor alpha (TNF α). Since then, additional studies have corroborated that maitake is a powerful, broad-spectrum cytokine inducer. In other words, maitake D fraction exhibits an antitumor effect on tumor-bearing mice through both enhanced cytotoxic activity and stimulation of macrophages, helping these white cells to live up to their fullest protective potential. This augmented activation of macrophages results in elevated production of interleukin-1 and thereby to activation of cytotoxic T lymphocytes, with the subsequent release of many additional cytokines.

Other basic research using an animal model shows that the D fraction of maitake mushroom can inhibit metastasis of cancer cells. In some cases, cancer metastases were reduced to less than one-tenth with its use.[37,44] Specifically, MM-164 liver cancer was injected into the left rear footpad of three groups of mice, and the footpad was cut off after 48 h. The first group was given normal feed. The second group received 20% maitake powder in their feed. The third group received one milligram per kilogram maitake D fraction with their feed. All three groups were then fed for another 30 days. The number of tumor metastases in the liver was counted by microscopic examination. Relative to a control group, metastases to the liver were prevented by 91% with maitake D fraction and by 81% with maitake crude powder at the doses administered.

Cancer researchers have studied the conditions necessary for cancer metastasis and discovered that tumor angiogenesis is a critical event.[1] Tumor angiogenesis is the proliferation of a network of blood vessels that penetrates into cancerous growths, supplying nutrients and oxygen and removing waste products. Tumor angiogenesis actually starts with cancerous tumor cells releasing molecules that send signals to surrounding normal host tissue. Through this signaling, certain genes in the host tissue that make proteins that encourage growth of new blood vessels are activated. Maitake's beta-glucans can prevent this process, at least to some extent, by stimulating tumor necrosis factor. Maitake D fraction was observed to affect angiogenesis *in vivo* and to enhance the proliferative and migratory capabilities of human vascular endothelial cell *in vitro*. D fraction also increased plasma vascular endothelial growth factor (VEGF) concentration significantly. Also, the production of VEGF and TNFα by the activated peritoneal macrophages was enhanced. Thus, consideration of all these results suggests that the antitumor activity of the D fraction not only works via the activation of the immunocompetent cells, but also possibly through inhibition of carcinoma angiogenesis induction.

30.6 APOPTOSIS

Normal or cancer cell death generally follows two distinct pathways: active apoptosis (programmed cell death) and passive necrosis. Apoptosis can be loosely defined as "cell suicide," although it is the highly organized biochemical process that is triggered by specific chemicals or biologicals.[45] Once the apoptotic machinery is turned on, cells "actively" prepare for suicide as a cascade of complex cellular events begins to take place. Only those cells programmed for death will die (or commit suicide) without disturbing adjacent cells. Generally, one of the major problems with cell death (particularly necrosis) is a secondary inflammation of adjacent and distant cells due to a release or spill of cytotoxic materials from dying cells as they rupture. However, during the apoptotic process, there is no sudden cell rupture or release of cytotoxic substances, so that all neighboring cells remain intact. This is the primary advantage of cells undergoing apoptosis.

In contrast, necrosis is considered "random cell murder," mediated by various extracellular elements and factors such as drugs, agents, radiation, microbes, etc., so that cells literally get killed or murdered. Compared with apoptosis, this is a passive, disorganized, "chaotic" process, involving cell rupture and a release of cytotoxic materials that will randomly exert their adverse effects on other healthy cells. Most of the current cancer treatments are aimed to induce necrosis in cancer cells. One can then understand why many patients suffer from various side effects such as hair loss, depression, nausea, dizziness, diarrhea, etc. under chemotherapeutic regimens. Instead, if regimens specifically triggering apoptosis in the targeted cancer cells become available, it could be a great relief for patients and their families. This is a major goal in improving the efficacy of the current cancer treatments while minimizing their side effects.

With the above comments in mind, D fraction has been reported to induce *apoptosis* in human prostatic cancer PC-3 cells *in vitro*.[46] D fraction alone as well as combination of D fraction and vitamin C caused over 90% apoptotic cell death in 24 h. Thus, D fraction may act as a potent

Control

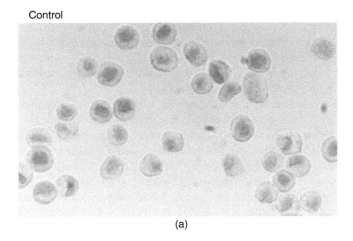

(a)

D-fraction-treated

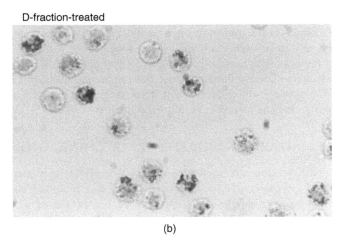

(b)

FIGURE 30.1 *In situ* hybridization (ISH) assay. Control (A) and D-fraction-treated (480 meg/ml) PC-3 cells (B) at 24 h were evaluated for apoptosis by the ISH assay. A greater than 90% (92 cells/100 cells counted) of D-fraction-treated cells were positively stained (B), indicating apoptotic cell death. In contrast, <10% (8/100) of controls showed artificial, nonspecific staining (A) irrelevant to apoptosis. [magnification: 200×] (From Files of Dr Konno)

apoptosis inducer for certain human cancers, and such a possibility deserves further investigation. In addition, small-scale clinical studies using D fraction on patients with prostate cancer are currently in progress, and we may have some useful data/results in the near future.

30.7 MAITAKE AND BETA-GLUCANS: RESEARCH ON HEALTH BENEFITS

In 1978, Mansell et al.[47,48] discussed the use of beta-glucans for immune modulation and as adjuvant therapy to control cancer. One of the most lethal cancers today is melanoma. Dr. Mansell and colleagues found that injection of beta-1,3-D-glucan into melanoma lesions caused successful regression. Examination of the involved site showed extreme macrophage activity during the time that they engulfed and destroyed tumor cells. Use of injections also appeared to reduce the risk of the spread of the cancer. The size of the large cancer lesions was markedly reduced in just 5 days, and in small lesions, resolution was complete.

A 1980 experimental study from the Department of Physiology at Tulane University School of Medicine, New Orleans, LA, also demonstrates that beta-glucans may inhibit the growth of these cancers.[49] This may have occurred via selective enhancement of apoptosis in the tumor cells.[50] Beta-glucan injections decreased tumor weight by about 70% in mice with melanoma and inhibited spread of the cancer to the lungs. Importantly, the researchers also reported that beta-glucans were effective in prolonging survival of mice with melanoma.

In other published studies, maitake D fraction has shown the potential to benefit persons suffering from cancers of the pancreas, brain, prostate, liver, lung, or breast. The end points assessed for this benefit were either (1) decreases or stabilization of tumor size, (2) reduction in the intensity of tumor markers, or (3) the prolongation of patients' expected life span by more than four-fold.[2,41] For example, Chinese researchers from Zhejiang Medical University and Zhejiang Cancer Hospital, Hangzhou, Zhejiang Province, China, used maitake extract on 63 cancer patients as adjuvant therapy with their chemotherapy and radiation treatments.[51] The patients took the maitake extract before meals, four times per day. The results were highly gratifying, with a 96% success rate against solid tumors and 91% among the leukemia patients. The researchers estimated the total effective rate for overall immune enhancement at 87%.

In a nonrandomized clinical study, 165 patients aged 25 to 65 and diagnosed with stage III–IV cancers were given maitake D fraction with crude maitake powder tablets alone or with chemotherapy.[2,41] Tumor regression or significant improvements in those receiving extract alone were observed in 11 of 15 breast cancer patients, 12 of 18 lung cancer patients, and 7 of 15 liver cancer patients. If taken with chemotherapy, these response rates improved by 12 to 28%. In several cases with both liver and lung cancer, the patients went from dangerous stage III status to more manageable stages. Virtually all patients reported improved quality of life.

30.8 EFFECTS OF MAITAKE ON SIDE EFFECTS OF CHEMOTHERAPY

The use of the maitake D fraction can ameliorate various side effects secondary to chemotherapy. Beneficial improvements were found in 90% of patients and included ameliorations of lost appetite, vomiting, nausea, and hair loss as well as increased counts of white blood cell; pain was reduced in 83%.[25] It is worth repeating that there are strong indications that while maitake extract contributes to tumor reduction without a high loss of white blood cells, it also reduces in many cases pain, hair loss, nausea, and other side effects usually associated with chemotherapy.

30.9 SUMMARY AND CONCLUSIONS

What makes maitake and its fractions so special as a cancer therapeutic agent compared with other natural agents used for similar purposes? For one thing, maitake has been studied extensively, allowing one to examine concrete data. Also, beta-glucans obtained from maitake mushrooms have unique, complex, and varied chemical structures that enhance their therapeutic benefits. A three-dimensional model of maitake's beta-1,3-D glucan shows it to be a helix with its 1,6 main chain having a greater degree of 1,3 branches.[6,12–14] It has been postulated that this greater degree of branching provides beta-glucans derived from maitake more chance to reach each immune cell for activation, resulting in greater potency. Many beta-glucans found in this natural medicine have a $(1 \rightarrow 6)$ branch in every other main-chain unit. Others have two $(1 \rightarrow 6)$ branches in every third main-chain unit. In studies on structure–activity relationships of beta-glucan-mediated immunopharmacological activity, it was found that some of the activities tested are influenced by the diversity of the beta-glucans. The maitake D fraction activates macrophages, natural killer cells, and other T cells to attack the tumor cells. While the beta-glucans in D fraction potentiate the activity of various mediators, mainly lymphokines and interleukin-1 and interleukin-2, other beta-glucan fractions obtained from maitake and those derived from other sources have completely different activities.[52] Although many more clinical studies are required before the true benefits of whole maitake powder and Fraction D can be established, the risk/benefits of the mushroom extract appear so favorable as to suggest strong consideration for its immediate therapeutic use for preventive measures or as an adjunct to more conventional therapies.

REFERENCES

1. Preuss, H.G. and Konno, S., *Maitake Magic*, Freedom Press, Topanga, CA, 2002.
2. Lieberman, S. and Babal, K., *Maitake, King of Mushrooms*, Keats Publishing, New Canaan, CT, 1997, pp. 20–21.
3. Borchers, A.T., Stern, J.S., Hackman, R.M., Keen, C.L., and Gershwin, M.E., Mushrooms, tumors, and immunity, *PSEBM*, 221, 281–293, 1999.
4. Jones, K., Maitake: a potent medicinal food, *Alternative Complementary Therapies*, (Dec.), 420–442, 1998.
5. Mizuno, T. and Zhuang, C., Maitake, *Grifola frondosa*: pharmacologic effects, *Food Rev. Int.*, 11, 135–149, 1995.
6. Ohno, N., Suzuki, I., Oikawa, S., Sato, K., Miyazaki, T., and Yadomae, T., Antitumor activity and structural characterization of glucans extracted from cultured fruit bodies of *Grifola frondosa*, *Chem. Pharm. Bull. (Tokyo)*, 32, 1142–1151, 1984.
7. Adachi, Y., Ohno, N., Ohsawa, M., Sato, K., Oikawa, S., and Kyadomae, T., Physio-chemical properties and antitumor activities of chemically modified derivatives of antitumor glucan "grifolan LE" from *Grifola frondosa*, *Chem. Pharm. (Tokyo)*, 37, 1838–1843, 1989.
8. Suzuki, I., Hashimoto, K., Oikawa, S., Sato, K., Wsawa, M., and Yadomae, T., Antitumor and immunomodulating activities of a beta-glucan obtained from liquid-cultured *Grifola frondosa*, *Chem. Pharm. Bull. (Tokyo)*, 37, 410–413, 1989.
9. Nanba, H., Hamaguchi, A., and Kuroda, H., The chemical structure of an antitumor polysaccharide in fruit bodies of *Grifola frondosa* (maitake), *Chem. Pharm. Bull. (Tokyo)*, 35, 1162, 1987.
10. Adachi, K., Nanba, H., and Kuroda, H., Potentiation of host-mediated antitumor activity in mice by beta-glucan obtained from *Grifola frondosa* (maitake), *Chem. Pharm. Bull. (Tokyo)*, 35, 35262–35270, 1987.
11. Ohno, N. et al., Effect of glucans on the antitumor activity of grifolan, *Chem. Pharm. Bull. (Tokyo)*, 34, 2149–2154, 1986.
12. Ohno, N., Hayashi, M., Iino, K., Suzuki, I., Oikawas, S., Sato, K., Suzuki, Y., and Yado-mae, T., Characterization of the antitumor glucan obtained from liquid-cultured *Grifola frondosa*, *Chem. Pharm. Bull. (Tokyo)*, 34, 1709–1715, 1986.
13. Iino, K., Ohno, N., Suzuki, I., Sato, K., Oikawa, S., and Yadomae, T., Structure-function relationship of antitumor beta-1,3-glucan obtained from matted mycelium of cultured *Grifola frondosa*, *Chem. Pharm. Bull. (Tokyo)*, 33, 4950–4956, 1985.
14. Ohno, N., Iino, K., Takeyama, T., Suzuki, I., Sato, K., Oikawa, S., and Yadomae, T., Structural characterization and antitumor activity of the extracts from matted mycelium of cultured *Grifola frondosa*, *Chem. Pharm. Bull. (Tokyo)*, 33, 3395–3401, 1985.
15. Ohno, N., Iino, K., Suzuki, I., Oikawa, S., and Yadomae, T., Neutral and acidic antitumor polysaccharides extracted from cultured fruit bodies of *Grifola frondosa*, *Chem. Pharm. Bull. (Tokyo)*, 33, 1181–1186, 1985.
16. Suzuki, I., Itani, T., Ohno, N., Oikawa, S., Sato, K., Miyazaki, T., and Yodamae, T., Anti-tumor activity of a polysaccharide fraction extracted from cultured fruiting bodies of *Grifola frondosa*, *J. Pharmacobiodyn.*, 7, 492–500, 1984.
17. Takeyama, T., Suzuki, I., Ohno, N., Oikawa, S., Sato, K., and Yadomae, T., Distribution of grifolan NMF-5N (I/B), a chemically modified antitumor beta-glucan in mice, *J. Pharmacobiodyn.*, 11, 381–385, 1988.
18. Takeyama, T., Suzuki, I., Ohno, N., Oikawa, S., Sato, K., Ohsawa, M., and Yadomae, T., Host-mediated antitumor effect of grifolan NMF-5N, a polysaccharide obtained from *Grifola frondosa*, *J. Pharmacobiodyn.*, 10, 644–651, 1987.
19. Ohno, N., Adachi, Y., Suzuki, I., Oikawa, S., Sato, K., Ohsawa, M., and Yadomae, T., Antitumor activity of a beta-1,3-glucan obtained from liquid cultured mycelium of *Grifola frondosa*, *J. Pharmacobiodyn.*, 9, 861–864, 1986.

20. Suzuki, I., Itani, T., Ohno, N., Oikawa, S., Sato, K., Miyazaki, T., and Yadomae, T., Effect of a polysaccharide fraction from *Grifola frondosa* on immune response in mice, *J. Pharmacobiodyn.*, 8, 217–226, 1985.

21. Nono, I., Ohsawa, M., Oikawa, S., and Yadomae, A.T., Modification of immunostimulating activities of grifolan by the treatment with $(1 \rightarrow 3)$-beta-D-glucanase, *J. Pharmacobiodyn.*, 12, 671–680, 1989.

22. Nono, I., Ohsawa, M., Oikawa, S., and Yadomae, A.T., Modulation of antitumor activity of grifolan by subsequent administration of $(1 \rightarrow 3)$-beta-D-glucanase *in vivo*, *J. Pharmacobiodyn.*, 12, 581–588, 1989.

23. Zhuang, C., Mizuno, T., Ito, H., and Shimura, K., Chemical modification and antitumor activity of polysaccharides from the mycelium of liquid-cultured *Grifola frondosa*, *Nippon Shokuhin Kogyo Gakkaishi*, 41, 733–740, 1994.

24. Hamura, J., Wagner, H., and Rollinghoff, M., (1-3) Glucans as a probe for T-cell specific immune adjuvants, II: enhanced *in vitro* generation of cytoxic T lymphocytes, *Cell. Immunol.*, 38, 328–333, 1978.

25. Maeda, Y., Chihara, G., and Ishimura, K., Unique increase of serum proteins and action of antitumour polysaccharides, *Nature*, 252, 250–251, 1974.

26. Brown, G.D. and Gordon, S., Immune recognition: a new receptor for beta-glucans, *Nature*, 413, 36–37, 2001.

27. *PDR for Herbal Medicines*, 2nd ed., Montvale, NJ, 2000.

28. Therapeutic Research Faculty, *Natural Medicines*, Comprehensive Data Base, Stockton, CA, 2000.

29. Suga, T., Shiio, T., Maeda, Y., and Chihara, G., Antitumor activity of Lentinen in murine syngenic and autochthonous hosts and its suppressive effect on 3-methylcholanthrene-induced carcinogenesis, *Cancer Res.*, 44, 5132–5137, 1984.

30. Manohar, V., Talpur, N., Echard, B.W., Lieberman, S., and Preuss, H.G., Effects of a water soluble extract of maitake mushroom on circulating glucose/insulin concentrations in KK mice, *Diabetes, Obesity Metab.*, 4, 43–48, 2002.

31. Adachi, K., Nanba, H., Otsuka, M., and Kuroda, H., Blood pressure-lowering activity present in the fruit body of *Grifola frondosa* (Maitake), I, *Chem. Pharm. Bull.*, 36, 1000–1006, 1988.

32. Nanba, H., Anti-hypertensive effect by the king of mushrooms, *Explore! Professionals*, 4, 17–19, 1993.

33. Talpur, N.A., Echard, B.W., Fan, A.Y., Jaffari, O., Bagchi, D., and Preuss, H.G., Antihyper-tensive and antidiabetic effects of whole maitake mushroom powder and its fractions in two rat strains, *Molecular Pharmacol. Biol.*, (in press).

34. Kabir, Y., Yamaguchi, M., and Kimura, S., Effect of shitake (*Lentinus edodes*) and maitake (*Grifola frondosa*) mushrooms on blood pressure and plasma lipids of spontaneously hypertensive rats, *J. Nutr. Vitaminol.*, 33, 341–346, 1987.

35. Kubo, K., Aoki, H., and Nanba, H., Anti-diabetic activity in the fruit body of *Grifola frondosa* (Maitake), *Biol. Pharm. Bull.*, 17, 1106–1110, 1994.

36. Kubo, K. and Nanba, H., The effect of maitake mushrooms on liver and serum lipids, *Alternative Ther.*, 2, 62–66, 1996.

37. Nanba, H. Activity of maitake D-fraction to inhibit carcinogenesis and metastasis, *Ann. N.Y. Acad. Sci.*, 768, 243–245, 1995.

38. Ohno, N., Suzuki, Y., Sato, K., Oikawa, S., and Yadomae, T., Effect of grifolan on the ascites form of Sarcoma 180, *Chem. Pharm. Bull. (Tokyo)*, 35, 2576–2580, 1987.

39. Suzuki, I., Takeyama, T., Ohno, N., Oikawa, S., Suzuki, Y., and Yadomae, T., Antitumor effect of polysaccharide grifolan NMF-5N on syngeneic tumor in mice, *J. Pharmacobiodyn.*, 10, 72–77, 1987.

40. Nanba, H., Maitake mushroom: immune therapy to prevent cancer growth and metastases, *Explore! Professionals*, 6, 1–3, 1995.

41. Nanba, H., Results of non-controlled clinical study for various cancer patients using Maitake D-fraction, *Explore! Professionals*, 6, 1995.

42. Adachi, Y., Okazaki, M., Ohno, N., and Yadomae, T., Enhancement of cytokine production by macrophages stimulated with $(1 \rightarrow 3)$-beta-D-glucan, grifolan (GRN), isolated from *Grifola frondosa*, *Biol. Pharm. Bull.*, 17, 1554–1560, 1994.

43. Okazaki, M., Adachi, Y., Ohno, N., and Yadomae, T., Structure-activity relationship of (1-3)-B-glucans in the induction of cytokine production from macrophages *in vitro*, *Biol. Pharm. Bull.*, 18, 1320–1327, 1995.

44. Sveinbjornsson, B., Rushfeldt, C., Seljelid, R., and Smedsrod, B., Inhibition of establishment and growth of mouse liver metastases after treatment with interferon gamma and beta-1-3-D-glucan, *Hepatology*, 27, 1241–1248, 1998.

45. Berges, R. et al., Programming events in the regulation of cell proliferation and death, *Clin. Chem.*, 39, 356–361, 1993.

46. Fullerton, S.A. et al., Induction of apoptosis in human prostatic cancer cells with β-glucan (maitake mushroom polysaccharide), *Mol. Urol.*, 4, 7–11, 2000.

47. Mansell, P.W., Ichinose, H., Reed, R.J., Krementz, E.T., McNamee, R., and DiLuzio, N.R., Macrophage-mediated destruction of human malignant cell *in vivo*, *J. Natl. Cancer Inst.*, 54, 571–580, 1975.

48. Mansell, P.W., DiLuzio, N.R., McNamee, R., Cowden, G., and Proctor, J.W., Recognition factors and nonspecific macrophage activation in the treatment of neoplastic diseases, *Ann. N.Y. Acad. Sci.*, 277, 20–44, 1976.

49. DiLuzio, N.R., McNamee, R.B., Williams, D.L., Gilbert, K.M., and Spanjers, M.A., Glucan induced inhibition of tumor growth and enhancement of survival in a variety of transplantable and spontaneous murine models, *Adv. Exp. Med. Biol.*, 121A, 269–290, 1980.

50. Konno, S., Maitake D-fraction: apoptosis inducer and immune enhancer, *Attern. Complement. Ther.*, (Apr.), 102–107, 2001.

51. Zhu, Y.P., Clinical study on antitumor and immunomodulating effects of Baolisheng [abstract O-20], in *International Programme and Abstracts: International Symposium on Production and Products of Lentinus Mushroom*, International Society for Mushroom Science, Committee on Science, Asian Region, Qingyuan County Government, Zhejiang Province, China, 1994.

52. Ohno, N., Asada, N., Adache, Y., and Yadomae, T., Enhancement of LPS triggered TNF-alpha (tumor necrosis factor-alpha) production by 1-3 beta-D-glucan in mice, *Biol. Pharm. Bull.*, 18, 126–133, 1995.

31 Taxol in Cancer Treatment and Chemoprevention

Sidney J. Stohs

CONTENTS

31.1 Historical Uses of Taxus (Yew) 519
31.2 The Historical Discovery of Taxol (Paclitaxel) 520
31.3 Clinical Applications of Taxol and Taxotere 521
    31.3.1 Taxol (Paclitaxel)...521
    31.3.2 Taxotere (Docetaxel) 522
31.4 Summary..523
References 523

31.1 HISTORICAL USES OF TAXUS (YEW)

Various species of the genus taxus have been long and widely used in the production of wood products and medicinal agents. *Taxus baccata* (common English yew) was used by the Druids in Great Britain as the source of wood for their temples,[1] while the branches were used to make long bows and were highly valued in archery.[2] The wood of *T. baccata* is dense, heavy, elastic, and very durable, and it has been also used to make handles of knives, backs of combs, and wood carvings.[3] Historically, poisonings have occurred among cattle and humans as the result of eating various parts of the yew. The seeds, leaves, and bark are toxic, but not the fruit pulp.[2]

Leaf decoctions of *T. baccata* have occasionally been used in folk medicine as an anthelmintic.[2] In homeopathy, tinctures of young shoots and berries have been used to treat a variety of conditions, including headache, neuralgia, cystitis, poor vision, gout, rheumatism, and various afflictions of the heart and kidneys.[1] Early chemical investigations on the leaves of *T. baccata* indicated the presence of alkaloids, sugar, glycosides, resins, and tannins.[4] The presence of terpenes, which are contained in Taxol™, was not reported. Aqueous extracts were demonstrated to exhibit a depressant effect on the central nervous system and potentiation of pentobarbital hypnosis while exhibiting no analgesic or anticonvulsant activity.[4]

Taxus brevifolia (Pacific or Western yew) grows along the western coast of the U.S. and Canada. Historically, the wood of *T. brevifolia* was used by natives of the northwest for making bows, paddles, and spear handles.[13] In folk medicine, decoctions of the leaves of *T. brevifolia* have been used to treat urinary tract and liver conditions, tonsillitis, epilepsy, rheumatism, and diphtheria.[5] Furthermore, preparations from this yew have been used for promoting menstruation, treating sunburn, and as an anthelmintic and abortifacient.[2,5] *Taxus brevifolia* bark was the initial source of the drug paclitaxel (Taxol), which will subsequently be discussed in detail.

Taxus canadensis (American yew) grows in eastern North America. Its seeds are poisonous, while decoctions of leaves have been used as a hypotensive and to treat rheumatism.[2,3] *Taxus cuspidata* (Japanese yew) is found in Japan and China, and its wood is very fine-grained. It is used for making furniture, sculptures, utensils, and wood carvings. Historically, the wood was also used for making arrows, while the heartwood was a source of a brown dye.[3] In folk medicine, leaf preparations have been used as an antidiabetic and abortifacient.[2] Preparations of *T. wallichiana* (Himalayan yew) have been used to treat coughs and colds.[2]

31.2 THE HISTORICAL DISCOVERY OF TAXOL (PACLITAXEL)

Conventional drug development normally follows an orderly process of preclinical studies using animals and cell cultures as well as phase I, phase II, and phase III studies, with the latter phase consisting of human clinical trials using large numbers of subjects. This orderly procedure does not usually lead to the development of complementary and alternative therapies, which traditionally and historically are based on anecdotal reports and word of mouth.[6] Furthermore, there is usually little isolation of active ingredients from crude products. Little, if any, preclinical testing is conducted in the development of complementary and alternative therapies, and double-blind clinical studies are rarely involved.

An exception to the normal evolutionary development of complementary and alternative therapies has been the discovery and development of Taxol (paclitaxel) from the bark of the Pacific yew tree (*T. brevifolia*), which is now widely used as a chemotherapeutic agent. The isolation, characterization, clinical, and preclinical studies involving Taxol have followed the conventional pathway to FDA approval and subsequent marketing.[6] Wall and Wani[7,8] have summarized the historical development from discovery to clinical use of Taxol.

In 1960, the National Cancer Institute (NCI) initiated a program whereby plant samples were randomly collected by the U.S. Department of Agriculture (USDA). Extracts of various plant parts were prepared by NCI and screened for antineoplastic activity. Among the plants collected were leaves, bark, twigs, and fruits of *T. brevifolia* in Washington state. Initial screenings indicated that extracts of *T. brevifolia* exhibited high cytotoxicity against 9KB cells, an index of potential activity against various neoplastic cells.

In 1964, Wall's laboratory at Research Triangle Institute, Research Triangle Park, NC, received initial samples of plant materials of *T. brevifolia* from NCI. Isolation of the cytotoxic principles were guided by using bioactivity determinations involving the inhibition of Walker WM solid tumor. By November 1966, Wall and Wani[7,8] had purified a fraction from this plant that exhibited exceedingly high cytotoxic activity. The 2-year isolation process involved standard ethanol extraction, partitioning of the ethanolic residue between water and chloroform, and finally countercurrent distribution.

The name *Taxol* was assigned to the compound purified from *T. brevifolia* based on the fact that the molecule contained hydroxyl groups and had a taxane nucleus. Studies on the structure of Taxol involved ultraviolet, infrared, and mass spectrometry, with full details of the structure being presented by Wani et al. in 1971.[9] The structure of Taxol (paclitaxel) is presented in Figure 31.1. Studies in the late 1960s indicated that Taxol had activity against L1210 and 1534 leukemia cells. Taxol was also shown to have activity against many solid tumors when assessed in a panel of tumor systems in 1975–1976, with initial preclinical development occurring in 1977.[7,8]

The determination of the mechanism of action of Taxol proved to be very important in the clinical development of this cytotoxic agent. Initial studies demonstrated that it was a mitotic spindle inhibitor.[10] However, subsequent studies demonstrated that the mechanism was unique because it stabilized microtubules and prevented their depolymerization back to tubulin, an effect opposite that of other antimitotic agents as colchicine, vincristine, vinblastine, and podophyllotoxin.[11,12]

Animal toxicity studies were conducted in 1982, while phase I clinical trials were completed in 1983–1984, and phase II clinical trials were completed in 1983–1986.[7,8] The NCI issued a Cooperative Research and Development Award (CRADA) to Bristol-Myers Squibb in 1991, which obtained a new drug application (NDA) in 1992.

Taxol was initially isolated from the bark of *T. brevifolia*, an exceedingly slow-growing plant. Taxol is present in very low amounts in the bark, and as a consequence, the short supply of Taxol presented a problem regarding general clinical use. This problem was solved by Bristol-Myers Squibb,

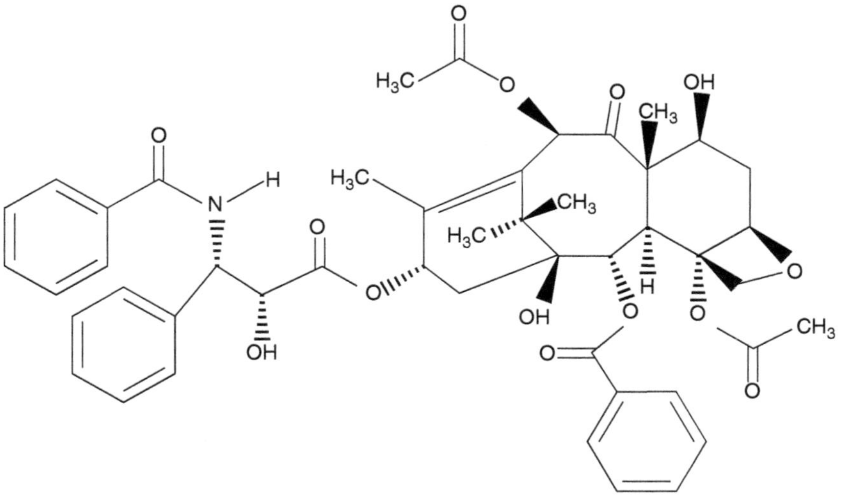

FIGURE 31.1 Taxol.

who figured out how to produce Taxol by semisynthesis from baccatin III or 10-deacetylbaccatin III, which occur in much higher quantities than Taxol in other taxus species, including the leaves of *T. baccata*, a renewable source.[7,8]

The total synthesis of Taxol was accomplished simultaneously by two research groups.[13,14] As is evident from the structure of Taxol (Figure 31.1), the large number of asymmetric carbon atoms results in a total synthesis that is exceedingly difficult. Although total synthesis will not lead to a practical source of supply of Taxol, valuable insight into the chemistry of Taxol was obtained. Furthermore, an analog of Taxol, Taxotere™, has been prepared and tested, and both Taxol and Taxotere (docetaxel) are used clinically.

Several studies have examined the abilities of plant tissue cultures to efficiently produce Taxol. Suspension cell cultures of *T. chinensis* are capable of producing Taxol.[15] Furthermore, protoplasts isolated from cultured cells of *T. cuspidita*, either in static culture or immobilized in agarose gel, effectively produce Taxol.[16] Thus, several systems exist for the potential economical production of Taxol.

31.3 CLINICAL APPLICATIONS OF TAXOL AND TAXOTERE

31.3.1 Taxol (Paclitaxel)

Taxol is considered as a first-line therapy in combination with cisplatin for ovarian carcinoma as well as non-small-cell lung cancer in patients who are not candidates for potentially curative radiation therapy or surgery.[17,18] It is also considered as a first-line therapy for refractory metastatic ovarian carcinoma and the treatment of breast cancer after failure of combination therapy for metastatic disease or relapse after adjuvant chemotherapy, as well as adjuvant treatment of node-positive breast cancer when administered sequentially with a standard doxorubicin-containing regimen. Furthermore, Taxol is used as a second-line treatment of AIDS-related Kaposi's sarcoma.[17,18] Taxol is also effective in treating patients with advanced head and neck carcinoma, although the results in this case do not appear to be better than those achieved with other standard treatment methods.[19] Taxol in combination with 4-hydroxytamoxifen have been shown to exhibit synergistic cytotoxic effects against estrogen-receptor-negative colon cancer and lung cancer cell lines.[20] This combination effectively remands proliferation of the colon cancer cells that were refractory to Taxol alone.

Taxol is an exceedingly potent drug with typical doses (i.v.) of 135 or 175 mg/m^2 every 3 weeks. Clinical studies have shown that the use of Taxol significantly increases response rate, time to progression, and survival time when compared with standard therapies that do not include Taxol.[17] Studies have compared expenditures associated with Taxol use for advanced non-small-cell lung cancer with other standard chemotherapeutic protocols, and the results have shown that Taxol is a cost-effective agent.[21]

Because Taxol is a cytotoxic agent, adverse events can be expected to occur in patients treated with this drug. Studies in patients with solid tumors receiving Taxol as a single agent have demonstrated that bone marrow suppression, including neutropenia and leukopenia, occur in approximately 90% of patients. Anemia also occurs in over 75% of patients treated with Taxol. Peripheral neuropathy as well as myalgia/arthralgia are observed in approximately 60% of patients, while alopecia occurs in over 80% of patients. Nausea and vomiting occur in approximately 50% of patients, while hypersensitivity reactions are evident in approximately 40% of subjects. Lower percentages of adverse events are associated with the liver and cardiovascular system.[17,18,22]

In addition to its use as an effective antineoplastic agent, several studies have demonstrated that Taxol is highly effective in preventing restenosis after angioplasty or vascular injury.[23–25]. The encapsulation of Taxol in polylactic acid microspheres within heparin-chitosan spheres has been shown to provide controlled release of Taxol and heparin for treatment of restenosis.[23] Furthermore, studies using stents coated with bioerodible polymer loaded with Taxol have been shown to be highly effective in reducing restenosis in human coronary arteries.[24,25] These initial studies suggest a number of related applications for the use of Taxol in preventing cell migration and proliferation.

31.3.2 Taxotere (Docetaxel)

Taxotere is a semisynthetic analog of Taxol (Figure 31.2). It is an antineoplastic agent that is used in the treatment of patients with advanced metastatic breast cancer or non-small-cell lung cancer after failure of prior chemotherapy. The drug is highly potent and is used in doses of 60 to 100 mg/m^2 (i.v.) every 3 weeks.[18,26] Taxotere has been shown to increase median survival, median time to progression, and overall response rate when used in combination with other selected chemotherapeutic agents in treating breast and lung cancers after failure of prior chemotherapy.[26] The incidence of adverse events in patients treated with Taxotere is similar to those observed for Taxol.[26]

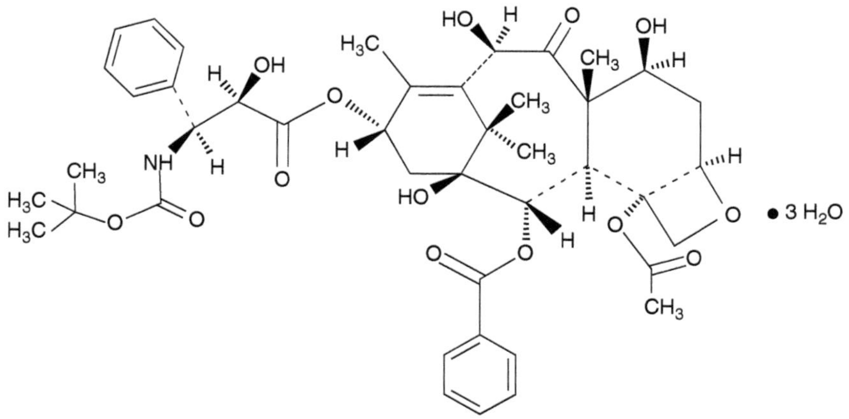

FIGURE 31.2 Taxotere.

31.4 SUMMARY

The development of Taxol as well as Taxotere as important chemotherapeutic agents represents an exceedingly interesting saga in the development of drugs derived from a natural product. Interestingly, preparations from Taxus spp. have been used for hundreds of years, although none of these uses reflected a common application for the treatment of cancers.

The unique mechanism of action of Taxol and its ability to act synergistically with other chemotherapeutic agents have resulted in the establishment of Taxol as a useful chemotherapeutic agent for selected cancers. Furthermore, the ability of Taxol to suppress cell migration and proliferation hold much promise for the use of Taxol for the prevention of restenosis as well as other conditions involving the inhibition of cell migration.

REFERENCES

1. Grieve, M., *A Modern Herbal,* Vol. 2, Hafner Publishing Co., Darien, CT, 1970.
2. Hocking, G.M., *A Dictionary of Natural Products*, Plexus Publishing, Medford, NJ, 1997.
3. Uphof, J.C.T., *Dictionary of Economic Plants*, 2nd ed., Weldon-Wesley, Codicote, U.K., 1968.
4. Vohora, S.B. and Kumar, I., Studies on *Taxus baccata*, 1: preliminary phytochemical and behavioral investigations, *Planta Medica*, 20, 100, 1971.
5. *Natural Medicines Comprehensive Data Base*, 5th ed., Therapeutic Research Faculty, Stockton, CA, 2003.
6. Harlan, Jr., W.R., New opportunities and proven approaches in complementary and alternative medicine research at the National Institutes of Health, *J. Altern. Compl. Med.*, 7, S53, 2001.
7. Wall, M.E. and Wani, M.C., Camptothecin and taxol: discovery to clinic (13th Bruce F. Caine Memorial Award lecture), *Cancer Res.*, 55, 753, 1995.
8. Wall, M.E. and Wani, M.C., Camptothecin and taxol: from discovery to clinic, *J. Ethnopharmacol.*, 51, 239, 1996.
9. Wani, M.E., Taylor, H.L., Wall, M.E., Coggon, P., and McPhail, A.T., Plant antitumor agents VI: the isolation and structure of taxol, a novel antileukemic and antitumor agent from *Taxus brevifolia, J. Am. Chem. Soc.*, 93, 2325, 1971.
10. Fuchs, D.A. and Johnson, R.K., Cytologic evidence that taxol, an antineoplastic agent from *Taxus brevifolia*, acts as a mitotic spindle poison, *Cancer Treat. Rep.*, 62, 1219, 1978.
11. Schiff, P.B., Fant, J., and Hortiz, S.B., Promotion of microtubule assembly *in vitro* by taxol, *Nature*, 22, 665, 1979.
12. Horwitz, S.B., Mechanism of action of taxol, *Trends Pharmacol. Sci.*, 13, 134, 1992.
13. Holton, R.A., Somoza, C., Kim, H.B., Liang, F., Biediger, R.J., Boatman, P.D., Nadizadeh, M., Suzuki, Y., Tao, C., Vu, P., Tang, S., Zhang, P., Murthi, K.K., Gentile, L.N., and Liu, J.H., First total synthesis of taxol, 1: functionalization of the B ring, *J. Am. Chem. Soc.*, 116, 1597, 1994.
14. Nicolaou, K.C., Yang, Z., Liu, J.H., Ueno, H., Nantermet, P.G., Guy, R.K., Clairborne, C.F., Ranaud, J., Couladouros, E.A., and Paulvannan, K., Total synthesis of taxol, *Nature*, 367, 630, 1994.
15. Choi, H.K., Kim, S.I., Song, J.Y., Son, J.S., Hong, S.S., Durzan, D.J., and Lee, H.J., Localization of paclitaxel in suspension cultures of *Taxus chinensis, J. Microbiol. Biotechnol.*, 11, 458, 2001.
16. Aoyagi, H., DiCosmo, F., and Tanaka, H., Efficient paclitaxel production using protoplasts isolated from cultured cells of *Taxus cuspidata, Planta Medica*, 68, 420, 2002.
17. Taxol, in *Physicians' Desk Reference*, 2002, p. 1129.
18. Adams, B.R. and Bence, A.K., Guide for the administration and use of cancer chemotherapeutic agents, 2003, *Pharm. Pract. News*, Dec. 31, 2002, p.
19. Grecula, J.C., Smith, R.E., Rhoades, C.A., Sharma, P., Agarwal, A., Zheang, H., Allen, J., Goldman, F.P., Young, D., and Schuller, D.E., Induction paclitaxel in previously untreated, resectable, advanced squamous cell carcinomas of head and neck: a phase II trial, *Cancer*, 89, 2587, 2000.
20. Gu, W.Z., Chen, Z., Tahir, S.K., Rosenberg, S.H., and Ng, S.C., Synergistic effect of paclitaxel 4-hydroxytamoxifen on estrogen receptor-negative colon cancer and lung cancer cell lines, *Anti-Cancer Drugs*, 10, 895, 1999.
21. Earle, C.C. and Evans, W.K., A comparison of the costs of paclitaxel and the best of supportive care in stage IV non-small-cell lung cancer, *Cancer Prev. Control*, 1, 282, 1997.

22. Finley, R.S. and Rowinsky, E.K., Patient care issues: the management of paclitaxel-related toxicities, *Ann. Pharmacother.*, 28, S27, 1994.

23. Chandy, T., Rao, G.H.R., Wilson, R.F., and Das, G.S., Development of poly (lactic acid)-chitosan co-matrix microspheres: controlled release of taxol-heparin for preventing restenosis, *Drug Deliv. J. Deliv. Targeting Ther. Agents*, 8, 77, 2001.

24. Drachman, D.E. and Rogers, C., Stent-based release of paclitaxel to prevent restenosis, *Z. Kardiol.*, 91, 41, 2002.

25. Park, S.J., Shimm, W.H., Ho, D.S., Raizner, A.E., Park, S.W., Hong, M.K., Lee, C.W., Choi, D., Jang, Y., and Lam, R., Weisman, N.J., and Minitz, G.S., Apaclitaxel-eluting stent for the prevention of coronary restenosis, *New England J. Med.*, 348, 1537, 2003.

26. Taxotere, in *Physicians' Desk Reference*, 2002, p. 778.

32 Lycopene and Cancer

Theresa Visarius, René Gysin, and Angelo Azzi

CONTENTS

32.1   Introduction 525
32.2   What Is Lycopene and Where Is It Available?...525
32.3   The Chemistry of Lycopene: Antioxidant and Other Properties 527
32.4   Bioavailability of Lycopene 528
    32.4.1 Absorption...528
    32.4.2 Tissue Distribution 529
    32.4.3 Lycopene Degradation 529
32.5   Molecular Basis of Lycopene Effects...530
    32.5.1 Lycopene the Antioxidant...530
    32.5.2 Lycopene in Signal Transduction 530
32.6   The combination of Lycopene with Other Molecules in Cancer Growth Inhibition 531
32.7   *In Vitro* Studies of Lycopene Effects on Cancer Cells................................532
32.8   Animal Studies of Lycopene Effects on Experimental Tumors....................533
32.9   Epidemiology of Lycopene and Relationship with Cancer Protection............534
32.10 Conclusions 536
Acknowledgment..536
References 536

32.1   INTRODUCTION

Interest in lycopene, a natural pigment biosynthesized by and accumulated in various fruits, vegetables, and plants, has been sparked by findings that indicate it may be beneficial in promoting the maintenance of health and prevention of disease. Although the health-promoting effects of lycopene have been known since the late 1950s,[1] evidence from recent epidemiological studies together with data gathered from animal and cell culture experiments provide strong evidence that an important physiological benefit supported by lycopene stems from its action as a chemoprotective and therapeutic agent against cancer. Clearly, in societies that are living longer, as proven by demographics of industrialized nations, a prime concern that parallels the extension of life expectancy is that the number of years lived in good health increases accordingly with the total number of years lived. Evidence suggests that there is a strong correlation between lycopene and cancer prevention, and this chapter reviews the scientific literature on this topic.

32.2   WHAT IS LYCOPENE AND WHERE IS IT AVAILABLE?

Lycopene, a carotenoid, is the pigment principally responsible for the deep-red color of ripe tomato but is also found in other plants, algae, microorganisms, and animals. The most established role for carotenoids in plants is to protect cells against photosensitization and to serve as light-absorbing pigments during photosynthesis.[2] The energy transfer from triplet excited species or

TABLE 32.1
Content of Lycopene in Different Foods

Food	Food Form	mg/100 g
Apricots	fresh	0.005
Apricots	canned, drained	0.1
Apricots	dried	0.9
Autumn olive*	fresh	15–54
Chili	processed	1.1–2.6
Grapefruit	pink, fresh	3.4
Guava	pink, fresh	5.4
Guava juice	pink, processed	3.3
Ketchup	processed	17.2
Papaya	red, fresh	2.0–5.3
Pizza sauce	canned	12.7
Pizza sauce	from pizza	32.9
Rosehip puree	canned	0.8
Salsa	processed	9.3
Spaghetti sauce	processed	17.5
Tomatoes	red, fresh	3.1–9.3
Tomatoes	whole, peeled, processed	11.2
Tomato juice	processed	7.8–10.8
Tomato soup	canned, condensed	4.0
Tomato paste	canned	30–55
Tomato puree	canned	16.7
Vegetable juice	processed	7.3–9.7
Watermelon	red, fresh	4.1

Source: Adapted from Khachik, F. et al., *Exp. Biol. Med. (Maywood)*, 227, 845–851, 2002; Nguyen, M.L. and Schwartz, S.J., *Proc. Soc. Exp. Biol. Med.*, 218, 101–105, 1998; Fordham, I.M. et al., *Hortscience*, 36, 1136–1137, 2001.

singlet molecular oxygen to carotenoids is very efficient, and lycopene is the most effective naturally occurring quencher of $^1O_2$.[3] The sequestration of carotenoids in plants is known to be largely dependent upon the carotenoid-associated proteins. Although these proteins are known to be important for carotenoid stability in plants, the basis for recognition and binding are poorly understood,[4] and their presence has not yet been described in animals. In humans, recent findings indicate that lycopene, and other carotenoids to some extent, may be accumulated in a tissue-specific manner and play an important role in health and resistance to degenerative conditions. Since mammals cannot synthesize carotenoids, they attain these micronutrients exclusively through their diets (Table 32.1).

In its natural sources, lycopene occurs essentially in the *trans* configuration. Thermal treatment and processing of lycopene-containing foodstuffs affects the chemical composition of lycopene. Isomerization converts all *trans* isomers to *cis* isomers due to additional energy input, resulting in a less stable, energy-rich form. Heat induces this isomerization, where the *cis* isomers increase with temperature and processing time. In general, the lycopene found in dehydrated and powdered tomatoes has poor stability unless the tomatoes were carefully processed and promptly placed in a hermetically sealed, inert atmosphere for storage. Here also, a significant increase in the *cis* isomers with a simultaneous decrease in the *trans* isomers is observed in tomato samples obtained from various dehydration methods. Frozen foods tend to exhibit excellent lycopene stability throughout their normal-temperature shelf-storage life.

Lycopene bioavailability is influenced by many factors. Since the bioavailability of lycopene *cis* isomers is higher than that of *trans* isomers, it follows that lycopene bioavailability from processed tomato products is higher than that from unprocessed fresh tomatoes. Further, the total lycopene content of processed foods is higher than that of untreated fruits and vegetables, in part due to the mechanical force used to break cell walls, which in turn weakens the bonding forces and affects the release of lycopene from the tissue matrix. This makes lycopene more accessible, thus enhancing the *cis* isomerization and, consequently, the overall bioavailability.[5]

Since lycopene levels in cultivated tomatoes are known to be generally low, genetically manipulating tomato to produce more lycopene in the fruit was a goal to enhance its nutrient value. Overexpression of phytoene synthase, from the bacterium *Erwinia uredovora (crtB)*, in tomato fruits induced an increase of total fruit carotenoids in primary transformants, where lycopene levels approximately doubled.[6] Also, tomato fruits transgenically engineered to produce higher levels of the polyamines spermine and spermidine during ripening exhibited 200 to 300% more lycopene than did the red fruit from the parental line, and thus were determined to have an improved nutritional quality.[7] The recommended daily allowance (RDA) for lycopene and the average daily intake have not yet been established. Preparations containing 5 to 20 mg of lycopene for daily intake are commercially available.

32.3 THE CHEMISTRY OF LYCOPENE: ANTIOXIDANT AND OTHER PROPERTIES

One hypothesis to explain the biological activity of carotenoids in disease prevention is based on their antioxidant ability, specifically, the quenching of $^1O_2$ and other oxidizing species, resulting in the protection of cells from oxidative damage.[8] Of the two broad classes that antioxidants fall into, namely preventive and chain-breaking, lycopene is classified as preventive, which is, a substance that intercepts oxidizing species in advance of damage rather than one that slows or stops an oxidative process after it begins by intercepting chain-carrying radicals. The physical quenching of $^1O_2$ by lycopene implies deactivation by transfer of excitation energy from $^1O_2$ to the lycopene molecule, yielding the triplet excited carotenoid ($^1O_2$ + lycopene $\rightarrow$ O_2 + lycopene*). The energy of the excited lycopene is dissipated through vibrational interactions to recover the ground state carotenoid. Lycopene remains intact in this process and is able to undergo further cycles of singlet oxygen deactivation. Chemical quenching contributes less than 0.05% to total $^1O_2$ quenching by carotenoids, but it is responsible for the eventual destruction of the molecule.

With its extended system of conjugated double bonds, lycopene contains a reactive electron-rich system that is susceptible to reactions with electrophilic compounds. As an acyclic carotenoid containing 11 conjugated double bonds, lycopene has the most potent $^1O_2$-quenching activity of the common carotenoids,[3] nearly twice that of β-carotene, and is known also to interact with reactive oxygen species such as hydrogen peroxide, peroxyl radicals, and nitrogen dioxide.[11–13] As opposed to other carotenoids, such as β-carotene, lycopene lacks a β-ionone ring, leaving it free of provitamin A activity, and because of its highly conjugated nature, it is particularly subject to both oxidative degradation and isomerization.

Chemical and physical factors known to degrade other carotenoids, including exposure to light, oxygen, elevated temperature, extremes in pH, and active surfaces, apply to lycopene as well. In the presence of oxygen, lycopene, and carotenoids in general, tend to autoxidize, a process that has been described as "bleaching".[14,15] At high oxygen levels, however, a carotenoid intermediate radical might add oxygen to form a peroxyl radical. Such an intermediate species could act as a pro-oxidant, initiating lipid peroxidation. Oxidation products of lycopene have been detected in human plasma[16] and possible pro-oxidant effects of carotenoids have been discussed in context with the adverse findings of long-term supplementation with β-carotene on the incidence of lung cancer in heavy smokers. Not all oxidation products are, however, detrimental. Carotenoids may decompose or be metabolized to retinoids with biological activities.[17] Acycloretinoic acid, an oxidation product of lycopene, as well as the parent compound itself stimulate gap-junctional communication. However, a much higher concentration of acycloretinoic acid is required to achieve an effect comparable with that of lycopene (1 μM vs. 0.1 μM, respectively), indicating the likelihood that lycopene, rather than its oxidation product, is the stimulant for improved gap-junctional communication.[18] In contrast, only small amounts of 4-oxoretinoic acid, a product of canthaxanthin decomposition, is sufficient to enhance gap-junctional communication in murine fibroblasts.[17]

Upon exposure to thermal energy, absorption of light, or involvement in specific chemical reactions, interconversion of lycopene isomers is known to take place. The *cis* isomers of lycopene, formed by rotation around any of its conjugated double bonds, have chemical and physical characteristics distinctly different from their all-*trans* counterparts. Some of the differences resulting from *trans*-to-*cis* isomerization include lower melting point, decreased color intensity, a shift in the lambda max, smaller extinction coefficients, and the appearance of a new maximum in the ultraviolet spectrum.[19,20]

The *in vivo* antioxidant behavior of lycopene depends on the concentration and localization in the actual target cells, tissues, or cellular compartments, as well as on many other factors.[21] Lycopene is highly hydrophobic and is, accordingly, most commonly located within cell membranes. Isomerization may alter the physicochemical relation between lycopene and subcellular structures, in turn allowing lycopene to interact with a greater variety of components within the cell and participate in reactions specific for subcellular compartments rather than remaining compartmentalized within the cell membrane.[22] Although compelling evidence suggests that the basis of lycopene action *in vivo* results rather from its nonantioxidant action, if lycopene does indeed function as an antioxidant *in vivo*, it could be involved in the regulation of transcription factors that are sensitive to the intracellular oxidant status. Alternatively, decreasing the extent of intracellular oxidative damage may prevent protein oxidation, thus reducing the risk of cancer-inducing events in the cells, as hypothesized in prostate cancer.[23]

32.4 BIOAVAILABILITY OF LYCOPENE

32.4.1 ABSORPTION

The presence of carotenoids in various human organs and tissues was reported as early as 1990,[24,25] and lycopene was found to be predominant in human liver, adrenals, adipose tissue, testes, and prostate.[26–30] Intake of tomato and tomato-based food products contributes to the absorption of lycopene as well as a wide range of other carotenoids found in human serum and tissues. An important determinant of carotenoid bioavailability is the matrix in which carotenoids are presented to the organism,[31] where consuming fat with carotenoids increases the efficiency of absorption.[27] To date, 25 carotenoids and 9 metabolites have been identified and characterized in the extracts from human serum and milk.[8] The absorption and distribution of [^{14}C]lycopene was studied in rats and rhesus monkeys following oral administration of labeled lycopene in olive oil supplemented with 1 mg α-tocopherol/mL. The peak accumulation of lycopene after a single gavage dose occurred between 4 to 8h and 8 to 48h in rat and monkey serum, respectively.[32] The concentration of lycopene in human plasma from healthy individuals is $\approx$1 μmol/L. With supplementation, this can double,[33] and lycopene appears to be equally bioavailable from tomato juice or from supplements.[34]

Cis-isomers of lycopene have been suggested to be better absorbed than the all-*trans* form due to the shorter length of the *cis* isomers, their greater solubility in mixed micelles, and the lower tendency of *cis* isomers to aggregate. Work with ferrets, a species that absorbs carotenoids intact, supports the hypotheses that *cis* isomers are substantially more bioavailable than all-*trans* lycopene.[35] Following a bolus oral dose, stomach and intestinal contents contained 6 to 18% *cis*-lycopene, whereas mesenteric lymph secretions contained 77% *cis* isomers. Further, *in vitro* studies suggest that *cis* isomers are more soluble in bile acid micelles and may be preferentially incorporated into chylomicrons.[20] The uptake of lycopene into intestinal mucosal cells is aided by the formation of bile acid micelles, a process stimulated by dietary fat.[27] The uptake of lycopene by the brush border membrane of the intestinal mucosal cell is thought to be by passive diffusion, and little is known about the intramucosal processing of lycopene. It remains to be elucidated whether lycopene is transported intracellularly by specific proteins or whether it migrates in lipid droplets. Lycopene exits the mucosal cell in chylomicrons, which are secreted via the mesenteric lymph system into the blood.

32.4.2 TISSUE DISTRIBUTION

Through the action of lipoprotein lipase on chylomicrons, lycopene and other carotenoids have the potential to be taken up passively by various tissues before clearance of chylomicron remnants by the liver via the chylomicron receptor. Carotenoids can accumulate in the liver or be repackaged into very-low-density lipoprotein (VLDL) and sent back into the blood. Uptake of carotenoids into tissues from VLDL and LDL occurs via the LDL receptor, and the tissues with the highest concentrations of carotenoids are known to have high LDL receptor activity.[20] Additionally, hormonal factors may also regulate the isomeric ratio of lycopene in tissues. Male F344 rats fed lycopene-containing diets achieved lycopene tissue concentrations and isomer patterns similar to those observed for humans, whereas castrated rats accumulated more total lycopene and more lycopene as *cis* isomers than intact rats, despite eating less total lycopene.[36]

Cis-isomers preferentially accumulate in tissues,[28,37–39] and all dietary carotenoids that are present in human serum may be accumulated in prostate. Lycopene and other major carotenoids have been shown to be present in higher concentrations in malignant prostate tissue relative to normal prostate tissue, although no apparent difference in the lycopene isomer ratio existed between benign and malignant tissues (all-*trans* lycopene 12 to 21% and *cis* isomers 79 to 88% of total lycopene).[28] A hypothesis to explain this is that since the metabolism of the prostate is regulated by the neuroendocrine axis and uptake of lycopene in the prostate is related to lipoprotein metabolism, the uptake is fairly nonspecific and will be greater in tissues that are more metabolically active and blood-vessel rich, as is the case in cancers. Interestingly, however, another study revealed significantly lower serum and tissue lycopene levels (44%, p = 0.04; 78%, p = 0.050, respectively) in cancer patients than in age-matched control subjects.[23]

32.4.3 LYCOPENE DEGRADATION

Numerous oxidation reactions of lycopene *in vitro* and *in vivo* have clearly shown that this compound is first oxidized at the 1,2- and the 5,6-position to form lycopene 1,2-epoxide and lycopene 5,6-epoxide, respectively. Although lycopene 1,2-epoxide was found to be quite stable, lycopene 5,6-epoxide was extremely unstable and underwent cyclization to give a mixture of 2,6 cyclolycopene–1,5-epoxides.[8] Lycopene epoxides and the rearrangement products of lycopene 5,6-epoxide have not been detected in human serum, although their corresponding cyclic diols, 2,6-cyclolycopene 1,5-diols, are present.[40] These diols may be formed from acidic or enzymatic ring opening of their respective epoxides. The origin of the metabolites of lycopene in human serum may be due to the presence of trace amounts of these compounds in tomato-based products. However, the concentration of 2,6-cyclolycopene–1,5-diols in raw tomatoes and tomato-based products is extremely low and most likely cannot account for their presence in human serum. The presence of low concentration of some of the lycopene metabolites in tomatoes and human serum could also be related to the physiological function of lycopene as a radical scavenger. Unfortunately, the fact that 2,6-cyclolycopene–1,5-diols and their precursor 2,6-cyclolycopene–1,5-epoxides are also found in tomato products at low concentrations makes it difficult to differentiate between the various processes that may ultimately be responsible for the presence of these compounds in human serum and tissues.[8]

32.5 MOLECULAR BASIS OF LYCOPENE EFFECTS

32.5.1 LYCOPENE THE ANTIOXIDANT

The ability of lycopene to act as an antioxidant and scavenger of free radicals is frequently cited as the most likely mechanism that could account for the hypothesized beneficial effects on human health.[22,41,42] Supporting this theory, protection against oxidative stress has been shown in parallel with a preferential destruction of lycopene relative to β-carotene in a study of human skin irradiated with ultraviolet (UV) light.[43] Further, the ability of carotenoids to act as antioxidants has been hypothesized as the mechanism underlying the protection of the human retina from photooxidation, a process that over time can result in the pathologies of age-related macular degeneration (AMD) and cataracts.[8]

The reported biological activities of lycopene are based on its potent antioxidant activity involving singlet-oxygen quenching and peroxyl-radical scavenging.[3] Free radicals are known to cause damage both in the structure and the function of cell membranes, DNA, and proteins, and such damage has, in turn, been linked to the onset of many degenerative diseases such as cancer, atherosclerosis, cataracts, AMD, as well as to premature aging. Consequently, the conclusion that lycopene protects via an antioxidant mechanism is, at first glance, logical. However, given that the concentration of lycopene in human serum ranges from 0.5 to 2 μM,[33] if all of the lycopene was consumed in antioxidative defense, one would not arrive at significant level of detoxification for most known oxidizing species important in physiology and medicine. Alternatively, if the antioxidant role of lycopene was restricted to that of quenching singlet oxygen (removing the excitation energy from $^1O_2$ without destroying the molecule), resulting in the inhibition of biologically deleterious effects, this would imply that $^1O_2$ is a signaling molecule, an argument for which no evidence presently exists. Rather, it may be more logical to envision lycopene as acting on the level of signal transduction.

32.5.2 LYCOPENE IN SIGNAL TRANSDUCTION

Dietary carotenoids protect human cells from damage[44] and suppress induced neoplastic transformation in model cell culture systems.[45] Specifically, the lycopene-induced prevention of breast and prostate cancer[46,47] may in part be achieved by lycopene-induced control of the cell cycle, as governed by an observed reduction in cyclin D1 levels and retention of p27(Kip1) in the cyclin E-cdk2 complexes,[48] leading to the inhibition of cdk2 activity. The end target of the catalytic cyclin-dependent kinase (CDK) activation is phosphorylation of the nuclear tumor-suppressor protein retinoblastoma during the G1 phase of the cell cycle, a process required for the continuation of the cell cycle. Thus, the root of the lycopene-induced inhibition of the cell cycle is suggested to be the decrease in retinoblastoma phosphorylation, resulting in an inhibition of the G1/S transition as observed in human breast cancer cells (MCF-7). Several mechanisms could be responsible for this effect, such as inhibition of p53 pathways, or disruption of growth factor signals known to induce p21 protein when cells reenter the cell cycle.[49,50]

There is also reason to suspect that the activity of lycopene may be in part analogous to that of tocotrienols: like tocotrienols, the end portions of lycopene have isoprenoid structure. This is not true for other carotenoids, which have cyclized tails. The well-known ability of tocotrienols to decrease serum cholesterol is rooted in the ability of these compounds to suppress the expression of HMG-CoA reductase, the rate-limiting enzyme for cholesterol synthesis that statin drugs inhibit allosterically by a posttranscriptional mechanism.[51,52] This effect presumably requires tocotrienols to interact with an as-yet-uncharacterized receptor, a receptor that is postulated to recognize other isoprenoids as well, based on the evidence that tocotrienols are not the only isoprenoids known to down-regulate HMG-CoA reductase.

Lycopene is also known to decrease the expression of HMG-CoA reductase in macrophage cell cultures, based on the results of a small clinical study where 60 mg lycopene daily was found to decrease serum LDL cholesterol.[53] Suppression of HMG-CoA reductase activity acts to retard cancer induction and slow cellular growth.[54,55] A main reason for this is that synthesis of the isoprenoid dolichol during G1 phase is crucial for cell surface expression of insulin-like growth factor I (IGF-I) receptors, and cells that cannot synthesize dolichol experience an effective deficiency of IGF-I activity.[56,57]

IGF-I is an important risk factor for numerous cancers,[58–60] and lycopene as well as the consumption of cooked tomatoes was found to be significantly inversely associated with IGF levels.[61] IGF-I mediates the effects of growth hormone, which is mainly secreted by liver but is also produced in several other tissues. IGF-I can act in an autocrine and paracrine manner to promote normal growth and malignant cellular proliferation.[62,63] *In vitro*, low micromolar concentrations of lycopene have slowed cancer cell proliferation by suppressing response to IGF-I.[64] There has been considerable progress recently in the understanding of the hormonal causes of prostate cancer. Particularly, a number of studies have shown a positive association between IGF-I and the incidence of prostate cancer, whereas the association was reversed after radical prostatectomy[65] and/or treatment with lycopene.[66]

Diminution of cell–cell interaction via gap junctions is associated with the neoplastic phenotype,[67] and lycopene is known to enhance gap-junctional communication in a manner unrelated to its antioxidant properties.[68] Lycopene simultaneously inhibits tumor cell proliferation and stimulates gap-junctional intercellular communication via up-regulation of connexin 43 expression, where lycopene, more potently than β-carotene, stimulated cell-to-cell communication in a concentration-dependent manner.[18,69]

32.6 THE COMBINATION OF LYCOPENE WITH OTHER MOLECULES IN CANCER GROWTH INHIBITION

Lycopene in combination with vitamin E or vitamin D has been reported to potentiate inhibition of cancer cell growth. A synergistic effect was observed when physiologically relevant concentrations of α-tocopherol and lycopene were studied in various prostate cancer cell lines.[70] Lycopene, in itself, was not a potent inhibitor of prostate carcinoma cell proliferation; however, up to 90% growth inhibition was reported with the α-tocopherol and lycopene combined treatment. This synergistic effect was not shared by β-tocopherol, ascorbic acid, or probucol, indicating a nonantioxidant mechanism to be the basis of the finding.

The combination of low concentrations of lycopene together with 1,25-dihydroxyvitamin D3 exhibited a synergistic effect on inhibiting the proliferation of promyelocytic leukemia cells and on their differentiation. Further, an additive effect on cell-cycle progression was reported, whereby these authors concluded with a recommendation for the inclusion of this carotenoid in the diet as a cancer-preventive measure.[71]

Combination of lycopene and the herbal medicine Sho-saiko-to (TJ-9), a crude extract of seven herbs — bupleurum root, pinellia tuber, scutellaria root, jujube fruit, ginseng root, glycyrrhiza root, and ginger rhizome — was studied in Long-Evans cinnamon rats. In this model of hepatocarcinogenesis, neither lycopene nor TJ-9 or the combination of the two resulted in protection,[72] although a later study showed the suppression of fibrogenesis by both lycopene and TJ-9, where the mechanism of action was suggested to be inhibition of stellate cell activity.[73]

Although no substantial evidence to date is available, many sources suggest the combination of lycopene with a plethora of 5-alpha-reductase inhibitors such as saw palmetto, riboflavin, zinc, and selenomethionine for the protection or therapy of prostate cancer. Despite the present fragmentary understanding, it is likely that novel useful combination therapies or anticancer prophylactic cocktails that include lycopene will be realized.

32.7 *IN VITRO* STUDIES OF LYCOPENE EFFECTS ON CANCER CELLS

The first study linking lycopene to inhibition of cancer cell growth was performed in rat glioma cells,[74] and the results indicated that both retinoids and carotenoids were active as antitumor agents against brain tumor cells in culture. Thereafter, a number of groups concluded that lycopene could regulate cancer cell growth by interfering with cell-cycle progression, thereby inhibiting proliferation (Table 32.2). Further, it was shown that lycopene, with a half-maximal inhibitory concentration (IC_{50}) of 1–2 μM, more effectively impaired growth of select cancer cell types than either α-carotene or β-carotene.[64]

Interestingly, many more lycopene-based studies have been conducted in animal models or in human subjects rather than in cell culture. The explanation for this may be seen as two-fold. Firstly, because of the high lipophilic nature of lycopene, it is most commonly solubilized in tetrahydrofuran, which, in itself, is toxic to cells. Second, since lycopene, a phytochemical, is a natural product and, in most studies, was either supplied to subjects in its natural form (tomato juice, etc.) rather than supplements or indirectly deduced from dietary questionnaires, ethical committee approval was relatively easy to obtain. Substantial evidence has been produced relating the consumption of lycopene to a decreased risk of acquiring degenerative diseases, such as certain kinds of cancers and cardiovascular disease. In addition, the usefulness of lycopene has, on several accounts, been shown in the treatment of cancers. Thus it becomes increasingly important to understand in detail the molecular mechanism of action of lycopene in various model cell systems.

Lycopene has been shown, thus far in two reports, to interfere in the mitogenic pathway of insulin growth factor I (IGF-I), as investigated in cell culture.[64,75] In summary, these studies revealed that the inhibitory effects of lycopene on cancer cell growth were not accompanied by apoptotic or necrotic cell death. Rather, it was found that low micromolar concentrations of lycopene slowed cancer cell proliferation by suppressing response to IGF-I, where lycopene treatment markedly reduced the IGF-I stimulation of tyrosine phosphorylation of insulin receptor substrate 1 and binding capacity of the AP-1 transcription complex. These effects were not associated with changes in the number or affinity of IGF-I receptors, but with an increase in membrane-associated IGF-binding proteins, which were previously shown in different cancer cells to negatively regulate IGF-I receptor activation. The inhibitory effect of lycopene on IGF signaling was associated with suppression of IGF-stimulated cell-cycle progression, and the conclusion was reached that that the inhibitory effects of lycopene on breast cancer cell growth were not due to the toxicity of the carotenoid but,

TABLE 32.2
In Vitro Studies Reporting the Actions of Lycopene in Various Cell Culture Systems

Lycopene Action Proposed Molecular Mechanism	Cell Type		Reference
Growth inhibition	Glioma	C-6	Wang[74]
Growth inhibition	Endometrial	Ishikawa	Levy[64]
	Breast	MCF-7	
	Lung	NCI-H226	
Growth inhibition	Prostate	LNCaP, DU-145, PC-3	Pastori[70]
Growth inhibition	Leukemia	HL-60	Amir[71]
Growth inhibition	Leukemia	HL-60	Nara,[76]
IGF-I pathway	Breast	MCF-7	Karas[75]
Gap junctional communication	Fibroblast	C3H 10T1/2	Zhang[67]
Gap junctional communication	Fibroblast	F9	Stahl[18]
Gap junctional communication	Oral cavity tumor	KB-1	Livny[69]
G1-S phase delay	Breast	MCF-7, T-47D	Nahum[48]
	Endometrial	ECC-1	

rather, to interference in IGF-I receptor signaling and cell-cycle progression where a G1/S transition delay was observed in various cancer cell lines.

Lycopene-induced inhibition of cancer cell cycle was elegantly investigated and recently reported in breast and endometrial cells.[48] The main finding was that reduction of cyclin D levels and retention of p27[KIP1] in the cyclin E-cdk2 complexes was responsible for the inhibitory effect of lycopene on cell-cycle progression. Specifically, a reduced cdk2 activity led to a reduced phosphorylation of the retinoblastoma protein, a main control protein of the G1 restriction point, leading to a delay in G1/S transition.

In accordance with an inhibition of cancer cell proliferation and increased cell differentiation, lycopene has also been described to stimulate gap-junctional communication by up-regulation of connexin 43. Connexin 43 is a key protein in the formation of gap junctions. It assembles to channels connecting the cytosol of neighboring cells for exchange of low-molecular-weight compounds. Cell–cell interaction via gap junctions is considered to be a key factor in tissue homeostasis, and its alteration is associated with the neoplastic phenotype. A first study showed in a 10T1/2 cell assay system that lycopene, and other carotenoids, were able to inhibit chemically induced neoplastic transformation, and this inhibition correlated with an enhanced gap-junctional communication.[67] In human fetal skin fibroblasts, it was later shown that very low concentrations of lycopene (0.1 μM) resulted in the stimulation of gap-junctional communication by increasing transcription and stabilization of connexin 43 mRNA,[18] and a third study validated this finding in human oral cavity tumor cells.[69] Lycopene strongly and dose dependently inhibited proliferation of KB-1 cancer cells, and treatment with 3 μM lycopene up-regulated transcription of connexin 43, leading to the enhanced gap-junctional communication.

Converse to most discovery processes involving the investigation of efficacy, potency, and mode of action of substances useful in inhibiting disease processes, more is known about the action of lycopene in whole organisms and tissues rather than in cells. Nevertheless, many ongoing *in vitro* investigations have been stimulated from the findings of the epidemiological studies, and the results are bound to enhance our knowledge of the mechanisms underlying the lycopene-induced control of cancer cell proliferation.

32.8 ANIMAL STUDIES OF LYCOPENE EFFECTS ON EXPERIMENTAL TUMORS

To date, only eight studies in animal models have been published on the cancer chemopreventive effects of lycopene or tomato carotenoids (Table 32.3), with the majority showing a protective effect of lycopene. A positive correlation between lycopene or tomato product intake and cancer

TABLE 32.3
Animal Studies of Lycopene Effects on Experimental Tumors

Cancer Type	Animal Model		Significance[a]	Reference
Lung	Mouse	B6C3F1	S	Kim,[77]
Urinary bladder	Rat	Fischer 344	NS	Okajima,[78]
Colon	Rat	F344/NSlc	S	Narisawa,[79]
Colon ACF	Rat	Sprague-Dawley	NS	Narisawa,[80]
Hepatic preneoplasia	Rat	weanling	S	Astorg,[81]
Mammary tumor	Mouse	SHN virgin	S	Nagasawa,[82]
Mammary tumor	Rat	DMBA-model	S	Sharoni[83]
Mammary tumor	Rat	NMU-model	NS	Cohen[84]

[a] S = significant; NS = not significant.

protection was reported in mouse lung, rat urinary bladder, rat colon cancer, against rat aberrant colon crypt formation, and in the rat hepatic preneoplasia model. The chemopreventive effects of lycopene and tomato products against mammary tumorigenesis were, however, less consistent.

In a high mammary tumor strain of SHN virgin mice, chronic treatment with lycopene significantly suppressed mammary tumor development.[82] This effect was associated with a decrease in the mammary gland activity of thymidylate synthetase, serum levels of free fatty acid, and prolactin. No deleterious side effects of lycopene were detected, and lycopene was suggested to be promising as a chemopreventive agent for mammary tumors. Two other groups used chemically induced tumor model systems. With the 7,12-dimethylbenz[a]anthracene (DMBA)-induced rat mammary tumor model system, one group compared the effect of lycopene-enriched tomato oleoresin on the initiation and progression of mammary tumors with that of β-carotene. Here, the lycopene-treated rats developed significantly fewer tumors, and the tumor area was smaller than that of the unsupplemented rats. Further, rats receiving β-carotene showed no protection against the development of mammary cancer.[83] The second study, which employed the N-methylnitrosourea (NMU)-induced rat mammary tumor model, provided contradictory findings, where neither pure lycopene nor lycopene in the form of a mixed carotenoid oleoresin exerted an inhibitory effect on tumor incidence, latency, multiplicity, volume, or total tumors per group compared with unsupplemented controls.[84] Unfortunately, no consensus can be drawn from the studies analyzing the effect of lycopene in breast cancer models. The animal studies published thus far are not comparable, since the lycopene preparations, doses, and routes of administration as well as animal type and model used were dissimilar.

The uptake and tissue-specific accumulation of lycopene was the subject of another study where male and female Fischer-344 rats were divided into groups fed various concentrations of lycopene over a 10-week period. Although 55% of the ingested lycopene was excreted, dose-dependent accumulation was observed in liver, prostate, lung, and mammary gland. This work first indicated specific, nonsimilar accumulation in a variety of target organs in addition to the dose dependency in both male and female rats.[85]

32.9 EPIDEMIOLOGY OF LYCOPENE AND RELATIONSHIP WITH CANCER PROTECTION

Nutritional factors are widely believed to be critical in carcinogenesis. Overwhelming evidence from epidemiological studies indicates that diets high in fruits and vegetables are associated with a lower risk of numerous cancers.[71,86,87] However, uncertainty exists concerning which components account for these benefits. Intake of tomatoes and tomato-based products and plasma levels of lycopene have been consistently associated with a lower risk of a variety of cancers.[87–89] Further, out of 72 epidemiological studies, 60 studies reported inverse associations between tomato intake or blood lycopene level and the risk of cancer at a defined anatomic site (Table 32.4). More than half of these inverse associations were statistically significant, and no study indicated that higher tomato consumption or blood lycopene level increased the risk of cancer at any of the investigated sites.

Two general types of study designs have been used to examine lycopene and tomato products in relation to cancer risk. One design has been based on a dietary questionnaire, used either to assess tomato products directly or to infer lycopene consumption; the second design relied on measuring the concentration of carotenoids in blood or tissue samples. Evidence for an inverse association between intake of tomatoes and tomato-based products or plasma levels of lycopene and a lower risk for cancer was strongest for cancers of the lung, stomach, and prostate gland and was suggestive for cancers of the cervix, breast, oral cavity, pancreas, colorectum, and esophagus. About half of studies comparing high to low intake of tomato products or lycopene showed a risk reduction of 30 to 40%.

Lung cancer is the leading cause of cancer death worldwide and is one of the cancer sites for which a benefit of fruits and vegetables has been most apparent. Initial findings led investigators to focus on β-carotene and to examine its relation to the risk of lung cancer in intervention trials.

TABLE 32.4
Summary of Total Epidemiological Studies Performed vs. Significant or Suggestive Studies Relating Lycopene Intake to Inhibition Cancer Progression

Cancer Site	Studies Performed	Number of Significant or Suggestive Studies
Lung and pleural	14	10
Stomach	12	10
Colorectal	5	4
Oral, laryngeal, pharyngeal	3	2
Pancreas	4	2
Prostate	17	10
Bladder	4	1
Cervix	5	3
Breast	8	2

Note: In most studies not supporting this association, the difference in serum lycopene concentration between patients with high plasma lycopene and low plasma lycopene was not as great as that observed in the suggestive or positive studies.

Unexpectedly, β-carotene supplementation was found to increase the risk for lung cancer in smokers. The Finnish Smoker's study and an American study called CARET[90] not only showed no protection, but also an increased risk of lung cancer in those participants who received β-carotene supplements in a dose of 25,000 IU (about the amount of β-carotene in a cup of carrots). Further, among subjects receiving β-carotene (n = 14,560), prostate cancer incidence was 23% higher and mortality was 15% higher compared with those not receiving it (n = 14,573).[91] Consumption of tomato or lycopene, however, in 10 of 14 epidemiological studies, suggested either a statistically significant or a suggestive association against lung cancer risk.[87] Apart from the epidemiological studies, an additional study indicated that higher prediagnostic dietary intake of tomatoes was associated with better survival from lung cancer after diagnosis.[92]

Stomach or gastric cancer is also one of the major causes of cancer deaths worldwide, although the risk has substantially decreased in economically developed countries. In 10 of 12 case-control studies, the results showed an inverse association between tomato or lycopene intake and the risk of gastric cancer. This protective correlation of tomato or lycopene was found in several populations, specifically in the U.S., Hawaiian Japanese, Japan, Israel, Italy, Spain, Poland, Belgium, and Sweden.[87] High intake of fruits and vegetables is well known to be inversely associated with gastric cancer, and the inverse correlation of tomato intake with gastric cancer has been among the most consistent and strongest, thus again suggesting a protective effect of tomatoes or lycopene consumption against cancer.

Prostate cancer is the most common noncutaneous cancer and is the second leading cause of death from malignancies.[93] The first clear connection between tomato products and a lower risk of prostate cancer was provided by data derived from the Health Professionals Follow-Up Study (HPFS) where the dietary habits of 47,356 men were followed. The ongoing prospective cohort study has indicated that lycopene-rich foods are inversely associated with the risk of prostate cancer, whereas the overall intake of fruits and vegetables was not related.[47,94] The most provocative observations linking the benefit of lycopene intake to the prevention or treatment of prostate cancer have recently been published by an American group.[66,95] Here, a clinical trial to investigate the biological and clinical effects of lycopene supplementation in patients with localized prostate cancer was performed with 26 men having newly diagnosed prostate cancer who were randomly assigned to receive a tomato oleoresin extract containing 30 mg of lycopene (n = 15) or no supplementation (n = 11) for 3 weeks before radical prostatectomy. After intervention, subjects in the intervention group had smaller tumors, less involvement of surgical margins and/or extra-prostatic tissues with cancer, and less diffuse involvement of the prostate by high-grade prostatic intraepithelial neoplasia, and mean plasma prostate-specific antigen levels were lower as compared with the control group.

Comparing differences in death rates resulting from prostate cancer between nations having nearly no deaths from this disease (China and Japan) and nations known to have a high risk of prostate cancer (North America, Western Europe, Australia) indicates that nutritional, environmental, and lifestyle factors may be key factors.[96] This is underscored by an observed increase in prostate cancer risk in migrants to the U.S. from nations having lower rates of this disease incidence. Further, the prevalence of latent prostate cancer in autopsy studies, being similar among geographic areas of the world that exhibit quite different rates of prostate cancer mortality, again suggests a prominent role for diet and nutrition in accelerating or inhibiting the process from latent forms of prostate cancer to aggressive tumors.

In summary, results from epidemiological studies are consistent for a variety of cancers across numerous diverse populations with the use of various study designs, prospective as well as retrospective ecologic, case-control dietary, prospective dietary, or blood-specimen based. Nevertheless, the safety and efficacy of lycopene supplementation still needs rigorous evaluation. Since the pharmacokinetic properties of lycopene supplementation in healthy persons and cancer patients are presently unclear, it is premature to recommend pharmacologic doses of lycopene with the goal of reducing the occurrence or inhibiting the progression of cancers.

32.10 CONCLUSIONS

From the literature reviewed in the present chapter, it appears that lycopene is an important compound in connection with the onset and the progression of cancer. Important epidemiological studies have traced the path that has been followed by the first clinical interventions and by the animal studies. Cellular mechanisms have appeared as well to show that the human and animal results are based on specific molecular events correlated with regulation of the cell cycle. An additional step in understanding the importance of lycopene has been the observed synergism between lycopene and α-tocopherol and vitamin D. Although the essential molecular basis of lycopene action on cancer cells has been described, several pieces are still missing from the puzzle of the complex mechanism of lycopene action. In particular, the specificity of lycopene protection will be a subject of importance in further studies. Another point that deserves further exploration is the synergistic action that some other compounds may exert in combination with lycopene. Finally, on the assumption that lycopene, due to the expression of its activity at very low concentrations, is not acting as an antioxidant, it will be important to identify which molecules specifically recognize lycopene and how lycopene-induced signaling regulates the cell cycle.

ACKNOWLEDGMENT

The studies reported in this article were supported by the Swiss National Science Foundation and the Foundation for Nutrition Research in Switzerland.

REFERENCES

1. Lingen, C., Ernster, L., and Lindberg, O., The promoting effect of lycopene on the non-specific resistance of animals, *Exp. Cell. Res.*, 16, 384–393, 1959.
2. Demmig-Adams, B., Gilmore, A.M., and Adams, III, W.W., Carotenoids 3: *in vivo* function of carotenoids in higher plants, *FASEB J.*, 10, 403–412, 1996.
3. Di Mascio, P., Kaiser, S., and Sies, H., Lycopene as the most efficient biological carotenoid singlet oxygen quencher, *Arch. Biochem. Biophys.*, 274, 532–538, 1989.

4. Vishnevetsky, M., Ovadis, M., and Vainstein, A., Carotenoid sequestration in plants: the role of carotenoid-associated proteins, *Trends Plant Sci.*, 4, 232–235, 1999.

5. Shi, J. and Le Maguer, M., Lycopene in tomatoes: chemical and physical properties affected by food processing, *Crit. Rev. Biotechnol.*, 20, 293–334, 2000.

6. Fraser, P.D., Romer, S., Shipton, C.A., Mills, P.B. et al., Evaluation of transgenic tomato plants expressing an additional phytoene synthase in a fruit-specific manner, *Proc. Natl. Acad. Sci. USA*, 99, 1092–1097, 2002.

7. Mehta, R.A., Cassol, T., Li, N., Ali, N. et al., Engineered polyamine accumulation in tomato enhances phytonutrient content, juice quality, and vine life, *Nat. Biotechnol.*, 20, 613–618, 2002.

8. Khachik, F., Carvalho, L., Bernstein, P.S., Muir, G.J. et al., Chemistry, distribution, and metabolism of tomato carotenoids and their impact on human health, *Exp. Biol. Med. (Maywood)*, 227, 845–851, 2002.

9. Nguyen, M.L. and Schwartz, S.J., Lycopene stability during food processing, *Proc. Soc. Exp. Biol. Med.*, 218, 101–105, 1998.

10. Fordham, I.M., Ba, C., Er, W., and Zimmerman, R.H., Fruit of autumn olive: a rich source of lycopene, *Hortscience*, 36, 1136–1137, 2001.

11. Burton, G.W. and Ingold, K.U., beta-Carotene: an unusual type of lipid antioxidant, *Science*, 224, 569–573, 1984.

12. Bohm, F., Tinkler, J.H., and Truscott, T.G., Carotenoids protect against cell membrane damage by the nitrogen dioxide radical, *Nat. Med.*, 1, 98–99, 1995.

13. Woodall, A.A., Lee, S.W., Weesie, R.J., Jackson, M.J. et al., Oxidation of carotenoids by free radicals: relationship between structure and reactivity, *Biochim. Biophys. Acta*, 1336, 33–42, 1997.

14. Rice-Evans, C.A., Sampson, J., Bramley, P.M., and Holloway, D.E., Why do we expect carotenoids to be antioxidants *in vivo*? *Free Radical Res.*, 26, 381–398, 1997.

15. Sies, H. and Stahl, W., Vitamins E and C, beta-carotene, and other carotenoids as antioxidants, *Am. J. Clin. Nutr.*, 62, 1315S–1321S, 1995.

16. Khachik, F., Beecher, G.R., Goli, M.B., and Lusby, W.R., Separation, identification, and quantification of carotenoids in fruits, vegetables and human plasma by high performance liquid chromatography, *Pure Appl. Chem.*, 71–80, 1991.

17. Hanusch, M., Stahl, W., Schulz, W.A., and Sies, H., Induction of gap junctional communication by 4-oxoretinoic acid generated from its precursor canthaxanthin, *Arch. Biochem. Biophys.*, 317, 423–428, 1995.

18. Stahl, W., von Laar, J., Martin, H.D., Emmerich, T. et al., Stimulation of gap junctional communication: comparison of acyclo-retinoic acid and lycopene, *Arch. Biochem. Biophys.*, 373, 271–274, 2000.

19. Gugger, E.T. and Erdman, Jr., J.W., Intracellular beta-carotene transport in bovine liver and intestine is not mediated by cytosolic proteins, *J. Nutr.*, 126, 1470–1474, 1996.

20. Boileau, T.W., Boileau, A.C., and Erdman, Jr., J.W., Bioavailability of all-*trans* and *cis*-isomers of lycopene, *Exp. Biol. Med. (Maywood)*, 227, 914–919, 2002.

21. van den Berg, H., Faulks, R., Fernando Granado, H., Hirschberg, J. et al., The potential for the improvement of carotenoid levels in foods and the likely systemic effects, *J. Sci. Food Agric.*, 80, 880–912, 2000.

22. Hadley, C.W., Miller, E.C., Schwartz, S.J., and Clinton, S.K., Tomatoes, lycopene, and prostate cancer: progress and promise, *Exp. Biol. Med. (Maywood)*, 227, 869–880, 2002.

23. Rao, A.V., Fleshner, N., and Agarwal, S., Serum and tissue lycopene and biomarkers of oxidation in prostate cancer patients: a case-control study, *Nutr. Cancer*, 33, 159–164, 1999.

24. Tanumihardjo, S.A., Furr, H.C., Amedee-Manesme, O., and Olson, J.A., Retinyl ester (vitamin A ester) and carotenoid composition in human liver, *Int. J. Vitam. Nutr. Res.*, 60, 307–313, 1990.

25. Kaplan, L.A., Lau, J.M., and Stein, E.A., Carotenoid composition, concentrations, and relationships in various human organs, *Clin. Physiol. Biochem.*, 8, 1–10, 1990.

26. Schmitz, H.H., Poor, C.L., Wellman, R.B., and Erdman, Jr., J.W., Concentrations of selected carotenoids and vitamin A in human liver, kidney and lung tissue, *J. Nutr.*, 121, 1613–1621, 1991.

27. Stahl, W. and Sies, H., Uptake of lycopene and its geometrical isomers is greater from heat-processed than from unprocessed tomato juice in humans, *J. Nutr.*, 122, 2161–2166, 1992.

28. Clinton, S.K., Emenhiser, C., Schwartz, S.J., Bostwick, D.G. et al., *cis–trans* Lycopene isomers, carotenoids, and retinol in the human prostate, *Cancer Epidemiol. Biomarkers Prev.*, 5, 823–833, 1996.

29. Stahl, W., Schwarz, W., Sundquist, A.R., and Sies, H., *cis–trans* Isomers of lycopene and beta-carotene in human serum and tissues, *Arch. Biochem. Biophys.*, 294, 173–177, 1992.

30. Freeman, V.L., Meydani, M., Yong, S., Pyle, J. et al., Prostatic levels of tocopherols, carotenoids, and retinol in relation to plasma levels and self-reported usual dietary intake, *Am. J. Epidemiol.*, 151, 109–118, 2000.

31. Castenmiller, J.J., West, C.E., Linssen, J.P., van Het Hof, K.H. et al., The food matrix of spinach is a limiting factor in determining the bioavailability of beta-carotene and to a lesser extent of lutein in humans, *J. Nutr.*, 129, 349–355, 1999.

32. Mathews-Roth, M.M., Welankiwar, S., Sehgal, P.K., Lausen, N.C. et al., Distribution of [14C]canthaxanthin and [14C]lycopene in rats and monkeys, *J. Nutr.*, 120, 1205–1213, 1990.

33. Olmedilla, B., Granado, F., Southon, S., Wright, A.J. et al., A European multicentre, placebo-controlled supplementation study with alpha-tocopherol, carotene-rich palm oil, lutein or lycopene: analysis of serum responses, *Clin. Sci. (Lond.)*, 102, 447–456, 2002.

34. Paetau, I., Khachik, F., Brown, E.D., Beecher, G.R. et al., Chronic ingestion of lycopene-rich tomato juice or lycopene supplements significantly increases plasma concentrations of lycopene and related tomato carotenoids in humans, *Am. J. Clin. Nutr.*, 68, 1187–1195, 1998.

35. Boileau, A.C., Merchen, N.R., Wasson, K., Atkinson, C.A. et al., *cis*-Lycopene is more bioavailable than *trans*-lycopene *in vitro* and *in vivo* in lymph-cannulated ferrets, *J. Nutr.*, 129, 1176–1181, 1999.

36. Boileau, T.W., Clinton, S.K., and Erdman, Jr., J.W., Tissue lycopene concentrations and isomer patterns are affected by androgen status and dietary lycopene concentration in male F344 rats, *J. Nutr.*, 130, 1613–1618, 2000.

37. Gartner, C., Stahl, W., and Sies, H., Lycopene is more bioavailable from tomato paste than from fresh tomatoes, *Am. J. Clin. Nutr.*, 66, 116–122, 1997.

38. Schierle, J., Bretzel, W., Buhler, I., Faccin, N. et al., Content and isomeric ratios of lycopene in food and human blood plasma, *Food Chem.*, 59, 459–465, 1997.

39. Yeum, K.J., Booth, S.L., Sadowski, J.A., Liu, C. et al., Human plasma carotenoid response to the ingestion of controlled diets high in fruits and vegetables, *Am. J. Clin. Nutr.*, 64, 594–602, 1996.

40. Khachik, F., Spangler, C.J., Smith, Jr., J.C., Canfield, L.M. et al., Identification, quantification, and relative concentrations of carotenoids and their metabolites in human milk and serum, *Anal. Chem.*, 69, 1873–1881, 1997.

41. Conn, P.F., Schlach, W., and Truscott, T.G., The singlet oxygen and carotenoid interaction, *J. Photochem. Photobiol.*, 11, 41–47, 1991.

42. Gerster, H., The potential role of lycopene for human health, *J. Am. Coll. Nutr.*, 16, 109–126, 1997.

43. Ribaya-Mercado, J.D., Garmyn, M., Gilchrest, B.A., and Russell, R.M., Skin lycopene is destroyed preferentially over beta-carotene during ultraviolet irradiation in humans, *J. Nutr.*, 125, 1854–1859, 1995.

44. Tinkler, J.H., Bohm, F., Schalch, W., and Truscott, T.G., Dietary carotenoids protect human cells from damage, *J. Photochem. Photobiol. B*, 26, 283–285, 1994.

45. Bertram, J.S., Pung, A., Churley, M., Kappock, T.D. et al., Diverse carotenoids protect against chemically induced neoplastic transformation, *Carcinogenesis*, 12, 671–678, 1991.

46. Zhang, S., Tang, G., Russell, R.M., Mayzel, K.A. et al., Measurement of retinoids and carotenoids in breast adipose tissue and a comparison of concentrations in breast cancer cases and control subjects, *Am. J. Clin. Nutr.*, 66, 626–632, 1997.

47. Giovannucci, E., Ascherio, A., Rimm, E.B., Stampfer, M.J. et al., Intake of carotenoids and retinol in relation to risk of prostate cancer, *J. Natl. Cancer Inst.*, 87, 1767–1776, 1995.

48. Nahum, A., Hirsch, K., Danilenko, M., Watts, C.K. et al., Lycopene inhibition of cell cycle progression in breast and endometrial cancer cells is associated with reduction in cyclin D levels and retention of p27(Kip1) in the cyclin E-cdk2 complexes, *Oncogene*, 20, 3428–3436, 2001.

49. Liu, Y., Martindale, J.L., Gorospe, M., and Holbrook, N.J., Regulation of p21WAF1/CIP1 expression through mitogen-activated protein kinase signaling pathway, *Cancer Res.*, 56, 31–35, 1996.

50. Macleod, K.F., Sherry, N., Hannon, G., Beach, D. et al., p53-Dependent and independent expression of p21 during cell growth, differentiation, and DNA damage, *Genes Dev.*, 9, 935–944, 1995.

51. Elson, C.E., Pfeffley, D.M., Hentosh, P., and Mo, H., Isoprenoid-mediated inhibition of mevalonate synthesis: potential application to cancer, *Proc. Soc. Exp. Biol. Med.*, 221, 294–311, 1999.

52. Parker, R.A., Pearce, B.C., Clark, R.W., Gordon, D.A. et al., Tocotrienols regulate cholesterol production in mammalian cells by post-transcriptional suppression of 3-hydroxy-3-methylglutaryl-coenzyme A reductase, *J. Biol. Chem.*, 268, 11230–11238, 1993.

53. Fuhrman, B., Elis, A., and Aviram, M., Hypocholesterolemic effect of lycopene and beta-carotene is related to suppression of cholesterol synthesis and augmentation of LDL receptor activity in macrophages, *Biochem. Biophys. Res. Commun.*, 233, 658–662, 1997.

54. Keyomarsi, K., Sandoval, L., Band, V., and Pardee, A.B., Synchronization of tumor and normal cells from G1 to multiple cell cycles by lovastatin, *Cancer Res.*, 51, 3602–3609, 1991.

55. Jakobisiak, M., Bruno, S., Skierski, J.S., and Darzynkiewicz, Z., Cell cycle-specific effects of lovastatin, *Proc. Natl. Acad. Sci. USA*, 88, 3628–3632, 1991.

56. Carlberg, M., Dricu, A., Blegen, H., Wang, M. et al., Mevalonic acid Is limiting for N-linked glycosylation and translocation of the insulin-like growth factor-1 receptor to the cell surface: evidence for a new link between 3-hydroxy-3-methylglutaryl-coenzyme A reductase and cell growth, *J. Biol. Chem.*, 271, 17453–17462, 1996.

57. Dricu, A., Wang, M., Hjertman, M., Malec, M. et al., Mevalonate-regulated mechanisms in cell growth control: role of dolichyl phosphate in expression of the insulin-like growth factor-1 receptor (IGF-1R) in comparison to Ras prenylation and expression of c-myc, *Glycobiology*, 7, 625–633, 1997.

58. Mantzoros, C.S., Tzonou, A., Signorello, L.B., Stampfer, M. et al., Insulin-like growth factor 1 in relation to prostate cancer and benign prostatic hyperplasia, *Br. J. Cancer*, 76, 1115–1118, 1997.

59. Hankinson, S.E., Willett, W.C., Colditz, G.A., Hunter, D.J. et al., Circulating concentrations of insulin-like growth factor-I and risk of breast cancer, *Lancet*, 351, 1393–1396, 1998.

60. Ma, J., Pollak, M.N., Giovannucci, E., Chan, J.M. et al., Prospective study of colorectal cancer risk in men and plasma levels of insulin-like growth factor (IGF)-I and IGF-binding protein-3, *J. Natl. Cancer Inst.*, 91, 620–625, 1999.

61. Mucci, L.A., Tamimi, R., Lagiou, P., Trichopoulou, A. et al., Are dietary influences on the risk of prostate cancer mediated through the insulin-like growth factor system? *BJU Int.* 87, 814–820, 2001.

62. LeRoith, D., Clemmons, D., Nissley, P., and Rechler, M.M., NIH conference: insulin-like growth factors in health and disease, *Ann. Intern. Med.*, 116, 854–862, 1992.

63. Daughaday, W.H., The possible autocrine/paracrine and endocrine roles of insulin-like growth factors of human tumors, *Endocrinology*, 127, 1–4, 1990.

64. Levy, J., Bosin, E., Feldman, B., Giat, Y. et al., Lycopene is a more potent inhibitor of human cancer cell proliferation than either alpha-carotene or beta-carotene, *Nutr. Cancer*, 24, 257–266, 1995.

65. Baffa, R., Reiss, K., El-Gabry, E.A., Sedor, J. et al., Low serum insulin-like growth factor 1 (IGF-1): a significant association with prostate cancer, *Tech. Urol.*, 6, 236–239, 2000.

66. Kucuk, O., Sarkar, F.H., Sakr, W., Djuric, Z. et al., Phase II randomized clinical trial of lycopene supplementation before radical prostatectomy, *Cancer Epidemiol. Biomarkers Prev.*, 10, 861–868, 2001.

67. Zhang, L.X., Cooney, R.V., and Bertram, J.S., Carotenoids enhance gap junctional communication and inhibit lipid peroxidation in C3H/10T1/2 cells: relationship to their cancer chemopreventive action, *Carcinogenesis*, 12, 2109–2114, 1991.

68. Zhang, L.X., Cooney, R.V., and Bertram, J.S., Carotenoids up-regulate connexin43 gene expression independent of their provitamin A or antioxidant properties, *Cancer Res.*, 52, 5707–5712, 1992.

69. Livny, O., Kaplan, I., Reifen, R., Polak-Charcon, S. et al., Lycopene inhibits proliferation and enhances gap-junction communication of KB-1 human oral tumor cells, *J. Nutr.*, 132, 3754–3759, 2002.

70. Pastori, M., Pfander, H., Boscoboinik, D., and Azzi, A., Lycopene in association with alpha-tocopherol inhibits at physiological concentrations proliferation of prostate carcinoma cells, *Biochem. Biophys. Res. Commun.*, 250, 582–585, 1998.

71. Amir, H., Karas, M., Giat, J., Danilenko, M. et al., Lycopene and 1,25-dihydroxyvitamin D3 cooperate in the inhibition of cell cycle progression and induction of differentiation in HL-60 leukemic cells, *Nutr. Cancer*, 33, 105–112, 1999.

72. Watanabe, S., Kitade, Y., Masaki, T., Nishioka, M. et al., Effects of lycopene and Sho-saiko-to on hepatocarcinogenesis in a rat model of spontaneous liver cancer, *Nutr. Cancer*, 39, 96–101, 2001.

73. Kitade, Y., Watanabe, S., Masaki, T., Nishioka, M. et al., Inhibition of liver fibrosis in LEC rats by a carotenoid, lycopene, or a herbal medicine, Sho-saiko-to, *Hepatol. Res.*, 22, 196–205, 2002.

74. Wang, C.J. and Lin, J.K., Inhibitory effects of carotenoids and retinoids on the *in vitro* growth of rat C-6 glioma cells, *Proc. Natl. Sci. Counc. Repub. China B*, 13, 176–183, 1989.

75. Karas, M., Amir, H., Fishman, D., Danilenko, M. et al., Lycopene interferes with cell cycle progression and insulin-like growth factor I signaling in mammary cancer cells, *Nutr. Cancer*, 36, 101–111, 2000.

76. Nara, E., Hayashi, H., Kotake, M., Miyashita, K. et al., Acyclic carotenoids and their oxidation mixtures inhibit the growth of HL-60 human promyelocytic leukemia cells, *Nutr Cancer,* 39, 273–283, 2001.

77. Kim, D.J., Takasuka, N., Kim, J.M., Sekine, K. et al., Chemoprevention by lycopene of mouse lung neoplasia after combined initiation treatment with DEN, MNU and DMH, *Cancer Lett.,* 120, 15–22, 1997.

78. Okajima, E., Tsutsumi, M., Ozono, S., Akai, H. et al., Inhibitory effect of tomato juice on rat urinary bladder carcinogenesis after N-butyl-N-(4-hydroxybutyl) nitrosamine initiation, *Jpn. J. Cancer Res.,* 89, 22–26, 1998.

79. Narisawa, T., Fukaura, Y., Hasebe, M., Nomura, S. et al., Prevention of N-methylnitrosourea-induced colon carcinogenesis in F344 rats by lycopene and tomato juice rich in lycopene, *Jpn. J. Cancer Res.,* 89, 1003–1008, 1998.

80. Narisawa, T., Fukaura, Y., Hasebe, M. et al., Inhibitory effect of natural carotenoids alphacarotene, beta-carotene, lycopene and lutein, on colonic aberrant crypt foci formation in rats, *Cancer Lett.,* 107, 137–142, 1996.

81. Astorg, P., Gradelet, S., Berges, R., Suschetet M., Dietary lycopene decreases the initiation of liver preneoplastic foci by diethylnitrosamine in the rat, *Nutr Cancer,* 29, 60–68, 1997.

82. Nagasawa, H., Mitamura, T., Sakamoto, S., and Yamamoto, K., Effects of lycopene on spontaneous mammary tumour development in SHN virgin mice, *Anticancer Res.,* 15, 1173–1178, 1995.

83. Sharoni, Y., Giron, E., Rise, M., and Levy, J., Effects of lycopene-enriched tomato oleoresin on 7,12-dimethyl-benz[a]anthracene-induced rat mammary tumors, *Cancer Detect. Prev.,* 21, 118–123, 1997.

84. Cohen, L.A., Zhao, Z., Pittman, B., and Khachik, F., Effect of dietary lycopene on N-methylnitrosourea-induced mammary tumorigenesis, *Nutr. Cancer,* 34, 153–159, 1999.

85. Zhao, Z., Khachik, F., Richie, Jr., J.P., and Cohen, L.A., Lycopene uptake and tissue disposition in male and female rats, *Proc. Soc. Exp. Biol. Med.,* 218, 109–114, 1998.

86. Block, G., Patterson, B., and Subar, A., Fruit, vegetables, and cancer prevention: a review of the epidemiological evidence, *Nutr. Cancer,* 18, 1–29, 1992.

87. Giovannucci, E., Tomatoes, tomato-based products, lycopene, and cancer: review of the epidemiologic literature, *J. Natl. Cancer Inst.,* 91, 317–331, 1999.

88. Khachik, F., Beecher, G.R., and Smith, Jr., J.C., Lutein, lycopene, and their oxidative metabolites in chemoprevention of cancer, *J. Cell. Biochem. Suppl.,* 22, 236–246, 1995.

89. Rao, A.V. and Agarwal, S., Role of antioxidant lycopene in cancer and heart disease, *J. Am. Coll. Nutr.,* 19, 563–569, 2000.

90. Omenn, G.S., Goodman, G., Thornquist, M., Grizzle, J. et al., The beta-carotene and retinol efficacy trial (CARET) for chemoprevention of lung cancer in high risk populations: smokers and asbestos-exposed workers, *Cancer Res.,* 54, 2038s–2043s, 1994.

91. Heinonen, O.P., Albanes, D., Virtamo, J., Taylor, P.R. et al., Prostate cancer and supplementation with alpha-tocopherol and beta-carotene: incidence and mortality in a controlled trial, *J. Natl. Cancer Inst.,* 90, 440–446, 1998.

92. Goodman, M.T., Hankin, J.H., Wilkens, L.R., and Kolonel, L.N., High-fat foods and the risk of lung cancer, *Epidemiology,* 3, 288–299, 1992.

93. Giovannucci, E., A review of epidemiologic studies of tomatoes, lycopene, and prostate cancer, *Exp. Biol. Med. (Maywood),* 227, 852–859, 2002.

94. Giovannucci, E., Rimm, E.B., Liu, Y., Stampfer, M.J. et al., A prospective study of tomato products, lycopene, and prostate cancer risk, *J. Natl. Cancer Inst.,* 94, 391–398, 2002.

95. Kucuk, O., Chemoprevention of prostate cancer, *Cancer Metastasis Rev.,* 21, 111–124, 2002.

96. Giovannucci, E. and Clinton, S.K., Tomatoes, lycopene, and prostate cancer, *Proc. Soc. Exp. Biol. Med.,* 218, 129–139, 1998.

33 NADH in Cancer Prevention and Therapy

George D. Birkmayer and Jiren Zhang

CONTENTS

33.1 Biological Functions of NADH 541
    33.1.1 NADH Is the Fuel for Cellular Energy Production...542
        33.1.1.1 NADH Increases the Mitochondrial Membrane Potential 542
        33.1.1.2 Extracellular NADH Increases Intracellular ATP
                Production in Heart Cells 542
    33.1.2 NADH Plays a Key Role in DNA and Cell Damage Repair...............................543
    33.1.3 NADH Stimulates Cellular Immune Functions 546
    33.1.4 NADH Is the Most Powerful Antioxidant 546
33.2 ENADA: The Stabilized Orally Absorbable Form of NADH.......................................547
    33.2.1 Bioavailability of ENADA–NADH..547
    33.2.2 ENADA–NADH: A Protector against Chemotoxicity
        and Radiation..548
    33.2.3 The Safety of ENADA–NADH..548
    33.2.4 ENADA–NADH as Therapeutic Concept
        for Certain Human Cancers 549
33.3 Case Studies 549
    33.3.1 Case 1 549
    33.3.2 Case 2 549
    33.3.3 Case 3 549
    33.3.4 Case 4 550
    33.3.5 Case 5 550
    33.3.6 Case 6 550
    33.3.7 Case 7 550
33.4 Mechanism of Action of NADH..551
Acknowlegdment..551
References 551

33.1 BIOLOGICAL FUNCTIONS OF NADH

NADH is the abbreviation for nicotinamide adenine dinucleotide hydride. NADH is also known under a number of other synonyms such as:

Diphosphopyridine nucleotide, reduced form
Adenine-D-ribose-phosphate-phosphate
D-ribose-nicotinamide, reduced form
Cozymase, reduced form

Coenzyme 1, reduced form
Codehydrogenase, reduced form
Nadide, disodium salt, reduced form

NADH is present in every living cell, where it catalyzes more than 1000 biochemical reactions. The most important biological functions of NADH are the following:

NADH is the cellular fuel for energy production
NADH plays a key role in DNA and cell damage repair
NADH stimulates cellular immune functions
NADH is the most potent antioxidant

33.1.1 NADH Is the Fuel for Cellular Energy Production

All living cells require energy to stay alive. Without energy, a cell dies because the energy production represents the essential prerequisites for every living cell.[1] How is energy produced in the cell? NADH reacts with oxygen in a cascade of biochemical reactions water and energy. This energy is stored in the form of the chemical compound adenosine triphosphate (ATP). NADH itself is produced from amino acids, sugars, and lipids via the citric acid cycle. One molecule of NADH yields three molecules of ATP, and the more NADH a cell has available, the more energy it can produce.[2] The amount of NADH a cell contains depends on the amount of energy it requires. Heart muscle cells, which have to contract themselves every second (86,400 times per day) for an entire lifetime, contain 90 μg of NADH per gram of tissue. Brain and muscle cells contain 50 μg.[3] One-third of all the energy produced by our body is used by our brain.

33.1.1.1 NADH Increases the Mitochondrial Membrane Potential

The British Nobel laureate Peter Mitchell postulated that energy in the mitochondria is formed by a gradient of electric charge between the outer and the inner side of the mitochondrial membrane. The higher the level of this electric potential, the more energy is produced. Researchers in China demonstrated that incubating cells with NADH leads to an increase in the mitochondrial membrane potential,[4] implying greater energy output.

33.1.1.2 Extracellular NADH Increases Intracellular ATP Production in Heart Cells

A recent study has shown that NADH can increase the biosynthesis of ATP inside the cell. Isolated single heart cells were incubated with NADH, and an increase of ATP inside the cell was found by two independent methods.[5] This observation provides convincing evidence that NADH can penetrate the cell membrane and increase the cellular energy level in the form of ATP. If the cell has more energy, it can live longer and can perform its functions better.

The consequences and implications of these findings are remarkable. Heart cells get more energy by NADH, hence their strength and capacity is higher. People with heart problems can benefit from NADH. After a heart attack, some areas in the heart may be damaged and hence not functioning but still be alive. If these cells are supplied with NADH, they may get more energy to repair the damage and become functional again.

The same principle may work in the brain. After a stroke, certain areas in the brain are not nurtured by blood, as the circulation is blocked. These brain parts may be still vital but not functioning. By offering them NADH, they get more energy and may regain their functionality. A number of anecdotal cases of stroke patients treated with NADH did show improvement of their symptoms even weeks after the event.

If NADH leads to an increase in energy in isolated heart cells, it should also work in other tissues such as the kidney, the liver, the pancreas, or the lung. It was some kind of dogma that NADH does not pass the cell membrane because it is too hydrophilic and too labile to penetrate into the intact cell. The study outlined has convincingly shown by two independent methods that NADH can increase ATP formation and energy production in isolated heart cells.[5] In doing so, NADH must penetrate the cell membrane to get to the point of action, the mitochondria. NADH is also taken up by cells lacking mitochondria such as erythrocytes. If you incubate human red blood cells with NADH, a decline of the extracellular NADH and an increase in ATP (= energy) in these cells is observed (Hallström et al., personal communication). The consumption of NADH by blood cells correlates (indirectly) to the level of ATP.[6] In other words, if blood cells consume a lot of NADH, the ATP level in these cells is low.

Highly conditioned athletes are assumed to have a high energy level in their muscles and blood cells. Hence, their blood cells consume only a low amount of NADH when incubated with it. Blood cells from elderly or sick people consume considerably more NADH than athletes. However, when athletes are tested after a marathon run or after long-distance cycling, their blood cells consume NADH in an amount comparable with that of old people. These observations were made with a newly developed and patented blood test known as ENMA (extracellular NADH metabolization assay).[6] It has an enormously broad application range.[7] It can be used not only for controlling the training performance of athletes, but also in the surveillance of patients in terms of energy recovery after a heart attack, a stroke, cancer treatment, or rehabilitation.

33.1.2 NADH Plays a Key Role in DNA and Cell Damage Repair

The DNA in the nucleus is well protected by histones and other macromolecules. Nevertheless, it can be damaged by exposure to various agents such as radiation, ultraviolet (UV) light, ozone, free radicals, carcinogens, and toxins such as cytostatic drugs, some of which are themselves carcinogenic. These potentially harmful agents can react with the chromosomes. If the DNA is affected and damaged by one of these agents, the genetic material will be altered. Replication of altered, defective DNA causes changed features in the newly divided cells provided that cell division can still occur. The greater the DNA damage, the more extensive are the alterations that can occur in cells and tissue.[8] Genetic damage is the biochemical basis for a number of chronic diseases such as cancer,[9,10] rheumatoid arthritis, immunodeficiencies, and arteriosclerosis.[9,10] Hence it is imperative that our genetic material remain unaltered in order to guarantee that any new progenitor cell developing after cell division is identical to its parent cells. If the DNA is altered by physical or chemical agents, the newly developing progenitor cells may be different from their mother cells and will not function in the originally programmed way.

In order to avoid the fatal consequences of DNA damage, mammalian cells have developed a system for repairing alterations of their genetic material. This so-called DNA repair system needs NADH to gain full functionality.[12,13] Therefore, the more NADH you have in your body, the better the DNA repair system functions, and the better you are protected from potentially developing diseases.

The exposure of cells to DNA-damaging reagents can trigger a wide range of cellular responses involved in the regulation of gene expression and cell-cycle progression, stimulation of DNA repair, and programmed cell death.[14,15] These processes are important for maintaining normal growth, antimutation, damage repair, and functional activity of cells. However, due to the unspecificity of chemotherapeutic drugs for the cancerous target cells, many normal cells get damaged as well, causing severe, sometimes fatal adverse reactions. The question is, how can normal cells be protected from the cytotoxic effects of chemotherapeutic agents? How can we stimulate the repair system and promote normal cellular responses after chemotherapy? The mechanism involved in repairing DNA-damaged cells exposed to cytostatics has been investigated in many clinical studies.[14–18] Whether the reduced form of the coenzyme nicotinamide adenine dinucleotide (NADH) can be used to protect cells from DNA damage has never been considered until recently. Previous studies in our laboratory have found that NADH can stimulate biosynthesis of endogenous cell factors and can rescue cells from apoptotic damage by triggering production of the bcl-2 oncogene proteins.[19]

The effect of NADH on DNA repair was investigated on PC12 cells damaged by doxorubicin. PC12 cells were incubated in medium with and without NADH before and after exposure to the DNA-damaging agent doxorubicin. The changes of the cell proliferation genes (c-myc, c-erb-2), the apoptosis inhibition gene bcl-2 and p53 (tumor suppressor gene), cell apoptosis inhibition gene bcl-2 and p53 (tumor suppressor gene), cell apoptosis gene (c-fos), and the proliferating cell nuclear antigen (PCNA) were investigated using a cytotoxicity assay and immunofluorescence flow cyto-metric analysis.

Doxorubicin induced DNA damage in PC12 cells by inhibiting the expression of the cell proliferation genes and by triggering apoptotic processes in the cells. This was shown by down-regulating the expression of c-erb-2, c-myc, bcl-2 and by up-regulating the expression of PCNA and c-fos of the PC12 cells.[20]

NADH not only increased the resistance of PC12 cells to the doxorubicin-induced DNA damage, but also partially repaired the damage. NADH promoted survival and differentiation by regulating the c-myc oncogene proteins. Furthermore, it supported the process of DNA repair by regulating the expression of p53 and bcl-2 on the PC12 cells damaged by doxorubicin. NADH also down-regulated expression of the cell apoptosis gene c-fos on the PC12 cells.

The expression of c-erb-2 oncogene proteins and PCNA on the PC12 cells did not show a significant change in the group of cells incubated with NADH in comparison with the group incubated with medium alone. In addition, no abnormal proliferation effect of NADH on PC12 cells was observed in these experiments.

As a consequence of these findings, NADH can be considered as a therapeutic adjunct for cancer patients to protect them against the general toxic effects of substances such as doxorubicin or cisplatin by stimulating the DNA-repair system and by promoting normal cellular biosynthetic responses after chemotherapy. NADH seems to exhibit a chemopreventive effect.

Drug-induced apoptosis is dependent on the balance between cell-cycle checkpoints and DNA-repair mechanisms. Doxorubicin is a DNA-damaging cytotoxic drug that accumulates in the nuclei of damaged cells. Increased accumulation of cellular doxorubicin is accompanied by apopto-sis.[14,15,21] Experiments indicate that the inhibition rate of PC12 cells correlates with the concentration of doxorubicin in medium and with time of exposure of the cells to the toxic environment. The cytotoxicity of doxorubicin for PC12 cells occurs not only in the phase of acute exposure but also in the lag phase.

Apoptosis induced by doxorubicin is accompanied by the down-regulation of the expression of the oncogene proteins c-erb-2 and c-myc, the anti-apoptotic gene proteins (bcl-2), p53 tumor suppressor protein, and up-regulation of the expression of PCNA[16] and c-fos.

These genetic changes occur not only in the early phase of the apoptosis induced by doxorubicin, but can also happen in the lag phase, when the damaged PC12 cells are incubated with new medium after removing the old doxorubicin-containing medium. DNA damage and activation of c-fos oncogene seem to be the major pathways of inducing apoptotic damage of PC12 cells. NADH can partially rescue cell activity of PC12 cells from DNA damage induced by doxorubicin. Cell damage repair is a complex biological process in which a number of reactions are involved.

NADH is an essential component of enzymes necessary for many metabolic reactions in the cell, including energy production. It plays a crucial role in triggering biological antioxidation and in regulating the expression of membrane glycoprotein receptors.[22,23] Previous studies have shown that NADH can rescue cells from apoptosis caused by inhibition of the mitochondrial respiratory chain induced by chemotherapeutic agents such as rotenone, and simultaneously can increase the production of endogenous biological factors necessary for proper functions. In addition, cell-cycle progression of PC12 cells is observed.[15,19]

When the apoptotic rate of PC12 cell was 82.2%, the rate of cells repaired by NADH was only 3.1%. After recovery incubation for 48 h, the expression of c-erb-2 oncogene proteins and PCNA on the PC12 cells did not show a significant increase in the group treated with NADH in comparison with the control. The change of c-erb-2 oncogene that occurs during the acute damage phase of the PC12 cells is difficult to be reversed by incubation with NADH or medium. However, the up-regulation of c-fos oncogene protein in the acute damage phase can be significantly down-regulated by incubation with NADH for 48 h. This suggests that NADH rescues PC12 from doxorubicin-induced damage not only by repairing the DNA, but also by increasing energy production within these cells.

Programmed cell death is an energy-dependent biochemically regulated process that is the result of the expression of a number of genes. The roles of several gene and gene families, such as Bcl-2/bax, p53, c-myc, c-jun, c-fos, considered to be critical for apoptosis have recently been described in different cell lines.[24,25] Many reports suggest that a rather complex genetic and molecular mechanism is involved in the process of apoptosis. It could also be triggered either by increased or by reduced gene expressions as well as by biochemical reactions not necessarily connected to altered gene expression.[14,24–26]

Observations from our studies provide evidence that complex molecular events are involved in the apoptotic process of PC12 cells induced by doxorubicin. After recovery incubation of PC12 cells with NADH for 48 h, the positive ratio and amount of c-erb-2 expressed on PC12 cells did not show an increase in comparison with the control with medium alone. The positive ratio of c-myc was not altered, but the amount of c-myc expressed on the vital PC12 cells was significantly up-regulated 47.7% and 52.9%, respectively, in comparison with the acute-damage phase and the group with medium alone. This suggests that regulating the expression of c-myc on PC12 cells may be involved in the DNA repair of PC12 cells damaged by doxorubicin. Although the exact function of c-myc remains largely unknown, its activation has been implicated in the induction of cell proliferation and differentiation. Some reports have also indicated that the c-myc oncogene protein acts as a sequence-specific factor that serves to regulate gene expression in normal cellular growth and differentiation and as a common intracellular transducer that promotes G0 to G1 transition. It may also be involved in the regulation of programmed cell death.[27,28–30]

In the processes of cell DNA damage repair, bcl-2 and p53 are two of the most important proteins encoded by the bcl-2 gene and p53 tumor suppressor gene. Wild-type p53 can suppress cell proliferation and slow DNA synthesis and block transition from G1 to S phase of the cell cycle.[30,31,32] The bcl-2 gene is a proto-oncogene and the most important inhibitor of apoptosis. Expression of bcl-2 may interfere with the apoptotic process mediated by the APO-1/Fas antigen and TNF receptor. Probably the ratio of bcl-2 and p53 determines how the cell responds to DNA-damaging agents. Current research indicates that expression of bcl-2 in pheochromocytoma cells is associated with that of the c-myc oncogene protein.[33–35] Overexpression of the proto-oncogene bcl-2 might block p53-induced apoptosis and inhibit p53 functional activity.[36] In our experiment, in which we investigated the effect of NADH on the recovery of PC12 cell from DNA-damage, the ratio of expression of p53 and bcl-2 on PC12 cells was down-regulated by 91.9 and 98.8% after exposure of the cells with doxorubicin. After recovery incubation of the cells in medium containing NADH for 48 h, the ratio of vital PC12 cells was up-regulated by 3.1%, and the p53 tumor suppressor protein expressed on the vital cells was down-regulated by 36.7%. However, the amount of bcl-2 expressed on the vital PC12 cells was found to be up-regulated by 12.7% in comparison with the control group (medium alone). These findings suggest that NADH can not only promote survival and differentiation of cells by regulating the c-myc oncogene protein, but also support the process of DNA repair by regulating the expression of p53 tumor-suppressor protein and proto-oncogene protein bcl-2 on the PC12 cells damaged by doxorubicin.

Cisplatin is one of the most frequently used drugs for chemotherapy of cancer. It damages the cell membrane, the mitochondria, and the nucleus not only of cancer cells, but from all normal noncancerous cells as well. The consequences are the so-called side effects of chemotherapy such as hair loss, gastrointestinal problems (vomiting, dizziness), etc. Preincubation of cells that might be damaged by cisplatin with NADH prevents the changes induced by cisplatin.[37] Based on these findings, cancer patients should protect themselves by taking NADH when receiving cisplatin, doxorubicin, or other cell-damaging cytostatic drugs. NADH is also involved in transcriptional pathways important for development, cell-cycle regulation, and transformation.

The corepressor CtBP (carboxy-terminal binding protein) binding to cellular and viral transcriptional repressors is regulated by the nicotinamide adenine dinucleotides NAD and NADH, with NADH being two to three orders of magnitude more effective.[38] The best-characterized target promoter for CtBP in mammalian cells is probably the E-cadherin gene.[39,40] Loss of E-cadherin expression in tumors correlates with metastasis, invasion, and poor clinical prognosis.[41,42] It has been shown that CtBP-mediated repression of the E-cadherin promoter is enhanced by hypoxia.[38] NADH may alleviate the hypoxic state by stimulating oxygen uptake into the cell. It has been shown that the oxygen uptake in the muscles of highly conditioned athletes increases after taking the stabilized, orally absorbable form of NADH.[42] In addition, NADH seems to be a sensor of blood flow needed in brain, muscle, and other tissues.[43] Increasing blood flow removes lactate and augments delivery of nutrients and oxygen for energy metabolism.

33.1.3 NADH STIMULATES CELLULAR IMMUNE FUNCTIONS

The cellular immune response in humans is based on the activities of the T lymphocytes, the B lymphocytes, and the macrophages. Macrophages have the capability for direct elimination of allogenic entities such as bacteria, viruses, and other foreign tissues. The first step in the elimination of bacteria is the perturbation of the plasma membrane of macrophages. As a consequence, the metabolic activity including oxygen consumption is markedly increased. Most of oxygen is converted to superoxide and hydrogen peroxide.[44] This phenomenon, known as "metabolic burst," appears to be the first and most critical step leading to the destruction of the invading foreign organism. During this metabolic burst and the cytotoxic activity induced in the macrophages, high amounts of NADH are needed and used. Hence, the immune-defense mechanism of white blood cells is fueled by NADH. Furthermore, it has been shown that NADH stimulates the biosynthesis of interleukin-6 (IL-6). Peripheral human blood leucocytes, when incubated with NADH, significantly stimulate the release of IL-6 in a dosage dependent manner.[45]

Besides a number of other functions, IL-6 has been reported to protect neurons from degeneration, although the mechanism has not yet been elucidated. If IL-6 protects neurons, it may protect other cells as well.[46,47]

33.1.4 NADH IS THE MOST POWERFUL ANTIOXIDANT

An antioxidant is a substance that acts against oxidation. The opposite of oxidation is reduction. Compounds with a high reduction potential exhibit a strong antioxidative power. NADH, the reduced form of coenzyme 1, has the highest reducing power as a single biological molecule. Only molecular hydrogen has a higher reduction potential, but this does not exist in living cells. Biological antioxidants are present in all living cells to protect the cell and its membrane from destruction by free radicals.[48] Free radicals are molecules with an unpaired electron. Hence they are extremely reactive. They interact with many compounds in human cells, in particular with the lipid-containing structures such as the cell membrane. In doing this, they violate the integrity of the cell wall, causing leakage and release of essential cellular components, usually resulting in cell death.[49] Free radicals have been shown to be involved in the development of cancer,[50] coronary heart disease, atherosclerosis, diabetes, neurodegenerative disorders, and autoimmune diseases.[51,52]

Free radicals are formed in human cells by agents knocking out electrons from a molecule. These agents can be x-rays or other forms of high-energy radiation, such as that used for radiotherapy of cancer. Small amounts of free radicals are also produced in normal cells by metabolic reactions. However, mammalian cells possess a defense system — called the "antioxidative protection shield" — to protect them from being irreversibly damaged.[53] The first and most important antioxidant component in this system is NADH, which has the highest reduction potential of any compound in the cells.[54] The thiobarbituric acid reactive species (TBARS) determination provides a measure of both free-radical formation and lipid peroxidation. In a study using spontaneous hypertensive rats (SHR), it was been found that the renal TBARS were significantly lower (1.9 nmol MDA/100 mg tissue) in the rats fed with 5 mg NADH orally as compared with the control animals (3.5 nmol MDA/100 mg tissue). MDA (malondialdehyde) is formed from the breakdown of polyunsaturated fatty acids. NADH also reduced total cholesterol and LDL cholesterol significantly as well as the blood pressure.[55] One of the conclusions the authors deduced from these findings is that NADH may be a useful agent for preventing and treating cardiovascular risk factors.

The antioxidative effect of NADH was also investigated in humans. When LDL cholesterol is oxidized *in vitro* induced by peroxyl radicals, NADH reveals an antioxidant effect identical to ascorbic acid during the first 90 min.[56] However, after 90 min, the effect of ascorbic acid ceases, whereas NADH continues to act antioxidatively. Hence, the antioxidative potency of NADH appears to last much longer than that of ascorbic acid. In a double-blind placebo-controlled study, 37 human subjects were given ENADA®–NADH (four tablets of 5-mg NADH) or placebo tablets for 4 weeks. NADH reduced malondialdehyde levels as well as the (oxidative stress-induced) carbonyl modification of proteins, particularly in smokers. A steady decrease of the initially elevated protein carbonyl modification levels of smokers was observed, ultimately approaching the levels of non-smokers within the study period of 4 weeks. This observation implies that ENADA–NADH may have a preventive or even curative effect on tissues damaged by cigarette smoking.

33.2 ENADA: THE STABILIZED ORALLY ABSORBABLE FORM OF NADH

NADH can be regarded as a biological form of hydrogen. Hence NADH is a very reactive compound, is very unstable, and becomes easily degraded by air, water, humidity, acids, and oxidizing agents such as sugars. Even in solid state, NADH reacts with lactose, the most common filler of tablets. In 1987, NADH was used intravenously (i.v.) for treatment of patients with Parkinson's disease (PD). The beneficial effects in improving the disability of the PD patients were remarkable.[57] The challenge was to transpose the intravenous form of NADH into an oral (tablet) form. After yearlong research, a galenic formulation was developed in which NADH was stable for at least 2 years, a prerequisite for registration as an ethical drug. For this special formulation of a stabilized orally absorbable form, one of the authors (G.B.) received worldwide patents.[58,59] The brand name for the patented, stabilized, orally absorbable form of NADH is ENADA. Numerous controlled clinical studies have been performed with ENADA since its development 1993.

33.2.1 Bioavailability of ENADA–NADH

When taken orally, the stabilized form of NADH is absorbed in the small intestine. Studies have shown that NADH passes the intestinal mucosa undegraded by passive diffusion.[60] In a further study, it was demonstrated that ENADA–NADH passes the blood–brain barrier. When rats were fed with two tablets ENADA–NADH (5 mg) an increase of the NADH level in the brain cortex was observed after 20 min of intake as measured by laser-induced fluorescence.[61]

Using a pulsed N-2 laser combined with a fiber-optic probe and photomultipliers, the NADH fluorescence was measured in the brain cortex of rats. After intraperitoneal application of NADH (50 mg/kg), an increase in the intensity of the cortical NADH fluorescence of about 18% was observed for approximately 30 min compared with the fluorescence intensity in the control group. Neither NAD+ (the oxidized form of NADH) nor nicotinamide (both at concentrations of 50 mg/kg) showed any effect on the NADH fluorescence in the cortex for the entire measurement period of 120 min.

Following oral application of NADH (two tablets of ENADA [5 mg NADH] = 51 mg/kg) the cortical fluorescence intensity was increased by about 20% compared with the control group.[61] The results of this study provide convincing evidence that NADH given orally increases the amount of

NADH in the brain. To achieve this, ENADA, the stabilized and orally absorbable form of NADH, had to have passed the blood–brain barrier.

33.2.2 ENADA–NADH: A PROTECTOR AGAINST CHEMOTOXICITY AND RADIATION

Cytostatic drugs such as cisplatin and doxorubicin are used in chemotherapy of cancer. These drugs trigger a wide range of cellular responses involved in the regulation of gene expression and cell-cycle progression and programmed cell death. As these cytostatic drugs are not specific to either cancerous or normal cells, the latter may be damaged during chemotherapy, causing severe, sometimes fatal, adverse reactions.

Studies from the authors have shown that NADH can stimulate the biosynthesis of endogenous cell factors and can rescue cells from apoptotic damage by triggering production of the bcl-2 oncogene proteins.[62] In the process of cell damage repair, bcl-2 and p53 are two of the most important proteins encoded by the bcl-2 gene and p53 tumor-suppressor gene. Wild-type p53 can suppress cell proliferation and can slow DNA synthesis and thus block transition from G1 to S phase of the cell cycle.[63] The bcl-2 gene is a proto-oncogene and the most important inhibitor of apoptosis. Current research indicates that expression of bcl-2 is associated with that of the c-myc oncogene protein.[64] In studies performed by the authors, it was found that doxorubicin down-regulated the expression of p53 and bcl-2 in PC12 cells by 91.9 and 98.8%, respectively. Incubation of the damaged cells by NADH promoted survival and differentiation by regulating the c-myc oncogene protein. NADH also supported the process of DNA repair by regulating the expression of the p53 tumor suppressor protein and the proto-oncogene protein bcl-2.[62] Similar results were obtained when cisplatin was used as cell-damaging agent.[37]

In the same publication, the effect of NADH on cells damaged by radiation was reported. When PC12 cells were exposed to radiation in a dose given routinely as radiotherapy of cancer, 90% of vital cells were damaged. Incubation of the damaged cells with NADH induced a repair process. More than half of the damaged cells could be repaired and gained full functionality (Zhang, personal communication).

A number of cytostatic drugs, including cisplatin, are carcinogenic. NADH is able to protect cells from the carcinogenic effects of these chemotherapeutic agents. Hence ENADA, the stabilized oral form of NADH, may present a safe, nontoxic, biological supplement for prevention of cancer.

33.2.3 THE SAFETY OF ENADA–NADH

The stabilized, orally absorbable form of NADH (ENADA) is a nutritional supplement available in the U.S. since 1995 and in the E.U. since 1997. Based on the patented formulation of this supplement, a number of clinical trials have been launched to prove scientifically that ENADA is effective. In order to get these studies started, an investigational new drug (IND) application was filed with the Food and Drug Administration (FDA). For FDA approval, it must be documented that ENADA (the stabilized oral form of NADH) is safe. For this reason, the maximum tolerated intravenous dose (MTD) of βNADH (reduced form of beta-nicotinamide adenine dinucleotide) in beagle dogs was elucidated. The maximum tolerated dose (MTD) of βNADH in dogs was found to be 500 mg NADH per kg of body weight per day. In other words, a dog weighing 10 kg will tolerate 5 g of NADH.[65] The oral form of NADH (ENADA) was also tested in beagle dogs, who were fed 150 mg/kg/day for 14 days. The drug was delivered in the form of 30 regular (5 mg) ENADA tablets filled in two gelatin capsules (15 ENADA tablets per capsule). This high dose was selected because it was considered to be the maximum amount that could be practically administered repeatedly over 14 days. All dogs survived the treatment, and no adverse reactions or side effects were observed. The dogs treated with ENADA showed no changes in comparison with the control animals regarding laboratory safety parameters and organ and tissue pathology. A dose of 150 mg/kg body weight means 1500 mg for a 10 kg beagle dog, and 1500 mg of ENADA corresponds to 300 5-mg ENADA tablets per day. This is a dose that beagle dogs tolerate without any side effect.[65]

In addition to the MTD findings, a study for potential chronic toxicity of ENADA was performed in rats, which were given one tablet of ENADA (5 mg NADH) per day for 26 weeks. No changes in laboratory parameters or in tissue and organ pathology were observed.[66] A dose of 5 mg for a rat weighing about 330 grams corresponds to 15 mg per kg body weight, or 1050 mg of NADH for a 70-kg human subject. A 1050-mg dose of NADH corresponds to 210 tablets of ENADA (5 mg NADH), which are tolerated without side effects when given for 26 weeks (6.5 months). Based on these safety data ENADA–NADH can be generally regarded as safe, and the FDA gave permission for two clinical trials in the U.S. about 2 weeks after application.

33.2.4 ENADA–NADH AS THERAPEUTIC CONCEPT FOR CERTAIN HUMAN CANCERS

NADH has been shown to inhibit the growth of murine fibrosarcoma and human laryngeal carcinoma cells *in vitro*.[67] Based on these findings and the various biological functions, the stabilized oral form of ENADA has been used as treatment for certain types of cancer.

33.3 CASE STUDIES

The following subsections describe a number of anecdotal cases in which ENADA–NADH was used as an anticancer therapy.

33.3.1 CASE 1

The first case involved a 48-year-old male suffering from a small-cell bronchial carcinoma. The diagnosis was made by MRT (magnetic resonance tomography) and verified by histopathological examination of biopsy specimen. The size of the tumor was 6 to 8 cm in diameter in September 2001 when the patient visited one of the authors (G.B.). The report of the University of Amsterdam indicated that the tumor was inoperable due to its localization very close to the mediastinum. The patient had received radiotherapy followed by chemotherapy before he came for a visit to one of the authors (G.B.). The patient was recommended to take four 5-mg tablets of ENADA per day. He was already taking selenium, vitamin C, and vitamin E. In January 2002, the size of the tumor was, as verified by MRT, to be the size of a cherry. The therapy with NADH was continued. In July 2002, an MRT report from the University of Amsterdam stated that no tumor was detectable.

33.3.2 CASE 2

A female, aged 63, underwent an operation for invasive duct carcinoma in August 1989. One year later, multiple liver and bone metastases were detected. Following four therapy cycles, according to the CMF diagram, further increase of liver and bone metastases were observed. Pain was reducible only with the strongest analgesics. Beginning in January 1991, the patient was treated with NADH (12.5 mg) administered intravenously three times a week. After four weeks of parenteral therapy, the patient was switched to an oral regimen, taking 5 mg of NADH every day. By April 1991 radiological detection showed metastasis regression. Some foci were greatly reduced in size, and some completely disappeared. The oral NADH therapy was continued. A check later in 1991 using CT scanning revealed a further marked regression of the liver metastases and the bone metastases were virtually undetectable. The patient was free from pain and no longer required analgesics. The serum concentration of CA15.3 dropped from 65.0 (January 1991) to 24 (August 1994).

33.3.3 CASE 3

A male, aged 59, had a colon carcinoma in 1987. In 1990, sonographic and radiological tests detected multiple liver metastases of cherry to plum size. Two chemotherapy cycles, Myleran or

Endoxan, were unsuccessful, and the liver foci increased in size. Therapy with NADH was started in December 1990, initially 12.5 mg intravenously three times a week. After four weeks, the therapy was changed to NADH orally, 5 mg, every day. In March 1991, sonographic detection showed a reduction in the size of the liver foci. In June 1991, a check by CT scanning and sonography revealed an almost complete disappearance of the metastases in the liver. The patient subjectively reported to feel extremely well. The tumor marker CEA was initially 110 in December 1990 and declined to 22 by November 1994.

33.3.4 Case 4

This case involved a female, aged 52. Three years after undergoing a quadrantectomy due to invasive scirrhous carcinoma of the breast, vertebral metastases were detected in January 1990. In April 1990, liver metastases were discovered by ultrasonics examinations. Therapy with Novaldex led to no regression of the metastases. There was also no response to a therapy cycle, according to the CMF diagram. Intravenous administration of NADH (12.5 mg every other week) was started in November 1990. After four weeks, the therapy was changed to NADH orally, 5 mg every day. Two months after the start of NADH therapy, there was a clear regression of liver metastases as well as disappearance/reduction of vertebral metastases. Liver metastases were greatly reduced or foci were no longer detectable. Tumor markers CEA and CA15.3 were 45 and 92, respectively, in April 1990. The last control in October 1994 showed CEA to be 14 and CA15.3 to 18.5.

33.3.5 Case 5

This case involved a male, aged 66. Parvicellular bronchial carcinoma was diagnosed in February 1990, and multiple foci in both pulmonary lobes were formed. Cytostatic therapy with methotrexate and Endoxan led to no regression. In October 1990, NADH was administered parenterally (10 mg intravenously) every other day. A radiographic check in 1991 revealed the remission of the neoplastic foci as regards to both number and size. NADH therapy was continued with 10 mg orally every day. A check in May 1991 by CT scanning confirmed a further reduction of tumor foci in both pulmonary lobes.

33.3.6 Case 6

This case involved a male, aged 72. In November 1990, a tumor mass in the liver (8 to 10 cm in diameter) was diagnosed. In summer 1993, multiple lung metastases of various sizes were found in a CT scan. The patient denied surgical intervention as well as chemo- or radiotherapy. Beginning in the spring of 1994, he took one tablet of NADH every day. Control examination by x-ray and computer tomography showed no increase of the lung metastases and a reduction of the liver mass, with indications of formation of necroses in the center of the tumor. The patient subjectively reported to feel well and had no pain. The lung-cancer-associated tumor marker CYFRA 21-1 was 35 before NADH therapy (April 94) and 21 in December 1994. The carcinoembryonic antigen CEA levels were measured to be 67 in April 94 and 28 in December 94.

33.3.7 Case 7

This case involved a female, aged 55. In February 1992, lymph-node metastases of a poorly differentiated mammary carcinoma were detected in the left neck region. The CA15.3 value was 37.0, the CEA level was 13.5, and the TPS was 145 in March 1992. The primary tumor could not be localized. The patient denied chemo- and radiotherapy. She was given 5 mg NADH every day. A year later, the previous palpable lymph-node metastases had disappeared. The tumor marker tests CA15.3, CEA, and TPS were 15.0, 8.0, and 95, respectively, in July 1994. Computer tomography and bone scan did not show any metastases (June 94).

33.4 MECHANISM OF ACTION OF NADH

For the time being, we can only speculate on the mechanism of action of NADH in stabilizing or reducing certain cancers. One possibility could be the function of NADH as a DNA-repairing agent. Cancer cells have a DNA that is altered from that of the original cells from which the carcinoma cells developed. If NADH is given to cancer patients, the content of NADH in the cancerous cell increases. The more NADH a cell has available the better the DNA repair system works, and the alteration of the genes may be reverted to normal.

Another possibility may be derived from the energy-increasing function of NADH. As mentioned earlier in this chapter, the intracellular level of ATP can be increased by incubating the cells with NADH. With more energy, cancerous cells increase their capacity of the biosynthesis of macromolecules, in particular proteins, glycoproteins, and glycolipids. These substances play a major role on the cell surface in regulating proliferation and differentiation. With more NADH and ATP in the cancerous cells, proliferation may be halted, and differentiation processes may be induced. These assumptions remain to be elucidated in further studies.

ACKNOWLEGDMENT

I want to thank Mrs. Strmljan Elfriede, M.A., for her assistance in preparing this manuscript.

REFERENCES

1. Alberts, B., Bray D., Lewis J., Raff, H., Roberts, K., and Watson, J.D., Energy conversion: mitochondria and chloroplasts, *Molecular Biology of the Cell* , 3rd ed., Garland Publishing, 1994, pp. 653–720.
2. Busheri, N., Jarrell, S.T., Lieberman, Sh., Mirdamadi-Zonozi, N., Birkmayer, G., and Preuss, H.G., Oral reduced B-nicotinamide adenine dinucleotide (NADH) affects blood pressure, lipid peroxidation, and lipid profile in hypertensive rats (SHR), *Geriatric Nephrol. Urol.*, 8, 95, 1998.
3. Klingenberg, M., Pyridinnucleotide und biologische Oxidation, Zur Bedeutung der freien Nukleotide, II, *Moosbacher Kolloquium*, Springer Verlag, Heidelberg, 1960, p. 82.
4. Xu, M., Zhang, J.-R., and Hui S., The cytoprotection of nicotinamide adenine dinucleotide (NADH) in the mitochondria regulation mechanism, *J. Tumor Marker Oncol.*, 17 (4), 167, 2002.
5. Pelzmann, B., Schaffer, P., Lang, P., Hallström, S., Nadlinger, K., Birkmayer, J., Reibnegger, G., Koidl, B., NADH-supplementation decreases pinacidil-primed $I_{K(ATP)}$ in ventricular cardiomyocytes via increase of intracellular ATP content, submitted for publication.
6. Nadlinger, K., Westerthaler, W., Storga-Tomic, D., and Birkmayer, J.G.D., Extracellular metabolisation of NADH by blood cells correlates with intracellular ATP levels, *Biochimica Biophysica Acta*, 1573, 177, 2002.
7. Enzyme-Based Assay for Determining Effects of Exogenous and Endogenous Factors on Cellular Energy production, U.S. Patent 6,248,552 B1, 2001.
8. Harris, C.C., Weston, A., Willey, J.C., Trivers, G.E., and Mann, D.L., Biochemical and molecular epidemiology of human cancer: indicators of carcinogen exposure, DNA damage, and genetic predisposition, *Environ. Health Perspect.*, 75, 109, 1987.
9. Demopoulos, H.B. et al., The possible role of free radical reactions in carcinogenesis, *J. Environ. Path. Tox.*, 3, 273, 1980.
10. Bankson, D.D., Kestin, M., and Rifai, N., Role of free radicals in cancer and atherosclerosis, *Clin. Lab. Med.*, 13, 463, 1993.
11. Halliwell, B., The role of oxygen radicals in human disease, with particular reference to the vascular system, *Haemostasis*, 23 (Suppl.), 118, 1993.
12. Ueda, K. and Hayaishi, O., ADP-ribosylation, *Ann. Rev. Biochem.*, 54, 73, 1985.
13. Satoh, M.S., Poirier, G.G., and Lindahl, T., NAD+ dependent repair of damaged DNA by human cell extracts, *Biol. Chem.*, 268 (8), 5480, 1993.

14. Wang, J.Y., Cellular responses to DNA damage, *Curr. Opin. Cell. Biol.*, 10 (2), 240, 1998.

15. Stecca, C. and Gerber, G.B, Adaptive response to DNA-damaging agents: a review of potential mechanisms, *Biochem. Pharmacol.*, 55 (7), 941, 1998.

16. Savio, M., Stivala, L.A., Bianchi, L., Tannini, V., and Prosperi, E., Involvement of the proliferation cell nuclear antigen (PCNA) in DNA repair induced by alkylating agents and oxidative damage in human fibroblasts, *Carcinogenesis*, 19, 591, 1998.

17. Anderson, C.W., Protein kinases and the response to DNA damage, *Semin. Cell. Biol.*, 5 (6), 427, 1994.

18. Fink, D., Aebi, S., and Howell, S.B., The role of DNA mismatch repair in drug resistance, *Clin. Cancer. Res.*, 4 (1), 1, 1998.

19. Birkmayer, G.J.D. and Birkmayer, W., Stimulation of endogenous L-dopa biosynthesis — a new principle for the therapy of Parkinson's disease: the clinical effect of nicotinamide adenine dinucleotide (NADH) and nicotinamide adenine dinucleotide (NADPH), *Acta Neurol. Scan.*, 126, 183, 1989.

20. Zhang, J.R., Vrecko, K., Nadlinger, K., Storga Tomic, D., Birkmayer, G.D., and Reibnegger, G., The reduced coenzyme nicotinamide adenine dinucleotide (NADH) repairs DNA damage of PC12 cells induced by doxorubicin, *J. Tumor Marker Oncol.*, 13 (4), 5, 1998.

21. Ramachandran, C., You, W., and Krishan, A., Bcl-2 and mdr-1 gene expression during doxorubicin-induced apoptosis in murine leukemic p388 and p388/R84 cells, *Anticancer Res.*, 17 (5A), 3369, 1997.

22. Marques, F. and Bicho, M.P., Activation of a NADH dehydrogenase in the human erythrocyte by beta-adrenergic agonists: possible involvement of a G protein in enzyme activation, *Biol. Signals*, 6 (2), 52, 1997.

23. Macaya, A., Apoptosis in the nervous system, *Rev. Neurol.*, 24 (135), 1356, 1996.

24. Huang, D.C., O'Reilly, L.A., Strasser, A., and Cory, S., The anti-apoptosis function of Bcl-2 can be genetically separated from its inhibitory effect on cell cycle entry, *EMBO J.*, 16 (15), 4628, 1997.

25. Lovschall, H., Kassem, M., and Mosekilde, L., Apoptosis: molecular aspects, *Nord. Med.*, 112 (8), 271, 1997.

26. Hopewell, R., Li, L., MacGregor, D., Nerlov, C., and Ziff, E.B., Regulation of cell proliferation and differentiation by myc, *J. Cell. Sci. Suppl.*, 19, 85, 1995.

27. Luscher, B. and Eisenman, R.N., New light on myc and myb, Part 1: myc, genes and development, 4, 2025, 1990.

28. Kangas, A., Nicholson, D.W., and Hittla, E., Involvement of CPP32/Caspase-3 in c-myc-induced apoptosis, *Oncogene*, 16 (3), 387, 1998.

29. Hughes, P.E., Alexi, T., and Schreiber, S.S., A role for tumour suppressor gene p53 in regulating neuronal apoptosis, *Neuroreport*, 8 (15), 5, 1997.

30. Blagosklonny, M.V. and Ei-Deiry, W.S., Acute overexpression of wt p53 facilitates anticancer drug-induced death of cancer and normal cell, *Int. J. Cancer*, 75 (6), 933, 1998.

31. Badwey, C.W. and Gerard, R.W., Production of superoxide and hydrogen peroxide by an NADH oxidase in guinea pig polymorphonuclear leukocytes, *J. Biol. Chem.*, 254, 11530, 1979.

32. Da-Gong, W., Colin, F.J., John, K.M., Kerry, V.P., Brew, A., Colin, F.J.R., and Keith, D.B., Expression of the apoptosis-suppressing gene bcl-2 in pheochromocytoma is associated with the expression of c-myc, *J. Clinical Endocrinol. Metab.*, 82 (6), 1949, 1997.

33. Katoh, S., Mitsui, Y., Kitani, K., and Suzuki, T., The rescuing effect of nerve growth factor is the result of up-regulation of bcl-2 in hyperoxia-induced apoptosis of a subclone of pheochromocytoma cells PC12h, *Neurosci. Lett.*, 232 (2), 71, 1997.

34. Birchall, M.A., Schock, E., Harmon, B.V., and Gobe, G., Apoptosis, mitosis, PCNA and bcl-2 in normal, leukoplakic and malignant epithelia of the human oral cavity: prospective, *in vivo* study, *Oral Oncol.*, 33 (6), 419, 1997.

35. Heidenreich, A., Schenkman, N.S., Sesterhenn, I.A., Mostofi, K.F., Moul, J.W., Srivastava, S., and Engelmann, U.H., Immunohistochemical and mutational analysis of the p53 tumor suppressor gene and the bcl-2 oncogene in primary testicular germ cell tumours, *APMIS*, 106 (1), 90, 1998.

36. Meng, X., Zhang, J.-R., and Li, P., The molecular mechanisms of nicotinamide adenine dinucleotide in inhibiting human liver cells from apoptosis induced by cisplatin, *J. Tumor Marker Oncol.*, 15 (2), 139, 2000.

37. Zhang, Q., Piston, D.W., and Goodman, R.H., Regulation of corepressor function by nuclear NADH, *Science*, 295, 1895, 2002.

38. Grooteclaes, M.L. and Frisch, S.M., *Oncogene*, 19, 3823, 2000.
39. Comijn, J. et al., *Mol. Cell.*, 7, 1267, 2001.
40. Meiners, S., Brinkmann, V., Naundorf, H., and Birchmeier, W., *Oncogene*, 16, 9, 1998.
41. Perl, A.K., Wilgenbus, P., Dahl, U., Semb, H., and Christofori, G., *Nature*, 392, 190, 1998.
42. Ido, Y., Chang, K., Woolsey, T.A., and Williamson, J.R., NADH: sensor of blood flow need in brain, muscle and other tissues, *FASEB J.*, 15, 1419, 2001.
43. Badwey, C.W. and Gerard, R.W., Production of superoxide and hydrogen peroxide by an NADH oxidase in guinea pig polymorphonuclear leukocytes, *J. Biol. Chem.*, 254, 11530, 1979.
44. Nadlinger, K., Birkmayer, J., Gebauer, F., and Kunze, R., Influence of reduced nicotinamide adenine dinucleotide on the production of interleukin-6 by peripheral human blood leukocytes, *Neuroimmunomodulation*, 9, 203, 2001.
45. Reyes, T.M., Fabry, Z., and Coe, C.L., Brain endothelial cell production of a neuroprotective cytokine, interleukin-6, in response to noxious stimuli, *Brain Res.*, 851, 215, 1999.
46. Loddick, S.A., Turnbull, A.V., and Rothwell, N.J., Cerebral interleukin-6 is neuroprotective during permanent focal cerebral ischemia in the rat, *J. Cerb. Blood Flow Metab.*, 18, 176, 1998.
47. Pryor, W.A., Free radical reactions and their importance in biochemical systems, *Fed. Proc.*, 32, 1862, 1973.
48. Tappel, A.L., Lipid peroxidation damage to cell components, *Fed. Proc.*, 32, 1870, 1973.
49. Halliwell, B. and Gutteridge, J.M.C., Oxygen toxicity, oxygen radicals, transition metals and disease, *Biochem. J.*, 219, 1, 1984.
50. Cranton, E.M. and Frackelton, J.P., Free radical pathology in age-associated diseases: treatment with EDTA chelation, nutrition and antioxidants, *J. Hol. Med.*, 6, 6,1984.
51. Halliwell, B. and Gutteridge, J.M.C., Role of free radicals and catalytic metal ions in human disease: an overview, *Methods Enzymol.*, 186, 1, 1990.
52. Demopoulos, H.B., Pietronigro, D.D., and Seligman, M.L., The development of secondary pathology with free radical reactions as a threshold mechanism, *J. Am. Coll. Tox.*, 2, 173, 1983.
53. Halliwell, B. and Gutteridge, J.M.C., *Free Radicals in Biology and Medicine, Clarendon Press*, Oxford, 1985.
54. Busheri, N., Taylor, J., Lieberman, S., Mirdamadi-Zonosi, N., Birkmayer, G., and Preuss, H.G., Oral NADH affects blood pressure, lipid peroxidation and lipid profile in spontaneously hypertensive rats, *J. Am. Coll. Nutr.*, 1997.
55. Reibnegger, G., Greilberger, J., Juergens, G., and Oettl, K., The antioxidative capacity of ENADA–NADH in humans, *ICMAN Proceedings* (2nd Int. Conf. on Mechanisms and Actions of Nutraceuticals), 37, 2002.
56. Birkmayer, G.J.D. and Birkmayer, W., Stimulation of the endogenous L-dopa biosynthesis — a new principle for the therapy of Parkinson's disease: the clinical effect of nicotinamide adenine dinucleotide (NADH) and nicotinamide adenine dinucleotidephosphate (NADPH), *Acta Neurol. Scan.*, 126, 183, 1989.
57. Birkmayer, J.G.D., Stable, ingestable and absorbable NADH and NADPH therapeutic compositions, U.S. Patent 5,332,727, 1994.
58. U.S. Patent EP 96 100 612 9.
59. U.S. Patent 5,654,288, 1997.
60. Mattern, C., Zur Entwicklung von stabilen Arzneiformen des Coenzyms NADH für die perorale und parenterale Applikation, Ph.D. thesis, Humboldt, Berlin, 1996.
61. Rex, A., Hentschke, M.-P., and Fink, H., Bioavailability of reduced nicotinamide-adenin-dinucleotide (NADH) in the central nervous system of the anaesthetized rat measured by laser-induced fluorescence spectroscopy, *Pharmacol. Toxicol.*, 90, 2002.
62. Zhang, J.R., Vrecko, K., Nadlinger, K., Storga Tomic, D., Birkmayer, G.D., and Reibnegger, G., The reduced coenzyme nicotinamide adenine dinucleotide (NADH) repairs DNA damage of PC12 cells induced by doxorubicin, *J. Tumor Marker Oncol.*, 13 (4), 5, 1998.
63. Blagosklonny, M.V. and Ei-Deiry, W.S., Acute overexpression of wt p53 facilitates anticancer drug-induced death of cancer and normal cell, *Int. J. Cancer*, 75 (6), 933, 1998.
64. Egle, A., Villunger, A., Marschitz, I., Kos, M., Hittmair, A., Lukas, P., Grunewald, K., and Greil, R., Expression of Apo-1/fas CD95, Bcl-2, Bax and bcl-x in myeloma cell lines: relationship between responsiveness to anti-Fas mab and p53 functional status, *Br. J. Haematol.*, 97 (2), 418, 1997.

65. Birkmayer, J.G.D., Nadlinger, K.F.R., and Hallström, S., On the safety of reduced nicotinamide adenine dinucleotide (NADH): the maximum tolerated dose (MTD) in dogs is 500 mg per kg, *J. Environmental Pathol. Toxicol. Oncol.*, 2003, in print.
66. Birkmayer, J.G.D. and Nadlinger, K., Safety of stabilized, orally absorbable, reduced nicotinamide adenine dinucleotide (NADH): a 26-week oral tablet administration of ENADA[(r)]/NADH for chronic toxicity study in rats, *Drugs Exptl. Clin. Res.*, 28 (5), 185, 2002.
67. Slade, N., Storga-Tomic, D. Birkmayer, G.D., Pavelic, K., and Pavelic, J., Effect of extracellular NADH on human tumor cell proliferation, *Anticancer Res.*, 19, 5355, 1999.

34 Astaxanthin and Cancer Chemoprevention

John E. Dore

CONTENTS

34.1 Introduction 555
34.2 Antioxidants and Cancer Prevention 556
    34.2.1 Fruits, Vegetables, and Carotenoids 556
    34.2.2 The β-Carotene Hypothesis 556
34.3 Dietary Carotenoids Other than β-Carotene 557
    34.3.1 Lycopene 558
    34.3.2 Lutein and Zeaxanthin 558
    34.3.3 α-Carotene and β-Cryptoxanthin 558
    34.3.4 Canthaxanthin, Astaxanthin, and Others 558
34.4 Properties of Astaxanthin 559
    34.4.1 Structure and Forms 559
    34.4.2 Antioxidant Potential 559
34.5 Astaxanthin as a Potential Cancer Preventative 561
    34.5.1 Cell Culture Studies 562
    34.5.2 Rodent Model Studies 562
    34.5.3 Possible Mechanisms of Action 563
        34.5.3.1 Antioxidation 563
        34.5.3.2 Immunomodulation 563
        34.5.3.3 Gene Regulation and Other Mechanisms 564
34.6 Safety and Metabolism of Dietary Astaxanthin 564
34.7 Conclusion 565
References 566

34.1 INTRODUCTION

There are clear links between human cancers and diet.[1,2] By some estimates, dietary risk factors rank higher than tobacco usage and much higher than pollution or occupational hazards in their association with cancer deaths.[3] In addition to avoidance of tobacco smoke and carcinogenic food items, regular intake of chemopreventive compounds is a promising approach for reducing cancer incidence.[3,4] A number of substances naturally occurring in foodstuffs, particularly antioxidant compounds in plant products, have shown promise as potential chemopreventive agents.[3–6] Among these phytonutrients, the yellow, orange, and red carotenoid pigments have recently sparked much interest. In epidemiological studies, vegetable and fruit consumption has consistently been associated with reduced incidence of various cancers,[5–7] and dietary carotenoid intake from these sources has similarly been correlated with reduced cancer risk.[8–10] However, several recent large-scale intervention trials failed to find any chemopreventive effect of long-term supplementation with β-carotene, the most abundant dietary carotenoid.[11–13] Several naturally occurring carotenoids other than β-carotene have exhibited anticancer activity[14–17] and are being considered further as potential chemopreventive agents. Among these carotenoids, the red pigment astaxanthin is of particular interest in health management due to its unique structural and chemical properties.[18–20] This chapter will review the evidence for anticarcinogenic behavior of selected carotenoids, with an emphasis on the chemopreventive activities of astaxanthin.

34.2 ANTIOXIDANTS AND CANCER PREVENTION

The higher eukaryotic aerobic organisms, including human beings, cannot exist without oxygen, yet oxygen represents a danger to their very existence due to its high reactivity. This fact has been termed the "paradox of aerobic life."[21] A number of reactive oxygen species are generated during normal aerobic metabolism, such as superoxide, hydrogen peroxide, and the hydroxyl radical.[22] In addition, singlet oxygen can be generated through photochemical events (such as in the skin and eyes), and lipid peroxidation can lead to peroxyl radical formation.[22] These oxidants collectively contribute to aging and degenerative diseases such as cancer and atherosclerosis through oxidation of DNA, proteins, and lipids.[21–23] Antioxidant compounds can decrease mutagenesis, and thus carcinogenesis, both by decreasing oxidative damage to DNA and by decreasing oxidant-stimulated cell division.[22] The human body maintains an array of endogenous antioxidants such as catalase and superoxide dismutase; however, exogenous dietary antioxidants such as ascorbic acid (vitamin C), α-tocopherol (vitamin E), and carotenoids play important roles in reducing oxidative damage as well,[21–23] and their serum levels have the potential to be manipulated.[23]

34.2.1 FRUITS, VEGETABLES, AND CAROTENOIDS

Human epidemiological studies have revealed a protective effect of vegetable and fruit consumption for cancers of the stomach, esophagus, lung, oral cavity and pharynx, bladder, endometrium, pancreas, colon and rectum, breast, cervix, ovary, and prostate.[24–26] A variety of compounds found in these foods have known bioactive mechanisms and are suspected as anticancer agents. These include vitamins C and E, flavonoids, isothiocyanates, phytosterols, selenium, folic acid, dietary fiber, protease inhibitors, isoflavones, indoles, carotenoids, and others.[1,25] The carotenoids are a group of approximately 600 naturally occurring pigments with diverse biological functions.[27] In plants and algae, carotenoids serve both photosynthetic and photoprotective roles; in animals, carotenoids are effective chain-breaking antioxidants and singlet oxygen quenchers, and some also serve as precursors for retinoids (vitamin A).[28] Some carotenoids also appear to have effects on cell communication and proliferation in animals.[29] Because animals cannot synthesize carotenoids *de novo*, they must obtain them from dietary sources.[30]

34.2.2 THE β-CAROTENE HYPOTHESIS

In a landmark 1981 paper, Peto and colleagues posed the provocative question, "Can dietary beta-carotene materially reduce human cancer rates?"[31] Their focus on this particular carotenoid was largely due to its known bioactivity (as provitamin A), emerging information on its antioxidant properties, and its abundance in common fruits and vegetables. These authors suggested that although an inverse correlation of dietary β-carotene intake and cancer incidence was evident, a genuine protective effect of β-carotene could not be verified without controlled trials.[31] Three large human-intervention trials were initiated to test the β-carotene hypothesis in the mid-1980s; the results from these trials were disappointing. Not only did β-carotene supplementation offer no significant protection from lung and other cancers, it actually increased lung cancer risk among smokers in two of the trials.[11–13]

It has been suggested that these negative results should not have been wholly unexpected. Rather than individual agents, the total diet and all its constituents need to be considered in determining nutrient factors related to cancer risk incidence.[32] A diet rich in fruits and vegetables provides a suite of phytonutrients, including some 40 to 50 carotenoids and their metabolites,[33] which may themselves have chemopreventive potential.[34,35] Biological antioxidants, including carotenoids and vitamins C and E, are known to act synergistically through radical repairing and other mechanisms.[36–40] An individual antioxidant, in high doses by itself, may yield undesirable effects not realized in combination with other antioxidants at normal biological doses.[41] In the case of β-carotene, although it normally functions as an antioxidant, it exhibits prooxidant effects at high concentration and especially at high oxygen tension.[42,43] Supplementation with high doses of this carotenoid therefore has the potential to enhance oxidation in the lungs, especially when radicals from tobacco smoke are present.[44,45] Thus, in considering the potential role of carotenoids in cancer prevention, we must not look at β-carotene as a supplement in isolation, but consider multiple dietary carotenoids and their various interactions within biological systems.

34.3 DIETARY CAROTENOIDS OTHER THAN β-CAROTENE

Despite the presence of 40 or more naturally occurring carotenoids in the human diet, only a handful of carotenoids are commonly detected in human plasma and tissues, along with several of their isomers and various metabolites.[33] The most common of these dietary carotenoids are three hydrocarbon carotenoids (carotenes): α-carotene, β-carotene, and lycopene; and three oxycarotenoids (xanthophylls): lutein, zeaxanthin, and β-cryptoxanthin.[33,46] Intake of these compounds is principally through consumption of fruits and vegetables; the xanthophyll astaxanthin, on the other hand, is obtained principally from seafood such as salmon and shrimp. Astaxanthin occurs in these animals naturally, but it also occurs in farmed fish, shellfish, and poultry as a result of its use as a feed additive.[47,48] Astaxanthin is therefore an occasional component of the human diet in most populations, but can be more significant in populations that regularly consume such foods.[49] Canthaxanthin, another potentially important xanthophyll, is also not generally considered a dietary carotenoid, but may be included in the human diet through its widespread use as a coloring agent in foods and animal feeds.[50,51] The structures of these eight important carotenoids are given in Figure 34.1.

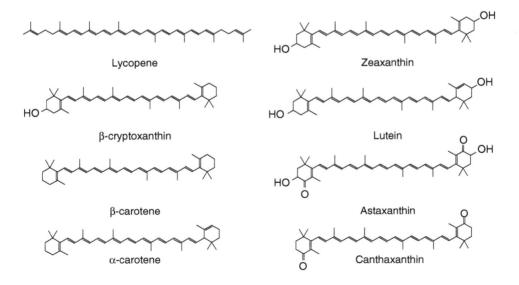

FIGURE 34.1 Chemical structures of eight important carotenoids in the human diet.

Among them, only α-carotene, β-carotene, and β-cryptoxanthin can be converted to vitamin A in humans.[51] Nevertheless, all of these dietary carotenoids have demonstrated some anticarcinogenic activity in animal experiments.[49,51–53]

34.3.1 Lycopene

Tomatoes and tomato-based products are the major dietary sources for the red carotenoid lycopene, although other plant sources exist, such as watermelon, grapefruit, and guava.[54] Lycopene is a very efficient biological singlet oxygen quencher[55] and has exhibited tumor-suppressive properties on animal and human cells *in vitro* and on mice *in vivo*.[56,57] Lycopene is found at high concentrations in the human prostate,[58] and epidemiological studies have revealed strong negative correlations between lycopene intake and prostate cancer risk,[26,59] and they have implicated lycopene as a factor in the prevention of several additional types of cancer and other human diseases.[60]

34.3.2 Lutein and Zeaxanthin

Lutein and zeaxanthin are yellow xanthophyll carotenoids common in green and yellow vegetables. Lutein is obtained primarily from leafy green vegetables such as spinach and kale, while orange peppers are rich in zeaxanthin.[49] These carotenoids accumulate in the macular region of the human retina, and are believed to play important roles in protecting the retina from photooxidative damage.[61–63] In cancer chemoprevention, a high intake of lutein and zeaxanthin has been correlated with a lower incidence of lung cancer in humans,[64,65] and lutein has exhibited antimutagenic effects *in vitro*.[66] Lutein has also demonstrated an ability to inhibit carcinogenesis in rat colons[67] and in the lungs of mice,[68] and it inhibits mammary tumor growth in mice[69] and in human cell cultures[70] by regulating apoptosis. Similarly, zeaxanthin has been shown to reduce the formation of liver tumors in mice.[71]

34.3.3 α-Carotene and β-Cryptoxanthin

Serum levels of the two other major carotenoids in the human diet, α-carotene and β-cryptoxanthin, have been inversely correlated with the incidence of human cervical cancer.[72] In addition, dietary intake of β-cryptoxanthin is associated with reduced risk for lung cancer.[65] Carrots and pumpkin are good sources of α-carotene, while β-cryptoxanthin is abundant in red bell peppers, papayas, and tangerines.[49,73] In studies with mice, α-carotene has been demonstrated to have a potent preventive action against lung, skin, and liver carcinogenesis.[14] Similarly, β-cryptoxanthin is effective at inhibiting skin tumor formation in mice.[68,71]

34.3.4 Canthaxanthin, Astaxanthin, and Others

Because it is not a significant dietary carotenoid, epidemiological data on canthaxanthin in disease prevention are lacking. However, it has exhibited potential anticancer properties *in vitro* and in animal models. Canthaxanthin can suppress proliferation of human colon cancer cells,[74] protect mouse embryo fibroblasts from transformation,[75] and protect mice from mammary and skin tumor development.[17,76] Canthaxanthin has also proved effective at inhibiting both oral and colon carcinogenesis in rats.[77,78] Although it is a potent antioxidant, the chemopreventive effects of canthaxanthin may also be related to its ability to up-regulate gene expression, resulting in enhanced gap-junctional cell-cell communication.[79,80] The chemopreventive effects of canthaxanthin may also be related to its ability to induce xenobiotic metabolizing enzymes, as has been demonstrated in the liver, lung, and kidney of rats.[81,82] Unfortunately, canthaxanthin overuse as a "sunless" tanning product has led to the appearance of crystalline deposits in the human retina.[83] Although these retinal inclusions are reversible[84] and appear to have no adverse effects,[83] their existence has prompted caution regarding intake of this carotenoid.

Several other naturally occurring carotenoids that are not considered significant in the human diet have shown potential as cancer chemopreventive agents. These include neoxanthin, fucoxanthin, phytofluene, ζ-carotene, phytoene, crocetin, capsanthin, peridinin, and astaxanthin.[52,53,85] The xanthophyll astaxanthin is a powerful antioxidant and has great potential for reducing human disease processes related to oxidative damage.[49] Therefore, it warrants a more detailed discussion as follows.

34.4 PROPERTIES OF ASTAXANTHIN

34.4.1 STRUCTURE AND FORMS

Like all carotenoids, astaxanthin (3,3′-dihydroxy-β,β-carotene-4,4′-dione) is derived from a central phytoene "backbone" of 40 carbon atoms linked by alternating single and double bonds. This structure is useful in energy transfer and dissipation and gives carotenoids their characteristic colors. As with all the dietary carotenoids except lycopene, the phytoene chain is terminated on either end by ionone rings. The presence of oxygen-containing functional groups on these rings classifies astaxanthin among the xanthophylls. These hydroxyl and keto groups allow astaxanthin to be esterified and also render it more polar than related carotenoids.[20] Astaxanthin has a number of geometric (Z) isomers, and also is optically active, having three possible stereoisomers.[47]

In nature, astaxanthin is usually found either conjugated to proteins (as in the flesh of salmon or in the lobster carapace) or esterified with fatty acids (as in *Haematococcus pluvialis* microalgae).[20] In contrast, synthetic astaxanthin is produced in the free form. Synthetic, algae-based, and yeast-based (from *Xanthophyllomyces dendrorhous*) astaxanthins are distinct in their stereoisomeric compositions as well.[48] Synthetic astaxanthin, as well as all three significant natural sources (*Haematococcus*, *Xanthophyllomyces*, and extracted crustacean shells), are used widely as feed additives.[48,86] Human dietary astaxanthin supplements derived from these three natural sources have also been marketed in recent years.[20,48]

34.4.2 ANTIOXIDANT POTENTIAL

Astaxanthin has demonstrated strong antioxidant behavior in a variety of *in vitro* studies. In organic solutions, astaxanthin is a potent quencher of singlet oxygen,[87–89] an effective inhibitor of peroxyl radical-dependent lipid peroxidation,[89–91] and an efficient peroxyl radical-trapping compound.[92,93] Both synthetic astaxanthin and a commercial *Haematococcus* algae extract were shown to be excellent scavengers of hydroxyl radicals and superoxide anions when introduced in DMSO to aqueous solutions (as shown in Figure 34.2).[94] These antioxidant properties of astaxanthin extend to model membrane systems and cultured animal cells. Astaxanthin and several other carotenoids inhibited peroxyl radical-mediated lipid peroxidation in liposomal[91,95] and microsomal[96–98] systems and in large unilamellar vesicles.[99] Similarly, astaxanthin was among the carotenoids found to be effective at quenching singlet oxygen[100] and at inhibiting photosensitized oxidation[101] in unilamellar liposomes. Astaxanthin was superior to β-carotene and lutein in its ability to protect rat kidney fibroblasts from ultraviolet-A (UVA) light-induced oxidative stress.[102] Astaxanthin also offered *in vitro* protection from chemically induced oxidation to cultured chicken embryo fibroblasts,[103] rat blood cells and mitochondria,[89] human lymphoid cells,[104] and human low-density lipoprotein (LDL).[105]

The antioxidant behavior of astaxanthin has been demonstrated *in vivo* as well. In *Haematococcus* algae, astaxanthin is accumulated as part of a stress response, and it is believed to protect cellular DNA from photodynamic damage.[106] This carotenoid also protects lipids from peroxidation in trout[107] and salmon.[108] In chicks, astaxanthin supplementation suppressed the formation of lipid peroxides in the plasma.[95] Significant biological antioxidant effects have been observed in vitamin E-deficient rats fed an astaxanthin-rich diet; these include protection of mitochondrial function[109] and inhibition of peroxidation of erythrocyte membranes.[89,109] In two independent studies, lipid peroxidation in the serum and liver of astaxanthin-fed rats treated with carbon tetrachloride was

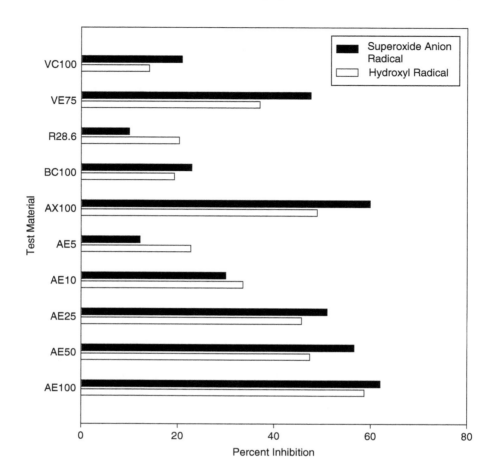

FIGURE 34.2 Radical quenching ability of astaxanthin and other compounds *in vitro*. The percentage inhibition of superoxide anion radical and hydroxyl radical-generated chemiluminescence is shown for the following test materials: VC100 = vitamin C at 100 mg l⁻¹; VE75 = vitamin E at 75 mg l⁻¹; R28.6 = all-*E* retinol at 28.6 mg l⁻¹; BC100 = β-carotene at 100 mg l⁻¹; AX100 = synthetic astaxanthin at 100 mg l⁻¹; AE5 through AE100 = algal extract (from *Haematococcus pluvialis*, containing 5% esterified astaxanthin) at 5 to 100 mg l⁻¹. (From Bagchi, D., Final Report to Cyanotech Corporation, Creighton University School of Health Sciences, Omaha, NE, 2001. With permission.)

significantly inhibited relative to rats fed a control diet.[97,110] Similar protection from peroxidation was afforded by astaxanthin to the serum, liver, kidney, spleen, and brain of rats exposed to cobalt-60 irradiation.[97] In an *ex vivo* study of human volunteers, dietary supplementation for 14 days with esterified astaxanthin extracted from krill significantly extended the lag time for chemically initiated LDL oxidation.[105] This effect appeared to be dose dependent: supplementation at 3.6, 14.4, or 21.6 mg astaxanthin per day produced significant differences from the control group, while 1.8 mg per day did not produce a significant effect.[105]

The interactions of carotenoids with free radicals are complex, depending on factors such as the structure of the carotenoid, the nature of the radical species, the composition of the surrounding matrix, the presence of other oxidants and antioxidants, and the concentrations of the radicals, carotenoids, and oxygen. All of these factors need to be taken into account to explain the uniquely effective antioxidant properties of astaxanthin. The radical quenching properties of carotenoids lie not only in the conjugated polyene chain, but in the functional groups as well.[111] The xanthophylls therefore have inherently different antioxidative properties than the carotenes. For example, astaxanthin and canthaxanthin are inherently poor antioxidants when compared with β-carotene in electron-transfer reactions with radicals,[112] yet the opposite is true in reactions that involve the formation of carotenoid–radical adducts.[113] Moreover, the overall antioxidant properties of carotenoids reflect not only their ability to scavenge radicals, but also on the reactivity of carotenoid radicals or carotenoid–radical adducts that are formed in the process of radical quenching.[114] Astaxanthin, for example, is the most difficult carotenoid to reduce to its radical cation;[115] the β-carotene radical cation, on the other hand, is more easily formed via electron transfer[112–114] and is itself long-lived and capable of oxidizing protein components such as tyrosine and cysteine.[115,116] In contrast, carotenoid–radical adducts formed with astaxanthin or canthaxanthin decay quickly to stable products.[113] Astaxanthin therefore has the advantage of being an effective radical quencher in some reactions, while not itself being converted into a damaging radical species in others. In addition, when compared with other carotenoids, the astaxanthin radical cation is the most easily reduced[117]; hence, if the astaxanthin radical cation should form, it can easily be converted back to the stable carotenoid via electron transfer from vitamin E, with which it reacts at a higher rate than do the other carotenoids.[112]

The position, concentration, and orientation of carotenoids within membranes may strongly influence both the structure and dynamics of the lipid bilayer and the antioxidant properties of the carotenoids in membrane systems.[118–120] Polar carotenoids such as zeaxanthin and astaxanthin may span the bilayer, where they tend to stabilize and rigidify the lipid membrane, while nonpolar carotenoids such as β-carotene are more likely to remain completely within the bilayer.[121–123] In the case of astaxanthin, intermolecular hydrogen bonds likely form with phospholipids in the membrane, anchoring the carotenoid molecule like a rivet. At the same time, intramolecular hydrogen bonding between the keto and hydroxyl groups of individual astaxanthin molecules can increase their hydrophobicity and thus keep them within the bilayer.[123] It has been suggested that roughly equal amounts of intra- and intermolecular hydrogen-bonded astaxanthin can exist simultaneously in a membrane, hence allowing for both scavenging of lipid peroxyl radicals within the membrane and interception of reactive oxygen species at the membrane surface.[123] Astaxanthin molecules spanning the bilayer may also be involved in a hypothesized mechanism in which they trap alkoxyl radicals within the hydrophobic core of the membrane and transport the unpaired electron up the polyene chain to the lipid–water interface, where it reacts with aqueous vitamin C to yield stable products in the lipid phase and an ascorbyl radical in the water phase.[124] Mechanisms such as these may explain the highly potent antiperoxidative activity of this carotenoid in lipid membranes.

The concentrations of carotenoids and the level of oxygen they are exposed to can also influence their antioxidant activities. At low oxygen partial pressures, diverse carotenoids effectively inhibit *in vitro* oxidation reactions, and their antioxidative abilities increase with increasing carotenoid concentration.[40,42] As oxygen levels are increased, however, their antioxidant potential typically decreases.[40,42] Certain carotenoids, notably β-carotene but also lycopene, exhibit unusual behavior; beyond a threshold carotenoid concentration, they actually decrease in antioxidant ability with increasing carotenoid concentration, and this effect is further exacerbated at high oxygen levels.[42,43,125,126] This prooxidant behavior of β-carotene appears to be related to its degradation products and their potential to be involved in radical chain reactions,[125] and may help to explain the unexpected increase in lung cancer deaths among smokers supplemented with this carotenoid.[41,45] The xanthophylls zeaxanthin, canthaxanthin, and especially astaxanthin are considered "pure" antioxidants because they exhibit little or no prooxidative behavior, even at high carotenoid concentration and high oxygen tension.[125,126]

34.5 ASTAXANTHIN AS A POTENTIAL CANCER PREVENTATIVE

Because astaxanthin has not typically been identified as a major carotenoid in human serum, information on its epidemiology in human health is lacking. Salmon, the principal dietary source of astaxanthin, is an important component of the traditional diets of Eskimos and certain coastal tribes in North America; these groups have shown unusually low prevalence of cancer.[127,128] This low cancer incidence has been attributed to the high levels in salmon of certain fatty acids, notably eicosapentenoic acid (EPA),[128] yet it is possible that astaxanthin has played a role in cancer chemoprevention among these peoples as well. Regardless, the existing data on the potential for astaxanthin to directly prevent cancer is limited to *in vitro* cell culture studies and *in vivo* studies with rodent models.

34.5.1 CELL CULTURE STUDIES

Methylcholanthrene-induced (Meth-A) mouse tumor cells grown in an astaxanthin-supplemented medium had reduced cell numbers and lower DNA synthesis rates 1 to 2 days postincubation than control cultures.[129] Similarly, astaxanthin inhibited murine mammary tumor cell proliferation by up to 40%, in a dose-dependent fashion, when included in the culture medium.[130] In addition, of eight carotenoids tested, astaxanthin was the most effective at inhibiting the invasion of rat ascites hepatoma cells in culture.[131] The growth of human cancer cell lines has also been inhibited by astaxanthin *in vitro*. Two human colon cancer cell lines were significantly less viable than control cultures after a 4-day incubation with astaxanthin, although a stronger effect was seen from α-carotene, β-carotene, or canthaxanthin.[74] Also, a weak effect of astaxanthin on human prostate cancer cell viability has been noted, but in this case, neoxanthin and fucoxanthin appeared to be much more effective.[85] On the other hand, significant inhibition of androgen-induced proliferation of human prostate cancer cells was recently demonstrated in the presence of either astaxanthin or lycopene.[132] Exposure to UVA radiation is believed to be the primary causative agent in skin tumor pathogenesis; both synthetic astaxanthin and an astaxanthin-rich algal extract gave significant protection from UVA-induced DNA damage to human skin fibroblasts, melanocytes, and intestinal CaCo-2 cells in culture.[133]

34.5.2 RODENT MODEL STUDIES

In studies with BALB/c mice, dietary astaxanthin inhibited the growth of transplanted Meth-A tumor cells in a dose-dependent fashion.[129] In a related study, Meth-A tumor cell growth was inhibited when dietary astaxanthin supplementation was started at 1 and 3 weeks prior to tumor inoculation, but not when supplementation was begun at the same time as tumor inoculation.[134] These results suggest that astaxanthin may inhibit tumor development in the early stages but not in the later stages of progression.[134] In other studies with mice, astaxanthin supplementation reduced transplanted mammary tumor growth[17] and suppressed spontaneous liver carcinogenesis.[71] Dietary consumption of egg yolks containing astaxanthin inhibited benzo(*a*)pyrene-induced mouse fore-stomach neoplasia[135] and sarcoma-180 cell-induced mouse ascites cancer.[136] In addition, dietary astaxanthin inhibited the accumulation of potentially tumor-promoting polyamines in the skin of vitamin A-deficient hairless mice after exposure to UVA and UVB irradiation.[137]

A series of studies on cancer chemoprevention by natural and synthetic substances in mice and rats revealed several carotenoids, including astaxanthin, as effective antitumor agents.[138] In one of these studies, dietary astaxanthin was found to significantly reduce both the incidence and prolif-eration of chemically induced urinary bladder cancer in mice.[139] In two related studies, the incidence and proliferation of chemically induced cancers of the oral cavity[78] and colon[77] were significantly reduced in astaxanthin-supplemented rats relative to control rats. Astaxanthin has shown effective-ness against the initiation of liver carcinogenesis in rats. An astaxanthin-supplemented diet reduced the number of DNA single-strand breaks and the number and size of liver preneoplastic foci induced in rats by aflatoxin B_1.[140,141] Dietary astaxanthin also reduced metastatic nodules and lipid peroxi-dation in the livers of rats treated with restraint stress.[142,143]

Although the above studies all point to potent anticarcinogenic effects of astaxanthin *in vivo*, a few studies have offered less compelling results. For example, in one study of chemically induced hepatocarcinogenesis in rats, dietary astaxanthin had no effect on the development of preneoplastic liver foci, while lycopene produced a significant reduction in foci.[144] Similarly, activation of *pim-1* gene expression (which is involved in regulating cell differentiation and apoptosis) was stimulated in lutein-fed but not in astaxanthin-fed mice.[145] Finally, one *in vivo* dietary astaxanthin study has reported negative results; dietary supplementation with either β-carotene or astaxanthin exacerbated carcinogenic expression in the skin of hairless mice after UV irradiation.[146]

34.5.3 POSSIBLE MECHANISMS OF ACTION

The proposed mechanisms of action in the cancer chemopreventive actions of carotenoids can be grouped into three major categories: carotenoids can act as potent biological antioxidants, as enhancers of immune system function, and as regulators of gene expression.[147] Astaxanthin is expected to function through each of these mechanisms in living systems.

34.5.3.1 Antioxidation

We have discussed above the potential for free radicals to initiate carcinogenesis and the unique antioxidative properties of astaxanthin against free radicals. Several recent examples testify to the effectiveness of astaxanthin in the prevention and treatment of oxidative cell and tissue damage *in vivo*. Dietary astaxanthin limits exercise-induced muscle damage in mice,[148] protects β-cell and renal function in diabetic mice,[149,150] and both retards and ameliorates retinal damage from photic injury in rats.[151] An algal extract containing astaxanthin was similarly found to attenuate selenite-induced cataract formation in the eyes of rat pups.[152]

Inflammation is believed to be a major contributor to carcinogenesis, through several mechanisms, including the production of free radicals by inflammatory cells.[153] Astaxanthin has been found effective at reducing the severity of several inflammatory conditions in rodents and humans. Gastric inflammation associated with infection by *Helicobacter pylori* bacteria was reduced in mice fed astaxanthin-containing algal meal[154] or algal cell extract.[155,156] Astaxanthin was also shown to have a dose-dependent ocular anti-inflammatory effect on lipopolysaccharide-induced uveitis in rats.[157] Two small, randomized, placebo-controlled trials were recently conducted on human volunteers to assess the effect of supplementation with an astaxanthin-rich algal extract on symptoms associated with the inflammatory diseases rheumatoid arthritis (RA) and carpal tunnel syndrome (CTS).[158,159] The results revealed that astaxanthin significantly relieved pain and improved performance in patients with RA[158]; the results on CTS patients were similar but statistically insignificant.[159] Although other mechanisms may be at work, the antioxidant properties of astaxanthin likely contribute to its ability to prevent or treat these various conditions, and thereby potentially reduce cancer risk.

34.5.3.2 Immunomodulation

It is well established that carotenoids can have an enhancing effect on immune function, and that such immunoenhancement may be manifested independently of their provitamin A activity or antioxidant potential.[160,161] Carotenoids appear to have specific immune functions that may enhance immunity to cancer cells.[160] Astaxanthin in particular has exhibited numerous immune-enhancing activities both *in vitro* and *in vivo*. In cell culture experiments, astaxanthin stimulated proliferation of mouse thymocytes and spleen cells, stimulated immunoglobulin production of murine spleen cells, and enhanced the release of interleukin-1α and tumor necrosis factor α (TNFα) from murine peritoneal adherent cells.[162] Similarly, production of antibodies in response to T-dependent antigens and other stimuli are enhanced by astaxanthin in mice *in vitro* and *in vivo*.[163–167] Astaxanthin also enhanced *in vitro* immunoglobulin production by human peripheral blood mononuclear cells in response to antigens.[168] Phytohemaglutinin-induced splenocyte proliferation and lymphocyte cytotoxic activity were stimulated in mice fed astaxanthin,[169] while dietary astaxanthin was able to delay symptoms of proteinuria and lymphadenopathy in autoimmune-prone mice.[170]

Similar immune responses in astaxanthin-fed mice have been noted when this carotenoid was used to reduce the inflammatory symptoms of *H. pylori* infections.[155,156] Moreover, immunoenhancement has been observed when astaxanthin was fed to tumor-inoculated mice. For example, Meth-A tumor-inoculated mice developed significantly higher cytotoxic T-lymphocyte activity and interferon-γ production by tumor-draining lymph node and spleen cells when fed an astaxanthin-supplemented diet relative to those fed a control diet; in parallel with these observations, a significant inhibition of tumor growth in the astaxanthin-fed mice was noted.[129,134] Taken together, these studies of the ability of astaxanthin to stimulate immune responses both *in vitro* and *in vivo* suggest that the immunoenhancing properties of this carotenoid may play an important role in its ability to function as a cancer chemopreventive agent.

34.5.3.3 Gene Regulation and Other Mechanisms

Other unexpected biological functions of carotenoids have been recently demonstrated that appear to be independent of their provitamin A and antioxidant activities.[79] Effective cell–cell communication through gap junctions is deficient in many human tumors, and its restoration tends to decrease tumor cell proliferation.[171] Several retinoids and carotenoids are now known to enhance gap-junctional communication between cells, and the enhancement by carotenoids is well correlated with their ability to inhibit neoplastic transformation in mouse embryo fibroblasts.[29,171,172] This stimulation of gap-junctional communication occurs as a result of a dose-dependent increase in the connexin 43 protein via up-regulation of the connexin 43 gene.[29,79,171] Interestingly, while β-carotene enhanced connexin 43 expression in murine fibroblasts, it did not do so in murine lung epithelial cells; this observation may at least in part explain why β-carotene is ineffective in chemoprevention of lung cancer.[173] It is not known if astaxanthin has an up-regulating effect on connexin 43, but the closely related carotenoid canthaxanthin has shown a strong stimulatory effect on gap-junctional communication between mouse embryo fibroblasts.[80,172]

Another regulatory function of carotenoids is the induction of xenobiotic-metabolizing enzymes (XME). By enhancing the production of these enzymes, carotenoids may help to prevent carcinogenesis by stimulating the detoxification of carcinogenic compounds. A number of studies have demonstrated such regulation by carotenoids, especially astaxanthin and canthaxanthin, in the liver of rats. Specific enzymes that were induced by astaxanthin and canthaxanthin included P4501A1 and 1A2, and CYP1A1 and 1A2, which are involved in the metabolism of such potential carcinogens as polycyclic aromatic hydrocarbons, aromatic amines, and aflatoxin.[81,140,141,174] These two xanthophylls also induced selected P450 enzymes in rat lung and kidney tissues, but not in the small intestine.[82] XME induction by astaxanthin is not only enzyme-specific and tissue-specific, but varies between species as well; different mechanisms appear to be at work in Swiss mice[175] and in human hepatocytes[176] than in rat liver.

Several additional regulatory mechanisms have been described involving astaxanthin that may underlie its anticarcinogenic effects. These include a regulatory influence of astaxanthin on transglutaminases in UV-irradiated hairless mice,[137] an inhibitory effect of astaxanthin and other carotenoids on metabolic activation of specific mutagens in bacteria,[177] and an induction of apoptosis by astaxanthin in murine mammary tumor cells.[130] Furthermore, inhibition of the enzyme 5α-reductase by astaxanthin may explain its antiproliferative effect on human prostate cancer cells,[178] and selective inhibition of DNA polymerases by astaxanthin and retinoids may result in reduced human gastric cancer cell growth.[179] Finally, direct blocking of nitric oxide synthase activity appears to be the mechanism by which astaxanthin reduces lipopolysaccharide-induced inflammation in rats.[157]

34.6 SAFETY AND METABOLISM OF DIETARY ASTAXANTHIN

Astaxanthin is not known to present any special health risk to humans. Astaxanthin is a natural, albeit minor, component of the human diet through consumption of salmon, trout, and various crustaceans, and it has been used as a dietary supplement at least since 1999.[20] The most common source of astaxanthin used in these supplements is an extract of *Haematococcus pluvialis* microalgae. Numerous acute and repeated-dose toxicity studies in mice, rats, and humans have demonstrated the lack of toxicity of the whole algal biomass.[180] Moreover, the extract has recently undergone a 13-week repeated-dose toxicity study in rats[181] as well as an 8-week randomized, double-blind, placebo-controlled clinical safety trial of 35 human volunteers;[182] no safety concerns were raised by either of these studies.

Despite the existing evidence attesting to the safety of dietary astaxanthin, little is known about the bioavailability and metabolism of this carotenoid in humans. Several steps are involved in the assimilation of carotenoids by mammals, including transfer from the food matrix, transfer to lipid micelles in the small intestine, uptake by intestinal mucosal cells, transport to the lymph system, and eventually, deposition of the carotenoid or its metabolites in specific tissues.[183,184] A number of factors can influence the progression of these steps, including the nature of the food matrix,[184,185] the structure of the carotenoid (including potential esterification and the nature of its isomeric composition),[183–187] the presence of other carotenoids,[184,188] and the amount and types of lipids in the diet.[189–191] Overall, human metabolism of astaxanthin should be somewhat similar to that of the other xanthophylls, but subtle differences are expected.

Astaxanthin absorption and metabolism has been fairly well researched in birds, crustaceans, and especially fish,[192] but only a handful of studies report on its uptake and metabolism in humans and other mammals. In rat hepatocytes, astaxanthin was metabolized into two racemic compounds: 3-hydroxy-4-oxo-β-ionone and its reduced form, 3-hydroxy-4-oxo-7,8-dihydro-β-ionone.[193] Both of these metabolites were also produced from astaxanthin in cultured human hepatocytes and in the plasma of human volunteers who ingested synthetic astaxanthin. However, in these systems, two additional metabolites, 3-hydroxy-4-oxo-β-ionol and 3-hydroxy-4-oxo-7,8-dihydro-β-ionol, were produced as well.[176] In terms of absorption, human volunteers ingesting a very large dose (100 mg) of synthetic astaxanthin readily incorporated this carotenoid into plasma lipoproteins to a considerable degree, reaching maximum plasma concentrations of astaxanthin in about 7 h.[194] All isomers of astaxanthin were incorporated, but there was a selective enrichment of the Z-isomers relative to all-E astaxanthin in the plasma.[194] The bioavailability of astaxanthin demonstrated in the above study was in contrast to the lack of astaxanthin detected in the plasma of human subjects who ingested an astaxanthin-containing salmon meal.[195] It is likely that the serum astaxanthin concentration achieved from this 500 g of salmon was below the detection limit of the assay, both because the salmon contained only 1.5 mg of astaxanthin, and because the salmon also contained canthaxanthin,[195] which could potentially have interfered with astaxanthin uptake.[188] The bioavailability of both free and esterified astaxanthin was also examined in healthy male volunteers who ingested a single 40-mg dose of this carotenoid in one of several different formulations; the results demonstrated an enhancement of astaxanthin bioavailability in humans when incorporated into lipid-based formulations.[196] It has been shown as well that the type of oil used influences astaxanthin bioavailability; in rats, astaxanthin assimilation was better when the carotenoid was introduced in olive oil than when it was introduced in corn oil.[191]

To date, no human bioavailability or metabolism studies have been reported that have utilized relevant dietary dosages of astaxanthin (4 to 12 mg daily is typically recommended by supplement manufacturers), nor has serum astaxanthin been tracked in humans undergoing longer-term (weeks to months) supplementation with this carotenoid.

34.7 CONCLUSION

A diet rich in fruits and vegetables is an important factor for the chemoprevention of a number of human cancers. Such a diet is rich in carotenoids, yet consumption of a wide variety of vegetables can have a greater bearing on the risk of specific cancers than intake of any specific carotenoid or total carotenoids.[197] The whole of the diet must be considered, including the various dietary carotenoids and other anticarcinogenic compounds.[198,199] It is becoming increasingly clear that relevant dietary dosages of a mixture of carotenoids are more likely to yield beneficial effects in cancer chemoprevention than high doses of a single carotenoid like β-carotene.[200]

Astaxanthin has exhibited potent antioxidant, immunomodulating, and enzyme-inducing properties, all of which suggest a potential role for this carotenoid in the prevention of cancer. Moreover, its unique structural properties and its lack of prooxidant activity make it a prime candidate for further investigation. More research is needed on the absorption and metabolism of this promising anticancer agent in humans and on its interactions with other carotenoids and vitamins in the human system.

REFERENCES

1. Papas, A.M., Diet and antioxidant status, in *Antioxidant Status, Diet, Nutrition, and Health*, Papas, A.M., Ed., CRC Press, Boca Raton, FL, 1998, chap. 5.
2. Chesson, A. and Collins, A., Assessment of the role of diet in cancer prevention, *Cancer Lett.*, 114, 237, 1997.
3. Lee, B.M. and Park, K.-K., Beneficial and adverse effects of chemopreventive agents, *Mutat. Res.*, 523–524, 265, 2003.
4. Sporn, M.B. and Suh, N., Chemoprevention: an essential approach to controlling cancer, *Nature Rev. Cancer*, 2, 537, 2002.
5. Wargovich, M.J., Experimental evidence for cancer preventive elements in food, *Cancer Lett.*, 114, 11, 1997.
6. Potter, J.D., Cancer prevention: epidemiology and experiment, *Cancer Lett.*, 114, 7, 1997.
7. Eastwood, M.A., Interaction of dietary antioxidants *in vivo*: how fruit and vegetables prevent disease? *Q. J. Med.*, 92, 527, 1999.
8. Zhang, S. et al., Dietary carotenoids and vitamins A, C, and E and risk of breast cancer, *J. Natl. Cancer Inst.*, 91, 547, 1999.
9. Holick, C.N. et al., Dietary carotenoids, serum β-carotene, and retinol and risk of lung cancer in the Alpha-Tocopherol, Beta-Carotene cohort study, *Am. J. Epidemiol.*, 156, 536, 2002.
10. Rock, C.L., Carotenoid update, *J. Am. Diet. Assoc.*, 103, 423, 2003.
11. The Alpha-Tocopherol, Beta Carotene Cancer Prevention Study Group, The effect of vitamin E and beta carotene on the incidence of lung cancer and other cancers in male smokers, *New England J. Med.*, 330, 1029, 1994.
12. Hennekens, C.H. et al., Lack of effect of long-term supplementation with beta carotene on the incidence of malignant neoplasms and cardiovascular disease, *New England J. Med.*, 334, 1145, 1996.
13. Omenn, G.S. et al., Effects of a combination of beta carotene and vitamin A on lung cancer and cardiovascular disease, *New England J. Med.*, 334, 1150, 1996.
14. Murakoshi, M. et al., Potent preventive action of α-carotene against carcinogenesis: spontaneous liver carcinogenesis and promoting stage of lung and skin carcinogenesis in mice are suppressed more effectively by α-carotene than by β-carotene, *Cancer Res.*, 52, 6583, 1992.
15. Levy, J. et al., Lycopene is a more potent inhibitor of human cancer cell proliferation than either α-carotene or β-carotene, *Nutr. Cancer*, 24, 257, 1995.
16. Park, J.S., Chew, B.P., and Wong, T.S., Dietary lutein from marigold extract inhibits mammary tumor development in BALB/c mice, *J. Nutr.*, 128, 1650, 1998.
17. Chew, B.P. et al., A comparison of the anticancer activities of dietary β-carotene, canthaxanthin and astaxanthin in mice *in vivo*, *Anticancer Res.*, 19, 1849, 1999.
18. Maher, T.J., Astaxanthin continuing education module, New Hope Institute of Retailing, Boulder, CO, 2000.
19. Naguib, Y., Pioneering astaxanthin, *Nutr. Sci. News*, 6, 58, 2001.
20. Guerin, M., Huntley, M.E., and Olaizola, M., *Haematococcus* astaxanthin: applications for human health and nutrition, *Trends Biotechnol.*, 21, 210, 2003.
21. Davies, K.J.A., Oxidative stress: the paradox of aerobic life, *Biochem. Soc. Symp.*, 61, 1, 1995.
22. Ames, B.N., Shigenaga, M.K., and Hagen, T.M., Oxidants, antioxidants, and the degenerative diseases of aging, *Proc. Natl. Acad. Sci. USA*, 90, 7915, 1993.

23. Ames, B.N. and Shigenaga, M.K., Oxidants are a major contributor to aging, *Ann. NY Acad. Sci.*, 663, 85, 1992.
24. Block, G., Patterson, B., and Subar, A., Fruit, vegetables, and cancer prevention: a review of the epidemiological evidence, *Nutr. Cancer*, 18, 1, 1992.
25. Steinmetz, K.A. and Potter, J.D., Vegetables, fruit, and cancer prevention: a review, *J. Am. Diet. Assoc.*, 96, 1027, 1996.
26. Giovannucci, E., Tomatoes, tomato-based products, lycopene, and cancer: review of the epidemiological literature, *J. Natl. Cancer Inst.*, 91, 317, 1999.
27. Krinsky, N.I., The antioxidant and biological properties of the carotenoids, *Ann. NY Acad. Sci.*, 854, 443, 1998.
28. Krinsky, N.I., Carotenoids in medicine, in *Carotenoids: Chemistry and Biology*, Krinsky, N.I., Ed., Plenum Press, New York, 1990, p. 279.
29. Wolf, G., Retinoids and carotenoids as inhibitors of carcinogenesis and inducers of cell-cell communication, *Nutr. Rev.*, 50, 270, 1992.
30. Schiedt, K., New aspects of carotenoid metabolism in animals, in *Carotenoids: Chemistry and Biology*, Krinsky, N.I., Ed., Plenum Press, New York, 1990, p. 247.
31. Peto, R., Doll, R., Buckley, J.D., and Sporn, M.B., Can dietary beta-carotene materially reduce human cancer rates? *Nature*, 290, 201, 1981.
32. Bland, J., The beta-carotene controversy in perspective, *J. Appl. Nutr.*, 48, 42, 1996.
33. Khachik, F., Askin, F.B., and Lai, K., Distribution, bioavailability, and metabolism of carotenoids in humans, in *Phytochemicals: A New Paradigm*, Bidlack, W.R. et al., Eds., Technomic Publishing Company, Lancaster, PA, 1998, chap. 5.
34. Khachik, F., Beecher, G.R., and Smith, Jr., J.C., Lutein, lycopene, and their oxidative metabolites in chemoprevention of cancer, *J. Cell. Biochem. Suppl.*, 22, 236, 1995.
35. King, T.J. et al., Metabolites of dietary carotenoids as potential cancer preventive agents, *Pure Appl. Chem.*, 69, 2135, 1997.
36. Palozza, P. and Krinsky, N.I., β-carotene and α-tocopherol are synergistic antioxidants, *Arch. Biochem. Biophys.*, 297, 184, 1992.
37. Böhm, F. et al., Carotenoids enhance vitamin E antioxidant efficiency, *J. Am. Chem. Soc.*, 119, 621, 1997.
38. Chen, H. and Tappel, A.L., Protection by vitamin E, selenium, trolox C, ascorbic acid palmitate, acetylcysteine, coenzyme Q, beta-carotene, canthaxanthin, and (+)-catechin against oxidative damage to liver slices measured by oxidized heme proteins, *Free Rad. Biol. Med.*, 16, 437, 1994.
39. Stahl, W. et al., Carotenoid mixtures protect multilamellar liposomes against oxidative damage: synergistic effects of lycopene and lutein, *FEBS Lett.*, 427, 305, 1998.
40. Young, A.J. and Lowe, G.M., Antioxidant and prooxidant properties of carotenoids, *Arch. Biochem. Biophys.*, 385, 20, 2001.
41. Paolini, M. et al., β-Carotene: a cancer chemopreventive agent or a co-carcinogen? *Mutat. Res.*, 543, 195, 2003.
42. Burton, G.W. and Ingold, K.U., β-Carotene: an unusual type of lipid antioxidant, *Science*, 224, 569, 1984.
43. Zhang, P. and Omaye, S.T., β-Carotene and protein oxidation: effects of ascorbic acid and α-tocopherol, *Toxicology*, 146, 37, 2000.
44. Crabtree, D.V. and Adler, A.J., Is β-carotene an antioxidant? *Med. Hypoth.*, 48, 183, 1997.
45. Zhang, P. and Omaye, S.T., Antioxidant and prooxidant roles for β-carotene, α-tocopherol and ascorbic acid in human lung cells, *Toxicol. In Vitro*, 15, 13, 2001.
46. Khachik, F. et al., Separation and identification of carotenoids and their oxidation products in the extracts of human plasma, *Anal. Chem.*, 64, 2111, 1992.
47. Bernhard, K., Synthetic astaxanthin: the route of a carotenoid from research to commercialisation, in *Carotenoids: Chemistry and Biology*, Krinsky, N.I., Ed., Plenum Press, New York, 1990, p. 337.
48. Lorenz, R.T. and Cysewski, G.R., Commercial potential for *Haematococcus microalgae* as a natural source of astaxanthin, *Trends Biotechnol.*, 18, 160, 2000.
49. Landrum, J.T., Bone, R.A., and Herrero, C., Astaxanthin, β-cryptoxanthin, lutein, and zeaxanthin, in *Phytochemicals in Nutrition and Health*, Meskin, M.S. et al., Eds., CRC Press, Boca Raton, FL, 2002, chap. 12.
50. Baker, R.T.M., Canthaxanthin in aquafeed applications: is there any risk? *Trends Food Sci. Technol.*, 12, 240, 2002.

51. Astorg, P., Food carotenoids and cancer prevention: an overview of current research, *Trends Food Sci. Technol.*, 8, 406, 1997.
52. Nishino, H., Cancer prevention by carotenoids, *Mutat. Res.*, 402, 159, 1998.
53. Nishino, H. et al., Carotenoids in cancer chemoprevention, *Cancer Metastasis Rev.*, 21, 257, 2002.
54. Nguyen, M.L. and Schwartz, S.J., Lycopene: chemical and biological properties, *Food Technol.*, 53, 38, 1999.
55. Di Mascio, P., Kaiser, S., and Sies, H., Lycopene as the most efficient biological carotenoid singlet oxygen quencher, *Arch. Biochem. Biophys.*, 274, 532, 1989.
56. Stahl, W. and Sies, H., Lycopene: a biologically important carotenoid for humans? *Arch. Biochem. Biophys.*, 336, 1, 1996.
57. Gerster, H., The potential role of lycopene for human health, *J. Am. Coll. Nutr.*, 16, 109, 1997.
58. Clinton, S.K. et al., *cis-trans* Lycopene isomers, carotenoids, and retinol in the human prostate, *Cancer Epidemiol. Biomarkers Prev.*, 5, 823, 1996.
59. Giovannucci, E. et al., Intake of carotenoids and retinol in relation to risk of prostate cancer, *J. Natl. Cancer Inst.*, 87, 1767, 1995.
60. Clinton, S.K., Lycopene: chemistry, biology, and implications for human health and disease, *Nutr. Rev.*, 56, 35, 1998.
61. Landrum, J.T. et al., A one year study of the macular pigment: the effect of 140 days of a lutein supplement, *Exp. Eye Res.*, 65, 57, 1997.
62. Sommerburg, O.G. et al., Lutein and zeaxanthin are associated with photoreceptors in the human retina, *Curr. Eye Res.*, 19, 491, 1999.
63. Landrum, J.T. and Bone, R.A., Lutein, zeaxanthin, and the macular pigment, *Arch. Biochem. Biophys.*, 385, 28, 2001.
64. Le Marchand, L. et al., An ecological study of diet and lung cancer in the South Pacific, *Int. J. Cancer*, 63, 18, 1995.
65. Voorrips, L.E. et al., A prospective cohort study on antioxidant and folate intake and male lung cancer risk, *Cancer Epidemiol. Biomarkers Prev.*, 9, 357, 2000.
66. de Mejía, E.G., Loarca-Piña, G., and Ramos-Gómez, M., Antimutagenicity of xanthophylls present in Aztec Marigold (*Tagetes erecta*) against 1-nitropyrene, *Mutat. Res.*, 389, 219, 1997.
67. Narisawa, T. et al., Inhibitory effects of natural carotenoids, α-carotene, β-carotene, lycopene and lutein, on colonic aberrant crypt foci formation in rats, *Cancer Lett.*, 107, 137, 1996.
68. Nishino, H. et al., Cancer prevention by carotenoids and curcumins, in *Phytochemicals as Bioactive Agents*, Bidlack, W.R. et al., Eds., Technomic Publishing Co., Lancaster, PA, 2000, chap. 9.
69. Brown, C.M. et al., Dietary lutein inhibits mouse mammary tumor growth by regulating angiogenesis and apoptosis, *FASEB J.*, 15, A954, 2001.
70. Sumantran, V.N. et al., Differential regulation of apoptosis in normal *versus* transformed mammary epithelium by lutein and retinoic acid, *Cancer Epidemiol. Biomarkers Prev.*, 9, 257, 2000.
71. Nishino, H. et al., Cancer prevention by carotenoids, *Pure Appl. Chem.*, 71, 2273, 1999.
72. Batieha, A.M. et al., Serum micronutrients and the subsequent risk of cervical cancer in a population-based nested case-control study, *Cancer Epidemiol. Biomarkers Prev.*, 2, 335, 1993.
73. Mangels, A.R. et al., Carotenoid content of fruits and vegetables: an evaluation of analytic data, *J. Am. Diet. Assoc.*, 93, 284, 1993.
74. Onogi, N. et al., Antiproliferative effect of carotenoids on human colon cancer cells without conversion to retinoic acid, *Nutr. Cancer*, 32, 20, 1998.
75. Bertram, J.S. et al., Diverse carotenoids protect against chemically induced neoplastic transformation, *Carcinogenesis*, 12, 671, 1991.
76. Mathews-Roth, M.M. and Krinsky, N.I., Carotenoid dose level and protection against UV-B induced skin tumors, *Photochem. Photobiol.*, 42, 35, 1985.
77. Tanaka, T. et al., Suppression of azoxymethane-induced rat colon carcinogenesis by dietary administration of naturally occurring xanthophylls astaxanthin and canthaxanthin during the postinitiation phase, *Carcinogenesis*, 16, 2957, 1995.
78. Tanaka, T. et al., Chemoprevention of rat oral carcinogenesis by naturally occurring xanthophylls, astaxanthin and canthaxanthin, *Cancer Res.*, 55, 4059, 1995.
79. Zhang, L.-X., Cooney, R.V., and Bertram, J.S., Carotenoids up-regulate Connexin43 gene expression independent of their provitamin A or antioxidant properties, *Cancer Res.*, 52, 5707, 1992.

80. Hanusch, M. et al., Induction of gap junctional communication by 4-oxoretinoic acid generated from its precursor canthaxanthin, *Arch. Biochem. Biophys.*, 317, 423, 1995.
81. Gradelet, S. et al., Effects of canthaxanthin, astaxanthin, lycopene and lutein on liver xenobiotic-metabolizing enzymes in the rat, *Xenobiotica*, 26, 49, 1996.
82. Jewell, C. and O'Brien, N.M., Effect of dietary supplementation with carotenoids on xenobiotic metabolizing enzymes in the liver, lung, kidney and small intestine of the rat, *Brit. J. Nutr.*, 81, 235, 1999.
83. Goralczyk, R. et al., Occurrence of birefringent retinal inclusions in cynomolgus monkeys after high doses of canthaxanthin, *Invest. Ophthalmol. Vis. Sci.*, 38, 741, 1997.
84. Leyon, H. et al., Reversibility of canthaxanthin deposits within the retina, *Acta Ophthalmol.*, 68, 607, 1990.
85. Kotake-Nara, E. et al., Carotenoids affect proliferation of human prostate cancer cells, *J. Nutr.*, 131, 3303, 2001.
86. Dore, J.E. and Cysewski, G.R., *Haematococcus* algae meal: a source of natural astaxanthin for aqua feeds, *Aqua Feed Int.*, 6, 22, 2003.
87. Di Mascio, P. et al., Carotenoids, tocopherols and thiols as biological singlet molecular oxygen quenchers, *Biochem. Soc. Trans.*, 18, 1054, 1990.
88. Shimidzu, N., Goto, M., and Miki, W., Carotenoids as singlet oxygen quenchers in marine organisms, *Fish. Sci.*, 62, 134, 1996.
89. Miki, W., Biological functions and activities of animal carotenoids, *Pure Appl. Chem.*, 63, 141, 1991.
90. Terao, J., Antioxidant activity of β-carotene-related carotenoids in solution, *Lipids*, 24, 659, 1989.
91. Woodall, A.A., Britton, G., and Jackson, M.J., Carotenoids and protection of phospholipids in solution or in liposomes against oxidation by peroxyl radicals: relationship between carotenoid structure and protective ability, *Biochim. Biophys. Acta*, 1336, 575, 1997.
92. Haila, K.M. et al., Carotenoid reaction with free radicals in acetone and toluene at different oxygen partial pressures: an ESR spin-trapping study of structure-activity relationships, *Z. Lebensm. Unters Forsch. A*, 204, 81, 1997.
93. Naguib, Y.M.A., Antioxidant activities of astaxanthin and related carotenoids, *J. Agric. Food Chem.*, 48, 1150, 2000.
94. Bagchi, D., Oxygen free radical scavenging abilities of vitamins C, E, β-carotene, pycnogenol, grape seed proanthocyanidin extract, astaxanthin and BioAstin® *in vitro*, Final Report to Cyanotech Corporation, Creighton University School of Health Sciences, Omaha, NE, 2001.
95. Lim, B.P. et al., Antioxidant activity of xanthophylls on peroxyl radical-mediated phospholipid peroxidation, *Biochim. Biophys. Acta*, 1126, 178, 1992.
96. Palozza, P. and Krinsky, N.I., Astaxanthin and canthaxanthin are potent antioxidants in a membrane model, *Arch. Biochem. Biophys.*, 297, 291, 1992.
97. Nishigaki, I. et al., Suppressive effect of astaxanthin on lipid peroxidation induced in rats, *J. Clin. Biochem. Nutr.*, 16, 161, 1994.
98. Nakagawa, K. et al., Inhibition by β-carotene and astaxanthin of NADPH-dependent microsomal phospholipid peroxidation, *J. Nutr. Sci. Vitaminol.*, 43, 345, 1997.
99. Rengel, D. et al., Exogenously incorporated ketocarotenoids in large unilamellar vesicles: protective activity against peroxidation, *Biochim. Biophys. Acta*, 1463, 179, 2000.
100. Cantrell, A. et al., Singlet oxygen quenching by dietary carotenoids in a model membrane environment, *Arch. Biochem. Biophys.*, 412, 47, 2003.
101. Oshima, S. et al., Inhibitory effect of β-carotene and astaxanthin on photosensitized oxidation of phospholipid bilayers, *J. Nutr. Sci. Vitaminol.*, 39, 607, 1993.
102. O'Connor, I. and O'Brien, N., Modulation of UVA light-induced oxidative stress by β-carotene, lutein and astaxanthin in cultured fibroblasts, *J. Dermatol. Sci.*, 16, 226, 1998.
103. Lawlor, S.M. and O'Brien, N.M., Astaxanthin: antioxidant effects in chicken embryo fibroblasts, *Nutr. Res.*, 15, 1695, 1995.
104. Tinkler, J.H. et al., Dietary carotenoids protect human cells from damage, *J. Photochem. Photobiol. B: Biol.*, 26, 283, 1994.
105. Iwamoto, T. et al., Inhibition of low-density lipoprotein oxidation by astaxanthin, *J. Atheroscler. Thromb.*, 7, 216, 2000.

106. Hagen, C., Braune, W., and Greulich, F., Functional aspects of secondary carotenoids in *Haematococcus lacustris* [Girod] Rostafinski (Volvocales) IV: protection from photodynamic damage, *J. Photochem. Photobiol. B: Biol.*, 20, 153, 1993.

107. Nakano, T. et al., Effect of astaxanthin rich red yeast (*Phaffia rhodozyma*) on oxidative stress in rainbow trout, *Biochim. Biophys. Acta*, 1426, 119, 1999.

108. Bell, J.G. et al., Depletion of α-tocopherol and astaxanthin in Atlantic salmon (*Salmo salar*) affects autooxidative defense and fatty acid metabolism, *J. Nutr.*, 130, 1800, 2000.

109. Kurashige, M. et al., Inhibition of oxidative injury of biological membranes by astaxanthin, *Physiol. Chem. Phys. Med. NMR*, 22, 27, 1990.

110. Kang, J.O., Kim, S.J., and Kim, H., Effect of astaxanthin on the hepatotoxicity, lipid peroxidation and antioxidative enzymes in the liver of CCl_4-treated rats, *Meth. Find. Exp. Clin. Pharmacol.*, 23, 79, 2001.

111. Di Mascio, P., Murphy, M.E., and Sies, H., Antioxidant defense systems: the role of carotenoids, tocopherols, and thiols, *Am. J. Clin. Nutr.*, 53, 194S, 1991.

112. Mortensen, A. and Skibsted, L.H., Relative stability of carotenoid radical cations and homologue tocopheroxyl radicals: a real time kinetic study of antioxidant hierarchy, *FEBS Lett.*, 417, 261, 1997.

113. Mortensen, A. et al., Comparative mechanisms and rates of free radical scavenging by carotenoid antioxidants, *FEBS Lett.*, 418, 91, 1997.

114. Miller, N.J. et al., Antioxidant activities of carotenes and xanthophylls, *FEBS Lett.*, 384, 240, 1996.

115. Mortensen, A., Skibsted, L.H., and Truscott, T.G., The interaction of dietary carotenoids with radical species, *Arch. Biochem. Biophys.*, 385, 13, 2001.

116. Burke, M. et al., One-electron reduction potentials of dietary carotenoid radical cations in aqueous micellar environments, *FEBS Lett.*, 500, 132, 2001.

117. Edge, R. et al., Relative one-electron reduction potentials of carotenoid radical cations and the interactions of carotenoids with the vitamin E radical cation, *J. Am. Chem. Soc.*, 120, 4087, 1998.

118. Fukuzawa, K. et al., Rate constants for quenching singlet oxygen and activities for inhibiting lipid peroxidation of carotenoids and α-tocopherol in liposomes, *Lipids*, 33, 751, 1998.

119. Barros, M.P. et al., Astaxanthin and peridinin inhibit oxidative damage in Fe^{2+}-loaded liposomes: scavenging oxyradicals or changing membrane permeability? *Biochem. Biophys. Res. Comm.*, 288, 225, 2001.

120. Socaciu, C. et al., Different ways to insert carotenoids into liposomes affect structure and dynamics of the bilayer differently, *Biophys. Chem.*, 99, 1, 2002.

121. Gabrielska, J. and Gruszecki, W.I., Zeaxanthin (dihydroxy-β-carotene) but not β-carotene rigidifies lipid membranes: a ^{1}H-NMR study of carotenoid-egg phosphatidylcholine liposomes, *Biochim. Biophys. Acta*, 1285, 167, 1996.

122. Sujak, A. et al., Lutein and zeaxanthin as protectors of lipid membranes against oxidative damage: the structural aspects, *Arch. Biochem. Biophys.*, 371, 301, 1999.

123. Goto, S. et al., Efficient radical trapping at the surface and inside the phospholipid membrane is responsible for highly potent antiperoxidative activity of the carotenoid astaxanthin, *Biochim. Biophys. Acta*, 1512, 251, 2001.

124. Jørgensen, K. and Skibsted, L.H., Carotenoid scavenging of radicals: effect of carotenoid structure and oxygen partial pressure on antioxidative activity, *Z. Lebensm. Unters Forsch.*, 196, 423, 1993.

125. Martin, H.-D. et al., Anti- and prooxidant properties of carotenoids, *J. Prakt. Chem.*, 341, 302, 1999.

126. Beutner, S. et al., Quantitative assessment of antioxidant properties of natural colorants and phytochemicals: carotenoids, flavonoids, phenols and indigoids: the role of β-carotene in antioxidant functions, *J. Sci. Food Agric.*, 81, 559, 2001.

127. Anonymous, Eskimo diets and diseases, *Lancet*, 1 (8334), 1139, 1983.

128. Bates, C. et al., Plasma essential fatty acids in pure and mixed race American Indians on and off a diet exceptionally rich in salmon, *Prostaglandins Leukot. Med.*, 17, 77, 1985.

129. Sun, S. et al., Anti-tumor activity of astaxanthin on Meth-A tumor cells and its mode of action, *FASEB J.*, 12, A966, 1998.

130. Kim, H.W., Park, J.S., and Chew, B.P., β-Carotene and astaxanthin inhibit mammary tumor cell growth and induce apoptosis in mice *in vitro*, *FASEB J.*, 15, A298, 2001.

131. Kozuki, Y., Miura, Y., and Yagasaki, K., Inhibitory effects of carotenoids on the invasion of rat ascites hepatoma cells in culture, *Cancer Lett.*, 151, 111, 2000.

132. Levy, J. et al., Lycopene and Astaxanthin Inhibit Human Prostate Cancer Cell Proliferation Induced by Androgens, presented at 13th Int. Carotenoid Symp., Honolulu, Jan. 2002, p. 135.

133. Lyons, N.M. and O'Brien, N.M., Modulatory effects of an algal extract containing astaxanthin on UVA-irradiated cells in culture, *J. Dermatol. Sci.*, 30, 73, 2002.

134. Jyonouchi, H. et al., Antitumor activity of astaxanthin and its mode of action, *Nutr. Cancer*, 36, 59, 2000.

135. Lee, S.H. et al., Inhibition of benzo(a)pyrene-induced mouse forestomach neoplasia by astaxanthin containing egg yolks, *Agric. Chem. Biotechnol.*, 40, 490, 1997.

136. Lee, S.H. et al., Inhibition of sarcoma-180 cell-induced mouse ascites cancer by astaxanthin-containing egg yolks, *J. Kor. Soc. Food Sci. Nutr.*, 27, 163, 1998.

137. Savouré, N. et al., Vitamin A status and metabolism of cutaneous polyamines in the hairless mouse after UV irradiation: action of β-carotene and astaxanthin, *Int. J. Vit. Nutr. Res.*, 65, 79, 1995.

138. Mori, H. et al., Chemoprevention by naturally occurring and synthetic agents in oral, liver, and large bowel carcinogenesis, *J. Cell. Biochem. Suppl.*, 27, 35, 1997.

139. Tanaka, T. et al., Chemoprevention of mouse urinary bladder carcinogenesis by the naturally occurring carotenoid astaxanthin, *Carcinogenesis*, 15, 15, 1994.

140. Gradelet, S. et al., Modulation of aflatoxin B1 carcinogenicity, genotoxicity and metabolism in rat liver by dietary carotenoids: evidence for a protective effect of CYP1A inducers, *Cancer Lett.*, 114, 221, 1997.

141. Gradelet, S. et al., Dietary carotenoids inhibit aflatoxin B_1-induced liver preneoplastic foci and DNA damage in the rat: role of the modulation of aflatoxin B_1 metabolism, *Carcinogenesis*, 19, 403, 1998.

142. Yang, Z. et al., Protective effect of astaxanthin on the promotion of cancer metastases in mice treated with restraint-stress, *J. Jpn. Soc. Nutr. Food Sci.*, 50, 423, 1997.

143. Kurihara, H. et al., Contribution of the antioxidative property of astaxanthin to its protective effect on the promotion of cancer metastasis in mice treated with restraint stress, *Life Sci.*, 70, 2509, 2002.

144. Astorg, P. et al., Dietary lycopene decreases the initiation of liver preneoplastic foci by diethylnitrosamine in the rat, *Nutr. Cancer*, 29, 60, 1997.

145. Park, J.S. et al., Dietary lutein but not astaxanthin or β-carotene increases *pim-1* gene expression in murine lymphocytes, *Nutr. Cancer*, 33, 206, 1999.

146. Black, H.S., Radical interception by carotenoids and effects on UV carcinogenesis, *Nutr. Cancer*, 31, 212, 1998.

147. Rousseau, E.J., Davison, A.J., and Dunn, B., Protection by β-carotene and related compounds against oxygen-mediated cytotoxicity and genotoxicity: implications for carcinogenesis and anticarcinogenesis, *Free Radical Biol. Med.*, 13, 407, 1992.

148. Aoi, W. et al., Astaxanthin limits exercise-induced skeletal and cardiac muscle damage in mice, *Antioxid. Redox Signal.*, 5, 139, 2003.

149. Uchiyama, K. et al., Beneficial effects of astaxanthin in type 2 diabetes model of db/db mouse, *Free Radical Biol. Med.*, 33, S211, 2002.

150. Uchiyama, K. et al., Astaxanthin protects beta-cells against glucose toxicity in diabetic db/db mice, *Redox Rep.*, 7, 290, 2002.

151. Tso, M.O.M. and Lam, T.-T., Method of retarding and ameliorating central nervous system and eye damage, U.S. Patent 5,527,533, 1996.

152. Wu, T.-H., Shah, P., and Maher, T.J., An astaxanthin-containing algal extract attenuates selenite-induced nuclear cataract formation in rat pups, *FASEB J.*, 16, A958, 2002.

153. Okada, F., Inflammation and free radicals in tumor development and progression, *Redox Rep.*, 7, 357, 2002.

154. Wang, X., Willén, R., and Wadström, T., Astaxanthin-rich algal meal and vitamin C inhibit *Helicobacter pylori* infection in BALB/cA mice, *Antimicrob. Agents Chemother.*, 44, 2452, 2000.

155. Bennedsen, M. et al., Treatment of *H. pylori* infected mice with antioxidant astaxanthin reduces gastric inflammation, bacterial load and modulates cytokine release by splenocytes, *Immunol. Lett.*, 70, 185, 1999.

156. Liu, B.H. and Lee, Y.K., Effect of total secondary carotenoids extracts from *Chlorococcum* sp. on *Helicobacter pylori*-infected BALB/c mice, *Int. Immunopharmacol.*, 3, 979, 2003.

157. Ohgami, K. et al., Effects of astaxanthin on lipopolysaccharide-induced inflammation *in vitro* and *in vivo*, *Invest. Ophthalmol. Vis. Sci.*, 44, 2694, 2003.

158. Nir, Y., Spiller, G., and Multz, C., Effect of an astaxanthin containing product on rheumatoid arthritis, *J. Am. Coll. Nutr.*, 21, 490, 2002.

159. Nir, Y., Spiller, G., and Multz, C., Effect of an astaxanthin containing product on carpal tunnel syndrome, *J. Am. Coll. Nutr.*, 21, 489, 2002.

160. Bendich, A., Carotenoids and the immune response, *J. Nutr.*, 119, 112, 1989.

161. Bendich, A., Carotenoids and the immune system, in *Carotenoids: Chemistry and Biology*, Krinsky, N.I., Ed., Plenum Press, New York, 1990, p. 323.

162. Okai, Y. and Higashi-Okai, K., Possible immunomodulating activities of carotenoids in *in vitro* cell culture experiments, *Int. J. Immunopharmacol.*, 18, 753, 1996.

163. Jyonouchi, H. et al., Studies of immunomodulating actions of carotenoids, I: effects of β-carotene and astaxanthin on murine lymphocyte functions and cell surface marker expression in *in vitro* culture system, *Nutr. Cancer*, 16, 93, 1991.

164. Jyonouchi, H., Zhang, L., and Tomita, Y., Studies of immunomodulating actions of carotenoids, II: Astaxanthin enhances *in vitro* antibody production to T-dependent antigens without facilitating polyclonal B-cell activation, *Nutr. Cancer*, 19, 269, 1993.

165. Jyonouchi, H. et al., Immunomodulating actions of carotenoids: enhancement of *in vivo* and *in vitro* antibody production to T-dependent antigens, *Nutr. Cancer*, 21, 47, 1994.

166. Jyonouchi, H. et al., Astaxanthin, a carotenoid without vitamin A activity, augments antibody responses in cultures including T-helper cell clones and suboptimal doses of antigen, *J. Nutr.*, 125, 2483, 1995.

167. Jyonouchi, H. et al., Effects of various carotenoids on cloned, effector-stage T-helper cell activity, *Nutr. Cancer*, 26, 313, 1996.

168. Jyonouchi, H., Sun, S., and Gross, M., Effect of carotenoids on *in vitro* immunoglobulin production by human peripheral blood mononuclear cells: astaxanthin, a carotenoid without vitamin A activity, enhances *in vitro* immunoglobulin production in response to a T-dependent stimulant and antigen, *Nutr. Cancer*, 23, 171, 1995.

169. Chew, B.P. et al., Dietary β-carotene and astaxanthin but not canthaxanthin stimulate splenocyte function in mice, *Anticancer Res.*, 19, 5223, 1999.

170. Tomita, Y. et al., Preventive action of carotenoids on the development of lymphadenopathy and proteinuria in MRL-*lpr/lpr* mice, *Autoimmunity*, 16, 95, 1993.

171. Bertram, J.S., Carotenoids and gene regulation, *Nutr. Rev.*, 57, 182, 1999.

172. Zhang, L.X., Cooney, R.V., and Bertram, J.S., Carotenoids enhance gap junctional communication and inhibit lipid peroxidation in C3H/10T1/2 cells: relationship to their cancer chemopreventive action, *Carcinogenesis*, 12, 2109, 1991.

173. Banoub, R.W., Fernstrom, M., and Ruch, R.J., Lack of growth inhibition or enhancement of gap junctional intercellular communication and connexin 43 expression by β-carotene in murine lung epithelial cells *in vitro*, *Cancer Lett.*, 108, 35, 1996.

174. Gradelet, S. et al., β-apo-8′-carotenal, but not β-carotene, is a strong inducer of liver cytochromes P4501A1 and 1A2 in rat, *Xenobiotica*, 26, 909, 1996.

175. Astorg, P. et al., Effects of provitamin A or non-provitamin A carotenoids on liver xenobiotic-metabolizing enzymes in mice, *Nutr. Cancer*, 27, 245, 1997.

176. Kistler, A. et al., Metabolism and CYP-inducer properties of astaxanthin in man and primary human hepatocytes, *Arch. Toxicol.*, 75, 665, 2002.

177. Rauscher, R., Edenharder, R., and Platt, K.L., *In vitro* antimutagenic and *in vivo* anticlastogenic effects of carotenoids and solvent extracts from fruits and vegetables rich in carotenoids, *Mutat. Res.*, 413, 129, 1998.

178. Anderson, M., Method of inhibiting 5α-reductase with astaxanthin, U.S. Patent 6,277,417, 2001.

179. Murakami, C. et al., Vitamin A-related compounds, all-*trans* retinal and retinoic acids, selectively inhibit activities of mammalian replicative DNA polymerases, *Biochim. Biophys. Acta*, 1574, 85, 2002.

180. Dore, J.E., Safety Profile: BioAstin® Natural Astaxanthin, technical bulletin Ax-072, Cyanotech Corp., Kailua-Kona, HI, 2002.

181. Ono, A. et al., A 13-week subchronic oral toxicity study of *Haematococcus* color in F344 rats, *Bull. Natl. Health Sci.*, 117, 91, 1999.

182. Spiller, G.A. and Dewell, A., Safety of an astaxanthin-rich *Haematococcus pluvialis* algal extract: a randomized clinical trial, *J. Med. Food*, 6, 51, 2003.

183. Furr, H.C. and Clark, R.M., Intestinal absorption and tissue distribution of carotenoids, *J. Nutr. Biochem.*, 8, 364, 1997.

184. Zaripheh, S. and Erdman, J.W., Factors that influence the bioavailability of xanthophylls, *J. Nutr.*, 132, 531S, 2002.

185. Castenmiller, J.J. et al., The food matrix of spinach is a limiting factor in determining the bioavailability of beta-carotene and to a lesser extent of lutein in humans, *J. Nutr.*, 129, 349, 1999.

186. Lavy, A., Ben Amotz, A., and Aviram, M., Preferential inhibition of LDL oxidation by the all-*trans* isomer of β-carotene in comparison with 9-*cis* β-carotene, *Eur. J. Clin. Chem. Clin. Biochem.*, 31, 83, 1993.

187. Gartner, C., Stahl, W., and Sies, H., Preferential increase in chylomicron levels of the xanthophylls lutein and zeaxanthin compared to beta-carotene in the human, *Int. J. Vitam. Nutr. Res.*, 66, 119, 1996.

188. Brown, E.D. et al., Vegetable concentrates interact with canthaxanthin to affect carotenoid bioavailability and superoxide dismutase activity but not immune response in rats, *Nutr. Res.*, 17, 989, 1997.

189. Roodenburg, A.J. et al., Amount of fat in the diet affects bioavailability of lutein esters but not of alpha-carotene, beta-carotene, and vitamin E in humans, *Am. J. Clin. Nutr.*, 71, 1187, 2000.

190. Clark, R.M. and Furr, H.C., Absorption of canthaxanthin by the rat is influenced by total lipid in the intestinal lumen, *Lipids*, 36, 473, 2001.

191. Clark, R.M. et al., A comparison of lycopene and astaxanthin absorption from corn oil and olive oil emulsions, *Lipids*, 35, 803, 2000.

192. Schiedt, K., Absorption and metabolism of carotenoids in birds, fish and crustaceans, in *Carotenoids, Vol. 3: Biosynthesis and Metabolism*, Britton, G., Liaaen-Jensen, S., and Pfander, H., Eds., Birkhauser, Basel, 1998, p. 285.

193. Wolz, E. et al., Characterization of metabolites of astaxanthin in primary cultures of rat hepatocytes, *Drug Metab. Dispos.*, 27, 456, 1999.

194. Osterlie, M., Bjerkeng, B., and Liaaen-Jensen, S., Plasma appearance and distribution of astaxanthin *E/Z* and *R/S* isomers in plasma lipoproteins of men after single dose administration of astaxanthin, *J. Nutr. Biochem.*, 11, 482, 2000.

195. Elmadfa, I. and Majchrzak, D., Absorption and transport of astaxanthin and canthaxanthin in humans after a salmon meal, *Ernährungs-Umschau*, 46, 173, 1999.

196. Odeberg, J.M. et al., Oral bioavailability of the antioxidant astaxanthin in humans is enhanced by incorporation of lipid based formulations, *Eur. J. Pharm. Sci.*, 19, 299, 2003.

197. Wright, M.E. et al., Dietary carotenoids, vegetables, and lung cancer risk in women: the Missouri women's health study (United States), *Cancer Causes Control*, 14, 85, 2003.

198. Goodman, G.E. et al., The association between lung and prostate cancer risk, and serum micronutrients: results and lessons learned from beta-carotene and retinol efficacy trial, *Cancer Epidemiol. Biomarkers Prev.*, 12, 518, 2003.

199. Le Marchand, L. et al., Vegetable consumption and lung cancer risk: a population-based case-control study in Hawaii, *J. Natl. Cancer Inst.*, 81, 1158, 1989.

200. Russell, R.M., Lycopene and lutein: the next steps to the mixed carotenoids, in *Nutraceuticals in Health and Disease Prevention*, Krämer, K., Hoppe, P.-P., and Packer, L., Eds., Marcel Dekker, New York, 2001, chap. 6.

35 Chemopreventive Effects of Selected Spice Ingredients

Young-Joon Surh, Hye-Kyung Na, and Hyong Joo Lee

CONTENTS

35.1 Introduction...575
    35.1.1 Dietary Prevention of Cancer: an Overview...575
    35.1.2 Spices...576
35.2 Chemopreventive Spices 576
    35.2.1 Turmeric..576
    35.2.2 Ginger and Related Rhizomes of Zingiberaceae Family 580
    35.2.3 Hot Red Pepper 581
    35.2.4 Saffron 582
    35.2.5 Allium Vegetables..583
        35.2.5.1 Garlic...584
        35.2.5.2 Onion...585
    35.2.6 Clove..587
    35.2.7 Rosemary 587
    35.2.8 Miscellaneous Spices..589
Acknowledgment...591
References 591

35.1 INTRODUCTION

35.1.1 DIETARY PREVENTION OF CANCER: AN OVERVIEW

Cancer is largely an avoidable or preventable disease. It is estimated that more than two-thirds of cancer-related deaths can be prevented through lifestyle modification, particularly our daily diet. Frequent consumption of diets rich in fruits and vegetables has been consistently shown to reduce the risk of several forms of human cancer. A major prevention strategy has been the "Five a Day for Better Health" program sponsored by the U.S. National Cancer Institute (NCI), encouraging the public to include more fruit and vegetables in their diet.[1]

A vast variety of phytochemicals present in our daily diet, including fruits, vegetables, grains, and seeds, have been found to possess substantial antimutagenic and anticarcinogenic activities. The chemopreventive effects of the majority of edible phytochemicals are often attributed to their antioxidative or anti-inflammatory activities.[2–4] Besides edible chemopreventives in vegetables and fruits, some phytochemicals derived from herbs and spices also have potential anticarcinogenic and antimutagenic activities as well as other beneficial health effects.[2,5,6] The use of herbal medications has been increasing worldwide. Spices are of particular interest in terms of both culinary and medicinal value. It is interesting to note that many spices contain antioxidant and anti-inflammatory ingredients, which may confer the chemopreventive activity.

35.1.2 SPICES

Spices — dried roots, bark, buds, fruits, seeds, or berries — are plant products used to enhance flavor, color, and palatability of foods and beverages. In contrast, herbs are mainly the fresh or dried leaves of aromatic plants. Since ancient times, spices have been an important part of our lives as exotic and aromatic enhancement to food and as folk medicine. Today, spice use is ubiquitous, although the frequency of use of individual spices varies widely from country to country.[7] In addition to their culinary value, spices have several health benefits, the most important of which may be reducing food-borne illnesses and food poisoning due to antibacterial, antifungal, and antiviral activity. Moreover, certain ingredients of spices have strong antioxidant and anti-inflammatory properties that may contribute to their chemopreventive potential.[2,8–10] Most spices contain dozens of such defensive phytochemicals, collectively called "phytoalexins," as secondary metabolites. This chapter covers chemopreventive potential of antioxidative and anti-inflammatory substances derived from representative spices (Figure 35.1). The biochemical and cellular events that are induced or modulated by selected chemopreventive spices are summarized in Table 35.1.

35.2 CHEMOPREVENTIVE SPICES

35.2.1 TURMERIC

Turmeric (*Curcuma longa* Linn.), which belongs to the ginger family (Zingiberaceae), is native to the tropics of Southeast Asia and India. It has been used for centuries as a natural food colorant, a preservative, or a spice as well as an herbal remedy. Numerous recent scientific investigations, as well as millennia of experience in folk medicine, reveal the breadth of therapeutic as well as flavoring potential of this spice. According to accumulated data in the literature, turmeric ameliorates inflammation, ulceration, platelet aggregation, lipid oxidation, and the growth of microbes. Moreover, the whole turmeric extract or the standardized preparations containing defined amounts of its ingredients have been shown to exert antitumorigenic effects by inhibiting both initiation and later steps (i.e., promotion and progression) of multistage carcinogenesis in various animal models and cultured cells.[11,12]

The ground, dried rhizome of turmeric contains phenolic compounds collectively named curcuminoids, such as curcumin, demethoxycurcumin, and *bis*-demethoxycurcumin, that provide yellow pigmentation of this plant. Besides their commercial use to color diverse food items, such as curry, cheese, pickles, and cakes, these curcuminoids have some interesting pharmacological and physiological activities, particularly anti-inflammatory and antioxidant activities. Of these curcuminoids, curcumin, a yellow pigment from *Curcuma longa*, is a major component of turmeric. Curcumin not only exhibits anti-inflammatory and antioxidant properties, but also substantial chemopreventive activities, protecting laboratory animals from the malignancies induced by a variety of chemical carcinogens.[1,4,13–15]

The chemopreventive properties of curcumin have been evaluated in numerous animal tumor models and cultured cells.[1,2,4,13–15] Topical application of curcumin inhibits chemically induced carcinogenesis in mouse skin, and oral administration of the compound suppressed oral, forestomach, duodenal, and colon carcinogenesis induced by various carcinogens.[14] Curcumin acts as both a blocking and a suppressing agent. Ciolino et al.[16] suggested that curcumin inhibited the activation of 7,12-dimethylbenz[a]anthracene (DMBA) and subsequent DNA adduction possibly by competing with the aryl hydrocarbon for the Ah receptor involved in expression of cytochrome P450 1A1 (CYP1A1), and by competitively inhibiting the enzyme in human mammary epithelial carcinoma MCF-7 cells.[16] Curcumin was found to inhibit benzo[a]pyrene (BaP)-induced forestomach cancer in mice[17], possibly by reducing the activity of hepatic CYP1A1 that activates BaP to the DNA-reactive diol epoxide, and by increasing levels of glutathione S-transferase (GSTs), particularly GSTP1-1, and epoxide hydrolase, which are important in detoxifying BaP diol epoxide.[18] Dietary supplementation of curcumin

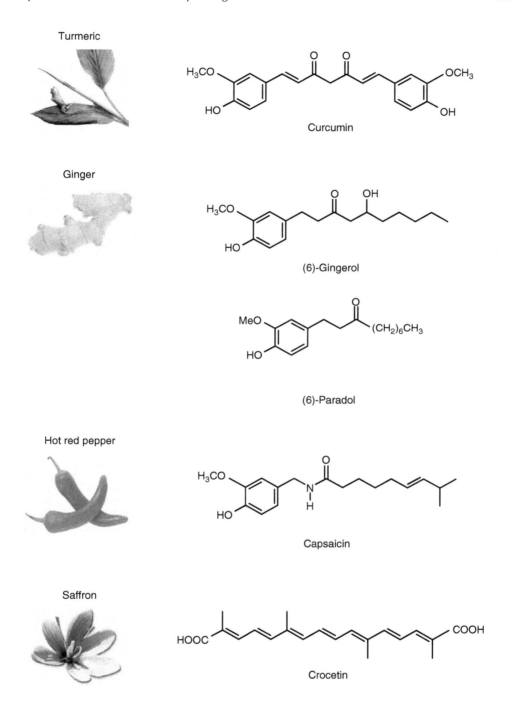

FIGURE 35.1 Examples of spices with chemopreventive potential and their principal ingredients.

(2%, w/v) to male ddY mice for 30 days significantly increased the activities of glutathione peroxidase, glutathione reductase, glucose-6-phosphate dehydrogenase, and catalase in the liver and kidney as compared with the corresponding control fed with a normal diet.[19] Parallel to these changes, curcumin administration resulted in a considerable enhancement in the activity of phase II detoxification enzymes, such as GST and NAD(P)H:quinone reductase (QR).

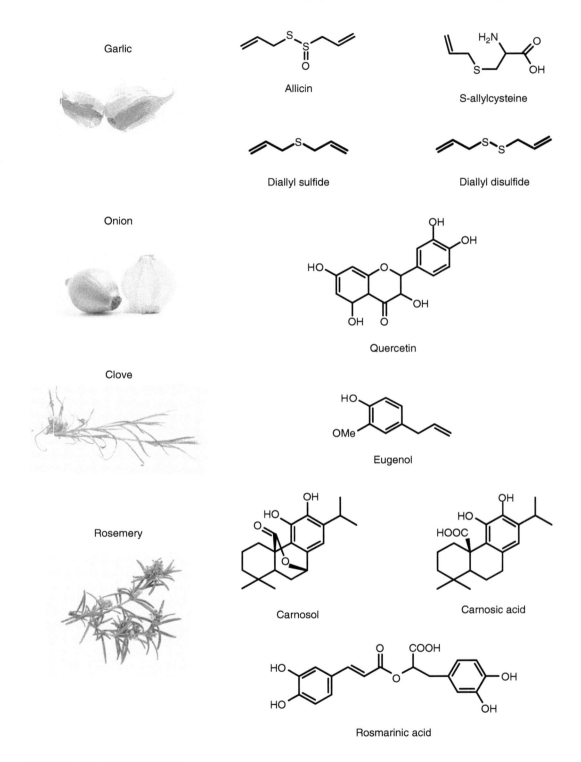

Garlic

Allicin

S-allylcysteine

Diallyl sulfide

Diallyl disulfide

Onion

Quercetin

Clove

Eugenol

Rosemery

Carnosol

Carnosic acid

Rosmarinic acid

FIGURE 35.1 (CONTINUED)

TABLE 35.1
Chemopreventive/Chemoprotective Effects of Representative Spices

Spices	Chemopreventive/Chemoprotective Effects
Turmeric	Inhibition of carcinogen activation and DNA binding
	Stimulation of carcinogen detoxification through induction of phase II enzymes
	Induction of apoptosis or differentiation in the cancerous cells
	Induction of cell cycle arrest
	Inhibition of angiogenesis, metastasis and invasion
	Inhibition of tumor promotion
Ginger	Inhibition of tumor promotion
	Inhibition of TPA-induced superoxide and hydrogen peroxide formation
	Inhibition of TPA-induced Epstein-Barr virus activation
	Induction of apoptosis and inhibition of transformation and metastasis
Hot red pepper	Inhibition of carcinogen activation and DNA binding
	Stimulation of carcinogen detoxification
	Suppression of tumor cell proliferation
	Induction of apoptosis
Saffron	Suppression of two-stage mouse skin carcinogenesis
	Inhibition of experimentally induced genotoxicity
	Inhibition of nucleic acid and protein synthesis and carcinogen-induced neoplastic transformation
	Inhibition of TPA-induced hydrogen peroxide production and myeloperoxidase activity
Garlic	Inhibition of lipid oxidation
	Augmentation of the GST activity and biosynthesis of cellular reduced glutathione
	Inhibition of the formation of carcinogen-DNA adducts
	Induction of apoptosis and cell cycle arrest in malignant cells
	Inhibition of neoplastic transformation and angiogenesis
Onion	Potentiation of cellular antioxidant capacity through up-regulation of γ-glutamylcysteine synthase
	Inhibition of DNA damage and growth of cancer cells mediated by perturbation of microtubule polymerization
	Inhibition of EGF-receptor signaling
	Induction of apoptosis and cell cycle arrest in cancerous cells
Clove	Inhibition of lipid peroxidation
	Anti-inflammatory, anti-mutagenic, and anti-carcinogenic effects
Rosemary	Antioxidant effect possibly by scavenging of •OH radical, singlet oxygen, and peroxynitrite
	Inhibition of metabolic activation of carcinogens
	Stimulation of carcinogen detoxification through induction of phase II enzyme
	Inhibition of tumor promotion
	Enhancement of the intracellular accumulation of chemotherapeutic agents through inhibition of the P-glycoprotein activity
	Induction of apoptosis and cell cycle arrest in tumor cells

However, data in the literature indicate that curcumin is more likely to possess tumor-suppressing properties and acts in the later stages of multistep carcinogenesis, interfering with cellular processes involved in tumor promotion and progression. Kinases, telomerase, cyclooxygenase 2 (COX-2), and transcription factors, including activator protein 1 (AP-1) and nuclear factor kappaB (NF-κB), are among the cellular targets.[1,13,18]

The inhibitory effect of curcumin when administered during the promotion and progression stages of carcinogenesis is associated with increased apoptosis, suggesting that stimulation of cancer cell death through apoptosis may be one of the plausible mechanisms by which curcumin blocks the process of carcinogenesis.[20] Curcumin has also been reported to induce cell-cycle arrest in a wide range of cell types, most frequently in the G1 or G2/M phase.

NF-κB plays a central role in cell survival and proliferation in many types of malignancies, including human multiple myeloma. In a recent study by Bharti et al.,[21] curcumin suppressed constitutively activated NF-κB in several types of human multiple myeloma cell lines. Curcumin down-regulated the expression of NF-κB-regulated gene products, including IκBα, Bcl-2, Bcl-x$_L$, cyclin D$_1$, and interleukin-6, which led to the suppression of proliferation and arrest of cells at the G1/S phase.[21] Curcumin also activated caspase-7 and caspase-9 and induced the cleavage of polyadenosine-5′-diphosphate-ribose polymerase.[21]

35.2.2 Ginger and Related Rhizomes of Zingiberaceae Family

The rhizome of ginger (*Zingiber officinale* Roscoe, Zingiberaceae) has been used not only as a food seasoning and a spice, but also as a useful component of oriental herbal medicine in which it is recommended for colds, fever, chills, rheumatism, motion sickness, and leprosy. Ginger has antiemetic, diuretic, anti-inflammatory, analgesic, carminative, stimulant, antioxidative, and antipyretic effects.[22]

In SHN virgin mice given hot-water extract of ginger (0.125%) in drinking water *ad libitum*, the spontaneous development of mammary tumors was significantly inhibited.[23] There were no apparent adverse effects observed following chronic administration of the ginger extract as estimated from body weight, food and water intake, and various plasma component levels. Preapplication of the ethanol extract of ginger significantly inhibited epidermal activities of COX, lipoxygenase, and ornithine decarboxylase (ODC) and ODC mRNA expression induced by the tumor promoter 12-O-tetradecanoylphorbol-13-acetate (TPA) in mouse skin.[24] The same treatment also afforded substantial inhibition of TPA-caused epidermal edema (56%) and hyperplasia (44%). In long-term tumor studies, topical application of the ginger extract 30 min prior to each TPA application after initiation with DMBA resulted in marked protection against skin tumor formation in SENCAR mice.[24] These findings provide clear evidence that ginger possesses antitumor-promoting effects and that the mechanism of such effects may involve inhibition of tumor promoter-caused cellular, biochemical, and molecular changes in mouse skin.

[6]-Gingerol, a principal pungent ingredient of ginger, strongly inhibited TPA-induced skin tumor promotion when topically applied onto dorsal skin of female ICR mice throughout the promotion stage.[25] [6]-Gingerol also suppressed ODC activity and tumor necrosis factor alpha (TNFα) production in TPA-stimulated mouse skin and lowered the TPA-induced superoxide production in differentiated human acute promyelocytic leukemia (HL-60) cells.[26] Likewise, [6]-paradol, a minor pungent ingredient of ginger but present in a larger quantity in the seeds of grains of paradise (*Aframomum meleguet* Roscoe, Zingiberaceae), inhibited TPA-induced ear edema and skin tumor promotion in mice and also suppressed TPA-stimulated superoxide production in HL-60 cells.[26] [6]-Paradol and its homologs exhibit inhibitory effects on TPA-induced hydrogen peroxide production and activities of myeloperoxidase and ODC as well as tumor promotion in mouse skin.[27] HL-60 cells treated with [6]-gingerol and [6]-paradol underwent apoptotic death.[28] Treatment of mouse epidermal JB6 cells with [6]-gingerol blocked epidermal growth factor (EGF)-induced anchorage-independent cell growth on the soft agar and AP-1 activation.[29]

When 6-week-old male F344 rats were given gingerol in the diet at a concentration of 0.02% for 3 weeks, the multiplicity of azoxymethane-induced intestinal neoplasms was significantly reduced compared with that of the animals treated with the carcinogen alone.[30] Pulmonary metastasis in mice implanted with B16F10 melanoma cells was abolished by gingerol.[31] The antimetastatic activity of [6]-gingerol was expressed through the host's antitumor immune functions, as the growth of B16F10 melanoma cells was not affected by this substance *in vitro*. The splenic CD8$^+$ T cells from mice treated with the compound showed inhibitory effects on pulmonary metastasis when these T cells were adoptively transferred to mice bearing B16F10 melanoma cells.

Zingiberaceae rhizomes commonly used in the Malaysian traditional medicine were screened for antitumor promoter activity using the short-term assay of inhibition of TPA-induced Epstein-Barr virus early antigen (EBV-EA) in Raji cells.[32] Rhizomes found to possess inhibitory activity toward TPA-induced EBV activation include: *Curcuma domestica*, *Curcuma xanthorrhiza*, *Kaempferia galanga*, *Zingiber cassumunar*, *Zingiber officinale*, and *Zingiber zerumbet*. These rhizome extracts capable of suppressing EBV activation had no cytotoxic effect in Raji cells. Zerumbone, a sesquiterpenoid from *Zingiber zerumbet* Smith, was found to be a potent inhibitor of TPA-induced EBV activation.[33] More recently, zerumbone has been reported to effectively suppress TPA-induced superoxide anion generation in differentiated HL-60 and in Chinese hamster ovary cells.[34] This α, β-unsaturated sesquiterpenoid inhibited the proliferation of several types of human colonic adenocarcinoma cell lines, while the growth of normal human dermal and colon fibroblasts was less affected.[34] The antiproliferative effects of zerumbone toward cancer cells appear to be associated with its ability to induce apoptotic cell death[34] or cell-cycle arrest.[35] Dietary administration of zerumbone caused significant reduction in the frequency of aberrant crypt foci and also in the expression of COX-2 and prostaglandins (PGs) in colonic mucosa.[36]

35.2.3 HOT RED PEPPER

Hot peppers belong botanically to the genus *Capsicum* (from the Greek *kapto*, "to bite") that includes cayenne, chili, jalapeno, green bell, paprika, etc. The dried fruits of capsicums have a medicinal value. Thus, regular consumption of red peppers may improve circulation, promote sweating, stimulate digestive secretions, increase appetite, relieve pain and inflammatory symptoms, etc. Hot peppers are healthy foods, and their nutritional values have been attributable to the relatively high content of carotenoids, vitamin C, and vitamin E, which are potent antioxidants. In contrast, the capsicums' medicinal value mainly comes from the alkaloid capsaicin, a major pungent ingredient of hot peppers. The U.S. Food and Drug Administration has approved phytomedicines containing 0.025 or 0.075% capsaicin in a cream base. Currently, capsaicin is sold as an over-the-counter ointment under such brand names as Zostrix and Capzacin-P (or Capzacin-HP for high potency) to ease a wide range of pains and aches associated with diabetic neuropathy, postmastectomy, arthritis, neurodermititis, etc.

The hot spicy foods have been suspected to damage the mucous membranes of gastrointestinal tract, and may cause or promote gastric cancer in humans. According to an epidemiological study conducted in Mexico City, people who eat heavily hot peppers are 17 times more likely to develop stomach cancer than the average.[37] A subsequent hospital-based case-control study has revealed the increased risk of gastric cancer among high-level consumers of capsaicin (90 to 250 mg of capsaicin per day, equivalent to approximately 9 to 25 jalapeno peppers per day) as compared with low-level consumers (0 to 29.9 mg of capsaicin per day, approximately zero to less than three jalapeno peppers per day) in Mexico.[38] The increased risk of gastric cancer observed among high-level capsaicin consumers was not related to the *Helicobacter pylori* status.[38] A prospective case-control study conducted in India also relates the higher stomach cancer risk to consumption of spicy food, including chili as well as high-temperature food.[39]

In contrast to the above findings, a different population study reported that people who ate cayenne peppers actually had lower rates of stomach cancer.[40] There is some evidence from laboratory as well as epidemiological studies to support the gastroprotective effects of red pepper. Researchers in Singapore have reported that hot chili pepper and its pungent ingredient capsaicin have preventive effects on ulceration of the digestive tract.[41–44] In support of this notion, Chinese residents living in Singapore have a high rate of ulcers compared with other ethnic populations in the same country who eat more peppers than Chinese people. Similarly, in the U.S., where the consumption of chili has doubled in the last two decades, stomach cancer rates have actually declined. Hot peppers or capsaicin, by stimulating the flow of digestive juices, may shield stomach lining against damage from acids and alcohol, and may even help heal them. Thus, intraesophageal application of a capsaicin-containing red pepper sauce (Tabasco) suspension had profound effects on upper gastrointestinal motility in healthy volunteers, which could improve clearance and protection of the esophagus and could lead to retention of the irritant in the stomach and faster transit through the small bowel.[45] Ethanol-induced oxidative and inflammatory damage in gastric mucosa of rats was ameliorated by intragastric administration of capsaicin.[46] The effect of capsaicin on experimental gastric ulcer was studied in 1-h pylorus-ligated rats. Capsaicin given at a low intragastric dose (0.1 μg kg⁻¹) protected against gastric mucosal injury evoked by i.g. administration of acidified aspirin, 96% ethanol, or 0.6 M HCl, while larger amounts of the compound (10 and 30 mg kg⁻¹) invariably aggravated the aspirin- and ethanol-induced gastric mucosal damage.[47]

Overall, the current scientific findings on health benefit versus risk of hot pepper consumption are contradictory, and the relationship between cayenne consumption and the increased risk of stomach cancer remains unclear. Although large quantities of hot peppers can severely irritate the stomach and esophagus, the amounts found in normal diets are more likely to protect against than cause cancer. In fact, hot pepper and hot pepper products are on the GRAS (generally recognized as safe) list defined by the U.S. Food and Drug Administration. Nevertheless, until more is known, not too much hot pepper should be taken, particularly by people with ulcers, heartburn, or gastritis.

It has been reported that capsaicin inhibits the growth of $H. pylori$ in a test-tube experiment.[48] In a prospective crossover study, groups of healthy $H. pylori$-infected adults were given tolerable amounts of fresh garlic or jalapeno peppers in their meal and subjected to the urea breath test to assess the status of the $H. pylori$ infection. Neither garlic nor jalapenos had any effect on $H. pylori$ in vivo,[49] which did not corroborate the findings from in vitro studies by other investigators.[48,50] It would also be worthwhile determining further whether long-term consumption of red pepper or capsaicin at nontoxic doses could protect gastric mucosa from oxidative or proinflammatory damage induced by $H. pylori$.

Capsaicin has been found to inhibit chemically induced carcinogenesis and mutagenesis in various animal models and cell culture systems.[4] Topical application of capsaicin inhibited TPA-induced activation of NF-κB and AP-1 in female ICR mice.[51] Similarly, TPA-induced DNA binding of NF-κB and AP-1 was suppressed in HL-60 cells in the presence of capsaicin. The modifying effects of dietary administration of capsaicin on azoxymethane-induced colon tumorigenesis were investigated in male F344 rats. Gavage with capsaicin significantly elevated phase II detoxification enzymes, such as GST and QR, in the liver and colon.[52] Capsaicin and rotenone given at a dose of 500 ppm for 4 weeks significantly inhibited aberrant crypt foci formation induced by azoxymethane. In a subsequent long-term study, dietary administration of capsaicin during the initiation phase significantly reduced the incidence of colonic adenocarcinoma.[52]

Interestingly, capsaicin has been found to preferentially suppress the growth of cancerous or transformed cells by inducing apoptosis, as recently reviewed by Surh.[53]

35.2.4 SAFFRON

Since ancient times, saffron (*Crocus sativus* L., Iridaceae) has been widely used as a spice for flavoring and coloring food preparations and also as a perfume or a dye. In addition, saffron has been used in folklore medicine. Egyptians used this medicinal plant for the treatment of diseases as recorded in the George Ebers papyrus (1500 B.C.).

Topical application of a saffron extract suppressed two-stage mouse skin carcinogenesis initiated with DMBA and promoted by croton oil.[54] Oral administration of the same extract (100 mg/kg body weight) also exhibited an inhibitory effect on 3-methylcholanthrene-induced soft tissue sarcomas in albino mice.[54] In another study, oral administration of 200 mg/kg body weight of saffron extract increased the life span of mice bearing sarcoma-180, Ehrlich ascites carcinoma, and Dalton's lymphoma ascites.[55] The growth of these tumors in vitro was inhibited by the saffron extract.

Liposome-encapsulated saffron effectively augmented its cytotoxic activity towards Sarcoma-180 and Ehrlich ascites carcinoma.[56] The intracellular levels of reduced glutathione and glutathione-related enzymes (e.g., GST and glutathione reductase) were elevated in Sarcoma-180 cells when incubated with saffron.[56]

Incubation of HeLa cells for 3 h with a concentrated extract of saffron prepared from the flowers of *Crocus sativus* resulted in significant inhibition of colony formation and cellular nucleic acid synthesis.[57] In another study, treatment of HeLa (cervical epithelioid carcinoma) cells with the saffron extract in combination with selenite increased the level of inhibition of colony formation and nucleic acid synthesis in comparison with cells that were treated with each of these agents alone.[58]

In order to compare the sensitivity of malignant and nonmalignant cells to saffron, Abdullaev and Frenkel[59] examined the effect of the saffron extract on macromolecular synthesis in three human cell lines, A549 cells (derived from a lung adenocarcinoma), WI-38 cells (normal lung fibroblasts), and VA-13 cells (WI-38 cells transformed *in vitro* by SV40 tumor virus). According to these studies, the malignant cells were more susceptible than the normal cells to the inhibitory effects of saffron on both DNA and RNA synthesis.[59] There was no effect on protein synthesis in any of the cells. Saffron extract has been shown to partially prevent the decrease in body weight, hemoglobin levels, and leukocyte contents and also to prolong the life span of cisplatin-treated mice almost three-fold.[60]

Saffron contains several pharmacologically active carotenoids (e.g., crocetin and crocin), picro-crocin, and safranal, which have the ability to inhibit the growth of different types of tumor cells.[61] Crocetin inhibited nucleic acid and protein synthesis in three malignant human cell lines, such as HeLa, A549, and VA13 cells.[62] Crocetin also had a dose-dependent inhibitory effect on DNA and RNA synthesis in isolated nuclei and suppressed the activity of purified RNA polymerase II.[62] Escribano and colleagues[63] have determined that the concentrations of saffron components inducing 50% cell growth inhibition (LD_{50}) in HeLa cells were 2.3 mg/ml for an ethanolic extract of saffron dry stigmas, 3 mM for crocin, 0.8 mM for safranal, and 3 mM for picrocrocin. Crocetin did not show any appreciable cytotoxic effect. Cells treated with crocin exhibited wide cytoplasmic vacuole-like areas, reduced cytoplasm, cell shrinkage, and pyknotic nuclei, indicative of apoptosis induction. Tarantilis et al.[64] compared the effects of carotenoids of *C. sativus* L. on the cell proliferation and differentiation of HL-60 cells with those of all-*trans* retinoic acid. The doses inducing 50% inhibition of cell growth were 0.12 μM for all-*trans* retinoic acid, 2 μM for crocin, and 2 μM for crocins. At 5 μM, all these compounds induced differentiation of HL-60 cells. A novel glucoconjugate isolated from corms of *C. sativus* L. exerted cytotoxic activity against different tumor cells.[65]

Besides its presence in the saffron, crocetin is also a major component in the fruit of *Gardenia jasminoides* Ellis, a Chinese herbal medicine. When pretreated to C3H10T1/2 cells at a nontoxic concentration (0.01 to 0.10 mM), crocetin significantly inhibited genotoxicity and DNA binding of BaP.[66] Crocetin also inhibited BaP-induced neoplastic transformation of C3H10T1/2 cells, which appeared to be associated with induction of GST activity.[66] Topical application of crocetin at 0.2 or 1.0 μmol with TPA (15 nmol) twice weekly for 20 weeks to female CD-1 mice previously initiated with BaP reduced the number of papillomas per mouse by 69 and 81%, respectively.[67] Preapplication of the same amount of crocetin also afforded significant protection against TPA-induced hyperplasia and edema in the ear. Topical application of crocetin inhibited TPA-mediated induction of epidermal ODC activity.[67] Pretreatment of mouse skin with various amounts of crocetin inhibited hydrogen peroxide production as well as myeloperoxidase activity induced by TPA. These findings indicate that crocetin possesses antitumor-promoting potential.[67]

35.2.5 ALLIUM VEGETABLES

Allium vegetables such as garlic, onions, leeks, and chives have been reported to protect against stomach, esophageal, and colorectal cancers, while evidence for a protective effect against cancer at other sites is still insufficient.[68] The chemopreventive effects of allium vegetables have been attributed to organosulfur compounds, mainly as allyl derivatives, which inhibit carcinogenesis in the forestomach, esophagus, colon, mammary gland, and lung of experimental animals.[68] Several mechanisms have been proposed to explain the chemopreventive effects of allyl derivatives. Organosulfur compounds have been found to modulate the activity of several xenobiotic metabolizing enzymes that activate or detoxify carcinogens and inhibit the formation of carcinogen DNA adducts in several target tissues. In addition, an antiproliferative activity has been observed in several tumor cell lines, which is possibly mediated by induction of apoptosis and alterations of the cell cycle. Organosulfur compounds derived from allium vegetables are thus potential chemopreventive agents. Limited population-based studies conducted in China have shown that frequent intake of allium vegetables is inversely associated with the risk for esophageal and stomach cancer[69] as well as prostate cancer.[70]

35.2.5.1 Garlic

Garlic (*Allium sativum*) has been used for both culinary and medicinal purposes. Recent studies have revealed that garlic is effective in preventing many chronic diseases, including cardiovascular disorders,[71] arthritis,[72] arteriosclerosis,[73] and cancer.[74,75] Garlic preparations are used as over-the-counter herbal medicines in the Western countries. The chemopreventive activity of garlic has been well documented.[74–77]

The major organosulfur compounds found in garlic are alliin (*S*-allyl-L-cysteine sulfoxide), allicin (thio-2-propene-1-sulfinic acid *S*-allyl ester), *S*-allylcysteine, diallyl sulfide, diallyl disulfide, and diallyl mercaptan.[74,75] While alliin, allicin, and *S*-allylcysteine are water soluble, diallyl sulfide, diallyl disulfide, and diallyl mercaptan are lipid soluble. In addition to characteristic odor and flavor, these organosulfur compounds possess distinct biological activities. Once garlic is crushed into garlic homogenate, the alliin that is originally present in garlic is converted into allicin by allinase. Thus, allicin is the major compound in garlic homogenate. Although alliin is also the principal bioactive compound in garlic powder, it is turned into allicin by allinase when garlic powder is rehydrated.[75] Allicin and allicin-containing garlic preparations have been found to retain a substantial antioxidant activity. Allicin irreversibly reduced oxidative activity of sulfhydryl (SH)-containing enzymes through modification of SH groups.[78] Garlic powder showed high hydroxyl radical-scavenging capacity.[79] When garlic powder tablets were given to humans for 2 weeks, the susceptibility of lipoproteins to oxidation significantly decreased.[80] Garlic leaf powder in rat diet effectively prevented lipid peroxidation and decreased the Mn-superoxide dismutase (SOD) and glutathione peroxidase activities associated with gentamycin nephrotoxicity.[80] When Sprague–Dawley rats (4 to 5 weeks old) were injected with azoxymethane (15 mg/kg body weight) and orally administered with 2% (w/v) of garlic, colonic aberrant crypt foci were reduced in garlic-treated groups.[81] In garlic-treated groups, the GST activity was elevated in both colon and liver, whereas considerable reduction in the lipid peroxidation was observed in liver as well as in colon in comparison with the carcinogen-only group. Moreover, there was a significant reduction in the bromodeoxyuridine (Brdu) labeling index with a concomitant increase in the apoptotic index in the colon of animals fed garlic extract. These results suggest that garlic has a protective effect on colon carcinogenesis, which is mediated by modulation of different biological pathways during multistep carcinogenesis.[74,81]

A number of reports have indicated the antiprollferative activity of allicin and allicin-containing garlic extract. Allicin and the water extract of garlic powder directly suppressed proliferation of human cancer cells.[82–84] In mammary and colon cancer cells, the extent of the decrease in glutathione levels correlated with the growth inhibitory activity of allicin, but the strong inhibition of cell growth by allicin is more likely due to cell-cycle arrest than induction of cell death, since apoptosis was not observed.[82] In leukemia HL-60 cells, allicin suppressed the cell proliferation by blocking cell-cycle arrest from S to G2/M phase transition and induced apoptosis, as was confirmed by DNA ladder formation.[84] In a hamster oral cancer model, aqueous garlic extract effectively decreased the expression of high-molecular-weight cytokeratins, which is a biomarker of oral cancer.[85]

Since allinase, responsible for converting alliin into allicin, is deactivated by heat and drying, alliin is the primary organosulfur constituent in heat-treated garlic. However, the antioxidant activity of alliin still remains unclear and ambiguous. When a garlic extract, in which alliin was a major active compound but allinase was inactivated was used, the antiproliferation of human tumor cells was not evident.[83] The growth of tumor cells was suppressed only when the extract was supplemented with garlic powder containing allinase, suggesting that the allicin primarily contributes to the antiproliferative activity retained in garlic. Similarly, the ability of garlic to inhibit binding of the carcinogen DMBA to rat mammary epithelial cell DNA was lost after microwave or oven heating, which caused inactivation of allinase.[86] Therefore, inactivation of allinase may abolish the antitumorigenic and antioxidant activity of garlic by blocking the conversion of alliin into allicin.

To remove unpleasant odor of garlic, odorless aged garlic extract is prepared by aqueous ethanol extraction of fresh garlic at room temperature for as long as 20 months. During this aging process, the content of allicin is considerably reduced because it is unstable and decomposes to form other organosulfur compounds. Water-soluble S-allylcysteine is the major compound in aged garlic extract but not in fresh garlic. S-allylcysteine has been reported to have strong antioxidant activity.[87–90] Administration of S-allylcysteine lowered the incidence of DMBA-induced hamster buccal pouch tumors, implying that this water-soluble compound may exert its chemopreventive effect by modulating lipid peroxidation, potentiating antioxidant activities, and inducing apoptosis.[87,91] However, S-allylcysteine was not effective in protecting against hepatocarcinogenesis and colon carcinogenesis, while S-methylcysteine and cysteine were protective.[92] While S-allylcysteine exhibited no apparent antiproliferative or proapoptotic activities, S-allylmercaptocysteine, a stable organosulfur compound of aged garlic extract, induced apoptosis in human leukemia,[93] colon cancer,[94,95] prostate cancer,[96] and breast cancer[96] cell lines. S-Allylcysteine protected against β-amyloid-induced apoptotic death in rat pheochromocytoma (PC12) cells.[97] A similar protection against β-amyloid-induced neuronal cell death was achieved with aged garlic extract.[97,98]

Garlic oil used for medicinal purposes is generally prepared by steam distillation, and its main allyl compounds consist of diallyl (57%), allyl methyl (37%), and dimethyl (6%) mono- to hexasulfides.[75] Allyl sulfides from garlic are known to contribute to elevation of cellular reduced glutathione, which protects cellular macromolecules (e.g., DNA, protein, membrane lipids) from damage caused by peroxides produced in aerobic metabolism. Garlic oil and its major bioactive constituents, diallyl sulfide, diallyl disulfide, and diallyl trisulfide, exhibited significant antioxidative activities by increasing the activity of glutathione reductase and decreasing the glutathione peroxidase activity in rats.[99,100] Diallyl sulfide has also been shown to possess antitumorigenic potential in various rodent tumor models. In a mouse skin tumor model, it was revealed that diallyl sulfide inhibited papillomagenesis by inducing apoptosis.[101] Diallyl sulfide inhibits the formation of DNA adducts and lipid peroxidation induced by a synthetic estrogen, diethylstilbestrol that is known to cause cancer in humans and animals.[102,103] The compound increased the life span of tumor-bearing mice and also suppressed angiogenesis.[104] Diallyl disulfide was found to inhibit aflatoxin B_1-initiated carcinogenesis in rat liver by modulating the enzymes involved in the metabolism of aflatoxin B_1, including GSTA5 and aflatoxin B_1 aldehyde reductase 1.[105] A similar chemopreventive activity of diallyl disulfide was observed in other chemically induced carcinogenesis animal models.[106,107] Some of these allyl sulfides have been reported to suppress the neoplastic cell transformation and growth through several mechanisms, including induction of apoptosis and regulation of cell-cycle progression.[95,96]

35.2.5.2 Onion

Onion (*Allium cepa* L., Liliaceae), a biennial plant, produces a large bulb in the first year of growth. Due to the high versatility and storability of onion, it has been used as an ingredient in many dishes and accepted by most cultures and races throughout the world.[108] Unlike garlic, which also belongs to allium plants, onion is abundant in flavonoids, representing one of the major sources of flavonoids in the human diet. The principal flavonoid found in a wide variety of onions is quercetin, either in free form or bound to glycosides. There have been numerous reports that quercetin, the major flavonol of onion, has strong antioxidant effects. Onion extract and quercetin showed significant inhibitory activity against lipid peroxidation induced by *tert*-butyl hydroperoxide in rat hepatocytes.[109] The strong antioxidant activity of quercetin appears to be exerted by the presence of orthocatechol group (3',4'-OH) in the B-ring.[110,111] Onion extract and quercetin increased the intracellular concentration of GSH[112,113] partly through up-regulation of the γ-glutamylcysteine synthetase.[113] Quercetin significantly inhibited the 4-hydroxy-2-nonenal-induced reactive oxygen species (ROS) production in rat liver epithelial RL34 cells.[114] DNA damage caused in TPA-stimulated monocytes was significantly attenuated by quercetin.[115] Quercetin also prevented renal dysfunction of rats caused by cyclosporine.[116] It is likely that quercetin, through its antioxidative action, prevented cyclosporine-induced ROS accumulation and, consequently, nephrotoxicity.[116] In this context, it is interesting to note that quercetin enhanced the expression of Mn-SOD localized in the mitochondrial matrix of rat kidneys.[117] Over the past decades, extensive studies have been conducted to evaluate the anticancer or chemopreventive properties of quercetin. Mechanisms underlying antitumorigenic effects of quercetin include down-regulation of the mutant form of p53,[118,119] G1-phase cell-cycle arrest,[120] tyrosine kinase inhibition,[121] induction of type II estrogen receptor,[122] inhibition of heat-shock protein,[123,124] suppression of Ras protein expression,[125] and inhibition of carcinogen-activating enzymes of the CYP1A family.[126,127] In a recent study, quercetin effectively inhibited cooking-oil-fume-induced DNA damage and repressed COX-2 mRNA and protein expression.[128]

Intestinal crypt cell proliferation was inhibited by quercetin *in vivo* at a low concentration.[129] It was demonstrated that quercetin inhibited the proliferation of the Ishiwaka endometrial cancer cell line by down-regulating the expression of epidermal growth factor (EGF) and cyclin D1.[130] It has been suggested that quercetin inhibits cancer cell proliferation by binding to tubulin, thereby resulting in perturbation of microtubule polymerization.[131] Quercetin also inhibited the proliferation of human mammary carcinoma (MCF-7) cells stimulated with environmental estrogens.[132] Quercetin blocked the EGF receptor-signaling pathway in the MiaPaCa-2 pancreatic cancer cell line, leading to the inhibition of cell proliferation and induction of apoptosis.[133] Quercetin inhibited the growth of pancreatic tumor in a nude mouse model, increased apoptosis by mitochondrial depolarization and cytochrome *c* release after caspase-3 activation, and inhibited NF-κB activation.[134] Enhancement of heat-induced inhibitory effects on cell growth was achieved by down-regulation of heat-shock protein 70 (hsp 70). Quercetin antagonized the hsp70 expression after heat treatment and induced apoptosis by increasing subG1 cells with lower levels of hsp70 in JCA-1 and LNcap prostate cancer cell lines.[135] In M14 and MNT1 human melanoma cells, quercetin induced apoptosis and reduced the expression of Hsp 70 at protein and mRNA levels.[136] Quercetin treated to MCF-7 human breast cancer cells blocked cell cycle progression by temporary M phase accumulation and successive G2 arrest and induction of apoptosis.[137] Quercetin also inhibited the invasiveness and migration of multiple myeloma B16-BL6 cells, which is likely to be associated with the cell-cycle arrest as well as the induction of apoptosis by decreasing the Bcl-2 expression.[138]

The association between onion consumption and the incidence of various carcinomas was investigated in the Netherlands Cohort Study. This study revealed that onion consumption is inversely related to the incidence of stomach cancer.[139] However, onion intake failed to reduce the risk of other cancers, such as breast, colon and rectum, and lung carcinomas.[140–142] Results from studies conducted in different provinces in China suggest that the consumption of onion is associated with reduced risk of developing stomach, esophageal, and brain cancers[69,143]

35.2.6 CLOVE

Cloves are the flower buds of the tree *Eugenia caryophyllata* (syn. *Syzygium aromaticum*). *E. caryophyllata* is a tropical tree originating from the Moluccas that is abundantly cultivated in Tanzania, Indonesia, Sri Lanka, and the Malagasy Republic.[144] Clove has been used in traditional Eastern medicine as a vermifuge, an antibacterial agent, as a treatment for toothache.[145] Eugenol (4-allyl-2-methoxyphenol), a terpene compound, is one of the major pharmacologically active compounds in clove essential oil,[144] which is also found in the cortex of *E. caryophyllata*.[145]

Eugenol has been widely used as a flavoring agent for food and cosmetic products and also as a component of zinc oxide eugenol cement in dentistry.[146] Eugenol as an antioxidant can trap peroxy radicals by donating its phenolic hydrogen atom[147] In an ascorbate- or H_2O_2/Fe^{2+}-induced lipid peroxidation system, eugenol significantly inhibited lipid peroxidation, which might be attributable to its free-radical scavenging ability.[148,149] The inhibitory activity of eugenol against the oxidation of low-density lipoproteins in humans was the highest among 13 phenolic compounds derived from essential oils.[150] The GST activity significantly increased in eugenol-treated rats.[151] Likewise, of the aqueous alcoholic extracts of commonly used spices, including garlic, ginger, onion, mint, cloves, cinnamon, and pepper, clove exhibited the highest inhibitory activity against oxidation of linoleic acid by soybean lipoxygenase.[152]

The eugenol-containing methanol extract of the cortex of *E. caryophyllata* was found to strongly inhibit the prostaglandin E_2 production in lipopolysaccharide (LPS)-activated murine macrophage RAW264.7 cells.[145] In addition, eugenol suppressed the expression of COX-2 in LPS-stimulated RAW264.7 cells[145] and HT-29 human colon cancer cells, and inhibited the proliferation of HT-29 cells.[145,153] Dietary eugenol attenuated carrageenan-induced edema in rats.[154] Eugenol inhibited the proliferation of HT-29 cells[145] and exhibited the cytotoxicity by inhibiting the DNA synthesis in both a salivary gland tumor cell line and normal human gingival fibroblasts.[155]

Eugenol has an antimutagenic activity[156–159] and suppressed chemically induced carcinogenesis in rodents, effects that appear to be associated with its induction of GST.[160] Similarly, eugenol and eugenol-rich oil attenuated murine papillomagenesis, possibly through potentiation of xenobiotic detoxification capacity, such as GST, sulfhydryl (-SH), and cytochrome b_5.[161–163] In contrast, eugenol provided minimal protection against DMBA-induced carcinogenesis in Swiss bare mice[164] and even promoted forestomach carcinogenesis in rats.[165] Therefore, additional studies are needed to further verify the chemopreventive effects of eugenol and eugenol-containing plant oils such as clove oil.

35.2.7 ROSEMARY

The dried leaves of rosemary (*Rosemarinus officinalis* L., Labiatae) are commonly used as spices and flavoring agents. Extracts of rosemary and sage have antioxidant effects that retard lipid peroxidation.[166–170] Rosemary also possesses anti-inflammatory and other pharmacological properties that confer therapeutic potential.[171] The ethanol extract of rosemary protected against oxidative DNA damage induced by H_2O_2- and visible light-excited methylene blue in colon cancer CaCo-2 cells and hamster lung V79 cells, possibly by scavenging of both hydroxyl radical and singlet oxygen.[172]

Several phenolic diterpenoids with antioxidant activities were isolated from rosemary leaves. These include carnosol, carnosic acid, rosmanol, isorosmanol, and epirosmanol. Carnosol showed potent antioxidative activity as revealed by scavenging α,α-diphenyl-β-picrylhydrazyl free radicals and protection of oxidative DNA damage.[173] In chronic inflammation, cytokines induce the production of nitric oxide, which is converted to DNA-damaging and carcinogenic peroxynitrite. Peroxynitrite is a cytotoxicant with strong oxidizing properties toward various cellular constituents, including sulfhydryls, lipids, amino acids, and nucleotides and can cause cell death, lipid peroxidation, carcinogenesis, and aging. Rosemary and sage showed the peroxynitrite scavenging effects.[174]

Extracts of *R. officinalis* L. have been reported to inhibit experimental carcinogenesis. Dietary supplementation of rosemary extract significantly reduced the mammary tumor incidence and also inhibited covalent binding of DMBA to mammary epithelial cell DNA.[175] Carnosol appears to be partly responsible for the antitumorigenic activity of rosemary. Intraperitoneal injection of carnosol as well as rosemary extract significantly decreased the number of DMBA-induced adenocarcinomas and the levels of DNA adducts in rat mammary gland compared with controls.[176]

When the rosemary extract was administered (i.g.) at 100 mg/kg/day for 5 consecutive days, the number and area of diethylnitrosamine-induced GST placental-form-positive (GST-P) hepatocellular foci were reduced in male F344 rats.[177] A methanol extract of the leaves of rosemary was evaluated for its effects on tumor initiation and promotion in mouse skin carcinogenesis. Topical application of the rosemary extract to mouse skin attenuated the covalent binding of BaP to epidermal DNA and inhibited tumor initiation by BaP and DMBA. Application of rosemary to mouse skin also inhibited TPA-induced ODC activity, inflammation, hyperplasia, and tumor promotion. Likewise, topical application of carnosol or ursolic acid isolated from rosemary inhibited TPA-induced ear inflammation, ODC activity, and tumor promotion.[178]

In an attempt to elucidate the mechanisms by which rosemary components block initiation of carcinogenesis, human bronchial epithelial cells were treated with the prototype procarcinogen BaP.[179] Whole rosemary extract or an equivalent concentration of its principal antioxidant constituents, carnosol and carnosic acid, inhibited DNA adduct formation by 80% after 6 h coincubation with 1.5 μM BaP. Under similar conditions, rosemary components reduced CYP1A1 mRNA expression by about 50% and inhibited the CYP1A1 activity by 70 to 90%.[179] Therefore, the observed suppression of BaP–DNA adduct formation by rosemary components may mostly result from their inhibition of the activation of BaP to its ultimate carcinogenic metabolites. Carnosol induced expression of GST, which is known to detoxify the ultimate electrophilic carcinogenic metabolite of BaP, including the bay-region diol epoxide. Moreover, expression of another important phase II detoxification enzyme QR was also up-regulated by carnosol in parallel with GST.[179] Therefore, rosemary components have the ability to decrease activation and facilitate detoxification of an important human carcinogen, thus identifying them as bifunctional chemopreventives. Similarly, the QR activity was elevated in a murine hepatoma cell line challenged with rosemary or other spices such as red pepper and basil.[180]

An extract of rosemary, when fed at concentrations of 0.3 and 0.6% for 4 weeks to female A/J mice, markedly enhanced GST and QR activities in the liver and stomach. However, diets supplemented with the rosemary extract failed to affect the activities of the same enzymes in the lung.[181] In another study, the effects of dietary intake and intraperitoneal (i.p.) administration of an extract of rosemary or its constituent carnosol on the activities of hepatic phase II detoxification enzymes were evaluated in female rats. The rosemary extract at concentrations from 0.25 to 1.0% in the diet resulted in a 3.5- to 4.5-fold increase in liver GST activity and a 3.3- to 4.0-fold increase in liver QR activity compared with controls. When the rosemary extract and carnosol were administered i.p., there was a significant increase in the activities of both enzymes in the liver.[182] More recently, the ability of rosemary to modulate CYP and phase II detoxification enzymes in rat liver was assessed by comparing the effects of dried-leaf extracts with different chemical compositions: essential oil containing monoterpenes, a dichloromethane extract containing phenolic diterpenes, and a water-soluble extract containing phenolic compounds such as rosmarinic acid and flavonoids. When male Wistar rats received the leaves or extracts of rosemary in their diet at 0.5% (w/w) for 2 weeks, essential oil selectively induced cytochrome P450s, particularly CYP2B.[183,184] Water-soluble extract enhanced both cytochrome P450s and detoxication enzymes. The dichloromethane extract of rosemary acted as a monofunctional inducer, inducing GST, QR, and UDP-glucuronosyltransferase.

Mace and colleagues[185] have examined the ability of carnosol and carnosic acid from rosemary as well as the synthetic dithiolethione, oltipraz, to block the formation of DNA adducts, and their effects on the expression of phase I and phase II enzymes. It was found that both rosemary extracts and oltipraz inhibited BaP- or aflatoxin B_1-induced DNA adduct formation by efficiently inhibiting CYP activities and inducing the expression of GST.[185] Treatment of female CD-1 mice with a 2% methanol extract of rosemary in AIN-76A diet for 3 weeks increased the liver microsomal 2-hydroxylation of estradiol and estrone by approximately 150%, increased their 6-hydroxylation by approximately 30%, and inhibited the 16α-hydroxylation of estradiol by approximately 50%.[186] The same treatment of rosemary also stimulated the liver microsomal glucuronidation of estradiol and estrone by 54 to 67% and 37 to 56%, respectively. In additional studies, feeding 2% rosemary diet to ovariectomized CD-1 mice for 3 weeks inhibited the uterotropic action of estradiol and estrone by 35 to 50% compared with animals fed a control diet.

P-glycoprotein (Pgp) is the transmembrane transport pump and causes the efflux of chemo-therapeutic agents from cells, which represents an important mechanism in multidrug resistance in tumors. An extract of rosemary was found to increase the intracellular accumulation of commonly used chemotherapeutic agents, such as doxorubicin and vinblastine, in drug-resistant MCF-7 human breast cancer cells that express Pgp.[187] Rosemary extract inhibited the efflux of the above anticancer drugs without affecting accumulation or efflux of doxorubicin in wild-type MCF-7 cells lacking Pgp. Treatment of drug-resistant cells with rosemary extract increased their sensitivity to doxorubicin, which was consistent with an increased intracellular accumulation of the drug. It appears that rosemary directly inhibits the Pgp activity by blocking the binding of drugs to Pgp.

Carnosol induced apoptosis in several acute lymphoblastic leukemia lines as characterized by loss of nuclear DNA, externalization of cell membrane phosphatidylserine, and depolarization of mitochondrial membranes.[188] Carnosol induced a substantial decrease in Bcl-2 in the cell population exhibiting a viable phenotype prior to detectable apoptotic changes in morphology.[188] In another study, carnosic acid with a strong antioxidant activity inhibited proliferation of HL-60 and U937 human myeloid leukemia cells without induction of apoptotic or necrotic cell death.[189] Growth arrest occurred concomitantly with a transient cell-cycle block in the G1 phase, which was accompanied by an increase in the levels of the universal cyclin-dependent kinase inhibitors $p21^{WAF1}$ and $p27^{Kip1}$. Carnosic acid alone caused only a marginal induction of differentiation. However, at low concentrations, this polyphenolic diterpene substantially augmented (100- to 1000-fold) the differentiating effects of 1,25-dihydroxyvitamin D_3 and all-*trans* retinoic acid. Furthermore, such combinations of carnosic acid (and any of these differentiation inducers) synergistically inhibited proliferation and cell-cycle progression.[189]

Intragastric administration of the ethanolic extract of rosemary (0.15 g/100 g body weight) to rats for 3 weeks produced the most pronounced protective effect against carbon tetrachloride-induced hepatotoxicity.[190] Pretreatment of mice for 7 days with the rosemary essential oil (1.1 mg/g body weight) significantly reduced the cyclophosphamide-induced mitodepression in the bone marrow cells. The potential hepatoprotective and antimutagenic activities of the rosemary ethanolic extract and essential oil, respectively, are attributed to the presence of relatively large amounts of phenolic compounds with high antioxidant activity.

35.2.8 MISCELLANEOUS SPICES

When various spices were screened for their ability to potentiate the typical carcinogen-detoxi-fying enzyme GST in Swiss mice, cumin seeds (*Cuminum cyminum* Linn.), basil leaves (*Ocimum sanctum* Linn.), and turmeric increased the enzyme activity by more than 78% in the stomach, liver, and esophagus.[191] GSH levels were also significantly elevated in all three organs by these plant products. These spices also significantly suppressed the chromosome aberrations caused by BaP in mouse bone-marrow cells. In a subsequent study, cumin seeds and basil leaves significantly decreased the incidence of both BaP-induced squamous cell carcinomas in the stomach of Swiss mice and 3′-methyl-4-dimethylaminoazobenzene-induced hepatomas in Wistar rats.[192]

Mace, which is the aril of the fruit of *Myristica fragrans* Houtt, has been used in Indonesian folk medicine as aromatic stomachics, analgesics, a medicine for rheumatism, etc. The methanol extract of mace exhibited an anti-inflammatory activity, as determined by carrageenan-induced edema in rats and acetic acid-induced vascular permeability in mice.[193] The chemopreventive effects of mace on 3-methylcholanthrene-induced cervical carcinogenesis[194] and DMBA-induced papillomagenesis in Swiss albino mice[195] were also reported. The major flavoring compounds in mace and nutmeg are myristicin and safrole.[196,197] Myristicin showed GST-inducing activities in the liver and small-intestinal mucosa of female A/J mice.[160] Reduction of myristicin yielded dihydromyristicin that retained the GST-inducing activity. When myristicin and dihydromyristicin were tested for their capability to inhibit BaP-induced tumor formation in female A/J mice, a 65% inhibition of the pulmonary tumor multiplicity was achieved by myristicin treatment, whereas dihydromyristicin caused little protection. In the forestomach, both myristicin and its hydrogenated analog exhibited antitumorigenic effects to a similar extent. Stimulation of GST activity by myristicin could be a major mechanism for its inhibition of BaP or other carcinogens that may be detoxified in the same manner. Treatment of mouse hepatoma Hepa-1c1c7 (Hepa-1) cells with myristicin increased *cyp1a1* transcription in a concentration-dependent manner.[198] The induction of *cyp1a1* gene expression by myristicin in Hepa-1 cells appears to occur in an Ah receptor-independent manner.

Bioassay-directed fractionation of celery seed oil from the plant *Apium graveolens* (Umbelliferae) led to the isolation of five natural products, including *d*-limonene, *p*-mentha-2,8-dien-1-ol, *p*-mentha-8(9)-en-1,2-diol, 3-*n*-butyl phthalide, and sedanolide, which exhibited high activities to induce the GST in the target tissues of female A/J mice.[199] When these compounds were further tested for their ability to inhibit BaP-induced tumors in the forestomach of mice, 3-*n*-butyl phthalide and sedanolide reduced the tumor incidence significantly. These results suggest that phthalides, as a class of bioactive natural products occurring in edible umbelliferous plants, may be effective chemopreventive agents.

Wasabi (*Eutrema wasabi* Maxim.) inhibited *N*-methylnitrosourea *N*-methyl-*N*′-nitro-*N*-nitrosoguanidine-induced rat gastric carcinogenesis.[200] Pretreatment of A/J mice with 6-methylthiohexyl isothiocyanate (6MHITC) isolated from *Wasabia japonica* (wasabi) for four consecutive days by gavage at a daily dose of 5 µmol significantly inhibited O^6-methylguanine formation in lungs at 4 h after the injection of 4-(methylnitrosamino)-1-(3-pyridyl)-1-butanone (NNK). In conjugation with this inhibitory effect, 6MHITC suppressed the increase in the proliferating cell nuclear antigen level as well as the ODC activity in the promotion stage of NNK-induced lung tumorigenesis. 6MHITC treatment did not lower the incidence of tumors, but it did significantly reduce the tumor multiplicity.[201] 6MHITC at nontoxic concentrations induced the cellular expression of QR as well as its activity in Hepa 1c1c7 cells through activation of antioxidant/electrophile-responsive elements.[202] 6MHITC also exhibited inhibitory effects on human platelet aggregation and GST activity in RL34 cells.[203,204] Similar effects were observed when 6MHITC was administered to rats or mice. The isothiocyanate moiety of 6MHITC is critical for its antiplatelet aggregation and anticancer activities because of its high reactivity with sulfhydryl groups in biomolecules.

The chloroform extract of mint (*Mentha cordifolia* Opiz.) showed antimutagenicity against tetracycline. An antimutagenic component was purified and identified by spectral analyses as 6,7-bis-(2,2-dimethoxyethene)-2,11-dimethoxy-2Z,4E,8E,10Z-dodecatetraendioic acid.[205] Perillyl alcohol, a naturally occurring monoterpene found in cherries and mint, is an inhibitor of farnesyltransferase.[206,207] In a mouse lung tumor bioassay, intraperitoneal injection of perillyl alcohol in 5-week-old male (C3H/HeJ X A/J) F1 hybrid mice prior to initiation with the carcinogen NNK reduced the tumor incidence and multiplicity [208].

ACKNOWLEDGMENT

This research was supported by Biogreen[21] Project, Republic of Korea. The authors wish to thank Ms. Dohee Kim and Mr. Jun-Wan Shin for their assistance in preparing the manuscript and illustration.

REFERENCES

1. Surh, Y.-J., Cancer chemoprevention by dietary phytochemicals, *Nature Rev. Cancer*, 3, 768, 2003.
2. Surh, Y.-J., Anti-tumor promoting potential of selected spice ingredients with antioxidative and anti-inflammatory activities: a short review, *Food Chem. Toxicol.*, 40, 1091, 2002.
3. Surh, Y.-J. et al., Molecular mechanisms underlying chemopreventive activities of anti-inflammatory phytochemicals: down-regulation of COX-2 and iNOS through suppression of NF-κ B activation, *Mutat. Res.*, 480–481, 243, 2001.
4. Surh, Y., Molecular mechanisms of chemopreventive effects of selected dietary and medicinal phenolic substances, *Mutat. Res.*, 428, 305, 1999.
5. Wargovich, M.J., Colon cancer chemoprevention with ginseng and other botanicals, *J. Korean Med. Sci.*, 16 (Suppl.), S81, 2001.
6. Wargovich, M.J. et al., Herbals, cancer prevention and health, *J. Nutr.*, 131, 3034S, 2001.
7. Fenner, L., The spices of life, *FDA Consumer*, 11–15, 1983.
8. Nakatani, N., Phenolic antioxidants from herbs and spices, *Biofactors*, 13, 141, 2000.
9. Rattan, S.I.S., Science behind spices: inhibition of platelet aggregation and prostaglandin synthesis, *Bioessays*, 8, 161, 1988.
10. Srivastava, K.C. and Mustafa, T., Prostaglandins leukotrienes and essential fatty acids, *Prostaglandins Leukot. Essent. Fatty Acids*, 38, 255, 1989.
11. Thapliyal, R. et al., Inhibition of nitrosodiethylamine-induced hepatocarcinogenesis by dietary turmeric in rats, *Toxicol. Lett.*, 139, 45, 2003.
12. Ammon, H.P. and Wahl, M.A., Pharmacology of *Curcuma longa*, *Planta Medica*, 57, 1, 1991.
13. Gescher, A.J., Sharma, R.A., and Steward, W.P., Cancer chemoprevention by dietary constituents: a tale of failure and promise, *Lancet. Oncol.*, 2, 371, 2001.
14. Conney, A.H. et al., Some perspectives on dietary inhibition of carcinogenesis: studies with curcumin and tea, *Proc. Soc. Exp. Biol. Med.*, 216, 234, 1997.
15. Nagabhushan, M. and Bhide, S.V., Curcumin as an inhibitor of cancer, *J. Am. Coll. Nutr.*, 11, 192, 1992.
16. Ciolino, H.P. et al., Effect of curcumin on the aryl hydrocarbon receptor and cytochrome P450 1A1 in MCF-7 human breast carcinoma cells, *Biochem. Pharmacol.*, 56, 197, 1998.
17. Singh, S.V. et al., Mechanism of inhibition of benzo[a]pyrene-induced forestomach cancer in mice by dietary curcumin, *Carcinogenesis*, 19, 1357, 1998.
18. Manson, M.M. et al., Blocking and suppressing mechanisms of chemoprevention by dietary constituents, *Toxicol. Lett.*, 112–113, 499, 2000.
19. Iqbal, M. et al., Dietary supplementation of curcumin enhances antioxidant and phase II metabolizing enzymes in ddY male mice: possible role in protection against chemical carcinogenesis and toxicity, *Pharmacol. Toxicol.*, 92, 33, 2003.
20. Reddy, B.S. and Rao, C.V., Novel approaches for colon cancer prevention by cyclooxygenase-2 inhibitors, *J. Environ. Pathol. Toxicol. Oncol.*, 21, 155, 2002.
21. Bharti, A.C. et al., Curcumin (diferuloylmethane) down-regulates the constitutive activation of nuclear factor-κB and IκBa kinase in human multiple myeloma cells, leading to suppression of proliferation and induction of apoptosis, *Blood*, 101, 1053, 2003.
22. Afzal, M. et al., Ginger: an ethnomedical, chemical and pharmacological review, *Drug Metabol. Drug Interact.*, 18, 159, 2001.
23. Nagasawa, H., Watanabe, K., and Inatomi, H., Effects of bitter melon (*Momordica charantia* l.) or ginger rhizome (*Zingiber officinale* Rosc.) on spontaneous mammary tumorigenesis in SHN mice, *Am. J. Chin. Med.*, 30, 195, 2002.
24. Katiyar, S.K., Agarwal, R., and Mukhtar, H., Inhibition of tumor promotion in SENCAR mouse skin by ethanol extract of *Zingiber officinale* rhizome, *Cancer Res.*, 56, 1023, 1996.

25. Park, K.K. et al., Inhibitory effects of [6]-gingerol, a major pungent principle of ginger, on phorbol ester-induced inflammation, epidermal ornithine decarboxylase activity and skin tumor promotion in ICR mice, *Cancer Lett.*, 129, 139, 1998.

26. Surh, Y.-J., Anti-tumor-promoting activities of selected pungent phenolic substance present in ginger, *J. Environ. Pathol. Toxicol. Oncol.*, 18, 131, 1999.

27. Chung, W.Y. et al., Antioxidative and antitumor promoting effects of [6]-paradol and its homologs, *Mutat. Res.*, 496, 199, 2001.

28. Lee, E. and Surh, Y.J., Induction of apoptosis in HL-60 cells by pungent vanilloids, [6]-gingerol and [6]-paradol, *Cancer Lett.*, 134, 163, 1998.

29. Bode, A.M. et al., Inhibition of epidermal growth factor-induced cell transformation and activator protein 1 activation by [6]-gingerol, *Cancer Res.*, 61, 850, 2001.

30. Yoshimi, N. et al., Modifying effects of fungal and herb metabolites on azoxymethane-induced intestinal carcinogenesis in rats, *Jpn. J. Cancer Res.*, 83, 1273, 1992.

31. Suzuki, F. et al., Keishi-ka-kei-to, a traditional Chinese herbal medicine, inhibits pulmonary metastasis of B16 melanoma, *Anticancer Res.*, 17, 873, 1997.

32. Vimala, S., Norhanom, A.W., and Yadav, M., Anti-tumour promoter activity in Malaysian ginger rhizobia used in traditional medicine, *Br. J. Cancer*, 80, 110, 1999.

33. Murakami, A. et al., Identification of zerumbone in *Zingiber zerumbet* Smith as a potent inhibitor of 12-*O*-tetradecanoylphorbol-13-acetate-induced Epstein-Barr virus activation, *Biosci. Biotechnol. Biochem.*, 63, 1811, 1999.

34. Murakami, A. et al., Zerumbone, a Southeast Asian ginger sesquiterpene, markedly suppresses free radical generation, proinflammatory protein production, and cancer cell proliferation accompanied by apoptosis: the alpha,beta-unsaturated carbonyl group is a prerequisite, *Carcinogenesis*, 23, 795, 2002.

35. Kirana, C. et al., Antitumor activity of extract of *Zingiber aromaticum* and its bioactive sesquiterpenoid zerumbone, *Nutr. Cancer*, 45, 218, 2003.

36. Tanaka, T. et al., Chemoprevention of azoxymethane-induced rat aberrant crypt foci by dietary zerumbone isolated from *Zingiber zerumbet*, *Life Sci.*, 69, 1935, 2001.

37. Lopez-Carrillo, L., Hernandez Avila, M., and Dubrow, R., Chili pepper consumption and gastric cancer in Mexico: a case-control study, *Am. J. Epidemiol.*, 139, 263, 1994.

38. Lopez-Carrillo, L. et al., Capsaicin consumption, *Helicobacter pylori* positivity and gastric cancer in Mexico, *Int. J. Cancer*, 106, 277, 2003.

39. Mathew, A. et al., Diet and stomach cancer: a case-control study in South India, *Eur. J. Cancer Prev.*, 9, 89, 2000.

40. Buiatti, E. et al., A case-control study of gastric cancer and diet in Italy, *Int. J. Cancer*, 44, 611, 1989.

41. Yeoh, K.G. et al., Chili protects against aspirin-induced gastroduodenal mucosal injury in humans, *Dig. Dis. Sci.*, 40, 580, 1995.

42. Kang, J.Y. et al., Effect of capsaicin and chilli on ethanol induced gastric mucosal injury in the rat, *Gut*, 36, 664, 1995.

43. Kang, J.Y., Teng, C.H., and Chen, F.C., Effect of capsaicin and cimetidine on the healing of acetic acid induced gastric ulceration in the rat, *Gut*, 38, 832, 1996.

44. Teng, C.H. et al., Protective action of capsaicin and chilli on haemorrhagic shock-induced gastric mucosal injury in the rat, *J. Gastroenterol. Hepatol.*, 13, 1007, 1998.

45. Gonzalez, R. et al., Effect of capsaicin-containing red pepper sauce suspension on upper gastrointestinal motility in healthy volunteers, *Dig. Dis. Sci.*, 43, 1165, 1998.

46. Park, J.S. et al., Capsaicin protects against ethanol-induced oxidative injury in the gastric mucosa of rats, *Life Sci.*, 67, 3087, 2000.

47. Abdel Salam, O.M., Mozsik, G., and Szolcsanyi, J., Studies on the effect of intragastric capsaicin on gastric ulcer and on the prostacyclin-induced cytoprotection in rats, *Pharmacol. Res.*, 32, 209, 1995.

48. Jones, N.L., Shabib, S., and Sherman, P.M., Capsaicin as an inhibitor of the growth of the gastric pathogen *Helicobacter pylori*, *FEMS Microbiol. Lett.*, 146, 223, 1997.

49. Graham, D.Y., Anderson, S.Y., and Lang, T., Garlic or jalapeno peppers for treatment of *Helicobacter pylori* infection, *Am. J. Gastroenterol.*, 94, 1200, 1999.

50. Sivam, G.P. et al., *Helicobacter pylori*: in vitro susceptibility to garlic (*Allium sativum*) extract, *Nutr. Cancer*, 27, 118, 1997.

51. Han, S.S. et al., Capsaicin suppresses phorbol ester-induced activation of NF-kB/Rel and AP-1 transcription factors in mouse epidermis, *Cancer Lett.*, 164, 119, 2001.

52. Yoshitani, S.I. et al., Chemoprevention of azoxymethane-induced rat colon carcinogenesis by dietary capsaicin and rotenone, *Int. J. Oncol.*, 19, 929, 2001.

53. Surh, Y.J., More than spice: capsaicin in hot chili peppers makes tumor cells commit suicide [comment], *J. Natl. Cancer Inst.*, 94, 1263, 2002.

54. Salomi, M.J., Nair, S.C., and Panikkar, K.R., Inhibitory effects of *Nigella sativa* and saffron (*Crocus sativus*) on chemical carcinogenesis in mice, *Nutr. Cancer*, 16, 67, 1991.

55. Nair, S.C., Pannikar, B., and Panikkar, K.R., Antitumour activity of saffron (*Crocus sativus*), *Cancer Lett.*, 57, 109, 1991.

56. Nair, S.C. et al., Effect of saffron on thymocyte proliferation, intracellular glutathione levels and its antitumor activity, *Biofactors*, 4, 51, 1992.

57. Abdullaev, F.I. and Frenkel, G.D., Effect of saffron on cell colony formation and cellular nucleic acid and protein synthesis, *Biofactors*, 3, 201, 1992.

58. Abdullaev, F.I. and Gonzalez de Mejia, E., Inhibition of colony formation of HeLa cells by naturally occurring and synthetic agents, *Biofactors*, 5, 133, 1995.

59. Abdullaev, F.I. and Frenkel, G.D., The effect of saffron on intracellular DNA, RNA and protein synthesis in malignant and non-malignant human cells, *Biofactors*, 4, 43, 1992.

60. Nair, S.C. et al., Modulatory effects of *Crocus sativus* and *Nigella sativa* extracts on cisplatin-induced toxicity in mice, *J. Ethnopharmacol.*, 31, 75, 1991.

61. Abdullaev, F.I., Cancer chemopreventive and tumoricidal properties of saffron (*Crocus sativus* L.), *Exp. Biol. Med.(Maywood)*, 227, 20, 2002.

62. Abdullaev, F.I., Inhibitory effect of crocetin on intracellular nucleic acid and protein synthesis in malignant cells, *Toxicol. Lett.*, 70, 243, 1994.

63. Escribano, J. et al., Crocin, safranal and picrocrocin from saffron (*Crocus sativus* L.) inhibit the growth of human cancer cells in vitro, *Cancer Lett.*, 100, 23, 1996.

64. Tarantilis, P.A. et al., Inhibition of growth and induction of differentiation of promyelocytic leukemia (HL-60) by carotenoids from *Crocus sativus* L., *Anticancer Res.*, 14, 1913, 1994.

65. Garcia-Olmo, D.C. et al., Effects of long-term treatment of colon adenocarcinoma with crocin, a carotenoid from saffron (*Crocus sativus* L.): an experimental study in the rat, *Nutr. Cancer*, 35, 120, 1999.

66. Chang, W.C. et al., Inhibitory effect of crocetin on benzo(a)pyrene genotoxicity and neoplastic transformation in C3H10T1/2 cells, *Anticancer Res.*, 16, 3603, 1996.

67. Wang, C.J. et al., Inhibition of tumor promotion in benzo[a]pyrene-initiated CD-1 mouse skin by crocetin, *Carcinogenesis*, 16, 187, 1995.

68. Bianchini, F. and Vainio, H., Allium vegetables and organosulfur compounds: do they help prevent cancer? *Environ. Health Perspect.*, 109, 893, 2001.

69. Gao, C.M. et al., Protective effect of allium vegetables against both esophageal and stomach cancer: a simultaneous case-referent study of a high-epidemic area in Jiangsu Province, China, *Jpn. J. Cancer Res.*, 90, 614, 1999.

70. Hsing, A.W. et al., Allium vegetables and risk of prostate cancer: a population-based study, *J. Natl. Cancer Inst.*, 94, 1648, 2002.

71. Brace, L.D., Cardiovascular benefits of garlic (*Allium sativum* L.), *J. Cardiovasc. Nurs.*, 16, 33, 2002.

72. Rahman, K., Garlic and aging: new insights into an old remedy, *Ageing Res. Rev.*, 2, 39, 2003.

73. Durak, I. et al., Effects of garlic extract on oxidant/antioxidant status and atherosclerotic plaque formation in rabbit aorta, *Nutr. Metab. Cardiovasc. Dis.*, 12, 141, 2002.

74. Thomson, M. and Ali, M., Garlic (*Allium sativum*): a review of its potential use as an anti-cancer agent, *Curr. Cancer Drug Targets*, 3, 67, 2003.

75. Banerjee, S.K., Mukherjee, P.K., and Maulik, S.K., Garlic as an antioxidant: the good, the bad and the ugly, *Phytother. Res.*, 17, 97, 2003.

76. Borek, C., Antioxidant health effects of aged garlic extract, *J. Nutr.*, 131, 1010S, 2001.

77. Das, S., Garlic: a natural source of cancer preventive compounds, *Asian Pac. J. Cancer Prev.*, 3, 305, 2002.

78. Rabinkov, A. et al., The mode of action of allicin: trapping of radicals and interaction with thiol containing proteins, *Biochim. Biophys. Acta*, 1379, 233, 1998.

79. Kourounakis, P.N. and Rekka, E.A., Effect on active oxygen species of alliin and *Allium sativum* (garlic) powder, *Res. Commun. Chem. Pathol. Pharmacol.*, 74, 249, 1991.

80. Phelps, S. and Harris, W.S., Garlic supplementation and lipoprotein oxidation susceptibility, *Lipids*, 28, 475, 1993.

81. Sengupta, A., Ghosh, S., and Das, S., Tomato and garlic can modulate azoxymethane-induced colon carcinogenesis in rats, *Eur. J. Cancer Prev.*, 12, 195, 2003.

82. Hirsch, K. et al., Effect of purified allicin, the major ingredient of freshly crushed garlic, on cancer cell proliferation, *Nutr. Cancer*, 38, 245, 2000.

83. Siegers, C.P. et al., The effects of garlic preparations against human tumor cell proliferation, *Phytomedicine*, 6, 7, 1999.

84. Zheng, S. et al., Initial study on naturally occurring products from traditional Chinese herbs and vegetables for chemoprevention, *J. Cell Biochem. Suppl.*, 27, 106, 1997.

85. Balasenthil, S., Rao, K.S., and Nagini, S., Altered cytokeratin expression during chemoprevention of experimental hamster buccal pouch carcinogenesis by garlic, *J. Oral Pathol. Med.*, 31, 142, 2002.

86. Song, K. and Milner, J.A., The influence of heating on the anticancer properties of garlic, *J. Nutr.*, 131, 1054S, 2001.

87. Balasenthil, S., Ramachandran, C.R., and Nagini, S., S-allylcysteine, a garlic constituent, inhibits 7,12-dimethylbenz[a]anthracene-induced hamster buccal pouch carcinogenesis, *Nutr. Cancer*, 40, 165, 2001.

88. Ide, N. and Lau, B.H., S-allylcysteine attenuates oxidative stress in endothelial cells, *Drug Dev. Ind. Pharm.*, 25, 619, 1999.

89. Kim, K.M. et al., Differential regulation of NO availability from macrophages and endothelial cells by the garlic component S-allyl cysteine, *Free Radical Biol. Med.*, 30, 747, 2001.

90. Wei, B.L. et al., In vitro anti-inflammatory effects of quercetin 3-O-methyl ether and other constituents from Rhamnus species, *Planta Medica*, 67, 745, 2001.

91. Balasenthil, S., Rao, K.S., and Nagini, S., Apoptosis induction by S-allylcysteine, a garlic constituent, during 7,12-dimethylbenz[a]anthracene-induced hamster buccal pouch carcinogenesis, *Cell Biochem. Funct.*, 20, 263, 2002.

92. Fukushima, S. et al., Suppression of chemical carcinogenesis by water-soluble organosulfur compounds, *J. Nutr.*, 131, 1049S, 2001.

93. Sigounas, G. et al., S-allylmercaptocysteine, a stable thioallyl compound, induces apoptosis in erythroleukemia cell lines, *Nutr. Cancer*, 28, 153, 1997.

94. Shirin, H. et al., Antiproliferative effects of S-allylmercaptocysteine on colon cancer cells when tested alone or in combination with sulindac sulfide, *Cancer Res.*, 61, 725, 2001.

95. Knowles, L.M. and Milner, J.A., Possible mechanism by which allyl sulfides suppress neoplastic cell proliferation, *J. Nutr.*, 131, 1061S, 2001.

96. Pinto, J.T. and Rivlin, R.S., Antiproliferative effects of allium derivatives from garlic, *J. Nutr.*, 131, 1058S, 2001.

97. Peng, Q., Buz'Zard, A.R., and Lau, B.H., Neuroprotective effect of garlic compounds in amyloid-beta peptide-induced apoptosis *in vitro*, *Med. Sci. Monit.*, 8, BR328, 2002.

98. Jackson, R. et al., Effect of aged garlic extract on caspase-3 activity, *in vitro*, *Nutr. Neurosci.*, 5, 287, 2002.

99. Shen, S.C. et al., *In vitro* and *in vivo* inhibitory activities of rutin, wogonin, and quercetin on lipopolysaccharide-induced nitric oxide and prostaglandin E_2 production, *Eur. J. Pharmacol.*, 446, 187, 2002.

100. Wu, C.C. et al., Effects of organosulfur compounds from garlic oil on the antioxidation system in rat liver and red blood cells, *Food Chem. Toxicol.*, 39, 563, 2001.

101. Arora, A. and Shukla, Y., Induction of apoptosis by diallyl sulfide in DMBA-induced mouse skin tumors, *Nutr. Cancer*, 44, 89, 2002.

102. Green, M. et al., Inhibition of DES-induced DNA adducts by diallyl sulfide: implications in liver cancer prevention, *Oncol. Rep.*, 10, 767, 2003.

103. Gued, L.R., Thomas, R.D., and Green, M., Diallyl sulfide inhibits diethylstilbestrol-induced lipid peroxidation in breast tissue of female ACI rats: implications in breast cancer prevention, *Oncol. Rep.*, 10, 739, 2003.

104. Shukla, Y., Arora, A., and Singh, A., Antitumorigenic potential of diallyl sulfide in Ehrlich ascites tumor bearing mice, *Biomed. Environ. Sci.*, 15, 41, 2002.

105. Guyonnet, D. et al., Mechanisms of protection against aflatoxin B_1 genotoxicity in rats treated by organosulfur compounds from garlic, *Carcinogenesis*, 23, 1335, 2002.

106. Guyonnet, D. et al., Liver subcellular fractions from rats treated by organosulfur compounds from Allium modulate mutagen activation, *Mutat. Res.*, 466, 17, 2000.

107. Guyonnet, D. et al., Antimutagenic activity of organosulfur compounds from Allium is associated with phase II enzyme induction, *Mutat. Res.*, 495, 135, 2001.

108. Griffiths, G. et al., Onions: a global benefit to health, *Phytother. Res.*, 16, 603, 2002.
109. Glasser, G. et al., Comparison of antioxidative capacities and inhibitory effects on cholesterol biosynthesis of quercetin and potential metabolites, *Phytomedicine*, 9, 33, 2002.
110. Mira, L. et al., Interactions of flavonoids with iron and copper ions: a mechanism for their antioxidant activity, *Free Radic. Res.*, 36, 1199, 2002.
111. Silva, M.M. et al., Structure-antioxidant activity relationships of flavonoids: a re-examination, *Free Radic. Res.*, 36, 1219, 2002.
112. Fiorani, M. et al., Quercetin prevents glutathione depletion induced by dehydroascorbic acid in rabbit red blood cells, *Free Radical Res.*, 34, 639, 2001.
113. Myhrstad, M.C. et al., Flavonoids increase the intracellular glutathione level by transactivation of the γ-glutamylcysteine synthetase catalytical subunit promoter, *Free Radical Biol. Med.*, 32, 386, 2002.
114. Feng, Q. et al., Anticarcinogenic antioxidants as inhibitors against intracellular oxidative stress, *Free Radic. Res.*, 35, 779, 2001.
115. Fabiani, R. et al., Antioxidants prevent the lymphocyte DNA damage induced by PMA-stimulated monocytes, *Nutr. Cancer*, 39, 284, 2001.
116. Satyanarayana, P.S., Singh, D., and Chopra, K., Quercetin, a bioflavonoid, protects against oxidative stress-related renal dysfunction by cyclosporine in rats, *Methods Find Exp. Clin. Pharmacol.*, 23, 175, 2001.
117. Shahed, A.R., Jones, E., and Shoskes, D., Quercetin and curcumin up-regulate antioxidant gene expression in rat kidney after ureteral obstruction or ischemia/reperfusion injury, *Transplant Proc.*, 33, 2988, 2001.
118. Avila, M.A. et al., Quercetin mediates the down-regulation of mutant p53 in the human breast cancer cell line MDA-MB468, *Cancer Res.*, 54, 2424, 1994.
119. Avila, M.A. et al., Quercetin as a modulator of the cellular neoplastic phenotype: effects on the expression of mutated H-ras and p53 in rodent and human cells, *Adv. Exp. Med. Biol.*, 401, 101, 1996.
120. Yoshida, M., Yamamoto, M., and Nikaido, T., Quercetin arrests human leukemic T-cells in late G1 phase of the cell cycle, *Cancer Res.*, 52, 6676, 1992.
121. Yokoo, T. and Kitamura, M., Unexpected protection of glomerular mesangial cells from oxidant-triggered apoptosis by bioflavonoid quercetin, *Am. J. Physiol.*, 273, F206, 1997.
122. Scambia, G. et al., Quercetin induces type-II estrogen-binding sites in estrogen-receptor-negative (MDA-MB231) and estrogen-receptor-positive (MCF-7) human breast-cancer cell lines, *Int. J. Cancer*, 54, 462, 1993.
123. Hansen, R.K. et al., Quercetin inhibits heat shock protein induction but not heat shock factor DNA-binding in human breast carcinoma cells, *Biochem. Biophys. Res. Commun.*, 239, 851, 1997.
124. Koishi, M. et al., Quercetin, an inhibitor of heat shock protein synthesis, inhibits the acquisition of thermotolerance in a human colon carcinoma cell line, *Jpn. J. Cancer Res.*, 83, 1216, 1992.
125. Ranelletti, F.O. et al., Quercetin inhibits p21-RAS expression in human colon cancer cell lines and in primary colorectal tumors, *Int. J. Cancer*, 85, 438, 2000.
126. Ciolino, H.P. and Yeh, G.C., Inhibition of aryl hydrocarbon-induced cytochrome P-450 1A1 enzyme activity and CYP1A1 expression by resveratrol, *Mol. Pharmacol.*, 56, 760, 1999.
127. Kang, Z.C., Tsai, S.J., and Lee, H., Quercetin inhibits benzo[a]pyrene-induced DNA adducts in human Hep G2 cells by altering cytochrome P-450 1A1 gene expression, *Nutr. Cancer*, 35, 175, 1999.
128. Lin, S.Y. et al., Protection by quercetin against cooking oil fumes-induced DNA damage in human lung adenocarcinoma CL-3 cells: role of COX-2, *Nutr. Cancer*, 44, 95, 2002.
129. Gee, J.M., Hara, H., and Johnson, I.T., Suppression of intestinal crypt cell proliferation and aberrant crypt foci by dietary quercetin in rats, *Nutr. Cancer*, 43, 193, 2002.
130. Kaneuchi, M. et al., Quercetin regulates growth of Ishikawa cells through the suppression of EGF and cyclin D1, *Int. J. Oncol.*, 22, 159, 2003.
131. Gupta, K. and Panda, D., Perturbation of microtubule polymerization by quercetin through tubulin binding: a novel mechanism of its antiproliferative activity, *Biochemistry*, 41, 13029, 2002.
132. Han, D.H. et al., Relationship between estrogen receptor-binding and estrogenic activities of environmental estrogens and suppression by flavonoids, *Biosci. Biotechnol. Biochem.*, 66, 1479, 2002.
133. Lee, L.T. et al., Blockade of the epidermal growth factor receptor tyrosine kinase activity by quercetin and luteolin leads to growth inhibition and apoptosis of pancreatic tumor cells, *Anticancer Res.*, 22, 1615, 2002.
134. Mouria, M. et al., Food-derived polyphenols inhibit pancreatic cancer growth through mitochondrial cytochrome C release and apoptosis, *Int. J. Cancer*, 98, 761, 2002.

135. Nakanoma, T. et al., Effects of quercetin on the heat-induced cytotoxicity of prostate cancer cells, *Int. J. Urol.*, 8, 623, 2001.

136. Piantelli, M. et al., Quercetin and tamoxifen sensitize human melanoma cells to hyperthermia, *Melanoma Res.*, 11, 469, 2001.

137. Choi, J.A. et al., Induction of cell cycle arrest and apoptosis in human breast cancer cells by quercetin, *Int. J. Oncol.*, 19, 837, 2001.

138. Zhang, X., Xu, Q., and Saiki, I., Quercetin inhibits the invasion and mobility of murine melanoma B16-BL6 cells through inducing apoptosis via decreasing Bcl-2 expression, *Clin. Exp. Metastasis*, 18, 415, 2000.

139. Dorant, E. et al., Consumption of onions and a reduced risk of stomach carcinoma, *Gastroenterology*, 110, 12, 1996.

140. Dorant, E., van den Brandt, P.A., and Goldbohm, R.A., A prospective cohort study on the relationship between onion and leek consumption, garlic supplement use and the risk of colorectal carcinoma in The Netherlands, *Carcinogenesis*, 17, 477, 1996.

141. Dorant, E., van den Brandt, P.A., and Goldbohm, R.A., Allium vegetable consumption, garlic supplement intake, and female breast carcinoma incidence, *Breast Cancer Res. Treat.*, 33, 163, 1995.

142. Dorant, E., van den Brandt, P.A., and Goldbohm, R.A., A prospective cohort study on Allium vegetable consumption, garlic supplement use, and the risk of lung carcinoma in The Netherlands, *Cancer Res.*, 54, 6148, 1994.

143. Hu, J. et al., Diet and brain cancer in adults: a case-control study in northeast China, *Int. J. Cancer*, 81, 20, 1999.

144. Zheng, G.Q., Kenney, P.M., and Lam, L.K., Sesquiterpenes from clove (*Eugenia caryophyllata*) as potential anticarcinogenic agents, *J. Nat. Prod.*, 55, 999, 1992.

145. Kim, S.S. et al., Eugenol suppresses cyclooxygenase-2 expression in lipopolysaccharide-stimulated mouse macrophage RAW264.7 cells, *Life Sci.*, 73, 337, 2003.

146. Huang, T.H., Lee, H., and Kao, C.T., Evaluation of the genotoxicity of zinc oxide eugenol-based, calcium hydroxide-based, and epoxy resin-based root canal sealers by comet assay, *J. Endod.*, 27, 744, 2001.

147. Fujisawa, S. et al., Antioxidant and prooxidant action of eugenol-related compounds and their cytotoxicity, *Toxicology*, 177, 39, 2002.

148. Reddy, A.C. and Lokesh, B.R., Studies on spice principles as antioxidants in the inhibition of lipid peroxidation of rat liver microsomes, *Mol. Cell. Biochem.*, 111, 117, 1992.

149. Nagababu, E. and Lakshmaiah, N., Inhibitory effect of eugenol on non-enzymatic lipid peroxidation in rat liver mitochondria, *Biochem. Pharmacol.*, 43, 2393, 1992.

150. Teissedre, P.L. and Waterhouse, A.L., Inhibition of oxidation of human low-density lipoproteins by phenolic substances in different essential oils varieties, *J. Agric. Food Chem.*, 48, 3801, 2000.

151. Vidhya, N. and Devaraj, S.N., Antioxidant effect of eugenol in rat intestine, *Indian J. Exp. Biol.*, 37, 1192, 1999.

152. Shobana, S. and Naidu, K.A., Antioxidant activity of selected Indian spices, *Prostaglandins Leukot. Essent. Fatty Acids*, 62, 107, 2000.

153. Hong, C.H. et al., Evaluation of natural products on inhibition of inducible cyclooxygenase (COX-2) and nitric oxide synthase (iNOS) in cultured mouse macrophage cells, *J. Ethnopharmacol.*, 83, 153, 2002.

154. Reddy, A.C. and Lokesh, B.R., Studies on anti-inflammatory activity of spice principles and dietary n-3 polyunsaturated fatty acids on carrageenan-induced inflammation in rats, *Ann. Nutr. Metab.*, 38, 349, 1994.

155. Atsumi, T. et al., Cytotoxicity and radical intensity of eugenol, isoeugenol or related dimers, *Anticancer Res.*, 20, 2519, 2000.

156. Rompelberg, C.J. et al., Effect of eugenol on the mutagenicity of benzo[a]pyrene and the formation of benzo[a]pyrene-DNA adducts in the lambda-lacZ-transgenic mouse, *Mutat. Res.*, 369, 87, 1996.

157. Rompelberg, C.J. et al., Antimutagenicity of eugenol in the rodent bone marrow micronucleus test, *Mutat. Res.*, 346, 69, 1995.

158. Sukumaran, K. and Kuttan, R., Inhibition of tobacco-induced mutagenesis by eugenol and plant extracts, *Mutat. Res.*, 343, 25, 1995.

159. Yokota, H., Hoshino, J., and Yuasa, A., Suppressed mutagenicity of benzo[a]pyrene by the liver S9 fraction and microsomes from eugenol-treated rats, *Mutat. Res.*, 172, 231, 1986.

160. Zheng, G.Q. et al., Inhibition of benzo[a]pyrene-induced tumorigenesis by myristicin, a volatile aroma constituent of parsley leaf oil, *Carcinogenesis*, 13, 1921, 1992.

161. Bhide, S.V. et al., Chemopreventive efficacy of a betel leaf extract against benzo[a]pyrene-induced forestomach tumors in mice, *J. Ethnopharmacol.*, 34, 207, 1991.

162. Singh, A., Singh, S.P., and Bamezai, R., Modulatory potential of clocimum oil on mouse skin papillomagenesis and the xenobiotic detoxication system, *Food Chem. Toxicol.*, 37, 663, 1999.

163. Sukumaran, K., Unnikrishnan, M.C., and Kuttan, R., Inhibition of tumour promotion in mice by eugenol, *Indian J. Physiol. Pharmacol.*, 38, 306, 1994.

164. Azuine, M.A., Amonkar, A.J., and Bhide, S.V., Chemopreventive efficacy of betel leaf extract and its constituents on 7,12-dimethylbenz(a)anthracene induced carcinogenesis and their effect on drug detoxification system in mouse skin, *Indian J. Exp. Biol.*, 29, 346, 1991.

165. Imaida, K. et al., Effects of naturally occurring antioxidants on combined 1,2-dimethylhydrazine- and 1-methyl-1-nitrosourea-initiated carcinogenesis in F344 male rats, *Cancer Lett.*, 55, 53, 1990.

166. Ho, C.-T. et al., Phytochemicals in tea and rosemary and their cancer-preventive properties, in Food Phytochemicals for Cancer Prevention II, Ho, C.-T., Osawa, T., Huang, M.-T., and Rosen, R.T., eds., American Chemical Society, Washington, DC, 1992, p. 2.

167. Nakatani, N., Natural antioxidants from spices, in *Phenolic Compounds in Food and Their Effects on Health II: Antioxidants and Cancer Prevention*, Huang, M.-T., Ho, C.-T., and Lee, C.Y., Eds., American Chemical Society, Washington, DC, 1992.

168. Aruoma, O.I. et al., An evaluation of the antioxidant and antiviral action of extracts of rosemary and Provencal herbs, *Food Chem. Toxicol.*, 34, 449, 1996.

169. Ho, C.T. et al., Chemistry and antioxidative factors in rosemary and sage, *Biofactors*, 13, 161, 2000.

170. Leal, P.F. et al., Functional properties of spice extracts obtained via supercritical fluid extraction, *J. Agric. Food Chem.*, 51, 2520, 2003.

171. al-Sereiti, M.R., Abu-Amer, K.M., and Sen, P., Pharmacology of rosemary (*Rosmarinus officinalis* Linn.) and its therapeutic potentials, *Indian J. Exp. Biol.*, 37, 124, 1999.

172. Slamenova, D. et al., Rosemary-stimulated reduction of DNA strand breaks and FPG-sensitive sites in mammalian cells treated with H_2O_2 or visible light-excited methylene blue, *Cancer Lett.*, 177, 145, 2002.

173. Lo, A.H. et al., Carnosol, an antioxidant in rosemary, suppresses inducible nitric oxide synthase through down-regulating nuclear factor-κB in mouse macrophages, *Carcinogenesis*, 23, 983, 2002.

174. Choi, H.R. et al., Peroxynitrite scavenging activity of herb extracts, *Phytother. Res.*, 16, 364, 2002.

175. Singletary, K.W. and Nelshoppen, J.M., Inhibition of 7,12-dimethylbenz[a]anthracene (DMBA)-induced mammary tumorigenesis and of *in vivo* formation of mammary DMBA-DNA adducts by rosemary extract, *Cancer Lett.*, 60, 169, 1991.

176. Singletary, K., MacDonald, C., and Wallig, M., Inhibition by rosemary and carnosol of 7,12-dimethylbenz[a]anthracene (DMBA)-induced rat mammary tumorigenesis and *in vivo* DMBA-DNA adduct formation, *Cancer Lett.*, 104, 43, 1996.

177. Kitano, M. et al., Chemopreventive effects of coumaperine from pepper on the initiation stage of chemical hepatocarcinogenesis in the rat, *Jpn. J. Cancer Res.*, 91, 674, 2000.

178. Huang, M.T. et al., Inhibition of skin tumorigenesis by rosemary and its constituents carnosol and ursolic acid, *Cancer Res.*, 54, 701, 1994.

179. Offord, E.A. et al., Rosemary components inhibit benzo[a]pyrene-induced genotoxicity in human bronchial cells, *Carcinogenesis*, 16, 2057, 1995.

180. Tawfiq, N. et al., Induction of the anti-carcinogenic enzyme quinone reductase by food extracts using murine hepatoma cells, *Eur. J. Cancer Prev.*, 3, 285, 1994.

181. Singletary, K.W. and Rokusek, J.T., Tissue-specific enhancement of xenobiotic detoxification enzymes in mice by dietary rosemary extract, *Plant Foods. Hum. Nutr.*, 50, 47, 1997.

182. Singletary, K.W., Rosemary extract and carnosol stimulate rat liver glutathione-S-transferase and quinone reductase activities, *Cancer Lett.*, 100, 139, 1996.

183. Debersac, P. et al., Effects of a water-soluble extract of rosemary and its purified component rosmarinic acid on xenobiotic-metabolizing enzymes in rat liver, *Food Chem. Toxicol.*, 39, 109, 2001.

184. Debersac, P. et al., Induction of cytochrome P450 and/or detoxication enzymes by various extracts of rosemary: description of specific patterns, *Food Chem. Toxicol.*, 39, 907, 2001.

185. Mace, K. et al., Development of *in vitro* models for cellular and molecular studies in toxicology and chemoprevention, *Arch. Toxicol. Suppl.*, 20, 227, 1998.

186. Zhu, B.T. et al., Dietary administration of an extract from rosemary leaves enhances the liver microsomal metabolism of endogenous estrogens and decreases their uterotropic action in CD-1 mice, *Carcinogenesis*, 19, 1821, 1998.

187. Plouzek, C.A. et al., Inhibition of P-glycoprotein activity and reversal of multidrug resistance *in vitro* by rosemary extract, *Eur. J. Cancer*, 35, 1541, 1999.

188. Dorrie, J., Sapala, K., and Zunino, S.J., Carnosol-induced apoptosis and downregulation of Bcl-2 in B-lineage leukemia cells, *Cancer Lett.*, 170, 33, 2001.

189. Steiner, M. et al., Carnosic acid inhibits proliferation and augments differentiation of human leukemic cells induced by 1,25-dihydroxyvitamin D3 and retinoic acid, *Nutr. Cancer*, 41, 135, 2001.

190. Fahim, F.A. et al., Allied studies on the effect of *Rosmarinus officinalis* L. on experimental hepatotoxicity and mutagenesis, *Int. J. Food Sci. Nutr.*, 50, 413, 1999.

191. Aruna, K. and Sivaramakrishnan, V.M., Plant products as protective agents against cancer, *Indian J. Exp. Biol.*, 28, 1008, 1990.

192. Aruna, K. and Sivaramakrishnan, V.M., Anticarcinogenic effects of some Indian plant products, *Food Chem. Toxicol.*, 30, 953, 1992.

193. Ozaki, Y. et al., Antiinflammatory effect of mace, aril of *Myristica fragrans* Houtt., and its active principles, *Jpn. J. Pharmacol.*, 49, 155, 1989.

194. Hussain, S.P. and Rao, A.R., Chemopreventive action of mace (*Myristica fragrans*, Houtt.) on methylcholanthrene-induced carcinogenesis in the uterine cervix in mice, *Cancer Lett.*, 56, 231, 1991.

195. Jannu, L.N., Hussain, S.P., and Rao, A.R., Chemopreventive action of mace (*Myristica fragrans*, Houtt.) on DMBA-induced papillomagenesis in the skin of mice, *Cancer Lett.*, 56, 59, 1991.

196. Fisher, C., Phenolic compounds in spices, in *Phenolic Compounds in Food and Their Effects on Health I: Analysis, Occurrence, and Chemistry*, Ho, C.-T., Lee, C.Y., and Huang, M.-T., Eds., American Chemical Society, Washington, DC, 1992.

197. Archer, A.W., Determination of safrole and myristicin in nutmeg and mace by high-performance liquid chromatography, *J. Chromatogr.*, 438, 117, 1988.

198. Jeong, H.G. et al., Murine Cyp1a-1 induction in mouse hepatoma Hepa-1C1C7 cells by myristicin, *Biochem. Biophys. Res. Commun.*, 233, 619, 1997.

199. Zheng, G.Q. et al., Chemoprevention of benzo[a]pyrene-induced forestomach cancer in mice by natural phthalides from celery seed oil, *Nutr. Cancer*, 19, 77, 1993.

200. Tanida, N. et al., Suppressive effect of wasabi (pungent Japanese spice) on gastric carcinogenesis induced by MNNG in rats, *Nutr. Cancer*, 16, 53, 1991.

201. Yano, T. et al., The effect of 6-methylthiohexyl isothiocyanate isolated from Wasabia japonica (wasabi) on 4-(methylnitrosamino)-1-(3-pyridyl)-1-buatnone-induced lung tumorigenesis in mice, *Cancer Lett.*, 155, 115, 2000.

202. Hou, D.X. et al., Transcriptional regulation of nicotinamide adenine dinucleotide phosphate: quinone oxidoreductase in murine hepatoma cells by 6-(methylsulfinyl)hexyl isothiocyanate, an active principle of wasabi (*Eutrema wasabi* Maxim.), *Cancer Lett.*, 161, 195, 2000.

203. Morimitsu, Y. et al., Antiplatelet and anticancer isothiocyanates in Japanese domestic horseradish, wasabi, *Mech. Ageing Dev.*, 116, 125, 2000.

204. Morimitsu, Y. et al., Antiplatelet and anticancer isothiocyanates in Japanese domestic horseradish, wasabi, *Biofactors*, 13, 271, 2000.

205. Villasenor, I.M., Echegoyen, D.E., and Angelada, J.S., A new antimutagen from *Mentha cordifolia* Opiz., *Mutat. Res.*, 515, 141, 2002.

206. Lantry, L.E. et al., Chemopreventive efficacy of promising farnesyltransferase inhibitors, *Exp. Lung Res.*, 26, 773, 2000.

207. Karlson, J. et al., Inhibition of tumor cell growth by monoterpenes *in vitro*: evidence of a Ras-independent mechanism of action, *Anticancer Drugs*, 7, 422, 1996.

208. Lantry, L.E. et al., Chemopreventive effect of perillyl alcohol on 4-(methylnitrosamino)-1-(3-pyridyl)-1-butanone induced tumorigenesis in (C3H/HeJ X A/J)F1 mouse lung, *J. Cell Biochem. Suppl.*, 27, 20, 1997.

36 Coenzyme Q_{10} and Neoplasia: Overview of Experimental and Clinical Evidence*

Emile G. Bliznakov, Raj K. Chopra and Hemmi N. Bhagavan

Illness is nothing else than a break of body's harmony that has the endogenous tendency to reconstitute itself, an endeavor that the physician can only support by appropriate measures.

Hippocrates, 460–377 BC

CONTENTS

36.1 The Integration of Cancer, Immune System, Aging, and Coenzyme Q_{10}..........................599
    36.1.1 The View from Afar 599
    36.1.2 Coenzyme Q, Mitochondria, and Bioenergetics.....................601
    36.1.3 Tumor–Host Interplay 608
        36.1.3.1 General Setting..608
        36.1.3.2 Coenzyme Q_{10}: Animal Studies 609
        36.1.3.3 Coenzyme Q_{10}: Clinical Studies 610
    36.1.4 The Burden of Aging 612
        36.1.4.1 Pandemic Relevance..612
        36.1.4.2 Coenzyme Q Implications 613
36.2 Interference between Coenzyme Q Biosynthesis and Drugs
Commonly Used in Clinical Practice 615
36.3 Coenzyme Q_{10}: Safety Considerations...617
36.4 Reflections and Concluding Thoughts 618
References 618

36.1 THE INTEGRATION OF CANCER, IMMUNE SYSTEM, AGING, AND COENZYME Q_{10}

36.1.1 THE VIEW FROM AFAR

Over millions of years of evolution, organisms have developed immune systems. The human immune system has evolved from a primitive state to a highly sophisticated, very complex, efficient, interrelated host-defense system. Immunological thinking was and still is considerably influenced by the phenomenon of protective immunity to infections. This feature was already well known to the ancient Greeks and has been impressively described by Thucydides in his report on the casualties during the Peloponnesian war (431–406 B.C.).

---

* Dedicated to the memory of Dr. Bliznakov who passed away September 2003

The idea that the immune responses are the body's principal defense mechanisms has deeply influenced cancer research during the past decades. This idea, although suggested by Paul Ehrlich in 1909, remained dormant until it was reformulated and crystallized as a general theory of "immunologic surveillance" by Frank McFarlan Burnet in 1959. An extensive review of the earlier publications on this subject was compiled by Stutman in 1975.[1] The main contention of immunologic surveillance is that the immune functions have evolved, in part, as a mechanism to prevent the emergence of cancer cells. Burnet[2] regarded neoplasia as "not self," and postulated that a major function of the immune system is to seek out and destroy new cancer cells as they arise. This revolutionary view of cancer as "nonself" gained wide acceptance, and substantial effort has gone into the development of practical applications. In a way, as suggested by Stutman,[1] an established advanced tumor would represent an actual failure of the immune surveillance mechanism. Furthermore, individuals with immune disorders have an increased likelihood of being diagnosed with cancer, and those with congenital immunodeficiencies develop cancer at 200 times the expected prevalence.[3]

The current definition of the immune system in general and immune surveillance in particular is, as expected, subject to persuasive genuine as well as fictitious criticism, based on biomedical, philosophical, and theological rationale, despite the vast practical accomplishments. A thought-provoking and an intriguing personal perception of several "classical" concepts ruling contemporary immunology is debated in a recently published review by Thiele,[4] where the "accepted" definitions of self and nonself discrimination, immune surveillance, the network theory, and others are discussed and criticized. In an earlier review, Fuchs and Matzinger[5] proposed an alternative assessment of immune surveillance. According to a new concept, the immune system reacts to disruption of tissue integrity, allowing its renewal; thus the immune system protects the integrity of tissues.[6]

The current doctrine governing the morphology and the function of the immune system is the subject of numerous comprehensive reviews. The immune system, through a complex interplay of highly specialized cell lines and seemingly endless number of newly defined soluble mediators, contends to ensure protection from the potentially harmful endogenous and exogenous insults that we encounter throughout our lifetimes. There are two major divisions of the immune system, namely the phylogenetically older innate and the acquired. The cellular components (macrophages, eosinophils, natural-killer lymphocytes) and the molecular components (complement, antibacterial peptides, diverse cytokines) of the innate immune system provide multiple signals for activation of the acquired immune system. The innate system has evolved to recognize common potentially pathogenic structures. The recognition of insulting agents by the acquired immune system involves specific interactions between antigen receptors on B and T cells and specific sites on foreign molecules (antigen determinants). The B and the T cells represent the decisive armament of the immune system. According to earlier classification, T cells represent the cellular immune response (operating in neoplasia), and the B-cell compartment is the source of all antibodies (IgG, IgM, IgA, IgD, and IgE). The major features of acquired immunity are specificity and memory.[7] However, an efficient immune response requires the synchronous operation of all components of the immune system. The critical effective players in neoplastic cell elimination, prevention of metastatic tumor cell dissemination, and immune response facilitation (also against viruses) are the natural killer (NK) cells.[8] They represent a segment of the lymphoid cell population accounting for only 10 to 20% of the peripheral blood lymphocytes.[9]

It is not unexpected that the immune system is not a totally autonomously existing entity. The interactions between the immune and the nervous systems at all levels, including the brain, pituitary, peripheral nervous system, and the immune cells, was the subject of an international conference in 1998 entitled "Neuroimmunomodulation: Molecular Aspects, Integrative Systems, and Clinical Advances," organized by the New York Academy of Sciences.[10] Anecdotal and, occasionally, hyperbolized reports on this interaction are also abundant. The interrelationship between the endocrine and the immune systems was well recognized much earlier.

It should be emphasized that the developing malignancy is not entirely defenseless and possesses mechanisms capable of overwhelming, overriding, or evading the host's attack, referred to as the "sneak-in".[8] Postulated mechanisms include changes in the immunogenicity of the tumor cells and alteration of the tumor-specific antigen; production of blocking antibodies "masking" the neoplastic cells; intervention by inhibitory T cells; or overwhelming the host immune functions by the tumor's rapid, destructive growth (blockade). Of special interest here is the profound immunosuppression associated with the progression of some viral infections, particularly retroviral, in animals (murine Friend leukemia virus) and in humans (HIV/AIDS virus).

In order to understand both the development and differentiation of normal tissues and to develop therapies for tissues that have exhibited abnormal growth patterns such as hyperplasia, dysplasia and neoplasia, it is imperative to comprehend the nature and the regulation of the normal cell cycle. Persuasive evidence indicates that the growth of tissues in both health and disease is a perfect balance between cell division and programmed cell death (apoptosis). Most often undefined, minute, generally long-lasting exogenous or endogenous impacts can tip this delicate balance in favor of the intruder, with the development of overt clinical neoplastic manifestations or in favor of the host, erroneously designated as spontaneous cancer regression. Established clinical events in the evolution of neoplastic disease include the loss of proliferative control, the failure to undergo apoptosis, the onset of neoangiogenesis, tissue remodeling, invasion of tumor cells into surrounding tissue, and finally, metastatic dissemination of tumor cells to distant sites.[11] A major promising future therapeutic aspect of neoplastic (and viral) disease progression will be the regulation of apoptosis. This and other fast-advancing areas in cancer research have led to four conferences on cancer during the past eight years under the auspices of the New York Academy of Sciences, covering subjects such as immune system, aging, genetics, prophylaxis, and therapy. The domain of clinical cancer research representing an integration of the natural defense mechanisms of the host, augmented by the use of pharmacotherapeutic agents including nutritional modalities, deserves high priority, in our opinion, in the development of new and more physiological approaches to cancer prevention and treatment.[12]

New data recently released by the American Cancer Society (February 2003) corroborate that for the year 2000, the U.S. leading causes of death (as percent of total deaths) still are heart disease (No. 1, 29.6%) and cancer (No. 2, 23.0%). Unlike the mortality from cardiovascular diseases, which is declining in people of all age groups, the mortality due to cancer, notwithstanding the declared "war against cancer," is still rising for older individuals (Figure 36.1). Approximately 50% of all neoplasms occur in the 12% of the population aged 65 or older, and this number will increase with the projected expansion of the older population.[12] Of particular interest is the swift increase in the incidence of certain neoplasms in older individuals that has become evident during the past 20 years (Figure 36.2). These neoplasms include nonmelanoma skin cancer, non-Hodgkin's lymphoma, and malignant brain tumors. This phenomenon cannot be ascribed solely to improved diagnostic techniques. According to a meritorious hypothesis that has both experimental and clinical support, older people may develop cancer earlier than younger individuals when exposed to the same dose of a carcinogen.[12] This is an important phenomenon in light of the current life-extension programs, further strengthening our position concerning the interplay between cancer, aging, and the immune system.

36.1.2 COENZYME Q, MITOCHONDRIA, AND BIOENERGETICS

All biochemical processes involve energy production, transformation, and utilization; thus the term *bioenergetics* could validly be applied to the whole of life sciences. Bioenergetics as a well-defined discipline rose to prominence in the 1950s. In this complex process, the source of energy is adenosine triphosphate (ATP), which is generated by oxidative phosphorylation in the inner membrane of the mitochondria. This evolved and adapted biological system converts energy in the most efficient manner for events such as biosynthetic processes, transport of molecules across cell membranes, and muscle contraction. Our knowledge has advanced dramatically during the past 15 years, as evidenced by the explosion of investigations and the elucidation of molecular

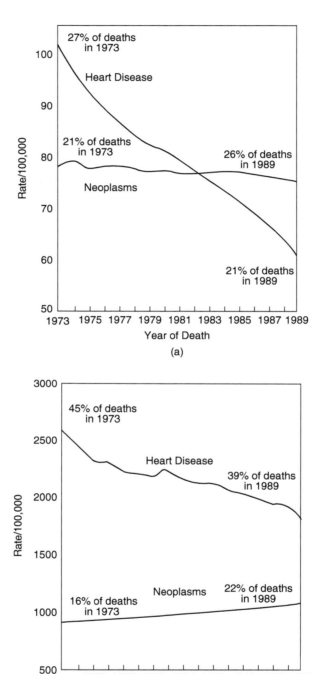

FIGURE 36.1 Variations in causes of death for younger (A, under 65 years) and older (B, over 65 years) individuals over time. (From Berlin, A., *Cancer Invest.*, 13, 540, 1995. Reproduced with permission.)

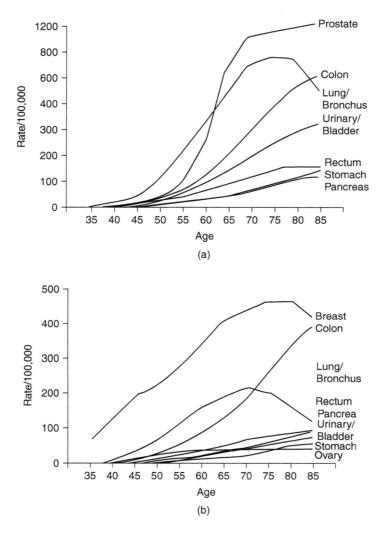

FIGURE 36.2 Age-related incidence in different neoplasms: A, male; B, female. (From Yanick, R. and Ries, L.A.[12] Reproduced with permission.)

mechanisms of mitochondrial physiology and pathology, bringing a better understanding of the role of mitochondria in health and disease. The mitochondrion was once thought to be stable and reliable, and its possible dysfunctions were not even considered. It is now revealed to be a primary intracellular structure operating close to its design limits and extremely prone to damage, with potential disastrous consequences for the organelle itself and for its host cell. Many disorders, particularly the chronic neurodegenerative and cardiovascular diseases, neoplasia, diabetes, and others, are now considered to involve, to greater or lesser extent, mitochondrial dysfunction and therefore impaired energy production. More than 75 disease states are now linked to this pathologic process.[13] The outburst of activity in this area necessitated the creation of a new nosological entity, "diseases of bioenergetics." A testimonial for this productive research effort is the Nobel prize awarded to Peter Mitchell in 1978 for his elucidation of the "Q-cycle" in the electron transport chain, and to Paul Boyer and John Walker in 1997 for their work on ATP synthase enzyme. A comprehensive review of the current advances in bioenergetics has been compiled by Nichols and Ferguson.[14]

It is now well established that mitochondria are critical targets in many, if not most, events of apoptotic cell death. The concept of a mitochondrial basis for eliminating tumor cells has been advanced since the early part of the last century, built upon general differences in metabolic control between neoplastic and cancer cells.[15] In 1956, Warburg postulated that mitochondrial oxidative phosphorylation was defective in tumor cells, and that the initial insult led to increase in glycolytic ATP production as the central event of cell transformation. He predicted that treatment producing mitochondrial injury of a general nature would strike a greater blow against cancer cells than against normal cells. Furthermore, aerobic glycolysis (the Warburg effect) in cancer cells may compete for ADP and phosphate, with oxidative phosphorylation resulting in a mitochondrial shift to a non-phosphorylating state. Another possibility suggested by Hockenberry[15] is that high rates of glycolysis may suppress respiration, known as the Crabtree effect.

ATP, the purine ubiquitously responsible for the storage and resupply of metabolic energy in biological systems, has sprung a new intriguing surprise on scientists working in this area. Recent studies indicate that ATP, in addition to its well-recognized intracellular functions, has unrealized diverse extracellular functions[16] such as:

- Neurotransmitter (or a cotransmitter) function in the central and the sympathetic nervous system, thus transmitting signals across the synapse
- Modulation of the immune system
- Trophic effects in the development and regeneration of the nervous system
- Inhibition of platelet aggregation
- Cytotoxic and apoptotic effects

It is premature at this time even to speculate on the paramount role of this new player in the fast-evolving area of research involving the link between disease, aging, and coenzyme Q, but the potential pharmacological and clinical implications are immense.

Coenzyme Q is well defined as a crucial component of the oxidative phosphorylation system in the mitochondria, where energy derived from the products of fatty acids and carbohydrates is converted into ATP to drive cellular machinery and biosynthetic processes (Figure 36.3). Coenzyme Q (also known as ubiquinone) was discovered by Fred Crane and his colleagues in 1957 in beef heart mitochondria.[17] Karl Folkers and his associates elucidated its chemical structure in 1958.[18] Coenzyme Q is composed of a homologous series of compounds differing in the length of the isoprene side chain, and coenzyme Q_{10}, the homologue present in humans and several other species, has 10 isoprene units with the structure 2,3-dimethoxy-5-methyl-6-decaprenyl-1,4-benzoquinone (Figure 36.4). Coenzyme Q is a ubiquitous, naturally occurring, fat-soluble molecule with characteristics similar to those of vitamins and is essential for the sustenance and functioning of most organisms.

Crane[19,20] has concisely summarized the currently recognized functions of coenzyme Q as:

- Needed for energy conversion (ATP production)
- An essential antioxidant
- Regenerates other antioxidants
- Stimulates cell growth and inhibits cell death
- Decreased biosynthesis may cause deficiency

The normal content of coenzyme Q in mitochondrial membranes has been reported to be below that required for kinetic saturation.[21] This finding strongly suggests that coenzyme Q may be the rate-limiting component in the respiratory chain, especially in the mitochondria of compromised or incapacitated tissues.

Coenzyme Q_{10} is of dual origin in humans, partly exogenously derived (food) and partly synthesized in the body. The dietary intake of coenzyme Q_{10} is likely to be much lower these days

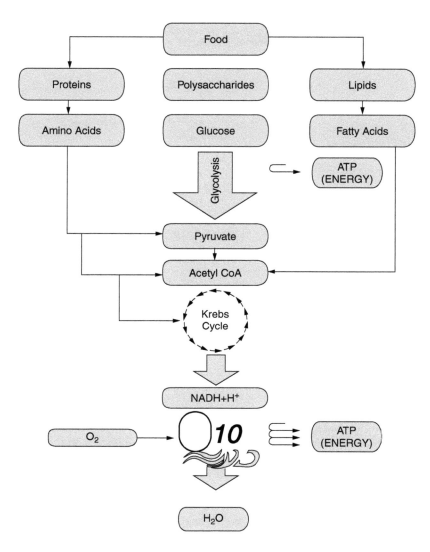

FIGURE 36.3 Role of coenzyme Q_{10} in mitochondrial ATP (energy) production. (From Littarru, G.P., *Energy and Defense*, Casa Editrice Scientifica Internazionale, Roma, Italy, 1995. Reproduced with permission.)

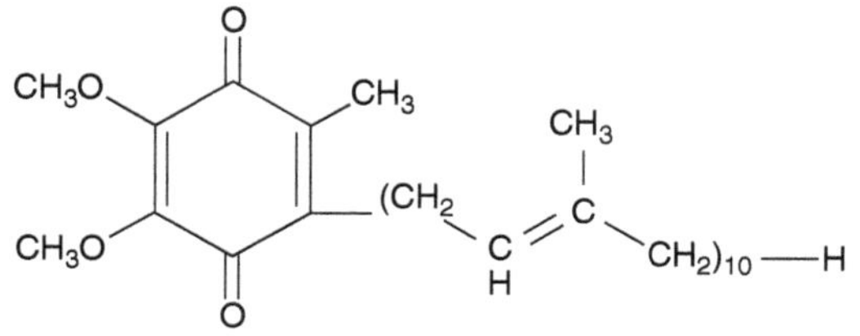

FIGURE 36.4 Structure of coenzyme Q_{10}.

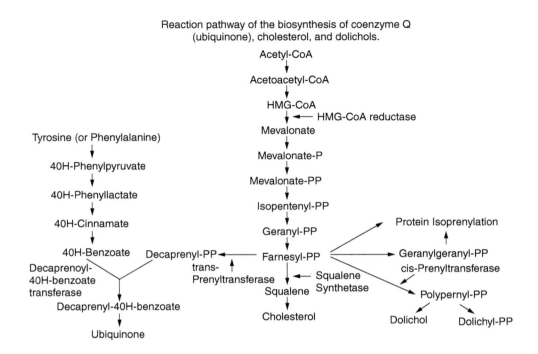

FIGURE 36.5 Reaction pathway of the biosynthesis of coenzyme Q_{10} (ubiquinone), cholesterol, and dolichols.

due to avoidance of foods high in fat, which for the most part also contain higher amounts of coenzyme Q_{10}.[22] This has an effect on other nutrients as well. During gastrointestinal uptake, dietary coenzyme Q_{10} is efficiently reduced to the antioxidant-active ubiquinol form that enters the circulation within the lipoproteins for potential uptake by tissues.[23] The endogenous synthesis of coenzyme Q_{10} is an elaborate process first described by Folkers[24] that depends on the availability of several vitamins and minerals (Figure 36.5). Deficiencies in any one of those building blocks will impair its biosynthesis and result in coenzyme Q_{10} deficiency, thus compromising coenzyme Q-dependent vital functions. Furthermore, the age-dependent decline in the coenzyme Q_{10} content of various tissues in humans and animals, to be discussed later, is postulated to be responsible, at least in part, for the "diseases of aging".[25] Bioenergetic degradation affects first and most intensely the cardiovascular and immune systems as well as the brain, tissues with the highest energy needs. As a consequence, a cascade of functional impairment begins in these systems followed, if not corrected, by overt clinical manifestations. Not surprisingly, cardiovascular, neoplastic, and neurodegenerative diseases are the most common causes of morbidity and mortality in the elderly. Furthermore, as a potent and versatile antioxidant, coenzyme Q blocks oxidative injuries to DNA, lipids, proteins, and other essential structures. This well-documented function prevents or retards the development of many diseases, particularly the "diseases of aging."

In this context, it is interesting to note that in his book on coenzyme Q_{10}, Littarru[26] devoted over 70% of the content to coenzyme Q_{10} prevention of oxidative damage. The rationale and the benefits of coenzyme Q_{10} and idebenone use in the treatment of respiratory chain diseases were reviewed recently.[27] Summing up the practical aspect, data from numerous experimental studies as well as clinical trials demonstrate the strong relationship between coenzyme Q_{10} deficiency, progression of many disease states ("diseases of bioenergetics"), and the beneficial effects of coenzyme Q_{10} supplementation.

Since 1960, the biochemistry, physiology, and clinical effectiveness of coenzyme Q_{10} have been presented at 17 specialized international symposia, with a total of almost 5000 published pages.

Most coenzyme Q$_{10}$ clinical research is focused on the large, heterogeneous group of cardiovascular diseases. The published results of 34 controlled clinical trials and several open-labeled and long-term studies were critically reviewed.[28] The evaluation reveals that out of 58 early and current trials involving 5727 patients with various forms of cardiovascular diseases, only three trials reported negative results (1.7%) experienced by 110 participants (1.9%). A testimonial for the heightened interest in the biomedical and clinical functions of coenzyme Q$_{10}$ is the recent increase in publications on this subject, particularly the two books edited by Kagan and Quinn[29] and Ebadi et al.[30] that contain comprehensive reviews covering all aspects of coenzyme Q.

One issue that needs to be addressed here and is rarely considered by most physicians is the question of bioavailability of coenzyme Q$_{10}$ products intended for oral use. Being somewhat fat-soluble (and insoluble in water), its absorption and bioavailability are very limited and depend on food fat intake to some extent. Commercially available preparations of coenzyme Q$_{10}$ in the U.S. and elsewhere are generally based on the pure crystalline material (as powder-filled capsules or tablets) and oil suspensions (in soft gel capsules), and these forms have limited bioavailability.[31] A recently introduced product that is "hydrosoluble" (called Q-Gel, Tishcon Corp., Westbury, NY) has been shown to have superior bioavailability (about three-fold higher than the other products) in clinical testing.[31-34] There have been numerous experimental and clinical studies using this product. Previous data based on animal experiments using coenzyme Q$_{10}$ in the powder form had not provided clear evidence that coenzyme Q$_{10}$ could be incorporated into tissue mitochondria. It is therefore noteworthy that in a recent study, Kwong et al.[35] demonstrated conclusively that in rats supplemented with Q-Gel in the diet, coenzyme Q$_{10}$ content of tissues and isolated mitochondria showed a significant increase along with an increase in the antioxidative potential and a decrease in protein oxidative damage. This study thus confirms not only the improved bioavailability of Q-Gel, as shown by its concentration in blood, but also the ability to increase the intracellular and mitochondrial uptake of coenzyme Q$_{10}$ in various tissues, including the brain.

The Food and Drug Administration (FDA) recently granted an "orphan drug" designation for UbiQ-Gel for the treatment of mitochondrial cytopathies. In Europe and in Japan, a synthetic analogue of coenzyme Q$_{10}$ called Idebenone (Takeda Chemical Industries, Osaka, Japan) is also available with claims that it has improved bioavailability, but it cannot be legally sold in the U.S. until it has been tested and approved as a "new drug" by the FDA. Another new development in this area is the introduction of stabilized ubiquinol, the reduced and the predominant form of coenzyme Q$_{10}$ in blood circulation (Q-Nol, GelTec/Tishcon Corp., Westbury, NY). Human and animal data have shown that Q-Nol has even higher bioavailability.[33,34] The clinical efficacy of both Q-Gel and Q-Nol are currently being evaluated in several clinical trials.

An intriguing subject is the production and availability of coenzyme Q$_{10}$ for commercial purposes. In the late 1960s and early 1970s, coenzyme Q$_{10}$ was available only in minute amounts for use as an analytical standard. Slowly, with increased demand, the production increased, first using beef heart as the source, then horse heart, and the product was marketed in the U.S. by the meat-packing industry and not by pharmaceutical companies. At that time, the Swiss giant Hoffmann-La Roche ventured that the limited demand for coenzyme Q$_{10}$ would eventually explode and, after enormous effort initiated in 1958, developed and patented a very long, complicated, and expensive synthetic method for its industrial production. The demand slowly increased, but at that time, Japanese scientists developed and patented a much simpler and cheaper method by way of fermentation. Then the price dropped from $1000 per gram to $17 to 20 per gram, and it is down to between $1 to 2 per gram today. Coenzyme Q$_{10}$ is produced exclusively in Japan, and extremely high-quality material is available in large quantities from Kaneka Corp., Osaka (yeast fermentation process, initiated in 1977), Nisshin Flour Milling Co., Tokyo (semisynthetic process using plant cell culture, initiated in 1974), and Mitsubishi Gas and Chemical Co. (bacterial fermentation process). The Nisshin process for industrial coenzyme Q$_{10}$ production qualifies the product as a phytopharmaceutical. Estimates put the combined annual production of coenzyme Q$_{10}$ by the three companies in Japan to between 150 and 200 metric tons.

36.1.3 Tumor–Host Interplay

36.1.3.1 General Setting

Tumorigenesis is a multistage process that has been classified into initiation, promotion, and progression phases, and each stage involves both genetic and epigenetic alterations, functional as well as morphological. These changes can be caused by chemical or physical factors and frequently involve the formation of reactive oxygen species (ROS) and other highly reactive molecules. These agents can damage DNA and other essential cellular components, modify gene expression, alter the cellular antioxidant defenses, and thus affect cell growth and differentiation.[36]

During metabolism of oxygen to water in the mitochondria, a small fraction of the oxygen is reductively converted into superoxide as a by-product. Superoxide may be further converted into various ROS, i.e., hydroxyl radical, hydrogen peroxide and others. However, the majority of the cells possess intracellular antioxidant defense mechanisms against the potentially harmful effects of the ROS. A second defense mechanism at the DNA region is the base excision and strand-break repair enzymes, which remove the damaged segments in order to maintain the integrity of the genome. However, deleterious genetic alterations, particularly poorly repaired oxidative damages, may accumulate in cells with age through errors in repair, replication, and recombination. Furthermore, the level of repair may decrease with the aging of the cell. Recent reviews have addressed the role of oxidative damage to the DNA[37] and its implication in carcinogenesis, atherosclerosis and AIDS,[38] and the inhibition of intracellular oxidative stress by anticarcinogenic antioxidants.[39]

Experimental and epidemiological data validating the concept that most tumors are associated with mutagens and mitogens suggest an empirical therapeutic approach, searching for agents that inhibit or reverse the cellular transformation elicited by these substances. Since 1987, over 1000 compounds have been screened for their anticarcinogenic potential under the National Cancer Institute Chemoprevention Testing Program, and a few such as tamoxifen, retinoic acid, vitamin E, selenium, and calcium have been further evaluated in large-scale intervention trials.[40]

Cancer chemoprevention is defined as the use of specific compounds or agents to prevent, inhibit, or reverse the progress of carcinogenesis.[40] Human cancer development requires 20 to 40 years or more in many major target tissues, and the scope of chemoprevention actually encompasses all phases of this process — from healthy subjects exposed to "normal" risk; to populations at intermediate risk due to environmental and lifestyle factors, genetic predisposition, and precancerous lesions; and then to previous cancer patients at high risk for other primaries. Clearly, the distinction between cancer prophylaxis and cancer treatment (primary, regrowth after treatment, or distant metastasis) is blurred or overlapping and thus is not practical. For this reason, we will assess the options for therapeutic intervention at all stages of tumorigenesis in this chapter.

The early clinical literature is replete with ample anecdotal verification of the potential role of the immune system in the surveillance, elimination, and the "cure" of cancer. The factors most commonly associated with the spontaneous regression have been fever, concurrent bacterial infections, administration of bacterial vaccines, or the removal of at least some portions of the tumor. The occasional regression that accompanied bacterial infections did not go unnoticed by physicians. This technique, referred as "nonspecific cancer immunotherapy," dates back to at least 1774, when a Parisian physician injected pus into the leg of a patient with inoperable breast cancer. As the induced infection worsened, the patient's cancer disappeared. More than a century later, Coley, noting the apparent beneficial effect of erysipelas infection on cancer regression, formulated a well-publicized preparation based on soluble toxins from erysipelas. More recently, local application of the tuberculosis BCG vaccine has proven to be effective therapy for certain cancers in randomized clinical trials. It is interesting to note here that in 1971, Laborit and his colleagues[41] reported an improvement in mouse Ehrlich ascites cancer by treatment with isolated mouse hepatic mitochondria. They speculated that the improvement resulted from an increase in the "defense system" activity. This study perhaps represents the first step in our understanding of the interrelationship between cancer, immune system, and coenzyme Q. Following a review of these findings, Fuchs and Matzinger[5] suggested that these "nonspecific" immune stimulants may be providing the necessary signals to stimulate local cells to initiate an immune response to tumor-specific antigens.

A comprehensive evaluation of current cancer chemotherapy status was outlined by Davis[42] and Agarwala.[43] A more specialized novel area of research with practical applicability is the effect of immune reactivity, expressed as an inflammatory process, on angiogenesis and thus on tumor growth.[44] It as appropriate to recall here that the direct effect of cholesterol on the immune system, more specifically on phagocytosis, was observed as early as 1914.[45] Furthermore, McMichael et al.[46] accumulated epidemiologic data suggesting that low serum cholesterol is associated with increased risk of cancer of the lung or colon. Klurfeld[45] concluded his review, stating emphatically that "cholesterol interactions with cells of the immune system are potentially some of the most important in the area of nutrition and immunity." This serious concern and the implications for cancer and other diseases are hotly denounced by today's anticholesterol crusaders.

36.1.3.2 Coenzyme Q_{10}: Animal Studies

In the early 1960s, we established our then-called "host defense program," representing a search for a physiologic stimulant of the immune system with applicability to humans. (The term *stimulant* was later replaced by the technically more precise term *modulator*). Many compounds were evaluated for activity based on a battery of animal tests, measuring their response upon stimulation of various parameters of the immune system. Satisfactory activity was first demonstrated by coenzyme Q_6 and then by coenzyme Q_{10}. For the first time, our results suggested the capability of the coenzyme Q group of compounds to stimulate *in vivo* two fundamental parameters of the immune system response: the phagocytic process in rats and the antibody production in mice, and the preliminary results were published in 1970.[47] Subsequently, studies were carried out using more complex test systems measuring the total host response to *Plasmodium berghei* infection in mice (causing a fulminate malaria infection) treated by coenzyme Q_{10} in combination with chloroquine.[48] This model exemplifies the effectiveness of a nonspecific stimulation of the immune system (coenzyme Q_{10}) plus a specific antimalarial drug. The combined treatment resulted in increased number of survivals, prolonged survival time, and reduced intraerythrocytic parasitemia. The involvement of the immune system in protozoan infections, particularly in malaria, has long been recognized but never fully demonstrated, and our study contributed significantly in enhancing our knowledge in this area.

Chemically induced tumors are still considered a good model for human neoplasia, and we evaluated the effectiveness of coenzyme Q_{10} in mice with dibenzpyrene-induced tumors. A second test system in this study was a retroviral infection in mice (Friend leukemia virus), a model used in the evaluation of new drugs for HIV/AIDS. Our findings and conclusions could be summarized as follows.[49]

Treatment with coenzyme Q_{10} reduced the percentage of mice with tumors, increased the number of survivors, and reduced tumor size in mice with tumors induced by 3,4,9,10-dibenzpyrene.

Treatment with coenzyme Q_{10} decreased splenomegaly and hepatomegaly (critical early symptoms in the progression of the infection) and increased the number of surviving mice infected with Friend leukemia virus.

In a joint program with Karl Folkers at the University of Texas in Austin, TX, we evaluated the changes in coenzyme Q_{10} levels in blood and other tissues following infection with murine Friend leukemia virus. The data revealed a significant coenzyme Q_{10} deficiency in blood and spleen (mitochondria) as the infection progressed. On day 20 after the infection, the coenzyme Q_{10} was reduced to 36% in blood and to 62% in spleen mitochondria (spleen being a major immune

TABLE 36.1
Coenzyme Q_{10} and Leukemia Experimental Infection in Mice

Treatment	Spleen wt. Body wt.	%
1. FLV infection (alone)	50.28	100.0
2. FLV + cyclophosphamide	47.61	94.7
3. FLV + cyclophosphamide + coenzyme Q_{10}	23.72	42.7
1. FLV infection (alone)	50.29	100.0
2. FLV + hydrocortisone	28.61	56.9
3. FLV + hydrocortisone + coenzyme Q_{10}	17.43	34.7

Note: FLV = Friend leukemia virus. Mice were treated on day 20 after infection.

Source: Bliznakov, E.G., in *Biomedical and Clinical Aspects of Coenzyme Q,* Folkers, K. and Yamamura, Y., Eds., Elsevier, Amsterdam, 1977, p. 73. With permission.

system organ). Our conclusion at that time was that the limiting factor in the resistance to retroviral infection in mice appeared to be the intracellular availability of coenzyme Q_{10}.

We presented a review of our earlier studies on experimental infections and neoplasia along with some new data at the International Symposium on Coenzyme Q_{10}, held at Lake Yamanaka, Japan, in September 1976.[50] These are summarized in Table 36.1. Consequently and in confirmation of our earlier published results, Folkers et al.[51] reported that coenzyme Q_{10} administration in patients increased the blood count of T4 lymphocytes and the blood level of IgG antibody. More recently, Barbieri et al.[52] demonstrated that coenzyme Q_{10} increased in a dose-dependent manner the antibody response to hepatitis B vaccine. In this single-blind, placebo-controlled and randomized clinical study, 21 volunteers per study group were supplemented with coenzyme Q_{10}, 90 mg/day or 180 mg/day, for 14 days prior and for 90 days after the vaccine administration. At day 60 after the vaccination, the antibody titers were increased by 135% for the 90-mg coenzyme Q_{10} and by 155% for the 180-mg coenzyme Q_{10} groups. It is interesting to note that the authors of this Swedish study offered an alternative and intriguing speculation. According to this group, the immunomodulating coenzyme Q_{10} activity is associated with its antioxidant effectiveness, which minimizes the oxidative stress and cell membrane turnover, thus improving antigen recognition and antibody production. They proposed that "the immunoenhancing coenzyme Q_{10} effect can be of certain importance for elderly people with lower endogenous production of coenzyme Q_{10}."

36.1.3.3 Coenzyme Q_{10}: Clinical Studies

The encouraging results from our extensive animal studies on the coenzyme Q–neoplasia link, discussed earlier, led many scientists to test further the validity of this link in limited clinical trials. Two reviews, one being short and popular and the other extensive and technical, were published recently on this still-neglected subject by the National Cancer Institute dated June 12, 2002.[53,54] The latter report traces the interest in coenzyme Q as a cancer therapeutic agent as early as 1961, when coenzyme Q_{10} deficiency was detected in patients with breast and other forms of cancer. A subsequent study showed a statistically significant relationship between the degree of coenzyme Q_{10} deficiency and the progression of breast cancer. Yet, these findings on the encouraging and potentially successful approach to cancer management have been largely overlooked and even forgotten over the years. Back in 1987, Eggens et al.[55] had reported that in humans with hepatocellular carcinoma, the coenzyme Q_{10} content in liver dropped to one-half that found in control tissue.

There has been some sporadic activity in this area during the past few years. Recently, Portakal et al.[56] published data on the relationship between oxidative stress and breast cancer development in 21 patients, and their results showed a decrease in the coenzyme Q$_{10}$ content in the tumor tissue and an increase in the oxidative stress as compared with that in the surrounding normal tissue. They construed that administration of coenzyme Q$_{10}$ may afford protection against increased oxidative stress and its consequences in breast tissue. In another study involving 28 untreated patients with lung and breast cancer, increased oxidative stress and reduced levels of lipid-soluble antioxidants, in particular reduced coenzyme Q$_{10}$, were found in blood and bone marrow plasma as compared with those of controls.[57] This, according to the authors, represents an imbalance between oxidant generation and antioxidant defense in favor of the former in cancer patients. In a study with 80 breast cancer patients, plasma coenzyme Q$_{10}$ level was found to be significantly lower, and this reduction was more pronounced in patients with larger tumor volume.[58] Similar results have been reported in 116 patients with myeloma and breast cancer.[59] Mikhail et al.[60] examined the association between plasma coenzyme Q$_{10}$ and α-tocopherol in cervical cancer patients and found that the mean plasma levels of both antioxidants were significantly lower than those in controls. They concluded that this reflected increased utilization of the antioxidants due to oxidative stress and that both coenzyme Q$_{10}$ and α-tocopherol might play a role in the pathogenesis of cervical neoplasia.

Three small-scale clinical trials with cancer patients undergoing standard treatment along with coenzyme Q$_{10}$ supplementation indicated a tumor suppressive effect.[61–63] A study by Lockwood et al.[61] in Denmark involved 32 breast cancer patients. In six of them (18.8%), signs of remission were reported following treatment with antioxidants that included coenzyme Q$_{10}$ at a relatively low dose (90 mg per day). All the patients also reported a decreased use of painkillers, absence of weight loss, and improved quality of life. The survival rate at 18 months was 100% against an expected rate of 87.5%, and at 24 months the survival rate was still 100%, whereas the expected rate was 81.2%. In a follow-up study, one new patient and one previously treated patient exhibiting cancer remission were given higher doses of coenzyme Q$_{10}$ (300 mg and 390 mg per day) for three to four months. Both patients showed complete regression of the tumors following this treatment. In another study by the same group,[62,63] three breast cancer patients were treated with a high dose of coenzyme Q$_{10}$ (390 mg per day) and followed for three to five years. Two showed complete regression of cancer and metastases, and the third had no microscopic evidence of cancer after mastectomy. One of the clinicians participating in this study stated ardently that he had seen around 200 cases of breast cancer a year, but he had never seen a spontaneous complete regression of 1.5- to 2.0-cm tumor, and had never seen a comparable regression following any conventional antitumor therapy in his 35 years of oncology practice.

In extension of the previously reported murine retroviral–coenzyme Q$_{10}$ link studies, two preliminary clinical observations deserve notice. In a blinded study, Bui et al.[64] found that plasma coenzyme Q$_{10}$ levels were significantly lower in HIV-positive patients than in healthy individuals and even lower in more advanced stages of the infection (level 4 or 5). In an early open-label study, coenzyme Q$_{10}$ (200 mg/day) increased the ratios of T lymphocytes (T4 helper to T8 suppressor cells) in AIDS patients.[65] This ratio is used as a sensitive indicator for evaluation of immunologic parameters and is depressed in HIV infections and in cancer. The explosion of AIDS infections in some parts of the world mandates a concentrated effort to evaluate the involvement of coenzyme Q$_{10}$ in the progression of this debilitating infection.

Admittedly, many of the clinical trials reported in the literature on coenzyme Q$_{10}$ and cancer link have design flaws, often accentuated by the critics. Another limiting factor in the earlier trials was the very low dose of coenzyme Q$_{10}$ employed along with too short a treatment period, both factors reflecting the insufficient supply of coenzyme Q$_{10}$ and the high price of the product at that time. Nevertheless, the results of these studies, as well as anecdotal reports of increased survival of patients with various types of cancer (pancreas, lung, colon, rectum, and prostate) treated with coenzyme Q$_{10}$,[63,66] suggest the therapeutic potential of coenzyme Q$_{10}$ in cancer therapy and also possibly in its prophylaxis, as demonstrated in the more extensive animal studies. Additional support for this optimistic outlook is provided by the simultaneous achievements in the neoplasia-related domain of research: immunology and aging, and their dependence on coenzyme Q$_{10}$ availability.

There are numerous reports on the efficacy of coenzyme Q_{10} as a protective agent in preventing cardiotoxicity in cancer patients treated with anthracycline drugs, as previously shown in animal studies. This was discussed in a short review published recently.[67] Published studies indicate that inclusion of coenzyme Q_{10} in the treatment regimen of cancer patients does not compromise the antitumor effect of Adriamycin™. On the other hand, an additive or even a synergistic therapeutic effect of combined treatment with anthracyclines and coenzyme Q_{10} in cancer patients could be expected. Treatment of cancer patients with doxorubicin and coenzyme Q_{10} was shown to lower the incidence of cardiac dysfunction.[68] This was followed by a study by Tsubaki et al.[69] showing that 18 of 25 patients (72%) with malignant lymphoma treated with Adriamycin and daunorubicin and 20 of 50 patients (40%) also treated with coenzyme Q_{10} developed abnormal ECGs. The coenzyme Q_{10}-treated patients also showed significant improvements in diarrhea and stomatitis. In a longer-term study, the same authors reported that while 11 of 12 patients (91.7%) treated with anthracyclines developed abnormal ECGs, only 11 of 17 (64.7%) also treated with coenzyme Q_{10} manifested ECG problems. A study by Karlsson et al.[70] established that patients on doxorubicin have lower concentrations of coenzyme Q_{10} in heart, skeletal muscle, and blood as compared with healthy subjects. The authors concluded that the lower the heart muscle coenzyme Q_{10}, the more impaired is the cardiac function. A randomized clinical trial with 10 cancer patients treated with Adriamycin and coenzyme Q_{10} (100 mg, twice daily) also demonstrated the ECG evidence for the protective effect of coenzyme Q_{10}.[71]

The results of numerous uncontrolled clinical trials, although often dismissed as not being rigorous enough, lend further support to the effectiveness and protective effect of coenzyme Q_{10} in cancer patients [reviewed in 66, 72]. The differences in the response to coenzyme Q_{10} in the various studies could be attributed to dosage, duration, and bioavailability factors.

More than a quarter century has passed since Karl Folkers postulated that coenzyme Q_{10} could have therapeutic potential for the treatment of cancer.[73] He argued then that, as coenzyme Q_{10} was essential for normal cell respiration and function, any deficiency in its availability or biosynthesis could disrupt normal cellular functions. This could lead to abnormal pattern of cell division that, in turn, might induce an oncogenic response. Today, it is unequivocal that additional clinical evidence evolving from larger, well-designed, state-of-the-art trials will bring us closer to the prophetic assertions of Karl Folkers. While the pharmaceutical industry has shown little interest in the past, further progress in this important area cannot be made without the participation of foundations, scientific organizations, academia, and the government.

36.1.4 THE BURDEN OF AGING

36.1.4.1 Pandemic Relevance

Another domain of scientific quest in which information from three areas of exploration sometimes collide and sometimes complement each other is the merger of immunologic, cancer, and aging research. The parallel increase in cancer risk with advancing age is well recognized,[74] as evident from the very low occurrence in the first decade of life, which rises to an incidence of about 30% in the seventh decade (Figure 36.2). In this regard, cancer may be considered a disease of the elderly, although from another point of view, the multistep nature of carcinogenesis indicates that the disease actually originates much earlier in life. Several pathophysiologic mechanisms common to all three areas of exploration have been proposed to explain the increased incidence such as genetic defects; exposure to environmental, physical, or chemical carcinogenic agents; and even infections. Additionally, many neoplastic alterations develop spontaneously as a result of accumulated random damage to genetic material and to other critical cellular structures, and the resulting functional dysfunctions not infrequently bear a resemblance to those associated with aging. For this and other reasons, cancer and other pathological states, i.e., cardiovascular and neurodegenerative diseases, are contemplated as "diseases of aging." In particular, ROS and oxidative stress with damage to critical biomolecules indicate another common mechanism underlying the progressive cellular functional insufficiency, characteristic of all three areas.

The replication of mitochondrial DNA mutations eventually leads to impairment of cellular function. The resulting ATP deficit and increased redox stress cause cell senescence and eventually cell death. Among the systems affected by this cellular disintegration are various compartments of the immune system. This overwhelming cellular functional erosion justifies proposed strategies aimed at reducing oxidative and other noxious tissue damages in an attempt to delay the aging process and to compress the time frame of chronic disease advent, thus enhancing the quality of life and ameliorating the social burden on the community.[12,75]

"Probably no subject so deeply interests human beings as that of the duration of life," wrote Raymond Pearl in his classic *The Biology of Death*. Ancient societies attributed to their gods, whether in Valhalla or on Mount Olympus, magical ways of postponing old age forever. Today's multifaceted research on aging based on scientific principles, rather than the magical *Fountanus Juventus* of Ponce de Leon and Columbus followers, began with the 1955 formulation of the "free radical theory" of aging by Denham Harman, a theory that is now supported by extensive experimental data.[76] His earlier studies had shown an effect of vitamin E on immune function, resulting in increased antibody-forming capacity of old mice by 189%.[77] As expected, some criticism is voiced on this now-accepted "theory of aging".[78,79]

Aging is both a complex scientific predicament as well as a human event of universal concern. A recent publication by the New York Academy of Sciences entitled "Towards Prolongation of the Healthy Life Span" was based on a conference to address this issue.[80] The genetic approach is a novel and a fashionable mode of exploration employed by gerontological science. The dichotomy of this is well conveyed by Martin,[81] who asks, "Do senescent phenotypes and their associated increased vulnerability to organismal death result from a relatively simple program of gene action or is it exceedingly complex, modulated by allelic variants at hundreds or thousands of loci?" The answer to this enigmatic question would provide a rationale (or the lack of it) for the development of interventions for the extension of our own life spans.

Overall, there is profusion of evidence that mitochondrial function and bioenergetic capacity deteriorate during normal aging, especially in postmitotic tissues such as brain, heart, and skeletal muscle.[82] Furthermore, Nicholls and Budd[83] recently reviewed mitochondrial function and dysfunction in cells and their relevance to aging and aging-related disorders. Additionally, Bruce Ames and his colleagues[84] have demonstrated that mitochondria from old rats can be "rejuvenated" and memory improved with a diet supplemented with acetyl-L-carnitine and α-lipoic acid, again confirming the participation of the mitochondrial energy and antioxidant processes in aging. In line with this and other similar observations, the Mitochondria Research Society recently published well-grounded recommendations for the design of clinical trials evaluating not single, but a combination of compounds that target two or more of the final common pathways of energy dysfunction.[85]

36.1.4.2 Coenzyme Q Implications

One segment of our host defense system program explored some of the modern aspects of aging and their links with the immune system and with some disease states emerging with accelerated frequency during aging. Our objective was not to discover the proverbial "fountain of youth," but to develop the means to delay or to ameliorate the clinical symptoms of the so-called diseases of aging, more specifically the diseases that accompany aging, and thus improve the quality of life.

The immunological senescence in mice and its reversal by coenzyme Q$_{10}$, further supporting its role as a potent immunomodulating agent, has been discussed previously.[86,87] In one set of experiments, old mice 16 to 18 months of age were treated with weekly injections of coenzyme Q$_{10}$. This resulted in a significant reduction in overall age-related mortality as compared with control mice of the same age (Figure 36.6). Furthermore, the mean survival time of the coenzyme

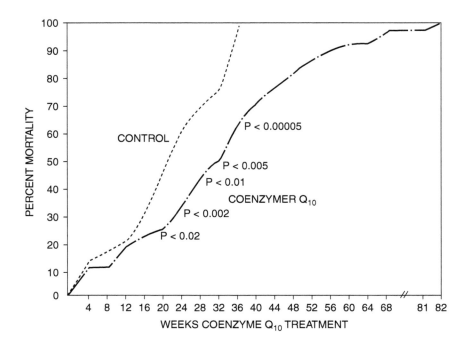

FIGURE 36.6 Modification of overall age-related mortality in old mice (CF1 female) by coenzyme Q_{10} treatment. Coenzyme Q_{10}: 50 µg/mouse (i.p.), administered as an emulsion weekly for 81 weeks starting at age 16 to 18 months (50 mice per group). (From Bliznakov, E.G., in *Biomedical and Clinical Aspects of Coenzyme Q*, Vol. 3, Folkers, K. and Yamamura Y., Eds., Elsevier/North-Holland Biomedical Press, Amsterdam, 1981, p. 311. Reproduced with permission.)

Q_{10}-treated group was significantly increased (Table 36.2). These findings have been presented at several conferences and were also recently reviewed.[87–90]

A well-known event associated with the decline of the immune function with aging is the involution of the thymus, regulated tightly by the brain and by the hormonal system. A predominant immunologic change observed in aging individuals is a decline in T-cell (thymus derived) function, affecting both subsets CD4 and CD8. Because of strong T-cell–B-cell interactions, this decline affects the function of both T and B cells, although the B cells themselves remain reasonably intact

TABLE 36.2
Survival of Old Mice Treated with Coenzyme Q_1

Treatment	Mean Survival Time after Treatment	
	Weeks	%
Control	20.0	100.0
Coenzyme Q_{10}	32.2	156.0 [a]

Note: Duration of coenzyme Q_{10} treatment was 81 weeks starting at age 16 to 18 months.

[a] $p < 0.001$

Source: Bliznakov, E.G., in *Biomedical and Clinical Aspects of Coenzyme Q*, Vol. 3, Folkers, K. and Yamamura Y., Eds., Elsevier/North-Holland Biomedical Press, Amsterdam, 1981, p. 311. With permission.

during aging. Furthermore, the process of T-cell maturation is determined by the degree of age involution of the thymus. The development of new therapies that can be used to enhance thymic regenerative capability in humans with T-cell-depletion states such as aging, cytotoxic chemotherapy, and infections (HIV/AIDS) are urgently needed.[91] Our studies have confirmed the age-dependent involution of the thymus. Old mice aged 24 months retained only 25% of the thymus weight as compared with young mice 10 weeks of age. In a recent issue of the popular *Health and Healing* newsletter (February 2000), it was reported that the thymus in infants is 70 g and shrinks to an incredible weight of just 3 g after age 50. Furthermore, in a joint study with Karl Folkers and his group, we extended the thymus involution study and established that the coenzyme Q$_{10}$ deficiency index of old mice at age 24 months reached 80%.[92]

Normal human cells reach senescence after dividing around 50 to 90 times in cell culture. Recent studies show that they have a "molecular clock" that informs them of their limited life span. This clock is the telomere, a structure at the end of each chromosome that shortens with each cell division and thus reminds a cell how many times it has divided before reaching the stage of "replicative senescence".[93] The involvement of telomeres not only in the aging process but also in cancer progression is evident but unexplored.

Following the publication of our aging studies, some groups of scientists ignored them, while others confirmed our then-unconventional concept. The oft-cited, now classical publication by Kalen et al.[94] verified that in humans 77 to 81 years old, the coenzyme Q$_{10}$ content of heart is only 43% of that present in subjects 19 to 21 years old. A similar pattern was observed in rats. In a small study, Fahy[95] was able to confirm our results, noting the extension of mean survival time in a group of mice receiving coenzyme Q$_{10}$ in their diet. In another publication, Coles and Harris,[96] as a part of a general survey, screened "promising single agents for anti-aging properties" using a long-lived strain of mice. At 39 months, 50% of the mice on coenzyme Q$_{10}$-supplemented diet were still alive compared with only 25% of the control animals. The average life span of the control mice was 30.8 months as opposed to 37.0 months for the coenzyme Q$_{10}$-supplemented group, a statistically significant difference. The "most spectacular" difference between the two groups, however, was the far greater level of activity in the coenzyme Q$_{10}$-treated mice. In a study by Lonnrot et al.[97] with rats and mice, treatment with coenzyme Q$_{10}$ did not prolong or shorten their life span. Yet, in a subsequent publication,[98] they reported that extended treatment of senescent rats with coenzyme Q$_{10}$ ameliorated the age-associated deterioration of the arterial vasodilatation, frequently the cause of increased mortality due to myocardial infarction and stroke. The authors stressed the possibility that dietary coenzyme Q$_{10}$ would protect arteries against the age-related pathological changes. In a recent article very pointedly entitled "Human aging and global function of coenzyme Q$_{10}$," Linnane et al.[99] concluded much more generally, "In this paper we asked the question whether coenzyme Q$_{10}$ can ameliorate the rate of tissue aging. The results suggest that this may indeed be possible." In a recently published communication, Wouters-Wesseling et al.[100] evaluated the effect of a "complete nutrition supplement" that also contained coenzyme Q$_{10}$ on the antibody response to influenza vaccine in elderly subjects. The author concluded that the nutritional supplement might have a beneficial effect on the antibody response in the elderly population.

36.2 INTERFERENCE BETWEEN COENZYME Q BIOSYNTHESIS AND DRUGS COMMONLY USED IN CLINICAL PRACTICE

The endogenous synthesis of coenzyme Q in the body is a very complex process requiring numerous substrates and cofactors. A number of commonly used drugs have been shown to cause coenzyme Q deficiency, either by inhibiting its synthesis or by interfering with its transport via blood circulation. This phenomenon has been described adequately during the past 20 years, but it is still ignored or not recognized by the majority of medical professionals and, most importantly, by the pharmaceutical industry. Furthermore, it is interesting to note that coenzyme Q deficiency is involved in the initiation and progression of the side effects of many drugs. Of particular importance in this group are the widely prescribed lipid-lowering drugs known as statins (HMG-CoA reductase inhibitors). Their use has been shown to cause significant reductions in coenzyme Q and ATP in blood. The reasons are obvious. Cholesterol and coenzyme Q share a common biosynthetic pathway, and an inhibition of cholesterol synthesis at the HMG-CoA reductase step results in an unintended inhibition of coenzyme Q, leading to a deficiency of coenzyme Q as well as other end products in the same pathway.

Coenzyme Q deficiency is implicated at least in part in the development of many of the side effects of statins, such as myopathies and rhabdomyolysis with renal failure, increased incidence of neoplasia, cataracts, peripheral neuropathies, and also some psychiatric disturbances. Despite this long list, the medical literature describes statins as "well tolerated" drugs, and their use in some cases is indiscriminate. *Health News* (September 2001), a publication of the Massachusetts Medical Society, recently sponsored a shocking editorial entitled, "A statin in every medicine cabinet." Another publication in the *American Journal of Cardiology*,[101] authored by its editor, is entitled, "Getting more people on statins." It is interesting that the popular magazines such as *U.S. News & World Report* (October 7, 2002) are more critical of this trend than the medical professionals. Notwithstanding this optimistic stand, the German pharmaceutical giant Bayer AG announced on August 6, 2001, the voluntary withdrawal from the market of their statin drug Baycol (Lipobay), also known as cerivastatin. The reason for this swift withdrawal, proclaimed in the German publications as the "Baycol Katastrophe," was the unacceptable high number of fatal side effects following treatment with this drug.

The multiplicity of effects resulting from treatment with statins has been the center of attention, frequently overstated, during the past few years. Referred to as pleiotropic effects, statins influence a wide range of physiologic functions. One of the newly described effects declared as a beneficial capability is immunosuppression.[102,103] Surely, it might be beneficial in patients with organ transplantation, but it could also explain the increased incidence of infections, neoplasia, and other side effects after long-term statin treatment, events well recognized by clinicians. Since the normal immune response provides surveillance against neoplastic transformation, it is not unexpected that the immunosuppressed host is at higher risk. Statistical evidence indicates that the chance of lymphoma developing in the immunosuppressed transplant recipient is increased 35-fold, with reticulum cell sarcoma occurring at 300 times the expected frequency.[104]

Of particular interest is the significant increase in the incidence of breast cancer in one of the clinical trials with statins (designated, regrettably, as CARE) that has been disregarded by most clinicians. Muldoon [*New York Times*, September 5, 2000] set forth this controversy as follows: "We have studies that show statins don't cause cancer within a five year period. Of course, neither does smoking." We have reviewed the literature on the relationship between statin treatment, side effects, and coenzyme Q deficiency, considering the medical[22,105,106] as well as the ethical aspects,[107] and we advocate the concomitant administration of coenzyme Q_{10} during statin treatment, especially in older patients, to prevent coenzyme Q_{10} deficiency and to support the impaired cellular bioenergetics. It should be stressed here that in addition to statins, there are numerous other clinically useful drugs that interfere with the biosynthesis of coenzyme Q_{10} or its transport. The list of drugs included in this roster contains 49 drugs.[108] Another name was just added to the list based on the observation that long-term corticosteroid administration (the routine treatment for autoimmune and allergic diseases) induces oxidative stress-mediated mitochondrial injury with oculoskeletal symptoms, including myopathy.[109]

Adriamycin is an antibiotic of the anthracycline family, and it is a major drug in cancer therapy. Its clinical application, however, has been limited by its often lethal cardiotoxicity in the form of cardiomyopathy. Ultrastructurally, this is characterized by necrosis and degeneration of the mitochondria, resulting in inhibition of respiratory activity. Adriamycin is accumulated within the macrophages, causing loss of cellular function. This contributes to its strong immunosuppressive

TABLE 36.3
Coenzyme Q$_{10}$ Treatment and Humoral Antibody Response in Mice

Treatment	Hemolytic Antibody Units	Percent
1. SRBC (alone)	127.5	100.0
2. SRBC + cyclophosphamide	80.1	62.8
3. SRBC + cyclophosphamide + coenzyme Q$_{10}$	113.3	88.9

Note: SRBC = sheep red blood cells.

Source: Bliznakov, E.G., in *Biomedical and Clinical Aspects of Coenzyme Q,* Folkers, K. and Yamamura, Y., Eds., Elsevier, Amsterdam, 1977, p. 73. With permission.

effect, associated with a significant reduction (over 50%) in thymus and spleen weights in mice. A single dose of this drug almost completely suppressed the hemolytic antibody production in mice, and coenzyme Q$_{10}$ administration led to a partial but significant reversal.[88] These findings justify consideration of coenzyme Q$_{10}$ as an adjunct to the current treatment regimen with Adriamycin in cancer patients. Similar results have been obtained with the use of cyclophosphamide, another well-known anticancer drug. Data from one study with mice are shown is Table 36.3.[50]

Cancer chemotherapy is certainly effective in the treatment of various forms of neoplasia. Yet, this effectiveness is generally associated with a profound immunosuppression that is well recognized by the medical professionals. It is for this reason that some of the same drugs are used in the treatment of autoimmune diseases, such as lupus erythematosus. Nevertheless, there was a publication in *Cancer Research* many years ago entitled, "Are the cancer drugs self-defeating?" This is a very relevant question that still has not been answered.

The foregoing discussion illustrates the importance of coenzyme Q as a critical component for the optimal function of the immune system at the mitochondrial level. Furthermore, we established the role of coenzyme Q as an active immunomodulating agent using animal models to evaluate various parameters of the immune system (phagocytic rate, circulating antibody levels, neoplasia, viral and parasitic infections). This capability is further enhanced if coenzyme Q$_{10}$ is administered in combination with other specific drugs, i.e., chemotherapy.

36.3 COENZYME Q$_{10}$: SAFETY CONSIDERATIONS

Coenzyme Q$_{10}$ has an excellent safety profile. No significant side effects have been observed in clinical testing involving thousands of patients on long-term high-dose coenzyme Q$_{10}$ treatment other than mild GI symptoms in a few cases. Parallel with our studies on the effectiveness of coenzyme Q$_{10}$, we carried out extensive toxicological evaluation, first as preclinical and then as FDA Phase I clinical trial in terminal cancer patients at the Yale Medical School (New Haven, CT) during 1972–1974, and our results on the lack of any significant adverse effects of coenzyme Q$_{10}$ have been confirmed by others. In one recent clinical trial in patients with Parkinson's disease, high-dose coenzyme Q$_{10}$ (up to 1200 mg a day for 16 months) was found to be safe and well tolerated.[110] Chronic toxicity testing using rats at doses up to 1200 mg per kg per day for 1 year has not revealed any adverse changes.[111] Thus, the safety of high-dose coenzyme Q$_{10}$ is well documented. Furthermore, the *Physicians Desk Reference* (1999, p. 3286) categorically states, "Adverse effects — none reported."

Another observation that is relevant to the clinical applicability of coenzyme Q$_{10}$ is that the enhancement of the immune functions established in our program did not result from hyperplastic alterations of spleen, liver, and other organs directly involved in immune system responsiveness. These would be considered as undesirable side effects. Accordingly, the described effects most likely result from an increased activity of existing cells via improved bioenergetic balance.

36.4 REFLECTIONS AND CONCLUDING THOUGHTS

Despite the many advances in both the basic and clinical areas of bioenergetics, and particularly the critical role of coenzyme Q_{10}, progress can be characterized as slow-paced and sometimes disorganized. There are several underlying reasons for this unfortunate situation. The most important ones are as follows:

The pharmaceutical industry continues to have a negative attitude, despite convincing evidence of the therapeutic potential of coenzyme Q_{10}, and it shows a lack of interest in developing products based on coenzyme Q_{10}. The obvious reason, and perhaps a pragmatic one, is that there is no patent protection. A secondary reason may be that the supply of the bulk material is totally dependent on a few producers in Japan, with no alternative sources.

The U.S. government agencies have provided limited financial support for the clinical evaluation of coenzyme Q_{10}, with the exception of three major recent clinical trials funded by the NIH on three neurodegenerative diseases.

Although our understanding of the origin and pathogenesis of neoplasia is far from complete, substantial knowledge has been gained with respect to possible mechanisms and potential therapeutic agents to justify large-scale clinical trials with coenzyme Q_{10} in selected forms of cancers such as prostate and breast, alone or in combination with other therapeutics, as discussed earlier.

A more philosophical point that needs to be emphasized is that coenzyme Q_{10} is a nutrient and not a drug, and therefore it should not be treated as such. Being labeled as a nutrient hinders its acceptance as a potentially useful therapeutic agent by the pharmaceutical industry as well as by mainstream physicians. In this context, we need only to look at the status of carnitine. It is an important nutrient essential for cellular functions like coenzyme Q_{10}, and it is available as a nutritional supplement. Since its therapeutic efficacy has been well established in certain genetic disorders and also in kidney disease, carnitine is also approved and marketed as a drug in the U.S. Coenzyme Q_{10} should be viewed exactly in the same light.

In addition to continuing to explore the efficacy of coenzyme Q_{10}, research is also warranted on the therapeutic potential of derivatives of coenzyme Q_{10}. It is possible that we might identify a molecule that is perhaps more versatile and potent.

Notwithstanding these paramount and rather grim obstacles, data presented in this review indicate that coenzyme Q_{10} is an appropriate candidate for future clinical applications in the "diseases of bioenergetics" and, more specifically, in disease states with impaired immune system function. The incorporation of coenzyme Q_{10} in our, as of now, confined armament in the "war against cancer" should be intensified and facilitated.

In closing, it is appropriate and timely to quote Arthur Schopenhauer (1788–1860), an authority on the philosophy of pessimism, who wrote prophetically, "All truth passes through three stages: First, it is ridiculed; second, it is violently opposed; and third, it is accepted as self-evident."

REFERENCES

1. Stutman, O., Immunodeficiency and cancer, in *Mechanisms of Tumor Immunity*, Green, I., Cohen, S., and McCluskey, R.T., Eds., Wiley Medical Publications, New York, 1977, p. 27.
2. Burnet, F.M., The concept of immunological surveillance, *Progr. Exp. Tumor Res.*, 13, 1, 1970.
3. Whiteside, T.L. and Herberman, R.B., Role of human natural killer cells in health and disease, *Clin. Diagn. Lab. Immunol.*, 1, 125, 1994.

4. Thiele, H.G., Contemplations of the paradigm of self and nonself discrimination and on other concepts ruling contemporary immunology, *Cell. Mol. Biol.*, 48, 221, 2002.
5. Fuchs, E. and Matzinger, P., Is cancer dangerous to the immune system? *Semin. Immunol.*, 8, 271, 1996.
6. Dembic, Z., Immune system protects integrity of tissues, *Mol. Immunol.*, 37, 563, 2000.
7. Robles, D.T. and Eisenberg, G.S., Immunology primer, *Endocrinol. Metab. Clin. N. Amer.*, 31, 261, 2002.
8. Long, E.O., Tumor cell recognition by natural killer cells, *Semin. Cancer Biol.*, 12, 57, 2002.
9. Moretta, L., Biassoni, R., Battino, C., Mingari, M. C., and Moretta, A., Natural killer cells: a mystery no more, *Scand. J. Immunol.*, 55, 229, 2002.
10. Neuroimmunomodulation: molecular aspects, integrative systems, and clinical advances, *Ann. NY. Acad. Sci.*, 840, 1998.
11. Herzig, M. and Christofori, G., Recent advances in cancer research: mouse models of tumorigenesis, *Biochem. Biophys. Acta*, 1602, 97, 2002.
12. Balducci, L. and Extermann, M., Cancer and aging: An evolving panorama, *Hematol. Oncol. Clin. N. Amer.*, 14, 1, 2000.
13. Moos, W.H., Anti-aging drug discovery and development, *Summit*, 33, 2002.
14. Nichols, D.G. and Ferguson, S.J., in *Bioenergetics 3,* Academic Press, San Diego, CA, 2001.
15. Hockenbery, D.M., A mitochondrial Achilles' heel in cancer? *Cancer Cell*, 2, 1, 2002.
16. Ghildyal, P. and Manchanda, R., Neurotransmission by ATP: new insights, novel mechanisms, *Ind. J. Biochem. Biophys.*, 39, 137, 2002.
17. Crane, F.L., Hatefi, Y., Lester, R.L., and Widmer, C., Isolation of a quinone from beef heart mitochondria, *Biochim. Biophys. Acta*, 25, 220, 1957.
18. Wolf, D.E., Hoffman, C.H., Trenner, N.R., Arison, B.H., Shunk, C.H., Linn, B.D., McPherson, J.F., and Folkers, K., Structure studies on the coenzyme Q group, *J. Am. Chem. Soc.*, 80, 4752, 1958.
19. Crane, F.L., New functions for coenzyme Q, *Protoplasma*, 213, 127, 2000.
20. Crane, F.L., Biochemical functions of coenzyme Q$_{10}$, *J. Am. Coll. Nutr.*, 20, 591, 2001.
21. Estornell, E., Fato, R., Castelluccio, C., Cavazzoni, M., Parenti-Castelli, G., and Lenaz, G., Saturation kinetics of coenzyme Q in NADH and succinate oxidation in beef heart mitochondria, *FEBS Lett.*, 311, 107, 1992.
22. Bliznakov, E.G. and Wilkins, D.J., Biochemical and clinical consequences of inhibiting coenzyme Q$_{10}$ biosynthesis by lipid-lowering HMG-CoA reductase inhibitors (statins): a critical overview, *Adv. Ther.*, 15, 218, 1998.
23. Stocker, R., Possible health benefits of coenzyme Q$_{10}$ (invited article), *Linus Pauling Inst. Newsl.*, (fall/winter), 4–5, 2002.
24. Folkers, K., Survey on the vitamin aspects of coenzyme Q, *Int. J. Vit. Nutr. Res.*, 39, 334, 1969.
25. Ernster L. and Dallner, G., Biochemical, physiological and medical aspects of ubiquinone function, *Biochim. Biophys. Acta*, 1271, 195, 1995.
26. Littarru, G.P., *Energy and Defense*, Casa Editrice Scientifica Internazionale, Roma, Italy, 1995.
27. Geromel, V., Darin, N., Chretien, D., Benit, P., DeLonlay, P., Rotig, A., Munnich, A., and Rustin, P., Coenzyme Q$_{10}$ and idebenone in the therapy of respiratory chain diseases: rationale and comparative benefits, *Mol. Genet. Metab.*, 77, 21, 2002.
28. Langsjoen, P.H. and Langsjoen, A.M., Overview of CoQ10 in cardiovascular diseases, *Biofactors*, 9, 273, 1999.
29. Kagan, V.E. and Quinn, P.J., *Coenzyme Q: Molecular Mechanisms in Health and Disease,* CRC Press, Boca Raton, FL, 2001.
30. Ebadi, M., Marwah, J., and Chopra, R.K., *Mitochondrial Ubiquinone (Coenzyme Q$_{10}$)*, Vol. I and II, Prominent Press, Scottsdale, AZ, 2001.
31. Chopra, R.K., Goldman, R., Sinatra, S.T., and Bhagavan, H.N., Relative bioavailability of coenzyme Q$_{10}$ formulations in human subjects, *Int. J. Vit. Nutr. Res.*, 68, 109, 1998.
32. Bhagavan, H.N., Chopra, R.K., and Sinatra, S.T., Absorption and bioavailability of coenzyme Q$_{10}$, in *Mitochondrial Ubiquinone (Coenzyme Q$_{10}$)*, Vol. I, Ebadi, M., Marwah, J., and Chopra, R.K., Eds., Prominent Press, Scottsdale, AZ, 2001, chap. 3.
33. Miles, M.V., Horn, P., Miles, L., Tang, P., Steele, P., and DeGraw, T., Bioequivalence of coenzyme Q$_{10}$ from over the counter supplements, *Nutr. Res.*, 22, 919, 2002.
34. Zaghloul, A., Gurley, B., Khan, M., Bhagavan, H., Chopra, R., and Reddy, I., Bioavailability assessments of oral coenzyme Q10 formulations in dogs, *Drug Develop. Ind. Pharm.*, 28, 1195, 2002.

35. Kwong, L.K., Kamzalov, S., Rebrin, I., Bayne, A-C.V., Jana, C.K., Morris, P., Forster, M.J., and Sohal, R.S., Effects of coenzyme Q_{10} administration on its tissue concentrations, mitochondrial oxidant generation, and oxidative stress in the rat, *Free Rad. Biol. Med.*, 33, 627, 2002.

36. Wei, Y.H. and Lee, H.C., Oxidative stress, mitochondrial DNA mutation, and impairment of antioxidant enzymes in aging, *Exp. Biol. Med.*, 227, 671, 2002.

37. Chatgilialoglu, C. and O'Neill, P., Free radicals associated with DNA damage, *Exp. Gerontol.*, 36, 1459, 2001.

38. Olinski, R., Gackowski, D., Folksinski, M., Rozalski, R., Roszkowski, K., and Jaruga, P., Oxidative DNA damage: assessment of the role in carcinogenesis, atherosclerosis, and acquired immunodeficiency syndrome, *Free Radical Biol. Med.*, 33, 192, 2002.

39. Feng, Q., Kumagai, T., Torii, Y., Nakamura, Y., Osawa, T., and Uchida, K., Anticarcinogenic antioxidants as inhibitors against intracellular oxidative stress, *Free Radical Res.*, 35, 779, 2001.

40. Kelloff, G.J., Crowell, J.A., Steele, V.E., Lubet, R.A., Boone, C.W., Malone, W.A., Hawk, E.T., Lieberman, R., Lawrence, J.A., Kopelovich, L., Ali, I., Viner, J.L., and Sigman, C.C., Progress in cancer chemoprevention, *Ann. N.Y. Acad. Sci.*, 889, 1, 1999.

41. Laborit, H., de Brux, J., and Thuret, F., Temporary arrest of ascitic cancer growth in mice by intraperitoneal injection of liver mitochondrial suspension, *Agressologie*, 12, 325, 1971.

42. Davis, I.D., An overview of cancer immunotherapy, *Immunol. Cell Biol.*, 78, 179, 2000.

43. Agarwala, S.S., New applications of cancer immunotherapy, *Semin. Oncol.*, 29, 1, 2002.

44. Davis, L.S., Sackler, M., Brezinschek, R.I., Lightfoot, E., Bailley, J.L., Oppenheimer-Marks, N., and Lipsky, P.E., Inflammation, immune reactivity, and angiogenesis in a severe combined immunodeficiency model of rheumatoid arthritis, *Am. J. Pathol.*, 160, 357, 2002.

45. Klurfeld, D.M., Cholesterol as an immunomodulator, in *Human Nutrition — A Comprehensive Treatise, Vol. 8: Nutrition and Immunology*, Klurfield, D.M., Ed., Plenum Press, New York, 1993, p. 79.

46. McMichael, A.J., Jensen, O.M., Parkin, D.M., and Zaridze, D.G., Dietary and endogenous cholesterol in human cancer, *Epidemiol. Rev.*, 6, 192, 1984.

47. Bliznakov, E.G., Casy, A., and Premuzic, E., Coenzyme Q: stimulants of the phagocytic activity in rats and immune response in mice, *Experientia*, 26, 953, 1970.

48. Bliznakov, E.G., Protective effect of reticuloendothelial system stimulants in combination with chloroquine on Plasmodium Berghei infection in mice, in *The Reticuloendothelial System and Immune Phenomenon*, DiLuzio, N.R., Ed., Plenum Press, New York, 1971, p. 315.

49. Bliznakov, E.G., Effect of stimulation of the host defense system by coenzyme Q_{10} on dibenzpyrene-induced tumors and infection with Friend leukemia virus in mice, *Proc. Natl. Acad. Sci.*, 70, 390, 1973.

50. Bliznakov, E.G., Coenzyme Q_{10} in experimental infections and neoplasia, in *Biomedical and Clinical Aspects of Coenzyme Q*, Folkers, K. and Yamamura, Y., Eds., Elsevier, Amsterdam, 1977, p. 73.

51. Folkers, K., Morita, M., and McRee, Jr., J., The activities of coenzyme Q_{10} and vitamin B6 for immune responses, *Biochem. Biophys. Res. Commun.*, 193, 88, 1993.

52. Barbieri, B., Lund, B., Lundstrom, B., and Scaglione, F., Coenzyme Q10 administration increases antibody titer in hepatitis B vaccinated volunteers: a single blind placebo-controlled and randomized clinical study, *Biofactors*, 9, 351, 1999.

53. NCI, Coenzyme Q_{10} (PDQ), National Cancer Institute, Bethesda, MD, 2002.

54. NCI, Coenzyme Q_{10}: Questions and Answers, National Cancer Institute, Bethesda, MD, 2002.

55. Eggens, I., Biosynthesis of sterols and dolichol in human hepatomas, *Acta Chem. Scand., B*, 41, 7, 1987.

56. Portakal, O., Ozkaya, O., Erden, I.M., Bozan, B., Kosan, M., and Savek, I., Coenzyme Q_{10} concentrations and antioxidant status in tissues of breast cancer patients, *Clin. Biochem.*, 33, 279, 2000.

57. de Cavanagh, E.M., Honegger, A.E., Hofer, E., Bordenave, R.H., Bullorsky, E.O., Chasseing, N.A., and Fraga, C., Higher oxidation and lower antioxidant levels in peripheral blood plasma and bone marrow plasma from advanced cancer patients, *Cancer*, 94, 3247, 2002.

58. Jolliet, P., Simon, N., Barre, J., Pons, J.Y., Boukef, M., Paniel, B.J., and Tillement, J.P., Plasma coenzyme Q_{10} concentrations in breast cancer: prognosis and therapeutic consequences, *Int. J. Clin. Pharmacol. Ther.*, 36, 506, 1998.

59. Folkers, K., Osterborg, A., Nylander, M., Morita, M., and Mellstedt, H., Activities of vitamin Q_{10} in animal models and a serious deficiency in patients with cancer, *Biochem. Biophys. Res. Commun.*, 234, 296, 1997.

60. Mikhail, M.S., Palan, P.R., and Romney, S.L., Coenzyme Q10 and alpha-tocopherol concentrations in cervical intraepithelial neoplasia and cervix cancer, *Obstet. Gynecol.*, 97, S3, 2001.

61. Lockwood, K., Moesgaard, S., Hanioka, T., and Folkers, K., Apparent partial remission of breast cancer in "high risk" patients supplemented with nutritional antioxidants, essential fatty acids and coenzyme Q$_{10}$, *Mol. Aspects Med.*, 15, S231, 1994.

62. Lockwood, K., Moesgaard. S., and Folkers, K., Partial and complete regression of breast cancer in patients in relation to dosage of coenzyme Q$_{10}$, *Biochem. Biophys. Res. Commun.*, 199, 1504, 1994.

63. Lockwood, K., Moesgaard, S., Yamamoto, T., and Folkers, K., Progress on therapy of breast cancer with vitamin Q$_{10}$ and the regression of metastases, *Biochem. Biophys. Res. Commun.*, 212, 172, 1995.

64. Bui, T., Xu, Y.D., and Woodhouse, C., Co-enzyme Q10 level in HIV-positive patients, *Pharm. Res.*, 13 (Suppl.), 462, 1996.

65. Folkers, K., Langsjoen, P., Nara, Y., Muratsu, K., Komorowski, J., Richardson, P.C., and Smith, T.H., Biochemical deficiencies of coenzyme Q10 in HIV-infection and exploratory treatment, *Biochem. Biophys. Res. Commun.*, 153, 888, 1988.

66. Folkers, K., Brown, R., Judy, W.V., and Morita, M., Survival of cancer patients on therapy with coenzyme Q$_{10}$, *Biochem. Biophys. Res. Commun.*, 192, 241, 1993.

67. Bliznakov, E.G., Cardiovascular diseases, oxidative stress and antioxidants: the decisive role of coenzyme Q10, *Cardiovasc. Res.*, 43. 248, 1999.

68. Cortes, E.P., Gupta, M., Chou, C., Amin, V.C., and Folkers, K., Adriamycin cardiotoxicity: early detection by systolic time interval and possible prevention by coenzyme Q10, *Cancer Treat. Rep.*, 62, 887, 1978.

69. Tsubaki, K., Horiuchi, A., Kitani, T., Taniguchi, N., Masaoka, T., Shibata, H., Yonezawa, T., Tsubakio. T., Kawagoe, H., and Shinohara, Y., Investigation of the preventive effect of CoQ$_{10}$ against the side-effects of anthracycline antineoplastic agents, Gan, *To Kagaku Ryoho*, 11, 1420, 1984.

70. Karlsson, I., Folkers, K., Astrom, H., Jansson, E., Pernow, B., Holmgren, A., Mellstedt, H., and Diamant, B., Effect of Adriamycin on heart and skeletal muscle coenzyme Q10 (CoQ10) in man, in *Biomedical and Clinical Aspects of Coenzyme Q*, Folkers, K. and Yamamura, Y., Eds., Elsevier, Amsterdam, 1986, p. 241.

71. Iarussi, D., Auricchio, U., and Agretto, A., Protective effect of coenzyme Q$_{10}$ on anthracyclines cardiotoxicity: control study in children with acute lymphoblastic leukemia and non-Hodgkin lymphoma, *Mol. Aspects Med.*, 15, S207, 1994.

72. Hodges, S., Hertz, N., Lockwood, K., and Lister, R., CoQ$_{10}$: could it have a role in cancer management? *Biofactors*, 9, 365, 1999.

73. Folkers, K., *Cancer Chemother. Rep.*, The potential of coenzyme Q$_{10}$ (NSC-140865) in cancer treatment, *Cancer Chemother. Rep.*, 4, 19, 1974.

74. Berlin, A., The conquest of cancer, *Cancer Invest.*, 13, 540, 1995.

75. Dreosti, I.E., Nutrition, cancer, and aging, *Ann. N.Y. Acad. Sci.*, 854, 371, 1998.

76. Harman, D., Role of antioxidant nutrients in aging: overview, *Age*, 18, 51, 1995.

77. Harman, D., Heidrick, M.L., and Eddy, D.E., Free radical theory of aging, *Clin. Res.*, 24, 110A, 1976.

78. Ramasarma, T., Some radical queries, *Toxicology*, 148, 85, 2000.

79. Sohal, R.S., Mockett, R.J., and Orr, W.C., Mechanism of aging: an appraisal of the oxidative stress hypothesis, *Free Rad. Biol. Med.*, 33, 575, 2002.

80. Towards prolongation of the healthy life span, *Ann. N.Y. Acad. Sci.*, 854, 1998.

81. Martin, G.M., Mechanisms of senescence-complicationists versus simplificationists, *Mech. Age Dev.*, 123, 65, 2002.

82. Meissner, C., Mohamed, S.A., von Wurmb, N., and Oehmichen, M., Aging and mitochondria, in *Research in Legal Medicine*, Vol. 27, Schmidt-Romhild, Lubeck, Germany, 2002, p. 325.

83. Nicholls, D.G. and Budd, S.L., Mitochondria and neuronal survival, *Physiol. Rev.*, 80, 315, 2000.

84. Liu, J., Atamna, H., Kuratsune, H., and Ames, B.N., Delaying brain mitochondrial decay and aging with mitochondrial antioxidants and metabolites, *Ann. N.Y. Acad. Sci.*, 959, 133, 2002.

85. Tarnopolsky, M., Practical issues in the design of studies evaluating therapy in mitochondrial diseases, *MitoMatters*, 1, 3, 2002.

86. Bliznakov, E.G., Immunological senescence in mice and its reversal by coenzyme Q10, *Mech. Age Dev.*, 7, 189, 1978.

87. Bliznakov, E.G., Control and reversal of the immunological senescence, U.S. Patent 4,156,718, May 29, 1979.

88. Bliznakov, E.G., Coenzyme Q, the immune system and aging, in *Biomedical and Clinical Aspects of Coenzyme Q*, Vol. 3, Folkers, K. and Yamamura Y., Eds., Elsevier/North-Holland Biomedical Press, Amsterdam, 1981, p. 311.

89. Bliznakov, E.G., Restoration of Impaired Immune Functions in Aged Mice by Coenzyme Q, paper presented at 4th Int. Congr. Immunol. Workshop, The Immunobiology of Aging, Paris, 3.8.04, July 1980.

90. Bliznakov, E.G., Aging, mitochondria, and coenzyme Q10: the neglected relationship, *Biochimie*, 81,1131, 1999.

91. Fry, T.J. and Mackhall, C.L., Current concepts of thymic aging, *Springer Semin. Immunopathol.*, 24, 7, 2002.

92. Bliznakov, E.G., Casey, A., Kishi, T., Kishi, H., and Folkers, K., Coenzyme Q deficiency in mice following infection with Friend leukemia virus, *Int. J. Vit. Nutr. Res.*, 45, 388, 1975.

93. Shay, J.W. and Wright, W.E., When telomeres matter, *Science*, 291, 839, 2001.

94. Kalen, A., Appelkvist, E.L., and Dallner, G., Age related changes in the lipid compositions of rat and human tissue, *Lipids*, 24, 579, 1989.

95. Fahy, G.M., Li extension benefits of CoQ10 (coenzyme Q10), *Anti-Aging News*, 3, 1, 1983.

96. Coles, L.S. and Harris, S.B., Coenzyme Q10 and lifespan extension, in *Advances in Anti-Aging Medicine,* Mary Ann Liebert, New York, 1996, p. 205.

97. Lonnrot, K., Holm, P., Lagerstedt, A., Huhtala, H., and Alho, H., The effect of lifelong ubiquinone Q10 supplementation on Q9 and Q10 tissue concentrations and life span of male rats and mice, *Biochem. Mol. Biol. Int.*, 44, 727, 1998.

98. Lonnrot, K., Porsti, I., Alho, H., Wu, X., Hervonen, A., and Tolvanen, J.P., Control of arterial tone after long-term coenzyme Q10 supplementation in senescent rats, *Brit. J. Pharmacol.*, 124, 1500, 1998.

99. Linnane, A.W., Zhang, C., Kopsidas, G., Kovalenko, S., Papakostopoulos, P., Eastwood, H., Graves, S., and Richardson, M., Human aging and global function of coenzyme Q10, *Ann. N.Y. Acad. Sci.*, 959, 396, 2002.

100. Wouters-Wesseling, W., Rozendaal, M., Snijder, M., Graus, Y., Rimmelzwaan, G., de Groot, L., and Bindels, J., Effect of a complete nutritional supplement on antibody response to influenza vaccine in elderly people, *J. Gerontol.*, 57, M563, 2002.

101. Roberts, W.C., Getting more people on statins, *Am. J. Cardiol.*, 90, 683, 2002.

102. Kwak, B., Mulhaupt, F., Myit, S., and Mach, F., Statins as newly recognized type of immunomodulators, *Nat. Med.*, 6, 1399, 2000.

103. Mach, F., Immunosuppressive effects of statins, *Atheroscler.*, Suppl. 3, 17, 2002.

104. Bellanty, J.A., Immunologically mediated disease, in *Immunology III*, Bellanty, J.A., Ed., W.B. Saunders, Philadelphia, 1985, p. 346.

105. Bliznakov, E.G., Lipid-lowering drugs (statins), cholesterol, and coenzyme Q10: the Baycol case — a modern Pandora's box, *Biomed. Pharmacother.*, 56, 56, 2002.

106. Bliznakov, E.G., Coenzyme Q10, lipid-lowering drugs (statins) and cholesterol: a present day Pandora's box, *J. Am. Nutraceut. Assoc.*, 5, 32, 2002.

107. Bliznakov, E.G., Medical ethics, *Lancet*, 356, 1522, 2000.

108. Pelton, R., LaValle, J.B., Hawkins, E.B., and Krinsky, D.L., *Drug-Induced Nutrient Depletion Handbook,* Lexi-Comp, Hudson, OH, 2000.

109. Mitsui, T., Umaki, Y., Nagasawa, M., Akaike, M., Aki, K., Azuma, H., Ozaki, S., Odomi, M., and Matsumoto, T., Mitochondrial damage in patients with long-term corticosteroid therapy: development of oculoskeletal symptoms similar to mitochondrial disease, *Acta Neuropathol.*, 104, 260, 2002.

110. Shults, C.W., Oakes, D., Kieburtz, K., Beal, M.F., Haas, R., Plumb, S., Juncos, J.L., Nutt, J., Shoulson, I., Carter, J., Kompoliti, K., Perlmutter, J.S., Reich, R., Stern, M., Watts, R.L., Kurlan, R., Molho, E., Harrison, M., Lew, M., and the Parkinson Study group, Effect of coenzyme Q10 in early Parkinson disease: evidence of slowing of the functional decline, *Arch. Neurol.*, 59, 1541, 2002.

111. Williams, K.D., Maneke, J.D., AbdelHameed, M., Hall, R.L., Palmer, T.E., Kitano, M., and Hidaka, T., 52-Week oral gavage chronic toxicity study with ubiquinone in rats with a 4-week recovery, *J. Agr. Food Chem.*, 47, 3756, 1999.

37 Probiotics in the Prevention of Cancer

Karen Madsen and John Walker

CONTENTS

37.1 Introduction...624
37.2 Colon Cancer...624
    37.2.1 Rates and Prevalence..624
    37.2.2 Pathogenesis..624
    37.2.3 Role of Diet...625
        37.2.3.1 Pro-Carcinogenic Substances.......................................625
        37.2.3.2 Anticarcinogenic Substances 626
37.3 Anatomy and Physiology of the Colon 626
    37.3.1 Anatomy 626
    37.3.2 Colonic Functions...627
    37.3.3 Role and Functions of Colonic Bacteria.....................................627
    37.3.4 Factors Affecting Colonic Microflora 628
37.4 Definitions: Probiotics, Prebiotics, Synbiotics..628
    37.4.1 Probiotics 628
    37.4.2 Prebiotics 628
    37.4.3 Synbiotics 629
37.5 Microbial Ecology of the Human Gut 629
    37.5.1 Survival of Probiotics...629
    37.5.2 Strain Selection...630
37.6 Role of Probiotics and Prebiotics in Prevention
    of Colorectal Cancer 630
    37.6.1 Mechanisms of Protection..630
        37.6.1.1 Inhibition of Mutagenic Metabolic Activity
            of Intestinal Microflora 630
        37.6.1.2 Suppression of Tumor Activities 631
        37.6.1.3 Enhancing the Host's Immune Response 631
        37.6.1.4 Production of Protective Metabolites 632
        37.6.1.5 Binding of Carcinogens 632
        37.6.1.6 Alteration of Physico-Chemical
            Conditions in Colon 632
37.7 Clinical Trials Using Probiotics in the Treatment
    of Other Cancers 633
    37.7.1 Uterine Cancer..633
    37.7.2 Bladder Cancer 633
37.8 Conclusions..633
Acknowledgment...634
References 634

37.1 INTRODUCTION

Nearly 100 years ago, the Russian Nobel prizewinner Elie Metchnikoff observed a high life expectancy in Bulgarian peasants who ate large amounts of fermented milk products. He believed that the reason for their longevity was their consumption of lactic acid bacteria (lactobaccilli and bifidobacteria) in fermented milk products. He subsequently published his landmark paper stressing the importance of intestinal lactobaccilli in maintaining human health (Metchnikoff, 1907). Recently, his ideas regarding lactic acid bacteria have seen a major resurgence in popularity. The term *probiotic* was popularized by Fuller (Fuller, 1989) and was initially used in reference to microflora used as supplements in livestock feed. Fuller's initial definition of probiotics as "a live microbial feed supplement which beneficially affects the host by improving its microbial balance," has been expanded to include effects such as host immunomodulation. The idea that commensal microflora could have beneficial effects on human health is based upon empirical observations and rational scientific investigation. Indeed, commensal microflora appear to be necessary for the establishment of oral tolerance to food antigens, as well as having an active role in the maintenance of an intestinal microbial barrier against pathogens. However, experimental evidence has also linked certain intestinal microflora with the development of colon cancer and, conversely, probiotic bacteria with having a protective effect against the development of colon cancer.

37.2 COLON CANCER

37.2.1 Rates and Prevalence

Colorectal cancer (CRC) is the third leading cause of cancer deaths in North America (Boyle and Langman, 2000). Colorectal cancer affects 6% of men and women by age 75, and about half of diagnosed individuals will die from this disease. The 5-year survival rate depends on the stage of disease, with a 90, 65, and 10% rate of survival for localized, regional, and distant cancers, respectively (Boyle and Langman, 2000). Only ≈30% of cases are diagnosed at an early stage.

Incidence rates of colorectal cancer are high in all economically developed countries and are increasing (Boyle and Langman, 2000). Both genetic and environmental factors have been identified as causes of colorectal cancer. The importance of environment is clear from epidemiological data showing that substantial geographic variation exists in the incidence of colon cancer. The Western world suffers from the highest rates, while Third World countries show a relatively low incidence rate. Striking increases in the incidence of colorectal cancer occurs in groups that have migrated from low- to high-incidence areas, once they assume the cultural habits of their adopted country (Potter, 1999). Table 37.1 lists the known risk factors associated with the development of colorectal cancer.

37.2.2 Pathogenesis

Most human colorectal cancers arise from colonic adenomas. Adenomas are dysplastic, nonmalignant masses in the colon, and they are typically observed 10 to 15 years prior to the onset of cancer in both sporadic and familiar colon cancers (Lynch and Hoops, 2002). Early carcinomas can be found in adenomatous polyps, and removal of polyps reduces the subsequent incidence of colon cancer in treated patients. The formation of adenomatous polyps in the colon occurs when regulatory mechanisms of epithelial renewal become disrupted. Under normal conditions, epithelial cells in the colon are constantly renewed in the crypt region and migrate up the villus, where they differentiate into mature enterocytes. Upon reaching the tip of the villus (3 to 5 days), the cells are sloughed off into the lumen and are replaced. In adenomas, the proliferative zone shifts to include the entire crypt and villus, and the resultant increase in the number of epithelial cells causes the irregular infolding and branching that are characteristic of adenomas.

Aberrant crypt foci (ACF) are thought to be the intermediate between normal colonic mucosa and the adenomatous polyp. Because they are considered to be the earliest precursor lesion in the

---

TABLE 37.1
Risk Factors for Colorectal Cancer

Genetic

Age (increased incidence after age 60)
Inflammatory bowel disease
Family history of colorectal cancer
Autosomal dominant trait (familiar adenomatous polyposis syndrome)
Personal history of gynecologic cancer

Environmental

High-fat, high-meat diet
Low-fiber diet
Low calcium
Physical inactivity
High alcohol intake
Obesity
Smoking

---

progression to colon cancer, they are being used as early biomarkers for neoplasia in both human and animal studies. It is estimated that 85% of colorectal cancers are associated with a mutation in the adenomatous polyposis carcinomatous (APC) gene, although mutations in other tumor-suppressor genes such as MCC, DCC, and p53, and in oncogenes such as K-ras can also occur (Lynch and Hoops, 2002).

37.2.3 ROLE OF DIET

More than 20 years ago, Doll and Peto (Doll and Peto, 1981), using international comparisons of exposure prevalences and disease rates, estimated that up to 90% of colonic cancers had a primary dietary contribution. Since then, numerous epidemiological and intervention studies have attempted to identify those components of the diet responsible, but to date it has not been possible to identify conclusively which substances are responsible (Bingham, 2000; Potter, 1999).

37.2.3.1 Pro-Carcinogenic Substances

The role of red meat as a risk factor for colon cancer has proven controversial. While earlier prospective studies showed inconsistent findings for an association between red meat intake and colon cancer risk, other prospective studies have demonstrated statistically significant associations (Norat et al., 2002; Willett et al., 1990). The reasons why red meat is frequently, but not always, associated with an increased risk of colon cancer are not clear. It has been hypothesized that those persons who consume meat with a heavily browned surface may have an increased risk, due to the production of mutagenic heterocyclic aromatic amines during frying, grilling, or broiling meat at high temperatures (Sinha et al., 2001). In addition, a recent meta-analysis implicated processed meats to be associated with an increased cancer risk, as opposed to fresh meat (Sandhu et al., 2001).

In addition to meat consumption, there is a correlation between a high intake of saturated or animal fat and an elevated risk of CRC (Potter, 1999; Zock, 2001). Excess caloric intake has also been shown to be a risk factor, possibly due to hyperinsulineamia associated with excess caloric intake and obesity. Finally, a high level of alcohol intake is also a risk factor for CRC (World Cancer Research Fund, 1997).

37.2.3.2 Anticarcinogenic Substances

Data from case-control studies implicates both diets rich in vegetables and overall high-fiber diets as being protective against the development of CRC, although prospective studies are overall less convincing (Bingham, 2000). High calcium intake has also been suggested to have a protective effect, possibly through binding cytotoxic bile and fatty acids (Kleibeuker, 1996). Consumption of large quantities of dairy products such as yogurt and fermented milk containing *Lactobacillus* or *Bifidobacterium* may be related to a lower incidence of colon cancer (Shahani and Ayebo, 1980). Epidemiological studies from Finland demonstrated that, despite a high fat intake, colon cancer incidence was lower in that country, possibly due to a high consumption of milk, yogurt, and other dairy products (Malhotra, 1977). In two population-based case-control studies of colon cancer, an inverse association was observed for yogurt (Peters et al., 1992) and cultured milk consumption (Young and Wolf, 1988), adjusted for potential confounding variables. However, two American prospective studies, the 1980–1988 follow-up of the Nurses Health Study and the 1986–1990 follow-up of the Health Professionals Follow-up Study did not provide evidence that intake of dairy products was associated with a decreased risk of colon cancer (Kampman et al., 1994).

37.3 ANATOMY AND PHYSIOLOGY OF THE COLON

37.3.1 ANATOMY

The colon is 4.5 to 5 ft long, with a diameter of 2.5 in. The colon can be functionally divided through the transverse colon into two parts: the right colon, consisting of the cecum and ascending colon; and the left, consisting of the descending colon, sigmoid colon, and rectum (Figure 37.1). Small intestinal contents enter into the colon through the ileocecal valve. The vermiform appendix extends from the base of the cecum and consists of a blind tube of lymph tissue. Two sphincters control movement through the colon: the ileocecal valve between the ileum and the colon and the O'Beirne sphincter that controls the movement of fecal material through the sigmoid colon into the rectum. Smooth muscle surrounds the anal canal, forming the internal anal sphincter, while striated muscle overlaps and forms the external anal sphincter. In the cecum and colon, the longitudinal muscle layer consists of three longitudinal bands called teniae coli. The circular muscles of the colon form the haustra.

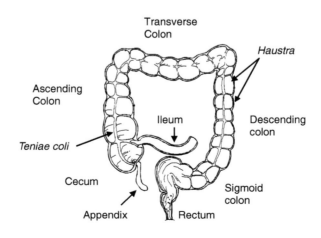

FIGURE 37.1

37.3.2 COLONIC FUNCTIONS

Primary functions of the right colon (cecum and ascending colon) include absorption of water and electrolytes and fermentation of undigested sugars. In contrast, the left colon (descending colon, sigmoid colon, and rectum) is primarily involved in storage and evacuation of fecal material. The colon is highly efficient at absorbing water. Under normal physiological conditions, approximately 1.5 l of fluid enters the colon daily, but only about 100 to 200 ml is excreted in the stool. The maximal absorptive capacity of the colon is approximately 4.5 l per day.

37.3.3 ROLE AND FUNCTIONS OF COLONIC BACTERIA

Bacteria are found throughout the entire intestine but are concentrated in the colon. The stomach and small intestine have a sparse microflora of approximately 10^5 colony forming units (CFU)/ml contents. In the ileum, microflora concentrations increase to 10^8 to 10^9, finally reaching 10^{11} to 10^{12} CFU in the colon. It is estimated that more than 400 to 500 bacterial species are inhabitants of the human colon, although five genera account for the majority of anaerobic bacteria: *Bacteroides*, *Eubacterium*, *Bifidobacterium*, *Peptostreptococcus*, and *Fusobacterium* (Simon and Gorbach, 1986) (Figure 37.2). The typical intestinal flora in humans eating a Western-style diet consists of the following organisms, in decreasing order of concentration: (1) bacteroides, eubacteria, peptostreptococci, and bifidobacteria (10^{10} to 10^{11} CFU/g feces); (2) enterobacteria and streptococci (10^8 to 10^9 CFU/g feces), with the former including mainly *Escherichia coli* and some klebsiella and proteus species; (3) lactobacilli (10^5 to 10^8 CFU/g); and (4) clostridia and staphylococci.

Various other facultative and aerobic bacterial strains are also found sporadically throughout the small and large intestine. Colonic microflora colonize the intestine within hours of birth, with the first organisms being enterobacteria and enterococci (Lejeune, 1984). In breast-fed infants, the bifidobacteria count then increases dramatically, while lactobacilli and bacteroides counts increase to a lesser amount, and the enterobacteria and enterococci counts decrease. The primary fecal flora in breast-fed infants consists of the genus *Bifidobacterium*, with *B. bifidum*, *B. infantis*, *B. breve*, and *B. longum* being the predominant strains. In formula-fed infants, the enterobacteria and enterococci counts remain stable, while bifidobacteria and bacteroides increase. Thus, while both breast-fed and formula-fed infants have a preponderance of bifidobacteria, formula-fed infants have other microorganisms at greater frequency and at higher levels. The introduction of solid food causes a major shift in bacterial strains colonizing the colon, with a rise of enterobacteria and enterococci, followed by colonization with *Bacteroides* spp., clostridia, and anaerobic streptococci. The composition

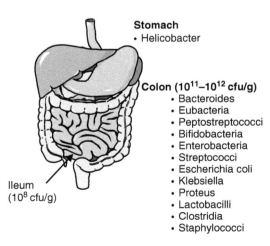

FIGURE 37.2

of human colonic microflora changes with advancing age. In weaned infants, children, and adults, the composition remains relatively stable.

37.3.4 FACTORS AFFECTING COLONIC MICROFLORA

Numerous environmental factors influence the intestinal microflora. Diet, infections, and the use of antibiotics or other drugs can significantly alter the types and numbers of bacterial species present. In addition, several bacteria, such as *Bacteroides*, *Enterobacter*, and *Enterococcus*, can release compounds that act as growth stimulators for other species, including bifidobacteria (Naidu et al., 1999). Thus, the levels of lactic acid bacteria in the colon may be regulated in part by the presence or absence of other bacterial species. Furthermore, factors such as gastric acidity, intestinal motility, immune status of the host, and the presence of adherence factors can alter colonization. Although ethnic origin and climate do not appear to be major determinants of intestinal microflora (Hill et al., 1971), geographic differences do exist in type of colonic microflora, although it is thought that differences are linked with the diets associated with different geographical regions.

37.4 DEFINITIONS: PROBIOTICS, PREBIOTICS, SYNBIOTICS

37.4.1 PROBIOTICS

Oral probiotics are living microorganisms that, upon ingestion, affect the host in a beneficial manner by modulating mucosal and systemic immunity, as well as improving nutritional and microbial balance in the intestinal tract (Fuller, 1989; Madsen, 2001). The main probiotic preparations currently on the market belong to a large group of bacteria designated as lactic acid bacteria (lactobacilli, streptococci, bifidobacteria) that are important and normal constituents of the human gastrointestinal microflora (Table 37.2). However, studies are also investigating potential probiotic roles of other microbes such as yeast (*Saccharomyces boulardii*), which are not normally found in the gastrointestinal tract (Goossens et al., 2003).

37.4.2 PREBIOTICS

Prebiotics are nondigestible food ingredients that beneficially affect the host by selectively stimulating the growth or activity of one or a limited number of probiotic bacterial species already established in the colon (Gibson and Roberfroid, 1995). Substances that act as prebiotics cannot

TABLE 37.2
Common Probiotics

Probiotics			
Lactobacilli	**Bifidobacteria**	**Streptococcus**	**Yeasts**
L. acidophilus	*B. bifidum*	*Lactis salivarius* subsp. *thermophilus*	*Saccharomyces boulardii*
L. casei	*B. infantis*		*S. cerevisiae*
L. delbrueckii subsp. *Bulgaricus*	*B. longum*		
L. reuteri	*B. thermophilum*		
L. brevis	*B. adolescents*		
L. cellobiosus			
L. curvatus			
L. fermentum			
L. plantarum			
L. rhamnosus			
L. salivarius			

be hydrolyzed or absorbed in the upper part of the gastrointestinal tract, but are instead available as substrates for the indigenous bacteria in the colon. Prebiotics act to modify the composition of the colonic microflora in such a way to allow the growth of health-promoting lactobacilli and bifidobacteria. Nondigestible oligosaccharides in general, and fructo-oligosaccharides in particular, are prebiotics. These low-molecular-weight carbohydrates are found naturally in onions, garlic, leeks, chicory, artichokes, and some cereals. Other oligosaccharides (raffinose and stachyose) are found in beans and peas. These molecules are also being produced industrially, and a number of new prebiotic preparations are being developed for the market. The beneficial effect of prebiotic intake on the host may result from either suppression of the growth of harmful microorganisms or the stimulation of favorable organisms (mainly lactobacilli and bifidobacteria).

37.4.3 SYNBIOTICS

The use of products in which a probiotic and a prebiotic are combined is also being investigated. The rationale for this combination is the belief that the presence of a prebiotic will help improve survival of the probiotic bacteria during the passage through the upper intestinal tract, resulting in a more efficient implantation in the colon. Moreover, it is believed that the prebiotic will also have a stimulating effect on the growth of both the exogenous (probiotic) and endogenous bacteria.

37.5 MICROBIAL ECOLOGY OF THE HUMAN GUT

37.5.1 SURVIVAL OF PROBIOTICS

It is estimated that more than 400 bacterial species are inhabitants of the human intestinal tract. However, among these, only 30 to 40 species constitute some 99% of the mass of intestinal flora (Figure 37.2). Although environmental factors and physiological interactions can modulate the distribution of the microflora, diet appears to be the major factor that regulates the frequency and concentration of individual species of microorganisms that colonize the gut.

Bacteria entering the mouth are washed with saliva into the stomach. Most bacteria are destroyed by gastric acid, resulting in a very sparse bacterial population in the upper small bowel, because only the most acid-resistant organisms survive transit through the stomach. The only known bacteria to survive and colonize the stomach is *Helicobacter*. The small intestine constitutes a zone of transition between the sparsely populated stomach and the luxuriant bacterial flora of the colon. In addition, bile acids, bicarbonate, lactozyme, mucins, peristaltic contractions, and antimicrobial peptides all contribute to the relative scarcity of bacterial colonization in the small intestine. In the colon, all available habitats are occupied by indigenous microorganisms. At least three major bacterial habitats have been described: the lumen of the gastrointestinal tract, the mucus gel that overlies the epithelium, and the adherence of bacteria to receptors on mucosal epithelial cells (Berg, 1996).

One of the major problems faced by supplemental oral probiotics is how to ensure survival of the microbe during the passage from mouth to the colon. Indeed, microbial strains used as probiotics must be both acid and bile resistant. Whether or not it is necessary for probiotic bacteria to have the ability to colonize the colon for long-term survival or benefit is not known. For example, common commercial strains such as *L. bulgaricus* and *L. acidophilus* are not adhesive in humans. However, convincing mucosal adhesiveness has been shown for *L. plantarum* strains 299 and 299V, *L. rhamnosus* strains GG and 271, and, recently, *L. acidophilus* strain LA1, *L. salivarius*, and *B. longum infantis*. It must be remembered that when using preparations of microbes unable to adhere to the colonic mucosa, continuous consumption is necessary to maintain any beneficial effects. However, even those strains such as *L. rhamnosus*, which does adhere to mucosa, gradually disappear by approximately 2 weeks after the end of administration of the bacteria (Alander et al., 1999).

37.5.2 STRAIN SELECTION

Probiotic bacteria differ on the basis of genus, species, and strains. Indeed, strains of the same species vary widely in traits such as expression of enzymes, types of inhibitors produced, carbohydrate fermentation patterns, resistance to acid and bile, ability to colonize the gastrointestinal tract, and clinical efficacy (Berg, 1996; Norat et al., 2002). Furthermore, the active principle has not always been associated with live bacteria, as some immune system modulation activities and macromolecular degradation have been linked to nonviable bacterial components such as enzyme activities or fermentation products (Huttner and Bevins, 1999).

For probiotic bacteria to be used in the treatment of human disease, they should meet the following criteria. First, the bacterial strains must be of human origin. Second, they need to exhibit nonpathogenic behavior and be resistant to pH, gastric juice, bile, and pancreatic secretions. It is believed, although unproven, that it is not the consumed dose that is relevant, but the dose reaching the target site (i.e., colon). Thus, it is conceivable that strains particularly suited to survive passage through the intestinal tract could be consumed in lower numbers than strains that are less able to survive passage. Finally, strains in use must exhibit beneficial functional activity when present in the gut lumen. The activity of individual strains of bacteria may be enhanced or, indeed, even eliminated by the presence of other strains present in the lumen.

37.6 ROLE OF PROBIOTICS AND PREBIOTICS IN PREVENTION OF COLORECTAL CANCER

The relationship between diet, colonic microflora, and the incidence of colorectal cancer is complex. Alterations in diet affect the growth of microflora, and particular bacterial species present in the colon can have either protective or detrimental effects on the host. Indeed, while certain luminal bacteria have been implicated in both the pathogenesis and etiology of colon cancer (Arimochi et al., 1997; Gallaher et al., 1996; Mallett and Rowland, 1990), studies performed in animal models of colon cancer have generally demonstrated a protective effect of probiotic bacteria against tumor development (Bolognani et al., 2001; Gallaher et al., 1996; Goldin et al., 1996; Hirayama and Rafter, 2000). In addition, a wide range of studies in rodent models on the effect of dietary prebiotics on cancer pathogenesis has consistently demonstrated a reduction in colonocyte DNA damage, aberrant crypt foci in the colon, number of tumors per animal, size and growth rate of tumors, mean survival time, and increase in life span (Pierre et al., 1997). Furthermore, in all animal studies where mixtures of probiotics and prebiotics were examined, the effect of the combination was greater than the sum of the two separately, suggesting a synergistic effect of symbiotic compounds. Based on these studies, considerable attention is being paid to dietary components that can influence the colonic microflora as well as supplemental probiotics as strategies for colon cancer prevention.

37.6.1 MECHANISMS OF PROTECTION

37.6.1.1 Inhibition of Mutagenic Metabolic Activity of Intestinal Microflora

Certain bacterial enzymes, including β-glucuronidase, azoreductase, nitroreductase, 7-α-dehydroxylase, and 7-α-dehydrogenase, have been implicated in the pathogenesis of colorectal cancer by converting precarcinogenic compounds in the colon into carcinogens (Arimochi et al., 1997; Hayatsu and Hayatsu, 1993; Mallett and Rowland, 1990). Toxic and potentially carcinogenic substances are frequently metabolized and conjugated in the liver before being excreted via bile into the small intestine. These compounds pass into the colon, where bacterial enzymes act to hydrolyze the conjugates and release the parent toxic compound into the colonic lumen. The presence and activity of these enzymes are lowest among lactic acid bacteria (lactobacilli and bifidobacteria). Thus — in that high levels of β-glucuronidase activity are associated with high-risk diets for colon cancer, and lower levels with low-risk diets — it has been suggested that by increasing levels of lactobacilli and bifidobacteria in the colon and decreasing the levels of other species, a corresponding decrease in the production of carcinogenic compounds would occur. Supporting this concept are studies in rodents, showing that the addition of lactobacilli to the diet significantly reduced the activity of β-glucuronidase in the colon (Rowland et al., 1998). In addition, lactobacilli have also been shown to reduce the specific activities of fecal enzymes in humans (Goldin and Gorbach, 1984). Goldin and Gorbach (Goldin and Gorbach, 1984) looked at the effect of feeding *Lactobacillus acidophilis* strains NCFM and N-2 on the activity of three bacterial enzymes — β-glucuronidase, nitroreductase, and azoreductase — and found that both *Lactobacillus* strains caused a significant reduction in all three enzymes after 10 days of feeding. However, the effect was reversed upon removal of the *Lactobacillus* from the diet, suggesting that continual consumption is necessary for the effect.

37.6.1.2 Suppression of Tumor Activities

The exposure of tumors to certain lactic acid bacteria results in suppression of tumor growth (Naidu et al., 1999). Antitumor activity of blastolysin, a preparation consisting of glycopeptide fragments of the cell wall of *L. bulgaricus*, was shown by Bogdanov et al. (Bogdanov et al., 1977). In mice, blastolysin suppressed tumor growth, including sarcoma S-180, leukemia P-388, plasmacytoma MOPC-315, adenocarcinoma AKATOL, melanosarcoma B-16, carcinoma LIC, and spontaneous tumors. Other studies have shown that a peptidoglycan isolated from *B. infantis* is effective in causing tumor regression in mice (Sekine et al., 1985). Biffi et al. (Biffi et al., 1997) examined the ability of milk fermented by five bacterial strains (*B. infantis*, *B. bifidum*, *B. animalis*, *L. acidophilus*, and *L. paracasei*) to inhibit the growth of the MCF-7 breast cancer cell line. Growth inhibition was induced by all fermented milks, with *B. infantis* and *L. acidophilus* being the most effective. The biological basis for the ability of lactic acid bacteria to suppress tumor growth may be related to either bacterial-induced reductions in neoplastic enzyme activity, bacterial-induced enhancement of colonic enzymatic activity, or to immunostimulation of the host.

37.6.1.3 Enhancing the Host's Immune Response

The ability of lactic acid bacteria to modulate various pathways of the mucosal and systemic immune system has been well documented (Kato et al., 1983; Matsuzaki, 1998; Matsuzaki et al., 1996; Norat et al., 2002). Indeed, numerous studies suggest that lactic acid bacteria play an important role and function as an integral component of the host's immunoprotective system by stimulating specific and nonspecific immune responses. Specific immunity may be induced by gut immune responses at the mucosal level. In the small intestine, M cells from Peyer's patches continually sample luminal antigens. Depending upon the antigen encountered, precursor T and B cells will undergo proliferation and differentiate, then migrate into the systemic circulation. In mice, Kato et al. (Kato et al., 1981) showed that intraperitoneal (i.p.) treatment with lactobacilli could prolong the life of mice implanted with sarcoma-180 or L1210 leukemia. This protective effect was reduced by concurrent treatment with the antimacrophage agent, carrageenan, and in T-cell deficient athymic nude mice, suggesting a macrophage-dependent antitumor effect. Further studies by Matsuzaki et al. (Matsuzaki et al., 1996) examining the effects of lactobacilli on antitumor activity and cytokine production in Meth A fibrosarcoma-bearing BALB/c mice found that the beneficial effect of the lactobacilli was inhibited by anti-CD3 and anti-CD8 antibodies, but not by anti-CD4 antibodies, suggesting that CD8+ T cells were involved. Intraperitoneal administration of *Lactobacillus casei* strain Shirota (LcS) has been shown to induce the production of several cytokines, including IFNγ, IL-1β, and TNFα. This increase in cytokine production was associated with an inhibition of tumor growth and increased survival (Matsuzaki, 1998).

37.6.1.4 Production of Protective Metabolites

There is some evidence to suggest that soluble compounds produced by lactic acid bacteria may act directly on tumor cells to inhibit their growth. Arimochi et al. (1997) showed (a) an inhibitory effect of *L. acidophilus* on azoxymethane-induced ACF formation in rat colon and (b) enhanced removal of O^6-methylguanine from the colonic mucosal DNA. This inhibitory effect was found in culture supernatants and not from bacterial cells. Furthermore, other studies have shown that dietary administration of lyophilized cultures of *B. longum* suppressed azoxymethane-induced colonic tumor development, along with a decrease in colonic mucosal cell proliferation and colonic mucosal and tumor ornithine decarboxylase and ras-p21 activities (Reddy, 1998).

37.6.1.5 Binding of Carcinogens

Lactic acid bacteria are able to bind and inactivate certain mutagenic compounds. Morotomi and Mutai (1986) showed that 22 strains of commensal bacteria were able to bind 3-amino-1,4-dimethyl-5H-pyrido[4,3-*b*] indole (Trp-P-1) and 3-amino-1-methyl-5H-pyrido[4,3-*b*] indole (Trp-P-1) in a pH-dependent fashion. Binding of these compounds subsequently significantly reduced their mutagenicity. Further work by Zhang and Ohta (1993) demonstrated that freeze-dried lactic acid bacteria reduced the absorption of Trp-1 from the small intestine in rats, which resulted in decreased levels of Trp-1 found in the blood. Pool-Zobel et al. (1996) studied the ability of *L. acidophilus*, *L. gasseri* P79, *L. confusus* DSM20196, S. *thermophilus* NCIM 50083, *Bifidobacterium breve*, and *Bifidobacterium longum* to prevent the induction of DNA damage by N-methyl-N'-nitro-N-nitrosoguanidine (MNNG) in rat colonic cells. Metabolically active *L. acidophilus*, *L. confusus*, *L. gasseri*, *B. longum*, and *B. breve*, as well as the peptidoglycan fraction and whole freeze-dried cells, prevented MNNG-induced DNA damage. However, cytoplasm and cell wall were inefficient. *In vitro* studies have demonstrated that several lactobacilli strains are able to bind numerous food-borne mutagens, including the amino acid pyrolysates 3-amino-1-4-dimethyl-5H-pyrido[4,3-*b*]indole, 3-amino-1-methyl-5H-pyrido[4,3-*d*]indole, and 2-amino-6-methyl-dipyrido[1,2-3',2'-*d*]imidazole (Thyagaraja and Hosono, 1994). However, no significant binding to other mutagens, such as aflatoxins, was observed.

Orrhage et al. (1994) studied the binding capacity of lactobacilli strains for mutagenic heterocyclic amines formed during cooking of protein-rich foods. The mutagens Trp-P-2, PhIP, IQ, and MeIQx were bound with markedly different efficiencies by the lactobacilli. Trp-2 was almost completely bound, and the binding was irreversible. The binding of PhIP reached approximately 50%, but IQ and MeIQx were only slightly bound. Again, pH was of major importance to the binding capacity of the probiotic bacterial strains. These results suggest that mutagenic compounds commonly found in the Western meat-rich diet can be bound to lactic acid bacteria found normally in the large intestine and subsequently inactivated.

37.6.1.6 Alteration of Physico-Chemical Conditions in Colon

In the colon, lactic acid bacteria break down undigested polysaccharides, resistant starches, and fiber, resulting in the formation of short-chain fatty acids (acetate, propionate, butyrate) as fermentation products. Increased production of short-chain fatty acids leads to a decrease in the luminal pH. A low pH in feces has been associated with a reduced incidence of colon cancer in various populations (Malhotra, 1977; Segal et al., 1995). Butyrate is known to modify gene expression in colonic epithelial cells, as well as directly enhancing cell proliferation in normal cells and suppressing proliferation in transformed cells. In addition, apoptosis is increased in transformed cells but inhibited in normal cells in the presence of butyrate (Hague et al., 1995; Hass et al., 1997; Marchetti et al., 1997). Butyrate can protect colonic cells from oxidative damage induced by hydrogen peroxide, and it increases glutathione transferase π in colonic cells (Treptow-van Lishaut et al., 1999). Glutathione transferase π, the most abundant glutathione transferase species in colon cells, is important for the detoxification of both electrophilic products and compounds associated with oxidative stress.

Decreased colonic luminal pH induced by supplemental probiotic bacteria will also inhibit the growth of other potentially pathogenic bacteria. For instance, consumption of fermented milk containing *L. acidophilus* has been shown to reduce the counts of fecal putrefactive bacteria, such as coliforms, and increase the levels of lactobacilli (Shahani and Ayebo, 1980), suggesting that supplemental probiotics may protect against cancer by suppressing the putrefactive/pathological organisms that may be involved in the production of tumor promoters and putative precarcinogens.

Dietary fat has been implicated as a risk factor for colon cancer, possibly due to the enhanced levels of bile acids that are released. Bacterial 7α-dehydroxylase acts on primary bile acids to release secondary bile acids, which have a cytotoxic effect on the colonic epithelium. *Lactobacillus acidophilus* administration has been shown to reduce the levels of soluble bile acids in feces (Pigeon, 2002).

Lactic acid bacteria have been shown to increase colonic NADPH-cytochrome P450 reductase activity (Pool-Zobel et al., 1996) and glutathione *S*-transferase levels (Challa et al., 1997), and to reduce hepatic uridine diphosphoglucuronyl transferase activity (Abdelali, 1995). All of these enzymes are involved in the metabolism of carcinogens in rats.

37.7 CLINICAL TRIALS USING PROBIOTICS IN THE TREATMENT OF OTHER CANCERS

37.7.1 UTERINE CANCER

A few clinical trials have been undertaken looking at the use of lactic acid bacteria as adjunctive therapy in the treatment of various types of cancer. The clinical efficacy of heat-killed LC9018 (*Lactobacillus*) in combination with radiation was evaluated in a randomized controlled trial on 61 patients with carcinoma of the uterine cervix of Stage IIB or III (Okawa et al., 1989). Supplementation with LC9018 enhanced the therapeutic effect of irradiation and also protected the patients from leukopenia during radiotherapy. The combination therapy also prolonged survival and the relapse-free interval compared with radiation alone.

37.7.2 BLADDER CANCER

Aso and Akazan (Aso and Akazan, 1992) conducted a randomized controlled trial with 58 patients with bladder cancer to investigate the safety and efficacy of Biolactis® powder (*L. casei*) in preventing recurrence after transurethral resection of the bladder tumor. Biolactis prolonged disease-free duration by 1.8 times (350 days) compared with that seen in the control group (195 days). Another study examined Biolactis in the prophylaxis of recurrence in three subgroups of patients with bladder cancer: (1) with primary multiple tumors, (2) with recurrent single tumors, and (3) with recurrent multiple tumors (Aso et al., 1995). Biolactis was effective in groups 1 and 2 but was ineffective in group 3. These studies suggest that oral administration of lactic acid bacteria may have a role as adjunctive therapy in the prevention of recurrence of bladder and uterine cancer. One hypothesis for the prevention or delay of bladder tumor development by lactic acid bacteria involves the ability of the bacteria to bind to mutagenic compounds in the intestine, thus decreasing their absorption and subsequent urinary excretion. Indeed, studies by Hayatsu (Hayatsu and Hayatsu, 1993) have shown that intake of freeze-dried *L. casei* for 3 weeks reduced the urinary excretion of mutagens after a test meal by 50% in six volunteers, compared with pretest values.

37.8 CONCLUSIONS

The use of probiotic strains of bacteria to prevent the development and progression of cancer is an interesting concept that holds promise. Carefully designed epidemiological studies are needed in the human population to corroborate the numerous experimental findings from animal models and cell culture studies. More work also needs to be done to investigate the mechanisms involved in cancer progression as well as the role of intestinal microflora in the process. In addition, identification of specific strains that demonstrate antitumor effects and the mechanisms by which these effects are mediated will aid in the design of effective therapeutic strategies.

ACKNOWLEDGMENT

Dr. Madsen is supported as an Alberta Heritage Foundation for Medical Research senior scholar. Crohn's and Colitis Foundation of Canada and the Canadian Institutes for Health Research are also acknowledged.

REFERENCES

Abdelali, H., Cassand, P., Soussotte, V., Daubeze, M., Bouley, C., and Narbonne, J.F., Effect of dairy products on initiation of precursor lesions of colon cancer in rats, *Nutr. Cancer*, 24, 121–132, 1995.

Alander, M., Satokari, R., Korpela, R., Saxelin, M., Vilpponen-Salmela, T., Mattila-Sandholm, T., and von Wright, A., Persistence of colonization of human colonic mucosa by a probiotic strain, *Lactobacillus rhamnosus* GG, after oral consumption, *Appl. Environ. Microbiol.*, 65, 351–354, 1999.

Arimochi, H., Kinouchi, T., Kataoka, K., Kuwahara, T., and Ohnishi, Y., Effect of intestinal bacteria on formation of azoxymethane-induced aberrant crypt foci in the rat colon, *Biochem. Biophys. Res. Commun.*, 238, 753–757, 1997.

Aso, Y. and Akazan, H., Prophylactic effect of a *Lactobacillus casei* preparation on the recurrence of superficial bladder cancer: BLP study group, *Urol. Int.*, 49, 125–129, 1992.

Aso, Y., Akazan, H., Kotake, T., Tsukamoto, T., Imai, K., and Naito, S., Preventive effect of a *Lactobacillus casei* preparation on the recurrence of superficial bladder cancer in a double-blind trial: BLP study group, *Eur. Urol.*, 27, 104–109, 1995.

Berg, R.D., The indigenous gastrointestinal microflora, *Trends Microbiol.*, 4, 430–435, 1996.

Biffi, A., Coradini, D., Larsen, R., Riva, L., and Di Fronzo, G., Antiproliferative effect of fermented milk on the growth of a human breast cancer cell line, *Nutr. Cancer*, 28, 93–99, 1997.

Bingham, S.A., Diet and colorectal cancer prevention, *Biochem. Soc. Trans.*, 28, 12–16, 2000.

Bogdanov, I.G., Velichkov, V.T., Gurevich, A.I., Dalev, P.G., and Kolosov, M.N., Antitumor effect of glycopeptides from the cell wall of *Lactobacillus bulgaricus*, *Biull. Eksp. Biol. Med.*, 84, 709–712, 1977.

Bolognani, F., Rumney, C.J., Pool-Zobel, B.L., and Rowland, I.R., Effect of lactobacilli, bifidobacteria and inulin on the formation of aberrant crypt foci in rats, *Eur. J. Nutr.*, 40, 293–300, 2001.

Boyle, P. and Langman, J.S., ABC of colorectal cancer: epidemiology, *BMJ*, 321, 805–808, 2000.

Challa, A., Rao, D.R., Chawan, C.B., and Shackelford, L., *Bifidobacterium longum* and lactulose suppress azoxymethane-induced colonic aberrant crypt foci in rats, *Carcinogenesis*, 18, 517–521, 1997.

Doll, R. and Peto, R., The causes of cancer: quantitative estimates of avoidable risks of cancer in the United States today, *J. Natl. Cancer Inst.*, 66, 1191–1308, 1981.

Eriyamremu, G.E. and Adamson, I., Early changes in energy metabolism in rats exposed to an acute level of deoxycholate and fed a Nigerian-like diet, *Ann. Nutr. Metab.*, 38, 174–183, 1994.

Fuller, R., Probiotics in man and animals, *J. Appl. Bacteriol.*, 66, 365–378, 1989.

Gallaher, D.D., Stallings, W.H., Blessing, L.L., Busta, F.F., and Brady, L.J., Probiotics, cecal microflora, and aberrant crypts in the rat colon, *J. Nutr.*, 126, 1362–1371, 1996.

Gibson, G.R. and Roberfroid, M.B., Dietary modulation of the human colonic microbiota: introducing the concept of prebiotics, *J. Nutr.*, 125, 1401–1412, 1995.

Goldin, B.R. and Gorbach, S.L., The effect of milk and *Lactobacillus* feeding on human intestinal bacterial enzyme activity, *Am. J. Clin. Nutr.*, 39, 756–761, 1984.

Goldin, B.R., Gualtieri, L.J., and Moore, R.P., The effect of *Lactobacillus* GG on the initiation and promotion of DMH-induced intestinal tumors in the rat, *Nutr. Cancer*, 25, 197–204, 1996.

Goossens, D., Jonkers, D., Stobberingh, E., Van den Bogaard, A., Russel, M., and Stockbrugger, R., Probiotics in gastroenterology: indications and future perspectives. Scand. J. Gastroenterol., Suppl., 239, 15–23, 2003.

Hague, A., Elder, D.J., Hicks, D.J., and Paraskeva, C., Apoptosis in colorectal tumour cells: induction by the short chain fatty acids butyrate, propionate and acetate and by the bile salt deoxycholate, *Int. J. Cancer*, 60, 400–406, 1995.

Hass, R., Busche, R., Luciano, L., Reale, E., and Engelhardt, W.V., Lack of butyrate is associated with induction of Bax and subsequent apoptosis in the proximal colon of guinea pig, *Gastroenterology*, 112, 875–881, 1997.

Hayatsu, H. and Hayatsu, T., Suppressing effect of *Lactobacillus casei* administration on the urinary mutagenicity arising from ingestion of fried ground beef in the human, *Cancer Lett.*, 73, 173–179, 1993.

Hill, M.J., Drasar, B.S., Hawksworth, G., Aries, V., Crowther, J.S., and Williams, R.E., Bacteria and aetiology of cancer of large bowel, *Lancet*, 1, 95–100, 1971.

Hirayama, K. and Rafter, J., The role of probiotic bacteria in cancer prevention, *Microbes Infect.*, 2, 681–686, 2000.

Huttner, K.M. and Bevins, C.L., Antimicrobial peptides as mediators of epithelial host defense, *Pediatr. Res.*, 45, 785–794, 1999.

Kampman, E., Giovannucci, E., van 't Veer, P., Rimm, E., Stampfer, M.J., Colditz, G.A., Kok, F.J., and Willett, W.C., Calcium, vitamin D, dairy foods, and the occurrence of colorectal adenomas among men and women in two prospective studies, *Am. J. Epidemiol.*, 139, 16–29, 1994.

Kato, I., Kobayashi, S., Yokokura, T., and Mutai, M., Antitumor activity of *Lactobacillus casei* in mice, *Gann*, 72, 517–523, 1981.

Kato, I., Yokokura, T., and Mutai, M., Macrophage activation by *Lactobacillus casei* in mice, Microbiol. Immunol., 27, 611–618, 1983.

Kleibeuker, J.H., Mulder, N.H., Cats, A., van der Meer, R., and de Vries, E.G., Calcium and colorectal epithelial cell proliferation, *Gut*, 39, 774–775, 1996.

Lejeune, C., Bourrillon, A., Boussougant, Y., and de Paillerets, F., Sequential development of the intestinal flora in newborn infants: a quantitative differential analysis, Dev. Pharmacol. Ther., 7, 138–143, 1984.

Lynch, J.P. and Hoops, T.C., The genetic pathogenesis of colorectal cancer, *Hematol. Oncol. Clin. North Am.*, 16, 775–810, 2002.

Madsen, K.L., The use of probiotics in gastrointestinal disease, *Can. J. Gastroenterol.*, 15, 817–822, 2001.

Malhotra, S.L., Dietary factors in a study of cancer colon from Cancer Registry, with special reference to the role of saliva, milk and fermented milk products and vegetable fibre, *Med. Hypotheses*, 3, 122–126, 1977.

Mallett, A.K. and Rowland, I.R., Bacterial enzymes: their role in the formation of mutagens and carcinogens in the intestine, *Dig. Dis.*, 8, 71–79, 1990.

Marchetti, C., Migliorati, G., Moraca, R., Riccardi, C., Nicoletti, I., Fabiani, R., Mastrandrea, V., and Morozzi, G., Deoxycholic acid and SCFA-induced apoptosis in the human tumor cell- line HT-29 and possible mechanisms, *Cancer Lett.*, 114, 97–99, 1997.

Matsuzaki, T., Immunomodulation by treatment with *Lactobacillus casei* strain Shirota, *Int. J. Food Microbiol.*, 41, 133–140, 1998.

Matsuzaki, T., Hashimoto, S., and Yokokura, T., Effects on antitumor activity and cytokine production in the thoracic cavity by intrapleural administration of *Lactobacillus casei* in tumor- bearing mice, *Med. Microbiol. Immunol. (Berl.)*, 185, 157–161, 1996.

Morotomi, M. and Mutai, M., *In vitro* binding of potent mutagenic pyrolysates to intestinal bacteria, *J. Natl. Cancer Inst.*, 77, 195–201, 1986.

Naidu, A.S., Bidlack, W.R., and Clemens, R.A., Probiotic spectra of lactic acid bacteria (LAB), *Crit. Rev. Food Sci. Nutr.*, 39, 13–126, 1999.

Norat, T., Lukanova, A., Ferrari, P., and Riboli, E., Meat consumption and colorectal cancer risk: an estimate of attributable and preventable fractions, *IARC Sci. Publ.*, 156, 223–225, 2002.

Okawa, T., Kita, M., Arai, T., Iida, K., Dokiya, T., Takegawa, Y., Hirokawa, Y., Yamazaki, K., and Hashimoto, S., Phase II randomized clinical trial of LC9018 concurrently used with radiation in the treatment of carcinoma of the uterine cervix: its effect on tumor reduction and histology, *Cancer*, 64, 1769–1776, 1989.

Orrhage, K., Sillerstrom, E., Gustafsson, J.A., Nord, C.E., and Rafter, J., Binding of mutagenic heterocyclic amines by intestinal and lactic acid bacteria, *Mutat. Res.*, 311, 239–248, 1994.

Peters, R.K., Pike, M.C., Garabrant, D., and Mack, T.M., Diet and colon cancer in Los Angeles County, California, *Cancer Causes Control*, 3, 457–473, 1992.

Pierre, F., Perrin, P., Champ, M., Bornet, F., Meflah, K., and Menanteau, J., Short-chain fructo-oligosaccharides reduce the occurrence of colon tumors and develop gut-associated lymphoid tissue in Min mice, *Cancer Res.*, 57, 225–228, 1997.

Pool-Zobel, B.L., Neudecker, C., Domizlaff, I., Ji, S., Schillinger, U., Rumney, C., Moretti, M., Vilarini, I., Scassellati-Sforzolini, R., and Rowland, I., *Lactobacillus*- and *Bifidobacterium*-mediated antigenotoxicity in the colon of rats, *Nutr. Cancer*, 26, 365–380, 1996.

Potter, J.D., Colorectal cancer: molecules and populations, *J. Natl. Cancer Inst.*, 91, 916–932, 1999.

Reddy, B.S., Prevention of colon cancer by pre- and probiotics: evidence from laboratory studies, *Br. J. Nutr.*, 80, S219–223, 1998.

Reddy, B.S., Mangat, S., Weisburger, J.H., and Wynder, E.L., Effect of high-risk diets for colon carcinogenesis on intestinal mucosal and bacterial beta-glucuronidase activity in F344 rats, *Cancer Res.*, 37, 3533–3536, 1977.

Rowland, I.R., Rumney, C.J., Coutts, J.T., and Lievense, L.C., Effect of *Bifidobacterium longum* and inulin on gut bacterial metabolism and carcinogen-induced aberrant crypt foci in rats, *Carcinogenesis*, 19, 281–285, 1998.

Sandhu, M.S., White, I.R., and McPherson, K., Systematic review of the prospective cohort studies on meat consumption and colorectal cancer risk: a meta-analytical approach, *Cancer Epidemiol. Biomarkers Prev.*, 10, 439–446, 2001.

Schiffrin, E.J., Rochat, F., Link-Amster, H., Aeschlimann, J.M., and Donnet-Hughes, A., Immunomodulation of human blood cells following the ingestion of lactic acid bacteria, *J. Dairy Sci.*, 78, 491–497, 1995.

Segal, I., Hassan, H., Walker, A.R., Becker, P., and Braganza, J., Fecal short chain fatty acids in South African urban Africans and whites, *Dis. Colon Rectum*, 38, 732–734, 1995.

Sekine, K., Toida, T., Saito, M., Kuboyama, M., Kawashima, T., and Hashimoto, Y., A new morphologically characterized cell wall preparation (whole peptidoglycan) from *Bifidobacterium infantis* with a higher efficacy on the regression of an established tumor in mice, *Cancer Res.*, 45, 1300–1307, 1985.

Shahani, K.M. and Ayebo, A.D., Role of dietary lactobacilli in gastrointestinal microecology, *Am. J. Clin. Nutr.*, 33, 2448–2457, 1980.

Simon, G.L. and Gorbach, S.L., The human intestinal microflora, *Dig. Dis. Sci.*, 31, 147S–162S, 1986.

Sinha, R., Kulldorff, M., Chow, W.H., Denobile, J., and Rothman, N., Dietary intake of heterocyclic amines, meat-derived mutagenic activity, and risk of colorectal adenomas, *Cancer Epidemiol. Biomarkers Prev.*, 10, 559–562, 2001.

Thyagaraja, N. and Hosono, A., Binding properties of lactic acid bacteria from "Idly" towards food-borne mutagens, *Food Chem. Toxicol.*, 32, 805–809, 1994.

Treptow-van Lishaut, S., Rechkemmer, G., Rowland, I., Dolara, P., and Pool-Zobel, B.L., The carbohydrate crystalean and colonic microflora modulate expression of glutathione S-transferase subunits in colon of rats, *Eur. J. Nutr.*, 38, 76–83, 1999.

Willett, W.C., Stampfer, M.J., Colditz, G.A., Rosner, B.A., and Speizer, F.E., Relation of meat, fat, and fiber intake to the risk of colon cancer in a prospective study among women, *N. Engl. J. Med.*, 323, 1664–1672, 1990.

World Cancer Research Fund, *Diet, Nutrition and the Prevention of Cancer: a Global Perspective*, World Cancer Research Fund/American Institute of Cancer Research, Washington, DC, 1997.

Young, T.B. and Wolf, D.A., Case-control study of proximal and distal colon cancer and diet in Wisconsin, *Int. J. Cancer*, 42, 167–175, 1988.

Zhang, X.B. and Ohta, Y., Microorganisms in the gastrointestinal tract of the rat prevent absorption of the mutagen-carcinogen 3-amino-1,4-dimethyl-5H-pyrido(4,3-b)indole, *Can. J. Microbiol.*, 39, 841–845, 1993.

Zock, P.L., Dietary fats and cancer, *Curr. Opin. Lipidol.*, 12, 5–10, 2001.

Section IV

Concluding Remarks

Commentary 1

Do Dietary Antioxidants Really Help Prevent or Treat Cancer?

Jeffrey Blumberg

Cancer has long been recognized as a chronic disease that can be influenced during its many stages from initiation through metastasis by nutritional factors, which can have an impact on its prevention and treatment. Early investigators proposed that cancer mortality resulted when the tumor depleted the patient of essential nutrients (Warren, 1932). This insight led subsequent generations of scientists to explore and characterize the interactions between tumors and host metabolism and to begin applying the principles of nutritional oncology (Heber et al., 1999). While many of the original studies examined abnormalities in carbohydrate and protein metabolism in cancer, it has only recently been recognized that a critical chemopreventive role may exist for essential micronutrients and phytochemicals, especially those with antioxidant properties (Prasad et al., 1998; Waladkhani and Clemens, 1998).

The significant reduction in cancer risk associated with increased consumption of plant foods, particularly fruits and vegetables, has often been attributed to their antioxidant ingredients, such as vitamins C and E, carotenoids, and flavonoids (Block et al., 1992; World Cancer Research Fund and American Institute for Cancer Research, 1997). There is a plausible basis for this hypothesis, as oxidant damage to DNA can be an early step in carcinogenesis. For example, hydroxyl radical attack upon DNA generates a series of modified purine and pyrimidine bases that are known to be mutagenic. The steady-state levels of oxidative base lesions in DNA of normal cells have been estimated at about 1 per 10^6 bases; this level is 100-fold higher than adducts induced by carcinogens like benzpyrene, implicating endogenous damage to DNA by reactive oxygen species as making an important contribution to the age-related development of cancer (Loft and Poulsen 1996). Oxidative damage to lipids and proteins may also lead indirectly to mutagenic effects. However, it is important to recognize that dietary antioxidants may possess other relevant anticancer bioactivities. For example, carotenoids promote gap-junctional communication (Yamasaki and Naus, 1996); flavonoids modulate phase I and II xenobiotic detoxification enzymes (Nijveldt et al., 2001); vitamin E inhibits protein kinase C activity (Studer et al., 1997); and ellagic acid reduces DNA–carcinogen adduct formation. Examining these actions as well as analyzing standard biomarkers of oxidative stress should be an essential feature of new research studies as a measure of exposure efficacy and to help determine mechanisms of action.

While the epidemiological evidence for a benefit of a range of dietary antioxidants has been consistently compelling, the outcome of the large-scale randomized chemoprevention trials employing antioxidant supplement preparations of vitamin C, vitamin E, β-carotene, and selenium has been mixed and somewhat disappointing. The Linxian General Population Trial in China showed a significant reduction in cancer mortality, particularly from stomach cancer, by daily use of a combination supplement of β-carotene, vitamin E, and selenium (Blot et al., 1993). A benefit was also observed in the Linxian Dysplasia Trial, where individuals receiving the supplement were more likely to have no esophageal dysplasia after 6 years than those receiving a placebo (Mark et al., 1994). However, the positive outcome of these studies was largely dismissed as not being directly applicable to Western societies, which tend to be well nourished and not deficient in multiple micronutrients, in contrast to the Linxian cohort. Nonetheless, the proof of principle for antioxidant efficacy seems clearly demonstrated in these trials, with remaining issues to be examined, such as better identifying responsive populations, characterizing more potent doses and combinations, determining the best stage of carcinogenesis for intervention, etc.

The Alpha-Tocopherol Beta-Carotene (ATBC) trial in Finland employed β-carotene with or without vitamin E, and the CARET study in the U.S. employed β-carotene plus retinol in cohorts at high risk for lung cancer due, in part, to lifelong heavy smoking, and both reported significantly increased incidence of lung cancer in the treated versus placebo groups (ATBC Cancer Prevention Study Group, 1994; Omenn et al., 1996). In contrast, no adverse effect on lung cancer was noted in either the healthy individuals or the subset of men who smoked in the Physicians' Health Study, who received a lower dose of β-carotene (Hennekens et al., 1996). The basis for the unexpected findings in the ATBC and CARET studies have been intensely debated and include (a) design issues such as β-carotene dose and form, nutrient selection, and subject eligibility criteria and (b) confounding factors such as complex interactions between β-carotene, cigarette smoke, and alcohol intake as well as overall dietary pattern and nutritional status (Blumberg and Block, 1994; Milbury and Blumberg, 2003). Nonetheless, it is noteworthy that participants with higher serum β-carotene at entry into the studies developed fewer lung cancers during the trial, even among those who received the active supplement, suggesting that longer-term intakes at lower doses might prove efficacious. Alternatively, serum β-carotene may merely reflect long-term consumption of total fruit and vegetable intake, with their constituent total carotenoids and other antioxidant phytochemicals. Interestingly, the low-dose vitamin E supplement in the ATBC trial significantly reduced prostate cancer incidence and mortality.

The need for long-term primary prevention trials with antioxidants in low-risk populations is self-evident, but only the ongoing Physicians Health Study II (Christen et al., 2000), Womens Health Study (Rexrode et al., 2000), SU.VI.MAX (Hercberg et al., 1998), and SELECT (Flesherner and Kucuk, 2001) meet these criteria. Preliminary reports from the SU.VI.MAX trial indicate a significant reduction in cancer incidence among men (Hercberg, 2003). However, even these trials employ principally middle-aged and older adults, and the supplement formulations are far from inclusive of all the major components of the antioxidant nutrient defense network, e.g., lacking α-lipoic acid, coenzyme Q_{10}, or any flavonoids or carotenoids other than β-carotene. Indeed, the apparent synergy between dietary antioxidants with regard to antiproliferative activity and other relevant mechanisms of action suggests both the complexity of this relationship and the challenge to formulating successful interventions (Chu et al., 2002; Sun et al., 2002). In fact, the limitations inherent in randomized chemoprevention trials, including their prohibitive cost, long duration, and assessment of only a few nutrients and single doses, suggest that recommendations today concerning antioxidants and cancer must be based on the totality of available research information and not this single research approach.

The anticancer actions of phytochemicals such as the allyl sulfides in garlic, isoflavones in soy, and monoterpenes from fruit skins may also be due, in part, to their antioxidant capacity. While many of these secondary metabolites have evolved to protect the plant from threats like fungus and insects, humans may have evolved to use them to help balance their own genetic–environmental interactions that, when perturbed, initiate and promote the carcinogenic process (Eaton and Konner, 1985; Harper, 2000). It is worth noting that only about 10% of Americans consume the minimum of five daily servings of fruits and vegetables recommended by the National Cancer Institute. While our genome was designed to complement a high intake of the extraordinary diversity of edible plant species, today our consumption is extraordinarily limited to a very small number of fruits, vegetables, cereals, and grains.

There is no doubt from the extensive body of epidemiological and experimental evidence that there is a causal link between diet and cancer. Dietary fat, excessive calories, obesity, and alcohol have been shown to increase cancer risk. Dietary fiber, fruits, vegetables, and certain vitamins and minerals common to plant foods appear protective against cancer. The enormous diversity of phytochemicals suggests that their mechanisms of action and dynamic interactions will be different for specific cancer stages and cancer sites. However, the lesson we have learned with other nutrient–disease relationships will certainly apply to the dietary antioxidants, i.e., no single nutrient will prove to be a "magic bullet" in the prevention or treatment of cancer. Our knowledge about the anticancer actions of antioxidants continues to grow through positive, null, and negative outcomes of research studies and suggests important applications to public health as well as to individual patients. These studies present an opportunity for the development of better dietary guidelines, engineered foods, and dietary supplements targeted to cancer prevention and treatment. In the meantime, there is a general consensus about diet to benefit overall health and reduce the risk for cancer: the most prudent approach for everyone is to adopt a low-saturated-fat, high-fiber diet that includes a variety of fruits, vegetables, and whole grains; to limit the consumption of alcohol; and to maintain a healthy body weight (Willett, 2002).

REFERENCES

Alpha-Tocopherol Beta-Carotene (ATBC) Cancer Prevention Study Group, Heinonen, O.P., Huttunen, J.K., and Albanes, D., The effect of vitamin E and beta-carotene on the incidence of lung cancer and other cancers in male smokers, *New England J. Med.*, 330, 1029–1035, 1994.

Block, G., Patterson, B., and Subar, A., Fruit, vegetables, and cancer prevention: a review of the epidemiological evidence, *Nutr. Cancer*, 18, 1–29, 1992.

Blot, W.J., Li, J.-Y., Taylor, P.R., Guo, W., Dawsey, W.M., Wnag, G.-Q., Yang, C.S., Zheng, S.-F., Gail, M.H., Li, G.-y., Yu, Y., Liu, B.-Q., Tangrea, J.A., Sun, H.-H., Liu, F., Fraumeni, J.F., Zhang, Y.-H., and Li, B., Nutrition intervention trials in Linxian, China: supplementation with specific vitamin/mineral combinations, cancer incidence, and disease-specific mortality in the general population, *J. Natl. Cancer Inst.*, 85, 1483–1492, 1993.

Blumberg, J.B., and Block, G., The alpha-tocopherol, beta-carotene cancer prevention study in Finland, *Nutr. Rev.*, 52, 242–245, 1994.

Christen, W.G., Gaziano, J.M., and Hennekens, C.H., Design of Physicians' Health Study II — a randomized trial of beta-carotene, vitamins E and C, and multivitamins, in prevention of cancer, cardiovascular disease, and eye disease, and review of results of completed trials, *Ann. Epidemiol.*, 10, 125–134, 2000.

Chu, Y.-F., Sun, J., Wu, X., and Liu, R.H., Antioxidant and antiproliferative activities of vegetables, *J. Agric. Food Chem.*, 50, 6910–6916, 2002.

Eaton, S.B. and Konner, Paleolithic nutrition: a consideration of its nature and current implications, *New England J. Med.*, 312, 283–289, 1985.

Fleshner, N.E. and Kucuk, O., Antioxidant dietary supplements: rationale and current status as chemopreventive agents for prostate cancer, *Urology*, 57, 90–94, 2001.

Harper, A.E., Ed., Physiologically active food components, *Am. J. Clin. Nutr.*, 65, 647S–743S, 2000.

Heber, D., Blackburn, G.L., and Go, V.L.W., Eds., *Nutritional Oncology*, Academic Press, San Diego, CA, 1999.

Hennekens, C.H., Buring, J.E., Manson, J.E., Stampfer, M., Rosner, B., Cook, N.R., Belanger, C., LaMotte, F., Gaziano, J.M., Ridker, P.M., Willett, W.C., and Peto, R., Lack of effect of long-term supplementation with beta carotene on the incidence of malignant neoplasms and cardiovascular disease, *New England J. Med.*, 334, 1145–1149, 1996.

Hercberg, S., http://www.nutraingredients.com/news/news.asp?id=7245, last updated June 26, 2003.

Hercberg, S., Galan, P., Preziosi, P., Roussel, A.M., Arnaud, J., Richard, M.J., Malvy, D., Paul-Dauphin, A., Briancon, S., and Favier, A., Background and rationale behind the SU.VI.MAX Study, a prevention trial using nutritional doses of a combination of antioxidant vitamins and minerals to reduce cardiovascular diseases and cancers: Supplementation en Vitamines et Mineraux AntioXydants Study, *Int. J. Vitamin Nutr. Res.*, 68, 3–20, 1998.

Loft, S. and Poulsen, H., Cancer risk and oxidative DNA damage in men, *J. Mol. Med.*, 74, 297–305, 1996.

Mark, S.D., Liu, S.-F., Li, J.-Y., Gail, M.H., Shen, Q., Dawsey, S.M., Liu, F., Taylor, P.R., Li, B., and Blot, W.J., The effect of vitamin and mineral supplementation on esophageal cytology: results from the Linxian Dysplasia Trial, *Int. J. Cancer*, 57, 162–166, 1994.

Milbury, P. and Blumberg, J.B., Dietary antioxidants: human studies overview, in *Oxidative Stress and Aging: Advances in Basic Science, Diagnostics, and Intervention*, Rodriguez, H. and Cutler, R.G., Eds., World Scientific Publishing Co., 2003, pp. 487–502.

Nijveldt, R.J., van Nood, E., van Hoorn, D.E.C. et al., Flavonoids: a review of probable mechanisms of action and potential applications, *Am. J. Clin. Nutr.*, 74, 418–425, 2001.

Omenn, G.S., Goodman, G.E., Thornquist, M.D., Balmes, J., Cullen, M.R., Glass, A., Keogh, J.P., Meyskens, F.L., Valanis, B., Williams, J.H., Barnhart, S., and Hammar, S., Effects of a combination of beta carotene and vitamin A on lung cancer and cardiovascular disease, *New England J. Med.*, 334, 11505, 1996.

Prasad, K.N., Cole, W., and Hoveland, P., Cancer prevention studies: past, present and future, *Nutrition*, 14, 197–210, 1998.

Rexrode, K.M., Lee, I.M., Cook, N.R., Hennekens, C.H., and Buring, J.E., Baseline characteristics of participants in the Women's Health Study, *J. Womens Health Gender-Based Med.*, 9, 19–27, 2000.

Studer, R.K., Craven, P.A., and DeRubertis, F.R., Antioxidant inhibition of protein kinase C-signaled increases in transforming growth factor-beta in mesangial cells, *Metabolism*, 46, 918–923, 1997.

Sun, J., Chu, Y.-F., Wu, X., and Liu, R.H., Antioxidant and antiproliferative activities of fruits, *J. Agric. Food Chem.*, 50, 7449–7454, 2002.

Waladkhani, A.R. and Clemens, M.R., Effect of dietary phytochemicals on cancer development, *Int. J. Mol. Med.*, 1, 747–753, 1998.

Warren, S., The immediate causes of death in cancer, *Am. J. Med. Sci.*, 184, 610–615, 1932.

Willett, W.C., Balancing life-style and genomics research for disease prevention, *Science*, 296, 695–698, 2002.

World Cancer Research Fund and American Institute for Cancer Research, *Food, Nutrition and the Prevention of Cancer: a Global Perspective*, American Institute for Cancer Research, Washington, DC, 1997.

Yamasaki, H. and Naus, C.C.G., Role of connexin genes in growth control, *Carcinogenesis*, 17, 1199–1206, 1996.

Commentary 2

Who Gets Cancer? Do Healthy Foods, Healthy Living, and Natural Antioxidants Really Help?

Jean Carper

Who gets cancer? About 1,334,100 Americans will be diagnosed with cancer in 2003 (not counting 1 million cases of basal and squamous skin cancers), according to the American Cancer Society. The leading cancer sites are: cancer of the prostate (220,900), breast (211,300), lung and bronchus (171,900), and colon and rectum (147,500). Cancer will kill an estimated 556,500 Americans in 2003, more than 1500 per day.

Yet, evidence suggests that at least one-third of the expected cancer deaths are tied to nutrition, physical inactivity, obesity, and other lifestyle factors, and another one-third to cigarette smoking. It is clear that fully two-thirds of those who die of cancer might have prevented or delayed their cancers by changing their diets, activity levels, and lifestyle.

It is a startling fact that the public is slow to accept that most cancer is not haphazard and capricious, a bolt from the blue, or a fulfillment of destiny. The truth is that cancer is a catastrophe, like a wreck in the road, that need not happen for most people.

Exciting discoveries, notably in the last three decades about how cancer is triggered at the cellular level and is stimulated to progress, have given rise to powerful diet and lifestyle antidotes that can help individuals avoid and control cancer.

Here are some of the scientific high points from my perspective as a medical and nutrition journalist and author:

- The breakthrough realization that chemicals, mainly antioxidants, in common foods could manipulate cancer initiation and progression in cells

  In 1979 I wrote a major article for the *Washington Post* on the pioneering work of Lee Wattenberg, University of Minnesota Medical School. Wattenberg coined the term "chemoprevention" in the 1970s, opening up a new area of cancer investigation by documenting that cruciferous vegetables and pure indoles, fed to rats and mice, acted as antidotes to the initiation and progression of cancer. His elegant work, showing how food chemicals blocked carcinogenic activity and boosted detoxification defenses at the cellular level, led to an explosion of research on the chemopreventive potential of antioxidants and other food chemicals that is still going strong.

- The understanding that foods contain a broad spectrum of antioxidants, their organization into categories and measurements of individual foods' anticancer potential

  At first, the scientific focus was on the antioxidant powers of vitamins E and C as well as beta carotene (in pill form and in orange and green fruits and vegetables). That narrow emphasis disappeared when research began to reveal equal or superior antioxidant and anticancer activity in carotenoids, such as lycopene, lutein, zeaxanthin, and alpha carotene, as well as isoflavones and flavonoids, such as catechins, quercetin, and anthocyanidins. Pushing knowledge further, the USDA developed tables listing the amount of various antioxidants in specific foods. Tufts University and other centers analyzed foods for their total ORAC (antioxidant potential), finding berries, spinach, and dried fruits at the top. It is much-needed evidence backing up the National Cancer Institute's advice to eat five to nine fruits and vegetables a day.

- The decline of the magic bullet theory; the rise of the antioxidant network theory.

  Several tests of antioxidant supplements as "magic bullets" to prevent cancer and other chronic diseases have been disappointing. High doses of beta carotene proved to be a lung-cancer promoter in current smokers and those exposed to asbestos. Vitamin E did not perform as expected. These results were controversial and sobering setbacks, perhaps signaling the decline of the idea that antioxidants perform like silver-bullet drugs. The work of researchers Lester Packer (University of Southern California) and Bruce Ames (University of California at Berkeley) has focused on the synergy of antioxidants as the secret to success, insisting that they are much more powerful together than alone. Packer calls it an antioxidant "network," in which the five most important collaborators are vitamin E, vitamin C, alpha lipoic acid, coenzyme Q_{10}, and glutathione. Their approach suggests that isolating and testing antioxidants as chemopreventives one by one in clinical trials may not be the most sensible approach. It may also explain why fruits and vegetables, with a wide variety of antioxidants, consistently emerge as reducers of cancer risk in epidemiological studies.

- The growing awareness that meat, notably red meat and cured meat, raises cancer risks that can be dramatically counteracted by antioxidants.

  In my profession, I talk with many researchers, and one message I feel has not received enough attention is the hazard of cancer from eating meat. Many think if they slash the fat in meat, they are home free. Yet, several studies show that regardless of fat content, carcinogenic heterocyclic amines (HCA) form on the surface of cooked meats, more so at prolonged high temperatures such as grilling, broiling, or frying. Studies also link high meat consumption, including fried and cured red meats, with colon and possibly breast cancers. However, studies at the Lawrence Livermore National Labs, Michigan State University, American Health Foundation, and the Cancer Research Center of Hawaii show that antioxidants can zap the formation of the carcinogens — for example, by adding vitamin E, tea, cherries, blueberries, garlic, or textured soy protein to hamburgers or marinating steaks and chicken with antioxidant herbs. Precooking meat and poultry in a microwave before grilling has reduced 90% of HCAs. Of course, it has been known for half a century that vitamin C helps block formation of nitrosamines arising from eating cured meats.

Scientific studies have demonstrated that much cancer is, at least partially, self-inflicted, as resistant as many may be to that notion, and that even genetic predisposition to cancer can be disarmed by a healthy lifestyle. What is left now is to further define and refine the interactions between cancer and lifestyle factors, including diet and antioxidants, to make this message so compelling that the public can no longer ignore it.

Concluding Remarks

This book has focused mainly on the potential of phytopharmaceuticals to aid in the war on cancer. As with drugs, there are no discernible magic bullets. Nevertheless, these "natural alternatives" can, alone or combined with conventional approaches, provide some benefits. In some cases, one can question the extent of the investigation on a particular ingredient. We expect this. The purpose of this book is to allow the reader to discern what is not known as much as what is known in any given area. Unlike drugs, the profit margins often do not allow for the performance of extensive studies using dietary supplements and nutraceuticals, i.e., randomized, double-blind, placebo-controlled studies performed at multicenters. We believe the information provided in this book will allow for the future development of projects designed to show the utility or lack of such for many phytopharmaceuticals.

Index

A

Aberrant crypt foci (ACF), 231–234, 241, 474, 581, 624–625
Abies concolor, 495
A549 cell line, 442, 583
Acetaldehyde (ACT), 92, 94
Acetaminophen, 313
Acetone (ACON), 92, 94
Acetylsalicylic acid (ASA), 362
Acid suppression therapy, 157
Actinidia polygama, 469
Action mechanisms of carcinogens. *See* Xenobiotics, chemical properties/biotransformations of
Activating factor-1 (AF-1), 110, 112
Activating transcription factor (ATF), 47
Activator protein-1 (AP-1)
    antioxidants, 494–495
    asbestos, 53–54
    curcumin, 355, 579
    inflammation, 86
    lung cancer, 47
    oxidative stress, 178–179
    peppers, hot red, 582
    reactive oxygen and nitrogen species, 494
    tea, 256
ActiVin. *See* Grape seed proanthocyanidin extract
Acute lymphocytic leukemia, 20
Acute myeloid leukemia (AML), 20
Acute radiation syndrome, 81
Adenocarcinomas, 67; *see also* Esophageal cancer; *specific adenocarcinoma*
Adenomatous polyposis carcinomatous (APC) gene, 625
Adenomatous polyps in the colon, 624
Adenosine diphosphate (ADP), 173
Adenosine triphosphate (ATP), 173, 337, 542–543, 601, 604, 613
Adhesion molecules, 355
Adriamycin, 257, 329, 332, 352, 371, 612, 616–617
Aflatoxin B, 399, 400, 585
Africa, 152, 154
AG1478 cell, 53
Age
    and chemical carcinogenesis, 342
    coenzyme Q$_{10}$, 610, 613–615
    at diagnosis, 9–10
    endothelial cell neoplasms, 207
    esophageal cancer, 152
    fruits and berries, 467
    gastric cancer, 143
    incidence and mortality rate, 7–11, 601–603
    obesity, 197
    pandemic relevance, 612–613
    prevalence, 10–11
Age-related macular degeneration (AMD), 530
Angiogenic myeloid metaplasia (AMM), 19
AIN-76 diet, 368, 589
Akt, 46, 69, 72
Albumin, 369
Alcohol
    curcumin and liver injury, 368–369
    esophageal cancer, 153, 154
Alkaline phosphatase, 370, 438, 456
Allicin, 584–585
Alliin, 584–585
Allium, 318
Allium vegetables, 583–584
Allopurinol, 93
S-allylcysteine, 254, 584–585
S-allylmercapto-L-cysteine, 254
Allylthio-pyrazine, 239
Aloe, 258, 259
Alpha-Tocopherol Beta-Carotene (ATBC) study, 286, 290–292, 640
Alternative medicine, consumer demand and increased research on, 205
ALX3 gene, 23
Alzheimer's disease, 367–368, 433
American Cancer Society, 14, 247, 249, 601
American College of Gastroenterology, 156
American Conference of Governmental Industrial Hygienists, 30
American Health Foundation, 644
American Joint Committee on Cancer, 147–148
American Medical Association (AMA), 251
American yew, 519–520
Ames, Bruce, 644
Aminocamptothecan, 262
Amino-3-methylimidazo-4,5-quinoline (IQ), 406–407
Amphiboles, 51
Amygdalin, 251
Androgen receptors and cofactors, 358–359
Androstenedione, 431
Ames testing, 30
Angiogenesis, 106, 108–110, 133, 356, 408–409, 512; *see also* antiangiogenic effects of dietary chemopreventive agents *under* Diet/nutrition
Angiostatin, 206
Angiotensin-converting enzyme (ACE), 369, 370
Animal studies, advantage of, 30–31
Anthocyanadines, 253
Anthocyanins, 467–468, 470, 471–472

Antioxidants
  activator protein-1, 494–495
  astaxanthin, 559–561, 563
  berries, 467, 468
  curcumin, 257, 351
  inhibition of carcinogenesis, 253
  lycopene, 527–528, 530
  metals/metalloids, 35
  NADH, 546–547
  nuclear factor-kB, 494–495
  overview, 556–557, 639–641, 643–644
  oxidative stress, 174, 180
  palm tocotrienols, 485–486
  paradoxical effects, 263
  phytoestrogens, 432–433
  plants, 85
  radiation, 83–84
  radiation therapy, adverse effects of, 220–221
  resveratrol, 454
  rosemary, 587
  smokeless tobacco, 93
Apigenin, 318, 324, 326
Apoptosis
  astaxanthin, 564
  curcumin, 352–353, 360, 579
  DNA damage, 91
  doxorubicin, 544, 545
  fruits and berries, 470
  lung cancer, 45–46
  metals/metalloids, 36–37
  mitochondria, 604
  mushrooms, maitake, 512–513
  oligomeric proanthocyanidins, 326, 332–337
  overview, 90–91
  PC12 cells, 545
  resveratrol, 257, 453–454
  rosemary, 589
  silica, 49–50
  smokeless tobacco, 98, 99
  tea, 410
  vitamins C/E and beta-carotene, 281–283
Arabinosylcytosine (ara-C), 22
Arachidonate, 362
Arachidonic acid, 127, 130, 406
Argentina, 32
Argon beam coagulation, 157, 160, 161
Arm lymphedema, radiation-induced, 219, 221
Arsenic, 29–32, 35, 37
Artemisia asiatica, 239
Arthritis and curcumin, 365–366
Aryl hydrocarbon hydroxylase (AHH), 90
Asbestos. *See under* Lung cancer
Ascorbate, 180
Ascorbic acid, 239
Ascorbyl radical, 272–273, 498
Asia
  curcumin, 257
  Eastern lifestyles and cancer risk, 435
  esophageal cancer, 152
  gastric cancer, 143
  phytochemicals, 239

  soy foods, 258
  tea, 390, 392
Aspirin, 125–129, 241, 357
Astaxanthin
  antioxidant potential/effects, 556–557, 559–561
  chemopreventive potential, 561–564
  safety and metabolism, 564–565
  structure and forms, 559
Atharva Veda, 248
Atherosclerosis, 433; *see also under* Curcumin
ATM cell, 45
Atomic bomb, Japan and the dropping of the, 81
Atrophic gastritis, 144
Australia, 536
Autocrine growth factors, 454
Autofluorescence, 157
Autoxidation. *See* Lipid peroxidation
Ayurveda, 248–249, 257, 263, 373
Ayurvedic Drugs in the Management of Cancer, 248
Azoreductase, 630, 631
Azoxymethane (AOM), 233, 337, 399, 473, 474

B

Bacille Calmette-Guerin (BCG) vaccine, 608
Bacteria, intestinal, 624, 627–628; *see also* Probiotics
Bacteroides, 628
Baicalin, 321
Bangladesh, 32
Barret, Norm, 155
Barrett's esophagus, 154, 155–157
Basic excision repair (BER), 43, 44
Basil leaves, 589–590
Batimistat, 206
Baycol, 616
Bayer AG, 616
B cells, 352, 600, 614
Bcl-2 genes, 91, 99–101, 545
Beer, Charles T., 261
Benign tumors, 4–5
Benzoapyrene (BAP), 90, 351
Berberis root, 251
Berries/berry extracts, 208–209, 467–469; *see also* Fruits
    and berries
Beryllium, 32
Beta-glucans. *See* Mushrooms, maitake
B16F10 melanoma cells, 580
Bhallatak, 248
Bilberry extract, 469–470
Biochanin A, 431
Bioenergetics, 601, 603–604, 606, 613
BioRad's Gel Analysis System, 99
Biotransformation. *See* Xenobiotics, chemical
    properties/biotransformations of
Bisdemethoxycurcumin, 376
Blackberries, 474
Black tea, 232–233; *see also* Tea polyphenols
Bladder cancer
  monocyte chemoattractant protein-1, 208
  probiotics, 633

tea, 413
vitamins C/E and beta-carotene, 288–289
Bleomycin, 160, 370
Blueberries, 474
B16 melanoma cell lines, 256
Body mass index (BMI), 197, 198–199, 289
Bowel injury, radiation-induced, 218
Boyer, Paul, 603
Brachial plexopathy, radiation-induced, 218–219
Brain cancer, 586
Brassinin, 242, 243
BRCA 1/2 genes/proteins, 19, 45, 124
Breast cancer
  BRCA, 19
  chance of getting, lifetime, 341
  coenzyme Q_{10}, 254, 610–611
  curcumin, 352
  estrogen, 108
  genistein, 113
  hormone-related, 433
  incidence rate, 7
  lipid peroxidation, 494
  lycopene, 234
  MCF-7/6 cells, 438–439
  meat consumption, 644
  monocyte chemoattractant protein-1, 208
  mortality rate, 7, 8
  mouse mammary gland organ culture assay, 240–241
  NADH, 550
  obesity, 200, 201
  oral contraceptives, 123–124
  oxidative stress, 610–611
  palm tocotrienols, 486–489
  paradoxical effects of natural products/supplements, 263
  phytoestrogens, 113, 442
  resveratrol, 451–452
  rosemary, 588
  soy foods, 238, 258, 435
  statins, 616
  tamoxifen, 428
  Taxol, 521
  tea, 256, 400, 415
  vitamins C/E and beta-carotene, 289–290
Bristol-Myers Squibb, 520–521
Bromode-oxyiridine (Brdu) labeling index, 584
Bromoflavone, 242, 243
Bronchoalveolar lavage (BAL) cells, 51
Bronchoalveolar lavage fluid (BALF), 370
Bronchogenic carcinomas, 42
Brusch, Charles, 252
BT20 cell line, 352
Buckthorn bark, 251
Burdock root, 251
Butyrate, 632

C

Cadmium, 29, 32–33, 35
Caffeic acid, 239
Cag A protein/Cag pathogenicity island, 165–166

Caisse, Rene, 252
Calcitonin, 20
Calcium, 608
Calcium glucarate (CGT), 330
Camellia sinensis. See Tea polyphenols
Camptotheca, 262
Camptotheca acuminata, 343
Camptothecins, 239, 343
Cancer Advisory Council (CA), 252
Cancer Chemotherapy National Service Center, U.S., 495
Cancer Research, 617
Cancer Research Center of Hawaii, 644
Canthaxanthin, 558, 561
Capsaicin, 239, 581–582
Carbon tetrachloride, 589
Carboplatin, 74
Carboxyethyl-hydroxychromans (CEHC), 484–485
Carboxy-terminal binding protein (CtBP), 546
Carcinogens, action mechanism of. *See* Xenobiotics,
    chemical properties/biotransformations of
Carcinomas, 5
Cardiovascular system
  Adriamycin, 616
  coenzyme Q_{10}, 612
  curcumin, 362–363, 368
  NADH, 542–543
  phytoestrogens, 436
  resveratrol, 456, 457
Carnosic acid, 588, 589
Carnosol, 588, 589
Alpha-carotene, 558
Beta-carotene, 153, 180, 337–338, 467, 526–527, 535; *see also* Astaxanthin; Vitamins C/E and beta-carotene as chemopreventive agents
Beta-Carotene and Retinol Efficacy Trial (CARET), 290–291, 640
Carotenoids, 239, 275, 282, 286, 555–559; *see also* Astaxanthin; Lycopene
Carpal tunnel syndrome, 563
Carrageenan, 631
Carrot juice, 281
Cascara, 251
Caski epithelial cancer cells, 488
Cassia quinquangulata, 242
Catalase, 176, 362, 433
Cataract formation and curcumin, 368
Catechins, 208, 338, 497; *see also* Pycnogenol; Tea polyphenols
CDC10 protein, 502
cDNA arrays, 356, 500
Celecoxib, 128–130, 133, 177
Celery seed, 590
Cellular differentiation, promotion of, 455
Cerivastatin, 616
Cervical cancer
  obesity, 200
  oxidative stress, 177
  vitamins C/E and beta-carotene, 292–293
Charaka Samhita, 248
Chemically induced dynamic nuclear polarization (CIDNP), 79–80

Chemical reactions and metals/metalloids, 29
Chemiluminescence (CL), 50–51, 93, 221
Chemoprevention. *See also* Diet/nutrition; Natural
    products/supplements; Phyto *listings; specific
    phytopharmaceutical/food type*
  age influencing, 342
  carotenoids, 555–556
  clinical trials, 342–343
  curcumin, 356–357
  defining terms, 391
  gender influencing, 342
  lifestyle change, 341
  metabolism, differences in carcinogenic, 341
  oncogenes and oncosuppressor genes, 341
  oral contraceptives, 123–124
  race/ethnicity influencing, 342
  research, need for more, 343–344
  resveratrol, 449–450
Chemotherapy
  coenzyme Q_{10}, 617
  conclusions/summary, 245
  defining terms, 311
  DNA methylation patterns, 22–23
  dysplasia, 311–312
  esophageal cancer, 160, 161
  gastric cancer, 149
  mushrooms, maitake, 514
  oligomeric proanthocyanidins, 329
  tea, 402
Chile, 32
China
  camptotheca, 262
  clinical trials, 342
  esophageal cancer, 154
  ginseng, 254
  kurarinone, 437
  mushrooms, maitake, 514
  onions, 586
  phytoestrogens, 443
  prostate cancer, 536
  tea, 390, 392
  yew, 260
Chk1/2 protein kinases, 45
Chlorodeoxyadenosine (CdA), 22
CHO-K1 cells, 452
Cholecystectomy (CCK), 155
Cholesterol, 360–361, 364, 375, 616
Chromatin, 34
Chromium, 29, 30, 33, 34, 36, 37
Chronic myeloid leukemia, 20
Chrysin, 324
C3H10T1/2 cells, 583
Cigarettes. *See* Smokeless tobacco; Tobacco
Cisplatin, 160, 161, 545–546, 548
Cis-retinoic acid, 313
c-Kit receptor tyrosine kinase, 63, 65–66, 68–69, 72, 74
Clinoptilolite-zeolite, 327
Cloves, 587
c-Met tyrosine kinase, 63–65, 67, 68, 70, 74
Coactivators and estrogen, 112
Cobalt, 29, 33, 35, 36

Cockayne's syndrome A (CSA), 43
Cocoa polyphenols, 471
Coenzyme Q_{10}
  bioavailability, 607
  breast cancer, 254
  conclusions/summary, 618
  immune system/aging/cancer, integration of, 600–615
  interface between biosynthesis and commonly used
    drugs, 615–617
  production and availability, 607
  reaction pathway of the biosynthesis, 606
  safety considerations, 617
  structure of, 605
Cognitive function and phytoestrogens, 436
Colectomy, 135
Colorectal cancer
  Abies concolor, 495
  anatomy/physiology of the colon, 626–628
  camptotheca, 262
  cloves, 587
  coenzyme Q_{10}, 611
  curcumin, 352, 356–357
  diet/nutrition, 625–626
  DNA methylation patterns, 19
  fruits and berries, 466
  ginger, 581
  grape seed proanthocyanidin extract, 473
  incidence rate, 7, 8
  lycopene, 234
  meat consumption, 644
  mistletoe, 250
  mortality rate, 7
  NADH, 549–550
  NSAIDs, 123–135
  obesity, 200
  oligomeric proanthocyanidins, 326
  oxidative stress, 465
  pathogenesis, 624–625
  rates and prevalence, 624, 625
  resveratrol, 257, 473
  Taxol, 521
  tea, 399, 413–414
  turmeric, 257
  vitamins C/E and beta-carotene, 287–288, 293–294
  worldwide, cases, 12
Columnar cells, 151
Combination therapy, 312
Comet assay, 276, 281
Computed tomography (CT), 147, 158
Con-A-bound fraction, 366
Confocal microscopy, laser scanning, 95–96
Connexin 26 gene, 19
Contraceptives, oral, 108, 123–124
Copper, 34
Costa Rica, 143
Cosuppressors and estrogen, 112
CpG (cytosine phosphoguanosine)/CpG islands. *See* DNA
  methylation patterns
Cranberries, 470
Crane, Fred, 604
C-reactive protein (CRP), 287

Creatine kinase (CK), 362, 368
Crk family, 68
Crocetin, 583
Cryptoxanthin, 286, 467, 558
Cumin seeds, 589–590
Curcuma xanthorrhiza, 360
Curcumin
  analogs, 376–377
  anticancer properties, 352–359, 576, 579–580
  atherosclerosis, 359–363
  clinical experience, 374–376
  colorectal cancer, 133
  conclusions/summary, 378–379
  degradation kinetics, 351
  metabolism, 373–374
  other effects
    Adriamycin-induced nephrotoxicity, 371
    Alzheimer's disease, 367–368
    arthritis, 365–366
    cataract formation, 368
    cyclosporine, 372
    diabetes, 363–364
    endotoxin shock, 372
    gallstone formation, 366
    HIV (human immunodeficiency virus), 367
    inflammatory bowel disease, 371
    liver injury, alcohol-induced, 368–369
    lung injury, drug-induced, 369–370
    multidrug resistance, 372
    multiple sclerosis, 366
    muscle regeneration, 364
    myocardial toxicity, 368
    pancreatitis, 372
    scarring, 371
    stress, 372
    wound healing, 365
  overview, 350–351
  P-form PST, 324
  sources of, 377–378
  stability of, 351
  toxicology/pharmacokinetics/effective dose, 375–376
Cyclic adenosine monophosphate (cAMP), 232
Cyclins/cyclin-dependent kinases (CDK), 209, 325–327, 356, 454, 473, 530, 586
Cyclooxygenase activity/inhibition (COXs)
  Alzheimer's disease, 368
  angiogenesis, 133
  Barrett's esophagus and esophageal cancer, 156
  cloves, 587
  curcumin, 355, 369, 579
  ginger, 580
  NSAIDs and colorectal cancer, 127–130, 134
  onions, 586
  oxidative stress, 176
  resveratrol, 232, 242–243, 256, 455
  target-organ specificity for chemopreventive agents, 241
  tea, 406
  vitamins C/E and beta-carotene, 283
Cyclophosphamide, 259, 589, 617
Cyclosporine, 372

CYP enzymes
  curcumin, 576
  fruits and berries, 473
  oligomeric proanthocyanidins, 323–324, 330, 331
  P450 cytochrome, 323–324
  resveratrol, 455
  rosemary, 588, 589
  tea, 406
Cystitis, radiation therapy and hemorrhagic, 218
Cytokeratins, 584
Cytokines
  asbestos, 53
  curcumin, 372
  estrogen, 113
  hemangiomas and breast cancer, 208–209
  mushrooms, maitake, 511
  oxidative stress, 176
  probiotics, 631
  radiation, 83
  rosemary, 587
  tea, 404
  vitamins C/E and beta-carotene, 283
Cytosine, 34
Cytotoxin association gene A (cag A), 164
Czech Republic, 12

D

Daidzein, 238–239, 258, 318, 321, 343, 431, 436
Daunorubicin, 612
Deguelin, 241, 242
Dehydrogenase, 630
Dehydroxylase, 630
Delayed-type hypersensitivity (DTH), 283
Deldman, Gabriel, 341
Demethoxycurcumin, 376
Denmark, 12, 197
Deoxy-5-azacytidine (DAC), 22
Deoxycytidine, 19, 22
Detoxification enzymes, 406–407
Dexamethasone, 365
Diabetes, 363–364
Diallyl disulfide (DADS), 254
Diallyl sulfide, 585
Dichloro-4-nitrophenol, 324
Diepoxybutane (DEB), 190–191
Diethylnitrosamine, 588
Diethylstilbestrol, 108
Diet/nutrition
  antiangiogenic effects of dietary chemopreventive agents, 205–209
  colorectal cancer, 625–626
  esophageal cancer, 153–154
  fat and cancer risk, dietary, 433, 435, 633, 641
  free radicals, 180
  gastric cancer, 144
  metals/metalloids, 29, 31, 33
  obesity, 198, 201–202
  overview, 575
  plants and antioxidants, 85

plant vs. high-fat/meat-based diets, 433, 435
    red meat and cancer risk, 625, 644
Dihydrocurcumin, 373
Dihydrodeoxyadenosine, 43
Dihydrodeoxyguanosine, 43
Dihydrotestosterone, 431
Dihydroxy-3,4,6-trimethoxyflavone, 239
Dimerization, 81
Dimethylaminoazobenzene, 590
Dimethylbenz-a-anthracene (DMBA) model
    animal models, 284
    beta-carotene, 338
    cloves, 587
    curcumin, 351, 576
    fruits and berries, 473, 474
    garlic, 585
    genistein, 113, 331
    ginger, 580
    lycopene, 534
    mace, 590
    overview, 241–242
    resveratrol, 455–456
    rosemary, 588
    tea, 394–396, 400
Dimethylhydrazine (DMH), 474
Dimethylnitrosamine (DEN), 341, 398
Diolepoxide-2 (BPDE-2), 407
Dioscorides, 253
Diosmin, 321, 337
Diphenyl-2-picrylhydrazyl (DPPH), 470
Disproportionation reaction, 81
Dithiolethione, 588
DNA (deoxyribonucleic acid)
    apoptosis, 91
    asbestos, 52
    coenzyme Q_{10}, 608
    double-strand breaks, 44–45
    estrogen, 107, 110–111
    free radicals, 82–83, 172, 343–344
    initiation and promotion, 486
    metals/metalloids, 34, 35
    8-methoxypsoralen, 188–189
    NADH, 543–546
    oligomeric proanthocyanidins, 324, 325, 330–331
    oxidative stress, 174–175, 177, 178
    Pycnogenol, 502
    reactive oxygen and nitrogen species, 493–494
    reactive oxygen species, 43
    repair process, 22, 43–44
    resveratrol, 257
    rosemary, 588–589
    signaling pathways, 44–45
    silica, 49
    smokeless tobacco, 97
    soy foods, 258
    tea, 408
    TNF-related apoptosis-inducing ligand, 352
    topoisomerase, 343
    vitamins C/E and beta-carotene, 275–281
    xenobiotic metabolism, 189–190
DNA (deoxyribonucleic acid) methylation patterns

chemotherapeutic response, 22–23
    conclusions/summary, 23–24
    DNA hypermethylation and tumorigenesis, 18–21
    DNA hypomethylation in tumor development, 21–22
    maintenance of, 18
    overview, 17–18
DNA methyltransferases (DNMT), 17, 18, 21
DNMT gene, 21, 23
Doxorubicin, 221, 313, 326, 332, 544, 612
Drug resistance and DNA methylation patterns, 17–18,
    22–23
Drugs, chemopreventive. *See* Phyto *listings; specific*
    phytopharmaceutical
Duodenum and tea, 399
Dysphagia and esophageal cancer, 153, 160–161
Dysplasia, 311–312

E

Ebers papyrus, 249, 582
Ecteinascidin, 332
E2F1 transcription factor, 46
Ehrlich, Paul, 600
Eicosapentaenoic acid (EPA), 562
Elastic light scattering, 157
Electron spin resonance (ESR), 79
Eli Lilly Corp., 261
Ellagic acid, 324, 474
Elmer 2000 gas chromatograph, 91
ENADA. *See under* NADH
Endocrine disrupters, 106
Endometrial cancer
    hormone-related, 433
    phytoestrogens, 442
Endoscopic ultrasound and esophageal cancer, 158–159
Endoscopy
    esophageal cancer, 156, 157, 160
    gastric cancer, 146, 149
    Helicobacter pylori, 167
Endothelial cell neoplasms, 112, 206–207, 432
Endotoxin shock and curcumin, 372
Energy production, cellular, 542
English yew, 519
Enterobacter, 628
Enterococcus, 628
Enterodiol, 113, 431
Enterolactone, 113, 431
Environmental Protection Agency (EPA), 30, 32
Eosinophils and oxidative stress, 176
Epicatechin gallate (ECG), 232; *see also* Tea polyphenols
Epidemiology of cancer, 3, 4, 6–14
Epidermal growth factor receptors (EGFR), 66–67, 72–74,
    109, 353–354
Epigallocatechin gallate (EGCG), 205–209, 308, 326–327;
    see also Tea
Epirubicin, 161
Epsilon-DNA adduct level, 494
Epstein-Barr virus (EBV), 581
Equol, 431
ERK1/2 enzymes, 53

Erwinia uredovora, 527
Esophageal cancer
  Barrett's esophagus, 155–157
  clinical presentation, 153
  conclusions/summary, 161
  cumin seeds, 589
  dysphagia, 160–161
  epidemiology, 151–152
  fruits and berries, 473
  geographical distribution, 152, 153
  grape seed proanthocyanidin extract, 473
  incidence, 151–153
  nitrosamines, 342
  obesity, 200
  onions, 586
  overview, 151
  risk factors, 153–155
  staging and preoperative evaluation, 157–159
  tea, 338, 398, 411–412
  treatment and prognosis, 159–160
  turmeric, 589
  vitamins C/E and beta-carotene, 290
Esophagectomy, 159
Essiac, 252
Estradiol, 105, 107, 110, 112, 429–432, 456
Estriol, 105
Estrogen. *See also* Phytoestrogens
  angiogenesis, 108–110
  coactivators and cosuppressors, 112
  conclusions/summary, 113–114
  genomic/nongenomic actions, 110–112
  obesity, 200
  overview, 105–106
  phytoestrogens, 112–113
  receptor, 112, 429
  response elements, 110–111, 452
  resveratrol, 451–452
  role of in tumorigenesis, 107–108
Estrone, 105, 431
Ethanobotany, 264
Ethoxyresorufin-O-deethylase (EROD), 406
Etoposide, 262
Eugenol, 587
Eukaryotic cells and DNA damage, 44
Eupatilin, 239
Europe, 197, 536
European Organization for the Research and Treatment of Cancer, 343
Excision repair, DNA damage and, 43–44
Exisulind, 132
Extracellular matrix (ECM), 327
Extracellular signal-regulated kinase (ERK), 177, 409

F

Familial adenomatous polyposis (FAP), 124, 126, 129, 130, 132–135, 494
Fanconi anemia, 189–192
Fas/FasL system, 50
Fat and cancer risk, dietary, 633, 641

Fc-gammaRIII, 501
Fears/misconceptions, overcoming common, 14
Fenton-type reactions, 35, 36
Ferrulic acid, 208
Fibroblast growth factors (FGFs), 109
Fibrosis after radiation therapy, 216–217
Finesteride, 313
Fisetin, 324
Flavonoids, 84, 106, 208, 260, 314–317, 339–340, 432, 471; *see also* Fruits and berries; Genistein; Grape seed proanthocyanidin extract; Natural products/supplements; Oligomeric proanthocyanidins; Phyto *listings;* Pycnogenol; Quercetin
Flavopiridol, 326
5-Fluorouracil, 160, 161, 259
Folkers, Karl, 604, 609, 612
Food and Drug Administration (FDA)
  aloe, 259
  camptotheca, 262
  capsaicin, 581
  celecoxib, 129, 130
  clinical trials, 342–343
  coenzyme Q_{10}, 607
  hoxsey, 251
  laetrile, 252
  mayapple, 262
  NADH, 548
  soy foods, 258
Food-derived products. *See* Diet/nutrition; Fruits and berries; Mushrooms, maitake; Palm tocotrienols; Phytochemical *listings;* Spices; Tea polyphenols
Formaldehyde (FA), 92, 94
France, 152, 197
Free radicals. *See also* Antioxidants; Lipid peroxidation; Oxidative/nitrosative stress; Reactive oxygen species
  astaxanthin, 560–561, 563
  balance, critical, 179–180
  in biological systems and the environment, 172–174
  conclusions/summary, 181
  defining terms, 80
  dimerization, 81
  disproportionation reaction, 81
  DNA, 343–344
  lead, 33
  lipid peroxidation, 486
  olefinic double bond, addition of a radical to an, 80
  overview, 171–172
  oxidizing agents, reaction with, 80
  as promoting agents, 486
  Pycnogenol, 496–497, 499
  radiation and the environment, 83–86
  research tools, 79–80
  smokeless tobacco, 91–92
  substitution homolytic bimolecular reactions, 80
  tea, 407–408
  treatment and protection strategies, 180, 181
  triphenylmethyl radical, 79
  unimolecular radical reactions, 80
  xenobiotic metabolism, 192
"Free radical theory of aging" (Harman), 613

Fruits and berries
    animal models, 473–474
    berries, major phytochemicals, 467–469
    carotenoids, 556
    conclusions/summary, 474–475
    epidemiology of consumption, 466–467
    minimum daily servings, 640
    overview, 465–466
    in vitro system studies, 469–473
Fucoxanthin, 338
Furocoumarins, 188–189

G

Galangin, 324
Gallbladder cancer and obesity, 200
Gallotannin, 471
Gallstone formation and curcumin, 366
Garlic, 253–254, 584–585
Gastric cancer
    astaxanthin, 564
    basil leaves, 589
    classification and prognosis, 147–148
    clinical features, 145
    conclusions/summary, 149
    cumin seeds, 589
    curcumin, 352
    diagnosis, 146–147
    DNA methylation patterns, 20
    etiology, 144–145
    grape seed proanthocyanidin extract, 471
    Helicobacter pylori, 163–169
        etiology, 144
        pathology, 145, 168
        precancerous lesions, 294
        vitamin C, 287
    incidence and epidemiology, 12, 143–144
    lycopene, 535
    metals/metalloids, 33
    mortality rates, 12, 144
    nitrosamines, 342
    oligomeric proanthocyanidins, 340
    onions, 586
    oxidative stress, 465
    pathology, 145
    peppers, hot red, 581
    tea, 256, 338, 398, 412–413
    therapy, 148–149
    turmeric, 589
    vitamins C/E and beta-carotene, 287, 290, 294
Gastroesophageal reflux disease (GERD), 154, 155, 157
Gastrointestinal (GI) tract, 128, 322
Gefitinib, 74
Geldanamycin, 74
Gender
    and chemical carcinogenesis, 342
    esophageal cancer, 152
    gastric cancer, 143
    incidence and mortality, 6–11, 247–248

obesity, 200
    prevalence, 10–11
Genetic effects, radiation-induced, 81
Genistein
    antioxidant efficacy, 432–433
    breast cancer, 113
    efficacy of, 238–239
    leguminous plants, 113
    MCF-7/6 cells, 331
    mechanism of action, 230
    overview, 231
    paradoxical but beneficial role, 343
    P-form PST, 324
    prostate cancer, 258
    protein kinases, 326
    5-alpha reductase, 431
Genomic mutations, 43
Genotoxicity and tea, 404–405
Ginger and related rhizomes of Zingiberaceae family, 580–581
Ginseng, 254–255
Gleason System, 6
Glomerular filtration rate (GFR), 371
Glucarolactone, 330
Glucose-6-phosphate/phosphatase, 364
Glucuronidase, 330, 630, 631
Glutamylcysteine, 474
Glutathione (GSH)
    curcumin, 362
    fruits and berries, 474
    garlic, 584
    lead, 33
    lipid peroxidation, 176
    phytoestrogens, 433
    Pycnogenol, 497
    reactive oxygen species, 35, 179, 180
    spices, 589
Glutathione-s-transferase (GST)
    cloves, 587
    curcumin, 368, 576, 577
    diepoxybutane, 190–191
    DNA methylation patterns, 21
    fruits and berries, 473
    genetic polymorphisms, 296
    probiotics, 632–633
    rosemary, 588
    tea, 406
Glycinoly-3-methoxyphenyl-1,6-heptadiene-3,5-dione, 377
Glycitein, 343
G2/M phase cell cycle, 45
Grading systems, 5–6
Grapefruit juice, 323, 330
Grape juice, 331
Grape seed proanthocyanidin extract (GSPE). *See also* Oligomeric proanthocyanidins
    angiogenesis, 470
    animal models, 473, 474
    antitoxic/cytotoxic effects, 313, 471
    free radicals, 312
    overview, 221–222
    smokeless tobacco, 99, 100, 313

Green tea, 205–206, 230, 255–256, 312, 326; *see also* Tea polyphenols
GST placental-form-positive (GST-P), 588
GTP in tea. *See* Tea polyphenols
Guanine, 34

H

Haber-Weiss type reactions, 35, 36, 173
HaCaT cell line, 498
Harman, Denham, 613
Hartwell, J., 263
HCT-15 cell line, 442
Head and neck squamous cell carcinoma (HNSCC), 326, 521
Health Professionals Follow-Up Study (HPFS), 288, 535
Heat shock/stress protein 90 (HSP90), 94
HeLa cells, 174, 282, 583
Helicase-like transcription factor (HLTF), 19
Helicobacter pylori
    astaxanthin, 563
    curcumin, 357
    esophageal cancer, 155
    oxidative stress, 175, 177
    peppers, hot red, 581, 582
    probiotics, 629
Helix pomatia lectin, 366
Hemangioendotheliomas (HE). *See* antiangiogenic effects of dietary chemopreventive agents *under* Diet/nutrition
Heparin-binding epidermal growth-factor-like growth factor (HB-EGF), 47
Hepatitis B, 177
Hepatocyte growth factor (HGF), 63–64
Hepatomas/hepatotoxicity, 20, 175–176, 259, 589
Herbal supplements. *See* Natural products/supplements
Hereditary nonpolyposis colorectal cancer (HNPCC), 124, 129, 132
HER2/neu expression, 353–354
Hesperidin, 321, 337
Heterocyclic amines (HCA), 405, 644
HGF/SF-Met, 67, 68
HIC-1 (hypermethylated in cancer) gene, 19
High-density lipoprotein (HDL), 360
High-pressure liquid chromatography (HPLC), 92, 241, 276, 482
Hippocrates, 253
HIV (human immunodeficiency virus), 367, 611
HL-60 cells, 256, 432, 442, 470, 580
HMG-CoA reductase inhibitors, 530–531
Hodgkin's disease, 10, 261
Hoffmann-LaRoche, 607
Homeopathic Pharmacopoeia of the United States, 250
Homologous recombination repair (HRR), 44, 45
Hormone replacement therapy (HRT), 113, 433, 435–436
Hot red peppers, 581–582
Hoxsey, 250–251
Human growth factor (HGF), 71–74
Human papillomavirus (HPV), 177
Human telomerase reverse transcriptase (hTERT), 20
Human umbilical vein endothelial cells (HUVEC), 409

Hungary, 12
Hycamtin, 262, 343
Hydrazinocurcumin, 377
Hydrogen peroxide (H$_2$O$_2$), 35, 179, 208, 470, 497
Hydroxphenyl-hepat-1,6-diene-3,5-dione, 376–377
Hydroxyeicosatetraenoic acids, 406
4-Hydroxyestradiol (4-OHE2), 455
Hydroxyl groups, 324, 429
Hydroxyl radical (HO)
    asbestos, 52
    DNA damage, 43, 639
    grape seed proanthocyanidin extract, 221
    ionizing radiation effects, mimicking, 35–36
    ischemia/perfusion, 173
    oxidation-reduction cycle, 35
    Pycnogenol, 496
    silica, 49
Hydroxy-3-methoxybenzoic acid methyl ester (HMBME), 376
4-Hydroxynonenal (4-HNE), 175
Hydroxy-2-nonenal, 368
Hydroxyproline, 370
Hyperbaric oxygen therapy (HBO), 217–219
Hypericum perforatum, 330
Hypoxanthine, 173

I

ICF (immunodeficiency/centromeric instability/facial abnormality) syndrome, 21
IkB protein family, 47–48
Image Pro Plus, 99
Immune system
    aloe, 259
    astaxanthin, 563–564
    coenzyme Q$_{10}$, 254, 616–617
    garlic, 254
    mushrooms, maitake, 510–511
    NADH, 546
    probiotics, 631
    statins, 616
    vitamins C/E and beta-carotene, 283
Incidence rates, 6–13, 247–248, 284–285, 433, 435, 643
India
    arsenic, 32
    Ayurveda, 248
    leukoplakia, 292
    tea, 390, 392
    tulsi, 259
    yew, 261
Indigenous peoples and natural products/supplements, 260–262
Indonesia, 590
Inflammation
    Alzheimer's disease, 367–368
    apoptosis, 50
    asbestos, 53
    astaxanthin, 563
    curcumin, 351, 366, 375
    free radicals, 174, 181
    kurarinone, 443

oligomeric proanthocyanidins, 312
oxidative stress, 176–177
paraquat, 370
phytoestrogens, 433, 434
redox-sensitive transcription factors, 86
rosemary, 587
tea, 403–404
vitamins C/E and beta-carotene, 287
Inflammatory bowel disease (IBD), 371
Insulin-like growth factor binding proteins (IGFBP), 488, 531, 532–533
Insulin-like growth factor (IGF-I/II), 46
Insulin resistance and obesity, 201
Interferon-alpha, 207
Interleukin 4, 18
Interleukin 6, 353, 433, 546
Interleukin 10, 404
Interleukin 12, 206, 366
International Agency for Research on Cancer, 30, 32, 199
International Symposium on Coenzyme Q_{10}, 610
Ipriflavone, 436
Ireland, 390
Iressa, 74
Irinotecan, 262, 343
Iron, 29, 34
Ischemia/reperfusion and free radicals, 173
Ishikawa-Var-I assay, 438, 439
Isoflavones, 113, 258, 312, 343, 429, 432, 435; *see also* Genistein; Oligomeric proanthocyanidins; Phytoestrogens
Isomerization, 526
Isoprenoid intermediates, 488–489
Isoproterenol, 362, 363
Isothiocyanates, 253
Italy, 197

J

J774A.1 cells, 92, 313, 471, 497
JB6 cells, 174
Janus kinase STAT pathway, 355, 366
Japan
    atomic bomb survivors, 81
    coenzyme Q_{10}, 254, 607
    gastric cancer, 12, 143, 147, 149
    kurarinone, 437
    prostate cancer, 536
    soy foods, 258
    tea, 390, 392, 414
Jun activated gene in CEF (JAC), 47
Jun/Fos complex, 47
c-Jun N terminal kinase-1 (JNK-1), 48, 53, 112, 209, 327, 355, 409
Jurkat-T cells, 22, 174

K

Kasabach-Merritt syndrome, 206, 207
Kempferol, 318, 324
Keratinocytes and smokeless tobacco, 98, 99

Kidney cancer
    metals/metalloids, 32, 33
    obesity, 200
    tea, 413
Korea, 12, 437
Korea Cancer Center Hospital, 255
Kuopio ischemic heart disease risk factor (KIHD), 466
Kurarinone, isolation/characterization of
    conclusions/summary, 443–444
    material/methods, 436–439
    results, 439–442

L

Lactate dehydrogenase (LDH), 92, 362, 364, 365, 368, 369
Lactic acid bacteria, 624; *see also* Probiotics
Lactobacillus, 629, 632, 633
Laetrile, 251–252
Lamina propria, 151
Laser scanning confocal microscopy, 95–96
Lead, 29, 33, 35
Lecithin, 239
Leptin, 198
Leukemias
    aloe, 259
    coenzyme Q_{10}, 609–610
    curcumin, 352
    defining terms, 5
    DNA methylation patterns, 20
    fruits and berries, 470
    garlic, 584
    mayapple, 262
    metals/metalloids, 32
    oligomeric proanthocyanidins, 326
    periwinkle, rosy, 261
    resveratrol, 456
Leukoplakia, 292
Lewis lung carcinoma (LLC) tumors, 256
L86-8276 (flavone), 326
Life expectancy and cancer deaths, 14
Lifestyle and cancer risk, 341, 416, 433, 435
Ligand binding domain (LBD), 112
Lignanes, 113, 429, 431
Lilium, 338
Limonene, 239
Lingonberry, 470
Linitis plastica (leather bottle stomach), 146
Linxian General Population Trial, 639–640
Lipid metabolism and curcumin, 361–362
Lipid peroxidation
    astaxanthin, 559–560
    breast cancer, 494
    curcumin, 375
    free radicals, 486
    fruits and berries, 474
    lycopene, 527
    onions, 586
    oxidative stress, 175–176
    palm tocotrienols, 485

permeability of cell membranes, 36
phytoestrogens, 433
reactive oxygen and nitrogen species, 494
rosemary, 587
smokeless tobacco, 91–93, 95
stages, three basic, 36
tulsi, 260
Lipopolysaccharide (LPS), 164
Lipoxygenase, 580
Liver cancer
  basil leaves, 589
  cumin seeds, 589
  fruits and berries, 469
  metals/metalloids, 32
  mushrooms, maitake, 512
  NADH, 549, 550
  obesity, 200
  tea, 399–400, 415
  turmeric, 589
Liver injury/weight and curcumin, 363–364,
  368–369
LNCaP cell line, 352
Loss of heterozygosity (LOH) and DNA methylation
  patterns, 23
Low-density lipoproteins (LDL), 231, 361, 364, 433, 485,
  529
Low-molecular-weight (LMW), 366
Lung cancer
  asbestos, 52–54
  curcumin, 358
  estrogen, 108
  fruits and berries, 473
  grape seed proanthocyanidin extract, 471
  incidence rate, 7, 8, 41
  lycopene, 234, 527, 534–535
  mayapple, 262
  metals/metalloids, 32, 33
  molecular events, basic, 43–48
  mortality rates, 7, 8, 41
  NADH, 549, 550
  prevalence, 11
  receptor tyrosine kinases, 64–74
  resveratrol, 256
  silica, 49–51
  Taxol, 521
  tea, 256, 338, 398, 402–403, 415
  vitamins C/E and beta-carotene, 285–287, 290–291
  worldwide, cases, 12
Lung injury, curcumin and drug-induced, 369–370
Lupus erythematous, 109
Lutein, 286, 289, 558
Luteolin, 318
Lycopene
  animal studies, 234, 533–534
  bioavailability, 526–529
  chemistry, 527–528
  combination with other molecules in growth inhibition,
    531
  content of different foods, 526
  epidemiology, 534–536
  fruits and berries, 467

mechanism of action, 230
molecular basis of effects, 530–531
observational studies, 276–277
overview, 525, 558
prooxidant behavior, 561
in vitro studies, 532–533
Lymphadenectomy, 148–149, 159
Lymphocyte ascorbate concentrations, 276, 277, 281
Lymphocytic leukemia, 259
Lymphokine activated killer (LAK) cells, 91
Lymphomas
  coenzyme Q_{10}, 612
  defining terms, 5
  genistein, 231
  Helicobacter pylori, 167, 168
  metals/metalloids, 32
  mortality rate, 8
  nuclear factor-kB, 48
  obesity, 200
  oligomeric proanthocyanidins, 326
Lysosomal hydrolases, 362

M

Mace, 590
Madhuyasti, 248
Magic bullet theory, decline of the, 644
Magnetic resonance imaging (MRI), 147, 549
Maize and esophageal cancer, 154
Malaysian Palm Oil Board (MPOB), 482
Malignant pleural mesothelioma (MPM), 66
Malignant tumors, 4–5; *see also specific type*
Malonaldehyde, 362
Malondialdehyde (MDA), 91, 92, 94, 175, 362
MALT (mucosa-associated lymphoid tissue) lymphoma,
  167, 168
Mandibular osteo-radionecrosis, 217–218
Mandulea sericea, 242
Manganese superoxide dismutase (MnSOD), 174, 176
Mannitol, 93
MAPKAP-2 kinase pathway, 112, 177
Matrix metalloproteinases 1/2 (MMP-1/2), 209, 217,
  327–330, 356, 358, 365, 409
Mayapple, 261–262, 264
M cells, 631
McFarlan, Frank, 600
MCF-7/6 cells, 174, 331, 352, 438, 442, 452, 453, 471,
  530, 586
MDA-MB-231 cell line, 67, 453
MDM2, 45
Meat consumption, 435, 625, 644
Meerwein reaction, 80
Memorial Sloan-Kettering Cancer Center
  (MSKCC), 252
Menopause and estrogens/phytoestrogens, 105–106, 428,
  435–436, 456, 457
Menstrual cycle and phytoestrogens, 432
Merkel cell carcinoma, 259
Mesotheliomas, 42
Metalloids. *See* Metals/metalloids

Metalloproteinase. *See* Matrix metalloproteinases 1/2
Metals/metalloids
     carcinogenic, 32–34
     epidemiology, 30–31
     free radicals, 173
     mechanisms of cancer causation, 34–37
     overview, 29–30
     radiation, 83
Metchnikoff, Elie, 624
8-Methoxypsoralen (8-MOP), 188–189
Methoxyresorufin demethylase (MROD), 405
Methylation, DNA. *See* DNA methylation patterns
Methylcholanthrene-induced (Meth-A) mouse tumor cells, 562, 564
Methylcytosine, 18, 83
Methyl nitrosamino-1-butanone (NNK), 90
Methylthiohexyl isothiocyanates (MHITC), 590
Met receptor. *See* receptor tyrosine kinases *under* Lung cancer
Mevalonate biosynthetic pathway, 488–489
MGMT gene, 22
Michigan State University, 644
Microflora, intestinal, 624, 627–628; *see also* Probiotics
Microsatellite instability (MSI), 127
Mint, 590
Mismatch repair (MMR), 22, 44
Mistletoe, 250
Mitchell, Peter, 603
Mitochondrial activity/physiology/pathology, 173, 542, 601, 603–604, 607, 613, 616–617
Mitogen-activated protein kinase (MAPK), 366, 371, 409, 494
Mitomycin C (MMC), 189–190
MLH-1 gene, 282
Mms2/Ubc13 enzyme, 44
MOCAP-induced neurotoxicity, 313
ModFit cell cycle analysis software, 99
Monkey's Bench mushroom family, 511
Monocyte chemoattractant protein-1 (MCP-1), 208–209, 470
Monomers, 492
Mortality rates, 6–14, 247, 602, 643
Mouse mammary gland organ culture (MMOC) assay, 240–243
mRNA expression, 586
MTT assay, 98, 352
Mucosal junction, 151
Multidrug resistance, 337, 372
Multiple myeloma, 200
Multiple sclerosis and curcumin, 366
Murine models of endothelial cell neoplasms, 206
Muscle regeneration and curcumin, 364
Muscularis propria, 151
Mushrooms, maitake, 509–515
Mutagenicity and tea, 404–405
Myelofibrosis, 19
Myeloperoxidase, 176, 362
Myocardial infarction, 362–363
Myricetin, 318, 324
Myths, persistent cancer, 14

N

N-acetyl-beta-D-glucosaminidase, 369, 370
N-acetylcysteine, 73, 206
NADH (nicotinamide adenine dinucleotide)
     biological functions of, 542–547
     case studies, 549–550
     curcumin, 363
     ENADA, 547–549
     ischemia/reperfusion, 173
     mechanism of action, 551
     mitomycin C, 190
NADPH (nicotinamide adenine dinucleotide phosphate oxidase)
     curcumin, 363, 577
     inflammation, 176
     ischemia/reperfusion, 173
     mitomycin C, 190
     probiotics, 633
     tea, 406
Napthoflavone, 322
Naringin, 321
Nasopharyngeal carcinomas, 20
National Cancer Institute (NCI)
     Chemoprevention Testing Program, 608
     classifying phytochemicals, 313
     clinical trials, 342, 343
     coenzyme Q_{10}, 610
     diet/nutrition, 575, 640
     drug discovery/development program, 230
     essiac, 252
     hoxsey, 251
     laetrile, 251, 252
     mayapple, 262
     natural products/supplements, 249
     resveratrol, 257
     Surveillance, Epidemiology, and End Results program, 4, 152
     Taxol, 520
     tea, 232–233
     yew, 261
National Institute for Occupational Safety and Health, 30, 32
National Institutes of Health (NIH), 205
National Toxicology Program, 30, 32
Native Americans, 260–262
Natural killer (NK) cells, 255, 511
Natural Product Alert (NAPRALERT), 239
Natural products/supplements; *see also* Phyto *listings; specific phytopharmaceutical/food type*
     Ayurveda, 248–249
     bioprospecting efforts, need to increase, 264
     conclusions/summary, 262–264
     cure, cancer, 250–252
     drugs developed from, 260–262
     prevention, cancer, 253–258
     radiation therapy, antioxidants as adjuncts to, 259–260
Netherlands, 12
Netherlands Cohort Study on Diet and Cancer, 412–413, 414, 586
N-ethyl-N-nitro-N-nitrosoguanidine (ENNG), 399

Neuroblastoma and DNA methylation patterns, 23
Neurodegenerative disorders, 436
Neutrophils, 174, 176, 370
Nickel, 33–36
Nitric oxide (NO)
  astaxanthin, 564
  curcumin, 356
  free radicals, 181
  inflammation, 176
  NSAIDs, 128–129
  Pycnogenol, 497, 499
  silica, 50–51
  wound healing, 365
Nitrilotriacetate (NTA), 498
Nitroreductase, 630, 631
Nitrosamines, 90, 342, 644
Nitrosation, inhibition of, 281
Nitrosative stress. *See* Oxidative/nitrosative stress
N-methyl-N-nitro-N-nitrosoguanidine (MNNG), 168
N-methyl-N-nitrosourea (MNU), 168
N-nitrosomethylbenzylamine (NMBA), 233
N-nitroso-methylbenzylamine (NMBzA), 398
N-nitroso-N-methylurea (NMU) model, 241–242
N-nitrosonornicotine (NNN), 90
Noble, Robert L., 261
non-Hodgkin's lymphoma, 8, 200
Nonhomologous end joining (NHEJ), 44, 45
Non-small-cell lung cancer (NSCLC), 63, 64, 66, 67, 74
Normal human oral keratinocytes (NHOK), 94–97, 99–101
Norway, 200
NSAIDs (nonsteroidal anti-inflammatory drugs), 180, 336,
  368; *see also under* Colorectal cancer
Nuclear factor-kB (NF-kB) proteins
  antioxidants, 494–495
  asbestos, 52–53
  curcumin, 355, 364, 371, 579, 580
  dietary chemopreventive strategies, 209
  E2F1 transcription factor, 46
  inflammation, 86, 433
  overview, 47–48
  oxidative stress, 178–179
  peppers, hot red, 582
  phytoestrogens, 433, 442
  Pycnogenol, 498
  reactive oxygen and nitrogen species, 494
  resveratrol, 256
  silica, 49
  tea, 256, 410–411
  turmeric, 257
Nuclear Regulatory Commission, U.S., 81
Nucleotide excision repair (NER), 43, 44
Nulliparity, 108
Nurses Health Study, 128, 134, 286, 289
Nutraceuticals, 205, 491
Nutrition. *See* Diet/nutrition

O

Obesity
  association of cancer with, 199–200
  conclusions/summary, 201–202
  esophageal cancer, 154
  estimating fat gain/loss, 198–199
  etiology, 198
  overview, 197–198
  pathogenesis behind associations, 200–201
Occupational Safety and Health Administration (OSHA),
  30
Olefinic double bond, radical added to, 80
Oligomeric proanthocyanidins (OCPs)
  absorption and bioavailability, 321–322
  cell-cycle analysis, 325–327
  chemical structure, 321
  conclusions/summary, 339–340
  DNA damage/repair, 330–331
  evidence for protective effect, 318–320
  execution of anticarcinogenic action, 314–317
  gastrointestinal metabolism, 322
  grape seed proanthocyanidin extract, 312–313
  metalloproteinases, 327–330
  miscellaneous mechanisms, 337–338
  overview, 313
  p450 cytochrome, 322–325
  programmed cell death, 332–337
  types/concentrations within plants and among species,
    318
  in vivo, their role, 320–322
Oltipraz, 588
Oncogenes having transforming capabilities, 341
Oncosuppressor genes, 341
Oncovine, 264
Onions, 585–586
Oolong tea, 232; *see also* Tea polyphenols
Optical coherence tomography, 157
Oral cancer, 584
Oral contraceptives (OC), 108, 123–124
Oral leukoplakia, 292
Orientin, 260
Ornithine decarboxylase (ODC)
  curcumin, 257
  evaluation of, 240
  fruits and berries, 474
  ginger, 580
  prostate cancer, 401
  rosemary, 588
Osteoporosis, 428, 436, 456, 457
Ovarian cancer
  camptotheca, 262
  epigallocatechin gallate, 209
  hormone-related, 433
  monocyte chemoattractant protein-1, 208
  mortality rate, 8
  obesity, 200
  oral contraceptives, 123
  Taxol, 521
Oxidations, radicals undergoing one-electron, 80
Oxidative/nitrosative stress
  aging, disease of the, 612–613
  biomolecules, modifications of, 174–177
  breast cancer, 610–611
  coenzyme Q_{10}, 606, 610–611
  colorectal cancer, 465

conclusions/summary, 181
dichotomous behavior of oxygen, 192
gastric cancer, 465
grape seed proanthocyanidin extract, 313
Helicobacter pylori, 168
oligomeric proanthocyanidins, 332
overview, 174
redox-dependent toxicity in Werner syndrome/Fanconi
    anemia, 189–192
signaling pathways/transcriptional factors, 177–179; *see
    also* Antioxidants; Free radicals; Hydroxyl radical;
    Lipid peroxidation; Reactive oxygen species
8-Oxoadenine, 276, 277
8-Oxoguanine, 276
Oxomate, 243
8-oxoGua, 83, 276, 277

P

Pacific yew, 260–261, 264, 519
Packer, Lester, 644
Paclitaxel. *See* Taxol
Palm tocotrienols, 481–489
Pancreatic cancer
    coenzyme Q_{10}, 611
    DNA methylation patterns, 20
    mortality rate, 7, 8
    obesity, 200
    tea, 400–401, 415
Pancreatitis, 372
Paraquat, 370
Paxillin, 67, 73
PC-cell derived growth factor (PCDGF), 454
PC12 cells, 544–545
PCNA (proliferating cell nuclear antigen), 44, 328–329,
    544–545
Penobscot Indians, 262, 264
Pentoxifylline, 220–221
Peppers, hot red, 581–582
Perilly alcohol, 590
Periwinkle, 261, 264
Perkin Elmer Lambda 6 spectrophotometer, 92
Peroxisome proliferator-activated receptor-γ (PPAR-γ),
    355, 369
Peroxynitrite (ONOO), 496, 497
Pesticides, 106
Peyer's patches, 631
P-form PST, 324
P-glycoprotein (Pgp), 329, 589
p15 gene, 20
p16 gene, 20
p21 gene, 209, 326, 327, 329, 356
p27 gene, 209, 326, 327, 329, 356
P38 gene, 177
p73 gene, 501
PI3K gene, 67–68, 71, 72
p16$^{INK4A/B}$ gene, 19
p15^{INK4B} gene, 18–19
p53 protein
    activator protein-1, 47

asbestos, 52
curcumin, 356
DNA damage response, 45
DNA methylation patterns, 19
DNA repair, 545
doxorubicin, 544, 545
esophageal cancer, 155–156
inactive in half of all cancers, 209
metals/metalloids, 36–37
nuclear factor-kB, 48
oligomeric proanthocyanidins, 326
oncosuppressor genes, 341
overview, 90
oxidative stress, 177, 178
Pycnogenol, 501
resveratrol, 454, 455
smokeless tobacco, 99–101
p57KIP2 gene, 20
P450 cytochrome
    astaxanthin, 564
    curcumin, 576
    fruits and berries, 473
    genistein, 331
    oligomeric proanthocyanidins, 322–325
    resveratrol, 256
    tea, 405
    xenobiotic metabolism, 192
pH and probiotics, 632, 633
Phellamurin, 321
Phenols, 314–317; *see also* Oligomeric proanthocyanidins
Phenylimidazo-4,5-b-pyridine (PhIP), 399
Phorbol ester-induced prostaglandin E2 (PGE2), 373
Phospholipid levels, 364
Photodynamic therapy (PDT), 157, 160
Physicians' Health Study (PHS), 290, 640
Phytoalexins, 576
Phytochemicals, development of selected, 229–234
Phytochemicals as potential cancer chemopreventive agents
    candidate to clinical trials, 243–244
    discovery of new agents, 239–243
    in foods, 238–239
    overview, 237–238
Phytoene, 338
Phytoestrogens. *See also* Kurarinone
    benefits, potential health, 433, 435–436
    biological effects, 112–113, 430–434
    classification, 429–431
    conclusions/summary, 442–444
    discovery of, 428
    estrogens, understanding actions of, 428
    overview, 106
    receptor, the estrogen, 429
Pine bark, 495–496; *see also* Pycnogenol
Piroxicam, 241
Plants and chemoprevention, 83–85, 239–240; *see also*
    Natural products/supplements; Phyto *listings;*
    specific phytopharmaceutical/food type
Plasminogen, 327
Plasmodium berghei, 609
Platelet aggregation and curcumin, 362
Platelet-derived growth factors (PDGFs), 109, 208

Podophyllin, 262
Poke root, 251
Polygonum cuspidatum, 450
Polymorphonuclear neutrophils (PMNs), 174
Polypectomy, 132
Polyphenols, 239, 253, 314–317, 437–438, 467; *see also*
      Natural products/supplements; Oligomeric
      proanthocyanidins; Phyto *listings;* Pycnogenol;
      Tea polyphenols
Polyphenon E, 233–234
Polyunsaturated fatty acids (PUFA), 273, 281, 494
Portugal, 143
Post-replication repair, DNA damage and, 44
Potassium, 251
Prebiotics, 628–629
Prevalence, 10–12, 624, 625
Prickly ash bark, 251
Primary cancer, 5
Proanthocyanidins, 467, 469–471; *see also* Oligomeric
      proanthocyanidins; Pycnogenol
Probiotics, 629–633
Proctitis, 218
Procyanidins, 495–496; *see also* Pycnogenol
Programmed cell death (PCD), 332–337
Prokaryotic cells and DNA damage, 44
Promyelocytic leukemia, 32
Prostaglandins
   cloves, 587
   curcumin, 373
   inflammation, 176
   NSAIDs, 127, 130, 131
   tea, 406
   vitamins C/E and beta-carotene, 283
Prostate cancer
   chance of getting, lifetime, 341
   coenzyme Q_{10}, 611
   curcumin, 358
   DNA methylation patterns, 21
   hormone-related, 435
   incidence rate, 7, 8
   lycopene, 531, 535–536
   metals/metalloids, 32
   mortality rate, 8
   oligomeric proanthocyanidins, 326
   phytoestrogens, 442
   5 alpha-reductase, 531
   soy foods, 258
   tea, 401–402, 414–415
   vitamins C/E and beta-carotene, 289, 291
Protein kinase C (PKC), 72, 96–97, 253, 282, 360,
      401–402, 488
Protein(s)
   metals/metalloids, 34
   oxidative stress, 176
   Pycnogenol, 498–499, 502
Protein Technologies, Inc., 231
Protocatechuic acid (PCA), 472
Psoralen, 323
PTEN gene, 46
PTI G-2535 (isoflavone), 231
Pulmonary artery endothelial cells (PAEC), 497

Pycnogenol
   antioxidant capacity, 496–498
   antioxidants as preventive factors, 494–495
   bioavailability, 492–493
   classifying, 312, 313
   conclusions/summary, 502–503
   exploration using new approaches, 500–502
   historical landmarks, 492
   monomers and flavonoids, 492
   overview, 337
   pine bark, 491–492
   procyanidins, 495–496
   proteins and enzyme inhibition, binding to, 498–499
   reactive oxygen and nitrogen species, 493–494
Pyrimidines, 83, 277

Q

"Q-cycle" in the electron transport chain, 603
Quercetin, 312, 318, 321, 324, 326, 337, 472, 586
Quinone, 240

R

Race/ethnicity
   and chemical carcinogenesis, 342
   esophageal cancer, 152–153
   gastric cancer, 143
   obesity, 200
   peppers, hot red, 581
Rad5 gene, 44
Rad18 gene, 44
Rad51 gene, 45
Rad6/UbcH1 gene, 44
Radiation, 81–86
Radiation therapy
   esophageal cancer, 153, 158, 160
   gastric cancer, 147
   late adverse effects, modulation of, 215–222
   NADH, 548
Raman spectroscopy, 157
Raspberries, 470, 473
Reactive nitrogen species (RNS), 272
Reactive oxygen and nitrogen species (RONS), 493–494
Reactive oxygen species (ROS). *See also* Free radicals;
      Hydroxyl radical (HO); Lipid peroxidation
   aging, disease of the, 612–613
   DNA damage, 43
   examples of, 172
   generated, reasons for being, 171–172
   glutathione, 35, 179, 180
   metals/metalloids, 35–36
   oligomeric proanthocyanidins, 331–332
   onions, 586
   plants, 84
   radiation, 83
   receptor tyrosine kinases, 73–74
   silica, 49, 50–51
   smokeless tobacco, 97

tea, 407
vitamin C, 272
xenobiotic metabolism, 190
Receptor tyrosine kinases (RTKs). *See under* Lung cancer
Recombinational repair, DNA damage and, 44
Rectal cancer
  coenzyme Q_{10}, 611
  fruits and berries, 466
  obesity, 200
  tea, 413
Red meat and cancer risk, 625, 644
Redox-dependent toxicity in Werner syndrome/Fanconi
    anemia, 189–192
Redox factor 1 (Ref-1), 53
Redox-sensitive transcription factors, 86
5 Alpha-reductase, 431
Red wine, 256
Relative risk, 30
Renal lesions and curcumin, 364
Research, need for more, 343–344
Research Triangle Institute, 520
Resection/surgery. *See specific treatment/disease*
Restriction landmark genome scanning (RLGS), 23
Resveratrol
  benefits of using, 456–457
  bioavailability, 456
  breast cancer, 451–455
  chemical structure, 450–451
  conclusions/summary, 457
  efficacy of, 242–243
  mechanism of action/molecular targets, 230
  in other cancers, 455–456
  overview, 232, 256–257, 450, 472–473
  pharmacological activities, 450
  in vivo, effect of, 456
Retinoic acid, 19–20, 608
Retinoids, 241, 313
Rheumatoid arthritis (RA), 563
Riboflavin, 154
RNA polymerase II (RNAPII), 43
Rochefort, H., 438
Rohitak, 248
Rosa rugosa, 469
Rosemary, 587–589
Rutin, 208, 321

S

Saffron, 582–583
St. Johns-wort (SJW), 323, 330
Salicylates, 473
Salmonella typhimurium-E assay, 233–234
Sandoz Ltd., 262
Sarcomas, 5
Scarring and curcumin, 371
SCRC1 gene, 112
Secondary cancer, 5
Selenium, 239, 608
Serine/threonine phosphorylation and receptor attenuation,
    72–73

Serpentines, 51
Sex-hormone-binding globulin (SHBG), 432
Shen Nung, 390
Sho-saiko-to (TJ-9), 531
Shustruta Samhita, 248
Siegel, Ronald K., 255
Signaling pathways and regulating cell proliferation
  astaxanthin, 564
  curcumin, 354–355, 366
  free radicals, 172
  lycopene, 530–531
  metals/metalloids, 36–37
  oligomeric proanthocyanidins, 325–327
  oxidative stress, 177–178
  tea, 401
  vitamins C/E and beta-carotene, 281–283
Silica. *See under* Lung cancer
Singapore, 581
Sitosterol, 467
SKBR3 cell line, 352
Skin cancer
  curcumin, 352
  grape seed proanthocyanidin extract, 473
  metals/metalloids, 32
  Pycnogenol, 495, 498
  tea, 256, 394–398
  vitamins C/E and beta-carotene, 291–292
SK-MEL-2 cell line, 442
SK-OV-3 cell line, 442
Small-cell lung cancer (SCLC), 63, 66, 67, 69, 74, 262
Small intestine tumors and tea, 399
Smokeless tobacco
  antioxidants, 93
  apoptosis, 98, 99
  conclusions/summary, 101
  DNA fragmentation, 97
  DNA ladder analysis, 97
  free-radical formation and lipid peroxidation, 91–92
  heat shock/stress protein 90, 94
  hepatic lipid peroxidation/urinary metabolites, 93–94
  lactate dehydrogenase and peritoneal macrophage cells,
      92
  laser scanning confocal microscopy, 95–96
  lipid peroxidation, 95
  MTT assay, 98
  normal human oral keratinocytes, 94–97, 99–101
  overview, 90–91
  p53 and Bcl-2 genes, 99–101
  preparation of extract, 91
  protein kinase C, 96–97
  superoxide anion production, 95
  urinary metabolites, excretion of, 92
Smoking. *See* Tobacco
Snuff dipping, 91
Socioeconomic status (SES) and gastric cancer, 143
Soft tissue radionecrosis, 218
Sophora flavescens. See Kurarinone
Sorbus sambucifolia, 469
Soviet Union, former, 143, 152
Soy foods/isoflavones, 231, 238–239, 258, 263, 312, 435,
    436

Spain, 197
Spices
    allium vegetables, 583–584
    basil leaves, 589–590
    celery seed, 590
    chemopreventive potential/ingredients, 577–578
    cloves, 587
    cumin seeds, 589–590
    curcumin, 576, 579–580
    effects of representative, 579
    garlic, 584–585
    ginger and related rhizomes of Zingiberaceae family,
        580–581
    mace, 590
    mint, 590
    onions, 585–586
    overview, 575–576
    peppers, hot red, 581–582
    rosemary, 587–589
    saffron, 582–583
    wasabi, 590
Squamous cell carcinomas, 67, 151, 152
Staging systems, 6, 147–148, 158
Statins, side effects of, 616
STAT3 pathway, 354–355, 366
Stem cell factor (SCF), 63, 65–66, 74
Stilling root, 251
Strawberries, 470, 473, 474
Stress and curcumin, 372
Stroke patients and NADH, 542
Substitution homolytic bimolecular reactions, 80
Sulfhydrl (SH) enzymes, 584
Sulforaphane, 243
Sulforhodamine-B (SRB) bioassay, 438–439
Sulformate, 243
Sulindac, 126, 131–132
SUMO gene, 44
Superoxide anions, 271, 432–433, 496
Superoxide dismutase (SOD), 93, 220, 362, 363, 433, 584
Superoxide radical, 35, 95
Supplements. See Natural products/supplements
Surgery/resection. See specific treatment/disease
Surveillance, Epidemiology, and End Results (SEER)
        program, 4, 152
Svoboda, Gordon, 261
Swabhavoparana, 248
Sweden, 197
Swedish Mammography Cohort, 289
SYK gene, 19
Synbiotics, 629
Syrian hamster embryo (SHE) cell transformation model,
        470

T

Taiwan, 32
Takayasu's arteritis, 109
Tamoxifen, 313, 428, 608
Tamra bhasma, 248
Target-organ specificity for chemopreventive agents, 241

Taxol
    clinical applications of Taxotere and, 521–522
    conclusions/summary, 523
    discovery of, 520–521
    lung cancer, 74
    side effects, 522
    yew, 261, 264, 517–520
Taxotere, 522
Taxus, 519–520
T cells. See also Immune system
    alternative assessment of immune surveillance, 600
    curcumin, 352, 355
    DNA methylation, 22
    DNA methylation patterns, 18, 22
    ginger, 580
    mushrooms, maitake, 511
    oxidative/nitrosative stress, 174
    quercetin, 326
    thymus, involution of the, 614–615
T47D cells, 452
Tea polyphenols
    anticarcinogenic effects: experimental studies, 394–403
    anti-inflammatory effects, 403–404
    black and oolong tea, 392–394
    chemistry, 393–394
    composition, 392, 393
    conclusions/summary, 415–416
    consumption, 391–392
    doxorubicin, 256, 326
    epidemiological studies, 411–415
    epigallocatechin gallate, 205–206, 255–256
    green tea, 392
    incidence and green tea consumption, 414
    initiation/promotion/progression, effective against,
        255–256
    mechanisms of biological effects, 404–411
    mechanisms/targets for selected agents, 230
    National Cancer Institute testing, 232–233
    overview, 390–391
    population-based studies, 338
    studies mixed results, 206
    types of tea, 232–233
    ultraviolet radiation, 255
Teniposide, 262
Testicular cancer, 262
Testosterone, 431
Tetradecanoylphorbol-13-acetate (TPA) model, 177, 406,
        470–472, 474; see also Dimethylbenz-a-anthracene
        model
Tetrahydrocurcumin, 373
Thalidomide, 133
Theaflavins, 232, 392–394
Thearubigins, 392–394
Thermal lasers, 157, 160, 161
Thiobarbituric acid reactive substances (TBARS)
    curcumin, 361, 363, 370
    NADH, 547
    palm tocotrienols, 486
    smokeless tobacco, 91, 93, 95
Thioredoxin, 180
Third world countries and colorectal cancer, 624

Thrombospondin-1, 109, 110
Thymine, 34, 83
Thymus, involution of the, 614–615
Tibet, 262
TNF-related apoptosis-inducing ligand (TRAIL), 352–353
TNM staging
    esophageal cancer, 158
    gastric cancer, 147–148
TNP-470 inhibitor, 206
Tobacco. *See also* Smokeless tobacco
    esophageal cancer, 153, 154
    gastric cancer, 145
    lycopene, 527, 535
    NADH, 547
    tea, 255, 399–400
    vitamins C/E and beta-carotene, 281, 286, 287, 290–291
Tobacco and Health Research Institute, 91
Tobacco-specific nitrosamines (TSNA), 90
Tocopherol-mediated peroxidation (TMP), 273
Tocopherols, 180, 208, 481–482; *see also* E *under* Vitamins; Palm tocotrienols; Vitamins C/E and beta-carotene as chemopreventive agents
Tocotrienols. *See* Palm tocotrienols
Tomatillo, 243; *see also* Lycopene
Topoisomerase, 240, 253, 343
Topotecan, 262, 343
Toxicity. *See* Xenobiotics, chemical properties/biotransformations of
TPA response element, 47
Transcription-coupled repair (TCR), 43
Transforming growth factor (TGF), 453, 454
Transgenic adenocarcinoma of the mouse prostate (TRAMP), 402
Trifolium pratense, 331
Trikatu, 373
Triphenylmethyl radical, 79
Trolox equivalents, 467
Trypan blue exclusion technique, 94–95
Tulsi, 258, 259–260
Tumor necrosis factor (TNF)
    astaxanthin, 563
    fruits and berries, 470
    ginger, 580
    mushrooms, maitake, 511
    Pycnogenol, 501
Turmeric, 257, 589; *see also* Curcumin
Tyrosine dephosphorylation, 336
Tyrosine kinase C, 253

U

Ubiquinone. *See* Coenzyme Q$_{10}$
U937 cell line, 500
Ultrasound and esophageal cancer, 158–159
Ultraviolet (UV) radiation
    astaxanthin, 559, 562, 563
    lycopene, 530
    ozone, stratospheric, 83
    Pycnogenol, 493, 498
    tea, 255, 396–397, 403–404

Umbelliferae family, 318
Unani, 263
Unimolecular radical reactions, 80
Urinary/bladder cancer
    incidence rate, 8
    metals/metalloids, 32
Urinary metabolites, smokeless tobacco and excretion of, 92, 93–94
Urokinase activity and green tea, 409
Ursolic acid, 260
Uterine cancer
    estrogen, 108
    probiotics, 633
    tea, 413
Uzbekistan, 292

V

Vaccinium smallii, 469
VA13 cells, 583
Vacuolating cytotoxin gene A (vacA), 164
Vascular endothelial growth factor (VEGF)
    angiogenesis, 133
    dietary chemopreventive strategies, 208
    estrogens, 109, 110
    fruits and berries, 470
    mushrooms, maitake, 512
    radiation therapy, adverse effects of, 220
    tea, 256
    vitamins C/E and beta-carotene, 291
Vascular injury and fibrosis, radiation therapy and, 216–217
Vascular smooth muscle cells (VSMC), 359–360
Vedas, 248
Vegetables/vegetarian diet, 229, 340, 466–467, 556, 575, 583–584, 626, 640; *see also* Fruits and berries
Velban, 264
Very low-density lipoproteins (VLDL), 364, 529
Vicenin, 260
Vimentin, 501
Vinblastine, 261, 264
Vincristine, 239, 264
Vindesine, 160
Virtual genome scan (VGS), 23
Viscum album, 336
Vitamins; *see also* Palm tocotrienols
    A, 313
    C
        esophageal cancer, 153
        gastric cancer, 287, 294
        metals/metalloids, 35
        nitrosamines, 644
        smokeless tobacco, 95, 97
    E
        esophageal cancer, 153
        immune function, 613
        intervention trials, 608
        lycopene, 531
        metals/metalloids, 35
        radiation therapy, adverse effects of, 220–221
        smokeless tobacco, 93, 95

Vitamins C/E and beta-carotene as chemopreventive agents
  antioxidant/other biological functions, 272–275
  conclusions/summary, 294–296
  evidence of chemoprevention, 284–294
  potential prevention mechanisms, 275–283

W

Waist-to-hip ratio (WHR), 200
Walker, John, 603
Wasabi, 590
Water extracts of green tea (WEGT), 397, 402–403
Wattenberg, Lee, 643
Werner syndrome (WS), 189–192
Wheat germ agglutinin (WGA), 366
Wine, red, 256
Womens Health Study, 640
Workplace exposures, 41–42; *see also* asbestos *and* silica
    under Lung cancer; Metals/metalloids
World Health Organization (WHO), 197, 248, 262
Wound healing and curcumin, 365

X

Xanthine oxidase pathway, 173
Xanthophylls, 560–561
Xenobiotic-metabolizing enzymes (XME), 564
Xenobiotics, chemical properties/biotransformations of,
    188–192
XF498 cell line, 442

Y

Yeast, estrogen-inducible, 438, 439
Yew, 260–261, 264, 519–520; *see also* Taxol

Z

Zapotin, 243
Zeaxanthin, 286, 289, 558
Zerumbone, 581
Zidovudine (AZT), 22
Zingiberaceae family, 318, 580–581